MODERN METHODS OF PLANT ANALYSIS

EDITED BY

K. PAECH   M. V. TRACEY

VOLUME IV

CONTRIBUTORS

G. BRAUNITZER - J. GLOVER - M.J.R. HEALY - E. HECKER - H. HELLMANN
R. HILL - E.C. HUMPHRIES - R.H. KENTEN - G. KORTUM
M. KORTUM-SEILER - K. PAECH - N.W. PIRIE - J. SMALL
R.L.M. SYNGE - F.R. WHATLEY

WITH 82 FIGURES

REPRINT 1980

SPRINGER-VERLAG
BERLIN · GÖTTINGEN · HEIDELBERG
1955

MODERNE METHODEN DER PFLANZENANALYSE

HERAUSGEGEBEN VON

K. PAECH   M. V. TRACEY

VIERTER BAND

BEARBEITET VON

G. BRAUNITZER · J. GLOVER · M.J.R. HEALY · E. HECKER · H. HELLMANN
R. HILL · E.C. HUMPHRIES · R.H. KENTEN · G. KORTUM
M. KORTUM-SEILER · K. PAECH · N.W. PIRIE · J. SMALL
R.L.M. SYNGE · F.R. WHATLEY

MIT 82 ABBILDUNGEN

REPRINT 1980

SPRINGER-VERLAG
BERLIN · GÖTTINGEN · HEIDELBERG
1955

ISBN-13: 978-3-642-64963-9   e-ISBN-13: 978-3-642-64961-5
DOI: 10.1007/978-3-642-64961-5

Inhaltsverzeichnis. — Contents.

Inhalt der übrigen Bände. — Contents of other Volumes.

Erster Band. — Volume I.

Zweiter Band. — Volume II.

Dritter Band. — Volume III.

Mitarbeiter von Band IV.   Contributors to Volume IV.

Professor HARRY G. ALBAUM, Department of Biology, Brooklyn College,
Brooklyn 10, N. Y. (USA).

ALLEN BENITEZ, Department of Plant Biology, Carnegie Institution of Washington,
Stanford University, Stanford, Calif. (USA).

Dr. B. T. CROMWELL, Dept. of Botany, The University, Hull,
(Great Britain).

Dr. J. DUCKWORTH, The Rowett Research Institute, Bucksburn,
Aberdeenshire (Great Britain).

Dr. E. F. HARTREE, Molteno Institute, Downing Street,
Cambridge (Great Britain).

Professor Dr. K. HASSE, Institut für organische Chemie d. T. H.,
Karlsruhe, Englerstraße 11.

Dr. E. JUCKER, c/o Sandoz AG.,
Basel (Schweiz).

Dr. R. MARKHAM, Molteno Institute, Downing Street,
Cambridge (Great Britain).

J. R. P. O'BRIEN, Dept. of Biochemistry, University Museum,
Oxford (Great Britain).

Dr. J. PACE, Research Association of British Flour Millers, Cereal Res. Station,
St. Albans, Herts. (Great Britain).

Sir R. A. PETERS, F. R. S., Institute of Animal Physiology,
Babraham Hall, Babraham, Cambs. (Great Britain).

N. W. PIRIE, F. R. S., Rothamsted Experimental Station,
Harpenden, Herts. (Great Britain).

Privatdozent Dr. Dr. P. SEIFERT, Medizin.-Diagnost. Institut,
Mannheim, P 7/19.

Professor James H. C. SMITH, Department of Plant Biology, Carnegie Institution
of Washington, Stanford University, Stanford, Calif. (USA).

Professor Dr. A. STOLL, c/o Sandoz AG.,
Basel (Schweiz).

Dr. F. M. STRONG, Department of Biochemistry, University of Wisconsin,
Madison 6, Wisc. (USA).

Dr. R. L. M. SYNGE, F. R. S., The Rowett Research Institute,
Bucksburn, Aberdeenshire (Great Britain).

Professor M. THOMAS, F. R. S., Department of Botany, King's College,
Newcastle-upon-Tyne 1 (Great Britain).

M. V. TRACEY, Rothamsted Experimental Station,
Harpenden, Herts. (Great Britain).

Professor Dr. Dr. E. WERLE,
München, Nußbaumstraße 20.

Peptides (Bound Amino Acids) and Free Amino Acids[1].

By

R. L. M. Synge.

The term "peptide" as distinct from "protein" has come to be reserved for compounds having molecular weight less than, at most, 10,000 or, if of higher molecular weight, having unusually simple amino acid composition. Natural compounds so far shown unequivocally to belong to this class do not exhibit denaturation in its usual sense or, particularly, coagulation by heat.

It should be made clear at the outset that, whilst tne amino acids found free in plant tissues are compounds of known or readily ascertainable chemical structure, this is not in general true of the "peptides"; peptides are usually only distinguished from proteins by the means used for separating them and therefore, in the present absence of adequate understanding of the chemical structure of both the proteins and the peptides found in nature the distinction is largely an empirical one. "Peptides" in the sense of much of the literature of plant analysis are no more than chemically bound amino acids found after applying a procedure expected to remove protein. The bonding of the amino acids may not necessarily involve peptide links (see below); indeed, where nothing but an increase of amino nitrogen after hydrolysis of such fractions has been demonstrated, even amino acids may not be concerned.

The least empirical procedures for separating peptides from proteins are dialysis and ultrafiltration with suitable membranes. Thus, there are good grounds for believing that no molecules having molecular weight greater than 10,000 can pass a cellophan membrane. The juice obtained from leaves by the ether-water and similar procedures (cf. CHIBNALL, 1939) has already undergone ultrafiltration through cell walls, and a notably high proportion of the nitrogen found therein passes freely through cellophan (e. g. SYNGE, 1951 b). DANIELSSON (1951, 1952) has used dialysis for studying the nitrogenous fractions of germinating and ripening peas.

Juices obtained after heat coagulation of proteins are likely also to be free from protein, although there are quite sufficient proteins that are not heat-coagulable (e. g. casein, gelatin) to emphasise the need for caution in interpreting fractions thus obtained (p. 24). Heat coagulation (and some protein precipitants) have the great advantage of quickly inactivating proteolytic enzymes. There is thus less chance that peptides and amino acids found in extracts of tissues so treated are artefacts of post-mortem proteolysis etc. Precipitation of protein by boiling has worked well in the hands of MACPHERSON (1952), who used mainly ensiled, already acid materials. 92--98% of the N extracted by him passed through cellophan. In most of the studies referred to below on peptide materials from seaweeds the raw material was initially extracted with boiling or hot water. Boiling has also been used in the isolation of seed "proteoses" (see below).

---

[1] Determination of total nitrogen cf. Vol. I.

Bathurst (1953), using aqueous extracts of freeze-dried ethanol-extracted grass, precipitated protein by acidification with acetic acid, and commented that this procedure precipitated the same amount of, or more N than conventional deproteinization with trichloroacetic acid. However, he did not subject the filtrates from acetic acid precipitation to treatment with trichloroacetic acid.

In general, there have been few critical studies (e. g. Neuberger and Sanger, 1942; Bisset, 1954) of yields of non-protein nitrogen after different deproteinizing procedures and often very little evidence has been adduced that the extracts obtained are free from protein. Students of plant peptides and amino-acids should note and emulate the critical studies by H. N. Christensen and colleagues on non-protein fractions of blood and other tissues (e. g. Christensen and Lynch, 1946a, b). The use of protein precipitants such as picric acid, sulphosalicylic acid, trichloroacetic acid, colloidal iron etc. is even more empirical than the procedures referred to above. Early work with such reagents has been reviewed by Rona and Strauss (1922). From this period the Stutzer treatment with copper hydroxide has survived by virtue of having been adopted as the official method for determining "true protein" (see Official Methods of Analysis of the Association of Official Agricultural Chemists. 7th Ed^n. (1950) p. 345. Washington: Association of Official Agricultural Chemists; Agric. Progress 20, 51 (1945).

The use of aqueous ethanol for obtaining protein-free extracts can be subjected to the same criticisms. It seems first to have been used to good effect for isolating non-protein nitrogenous fractions by Osborne, Vickery and their colleagues (Osborne, Wakeman and Leavenworth, 1921, 1922; Vickery, 1924a, b 1925a, b; Vickery and Leavenworth, 1925; Vickery and Vinson, 1925). In recent years it has achieved considerable popularity after its use by Dent, Stepka and Steward (1947) in preparing an extract of potato for chromatography of free amino acids. The slight extraction of inorganic salts by ethanol is an attraction of the procedure. Although Bathurst (1953) has claimed that peptide and protein material are absent from extracts of freeze-dried grasses made with 80% ethanol (w/v or v/v not specified), it would be well to subject each case to individual study in view of the very different results, especially under alkaline conditions, found by Osborne, Wakeman and Leavenworth (1921, 1922).

In all these separation procedures it should be noted that no control is exercised over the possible retention on the protein of smaller non-protein molecules by adsorption, ion exchange etc. Thus peptides containing residues with basic, acidic, aromatic or higher paraffinic side chains would be expected to be selectively retained on protein, and this may explain the rather simple amino acid composition of the peptide-like materials so far detected in ordinary tissues (see Synge, 1953b). Such effects with peptides and amino acids are likely, however, not to give rise to such serious difficulties as those found in extracting large molecules from plant tissues, referred to elsewhere in this work (Vol. I, p. 27). Astrup and Øhlenschläger (1948) noticed interesting differences in the power of different extracting agents to remove glutathione from yeast.

In agricultural analysis the difference between total N ("crude protein") and N of "true protein" (determined by the Stutzer method referred to above) is often called "amide N". As Chibnall (1939, p. 19) has pointed out, this is not because at any time people have believed that only the amides glutamine and asparagine are present in this fraction; other amino acids have been known from the first studies to be present. It is simply that modern chemical nomenclature has been slow to penetrate agricultural analysis and the old name amido-acid (Amidosäure), often shortened to amide, has persisted. These terminological changes should be remembered also when studying the older literature. It has long been known that the amino-acid composition of the non-protein fraction is less favourable for nutrition, at least of monogastric (non-ruminant) animals, than that of ordinary proteins (see section on amino acids below). The convention of regarding its N as having half the value of the N of "true protein" has thus had some justification in practice.

STREET (1949) has briefly reviewed techniques for determination of some of the main nitrogenous constituents of plants.

A. Peptides (Bound Amino Acids).

It seems clear that among the nitrogenous compounds found in plant extracts deproteinized by the methods referred to above peptides and bound amino acids are not usually present in large amount compared with free amino acids. However, there do not seem to be many cases where the former have been searched for by valid analytical means and not found. The many negative reports based on paper chromatograms coloured with ninhydrin can be ignored since many peptides (especially cyclic peptides) give little or no colour with ninhydrin. (In this connection the reaction of RYDON and SMITH (1952) seems likely to prove very useful for visualizing peptide material on filter-paper chromatograms and in spot tests. It should be noted, further, that peptides containing more than a few amino acid residues per molecule may not give good spots, but only diffuse streaks, on filter-paper chromatography.) However, small open-chain peptides would almost certainly, if commonly present in plant tissues, have been detected, so there are grounds for supposing that such substances, other than glutathione (see below) are usually absent.

It seems useless, in the present state of knowledge of plant peptides, to discuss in detail isolative procedures. Such procedures for peptides in general, as well as methods for further study of their chemical structure, have been well reviewed by SANGER, 1952 (see also BOULANGER and BISERTE, 1952). There is every reason for supposing that plant peptides will prove just as difficult to purify as peptides from other sources, each starting material, as well as each product to be isolated, setting a separate problem.

If isolation or separation is not required, analysis of a protein-free extract for peptide or chemically bound amino acid is comparatively simple. The increase on hydrolysis of free amino acids, determined by the ninhydrin-CO_2 procedure, gives a useful and specific measure of bound amino acids. Changes in other quantities (amino N, titratable basic and acidic groups, extent of Cu complex formation), though less specific, are useful on occasion. The most informative method of analysis, however, is to determine individual free amino acids present in an extract before and after hydrolysis by any of the *specific* methods for amino acids (see below). In applying such procedures account should be taken of the presence of amides.

The variations in "peptide N", determined by one or other of the less specific means just mentioned, or by still cruder difference procedures, have been studied by plant physiologists in a number of materials subjected to varying conditions. While such studies have been useful in throwing light on N metabolism, the variety of methods used prevents comparison of the results of different workers and it would be misleading to draw up comparative tables of total peptide present in different plant materials.

Special caution should be exercised in interpreting such data when there is a possibility of the MAILLARD reaction between sugars and amines having proceeded. This has been the subject of a number of studies in recent years. It proceeds fastest in concentrated solution and at higher temperatures and is likely therefore to have occurred in any dried or stored whole plant tissue. The reaction proceeds with loss of amino N and there is at least partial regeneration of amines involved (including amino acids) on acid hydrolysis (cf. Lea, 1950; Lea and Hannan, 1950; Friedman and Kline, 1950a, b; Gottschalk and Partridge, 1950; Hannan and Lea, 1951; Täufel and Iwainsky, 1952). Vickery, Pucher, Wakeman and Leavenworth (1937) (pp. 780—785) noted loss of amino N on drying tobacco leaves; the lost amino N probably reappeared in their "peptide" fraction on acid hydrolysis. In the light of more recent knowledge, these effects are probably attributable to the Maillard reaction. It has in fact been studied in direct connection with tobacco curing (Bolgunov and Pokhno, 1950; Mashkovtsev, 1950, 1951).

Since substances giving at least the colour reactions of hexosamines are formed in the same reaction (see also Horowitz, Ikawa and Fling, 1950; Immers and Vasseur, 1950; Heyns and Koch, 1952), it is reasonable to regard the hexosamine reaction observed in pineapple leaves by Sideris, Young and Krauss (1938) as attributable to the same cause; especially in view of the increase in the reaction in older leaves and when N was supplied to the plants as NH_3 rather than nitrate.

Purines and purine derivatives yield glycine under the conditions of protein hydrolysis, and this additional source of artefact "bound amino acid" should be borne in mind (Markham and Smith, 1949).

Pyrrolidone carboxylic acid, which yields glutamic acid on acid hydrolysis, is formed readily from glutamine in neutral solution at high temperatures and in weakly acid solution at low temperatures. It is thus another artefact source of amino acid. Analytical methods for pyrrolidone carboxylic acid were reviewed by Ellfolk and Synge (1955), who concluded that it was absent from freshly prepared grass extracts.

The general occurrence of peptides in nature has been reviewed by Synge (1949, 1953b) and Fromageot (1953). Since few generalisations about analytical procedure can justifiably be made at the present stage of understanding, the most helpful approach seems to be to give a documented outline of observations so far made on the occurrence of peptides and bound amino acids in plants and plant products.

I. Higher Plants.

Studies of substances in seeds responsible for allergic reactions in humans have revealed an interesting group of compounds. Schloss (1912) studied materials from almonds and oats capable of diffusing through collodion or parchment membranes and more detailed observations on a variety of materials were made by Wells and Osborne (1915) who classified the substances as "natural proteoses" and gave some additional references. Spies and colleagues have continued such studies in more recent years and have shown in a series of papers (for references see Spies, Coulson, Chambers, Bernton, Stevens and Shimp, 1951) that a large number of seeds yield materials soluble in boiling water and in 25% (but not in 75%) ethanol, not precipitable by lead acetate and partially diffusing through cellophan. These materials contain some carbohydrate and a wide variety of amino acids and possess allergic activity. (See also Reeves and Guthrie, 1950; Grabar and Koutseff, 1934.) Basic material having rather low molecular weight and some antibiotic activity has been isolated from wheat flour (Balls, Hale and Harris, 1942; Balls and Harris, 1944). Somewhat similar materials isolated from pollens have likewise been described.

The earlier literature has been reviewed by NEWELL, 1942 (see also ROCKWELL, 1943a, b; LOVELESS, 1949; STONE, HARKAVY and BROOKS, 1947; ABRAMSON, MOORE and GETTNER, 1941, 1942; WODEHOUSE and COCA, 1946; PERLMAN, 1951; DANKNER et al., 1951).

In germinating seeds, peptides of low molecular weight giving a colour reaction with ninhydrin may be present. This is suggested by qualitative filter-paper chromatography with wheat (DEVAY, 1952) and barley (including malt and wort) (LJUNGDAHL and SANDEGREN, 1950; BISERTE and SCRIBAN, 1950, 1951). BISERTE and SCRIBAN (1952) used ion-exchange resins for fractionating peptides of wort, and obtained fractions embodying large proportions of dicarboxylic amino acid residues. They also observed chemically bound β-alanine and γ-aminobutyric acid. FROMAGEOT, JUTISZ and TESSIER (1949) obtained some indirect evidence for peptides in corn steep liquor.

Studies of deproteinized alfalfa juice by VICKERY and colleagues (VICKERY, 1924a, b; 1925a, b; VICKERY and LEAVENWORTH, 1925; VICKERY and VINSON, 1925), whilst not yielding any well-defined peptide material, gave abundant evidence for the occurrence of chemically bound amino acids, especially basic ones. Some were precipitated by lead acetate, other escaped precipitation. SYNGE (1951b) studied the ionophoretic behaviour of diffusible "bound amino acids" of the juice of rye grass. Small amounts of basic and acidic material were separated, but the bulk of the material, amounting to about 5% of the diffusible non-protein nitrogen, exhibited no charge at the pH's studied and yielded only a few different amino acids on hydrolysis. Countercurrent distribution with phenol-water separated the material from much contaminating carbohydrate and further fractionation was achieved by partition chromatography using kieselguhr (SYNGE and WOOD, 1954). VIRTANEN and MIETTINEN (1953) have studied bound amino acids in peas and described somewhat similar results using ion-exchange resins and charcoal for fractionation. They found chemically bound β-alanine and γ-aminobutyric acid and suggested that compounds of amino acids with sugars may be involved.

Peptide-like material is extractable from filter paper (WYNN, 1949; HANES, HIRD and ISHERWOOD, 1952). GOVINDARAJAN and SREENIVASAYA (1950) obtained evidence of bound amino acids in mangoes.

A monoacetyl derivative of L-ornithine was isolated from *Corydalis ochotensis* by MANSKE (1937). A yield of 10% was obtained from the taproots. Later MANSKE (1946) obtained the same compounds from *Corydalis cornuta*. NIERENSTEIN (1914) isolated laevorotatory crystalline material from galls caused by *Cynips calcis* on *Quercus aegilops*, which may have been N-galloyl-L-leucine (cf. NIERENSTEIN, 1915).

In work on the toxic principle of mistletoe *(Viscum album)*, WINTERFELD and BIJL (1948) and WINTERFELD and RINK (1949) isolated material which was pharmacologically active. This was said to yield on hydrolysis cysteine, arginine and serine, which appeared to be linked with glucuronic acid and phosphate residues and with a hydrogenated naphthalene derivative.

HASSALL, REYLE and FENG (1954) briefly reported the isolation from seeds of the fruit of *Blighia sapida* by chromatographic means of two compounds, apparently peptides, which they called hypoglycin A and B. These were toxic in animals, producing hypoglycaemia, and may be responsible for toxicity of the fruit.

II. Algae.

There is now a considerable literature on peptides obtainable from marine algae by extraction of the tissues with hot water or aqueous ethanol. HAAS and

Hill (1931) used clarification with lead acetate followed by precipitation with mercuric acetate to isolate 0.18% of a substance, yielding on hydrolysis much glutamic acid, from *Pelvetia canaliculata*. From amino-N figures they concluded it was an octapeptide. A similar procedure applied to *Corallina officinalis* gave 0.8% of amorphous material, judged in the same way to be a pentapeptide, which yielded on hydrolysis an unspecified quantity of aspartic acid (Haas and Hill, 1933a). Some metabolic conclusions were drawn (Haas and Hill, 1933b). The same procedure gave similar yields of amorphous material from a number of other seaweeds; alanine, arginine, histidine, aspartic acid, glutamic acid and phenylalanine were detected (Haas, Hill and Russell-Wells, 1938; Haas, 1950).

Applying a similar procedure to *Eisenia bicyclis*, Ohira (1939) obtained 0.8% of a crystalline product which yielded on hydrolysis 2 moles of glutamic acid, one of alanine and one of ammonia. This was called "eisenin". Subsequent chemical studies led Ohira (1940a, b, 1942) to formulate eisenin as pyrrolidonoyl-glutaminylalanine. Tazawa (1949) contested this formulation.

Dekker, Stone and Fruton (1949) followed Haas' procedure (full experimental details given) and obtained from *Pelvetia fastigiata* a small yield (0.05%) of material for which they produced convincing evidence for the structure L-pyrrolidonoyl-L-glutaminyl-L-glutamine. They thought it likely that the pyrrolidonoyl residue was formed from a glutaminyl residue in the course of isolation. Channing and Young (1953) attempted to isolate peptides extractable from *Pelvetia canaliculata* with cold aqueous alkali as N-2-bromo-4:6-dinitrophenyl derivatives. After substitution, precipitation with mercury was followed by countercurrent distribution. Glutamic acid, aspartic acid, glycine, alanine, serine and histidine were recognized in hydrolysates of the resulting fractions, which were not, however, characterized in any detailed way.

Fogg (1952) studied "peptide" formation in the culture medium of the nitrogen-fixing blue-green alga *Anabaena cylindrica*. Various amino acids were liberated on acid hydrolysis of the medium, but no evidence was presented for assessing the molecular weight of the substances present.

III. Fungi.

A variety of peptide-like substances have been isolated from fungi. The majority of these possess some pharmacological activity; search for the active principle was the reason for their discovery. It is remarkable how many of them contain unusual amino acid residues and unusual ring systems.

Amanita spp. In continuation of studies using liquid-liquid extraction and adsorption chromatography (Lynen and U. Wieland, 1938; H. Wieland and Witkop, 1940; H. Wieland and Hallermeyer, 1941) Th. Wieland and colleagues have used some more recent methods for isolating and characterizing the toxic peptide-like constituents of *Amanita phalloides*. Th. Wieland (1949a) found filter-paper chromatography ineffective, but using filter-paper electrophoresis and countercurrent distribution with phenol-chloroform-water systems isolated a third component, β-amanitin, as well as the previously characterized α-amanitin and phalloidin. Th. Wieland and Schmidt (1952) later reported useful solvent systems for paper chromatography of these substances, and were able to use the same systems with cellulose-powder chromatograms for preparative work. They were also able to demonstrate the absence of these toxic compounds from the closely related non-toxic *Amanita mappa*, from which a different peptide, mappin, was isolated by similar means. Phalloidin yields on hydrolysis L-cystine, L-alanine, L-hydroxytryptophan, L-*allo*hydroxyproline and threonine. The amanitins yield aspartic acid, glycine, cysteine and a hydroxyproline isomer. Mappin, which is S-free, yields aspartic acid, serine, glycine, alanine, a hydroxyproline and other not yet identified constituents. (See also Šorm and Keil, 1951.)

Fusarium spp. Compounds showing antibacterial activity have been isolated from a number of *Fusarium* spp. and were found on hydrolysis to yield α-N-methylamino acids and α-hydroxy acids. There is evidence for peptide and ester linkages of the residues both being present, but there is not yet final agreement as to the structures (Plattner and Nager, 1948 a, b, c; Plattner, Nager and Boller, 1948; Cook, Cox and Farmer, 1949). Plattner and Nager (1948 a) gave directions for chromatographic identification of various N-methylamino acids, and studied their distribution in extracts from a number of *Fusarium* spp.

Fusarium lycopersici yields lycomarasmin, which seems to be the active principle by which the fungus causes wilting of tomato plants. On hydrolysis, approx. equimolar amounts of glycine, aspartic acid, ammonia and pyruvic acid are obtained from lycomarasmin. Woolley (1948) has produced strong evidence for the formulation:

$$CH_2{-}CO{-}NH_2 \qquad\qquad CH_3$$
$$|\qquad\qquad\qquad\qquad\qquad |$$
$$CH{-}NH{-}CO{-}CH_2{-}NH{-}C{-}OH$$
$$|\qquad\qquad\qquad\qquad\qquad |$$
$$COOH \qquad\qquad\qquad\qquad COOH$$

(cf. Plattner, Clauson-Kaas, Boller and Nager, 1948). The molecular weight has been established by X-ray crystallography and diffusion measurements (Plattner, Günthard and Boller, 1952).

Aspergillus and *Penicillium* spp. Aspergillic acid, isolated from culture filtrates of *Aspergillus flavus*, bears a close chemical relationship to the anhydride of leucine-isoleucine (Dutcher and Wintersteiner, 1944; Dutcher, 1947; Dunn, Gallagher, Newbold and Spring, 1949; Dunn, Newbold and Spring, 1949; Newbold, Sharp and Spring, 1951 a, b; Gallagher, Newbold, Sharp and Spring, 1952). In this connection the isolation of L-leucine-L-proline anhydride by Johnson, Jackson and Eble (1951) by organic-solvent extraction of culture filtrates of *Streptomyces* spp. and *Aspergillus fumigatus* is interesting.

Only passing reference can be made to the very extensive work on the natural occurrence, structure and metabolism of the penicillins, which may be regarded as containing residues of L-α-aminomalonaldehydic acid and D-penicillamine (β-thiolvaline), although biosynthetically they probably arise by condensation of L-cysteine with a degradation product from valine (Arnstein and Grant, 1954).

Fumaryl-DL-alanine has been isolated from *Penicillium resticulosum* by Birkinshaw, Raistrick and Smith (1942).

Ergot Alkaloids. Whilst these are discussed elsewhere (Cromwell, p. 409) it should be noted that the majority of them contain bound amino acid residues which are set free on hydrolysis, and that it is differences in the amino acid residues which determine the differences between the ergot alkaloids. Stoll, Hofmann and Petrzilka (1951) (cf. earlier papers in series, also Stoll and Petrzilka, 1952) have shown a number of interesting features in the complicated ring system into which the amino acid residues are condensed: (a) an α-hydroxy-α-amino acid residue is present; (b) a "cyclol" structure is present; (c) the proline residue has its carboxyl group in the ortho form, linked partly in ester and partly in amide linkage and presenting an extra centre of asymmetry which is lost on hydrolysis; (d) the configuration at the α-C atom of the proline residue is almost certainly L-, the D-proline found in hydrolysates of the alkaloids arising by inversion.

Yeasts. Evidence for peptides in yeasts has been obtained by LJUNGDAHL and SANDEGREN (1950), MIETTINEN (1951), LINDAN and WORK (1951), COLEMAN (1951) and REINDEL and HOPPE (1952). See also below under glutathione and pteroic acid derivatives etc.

IV. Glutathione.

This open-chain tripeptide is of very wide distribution throughout the plant and animal kingdoms. The earlier literature on distribution and analytical methods can be found in BEILSTEIN (zweites Ergänzungswerk Bd. IV, S. 931). Many of the analytical methods proposed and used are completely non-specific, being for reducing or thiol groups. They may, however, be useful on occasion for setting upper limits to the amount of glutathione present in an extract.

Isolation of glutathione was greatly facilitated when HOPKINS (1929) first used Cu_2O for precipitation of the cuprous mercaptide, and large yields are readily obtained from yeast in this way. PIRIE (1930) improved the procedure by using ethanol, ether and sulphuric acid for preparing the initial extract (see also ASTRUP and ØHLENSCHLÄGER, 1948; SCHROEDER, COLLIER and WOODWARD, 1939).

Several isolations from plant material using the cuprous mercaptide procedure have been reported. HOPKINS and MORGAN (1943) had difficulty applying the simple procedure to trichloroacetic acid extracts of peas, but succeeded using sulphuric acid extracts that had been half-saturated with ammonium sulphate. However they later found (1945) deproteinization of sulphuric acid extracts to be unnecessary, and further isolated glutathione from broad beans *(Vicia faba)* and vegetable marrow *(Cucurbita pepo)*. GUTHRIE (1932) isolated glutathione from potato tubers that had been treated with ethylene chlorohydrin. REEVES and GUTHRIE (1950) obtained 43 mg. glutathione from 9 kg. of shelled peanut kernels, and established chromatographically the purity of their product.

JANSEN and JANG (1952) isolated glutathione from orange juice as the *S*-benzyl derivative after mercury precipitation. SULLIVAN and HOWE (1937) isolated glutathione from wheat germ using various metal precipitations. Precipitation as the cadmium complex has also been used (BINET and WELLERS, 1947).

Of specific non-isolative analytical procedures the method of WOODWARD, 1935 (cf. SCHROEDER and WOODWARD, 1939) has proved useful in plant analysis. It depends on the action of glutathione as coenzyme for glyoxalase, and seems highly specific. SCHROEDER and WOODWARD analysed potato, apple and corn sprouts in this way, and the method has been used by LAWRENCE (1950) for a study of glutathione in peas subjected to various conditions. SHAMSHIKOVA and IOFFE (1947) describe steps necessary to prevent interference· by cysteine. See also ENNOR (1939).

Filter-paper chromatography is useful for the specific detection of glutathione, and the most interesting example of its use hitherto is by STEWARD, THOMPSON, MILLAR, THOMAS and HENDRICKS (1951) who studied juice from alfalfa plants to which $^{35}SO_4''$ had been supplied. The spot corresponding to glutathione coincided whether revealed by ninhydrin or by radioautography. See also LINDAN and WORK (1951); HASHIZUME (1951/52).

V. Pteroic Acid Derivatives etc.

Substances possessing the biological activities of pteroylglutamic acids are widely distributed. The original "folic acid" preparations were concentrated from spinach leaves, but no pure compounds were isolated from this source. "Pteroylheptaglutamic acid" has been isolated from yeast, and seems to be related to a compound of *p*-aminobenzoic acid with several residues of glutamic acid isolated from yeast by RATNER, BLANCHARD and GREEN (1946). The whole subject was thoroughly reviewed by JUKES and STOKSTAD (1948). See also STRONG (pp. 651—656).

Conjugated β-alanine found in plant products (e. g. SYNGE, 1951b) may be present in pantothenic acid, coenzyme A and related compounds of wide distribution in nature (cf. pp. 624 ff.).

Progress in these fields has been surveyed regularly in recent years in the "Annual Review of Biochemistry".

B. Amino Acids.

It is not intended in this section to deal with the analysis for amino acids of proteins, peptides etc. as this has been reviewed at length elsewhere, and it will suffice to refer to a number of such reviews. Some of the special considerations applying in amino acid analysis of plant proteins (hydrolysis conditions, etc.) are discussed by PIRIE (pp. 29—32) and representative results of such analyses are tabulated. In general, the purer the starting material, the simpler it is to analyse for amino acids; the lower the molecular weight, the less analytical accuracy is needed for arriving at stoichiometric relations between the amino acid residues present. Detailed treatment here of the voluminous literature of amino acid analysis would be the less suitable because at the present time it seems as if the chromatographic methods developed by S. MOORE and W. H. STEIN are coming to hold much the same position in this field as the elementary microanalytical procedures of PREGL hold in organic chemistry as a whole. While there will always be scope for less laborious methods designed to determine individual amino acids or for application when long series of less accurate analyses have to be carried out, it is clear that most of the methods so far proposed are already obsolete.

MITCHELL and HAMILTON (1929) and WINTERSTEIN (1933; cf. TAUBÖCK and WINTERSTEIN, 1933) comprehensively reviewed the methods available up to about that time, and the next fifteen years were covered by MARTIN and SYNGE (1945) and SNELL (1945) (microbiological assay). For more recent developments, the following may be consulted:

BALSTON and TALBOT (1952) (paper and cellulose chromatography); BARTON-WRIGHT (1952) (microbiological assay); BLOCK and BOLLING (1951); DUNN (1949) (microbiological assay); GALE (1946a, 1948) (decarboxylases); HAIS and MACEK (1954); JUTISZ (1952) (paper chromatography); LEDERER and LEDERER (1953) (chromatography); MARDASHEV (1950) (microbiological and enzymic methods); PASKHINA (1950) (partition chromatography); TRISTRAM (1949, 1953); TURBA (1954) (chromatography etc.); WERLE (1951) (decarboxylases); TH. WIELAND (1949b).

See also "Conference on Amino Acid Analysis of Proteins". Ann. N. Y. Acad. Sci. 47, 57 (1946). New developments in amino acid analysis have been noted, almost annually, in the "Annual Review of Biochemistry" since its inception.

The particular concern of the present chapter is the analysis of plant materials for free amino acids. Reference to the review by VICKERY and SCHMIDT (1931) on the history of the discovery of the amino acids shows what a high proportion of the amino acids present in proteins were at an early stage demonstrated to occur free in plants (see also WEHMER and HADDERS, 1933). Nevertheless the study of these free amino acids lagged behind amino acid analysis of proteins and it is only in recent years that a reasonably objective picture of amino acid distribution has been reestablished. It is convenient to consider (I) what amino acids have been found free in plants; (II) general methods of analysis; (III) special methods for determining particular free amino acids that are or may prove specially useful with plant extracts; (IV) some representative studies of free amino acids in plant materials.

I. Amino Acids Found in Plants.

The following 20 amino acids (for the most part having L configuration at α-C) are generally found as protein components throughout the whole range of living organisms: glycine, alanine, valine, leucine, isoleucine, serine, threonine, cystine-cysteine, methionine, aspartic acid, glutamic acid, asparagine, glutamine, arginine, lysine, histidine, proline, phenylalanine, tyrosine and tryptophan. These all likewise have been found free in plant material, although concentrations may in particular cases be so low as to give negative results by ordinary means of detection. There are very few examples of amino acids other than the above being incorporated in proteins. As far as plants are concerned, the following claims have been made and not so far refuted:— *Iodogorgoic acid* (diiodotyrosine) in seaweed protein (Roche and Lafon, 1949) [Ericson and Sjöstrom (1952) and Channing and Young (1953) were not able to demonstrate its presence]; *hydroxyproline* in sugar beet protein (Sisakyan, Bezinger and Kuvaeva, 1951; Vavruch, 1952) and pollen (Auclair and Jamieson, 1948); a dehydrogenated form of hydroxyproline, "*oxyminaline*", in pectases of *Penicillium* and *Aspergillus* (Minagawa, 1945); *α-aminoadipic acid* in seed protein of maize (Windsor, 1951); *sarcosine* in groundnut seed protein (Haworth, MacGillivray and Peacock, 1951); Marquardt and Vogg (1952) claimed to have detected *lanthionine* and *taurine* in hydrolysates of pollen.

Horn and Jones (1941) gave evidence for the incorporation of Se into the proteins of wheat grown on seleniferous soils, and gave references to earlier work. This presumably occurs by substitution of Se for S in some amino acid residues, and confirmatory evidence was obtained by studying the non-protein fraction of *Astragalus pectinatus* (see below). *αε-Diaminopimelic acid*, a widely distributed constituent of bacterial proteins, seems to be absent from fungi and higher plants so far studied (Work and Dewey, 1953; Synge, 1953a). Wingo and Davis (1949) have studied a number of seed globulins and consider it unlikely that *citrulline* residues are present (cf. Martin and Synge, 1945).

A number of further kinds of amino-acid residue found in compounds of lower molecular weight than proteins have been referred to above (pp. 5—6). There are, however, also a considerable number of amino acids which have been found free in plant material and which are normally absent from proteins. Such reports are particularly frequent for the order *Leguminosae*, and perhaps reflect, apart from the obvious bias due to the economic importance of the order, a genuine peculiarity of metabolism.

γ-Aminobutyric acid has been recognized chromatographically in nearly all plant tissues investigated, and often represents a substantial fraction of the non-protein nitrogen. Actual isolations have been reported from yeast by Reed (1950), from red beetroot by Westall (1950a, b), from sugar beet by Brown (1951) and from rye grass by Synge (1951a). Other data on its occurrence were reviewed by Synge (1951a, b).

β-Alanine has also frequently been recognized chromatographically as well as by microbiological tests. It usually occurs in smaller amounts than does γ-aminobutyric acid. No bulk isolation seems yet to have been reported (see, however, Westall, 1950a, b; Hulme and Arthington, 1950).

α-*Aminobutyric acid* has likewise often been asserted to be present on the basis of otherwise unsupported chromatographic evidence [cf. CHRISTIANSEN and THIMANN (1950) who also made similar claims in respect of *hydroxylysine* and *norvaline*].

3:4-*Dihydroxyphenylalanine* has been found free in various beans (for early references see VICKERY and SCHMIDT, 1931). It was isolated from seeds of *Mucuna pruriens* by DAMODARAN and RAMASWAMY (1937; cf. SARKAR, 1945) and of *Mucuna capitata* by YOSHIDA (1945). A detailed account of the preparation from velvet beans has been given by SEALOCK (1949). SBOROV, PETERS and ARNOW (1942) failed after careful search to find dihydroxyphenylalanine in hydrolysates of velvet-bean and other plant proteins. JOSLYN and STEPKA (1949) searched for it chromatographically in various fruits and did not find it, although their method was shown to be effective. KOLOUŠEK and COULSON (1952) detected it chromatographically in hay.

Surinamine (*N*-methyl-L-tyrosine) occurs in the bark of various leguminous trees. All the data on its natural occurrence seem to be before 1910 [see BEILSTEIN (Hauptwerk) 14, 612].

Sarcosine was recognised by LINKO, ALFTHAN, MIETTINEN and VIRTANEN (1953) among the free amino acids of reindeer moss *(Cladonia silvatica)*.

γ-*Methyleneglutamic acid*, γ-*methyleneglutamine and* γ-*amino-*α-*methylenebutyric acid* have been found in the tissue juices of groundnut plants *(Arachis hypogaea)* by DONE and FOWDEN (1952) and FOWDEN and DONE (1953). In the second paper they mention reports of similar compounds occurring in liliaceous plants.

L-*Pipecolinic acid* (piperidine-2-carboxylic acid, the first homologue of proline) was isolated from an ethanolic extract of leaves of *Trifolium repens* by MORRISON (1951, 1953) and he presented evidence for its widespread occurrence in other plants. ZACHARIUS, THOMPSON and STEWARD (1952) isolated the same substance from beans, and gave evidence for its occurrence in other plants. Further evidence of such occurrence, with notes on its chromatographic behaviour, has been given by HARRIS and POLLOCK (1952). HULME and ARTHINGTON (1952) isolated it (and another so far unidentified compound) from apples. See also KOLOUŠEK (1953); PHILLIPS (1953).

Baikiaine was isolated from extracts of the heartwood of the Rhodesian leguminous tree *Baikiaea plurijuga* and shown to be L-1:2:3:6-tetrahydro-pyridine-2-carboxylic acid by KING, KING and WARWICK (1950). This is a simple dehydrogenation product of L-pipecolinic acid. See also KING and KING (1953); CARRUTHERS and FARMER (1953).

Hydroxyproline. JOSLYN and STEPKA (1949) produced chromatographic evidence for its occurrence in prunes, and GIRI, GOPALKRISHNAN, RADHAKRISHNAN and VAIDYANATHAN (1952) claimed that 1.7—3.0% was present in dried leaves of *Santalum album* (sandal). They used paper chromatography for identification and determined the amount present by a specific colorimetric procedure. In later work (RADHAKRISHNAN and GIRI, 1954) the material was isolated and found to be *allo*hydroxy-L-proline. See also AUCLAIR and JAMIESON (1948); IZZAT et al. (1953).

Homoserine (γ-hydroxy-α-aminobutyric acid). VIRTANEN and MIETTINEN (1953) refer to the isolation of this amino acid from *Pisum sativum* by MIETTINEN *et al.* (1953). See also VIRTANEN, BERG and KARI (1953); BERG, KARI, ALFTHAN and VIRTANEN (1954).

Mimosine. A substance giving an intense colour reaction with $FeCl_3$ and crystallizing from cut shoots of *Mimosa pudica* and *Leucaena glauca* has been known since 1890. RENZ (1936) isolated on a larger scale and characterized the material in more detail, giving references to the earlier literature. Further study led various workers independently to conclude that preparations from *Mimosa* and *Leucaena* are identical and have the formula

$$O=C \begin{matrix} \overset{H}{\overset{O}{\underset{}{|}}} \\ C=CH \\ \diagdown \\ \diagup \\ C=CH \\ \underset{H}{|} \end{matrix} N-CH_2-CHNH_2-COOH$$

(literature reviewed by KLEIPOOL and WIBAUT, 1950; see also earlier papers in this series and KOSTERMANS, 1946a, b, 1947; BICKEL, 1944, 1948 and earlier papers in series; MASCRÉ and OTTENWAELDER, 1941; ROBINSON, 1951; WIBAUT and SCHUHMACHER, 1952). BONTING and SCHEPMAN (1950) studied the polarographic behaviour of mimosine. MATSUMOTO and SHERMAN (1951) described a colorimetric method for determining mimosine. MATSUMOTO, SMITH and SHERMAN (1951) described the effect of various heat treatments on the mimosine content and toxicity of *Leucaena glauca*.

Selenium-containing amino acids. HORN and JONES (1941) isolated from *Astragalus pectinatus* grown on Se-rich soil crystalline fractions which they considered to be mixed crystals of cystathionine with its Se homologue. Synthetic studies related to this problem have been undertaken by PAINTER (1947) and KLOSTERMAN and PAINTER (1947). See also STEKOL (1942) and KLUG and PETERSEN (1949). Although there is considerable evidence for cystathionine being an intermediate of amino acid metabolism, it does not seem yet to have been isolated from plant material by other workers.

Djenkolic acid ($HOOC \cdot CHNH_2 \cdot CH_2 \cdot S \cdot CH_2 \cdot CHNH_2 \cdot COOH$) was isolated by VAN VEEN and HYMAN (1935) from djenkol beans *(Pithecolobium lobatum)* after low-temperature alkali treatment and from the urine of persons poisoned by these beans. The structure suggested by them was confirmed synthetically by DU VIGNEAUD and PATTERSON (1936) and ARMSTRONG and DU VIGNEAUD (1947). Djenkolic acid was shown by VAN VEEN and LATUASAN (1949) to occur free in the plant (see also VAN VEEN and HYMAN, 1933; HYMAN and VAN VEEN, 1936; VAN VEEN, 1938).

S-Methylated methionine has been isolated from cabbage juice by MACRORIE et al. (1954).

Alliine ($HOOC \cdot CHNH_2 \cdot CH_2 \cdot SO \cdot CH_2 \cdot CH{:}CH_2$) occurs free in garlic, and is responsible for the characteristic odoriferous substances of garlic, which are liberated from it by enzyme action. STOLL and SEEBECK (1951) have reviewed the subject in detail (cf. STOLL and JUCKER pp. 696,698). See also GORYACHENKOVA (1952); VILKKI (1954).

Substituted glutamic acid. ROBERTS and WOOD (1951) obtained chromatographic evidence for the abundant occurrence in tea leaf and stem of a supposed mono-ester of glutamic acid. It is not clear if this is the same as "theanine", isolated by SAKATO (1950) in a yield of approx. 1% from tea and suggested by him to be the γ-ethylamide of L-glutamic acid. This was confirmed by synthesis (SAKATO, HASHIZUME and KISHIMOTO, 1950). See also HASHIZUME (1951—1952). STARK, GOODBAN and OWENS (1950) reported the presence of a methylene derivative of pyrrolidonecarboxylic acid in beet sugar diffusion juice, which was probably an artefact due to added formaldehyde.

α-Aminoadipic acid has been found free in a number of plants by BERG et al. (1954).

Ornithine. Although this, from metabolic evidence, is certainly an important intermediate in plant metabolism, reports of its presence in plants are few (e. g.

PHILPOTT, 1948; VAVRUCH, 1952; VIRTANEN and MIETTINEN, 1952; MIETTINEN and VIRTANEN, 1952) and are mainly based on evidence from two-dimensional filter-paper chromatograms. These should be accepted with caution where electrolytic desalting has been used for preparing the samples, as this can cause ornithine to be formed from arginine (STEIN and MOORE, 1951). Ornithine has been found in large quantity in manioc flour as usually processed for human consumption; it is considered to arise from arginine (probably free) by the action of arginase during the processing (CLOSE, ADRIAENS, MOORE and BIGWOOD, 1953).

Citrulline (δ-carbamyl-ornithine) was isolated from press-juice of water melons by WADA (1930) and has since been generally recognized as an important intermediate in the metabolism of urea. There have been various reports on its possible occurrence in plant juices based on two-dimensional filter paper chromatograms, on which it is liable to be confused with histidine and methionine sulphoxide. The amounts noted have usually been small. However, VIRTANEN and MIETTINEN (1952) and MIETTINEN and VIRTANEN (1952) found citrulline to be the most abundant free amino acid in roots, root-nodules and other parts of *Alnus* spp. and isolated a specimen as the Cu derivative. ARCHIBALD (1944) described a procedure for determining citrulline.

Canavanine [HOOC · CHNH$_2$ · CH$_2$ · CH$_2$ · O · NH · C(NH) · NH$_2$] was isolated from jack beans *(Canavalia ensiformis)* by KITAGAWA and colleagues whose structural studies led to final establishment of the structure through synthesis (KITAGAWA and TOMITA, 1929; KITAGAWA and TOMIYAMA, 1929; KITAGAWA and YAMADA, 1932a, b; KITAGAWA and MONOBE, 1933a, b; KITAGAWA, SAWADA and HOSOKI, 1935; KITAGAWA and TAKANI, 1935, 1936; KITAGAWA, 1936a, b, 1937; KITAGAWA and WADA, 1935; KITAGAWA and TSUKAMOTO, 1937; KITAGAWA and EGUCHI, 1938). See also GULLAND and MORRIS (1935); TOMIYAMA (1935); BOREK and CLARKE (1938). Canavanine was incorrectly stated in reviews by DUNN (1938) and by MARTIN and SYNGE (1945) to have been obtained from soya beans. DAMODARAN and NARAYANAN (1939) isolated canavanine from *Canavalia obtusifolia*. CADDEN (1940) described a simplified isolation from jack-bean meal and gave data on the optical rotation in acid and alkali. MÜLLER (1941) described precipitation of canavanine with nitranilic acid. ARCHIBALD and HAMILTON (1943) described the removal of canavanine from preparations of jack bean urease. ARCHIBALD (1946) described a colorimetric method for the terdemination of canavanine and FEARON (1946) a colour reaction (cf. TRACEY pp. 126, 131).

A guanidino-substituted form of arginine *("arginosuccinic acid")* was isolated by WALKER (1952) on incubating dried *Chlorella* cells with fumarate or malate. This may be a normal intermediate of metabolism accumulating under un-physiological conditions. See also WALKER and MYERS (1953); WALKER (1953).

II. Determination of Free Amino Acids: General Methods.
1. Aggregate Amino Acids.

Of the methods for determining the aggregate of free amino and imino acids, none can be expected to be completely satisfactory with crude extracts of plant material. The various acid- and base-titrimetric procedures will respond to carboxyl groups of plant acids and other compounds and to phenolic groups of glycosides etc. The formol titration will be fairly specific for amino groups, but will not distinguish between amino acids and other amines, and the same is true for the VAN SLYKE amino-N determination with nitrous acid, though here there are differences in the reaction times for amino groups α- to carboxyl and other amino groups. There is also interference in this reaction by the amide group of glutamine, and glycine and glutathione give anomalous results. The nitrous acid procedure gives negative results with proline and other secondary amines (Peters and Van Slyke, 1932; cf. Chibnall and Westall, 1932).

Methods based on complex formation with copper such as that of Pope and Stevens (1939) have their use for rough quantitative work, as the formation of Cu complexes is fairly specific for free amino acids (e. g. Barton-Wright, 1949). Uncertainty as to the behaviour in such conditions of the peptides and bound amino acids is a complication (cf. Kerkkonen, 1948), but these seem usually to be present in low concentration compared with free amino acids. Interference by other types of compound forming complexes with Cu and interference by other compounds sensitive to iodine, if iodimetry is used for detecting the dissolved Cu, should be borne in mind. γ-Aminobutyric acid does not form a complex with Cu but β-alanine does so to some extent.

The very high specificity of the ninhydrin-CO_2 procedure (Van Slyke, Dillon, MacFadyen and Hamilton, 1941; Van Slyke, MacFadyen and Hamilton, 1941) makes it particularly attractive for aggregate determination of free amino acids. In spite of this, there has been an unaccountable reluctance to apply it in biochemical studies generally and in plant analysis particularly. It has the inherent disadvantage of giving 2 moles of CO_2 with aspartic acid, but otherwise measures free α-amino acids with reasonable accuracy (glutathione included). The method has been used by Neuberger and Sanger (1942) with potato extracts (these workers also used it for determining glutamine — see below) by Synge (1951b) with grass juice (some trouble with blanks was reported) and by Macpherson (1952) with silage. The somewhat similar reaction with chloramine-T has been studied (with amino acids at least partially purified) by Kemble and Macpherson (1954a, b, c). This can be carried out in Warburg manometers at lower temperatures than are practicable with the ninhydrin-CO_2 reaction.

γ-Aminobutyric acid is not determined by these procedures, but β-alanine reacts partially.

The colour reaction of amino acids with ninhydrin (Moore and Stein, 1948) has been used occasionally for rough colorimetric determination of amino acids in plant extracts, but has the disadvantage that the different amino acids give differing amounts of colour. Neither is it specific for amino acids, being given by amines generally.

2. Individual Amino Acids.

All the procedures discussed in the previous section must be regarded as labour-saving but inferior substitutes for a full chromatographic analysis of the amino acids present. No fully satisfactory procedure with plant extracts has yet been described, but the application of the methods recently developed by S. Moore and W. H. Stein to urine, which presents a problem not dissimilar, by Stein (1953) is surely a landmark in the study of amino acids in biological fluids. The procedure described gave amino acid zones in reproducible positions on chromatography of unfractionated urine. Quantitative determination of the amino acids before and after acid hydrolysis of the urine likewise gave a quantitative measure of the bound amino acids present. None of the zones attributed to free amino acids decreased appreciably on hydrolysis (excepting of course those of glutamine and asparagine, which were hydrolysed).

There is clearly scope for simplification of the long procedure of collecting fractions and separately performing a colorimetric determination on each. Kemble and Macpherson (1954a, b, c) describe a method for eluting amino acid zones from paper and subsequent quantitative determination, using mixtures previously simplified by electrical transport procedures (cf. SYNGE, 1951b and section on electrical transport methods, Vol. I p. 55).

Most other attempts to perform quantitative estimations of amino acids separated on paper chromatograms have been at best less accurate but paper chromatography is very useful for qualitative and rough quantitative work. Numerous minor modifications of the two-dimensional procedure of CONSDEN, GORDON and MARTIN (1944) have been proposed. This was first used effectively with plant material by DENT, STEPKA and STEWARD (1947) and early results with the method have been reviewed by STEWARD and THOMPSON (1950). The method has been useful to give a general picture of the free amino acids of a wide variety of plant materials, and has been instrumental in revealing a number of the unusual amino acids already referred to. Several examples of its use are also referred to in the section below on results of analyses for free amino acids.

The methods for determining amino acids as their N-2:4-dinitrophenyl derivatives are of great potential value for rapid chromatographic determination of free amino acids, as the substitution with 2:4-dinitro-fluorobenzene proceeds under mild conditions ($NaHCO_3$ in aqueous ethanol) and the substituted amino acids are then readily extracted into ether before chromatography. The resulting zones, being coloured, are easily separated and assayed colorimetrically [SANGER, 1945; PORTER and SANGER, 1948; BLACKBURN, 1949; MIDDLEBROOK, 1951; CALLOW and WORK, 1952; KROL, 1952; PERRONE, 1951; KNESSL, KEIL, MALÝ and SORM, 1951; BLACKBURN and LOWTHER, 1951; MILLS, 1952; GREEN and KAY, 1952 (includes some additional references)].

Where ultramicro quantities of amino acid have to be determined, or where a detailed comparison of natural with synthetic material is desired, the methods involving chromatography of p-iodobenzenesulphonyl derivatives differentially labelled with ^{131}I and ^{35}S can be of value (KESTON, UDENFRIEND and LEVY, 1950; UDENFRIEND, 1950).

It is probably unwise to use microbiological assay with unhydrolysed plant extracts, since bound amino acids can influence the behaviour of the micro-organism used and it is difficult to control for the presence or absence of other stimulating or inhibitory materials. Results on acid hydrolysates of plant extracts can be regarded as more reliable, though not as satisfactory as results with hydrolysates of purified proteins.

If it is desired not merely to determine free amino acids but to isolate them from plant material on a larger scale (e. g. for degradative chemical study of the positions of labelling by isotopes or for use in further studies of biosynthetically labelled amino acids) displacement chromatography on ion-exchange resins can be of great value. A series of methods, developed by S. M. PARTRIDGE and colleagues, has been so applied as to give good yields of nearly all the common amino acids (see PARTRIDGE and BRIMLEY, 1952). WESTALL (1950a, b) described the use of such methods for a fairly large-scale isolation of glutamine, γ-amino-butyric acid and betaine from an extract of red beetroots (see also MORRISON, 1953). It is also, of course, possible for isolative purposes to increase the scale of any of the chromatographic methods referred to above. The liquid-liquid chromatographic methods also have their counterpart in countercurrent distribution methods, which can be applied on a fairly large scale if suitable apparatus is available (e. g. CRAIG, HAUSMANN and WEISIGER, 1952; cf. HECKER Vol. I, p. 66).

III. Special Methods for Free Amino Acids.

Only a few of the various special methods that have been proposed for determinations of individual amino acids in protein hydrolysates, etc. can be recommended for use with unfractionated plant extracts. Thus, many of the methods proposed for tyrosine are general reactions for phenols; for cystine, general reactions for potential thiol groups. Among methods that can be recommended, it will be noticed that those depending on measuring reaction products after allowing a specific enzyme to act on the extract have especial prominence, and it seems likely that such methods will be developed in increasing numbers during the coming years. All the methods described in this section are likely to be of most use for doing series of determinations of an individual amino acid; if several amino acids are to be determined on the same material, the chromatographic methods already referred to will prove more useful. It should further be noted that many of the methods described below give particular atoms of the amino acid to be determined in the form of simple degradation products which can be of use in metabolic studies using isotopic labelling.

1. Use of Aldehydes Liberated in Reaction with Ninhydrin.

Of the common amino acids, only glycine gives formaldehyde in this reaction. Alexander, Landwehr and Seligman (1945; cf. MacFadyen, 1945) used a specific colorimetric method for determining formaldehyde in distillates from the reaction mixtures. The procedure was criticised by Krueger (1949) on the grounds of interference by other amino acids, but Christensen, Riggs and Ray (1951) have shown that with proper precautions such interference is not serious in ordinary tissues, referring to previous work by H. N. Christensen and colleagues with tissue extracts where the method gave valid results. See also Michel, Michel and Bozzi-Tichadou (1947).

Virtanen and Rautanen (1946, 1947; cf. Roine and Rautanen, 1947) determined the aggregate of volatile aldehydes arising from alanine, valine, leucine, isoleucine, phenylalanine and methionine. The gas-liquid chromatographic procedures since developed (James and Martin, 1954) seem now to offer the possibility of rapidly determining these aldehydes individually.

2. Estimation of Serine and Threonine with Periodate.

Rees (1946) improved the available methods and reviewed the earlier literature on these procedures. If hydroxylysine, hexosamines etc. are absent, the ammonia evolved on treatment with periodate should be a measure of (serine + threonine) present, although seryl and threonyl peptides (in which the amino group of these residues is free) will also react. As sugars will generally be present in the mixtures analysed, formaldehyde cannot be used to estimate serine; the yield of acetaldehyde should, however, give a measure of threonine present, provided that free methylpentoses and similar compounds are absent. Diminution of the yield of CO_2 with ninhydrin or chloramine T after treatment with periodate (Kemble and Macpherson, 1954) can serve as a useful confirmation that free α-amino acids are being determined.

3. Enzymic Methods.

Arginine can conveniently be determined by conversion to urea with liver arginase, the urea being determined with jackbean urease (see Urea, p. 119) (Hunter and Dauphinee, 1930; Hunter and Pettigrew, 1937). The arginase reacts with free arginine or arginine in peptide linkage having a free carboxyl group and is otherwise very specific. It does, however, form urea from canavanine, and this must be removed from the urease preparation (Archibald and Hamilton, 1943).

GALE (1945, 1946b) described procedures for the specific determination of *lysine, ornithine, tyrosine, histidine, glutamic acid* and *arginine* using bacterial decarboxylases. The procedure for glutamic acid was adapted by KREBS (1948; cf. HUGHES, 1949; KREBS, BOULANGER and OSTEUX, 1950; FOSS, 1953), making use of a specific glutaminase also present in the bacteria, for parallel determination of *glutamic acid* and of *glutamine*. This glutaminase seems to be more specific than that previously available (ARCHIBALD, 1945). Later KREBS (1950) was able to use the same enzyme system in the presence of α-oxoglutarate, transaminase and (for asparagine) asparaginase to determine *aspartic acid* and *asparagine* and reported data on both dicarboxylic amino acids and their amides in peas. MARDASHEV and GLADKOVA (1948) described a decarboxylase specific for *aspartic acid* (see also OKUNUKI, SAKURAI and MATSUDA, 1952; ASTRUP and MUNKVAD, 1950; MEISTER, SOBER and TICE, 1951). HASSE and SCHUMACHER (1952) used decarboxylase from *Raphanus sativus* for determining *glutamic acid*.

4. Glutamic and Aspartic Acids.

Since these are the only common amino acids possessing a net negative charge at pH 6—7 they may readily be isolated by electrical transport methods (see Vol.1, also THEORELL and ÅKESON, 1942; SYNGE, 1951b) or by chromatography on anion-exchange columns (see WIELAND and WIRTH, 1943; CONSDEN, GORDON and MARTIN, 1948; DRAKE, 1947; KRETOVICH and BUNDEL, 1948, 1950). The latter group of methods permits simultaneous separation of aspartic and glutamic acids on account of their different ionization constants. The ninhydrin-CO_2 procedure is probably the best method for determining the isolated materials, since plant acids (other than ascorbic acid), acidic peptides and acyl amino acids will not interfere. However, interference by glutathione, which yields CO_2 with ninhydrin, must be allowed for. Since aspartic acid yields two moles of CO_2, it may sometimes be justifiable to determine N and ninhydrin-CO_2 on such acidic fractions and calculate the amounts of glutamic and aspartic acids present.

5. Glutamine and Asparagine.

Total NH_3 liberated on mild acid hydrolysis can serve as a measure of these amino acids (see VICKERY, PUCHER, CLARK, CHIBNALL and WESTALL, 1935) for rough work, but should not be relied on since it will be liberated also from urea and bound amino acid fractions (e. g. SYNGE, 1951b). ARCHIBALD (1945) reviewed methods for determining glutamine available at that time. Since then, the enzymic methods mentioned above have been most widely used, although SORGATO and SORENZO (1950) followed earlier workers and extracted the pyrrolidonecarboxylic acid formed by heating plant tissue extracts. Asparagine was usually determined by difference between glutamine and "total amide", although this was known to be unreliable, until the enzymic methods become available. BUTLER (1951) subjected grass juice to preliminary chromatography on paper with phenol-water and then determined amide N set free on acid hydrolysis of the glutamine and asparagine zones.

IV. Some Results of Analyses for Free Amino Acids in Plant Materials.

In spite of the great interest in free amino acids of plant tissues and the rapid development and improvement of methods in recent years, there are very few quantitative general analyses yet available, and such as have been published are not expressed on comparable bases and are on extracts prepared by a variety of different procedures. However, it is already clear that in general there is a predominance of the amino acids not nutritionally essential for most animals. These are known to be closely linked with the metabolism of carbohydrate. They include glycine, serine, alanine, aspartic acid, glutamic acid, asparagine, glutamine and γ-aminobutyric acid, which usually represent a far higher proportion of the total free amino acids than do the corresponding residues of the total amino acid residues in ordinary proteins. Nevertheless, there are great variations in the relative proportions of the free amino acids both between species, between different parts of the same plant and with the physiological state of the plant.

Below are given some references which may be useful as an orientation. See also STEWARD and THOMPSON (1950) and reviews referred to by them.

Potato tubers. NEUBERGER and SANGER, 1942; DENT, STEPKA and STEWARD, 1947; CHICK and SLACK, 1949; PAYNE, FULTS and HAY, 1951.

Root nodules of leguminous plants. HUNT, 1951; ZELITCH, WILSON and BURRIS, 1952.

Root nodules of alder. MIETTINEN and VIRTANEN, 1952; VIRTANEN and MIETTINEN, 1953.

Grasses. RABIDEAU and EDWARDS, 1951; SYNGE, 1951b; BATHURST, 1953; KEMBLE and MACPHERSON, 1954c; PRATT and WIGGINS, 1949 (sugar cane juice).

Meristematic tissue of a variety of plants. ALLSOPP, 1948.

Tobacco leaves. PEARSE and NOVELLIE, 1953; COMMONER and NEHARI, 1953.

Lucerne (alfalfa) leaves. STEWARD et al., 1951; VICKERY, 1924a, b, 1925a, b; VICKERY and LEAVENWORTH, 1925; VICKERY and VINSON, 1925.

Pea plant extracts. AUCLAIR and MALTAIS, 1952; CHRISTIANSEN and THIMANN, 1950; RAUTANEN, 1948; KREBS, 1950.

Clover leaves. MORRISON, 1953.

Cereal seeds. SULLIVAN and PAYNE, 1951; BISERTE and SCRIBAN, 1950 (germinating barley); DE VAY, 1952.

Beet-sugar molasses. SCHNEIDER et al., 1952.

Fruits. JOSLYN and STEPKA, 1949; HULME and ARTHINGTON, 1950, 1952; CASIMIR, JAKOVLIV and JAKOVLIV, 1952; VALAISE and DUPONT, 1951.

Freshwater algae. FOWDEN, 1951.

Marine algae. ERICSON and SJÖSTRÖM, 1952; CHANNING and YOUNG, 1953.

Yeast. LINDAN and WORK, 1951; MIETTINEN, 1951; REINDEL and HOPPE, 1952.

Ergot. KAWATANI, KATAYAGANI and KIYOOKA, 1951.

References.

ABRAMSON, H. A., D. H. MOORE and H. H. GETTNER: Proc. Soc. Exp. Biol. Med., N. Y. 46, 153 (1941); J. Phys. Chem. 46, 192 (1942). — ALEXANDER, B., G. LANDWEHR and A. M. SELIGMAN: J. Biol. Chem. 160, 51 (1945). — ALLSOPP, A.: Nature (Lond.) 161, 833 (1948). — ARCHIBALD, R. M.: J. Biol. Chem. 156, 121 (1944); Chem. Revs. 37, 161 (1945); J. Biol. Chem. 165, 169 (1946). — ARCHIBALD, R. M., and P. B. HAMILTON: J. Biol. Chem. 150, 155 (1943). — ARMSTRONG, M. D., and V. DU VIGNEAUD: J. Biol. Chem. 168, 373 (1947). — ARNSTEIN, H. R. V., and P. T. GRANT: Biochem. J. 57, 353, 360 (1954). — ASTRUP, P., and I. MUNKVAD: Scand. J. Clin. Lab. Invest. 2, 133 (1950). — ASTRUP, T., and V. ØHLENSCHLÄGER: Biochem. J. 42, 211 (1948). — AUCLAIR, J. L., and C. A. JAMIESON: Science (Lancaster, Pa). 108, 357 (1948). — AUCLAIR, J. L., and J. B. MALTAIS: Nature (Lond.) 170, 1114 (1952).

BALLS, A. K., W. S. HALE and T. H. HARRIS: Cereal Chem. 19, 279, 840 (1942). — BALLS, A. K., and T. H. HARRIS: Cereal Chem. 21, 74 (1944). — BALSTON, J. N., and B. E. TALBOT: Guide to filter paper and cellulose powder chromatography. London and Maidstone: Reeve Angel & Balston 1952. — BARTON-WRIGHT, E. C.: Biochim. Biophys. Acta 3, 679 (1949); The Microbiological Assay of the Vitamin B-Complex and Amino Acids. London: Pitman 1952. — BATHURST, N. O.: J. Sci. Food Agric. 4, 221 (1953). — BERG, A. M., S. KARI, M. ALFT-HAN and A. I. VIRTANEN: Acta Chem. Scand. 8, 358 (1954). — BICKEL, A. F.: Bijdrage tot de Structuur van Leucaenol uit *Leucaena glauca* Bentham. Amsterdam: H. J. Paris. Through Chem. Abs. 38, 2342 (1944). — BICKEL, A. F.: J. Amer. Chem. Soc. 70, 328 (1948). — BINET, L., and G. WELLERS: Compt. rend. Acad. Sci. (Paris) 224, 870 (1947). — BIRKINSHAW, J. H., H. RAISTRICK and G. SMITH: Biochem. J. 36, 829 (1942). — BISERTE, G., and R. SCRIBAN: Bull. Soc. Chim. biol. (Paris) 32, 959 (1950); 33, 114 (1951); 34, 350 (1952). — BISSET, S. K.:

Biochem. J. **58**, 225 (1954). — BLACKBURN, S.: Biochem. J. **45**, 579 (1949). — BLACKBURN, S., and A. G. LOWTHER: Biochem. J. **48**, 126 (1951). — BLOCK, R. J., and D. BOLLING: The amino acid composition of proteins and foods: analytical methods and results. 2nd. edn. Springfield, Ill.: C. C. Thomas 1951. — BOLGUNOV, G. P., and M. T. POKHNO: Biokhimiya, **15**, 67 (1950). — BONTING, S. L. JR., and F. R. SCHEPMAN: Rec. Trav. chim. **69**, 1007 (1950). — BOREK, E., and H. T. CLARKE: J. Biol. Chem. **125**, 479 (1938). — BOULANGER, P., and G. BISERTE: Bull. Soc. chim. Fr. **1952**, 830. — BROWN, R. J.: Industr. Engng. Chem. **43**, 610 (1951). — BUTLER, G. W.: Anal. Chem. **23**, 1300 (1951).

CADDEN, J. F.: Proc. Soc. Exp. Biol. Med. (New York) **45**, 224 (1940). — CALLOW, R. K., and T. S. WORK: Biochem. J. **51**, 558 (1952). — CARRUTHERS, W. R., and R. H. FARMER: Chem. and Ind. **1953**, 641. — CASIMIR, J., G. JAKOVLIV and G. JAKOVLIV: Inds. agr. et aliment. (Paris) **69**, 787 (1952) (Through Chem. Abs. 47, 3488). — CHANNING, D. M., and G. T. YOUNG: J. Chem. Soc. **1953**, 2481. — CHIBNALL, A. C.: Protein metabolism in the plant. New Haven: Yale University Press 1939. — CHIBNALL, A. C., and R. G. WESTALL: Biochem. J. **26**, 122 (1932). — CHICK, H., and E. B. SLACK: Biochem. J. **45**, 211 (1949). — CHRISTENSEN, H. N., and E. L. LYNCH: (a) J. Biol. Chem. **163**, 741 (1946); (b) J. Biol. Chem. **166**, 87 (1946). — CHRISTENSEN, H. N., T. R. RIGGS and N. E. RAY: Anal. Chem. **23**, 1521 (1951). — CHRISTIANSEN, G. S., and K. V. THIMANN: Arch. Biochem. **28**, 117 (1950). — CLOSE, J., E. L. ADRIAENS, S. MOORE and E. J. BIGWOOD: Bull. Soc. Chim. biol. **35**, 985 (1953). — COLEMAN, I. W.: Can. J. Med. Sci. **29**, 151 (1951). — COMMONER, B., and V. NEHARI: J. Gen. Physiol. **36**, 791 (1953). — CONSDEN, R., A. H. GORDON and A. J. P. MARTIN: Biochem. J. **38**, 224 (1944); Biochem. J. **42**, 443 (1948). — COOK, A. H., S. F. COX and T. H. FARMER: J. Chem. Soc. **1949**, 1022. — CRAIG, L. C., W. HAUSMANN and J. R. WEISIGER: J. Biol. Chem. **199**, 865 (1952).

DAMODARAN, M., and K. G. A. NARAYANAN: Biochem. J. **33**, 1740 (1939). — DAMODARAN, M., and R. RAMASWAMY: Biochem. J. **31**, 2149 (1937). — DANIELSSON, C. E.: Acta Chem. Scand. **5**, 541 (1951); **6**, 149 (1952). — DANKNER, A., A. C. BUCANTZ, M. C. JOHNSON and H. L. ALEXANDER: J. Allergy **22**, 437 (1951); Through Chem. Abs. 47, 10630 (1953). — DEKKER, C. A., D. STONE and J. S. FRUTON: J. Biol. Chem. **181**, 719 (1949). — DENT, C. E., W. STEPKA and F. C. STEWARD: Nature (Lond.) **160**, 682 (1947). — DEVAY, J. E.: Cereal Chem. **29**, 309 (1952). — DONE, J., and L. FOWDEN: Biochem. J. **51**, 451 (1952). — DRAKE, B.: Nature (Lond.) **160**, 602 (1947). — DUNN, G., J. J. GALLAGHER, G. T. NEWBOLD and F. S. SPRING: J. Chem. Soc. **1949**, S. 126. — DUNN, G., G. T. NEWBOLD and F. S. SPRING: J. Chem. Soc. **1949**, S. 131. — DUNN, M. S.: The Chemistry of the Amino Acid and Proteins ed. C. L. A. Schmidt p. 21. Springfield, Ill.: C. C. Thomas 1938; Physiol. Revs. **29**, 219 (1949). — DUTCHER, J. D.: J. Biol. Chem. **171**, 321, 341 (1947). — DUTCHER, J. D., and O. WINTERSTEINER: J. Biol. Chem. **155**, 359 (1944).

ELLFOLK, N., and R. L. M. SYNGE: Biochem. J. **59**, 523 (1955). — ENNOR, A. H.: Austral. J. Exp. Biol. Med. Sci. **17**, 157 (1939). — ERICSON, L. E., and A. G. M. SJÖSTRÖM: Acta Chem. Scand. **6**, 805 (1952).

FEARON, W. R.: Analyst **71**, 562 (1946). — FOGG, G. E.: Proc. Roy. Soc. Lond. B, **139**, 372 (1952). — FOSS, O. P.: Scand. J. Clin. Lab. Invest. **5**, 334 (1953); Through Chem. Abs. 48, 5255 (1954). — FOWDEN, L.: Nature (Lond.) **167**, 1030 (1951). — FOWDEN, L., and J. DONE: Biochem. J. **55**, 548 (1953). — FRIEDMAN, L., and O. L. KLINE: (a) J. Nutr. **40**, 295 (1950); (b) J. Biol. Chem. **184**, 599 (1950). — FROMAGEOT, C.: Adv. Prot. Chem. **8**, 1 (1953). — FROMAGEOT, C., M. JUTISZ and P. TESSIER: Bull. Soc. Chim. Biol. **31**, 689 (1949).

GALE, E. F.: Biochem. J. **39**, 46 (1945); (a) Adv. Enzymol. **6**, 1 (1946); (b) Nature (Lond.) **157**, 265 (1946); Eiweiß-Forschg. **1**, 145 (1948). — GALLAGHER, J. J., G. T. NEWBOLD, W. SHARP and F. S. SPRING: J. Chem. Soc. **1952**, 4870. — GIRI, K. V., K. S. GOPALKRISHNAN, A. N. RADHAKRISHNAN and C. S. VAIDYANATHAN: Nature (Lond.) **170**, 579 (1952). — G RYACHENKOVA, E. V.: Doklady Akad. Nauk SSSR. **87**, 457 (1952); Through Chem. Abs. 47, 4928 (1953). — GOTTSCHALK, A., and S. M. PARTRIDGE: Nature (Lond.) **165**, 684 (1950). — GOVINDARAJAN, V. S., and M. SREENIVASAYA: Current Sci. (India) **19**, 234 (1950). — GRABAR, P., and A. KOUTSEFF: C. r. Soc. Biol. (Paris) **117**, 700 (1934). — GREEN, F. C., and L. M. KAY: Anal. Chem. **24**, 726 (1952). — GULLAND, J. M., and C. J. O. R. MORRIS: J. Chem. Soc. **1935**, 763. — GUTHRIE, J. D.: J. Amer. Chem. Soc. **54**, 2566 (1932).

HAAS, P.: Biochem. J. **46**, 503 (1950). — HAAS, P., and T. G. HILL: Biochem. J. **25**, 1472 (1931); (a) Biochem. J. **27**, 1801 (1933); (b) Ann. Botany **47**, 55 (1933). — HAAS, P., T. G. HILL and B. RUSSELL-WELLS: Biochem. J. **32**, 2129 (1938). — HAIS, I. M., and K. MACEK: Papírová Chromatografie. Praha: Nakládelství Československé Akademie Věd, 1954. — HANES, C. S., F. J. R. HIRD and F. A. ISHERWOOD: Biochem. J. **51**, 25 (1952). — HANNAN, R. S., and C. H. LEA: Nature (Lond.) **168**, 744 (1951). — HARRIS, G., and J. R. A. POLLOCK: Chem. and Ind. **1952**, 931. — HASHIZUME, T.: J. Agr. Chem. Soc. (Japan) **25**, 25, 129, 508 (1951/52); Through Chem. Abs. 47, 6065 (1953). — HASSALL, C. H., K. REYLE and P. FENG: Nature (Lond.) **173**, 356 (1954). — HASSE, K., and H. W. SCHUMACHER: Z. anal. Chem. **137**, 433 (1953). —

Haworth, R. D., R. MacGillivray and D. H. Peacock: Nature (Lond.) 167, 1068 (1951). — Heyns, K., and W. Koch: Z. Naturforschg. 7 B, 486 (1952). — Hopkins, F. G.: J. Biol. Chem. 84, 269 (1929). — Hopkins, F. G., and E. J. Morgan: Nature (Lond.) 152, 288 (1943); Biochem. J. 39, 320 (1945). — Horn, M. J., and D. B. Jones: J. Biol. Chem. 139, 649 (1941). — Horowitz, H. N., M. Ikawa and M. Fling: Arch. Biochem. 25, 226 (1950). — Hughes, D. E.: Biochem. J. 45, 325 (1949). — Hulme, A. C., and W. Arthington: Nature (Lond.) 165, 716 (1950); 170, 659 (1952). — Hunt, G. E.: Amer. J. Botany 38, 452 (1951). — Hunter, A., and J. A. Dauphinee: J. Biol. Chem. 85, 627 (1930). — Hunter, A., and J. B. Pettigrew: Enzymologia 1, 341 (1937). — Hyman, A. J., and A. G. van Veen: Geneesk. Tijdschr. Ned.-Ind. 76, 840 (1936).

Immers, J., and E. Vasseur: Nature (Lond.) 165, 898 (1950). — Izzat, I., F. Penaranda, M. Robert and M. Polonovski: Bull. Soc. Chim. biol. 35, 266 (1953).

James, A. T., and A. J. P. Martin: Biochem. J. 57, v (1954). — Jansen, E. F., and R. Jang: Arch. Biochem. Biophys. 40, 358 (1952). — Johnson, J. L., W. G. Jackson and T. E. Eble: J. Amer. Chem. Soc. 73, 2947 (1951). — Joslyn, M. A., and W. Stepka: Food Res. 14, 459 (1949). — Jukes, T. H., and E. L. R. Stokstad: Physiol. Revs. 28, 51 (1948). — Jutisz, M.: Bull. Soc. chim. fr. 1952, 821.

Kawatani, T., M. Katayanagi and S. Kiyooka: J. Pharm. Soc. Japan 71, 385, 389 (1951); Through Chem. Abs. 45, 8207 (1951). — Kemble, A. R., and H. T. Macpherson: (a) Biochem. J. 56, 548 (1954).; (b) Biochem. J. 58, 44 (1954); (c) Biochem. J. 58, 46 (1954). — Kerkkonen, H. K.: Acta Chem. Scand. 2, 518 (1948). — Keston, A. S., S. Udenfriend and M. Levy: J. Amer. Chem. Soc. 72, 748 (1950). — King, F. E., and T. J. King: Chem. and Ind. 1953, 489. — King, F. E., T. J. King and A. J. Warwick: J. Chem. Soc. 1950, 3590. — Kitagawa, M.: (a) J. Biochem. (Japan) 24, 107 (1936); (b) J. Agric. Chem. Soc. (Japan) 12, 871 (1936); J. Biochem. (Japan) 25, 23 (1937). — Kitagawa, M., and Y. Eguchi: J. Agric. Chem. Soc. Japan 14, 525 (1938); Bull. Agric. Chem. Soc. (Japan) 14, 43 (1938). — Kitagawa, M., and S. Monobe: (a) J. Biochem. (Japan) 18, 333 (1933); (b) J. Agric. Chem. Soc. (Japan) 9, 845 (1933). — Kitagawa, M., K. Sawada and Y. Hosoki: J. Agric. Chem. Soc. (Japan) 11, 539 (1935). — Kitagawa, M., and A. Takani: J. Agric. Chem. Soc. (Japan) 11, 1077 (1935); Bull. Agric. Chem. Soc. (Japan) 11, 170 (1935); J. Biochem. (Japan) 23, 181 (1936). — Kitagawa, M., and T. Tomita: Proc. Imp. Acad. (Japan) 5, 380 (1929). — Kitagawa, M., and T. Tomiyama: J. Biochem. (Japan) 11, 265 (1929). — Kitagawa, M., and J. Tsukamoto: J. Biochem. (Japan) 26, 373 (1937). — Kitagawa, M., and M. Wada: J. Agric. Chem. Soc. (Japan) 11, 1083 (1935); Bull. Agric. Chem. Soc. (Japan) 11, 171 (1935). — Kitagawa, M., and H. Yamada: (a) J. Biochem. (Japan) 16, 339 (1932); (b) J. Agric. Chem. Soc. (Japan) 8, 1201 (1932). — Kleipool, R. J. C., and J. P. Wibaut: Rec. Trav. chim. 69, 37 (1950). — Klosterman, H. J., and E. P. Painter: J. Amer. Chem. Soc. 69, 2009 (1947). — Klug, H. L., and D. F. Petersen: Proc. S. Dakota Acad. Sci. 28, 87 (1949). — Knessl, O., B. Keil, A. Malý and F. Šorm: Coll. Czechoslov. Chem. Communs. 15, 918 (1951). — Koloušek, J.: Chem. Listy 47, 473 (1953); Through Chem. Abs. 48, 3633 (1954). — Koloušek, J., and C. B. Coulson: Chem. Listy 46, 186 (1952); [Through Chem. Abs. 46, 11501 (1952)]. — Kostermans, D.: (a) Rec. Trav. Chim. 65, 319 (1946); (b) Natuurw. Tijdschr. 102, 183 (1946); Rec. Trav. Chim. 66, 93 (1947). — Krebs, H. A.: Biochem. J. 43, 51 (1948); 47, 605 (1950). — Krebs, H. A., P. Boulanger and R. Osteux: Bull. Soc. Chim. biol. 32, 851 (1950). — Kretovich, V. L., and A. A. Bundel: Doklady Akad. Nauk SSSR. 61, 861 (1948); 73, 137 (1950). — Krol, S.: Biochem. J. 52, 227 (1952). — Krueger, R.: Helv. Chim. Acta 32, 238 (1949).

Lawrence, J. M.: Arch. Biochem. 27, 1 (1950). — Lea, C. H.: Chem. and Ind. 1950, 155. — Lea, C. H., and R. S. Hannan: Nature (Lond.) 165, 438 (1950). — Lederer, E., and M. Lederer: Chromatography. Amsterdam: Elsevier 1953. — Lindan, O., and E. Work: Biochem. J. 48, 337 (1951). — Linko, P., M. Alfthan, J. K. Miettinen and A. I. Virtanen: Acta Chem. Scand. 7, 1310 (1953). — Ljungdahl, L., and E. Sandegren: Acta Chem. Scand. 4, 1150 (1950). — Loveless, M. H.: 1st. Int. Congr. Biochem. Abs. Commun. p. 451. Cambridge 1949. — Lynen, F., and U. Wieland: Liebigs Ann. 533, 93 (1938).

MacFadyen, D. A.: J. Biol. Chem. 158, 107 (1945). — Macpherson, H. T.: J. Sci. Food Agric. 3, 362 (1952). — MacRobie, R. A., G. L. Sutherland, M. S. Lewis, A. D. Barton, M. R. Glazener and W. Shive: J. Amer. Chem. Soc. 76, 115 (1954). — Manske, R. H. F.: Canad. J. Res. 15 B, 84 (1937); 24 B, 66 (1946). — Mardashev, S. R.: Uspekhi biol. Khim. 1, 281 (1950). — Mardashev, S. R., and V. N. Gladkova: Biokhimiya 13, 315 (1948). — Markham, R., and J. D. Smith: Nature (Lond.) 164, 1052 (1949). — Marquardt, P., and G. Vogg: Arzneimittel-Forschg. 2, 267 (1952). — Martin, A. J. P., and R. L. M. Synge: Advanc. Protein Chem. 2, 1 (1945). — Mascré, M., and A. Ottenwaelder: Bull. Sci. pharmacol. 48, 65 (1941). — Mashkovtsev, M. F.: Biokhimiya 15, 528 (1950); 16, 615 (1951). — Matsumoto, H., and G. D. Sherman: Arch. Biochem. 33, 195 (1951). — Matsumoto, H., E. G. Smith and G. D. Sherman: Arch. Biochem. 33, 201

(1951). — MEISTER, A., H. A. SOBER and S. V. TICE: J. Biol. Chem. 189, 577 (1951). — MICHEL, R., O. MICHEL and M. BOZZI-TICHADOU: Bull. Soc. chim. Biol. 29, 881 (1947). — MIDDLEBROOK, W. R.: Biochim. Biophys. Acta 7, 547 (1951). — MIETTINEN, J. K.: Acta Chem. Scand. 5, 962 (1951). — MIETTINEN, J. K., S. KARI, T. MOISIO, M. ALFTHAN and A. I. VIRTANEN: Suomen Kemistilehti B, 26, 26 (1953). — MIETTINEN, J. K., and A. I. VIRTANEN: Physiol. Plantarum 5, 540 (1952). — MILLS, G. L.: Biochem. J. 50, 707 (1952). — MINAGAWA, T.: Proc. Imp. Acad. (Tokyo) 21, 33, 37 (1945); Through Chem. Abs. 47, 151 (1953). — MITCHELL, H. H., and T. S. HAMILTON: The Biochemistry of the Amino Acids. New York: Chemical Catalog Co. 1929. — MOORE, S., and W. H. STEIN: J. Biol. Chem. 176, 367 (1948). — MORRISON, R. I.: Biochem. J. 50, xiv (1951).; 53, 474 (1953). — MÜLLER, E.: Hoppe-Seylers Z. 268, 245 (1941).

NEUBERGER, A., and F. SANGER: Biochem. J. 36, 662 (1942). — NEWBOLD, G. T., W. SHARP and F. S. SPRING: (a) Chem. and Ind. 1951, 651; (b) J. Chem. Soc. 1951, 2679. — NEWELL, J. M.: J. Allergy 13, 177 (1942). — NIERENSTEIN, M.: Z. physiol. Chem. 92, 53 (1914); Biochem. J. 9, 240 (1915).

OHIRA, T.: J. Agric. Chem. Soc. (Japan) 15, 370 (1939); Bull. Agric. Chem. Soc. (Japan) 15, 78 (1939); (a) J. Agric. Chem. Soc. (Japan) 16, 1 (1940); Bull. Agric. Chem. Soc. (Japan) 16, 10 (1940); (b) J. Agric. Chem. Soc. (Japan) 16, 293 (1940); Bull. Agric. Chem. Soc. (Japan) 16, 71 (1940); J. Agric. Chem. Soc. (Japan) 18, 915 (1942); Bull. Agric. Chem. Soc. (Japan) 18, 74 (1942). — OKUNUKI, K., S. SAKURAI and G. MATSUDA: Proc. Japan Acad. 28, 347 (1952); Through Chem. Abs. 47, 3357 (1953). — OSBORNE, T. B., A. J. WAKEMAN and C. S. LEAVENWORTH: J. Biol. Chem. 49, 63 (1921); 53, 411 (1922).

PAINTER, E. P.: J. Amer. Chem. Soc. 69, 229, 232 (1947). — PARTRIDGE, S. M., and R. C. BRIMLEY: Biochem. J. 51, 628 (1952). — PASKHINA, T. S.: Uspekhi biol. Khim. 1, 307 (1950). — PAYNE, M. G., J. L. FULTS and R. J. HAY: Science (Lancaster, Pa.) 114, 204 (1951). — PEARSE, H. L., and L. NOVELLIE: J. Sci. Food Agric. 4, 108 (1953). — PERLMAN, E.: Bull. N. Y. Acad. Med. 27, 586 (1951). — PERRONE, J. C.: Nature (Lond.) 167, 513 (1951). — PETERS, J. P., and D. D. VAN SLYKE: Quantitative Clinical Chemistry Vol. 2, p. 385. London: Baillière, Tindall & Cox. 1932. — PHILLIPS, D. M.: Chem. and Ind. 1953, 127. — PHILPOTT, M. W.: Report of the Rubber Research Institute of Malaya (1945/48) 1948, 191; Through Chem. Abs. 45, 9291 (1951). — PIRIE, N. W.: Biochem. J. 24, 51 (1930). — PLATTNER, PL. A., N. CLAUSON-KAAS, A. BOLLER and U. NAGER: Helv. Chim. Acta 31, 860 (1948). — PLATTNER, PL. A., Hs. H. GÜNTHARD and A. BOLLER: Helv. Chim. Acta 35, 999 (1952). — PLATTNER, PL. A., and U. NAGER: (a) Helv. Chim. Acta 31, 665 (1948); (b) Helv. Chim. Acta 31, 2192 (1948); (c) Helv. Chim. Acta 31, 2203 (1948). — PLATTNER, PL. A., U. NAGER and A. BOLLER: Helv. Chim. Acta 31, 594 (1948). — POPE, C. G., and M. F. STEVENS: Biochem. J. 33, 1070 (1939). — PORTER, R. R., and F. SANGER: Biochem. J. 42, 287 (1948). — PRATT, O. E., and L. F. WIGGINS: Proc. Brit. W. Indies Sugar Technol. 1949, 29.

RABIDEAU, G. S., and M. B. EDWARDS:Plant Physiol. 26, 798 (1951). — RADHAKRISHNAN, A. N., and K. V. GIRI: Biochem. J. 58, 57 (1954). — RATNER, S., M. BLANCHARD and D. E. GREEN: J. Biol. Chem. 164, 691 (1946). — RAUTANEN, N.: Acta Chem. Scand. 2, 127 (1948). — REED, L. J.: J. Biol. Chem. 183, 451 (1950). — REES, M. W.: Biochem. J. 40, 632 (1946). — REEVES, W. A., and J. D. GUTHRIE: Arch. Biochem. 26, 316 (1950). — REINDEL, F., and W. HOPPE: Chem. Ber. 85, 716 (1952). — RENZ, J.: Hoppe-Seylers Z. 244, 153 (1936). — ROBERTS, E. A. H., and D. J. WOOD: Current Sci. (India) 20, 151 (1951). — ROBINSON, R.: Proc. Roy. Soc. London B 138, 1 (1951). — ROCHE, J., and M. LAFON: Compt. r. Acad Sci. (Paris) 229, 481 (1949). — ROCKWELL, G. E.: (a) Ohio State Med. J. 39, 128 (1943); (b) Ann. Allergy 1, 43 (1943). — ROINE, P., and N. RAUTANEN: Acta Chem. Scand. 1, 854 (1947). — RONA, P., and E. STRAUSS: Handbuch der biologischen Arbeitsmethoden ed. E. ABDERHALDEN I. 8, p. 715. Berlin und Wien: Urban & Schwarzenberg 1922. — RYDON, H. N., and P. W. G. SMITH: Nature (Lond.) 169, 922 (1952).

SAKATO, Y.: J. Agr. Chem. Soc. (Japan) 23, 262 (1950); Through Chem. Abs. 45, 3528 (1951). — SAKATO, Y., T. HASHIZUME and Y. KISHIMOTO: J. Agr. Chem. Soc. (Japan) 23, 269 (1950); Through Chem. Abs. 45, 3528 (1951). — SANGER, F.: Biochem. J. 39, 507 (1945); Advanc. Protein Chem. 7, 1 (1952). — SARKAR, S. N.: Ann. Biochem. Exp. Med. 5, 39 (1945). — SBOROV, A. M., L. PETERS and L. E. ARNOW: Proc. Soc. Exp. Biol. Med. (New York) 49, 698 (1942). — SCHLOSS, O. M.: Amer. J. Diseases Children 3, 341 (1912). — SCHNEIDER, F., E. REINEFELD, H. GROVE, H. MÜLLER and B. ZENKER: Zucker-Beihefte, No. 6, 79 (1952). — SCHROEDER, E. F., V. COLLIER JR. and G. E. WOODWARD: Biochem. J. 33, 1180 (1939). — SCHROEDER, E. F., and G. E. WOODWARD: J. Biol. Chem. 129, 283 (1939). — SEALOCK, R. R.: Biochem. Preps. 1, 25 (1949). — SHAMSHIKOVA, G. A., and A. L. IOFFE: Biokhimiya 12, 437 (1947). — SIDERIS, C. P., H. Y. YOUNG and B. H. KRAUSS: J. Biol. Chem. 126, 233 (1938). — SISAKYAN, N. M., E. N. BEZINGER and E. B. KUVAEVA: Biokhimiya 16, 358 (1951). — SNELL, E. E.: Advanc. Protein Chem. 2, 85 (1945). — SORGATO, I., and B. SORENZO: Ind. saccar. ital. 48, 330 (1950); Through Chem. Abs. 45, 3901 (1951). — ŠORM, F., and B. KEIL: Coll.

Czechoslov. Chem. Communs. **16**, 366 (1951). — Spies, J. R., E. J. Coulson, D. C. Chambers, H. S. Bernton, H. Stevens and J. H. Shimp: J. Amer. Chem. Soc. **73**, 3995 (1951). — Stark, J. B., A. E. Goodban and H. S. Owens: Proc. Amer. Soc. Sugar Beet Technol. **6**, 578 (1950). — Stein, W. H.: J. Biol. Chem. **201**, 45 (1953). — Stein, W. H., and S. Moore: J. Biol. Chem. **190**, 103 (1951). — Stekol, J. A.: J. Amer. chem. Soc. **64**, 1742 (1942). — Steward, F. C., and J. F. Thompson: Ann. Rev. Plant Physiol. **1**, 233 (1950). — Steward, F. C., J. F. Thompson, F. K. Millar, M. D. Thomas and R. H. Hendricks: Plant Physiol. **26**, 123 (1951). — Stoll, A., A. Hofmann and Th. Petrzilka: Helv. Chim. Acta **34**, 1544 (1951). — Stoll, A., and Th. Petrzilka: Helv. chim. Acta **35**, 589 (1952). — Stoll, A., and E. Seebeck: Adv. Enzymol. **11**, 377 (1951). — Stone, G. C. H., J. Harkavy and G. Brooks: Ann. Allergy **5**, 546 (1947). — Street, H. E.: New Phytologist **48**, 84 (1949). — Sullivan, B., and M. Howe: J. Amer. chem. Soc. **59**, 2742 (1937). — Sullivan, B., and W. E. Payne: Cereal Chem. **28**, 340 (1951). — Synge, R. L. M.: Quart. Revs. Chem. Soc. **3**, 245 (1949); (a) Biochem. J. **48**, 429 (1951); (b) Biochem. J. **49**, 642 (1951); (a) J. gen. Microbiol. **9**, 407 (1953); (b) The chemical structure of proteins, ed. G. E. W. Wolstenholme and M. P. Cameron, p. 43. London: Churchill, 1953. — Synge, R. L. M., and J. C. Wood: Biochem. J. **56**, xix (1954).

Täufel, K., and H. Iwainsky: Biochem. Z. **323**, 299 (1952). — Tauböck, K., and A. Winterstein: Handbuch der Pflanzenanalyse. ed. G. Klein Bd. **3**, 190. Wien: Springer 1931/33. — Tazawa, Y.: Acta phytochim. (Japan) **15**, 141 (1949). — Theorell, H., and Å. Åkeson: Ark. Kemi Min. Geol. **16** A no. 8 (1942). — Tomiyama, T.: J. biol. Chem. **111**, 45 (1935). — Tristram, G. R.: Advanc. Protein Chem. **5**, 83 (1949); The Proteins ed. H. Neurath and K. Bailey IA, 181. New York: Academic Press 1953. — Turba, F.: Chromatographische Methoden in der Protein-Chemie. Berlin-Göttingen-Heidelberg: Springer 1954.

Udenfriend, S.: J. biol. Chem. **187**, 65 (1950).

Valaise, H., and G. Dupont: Inds. agr. et aliment. **68**, 245 (1951); Through Chem. Abs. **46**, 4731 (1952). — Van Slyke, D. D., R. T. Dillon, D. A. MacFadyen and P. Hamilton: J. biol. Chem. **141**, 627 (1941). — Van Slyke, D. D., D. A. MacFadyen and P. Hamilton: J. biol. Chem. **141**, 671 (1941). — Vavruch, I.: Sugar, **47**, 35 (1952). — van Veen, A. G.: Geneesk. Tijdschr. Ned.-Ind. **78**, 2619 (1938). — van Veen, A. G., and A. J. Hyman: Geneesk. Tijdschr. Ned.-Ind. **73**, 991 (1933); Rec. Trav. chim. **54**, 493 (1935). — van Veen, A. G., and H. E. Latuasan: Chronica Naturae **105**, 288 (1949). — Vickery, H. B.: (a) J. Biol. Chem. **60**, 647 (1924); (b) J. Biol. Chem. **61**, 117 (1924); (a) J. Biol. Chem. **65**, 81 (1925); (b) J. Biol. Chem. **65**, 657 (1925). — Vickery, H. B., and C. S. Leavenworth: J. Biol. Chem. **63**, 579 (1925). — Vickery, H. B., G. W. Pucher, H. E. Clark, A. C. Chibnall and R. G. Westall: Biochem. J. **29**, 2710 (1935). — Vickery, H. B., G. W. Pucher, A. J. Wakeman and C. S. Leavenworth: Bull. Conn. Agric. Exp. Sta. no. 399 (1937). — Vickery, H. B., and C. L. A. Schmidt: Chem. Revs. **9**, 169 (1931). — Vickery, H. B., and C. G. Vinson: J. Biol. Chem. **65**, 91 (1925). — du Vigneaud, V., and W. I. Patterson: J. Biol. Chem. **114**, 533 (1936). — Vilkki, P.: Suomen Kemistilehti, B, **27**, 21 (1954). — Virtanen, A. I., A. M. Berg and S. Kari: Acta Chem. Scand. **7**, 1423 (1953). — Virtanen, A. I., and J. K. Miettinen: Nature (Lond.) **170**, 283 (1952); Biochim. Biophys. Acta **12**, 181 (1953). — Virtanen, A. I., and N. Rautanen: Suomen Kemistilehti, B, **19**, 56 (1946); Biochem. J. **41**, 101 (1947).

Wada, M.: Biochem. Z. **224**, 420 (1930). — Walker, J. B.: Proc. nat. Acad. Sci. **38**, 561 (1952); J. Biol. Chem. **204**, 139 (1953). — Walker, J. B., and J. Myers: J. Biol. Chem. **203**, 143 (1953). — Wehmer, C., and M. Hadders: Handbuch der Pflanzenanalyse, ed. G. Klein Bd. **3**, 80, 222. Wien: Julius Springer 1933. — Wells, H. G., and T. B. Osborne: J. Infect. Diseases **17**, 259 (1915). — Werle, E.: Angew. Chem. **63**, 550 (1951). — Westall, R. G.: (a) Nature (Lond.) **165**, 717 (1950); (b) J. Sci. Food Agric. **1**, 191 (1950). — Wibaut, J. P., and J. P. Schuhmacher: Rec. Trav. Chim. **71**, 1017 (1952). — Wieland, H., and R. Hallermayer: Liebigs Ann. **548**, 1 (1941). — Wieland, H., and B. Witkop: Liebigs Ann. **543**, 171 (1940). — Wieland, Th.: (a) Liebigs Ann. **564**, 152 (1949); (b) Fortschr. chem. Forsch. **1**, 211 (1949). — Wieland, Th., and G. Schmidt: Liebigs Ann. **577**, 215 (1952). — Wieland, Th., and L. Wirth: Ber. dtsch. chem. Ges. **76**, 823 (1943). — Windsor, E.: J. Biol. Chem. **192**, 595 (1951). — Wingo, W. J., and O. L. Davis: Proc. Soc. Exp. Biol. Med. **72**, 415 (1949). — Winterfeld, K., and L. H. Bijl: Liebigs Ann. **561**, 107 (1948). — Winterfeld, K., and M. Rink: Liebigs Ann. **561**, 186 (1949). — Winterstein, A.: Handbuch der Pflanzenanalyse, ed. G. Klein Bd. **3**, 1. Wien: Julius Springer 1933. — Wodehouse, R. P., and A. F. Coca: Ann. Allergy **4**, 58 (1946). — Woodward, G. E.: J. Biol. Chem. **109**, 1 (1935). — Woolley, D. W.: J. Biol. Chem. **176**, 1291, 1299 (1948). — Work, E., and D. L. Dewey: J. Gen. Microbiol. **9**, 394 (1953). — Wynn, V.: Nature (Lond.) **164**, 445 (1949).

Yoshida, T.: Tohoku J. Exp. Med. **48**, 27 (1945); [Through Chem. Abs. **42**, 7384 (1948)].

Zacharius, R. M., J. F. Thompson and F. C. Steward: J. Amer. Chem. Soc. **74**, 2949 (1952). — Zelitch, I., P. W. Wilson and R. H. Burris: Plant Physiol. **27**, 1 (1952).

Proteins[1]

By

N. W. Pirie.

Proteins are not a rigidly defined category of substances. One extreme definition would restrict the term to substances composed entirely of amino acids held together by peptide links. That might be too narrow because many generally accepted proteins carry prosthetic groups which are not made up from amino acids; there is indeed no evidence that chlorophyll and nucleic acid have ever been found in significant quantity in a plant except in combination with proteins. Furthermore it is not always possible to demonstrate that peptide bonds are responsible for holding the structure together. At the other extreme lies the common practice of looking on protein as the only form of combination in which nitrogen occurs in plant material and getting the alleged protein content by multiplying the nitrogen content by some such factor as 6.25. This practice includes free amino-acids, amides, purines, and even some of the nitrate, in the protein category and must be roundly condemned. When the figure given for protein content depends on a nitrogen determination its trustworthyness clearly depends on the completeness with which other forms of nitrogen have been eliminated. The assumption that they have been eliminated is more commonly made than shown to be valid and it is possible, or even probable, that there are several forms of non-protein-nitrogen that have not yet been recognised because of the attraction of the simplifying assumption.

A trustworthy method for determining protein depends more on the separation of substances that would interfere with the final analytical method than on the precision of that method. Such separations depend on the general properties of proteins and those that are relevant may therefore be considered briefly.

A. Methods for Separating Protein from Non-Protein Nitrogenous Material.

By definition proteins are substances of large molecular weight and 10,000 may arbitrarily be taken as the lower limit. The trypsin inhibitor from soya bean has this molecular weight (FRAENKEL-CONRAT, BEAN, DUCAY and OLCOTT, 1952) so that its status is equivocal. By dialysis or ultrafiltration it is often possible to get sharp separation; materials that have passed through a membrane are likely to be free from protein and conversely those that have not will have been freed from some non-protein material. But dialysis can only remove soluble materials that are not attached to or associated with the indiffusible substances and it is ineffective with materials of molecular weight 1,000—10,000 because these diffuse slowly. It is not therefore to be relied on.

Some proteins, notably the prolamines from cereal seeds, dissolve in aqueous alcohol but no protein that is substantially free from non-amino acid material is

---

[1] Determination of total nitrogen see Vol. I. Seed proteins see p. 69.

known to be soluble in the fat solvents that are immiscible with water. The ether extract from some materials does however contain peptide:lipid complexes (BALLS, 1942) and the possibility that proteins may be similarly linked must be borne in mind. In general however solvent extraction is a trustworthy method of separating lipides from proteins. The nitrogen-containing lipides, which are obviously the ones that interfere most with protein determinations based on nitrogen determination, can often only be removed completely if the material is made acid before extraction (LOVERN, 1942; HOLDEN, 1952).

Many proteins denature and coagulate on heating so that much non-protein material can be removed by extracting the tissue or preparation with boiling water. This is a standard procedure incorporated into many of the methods for determining what is sometimes called the "true protein" but it also cannot be wholly relied on. Some proteins are soluble in boiling water; these are not sufficiently widespread to be a frequent cause of error. The large group of substances either insoluble in water under all conditions or kept out of solution by association with protein or other tissue components is more important. Thus the nucleic acids are soluble at neutrality when free, but in many nucleoproteins the nucleic acid remains associated with the denatured protein after heating (c. f. PIRIE, 1954); similarly haematins and other nitrogenous substances sometimes remain with the coagulated protein. BONDI and MEYER (1948) claim that part of the nitrogen in the leaves of young plants is in condensed ring systems that separate analytically along with the lignin fraction. Substances such as this will therefore be included in a "true protein" determination.

Several substances precipitate most proteins from solution and of them alcohol, acetone, tannic acid and trichloroacetic acid are most often used analytically. The last is the most satisfactory for very few proteins are soluble in 20 to 50 g./l. solutions of it whereas many other tissue constituents, including the polysaccharides, are. The trichloroacetic acid precipitate, like that got on boiling, may retain most of the nucleic acid and haematin but the two types of precipitation are by no means always the same. Thus 10—20% more N appears in the trichloroacetic acid precipitate from some leaf extracts than in heat coagulums from the same extracts (SLADE and BIRKINSHAW, 1939; PIRIE, unpublished). The nature of this extra N has not been established.

For many purposes it is not necessary to take these factors into consideration; the amount of non-protein nitrogen may be insignificant; the analysis may only be a prelude to feeding experiments in which some at least of the non-protein material may be of as much value as the protein; it may be possible to determine the concentration of a substance that is known to interfere and deduct this value from the "true protein". For other purposes it may be necessary either to determine the protein by an unequivocal method or to make a quantitative extraction and so free it from interfering material. Extraction, where it is possible, is very much to be preferred but its difficulties have already been stressed in an earlier section (Vol. I, p.26). When the object of the extraction is purely analytical, so that it is immaterial that the protein is no longer in its original state, there are several extra possibilities. Protein can be separated from finely ground leaves by repeated extraction with water or with dilute alkali, and residual protein that does not come away with dilute aqueous solvents will often dissolve in formic acid or formamide or in strong solutions of urea, guanidine, guanidine nitrate, pyridine, phenol, thiocyanates, strong mineral acids, etc. From these solvents the protein, now somewhat modified, can be precipitated by dilution, dialysis or the addition of another solvent such as alcohol or ether. The disadvantage of this method is that the powerful protein solvents are also often solvents for other tissue components, so that extracts made in this way, while they may be free from some interfering substances, are not necessarily free from all of them. Treatment with agents in this category also carries with it the risk that amino acid or peptide fragments may be split off from the protein.

B. Methods for Determining Protein.

1. Determination of Total Dry Matter.

The concentration of a solution can be most conveniently measured by drying a suitable volume, by the methods described elsewhere (Vol. 1, p. 49) and weighing. If the solution is known to contain only protein this gives the protein content and the method has the added merit that the dry material can be used for other analyses. The difficulties involved in weighing small amounts of material lead many people to use volumetric or colorimetric methods even when this circuitous approach is not strictly necessary. A method that depends on the determination of dry matter clearly includes the prosthetic group, if any, along with the protein. Other methods may not do so and it is therefore important to be sure, if comparisons are being made, that the methods used are in fact comparable.

2. Determination of Nitrogen.

When a protein solution contains salts, sugars etc. the determination of dry weight is inapplicable and that of nitrogen usual. If amino acids or ammonium salts are present the protein must first be separated from them and for this the standard procedure is precipitation with trichloroacetic acid, but precipitation has few other merits. As already pointed out it does not, as a rule, remove prosthetic groups and some of these, for example nucleic acid and chlorophyll, contain nitrogen so that the final figure will still include some non-protein nitrogen. Unless it is clear that there is some advantage in precipitation, the step may well be omitted. Trichloroacetic acid, like other acids, may bring about partial hydrolysis with loss of particular amino acids; the extent of this is generally small but it may cause error if the protein is to be used later for the determination of amino-acid distribution.

The widespread use of figures for total nitrogen in assessing the value of vegetable materials in animal feeding (see p. 106) has been an important incentive in the development of trustworthy methods of nitrogen analysis. Methods based on that of KJELDAHL have now become almost universal but there have been many variations in the experimental procedure since it was first published in 1883. The history of these variations has been given in admirable detail by VICKERY (1946 a and b). CHIBNALL, REES and WILLIAMS (1943) point out some of the uncertainties that have been introduced by the use of different methods in different laboratories and KIRK (1947) surveys the whole subject. The incineration procedure of CHIBNALL, REES and WILLIAMS may be quoted verbatim:—

"The sample for analysis, which should not contain more than 10—15 mg. N, is weighed out in a small glass flat-bottomed cup, stoppered if there be need to dry to constant weight. After weighing, the cup is dropped carefully into a long-necked 250 ml. KJELDAHL flask in such a way that no protein is scattered; about 5 ml. H_2O followed by 20 ml. N-free H_2SO_4 are added and the mixture boiled gently on the stand until charring commences. The flask is then removed from the stand, cooled slightly and 5 g. catalyst mixture added. The latter is prepared by grinding to a fine powder 80 g. K_2SO_4, 20 g. $CuSO_4 \cdot 5H_2O$ and 0.34 g. Na selenate. The flask is then returned to the stand and the contents gently boiled. The digest clears in 15—30 min. and the heating is continued fairly vigorously for another 8 hr. or, if convenient, overnight. The cooled digest is finally made up to standard volume and an aliquot withdrawn for NH_3 estimation. It is essential, of course, that control blanks for the complete procedure be carried out at frequent intervals. In the case of protein hydrolysates an aliquot containing about 0.1 mg. N is withdrawn and incinerated in the usual micro-Kjeldahl tube, using 2 ml. acid and 0.5 g. catalyst mixture. The procedure then follows that given above, except that the whole digest is used for the NH_3 estimation.''

The merits of ammonia determination by titration and Nesslerisation are considered in the section on Mineral Components and Ash Analysis (Vol. I).

From the nitrogen content, the protein content of a material is calculated by the use of a factor but this factor should be determined for the protein that is present or predominant in the material being analysed. Published values for the nitrogen contents of purified proteins range from 14 to 19% and the use of the traditional factor, 6.25, depends on the assumption that all contain 16%. Chibnall (1939) has suggested that in plant materials 16.7% is a more usual value so that, if an attempt were being made to use one factor, it should be 6.0.

3. Determination of Amino Acids.

Carbon and hydrogen determinations have been used in protein analysis but they do not seem to offer any advantage and are particularly subject to interference from other materials. The determination of amino acids or of characteristic components of them such as amino or carboxyl groups adjacent to one another is much to be preferred. Two or three amino acids, e. g. valine and glutamic acid, have been found in every protein so far examined; we could therefore use the absence of one of them from any material as provisional evidence that it does not contain protein. But the recognition of an amino acid in an insoluble or indiffusible fraction does not necessarily demonstrate the presence of protein, for amino acids are formed when hot acids act on other types of substance, for example, some of the alkaloids, purines and vitamins. It is nevertheless traditional to use colour reactions for amino acids as if they were tests for protein and those generally tested for — arginine by the Sakaguchi reaction, tyrosine and phenyl-alanine by Millon's reaction, histidine by the diazo reaction and tryptophan by several reactions — are sufficiently widespread for the assumption that their presence or absence is an indication of the presence or absence of protein to be misleading only infrequently. The main objection to the use of analyses for a particular amino acid in the determination of protein, is obviously the great extent to which proteins vary in their amino acid composition. Because of this, the amount of a particular amino acid in a material only has a bearing on the protein content when the composition of the constituent proteins is known. There are a few amino acids e. g. diaminopimelic acid which occurs is several species of bacterium (Work, 1951), that are such infrequent constituents of proteins that their presence in a hydrolysate may be used as tentative evidence for the presence of a particular protein but this method has no general applicability.

A few amino acids can be determined on intact proteins but in general hydrolysis is necessary first; the difficulties encountered in hydrolysis and the results of amino-acid determinations are considered in a later section (p. 29).

4. Determination of the Peptide Bond.

The determination of peptide bonds is a more general method for determining protein, and the *biuret reaction* is the best known example of it. In this reaction the colour of a dilute alkaline solution containing a copper salt is greatly intensified by the presence of protein and some other substances such as polypeptides and biuret itself. For quantitative work the copper in the solution is stabilised by adding complex-formers such as ethylene glycol or tartarate. The precise molecular configuration responsible for the action is not known and there are slight variations in the intensity and tint of the colour given by different proteins (GORNALL, BARDAWILL and DAVID, 1949) so that, without calibration, the method cannot be used as an absolute method of protein determination but the variations are small enough to make the method suitable for many purposes. The reaction has several defects. It is not very sensitive; about 1 mg. of protein is needed to get a satisfactory colour. It is only applicable to materials that are either in solution or can be dissolved in fluids with a composition comparable to the biuret reagent. There is the obvious interference by other colours in the protein preparation. This last is a serious objection because many materials of vegetable origin have a brown colour and this is intensified in the alkaline conditions of the reaction. Some coloured contaminants and interfering substances such as polypeptides are removed by preliminary precipitation of the protein by agents such as trichloroacetic acid and by extraction with fat solvents but so much colour often remains so to make the method unuseable.

Alternatively, the recognition that peptide bonds are being hydrolysed can be used as a method for determining proteins. This is most satisfactory if the hydrolysis is brought about enzymically. Clearly, if the enzyme preparation is known to contain only protease, any breakdown that it brings about is evidence for the presence of protein in the material being changed. That is a tautology. Enzyme preparations are not in general so specific. It is therefore more satisfactory to demonstrate the appearance of groups that are unlikely to have been derived from anything except protein; thus methods based on the *appearance of amino groups* are preferable to those based on the appearance of carboxyl unless, as with the methods based on the evolution of CO_2 on oxidation with ninhydrin, only a carboxyl adjacent to an amino group reacts. The determination of NH_2 groups by measuring the evolution of N_2 after reaction with nitrous acid is described by KEMEN (Vol. I). It is generally trustworthy although there are several substances which do not contain NH_2-, e. g. alcohol and phenols, which evolve gas. Blank determinations eliminate interference from such substances if they are preformed but it would be more difficult to correct the value if such substances were formed during an action designed to liberate amino groups and this could be a consequence of esterase or pectin esterase action. POPE and STEVENS (1939) found that amino-acids and peptides formed soluble copper complexes when mixed with a suspension of copper phosphate. This phenomenon has been used for assaying proteolytic enzymes (e. g. TRACEY, 1948); it could also be used to estimate the original protein content of a system to which excess protease had been added. Clearly this type of evidence is only valid when an action is found to take place. Many proteins are not attacked by proteases and some are attacked incompletely; as mentioned in the section on Grinding (Vol. I, p. 27) this may sometimes be a consequence of the pretreatment to which the material has been exposed.

The determination of amino groups with formaldehyde, the so called *formol titration*, is widely used but very misleading. Accurate results depend on a knowledge of the original pK of the amino group and the pK after sufficient formaldehyde has been added to raise the concentration to about 10%. The fluid is adjusted to a pH, generally in the neighbourhood of the original pK, that lies beyond the buffering range of the formaldehyde-NH_2 complex. Alkali is then added to restore the original pH, but one amino group does not require one equivalent of alkali for this, but only that proportion of an equivalent that corresponds

to the proportion of the amino group left untitrated at the first adjustment i. e.
to half if the original adjustment was exactly to the amino pK and if sufficient
formaldehyde was added. A further defect of the method is that the pH of some
substances e. g. reduced glutathione (Pirie and Pinhey, 1929) increases when
formaldehyde is added to an alkaline solution and so could partially compensate
for the fall due to amino groups. The method has been applied to intact proteins
as well as to amino acids but, as Kirk (1947) wisely remarked, "The formol
titration applied to protein analysis appears to offer some of the best illustrations
of the analytical confusion often to be found in the biochemical literature".

The *reaction of ninhydrin* (triketohydrindene) with free amino acids is relatively
specific and results in the formation of a coloured derivative of the ninhydrin,
CO_2, an aldehyde with one carbon atom less than the amino acid, and ammonia.
Estimation methods have been published depending on all four. Thus MacFadyen
(1944) determined the ammonia. Colorimetry is more difficult because the amount
and tint of the colour varies with the amino acid so that particular attention to
detail is needed when the colorimetric method is used but methods have been
described by Moore and Stein (1954) and by Cocking and Yemm (1954).
The determination of the aldehyde has some advantages (c. f. Virtanen and
Laine, 1940) but CO_2 determination is both simpler and more general. Van
Slyke, Dillon, MacFadyen and Hamilton (1941) have dealt in critical detail
with the manometric method, and titrimetric methods have been described by
van Slyke, MacFadyen and Hamilton (1941) and Christensen, West and
Dimick (1941). When aldehyde is determined differences arise between proteins according to the proportion in which the different amino acids occur in
them. Otherwise these methods do not distinguish between the amino acids
sufficiently for differences in composition to affect the results and this is an
advantage when total protein is being determined. The position is quite different when the individual amino acids are being determined and the factors that
then arise will be considered later (p. 29).

5. Physical Methods.

The concentration of protein in solutions that are too dilute to be amenable
to any of these methods can sometimes be determined by the measurement of
some physical property. Density, refractive index and absorbtion spectrum are
the three most commonly used. The first two are clearly only applicable to
solutions containing protein only, or to solutions of protein in a solution of known
properties. They have been extensively applied to such fluids as serum, in which
Moore and van Slyke (1930) found that a change of 1 g. per litre in the protein
concentration caused a change of 0.00029 in the *density*, but less often to plant
proteins. Wolf (1942), however, used refractive index measurements to determine
the protein in potato juice. *Refractive index* measurements are regularly used to
measure differences in the concentration of proteins in different parts of the tube
during electrophoresis or ultracentrifugation and the methods adapted to this
type of measurement are described in the handbooks on these processes.

Measurements of *light absorption* are an obvious method for determining the
concentration of coloured proteins such as the chlorophyll-protein complexes, the
copper or iron containing enzymes and the coloured algal proteins. By virtue of
the aromatic rings in several amino acids, all proteins absorb light in the ultraviolet and measurements at 290 mμ or some similar wave length are often taken
as a measure of the protein content of a solution or even of a histological preparation. Provided the absorption spectrum of all the other components of the
system have been measured, the method is convenient and reliable, but unless it

is clear that these control measurements have been made, results based on ultraviolet absorption should be treated with reserve. The strong absorption at 260 mμ by ascorbic acid, for example, is often overlooked (MOHLER and LOHR, 1938). The general precautions to be taken in absorption spectrum measurement are dealt with elsewhere.

6. Methods Depending on Specific Activities.

Proteins are generally of interest because they have a specific activity e. g. they are enzymes, allergens or viruses, and many more are antigenic so that it is in principle possible to make a specific antiserum with which to determine the titre of protein in a solution. Methods based on these properties have the advantage, not only that interference by non-protein materials is uncommon, but also that they differentiate between the specific protein and others in the system. This specificity is, however, not always so great as is assumed and the activity measured is often the property of a category of materials rather than of an individual substance. Proteins, which can be differentiated by other means, may share a serological or enzyme activity (c. f. PIRIE, 1940, 1950a, 1952a, COLVIN, SMITH and COOK, 1954.

In the other parts of this section on Proteins the assumption will be made that all enzymes are proteins and that the determination of enzyme activity is tantamount to the determination of a specific protein. This assumption is not necessarily justified; partly because there may be non-protein enzymes, although none have been demonstrated unequivocally so far, and partly because the rate of enzyme action in a tissue is affected by factors such as the accessibility of the enzyme to substrate as much as by the actual quantity of enzyme present. The methods used in determining enzymes are too diverse to be discussed here and many are described elsewhere in the book. An account of the use of serological methods is given in the section on Plant viruses (p. 46) and is also dealt with in several monographs e. g. WILSON and MILES (1955) and BAWDEN (1950). The latter also describes the technique of infectivity measurements and the general principles underlying biossay are given in another section (HEALY, Vol. 1.).

C. The Amino Acid Composition of Tissues and Isolated Proteins.

There are two aspects to a discussion of the amino acid composition of proteins: the techniques used in the analysis and the motives that prompted it. It is only in the light of these motives that the adequacy of the methods of analysis can be assessed. The determination of the composition of a pure protein has obvious general scientific value and need not be discussed further; the difficulties arise with determinations on protein mixtures because these are done for practical reasons and generally to assess the protein or tissue as a feeding stuff (c. f. DUCKWORTH, p. 106).

Results in the older literature are of doubtful validity mainly because the initial hydrolysates were unsatisfactory. When enzymes are used as the hydrolytic agent, difficulties arise primarily from a failure of the action to go to completion; with acids they arise from destruction of some of the amino acids and from complex formation between them and other substances in the tissue. These factors arise in a more acute form with vegetable proteins than with proteins made from other types of organism because part of the protein is often encased in, or associated with, materials such as cellulose and lignin which resist the usual enzyme preparations. There is also, commonly, a high ratio of carbohydrate to protein and this increases the extent to which amino acids are lost by combination as "humin". LUGG (1939, 1949) and BLOCK and BOLLING (1951) have discussed in detail the losses of amino acids under normal conditions of hydrolysis and the precautions that can be taken to minimise loss. The greatest difficulties arise when unfractionated dried tissue is hydrolysed but even extracted proteins, especially leaf proteins, generally carry sufficient impurity to cause low yields. The conditions of hydrolysis can therefore, affect greatly the apparent composition of the protein; thus WAITE, FENSOM and LOVETT (1953) trebled the apparent amount of lysine in the protein extracted from senescent cocksfoot grass by hydrolysing in a sealed tube at 105° with 6 N HCl rather than with 2 N HCl under reflux. Recently DUSTIN, CZAJKOWSKA, MOORE and BIGWOOD (1953) found that there was no perceptible loss when a mixture of 15 amino acids was heated with 100 times its weight of starch for 24 hours under reflux with 6 N HCl when a very large volume of acid was used. The same methods were applied by DUSTIN, SCHRAM, MOORE and BIGWOOD (1953) to hay and other similar materials and they got nearly quantitative recovery of the N as amino acids when 4 ml. of acid was used for each mg. of protein in the material. It seems likely that this technique will get rid of most, if not all, the difficulties encountered in the analysis of un-fractionated tissues.

To avoid the interference from other tissue constituents the analyses are often done on extracted proteins. This is a valid technique only when the extraction is nearly complete. Sometimes however, figures purporting to be a guide to the nutritional value of a crop have been published although they are based on the analysis of a protein fraction which made up as little as 15% of the whole. At the same time the claim may be made that this protein is representative. It is difficult to see how evidence that it is representative could be got; for if trustworthy analyses could have been done on the whole tissue there would have been no need for the preliminary extraction and it there are no such figures the claim is a pure assertion and an improbable one at that. The evidence that proteins prepared in different ways come from different parts of the leaf in discussed later (p. 36, 41); essentially the same principles apply to most other tissues. Few analyses of a series of fractions from the same tissue have so far been made so that it is still not possible to assess the practical importance of this obvious theoretical objection to the use of analyses on a fraction as a guide to the nutritional value of the whole. TIMM (1942) found that glutamic acid, histidine and lysine made up 9.2, 1.6 and 5.4% of the protein from spinach "cytoplasm" but 8.4, 1.8 and 4.7% of the protein from chloroplasts. More extensive analyses on barley (YEMM and FOLKES, 1953) are set out in the table; the most significant difference here between cytoplasm protein and the mixed leaf protein is the lower lysine content of the latter. Here, therefore, the objection is theoretical rather than practical. Presumably this arises because each "protein" is a mixture of so many different individual proteins that an average amino acid composition is reached. The assumption that this will always happen when a non-representative protein sample is separated from a tissue is tantamount to assuming that different tissues also will have the same average amino acid composition and if that is assumed there is no need to do the analyses.

In most of the papers quoted in this section consideration is given to the precautions that have to be observed when determining amino acids in protein hydrolysates. The matter has been considered in detail by MARTIN and SYNGE (1945), and TRISTRAM (1949, 1953), BLOCK and BOLLING (1951). Analyses on six proteins or mixtures of proteins are collected in the Table 1 to illustrate the extent of variation in plant protein composition. For this purpose only analyses that account for nearly the whole of the protein are significant because of the uncertainty, when analyses only account for 70 to 80% of the protein, about which amino acids the loss has fallen on. Analyses of preparations of tobacco mosaic virus strains are set out in table 4 in the section on the virus infected leaf (p. 46) and there are more figures in the section on Seed Proteins (p. 69). Many incomplete amino acid determinations are scattered through the literature and some of the values are collected together in review articles e. g. TRISTRAM (1949, 1953), BLOCK and MITCHELL (1946—1947) and BLOCK and BOLLING (1951). The

Table 1. *Amino Acid Composition of Proteins and Mixtures of Proteins.*

Protein Analysed	Edestin	Zein	Gliadin	Protein from *Chlorella*	Protein from Barley Leaf	
					Mixed	Cytoplasmic
% Nitrogen in ash and moisture free protein Amino Acid with Molecular Weight	18.55	16.2	17.66	11.5	14.2	14.8
Alanine　　89.1	3.6	10.2	1.9	7.7	8.0	7.9
Arginine　174.15	29.0	3.4	5.0	15.8	15.0	15.0
Aspartic Acid　133.1	6.8	3.0	0.8	6.4	6.6	6.5
Cysteine and Cystine　121.1	0.9	0.6	1.7	0.2	0.9	0.9
Glutamic Acid　147.1	10.6	15.8	24.7	7.8	8.0	8.1
Glycine　75.05	5.1	—	—	6.2	6.5	6.7
Histidine　155.1	4.2	2.2	2.9	3.3	4.0	4.1
Isoleucine　131.1	4.3	} 14.8	7.2	{ 3.5	4.7	4.2
Leucine　131.1	2.7			6.1	5.8	5.9
Lysine　146.1	2.5	0.0	0.7	10.2	6.5	7.1
Methionine　149.2	1.2	1.4	0.9	1.4	1.3	1.4
Phenylalanine　165.1	2.5	3.1	3.1	2.8	3.2	3.0
Proline　115.1	2.8	8.0	9.2	7.2	4.0	4.1
Serine　105.1	4.5	5.8	3.7	3.3	3.3	3.5
Threonine　119.1	2.4	2.5	1.4	2.9	4.0	4.0
Tryptophan　204.1	1.1	0.1	0.5	2.1	1.8	2.0
Tyrosine　181.1	1.8	2.5	1.4	2.8	2.2	2.3
Valine　117.1	3.7	2.6	1.8	5.5	5.0	4.9
Amide N.　—	9.6	18.4	25.4	6.1	4.75	4.8
Total　—	99.3	94.4	92.3	101.3	95.55	96.3

　　The figures for edestin, zein and gliadin given in this Table are recalculated from those given by TRISTRAM (1949) and BLOCK and BOLLING (1951) so as to express the amount of nitrogen in each amino acid as a percentage of the total nitrogen of the protein. The nitrogen content of the protein is also given and, for convenience, the molecular weight of the amino acid. By calculation the yield of amino acid to be expected on hydrolysis can therefore be got and also the amount of each amino acid residue that goes to make up 100 g of protein. This last is the quantity used in Table 4 (p. 55). Figures for *Chlorella* protein are from FOWDEN (1952) and those for the barley leaf proteins are from YEMM and FOLKES (1953).

figures in the table for the three purified proteins have been taken from TRISTRAM (1949) and from BLOCK and BOLLING (1951). They are based on many different types of determination depending on isolation of the amino acid or a derivative, colorimetry, microbiological assay or chromatographic fractionation followed by ninhydrin or amino nitrogen determinations. The uncertainty of these methods varies from ± 10% for microbiological assay to ± 3% for the rest. In the light of this the nearness of approach of the sums of the N accounted for to 100% is satisfactory. With these materials the methods used are almost certainly valid; this is by no means always so and any method that does not depend on the isolation of a characteristic derivative of the amino acid, or on a clearly defined chromatographic peak, must be examined critically when it is applied to a new type of material. Caution is especially needed when colorimetric methods are used or methods depending on the action of a single group in the amino acid. The literature contains many remarkable claims made because the need for caution had been overlooked. Thus MILES and PIRIE (1939) commented adversely on the claim by PENNELL and HUDDLESON (1937) that an antigen from *Brucella melitensis* contained 19% of tryptophan; BLOCK and BOLLING (1951) recommend similar reserve towards the suggestion by MAZUR and CLARKE (1942) that the protein in *Macrocystis* contains 17% of methionine.

From the figures few generalisations can be drawn. Some plant proteins, like some animal proteins, are unusually rich in one amino acid and in consequence are poor sources of the others; thus the glutamic acid content of gliadin is as exceptional as the glycine content of collagen or the arginine content of clupein. Similarly the absence of lysine from zein can be paralleled by the absence of tryptophan from collagen and from the ribonuclease made from mammalian pancreas. These anomalies appear in general in pure proteins or in mixtures containing only a small number of components. The attempts that have been made to interpret the enzyme or other activities of a protein in terms of its composition are discussed critically by TRISTRAM (1953). They still seem unsatisfactory and are of little relevance here because figures for animal proteins have been used almost exclusively.

The old distinction between "first" and "second class proteins" was made on the basis of feeding experiments on animals; the former tended to be of animal and the latter of vegetable origin. But the animal proteins that were considered "first class" are either very complex protein mixtures such as muscle and liver or else they are less complex mixtures such as milk or egg proteins that are the main protein source for young animals. It is reasonable to suppose that, in the latter, evolutionary selection has operated in favour of a balanced distribution of amino acids. Those plant proteins which are also complex mixtures, and of these the proteins from leaf and *Chlorella* are the most fully analysed examples, have an amino acid distribution comparable to that of the "first class" mixed proteins of animal origin and there is no *a priori* reason to think that they will be of less value nutritionally. The point has not yet been settled experimentally but DEAN (1953) has published a useful survey of the nutritional adequacy of other plant proteins and of the possibilities of using them to supplement one another.

Another reason for interest in the composition of the bulk proteins from various species of plants is the idea that it may contribute to an understanding of phylogeny. This underlies the work of MAZUR and CLARKE (1938, 1942) on the algae and the review by LUGG (1949). The subject is of obvious importance but it seems unlikely that it can be studied by the methods used hitherto.

We may assume, in the light of the discussion in the preceding paragraphs, that it is, or soon will be, possible to separate so large a proportion of the protein in a tissue that the preparation may be taken as typical. Alternatively, methods may be developed whereby trustworthy analyses can be carried out on whole tissues. The amino acid content of a tissue will then not be in dispute but, for various reasons, it may be difficult to interpret. First, it may not be characteristic of the species but depend rather on the physiological state of the specimens used. Thus TIMM (1942) and YEMM and FOLKES (1953) find a small but significant difference between the compositions of chloroplast and cytoplasm protein; cultural conditions affect the ratio in which these two broad protein mixtures occur in a leaf and so must affect the composition of the bulk protein. Similarly the state of nutrition affects the composition of the seed proteins of many plants

e. g. rice (KIK, 1941). Variations of this type are commented on in the next section (p. 69); they could probably be most easily studied with species, like *Chlorella*, in which great variations in composition are possible. The analyses set out in Table 1 were made on protein from cells containing 4.3% N. By varying the conditions of growth the N content can be varied from 1.4 to 9.3% (MILNER, 1953); analyses on proteins made from cells with these diverse protein contents would be of great interest. The degree of ripeness of a tissue also affects the ratio in which proteins occur in it. DANIELSSON (1952) finds that the ratio of vicilin to legumin falls as pea seeds ripen; the latter contains 1.3% of tryptophan and the former 0.4 so that the tryptophan content of total pea protein varies with the ripeness of the seed.

Secondly, the bulk proteins from closely related species may differ significantly in composition. Thus ARMSTRONG (1951) found 2.3% of tyrosine in preflowering timothy grass protein and 3.95% in protein from cocksfoot taken at the same stage. He determined six amino acids in fifteen species and, although the other differences were not as large as the one quoted, his figures offer no support for the idea that leaf proteins have the same composition in related plants.

Thirdly, much of the protein in a tissue is made up of enzymes; it may be that nearly all of it is. A clearly established difference in the composition of protein taken from two species in the same physiological and nutritional state may be due to differences in the ratios in which different enzymes or other proteins occur, or to differences in the composition of the same enzyme when derived from different sources, or to both factors. Before a phylogenetic interpretation of a difference becomes significant it is necessary to find out to which factor it is due. A clearly established difference in the amino acid composition of an enzyme, filling the same role in two species, would be of more significance than a difference in the composition of the bulk proteins. It may, on the other hand, turn out that each enzyme has a relatively constant composition in a range of species but that different enzymes in these species differ. Differences in the composition of the bulk proteins will then reflect a difference in the ratios in which the enzymes or other proteins occur in the different species, but this could have been more satisfactorily established by determining the proteins instead of the amino acids. It will be of the greatest interest if it should turn out that the enzymes in certain plant families tend to be either rich or deficient in certain amino acids and these differences may be interpretable phylogenetically but they will not be established by the methods of analyses adopted so far.

D. Leaf Proteins.

I. The Normal Mature Leaf.

1. Effects of the Physiological State of the Leaf on its Protein.

Most of the published information about leaf protein is based on no more than a measurement of the total nitrogen content of the leaf. When small changes are being described it is therefore possible that the change does not take place in anything properly called protein, but, in most leaves, protein is certainly the predominant form of nitrogen so that large changes must reflect changes in the protein. Most of the studies have been concerned with the value as fodder of different plant species and of the same plant species grown under different cultural conditions and cut at different stages of growth. Physiological investigations of the translocation of protein from the leaf before abscission, or from leaf to seed, have accounted for many of the remainder.

From these analyses a few conclusions can be drawn. In young leaves growing under good conditions as much as 40% of the dry matter may be protein and 30% is fairly common. As the leaf matures the protein content falls. This may be caused by a change in the ratio of fibrous tissue to tissue rich in plastids but it is often caused by nutritional deficiencies. Thus Wadleigh (1944) and Hall (1951) find that the leaf of mature but well manured cotton contains 3.2—3.8% of nitrogen whereas on field grown plants the percentage of nitrogen often falls below 1% (Jewitt, 1949). The grasses, on the other hand, can only be maintained at a high nitrogen content by regular cutting as well as liberal manuring so that the material handled will contain a high proportion of blade rather than stalk. This is one of the central problems of grassland husbandry and much has been written about it e. g. Watson (1939) and Holmes (1951).

The physiological and metabolic behaviour of protein in the leaves and other parts of plants has been discussed in several recent reviews Chibnall (1939), Wood (1945), McKee (1949), Lugg (1949) and Street (1949). There is, therefore, no need to survey the field again but some arbitrarily selected examples will be used to illustrate a general theme. The extreme physiological plasticity of plants makes bald statements about their analytical composition or about the extractability of their components, of little value. Experimental results will only be replicated if plants grown under the same conditions are used. Thus Crook and Holden (1948) found that the extractability of the nitrogen in tobacco leaf increases with the nitrogen content and Holden and Tracey (1948) found that variation in the fertiliser treatment affected the elementary and the enzymic composition of the leaf and also the distribution of the enzymes between the different leaf fractions. The effects of different forms of nitrogen manuring on the amounts of carbonic anhydrase, catalase and peroxidase in tobacco leaves were measured by Tombesi and Cervigni (1952); the last was affected very little but nitrogen manuring increased the first two. In a similar but more elaborate study Nason, Oldewurtel and Propst (1952) varied the micronutrient elements on which tomato plants were grown and found that, in general, a deficiency increased the enzyme content if this is expressed as a function of the total leaf protein except when the enzyme contains the element in which the plant has been made deficient. They examined six enzymes, polyphenol oxidase, peroxidase, diaphorase, ascorbic oxidase, lactic dehydrogenase and glycollic dehydrogenase, and discuss critically the manner in which the deficiency may act.

The intensity of light in which a plant has been grown affects the protein content, Nightingale (1927), by restricting the daily period of illumination of somewhat nitrogen deficient tomato plants raised the nitrogen content of their leaves from 1.7% of the dry matter to 2.47. By growing well manured tobacco plants in the shade, Bawden and Roberts (1947) raised the percentage of nitrogen in the dry matter of the leaves from 5 to 9.8. In neither of these experiments was the change in protein content measured but similar increases have been found in the chlorophyll content of shaded leaves (c. f. Olsen, 1942) and this may reasonably be taken as evidence for a change in the protein with which the chlorophyll is normally combined (see p. 40). The changes that follow the illumination of etiolated barley seedlings have been followed by Tolbert and Burris (1950) who find that the activity of glycolic oxidase increases even more rapidly than the chlorophyll content but that changes in other enzymes are less marked. These changes took place in a few hours but it is not certain that they are the consequence of a total increase in the enzyme, which is presumably a protein, rather than of an alteration in its accessibility. There have been frequent claims for a diurnal variation in the protein content of leaves, but its reality does not seem to have been established with certainty (cf. DENNY, 1932). Differences in the feeding value of Spring and Autumn forage, with apparently the same botanical and chemical compositions, have also been claimed and these differences are sometimes attributed to differences in the ratio in which the different leaf proteins occur in leaves in differing physiological states. This is one possibility and it deserves careful investigation but it is also possible that the effect is due to entirely different factors such as variation in the carbohydrate or oestrogen content of the leaves.

As a final example of the effects of the conditions to which the leaf is exposed, on the protein in it, the changes brought about by starvation and wilting may be mentioned. In general, shortage of water leads to protein hydrolysis (PETRIE and WOOD, 1938; PETRIE, 1943). The course of disappearance of protein in starved barley leaves has been followed by YEMM (1950) and in oat leaves by CRUICKSHANK and WOOD (1945). On prolonged starvation 80% of the protein may be either oxidised or hydrolysed. Few studies have dealt with the differential rate of loss of the different proteins in the leaf but WOOD, CRUICKSHANK and KUCHEL (1943) observed differences between Sudan Grass and Kikuyu grass; in the former the chloroplast protein disappeared more rapidly than the cytoplasmic protein whereas they disappeared at equal rates in the latter. In tobacco leaves the phosphatase, invertase and peroxidase content does not diminish proportionately to the diminution in total protein (AXELROD and JAGENDORF, 1951). An examination of the ratios in which the different enzymes occur in leaves kept under different physiological conditions would throw much light on the nature of leaf protein and on the question whether there is anything in the leaf that could properly be called "storage protein" or whether all the protein is made up of enzymes. The immediate issue however, is that leaves that have been taken from the plant and exposed to adverse conditions for only a few hours may show a distribution of nitrogen significantly different from that in the leaves when they were picked. The extent of these metabolic changes can be diminished by storing the leaves at 0° but actual freezing of the tissue leads to the irreversible coagulation of some of the protein in many leaves. This can be an advantage if preparations of the other proteins are being made and the technique is often used, but the assumption should never be made that proteins isolated from a leaf that has been frozen are typical of those originally present; they may indeed be artefacts.

In brief, the physiological plasticity of plants is so great that it is only when two species have been examined after cultivation under a wide range of different conditions that differences in their enzymic or analytical composition become significant.

2. Bulk Preparations of Leaf Protein.

ROUELLE (1773) separated "La matiere glutineuse" by heat coagulation of the juice expressed from minced hemlock leaves and he recognised its resemblance to animal proteins, "la partie caséeuse du lait". Much of the work that has been done since, has made use of essentially the same method and effort has been expended on increasing to the greatest extent possible the percentage extraction of protein. The methods all involve fine grinding and the difficulties involved in this process have been commented on elsewhere in this book (Vol. 1, p. 26). The object is to separate the protein, in a still unhydrolysed form, from the fibrous, carbohydrate-rich parts of the leaf. As grinding proceeds, a smaller proportion of the nitrogen remains associated with that part of the fibre which is still sufficiently coarse to be retained on a cloth or sedimented at ca. 1,000 g. But by the time extraction is nearing completion, much of the fibre has been so comminuted that it also appears in the "protein" fraction. If therefore, protein preparations are made from successive extracts after different intensities of grinding the nitrogen content of the preparation falls steadily so that the final fractions should only be called protein preparations after the finely divided fibre has been removed by some other means.

We must assume that the different proteins of the leaf are not extracted from the matrix with the same readiness. From this it follows that methods which give different percentage extractions will extract a different mixture of components, so that the amino acid composition and other properties of the preparations may well be different. It has seldom been possible to get unequivocal evidence about the region in the cell from which a leaf protein originated. Histological study of the leaf residues left after some of the usual grinding procedures shows that some cells have not been torn open and that the extent to which the chloroplasts are destroyed varies with species and physiological state. Thus, with some species a bulk preparation will contain components derived from all parts of the leaf, while with other species the leaf elements are robust so that it is relatively easy to separate them first and then study in isolation the proteins derived from them. The effects of factors such as these on the interpretation of analytical figures got on bulk preparations of tissue proteins, have already been discussed.

The factors governing completeness of extraction are by no means clearly understood. Fine grinding is obviously an advantage unless it is carried so far that the difficulties already mentioned obtrude. Slightly alkaline conditions promote extraction but may damage the protein that is being extracted. The addition of agents such as ether which damage the osmotic control exerted by the cell membrane has the merit (Chibnall, 1923, 1939) that it allows the separation of low molecular weight material before the protein is liberated from the cells, but it may impede the subsequent dispersal of the leaf protein. Chibnall found that this defect was avoided to some extent by using the fluid that had been expressed from an earlier batch of leaves and Vickery (1945) suggests that this phenomenon is due to the maintenance of an optimum ionic environment during the treatment. It seems probable, however, that there is some specificity in the effect because Crook (1946) has stressed the depressant effect of salts on protein extraction. He examined various buffers in which leaves could be ground and concluded that the most convenient one that did not contain nitrogen, and so did not complicate subsequent nitrogen determination, was o-chlorophenol. But he found that extraction was better if no buffer was added and the pH value was adjusted by the cautious addition of NaOH to whatever alkaline value is considered safe. The adjustment of the pH value of a suspension of ground leaf is not, however, simple. Bawden and Pirie (1944) found that the pH value of a tomato suspension returned repeatedly to about 6 after adjustment to 7.6 and Holden (1945) showed that this was due mainly to the action of pectin esterase and that a stable pH value could only be reached after the enzyme in the leaf fibre had completed the hydrolysis of all the pectic methyl ester. This behaviour is common, though some leaves, notably from the Cucurbitaceae (Holden, 1948), do not show an acidwards drift but stay alkaline throughout. Holden concluded that this phenomenon is due to a reaction between calcium carbonate, phosphate and pectin in these leaves and was able to simulate the effect by grinding acid drifting fibre with calcium carbonate. Table 2 is from Crook and Holden (1948); it lists the species found to fall into the two categories and gives the percentage extraction of nitrogen found on an extensive range of plants after passage through a domestic meat mincer and a roller mill.

Much work has been done on the amino acid composition of these bulk preparations in which a high yield of protein is looked on as of more importance than knowledge about the part of the leaf from which the protein came, but it is clear that the figures do not contribute to our general understanding of protein composition because they relate to a mixture of proteins and the ratio in which different components occur will depend on the precise conditions of extraction. and on its completeness. These factors have already been discussed (p. 30). In several countries an increasing interest is now being taken in leaf protein preparations as a source of animal and human food (cf. PIRIE, 1942, 1952b; BICKOFF, BEVENUE and WILLIAMS, 1947; SULLIVAN, 1943). An increased interest in analytical methods adapted to them is therefore probable.

Table 2. *Extraction of Nitrogen and Dry Matter from Leaves of Various Plant Species.*

Family	Latin name	Common name	Dry matter (% of wet wt.)	Nitrogen % Dry matter	Percentage of N extracted
(a) Plants whose fibres require neutralization.					
Compositae	*Senecio vulgaris*	Groundsel	8.2	4.68	92
	Chrysanthemum hortorum	Chrysanthemum	18.6	2.06	62
Cruciferae	*Brassica oleracea*	Cabbage	9.7	4.47	85.5
Rosaceae	*Fragaria vesca, var. Mayo*	Strawberry	33.0	2.84	71
	var. Royal Sovereign	Strawberry	36.1	1.57	39.5
	Prunus laurocerasus	Laurel	28.8	1.75	52
Cucurbitaceae	*Ecballium elaterium*	Squirting cucumber	11.8	4.52	90
Leguminosae	*Phaseolus vulgaris*	Bean	14.1	2.80	73
Equisetaceae	*Equisetum sp.*	Horsetail	17.1	3.23	59
Gramineae	*Triticum vulgare*	Wheat	11.6	4.46	92
	Dactylis glomerata	Cocksfoot	27.0	2.49	69
Solanaceae	*Physalis alkekengi*	Winter cherry	23.7	4.03	64
	Nicotiana glutinosa	—	9.8	2.76	82
	Nicotiana tabacum	Tobacco	9.5	4.35	93.5
	Datura stramonium	Thorn apple	13.4	4.66	93
	Lycopersicum esculentum	Tomato	13.5	3.26	83
Buxaceae	*Buxus sempervirens*	Box	52.1	3.29	83
Moraceae	*Ficus carica*	Fig	20.2	4.52	87.5
(b) Self-neutralizers.					
Compositae	*Helianthus annuus*	Sunflower	17.6	3.24	56
	H. tuberosus	Artichoke	18.9	2.49	44.5
Boraginaceae	*Symphytum officinale*	Comfrey	13.2	4.14	52.5
Urticaceae	*Urtica dioica*	Nettle	21.8	3.75	41.5
Cucurbitaceae	*Cucurbita ovifera*	Marrow	14.7	4.66	75
	C. ovifera	Marrow	13.2	2.95	69
	Cucumis sativus	Cucumber	9.8	4.38	73
	Bryonia dioica	Bryony	11.1	4.60	88.5
	B. dioica	Bryony	11.5	6.21	89
	Lagenaria leucantha	Calabash	11.4	4.21	55
	Thladiantha sp.	—	13.9	3.17	77
	Cyclanthera explodens	—	12.0	4.85	87

3. Proteins from Mechanically Isolated Fractions.

The leaves of a few plant species carry glands or other organs from which a fluid either exudes normally or comes out on damage. The best known of the latter group is the nettle sting and it was at one time thought that the fluid coming out of the sting contained an enzyme comparable to those in insect stings. EMMELIN and FELDBERG (1947) have however shown that there is no need to postulate an enzyme; they find sufficient acetylcholine, histamine and another substance apparently of low molecular weight to account for the sting. The digestive fluids

of the insectivorous plants are the only leaf secretion on which much work has
been done; Darwin recognised proteases in them and this has often been confirmed. Clear evidence that the enzyme was actually in the secretion and was not
due to bacteria was given by Stern and Stern (1932) working with *Nepenthes*
and by Holter and Linderstrøm-Lang (1933) with *Drosera*.

Most work on leaf proteins has however been done on extracts and interpretation is by no means simple because, as soon as the cells are damaged, substances which did not have access to one another in the intact cell come into
contact and a series of changes start. It is possible that some of these are of
physiological interest in that they are normal actions speeded up or proceeding
in an environment that does not favour a back-reaction that had hitherto maintained an equilibrium state; but it is probable that most are unphysiological. It
is usual to bring about the first stages of the fractionation of leaf components by
filtering or centrifuging and these operations should be carried out as quickly as
possible and at as low a temperature as possible because many components
become relatively stable once they have been separated from some other components of the leaf. Thus Arnon and Whatley (1949) observed that the photo-
lytic activity of chloroplast fragments was retained for longer if they were centrifuged immediately from the leaf extract and Pirie (1950 b) got better yields
of leaf nucleoprotein if it was isolated quickly. It is a matter of general observation
that there is precipitation of material containing protein and carbohydrate for
several days from leaf sap or extracts from leaves. Precipitation is generally more
rapid if old leaves have been used but the final quantity is larger the younger
the leaf.

a) Chloroplast Protein.

Many methods have been used to prepare chloroplasts and their fragments;
the choice depends on the use to which the preparation is to be put and the species
from which it is being made. Menke (1938) and Neish (1939) removed coarse
particles such as starch grains and tissue fragments, the former then separated
the chloroplasts by differential centrifugation while the latter flocculated them
by adding enough $CaCl_2$ to bring the concentration to 0.1 M. White et al. (1948)
found that the chloroplasts in some leaves were so robust that they withstood
incubation with the enzymes produced by *Clostridium roseum*, under conditions
that disintegrated the leaf structure, so that they could be separated afterwards
from the digest. These processes give products that have undergone some modification so that they are not suitable for studies on photosynthesis. By grinding
the leaf in 0.5 M sucrose, Granick (1938) protected the chloroplast from damage
by exposure to the low osmotic pressure of the leaf sap and made preparations
that are in a more nearly normal state.

Similar preparations have been made in many different laboratories with
minor variations in technique. When carefully made the main constituent is
indubitably chloroplast but Weier and Stocking (1952) have pointed out that
there is generally significant contamination by cell nuclei and by nuclear fragments,
and that it is difficult to remove them completely. Contamination of this type
has probably little effect on analytical composition but in some work the material
that is called "chloroplasts" has been grossly contaminated by other leaf protein.
This happens when chloroplasts are not separated by relatively discriminatory
methods such as differential centrifugation or precipitation by low concentrations
of ammonium sulphate or calcium salts but by more general methods of protein
precipitation such as stronger ammonium sulphate or acidification. Before
considering the conclusions come to, the technique of preparation used should be
considered critically.

HILL and SCARISBRICK (1940), working with chickweed, found varietal differences in the stability of the system bringing about the liberation of O_2 when illuminated in the presence of reducing agents (the HILL reaction) but all lost activity nearly completely in 4 hours; this was the time for which HANSON (1941) found that chloroplasts from a grass retained their characteristic bright appearance under the microscope. Activity is retained for longer if the preparations are kept cool and there are some scattered observations on protective actions by substances such as methanol and propylene glycol. In general, however, preparations have been used immediately when photosynthesis is being studied.

Between 30 and 50% of the protein in a leaf is present in the chloroplasts and they contain between 30 and 50% of protein. It is to be expected therefore that much of the enzyme activity would be located in them and this expectation is borne out. A note of caution has been introduced by HAGEN and JONES (1952) who found, as others have done, that almost all the catalase of the leaf was associated with the chloroplasts when these are separated in the normal way, but that if the pH value was altered, from the normal value of 5, to 3.3 or 5.6, much of the catalase dissociated. HOLDEN and KENTEN (unpublished) find that the partitioning of catalase between the chloroplast preparation and the fluid is affected by very slight changes in the acidity and varies with different plant species. The acidity to which the chloroplast is exposed in the cell is not known, it is therefore uncertain whether catalase is a genuine component or not. In a similar way WARBURG and LÜTTGENS (1946) found that only part of the polyphenol oxidase was attached to the chloroplasts but ARNON (1949) gives reasons for thinking that it is all attached and showed that it was not readily dissociated even by fine grinding. Doubts of the same sort have been expressed by SISAKYAN and KOBYAKOVA (1948). Many of the other enzymes that must be postulated in chloroplasts because they are able to bring about the HILL reaction are considered elsewhere but, as a further example of the phenomena emphasised at the beginning of this section, it is worth recording that HILL and SCARISBRICK (1940) found that the activity of preparations made from chickweed picked in the middle of a sunny afternoon had fallen to zero. Phosphorylase (YIN, 1948) and cytochrome oxidase (McCLENDON, 1953) are among the other enzymes whose presence in chloroplasts seems to have been clearly established. General information on the composition of chloroplasts is surveyed by RABINOWITCH (1945, 1950). Many measurements of the ratio of chlorophyll to protein have been made; in almost all preparations it has been in the range 1:4.3 to 1:6.3. This constancy has stimulated attempts at interpretation on a molecular basis but these seem to be premature until there is some evidence about the proportion of the chloroplast protein that is made up of enzymes and which therefore may not be associated with the chlorophyll.

Chloroplasts are easily fragmented by ageing, shaking with glass beads, exposure to ultrasonic vibration or running at high pressure through a needle valve (MILNER, LAWRENCE and FRENCH, 1950). The enzyme activities are retained after some fragmentation and those responsible for the HILL reaction have been studied in some detail. Thus THOMAS, BLAAUW and DUYSENS (1953) have followed the manner in which activity diminishes with particle size and the reappearance of activity on reaggregation has been described by MILNER, KOENIG and LAWRENCE (1950). Examination, both by the light and the electron microscope, reveals dense particles, the grana, inside chloroplasts; the literature on these and other aspects of chloroplast structure is reviewed by STRAUS (1953). There is evidence (GRANICK and PORTER, 1947) that the particles in fragmented preparations are grana or grana aggregates but the point has not been definitely settled and no characteristic subunits with constant properties have as yet been made from chloroplasts. The study is, however, being actively pursued in many laboratories because of its importance in an understanding of the mechanism of photosynthesis. (cf. HOLT and FRENCH, 1949).

The chlorophylls are the most characteristic components of chloroplasts. They are treated in another section (p. 142), but evidence is accumulating that tends to confirm LUBIMENKO's (1927) opinion that chlorophyll does not exist free but as a protein complex, and this necessitates some consideration here of the complex. There are two main lines of evidence in favour of the complex; chlorophyll is not extracted from leaves or suspensions of chloroplast fragments, by solvents immiscible with water, unless a treatment that would denature proteins has been used; the bands in the absorbtion spectrum of isolated chlorophyll fall at slightly shorter wave lengths than those of chlorophyll in the leaf or in chloroplast fragments and this shift has not been simulated by the variations in the state of dispersion or association that have been tried so far (SMITH, 1941a; RODRIGO, 1953). The various treatments commonly used as first steps in the isolation of chlorophyll — heating, drying, extraction with alcohol or acetone etc. — break the linkage with protein but by other treatments a water soluble chlorophyll: protein complex can be prepared. SMITH (1941a and b) extracted fresh moist preparations made from leaves with solutions of highly surface active substances such as digitonin, bile salts and sodium dodecyl sulphate. The last is especially effective in breaking up the chloroplasts and giving clear chromoprotein solutions, in these the chlorophyll is converted, by removal of the magnesium, to pheophytin in acid solution but it remains intact if kept neutral. After this treatment the complex can still be sedimented on the ultracentrifuge (SMITH and PICKELS, 1941). TAKASHIMA (1952) dispersed clover grana with picoline and then crystallised out a chlorophyll:protein complex by adding dioxane; similar crystalline complexes have been made from several species including pteridophytes and algae (SHERRATT and EVANS, unpublished). There is no evidence that the chlorophyll was associated with this particular protein in the intact chloroplast and the crystallisable association may well be an artefact. No properly characterised free proteins have been made from the complex, for the protein is either denatured during the dissociation or, if dissociation is brought about by detergents, it may remain soluble but is very unstable.

From time to time the claim is made that plastids contain nucleoprotein. It is intrinsically probable that they do so and the evidence that nucleoprotein was present in the preparations examined is good, but it is less certain that these preparations were not contaminated with other components of the leaf. JAGENDORF and WILDMAN (1954) conclude that the nucleic acid content of carefully washed tobacco chloroplasts can be as low as 0.3% of the dry matter. Among the other proteins found in chloroplast preparations one is of exceptional interest because of the part that it may play in photosynthesis. HILL and SCARISBRICK (1951) and DAVENPORT and HILL (1952) describe the preparation from leaves of cytochrome f, a haem:protein complex similar to cytochrome c. DAVENPORT (1952) gives reasons for thinking that it is concentrated in the chloroplasts. Unwilted parsley leaves are ground in cold ammoniacal ethanol and then the cytochrome f is isolated from the extract. Only traces could be prepared when the same method was applied to chloroplasts made by the usual methods of grinding and centrifuging; this is taken as evidence that chloroplasts made in this way have in fact undergone modification and HILL and SCARISBRICK describe a method for making chloroplasts that still appear to contain the cytochrome f.

The work on chloroplasts illustrates very clearly the most common difficulty in studies of this kind. A fragile tissue component, subjected to only gentle procedures of fractionation, must always be treated with suspicion and the possibility borne in mind that its apparent constituents are the results of mechanical mixture or secondary aquisition. If, on the other hand, it is subjected to more rigorous purification there is the equally great risk that constituents will be lost or damaged during the fractionation. When the quantities are large, as with chlorophyll, the issue can become simple but more often all that can be hoped for is a balancing of probabilities.

b) Protein in the Organelles.

At one time it was sufficient to distinguish between leaf proteins derived from the chloroplasts and those derived from the cytoplasm, but in the last 10—15 years there has been such an increase in our knowledge of microscopic and sub-microscopic cell components that it is now necessary to consider the proteins in them as another category. The old information about these particles (cf. GUILLIERMOND, 1941; NEWCOMER, 1940) was mainly derived from histology; chemical study was impeded by their fragility. Interest in the chemistry of mitochondria, microsomes and other particles springs largely from their ability to bring about many integrated enzyme actions. This suggests that the enzymes in them are organised in a spatial array and has led to the use of the term "organelle" to cover the whole group. The localisation of enzymes in the cell and in fractions made from it has been reviewed in detail by GODDARD and STAFFORD (1954). Interest is also taken in the group because particles of this type are common contaminants of virus preparations.

The preparations used are made by centrifuging leaf extracts at 10,000 to 60,000 g. after removing chloroplasts and chloroplast fragments by centrifuging at lower speeds. Sometimes the product is referred to as the "particulate" fraction or even as a "particulate". This usage must be roundly condemned. The reality of atoms is now generally conceded and from this it necessarily follows that everything is particulate. It is meaningless to distinguish between particulate and non-particulate materials; all are particulate but they differ in size and the use of a general term obscures the importance of describing clearly the observations that have been made on the particles in a preparation so that the size range in which they fall can be deduced. There is general agreement that the traditional simple dichotomies such as those between a molecule and a particle or between a solution and a suspension are seldom either real or useful. They may be valuable when substances at the extremes of the range are being contrasted, but in the middle of the range, and this is the part with which we are now concerned, only confusion results from the assumption that they are applicable. It would be possible to lay down arbitrarily the intensity of centrifugation needed to sediment the limiting particle to which the term "particulate" is to be applied but this prejudging of the issue would offer no advantages and would only serve to discourage the complete description of the properties of the preparation.

Comment has already been made on the difficulty of making an extract in which materials will be in the state in which they existed in the intact leaf. This difficulty arises in its most acute form in the study of the organelles. Some can be seen with the microscope or they can be inferred from other properties such as the scattering of infra-red light (LEPESCHKIN, 1942), or the absorption of X-rays at discrete points that BARCLAY and LEATHERDALE (1948) observed, but there is no certainty that the material isolated on the centrifuge is even derived from, let alone identical in properties with, the structure that had been seen. Histological

staining reactions have been applied to the different fractions and the particles
sedimenting at 10.000—16,000 g. in 20 min. from a leaf extract are found to stain
with Janus Green B. This is generally accepted as a stain for mitochondria, and
these are probably an important component of preparations made in this way,
but contamination by chloroplast fragments and other material is to be expected.
There is little evidence about their analytical composition. McClendon (1952)
has measured the ratios between total nitrogen, chlorophyll, ribonucleic acid
phosphorus, and desoxyribonucleic acid phosphorus on a succession of centrifugal
fractions from tobacco leaves, but these ratios are not related to a determination
of dry matter. It is clear that the mitochondrial fraction is richer in ribonucleic
acid and poorer in chlorophyll and desoxyribonucleic acid than the others but it
is not clear what proportion of the whole preparation has been accounted for.

The enzyme activity of mitochondrial fractions has been examined in some
detail. Bhagvat and Hill (1951) measured the cytochrome oxidase activity of
sediments made at 15,000 g. from a variety of leaves, shoots and buds and found
that the enzyme from these sources resembled the enzyme from animal sources
in being concentrated in that fraction. This had been inferred by du Buy, Woods
and Lackey (1950) and has been amply confirmed by others. Several other
oxidative enzyme systems have been found in this fraction and the conditions
under which different enzyme activities are retained by the preparations have
been considered in detail by Millerd, Bonner, Axelrod and Bandurski (1951)
and by McClendon (1953). The first of these papers also contains evidence that
mitochondria can catalyse some phosphorylative reactions. Chantrenne (1947),
working with animal tissues, has argued that there is a spectrum of particles in
an extract varying in size, composition and enzyme activity and that the restriction
of the term "mitochondria" to one particular size is arbitrary. A similar state of
affairs has not been shown to exist in plant extracts but it is probable that it
does so.

Weier and Stocking (1952) remarked wisely that the cytoplasm was any
part of a cell in which the research worker was not at the time interested. This
attitude of mind resulted in a disregard for the parts of a leaf extract that did not
sediment at 15,000 g. because the oxidative enzymes were mainly found in that
sediment. An interest in the contaminants of virus preparations, however, led
Pirie (1950b) to examine the protein sedimenting from leaf extracts at up to
100,000 g. This fraction carried most of the ribonucleic acid of the leaf and Pirie
suggested that it was made up, in part at least, of microsomes and was the precursor
from which a nucleoprotein of much smaller particle size had been derived by
Wildman, Campbell and Bonner (1949). This view is contested by Wildman
and Jagendorf (1952) who look on the material as an artefact made by the
coagulation of low molecular weight proteins by acids liberated from vacuoles
or some other site in the leaf cell. This does not seem likely because the addition
of citrate and other buffers increases the yield slightly and variations in pH in
the range 5.0 to 7.5 were found not to affect it. The possibility that the nucleoprotein is an artefact and not the microsome fraction cannot, however, be excluded
and was indeed suggested in the first place (Pirie, 1950b). The disagreement
illustrates clearly a basic difficulty in this kind of work. Leaf sap is not a physiological fluid and cell components are not exposed to it until the cell is broken;
furthermore cells of many different types and compositions get mixed together
during grinding so that there are ample opportunities for the modification of
proteins. Among the widely distributed leaf components the tannins may be
mentioned as agents that can alter proteins in this way. Bawden and Kleczkowski
(1945) have suggested that precipitation by tannins is responsible for the absence

of virus from extracts of virus-infected strawberry leaves and WAYGOOD and CLENDENNING (1951) suggest that carbonic anhydrase is lost from some leaf extracts in a similar way. If steps are taken to minimise these changes the risk is introduced that artefacts of another type will be made. This may be unavoidable and it suggests that unequivocal evidence that any particle is still in the state in which it existed in the cell may be hard to get.

Whatever its origin, the nucleoprotein that PIRIE (1950 b) found in concentrations as high as 5 g. per litre in sap from the young leaves of tobacco and several other species, is likely to be a component of microsome preparations and it has some interesting properties. It is heterogeneous on the ultracentrifuge and on repeated centrifugation dissociates gradually into a slowly sedimenting protein that may contain 40% of nucleic acid and a nucleoprotein that flocculates on standing for a few hours and contains only 10—15%. Each product contains ribonuclease and the whole microsome fraction brings out about one fifth of the ribonuclease activity of young tobacco leaves. This partition of the enzyme raises again the question, discussed in the section on chloroplasts, of the original position of the enzyme. Here a decision is perhaps easier because the nucleoprotein is a substrate for ribonuclease of either plant or animal origin so that the obvious interpretation is that the enzyme-substrate complex is being separated before it has time to undergo reaction; later there is decomposition of the nucleic acid and separation of the enzyme. The early phases of the action of ribonuclease on the nucleoprotein are accompanied by the precipitation of a protein which still contains some nucleic acid, this is lost on further incubation. It is only possible to prepare the nucleoprotein because these actions are slow especially after the nucleoprotein has been separated from the other components of sap and it is likely that secondary changes explain the very low yields that were got from old leaves. This explanation would account for the apparent discrepancy between the results of HOLDEN (1952) and PIRIE (1950 b) for the former found that even in old leaves there was reason to think that 30% of the total phosphorus was present as ribonucleic acid whereas the latter got negligible yields of nucleoprotein from them.

c) Protein in Extracts Freed from Particles Visible in the Microscope.

Protein is not readily extracted from leaves with a dry texture nor from leaves that contain large amounts of tannin or of acid. This limits the range of species studied and convenience of growing the plants limits the choice still more. The greater part of the published work deals with proteins from spinach, tobacco, tomato, pea, various beans and the various grasses or the mixtures of grass and other plants generally used as forage. Most of the protein in neutral extracts from the fresh leaves of these plants remains in solution when the extracts are centrifuged sufficiently to sediment chloroplasts and the organelles — or if they are filtered through a layer of paper pulp. If the technique used for disintegrating the leaf is efficient the clarified sap from young and well fertilised leaves may contain 50 g. of protein per litre. This is generally called cytoplasmic protein and it is probable that much of it existed *in vivo* in the apparently structureless part of the cell. But, as has been pointed out in earlier sections, difficulties arise when we consider each definable component Lof the protein mixture. Although many enzymes are found predominantly in the clarified sap, part of the activity may be carried by the chloroplasts, organelles, or residual fibre, and the ratio of the two forms of activity depends on the physiological state of the leaf, and on the precise conditions used during the extraction. It is possible that this is the actual state of affairs in the leaf and that the control of its enzyme actions depends in part. on the varying extent to which the enzymes are anchored to or dissociated from leaf structures but it is also possible that the state of dissociation or anchoring is an artefact and is produced by the processes of extraction. No unequivocal evidence about the original state of any of the characteristic cytoplasmic enzymes has so far been produced.

There are two main obstacles to the separation of homogeneous proteins from leaf sap; there are so many different enzymes present that the proportion of the whole protein that exists in the form of each protein is very small, and the proteins resemble each other in their physical properties rather closely. Thus Wildman and Bonner (1947) got a main peak, which accounted for 75% of the protein present, when they examined electrophoretically the product made by passing spinach leaves through a colloid mill and then freeze drying. They recognised that the grinding technique dispersed the chloroplasts so that some chloroplast protein would be added to the mixture, and freezing has been found to bring about changes in other samples of leaf protein. The main peak carried both auxin and phosphatase activity and Wildman and Bonner point out that the minimum molecular weight of the protein, calculated from the auxin content, is 45 times that calculated from osmotic pressure measurements. Partly purified sap phosphatase preparations have 200 times the activity, per mg. of protein nitrogen, that the whole cytoplasmic protein had (Holden, unpublished). It seems likely, therefore, that these activities are carried by components in a mixture of similar proteins and that the observed homogeneity is caused by the electrophoretic uniformity of the normal proteins rather than by the presence of a single component making up 75% of the leaf protein (see section on Electrical Transport, Vol. 1, p. 55).

The uniformity does not extend to all the physical properties of the proteins. Singer, Eggman, Campbell and Wildman (1952) examined by ultracentrifugation tobacco leaf extracts made on a colloid mill and found more peaks in the ultracentrifuge than in the electrophoretic pattern. Sedimentation constants ranged from 25 S to 4 S. The former was associated with protein that precipitated if the extracts were dialysed at pH 6. These observations illustrate yet another difficulty encountered in handling the leaf proteins: the instability of the initial extract. In the section on organelles, the instability of the leaf nucleoprotein in the presence of the ribonuclease and salts from sap was mentioned. The other proteins seem to be similarly unstable and may be precipitated or in other ways altered by ageing in the extract or by various simple treatments. On dialysis against water, a brown precipitate separates containing predominantly protein and carbohydrate. The extent of this separation is diminished if the extract is made alkaline before dialysis or if reducing agents such as cysteine or ascorbic acid are added, these agents are especially effective if added during the initial grinding. Precipitation is also diminished if the dialysis is carried out against 0.02 to 0.1 M salts instead of water and potassium salts are particularly effective. This precipitation is characteristic of extracts made by methods that cause only slight damage to the leaves e. g. grinding in a domestic meat mincer, rather than the more thorough methods of grinding e. g. roller mill or colloid mill. The extent to which these changes are due to enzyme action, to the combination of materials that were kept separate in the leaf before mincing and to slow denaturation or other similar changes in the protein is by no means clear. The phenomenon has not been studied in any detail because, in most investigations of leaf proteins, it is a convenience. The unstable protein does not have associated with it any of the viruses or enzymes that have so far been investigated so that their purification is made easier if the sap is allowed to "mature" sufficiently for this protein fraction to be removed.

Various other procedures remove what is probably the same protein fraction and they have often been used for the preliminary clarification of leaf extracts. The most successful are acidification to pH 5.0 or 4.5, the addition of from one twentieth to one third of a volume of alcohol to the ice cold extract, the addition of K_2HPO_4 or Na_2HPO_4 and heating to 50—55°. By suitable combination of these processes 80—90% of the protein may be removed from the original sap with little fall in the activity of many enzymes. Finally, the processes of fractionation alter the properties of much of the leaf protein in such a way as to make it easier to remove. Thus the supernatant fluid made by ultracentrifuging fresh tobacco sap gives a precipitate with ammonium sulphate which is soluble in water and from this solution much of the protein can be sedimented by further ultracentrifugation under the same conditions as before.

It is convenient to carry out the initial extraction of certain enzymes under conditions designed to avoid bringing out the main bulk of the protein. Thus BUNTING and JAMES (1941) found that much of the carboxylase came out in the juice that could be pressed from frozen but unminced barley leaves. KENTEN and MANN (1952) purified α-hydroxy acid oxidase after extracting it from the outside leaves of brussels sprouts that had been dried at 37°. DAVENPORT and HILL (1952) ground parsley in sufficient cold ammoniacal alcohol to give 50% alcohol in the final solution and so extracted cytochrome f with fewer contaminants that interfere with subsequent purification. The use of acetone powders is traditional both with animal and plant tissue; by this technique most of the tissue protein is made insoluble so that is does not come out in later aqueous extracts and the tissue is made sufficiently brittle by the acetone for grinding to be easy. All these pretreatments damage the structures in the leaf to such an extent that by the study of the subsequent extracts it is no longer possible to get any evidence about the original mode of occurrence of the proteins that are being extracted.

So far no enzyme preparation has been made from leaves that satisfies any of the normally used criteria of homogeneity. GALSTON, BONNICHSEN and ARNON (1951) crystallised a catalase preparation (see also p. 235) from spinach, this had properties in many ways similar to those of preparations from blood but the crystals were always accompanied by some amorphous material. Several enzyme preparations have been made in which the ratio of enzyme activity to protein nitrogen is 300 to 800 times that of the original sap and from these preparations tentative conclusions have been drawn about the probable nature of the enzyme. From among this group the purification of glycolic acid oxidase (ZELITCH and OCHOA, 1953), hydroxy acid oxidase (TOLBERT, CLAGETT and BURRIS, 1949), ribonuclease (HOLDEN and PIRIE, unpublished) and peroxidase (KENTEN, unpublished) may be mentioned.

The study of leaf cytoplasmic protein appears to raise an issue similar to that discussed briefly in the section on latex (p. 59). The peculiarity in latex is the presence of a large amount of enzyme for which no apparent use could be found; here there is a large amount of protein with which no activity at all has so far been associated. The same tentative explanation seems probable — that this protein is made up of enzymes for which we do not yet know the correct substrate. There is no logical objection to such an explanation because we have probably only examined a tiny proportion of the whole number of enzymes present in the leaf but the attribution of unknown enzyme activity to such a large proportion of the total protein, and that, protein of a somewhat uniform type, raises difficulties. NORTHROP (1948) has suggested that part of the protein of the cell is an undifferentiated raw material from which enzymes and other specific proteins are made by simple changes comparable to the change that turns trypsinogen into trypsin. There does not seem to be any positive evidence either for or against this idea but it would offer a simple explanation for at least the uniformity of this part of the protein of the leaf.

There are many other possibilities and some of them are no more unsatisfactory than the two that have been mentioned; they should not, therefore, be overlooked when the interpretation of new phenomena is being considered. There is nothing approximating to a protein balance sheet for any tissue; even in the most intensively studied there is a wide gap between the sum of the known components and the observed protein content. Furthermore there is no means of knowing whether a proportion of the normal protein is dispensible. Some of the apparently normal proteins may be as anomalous as viruses. They may stimulate synthesis in a comparable way and may be produced by the cell as a consequence of an originally external stimulus. Such a substance, doing no good but little harm, would not necessarily be selected against (cf. Pirie, 1952a). Viruses are seldom deliberately sought. They are recognised only when the symptoms caused are so obtrusive that their existence can no longer be ignored so that attention gets focussed on extreme manifestations. There may be a rich vein of trivial and transmissible changes awaiting exploration. On this view much of the protein is innocuous but has no function.

Many substances begin to precipitate from sap as it ages; this raises the possibility that the fresh cytoplasmic protein acts as a solvent and loses this property as it begins to denature. On this view it would resemble the albumin in blood, part of the function of which is thought to be the maintenance in solution of some water insoluble substances such as steroids. Most plants are naturally immune to attack by a wide range of micro-organisms and there is an extensive literature on the antibacterial action of plant extracts (cf. Gäumann, 1950). Part of this resistance may reside in the cytoplasmic protein, that is, it would have a function comparable to that of blood serum globulin. The usual interpretation offered for any biological substance with no obvious function is that it is a reserve; this has a certain plausibility when we consider the cytoplasmic protein because its amount varies with the state of nutrition of the plant whereas many of the enzymes that are characteristic of the mature leaf do not seem to be depleted when the leaf is starved or ages. Finally, there is the possibility that there is very little cytoplasmic protein *in vivo* but that much of it is pulled off from the cell wall, the chloroplasts, or the organelles during the process of making the extract.

II. The Proteins of Virus Infected Leaves.

There is no rigid definition of a perfectly normal and healthy plant. Every variation in the nature and composition of the environment can, if carried to extremes, produce signs that are generally accepted as abnormal and these must be expected to modify at least the proportions in which proteins occur in the plant. They may do more than this. Thus plants growing on seleniferous soil make proteins containing selenium and nucleoprotein synthesised in tobacco plants simultaneously infected with tobacco mosaic virus and treated with an 8-azaguanine derivative or with thiouracil contains these abnormal components (Matthews, 1953; Jeener and Rosseels, 1953). These observations open up interesting possibilities for the study of the chemical mechanism of some forms of poisoning and, incidentally, of protein synthesis, but only a beginning has so far been made.

Definition becomes easier when we consider infected plants but even here it is not perfectly simple. A legume is "normal" when it is infected with the *Rhizobium* causing root nodules and there are many other similar symbionts in other plants. Some bacteria and funguses, however, cause characteristic disease symptoms from which the normal plant is free. These probably reflect great changes in the ratios in which different normal proteins occur, and changes in the enzymic make up of the plant have been described, but no characteristic new enzyme is known to be formed as a result of this type of infection. Infection may produce substances which make the plant immune to further attack or to attack by related organisms, and extracts from plants suffering from bacterial infection have been found to agglutinate bacteria or even to be bactericidal. These phenomena are presumably sometimes due to the appearance of specific proteins but no proteins with these properties have so far been isolated. The extensive literature on this subject is fully reviewed by GÄUMANN (1950).

The plant carrying a bacterial or fungal infection is generally looked upon essentially as a culture medium on which the pathogen grows and from which, in principle at least, it can be distinguished. The virus infected plant is in a different position. The virus is not known to have any independent powers of multiplication but it affects the host in such a way that more virus is made. This is not, as in the case of selenium poisoning, because the agent is built into new protein but because it modifies the subsequent course of protein synthesis in the infected plant. Put simply, virus infection appears to be an induced aberration of nucleoprotein synthesis of such a type that one result of the aberration is the production of more virus (PIRIE, 1950a; BAWDEN and PIRIE, 1953). The multiplication of the virus is thus most intimately involved with the metabolism of the host and for this reason a brief account of the properties of viruses forms a logical part of an article about plant proteins whereas an account of the properties of organisms that could grow on a culture medium would not.

The proteins associated with virus infections have been very much more fully investigated and characterised than those present in uninfected leaves. Various explanations can be given for this anomaly. One is that more effort has been expended on the study because it holds out the possibility of devising prophylactic or curative measures based on knowledge of the properties of viruses. But a more important reason is probably the large size of the particles that make up virus preparations. By ultracentrifugation it is, therefore, easy to separate the proteins carrying infectivity from almost all the rest of the protein in a leaf extract and a centrifuge able to give at least 20,000 g. is essential if work on the chemical properties of viruses is being undertaken. Those viruses that have been well characterised are a uniform group; all are ribonucleoproteins and all are more stable *in vitro* than the generality of leaf proteins or of viruses attacking other hosts. It is easy to overrate the importance of these generalisations. A hundred or more plant diseases are attributed to viruses and from plants with many of these diseases infective extracts have been made. But only about twenty have yielded infective preparations about whose chemical properties there is any unanimity and in which it seems likely that the chemical properties are related to those of the virus. Unsuccessful attempts have been made to study many of the others and it is reasonable to suggest that one reason for the failure is that the viruses in these others have a different type of chemical and physical property. The processes of separation from the tissue and subsequent fractionation are highly selective in a chemical and physical sense and it is to be expected that those viruses that have proved amenable to isolation by a rather uniform group of techniques should share some rather uniform properties.

The disease symptoms caused by the viruses that have been purified cover nearly the whole known range including necroses, mosaics, mottles, and enations (i. e. leafy frills). The viruses are transmitted in many different ways. It is therefore unlikely that the chemical uniformity of the group of thoroughly investigated viruses reflects any underlying biological unity between them. It is, however, characteristic of all of them that they multiply in tissues, such as the leaf, in which cell division is no longer taking place. BAWDEN and PIRIE (1952) have suggested that it is possible that this may be correlated with the universal appearance of ribonucleic acid in them because it is reasonable to assume that deoxyribonucleic acid metabolism plays a smaller part in the metabolism of cells in which no chromosome multiplication is going on than it does in the actively dividing cells in which animal and bacterial viruses are generally studied. It there is substance in this suggestion, the plant viruses that affect predominantly tissues that are undergoing cell division, or that promote cell division as a consequence of infection, would be more likely to contain deoxyribonucleic acid.

This is not the place to discuss whether the characteristic proteins that have been isolated from plants with different virus diseases are indeed the viruses; the issue has been discussed abundantly elsewhere (BAWDEN, 1950; PIRIE, 1950a; BAWDEN and PIRIE, 1952). It seems probable that no preparation consists exclusively of virus particles but that all the particles in it, whether infective or not, resemble each other closely. There is no evidence on the degree of difference between particles that is necessary to cause a difference in their infectivity; with particles as large as plant viruses there is no reason to assume that it would be analytically perceptible. Even the smallest particles for which infectivity is claimed contain at least 10,000 amino acids and any one of them may be essential. This fact makes irrelevant all the evidence, designed to show that preparations are homogenous, which is based on constancy of analytical, ultracentrifugal or electrophoretic properties. The preparations have, however, an interest as plant proteins because some of them are as uniform in their properties as any of the enzymes or other proteins that have been isolated from plants and are, therefore, as likely as the others to be pure.

1. The Preparation of Virus-containing Extracts.

Some information about the properties of viruses can be got by examining the infected tissue under the microscope and some can be got by different forms of analysis performed directly on the infected tissue. Generally, however, it is necessary to extract the virus from the tissue before it can be studied. The difficulties inherent in this process have been discussed in an earlier section; they are abundantly illustrated here.

The efficiency of different methods of extracting virus can be most conveniently studied with tobacco mosaic virus (TMV) because no other virus makes up such a large proportion of the leaf and it is so resistant to enzyme attack that the amount remaining in a tissue fraction can be found by dissolving most of the fraction with a suitable enzyme mixture. BAWDEN and PIRIE (1946) measured the amount of virus liberated from leaves by coarse grinding followed by finer grinding, by incubation with a commercial trypsin preparation, and with the mixed enzymes from snail's crops. Fine grinding in a roller mill (Vol. 1, p. 28) liberated more virus than the commonly used coarse grinding in a domestic mincer; but even after this, more virus could be liberated by enzyme incubation and the snail enzymes were more effective than trypsin. There was, however, evidence that fine grinding was either destroying some virus or irreversibly anchoring it to the fibre because there was better extraction when enzyme incubation preceded grinding instead of following it. LIMASSET, CORNUET and GENDRON (1950) dispute these conclusions and claim complete liberation if the leaves are ground on a high speed mill. The basis for this claim is their failure to get significant virus liberation on incubating with enzymes the fibre residue left after milling, and also their failure to get out more virus by two incubations with enzyme than by milling. The published results do not, however, support the claim. BAWDEN and PIRIE found that fine grinding makes any virus that it does not extract unextractable by other means; the first piece of evidence is, therefore, irrelevant. The figures published by LIMASSET et al. show that more virus comes out on the second incubation with snail enzymes than on the first which suggests that they were working under conditions in which the enzyme was largely inhibited (perhaps by Ca), unfortunately they did not incubate a third time and unless a third incubation is shown to release no more virus there is clearly no reason to think that liberation is complete. Furthermore, samples of the same fibre which had only been milled, were not compared with samples which had been incubated with enzymes before milling. COMMONER, MERCER, MERRILL and ZIMMER (1950) and TAKAHASHI (1951) also claim complete liberation of virus in a glass "homogenizer" because their yields per gram dry weight of infected tissue are of the same order as the 100 mg. per g. of leaf dry matter found by BAWDEN and PIRIE. This cannot be accepted as evidence for complete liberation because BAWDEN and PIRIE never suggested that this was the maximum attainable virus content and there was already evidence at that date, which has been supplemented by the work of BAWDEN and KASSANIS'(1950), that by suitable choice of virus strain and fertiliser treatment higher virus contents can be reached. It seems, therefore, that any claim that all the virus has been separated from a leaf should be treated with reserve unless it is supported by evidence got from exhaustive enzyme digestion. There is no reason to think that a purely mechanical separation of virus from leaf fibre cannot be made because, with TMV, it is only the difficulty of extraction that suggests the presence of a chemical link. All that is maintained here is that there is still no reason to think that this mechanical separation has been achieved.

The small amount of work that has been done with other viruses leads to similar conclusions. With both tomato bushy stunt virus (BAWDEN and PIRIE, 1944) and potato X-virus (BAWDEN and CROOK, 1947) there is clear evidence that the process of fine grinding destroys virus or makes it unextractable from the fibre even by enzyme digestion. The nature of the action is not clear but, with bushy stunt, it can be prevented by incubating the fibre with commercial trypsin preparations before grinding and this procedure gives the highest yields of virus. A few scattered observations suggest that this is a general phenomenon and that the usual techniques of mincing cannot be relied on to extract more than a fraction of the total virus in a leaf.

2. The Determination of Virus in Extracts.

In spite of the uncertainty that there is about what proportion of the total virus appears in any particular type of extract, it is still worth while to consider the methods that can be used to assay the virus in extracts because it is likely that the same method will bring out approximately the same proportion of the virus from leaves of differing total virus content. Also, information about the easily extractible virus in the leaf is of value even although information about the total virus content would be preferable.

The methods that can be used clearly depend on what we choose to include in the category "virus". If we operate restrictively and only include infective material, then no chemical or serological method is valid, and all information must be derived from infectivity measurements. If we extend the category a little more widely and include material that carries the antigenic specificity of the virus although it may not be infective, then the measurement of serological precipitation end point can be used. A further extension of the category is necessary before a method depending on chemical analysis can be considered valid.

a) Serological Determination.

Serological methods are simple and take very little material. Antiserum is made by injecting an animal, usually a rabbit, intravenously three or four times at weekly intervals with a suitable quantity of the antigen; 1 mg. is generally sufficient for each injection. The more highly purified the antigen the greater the specificity of the serum but the normal leaf proteins are such feeble antigens compared to the viruses that they seldom evoke enough antibody formation to cause any cross reactions even when crude preparations are used for the immunisation. The serum is stable for many months if kept at 0° and can be used diluted 1:20 to 1:200 according to the efficiency of the immunisation. Suitably diluted antiserum is mixed with an equal volume of a solution of virus in saline and observed after a few hours at 50°. It is convenient to have 1 ml. of each solution and use a tube about 7 mm. wide which is half immersed in a water bath; convection then keeps the fluid continually mixed and promotes precipitation. Tubes are set up containing different concentrations of a standard purified preparation of the virus being assayed and compared with a range of dilutions of the extract being tested. The precise end point depends on the virus that is being used and on the quality of the serum, the time for which precipitation is allowed to proceed and the visual acuity of the observer. If all these are kept constant it is reproducible within 30%. Some of the strains of TMV will precipitate to an end point of 0.1 mg. per litre but 1 mg. per litre is more usual and this value is also given by strains of potato virus X. The physical properties of solutions and the appearance of dried preparations in the electron microscope demonstrate that these viruses have elongated particles. Other viruses, which appear to have approximately spherical particles, generally do not precipitate at such great dilutions: 10 mg. per litre is the limit for broad-bean mottle virus and 2—3 mg. per litre for tomato bushy stunt virus and some of the tobacco necrosis virus strains. Two types of precipitate are readily distinguished; that given by the elongated viruses has the open, fluffy texture that is characteristic of the "flagellar" antigens of bacteriology whereas the others give the more compact and granular precipitate that is called "somatic".

Several viruses, notably TMV, can exist in different states of aggregation and this affects the serological behaviour. This is important because different types of particle are predominant in extracts made in different ways (BAWDEN and PIRIE, 1945b); to get consistent results all must be brought to the same state before testing. This is most easily done by heating at 60° for a few minutes in the presence of 0.01 to 0.1 M buffer at pH 5.5. The end point of precipitation then seems to be a trustworthy index of the virus content. In the presence of the other components generally present in unfractionated leaf extracts the measurement of the weight, or nitrogen content, of the precipitate separating when virus and antiserum are mixed in equivalent proportion is not a trustworthy index because there is extensive co-precipitation. This widely used method, while it may be suitable for the study of other systems, such as purified virus preparations, is, therefore, inapplicable here.

b) Physico-chemical Determination.

In spite of the convenience of serological analysis much effort has been expended on the design of alternative chemical methods. These depend, in essence, on the determination of the protein or nucleic acid that remains in the system after impurities have been removed by some fractionation procedure. They all, therefore, have the defects already discussed when the validity of total nitrogen determination as a measure of protein content was being considered and also the added defect that discrimination is now being attempted between proteins. The thoroughly studied viruses have, as has already been mentioned, a close chemical uniformity and this might be expected to simplify the problem of chemical determination because a method which works for one should work for several others. In fact, chemical determination has only been seriously attempted with TMV because it is the only one that reaches analytically satisfactory levels. Thus TMV and its strains may account for more than 10% of the dry matter of a leaf, potato virus X may get near this value but tomato bushy stunt, turnip yellow mosaic, broad-bean mottle and southern bean mosaic reach only a third to a tenth of it. Many other viruses are not known to accumulate to an extent greater than 0.1% of the dry matter of the leaf. For the assay of these viruses, therefore, only serological methods are likely to be useful.

In Table 3 an outline of a representative group of the methods that have been used for the chemical determination of TMV is set out. As emphasised in an earlier section, there is no reason to think that there is complete liberation of virus with any of these grinding procedures but the percentage liberation may

Table 3. *Methods for the Chemical Determination of TMV.*

Conditions of grinding	Fractionation procedure	Method of assay	Authors
1. Frozen in pH 7 phosphate	Discard ppt. at pH 4.2 and fluid at pH 3.4	Kjeldahl	HILLS and McKINNEY (1942)
2. Unfrozen in pH 6.9 maleate	Freeze, thaw and dialyse	Planimeter on electrophoresis diagram	WILDMANN, CHEO and BONNER (1949)
3. Unfrozen in 0.5 M sucrose with phosphate	Discard ppt. at pH 4.2 and fluid at pH 3.4	Folin phenol determination	COMMONER, MERCER, MERRIL ZIMMER, 1950
4. Unfrozen	Dialyse at 40° and pH 5 in presence of chloroform	Spectrophotometry at 260 mμ	KUTSKY and RAWLINS (1950)
5. Unfrozen in pH 7 maleate	Freeze, thaw and ultracentrifuge twice	Kjeldahl	MENEGHINI and DELWICHE (1951)
6. Frozen with some phosphate	Ultracentrifuge twice	Spectrophotometry at 260 mμ	TAKAHASHI (1951)
7. Frozen	Heat at 60° for 15 m. and ultracentrifuge	Spectrophotometry at 260 mμ	SCHLEGEL and RAWLINS (1953)
8. Frozen	Shake with chloroform and amyl alcohol and ultracentrifuge	Spectrophotometry at 260 mμ	SCHNEIDER (1953)

well be reproducible. When the extract is only to be used for chemical analysis, the preliminary freezing (methods 1, 6, 7, 8) is an advantage, it is not so efficient for removing non-virus protein as freezing after grinding but it makes the grinding easier and more complete.

Methods 1 and 3 depend on there being little normal leaf protein that precipitates at pH 3.4 while being soluble at pH 4.2 and on the virus being quantitatively precipitated at pH 3.4. The second characteristic may be relied on; the first is more uncertain. Little of the virus gets carried down on the pH 4.2 precipitate when the leaves have a high virus content but the risk of loss increases as the virus content diminishes, conversely the pH 3.4 precipitate from plants that have been long infected and have a high virus content contains much material that becomes unsedimentable on the ultracentrifuge after incubation with trypsin. There seems little advantage in extending the category "virus" to include this. In methods 4, 7 and 8 the normal leaf protein is removed by making use of its ready denaturation or destruction and, because all three methods finish with spectrophotometry at 260 mμ it is particularly the normal nucleoproteins that have to be removed. It follows from what is known about the properties of the main component of the normal extractable nucleoprotein (Pirie, 1950b) that these treatments are likely to be effective. If errors arise they are likely to be caused by incomplete removal of normal purines and pyrimidines on dialysis, which would lead to high values in method 4, and to losses from the ultracentrifuge pellet, which would lead to low values in methods 5, 6, 7 and 8. While an ultra-centrifuge is slowing down and while it is being opened and the tubes removed, the pellets begin to soften and go into solution. This is easily seen if the pellet is watched carefully as the last drops of supernatant fluid are poured away from it. This loss can be avoided if the supernatant is carefully sucked away leaving the small dense layer of redissolved pellet undisturbed, but if this is done the efficiency with which interfering materials are removed by ultracentrifugation is greatly diminished. The loss depends on the exposed area of the surface of the pellet and is, therefore, proportionately greater, the smaller the pellet. In method 2 less attention is paid to the removal of non-virus protein because the specificity of the electrophoretic measurement permits greater discrimination between components in the final stage of the determination.

Lojkin and Beale (1944) pointed out the convenience of the Folin phenol reagent, used on unhydrolysed virus samples, for the determination of amounts of TMV inconveniently small for determination by Kjeldahl and this procedure is used in method 3. The method is trustworthy so long as it is certain that the same strain of virus is being used throughout, but different strains give significantly different amounts of colour under the same conditions. No comment is needed on methods 1 and 4 which depend on nitrogen determination but difficulties arise in the spectrophotometric methods (4, 6, 7 and 8). If extracts from young infected leaves are given a suitable pretreatment, e. g. heating to 60°, colourless virus preparations are easily made from them and the absorbtion spectrum shows the characteristic maximum at 260 mμ (Bawden and Pirie, 1937). The amount of normal protein that is likely to contaminate such a preparation would have little effect on a measurement of absorbtion at 260 mμ because of the feebler absorption here of proteins that do not contain nucleic acid. When older leaves are used, especially from Turkish tobacco plants, the preparations are brown and can only be made colourless by more elaborate fractionation procedures; incubation with trypsin is generally effective. The brown contaminant absorbs in the ultra-violet as well as in the visible part of the spectrum so that the preparations made from leaves in this state would appear to have too high a virus content. It is likely that virus coming from necrotic areas of a leaf will be accompanied by similar substances. Thus Best (1944) finds that virus infection increases greatly the amount of scopoletin in the leaf; this absorbs in the ultraviolet but the extent to which it accompanied TMV during the early stages of purification has not been studied.

The chemical determination of viruses would be simplified if there were some characteristic component that could be determined after total hydrolysis of the leaf. Martin, Balls and McKinney (1939) made use of the resistance of TMV to proteolytic enzymes; their figures for the virus content are probably too high because there are other nitrogen compounds in normal leaves which resist proteases and there is no reason to assume that this fraction is unaffected by virus infection.

A similar difficulty arises in attempts to use the different states of combination of nucleic acid in TMV and normal leaf protein. The amino acid composition of TMV differs significantly from that of the healthy leaf and of fractions made from it; LUGG and BEST (1945) have discussed the possibility of using these differences as a means of estimating the total virus content of the leaf and conclude that the differences are too small to be useful. The results on different amino acids even lead to different conclusions (cf. BAWDEN and PIRIE, 1946). Thus LUGG and BEST found 1.41 to 1.51% of methionine in the protein from different fractions of healthy leaf and 1.13 to 1.34% in those from infected leaves; this fall is reasonable because TMV does not contain methionine. There was, however, also a slight fall in the tryptophan content although TMV contains 2.1% of tryptophan which is more than the percentage found in any fraction from either healthy or infected leaf. Presumably, therefore the process of infection depresses synthesis of a normal high-tryptophan protein. That virus infection should affect the proportions in which the leaf proteins occur is probable *a priore* and the agreement between expectation and observation suggests that amino acid analysis is not likely to be a successful method of virus determination.

3. The Properties of Virus Preparations.

Only two viruses, tomato bushy stunt and TMV, will be considered in any detail, partly because they are sufficient to illustrate the general nature of the phenomena encountered and partly because few of the other viruses have been studied in a sufficient number of laboratories for there to be an agreed body of information about them. As a matter of general comment, stress must be laid on the frequency with which careful study has shown that infection leads to the formation of two or more distinguishable anomalous proteins. With the Rothamsted strain of tobacco necrosis virus there are two serologically related nucleoproteins (BAWDEN and PIRIE, 1945 a, 1950) and, with turnip yellow mosaic virus, MARKHAM and SMITH (1949) find a nucleoprotein and also what is apparently the same protein free from nucleic acid. Similar phenomena with TMV will be discussed later. There is no clear evidence about the relationship between these different products. The link between protein and nucleic acid in turnip yellow mosaic virus is exceptionally fragile so that the obvious interpretation of the anomalous free protein is that it is derived by the breakdown of the virus during isolation but MARKHAM and SMITH present evidence that makes this interpretation unlikely. The possibility that the different products are stages in either the synthesis or the decomposition of the virus has often been mooted but not demonstrated unequivocally. It need not be discussed here because all that is needed for the present purpose is the realisation that the diverse products are there.

a) Tomato Bushy Stunt Virus.

Preparations made by simply ultracentrifuging the sap from infected tomato, *Datura stramonium*, or *Nicotiana glutinosa* plants are pale green and slightly contaminated with normal leaf protein; those made by precipitation with ammonium sulphate under slightly acid conditions are colourless and there is no reason to suspect contamination by any material not closely related to the virus (BAWDEN and PIRIE, 1938, 1943 a). The virus is soluble in water and dilute salt solutions over the whole pH range and crystallises readily as rhombic dodecahedra when precipitated slowly by the addition of several salts including ammonium sulphate (PIRIE, 1945). The crystal form, appearance under the electron microscope, and absence of any anomalies in the physical character of the solutions or of the pellets thrown down on ultracentrifugation all suggest that the particle is spherical.

The sedimentation constant is $S_w^{20} = 132 \times 10^{-13}$ cm. sec.$^{-1}$ dyne^{-1} and this suggests that the hyd:ated particle has a "molecular weight" of 18,000,000 (or 18 Md.). The weight of the dry particle would then be about 10 Md. (MARKHAM, SMITH and LEA, 1942).

Thoroughly purified preparations contain 1.5% of phosphorus and the carbohydrate content and yield of nucleic acid after disruption with alkali suggest that all this is present as nucleic acid. Less rigorously purified products contain an excess of carbohydrate. The nature of the excess carbohydrate is not known but there is chromatographic evidence (MARKHAM and SMITH, 1951) that the sugar in the nucleic acid is ribose. No complete analysis of the amino acid composition has been published.

Solutions of purified virus are stable for many months or even years if kept cold; they gradually lose infectivity but retain their other properties including serological activity. It is characteristic of most viruses that infectivity can be destroyed by ageing or treatment with a variety of agents (H_2O_2, HNO_2, $HCHO$, ultraviolet light etc.) before any other change is observed; this makes it impossible to be sure that every particle in even the most infective preparation is infective because deliberately made mixtures of infective and non-infective material have so far proved inseparable. Bushy stunt virus is denatured if the salt-free isoelectric solution is frozen (BAWDEN and PIRIE, 1943 b) but it is protected by the other components if sap is frozen.

All published work on bushy stunt virus appears to relate to one strain and it is clear that the main anomalous component of sap, made by coarse grinding in a domestic mincer, has the properties described above. By more intimate grinding a further quantity of virus is extracted, part of which is associated with chromoprotein (BAWDEN and PIRIE, 1944); there is reason to think that this is an artefact of grinding. A small part of the virus in mincer extracts is associated with other protein and it precipitates at pH 4, part is associated with solanin and it precipitates at pH 8. Each retains these properties on repeated ultracentrifugation. They have not been studied in any detail but are worth mention as another example of the variety of products that can result from an infection.

b) Tobacco Mosaic Virus.

The position is more complex with this virus because it is clear that infection leads regularly to the production in the plant of several different proteins. There is disagreement about the relationship between these proteins and about the extent to which it is useful to include all of them in the category of "virus".

A typical preparation contains rods about 300 mμ long which sediment easily on the ultracentrifuge (S_w^{20} = about 200×10^{-13} cm. sec.$^{-1}$ dyne^{-1}) and orientate when the solution is stirred or shaken so that a bright pattern is seen when the solution is examined between crossed polarisers. The properties have often been set out in detail e. g. PIRIE (1945); BAWDEN (1950). There is good reason to think that material with these properties exists in the infected cell *in vivo*, but there is also reason to think that much of the material showing these properties in purified preparations is an artefact derived by the aggregation of less anisometric material. Preparations not only contain material of a virus-like nature but originating from material that had been in a different state, they also contain material carrying the serological specificity of the host plant (BEALE and LOJKIN, 1944). This is a very interesting type of contamination and the interpretations that can be put on it are discussed elsewhere (PIRIE, 1953).

Besides the typical preparation, other nucleoproteins, that do not appear to be present in the normal leaf, can be made from plants infected with TMV (BAWDEN and Pirie, 1945 b). These react specifically with antiserum made against TMV, and are either nucleoproteins containing about 5% of nucleic acid and showing the physical properties characteristic of a suspension of highly anisometric rods, or else they can be turned into this form by incubation, especially by incubation with trypsin. It is also claimed (Takahashi and Ishii, 1953; Jeener and Lemoine, 1953) that another anomalous protein is present, this is free from nucleic acid but can be aggregated to rods similar to those of TMV and reacts with anti-TMV antiserum. These claims have not yet been put on a fully satisfactory quantitative basis; that is to say, analytical figures and absorption spectra have not been published for the actual preparation giving a specified serological reaction. There is, therefore, still the possibility that some of the results could be due to the aggregation of small amounts of serologically active nucleoprotein.

The greater part of the anomalous nucleoprotein is either in the typical form or gets into it during the course of isolation. It is soluble in water and dilute salt solutions at neutrality but shows a zone of precipitation on acidification. The region of precipitation varies slightly with the strain of virus and the amount of salt present, in the absence of salt the pH of maximum precipitation is about 4.2 and in presence of dilute salt solutions it is about 3.3. As the salt concentration is increased the solution gets a more intense sheen and becomes increasingly thixotropic and viscous. With ammonium sulphate, at concentrations between 0.3 and 0.4 of saturation, the system breaks up on stirring into a suspension of fibres which have sometimes loosely been called "crystals". The concentration of ammonium sulphate at which this happens depends on the pH, virus concentration, previous history of the virus solution, and the intensity of stirring. Several methods of making purified preparations have been published and they depend on a suitable combination of ultracentrifugation and precipitation by acid and

Table 4. *Amino Acid Composition of Strains of Tobacco Mosaic Virus.*

Amino acid	Strain of Virus							
	TMV	B 2	B 2 A	B 3	B 4	S 1	S 2	S 3
Alanine	5.9	6.1	7.7	7.7	6.9	4.7	4.5	5.3
Arginine	8.7	8.7	9.8	8.4	9.0	9.0	8.6	9.4
Aspartic Acid	10.3	12.3	9.7	10.5	12.6	11.7	11.7	11.2
Cysteine	0.1	0.7	0.6	0.7	0.6	0.6	0.7	0.7
Glutamic Acid	9.7	10.6	11.4	9.7	10.9	10.0	10.0	10.5
Glycine	1.9	2.2	2.5	2.1	1.9	1.8	1.8	1.9
Histidine	0.0	0.1	0.0	0.0	0.0	0.0	0.0	0.0
Isoleucine	5.1	4.5	4.9	5.7	5.7	6.3	6.0	6.2
Leucine	6.9	7.0	7.0	7.3	6.9	7.3	7.4	7.5
Lysine	1.2	1.5	1.8	1.3	1.3	1.3	1.3	1.3
Methionine	0.1	0.1	0.1	0.1	0.1	0.0	0.0	0.0
Phenylalanine	7.3	6.7	7.1	7.0	7.4	6.2	5.9	6.2
Proline	4.6	4.6	4.8	5.0	4.7	4.5	4.1	4.4
Serine	7.5	8.8	8.1	8.8	9.0	7.0	8.8	9.0
Threonine	10.1	8.4	9.4	11.0	12.2	10.8	11.0	12.0
Tyrptophan	1.7	1.8	1.8	1.8	1.7	1.7	1.7	1.6
Tyrosine	3.3	3.1	3.8	3.5	3.4	3.2	3.5	3.8
Valine	9.2	8.9	9.9	10.2	10.7	9.8	9.8	9.6
Sum	94.3	96.0	100.1	100.6	105.0	96.8	97.0	100.6

The figures in this Table have been recalculated from those of Black and Knight (1953) by multiplying by $\dfrac{\text{Molecular weight of amino acid-18.}}{\text{Molecular weight of amino acid}}$

They therefore give the weight of each amino acid, less the elements of water, found in 100 g. of virus preparation. If the virus strains contain 5% of nucleic acid their sum should clearly be 95.

salts. Like some other materials made up of very anisometric particles, TMV solutions separate into two immiscible phases at a concentration that depends on the state of aggregation of the component rods. This generally happens between 2 and 3%, the lower phase is the more concentrated and it is liquid crystalline and contains a smaller proportion of impurity than the upper phase.

Purified preparations made up exclusively of particles more than 200 mμ long contain 0.5—0.6% of phosphorus all in the form of ribonucleic acid; preparations containing shorter particles have a variable composition but it is uncertain to what extent this is due to their being simple mechanical mixtures of unaggregated virus and sedimentable parts of the normal leaf protein. The nucleic acid is sufficiently firmly attached to the protein to accompany it during sedimentation on the ultracentrifuge over a wide pH range and in different salt solutions but it is easily separated by heating or treatment with acetic acid or pyridine (BAWDEN and PIRIE,1937), by urea and a group of other organic substances (BAWDEN and PIRIE,1940) and by strontium nitrate (PIRIE,1954). The protein moiety has been the subject of extensive amino acid determinations which have not always given consistent results. With the improvement in analytical methods in recent years there has been better agreement and Table 4 is based or the most recent figures published by BLACK and KNIGHT (1953). The different B and S strains were derived from the parent TMV and are thought to be the results of successive one-step mutations accompanied by a change in the proportion of one amino acid for each step.

The nitrogen contents of the virus preparations analysed were not stated so that it is not possible to express the results in the same form as in Table 1 (p. 31). Instead, the percentage of each amino acid, as given by BLACK and KNIGHT (1953), is multiplied by $\dfrac{\text{Molecular wt} - 18}{\text{Molecular weight}}$

The figures are therefore the weights of each amino acid minus the elements of water and, for any protein that is so large that the proportion of the amino acids in terminal positions is negligible, their sum should be 100 for a protein without prosthetic groups. All these strains of TMV presumably contain 5—6% of nucleic acid so the sum of the figures in the Table 4 should be 94—95. Until we know why so many come to more than this it is not clear what weight can be put on the apparent differences in composition.

This work is a useful beginning to the study of the essential differences between virus strains but it needs to be supplemented by analyses on batches of virus isolated from different host species and from hosts in differing nutritional condition. These may well alter the amino acid composition of the virus for, as mentioned earlier, a particle sufficiently like TMV to accompany it through the processes of purification, but containing azaguanine, is built up when virus synthesis proceeds in plants poisoned with a related substance. There is no reason to assume, *a priore*, that the amino acids in the protein moiety will be more completely insulated from changes in the host than the components of the nucleic acid. On the other hand a significant difference in the biological properties of a virus particle could be the consequence of the substitution of a single amino acid by another or even of a rearrangement. Neither change would be perceptible by the analytical methods used so far.

III. The Immature Leaf and Other Young Tissues.

Many tissues in the early stages of development offer peculiar advantages for biochemical work. There is active cell division in them, whereas there is none in seeds and most parts of the leaf, and little in tubers and the other underground parts that have been extensively studied. With this goes a high metabolic rate. The enzyme activity of these tissues is often high and the ratios in which different enzymes occur may be quite different from that in mature tissue; it may therefore be essential to use them as starting material for certain enzyme preparations. In the same way the absence of chlorophyll and the common absence of substances such as tannins, that can complicate separations, is an advantage. The individual cells are, in general, smaller than those in mature tissue so that a larger proportion of the dry matter is likely to be derived from the nuclear apparatus; this expectation is borne out by the high deoxynucleic acid content. Seedlings have two other practical advantages; they can be grown quickly with little equipment at any time of year, so that experiments do not have to be planned months in advance, and they are generally soft or brittle so that grinding is exceptionally easy.

Only scattered observations have been made on buds and other types of immature leaf and most of them have been incidental to work on the nucleic acids. Brussels sprouts are sometimes used as a source of enzymes and MARSH and GODDARD (1939) found differences between the terminal oxidases of immature and mature carrot leaves. Only in the former was oxygen uptake inhibited by cyanide and azide. Most work has been done on seedlings and the age of the seedling clearly affects the ratios in which immature leaf, cotyledon and root occur. Sometimes the parts are separated but in the published work it is not always clear whether this has been done. It is important to use seedlings at the right stage of development, not only because of this variation in ratio but also because enzyme activity varies greatly with age. Sometimes it falls steadily; this happens with the ability of pea seedlings to convert fructose diphosphate to phosphoglyceric acid (TEWFIK and STUMPF, 1951). More often there is a period of maximum activity. KENTEN and MANN (1951) found that the amine oxidase activity of pea seedlings was highest at the tenth day and barely perceptible by the 30th, GOKSÖYR, BOERI and BONNICHSEN (1953) found a maximum on the second day for alcohol dehydrogenase and on the sixth for catalase.

Seedlings have been ground by the methods used with other tissues so that the distribution of enzymes among the fractions isolated by differential centrifuging can be studied. The results are comparable with those on other tissues; protein is found in all the fractions and there is very definite concentration of certain enzymes in some of them. The difficulties involved in drawing conclusions from such measurements of enzyme distribution have already been emphasised in the sections on Grinding and on the Proteins of Organelles (Vol. 1, p. 26 and this Vol. p. 41) and need not be stated again. STAFFORD (1951) has discussed these in the course of setting out her very thorough investigation of the distribution of enzymes in extracts from pea seedlings grown for three days. Cytochrome oxidase and succinic dehydrogenase were associated with the readily sedimentable particles whereas ascorbic oxidase, amylase and phosphorylase were mainly in the supernatant fluid even after 30 min. at 60,000 g. Many other enzyme systems have been found associated with the organelles from seedlings. One of the most interesting is the enzyme from mung beans that brings about oxidative phosphorylation (BONNER and MILLERD, 1953). The proteins of the very young shoots of barley and other seeds are discussed in the section on Seed Proteins (PACE, p. 69).

Work on seedlings is likely to increase rapidly for the reasons already given and also because of the interest that is always associated with a developmental process. We have here a system into which only water and oxygen need go and in which the sequential development of structure and of the processes of synthesis can be followed. But it is the mature leaf that is of primary importance in agriculture and the sciences that spring from it and it will be a pity if work on seedlings, which differ significantly from the mature leaf in composition and metabolism, should divert too much attention from it.

E. Proteins from Miscellaneous Parts of the Plant.

It might be logical to consider the proteins, and the methods used in their study, of all the tissues connected with reproduction together. But so much more work has been done on the seed proteins than on those of any of the other tissues connected with this function that it is convenient to treat them separately (see Pace, p. 69). The remaining tissues are united, not only by their function, but also by having been but little investigated. Enzymes have been demonstrated in all the tissues, total nitrogen determinations have been made on most of the materials that are used as food by man or domestic animals and the part that the pollens play in allergic states has led to some investigation of pollen protein. But for the most part this is an undeveloped field.

1. Flowers.

The only flowers used in bulk as foods rather than condiments are the immature masses of flower buds in cauliflowers and broccoli. The nitrogen content of selected parts may be as high as 6%; this is largely protein nitrogen. Several enzyme preparations have been made by grinding and extracting cauliflower heads but the only claim for the preparation of a protein with characteristic properties relates to a crystalline nucleoprotein carrying ascorbic oxidase activity (Tadokoro and Takasugi, 1939, 1941) and this has not yet been confirmed. Protein preparations have been made from the petals of a few flowers e. g. *Wisteria* (Sinano, 1939), by extraction with dilute alkali and precipitation with acid but they have not been characterised. A flavo-protein has been recognised in extracts of the spadix of *Arum* (James and Beevers, 1950); this tissue has an exceptionally high metabolic rate and its temperature may be 20° above that of the surroundings.

2. Fruits.

Total nitrogen has been determined on most of the fruits that are normally eaten and the presence of a wide range of enzymes has been demonstrated in them. Nitsch (1953) reviews the extensive literature on the changes in enzyme activity and protein content during the growth and ripening of fruits. The fruit resembles other parts of the plant in that its enzyme content may depend more on its age and on the cultural conditions than on the variety of plant used. Because of the economic importance of fruit storage, much work has been done on the changes that go on in fruit kept under different conditions. In general ageing is accompanied by proteolysis but Turner (1949) has shown that the apple synthesises protein for many months at 0°. It has proved difficult to extract protein from many fruits in an unaltered form and extraction with dilute alkali has generally been resorted to. The separation of protein from orange pulp by extraction with 1% NaOH in 50% ethanol (Sinclair, Bartholomew and Nedvidek, 1935) is typical of this type of preparation.

Hulme (1946) suggested that with acid fruits, such as apples, the normal processes of grinding might bring the protein into contact with acids from which it was separated *in vivo*. He therefore froze the fruit and ground it finely while still frozen, this cold powder was then stirred into a large volume of borate or phosphate buffer so that the protein was only exposed momentarily to acid while in solution. By this means he (Hulme, 1951) has made preparations containing up to 50% of protein. These preparations are not in a state in which the protein mixture can readily be fractionated but they are a definite advance on those made hitherto.

Although many fruits have a low protein content, and what protein there is is not easy to bring out in an extract, there are some from which relatively pure enzyme preparations can be made in quantity. It is well known that some fruits are powerfully proteolytic and extracts from them have been used for softening meat both in primitive and modern cookery. Of these papain is the best known and has been most widely studied. It is the name for a group of enzymes found in the latex of unripe papaya and from it two crystalline preparations have been made with distinct properties and enzyme activities (BALLS and LINEWEAVER, 1939; JANSEN and BALLS, 1941). Protease preparations have also been made from pineapple and from the fruits of *Bromelia pinguin* (ASENJO and FERNANDEZ, 1942), *Pileus mexicanus* (CASTANEDA, BALCAZAR and GAVARRON, 1942), *Tabernae-montana grandiflora* (JAFFE, 1943) and *Solanum elaeagnifolium* (GREENBERG and WINNICK, 1940). DUNN and DAWSON (1951) have crystallised a copper-containing protein which carries ascorbic oxidase activity from the extract of the rind from squash. The yield was only about 0.2 mg. per litre of rind pulp; the other fruits gave much higher yields of protein but with some there was little or no evidence that the protein was homogeneous.

3. Pollen.

In a series of papers SOSA-BOURDOUIL has reported the analytical composition of pollens; the nitrogen contents range from 3 to 7% and she (1939) finds that different samples of pollen from the same species have the same nitrogen content and that this, at least in the *Ranunculaceae*, can be used as a basis for classification. Because of the importance of pollen as a food for bees several analyses of mixed pollens have been made (e. g. VIVINO and PALMER, 1944) and these also suggest that the average protein content is 25% of the dry matter. Pollens have been found to bring about a number of enzyme actions (OKUNUKI, 1939) but active protein preparations have not apparently yet been separated. Histological studies on pollen show, as would be expected from their importance in reproduction, a high concentration of nucleic acid (e. g. PAINTER, 1943) and a nucleoprotein has been isolated (v. EULER, HELLER and HOGBERG, 1949).

Most of the work on the isolation of proteins from pollen has had as its object the preparation of test or immunising antigens for use in hay fever. Several proteins are present in pollen extracts. AUGUSTIN (1953) separated from grass pollen an albumin fraction which gave only skin reactions and a globulin fraction which also precipitated specifically with antisera. The latter could be fractionated further into a number of components all active. Ragweed pollen extracts have been fractionated with somewhat similar results in several different institutes. The main allergen is a heat coagulable protein that precipitates with antisera but skin-reacting substances of uncertain nature and apparently smaller molecular weight are also present. In this work quantitative data have seldom been given but BROWN and BENOTTI (1943) report yields of soluble protein amounting to 1.7 mg. of nitrogen per gram of defatted pollen.

4. Proteins in Latex.

In many families there is a system of latex cells or vessels which pervades all parts of the plant. When this system is cut a fluid, generally milky, exudes; this is the latex and it is peculiar among the fluids that have been discussed in connection with plant proteins in that it is natural and is not an artificial mixture made during the process of extraction. There are great differences in composition between the latexes of different species; some, as is well known, contain predominantly rubber; others have a high nitrogen content. Thus ULTÉE (1923) found 14.6% of nitrogen in the dry matter of *Ficus callosa* latex and much of this appears to be protein. Only two types of latex protein have been studied in detail; the protein that accompanies rubber through the first stages of refining and the proteases from a number of different plants.

The protein in the latex of *Hevea brasiliensis* is of great technological importance because many of the methods of coagulating the rubber particles depend on its properties. After coagulation much of the protein is associated with the rubber but some remains in the serum and the question whether there are two distinct proteins or one protein which is only partly precipitated has been vigorously discussed. Most of the protein can be dissociated from the rubber particles by replacing it with ammonium oleate (VERGHESE, 1948) and the protein in the serum can be precipitated by the usual agents such as ammonium sulphate (BONDY and FREUNDLICH, 1938). TRISTRAM (1941) found no difference in the amino acid composition of protein isolated from crepe rubber and that extracted by borate buffer from dried whole latex; the former is predominantly the protein of the serum so that his observation does not support the idea of two proteins but it does not refute it. Several enzymes have, however, been recognised in latex (cf. DE HAAN-HOMANS, 1950) so that it seems only reasonable to conclude that the protein is a mixture.

HOMER knew that the juice from fig leaves would clot milk, and juices and extracts from many plant species have been used for centuries in cheese making in different parts of the world. The action is generally due to the presence of a protease but it is sometimes due to a heat stable component. Thus the active extract of *Galium verum*, which was at one time the plant most commonly used for cheese making in Europe and whose popular name in most European languages implies this, is made by boiling. Slight protease activity is found in most leaves (TRACEY, 1948) but in some it is powerful and in them the activity is greatest in the latex. From fig latex a crystalline protein which carries protease activity can be made very simply. WALTI (1938) clarified the latex, adjusted the pH to 5 and allowed crystallisation to proceed at 5°. The purified enzyme is available commercially and a beginning has been made (WOOD, 1942) in the study of the variation of enzyme content with cultural conditions. The latex from papaya has already been mentioned and that from *Asclepias speciosa* contains enough protease to be of possible economic importance. Here also the activity is carried by a crystallisable protein (CARPENTER and LOVELACE, 1943). The first steps towards the purification of the proteases from other latexes have been taken with several plants e. g. *Euphorbia lathyris*, *Asclepias mexicana* and *Hura crepitans*, but these preparations have not, so far, even been crystallised.

The high concentration of protease in the latex of so many species invites speculation about the possible function of the enzyme. If we assume that the protein is acting as an enzyme which catalyses *in vivo* an action similar to that by which we investigate it *in vitro*, it is very difficult to understand the capricious distribution of enzyme activity. Thus, although all leaves with powerful protease activity contain latex, not all leaves with latex are strongly proteolytic. The latex of *Hevea brasiliensis* for example, is not. Leaves with the usual feeble proteolytic activity seem to be at no disadvantage and it seems unlikely that the feebleness is due to a masking of the activity because TRACEY (1948) has shown that the amount of protease in tobacco is sufficient to account for the rate at which protein-splitting is known to take place. The same type of difficulty arises in interpreting the protease in pineapple or the urease in soya bean; other similar fruits and beans function as well with very much less enzyme. Some have argued that the enzyme has no function but evolved accidentally and survives because it is doing no harm. This is possible and such an origin may well explain the presence of enzymes and other substances when they make up only a small proportion of the whole weight of the tissue (cf. PIRIE, 1952 a) but it does not seem likely that any large scale useless diversion of metabolites would survive. An alternative explanation is that these proteins are storage or reserve substances and that their enzymic activity is an accident of which no use is made by the plant. Again this is possible but there is no evidence favouring it. A more attractive suggestion is that they function as enzymes in the plant but that they catalyse a different action from the one that we are using for their detection. On this theory all comparable species and organs would manifest enzyme activity at a similar level if we knew what reaction to investigate. The peculiarity of the enzymes mentioned here would then be that in these particular species the protein has a less restricted specificity and can bring about another action as well. This latter action would be the one by which we have hitherto recognised these enzymes but it may have no role or only a trivial role *in vivo*. The weakness of this suggestion is that it is not easy to see what the normal action might be. The most obvious possibility is that it is some phase of protein synthesis. If this were the correct interpretation it would mean that, whereas all tissues contain enzymes that can bring about protein synthesis if we use the correct substrates and maintain the correct conditions, it is only in certain tissues that these enzymes are also able to cause proteolysis. Such an interpretation is far from satisfactory and the whole subject is so important that it would amply repay a detailed investigation of the manner in which specific enzymes accumulate during growth and their variation with changes in the physiological state of the tissue.

5. Proteins in Roots, Tubers and Other Underground Parts of the Plant.

Little is known about the protein content or the properties of the proteins in the ordinary roots with which plants anchor themselves and absorb nutrients and what is known has been learnt mainly by histochemical methods. These show extremely active metabolism at the growing tip where cell division is taking place; many enzymes have been recognised at this site and some have been partially purified. The transparency of some roots makes them particularly suitable for spectroscopic study; LUNDEGARDH (1952) has followed in this way some enzyme actions in wheat roots *in vivo*.

The proteins in the hypertrophic parts of roots and in the tubers that are used for food have been more extensively studied. Most work has been done on the potato tuber but that of MORRIS, WEAST and LINEWEAVER (1945) on carrot must be mentioned because they showed clearly that peroxidase and some other enzymes were concentrated in the outer layers and they considered the uncertainties that may be introduced into enzyme assays on tissue suspensions by variation in the state of adsorption, rather than in the total quantity, of enzyme. Work on potato tuber proteins has had two main objects; the important part that the potato plays in nutrition has led to many separate investigations of the nutritive value of potato protein and an interest in the active metabolism of the tuber has led to the purification of several of its enzymes. This separation of interest has had the unfortunate result that those interested in nutrition appear to assume, and sometimes even state the point explicitly, that there is only one protein in the tuber whereas the work of the second group has always made it probable that there are many. The importance of the potato as a food has however, led to an unusually complete study of the effects of variety, age and cultural treatment, on the protein content.

STREET, KENYON and WATSON (1946) determined the weights of the different parts of King Edward potatoes during the growing season and the distribution of the different forms of nitrogen in each part. NEUBERGER and SANGER (1942) found that the protein content of ten different potato varieties ranged from 0.66 to 1.2 percent of the fresh weight although the potatoes were all grown under the same conditions; many studies have shown that differences in manurial treatment cause significant variations in tuber composition. Precise results are, however, difficult to get because the individual tubers from the same plant, even when they are the same size, do not always have the same composition (Goldthwaite, 1925). Under the usual conditions of storage the protein content remains approximately constant for several months (Crook and Watson, 1948), but changes begin as soon as the tuber is warmed. There is some evidence (Levitt, 1952) that at first there is protein synthesis. As the development of the new plant from the "seed" potato proceeds, protein disappears, though slowly, and even when the weight of the above ground part of the new plant reaches 285 g.. one third of the protein still remains in the old tuber (Street et al., 1946).

Protein is not uniformly distributed inside the tuber. The protein content, expressed as percentage of the dry matter, is highest in and immediately below the skin and falls in the successive concentric zones towards the centre (Glynne and Jackson, 1919; Neuberger and Sanger, 1942). The protein in the skin is less readily extracted than that in the other parts. For these reasons it is important to distinguish between results on whole and peeled potatoes.

Essentially the same method has been used to prepare bulk samples of potato protein from the early studies of Osborne and Campbell (1896) and Kiesel, Belozersky, Agatov, Biwschich and Pawlowa (1934) down to those of Saito and Watanabe (1951). The fresh tuber is grated and extracted with water or 2% NaCl solution, from the clarified extract, protein is precipitated by salting out or heat coagulation. Determinations of the amino acid composition of this bulk protein have been somewhat discordant, perhaps because of the different ratios in which the component proteins occur in different preparations, but there appears to be agreement that its principle nutritional inadequacy is deficiency in the sulphur containing amino acids. After repeated extraction Neuberger and Sanger (1942) found that 8.5% of the total nitrogen of the whole tuber remained insoluble; they compared different methods of precipitating the soluble protein from extracts and found that all gave approximately the same results.

Several different procedures have been used to fractionate extracts made in this way; from among them three studies may be mentioned. Groot, Jansen, Kentie, Oosterhuis and Trap (1947) separated the protein soluble in water from that soluble only in dilute salt and they also used differential precipitation with a quarternary ammonium soap. Chick and Slack (1949) separated the acid precipitable protein from the protein that could afterwards be coagulated by heat; they found that the latter predominated in freshly dug tubers and the former in those that had been stored for a year. Levitt (1952), in a more detailed study, fractionated extracts from freeze dried potatoes by ultracentrifugation at up to 43,000 g. The freezing and drying may well have brought about changes (cf. Vol. I, p. 53) but he found that as much as one tenth of the protein may be present in particles sedimenting in the manner of mitochondria or microsomes and that it is these fractions that change most during the breaking of dormancy.

Potatoes have been used as the source of other purified enzyme preparations. Thus Kubowitz (1937, 1938) prepared polyphenol oxidase, and Kalckar (1944) adenylpyrophosphatase, from them. The preparation were as active, or more active, than those from other sources but neither was claimed as pure.

Although about twenty different enzyme activities have been recognised in potato tubers only the enzymes associated with the synthesis of starch have been highly purified. In the early phases of this work conventional methods of precipitation with ammonium sulphate and lead acetate were used; the products were clearly inhomogeneous electrophoretically. GILBERT and PATRICK (1952) separated fractions at —5° C by adding ethanol in the presence of low concentrations of citrate and, after precipitating the fraction soluble in 11% v/v alcohol but precipitated by 14.9%, they were able to crystallise Q enzyme from it with ammonium sulphate. A litre of the original potato juice probably contains about 16 mg. of this protein. Q enzyme is responsible for the synthesis of the branched, amylopectin, starch chain and they found that it was no longer present after March when sprouting begins. Phosphorylase, which builds up the unbranched 1:4 chain, could not be crystallised by this method and it was not even possible to get repeatable results on its precipitation because of interference by other substances which varied with the variety of potato and the season. This difficulty was avoided by BAUM and GILBERT (1953) who first adsorbed both enzymes on to starch. This step removed the interfering substances so that both enzymes could be crystallised separately after elution from the starch. A crystalline preparation has also been made by FISHER and HILPERT (1953). Other enzymes which affect the branching of the chain have also been prepared but not in such a high state of purity.

Horseradish is the traditional source of peroxidase and it remains the most satisfactory (cf. p. 231). THEORELL (1942) made a crystalline enzyme preparation from this source by repeated precipitation with ammonium sulphate and alcohol, and removal of impurities by electrophoresis. From the haemin content, THEORELL and MAEHLY (1950) calculate a molecular weight of 44,000; this is the same as that found by measuring rates of diffusion and sedimentation on the ultracentrifuge. A simpler method of purification is given by KEILIN and HARTREE (1951) and the method has been still further simplified by KENTEN and MANN (1954). The first seems to be homogeneous when examined on the ultracentrifuge but it does not apparently crystallise; one fraction of the second crystallises but has not been examined centrifugally. Several other roots are also rich in peroxidase and highly purified preparations have been made from the sweet potato (KONDO and MORITA, 1952).

The characteristic reddish colour of legume root nodules attracted attention many years ago and the recognition that the colour was due to a protein resembling muscle haemoglobin (KUBO, 1939; KEILIN and WANG, 1945) has led many to investigate its function in nitrogen fixation in the nodules (see also p. 239). The pigment is not made by either the legume root or the *Rhizobium* alone but only when there is symbiosis and a nodule is formed; even then it is only made if the nodule is effective, that is, is able to fix nitrogen. It is confined to those cells in the nodule that contain bacteria and it is easily extracted by grinding with water (SMITH, 1949). There are no anomalies in the distribution of amino acids in it (WILSON, 1952).

ELLFOLK and VIRTANEN (1950, 1952) have studied the methaemoglobin in oxidised preparations made from soya bean nodules and find two components distinguishable by their electrophoretic mobility but not by ultracentrifugation. The molecular weight is about 20,000 and they contain 0.26—0.27% of iron.

There is general agreement that this haemoglobin plays a part in nitrogen fixation because it is present in all those legume nodules that fix nitrogen and absent from those that do not, furthermore, fixation is inhibited by agents such as carbon monoxide which combine with haemoglobin. There is not sufficient haemoglobin for it to act as a significant oxygen store and SMITH (1949) finds no difference in the rate of oxygen uptake by effective and ineffective nodules nor is the uptake by effective nodules diminished when sufficient carbon monoxide is introduced to combine with nearly all the haemoglobin. It is unlikely therefore, that it plays a role similar to that played by haemoglobin in oxygen transport in vertebrates and such invertebrates as contain haemoglobin. It is worth emphasising that it does not occur in comparable quantities in the root nodules of the non-leguminous plants e. g. alder and bog myrtle, which fix nitrogen, nor in the nitrogen fixing microorganisms such as *Nostoc, Azotobacter* and *Clostridium pasteurianum*. So far as is known haemoglobin occurs in no part of any higher plant though it is present in yeast (KEILIN and TISSIÈRES, 1954). No other substance has been found in a higher plant that can be oxygenated in the same way as haemoglobin. There seem, therefore, to be many clues for interpreting its mode of action but no satisfactory suggestion has yet been made. *In vitro* haemoglobin catalyses the decomposition of hydroxylamine to nitrogen and the obvious possibility is that *in vivo* this action is reversed but KEILIN and SMITH (1947) have discussed the difficulties that arise if this simple idea is adopted.

The other proteins of the root nodules have received little attention but ZELITCH, WILSON and BURRIS (1952) find a normal distribution of amino acids in that part of soya bean nodules that is soluble after grinding in 4 N HCl.

F. Proteins in Algae.

Seaweed has often been suggested as a cattle fodder and many analyses of the various species have been published. The N content varies from about 3 to 7 % of the dry and ash-free matter (STEWART and WOODWARD, 1953; MILNER, 1953) and most of this N appears to be protein, but few attempts have been made to isolate it and in the unisolated state, in ground whole weed, it is not well used by non-ruminant animals (RINGEN, 1939). Recently interest has been growing in the possibility of culturing unicellular algae such as *Chlorella* as sources of food or fodder. The N content can vary from 1.4 to 9.3 % of the dry and ash-free material according to the conditions of growth. FOWDEN (1952), using a batch containing ·4.3% N, extracted most of the protein by grinding the suspension in a needle valve and then treating it with an ether : ethanol mixture. His amino acid analysis of the bulk protein accounted satisfactorily for all the N present and it is given, along with some other analyses of plant proteins, on p. 31. As might be expected from the fact that it is a mixture of a great many different proteins no anomalies appear in the distribution of amino acids in it.

The two most fully characterised types of algal protein are phycoerythrin and phycocyanin, they are highly coloured and all recent work, e. g. EMERSON and LEWIS (1942) and FRENCH and YOUNG (1952), tends to confirm ENGLEMANN's original (1883, 1884) conclusion that they are an important part of the light absorbing mechanism used in photosynthesis. Proteins falling into these categories and with slightly different properties can be extracted from many species by drying, grinding and extracting with water (BOUILLENNE-WALRAUD and DELARGE, 1937) by allowing the tissue to putrefy slightly (SVEDBERG and KATSURAI, 1929) or even by leaving marine species in unaerated sea water (HAXO and BLINKS, 1950) or in fresh water. From these extracts the proteins can be crystallised by the addition of ammonium sulphate (SVEDBERG and KATSURAI, 1929; KRASNOVSKII, EVSTIGNEEV, BRIN and GAVRILOVA, 1952). Preparations made in these ways appear homogeneous by electrophoresis and on the ultracentrifuge and the sedimentation constant suggests a molecular weight of about 300,000. After fission with acid or after enzymic destruction of the protein the polypyrrole components, or phycobilins, can be prepared; their relations to the bile pigments have been discussed in detail by LEMBERG (1930) and by LEMBERG and BADER (1933).

References.

ARMSTRONG, R. H.: J. Sci. Food Agric. 2, 166 (1951). — ARNON, D. I.: Plant Physiol. 24, 1 (1949). — ARNON, D. I., and F. R. WHATLEY: Arch. Biochem. 23, 141 (1949). —ASENJO, C. F., and M. C. C. FERNANDEZ: Science 95, 48 (1942). — AUGUSTIN, R.: Biochem. J. (proc.) 54, 10 (1953). — AXELROD, B., and A. T. JAGENDORF: Plant Physiol. 26, 406 (1951.)

BALLS, A. K.: J. Wash. Acad. Sci. 32, 132 (1942). — BALLS, A. K., and H. LINEWEAVER: J. Biol. Chem. 130, 669 (1939). — BARCLAY, A. E., and D. LEATHERDALE: Brit. J. Radiol. 21, 544 (1948). — BAUM, H., and G. A. GILBERT: Nature (Lond.) 171, 983 (1953). — BAWDEN, F. C.: "Plant Viruses and Plant Diseases". Chronica Botanica. Waltham Mass. USA. 1950. — BAWDEN, F. C., and E. M. CROOK: Brit. J. Exp. Path. 28, 403 (1947). — BAWDEN, F. C., and B. KASSANIS: Ann. Appl. Biol. 37, 215 (1950). — BAWDEN, F. C., and A. KLECZKOWSKI: J. Pomol. 21, 2 (1945). — BAWDEN, F. C., and N. W. PIRIE: Proc. Roy. Soc. B. 123, 274 (1937); Brit. J. Exp. Path. 19, 251 (1938); Biochem. J. 34, 1258 and 1278 (1940); (a) Biochem. J. 37, 66 (1943); (b) Biochem. J. 37, 70 (1943); Brit. J. Exp. Path. 25, 68 (1944); (a) Brit. J. Exp. Path. 26, 277 (1945); (b) Brit. J. Exp. Path. 26, 294 (1945); 27, 81 (1946); J. gen. Microbiol. 4, 464 (1950); Ann. Rev. Plant Physiol. 3, 171 (1952). — BAWDEN, F. C., and N. W. PIRIE: in "The Nature of Virus Multiplication". Cambridge: Ed. P. Fildes and W. E. van Heyningen 1953. — BAWDEN, F. C., and F. M. ROBERTS: Ann. Appl. Biol. 34, 286 (1947). — BEALE, H. P., and M. E. LOJKIN: Contrib. Boyce Thompson Inst. 13, 385 (1944). — BEST, R. J.: Austr. J. Exp. Biol. Med. Sci. 22, 251 (1944). — BHAGVAT, K., and R. HILL: New Phytol. 50, 112 (1951). — BICKOFF, E. M., A. BEVENUE and K. T. WILLIAMS: Chemurgic Digest. 6, 215 (1947). — BLACK, F. L., and C. A. KNIGHT: J. Biol. Chem. 202, 51 (1953). — BLOCK, R. J., and D. BOLLING: "The Amino Acid composition of Proteins and Foods". Thomas, Springfield, Illinois 1951. — BLOCK, R. J., and H. H. MITCHELL: Nut. Abs. Rev. 16, 249 (1946/47). — BONDI, A., and H. MEYER: Biochem. J. 48, 248 (1948). — BONDY, C., and H. FREUNDLICH: C. R. Lab. Carlsberg 22, 89 (1938). — BONNER, J., and A. MILLERD: Arch. Biochem. 42, 135 (1953). — BOUILLENNE-WALRAUD, M., and L. DELARGE: Arch. Inst. Bot. Liége 14, 5 (1937). — BROWN, E. A., and N. BENOTTI: Ann. Allergy 1, 150 (1943). — BUNTING, A. H., and W. O. JAMES: New Phytol. 40, 262 (1941).

CARPENTER, D. C., and F. E. LOVELACE: J. Amer. Chem. Soc. 65, 2364 (1943). — CASTANEDA, M, M. R. BALCAZAR and F. F. GAVARRON: Science 96, 365 (1942). — CHANTRENNE, H.: Biochim. Biophys. Acta 1, 437 (1947). — CHIBNALL, A. C.: J. Biol. Chem. 55, 333 (1923); "Protein Metabolism in the Plant". Yale University Press 1939. — CHIBNALL, A. C., M. W. REES and E. F. WILLIAMS: Biochem. J. 37, 354 (1943). — CHICK, H., and E. B. SLACK: Biochem. J. 45, 211 (1949). — CHRISTENSEN, E. C., E. S. WEST and K. P. DIMICK: J. Biol. Chem. 137, 733 (1941). — COCKING, E. C., and W. YEMM: Biochem. J. (Proc.) 58, xii (1954). — COLVIN, J. R., D. B. SMITH and W. H. COOK: Chem. Rev. 54, 687 (1954). — COMMONER, B., F. L. MERCER, P. MERRILL and A. J. ZIMMER: Arch. Biochem. 27, 271 (1950). — CROOK, E. M.: Biochem. J. 40, 197 (1946). — CROOK, E. M., and M. HOLDEN: Biochem. J. 43, 181 (1948). — CROOK, E. M., and D. J. WATSON: J. Agric. Sci. 38, 440 (1948). — CRUICKSHANK, D. H., and J. G. WOOD: Austr. J. Exp. Biol. Med. Sci 23, 243 (1945).

DANIELSSON, C.-E.: Acta Chem. Scand. 6, 149 (1952). — DAVENPORT, H. E.: Nature (Lond.) 170, 1112 (1952). — DAVENPORT, H. E., and R. HILL: Proc. Roy. Soc. B. 139, 327 (1952). — DEEN, R. F. A.: "Plant Proteins in Child Feeding". Med. Res. Co. Special Rep. 297. London: H.M.S.O. 1953. — DE HAAN-HOMANS, L. N. S.: Trans. Instn. Rubb. Ind. 25, 346 (1950). — DENNY, F. E.: Contrib. Boyce Thompson Inst. 4. 65 (1932). — DU BUY, H. G., M. W. WOODS and M. D. LACKEY: Science 111, 572 (1950). — DUNN, F. J., and C. R. DAWSON: J. Biol. Chem. 189, 485 (1951). — DUSTIN, J. P., C. CZAJKOWSKA, S. MOORE and E. J. BIGWOOD: Anal. Chem. Acta 9, 256 (1953). — DUSTIN, J. P., E. SCHRAM, S. MOORE and E. J. BIGWOOD: Bull. Soc. chim. Biol. 35, 1137 (1953).

ELLFOLK, N., and A. I. VIRTANEN: Acta Chem. Scand. 4, 1014 (1950); 6, 411 (1952). — EMERSON, R., and C. M. LEWIS: J. Gen. Physiol. 25, 579 (1942). — EMMELIN, N., and W. FELDBERG: J. Physiol. 106, 440 (1947). — ENGLEMANN, T. W.: Z. Botanik 41, 1 (1883); 42, 81 (1884). — VON EULER, H. HELLER and K. G. HOGBERG: Ark. Kemi. Min. Geol. 26 A, (21) 1 (1949). —

FISHER, E. H., and H. M. HILPERT: Experientia 9, 176 (1953). — FOWDEN, L.: Biochem. J. 50, 355 (1952). — FRAENKEL-CONRAT, H., R. C. BEAN, E. D. DUCAY and H. S. OLCOTT: Arch Biochem. 37, 393 (1952). — FRENCH, C. S., and V. K. YOUNG: J. gen. Physiol. 35, 873 (1952).

GALSTON, A. W., R. K. BONNICHSEN and D. I. ARNON: Acta Chem. Scand. 5, 781 (1951). — GÄUMANN, E.: "Principles of Plant Infection". London: Crosby Lockwood 1950; „Pflanzliche Infektionslehre". Lehrbuch der allgem. Pflanzenpathologie. 2. Aufl. Basel 1951. — GILBERT, G. A., and A. D. PATRICK: Biochem. J. 51, 181 (1952). — GLYNNE, M. D., and V. G. JACKSON: J. Agric. Sci. 9, 239 (1919). — GODDARD, D. R., and H. A. STAFFORD: Ann. Rev. Plant

Physiol. **5**, 115 (1954). — Goksöyr, J., E. Boeri and R. K. Bonnichsen: Acta Chem Scand. **7**, 657 (1953). — Goldthwaite, N. E.: Colorado Expt. Sta. Bull. **1925**, 296. — Gornall, A. G., C. J. Bardawill and M. M. David: J. Biol. Chem. **177**, 751 (1949). — Granick, S.: Amer. J. Bot. **25**, 558 (1938). — Granick, S., and K. R. Porter: Amer. J. Bot. **34**, 545 (1947). — Greenberg, D. M., and T. Winnick: J. Biol. Chem. **135**, 761 (1940). — Groot, E. H., J. W. Jansen, A. Kentie, H. K. Oosterhuis and H. J. L. Trap: Biochim. Biophys. Acta **1**, 410 (1947). — Guilliermond: "The Cytoplasm of the Plant Cell". Waltham. Mass. USA. 1941.

Hagen, C. E., and V. V. Jones: Bot. Gaz. **114**, 130 (1952). — Hall, V. L.: Plant Physiol. **26**, 677 (1951). — Hanson, E. A.: Aust. J. Exp. Biol. Med. Sci. **19**, 157 (1941). — Haxo, F. T., and L. R. Blinks: J. Gen. Physiol. **33**, 389 (1950). — Hill, R., and R. Scarisbrick: Proc. Roy. Soc. B **129**, 238 (1940). New. Phytol. **50**, 98 (1951). — Hills, C. H., and H. H. McKinney: Phytopathology **32**, 435 (1942). — Holden, M.: Biochem. J. **39**, 172 (1945). — Holden, M.: Biochem. J. **42**, 332 (1948); **51**, 433 (1952). — Holden, M., and M. V. Tracey: Biochem. J. **43**, 147 (1948). — Holmes, W.: J. Agric. Sci. **41**, 64 (1951). — Holt, A. S., and C. S. French: In "Photosynthesis in Plants". p. 277. Iowa State College Press. 1949. — Holter, H., and K. Linderstrøm-Lang: Hoppe-Seylers Z. **214**, 233 (1933). — Hulme, A. C.: Nature (Lond.) **158**, 58 and 588 (1946); J. Sci. Food. Agric. **2**, 160 (1951).

Jaffee, W. G.: Rev. Brasil. Biol. **3**, 149 (1943). — Jagendorf, A. T., and S. G. Wildman: Plant Physiol. **29**, 270 (1954). — James, W. O., and H. Beevers: New Phytol. **49**, 353 (1950). — Jansen, E. F., and A. K. Balls: J. Biol. Chem. **137**, 459 (1941). — Jeener, R., and P. Lemoine: Nature (Lond.) **171**, 935 (1953). — Jeener, R., and J. Rosseels: Biochim. Biophys. Acta **11**, 438 (1953). — Jewitt, T. N.: J. Agric. Sci. **35**, 264 (1949).

Kalckar, H. M.: J. Biol. Chem. **158**, 355 (1944). — Keilin, D., and E. F. Hartree: Biochem. J. **49**, 88 (1951). — Keilin, D., and J. D. Smith: Nature (Lond.) **159**, 692 (1947). — Keilin, D., and A. Tissieres: Biochem. J. (Proc.) **57**, xxix (1954). — Keilin, D., and Y. L. Wang: Nature (Lond.) **155**, 227 (1945). — Kenten, R. H., and P. J. G. Mann: Biochem. J. **50**, 360 (1951); **52**, 125 (1952); **57**, 347 (1954). — Kiesel, A., A. Belozersky, P. Agatov, N. Biwschich and M. Pawlowa: Hoppe-Seylers Z. **226**, 73 (1934). — Kik, M. C.: Cereal Chem. **18**, 349 (1941). — Kirk, P. L.: Advanc. Protein Chem. **3**, 139 (1947). — Kondo, K., and Y. Morita: Bull. Res. Inst. Food Sci. Kyoto Univ. **10**, 33 (1952). — Krasnovskii, A. A., V. B. Evstigneev, G. P. Brin and V. A. Gavrilova: Doklady Akad. Nauk. USSR. **82**, 947 (1952). — Kubo, H.: Acta Phytochim. (Japan) **11**, 193 (1939). — Kubowitz, F.: Biochem. Z. **292**, 221 (1937); **293**, 308 (1937; **299**, 32 (1938). — Kutsky, R. J., and T. E. Rawlins: J. Biol. Chem. **60**, 763 (1950).

Lemberg, R.: Ann. Chem. **477**, 195 (1930). — Lemberg, R., and G. Bader: Naturwiss. **21**, 206 (1933). — Lepeschkin, W. W.: Protoplasma **37**, 25 (1942/43). — Levitt, J.: Physiol. Plant **5**, 470 (1952). — Limasset, P., P. Cornuet and Y. Gendron: Ann. des Epiphyties **1**, 1 (1950). — Lojkin, M. E., and H. P. Beale: Contrib. Boyce Thompson Inst. **13**, 337 (1944). — Lovern, J. A.: DSIR Food Investigation Spec. Rep. No. 52 (1942). — Lubimenko, V.: Rev. gén. Bot. **39**, 547 (1927). — Lugg, J. W. H.: Appendix to "Protein Metabolism in the Plant" by A. C. Chibnall. Yale University Press 1939. — Lugg, J. W. H.: Advanc. Protein Chem. **5**, 229 (1949). — Lugg, J. W. H., and R. J. Best: Aust. J. Exp. Biol. Med. Sci. **23**, 235 (1945). — Lundegårdh, H.: Ark. Kemi **5**, 97 (1952).

McClendon, J. H.: Amer. J. Bot. **39**, 275 (1952); **40**, 960 (1953). — Macfadyen, D. A.: J. Biol. Chem. **158**, 507 (1944). — McKee, H. S.: New Phytol. **48**, 1 (1949). — Markham, R., and J. D. Smith: Biochem. J. **49**, 401 (1951). — Markham, R., K. M. Smith and D. E. Lea: Parasitolgy **34**, 315 (1942). — Markham, R., and K. M. Smith: Parasitology **39**, 330 (1949). — Marsh, P. B., and D. R. Goddard: Amer. J. Bot. **26**, 724 (1939). — Martin, A. J. P., and R. L. M. Synge: Advanc. Protein Chem. **2**, 1 (1945). — Martin, L. F., A. K. Balls and H. H. McKinney: J. Biol. Chem. **130**, 687 (1939). — Matthews, R. E. F.: Nature (Lond.) **171**, 1065 (1953). — Mazur, A., and H. T. Clarke: J. Biol. Chem. **123**, 729 (1938). — Mazur, A., and H. T. Clarke: J. Biol. Chem. **143**, 39 (1942). — Meneghini, M., and C. C. Delwiche: J. Biol. Chem. **189**, 177 (1951). — Menke, W.: Z. Botanik **32**, 273 (1938). — Miles, A. A., and N. W. Pirie: Brit. J. Exp. Path. **20**, 278 (1939). — Millerd, A., J. Bonner, B. Axelrod and R. Bandurski: Proc. Nat. Acad. Sci. Wash. **37**, 855 (1951). — Milner, H. W.: In "Algal Culture from Laboratory to Pilot Plant". Ed. J. S. Burlew. Carnegie Inst. of Washington Publ. 600, 1953. — Milner, H. W., M. L. G. Koenig and N. S. Lawrence: Arch. Biochem. **28**, 185 (1950). — Milner, H. W., N. S. Lawrence and C. S. French: Science (Lancaster, Pa.) **111**, 633 (1950). — Mohler, H., and H. Lohr: Helv. Chim. Acta **21**, 485 (1938). — Moore, N. S., and D. D. von Slyke: J. Clin. Invest. **8**, 337 (1930). — Moore S., and W. H. Stein: J. Biol. Chem. **211**, 907 (1955). — Morris, H. J., C. A. Weast and H. Lineweaver: Bot. Gaz. **107**, 362 (1945).

Nason, A., H. A. Oldewurtel and L. M. Propst: Arch. Biochem. **38**, 11 (1952). —

NEISH, A. C.: Biochem. J. **38**, 293 (1939). — NEUBERGER, A., and F. SANGER: Biochem. J. **36**, 662 (1942). — NEWCOMER, E. H.: Bot. Rev. **6**, 85 (1940). — NIGHTINGALE, G. T.: Univ. Wisconsin Agric. Exp. Sta. Res. Bull. 74 (1927). — NITSCH, J. P.: Ann. Rev. Plant. Physiol 4, 199 (1953). — NORTHROP, J. H.: In "Crystalline Enzymes" ed. J. H. NORTHROP, M. KUNITZ and R. M. HERRIOTT. Columbia (New York) 1948. —

OKUNUKI, K.: Acta Phytochim. (Tokio) 11, 27, 65, 249 (1939). — OLSEN, C.: C. R. Lab. Carlsberg 24, 99 (1942). — OSBORNE, T. B., and G. F. CAMPBELL: J. Amer. Chem. Soc. 18, 575 (1896).

PAINTER, T. S.: Bot. Gaz. 105, 58 (1943). — PENNELL, R. B., and I. F. HUDDLESON: Mich. State Coll. Agric. Exp. Sta. Tech. Bull. No. 156 (1937). — PETRIE, A. H. K.: Biol. Rev. 18, 105 (1943). — PETRIE, A. H. K., and J. G. WOOD: Ann. Bot. (Lond.) N. S. **2**, 33, 88 (1938). — PIRIE, N. W.: Biol. Rev. 15, 377 (1940); Chem. and Industr. 61, 45 (1942); Advances in Enzymology 5, 1 (1945); (a) Nature (Lond.) **166**, 495 (1950); (b) Biochem. J. 47, 614 (1950); (a) Brit. J. Philosophy Science 2, 269 (1952); (b) World Crops 4, 374 (1952); 6 th. Int. Congr. Microbiol. Rome 1953; Biochem. J. **56**, 83 (1954). — PIRIE, N. W., and K. G. PINHEY: J. Biol. Chem. 84, 321 (1929). — POPE, C. G., and M. F. STEVENS: Biochem. J. **33**, 1070 (1939).

RABINOWITCH, E. I.: "Photosynthesis and Related Processes". Vol.I. New York: Interscience Publ. 1945.; "Photosynthesis and Related Processes". Vol. II. New York: Interscience Publ. 1950. — RINGEN, J.: Meld. Norges Landbrukshøgskole 19, 451 (1939). — RODRIGO, F. A.: Biochim. Biophys. Acta 10, 342 (1953). — ROUELLE, H. M.: J. de médicine, chirurgie, pharmacie etc. 40, 59 (1773).

SAITO, T., and S. WATANABE: J. Agric. Chem. Soc. (Japan) 25, 115 (1951). — SCHLEGEL, D. E., and T. E. RAWLINS: Phytopathology **43**, 89 (1953). — SCHNEIDER, I. R.: Science (Lancaster, Pa.) 117, 30 (1953). — SINANO, S.: Bull. Agric. Chem. Soc. (Japan) 15, 118 (1939). — SINCLAIR, W. B., E. T. BARTHOLOMEW and R. D. NEDVIDEK: J. Agric. Res. 50, 173 (1935). — SINGER, S. J., L. EGGMAN, J. M. CAMPBELL and S. G. WILDMAN: J. Biol. Chem. 197, 233 (1952). — SISAKYAN, N. M., and A. M. KOBYAKOVA: Biokhimiya 1**3**, 88 (1948). — SLADE, R. E., and J. H. BIRKINSHAW: British Patent 511525 (1939). — SMITH, E.L.: (a) J. Gen. Physiol. 24, 565 (1941); (b) J. Gen. Physiol. 24, 583 (1941). — SMITH, E. L., and E. G. PICKELS: J. Gen. Physiol. 24, 753 (1941). — SMITH, J. D.: Biochem. J. 44, 585, 591 (1949). — SOSA-BOURDOUIL, C.: C. R. Acad. Sci. (Paris) **208**, 536 (1939). — STAFFORD, H. A.: Physiol. Plant. 4, 696 (1951). — STERN, K. G., and E. STERN: Biochem. J. **252**, 81 (1932). — STEWART, A. B., and F. N. WOODWARD: "A Survey of Agricultural, Forestry and Fishery. Products in the United Kingdom and their Utilisation". London: H.M.S.O. 1953. — STRAUS, W.: Bot. Rev. 19, 147 (1953). — STREET, H. E.: Advanc. Enzymol. **9**, 391 (1949). — STREET, H. E., A. E. KENYON and G. M. WATSON: Ann. Appl. Biol. **33**, 1 (1946). — SULLIVAN, J. T.: Science (Lancaster, Pa.) **98**, 363 (1943). — SVEDBERG, T., and T. KATSURAI: J, Amer. Chem. Soc. **51**, 3573 (1929).

TADOKORO, T., and N. TAKASUGI: J. Chem. Soc. (Japan) **60**, 188 (1939). — TADOKORO, T., and N. TAKASUGI: J. Chem. Soc. (Japan) **62**, 1018 (1941). — TAKAHASHI, W. N.: Phytopathology 41, 903 (1951). — TAKAHASHI, W. N., and M. ISHII: Amer. J. Bot. 40, 85 (1953). — TAKASHIMA, S.: Nature (Lond.) **169**, 182 (1952). — TEWFICK, S., and P. K. STUMPF: J. Biol. Chem. **192**, 519 (1951). — THEORELL, H.: Ark. Kemi Min. Geol. 16A, No. 2 (1942). — THEORELL, H., and A. MAEHLY: Acta Chem. Scand. 4, 422 (1950). — THOMAS, J. B., O. H. BLAUW and L. N. M. DUYSENS: Biochim. Biophys. Acta 10, 230 (1953). — TIMM, E.: Z. Botanik **38**, 1 (1942). — TOLBERT, N. E., and R. H. BURRIS: J. biol. Chem. 186, 791 (1950). — TOLBERT, N. E., C. O. CLAGETT and R. H. BURRIS: J. Biol. Chem. 181, 905 (1949). — TOMBESI, L., and T. CERVIGNI: Tobacco 56, 6 (1952). — TRACEY, M. V.: Biochem. J. **42**, 281 (1948). — TRISTRAM, G. R.: Biochem. J. **35**, 413 (1941); Advanc. Protein Chem. **5**, 83 (1949); In "The Proteins." ed. H. NEURATH and K. BAILEY. Vol. I. New York: Academic Press 1953. — TURNER, J. F.: Aust. J. Sci. Res. B. **2**, 138 (1949).

ULTÉE, A. J.: Bull. jardin botan. Buitenzorg **5**, 245 (1923).

VAN SLYKE, D. D., R. T. DILLON, D. A. MacFADYEN and P. HAMILTON: J. Biol. Chem. 141, 627 (1941). — VAN SLYKE, D. D., D. A. MacFADYEN and P. HAMILTON: J. Biol. Chem. 141, 671 (1941). — VERGHESE, G. T.: Trans. Inst. Rubb. Ind. 24, 138 (1948). — VICKERY, H.B.: Physiol. Rev. **25**, 347 (1945); (a) J. Assoc. Off. Agric. Chemistr. **29**, 358 (1946); (b) Yale J. Biol. Med. 18, 473 (1946). — VIRTANEN, A. I., and T. LAINE: Hoppe-Seylers Z. **256**, 193 (1940). — VIVINO, A. E., and L. S. PALMER: Arch. Biochem. 4, 129 (1944).

WADLEIGH, C. H.: Arkansas Agric. Exp. Sta. Bull. 446 (1944) — WAITE, R., A. FENSOM and S. LOVETT: J. Sci. Food Agric. 4, 28 (1953). — WALTI, A.: J. Amer. Chem. Soc. **60**, 693 (1938). — WARBURG, O., and W. LÜTTGENS: Biochimiya 11, 303 (1946). — WATSON, S. J.: "The Science and Practice of Conservation." London: 1939. — WAYGOOD, E. R., and K. A. CLENDENNING: Science (Lancaster, Pa.) 11**3**, 177 (1951). — WEIER, T. E., and C. R. STOCKING: Amer. J. Bot. **39**, 720 (1952). — WHITE, J. R., L. WEIL, J. DAGHSKI, E. S. DELLA

MONICA and J. J. WILLAMAN: Industr. Eng. Chem. **40**, 293 (1948). — WILDMAN, S. G., and J. BONNER: Arch. Biochem. **14**, 381 (1947). — WILDMAN, S. G., J. CAMPBELL and J. BONNER: Arch. Biochem. **24**, 9 (1949). — WILDMAN, S. G., C. C. CHEO and J. BONNER: J. Biol. Chem. **180**, 985 (1949). — WILDMAN, S. G., and A. T. JAGENDORF: Ann. Rev. Plant Physiol. **3**, 131 (1952). — WILSON, G. S., and A. A. MILES: "Principles of Bacteriology and Immunology". London: Arnold 1955. — WILSON, P. W.: Advanc. Enzymol. **13**, 345 (1952). — WOOD, A.: Fruit Prod. J. Amer. Vinegar Industr. **21**, 308 (1942). — WOOD, J. G.: Ann. Rev. Biochem. **14**, 665 (1945). — WOOD, J. G., D. H. CRUICKSHANK and R. H. KUCHEL: Aust. J. Exp. Biol. Med. Sci. **21**, 37 (1943). — WOLF, O.: Z. Spiritusind. **65**, 87 (1942). — WORK, E.: Biochem. J. **49**, 17 (1951).

YEMM, E. W.: Proc. Roy. Soc. B **136**, 632 (1950). — YEMM, E. W., and B. F. FOLKES: Biochem. J. **55**, 700 (1953). — YIN, H. C.: Nature (Lond.) **162**, 928 (1948).

ZELITCH, I., and S. P. OCHOA: J. Biol. Chem. **201**, 707 (1953). — ZELITCH, I., P. W. WILSON and R. H. BURRIS: Plant Physiol. **27**, 1 (1952).

Seed Proteins.

By

J. Pace.

With 1 Figure.

A. Some General Considerations.

The heroic age of the study of seed proteins is the period of about twenty five years, from 1890, when Osborne and his collaborators made their extensive investigations. During this time fractions from seeds provided one of the most profitable sources of material for the development and application of the classical techniques of protein chemistry. As Lugg (1949) has noted: "For many years the only substantial information about plant proteins as substances pertained to those proteins which had been extracted in almost pure, and sometimes almost individual, state from seeds." This information was in large measure due to Osborne's sustained and patient work. It became clear, from his investigations of many types of seeds, that while from some of them it was comparatively easy to obtain protein fractions which were sufficiently free from extraneous matter to deposit in crystalline form, yet with others, notably the cereal grains, the protein was extremely intractable material. Indeed, the triumphs and the miseries of the study of seed proteins are reflected historically; Schmiedeberg's (1877) early preparation of crystalline protein material from the brazil nut may·be contrasted with the fact that crude wheat gluten is still prepared essentially by Beccari's (1728) technique, reported to the Bologna Academy in 1745, of gently washing away, in a stream of water, the bulk of the starch from wheaten flour.

I. The Major Seed Proteins.

The achievements of Osborne and his school using the now classical methods of fractionation, have been consolidated, extended and in some cases refined. But, in general, attempts to apply more modern techniques to the specific and difficult problems presented by many of the seed proteins are yet disproportionately few in number in relation to the decisive importance of seeds as a source of food, directly and indirectly, for man and animal. This point has been emphasised by Vickery (1945), who also considers that the time is ripe for a renewed attack upon the characterization of seed proteins with the techniques which are now available. Meanwhile Osborne's categories of globulins and albumins frequently remain as the best characterized examples of protein fractions from seeds and his methods, with minor modifications, continue to be used for their isolation. For example, in a recent study of flax seeds by Vassel and Nesbitt (1945) it is observed that no attempt to fractionate the protein constituents has been made since Osborne's publication, (1892). Osborne's criteria for the homogeneity of his preparations were usually based on the solubility relations in different salt solutions and other solvents, on the elementary composition and, in some instances, on the content of certain amino acids. Those examples which were obtained in the crystalline state, usually the globulin fractions, were, naturally at that time, considered to be homogeneous. But judged by modern criteria (Pirie, 1940), it is doubtful if more than a few proteins prepared from seeds can be considered to be adequately characterized or homogeneous.

Vickery (1945) has also pointed out that in the great majority of instances only a proportion of the total protein of a seed has been reasonably well characterized. This leads to a defective knowledge of the nutritional value of the total protein of the seed or of those fractions of the total which it has not been possible to characterize. This is important on the one hand because of the well known supplementary action of different proteins in promoting growth in animals, and on the other hand because of the possibility that certain seeds may contain protein fractions which are deleterious or toxic. Some of the seeds from leguminous plants, for example, contain protein fractions which may be harmful when ingested under certain conditions.

It may therefore not be out of place to consider briefly whether or not some reorientation of approach is possible. Some significance may perhaps be attached to the observation that much of the literature, on the seed proteins which have been moderately well characterized, deals with the globulin fractions from the seeds of dicotyledonous plants. This tendency to concentrate on the examination of those protein fractions which may most easily be obtained in a reasonably clean state has obviously desirable features, but it may at the same time have led to a rather mechanical approach to the complex problem of the total protein constituents. For it indicates an attitude of regarding a ground seed meal as primarily no more than a convenient moderately crude source of a particular protein fraction. As a partial corrective to this attitude it seems necessary to emphasise, platitudinous though it is, that seeds are viable organisms, in a resting condition, as we normally know them, only because their moisture content is below a certain critical level. It follows that they may be expected to contain a wide range of proteins of diverse character and function. Secondly, some consideration should be given to the structure of seed, with the possibility in mind that the distribution and type of protein in the different anatomical parts of the seed may be different. These considerations perhaps apply mainly to the seeds of the monocotyledons, and they were plainly recognized by Osborne (1907) in his studies of the wheat grain. A restatement of them may, however, be useful, particularly as the modern techniques of paper chromatography, ionophoresis, partition chromatography and ion-exchange columns need quite small quantities of material, and therefore, in conjunction with microdissection, open a prospect of examining proteins from particular parts of seeds in a more detailed and precise way than has been possible hitherto.

1. Localisation within Seed Structures.

The gross structural features of seeds differ in the two main botanical groups of the higher flowering plants — the monocotyledons and the dicotyledons. In the monocotyledons, the group to which the cereal grains belong, the main anatomical parts are the endosperm (generally regarded as the storage organ), the embryo, including the scutellum, and the outer seed coats. The aleurone layer of cells, important qualitatively for the character of its protein, is considered botanically to be part of the endosperm. The relation of these parts is illustrated for the wheat grain in Figure 1. In most dicotyledonous seeds, the outer coat surrounds two cotyledons between which, and attached to both, lies the embryo. The cotyledons are sometimes considered as part of the embryo.

Monocotyledons. The variation in the character of the protein in different anatomical parts of a monocotyledonous seed may be illustrated in a general way by the example of the wheat grain. The ungerminated grain contains a number of enzymes and presumably many of these are specific proteins. Their distribution in the grain is considered later, but it may be noted here that many of them are located in the germ and outer parts of the berry. The germ is also comparatively rich in nucleic acids, some of which are conjugated with proteins. Little or no information is available on how the total protein of the germ is distributed between the scutellum part and the rest of the embryo. The bulk of the protein of the wheat grain is contained in the endosperm. But within the endosperm itself there is evidence of a gradation both in quantity and in type. The outer layers of the endosperm are relatively rich in protein, compared with the central part. HINTON's (1947) data on nitrogen distribution, given in Table 1, indicate this trend. For, so far as the endosperm is concerned, variations in non-protein nitrogen in the different parts are almost certainly insufficient to change significantly the assumption that the nitrogen figures reflect, in a general way, the gradation in protein content.

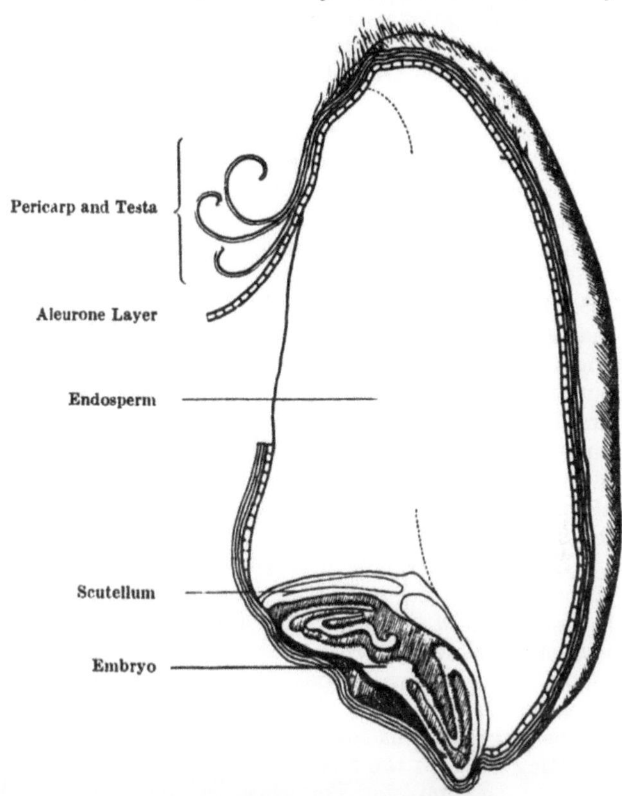

Fig. 1. Diagram of Wheat Grain Structure.

Table 1. *Nitrogen Contents of Dissected Fractions of the Wheat Grain (Triticum vulgare, variety Vilmoren 27). [HINTON 1947).]*

	Proportion of the grain %	Nitrogen content %
Whole grain	—	1.40
Pericarp and testa	8.0	0.70
Aleurone layer	7.0	3.15
Embryo	0.94	5.33
Scutellum	1.5	4.27
Endosperm 1 (outer layer, 150 μ in depth)	12.5	2.20
Endosperm 2 (second layer, 150 μ in depth)	12.5	1.40
Endosperm 3 (inner layer, 300 μ in depth)	57.5	1.00

There is also evidence from nutritional studies, amino acid analyses (BARTON-WRIGHT and MORAN, 1946), and enzyme distribution that the character of the protein in the outer layers is different. The endosperm also contains lipoprotein complexes, and it may also contain small amounts of nucleoproteins. It is still uncertain whether or not any of the endosperm protein is conjugated with carbohydrate. The outer seed coats contain some protein material, but this has hardly been investigated. While the broad differences between endosperm protein and protein from germ are well known, and will be referred to in detail later, uncertainty on some points remains, because the fractions used as source material to represent anatomical parts have usually been products derived from the operations of flour milling. Roller mills and their ancillary sifting and separating devices do effect a remarkable separation of parts of the wheat berry, but there are certain limitations to be noted. Commercial wheat germ contains a proportion of the outer seed coats and frequently some of the outer layers of the endosperm, while flour of only 70% extraction usually contains a little of the embryo. OSBORNE used flour as a source material for his studies of the endosperm protein, but noted that a proportion of the small amount of water soluble albumin, his leucosin, might arise from traces of germ in the flour. More precise work on the distribution of protein in the berry is needed, and for this it appears essential in the first instance to adopt micro-dissection by hand. This is laborious, but has been done by HINTON (1947) in his studies of the distribution of some of the members of the B group of vitamins. His careful dissection of the embryo, scutellum and aleurone layer would seem to be a procedure which might well be used in future investigations of the protein.

That such an approach may be fruitful, with seeds from the monocotyledons, is indicated by the interesting results recently obtained by DANIELSSON (1949) in his studies of the behaviour of globulin fractions in the ultracentrifuge. DANIELSSON, following earlier work in the Uppsala laboratory, has studied the seed globulins (a) of nine species in the family *Gramineae*, including the important cereal grains, and (b) of thirty-four different species in the family *Leguminosae*. In the *Gramineae* his results with wheat and barley indicate the value of differentiating between the anatomical parts, even though the anatomical separation may not be complete. For wheat he relied on the milling separations of flour, germ and bran as his source materials; with barley two fractions were used, the embryos were excised manually with a razor blade, and the material remaining comprised the second fraction. Earlier work by QUENSEL (1942) with globulin fractions from whole barley meal, obtained by fractional precipitation with ammonium sulphate from sodium chloride (phosphate buffered) extracts, had shown that, judged by sedimentation constants from measurements in the ultracentrifuge, there are four well defined globulins described as α, β, γ, δ, in order of increasing molecular weight. DANIELSSON now finds that the α component occurs principally in the residual fraction (grain minus embryo), the very small amount found in the embryo may be due to incomplete removal of endosperm or bran; the γ component is found only in the embryo and is not detectable in the rest of the grain. In wheat the α and γ components are present and as these have the same sedimentation constant as α and γ from barley DANIELSSON, perhaps prematurely, considers them as identical in the two grains. Be that as it may, the distribution of the α and γ components in the wheat grain is uneven, as it is in barley. The γ component only is found in the embryo, while the flour fraction contains considerably more of the α component than the γ and conversely the bran contains far more of γ than of α. BARLEY is unique, among the nine species of Gramineae studied, in possessing the β and δ components. Wheat, oats, rye, and maize all contain the α and γ components and in each species the sedimentation constants for the particular component are the same. In three grass seeds, *Paniceum miliaceum*, *Phleum pratense* and *Festuca pratensis*, only the γ component is observed, but this may be due, as DANIELSSON comments, to the fact that these seeds contain very little endosperm protein and therefore the embryo, containing the γ component, provides the major contribution to the total globulin which is extractable. Evidence from ultracentrifuge studies alone is insufficient to establish that the globulin fractions are individual protein species, but DANIELSSON's work marks a considerable advance in the knowledge of protein extractable by salt solution from the cereal grains, and definitely establishes a marked differentiation in particle size between material obtained from endosperm and that from germ. The estimated molecular weight for the α component (from barley) is 29,000, while that of the γ component (from wheat) is 210,000.

For many of the monocotyledonous seeds, however, the globulin fraction is but a small part of the total protein. In the wheat grain, for example, it is only about 6% of the total protein and most of it is located in the germ. At equilibrium moisture with air the germ is about 2.5% of the weight of the grain, the aleurone layer 6.5%, the endosperm about 83% and the outer seed coats about 8%. Despite some imprecision due to total protein content being estimated from nitrogen determinations, so that the figure obtained includes an uncertain proportion of nonprotein nitrogenous material, it may reasonably be assumed that over 80% of the total protein of the wheat grain is contained in the endosperm. Apart from the aleurone layer, the protein of the endosperm forms an amorphous matrix in which the starch grains are embedded. The aleurone layer, which as noted above, is relatively more rich in protein than the rest of the endosperm, contains well defined granules, but their composition appears still to be uncertain. The protein content of the endosperm, which varies with the variety of wheat, the geographical area in which the wheat is grown and the cultural conditions during growth, may range from 7—15%, reckoned on the air-dried grain. Quantitatively, therefore, the endosperm is the main protein depot of wheat and other cereal grains. And it is the endosperm protein which, despite intensive investigation, has proved to be so difficult to characterize, and for which it is still necessary to use OSBORNE's nomenclature for the fractions derived from it by methods based essentially on his procedure.

Dicotyledons. In the dicotyledonous seeds, a higher proportion of the total protein is more tractable than in the cereal grains. The main protein reserve is contained in the cotyledons and frequently a large proportion of it may readily be extracted with salt solutions. These globulins are considered to be the main storage proteins, and from the oil seeds especially they may be obtained in crystalline form. They are relatively stable substances having, in general, reproducible properties and can be put through cycles of solution and recrystallization with comparatively slight denaturation or modification of many of their physical properties. As early as 1855 HARTIG, by examining parts of dicotyledonous seeds under the microscope, concluded that some of the protein was present in crystalline form. In the cotyledons of some of the oil-bearing seeds, crystals, predominently protein, may frequently be observed. It is probable that this crystalline material contributes materially to the protein which may be artificially crystallized from the globulin fraction. Although much of the main depot protein of the dicotyledonous seeds may be readily extracted in a comparatively unmodified state, yet the same considerations apply to these seeds, although to a lesser extent, as those, noted above, to the monocotyledons. Apart from the main globulin fractions there may be expected to be protein material of different character derived from the embryo as well, possibly, as cytoplasmic protein and some deriving from the seed coats. Little appears to be known about the relative proportions in which these different protein components may occur.

2. The Status of Protein Fractions from Cereal Grains.

Some instances of how these general factors apply in the examination of seed proteins may now be considered. It is perhaps most useful to continue with the wheat grain as an example of a monocotyledonous seed, as it was the subject of Osborne's most detailed investigation and as it has been widely investigated ever since. Wheat endosperm may be finely ground and therefore easily prepared in a state which is highly satisfactory from the point of view of having a fine powder as a starting material for extracts. In other respects the fine flour is less promising. As a crude protein source it is poor material. Although the moisture content is no more than about 12% the total protein content is only of the order of 10%. It contains starch, oligosaccharides, pentosans, polyuronides, glycerides and phospholipids, amino acids, steroids, pigments and traces of other substances. It is not possible to extract all the fat from it with petroleum ether or ethyl ether. But its main characteristic is perhaps its behaviour when wetted with water, as it is the ease with which a dough is formed, and the physical properties of that dough, which have given the wheat grain pride of place as a bread making grain.

Most of the starch grains may be removed by continuously washing the dough in a slow stream of water and this leaves the bulk of the endosperm protein retained in a tough elastic mass — the gluten. It is the insolubility of most of the endosperm protein in neutral aqueous solutions that has made the characterization of that protein so uncertain. Gluten, even after prolonged kneading and washing, may still contain some 20—25% of non-protein constituents. These are mainly starch grains, fat and phospholipids which are tenaciously retained, and steroids.

Osborne's researches on the proteins of the wheat grain led him to conclude: "Thus five unquestionably different forms of protein, differing in composition, solubility and physical character, can be isolated from the wheat kernel. Whether each of these is itself a chemical individual or a mixture of two or more very similar substances, can not at present be asserted. All that can be said is that it has not yet been possible to separate them into fractions, the properties of which indicate a mixture" (1907).

Osborne's division of wheat proteins into five fractions depended primarily on differences between them in solubility relations. Leucosin, the water-soluble protein representing 0.3—0.4% of the wheat grain and the globulin fraction, about 0.6% of the grain, were shown to be contained mainly in the germ. The main (endosperm) proteins were two in number; gliadin, insoluble in neutral aqueous solutions but characterized by ready solubility in neutral 70% ethyl alcohol, and glutenin, insoluble in neutral aqueous solutions and in aqueous alcohol. The fifth fraction was material with the character of a proteose and this was located chiefly in the germ. Gliadin formed from 40 to 50% and glutenin about 40% of the total protein of the grain. The intermixture of gliadin and glutenin was responsible for the peculiar properties of gluten. For some considerable time gliadins and glutenins were regarded as well-defined proteins distinctive to the cereal grains. But as Lugg (1949) has pointed out: " . . . if gliadins are defined as 'proteins fairly soluble in slightly diluted ethanol' they are not confined exclusively to cereal seeds not even to the plant world." Furthermore, in the opinion of the present writer, it is certainly no longer possible to regard gliadin and gluten or the globulin as other mixtures of individual protein particles or complexes. Less attention has been given to leucosin and the proteose fraction, but, as we have noted, it seems highly probable that these are also mixtures.

There is a considerable literature dealing with the physical and chemical properties of gluten preparations, including attempts at fractionation, sometimes by methods so drastic that some modification of the original protein material must have occurred. It is not possible or desirable to review this literature here, except to refer briefly to some of the papers which have provided evidence which, in the writer's opinion, indicates that at the present time gliadin and glutenin should only be regarded as convenient historical names which describe fractions empirically attained. From the same flour, with a strictly standard procedure and by keeping carefully to the same times of extraction, temperatures and other operating details, it is possible to obtain fractions which are similar in many properties and useful for certain purposes. But this position does not warrant the description of the fractions as protein entities. At the outset there would seem to be grounds for serious doubt about the nature of glutenin. This fraction is, essentially, the material which remains after extraction with water or neutral salt solutions followed by aqueous ethanol. It seems at least possible therefore that glutenin may contain some of the original protein material of the endosperm which has been denatured or dissociated by the use of ethanol to extract the gliadin. Indeed the relation of the gluten proteins as a whole to the state of the proteins as they exist in the endosperm remains completely uncertain. This relation may in part be connected with whether or not, in the native endosperm, they are conjugated with lipoid, carbohydrate or other substances, and this is at present unknown. As an approach to this question of fundamental interest it might be worth investigating the possibility of obtaining thin sections of endosperm suitable for X-ray diffraction studies.

The view that gluten consisted essentially of two proteins began to be questioned in the first instance as a result of attempts to refine existing methods for the determination of flour proteins. It was noted that there were large variations in the quantity of protein extracted at different concentrations of ethanol, in aqueous ethanol mixtures, and that the solubility of gliadin in such solvents depended on the particular pretreatment which the preparation had received. The precarious nature of the system of classification was illustrated by the observations of GORTNER, HOFFMAN and SINCLAIR (1928). They made a careful study of the extraction of protein from flour with a series of neutral salts. Using dilute solutions, at equivalent concentrations, of potassium chloride, bromide, fluoride, iodide, they found the amount of protein extracted varied widely with the different salts. They noted that such behaviour is similar to the peptisation of material in the colloidal state, and LUGG (1949) has observed that it has disturbing implications, for the classification of the proteins on arbitary solubility relations. McCALLA and ROSE (1935) found that most of the gluten proteins may be dispersed in 10% sodium salicylate solution. With such dispersions they obtained a number of fractions by addition of $MgSO_4$ up to a level equivalent to about half saturation, when practically all the protein precipitated. The various fractions were then analysed for amide nitrogen and for arginine. The amide nitrogen increased in the fractions obtained with increasing amounts of ammonium sulphate up to a point corresponding to the precipitation of about 80% of the protein, and then, in the remaining fractions, decreased sharply. In corresponding fractions the arginine content decreased, then increased. While allowance should be made for the possibilities of modification, aggregation and co-precipitation effects, these experiments do indicate that gluten is a complex protein system. The authors own conclusion was that gluten may be regarded as a single protein complex made up of components differing progressively and systematically in both physical and chemical properties. In a later study McCalla and Gralen (1942) examined dispersions of gluten in sodium salicylate, measuring both sedimentation rates in the ultracentrifuge and diffusion rates. The proportion of total gluten protein molecularly dispersed increased as the concentration sodium salicylate was raised (up to 12%). They found that the most soluble 25% of the gluten was all molecularly dispersed but was definitely inhomogeneous. None of the other fractions was entirely molecularly dispersed. The diffusion curves showed that the different fractions, differentiated on the basis of solubility, were inhomogeneous, partly due to aggregation and partly to differences in the proportion and properties of the gluten in a truly molecular state of dispersion. McCalla and Gralen conclude that their results "support the hypothesis that gluten is a protein system showing progressive and regular changes in properties with change in solubility".

Because of its solubility in aqueous ethanol, gliadin may be obtained from gluten in a relatively clean state. But there is a substantial amount of evidence indicating that the best preparations are inhomogeneous. In a careful investigation Haugaard and Johnson (1930) working in Linderstrøm-Lang's laboratory, applied to gliadin solutions the methods of thermal fractionation, using temperatures down to —11° C. Their different fractions had approximately the same nitrogen content of about 17.65%, but differed in viscosity and in optical rotation. Tryptophan and tyrosine contents tended to differ between the most and least soluble fractions. Krejci and Svedberg (1935) examined fractions, prepared by the methods of Haugaard and Johnson, in the ultracentrifuge, and found a range of particles of different molecular weight, with the most soluble fraction containing the most homogeneous material with the lowest particle weights. Burk (1938) from osmotic pressure measurements concluded that gliadin is inhomogenous and the same conclusion may be drawn from Lamm and Polson's (1936) refractometric studies. Schwert, Putnam and Briggs (1944) worked on the technically difficult problem of applying electrophoretic methods to the study of gliadin preparations. They obtained convincing evidence, using a range of conditions with respect to pH, type of buffer, ionic strength and protein concentration, that gliadin is not an electrophoretically homogeneous protein. The electrophoretic patterns indicated that the components of gliadin do not migrate as independent entities under any of the conditions which were used. Under all conditions there was evidence of component interaction so that all the electrophoretic fractions were regarded as mixtures of components.

That the main proteins of the wheat grain still remain uncharacterized is certain. Whether or not there is a mixture of a relatively few closely related species which in different solvents may dissociate, aggregate or form complexes, or whether there is an array of individual components is a question which cannot at present be answered. The position appears to be similar with the main proteins of the other cereal seeds, rye, barley and oats.

3. The Status of Protein Fractions from Dicotyledonous Seeds.

With many of the dicotyledonous seeds the position is undoubtedly less complex than with members of the *Gramineae*. The oil-bearing seeds in particular have long been known to yield a good proportion of their total protein in a relatively well characterized state. Yet even with some of these the individual nature of proteins which were regarded as representative is now suspect. This may be illustrated with the example of the peanut. Following the work of Johns and

Jones (1916) the total protein of the peanut was considered to be made up of two globulins, arachin and conarachin, together with a small amount of heat coagulable albumin. Arachin and conarachin together account for about 95% of the total protein. Arachin alone represents about two-thirds of the total protein and, as it is easily isolated, has been frequently investigated as a typical seed globulin representative of the peanut. The work of Irving, Fontaine and Warner (1945) indicates that this conception must now be revised. They obtained an extract of defatted peanut meal containing at least 98% of the protein, and examined it electrophoretically. The electrophoretic analysis showed the presence of two major components A and B in the ratio of about 7:1 which together made up about 87% of the protein in the extract. The remaining 13% of the protein in the extract is considered to consist of at least two minor components. Arachin and conarachin were then prepared from the same meal and examined by electrophoretic analysis. Arachin was found to contain about 76% of component A and 24% of component B and none of the minor components. Conarachin contained about 80% of component A, 20% of the minor components and apparently none of component B.

There may be some uncertainty about these data because of the rather alkaline (pH 9.25) buffer used for extraction and analysis, but on the whole it may be concluded that arachin and conarachin are not protein entities. Furthermore Irving, Fontaine, and Warner themselves point out that it is uncertain whether the electrophoretically observed components are simple proteins. They consider that there is a possibility that one or more of the components may be a conjugated protein.

From the work described by Vickery (1945) it also would appear that the glycinin of the soyabean also can no longer be regarded as a characteristic entity. Following a description of the experiments carried out in his laboratory Vickery concludes that: "Obviously the total protein of soyabeans represents a complex mixture of components."

II. Other Seed Proteins.

1. Enzymes.

Up the present time all well-characterized enzymes have been found to be proteins or to contain a specific protein component. The future may very well show that this is not universally true, but at present it is justifiable to regard enzymes as essentially protein substances. Many of the known enzymes are found in seeds, and it follows that some of the components in the mixture of proteins contained in seeds are specific enzyme proteins. This consideration alone indicates the limitations imposed by an excessive emphasis on a few protein fractions, even if these apparently account for most of the total protein. This is particularly so if these main fractions are regarded, by implication and without experimental evidence, as catalytically inert. For, occasionally, some enzyme activity may be associated with the main fractions, and it would be useful to examine this possibility since the specific catalytic activity of a protein preparation is often one of the more sensitive criteria of its homogeneity. Such investigations, in parallel with conventional fractionations of seed proteins, are infrequently made. Isolated examples are to be found in the observations of Danielsson and Sandgren (1947) that β-amylase accumulates as the albumin fraction from barley seeds is fractionated, and those of Barré (1951) where it is shown that glucosidase and phosphatase activities are associated with one of five protein fractions prepared from sweet almonds. A particularly interesting example is the demonstration by VENNESLAND and FELSHER (1946) that the crystalline globulins of squash and pumpkin seeds display a thermolabile oxalacetate carboxylase activity. These globulins are normally regarded as the chief reserve protein of the seeds, and it seems unlikely that the whole of the crystalline protein is a homogeneous specific enzyme protein, particularly since in other tissues oxalacetic carboxylases are but a small proportion of the total dry material. In other words, there is at least a possibility that some crystalline protein fractions from seeds may contain a number of enzyme proteins, though these may be present in extremely small amounts. Conversely, where crystalline enzyme preparations have been obtained without much fractionation their homogeneity may be open to question as more refined techniques of separation become available.

Detailed studies of this character are uncommon, probably because the activity of many enzymes in the resting seed is of a much lower order than when the seeds germinate. Particular enzymes may not be present in the resting seed and only appear by synthesis during the processes of germination; or they may be present in inactive form as precursors or lacking activators or coenzyme moieties; or they may be present in barely detectable amounts confined to particular loci in the seed. In many instances therefore investigators, needing a good source of a particular enzyme, have naturally used seeds which have been allowed to germinate. Yet a wide range of enzymes has been demonstrated in resting seeds,

Table 2.

Enzyme	Seed	Reference
β-Amylase	Cereal grains	KNEEN, SANDSTEDT and HOLLENBECK (1943)
	Soyabean	NEWTON, FARLEY and NAYLOR (1943)
Phosphorylase	Pea	HANES (1940)
Phosphohexokinase	Maize, pea, oat, tomato, radish, soyabean, broad bean	AXELROD, SALTMAN, BANDURSKI and BAKER (1952)
α-Carboxylase	Wheat	SINGER and PENSKY (1952)
	Jack bean, soyabean	COHEN (1946)
	Oat	ZEIJLEMEKER (1939)
β-Carboxylase	Wheat, lentil, pea	VENNESLAND (1949)
Malease	Maize	SACKS and JENSEN (1951)
Thioglucosidase	*Sinapis alba*	NEUBERG and WAGNER (1926)
Aldolase	Pea, squash	STUMPF (1948)
Aconitase	Cucumber, pea, soyabean	OCHOA (1951)
Lipase	Cereal grains	PEERS and MARTIN (1953) PETT (1935)
	Castor bean	KELSEY (1939)
Phytase	Cereal grains	PEERS (1953), ALBAUM and UMBREIT (1943)
Phosphoamidase	Rice	BREDERECK and GEYER (1939)
Lipoxidase	Soyabean	THEORELL, HOLMAN and AKESON (1947)
Transaminase	Wheat	LEONARD and BURRIS (1947)
	Oat	ALBAUM and COHEN (1943)
Amino acid decarboxylase	Pea, maize, sunflower	SCHALES, MIMS and SCHALES (1946)
Urease	Soyabean	SUMNER (1926)
	Jack bean	KIRK (1933)
Cytochrome oxidase	Wheat	BROWN and GODDARD (1941)
Formic acid dehydrogenase	Pea	ADLER and SREENIVASAYA (1937)
Glutathione reductase	Pea	MAPSON and GODDARD (1951)
	Wheat	CONN and VENNESLAND (1951)

which in some instances are among the best sources. Frequently, but not exclusively, they are associated with the embryonic tissue, so that in the monocotyledons the germ is the main site of activity while in the dicotyledons the whole meal, as in the case of the pea, contains a diverse range of enzyme systems.

In Table 2, which is illustrative, rather than exhaustive, examples are given of some of the enzymes which have been investigated in resting seeds.

The distribution of enzymes in the different parts of the monocotyledonous seeds has received relatively little attention. In the case of the cereal grains a knowledge of the distribution is of some practical importance. From a more fundamental point of view it is of interest to know whether all the reserve proteins of the endosperm are so enzymically inert as this designation implies. ENGEL (1947), using histological methods, has examined the location of a number of enzymes in the wheat grain. This distribution, ENGEL and BRETSCHNEIDER (1947) is as follows:

Proteinase: No activity in the pericarp, testa and endosperm other than the aleurone layer. The chief site is the aleurone layer, which contains per unit weight 6—7 times as much as the germ.

Dipeptidase: As with the proteinase, except that the germ contains about 3 times, per unit weight, as much as the aleurone layer.

Amylase: No activity in the pericarp and testa or aleurone layer. The highest activity is found in the part of the endosperm adjoining the aleurone cells and that adjoining the scutellum. The scutellum has some activity but the other part of the germ has none.

A complete picture of the distribution of an enzyme in a resting seed appears only to be available for wheat phytase, which PEERS (1953) has investigated using hand-dissected grains. Table 3 is taken from his paper.

An example of an extremely uneven distribution of an enzyme is the lipase of oats. This is located almost entirely in the pericarp of the groat, PEERS and MARTIN (1953).

Some seeds are relatively rich in members of the B group of vitamins. Of these, nicotinamide, thiamine, riboflavin, are components of the coenzyme groups

Table 3. *Distribution of Phytase in the Wheat Grain (Cappelle Deprez).* [PEERS (1953).]

Fraction	% by wt. of whole grain	Phytase activity (μg. P/hr./mg. dry wt.)
Wholemeal	100	3.41
Endosperm	82.5	1.29
Germ	1.0	9.07
Scutellum	1.5	31.8
Epidermal layers . .	4.5	1.32
Testa cross-layer . .	3.5	4.34
Aleurone	7.0	17.7

of certain enzymes, while pyrodoxine in the form of pyrodoxal phosphate is a coenzyme for amino acid decarboxylases and transaminases, and pantothenic acid is a component of a coenzyme group, coenzyme A, involved in biological acetylation. How far these components are associated with their specific enzyme proteins in the resting seed is at present unknown. It probably varies in different types of seeds. Thus for example COHEN (1946) has found that the jack bean has a potent carboxylase system, but the soyabean is inactive in this respect, although the thiamine contents of both beans are approximately the same. In wheat germ some of the thiamine appears to be associated with specific protein components as it contains an active carboxylase system, SINGER and PENSKY (1952). The other members of the B group in cereals are "bound" in some way, since digestion enzymatically or by acid, is needed to make them available for assay. Whether or not this linkage is associated with specific proteins is at present unknown. The distribution of these vitamins in the wheat grain is now well established by the work of HINTON and co-workers (1952, 1953) and it may be relevant to note it here, Table 4, as future work may establish whether or not there is an association with specific proteins.

Table 4. *The Distribution of Vitamins in the Wheat Grain.* [Adapted from HINTON, PEERS and SHAW (1953).]

Fraction	% by wt. of whole grain (for variety Thatcher)	μg./gm.			
		Thiamine	Riboflavin	Pantothenic acid	Nicotinic acid
Pericarp testa. . .	8.2	0.6	—	7.8	25.7
Aleurone	6.7	16.5	10.0	45.1	741
Scutellum	2.0	156.0	12.9	14.1	38.2
Embryo	1.6	8.4	13.7	17.1	38.5
Endosperm . . .	81.5	0.5	0.6	3.9	42.5

Riboflavin, pantothenic acid and nicotinic acid determinations were made with var. Thatcher. The thiamine data are for var. Vilmorin 27.

This limited consideration of seed enzymes illustrates more specifically some of the general observations made earlier.

For the monocotyledons the example of the wheat grain indicates that the endosperm protein is not uniformly inert, nor identical in composition, and that certain parts of it contain specific enzyme proteins. The regions of greatest activity for those enzymes which have been investigated are the aleurone layer and the parts closely adjacent to it. The protein of the central part of the endosperm does appear to be relatively inert. The scutellum and the embryo proper are together a good source of many enzymes and are therefore the sites of a great variety of protein components.

Meals prepared from some of the dicotyledonous seeds are also well endowed with enzyme proteins. With material from both groups many of the enzyme proteins may be extracted with water, neutral salt solutions or buffer solutions. When such extracts are used for the preparation of protein fractions there is therefore the probability that initially many protein species are present. From a quantitative point of view some components may be present in extremely small amounts. But taken together they may make up an appreciable proportion of the protein. With relatively simple fractionation procedures it would then seem advisable to bear in mind that in a complex mixture co-precipitation, adsorption and mutual interaction may occur and that crystallinity by itself is no guarantee that a single component has been isolated. These possibilities appear to be less with the main endosperm proteins of the monocotyledons as the flour or meal is usually extracted first with water or dilute salt solutions. But the insoluble protein mass offers an extensive adsorbing surface which may retain soluble proteins to a variable degree depending on the precise mode of extraction and washing which has been employed. Among the dicotyledons pea meal, for example, is an established source of many enzymes. The complexity of its total protein, which this fact indicates, may remain unappreciated if the sights are never lifted from the three main targets of leguminin, vicilin, and legumelin. And this may be also true for other dicotyledonous seeds.

2. Inhibitors and Toxic Proteins.

It has been known for many years that some plant protein fractions e. g. ricin, crotin and abrin, are toxic. Some of these also agglutinate the red blood cells of some species of animals.

Nutritional studies and the practical feeding of animals have also shown that the raw seeds of some leguminous plants are either poor promoters of growth or positively harmful. The development of certain diseases of man, lathyrism and cicerism, where the diet contains a large proportion of certain legume seeds, may partly be due to a deficiency of methionine. But other deleterious factors are involved when the raw seeds of some legumes are fed to animals.

In 1917 OSBORNE and MENDEL (1917a) observed that when soyabeans were autoclaved their nutritive value was improved. Since then a similar effect has been observed with other legume seeds; the common bean, OSBORNE and MENDEL (1917b), the lima bean, FINKS and JOHNS (1921), the Adsuki bean, JOHNS and FINKS (1921), the velvet bean, SURE and READ (1921), and the jack bean ORRU and DEMIEL (1941). A considerable impetus was given to the study of this effect when HAM and SANDSTEDT (1944) found that extracts of raw soyabeans, made with dilute acid, inhibit tryptic digestion *in vitro*. The activity of the inhibitory material was lost when either the meal or the extract were autoclaved. HAM, SANDSTEDT and MUSSEHL (1945) subsequently suggested that the presence of this trypsin inhibitor was responsible for the lower nutritive value of the bean when raw than when autoclaved. The chief trypsin inhibitor was soon shown to be a protein and was prepared in crystalline form by KUNITZ (1945). As other evidence accumulated, it appeared that the improvement in nutritive value when the seeds are autoclaved could not be entirely due to the destruction of the labile trypsin inhibitor. In the first place, it was observed (WESTFALL, BOSSHARDT and BARNES, 1948), that a fraction from soyabeans inhibited the growth of rats when these were fed a diet which did not contain intact protein but enzyme-hydrolysed casein. Secondly, BORCHERS, ACKERSON and MUSSEHL (1948) found that a purified soyabean trypsin inhibitor preparation did not inhibit the growth of rats and chicks. Furthermore, other legume seeds were found to contain trypsin inhibitors and BORCHERS and ACKERSON (1950) investigated the effects of autoclaving, on the nutritive value, in parallel with tests for the presence of trypsin inhibitors in the raw seeds. Their results are given in Table 5 which is reproduced from their paper.

Table 5. *Trypsin Inhibitors in Raw Autoclaved Seeds.* [BORCHERS and ACKERSON (1950).]

Improved by autoclaving		Not improved by autoclaving	
Inhibitor Positive	Inhibitor Negative	Inhibitor Positive	Inhibitor Negative
Velvet bean (*Stizolobium spec.*)	Jack bean (*Canavalia ensiformis*)	Peanut (*Arachis hypogaea*)	Guar (*Cyamopsis tetragonaloba*)
Lima bean (*Phaseolus lunatus*)	Lentil (*Lens esculenta*)	Partridge pea	Common pea (*Pisum sativum*)
Common bean (*Phaseolus vulgaris*)	Horse bean (*Vicia faba*)	Chick pea (*Cicer sp.*)	Common vetch (*Vicia sp.*)
Soya bean (*Glycina max.*)		Sweet pea (*Lathyrus sp.*)	
Black-eyed bean		Lespedeza (*Lespedeza sp.*)	
		Mung bean (*Phaseolus aureus*)	

There is thus no correlation between the effect of autoclaving on the nutritive value and the presence or absence of trypsin inhibitors. This suggests that in some instances other heat-labile deleterious material may be present, and this view has recently been strengthened by the preparation of a toxic protein fraction, soyin from the soyabean by LIENER (1953). Two other comments on the data in the table may be made. It would seem worth while to examine some of the other seeds in the "improved" group for the presence of toxis protein fractions. In this connexion KLOSE, GREAVES and FEVOLD (1948) have reported that a preparation from the lima bean inhibits the growth of rats on a diet containing acid-hydrolysed casein in lieu of protein. Secondly, it should be noted that the material was autoclaved at its normal low moisture content for about thirty minutes at 15 lbs. pressure. Proteins are in general much more stable to heat at low moisture content. There is therefore a possibility, it may be slight, that some of the members of the unimproved group may contain deleterious proteins which are more stable to heat than those contained in the members of the improved group.

A property such as toxicity or a specific inhibitory effect on an enzyme is an attribute which is helpful in fractionating proteins and the soyabean trypsin inhibitor and ricin, both prepared in crystalline form by KUNITZ (1945) and KUNITZ and MACDONALD (1948) are perhaps better characterized than some other crystalline seed proteins. The value of a specific property in fractionating even crystalline protein material may be illustrated by the example of the trypsin inhibitor from the lima bean. TAUBER, KERSHAW and WRIGHT (1949) obtained a crystalline protein fraction which had the trypsin inhibitor activity. When FRAENKEL-CONRAT, BEAN, DUCAY and OLCOTT (1952) repeated the fractionation they found that by a slight modification of the recrystallization procedure the inhibitory activity could be removed from the crystalline material, which was then inert, while the activity was retained in the mother liquor. They have in turn crystallized the inhibitor, which is a low molecular weight (10,000) protein. It is much more stable to heat than the soyabean inhibitor.

KABAT, HEIDELBERGER and BEZER (1947) from immunochemical and toxicological studies have suggested that both toxic and non-toxic ricin may exist together in the castor bean. This view is based on their observation that a toxic protein fraction from the bean was electrophoretically, ultracentrifugally and immunochemically homogeneous and, judged by these criteria, identical with samples of crystalline ricin. But the protein fraction was only about two-thirds as toxic as the crystalline product. They cite this study as an example of the value of quantitative immunochemical methods, used together with other techniques, as an aid in the characterization of proteins. It may be recalled that OSBORNE was a pioneer in this field also, with his anaphylaxis studies with seed proteins. From the experiments with WELLS, cf. WELLS and OSBORNE (1911, 1913 and 1916), he concluded that these methods. "Indicated the almost infinite variety of vegetable proteins and the occurrence of possible identical proteins only in very closely related species." These experiments, for example, showed the close similarity of the legumin from the pea with that of the vetch and the gliadin of wheat with that of rye.

The mode of action of toxic protein fractions from seeds is at present unknown. A fundamental question here is whether the toxicity is a property of the protein itself or whether it is due to some other substance so closely associated with the protein that it has defied separation by the usual fractionation techniques. The evidence at present suggests that it is in fact a property of the protein. If this be granted then the mechanism of how the protein exerts its toxic effect is still obscure. It would still appear to be uncertain, for example, whether or not the toxic seed proteins have enzymic activity, as do some of the bacterial toxins. MOULÉ (1951), cf. LE BRETON and MOULÉ (1949) has prepared samples of highly toxic ricin which she states have proteolytic activity. This finding inevitably raises the question of whether the enzymic activity and the toxic property can be attributed to the same protein molecule.

In the case of a seed protein made toxic artifically, by treatment of wheat protein with agene, the toxicity is due to modification of some of the methionine residues to methionine sulphoximine which acts as an anti-metabolite to methionine and glutamine (HEATHCOTE, 1949; PACE and McDERMOTT, 1952).

Toxic proteins and anti-enzyme proteins have both an intrinsic and a general biological interest, and there seems to be a reasonable basis for further examination of other legume proteins for the presence of components with such distinguishing properties.

III. Amino Acid Composition.

OSBORNE's studies established that proteins similar in physical properties and general characteristics are found in seeds which are botanically closely related, but the proteins from uncorrelated groups may have markedly different properties. He noted (1924) that this is shown most strikingly with the cereal grains whose reserve proteins are generally similar, but are sharply differentiated in general properties from the main proteins derived from other groups of seeds.

These relationships, apparent in the general properties of the proteins, were more precisely delineated by the amino acid analyses reported in 1903, OSBORNE and HARRIS (1903). Restricted in scope and, to some extent, in technique as these inevitably were at that time, they established some of the salient points of resemblance and differences in amino acid distribution in proteins from different groups of seeds. The prolamines from cereals were shown to contain little or no lysine, relatively little of the other basic amino acids, but to be outstandingly rich in glutamic acid, proline and amide nitrogen. These features decisively differentiated the main cereal proteins from those, for example, of the group of leguminous seeds. Within this latter group botanical relationships were also reflected in the amino acid analyses of the proteins so that, for example, OSBORNE (1924) commented: "the proteins from different species of *Lupinus* present common characteristics but differ from the proteins of the other leguminous seeds. The proteins of *Lupinus*, however, resemble proteins of other leguminous seeds more closely than they do those of non-leguminous seeds."

Table 6. *Basic Amino Acids in Proteins.* [Adapted from OSBORNE (1924).]

Protein	Source	Percentage of the Protein		
		Histidine	Arginine	Lysine
Globulin	squash seed	2.42	14.44	1.99
Excelsin	para-nut	2.50	14.29	1.64
Edestin	hemp seed	2.19	14.17	1.65
Globulin	cotton seed	3.46	13.51	2.06
Globulin	castor bean	2.74	13.19	1.54
Amandin	almond	1.87	12.16	0.72
Legumin	pea	1.69	11.73	4.98
Legumin	vetch	2.94	11.06	3.70
Conglutin-α	lupin	2.51	10.93	2.74
Vicilin	pea	2.17	8.91	5.40
Glycinin	soyabean	2.10	7.69	3.39
Vignin	cow pea	3.08	7.20	4.31
Glutelin	maize	3.00	7.06	2.93
Leucosin	wheat	2.83	5.94	2.75
Legumelin	pea	2.27	5.45	3.03
Legumelin	soyabean	2.04	5.35	4.91
Phaseolin	kidney bean	2.62	4.87	4.58
Glutenin	wheat	1.76	4.72	1.92
Gliadin	wheat	0.58	3.16	0.63
Gliadin	rye	0.39	2.22	0.00(?)
Hordein	barley	1.28	2.16	0.00(?)
Zein	maize	0.82	1.55	0.00

These differences and relationships are illustrated in Table 6, from OSBORNE's determinations of the basic amino acids in proteins from various types of seeds.

A noteworthy feature of these data is the wide range in the arginine content of the different proteins, from the high figure for the crystalline globulins to the low of the cereal prolamines. In this respect the proteins of the leguminous seeds occupy an intermediate position between these extremes. But within this group are interesting examples of the variations referred to be OSBORNE. For example, the legumin of the pea and vetch have a similar arginine content but this is higher than in another similar pair, glycinin of the soyabean and viginin of the cow pea.

More recent work has confirmed the similarity of the amino acid composition of crystalline globulins from the seeds of closely related plants. Globulins from different genera may however sometimes be distinguished by differences, often slight, in their content of certain amino acids. Crystalline globulins from the seeds of several species of *Cucurbitaceae*, hemp and tobacco have been compared in this respect by VICKERY, SMITH, HUBBELL and NOLAN (1941) and by SMITH and co-workers (1946, 1947, 1948). SMITH and GREENE (1947) have summarized data on the amino acid composition of the seed globulins for which comparative values are available (globulins from hemp, pumpkin, squash, watermelon, cucumber and tobacco seed.)

Table 7. *Arginine Content of Crystalline Globulins (as percentage of the dry ash-free protein).* [VICKERY, SMITH, HUBBELL and NOLAN (1941).]

Popular Name	Species	Arginine %
Squash	*Cucurbita mosch.*	16.17
Squash	*Cucurbita pepo*	16.24
Squash	*Cucurbita maxima*	16.21
Watermelon	*Citrullis vulgaris* var. Stone Mountain	17.91
Watermelon	*Citrullis vulgaris* var. Kleckley Sweet	17.90
Cantaloupe	*Cucumis melo*	16.58
Cucumber	*Cucumis sativus*	15.84
Hemp		16.76
Tobacco		16.09

They point out that the chief characteristic of all these globulins is their high arginine content (exemplified in Table 7, from VICKERY, SMITH, HUBBELL and NOLAN, 1941) and comment: "The variations among all of these proteins are minor in character. Even tobacco seed globulin and edestin which belong to different botanical families are not remarkably outstanding in comparison with the members of the Cucurbitaceae. Nevertheless, all of the different genera can be distinguished from each other by the results. The pumpkin and squash seed globulins do not reflect any differences which are significantly greater than the experimental error involved in microbiological assays."

A more varied ranged of protein fractions from seeds has been analysed by HORN, JONES and BLUM (1950), for the content of those amino acids shown by ROSE (1937), to be essential for the growth of the weanling rat.

Most of the data on amino acid composition refer to fractions isolated from the total protein of seeds. For nutritional reasons especially it is important to know the amino acid composition of the seed protein as a whole and, in the case of the cereal grains the amino acid composition of the proteins in the different anatomical parts of the grain. As it is usually not possible to extract the whole of the protein in a reasonably clean state analyses are made using the hydrolysate of whole seed meals or the meals from parts of seeds. This procedure inevitably involves hydrolysis of the proteins in the presence of large amounts of carbohydrate or lipoids and the data are therefore less reliable than those obtained with relatively clean protein fractions. In the analysis of different parts of the cereal grains further complications arise. Thus in the analysis of the meal from

Table 8. *Content of Amino Acids in Various Seed Proteins. (Calculated on a moisture-ash-free basis.)* [HORN, JONES and BLUM (1950).]

	Arachin	Coconut Globulin	Cona-rachin	Cotton seed Globulin	Edestin	Glycinin	Peanut Total Globulins	Phaseo-lin Navy bean	Wheat bran Globulin	Zein
Nitrogen . . .	18.30	17.42	18.20	18.0	18.55	17.30	18.01	16.07	17.76	16.00
Arginine . . .	13.5	16.73	16.53	14.72	16.51	7.94	14.16	5.97	13.30	1.95
Histidine . . .	2.16	1.52	2.05	3.38	2.50	1.97	1.65	2.24	1.86	.0.76
Isoleucine . .	4.55	4.43	3.98	4.62	5.54	5.69	4.28	6.69	3.82	.5.03
Leucine . . .	7.61	7.18	6.61	7.06	7.50	8.07	7.45	10.5	6.35	21.10
Lysine	2.72	4.37	4.69	4.15	3.23	6.90	3.50	7.20	4.15	0.21
Methionine . .	0.24	2.02	2.13	1.40	1.12	1.15	1.0	1.12	1.04	1.41
Phenylalanine .	6.96	5.10	4.32	8.13	5.43	5.82	5.75	8.04	4.22	7.30
Threonine . .	2.89	4.06	1.93	3.96	4.34	3.0	3.24	4.16	3.50	2.62
Tryptophane .	0.90	0.66	1.20	1.29	1.31	1.05	0.92	0.50	0.80	0.03
Valine	4.85	5.92	3.68	6.05	6.39	4.56	5.00	6.00	6.56	3.98

inner endosperm, the protein is hydrolysed in the presence of a large amount of carbohydrate and a relatively small amount of lipoids while with germ the carbohydrate present is negligible but there is a high proportion of lipoids. It is desirable that the improvement, referred to elsewhere in this volume, which has been made in the hydrolytic procedure should be applied in future investigations of the amino acid composition of seed meals. This procedure has recently been used in the analysis of the amino acid content of barley meal, using the MOORE and STEIN method of buffered resin columns, by DUSTIN, SCHRAM, MOORE and BIGWOOD (1953). Meanwhile the data which are available suffice to illustrate the main differences in the composition of the total proteins of different types of seeds and of different parts of the cereal grains. The validity of the general picture shown by the analyses has in many instances received confirmation from feeding tests with animals.

The distribution of the essential amino acids in the proteins in different parts of the wheat grain may be illustrated (Table 9) by the data of BARTON-WRIGHT and Moran (1946) who examined different milling fractions corresponding approximately to the different anatomical parts of the grain.

Table 9. *Distribution of Essential Amino Acids in Wheat Protein. (Calculated to 16% N on moisture- and ash-free basis.)* [BARTON-WRIGHT and MORAN (1946).]

Amino Acid	Inner Endosperm	Outer Endosperm	Bran	Germ	Whole Wheat
Arginine	2.92	4.50	7.53	6.20	3.80
Cystine	1.55	1.90	1.50	1.37	1.74
Histidine	1.65	1.74	1.68	3.03	1.65
Isoleucine	7.02	6.56	4.50	5.23	6.97
Leucine	9.14	7.98	6.52	7.33	8.27
Lysine	1.92	2.60	3.87	5.44	2.80
Methionine . . .	1.12	1.40	1.09	1.28	1.32
Phenylalanine . .	3.95	3.43	2.45	2.47	3.68
Threonine	2.56	2.72	2.85	6.28	2.78
Tryptophane . . .	0.93	1.12	1.83	0.90	1.03
Valine	3.65	4.02	4.10	4.20	4.00

While future work may show that some of these values need revision they adequately demonstrate the changing character of the composition of the proteins from the inside to the periphery of the wheat grain. The nutritional quality of the proteins of the aleurone layer (mostly in the bran fraction in Table 9) and the germ is better than that of the endosperm. The proteins of the aleurone and the germ approach in their content and balance of the essential amino acids the level of nutritional quality of some of the best animal proteins. Compared with the endosperm proteins the most notable improvement is the much higher lysine content and an appreciably higher content of valine, tryptophane and methionine. As these anatomical parts only contribute a small fraction to the total proteins of the grain they cannot completely compensate for the unbalanced character of the endosperm proteins. For this reason the chief nutritional defect of wheat protein as a whole is the inadequate content of lysine. Similar considerations apply to a greater or less degree to the other cereal grains. Methionine is the limiting essential amino acid in most of the legume seeds.

LUGG (1949) has compiled data on the amino acid composition of the bulk proteins of sunflower seed, soyabean, maize and wheat. Most of the considerable data on the subject are unsatisfactory except for indicating broad differences between seeds of different types. DUSTIN, SCHRAM, MOORE and BIGWOOD (1953) have recently published figures for barley meal which are probably the best available for a whole seed meal. It is to be hoped that further studies using a similar technique, will now be made with other seeds.

The protein content of seeds of a given species frequently varies with the particular variety. Further, with a given variety in the case of the cereal grains for example, the cultural conditions and particularly the nitrogeneous treatment influence the protein content. How far the changes in protein content in different varieties and under different cultural conditions are reflected in different proportions of the individual protein components and hence in a different amino acid composition for the proteins as a whole is still uncertain. The changes may be no greater than the inevitable errors in the analytical determinations. In this connexion it must be noted that in amino acid analyses of whole meals frequently only about 75—85% of the total nitrogen is recovered. This important question has recently been investigated by SAUBERLICH, CHANG and SALMON (1953) who have studied the amino acid composition and protein content of maize as related to variety and nitrogen fertilization. The protein contents ranged from 6.8% to 12%. The samples on a low nitrogen treatment had a protein content ranging from 6.8 to 8.2% while those on a high nitrogen treatment ranged from 9.3 to 12%. Eighteen amino acids were determined in the hydrolysates and these accounted for about 85% of the KJELDAHL nitrogen. The percentage of some amino acids, leucine, alanine, phenylalanine and proline, in the total protein increased as the percentage protein increased. The percentage of others, arginine, glycine, lysine and tryptophane, in the total protein decreased as the percentage of protein increased. The remainder tended to stay relatively constant.

Further work on this question using recent techniques of hydrolysis and estimation would be most valuable.

B. Procedures.

I. Choice and Preparation of Initial Material.

It is preferable to start with seeds of known history and origin. The seeds should be in sound condition; ungerminated, free from mould or other microbiological infection, and from infestation by insects. The sample used should be examined for the presence of "foreign" seeds and for rodent excreta. Impurities of these kinds may often be detected by the use of a graded set of sieves or by gentle aspiration with a current of air. If the seeds are obtained from a commercial source it is important to ascertain that they have not been treated with fumigants, dressings or other chemical preservatives, or subjected to methods of drying which may possibly have damaged them. With the seeds which are of commercial importance all these hazards may exist. It is useful to know the age of the seeds, as oxidative and other changes may proceed, in some instances, even at their normally low moisture content. VICKERY, SMITH and NOLAN (1952), however state that for the preparation of globulins "seeds of low germination are as satisfactory as fresh viable seed". With the cereal grains the protein content and the behaviour of the ground-up material towards different solvent extractants varies markedly with different varieties and with the cultural conditions which prevailed during the growth of the parent plant.

It has already been emphasised that, in appropriate cases, attention should be paid to the anatomical parts of the seed. Whenever it is feasible the outer seed coats or integuments should be removed and if required, enamined separately. Whether or not it is possible to excise the embryo from the endosperm will depend on the type of seed which is being investigated.

The grinding of seeds, or their anatomical parts, to prepare a meal suitable for extraction generally presents less difficulties than similar operations with the leaves and other plant tissues which are discussed in detail elsewhere. Roller or hammer mills are usually adequate and for small samples a coffee bean grinding machine may suffice. The main risks are contamination with metal and over-heating-particularly in the case of the hammer mill.

The age of a seed meal may be important in relation to the extraction of its protein components and to their character and behaviour when dispersed and fractionated. JONES and GERSDOFF (1938) found that when soyabean meal ages the amount of nitrogen extractable with neutral salt solutions decreases and a similar effect was observed by SMITH and CIRCLE (1938) when investigating the proportion of the total nitrogen which is extracted by water. Some character-istics of the endosperm proteins of wheat are changed when flour ages. These are reflected in changes in the physical properties of the gluten, which are desirable in the doughing and baking processes of bread making, and moderately aged flour is, for these reasons, considered to be better for bread making than fresh flour. The nature of these changes in seed meals during aging is still largely unknown. In part they may be due to oxidative changes — the protein possibly being indirectly affected through a primary oxidation or peroxidation of unsatur-ated fatty acids (MORAN, PACE and McDERMOTT, 1953). In those meals containing reducing sugars some of the proteins may be modified by occurrence of the MAILLARD Reaction.

With some types of seed meals it is advantageous to carry out a preliminary extraction with a fat solvent or solvents. This is often helpful where subsequent protein extractions are to be made with aqueous solutions since it reduces, but does not eliminate entirely, the formation of lipoid-protein dispersions which are frequently difficult to clarify and fractionate. Low boiling petroleum ether is the most usual solvent. It should be noted however that this solvent does not remove all the lipoid material from seed meals and secondly that it may remove protein or peptide substances associated with the dissolved fat. If flour from the endosperm of wheat, for example, is extracted with petroleum ether at room temperature the clear solution obtained contains protein or high molecular weight peptides, cf., BALLS, HALE and HARRIS (1942). More lipoid may sometimes be removed if petroleum ether extraction is followed by alcohol. But this is an operation to be adopted with some caution, and its value assessed in relation to the particular operational requirement. It may cause some denaturation of the residual protein and with the cereal grains, and probably with some seeds

of other families, it dissolves out a fraction of the protein. For some purposes
however alcohol, acetone or benzene extractions may be a useful preliminary step.
Thus BROWN and GODDARD (1941) in the isolation of cytochrome oxidase from
wheat germ extracted the material with benzene and acetone while SINGER and
PENSKY (1952) defatted wheat germ with acetone in their isolation of α-carboxylase. BASS and SAUNDERS (1932), in the preparation of crystalline globulins
from citrus seeds, used benzene to defat the ground seed meal.

II. Extraction.

1. Extraction with Neutral Salt Solutions.

The choice of method for the extraction or dispersion of the protein contained
in seed meal, depends on how necessary it is, in a particular investigation, to
obtain the protein with the minimum of change from its native state. With meal
from dicotyledonous seeds water and neutral salt solutions extract a good proportion of the protein and with these mild conditions denaturation and modification of the protein is usually slight. But some denaturation may occur even
in neutral salt solution, as MACHEBOEUF and TAYEAU (1942) have noted in a
study of the effect of salt concentration and pH on the extraction of proteins from
peanut meal. Neutral aqueous solutions frequently disperse, in "association"
with the protein, lipoid and carbohydrate material as well as those water-soluble,
low molecular weight substances which may be removed by dialysis. This
dispersion of lipoid material may occur even when the meal has been previously
extracted with a solvent such as petroleum ether. This sometimes occurs with
wheaten flour as well as with meals from dicotyledonous seeds. MACHEBOEUF
and TAYEAU (1942) found that with saline extracts of defatted peanut meal
a large part of the lipoid and carbohydrate material associated with the proteins
remains in the mother liquor when the extracts are acidified to about pH 4.4 when
most of the protein precipitates. The precipitate still contains phospholipoid
and carbohydrate impurities. If it is extracted with boiling alcohol and then
with ether the coagulum still remains rich in phosphorus. SMILEY and SMITH
(1946), in a study of the preparation and nitrogen content of soyabean protein,
have found that the nitrogen content of different glycinin preparations, varies
even when they are made from the same source of material. They consider
that this variation is due to contamination with material which appears to be
mainly phospho-lipoid. The impurity is largely eliminated if ethanol is used as
the solvent for the preliminary extraction of the oil.

Other phosphorus compounds may be associated with the proteins in extracts
of seed meals and, in some circumstances, be retained with protein precipitates
derived from the extracts. A considerable proportion of the phosphate in many
seeds is present in the form of phytic acid. FONTAINE, PONS and IRVING (1946),
who have studied some of the relationships between phytic acid and protein
in a number of seed meals, have concluded that phytic acid may be a major
impurity in seed protein preparations. They suggest that the phytate ion may
be responsible for the cloudiness of some extracts, and have established, with
peanut, cottonseed and soyabean meals, that the naturally occurring phytic
acid is responsible for the suppression of the solubility of the seed meal proteins
at pH values below their isoelectric points. At these pH values they demonstrated
that phytic acid forms a complex with the proteins of peanut and cottonseed
meals. In a discussion of the results they state: "Since phytic acid can be considered to occur as an impurity in most seed protein preparations, it is reasonable
to suppose that its presence would influence the crystallization and denaturation

behaviour of these proteins and also influence the results of investigations that involve the measurement of the electrophoretic mobility, viscosity or action of proteolytic enzymes on solutions of these proteins." The significance of the experimental observations of FONTAINE, PONS and IRVING is emphasised by BOURDILLON's (1951) isolation, from the bean *Phaseolus vulgaris*, of a crystalline protein-phytic acid complex. This was crystallized at pH 4.0 in the form of anisotropic dodecahedra. The complex could be dissociated by prolonged dialysis in a slightly alkaline medium. The protein alone was crystallized, at pH 5.3, in the form of isotropic bisphenoids and quadrangular bipyramids. BOURDILLON suggests that the complex exists as such in the seed and is not an artifact arising during the preparation. Other phosphorus compounds may form complexes with seed proteins; COURTOIS and BARRÉ (1949) have shown that a protein fraction from almonds interacts with orthophosphoric and pyrophosphoric acids and with a number of phosphoric esters.

Neutral salt solutions do not extract all the protein and more drastic methods are required if a higher proportion is to be dispersed. With these there is a risk of denaturing or changing in other ways the native state of the proteins. For some purposes, e. g. the utilization of the protein for industrial purposes or for some types of feeding stuffs, this may not matter. But for investigations where it is important to work with material modified as little as possible from its native condition dispersing agents such as alkali, detergents and organic reagents are obviously to be avoided unless preliminary tests have clearly demonstrated that they are unobjectionable. Usually in fact the aim of extracting most of the proteins is incompatible with the object of obtaining it in approximately the native state. Some proteins appear to be sensitive, more perhaps than is usually appreciated, to atmospheric oxygen. HAUGAARD (1946) for example, has shown that enzymes containing essential —SH groups are inactivated by oxygen. Conditions which cause frothing may therefore, in the presence of air (apart from surface denaturation effects) lead to oxidative changes. We have observed (unpublished work) in experiments on the fractionation of wheat gliadin and examination of the fractions by paper chromatography and ionophoresis that different patterns are obtained when the fractionation is carried out entirely in an atmosphere of nitrogen compared with the usual procedure in air. The susceptibility of sulphydryl groups to oxidative change is well known. Under some conditions with wheat proteins, some of the methionine residues also appear to be easily oxidized to the sulphoxide.

2. Extraction with Alkaline Solutions.

It is found, with many seed meals, that the most efficient of the usual agents, for its purpose of dispersing a maximum amount of protein are solutions of the caustic alkalis. But it is unlikely that these are without effect on the more labile protein components. Although KLIMENKO (1950) has shown that 0.1—0.2% NaOH is the most efficient dispersing agent for the proteins of peas and wheaten flour he states that it is without effect upon the arginine and cystine residues. Other tests however might be expected to have shown some modification of the proteins. A detailed investigation of the conditions which favour the optimum dispersion of the proteins of the soyabean has been made by SMITH and CIRCLE (1938). They used defatted (petroleum ether) meal which had been finely ground in a Wiley mill. The amount of nitrogen extracted with a number of organic and inorganic acids was no greater than that obtained with water which extracted 84% of the total nitrogen at 26° C. It was possible to extract up to 95% of the total nitrogen with solutions of alkali used at about 0.8 milli-equivalents of the base per gm. of meal. SMITH and CIRCLE were concerned with optimum extraction of the protein and it is uncertain how far the alkaline extractants, clearly suitable for their immediate purpose, may have modified the protein.

Mild alkaline extraction, by the use of an ammonia-ammonium chloride buffer has been used by IRVING, FONTAINE and WARNER (1945) to extract about 98% of the nitrogen of peanut meal. This is an example where a very high proportion of the protein of a seed meal has been extracted, under relatively mild conditions, and then examined by electrophoretic analysis. It is thus of considerable general interest in relation to the study of the proteins from the meals of some of the dicotyledonous seeds, and the procedure of IRVING, FONTAINE and WARNER may therefore usefully be quoted here:

"White skinned peanuts were mechanically shelled and the shell fragments carefully removed. Unblanched meats were mechanically crushed and flaked and flakes extracted with petroleum ether (30—60° C) at room temperature. Solvent removed by aeration and the air dried meal ground in a hammer mill . . ."

"One gram of peanut meal was extracted successively in the cold (1—3°C) with five portions of ammonia buffer (0.2 M NH$_3$ and 0.1 M HCl, pH 9.26 at 25°, 0.1 ionic strength) by alternately stirring and centrifuging and the combined extracts were diluted with buffer to 50 ml. An alkaline buffer was used in order to obtain maximum extraction of the nitrogen in peanut meal and a sufficiently high protein concentration (0.8%) for accurate electrophoretic analysis. The combined extracts contained an average of 98% of the total nitrogen in the meal.."

"The extract was dialyzed in a cellophane membrane against 0.5 litres of ammonia buffer in the cold for 2—4 hours. The dialysis was continued in 1.8 litres of fresh buffer for an additional 14—16 hours and the dialyzed solution was then centrifuged at high speed for 15 minutes (relative centrifugal force of 3,500 × gravity) to remove a minute amount of precipitate which formed during dialysis. The clear dialyzed solution contained approximately 0.9% protein or approximately 98% of the protein of the meal . . ."

C. Experimental Procedures for the Preparation of Different Types of Protein Fractions from Seeds.

In Section A some of the problems involved in the investigation of seed proteins have been discussed in general terms. The methods which are used are now illustrated in greater detail by giving some examples of preparation of different types of proteins. Some of these are subject to the limitations which have been emphasised previously. As representative procedures, however, they indicate the position from which further advances may be made.

I. Globulin Fractions.

1. From Oil-bearing Seeds.

The globulins from the oil-bearing seeds are readily obtained in crystalline form. Characteristic examples are obtained, for example, from hemp seed, castor seed, brazil nut, BAILEY (1942a); many species of the *Cucurbitaceae*, VICKERY, SMITH, HUBBELL and NOLAN (1941) and CHENG-FA WANG (1940); citrus seeds, BASS and SAUNDERS (1932). Edestin, from hemp seed, has for some time been considered a good example of a well characterized protein, and therefore suitable for careful studies of amino acid composition, cf. CHIBNALL (1942), and TRISTRAM (1946) and for terminal amino acid assay by SANGER and PORTER (1947). The problems associated with establishing the homogeneity of the crystalline globulins are most aptly described in BAILEY's comments on edestin, in his paper dealing with the denaturation of this globulin by acid (1942b): "It is important at the outset to stress that neither edestin nor the majority of other plant proteins can certainly be considered as completely homogeneous with respect both to particle weight and to electrical charge ... In most cases homogeneity has been established in the ultracentrifuge but has not been supplemented by electrophoresis or solubility experiments. That edestin is monodisperse in the ultracentrifuge was established by SVEDBERG and STAMM (1929). Electrophoretic experiments are rendered almost impossible by the extreme insolubility of edestin at low temperatures and low ionic strengths. The solubility of a "standard" edestin preparation however was studied by the author ... and was found to be greatly dependent upon the solvent-solute ratio, both in polar and dipolar solvents. This does not necessarily indicate that edestin is not homogeneous ... A simple phase-rule solubility however has generally been demonstrated only in the case of proteins with particle weights below 70,000. For large molecules like those of edestin with a particle weight above 300,000 we may find that the laws governing the solubility of colloidal bodies hold more nearly."

Details of the preparation of some characteristic crystalline globulins are now given. These descriptions are given in the words of the quoted authors.

a) Crystalline Globulins from Cucurbit Seeds.

Seeds of pumpkins, squashes, marrows. Seeds of watermelon, cantaloupe or cucumber may be used (VICKERY, SMITH, HUBBELL and NOLAN, 1941; VICKERY, SMITH and NOLAN, 1952). "Four kilograms of air-dried seeds are ground in a meat grinder, first with the coarsest cutting plate and then with the finest. The seed should be fed into the grinder continuously in small amounts. The too rapid addition of seed will clog the grinder. The meal is mixed without unnecessary delay into 12 litres of warm (30—35°) 10% sodium chloride solution. Filter paper clippings are then incorporated into the thin suspension in sufficient amount to form a moist mass that can be molded into square, flat cakes that are individually wrapped in stout canvas press-cloth (drilling) without mitering the corners. The cakes are pressed in piles of three or four at a time between steel plates in a hydraulic press. Pressure is applied gently at first until the extract no longer runs freely. The full pressure is then slowly applied. With experience, conditions are soon found that do not lead to rupture of the press-cloth. It is sometimes desirable to rearrange and straighten the cakes in the press before the final high pressure is exerted.

The cakes are broken up and moistened with about 8 litres of warm salt solution, and the material is wrapped and pressed a second time. The turbid extract (about 16 litres) is placed in enamelware pails and rapidly brought to a temperature of 75° in a water bath to coagulate the small quantities of albumin-like proteins present. It is then transferred to large flasks or glass jars and allowed to stand overnight for the emulsion of fat and coagulated protein to rise. The still turbid solution is siphoned from beneath the emulsion layer and is filtered through a thick (about 5 cm.) layer of filter paper pulp packed hard on a BÜCHNER funnel (a commercial filter aid may also be employed) and washed thoroughly with water and then once with 10% sodium chloride solution. The vacuum should be applied slowly and should not be greater than about 13 cm. of mercury as, otherwise, the filter may clog or may yield a turbid filtrate. Time and material are saved by proceeding slowly at this point.

The fatty emulsion is centrifuged, and as much aqueous phase as possible is removed and added to the filter. When all has passed through, the clear filtrate is removed and the rest of the emulsion, diluted with a little salt solution used to wash the flask and centrifuge bottles, is added to the filter. The filter is finally washed with a little salt solution, and the filtrate, which may be turbid, is filtered clear on a small paper pulp filter.

The pale yellow filtrate should be perfectly clear, but it possesses a strong TYNDALL effect. It is warmed to 60° and diluted with 4 volumes of water at 60° and is then chilled overnight in a cold room or placed out of doors in cold weather. The supernatant solution is removed, and the precipitated protein is collected on folded soft filter papers, if recrystallization is to be carried out, or, otherwise, on hardened filter paper in a BÜCHNER funnel. For recrystallization, the soft filter papers containing the protein are reduced to pulp with the aid of sufficient 10% salt solution, and the solution is diluted with salt solution to about two-thirds of the original volume of the extract. This is filtered clear on a pulp filter as before and again diluted at 60° with 4 volumes of water at 60°. The recrystallization process may be repeated as many times as desired.

The protein, after being collected on hardened filter paper, is washed with 30% alcohol; in large-scale operations, it is convenient to collect successive lots under this solvent. Washing is carried out by suspending the preparation repeatedly in 30% alcohol until the filtrate gives only a negligible test for chlorides. The protein is finally washed twice with 95% alcohol and, while still slightly moist, is granulated by being brushed through a fine sieve. It is then spread in a thin layer on pans and allowed to dry and equilibrate with atmospheric moisture for a day or two.

Properties and Yield of Product. The dry white powder obtained seldom retains its crystalline form undamaged after the washing operation, although inspection under the microscope before this is done usually shows the protein to be wholly crystalline (regular octahedra). Nevertheless, it should be completely soluble to a clear solution in 10% salt solution. Samples of the crystalline material may be retained under 2% salt solution if protected with toluene and kept in the refrigerator. The yield from 4 kg. of *C. pepo* seeds is about 500 gm. or from 10—12% of the weight of the air-dried meal. Most of the other cucurbit seeds give similar yields, although cucumber seeds have given yields as high as 15%. The protein from *C. pepo* contains 18.42% of nitrogen, 0.99% of sulfur, corrected for ash and moisture, and 0.42% of ash ...

Comment. The present procedure is modified from the classic methods of OSBORNE. Extraction of the meal with fat solvents before treatment with salt solution facilitates the clarification of the protein extract but otherwise offers no substantial advantage either in yield or in quality of the product. The heat-coagulation step assists materially in the filtration, and no adverse effect of the high temperature upon the properties of the globulin has been noted. However, if highly accurate physico-chemical measurements upon the isolated protein are to be carried out, it would be advisable to omit this step until a demonstration has been obtained that it is entirely safe. It is a convenience only, not a necessity, and adequate purification of the globulin can be obtained by repeated recrystallization. Under circumstances where exposure of the seed globulin solution to any tempreature higher than room temperature is objectionable, the extraction should be carried out with even more liberal quantities of salt solution and the filtered extract should be subjected to dialysis to separate the protein rather than to the dilution process described."

The preparation of crystalline globulins from the seeds of watermelon, gourd, pumpkin and loofah has also been described by CHENG-FA WANG (1940).

b) Crystalline Globulins from Other Oil Seeds.

"Preparation of some seed globulins in the microcrystalline state." [BAILEY (1942 a).]

Edestin from Hemp Seed. 500 g. of freshly ground seeds were mixed with 2 litres of 5% NaCl and after raising the temperature slowly to 50—55° in a water bath, the mixture was agitated at this temperature for an hour, strained through calico and filtered clear through a three inch layer of filter pulp which had been saturated with hot 5% NaCl. During filtration the temperature was not allowed to drop below 35°; after filtration the residual liquid in the pad was displaced with hot salt solution and the combined filtrates were cooled slowly to room temperature and then to 5°. The crystalline sediment was now separated from the mother liquor by centrifuging and recrystallized twice from 5% NaCl. The protein was stored in paste form at 0° in presence of a little toluene, or the dry crystals were prepared by washing free from chloride with distilled water, dehydrating in alcohol and ethyl ether and finally in a current of warm air.

Castor Seed Globulin. After removal of the testa the procedure was similar to that for edestin. The globulin generally separated in the form of spheroids. A perfect crystalline preparation could always be obtained by shaking an excess of the spheroids at 26—28° with alcoholic NaCl (1 volume absolute alcohol diluted with 4 volumes of 5% NaCl solution) filtering and cooling.

Excelsin from the Brazil Nut. Kernels ground and partially defatted in the BÜCHNER press at 300 atmospheric pressure. Residue extracted as described in the case of hemp seed, using 400 ml. of salt solution for 100 g. of defatted kernel. Solubility of excelsin in aqueous salt solution is too high to permit of crystallization by cooling, but the typical micro-hexagonal plates were readily obtained by rapid dialysis in cellophane membranes against distilled water. Recrystallization was usually effected from alcoholic salt solution, as described in the case of castor seed globulin."

2. From a Seed of Low Oil Content.

"The globulins of the Jack bean *(Canavalia ensiformis)*". (SUMNER and GRAHAM, 1925; cf. SUMNER, 1919, and SUMNER and HOWELL, 1936.)

Concanavalin B. This globulin is present in the jack bean in very small amount. We have prepared it as a by-product in our method for the purification of urease. One mixes 2 liters of 30 to 35% alcohol with 1 kilo of meal for $^1/_2$ hrs. The extract is pressed out from cheese-cloth in a tincture press, centrifuged as clear as possible, and allowed to stand in a liter graduate overnight at —10° C. The supernatant liquid is then syphoned off and the precipitate is centrifuged off while cold in cold centrifuge tubes and holders. Upon stirring up the precipitate with a little dilute neutral phosphate solution the needles of concanavalin B separate out quantitatively within a short time and are centrifuged off and washed with 2% salt solution. The needles are recrystallized twice by dissolving in 10% sodium chloride solution and dialyzing. The second time the dialysis is continued until no more chloride is given off. The product is dehydrated with increasing concentration of alcohol, with absolute alcohol, and absolute ether, and is dried in a vacuum over sulfuric acid, then is dried in a hot water oven, and finally in a hot air oven at 105° C for an hour. The needles should not remain too long in water during the first part of the purification as they are likely to become denatured and cannot then be redissolved in salt solution.

Concanavalin A. The press-cake from which urease and concanavalin B were prepared is washed once with 30% alcohol and pressed out again in the tincture press. This serves to remove sediment that is difficult to filter later on. The press-cake is now mixed up with 2 litres of 2% aqueous sodium chloride solution and again pressed out. The extract is mixed with some toluene and filtered into a dialyzing membrane and the material is dialyzed against 0.001 N acetic acid until the globulins have precipitated completely. We have found that our dialysis parchment contains a considerable amount of a soluble carbohydrate and have removed this by washing the parchment many times in warm water before use. The precipitate by dialysis consists of canavalin, concanavalin A and a small amount of concanavalin B. The precipitate is centrifuged off and washed with water. The canavalin and concanavalin B are now dissolved by the addition of 10% salt solution. Concanavalin A is not dissolved and is centrifuged off and washed several times with 10% salt solution and once with distilled water, then it is centrifuged down as hard as possible. The bisphenoid crystals are now dissolved by stirring each 125 ml. of moist precipitate with 100 gm. of solid sucrose and 60 ml. of neutral phosphate solution, containing 6.8% K_2HPO_4 and 4.25% KH_2PO_4. The solution is treated with neutral phosphate solution, alcohol to 30%, and solid sodium chloride to precipitate the pentosan. After standing for 24 to 48 hours it is filtered and the filtrate is dialyzed until practically free from salt. The crystals are washed until salt-free, dehydrated with alcohol and ether, and dried as described under concanavalin B.

Canavalin. The salt solution of canavalin and concanavalin B, from which concanavalin A has been separated, is dialyzed until the globulins have precipitated. The precipitate is centrifuged off and washed with distilled water and dissolved in 2% sodium chloride. The needles will not dissolve and are centrifuged off and purified as described under concanavalin B. The solution of canavalin is filtered and dialyzed until precipitated, again dissolved in salt solution, and treated with excess of phosphate and sodium chloride and alcohol to 30%. After 24 to 48 hours it is filtered clear and again dialyzed. This time the precipitated globulin is centrifuged off and washed with 30% alcohol. It is then dehydrated with alcohol, washed with ether, and dried as had been already described."

3. Globulins of the *Gramineae*.

These have not been so well characterized as the oil seed globulins, but some advance has recently been made by workers in Svedberg's laboratory, cf. Quensel (1942) and Danielsson (1949). Before noting their methods a description of one of Osborne's procedures is of both historical and general interest.

The Globulin of Wheat Flour. (Osborne, 1907.) "Beside the proteins soluble in water, 10% sodium-chloride brine extracts from ground wheat kernels a globulin which is present in the seed in small quantity. Ten kilograms of "straight flour" were extracted with 34 litres of 10% NaCl solution by suspending the flour in the liquid, stirring frequently, and then allowing the whole to stand at rest overnight. The extract, separated from the flour and filtered as clear as possible, had a very slight acid reaction, was of a pink color, very viscid consistence, and formed about one-half of the solution added to the flour. This was saturated with ammonium sulphate, and the resulting precipitate filtered off and dissolved as far as possible in 4 litres of 10% NaCl solution. The exceedingly viscid solution was filtered with difficulty, placed in a dialyzer, and left in a stream of running water until the chlorides were removed. As the salts dialyzed out the globulin gradually separated in minute particles, the larger of these being evidently spheroidal in form. This precipitate weighed 5.8 grams . . .

Dilution of the solution of the globulin in 10% sodium-chloride brine precipitated the protein. Saturation with sodium chloride gave no precipitate. Saturation with magnesium sulphate, and also with ammonium sulphate, completely precipitated the globulin.''

As noted in Section A (p. 72), DANIELSSON has examined by the use of the ultracentrifuge, the differentiation of globulins in different anatomical parts of seeds of some of the *Gramineae*. The following description of the general method of preparing the fractions is derived from the details given by SAVERBORN, DANIELSSON and SVEDBERG (1944) and DANIELSSON (1949).

Globulin Fractions from Seeds of the *Gramineae*. (SAVERBORN, DANIELSSON and SVEDBERG, 1944; DANIELSSON, 1949, and JENSEN, 1952.) Seeds, or parts of seeds, were ground to a meal. The meal was extracted with M NaCl buffered to pH 7.0 with mono- and disodium phosphate of a total concentration of 0.1 M. Four parts of the extracting solution were taken to one part of meal. The extractions were made by vigorously stirring the mixture for one night at about 5° C. The extract was then separated by centrifuging and filtering. Fractions were obtained from the extract by fractional precipitation with solid ammonium sulphate. Three precipitates were obtained at 15% saturation, 40% saturation and 70% saturation, with $(NH_4)_2SO_4$, respectively. (100% saturation given as 760 gm. $(NH_4)_2SO_4$ per litre.) The precipitates were dissolved in a buffer of composition:— 0.2 M NaCl, 0.03 M NaH$_2$PO$_4$, 0.02 M NaH$_2$PO$_4$, pH 7.0, and dialyzed against distilled water to separate the globulins from albumins, material of low molecular weight and polydisperse material. Further repeating the cycle of solution in saline buffer and dialysis or reprecipitation with the appropriate concentration of $(NH_4)_2SO_4$.

In this way from barley meal fractions were obtained containing varying amounts of four globulins called α, β, γ and δ in order of increasing molecular weight. The components were distributed among the different fractions in the following way: (1) 15% saturation with $(NH_4)_2SO_4$, β-rich fraction; (2) 40% saturation, polydisperse fraction; (3) 70% saturation, α, γ and δ-rich fraction.

The distribution of these components in different parts of the grain has been described in Section A (p. 72). DANIELSSON's (1949) ultracentrifuge data for fractions which are observed in different members of the *Gramineae* are given in Table 10.

Table 10. *Ultracentrifuge Data for Globulin Fractions of Various Graminean Seeds.*
[DANIELSSON (1949).]

Species	Components				Sedimentation Constants (S)			
	α	β	γ	δ	α	β	γ	δ
Hordeum vulgare	α	β	γ	δ	2.5	6.2	8.3	12.0
Secale cereale	α	—	γ	—	2.6	—	8.2	—
Triticum vulgare	α	—	γ	—	2.5	—	8.2	—
Avena sativa	α	—	γ	—	2.6	—	8.1	—
Zea May	α	—	γ	—	2.6	—	8.2	—
Panicums milliaceum	—	—	γ	—	—	—	8.5	—
Phleum pratense	—	—	γ	—	—	—	8.5	—
Festuca pratensis	—	—	γ	—	—	—	8.3	—
Festuca rubra	α	—	—	—	2.4	—	—	—

4. Globulins of the *Leguminosae*. (DANIELSSON, 1946.)

By a procedure similar to that described in 3., but precipitating only at one $(NH_4)_2SO_4$ concentration, 70% saturation, DANIELSSON obtained from pea meal, after drying the final precipitate formed during dialysis *in vacuo* over CaSO$_4$, a powder containing two main globulins considered to be identical with OSBORNE's vicilin and legumin. DANIELSSON (1949) gives the following method for the separation of these globulins from each other:

"A dried preparation, consisting of a mixture of approximately equal portions of the two components is treated with a large excess of buffer of pH 4.5, with stirring. Acetate buffers containing 0.2 M NaCl have been used. During this procedure the light component dissolves while the other does not. After 24 hours the heavier component is centrifuged off and dissolved in standard buffer (0.2 M NaCl; 0.03 M Na$_2$HPO$_4$; 0.02 M NaH$_2$PO$_4$; pH 7.0). This solution is dialyzed against water. The centrifugate from the above centrifugation contains the lighter component and this solution is also dialyzed. After 1—2 days all the globulin has precipitated, and the two precipitates are dried in the cold . . .

By repeated use of this method the two components have been completely separated from each other."

By examination in the ultracentrifuge the molecular weight of vicilin was found to be 186,000 and that of legumin 331,000. DANIELSSON has examined the seed globulins from 34 different species of the family *Leguminosae* and he states: "Vicilin and legumin were found in all, except in some *Acacia* species and in *Trifolium repens*".

II. Prolamine from a Cereal Grain.

The classical example is gliadin, the protein fraction of wheat flour soluble in ethanol diluted with water. OSBORNE (1907) describes several procedures either by direct extraction of the flour or by extraction of the gluten washed out from the flour. More satisfactory preparations are obtained if gluten rather than the whole flour is used. Many modifications of OSBORNE's procedure have been made. In some of these a greater or less degree of fractionation may occur. Preparations are only strictly comparable it made in precisely the same way. DILL and ALSBERG's (1925) procedure is now described. Preparations using this method, have been used by HAUGAARD and JOHNSON (1930) in their investigation of the thermal fractionation of gliadin and fractions similar to those obtained by HAUGAARD and JOHNSON have been examined in the ultracentrifuge by KREJCI and SVEDBERG (1935).

We have already commented on the risks which may attend shaking in air.

Preparation of Wheat Gliadin. (DILL and ALSBERG, 1925.) "Preparation of the gluten, extraction therefrom of the gliadin with dilute alcohol, and concentration of the filtered extract under dimished pressure below 50° C were carried out essentially as described by OSBORNE and HARRIS (1903). The concentrated syrup was then poured into about 5 volumes of 1% aqueous sodium chloride solution contained in a 2500 ml. wide mouth bottle. Vigorous shaking for 1 or 2 minutes precipitated the gliadin as a foam which presents a large surface and is of such a nature that the bottle may be inverted and the wash water drained out. Three vigorous shakings with 1 liter portions of 0.1% sodium chloride solutions followed.

The wet foamy precipitate was dissolved by the addition of a suitable amount of 95% alcohol (the percentage of alcohol in this and subsequent cases is percentage by volume) and from the volume of the solution obtained, it was possible to adjust the final solvent to a concentration of approximately 70% of alcohol. This solution was held in the refrigerator at about 5° C for 24 hours. Much of the gliadin had then separated as a honey-like layer. It appeared to carry down with it most of the lipoids and suspenoid impurities which are difficult, if not impossible, to remove by filtration. The decanted supernatant liquid was turbid, but became water-clear when warmed to room temperature. The honey-like residue was dissolved in warm 70% alcohol and a second separation was carried through in the same manner. A third separation did not succeed.

This method of purifying gliadin is advantageous in that tedious filtration through filter paper pulp is unnecessary. In fact it was found that the gluten extract could be carried through this separation without filtration. Such a separation can only be achieved at the expense of some gliadin and hence is not applicable when a quantitative yield is desired. This method is possibly adaptable to the preparation of other proteins; almost certainly to the preparation of other prolamins.

The clear gliadin solutions from the first and second separations were combined and concentrated under diminished pressure. The gliadin was precipitated and washed in salt solution as described before except that sodium chloride was replaced by lithium chloride. The use of lithium chloride in this and subsequent precipitations renders the removal of electrolytes during the final dehydration with alcohol and ether more nearly attainable.

The wet gliadin was dissolved as before in the minimum quantity of strong alcohol. It was then precipitated by pouring it in a fine stream into 4 volumes of absolute alcohol, containing 0.025% of lithium chloride. Vigorous shaking promoted precipitation. The precipitate was dissolved in about 500 ml. of warm 60% alcohol and precipitated as before. This process was repeated a third time.

The final precipitate was dehydrated twice with absolute alcohol, ground in a mortar, again dehydrated with alcohol, and finally three times with dry aldehyde-free ether. It was then dried at 40° C and 2 mm. pressure . . .

Three other gliadin preparations which did not form clear solutions in dilute alcohol at room temperature had been obtained by a different method. This earlier method differed in an important detail from the method just described. It is necessary, in order to make clear the significance of this detail, to describe some of the conditions under which gliadin may be denatured by alcohol. OSBORNE (1924) reported that under certain conditions zein may be denatured by alcohol. In the course of the present investigation it became evident that gliadin also may be denatured by alcohol. After standing for some time in contact with 75—85% alcohol it becomes irreversibly altered. The conditions are difficult to reproduce for the conditions of the gliadin, i. e., whether anhydrous or moist, and the temperature are important variables. Under some conditions even 70% alcohol may denature gliadin. If dry gliadin is covered with 70% alcohol, it becomes solvated, forming a concentrated, clear viscous solution below the larger part of the solvent. If this is then allowed to stand quietly for 2 or 3 days, part of the gliadin will be altered and will not dissolve in any concentration of alcohol at room temperature.

In making these three first preparations, an insufficient amount of absolute alcohol was used for precipitation. The gliadin came down as a viscous translucent mass rather than in the form of opaque white flakes. This translucent mass dissolved only slowly in 70% alcohol. Here is where denaturing occurred. The difficulty may be avoided by using a large excess of absolute alcohol and by redissolving in 60% rather than in 70% alcohol. The use of 60% alcohol is advisable because the strong alcohol held by the precipitate raises the concentration of alcohol in the solvent . . .''

III. Enzymes, Inhibitors and Toxic Proteins.

Methods of fractionation in these cases are generally more complicated than those used for the preparation of main protein fractions. They illustrate the more exacting procedures required for the isolation of the (quantitatively) minor constituents in the multi-component protein system contained in certain seeds.

1. Enzymes.

Soyabean Lipoxidase. The preparation of crystalline soyabean lipoxidase according to HOLMAN and BERGSTRÖM (1951) is as follows:—

"Fat-free, low-temperature extracted soyabean flour (15 kg.) was suspended in 100 litres 0.1 M acetate buffer at pH 4.5. Centrifugation removed the insoluble material and the extract was adjusted to pH 6.7 with NH_3. Five volumes 20% barium acetate, 10 volumes acetone, and 2 volumes of basic lead acetate were added per 100 volumes of extract. The inactive precipitate was removed in a large separator. The inactive precipitate produced by addition of 25 g. $(NH_4)_2SO_4$ per 100 ml. extract was allowed to settle and the supernatant fluid was decanted. More $(NH_4)_2SO_4$ was added to bring its concentration to 40 g./100 ml. and the precipitate containing the activity was recovered by centrifugation. The solid was redissolved in a small quantity of water and heated to 63° C for 5 minutes to coagulate inactive albumins. The supernatant liquid (11.4 million units) was fractionated again with $(NH_4)_2SO_4$, and that which precipitated between 35 and 50% saturation retained. This was fractionated with ethanol at 0° C in 0.02 M phosphate buffer at pH 5.5 and the initial precipitate deposited below 12% alcohol contained the activity. Fractionation with $(NH_4)_2SO_4$ between 50 and 60% saturation yielded a product containing 1.8 million units. Electrophoresis in the large Tiselius apparatus yielded 0.34 million units of lipoxidase at 820 units per mg. This was concentrated and then dialyzed against $(NH_4)_2SO_4$, the concentration of the latter being slowly increased. The lipoxidase crystallized out as colorless plates or sheaves. The crystals were washed with slightly less concentrated $(NH_4)_2SO_4$ to remove the small amount of amorphous material, leaving crystals assaying at 850 units per mg. This preparation is a 150-fold concentration of the dry matter of a water extract. Isolation of the enzyme has been found not strictly reproducible with different batches of soyabeans and therefore pilot experiments and continual assays of fractions is necessary. The spectrophotometric assay was used on all fractions in the described isolation of the enzyme."

A description of the earlier experiments, leading to this isolation, is given by THEORELL, HOLMAN and AKESON (1947). These experiments are of interest in providing another example of the contamination of seed protein fractions with phytic acid.

Wheat Germ α-carboxylase. (SINGER and PENSKY, 1952.)

"(i) *Acetone Powder.* 1 kilo of wheat germ is defatted with 10 volumes of reagent grade acetone at —10°. A portion of the acetone is used to homogenize the germ in a Waring blendor, and the resulting suspension is poured into the rest of the acetone. After 10 minutes stirring, the suspension is filtered rapidly, and the washing and filtration are repeated. While still moist, the defatted germ is placed in desiccators and dried *in vacuo* over H_2SO_4 in the cold.

(ii) *Extraction.* The following procedure is workable on a 250 to 2500 gm. scale. Unless otherwise indicated, it is carried out at 5—7°. All pH values are measured at the temperature at which the step is carried out.

1 kilo of acetone powder is extracted with 5 litres of H_2O by vigorous stirring for 15 minutes, followed by centrifugation for 30 to 45 minutes at 2600 r. p. m. The cloudy supernatant (2 liters) contains all of the enzyme (750,000 to 780,000 units; activity ratio = 2.1, based on dry weight).

(iii) *Isoelectric Precipitation.* To the water extract 1 N acetic acid is added under vigorous stirring to exactly pH 5.20 at 0—4°. The suspension is centrifuged immediately (2600 r. p. m. for 30 minutes), yielding an almost clear, yellow supernatant. After chilling the solution to about 3°, the enzyme is precipitated from the supernatant by further addition of 1 N acetic acid to exactly pH 4.90. The suspension is kept overnight at 0° to complete the precipitation. The enzyme is quite stable at this stage, since it is insoluble at pH 4.90, whereas at pH 5.20 it is unstable. The suspension at pH 4.90 is centrifuged for 30 minutes at 2600 r. p. m. The precipitate is re-suspended in 900 ml. of 0.1 M succinate buffer, pH 6.0, with the aid of glass homogenizers and stirred for 20 minutes. Centrifugation for 30 minutes in an angle head centrifuge at 5000 r. p. m. yields a clear, yellow supernatant, containing 530,000 to 612,000 units ($D_{280}^{1\%}$ = 11.2; activity ratio = 77 to 99).

(iv) *Alcohol Fractionation.* The enzyme solution is chilled to 0° and adjusted to pH 5.50, with 1 N acetic acid. Ethanol (95 per cent, —15°) is added dropwise, with stirring, to a final concentration of 15 per cent by volume, while the enzyme solution is maintained at 0° to —1°. The stirring is continued for 15 minutes after the addition of alcohol and the precipitated enzyme is separated by centrifugation for 45 minutes at 2500 r. p. m. at 0° to —1°. The moist precipitate is immediately dried in a desiccator at 0° over H_2SO_4 (high vacuum). The resulting powder is stable for months in the cold. About 530 mg. of powder, containing 490,000 to 560,000 units, are obtained per kilo of acetone powder. Activity ratio = 1365 to 1510 on the basis of $D_{280}^{1\%}$ = 11.0.

(v) *$(NH_4)_2SO_4$ Precipitation and Dialysis.* The dry enzyme is dissolved in 0.05 M tris (hydroxymethyl) aminomethane (THA) buffer, pH 7.7, at 6°, 20 ml. of buffer being used for each 100 mg. of powder. The insoluble part is removed by centrifugation. The solution is adjusted to pH 6.0 with 1 N acetic acid at 0° and saturated $(NH_4)_2SO_4$ is added dropwise to give 0.33 saturation. After 15 minutes stirring the precipitated enzyme is centrifuged (45 minutes at 5000 r. p. m.). The precipitate is resuspended in about 10 ml. of H_2O for each kilo of acetone powder used and dialyzed for 24 to 48 hours against glass-distilled water. The enzyme precipitates quantitatively during dialysis. After a brief centrifugation at 5000 r. p. m., the precipitated enzyme is dissolved in 0.1 M imidazole buffer, pH 6.8, at 5° (10 ml. of buffer for each 100 mg. of alcohol-precipitated enzyme). Insoluble protein particles are removed by centrifugation at 18,000 r. p. m. In the best preparations, the clear, colorless imidazole extract contains 560,000 units of enzyme for each kilo of acetone powder used; activity ratio = 5600, based on $D_{280}^{1\%}$ = 10.7. The over-all yield is, therefore, 75%, and the losses are due mainly to occlusion of the soluble enzyme in the voluminous precipitates. The purification is about 2700-fold compared with the first extract (activity ratio = 2.1) and 11,000-fold compared with the acetone powder (activity ratio = 0.5). At this stage of purity the enzyme appeared to be homogeneous in the Tiselius apparatus, but ultracentrifugal examination revealed the presence of about 10% impurity. Further chemical fractionation failed to increase the activity ratio significantly, however, the impurity may be removed by differential ultracentrifugation.

Occasionally, the imidazole extract has a lower activity ratio (4000 to 4300) and a total yield of about 400,000 units. The purity of such preparations can be raised by treatment with $Ca_3(PO_4)_2$ gel. In a typical experiment 144 mg. of protein in 10 ml. of imidazole buffer (based on $D_{280}^{1\%}$ = 10.7), at an activity ratio of 4000, was treated dropwise with 70 mg. of $Ca_3(PO_4)_2$ gel, until 28% of the initial activity and 41% of the protein were adsorbed; the purity in the inadsorbed portion was raised to an activity ratio of 4850.

Concentrated solutions of the enzyme in imidazole buffer, pH 6.8, are stable for about a week at 5°. If precipitated with solid $(NH_4)_2SO_4$ (18.6 gm. per 100 ml. of solution) or by exhaustive dialysis against H_2O, the enzyme retains its activity for several weeks in the cold. Freezing or lyophilizing of enzyme solutions causes

7*

considerable inactivation. The best way to store the purified preparations for prolonged periods is in the form of the alcohol-precipitated dry powder.''

Pea Aldolase. (STUMPF, 1948.) ''For the most part pea seeds (Dwarf Telephone) were employed as the source for the enzyme, though squash seeds were also found suitable. The enzyme was readily extracted from pea seeds, which had been soaked for 12 hours in distilled water at 2°, with 0.1% potassium carbonate, as the extracting solvent. The extract was then subjected to (1) ammonium sulfate fractionation, (2) isoelectric precipitation with dilute acetic acid at pH 5.5, by which procedure much inert protein was precipitated, while aldolase remained in solution, and finally (3) acetone fractions followed by dialysis and a final isoelectric precipitation. The details of the purification procedure are summarized below:—

Table 11.

Fraction	Purification Procedure	Total units	Units per mg. protein
Extract		460	0.06
I	Neutral saturated $(NH_4)_2SO_4$ to 35% saturation	43	0.007
II	Neutral saturated $(NH_4)_2SO_4$ to 70% saturation	341	0.11
	II dialyzed 5 hrs. against distilled water at 50°		
IIa	Neutral saturated $(NH_4)_2SO_4$ to 40% saturation	14	0.03
IIb	Neutral saturated $(NH_4)_2SO_4$ to 51% saturation	98	0.14
IIc	Neutral saturated $(NH_4)_2SO_4$ to 61% saturation	165	0.19
IId	Neutral saturated $(NH_4)_2SO_4$ to 70% saturation	34	0.07
	IIc dialyzed overnight against distilled water		
IIc, residue		6	0.04
IIc, supernatant (A)	Ppt. with 1% acetic acid to pH 5.5	155	0.25
A	Acetone added to 55%	5	0.08
B	Acetone added to 62%	135	0.32
C	Acetone added to 72%	2	
	B dialyzed overnight		
B, residue		8	0.11
B, supernatant (Fraction Bs)	Ppt. with 1% acetic acid to pH 5.5	97	5.50

2. Inhibitors.

Crystalline Soyabean Trypsin Inhibitor. (KUNITZ, 1945.) ''. . . The details of the method of isolation are as follows:—

(i) *Washing with 80 per Cent Alcohol.* 1000 gm. cold-processed, defatted soyabean meal is added to a mixture of 2400 ml. 95% alcohol cooled to 5° C and of 450 ml. distilled water. The suspension is stirred well and left at 20—25° C for 30 minutes. It is then filtered with suction on a 32 cm. BÜCHNER funnel through filter cloth. The filtrate is rejected.

(ii) *Extraction in 0.25 N Sulfuric Acid.* The semidry meal is resuspended in 5000 ml. 0.25 N sulfuric acid (7 ml. concentrated H_2SO_4 per liter of water) at 20—25° C and left for 1 hour at room temperature with occasional stirring. The suspension is refiltered with suction on the same filter cloth. The meal residue is rejected.

(iii) *Removal of Inert Protein by Means of Bentonite.* 20 gm. of a stock mixture of one part by weight of bentonite (U. S. Powder, Amend Drug and Chemical Co., New York) and an equal part of hyflo super cel (Johns-Manville Corporation) is added to the acid filtrate and stirred for 10 minutes. The suspension is filtered with suction on 32 cm. No. 303 E & D filterpaper (The Eaton-Dikeman Co., Mt. Holly Springs, Pennsylvania). The residue on the paper is washed twice with portions of 125 ml. of water. The residue is rejected.

(iv) *Adsorption of the Inhibitor on Bentonite.* 100 gm. of the stock of bentonite super cell mixture is added gradually to the combined filtrate and washings from step (iii). The solution is stirred gently while the bentonite mixture is added and the stirring is continued for 10 minutes. The suspension is filtered and the residue is washed, as described in step (iii). The filtrate and washings are rejected.

(v) *Elution with Pyridine and Dialysis.* The bentonite residue is stirred up with 270 ml. of water. At this stage the suspension can be stored in the refrigerator overnight. The suspension is warmed to 25° C and 30 ml. of pyridine (Stock No. 214, Eastman Kodak Co.) is added with stirring. The thick suspension is filtered with suction on 24 cm. No. 303 E & D filter paper in a hood. The filtration generally requires several hours. The residue on the funnel is washed once with 200 ml. 5 per cent pyridine in water. The combined filtrate and washing is dialyzed overnight in 12 inch long cellophane tubings placed in a tall jar with running tap water, in the hood, if possible.

(vi) *Removal of Inert Material at pH* 5.3. The dialyzed solution, free of any gummy residue adhering to the dialysis tubes, is adjusted to pH 5.3 with the aid of about 2 ml. of 1 N HCl. (The pH value is tested by the drop method on a plate using 0.1 M acetate buffer as standards and 0.01 per cent methyl red as an indicator.) 4 gm. of bentonite-super cell mixture is stirred into the solution which is then filtered with suction on a 15 cm. No. 303 E & D filter paper and the residue is washed several times with 15 ml. portions of water. The washings, if not clear, are refiltered. The residue is rejected.

(vii) *First Precipitation of the Inhibitor at pH* 4.65. The combined filtrate and washings of step (vi) is cooled to 5° C and then titrated with 1 N HCl to pH 4.65 (tested carefully with 0.05 per cent bromcresol green on a drop plate). A heavy precipitate is formed which is filtered off at 5—8° C on 15 cm. No. 303 E & D filter paper on a BÜCHNER funnel without suction. The filtration is completed with very light suction. Weight of filter cake 10 to 12 gm. The filtrate is rejected.

(viii) *Second Precipitation at pH* 4.65. The filter cake is suspended in 100 ml. of water cooled to 5° C, the water being added gradually to the precipitate and incorporated thoroughly with a porcelain spatula. 1 N NaOH is added drop by drop with stirring until the precipitate is dissolved. Care should be taken, however, not to raise the pH of the solution above 6.4. The clear solution is warmed to 25° C and titrated slowly with 1 N HCl until a slight permanent precipitate is formed. Two gm. of standard super cel is stirred into the solution which is then filtered with suction on 7 to 11 cm. No. 303 E & D filter paper. The residue on the paper is washed with several milli-liters of water. The combined filtrate and washing is cooled to 5° C and titrated to pH 4.65. It is filtered with light suction on 15 to 18 cm. No. 612 E & D filter paper at 5—8° C. Yield about 8 to 10 gm. of filter cake which is stored in refrigerator. The filtrate is rejected.

(ix) *Crystallization.* The filter cake of step (viii) (about 10 gm.) is ground up to a uniform suspension with 10 ml. of cold water and then warmed to about 35° C. 0.5 N NaOH is added drop by drop with careful stirring until the precipitate is almost completely dissolved and the pH of the solution is about 5.2. The clear solution is decanted into a 5 ml. centrifuge tube. Any residue in the beaker is stirred up with 1 to 2 ml. of cold water, dissolved with the aid of a drop of 0.1 N NaOH, and added to the main bulk of solution in the centrifuge tube which is then placed at 35—37° C for crystallization. A heavy sediment of crystals is obtained within 5 to 6 hours. Inoculation with a few crystals greatly facilitates the process of crystallization. The suspension is centrifuged for 10 minutes at about 3000 r. p. m. The residue is stored in the refrigerator while the supernatant liquid is either stored or, if time permits, titrated with a few drops of 0.2 N HCl or pH 5.1 at 36—37° C, inoculated, and left at that temperature. Another crop of crystals is gradually formed which is centrifuged off after several hours and added to the first crop of crystals. The supernatant liquid is rejected.

It is preferable to begin step (ix) in the morning, so as to be able to centrifuge before the end of the day. The crystals, as well as supernatant solutions, should be stored overnight in the refrigerator. It is advisable to begin the crystallization with at least 50 gm. of amorphous precipitate collected from several preparations.

(x) *Recrystallization.* The combined crystal residues of step (ix) (about 7 ml.) are stirred up with twice the volume of cold water and titrated with 0.5 N NaOH to clearing, the final pH being 6.0. The clear solution is warmed to 35° C and titrated with 0.5 N HCl to pH 5.1 when a slight permanent precipitate is formed. The solution is mixed with 2 gm. standard super cel and filtered clear with suction on a small No. 303 E & D filter paper. The filtrate is inoculated and left at 36 to 37° C. A heavy suspension of crystals forms gradually and is centrifuged after 5 to 6 hours. The residue is stored at 5° C. The pH of the supernatant liquid is readjusted with 1 to 2 drops of 0.2 N HCl to pH 5.1, inoculated, and left for several hours longer at 36—37° C when another crop of crystals is formed which is centrifuged off and added to the first crop. The final supernatant solution may yield still more crystals by cooling it to 5° C and then adding one quarter of its volume of cold 95 per cent alcohol as described in the following section.

(xi) *Crystallization in Dilute Alcohol.* Centrifuged crystals are stirred up with five times the volume of cold water, and 0.5 N NaOH is added drop by drop until the crystals are all dissolved. The pH of the solution is not allowed, however, to rise about 6.6. The clear solution is titrated with 0.2 N HCl to pH 5.2. Any precipitate formed is filtered off with suction on No. 303 filter paper with the aid of 4 gm. of standard super cel per 100 ml. of solution. The residue on the funnel is washed with several milliliters of water. The volume of the filtrate and washings is measured and the solution is then cooled in an ice-water bath to about 5° C. A quarter of its volume of 95 per cent alcohol, cooled to 5° C, is added slowly to the cold solution. A heavy precipitate is formed. The pH of the mixture is adjusted with 0.2 N HCl to 5.0 and left at 30° C. The amorphous precipitate changes within 2 hours into well formed hexagonal and rhomboid crystals and plates which settle rapidly to the bottom of the vessel. The supernatant solution is decanted every hour, adjusted with 0.2 N, or more dilute HCl to pH 5.0, and returned to the original vessel containing the settled crystals. This is continued for several hours until no formation of precipitate is noticed on adjusting the pH of the supernatant solution to 5.0. The crystallization mixture is then allowed to stand at 30° C for 30 minutes longer and then filtered with suction on hardened paper, washed on the funnel several times with cold acetone, and allowed to dry in the room for 24 hours. It is stored in the refrigerator.

(xii) *Recrystallization in Alcohol.* The dry crystals are suspended in thirty times their weight of cold water, allowed to soak for 5 to 10 minutes, and then treated exactly as in step (xi).

Yield. The yield of trypsin inhibitor crystals varies considerably with the stock of soyabean meal used. Nutrisoy XXX generally yields about 1 gm. of four times crystallized inhibitor per 100 gm. of meal.''

For the preparation of a crystalline trypsin inhibitor from the lima bean see FRAENKEL-CONRAT, BEAN, DUCAY and OLCOTT (1952).

3. Toxic Proteins.

Crystalline Ricin. (KUNITZ and McDONALD, 1948.) As starting material an aqueous extract of castor bean meal, precipitated by saturation with sodium sulphate. The precipitate is filtered off and dried.

"(i) *Crystallization in the Presence of Sodium Sulphate.* A solution of crude Na_2SO_4-ricin (as above) in water yields crystals of toxic protein when stored for several weeks at about 5° C. The details are as follows:

10 gm. of dry powder of Na_2SO_4-ricin is suspended in 30—40 ml. H_2O. It is best to add the powder slowly to the measured amount of water so as to allow the powder to become wet gradually. The mixture is stored until uniformly dispersed. It is then filtered clear on fluted paper. The filtration, if slow, is allowed to proceed overnight in the cold room at about 5° C. The solution is adjusted with 1 M sodium hydroxide or HCl to pH 6.8 and stored at 5° C. A slight precipitate of rosettes of very fine needles generally appears within 10 days or more. The bulk of the precipitate increases gradually until it reaches a maximum in 6 to 8 weeks. Crystals are filtered or centrifuged. The yield is about 0.5 gm.

Recrystallization. The crystalline precipitate is suspended in a volume of water equal to about $1/_4$th of the volume of water used for the first crystallization. Enough 1 M HCl is added slowly with stirring until the crystals dissolve. The solution is filtered clear on fluted paper and the paper is washed with a small amount of 0.01 M HCl. The filtrate and washings are titrated to pH 6.8 with 1 M sodium hydroxide but the addition of sodium hydroxide is interrupted at the first appearance of turbidity even if pH 6.8 is not reached. The solution is stored at 5° C. A heavy crop of crystals generally is formed within 24 hours and the crystallization is complete in 2 or 3 days.

(ii) *Crystallization in the Presence of Ammonium Sulphate.* 10 gm. of dry Na_2SO_4-ricin powder is stirred up with 30 ml. H_2O. The mixture is filtered on fluted paper; the residue in the paper is washed with about 10 ml. H_2O. Enough solid ammonium sulphate is added to the combined filtrate and washings to bring the solution to 0.8 saturation (5.6 gm. per 10 ml.). The precipitate formed is filtered with suction. The filter cake is weighed and then dissolved in water in proportion of 1.3 ml. of H_2O to 1 gm. filter cake. Saturated $(NH_4)_2SO_4$ is added slowly until a slight turbidity appears which is removed by filtration through folded paper. The filtrate is titrated to pH 6.8 with 1 M sodium hydroxide. The solution is left for 2 to 3 days at 5° C. A heavy precipitate gradually forms. The suspension is centrifuged at a temperature not higher than 10° C. The residue is dissolved in about 3 ml. of H_2O and stored at 5° C. A heavy amorphous precipitate forms in a few hours. The amorphous precipitate gradually changes into fine needles. Suspension centrifuged or filtered after 2 to 3 weeks."

Recrystallization is carried out as described above.

Properties. "Ultracentrifuge and electrophoresis tests show the crystalline protein becomes fairly homogeneous after 3 or 4 crystallizations. This is also confirmed by toxicity measurements. Solubility tests, however, indicate the presence of more than one protein component in the crystalline material, possibly in the form of a solid solution."

Soyin, a Toxic Protein from the Soyabean. (Liener, 1953.) ". . . Fifteen hundred grams of defatted soyabean flour were suspended in 18 l. of distilled water at room temperature. With constant stirring the pH was adjusted to 6.7 with 5 N NaOH, and the stirring was continued for an additional hour. The suspension was acidified with 6 N HCl to pH 4.6, and the curd was allowed to settle overnight at 4° C. Most of the clear supernatant fluid could be siphoned off, and the remainder of the water extract was removed by filtration on 10-inch Büchner funnels with the aid of suction. To each 10 l. of water extract were gradually added, while stirring, 3 kg. of $(NH_4)_2SO_4$. The precipitate which formed was removed by filtration and discarded, and 1 kg. $(NH_4)_2SO_4$ was added to the clear yellow filtrate. After allowing the precipitate to settle overnight most of the supernatant fluid was siphoned off, and the precipitate was centrifuged free of the remaining liquor. Sufficient water was added to the precipitate to form a slurry (about 150 ml.), which was dialyzed against distilled water at 4° C for 36 hours. Any insoluble material which precipitated was removed by centrifugation after dialysis. The pH of the solution was adjusted to 4.6 by the dropwise addition of 70 gm. $(NH_4)_2SO_4$, the precipitate was centrifuged from the solution and re-dissolved in 70 ml. $M/20$ phosphate buffer at pH 6.1. The buffered solution was transferred to a mechanical dialyzer and dialyzed against 60% ethanol at —5 to —10° C for 24 hours. The precipitate formed within the dialysis sack was centrifuged down at —5° C, dissolved in about 25 ml. distilled water, and dialyzed overnight against a large volume of distilled water at 4° C. After centrifuging off any insoluble material that remained, the solution was dried *in vacuo* from the frozen state.

This procedure yielded an average of 2 gm. of material per kilogram of soyabean flour . . .''

References.

ADLER, E., and M. SREENIVASAYA: Hoppe-Seylers Z. **249**, 24 (1937). — ALBAUM, H., and W. UMBREIT: Amer. J. Bot. **80**, 553 (1943). — ALBAUM, H. G., and P. P. COHEN: J. Biol. Chem. **149**, 19 (1943). — AXELROD, B., P. SALTMAN, R. S. BANDURSKI and R. S. BAKER: J. Biol. Chem. **197**, 89 (1952). —

BAILEY, K.: (a) Trans. Faraday Soc. **38**, 186 (1942); (b) Biochem. J. **36**, 140 (1942). — BALLS, A. K., W. S. HALE and T. H. HARRIS: Cereal Chem. **19**, 279, 840 (1942). — BARRE, R.: Bull. Soc. Chim. biol. (Paris) **33**, 1473 (1951). — BARTON-WRIGHT, E. C., and T. MORAN: Analyst **71**, 278 (1946). — BASS, A. A., and F. SAUNDERS: Proc. Soc. Exp. Biol. (New York) **30**, 445 (1932). — BECCARI, A. A.: De Bononiensi Scientarium et Artium Instituto et que Academia Commentarii, II, Pt. 1, 122 (1745). — BORCHERS, R., and C. W. ACKERSON: J. Nutrit. **41**, 339 (1950). — BORCHERS, R., C. W. ACKERSON and F. E. MUSSEHL: Arch. Biochem. **19**, 317 (1948). — BOURDILLON, J.: J. Biol. Chem. **189**, 65 (1951). — BREDERECK, H., and E. GEYER: Hoppe-Seylers Z. **254**, 223 (1938). — BROWN, A. N., and D. R. GODDARD: Amer. J. Bot. **28**, 319 (1941). — BURK, N. F.: J. Biol. Chem. **124**, 49 (1938).

CHIBNALL, A. C.: Proc. Roy. Soc. London B **131**, 136 (1942). — COHEN, P. P.: J. Biol. Chem. **164**, 685 (1946). — CONN, E. E., and B. VENNESLAND: J. Biol. Chem. **192**, 231 (1951). — COURTOIS, J., and R. BARRE: Bull. Soc. Chim. biol. (Paris) **31**, 740 (1949).

DANIELSSON, C. E.: Biochem. J. **44**, 387 (1949). — DANIELSSON, C. E., and E. SANDGREN: Acta Chem. Scand. **1**, 917 (1947). — DILL, D. B., and C. L. ALSBERG: J. Biol. Chem. **65**, 279 (1925). — DUSTIN, J. P., E. SCHRAM, S. MOORE and E. J. BIGWOOD: Bull. Soc. Chim. biol. (Paris) **35**, 1137 (1953).

ENGEL, C.: Biochim. Biophys. Acta **1**, 43 (1947). — ENGEL, C., and L. H. BRETSCHNEIDER: Biochim. Biophys. Acta **1**, 357 (1947).

FINKS, A. J., and C. O. JOHNS: Amer. J. Physiol. **56**, 205 (1921). — FONTAINE, T. D., W. A. PONS and G. W. IRVING: J. Biol. Chem. **164**, 487 (1946). — FRAENKEL-CONRAT, H., R. C. BEAN, E. D. DUCAY and H. S. OLCOTT: Arch. Biochem. **37**, 393 (1952).

GORTNER, R. A., W. F. HOFFMANN and W. B. SINCLAIR: Colloid Symp. Monogr. **5**, 179 (1928).

HAM, W. E., and R. M. SANDSTEDT: J. Biol. Chem. **154**, 505 (1944). — HAM, W. E., R. M. SANDSTEDT and F. E. MUSSEHL: J. Biol. Chem. **161**, 635 (1945). — HANES, C. S.: Proc. Roy. Soc. London B **128**, 421 (1940). — HARTIG, T.: Bot. Ztg. **13**, 881 (1885). — HAUGAARD, G., and A. H. JOHNSON: C. r. Lab. Carlsberg **18**, (2) 1 (1930). — HAUGAARD, N.: J. Biol. Chem. **164**, 265 (1946). — HEATHCOTE, J. G.: Lancet (London) **257**, 1130 (1949). — HEATHCOTE, J. G., J. J. C. HINTON and B. SHAW: Proc. Roy. Soc. London B **139**, 276 (1952). — HINTON, J. J. C.: Proc. Roy. Soc. London B **134**, 418 (1947). — HINTON, J. J. C., F. G. PEERS and B. SHAW: Nature (London) **172**, 993 (1953). — HOLMAN, R. T., and S. BERGSTRÖM: The Enzymes (ed. J. B. SUMNER and K. MYRBÄCK) II, (1) 559. New York: Academic Press 1951. — HORN, M. J., D. B. JONES and A. E. BLUM: Misc. Publ. U.S. Dep. Agric. No. 696 (1950).

IRVING, G. W, T. D. FONTAINE and R. C. WARNER: Arch. Biochem. **7**, 475 (1945).

JENSEN, R.: Acta Chem. Scand **6**, 771 (1952). — JOHNS, C. O., and A. J. FINKS: Amer. J. Physiol. **56**, 208 (1921). — JOHNS, C. O., and D. B. JONES: J. Biol. Chem. **28**, 77 (1916). — JONES, D. B., and C. E. F. GERSDOFF: J. Amer. Chem. Soc. **60**, 723 (1938).

KABAT, E. A., M. HEIDELBERGER and A. E. BEZER: J. Biol. Chem. **168**, 629 (1947). — KELSEY, F. E.: J. Biol. Chem. **130**, 187, 195, 199 (1939). — KIRK, J. S.: J. Biol. Chem. **100**, 667 (1933). — KLIMENKO, V. G.: Biokhimiya **15**, 186 (1950). — KLOSE, A. A., J. D. GREAVES and H. L. FEVOLD: Science (Lancaster, Pa.) **108**, 88 (1948). — KNEEN, E., R. M. SANDSTEDT and C. M. HOLLENBECK: Cereal Chem. **20**, 399 (1943). — KREJCI, L., and T. SVEDBERG: J. Amer. Chem. Soc. **57**, 946 (1935). — KUNITZ, M.: J. Gen. Physiol. **29**, 149 (1945). — KUNITZ, M., and M. R. McDONALD: J. Gen. Physiol. **32**, 25 (1948).

LAMM, O., and A. POLSON: Biochem. J. **30**, 528 (1936). — LE BRETON, E., and Y. MOULE: Bull. Soc. Chim. biol (Paris) **31**, 94 (1949). — LEONARD, M. J. K., and R. H. BURRIS: J. Biol. Chem. **170**, 701 (1947). — LIENER, I. E.: J. Nutr. **49**, 527 (1953). — LUGG, J. W. H.: Advanc. Protein Chem. **5**, 229 (1949).

McCALLA, A. G., and N. GRALEN: Canad. J. Res. **20** C, 130 (1942). — McCALLA, A. G., and R. C. ROSE: Canad. J. Res. **12**, 346 (1935). — MACHEBOEUF, M., and F. TAYEAU: Bull. Soc. Chim. biol. (Paris) **24**, 260, 268, 273 (1942). — MAPSON, L. W., and D. R. GODDARD: Nature (London) **167**, 975 (1951). — MORAN, T., J. PACE and E. E. McDERMOTT: Nature (London) **171**, 103 (1953). — MOULE, Y.: Bull. Soc. Chim. biol. (Paris) **33**, 1461 (1951).

NEUBERG, C., and J. WAGNER: Biochem. Z. **174**, 457 (1926). — NEWTON, J. M., F. F. FARLEY and F. M. NAYLOR: Cereal Chem. **20**, 23 (1943).

OCHOA, S.: The Enzymes (ed. J. B. SUMNER and K. MYRBÄCK) I, (2) 1217. New York: Academic Press 1951. — ORRU, A., and C. V. DEMIEL: Quad. Nutrit. **7**, 273 (1941); Chem. Zbl. **2**, 631 (1941). — OSBORNE, T. B.: Amer. Chem. J. **14**, 629 (1892); The Proteins of the Wheat Kernel. Washington: D. C., Carnegie Institution of Washington 1907; The Vegetable Proteins, 2 nd. ed. London: Longmans, Green & Co. 1924. — OSBORNE, T. B., and I. F. HARRIS: J. Amer. Chem. Soc. **25**, 323 (1903). — OSBORNE, T. B., and L. B. MENDEL: (a) J. Biol. Chem. **32**, 369 (1917); (b) Yearb. Carneg. Instn. **16**, 329 (1917).

PACE, J., and E. E. McDERMOTT: Nature (London) **169**, 415 (1952). — PEERS, F. G.: Biochem. J. **53**, 102 (1953). — PEERS, F. G., and H. F. MARTIN: Biochem. J. **55**, 523 (1953). — PETT, L.: Biochem J. **29**, 1902 (1935). — PIRIE, N. W.: Biol. Rev. **15**, 377 (1940).

QUENSEL, O.: Dissertation. Univ. Uppsala, Sweden 1942.

ROSE, W. C.: Science (Lancaster, Pa.) **86**, 298 (1937).

SACKS, W., and C. O. JENSEN: J. Biol. Chem. **192**, 231 (1951). — SANGER, F., and R. PORTER: XIth. Int. Congr. Chem. 1947. — SAUBERLICH, H. E., W.-Y. CHANG and W. D. SALMON: J. Nutr. **51**, 241 (1953). — SAVERBORN, S., C. E. DANIELSSON and T. SVEDBERG: Svensk kem. Tidskr. **56**, 75 (1944). — SCHALES, O., V. MIMS and S. S. SCHALES: Arch. Biochem. **10**, 455 (1946). — SCHMIEDEBERG, O.: Hoppe-Seylers Z. 1, 205 (1877). — SCHWERT, G. W., F. W. PUTNAM and D. R. BRIGGS: Arch. Biochem. 4, 371 (1944). — SINGER, T. P., and J. PENSKY: J. Biol. Chem. **196**, 375 (1952). — SMILEY, W. G., and A. K. SMITH: Cereal Chem. **23**, 288 (1946). — SMITH, A. K., and S. J. CIRCLE: Industr. Engng. Chem. (Anal) **30**, 1414 (1938). — SMITH, E. L., and R. D. GREENE: J. Biol. Chem. **167**, 833 (1947); **172**, 111 (1948). — SMITH, E. L., R. D. GREENE and E. BARTNER: J. Biol. Chem. **164**, 159 (1946). — STUMPF, P. K.: J. Biol. Chem. **176**, 233 (1948). — SUMNER, J. B.: J. Biol. Chem. **37**, 137 (1919); 70, 97 (1926). — SUMNER, J. B., and V. A. GRAHAM: J. Biol. Chem. **64**, 257 (1925). — SUMNER, J. B., and S. F. HOWELL: J. Biol. Chem. **113**, 607 (1936). — SURE, B., and J. W. READ: J. Agric. Res. **22**, 5 (1921). — SVEDBERG, T., and A. J. STAMM: J. Amer. Chem. Soc. **51**, 2170 (1929).

TAUBER, H., K. B. KERSHAW and R. D. WRIGHT: J. Biol. Chem. **179**, 1155 (1949). — THEORELL, H., R. T. HOLMAN and A. AKESON: Acta Chem. Scand. 1, 571 (1947). — TRISTRAM, G. R.: Biochem. J. **40**, 721 (1946).

VASSEL, B., and L. L. NESBITT: J. Biol. Chem. **159**, 571 (1945). — VENNESLAND, B.: J. Biol. Chem. **178**, 591 (1949). — VENNESLAND, B., and R. Z. FELSHER: Arch. Biochem. **11**, 279 (1946). — VICKERY, H. B.: Physiol. Rev. **25**, 347 (1945). — VICKERY, H. B., E. L. SMITH, R. B. HUBBELL and L. S. NOLAN: J. Biol. Chem. **140**, 613 (1941). — VICKERY, H. B., E. L. SMITH and L. S. NOLAN: Biochem. Preparations 2, 5 (1952).

WANG, C.-F.: Chin. J. Physiol. 15, 231, 236 (1940). — WELLS, H. G., and T. B. OSBORNE: J. Infect. Dis. 8, 66 (1911); **12**, 341 (1913); **19**, 183 (1916). — WESTFALL, R. J., D. K. BOSSHARDT and R. H. BARNES: Proc. Soc. Exp. Biol. (New York) **68**, 498 (1948).

ZEIJLEMEKER, F. C. J.: Proc. Acad. Sci. (Amst.) **42**, 187 (1939).

Methods of Determining the Nutritive Value
of Proteins.

By

J. Duckworth.

The great importance of accurately measuring the nutritive value of plant proteins has long been recognized. As early as 1918 Osborne and Mendel pointed out that the world food crisis then prevailing called for a focussing of attention on the comparative feeding value of the proteins in the various cereals eaten by man and his domesticated animals. With penetrating insight they relegated variations in other nutrients to a level of secondary importance, in such comparisons, on the grounds that cereals resemble each other closely in their content of starch and that they are sufficiently alike in possessing or lacking individual minerals and vitamins.

Increased understanding of the nature of the world's food problems and the expansion of our knowledge of food values do not call for a drastic revision of this view of Osborne and Mendel. Even though the scope of the generalization must now be extended to include many other food materials besides cereals the essential validity remains. Energy can be provided in the diet by carbohydrates and fats in relative proportions that can be varied over a wide range for most animals, and its provision raises few problems beyond obtaining sufficient at low cost. Shortages of inorganic nutrients in diets are easily and inexpensively remedied by adding mineral salts. Vitamin deficiencies in diets are easily prevented, either by including naturally rich sources of them in the diet or, where this is not practicable, by adding vitamin concentrates that are becoming increasingly more plentiful as by-products of the expanding fermentation industry. But the provision of sufficient protein of adequate nutritive value remains the central dietary problem in many parts of the world.

Cereals, potatoes, cassava and fodder beet are the prime sources of energy for man and his non-ruminant livestock. These energy sources have this in common: the ratio of protein to energy in them is too low for most physiological states of animals and a protein-rich supplement must be added to make up the deficit. During the period of rapid growth in early life, during reproduction (both pregnancy and egg laying) and during lactation a high level of supplementing of the common energy sources with protein-rich foods is essential if a balanced diet is to be made. Only partial exceptions to this need for supplementing are found, as in cases where an arid climate favours the growing of cereals of unusually high protein content, and where poultry, allowed to forage for themselves, can prey upon the lesser fauna as a protein source. However, under most practical conditions of feeding the provisions of sufficient protein involves supplementing a low-protein energy source with a high-protein material, and thus the evaluation of proteins from the outset is not only of the intrinsic nutritive value of individual proteins nor only of the mixture of proteins in each foodstuff but also must include a measurement of the supplementary relationships between the proteins of one foodstuff and those of another.

Since RUBNER first observed that proteins differ greatly in their nutritive values — and hence in their supplementary values — the causes of these differences have been sought by many students of nutrition. It is now clear that most of these differences between proteins can be accounted for by differences in their amino acid composition, in other words it is generally true that proteins rich in amino acids for which the animal has an indispensable need are superior in feeding value to those poor in these amino acids. But there remain other differences not to be explained by simple differences in the relative amounts of amino acids in feeding stuffs of contrasting nutritive value.

These differences are still imperfectly understood, but can be grouped in two classes. In the first of these the reduction in feeding value below what would be expected from amino acid composition can be attributed to the action of substances other than protein in the material fed; such cases arise when a natural feeding stuff is tested directly in animal feeding trials, but not when an isolated protein is fed. In the second class are those cases where the difference lies in the protein itself, sometimes as a consequence of changes induced during commercial processing and in such cases affecting digestibility of the protein, availability of essential amino acids in metabolism, or *rate* at which essential amino acids are liberated in digestion, and sometimes as a consequence of the amino acids being in an unbalanced ratio for a particular species of test animal. Examples of these departures of feeding response from predicted values are given below.

[The term "protein" is used throughout to refer to the value obtained by multiplying the total nitrogen content of food stuffs by the conventional factor (usually 6.25 in nutritional studies). The various methods can be applied to evaluating pure and impure proteins, peptides, and mixtures of free amino acids.]

A. Evaluation for Ruminants.

It is premature to attempt a definition of the conditions essential for the accurate evaluation of plant proteins for ruminant animals. Although no serious technical difficulty faces the experimenter wishing to assess the effect of providing different kinds and amounts of nitrogenous substances in the diet on growth, pregnancy or lactation the extent to which his results can be built into a meaningful generalization about the comparative dietary values of nitrogen sources and levels of feeding is necessarily limited. As with non-ruminants the problem appears in its simplest form when evaluating a protein source for growth but even in this case — uncomplicated by the large drafts that may be made on body reserves of protein in studies of the value of dietary nitrogen for pregnancy and lactation — the fundamental difficulty is unresolved. This is the impossibility of predicting the kind and extent of transformations that the dietary nitrogen will undergo in the rumen, changes that will determine the nutritive value of the nitrogenous material subsequently available to the host animal. These transformations are subject to influences not solely related to the protein of the diet.

Certain protein characteristics have an influence on rumen transformations of protein, solubility and susceptibility to proteolysis being important. McDONALD (1952) investigated the accumulation of ammonia in rumen liquor, as an index of protein degradation, after feeding different proteins. When zein was fed, as an example of a protein of low solubility and high resistance to hydrolysis, he found no accumulation of ammonia in rumen liquor. When casein and gelatin were fed, as examples of proteins of high solubility and susceptibility to proteolysis there was rapid and high accumulation of ammonia. The rate and level of ammonia accumulation was between these extremes when meadow hay was fed.

The fate of this ammonia is conjectural. Information bearing on the problem is accumulating in those studies of the overall nitrogen metabolism of the ruminant that are aimed at finding the conditions under which simple nitrogen compounds — urea and ammonium salts — can be used by rumen microorganisms for synthesizing microbial protein which, on passage through the next portion of the alimentary tract, yields amino acids for use by the host. It is already clear that the type of carbohydrate provided by the diet has a decisive influence on the degree to which the ammonia, as fed or as derived from urea, will be utilized by the flora of the rumen for synthesis of microbial protein. If the carbohydrate of the diet is largely starch the conversion of ammonia to protein is favoured, presumably because the rate at which the starch is microbially digested to yield energy parallels the rate at which protein can be synthesized from ammonia. If the dietary carbohydrate is chiefly cellulose the rate of liberation of energy is too slow for efficient conversion of large amounts of ammonia to microbial protein. In contrast to this, sugars disappear from the rumen too rapidly for their energy to be available for sufficiently long periods and again conversion of ammonia to protein is slight.

These interrelationships between the susceptibility of dietary protein to degradation in the rumen and the availability of dietary energy on the utilization of the amino acids and ammonia produced explain, in part, the variations in Biological Value of proteins in ruminant feeding to which attention has been called by Lofgreen, Loosli and Maynard (1947, 1951). Until more is known about transformations of dietary nitrogen in the rumen great uncertainty must surround the evaluation of protein sources for the ruminant.

B. Evaluation for Non-Ruminants.

I. Techniques Involving Chemical Analysis, or Microbiological or Enzymatic Assay.

Determination of Amino Acid Composition of Protein.

Since the value of a protein to the non-ruminant animal is an expression of the extent to which the animal can extract and put to metabolic use its constituent amino acids the determination of amino acid content is of paramount importance. Indeed the main advances made during the last half century in the understanding of protein values have rested, fundamentally, on this basis.

In spite of this the contribution that amino acid analysis can make to assessing the nutritive value of a specific feeding stuff is often meagre. This is because analytical errors may be large in amino acid estimations and because the information provided by simple analysis (as distinct from analysis of amino acids released during *in vitro* enzymic hydrolysis, described below) gives no measure of the extent to which the amino acids are set free in digestion or metabolized after absorption.

Block and Bolling (1951) have compiled a comprehensive collection of methods of amino acid analysis and include chemical, chromatographic, microbiological and enzymatic techniques[1]. They emphasize that one of the main sources of error is the loss, by partial destruction, of amino acids during the hydrolysis that is carried out, in the majority of methods, preparatory to analysis or assay. This source of error combined with the errors inherent in the subsequent measurement of individual amino acids are responsible for the wide range of values recorded in the literature for the amino acid content of feeding stuffs.

---

[1] see also Synge, p. 1; Pirie, p. 23 ; Pace, p. 69;

An investigation of the reproducibility of results for the amino acid composition of beef muscle, casein, whole egg proteins, egg albumin, groundnut flour and wheat gluten has been reported by the Bureau of Biological Research of Rutger's University (1950). Samples of a single uniform batch of each material were distributed to a series of collaborating laboratories where they were hydrolysed and analysed according to methods individually favoured. The average values, together with the corresponding highest and lowest values, for 13 amino acids are set out in Table 1. Large differences were found between laboratories and it is a particularly serious drawback, from a nutritional point of view, that the variation is particularly wide in those amino acids — methionine, cystine and lysine — that are most likely to be limiting in diets constructed solely from plant products. With the introduction of improved techniques discrepancies such as these will disappear, but for the present caution should be exercised in using such analytical data in nutritional interpretations.

Contrary to expectation only a small part of the discrepancies could be accounted for by differences between laboratories in the extent of amino acid destruction taking place during hydrolysis preparatory to analysis or assay. Deviations of a similar order were found when collaborating laboratories analysed hydrolysates that had been prepared from the 6 test materials at the central laboratory (Rutger's University). Apparently the losses during hydrolysis, emphasized as a source of error by BLOCK and BOLLING, are fairly uniform from one laboratory to another thus influencing absolute rather than relative values.

Table 1. *Comparative Values for Different Amino Acids Found in Collaborative Trials at Various Laboratories.* (Values in mg. amino acid/100 mg. N.)

Amino acid	Number of laboratories participating	Groundnut flour			Wheat gluten		
		Mean	Low	High	Mean	Low	High
Cystine . . .	5	9.6	5.2	20.6	13.9	10.5	16.6
Methionine. .	12	5.4	2.6	6.6	10.2	8.3	12.5
Lysine . . .	10	23.2	15.4	30.6	12.5	9.2	21.1
Arginine . .	10	70.5	53.2	83.1	22.5	19.3	25.3
Aspartic . .	3	78.1	70.4	89.1	28.9	21.2	24.3
Histidine . .	10	13.6	12.2	17.3	12.4	10.5	14.0
Isoleucine . .	10	25.5	21.3	31.2	27.4	24.1	37.0
Leucine . . .	10	38.9	35.0	43.1	42.6	34.0	47.2
Phenylalanine	11	30.1	21.8	34.2	32.0	28.7	35.7
Threonine . .	10	16.3	14.6	18.7	16.9	15.8	19.4
Tryptophan .	11	6.4	3.8	9.2	5.8	2.3	8.2
Tyrosine . .	5	20.0	14.1	25.8	17.5	11.1	20.4
Valine. . . .	10	27.4	22.8	30.2	25.6	22.2	29.1

The same group of collaborating workers also carried out parallel measurements of the nutritive value of the protein in the 6 test materials using, according to individual preference, various biological methods of assay: growth of young animals, nitrogen balance in adult animals, protein repletion in adult animals after single or double depletion, reproduction and lactation trials. The order of rating of the nutritive values of the 6 test proteins in these biological assays corresponded, in general, with the experience of earlier research. Some of the deviations from expectation were associated with the particular physiological function used as the criterion of measurement and revealed valuable information about the utilization of dietary proteins for contrasting physiological purposes.

Two important conclusions were drawn from the work as a whole:

1. "Protein efficiency measurement in the growing rat is probably the most satisfactory method of establishing the nutritive value of proteins for growth.

However, considerable standardization of the method must be established before results from one laboratory can be considered valid and reproducible in other laboratories.''

2. ''. . . the nutritive value of proteins cannot be revealed by a restricted oversimplified analysis.''

Anderson and Williams (1951) used the growth response of the proteolytic ciliate *Tetrahymena geleii* W. to measure the Biological Value of intact proteins. Although the growth rates did not reflect accepted nutritive values of the proteins tested the quantitative nature of the response to graded amounts of each protein suggest that this type of microbiological test might be developed for the assay of intact proteins.

Melnick and Oser (1949) have devised a method based on the *in vitro* digestion of proteins with pancreatin in which the rate of hydrolysis is determined. In addition they determined the rate at which certain essential amino acids became microbiologically available during the process of *in vitro* digestion. As they state, food products may be altered beneficially or detrimentally by heat processing without any change in amino acid composition or digestibility, but with a profound effect on Biological Value and on rate of digestion by pancreatin *in vitro*. They conclude: ''for optimal utilization of food proteins all essential amino acids must not only be available for absorption but must be liberated during digestion *in vivo* at rates permitting mutual supplementation''. Extending this approach Halevy and Grossowicz (1953) have reported a method of testing such pancreatin hydrolysates for the 10 essential amino acids, using a single test organism: *Streptococcus faecalis*.

Chemical Score Method of Evaluation. A method of allocating an index value to a protein as a measure of nutritive worth has been proposed by Mitchell and Block (1946). Whole egg proteins are taken as the standard, in preference to milk proteins, in view of their superiority in digestibility and metabolic utilization. The test protein is analysed for its content of essential amino acids and the percentage deviation of each amino acid from the corresponding value for the same amino acid in whole egg proteins is calculated. ''Thus the amino acid limiting the nutritive value for maintenance and growth of the albino rat for any particular food protein would be that amino acid present in the least amount with reference to whole egg proteins, i. e. the amino acid with the greatest percentage deficit. Thus, the limiting amino acid in whole wheat proteins is lysine (—63%), that in blood fibrin is isoleucine (—37%).'' (Block and Mitchell, 1946—1947.)

Beyond providing a simple, not too accurate, and largely uniformative method of expressing analytical data in nutritional terms it is difficult to see much advantage in the proposed method. With the persisting uncertainty about the accuracy of amino acid analysis very little is to be gained by expressing the difference between two values, each subject to error, as a percentage of one of them.

II. Techniques Involving the Use of Animals.

There continues to be considerable disagreement between experimenters on the validity of different methods of using animals for measuring the nutritive value of proteins. Most of the disagreement is resolved once it is clearly understood that the response on animal yields is a measure of the use it has been able to make, according to its physiological state, of the protein intake that the experimenter has permitted and, further, that this response may have been modified detrimentally by the presence of deleterious substances or the absence of unknown but essential micro-nutrients.

Regarding the physiological state selected as the criterion for assessing the nutritive value of the test protein, it is now known that the order in which a series of proteins fall for growth is not necessarily the order in which they will fall for some other physiological states — haemoglobin regeneration or milk production — in the same species. In other words the amino acid that is limiting in a given protein for growth in the mouse is not necessarily the one that is limiting in it for egg production in the fowl.

As far as permitted protein intake is concerned, it is widely recognized, and indeed it is common feeding practice for both man and animals, that the inferiority of plant proteins in essential amino acids can be overcome by feeding these proteins at higher levels. By increasing the intake so that the amount of the limiting amino acid consumed is sufficient to meet the physiological need normal growth or production can be sustained. In tests designed to measure the nutritive value of proteins it is essential that the level at which the protein is included in the diet should be well below the level at which the inferiority of poor proteins is masked by overfeeding. Although there can be no logical justification for any particular level the selected level should be below that for supporting normal performance even if the test protein is equal in nutritive value to those of proved high value (e. g. egg proteins, milk proteins, fish proteins). In the case of whole egg proteins, 12% in a semi-synthetic diet meets the requirement for growth in the rat (MITCHELL and BEADLES, 1952).

Whether proteins are to be tested alone or in combination with other proteins depends on the aim of the experimenter. In those experiments where a single protein or foodstuff is the sole source of amino acids for the animal the information yielded has limited application in those practical circumstances where a mixture of proteins are fed and where supplementary relationships prevail, one protein remedying the shortcomings of another. If the purpose of the investigation is to study the effect on nutritive value of a variable in a manufacturing process, such as temperature, determination of the Biological Value, as described below, will suffice to give the information needed. On the other hand if it is important to know whether the damage caused to nutritive value is important in practice a test such as the measurement of the Gross Protein Value should be made. This test is designed to determine whether supplementary actions between the test protein and the other proteins with which it will be fed in practice will be sufficient to offset the damage to the test protein.

An important factor overlooked in some trials is the metabolizable energy content of the diets as a whole and of the particular material under test. This matter rarely arises when using so-called semi-synthetic diets in which energy is derived from starch, dextrin, sucrose or glucose and a purified fat such as cottonseed oil. However, when using diets that resemble practical rations wide differences in the content of energy can be found, diets based on maize often containing 20% more metabolizable energy than a corresponding diet composed of a mixture of other cereals and miller's offals. Since an animal usually consumes the amount of a diet that will satisfy its needs for calories, that is to a certain energy level, the quantity of food eaten when a low energy ration is freely offered is greater than when a high energy ration is given. When the percentage of protein is arbitrarily fixed at a level somewhat below requirement for a high energy diet this same level of protein may be close to or above requirement for a low energy diet because of the greater total weight of food eaten.

Toxic substances in test materials can be responsible for errors in evaluation. Familiar examples are: the gossypol of some cotton seed products; the antitryptic substance and the toxic protein, soyin, of raw soyabeans and of inadequately heat-treated soyabean meal (Liener, 1953); the saponin-like substance in lucerne, and perhaps in other plant materials (Hansen, Scott, Larson, Nelson and Krichevsky, 1953).

The extent to which amino acid imbalance can reduce performance — as distinct from the restriction of performance that arises from the shortage of an essential amino acid — and thus lead to error in the interpretation of results from growth tests designed to measure the nutritive value of proteins is not known. Lyman and Elvehjem (1951) reported that additions of gelatin with L-cystine or DL-methionine to diets containing casein caused marked inhibition of growth in the rat. Moreover the inhibition was greater when dextrin was used as the energy source in the diet than when sucrose was used. Further, the inhibition could be prevented by increasing the intake of nicotinic acid.

The following tests are designed mainly to measure the nutritive value of proteins for growth, but with appropriate modifications they can be adapted to evaluate proteins for other purposes provided that the conditions essential for a valid test are observed.

1. Simple Substitution Technique.

In trials of this type a standard practical ration known to be favoured for its reliability under practical conditions of feeding is used as the control ration. To measure the value of the protein source under test other rations are devised in which part or all of the protein concentrate in the standard ration is replaced by the test material. Usually, but not always, the amount of the test material included is such that the weight of protein it contributes at each level of substitution is equal to the weight of protein replaced. The test rations and the standard ration, as control, are then fed to comparable groups of animals and their performance — as growth, milk production, egg production, reproductive performance — measured. The value of the test material as a source of protein is judged from the results obtained, using performance on the standard ration as the basis of comparison.

Advocates of this type of test defend it on the grounds that the results obtained are directly and immediately applicable in practice. Such contentions are sometimes true, but only fortuitously. The scientific information yielded by simple substitution trials is very meagre.

Limitations on the extent to which results from this kind of trial can be interpreted are imposed by the nature of the standard ration. Because it is designed for use in practice on the farm it will be provided with a level of protein that is higher than the minimum needed by the class of stock to which it is to be fed. This extra allowance of protein — or margin of safety — is provided to protect the animal against risks of protein inadequacy arising from cereals of unusually low protein content and from protein concentrates in which the nutritive value of the protein has been impaired by errors in processing. In simple substitution experiments it is not possible to determine the extent to which the value attributed to the test protein rests on an exploitation of this margin of safety. In other words, where a standard ration contains a margin of safety of 20% in its protein content the replacement of this 20% by a test protein of no nutritive value will not affect performance, and the conclusion drawn that the test protein added is nutritionally equal to the protein replaced.

2. Successive Substitution Technique, with Nitrogen Balance Trials.

This technique was developed by Woodman and Evans in a series of studies of the value of protein-rich concentrates as protein supplements to mixtures of barley and miller's offals for the pig. Correlating the results of their past experiments they have laid down seven conditions that must be satisfied if valid estimates are to be made of the supplementary relationships among proteins for normal growth of the bacon pig (WOODMAN and EVANS, 1951). These conditions are:

(1) The comparison must be made during the growing period from weaning to 90 lb. live-weight. Beyond this weight the protein contained in practical barley-fine bran or barley-middlings mixtures is sufficient for the pig's needs, and additional protein is without effect on either growth or nitrogen retention.

(2) The diets compared should be equal, or almost equal, in their content of "total digestible nutrients" as computed by means of coefficients of digestibility determined in digestibility trials with pigs.

(3) The daily feed allowance must be related to live-weight and the feed itself must be sufficiently palatable to ensure that the whole of the daily feed allowance is eaten.

(4) The content of the various minerals and vitamins must be adequate to meet the requirements for normal growth.

(5) The technique of individual feeding must be used, and the experimental design must be such that the results can be analysed statistically.

(6) The standard ration, against which others are to be compared, should contain the minimum amount of protein needed for the maximum rate of growth, compatible with the available supply of net energy. Such a condition is met when a barley meal-middlings or barley meal-fine bran mixture, containing a small amount of lucerne meal and minerals, is supplemented with 7% of fish meal.

(7) In the experimental ration containing the protein supplement under test, the percentage of digestible crude protein should be at the same level as in the standard ration. In addition, the amount of digestible crude protein provided by the protein supplement under test should be equal to the amount of digestible crude protein provided by the 7% of fish meal in the standard ration.

The initial trial, under these conditions, gives a measure of the supplementary value of the test protein concentrate relative to fish meal protein. In subsequent tests the first six conditions are again observed but in place of the seventh successively larger amounts of the test material are given, thus increasing the amount of digestible crude protein. By these growth trials and by determination of nitrogen retention in metabolism trials the amount of the test protein necessary to equal the fish meal standard is determined.

This technique yields a valid estimate of the supplementary value of the test protein concentrate while at the same time yielding information of value for animal feeding.

3. Gross Protein Value.

This method, which is a development of early methods of a similar character, was devised by HEIMAN, CARVER and COOK (1939). It measures the value of protein sources to supplement the proteins of common energy sources in poultry rations. The main advantage gained by making an assessment of the supplementary protein source in the presence of the other proteins with which it will be fed in practice is that the value obtained has a more direct application in practical feeding than the values yielded by techniques that test each protein alone. The test is carried out with levels of protein feeding that are below requirement so that the confusing effects of "margins of safety" are avoided.

The test diets are made up as follows:

1. The control diet is made up containing those cereals, cereal offals, vitamin carriers and minerals with which the test protein will normally be fed. The cereals and cereal offals are included in the same ratio to each other as in a practical ration but the total amounts of these are only included at about 70% of the normal level, the remainder of the energy being supplied by sucrose. The ratio of cereals and cereal offals to sucrose is adjusted, according to protein contents of the foodstuffs used, so that the control ration finally contains 8% of protein.

2. *The reference diet* is made by replacing part of the sucrose with sufficient casein to add 3% of protein as a supplement to the 8% of cereal origin.

3. *The test ration* is made by replacing part of the sucrose with sufficient of the test protein source to provide 3% of protein to supplement the 8% of protein of cereal origin. The diet, like the reference diet, then contains 11% of protein.

Chicks are fed on the control ration for 2 weeks to deplete their protein reserves. They are then divided into 6 groups, 2 continue on the control ration, 2 are fed on the reference ration, and 2 are fed on the test ration. The feeding trial continues for a further 2 weeks, food consumption being measured. The Gross Protein Value is calculated by expressing the growth per g. of test protein consumed as a percentage of the growth per g. of casein protein consumed, both measured above the growth achieved on the control ration.

4. Biological Value of Protein.

This is the most valuable single method of protein evaluation, judged by the amount of information yielded to the nutritionist in an individual trial. Originally devised by Thomas (1909) it has been developed and its accuracy improved by Mitchell over a period of about 20 years, eventually to reach a stage where there is excellent reproducibility of results (Mitchell, Hamilton, Beadles and Simpson, 1945).

Growing rats are fed on a low protein diet (4% of laboratory-prepared whole egg protein) the nitrogen of which is completely digested and absorbed. Urinary and faecal outputs of nitrogen are determined. The whole egg protein is removed from the diet and replaced by the test protein at a level below requirement, and the urinary and faecal outputs of nitrogen are again determined.

The faecal nitrogen excreted when the egg protein diet is fed is the so-called "metabolic nitrogen" derived from nitrogenous secretions into the alimentary tract and fragments of intestinal epithelium (these will have been converted in part to bacterial protein in the lower bowel). The total output of metabolic nitrogen of faeces varies with the weight of diet consumed but will be increased if the roughage content of the diet is raised, presumably as a consequence of the greater mass of material passing through the intestine. When this excretion of metabolic nitrogen, calculated per g. of diet eaten, is subtracted from the faecal output of nitrogen during the period when the diet containing the test protein was fed the true digestibility of the test protein is determined. The remainder of the ingested test protein nitrogen is the absorbed fraction.

The output of nitrogen in urine during the period when the egg protein diet is fed measures the catabolism of the body tissue. This value subtracted from the output of urinary nitrogen when the test protein is fed yields a measure of the extent to which the absorbed nitrogen of the test protein has been wasted in metabolism. The rest of the absorbed nitrogen has been retained in the body and this quantity expressed as a percentage of the absorbed nitrogen of the test protein is its Biological Value.

The information provided by this method of evaluation allows discrimination between those cases in which poor nutritive value of a test protein arises from low digestibility and those in which it arises from an inferior content, or availability in metabolism, of essential amino acids in nitrogen that is well absorbed.

That the results obtained are highly reproducible is shown in a comparison of two series of estimations made by MITCHELL et al. (1945) in a study of autoclaved soyabean oilmeal carried out with a 6 year interval between series. The two mean values for true digestibility were 83.1% and 84.4% and for Biological Value were 67.0% and 67.5%.

5. Net Protein Value.

This value is obtained by multiplying together the true digestibility coefficient of a protein and its Biological Value. From a nutritional point of view this method of attributing a numerical value to a protein is superior to the Biological Value alone since it takes into consideration both the digestibility and the nutritive value of the absorbed amino acids.

6. Protein Efficiency Ratio.

This ratio was first introduced by OSBORNE, MENDEL and FERRY (1919) and is the gain in weight per g. of protein eaten. It was recognized from the beginning that the level of protein in the diet has a large effect on the ratio and the level of inclusion of protein is usually 10%.

In trials where the protein efficiency ratio is to be measured the test and standard proteins are fed at the same levels in all diets, the animals are usually allowed to eat to appetite and the amount of diet eaten during the test period is determined. HEGSTED and WORCESTER (1947) have shown that there is a very high correlation between the gain in weight on the diets and the protein efficiency ratios. Consequently the measurement of food eaten and the expression of gain in ratio to weight of protein eaten add little to the information provided by weight gains alone. In other words little more can be deduced from such calculated values than from the actual gains in weight, as far as grading a series of proteins for nutritive value is concerned.

SHERWOOD and WELDON (1953) also found no advantage in using protein efficiency ratios in place of gain in body weight for any of four methods of feeding. In order of sensitivity (measured by the ratio of each difference to its standard error) they ranked the different feeding techniques in the following order: 1. *ad libitum* feeding; 2. adjusted protein feeding (the daily supply of protein being varied in relation to the increasing body weight of the test animal); 3. constant protein feeding (0.5 g. protein per rat daily); 4. pair-feeding.

Some light has been cast on the problem of the relation of food intake to nutritive value of protein by HEGSTED and HAFFENREFFER (1949). They suggest that the food intake of animals is controlled by some unknown appetite factor at a relatively constant proportion above basal metabolism. If the food consumed is nutritionally adequate the animal grows, basal metabolism increases, and the food intake is progressively greater on succeeding days. If the food eaten is nutritionally inadequate the animal's growth is subnormal and the unwanted calories, above the needs for basal metabolism and the limited growth that has been made, are consumed probably in activity.

7. Carcase Analysis Technique.

The proportion of dietary nitrogen that is retained as body nitrogen is the absolute measure of nitrogen utilization. A comparison of this method with others, using the rat, was made by BARNES, MAACK, KNIGHTS and BURR (1945). Four protein sources were tested: spray-dried whole egg proteins, a well heat-treated soyflour (No. 1), a poorly heat-treated soyflour (No. 2), and a wheat gluten.

8*

The test substances were fed at different levels (ranging from 4 to 40%), nitrogen digestibility determinations were made, and carcases were analysed. Some of the results are set out in Table 2.

One of the findings of this study, with a bearing on the controversy about the relative merits of *ad libitum* feeding and the form of resticted feeding known as "pair-feeding", was that when the diet contains 10% of protein the percentages of protein in the carcase were more uniform in animals fed *ad libitum* (maximum range 0.7%) than in animals pair-fed (maximum range 1.9%). Pair-feeding caused a marked increase in the apparent digestibility of protein, and thus reduced those differences in growth-promoting value between proteins that arise from this source.

As Osborne et al. (1919) pointed out the level at which the maximum value is obtained for the protein efficiency ratio (gain per g. protein eaten) varies from one protein to another. The 10% level of inclusion in the diet is close to the point where the maximum protein efficiency ratio for whole egg proteins is obtained, but for soyflour proteins it is between 10% and 15% and for wheat gluten over 20%. Consequently when the 10% level is used for all the test proteins in a feeding trial the high quality protein used as the standard will be tested at the level where there is maximal utilization. But in the case of plant proteins their utilization will have been tested on the ascending slope of the efficiency curve, in some cases at levels well below the point of maximum efficiency. Differences will thus be exaggerated.

Table 2. *Comparison of Methods of Expressing the Growth-promoting Quality of proteins.*
(Whole egg protein arbitrarily set at 100 and other proteins assigned relative values.)

Protein source	Nutritive value ratios (see footnotes)					
	1	2	3	4	5	6
Whole egg	100	100	100	100	100	100
Soyflour No. 1	69	53	55	59	56	57
Soyflour No. 2	46	40	47	48	55	55
Wheat gluten	15	8	21	20	16	19

Column 1: Grams gain in body weight per gram of protein consumed; 10% protein pair-fed.

Column 2: Grams gain in body weight per gram of protein consumed; 10% protein, fed *ad libitum*.

Column 3: Grams gain in body weight per gram of protein consumed. Maximal ratios by *ad libitum* feeding.

Column 4: Grams gain in carcass protein per gram of protein absorbed; 10% protein pair-fed.

Column 5: Grams gain in carcass protein per gram of protein absorbed. Ratios established with 38 g of protein absorbed per rat in 42 days by *ad libitum* feeding.

Column 6: Grams gain in carcass protein per gram of protein absorbed. Maximal ratios by *ad libitum* feeding.

The values in Table 2 show how estimates of protein efficiency ratio change according to methods of feeding and of calculation. The authors consider that the techniques yielding the values in Columns 5 and 6 could be adapted to measure the maximum growth promoting capacity of dietary proteins (although it is difficult to see why the known nutritive difference between the proteins of properly and improperly heat-treated soyflours was not detected, digestibilities of the two being found to be almost identical). The corresponding values obtained by other procedures of experimentation and calculation differ from these values, sometimes markedly. The differences in protein efficiency found on pair-feeding at the 10% level (Column 1) are reduced when calculations are made on the basis of nitrogen retention (Column 4). Protein efficiency calculated from the results of *ad libitum* feeding at the 10% level (Column 2) penalizes the poorer proteins, each of which has a lower value relative to egg proteins than is given by any other procedure, an exaggeration of differences that is well-known to characterize comparisons by this method. The authors rightly emphasize that "the common parctice of measuring gain in body weight as a function of protein consumption with diets fed *ad libitum* and containing an arbitrarily selected amount of protein may be useful in a preliminary rating of proteins, but should not be given wide application without a full understanding of its limitations."

BOSSHARDT, YDSE, AYRES and BARNES (1946) developed a method of carcase analysis using the mouse which, in spite of certain disadvantages is useful for rapid tests of proteins that are available in only small amounts. They standardized the pre-test treatment of the animals, the duration of the test and the level of feeding of the test substance. They emphasized the importance of the length of the test period since in the case of proteins of high nutritive value the animals grow rapidly and quickly reach the stage where the growth rate decreases while the food consumption increases. This causes steadily declining values for protein efficiency ratios. In the case of proteins of poor nutritive value the growth rate tends to remain constant and the protein efficiency ratio remains steady. Differences between the ratios thus diminish. These observations are of particular importance because of their bearing on the general problem of design of experiments. It is often assumed that lengthy experiments with animals must inevitably be superior to short term experiments. This is not true in circumstances where the changes in the animal with increasing age themselves influence the outcome of the experiment.

8. Protein Repletion Techniques.

Methods of assessing the value of proteins and protein hydrolysates for the repletion of specific body proteins after experimental depletion have been reviewed by FROST, by SILBER and PORTER, and by CHOW (see ALBANESE, 1950). Broadly speaking the ratings given to proteins by these methods are similar to those given by conventional feeding techniques, but it is doubtful whether they can be used with advantage in most studies directed to solving nutritional problems. As SILBER and PORTER point out individual proteins may be superior for promoting regeneration of plasma protein but inferior for regeneration of liver protein. Consequently it should be expected that the results provided by repletion tests could be interpreted only with difficulty, if at all, when comparing different proteins as sources of amino acids for growth, lactation and egg production.

References.

ALBANESE, A. A.: Protein and Amino Acid Requirements of Mammals. New York: Academic Press Inc. 1950. — ANDERSON, M. E., and H. H. WILLIAMS: J. Nutrit. **44**, 335 (1951).

BARNES, R. H., J. E. MAACK, M. J. KNIGHTS and G. O. BURR: Cereal Chem. **22**, 273 (1945). — BLOCK, R. J., and D. BOLLING: The Amino Acid Composition of Proteins and Foods. 2nd Edition. Springfield, Ill.: C. C. Thomas 1951. — BLOCK, R. J., and H. H. MITCHELL: Nutrit. Abstr. Rev. **16**, 249 (1946/47). — BOSSHARDT, D. K., L. C. YDSE, M. M. AYRES and R. H. BARNES: J. Nutrit **81**, 23 (1946). — Bureau of Biological Research, Rutger's University: Report on Co-operative Determinations of the Amino Acid Content and of the Nutritive Value of six Selected Protein Food Sources 1950.

Halevy, S., and N. Grossowicz: Proc. Soc. Exp. Biol. Med. 82, 567 (1953). — Hansen, R. G., H. M. Scott, B. L. Larson, T. S. Nelson and P. Krichevsky: J. Nutrit. 49, 453 (1953). — Hegsted, D. M., and V. K. Haffenreffer: Amer. J. Physiol. 157, 141 (1949). — Hegsted, D. M., and J. Worcester: J. Nutrit. 33, 685 (1947). — Heiman, V., J. S. Carver and J. W. Cook: Poultry Sci. 18, 464 (1939).

Liener, I. E.: J. Nutrit. 49, 527 (1953). — Lofgreen, G. P., J. K. Loosli and L. A. Maynard: J. Animal Sci. 6, 343 (1947); 10, 171 (1951). — Lyman, R. L., and C. A. Elvehjem: J. Nutrit. 45, 101 (1951).

McDonald, I. W.: Biochem. J. 51, 86 (1952). — Melnick, D., and B. L. Oser: Food Technol. 3, 57 (1949). — Mitchell, H. H., and J. R. Beadles: J. Nutrit. 47, 133 (1952). — Mitchell, H. H., and R. J. Block: J. Biol. Chem. 163, 599 (1946). — Mitchell, H. H., T. S. Hamilton, J. R. Beadles and F. Simpson: J. Nutrit. 29, 13 (1945).

Osborne, T. B., L. B. Mendel and E. L. Ferry: J. Biol. Chem. 37, 223 (1919).

Sherwood, F. W., and V. Weldon: J. Nutrit. 49, 153 (1953).

Thomas, K.: Arch. Anat. Physiol., Physiol. Abt. 1909, 219.

Woodman, H. E., and R. E. Evans: J. Agric. Sci. 41, 102 (1951).

Urea and Ureides.

By

M. V. Tracey.

The necessity for the inclusion of a section on this group of compounds was unfortunately realised late by the editors. As a consequence it proved impossible to secure a contribution from a recognised authority in the field. Since however it seems to the editors that it is a subject that has been unjustly neglected — particularly in Britain and the United States — it is hoped that this review will serve the purpose of directing interest towards this fascinating and almost certainly rewarding field of research.

In 1933 WEHMER and HADDERS reported that urea had been detected in a very large number of plants representative of 31 natural families. Today it seems possible that urea may seldom, if ever, occur in a free state in vascular plants. Earlier reports of its widespread distribution are to be attributed to the common occurrence of urea precursors in plants and to their lability in response to changes in their physical environment and to their susceptibility to enzymic hydrolysis.

Ureides may be of the greatest quantitative importance in a consideration of the nitrogenous constituents of plants and plant extracts. The sap exuding from the severed roots of *Acer* sp. in spring may contain 99% or more of its nitrogen in the form of allantoin and allantoic acid and these ureides may account for a third of the total nitrogen of the roots and young branches (MOTHES and ENGELBRECHT, 1952). Similarly they may account for a quarter of the total nitrogen of the white leaves of *Acer Negundo* and for 45% of the total nitrogen of the inflorescence axes of *Acer pseudoplatanus* (ÉCHEVIN, BRUNEL and SARTORIUS, 1940). That they are not restricted to the *Aceraceae* may be seen from Table 3 at the end of this contribution (p. 133). The practical importance of the occurrence of the ureides in plants lies not only in the nutritional aspect but also in the field of crop preservation. During the harvesting and drying of crops ureides may be decomposed with the production of urea which in turn is hydrolysed by urease to ammonia with a concequent loss of nitrogen (REIFER and MELVILLE, 1949). The intrinsic interest of the widespread occurrence of ureides in plants is heightened by the conclusion that they must play an important role in the nitrogen metabolism of some plant families (BRUNEL and CAPELLE, 1947; MOTHES and ENGELBRECHT, 1952). Their importance seems to have been insufficiently appreciated so far (a recent text on plant biochemistry devotes 0.1% of the section "Metabolism of Nitrogenous Compounds" to allantoin and allantoic acid). The metabolic importance of the ureides is not the subject of this article and little work has been done on it. A few scattered observations in addition to those mentioned above indicate, however, that it is a subject worthy of investigation. It has been observed that glutamine increases in concentration in the sugar beet and then decreases and disappears — at this point it seems to be replaced by allantoin (RAVENNA and NUCCORINI, 1928a, b); that urea nitrogen (possibly derived from ureides) in a number of crop plants is of the same level as glutamine nitrogen (Reifer and Melville, 1949); and that leguminous seeds during growth may contain 16% of their total nitrogen as urea, (Damodaran and Venkatesan, 1948) much of which may have been formed from allantoic acid during the heating used in deproteinising the extracts. Owing to the scattered nature of the literature, which has not been adequately surveyed recently, more attention will be paid to non-analytical aspects than is usual in the remainder of this book.

A. The Occurrence of Urea and its Precursors in Plants.

So far as is known there are three degradative paths by which urea may be formed in plants and a possible means by which it may be used in the synthesis of ureides. The paths are (1) purines → uric acid → allantoin → allantoic acid → urea + glyoxylic acid; (2) arginine → urea + ornithine; and (3) canavanine → canaline + urea. Table 1 shows in summary form the dates at which these substances and their related enzymes were first identified in plants. It is hoped that this table may assist the reader in interpreting some of the earlier experimental results.

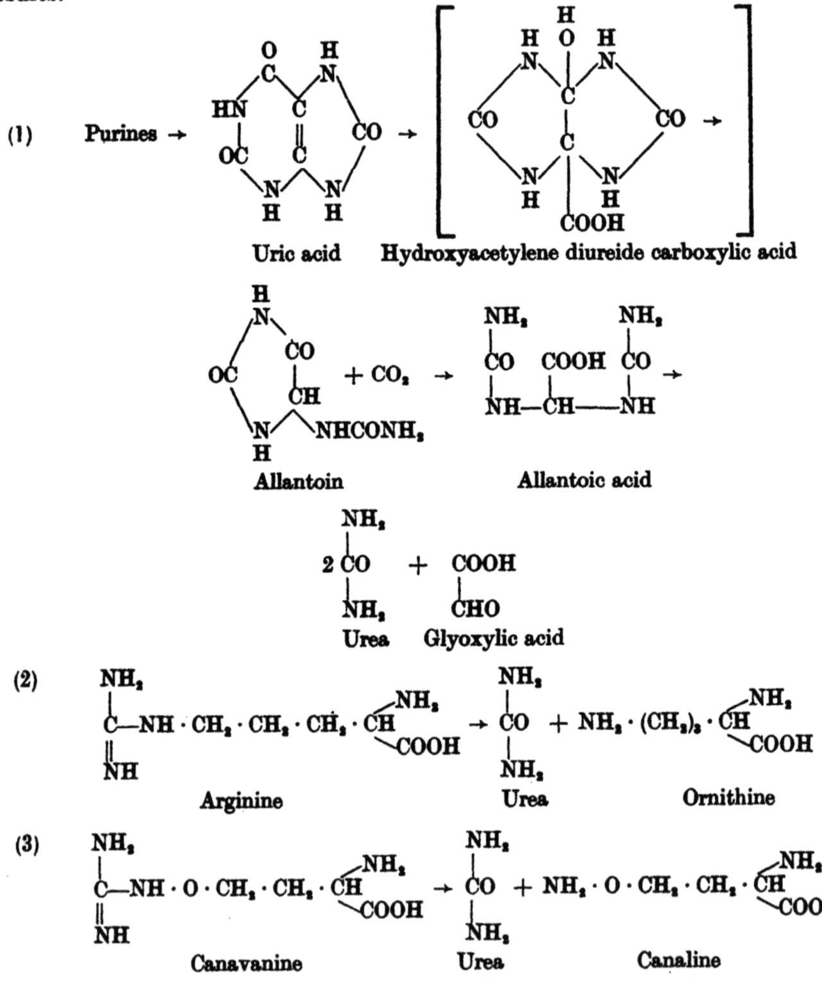

(1) Purines →

Uric acid   Hydroxyacetylene diureide carboxylic acid

Allantoin   Allantoic acid

Urea   Glyoxylic acid

(2)   Arginine   Urea   Ornithine

(3)   Canavanine   Urea   Canaline

Main Paths of Urea Formation in Vascular Plants.

Table 1. *Dates of the Discovery of Urea Precursors in Plants and of Enzymes effecting their Breakdown.* (References are given in the text.)

Enzyme	Year	Plant	Year	Enzyme	Plant
Urea	1903	fungi	1876	Urease	Bacteria
	1912	vasc. plants	1881	Allantoin	vasc. plants
Urease	1876	bacteria	1886	Arginine	vasc. plants
	1904	fungi	1903	Urea	fungi
			1904	Urease	fungi
	1909	vasc. plants		Arginase	yeast
Allantoic Acid	1926	vasc. plants	1909	Urease	vasc. plants
Allantoicase	1936	fungi		Arginase	vasc. plants
	1937	vasc. plants		Urea	vasc. plants
Allantoin	1881	vasc. plants	1912		
Allantoinase	1929	vasc. plants	1921	Uricase[1]	vasc. plants
Uric Acid	1928	fungi	1926	Allantoic acid	vasc. plants
	1932	vasc. plants	1928	Uric acid	fungi
Uricase	(1921)[1]	vasc. plants	1929	Allantoinase	vasc. plants
	1929[2]	vasc. plants		Uricase[2]	vasc. plants
Arginine	1886	vasc. plants		Canavanine	vasc. plants
Arginase	1904	yeast		(Canavanase	liver)
	1909	vasc. plants	1932	Uric acid	vasc. plants
Canavanine	1929	vasc. plants	1936	Allantoicase	fungi
(Canavanase	1929	liver)	1937	Allantoicase	vasc. plants
Canavanase = Arginase	1938		1938	Canavanase = Arginase	

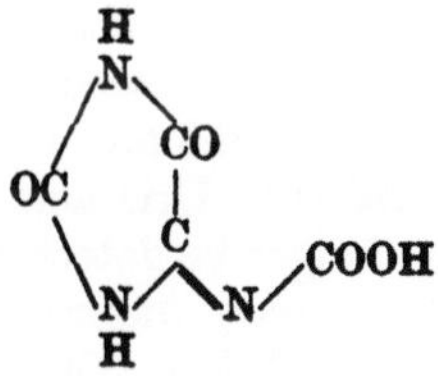

Allantoxanic acid

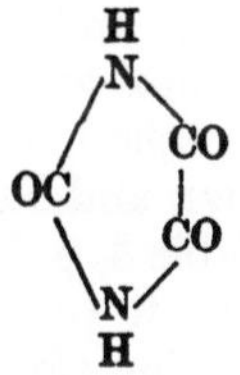

Parabanic acid

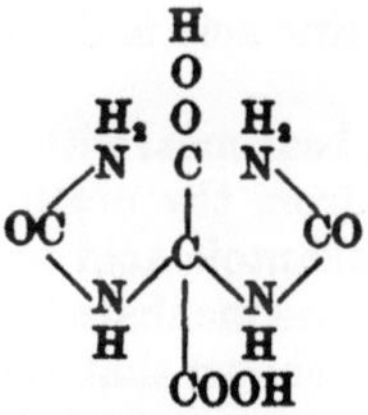

Uroxanic acid

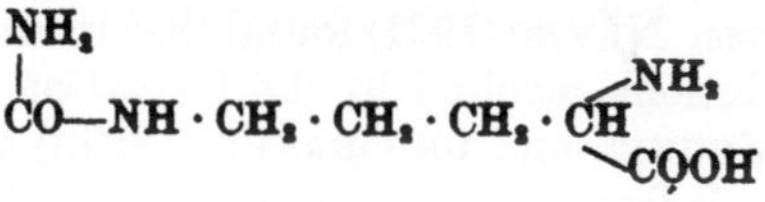

$COOH \cdot CO \cdot NH_2$

Oxamic acid[3]

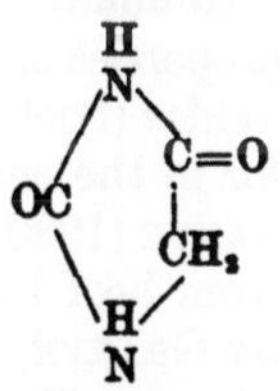

Citrulline

Hydantoin[4]

Compounds that may Prove to be Catabolic Products of Allantoin or to be Urea Precursors.

[1] Uric acid → ammonia.
[2] Uric acid → allantoin.
[3] Occurs in sugar beet, KMÍNEK (1936).
[4] Occurs in sugar beet, VON LIPPMANN (1896).

There is evidence that urea and glyoxylic acid may be condensed to form allantoin in some fungi (BRUNEL-CAPELLE, 1952) or allantoic acid (BRUNEL and BRUNEL-CAPELLE, 1951) but as yet there is none to indicate that a similar process occurs in vascular plants (MOTHES and ENGELBRECHT, 1953). Recent work has also indicated catabolism of allantoin may occur in fungi via an allantoin oxidase system with the possible formation of urea and oxalic or parabanic acids or, perhaps more likely, allantoxanic acid and ammonia (FRANKE, TAHA and KRIEG, 1952).

I. Intermediates in Purine Catabolism.

Between 1929 and 1937 FOSSE and his fellow workers found uric acid, uricase, allantoinase, allantoic acid and allantoicase in vascular plants for the first time. Allantoin was already known as a plant constituent. It seemed therefore that the catabolism of uric acid could account for the presence of ureides in plants and, together with the breakdown of arginine, for the urea that appeared in extracts. It later became apparent that the breakdown of purines could not possibly account for the relatively enormous amounts of ureides present in some plant tissues (BRUNEL and ÉCHEVIN, 1938; ÉCHEVIN, BRUNEL and SARTORIUS, 1940).

1. Uric Acid and Uricase.

Uric acid was first reported in plants by SUMI (1928) in the spores of *Aspergillus oryzae* (0.6% of the air dry wt.). FOSSE, DE GRAEVE and THOMAS (1932a) detected it in a number of plant seeds and isolated it in unstated yield from the seeds of *Melilotus officinalis*, and later from the seeds of *Trifolium sativum* and *Faba vulgaris* (1932b). Total uric acid values in mg./kg. dry seed, determined by an undescribed micro-method, were in the range 30—250 mg. for a number of seeds (FOSSE, DE GRAEVE and THOMAS, 1932b, 1933b). Uric acid cannot be detected in many seeds (Table 3, p. 133) and may no longer be detectable in seedlings from seeds that contain it (DE GRAEVE, 1937) or may diminish in amount during growth of the seedling (ÉCHEVIN and BRUNEL, 1937). The presence of uric acid in a number of seeds was confirmed by MICHLIN and IVANOV (1936). The reliability of the methods used in the detection and estimation of uric acid is discussed in a later section.

Uricase. NĚMEC (1921) found that incubation of soya bean meal with potassium urate solutions resulted in the formation of ammonia from the urate. In 1929 FOSSE, BRUNEL and DE GRAEVE (1929b) showed that allantoic acid was formed when meals from the seeds of sixteen species of legume were incubated with urate solutions. Since FOSSE and BRUNEL (1929a) had earlier shown that ten of these seeds contained allantoinase they concluded that uricase was responsible for the change of uric acid to allantoin and that allantoic acid was then formed by the allantoinase. The properties of plant uricase are of some importance in the enzymic analysis of plant ureides (Section B). Uricase seems to occur only in low concentrations in seeds, for in the enzymic estimation of uric acid described by FOSSE, BRUNEL and DE GRAEVE (1930) that present in soya bean meal was supplemented by a preparation from beef kidney. Some indication of its activity is given by FOSSE, BRUNEL, DE GRAEVE, THOMAS and SARAZIN (1930c). Plant uricase is said to be most active at pH 8.9 (BRUNEL and CAPELLE, 1947) though it may be used at pH 7.6 (BRUNEL and ÉCHEVIN, 1938). It is not clear whether the optimum pH given is based on experiments with plant or animal uricase. As with animal uricase the enzyme is completely inhibited by small concentrations of cyanide, e. g. 0.001% KCN. In the presence of allantoin sixty to eighty times as high a cyanide concentration may be necessary (FOSSE, BRUNEL, DE GRAEVE, THOMAS and SARAZIN, 1930c). After inhibition with KCN soya meal is still capable of the decarboxylation of hydroxyacetylene diureide carboxylic acid to allantoin (FOSSE, DE GRAEVE and THOMAS, 1933c). The uricase of *Soya hispida* seeds may also be inactivated by heat without affecting allantoinase or urease. Effective methods are heating a water suspension of the seed meal to 78° C for thirty minutes or heating the dry meal at 82° C for fifty hours (FOSSE, BRUNEL, DE GRAEVE, THOMAS and SARAZIN, 1930a). The known distribution of uricase is summarised in Table 3 (p. 133).

2. Allantoin and Allantoinase.

Allantoin has a long and respectable history as a plant component. It was isolated by SCHULZE and BARBIERI (1881) in a yield of 0.5—1.0% of the dry matter from sprouting shoots of *Platanus orientalis* (it accounted for 5—10% of the total nitrogen). SCHULZE and BOSSHARD (1885) later isolated it from other Aceraceae but were unsuccessful with a number of other trees. PURUCKER (1932)[1] after reviewing the field concluded that allantoin was distributed in at least nine families of vascular plants. The work of RAVENNA and NUCCORINI (1928a) on sugar beet should be added to his list which then covers all published work apart from the French. In 1926 FOSSE (1926a) found that allantoin and a ureide other than allantoin (later shown to be allantoic acid) were present in the saps of *Phaseolus vulgaris* and *Acer pseudoplatanus*. Since allantoic acid could arise from either allantoin (known to be present) or from uroxanic acid a search was made for the latter but it appeared to be absent (FOSSE, 1926c; FOSSE and HIEULLE, 1927, 1928). Allantoin was subsequently shown to occur in a large number of plants (Table 3, p. 133). Analysis of the seeds of fifty plants from eleven families showed allantoin contents in the range 20—90 mg./kg. in twenty species, 100—410 in another twenty, 500—880 in eight and the highest values of 1,700 and 3,300 mg./kg. in *Phaseolus Mungo* and *Dolichos sinensis* (FOSSE, BRUNEL and THOMAS, 1931). Allantoin contents may increase on germination, thus *Glycine hispida* seeds containing 14 mg. allantoin/kg. dry matter gave sixteen-day seedlings containing 1,800 mg./kg. dry matter (BRUNEL and ÉCHEVIN, 1938).

Optically Active Allantoin. FOSSE, THOMAS and DE GRAEVE (1934a) observed that the action of soya bean meal on a solution of allantoin resulted in the appearance of optical activity in the previously inactive solution. The substance responsible was isolated and proved to be L-allantoin ($[\alpha]_D$ —92° 24'). It was found to racemise rapidly in solution with the consequence that if enzymic action is sufficiently prolonged all the allantoin in solution is hydrolysed to allantoic acid. FOSSE, THOMAS and DE GRAEVE (1934c) then isolated allantoin from *Platanus orientalis* leaves taking precautions to avoid racemisation. They succeeded in getting a sample of D-allantoin ($[\alpha]_D$ +92°) and it was later isolated from calf urine (THOMAS and DE GRAEVE, 1934). Racemisation is rapid in alkaline solutions but slow in dilute acid (FOSSE, THOMAS and DE GRAEVE, 1934b). In the racemic form which is normally prepared allantoin is sparingly soluble in dilute acetic acid and on this property was based in part the claims for specificity in the histochemical tests for urea carried out by KLEIN, TAUBÖCK and LINSER (1930), KLEIN and TAUBÖCK (1931). Xanthydrol was used to precipitate urea extracted from plant sections with dilute acetic acid. D-Allantoin is however readily soluble in dilute acetic acid and forms a precipitate with xanthydrol. If in fact D-allantoin normally occurs in plants its different solubility from the racemic form does much, together with its instability, to explain the former impression of the widespread distribution of urea in plants. Allantoin is converted to allantoic acid at

---

[1] The results of PURUCKER's own analyses though described as for allantoin are, owing to the method used, for allantoin + allantoic acid.

alkaline pHs. In the presence of seed meals from a number of plants at pHs in the region of 7—9 the conversion is much more rapid than in their absence. Boiling the seed meals removes their activity (Fosse and Brunel, 1929). Table 3 summarises the known distribution of allantoinase in vascular plants. It also occurs in fungi and animals. A recent report states that the pH optimum is in the region 7.3—7.6 (Brunel and Capelle, 1947).

3. Allantoic Acid and Allantoicase.

Allantoic acid, the product of the chemical or enzymic hydrolysis of allantoin, was first identified in nature by Fosse (1926b). Five minutes heating at 100° of saps containing allantoin resulted in the appearance of urea and a substance later shown to be allantoic acid (Fosse, 1926c) by examination of the lead salt and xanthyl derivative. It occurs in small amounts in some seeds (Table 3, p. 133) and increases rapidly in amount after germination — thus in *Melilotus officinalis* it increased from 97 mg./kg. dry weight in the seed to 2.7 g./kg. dry weight at 15 days and 6.85 g./kg. at 35 days, an increase of seventy fold (Fosse, de Graeve and Thomas, 1933a). Little detailed study of its distribution in plants has been carried out but at present it appears to be typical of young rapidly growing organs (buds, flowers, developing fruits and seeds) and is also present in remarkable amounts in the parts of variegated leaves that are free from chlorophyll (allantoic acid nitrogen $= 14\%$ of total nitrogen in white parts of *Acer Negundo* leaves, only traces in green portions; Molliard, Échevin and Brunel, 1938). Allantoic acid often occurs in a concentration similar to that of allantoin but may occur alone in the maple (Mothes and Engelbrecht, 1952). During the germination of *Glycine hispida* it increases in amount per plant remaining at levels about 3/2 that of allantoin (Brunel and Échevin, 1938, see p. 123).

Allantoicase. The enzyme hydrolysing allantoic acid to urea and glyoxylic acid was first discovered by Brunel (1936) in the fungi *Aspergillus niger* and *A. oryzae* and in 1938 it was found in *Glycine hispida* seedlings (Brunel and Échevin, 1938). Its presence has not been reported in any other plant (Table 3, p. 133). The reason may be the inherent difficulties in detecting its presence in crude extracts unless special precautions are taken. Brunel-Capelle (1950a) pointed out that of the two products formed by its action, urea and glyoxylic acid, the first was not a good index since it is not easy to distinguish in small amounts from allantoic acid if the formation of xanthyl derivatives is used and also because the extracts often contain considerable amounts of urease. Theoretically ammonia production in the presence of urease could be used but this involves an exact knowledge of other systems present capable of liberating ammonia. Glyoxylic acid has the disadvantage of being rapidly destroyed in plant tissue extracts and indeed in fungal extracts the total amount of glyoxylic acid apparently formed from allantoic acid begins to decrease before the action of allantoicase is complete. If, however, a suitable "trapping" substance is added the glyoxylic acid may accumulate in the form of a stable derivative and she showed that phenylhydrazine was suitable for this purpose. It was shown that extracts from 24 species of Leguminosae contained either enzymic or non-enzymic (stable to heating at 100°, 30 mins.) mechanisms causing glyoxylic acid to disappear from solution and that these might occur together (Brunel-Capelle, 1950b). It is implied that in fact allantoicase is present in these plants.

4. Urea and Urease.

There is an abundance of statements that urea occurs in green plants. That urea occurs in any significant amounts in them is however a matter of conjecture at present. It will be obvious from the preceding section that evidence for the presence of urea in plants before the discovery that allantoic acid is a plant constituent must be suspect and the same must be true of subsequent work in which the authors seem unaware that allantoic acid may have been present. Unfortunately awareness that allantoic acid is both widely distributed in plants and extremely easily converted to urea seems to have been completely restricted to the French workers until the work of MOTHES and ENGELBRECHT in 1952. Thus the most recent workers on urea in plants (DAMODARAN and VENKATESAN, 1948; REIFER and MELVILLE, 1949) are forced to postulate the presence of an unknown urea precursor other than arginine in the plants they used though the French had shown that many contained allantoin and it is probable that allantoic acid was also present. If evidence for the occurrence of urea in vascular plants is restricted to that obtained by workers aware of the possible presence of allantoic acid and allantoin then all that remains is a statement by FOSSE (1939) that he repeated some of his work published in 1912 and 1913 and isolated the xanthyl derivative of urea from plant juices that had been kept at an acid pH and not heated. No yields or other details are given. BONNET (1929) may have been aware of the possible origin of urea from allantoic acid but the amounts of urea he found in lupin and lentil seeds and seedlings were small (2—20 mg./100 g. seed). He used the xanthydrol precipitation method for estimation; the precipitate got from the aliquots used often weighed less than a mg. and the specificity of the method in the presence of other ureides is open to doubt. In his examination of the flowers, fruits and seeds of 87 species of Leguminosae BRUNEL (1952) reports that urea was not once detected though this family had previously been regarded as particularly rich in urea. The evidence that urea precursors may give rise to urea in plant extracts is overwhelming and was first put forward by FOSSE (1926a) and by KLEIN (1928). There was some difference of opinion between the two as to the nature of the precursor for though KLEIN at first (1928) concluded that urea was combined with an aldehyde he later suggested that it originated mainly from arginine and arginine-like compounds (KLEIN and TAUBÖCK, 1932a).

There is no doubt that urea occurs in the fungi — indeed the first report that it occurred in plants was that of BAMBERGER and LANDSIEDL (1903) who isolated it in a yield of up to 3.5% of the dry matter from *Lycoperdon giganteum* Batsch (= *L. bovista* L.). Much of the literature is summarised by BRUNEL and CAPELLE (1947). Urea may be found in very large amounts — 13% of the dry matter of *Psalliota pratensis* (IVANOFF, 1923), 12% of the dry matter of *Lycoperdon giganteum* (TRACEY unpublished) for example. In the latter example the urea N accounted for 44% of the total N.

Urease. MUSCULUS (1876) was the first to discover urease; he used cell free extracts of bacteria. In 1904 it was found by SHIBATA in *Aspergillus niger* and in 1909 it was found by TAKEUCHI in *Glycine soja* seeds and seedlings and in the seeds of a number of other plants. He also observed that urease will act on biuret. This observation was forgotten and urease was long considered as mono-specific until TAKEUCHI's observation was confirmed by SHAW and KISTIAKOWSKY (1950). Urease is widely distributed in plants (KIESEL, 1911, 1927; FOSSE, 1914, 1916; DAMODARAN and SIVARAMAKRISHNAN, 1937; BRUNEL, 1952) particularly in the Cucurbitaceae and Leguminosae. The most active source is the jack or sword bean *(Canavalia ensiformis)* as shown by ANNETT (1914). It may contain 0.13% of urease while the soy bean contains about 0.01% (SUMNER, 1951).

II. Arginine as a Urea Precursor.

Arginine was discovered by SCHULZE and STEIGER in 1886. They isolated it from etiolated lupin seedlings in a yield of 3—4% of the dry matter. Later it was shown to occur in protein hydrolysates (Hedin, 1895). Arginase, the enzyme hydrolysing arginine to ornithine and urea, was found in yeast by Shiga (1904) and in vascular plants by Kiesel in 1909. The probable co-existence of enzyme and substrate in plants implies that their interaction is a possible source of urea and Klein and Tauböck (1932a, b) showed that in some plants the urea found in extracts was approximately accounted for by the amount of arginine known to be broken down. In others, all the urea could not be accounted for in this way. The possible complexity of the origin of urea in plant extracts is indicated by the fact that one of the plants in which arginine appeared to account for urea formation was *Canavalia ensiformis*. Seeds of this plant (and possibly the seedlings) are known to contain allantoin and also canavanine which is an alternative substrate for arginase which liberates urea from it (Section III). Damodaran and Venkatesan (1948) and Reifer and Melville (1949) have each unsuccessfully attempted to account for urea found in plant extracts in terms of formation from arginine and each have concluded that some other urea precursor must be invoked. It is of interest that the two plants used by Damodaran and Venkatesan were *Phaseolus Mungo* and *Dolichos biflorus*. Fosse, Brunel and Thomas (1931) had reported that the highest allantoin levels found in an examination of the seeds of fifty species were in *Phaseolus Mungo* and *Dolichos sinensis* (*D. biflorus* was not examined). It seems therefore plausible that allantoin or allantoic acid represents the precursor postulated by Damodaran and Venkatesan. Similarly allantoin has been reported (Table 3, p. 133) as being present in all the crop plants used by Reifer and Melville.

III. Canavanine as a Urea Precursor.

Canavanine was isolated from *Canavalia ensiformis* by Kitagawa and Tomita (1929); it is also present in *Canavalia obtusifolia* seeds (2% of fat free meal) and in *Canavalia lineata* (see Synge, p. 13, for detailed references). It is split enzymically by preparations from liver and other sources to canaline and urea and at first it was thought that a specific enzyme, not arginase, was responsible (Kitagawa and Tomita, 1929; Kitagawa and Tomiyama, 1929). Later it was shown that canavanine is in fact a substrate for arginase (Kitagawa and Eguchi, 1938; Damodaran and Narayanan, 1940). Unless canavanine is shown to be more widespread in plants its importance as a urea precursor must remain slight[1]. It is however of importance in that the simultaneous presence of canavanine, arginase and urease in the jack bean means that though this bean has the highest known urease content the urease must be freed from both canavanine and arginase (and possibly arginine) if high and variable blanks in the determination of urea by its use are to be avoided.

IV. Citrulline and Hydantoin.

Citrulline has been isolated from the juice of water melons and from the roots of *Alnus* sp. (see Synge, p. 13) and may occur in other plants. Srb and Horowitz (1944) showed that the ornithine cycle by which urea is synthesised in animals occurs in *Neurospora*. It is possible therefore that by entering the ornithine cycle citrulline might give rise to urea in plants but there is no evidence that this is so.

Hydantoin was isolated and identified from the juice of etiolated sugar beet sprouts by von Lippmann (1896). He says that this occurred only on a single occasion. Nothing more is known of the occurrence of this compound in plants.

---

[1] The probable presence of canavanine in 3 species of Leguminosae has been shown by the nitroprusside colour reaction. (p. 132). Canavanine was isolated from seeds of *Colutea aborescens* (*Leguminosae*). No colour reaction was given by extracts from species belonging to 20 other plant families (Fearon and Bell, 1955).

V. Thiourea.

KLEIN and FARKASS (1930) reported the isolation of thiourea from laburnum seeds. WEHMER and HADDERS (1933) reported that it is distributed in various organs of *Laburnum anagyroides* but not in other *Laburnum* sp. or in other plants. WINKLER (1949) reported that it does not occur in peaches but that added thiourea is oxidised by them to formamidine sulphonic acid (thiourea trioxide). OVCHAROV (1937) found that thiourea is synthesised by *Verticillium albo-atrum* and *Botrytis cinerea* when growing on synthetic media with asparagine or ammonia as sole N source but not if the N source is nitrate. He says further that plants affected by rust contained more thiourea than healthy plants. Qualitative tests were made for thiourea using its absorption in U.V. light [this paper has only been seen in abstract (OVCHAROV, 1938)]. If thiourea is synthesised by some fungal plant pathogens then it is a possible constituent of diseased plant tissues.

B. Some Analytical Principles.

Well tried, widely used methods for the estimation of most of the urea precursors dealt with in Section A do not exist. It is probable therefore that some discussion of the principles underlying their estimation, together with references to methods that have been used in this and other fields, will be of more value than a necessarily arbitrary selection of a few methods that have been insufficiently used, or if widely used, have been restricted to animal tissues.

I. Uric Acid.

Only two methods have so far been described as being used for the determination of uric acid in plants. FOSSE and his collaborators probably used an enzymic method. No detailed description of the method as applied to plant material has been found. In a paper giving the uric acid content of a number of seeds (FOSSE, DE GRAEVE and THOMAS, 1932b) reference is made to "une nouvelle methode de microdosage" which is not further described. In FOSSE's book (1939) the word "spectrophotometrique" is added but no other details are given. It may be surmised on the basis of other work published by this group (FOSSE, BRUNEL and DE GRAEVE, 1929a, 1930) that the principle of the method used involved the conversion of uric acid to allantoin by kidney uricase, of allantoin to allantoic acid by soya meal (perhaps freed from uricase activity), and of allantoic acid to glyoxylic acid and urea by acid hydrolysis. Glyoxylic acid would then be determined after formation of the deeply coloured product with phenylhydrazine. A control from which kidney uricase was omitted would correct for pre-existing allantoin and allantoic acid in the material analysed and in the soya bean meal. Inactivation of the kidney uricase before addition of uricase-free soya meal (which may contain uric acid) would eliminate interference from this source. Directions for the preparation of uricase-free soya meal are given by FOSSE, BRUNEL, DE GRAEVE, THOMAS and SARAZIN, 1930a, b). MICHLIN and IVANOV (1936) used a method involving the reduction of FOLIN's reagent for uric acid. It is known that this method may give erroneous results when used on animal material unless special precautions are taken (PETERS and VAN SLYKE, 1932).

It is probable that future investigators will make use either of paper chromatography (see MARKHAM, p. 246, for methods for the chromatography of purines) or of one of those methods depending on the destruction of uric acid by uricase and the measurement of some consequent change in the medium. KALCKAR (1947), for example, has described a specific method for uric acid in which the purine is destroyed by uricase and the change in U.V. absorption at 290 mμ gives a measure of the uric acid originally present. A method of the same nature, but perhaps a little less specific, is that of Blauch and Koch (1939), later modified by Block and Geib (1947), in which change in colour production after uricase treatment is measured using a complex tungstic acid as the colour producing agent.

II. Allantoin and Allantoic Acid.

An estimate of the concentration of allantoic acid alone will seldom be wanted and an accurate value for allantoin requires the simultaneous determination of allantoic acid. The two will accordingly be dealt with together. The methods described depend on the hydrolysis of allantoic acid to urea and glyoxylic acid and of allantoin to the same products *via* allantoic acid. The glyoxylic acid formed is then estimated by the Schryver (1910) reagent for formaldehyde. Obviously there are two possible sources of interference. If the hydrolysis of allantoin to allantoic acid is enzymic then either uricase or uric acid must be known to be absent from the system; glyoxylic acid must be absent, destroyed or separately estimated. The estimation of glyoxylic acid rather than urea, the other end product of hydrolysis, avoids a multitude of difficulties such as the presence of urea, urease, other urea precursors, or systems producing ammonia.

Hydrolysis of allantoin to allantoic acid by alkali was first used by Fosse and Bossuyt (1929) and this was followed by acid hydrolysis of the allantoic acid and the gravimetric determination of urea with a confirmatory qualitative colour test for glyoxylic acid. The immediately subsequent discovery of allantoinase (Fosse and Brunel, 1929a) led to its use in place of alkaline hydrolysis (Fosse, Brunel and de Graeve, 1929a). Since the enzyme source used, soya meal, contained urease, urea present or derived from precursors did not interfere as it was destroyed before the subsequent acid hydrolysis. The final determination of urea, after acid hydrolysis, had however to be gravimetric as the use of urease would lead to interference by any ammonia forming substances labile to acid hydrolysis. The discovery that uricase was also present in the soya meal (Fosse, Brunel and de Graeve, 1929b) led to the development of methods of inactivating the uricase without destroying allantoinase. Either heating the meal under controlled conditions or conducting the hydrolysis in the presence of KCN were found effective (Fosse, Brunel, de Graeve, Thomas and Sarazin, 1930a, b). The use of 0.8 g. KCN/litre in the hydrolysis is probably the most convenient method. Young and Conway (1942) examined one of the early variations of the Fosse methods and concluded that, with urine, enzymic hydrolysis of allantoin was unnecessary for the process was as fast in the absence of the enzyme at the pH they found necessary for complete hydrolysis (pH 12). Since Brunel and Echevin (1938) used a pH of 7.3 it is probable that Young and Conway used an unsuitable variety of soya bean; Brunel and Capelle (1947) have noted that the allantoinase content of soya beans is very variable. The alkaline hydrolysis method of Young and Conway is rapid and by modifying the details of the final estimation of glyoxylic acid they increased its accuracy. Alkaline hydrolysis, however, is often unsuitable with plant extracts as it may cause considerable darkening of the liquid which may make the final colorimetric estimation difficult or impossible (Mothes and Engelbrecht, 1952).

It seems likely, therefore, that any future agreed and reliable method for the determination of allantoin and allantoic acid will depend on the following steps.

1) Hydrolysis of allantoin to allantoic acid. If alkaline hydrolysis darkens the extract too much enzymic hydrolysis must be used. Soya bean samples should

first be examined for allantoinase activity. One gram of a suitable meal in one per cent suspension will hydrolyse at least 2.5 mg. allantoin (at a concentration of 100 mg./litre) in an hour at pH 7.3 and 40° (BRUNEL and CAPELLE, 1947). Directions for concentrating the enzyme extracted from suitable meals are given by these authors. The allantoinase is then used in the presence of sufficient KCN (0.8 g./litre) to inhibit any uricase activity.

2) Total allantoic acid (that formed from allantoin and that already present) is then determined with a parallel determination on material that has not been subjected to hydrolysis. The first step is the hydrolysis to urea and glyoxylic acid by 0.05 N HCl for 2 mins. at 100°. The hydrolysed samples are immediately cooled in ice water and one fifth volume of 0.33% freshly prepared phenyl-hydrazine hydrochloride solution added. Condensation with the glyoxylic acid is then allowed to take place for 15 mins. in a water bath at 30° C (the low temperature of condensation reduces interference by reducing sugars). The samples are then cooled again to incipient freezing in an ice-salt bath and the condensed product oxidised with ferricyanide in 3 N HCl. 0.6 vols of conc. HCl at —10° C are added followed by 0.2 vols of 1.67% freshly prepared potassium ferricyanide solution in water. After mixing 30 mins. is allowed for colour development. Standards and suitable blanks are run concurrently (YOUNG, MACPHERSON, WENTWORTH and HAWKINS, 1944).

3) The estimations described are aided by first drying the material and then extracting with 2.5% trichloracetic acid. If it is desired to estimate pre-formed glyoxylic acid drying should be cautious and the extraction should be carried out at pH 5 to minimise destruction of the acid (MOTHES and ENGELBRECHT, 1952). Glyoxylic acid is not a common acid in plants in significant quantities and is very labile.

III. Urea.

Success in the accurate and meaningful estimation of free urea in a plant tissue or extract will depend to a very large extent on the care with which the material is prepared for analysis and on an appreciation of the manifold possibilities of the formation of urea not originally present in a free form during the analysis. Much has been said in other sections of this work (PAECH in Vol. I p. 1, PIRIE in Vol. I p. 26 and this Vol., p. 38 BELL in Vol. II p. 2) of the importance of avoiding enzymic processes in the preparation of plant material for analysis and it will not be repeated here. The most satisfactory method for the determination of urea is, however, based on its enzymic transformation to ammonia and carbon dioxide. It is therefore of the greatest importance that either a pure urease (using pure in the sense that no other enzymic acitivity is detectable) or a urease preparation free from a number of specific activities is used.

Urease is one of the easiest enzymes to purify to a stage at which crystals are formed. This may give rise to a sense of false confidence, HELLERMAN and PERKINS (1935) reported that crystalline urease had arginine as an alternative substrate through they subsequently found that crystallising jack bean urease three times instead of once removed all arginase activity (STOCK, PERKINS and HELLERMAN, 1938). Similarly ARCHIBALD and HAMILTON (1943), showed that jack bean urease purified by the method of VAN SLYKE and CULLEN (1914), which is in common use, contains appreciable amounts of canavanine. Thus it is possible for a urease preparation sold or described as "purified" to contain both the enzyme (arginase) and the substrate (canavanine) necessary for the formation of urea. This presence in the urease preparation of a system capable of urea production is unimportant in itself since it may be allowed for by the use of suitable blanks but obviously if the material to be analysed contains either arginine or canavanine erroneous results will be got. It is known that jack bean contains allantoin and it is probable that the enzymes necessary for its conversion to urea are also present. After reports that the use of jack bean urease resulted in erroneous high urea values for animal tissues DAMODARAN and SIVARAMA-KRISHNAN (1937) surveyed a number of other seeds as urease sources. Jack bean *(Canavalia ensiformis)*, *Canavalia obtusifolia*, soya bean *(Glycine hispida)*, snake gourd *(Trichosanthes anguina)*, colocynth *(Citrullus colocynthis)*, common gourd *(Cucurbita maxima)* and horse-gram *(Dolichos biflorus)* all were good sources of urease and all gave "extra urea" when used in blood urea determinations. The seeds of the water melon *(Citrullus vulgaris)* though superior to soya bean in urease content gave rise to no "extra urea" (a meal of these seeds is now commercially available). In a later paper (DAMODARAN and NARAYANAN, 1940) it was shown that the seeds of the two *Canavalia* species contain arginase (both also contain canavanine). A number of other seeds giving rise to "extra urea" did not contain arginase and therefore some other enzyme or enzyme substrate mixture must have been responsible. Reference to Table 3 (p. 133) will show that of the seeds leading to "extra urea" production those of soya bean and common gourd are known to contain allantoin.

There are a number of methods that have been suggested for the determination of urea by the use of urease but of them all the CONWAY diffusion method is probably the best adapted for use with plant material. Details of the principle on which the method is based are given by KENTENl. (VoI, I of this handbook and by CONWAY (1950). A solution of boric acid containing a suitable indicator is placed in the central compartment of a diffusion unit and a sample of the urea containing extract in the outer compartment. The lid is greased and placed in position and, with the lid slightly displaced, urease solution is added to the sample (the urease used should be dialysed to remove amino acids). The lid is closed and the urease solution allowed to act for a suitable time (with active solutions 15 mins may be sufficient). The sample should preferably be not more than 1 ml. in volume and contain $10—100 \mu g$. urea N (these amounts may be modified considerably if desirable: CONWAY, 1950). The pH of the extract-enzyme mixture should lie between 7 and 8. This is conveniently assured by the addition of buffer to the enzyme solution but the final salt concentration should be less than $M/28$ if the reaction is not to be unduly slowed. After the action of the urease is complete 1 ml. of saturated potassium carbonate is added to the outer compartment, the lid closed, and after mixing diffusion is allowed to occur of the ammonia into the acid. This process is complete in one to two hours at room temperature depending on the volumes of reagents used. The acid in the central compartment is then titrated back with standard acid or may be removed and Nesslerised. When using the method on blood and tissue extracts CONWAY (1950) found no evidence of interference by the arginine-arginase system. If fungal extracts, known or suspected to contain free amino sugars, are being used it is advisable to carry out the final diffusion of ammonia at $5°$ C (TRACEY, 1952).

There exist a number of colorimetric methods for the determination of urea. The most recent of these are based on the carbamido diacetyl reaction of FEARON (1939). The test is positive for urea and for a number of substituted urea including allantoin, allantoic acid and citrulline. The method can, of course, be modified to give greater or less specificity for any particular substance (KAWERAU, 1946) but does not seem as well adapted as the enzymic methods for work with plants.

IV. Arginine, Canavanine, Citrulline and other Compounds.

Table 2 lists the results of the application of a number of colorimetric tests to compounds, related to urea, that are known to occur in plants or are closely related to those that do.

Table 2.

Reacting group	Diacetylcarbamido reaction (FEARON) $R \cdot NH \cdot CO \cdot NH_2$ (sometimes $R' \cdot NH \cdot CO \cdot NH \cdot R'$)	α-Naphthol-hypobromite reaction (SAKAGUCHI) $NH_2 \cdot \overset{\overset{NH}{\|}}{C} \cdot NH \cdot C$	Nitroprusside reaction (KITAGAWA) $NH_2 \cdot \overset{\overset{NH}{\|}}{C} \cdot NH \cdot O \cdot R$	Alkaline picrate (JAFFE)
Urea	+			
Allantoic Acid	+			
Allantoin	+			
Uric Acid	−			
Arginine	−	+	−	
Canavanine		−	red	
Citrulline	+			
Hydantoin	−			+
Thiourea	−		blue	
Creatinine	−		−	+
Creatine[1]	−			+
Glycocyamine[1]	−	+	−	
Hydantoic Acid[1]	+			
Uroxamic Acid[1]	−			

1. The Diacetyl Carbamido Test.

The reaction between diacetyl monoxime and substituted ureas in acid solution was investigated by FEARON (1939). The colours given are deepened after oxidation. ARCHIBALD (1944) developed the observations of FEARON and was able to use the reaction to determine citrulline and allantoin in mixtures. Urea is first removed from the sample by incubation with urease that has had impurities of low molecular weight removed from it by dialysis. After destruction of the urea the sample is dialysed against a suitable small volume of dilute sulphuric acid. This inactivates the urease and any other enzymes present in the urease preparation. The remainder of the assay is carried out on the dialysate which is, of course, free from protein — FEARON had shown that many proteins give a colour with the diacetyl carbamido reaction. The dialysate may contain allantoin, allantoic acid and citrulline all of which react to give a colour. Citrulline is removed from neutralised aliquot of the solution by passage through a column of Amberlite IR 100, allantoin and allantoic acid not being adsorbed. One volume of dialysate is then heated for ten minutes at 100° with 0.5 vols of sulphuric-phosphoric acid (1:3 of the conc. acids by volume) and a sixteenth volume of 3% diacetyl monoxime in water. After cooling for ten minutes the colour intensities are determined using light at 490 mμ. It is advisable to exclude light from the tubes from the beginning of heating until the final reading in the colorimeter. At the same time samples of the material that has passed through the resin and citrulline standards are analysed. Subtraction of the apparent citrulline values for the material that has passed through the resin from the values for the intact dialysate gives a measure of the amount of citrulline present. Archibald also gives details for the determination of allantoin in blood plasma in the presence of citrulline but the presence in plant extracts of allantoic acid and the availability

---

[1] Not known to occur in plants.

of other methods probably make this method inapplicable to plant extracts. KAWERAU (1946) has introduced further modifications of the method proposed by ARCHIBALD.

2. The α-Naphthol-Hypobromite Reaction (SAKAGUCHI-WEBER).

SAKAGUCHI (1925) showed that if α-naphthol were added to a strongly alkaline solution of arginine and followed by a small amount of hypochlorite a red colour was rapidly formed. Of the substances known to react only arginine is a known plant constituent. Glycocyamine (guanidinoacetic acid) also reacts but has not been identified in plants. Glycocyamine is a precursor of creatine and creatinine in animals and it has been shown that etiolated wheat seedlings are capable of creatine synthesis from added glycocyamine (BARRENSCHEEN and PANY, 1942). Creatinine has been isolated in small amounts from etiolated seedlings of *Lupinus luteus* (TOKAREWA, 1926). It is possible therefore that glycocyamine may occur in plants.

The SAKAGUCHI reaction has been developed as a quantitative method for the determination of both glycocyamine and arginine by DUBNOFF and BORSOOK, 1941; DUBNOFF, 1941). Passage of the solution to be analysed through a base exchange column (Permutit "after FOLIN") results in the retention of arginine while the glycocyamine passes through. Any remaining glycocyamine is washed out with 0.3% NaCl solution. The arginine is then eluted with 10% NaCl solution. The SAKAGUCHI reaction is then performed on aliquots of 2 ml. by adding 0.5 ml. of urea-α-naphthol solution (0.2% α-naphthol solution in absolute alcohol diluted with 4 vols 10% urea in water before use). After standing 2 mins 0.2 ml. of NaOBr solution (0.66 ml. Br_2 in 100 ml. 5% NaOH) are added. All reagents and samples are cooled to 0° before carrying out the reaction. After mixing and standing 20 mins at 0° the solutions are warmed quickly to room temperature and any bubbles of gas removed by tapping. The colour intensity is then determined in a photoelectric colorimeter using a filter with maximum transmission in the region of 525 mμ. The method is suitable for quantities of 5—40 μg. arginine in the sample.

3. Nitroprusside Reaction.

The only compounds that give a red colour with nitroprusside at pH 7.2 under the conditions to be described are hydroxyguanidine derivatives. Canavanine is the only one known to occur naturally. Thiourea and other thio compounds give a blue colour. The reagent used is not nitroprusside but a complex ion derived from it by irradiation in neutral solution (KITAGAWA and YAMADA 1932; KITAGAWA and TAKANI, 1936) or the addition of an oxidising agent (ARCHIBALD, 1946 discusses the nature of the active ion or ions responsible). 0.5 ml. of 20% K_2CO_3 and 0.4 ml. of 30% H_2O_2 are mixed with 10 ml. of 2% sodium nitroprusside solution (prepared weekly and kept at 0—4° C). The reagent should be prepared daily. 0.5 ml. of M phosphate buffer pH 7.2 and 0.5 ml. of the reagent are added to 1 ml. neutral samples and standards of canavanine (0—250 μg.). After mixing the samples are kept in the dark for two hours. If desired up to 10 ml. of water may be added after the period of colour development to enable larger vessels to be used in reading the colour developed. A filter with maximum transmission at 520 mμ is used. Some plant extracts contain material that inhibits colour development to a varying extent. The original paper should be consulted for details.

An essentially similar reagent has been developed for the determination of thiourea in fruit extracts by means of the blue colour developed (Ass. Off Agric. Chemists., 1948).

Table 3. *Occurrence of Urea and Ureides and the Corresponding Enzymes in the Plant Kingdom.*
(The numbers in brackets refer to the references at the end of the table.)

	Uric Acid	Uricase	Allantoin	Allantoinase	Allantoic Acid	Allantoicase	Urea	Urease
Acanthaceae								
Blepharis edulis			S+ (29)					
Aceraceae								
Acer campestris	S+ [12)		P					
circinnatum			Bark+ (25)					
negundo			L+(24,38)(44)		L+(24,38)(44)		LO (38)	
palmatum			Bark+ (25)					
pennsylvanicum			Bark+ (25)					
platanoides			F+(25)(27) Y+(44)		F+ (27) Y+(44)			
pseudoplatanus	S+ [14)	S+ (14)	P,S+ (14) F+ (21,24) Y+(44)	S+ (14)	S+(14)L+(44) L+(2—5) F+(21,24)	FO (21)	FO (21)	F+ (21)
saccharophorum			Bark+ (25)					
saccharum			Sap ?0 (45)					
Betulaceae								
Corylus avellana					L+ (28)			
Boraginaceae								
Anchusa officinalis			P					
Borrago officinalis			P					
Cordia atrofusca			P					
excelsa			P					
Symphytum officinale			P, L + R + (27, 44)		L+R+(27,44)			
Caryophyllaceae								
Agrostemma githago	YO (24)	YO (24, 17), SO (17)	Y+ (17)	Y+ SO (17)	Y+, SO (17)	YO, SO (17)	YO (17)	Y+SO (17)
Chenopodiaceae								
Anabasis aretiodes			P					
Atriplex hortensis				S+ (9)				
Beta vulgaris			P, S+ (11), R+ (30,32,36)	S+ (9)				
Spinacia oleracea			S+ (11)	S+ (9)				

Table 3. (Continued.)

	Uric Acid	Uricase	Allantoin	Allantoinase	Allantoic Acid	Allantoicase	Urea	Urease
Compositae								
Lactuca sativa			S+ (*11*)					
Scorzonera hispanica			Y+ (*27*)		Y+ (*27*)			
Tanacetum vulgare				S+ (*9*)				
Cruciferae								
Brassica napus		RO (*10*)	S+ (*11*), R+ (*10*)	RO (*10*)	R+ (*10*)			
oleracea		FO (*10*)	S+ (*11*) F+ (*10*)	F+ (*10*)	F+ (*10*)			
Lepidium sativum	S+ (*12*)		S+ (*11*)					
Raphanus sativus			RO (*22*) S+ (*11*),	RO (*22*)	RO (*22*)	RO (*22*)		
Sinapis nigra			S+ (*11*)					
Cucurbitaceae								
Cucumis sativus			S+ (*11*)	S+ (*9*)				
Cucurbita maxima			S+ (*11*)	S+ (*9*)				
Euphorbiaceae								
Ricinus communis	S+ (*12*)							
Fagaceae								
Quercus robur			F + L + (*27*)		F + L + (*27*)			
Geraniaceae								
Geranium microrhiza			L+ (*39*)					
phaeum			L+ (*39*)					
pratense			L+ (*39*)					
Pelargonium peltatum			L+ (*39*)					
zonale			L+ (*38*)		L+ (*38*)		LO (*38*)	
Gramineae								
Avena sativa			S+ (*11*)					
Hordeum vulgare	SO (*14*)	SO (*10, 14*)	S+ (*10,11,14*)	SO (*10, 14*)	SO (*10, 14*)			
Oryza sativa			S+ P(*11*)					
Secale cereale			S+ (*31*)					
Sorghum halepense	S+ (*12, 43*)							
vulgare			S+ (*11*)					

Table 3. (Continued.)

	Uric Acid	Uricase	Allantoin	Allantoinase	Allantoic Acid	Allantoicase	Urea	Urease
Triticum sativum	YO (*16*), S+ (*14, 16*)	SO (*14*), S+ (*10*)	S+P (*10, 11, 14*)	SO (*10, 14*)	SO (*10*), S+ (*14*)			
Zea mays	SO (*14, 16*)	S+ (*10, 14*)	S+ (*10,11,14*) Y+ (*27*)	SO (*10, 14*)	SO (*10, 14*) Y+ (*27*)			
Hippocastanaceae								
Aesculus flava			P, branches + (*26*)					
hippocastanum			P, branches + (*26*) F + L + (*27*)		F + L + (*27*)			
Labiatae								
Stachys sylvatica			P					
Liliaceae								
Allium cepa			S+ (*11*)					
porrum			S+ (*11*)					
Leguminosae								
Acacia longifolia				S+ (*7*)				
Anthyllis hermaniae				S + (*7*)				
Astragalus glycyphylloides				S+ (*7*)				
Canavalia ensiformis			S+ (*11*)					
Cercis siliquastrum				S+ (*7*)				
Cicer arietinum	SO (*14*)	S+ (*8, 14*)	S+ (*10,11,14*)	S+ (*7, 14*)	SO (*10, 14*)			
Colutea arborescens				S+ (*7*)				
Coronilla varia	S+ (*12*)		S+ (*11*)	S+ (*7*)				
Cytisus alschingeri				S+ (*7*)				
Dolichos lablab		S+ (*8*)						
sinensis			S+ (*11*)					
Faba vulgaris	S+ (*12,14,43*)	S+ (*8, 14*)	SO (*14*), S+ (*11*)	S+ (*14*)	SO (*14*)			
Genista scoparia	S+ (*14*)	S+ (*8, 14*)	S+ (*11, 14*)	S+ (*14*)	S+ (*14*)			
tinctoria				S+ (*7*)				
Glycine hispida	S+ (*14*)	S+ (*8, 14, 34, 37*),Y + (*20*)	S+ (*10, 11, 14,37*),Y+(*23*)	S+ (*6, 14, 35, 37*),Y+ (*21, 23*)	SO (*10*, S+ (*14*), Y+ (*23*)	Y+ (*20, 23*)	SO, YO (*20*)	S+ Y+, (*20*)

Table 3. (Continued.)

	Uric Acid	Uricase	Allantoin	Allantoinase	Allantoic Acid	Allantoicase	Urea	Urease
Leguminosae (cont.)								
Glycine soya	S+ (*12, 43*)	S+ (*8*)		S+ (*7*)				
Lathyrus latifolius	SO (*14*)	S+ (*14*)	S+ (*11, 14*)	S+ (*7, 14*)	S+ (*14*)			
niger				S+ (*7*)				
odoratus				S+ (*7*)				
pratensis				S+ (*7*)				
sylvestris		S+ (*8*)						
Lens esculenta			SOYO (*15,41*)	S+ (*7*)	S+,Y+ (*41*)			
Lotus corniculatus			S+ (*11*)					
tetragonolobus				S+ (*7*)				
Lupinus albus	S+ (*12,14,19*)	SO (*14, 19*), Y+ (*19*)	S+ (*11,14,19*), Y+ (*19*)	S+ (*7,14,19*), Y+ (*19*)	SO(*14*), S+ (*19*), Y+ (*19*)	SO, YO (*19*)	SO, YO (*19*)	S+, Y+ (*19*)
luteus			SO,YO(*15,41*)		S+,Y+(*15, 41*)		S+, Y+ (*15*)	
Medicago minima				S+ (*7*)				
Melilotus alba		S+ (*8*)		S+ (*7*)				
officinalis	S+ (*12,14,18*), YO (*18*)	S+ (*8, 14*)	S+ (*11, 14*) Y+ (*27*)	S+ (*7,14*)	SO(*14*), S+ (*13*), Y+ (*13,27*)			
Mimosa pudica				S+ (*7*), +(*40*)	+ (*40*)		O (*40*)	S+ (*40*)
Ononis fruticosa				S+ (*7*)				
Phaseolus lunatus		S+ (*8*)	S+ (*11*)	S+ (*6*)				
multiflorus		S+ (*8*)	P	S+ (*7*)				
mungo		S+ (*8*)	S+ (*11*)	S+ (*7*)				
vulgaris	SO (*14*)	S+ (*8, 14*)	P, S+ (*10,11, 14*), L+ (*1*)	S+ (*6, 14*)	S+ (*10, 14*), L+ (*1*)			
Pisum sativum	SO (*14*)	S+ (*8, 14*), Y+ (*10*)	P, S+ (*10,11, 14*), Y+ (*10*)	S+ (*7, 14*)	SO(*10*)S+(*14*) Y+ (*10*)			
Scorpiurus sulcata				S+ (*7*)				
Spartium junceum		S+ (*8*)						
Trifolium montanum				S+ (*7*)				
pratense			Y+ (*27*)			Y+ (*27*)		
procumbens				S+ (*7*)				
sativum	S+ (*12, 14, 16, 18*)							
Trigonella polycerata	YO(*18*),Y+ (*16*) + (*43*)	S+ (*14*)	S+ (*14*)	S+(*14*), S+ (*7*)	S+ (*13, 14*), Y+ (*13*)			

Table 3. (Continued.)

	Uric Acid	Uricase	Allantoin	Allantoinase	Allantoic Acid	Allantoicase	Urea	Urease
Leguminosae (cont.)[1]								
Vicia hirsuta				S+ (7)				
lutea				S+ (7)				
pisiformis				S+ (7)				
sativa	S+ (14)	S+ (14)	S+ (14)	S+ (14)	SO (14)			
sylvatica				S+ (7)				
Wistaria sinensis			F+ (24)		F+ (24)			
Liliaceae				+ (42)				
Malvaceae								
Althaea rosea				S+ (9)				
Orchidaceae				+ (42)				
Platanaceae								
Platanus acerifolia orientalis			F + L + (27) P		F + L + (27)			
Polygonaceae								
Fagopyrum esculentum			S+ (11)					
Rosaceae								
Fragaria indica				S+ (9)				
Geum sylvaticum				S+ (9)				
Persica vulgaris				S+ (9)				
Potentilla recta				S+ (9)				
Prunus domestica				S+ (9)				
laurocerasus				S+ (9)				
Pyrus malus							L + R + (46)	

[1] The flowers, fruits and seeds of a further 87 Leguminosae native to or introduced into Indo China have been examined (40). In none was urea found. The sub-family Mimosoideae (13 species + *Mimosa pudica* — see table) was characterised by the general occurrence of allantoinase, a low level of allantoic acid (often only detectable in the young fruit), and a low urease activity in the seeds. The Caesalpinioideae were similar to the Mimosoideae save that urease was absent from the flowers and fruit and appeared in traces in the seeds of only six species of the total seventeen examined. Allantoinase was widely distributed in the Papilionatae (56 species) being found in the flowers and accumulating in the seeds. Allantoic acid appeared in the flowers, accumulated in the fruits and disappeared with the ripening of the seeds. Its level, as would be expected, was higher in those species with a high allantoinase content. The urease content of the seeds was variable.

Table 3. (Continued.)

	Uric Acid	Uricase	Allantoin	Allantoinase	Allantoic Acid	Allantoicase	Urea	Urease
Rutaceae								
Ruta graveolens				S+ *(9)*				
Solanaceae								
Datura metel			**P**					
Lycopersicum esculentum			S+ *(11)*					
Nicotiana tabacum			**P,** S+ *(11)*	S+ *(9)*				
Solanum tuberosum			**P**					
Tiliaceae								
Tilia sylvestris			YO *(24)*					
Umbelliferae								
Carum petroselinum			S+ *(11)*					
Daucus carota			S+ *(11)*	S+ *(9)*				
Foeniculum dulce				S+ *(9)*				
Urticaceae								
Cannabis sativa			S+ *(11)*	S+ *(9)*				
Humulus lupulus				S+ *(9)*				
Urtica dioica	S+ *(14)*	SO *(14)*	S+ *(11, 14)*	SO*(14)*,S+*(9)*	S+ *(14)*			
Vitaceae								
Ampelopsis hederacea			YO *(24)*					

O = not detectable, + present; S = seed; Y = young leaves or seedlings; F = flowers or inflorescences; L = leaves; R = roots or rhizomes. P = early work (mostly isolative) summarised by PURUCKER *(39)*. Figures in *(27)* and *(44)*, except those for *Acer platanoides* refer to allantoin and for allantoic acid; the same is true for some of PURUCKER's figures *(39)*.

References to Table 3.

(1) Fosse, R.: Compt. rend. Acad. Sci. (Paris) 182, 869 (1926).
(2) Fosse, R., and A. Hieulle: Loc. cit. 184, 1596 (1927).
(3) Fosse, R., and V. Bossuyt: Loc. cit. 185, 308 (1927).
(4) Fosse, R., and V. Bossuyt: Bull. soc. Chim. biol. 10, 308 (1928).
(5) Fosse, R., and V. Bossuyt: Loc. cit. 10, 313 (1928).
(6) Fosse, R., and A. Brunel: Compt. rend. Acad. Sci. (Paris) 188, 426 (1929).
(7) Fosse, R., and A. Brunel: Compt. rend. Acad. Sci. (Paris) 188, 426 (1929).
(8) Fosse, R., A. Brunel and P. de Graeve: Loc. cit. 189, 213 (1929).
(9) Fosse, R., A. Brunel and P. de Graeve: Loc. cit. 189, 716 (1929).
(10) Fosse, R., A. Brunel, P. de Graeve, P. E. Thomas and J. Sarazin: Loc. cit. 191, 1153 (1930).
(11) Fosse, R., A. Brunel and P. E. Thomas: Loc. cit. 193, 7 (1931).
(12) Fosse, R., P. de Graeve and P. E. Thomas: Loc. cit. 195, 1198 (1932).
(13) Fosse, R., P. de Graeve and P. E. Thomas: Loc. cit. 196, 883 (1933).
(14) Fosse, R., P. de Graeve and P. E. Thomas: Loc. cit. 196, 1264 (1933).
(15) Bonnet, R.: Loc. cit. 189, 373 (1929).
(16) de Graeve, P.: Loc. cit. 204, 798 (1937).
(17) Brunel, A., and R. Échevin: Loc. cit. 205, 81 (1937).
(18) de Graeve, P.: Loc. cit. 204, 445 (1937).
(19) Échevin, R., and A. Brunel: Loc. cit. 204, 881 (1937).
(20) Échevin, R., and A. Brunel: Loc. cit. 205, 294 (1937).
(21) Brunel, A., and R. Échevin: Loc. cit. 207, 592 (1938).
(22) Brunel, A., and R. Échevin: Loc. cit. 208, 1043 (1939).
(23) Échevin, R., and A. Brunel: Loc. cit. 208, 826 (1939).
(24) Échevin, R., A. Brunel and I. Sartorius: Loc. cit. 211, 71 (1940).
(25) Plouvier, V.: Loc. cit. 227, 225 (1948).
(26) Plouvier, V.: Loc. cit. 228, 1886 (1949).
(27) Mothes, K., and L. Engelbrecht: Flora 139, 586 (1952).

(28) Leroux, L.: Compt. rend. Acad. Sci. (Paris) 205, 172 (1937).
(29) Lal, J. B.: J. Ind. Chem. Soc. 13, 109 (1936).
(30) Kmínek, M.: Chem. Abstr. 30, 7899 (1936).
(31) Ihde, A. J., and H. A. Schuette: J. Amer. Chem. Soc. 63, 2486 (1941).
(32) Ravenna, C., and R. Nuccorini: Ann. Chim. Applic. 18, 509 (1928).
(33) Macalister, C. J.: Brit. Med. J. 1, 10 (1912).
(34) Němec, A.: Biochem. Z. 112, 286 (1921).
(35) Ro, K.: J. Biochem. (Japan) 14, 405 (1932).
(36) Stanek, V., and M. Kmínek: Listy Cukrovar 53, 403 (1935).
(37) Sosa-Bourdouil, C., A. Brunel and A. Sosa: Compt. rend. Acad. Sci. (Paris) 212, 1049 (1941).
(38) Molliard, M., R. Échevin and A. Brunel: Loc. cit. 207, 1021 (1938).
(39) Purucker, H.: Planta 16, 277 (1932).
(40) Brunel, A.: Compt. rend. Acad. Sci. (Paris) 234, 1470 (1952).
(41) Bonnet, R.: Bull. Soc. chim. biol. 11, 1025 (1929).
(42) Brunel, A., and G. Capelle: Bull. Soc. chim. biol. 29, 427 (1947).
(43) Michlin, D., and N. N. Ivanov: Planta 25, 54 (1936).
(44) Mothes, K., and L. Engelbrecht: Hoppe Seylers Z. 295, 387 (1953).
(45) Pollard, J. K. and T. Sproston: Plant Physiol. 29, 360 (1954).
(46) Oland, K.: Physiologia Plantarum 7, 463 (1954).

4. Jaffe Reaction for Hydantoin and Creatinine.

It has long been known that creatinine when treated with alkaline picrate gives an orange colour. Heating creatine with 0.04 M picric acid for 35 mins at 100° results in an 80% conversion[1] to creatinine. Bernheim and Bernheim (1946) reported that the Jaffe reaction for creatinine is also positive for hydantoin though not for hydantoic acid.

Bonsnes and Taussky (1945) recommend the following procedure for creatinine. Add 1 ml. of 0.04 M picric and 1 ml. of 0.75 N NaOH to a sample containing

---

[1] An improved method with 100% conversion has now appeared (Taussky. 1954).

up to 50 μg. creatinine. Allow the colour to develop for 15 min. If creatine is present heat for 45 mins at 100° before adding the NaOH. Subtract from this value the creatinine value and multiply the difference by 1.25 to allow for incomplete conversion of creatine to creatinine. This method has been used by BERNHEIM and BERNHEIM, 1946) for the estimation of hydantoin. It has not been used on plant materials and may require modification.

References.

ANNETT, H. E.: Biochem. J. 8, 449 (1914). — ARCHIBALD, R. M.: J. Biol. Chem. 156, 121 (1944); 165, 169 (1946). — ARCHIBALD, R. M., and P. B. HAMILTON: J. Biol. Chem. 150, 155 (1943). — Ass. Off. Agric. Chemists J. 31, 100 (1948).

BAMBERGER, M., and A. LANDSIEDL: Monats. Chem. 24, 218 (1903). — BARRENSCHEEN, H. K., and J. PANY: Biochem. Z. 310, 344 (1942). — BERNHEIM, F., and M. L. C. BERNHEIM: J. Biol. Chem. 163, 683 (1946). — BLAUCH, M. G., and F. C. KOCH: J. Biol. Chem. 130, 443 (1939). — BLOCK, W. D., and N. C. GEIB: J. Biol. Chem. 168, 747 (1947). — BONNET, R.: Bull. Soc. Chim. Biol. 11, 1025 (1929). — BONSNES, R. W., and H. H. TAUSSKY: J. Biol. Chem. 158, 581 (1945). — BRUNEL, A.: Thèse Doct. Sci., Paris 1936; Compt. rend. Acad. Sci., Paris 234, 1470 (1952). — BRUNEL, A., and G. BRUNEL-CAPELLE: Compt. rend. Acad. Sci. (Paris) 232, 1130 (1951); Bull. Soc. Chim. Biol. 29, 427 (1947). — BRUNEL, A., and R. ÉCHEVIN: Rev. gén. Bot. 50, 73 (1938). — BRUNEL- CAPELLE, G.: (a) Compt. rend. Acad. Sci. (Paris) 230, 1979 (1950); (b) Compt. rend. Acad. Sci. (Paris) 230, 2224 (1950); 234, 1466 (1952).

CONWAY, E. J.: Microdiffusion Analysis and Volumetric Error, 3rd. edn. London: Crosby Lockwood 1950.

DAMODARAN, M., and K. G. A. NARAYANAN: Biochem. J. 34, 1449 (1940). — DAMODARAN, M., and P. M. SIVARAMAKRISHNAN: Biochem. J. 31, 1041 (1937). — DAMODARAN, M., and T. R. VENKATESAN: Proc. Ind. Acad. Sci. 27 B, 26 (1948). — DUBNOFF, J. W.: J. Biol. Chem. 141, 711 (1941). — DUBNOFF, J. W., and H. BORSOOK: J. Biol. Chem. 138, 381 (1941).

ÉCHEVIN, R., and A. BRUNEL: Compt. rend. Acad. Sci. (Paris) 204, 881 (1937). — ÉCHEVIN, R., A. BRUNEL and I. SARTORIUS: Compt. rend. Acad. Sci. (Paris) 211, 71 (1940).

FEARON, W. R.: Biochem. J. 33, 902 (1939). FEARON, W. R. and E. A. BELL: Biochem. J. 59, 221 (1955) — FOSSE, R.: Compt. rend. Acad. Sci. (Paris) 155, 851 (1912); 156, 567, 1938 (1913); 158, 1374 (1914); Ann. Chim. (Ser. 9) 6, 13, 155 (1916); (a) Compt. rend. Acad. Sci. (Paris) 182, 175 (1926); (b) Compt. rend. Acad. Sci. (Paris) 182, 869 (1926); (c) Compt. rend. Acad. Sci. (Paris) 183, 1114 (1926); "Uréogénese et Métabolisme de l'Azote Purique chez les Végétaux". Paris: Gauthier-Villars 1939. — FOSSE, R., and V. BOSSUYT: Compt. rend. Acad. Sci (Paris) 188, 106 (1929). — FOSSE, R., and A. BRUNEL: Compt. rend. Acad. Sci. (Paris) 188, 426 (1929). — FOSSE, R., A. BRUNEL and P. DE GRAEVE: (a) Compt. rend. Acad. Sci. (Paris) 188, 1418 (1929); (b) Compt. rend. Acad. Sci. 189, 213 (1929); 190, 693 (1930). — FOSSE, R., A. BRUNEL and P. E. THOMAS: Compt. rend. Acad. Sci. (Paris) 193, 7 (1931). — FOSSE, R., A. BRUNEL, P. DE GRAEVE, P. E. THOMAS and J. SARAZIN: (a) Compt. rend. Acad. Sci. (Paris) 191, 1025 (1930); (b) Compt. rend. Acad. Sci. (Paris) 191, 1153 (1930); (c) Compt. rend. Acad. Sci. (Paris) 191, 1388 (1930).— FOSSE, R., P. DE GRAEVE and P. E. THOMAS: (a) Compt. rend. Acad. Sci. (Paris) 194, 1408 (1932); (b) Compt. rend. Acad. Sci. (Paris) 195, 1198 (1932); (a) Compt. rend. Acad. Sci. (Paris) 196, 883 (1933); (b) Compt. rend. Acad. Sci. (Paris) 196, 1264 (1933); (c) Compt. rend. Acad. Sci. (Paris) 197, 370 (1933). — FOSSE, R., and A. HIEULLE: Compt. rend. Acad. Sci. (Paris) 184, 1596 (1927); Bull. Soc. Chim. Biol. 10, 308 (1928). — FOSSE, R., P. E. THOMAS and P. DE GRAEVE: (a) Compt. rend. Acad. Sci. (Paris) 198, 689 (1934); (b) Compt. rend. Acad. Sci. (Paris) 198, 1374 (1934); (c) Compt. rend. Acad. Sci. (Paris) 198, 1953 (1934). — FRANKE, W., E.—E. M. TAHA and L. KRIEG: Arch. f. Mikrobiol. 17, 255 (1952).

DE GRAEVE, P.: Compt. rend. Acad. Sci. (Paris) 204, 445, 798 (1937).

HEDIN, S. G.: Hoppe-Seylers Z. 20, 186 (1895). — HELLERMAN, L., and M. E. PERKINS: J. Biol. Chem. 112, 175 (1935).

IVANOV, N. N.: Biochem. Z. 143, 62 (1923).

KALCKAR, H. M.: J. Biol. Chem. 167, 429 (1947). — KAWERAU, E.: Sci. Proc. Roy. Dublin Soc. 24, 63 (1946). — KIESEL, A.: Hoppe-Seylers Z. 60, 460 (1909); 75, 169 (1911); Ergebn. Biol. 2, 257 (1927). — KITAGAWA, M., and S. EGUCHI: J. Agr. Chem. Soc. Jap. 14, 525 (1938).— KITAGAWA, M., and A. TAKANI: J. Biochem. Jap. 23, 181 (1936). — KITAGAWA, M., and T. TOMITA: Proc. Imp. Acad. Jap. 5, 380 (1929). — KITAGAWA, M., and T. TOMIYAMA: J. Biochem. Jap. 11, 265 (1929). — KITAGAWA, M., and H. YAMADA: J. Biochem. Jap. 16 339 (1932). — KLEIN, G.: Z. Pfl.-Ernähr. Düng. 12A, 390 (1928). — KLEIN, G., and E. FARKASS: Oesterr. Bot. Z. 79, 107 (1930).— KLEIN, G., and K. TAUBÖCK: Jahrb. wiss. Bot.

74, 429 (1931); (a) Biochem. Z. 251, 10 (1932); (b) Biochem. Z. 255, 278 (1932). — KLEIN, G., K. TAUBÖCK and H. LINSER: Jahrb. wiss. Bot. 78, 193 (1930). — KMÍNEK, M.: Chem. Abst. 30, 7899 (1936).

VON LIPPMANN, E. O.: Ber. dtsch. chem. Ges. 29, 2645 (1896).

MICHLIN, D., and N. N. IVANOV: Planta 25, 59 (1936). — MOLLIARD, M., R. ÉCHEVIN and A. BRUNEL: Compt. rend. Acad. Sci. (Paris) 207, 1021 (1938). — MOTHES, K., and L. ENGELBRECHT: Flora 139, 586 (1952). — MOTHES, K., and L. ENGELBRECHT: Hoppe-Seylers Z. 295, 387 (1953). — MUSCULUS, F.: Compt. rend. Acad. Sci. (Paris) 82, 333 (1876).

NĚMEC, A.: Biochem. Z. 112, 286 (1921).

OVCHAROV, K. E.: Compt. rend. Acad. Sci. (U.R.S.S.) 16, 461 (1937); Chem. Abst. 32, 1745 (1938).

PETERS, J. P., and D. D. VAN SLYKE: Quantitative Clinical Chemistry, Vol. II. London: Ballière, Tindall & Cox. 1932. — PURUCKER, H.: Planta 16, 277 (1932).

REIFER, I., and J. MELVILLE: J. Biol. Chem. 178, 715 (1949). — RAVENNA, C., and R. NUCCORINI: (a) Boll. inst. agr. Pisa 4, 159 (1928); (b) Ann. Chim. applicata 18, 509 (1928).

SAKAGUCHI, S.: J. Biochem. Japan 5, 13, 133 (1925). — SCHRYVER, S. B.: Proc. Roy. Soc. (London) B, 82 226 (1910). — SCHULZE, E., and J. BARBIERI: Ber. dtsch. chem. Ges. 14, 1602 (1881). — SCHULZE, E., and E. BOSSHARD: Hoppe-Seylers Z. 9, 420 (1885). — SCHULZE, E., and E. STEIGER: Ber. dtsch. chem. Ges. 19, 1177 (1886). — SHAW, W. H. R., and C. B. KISTIAKOWSKY: J. Amer. Chem. Soc. 72, 634 (1950). — SHIBATA, K.: Beitr. chem. Physiol. Path. 5, 384 (1904). — SHIGA, K.: Hoppe-Seylers Z. 42, 505 (1904). — SRB, A. M., and N. H. HOROWITZ: J. Biol. Chem. 154, 129 (1944). — STOCK, C. C., M. E. PERKINS and L. HELLERMAN: J. Biol. Chem. 112, 175 (1935). — SUMI, M.: Biochem. Z. 195, 161 (1928). — SUMNER, J. B.: In "The Enzymes", Vol. I, p. 873. New York: J. B. SUMNER & K. MYRBÄCK. Academic Press 1951.

TAKEUCHI, T.: J. coll. agr. Tokio 1, 1 (1909). — TAUSSKY, H. H.: J. Biol. Chem. 208, 853 (1954). — THOMAS, P. E., and P. DE GRAEVE: Compt. rend. Acad. Sci. (Paris) 198, 2205 (1934). — TRACEY, M. V.: Biochem. J. 52, 265 (1952). — TOKAREWA, A.: Z. physiol. Chem. 158, 28 (1926).

VAN SLYKE, D. D., and G. E. CULLEN: J. Biol. Chem. 19, 211 (1914).

WEHMER, C., and M. HADDERS: In KLEINS „Handbuch der Pflanzenanalyse", Bd. 4, S. 222. Berlin: Julius Springer 1933. — WINKLER, W. O.: J. Ass. Off. Agr. Chem. 32, 430 (1949).

YOUNG, E. G., and C. F. CONWAY: J. Biol. Chem. 142, 839 (1942). — YOUNG, E. G., C. C. MACPHERSON, H. P. WENTWORTH and W. W. HAWKINS: J. Biol. Chem. 152, 245 (1944).

Chlorophylls: Analysis in Plant Materials.

By

James H. C. Smith and Allen Benitez.

With 10 Figures.

Chlorophylls are the green pigments in plants some of which, and perhaps all, participate in photosynthesis. An understanding of their nature, genesis, transformation, and function in plants requires that they and their near chemical relatives be recognized qualitatively and oftentimes be determined quantitatively. In addition to the use of analytical procedures for the identification and determination of the chlorophylls as participants in plant physiological processes, these methods may be used in various other fields of science and technology some of which are genetics, plant nutrition, certain aspects of pure chemistry, crop production of land and sea, medicine, pharmaceuticals, food technology, and paleontology.

Because of the wide application of the analytical methods for the chlorophylls and their derivatives, it is well to review and revise the methods of analysis from time to time, and it is with a deep sense of indebtedness to previous reviews and to particular articles cited in this review, that this compilation of methods is made.

Since the review by TREIBS (1932), older methods of analysis have been improved and novel methods have been developed. The chlorophylls then known have been more adequately characterized and new chlorophylls have been discovered. Accordingly, it is the object of this review to integrate the new with the old in such a manner as to give as reliable and up-to-date procedures as possible for the preparation, identification. and estimation of the various chlorophylls and some of their near relatives.

The various naturally occurring chlorophylls to be treated here, which have been reported and described more or less in detail, are: chlorophylls a, b, c, d and e, protochlorophyll, bacteriochlorophyll, and bacterioviridin. Some of these, notably chlorophylls a and b, have been rather thoroughly characterized both in physical and chemical properties, whereas others, such as chlorophyll e, have been characterized by little more than a name and a qualitative absorption spectrum.

The chlorophylls are not stable crystalline compounds of constant properties which are readily obtainable for standardization purposes. Even the pure chlorophylls prepared in various laboratories are reported to have different spectroscopic properties. Whether the divergences have arisen from differences in the spectroscopic equipment used, from impurities in solvents or pigments, or from nonuniformity of isomeric composition of the pigment (cf. MANNING and STRAIN, 1943) is not known. At present it would seem advisable for each analyst, himself, to prepare samples of the highest purity possible and to determine the analytical constants under the conditions of analysis used. For this reason, methods for preparing purified. samples of the chlorophylls will be given.

A. Chlorophylls *a* and *b*.

I. Structure and Distribution.

Chlorophyll *a*. The structural and empirical formulas of chlorophyll *a* are shown below (FISCHER and STERN, 1943, p. 48; STOLL and WIEDEMANN, 1952). The calculated molecular weight is 893.48[1]. The calculated elementary composition is

$$C, 73.93; H, 8.12; O, 8.95; N, 6.27; Mg, 2.72$$

(cf. STOLL and WIEDEMANN, 1933a, b; WINTERSTEIN and STEIN, 1933).

Chlorophyll *a* occurs in all photosynthetic organisms which are known to give off oxygen in photosynthesis. It is found in the higher plants, green, red, and brown algae, diatoms, etc.

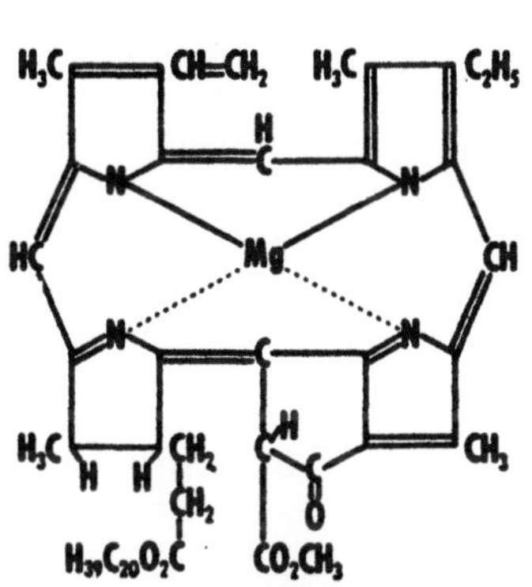

$C_{55}H_{72}O_5N_4Mg$
CHLOROPHYLL *a*

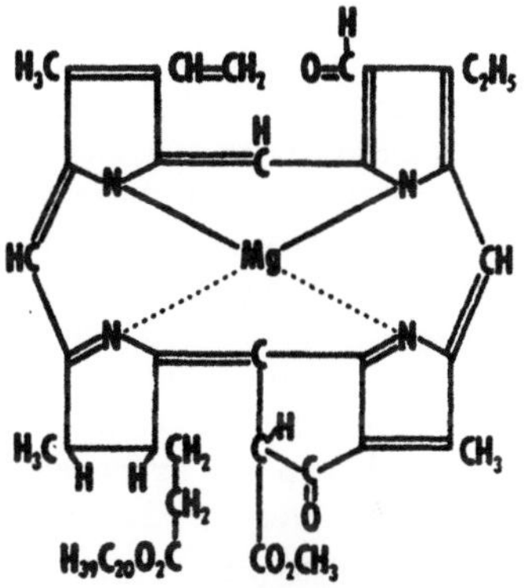

$C_{55}H_{70}O_6N_4Mg$
CHLOROPHYLL *b*

Chlorophyll *b*. The structural and empirical formulas of chlorophyll *b* are given above (FISCHER and STERN, 1943, p. 242; STOLL and WIEDEMANN, 1952).

The calculated molecular weight is 907.46. The calculated elementary composition is

$$C, 72.97; H, 7.78; O, 10.58; N, 6.17; Mg, 2.68$$

(cf. STOLL and WIEDEMANN, 1933a, b; WINTERSTEIN and STEIN, 1933).

Chlorophyll *b* occurs normally in all the higher land plants and green algae. It appears to be absent from the red and brown algae, diatoms, etc.

II. Preparation.

Methods for preparing pure chlorophyll *a* and chlorophyll *b* have been published by a number of workers. Partition methods have been used by WILLSTÄTTER and STOLL (1913). Chromatographic methods have been used by WINTERSTEIN and STEIN (1933), MACKINNEY (1940), ZSCHEILE and COMAR (1941), STRAIN and MANNING (1942), HARRIS and ZSCHEILE (1943), and by Dr. V. M. KOSKI in our own laboratory (1950).

In this section, two methods will be described for the preparation of the chromatographically separated chlorophylls *a* and *b*: The method of STOLL and

---

[1] The atomic weights used in the calculation of the molecular weights and the composition of the various pigments are: C, 12.01; H, 1.008; O, 16.00; N, 14.008; Mg, 24.32.

Wiedemann (1952) for obtaining the solid chlorophylls from dried leaves, for the description of which we are greatly indebted to Professor Stoll; and the method of the writers for obtaining solutions of the spectroscopically pure pigments from fresh leaves.

Pure Chlorophylls *a* and *b* from Dried Leaves. Stoll and Wiedemann have used the following method for the preparation of the pure chlorophyll components.

Crude Chlorophyll. Five hundred grams of carefully dried stinging nettles are ground to a medium fine powder and placed in a Büchner funnel of approximately 30 cm. diameter. The dry powder is sucked down firmly to give as even a layer as possible and then extracted with 1.5 litres 80% acetone poured over it in small portions. The dark green extract is stirred with 300 grams talc and 1.5 liters distilled water are then added. One hundred grams talc moistened with water are placed in a second Büchner funnel of approximately 20 to 25 cm. diameter, so as to form as even a layer as possible, and the green talc precipitate is then filtered onto this as soon as it begins to separate. After sucking off the yellowish brown mother liquor, the green talc layer is washed with 1.5 liters 60% acetone. Washing is then continued with 1 liter 0.01 N caustic soda to which approximately 0.5 gram sodium bisulfite has been added and immediately afterwards with distilled water until the filtrate is neutral in reaction. The talc layer is then dried rapidly by strong suction. The dry talc layer is extracted as exhaustively as possible with anhydrous ether and the deep green ether solution is dried with anhydrous sodium sulfate, filtered, and then concentrated under slightly reduced pressure at room temperature to approximately 20 ml. The remaining ether is then driven off by repeated addition of small quantities of petroleum ether and further, very gradual concentration until the chlorophyll separates out. If very pure solvents are used, it is permissible to obtain the crude chlorophyll by driving off the petroleum ether completely. The yield is almost quantitative and is usually around 6 grams from 100 grams of powdered leaves, but varies according to the quality of the leaves.

The operations described so far must only be carried out in a dim light and the subsequent operations must be performed as far as possible in the absence of light and oxygen.

Chromatographically Pure Chlorophyll Components. Powdered sugar (sucrose) of good commercial quality is dried, if necessary, and then rubbed through a sieve No. 50 (perforation size ca. 0.2 sq. mm.). The necessary solvents — thiophene-free benzene, petroleum ether, and ether — are purified by the addition of drying agents or caustic potash and fractionation through a column until they are anhydrous and free from acid and alkali. Care must also be taken that the ether does not contain any traces of peroxide. The filter papers to be used must first be extracted thoroughly with pure ether and should afterwards be stored under highly purified ether in the dark.

When all these materials are ready, a series of chromatographic columns each having an effective length of 50 cm. (upper rim to sintered glass disk) and a diameter of 9 to 10 cm. are prepared in the following way: The outlets are closed and each column is filled with a suspension of 1.3 kg. of the prepared sugar in 2.5 liters of a mixture of equal parts of the previously purified benzene and petroleum ether. When the columns have settled (with the outlets open) and immediately after the supernatant solvent has entered the columns, a solution containing 500 mg. crude chlorophyll in 250 ml. of the same solvent mixture (equal parts of benzene and petroleum ether) is passed through each column. The chromatograms are developed using the same solvent mixture. When the zones have formed, their progress down the column is stopped by the addition of pure petroleum ether and finally all benzene remaining in the column is displaced by passing through further quantities of petroleum ether. When this has been accomplished the columns are freed from excess solvent by brief suction and the next procedure is then commenced immediately.

At this point it may be mentioned that the preparation of the column in the manner described above, using a suspension of sugar, enables the troublesome irregularities in the line of demarcation between the zones to be largely prevented. A further consequence of the use of 1:1 mixture of benzene and petroleum ether is that the first fractions, containing the magnesium-free components, are largely removed during the passage through the column. Nevertheless, even with this method, a quantitative separation of the chlorophylls in a single operation can only be achieved with certainty if the quantity of chlorophyll per column is reduced still further. However, this would be very uneconomical. As will be shown below, it is possible to work very satisfactorily with the given quantities of chlorophyll and only a relatively small mixed fraction is obtained. This is collected from several columns and passed through a further column, and in this way an almost complete separation of the starting material into the chromatographically pure components can be achieved with little additional labor.

The procedure adopted is as follows: By removing from the columns the zones exhibiting a blue-green color the greater part of the chlorophyll *a* is obtained in the highest state of purity attainable; and by removing the lower portions of the zones with a yellow-green color at least half the quantity of chlorophyll *b* is obtained in a pure state. The intermediate layers, in which there is some overlapping of the two chlorophyll zones, and which contain more *b* than *a* are collected from several columns and separation effected as rapidly as possible by passage through a further column. The upper portions of the yellow-green bands, recognizable by their rather more dirty coloration, are rejected, as they contained only a little intact chlorophyll *b*.

After dismantling the columns and separating into segments, each of the usable fractions is suspended in approximately one liter of ether (per column). After stirring to insure distribution of the sugar, the suspension is immediately filtered through a previously prepared filter paper. The sugar remaining on the filter is washed with highly purified ether and the washing combined with the main filtrate. Immediately afterwards, the combined ether solutions are concentrated to a small volume by evaporation at 20 to 25° under slightly reduced pressure, and the remainder of the solvent is then removed in desiccators after transferring the concentrates to a number of small flasks.

In this way, the chromatographically pure components are obtained direct without the necessity for fractionation by solution and the products are stable for years if stored in the dark out of contact with air. The approximate yields are: chlorophyll *a*, 80%; chlorophyll *b*, 70%.

Ether Solutions of Chlorophylls *a* and *b* for Spectrophotometry. The writers have separated the chlorophyll components *a* and *b* from fresh leaves by the following procedure: Fifty grams of grape leaves were washed, freed from extraneous water, cut into strips about 1 cm. wide, and disintegrated in a Waring Blendor for 3 min. with 200 ml. of acetone containing about 1 g. of magnesium carbonate. The green suspension was centrifuged and the supernatant extract added to 100 ml. of petroleum ether, b. p. 50—70° C. The petroleum ether was washed once with 200 ml. of saturated magnesium carbonate solution by MACKINNEY's method (1940). This was accomplished by streaming the petroleum ether solution up through the magnesium carbonate solution contained in a separatory funnel. The petroleum ether was poured through a long-stemmed funnel which emptied deep in the aqueous phase. A second extract, prepared in the same way, was added to the same petroleum ether and washed in like manner. The sediment from the two lots of leaves was extracted with 100 ml. of acetone. The extract, after being centrifuged, was added to the same petroleum ether. The petroleum ether was washed successively with 200 ml. of saturated magnesium carbonate solution, two portions of 200 ml. of 70% methanol, and with three 200 ml. portions of saturated magnesium carbonate solution. All these operations were carried out in a dimly lighted coldroom at 4 to 6° C.

The petroleum ether, after being treated with a small quantity of anhydrous sodium sulfate was decanted onto a column of confectioner's sugar (powdered sucrose containing 3% starch), 45×100 mm. The adsorption columns used by the writers were prepared by pouring the dry sugar in small lots into a special glass tube and, after each addition, pressing the powder to a compact mass with a metal tamper. The diameter of the tamper should be only slightly smaller than the diameter of the glass tube (cf. Strain, 1942, pp. 41—42). The sodium sulfate was rinsed with petroleum ether and the rinsings poured onto the same adsorption column. The column was washed with about 100 ml. or petroleum ether, b. p. 50 to 70° C, until the washings were nearly colorless. This treatment removed substances that were strong eluents of the pigments. The green filtrate was poured onto a powdered sugar adsorption column, 65×390 mm., and the chromatogram developed with about 600 ml. of petroleum ether, b. p. 50 to 70° C.

The olive green chlorophyll b zone was dug out, packed into an elution cylinder, and eluted with about 175 ml. of acetone. If an appreciable amount of chlorophyll b was held in the first column this was eluted with acetone and added to the main bulk of the chlorophyll b eluate.

The blue chlorophyll a zone was dug out and eluted with acetone in the same manner as the chlorophyll b.

Chlorophyll b. To the chlorophyll b eluate was added 20 ml. of petroleum ether, b. p. 50 to 70° C. This mixture was washed with saturated magnesium carbonate solution, first with 50 ml., and then three times with 200 ml. each. The free water was removed with anhydrous sodium sulfate and the petroleum ether solution, including the petroleum ether washings of the sodium sulfate, was put onto a powdered sugar column 65×200 mm. The column was washed with 300 ml. of petroleum ether, b. p. 50 to 70° C.

The chlorophyll b and chlorophyll a zones were dug out separately, each as free from other pigments as possible, and eluted with acetone. The chlorophyll b eluate was worked up to recover pure chlorophyll b, and the chlorophyll a eluate was added to the main bulk of the chlorophyll a solution.

The chlorophyll b eluate was concentrated to about 10 ml. under reduced pressure, the remaining acetone expelled by treating with 25 ml. of petroleum ether, b. p. 50 to 70° C, and evaporating the liquid under reduced pressure until the volume of the solution was about 10 ml. To this concentrate was added 50 ml. of petroleum ether, b. p. 50 to 70° C, and the solution poured onto an adsorption column of powdered sugar 65×150 mm. The column was washed successively with 50 ml. of petroleum ether, 200 ml. of 25% benzene in petroleum ether, 100 ml. of 30% benzene, and 300 ml. of 40% benzene (Reagent grade). This washing procedure with benzene-petroleum ether may be varied to get the most effective separations. In some cases 40% benzene may be so high a concentration as to elute the pigments, in which case a lower concentration should be used. This treatment washes out chlorophyll a, pheophytin, and yellow pigments.

The chlorophyll b zone was dug out and eluted with acetone. The acetone eluate was concentrated to about 10 ml. and the remaining acetone flushed out by adding 25 ml. of petroleum ether, and concentrating under reduced pressure. To the 10 ml. remaining was added 50 ml. of petroleum ether and the adsorption resumed on a column of powdered sugar, 65 × 150 mm. The chromatogram was developed by washing with 300 ml. of 5% acetone in petroleum ether. The yellow zone, above the chlorophyll *b* zone, was discarded and the chlorophyll *b* zone dug out and eluted with acetone in the usual manner. Again the eluate was concentrated to 10 ml. and the acetone flushed out with petroleum ether. The concentrated solution was taken up in 50 ml. of petroleum ether and put onto a column of powdered sugar, 65 × 150 mm. The chromatogram was developed with 300 ml. of 20% ether in petroleum ether, b. p. 40 to 50° C. The column was then washed with 200 ml. of petroleum ether, b. p. < 40° C, the column sucked dry, and the chlorophyll *b* zone dug out and eluted with ether. The ether eluate was diluted to 50 ml. and this solution used for the analytical determinations.

Chlorophyll a. The combined chlorophyll *a* eluates were concentrated to about 10 ml. under reduced pressure at a temperature of about 30° C. The remainder of the solution was treated with 20 ml. of petroleum ether, b. p. 50 to 70° C, and concentrated again to about 10 ml. To the concentrate was added 5 ml. of benzene and 50 ml. of petroleum ether and the solution poured onto a sugar column 65 × 270 mm. A precipitate may be formed at this point if the concentration is high and if too long a time intervenes between dissolving the pigment and putting it onto the column. The column was washed with 50 ml. of petroleum ether and the chromatogram developed with petroleum ether containing various percentages of benzene (Reagent grade): 100 ml. of 25% benzene, 200 ml. of 30%, and 100 ml. of 40% benzene. The column was sucked dry. The pigment zones above the blue chlorophyll *a* zone were dug out and discarded and the chlorophyll *a* zone recovered free of other zones and eluted with redistilled acetono, about 100 to 150 ml. The acetone solution was concentrated to 10 ml. and swept free of acetone as already described for chlorophyll *b*. This concentrate after adding 5 ml. of washed and redistilled ether, and 50 ml. of petroleum ether, b. p. 50 to 70° C, was poured onto a sugar column 65 × 250 mm. The chromatogram was developed by washing successively with 50 ml. of petroleum ether, 200 ml. of 10% ether in petroleum ether (b. p. 50 to 70° C), 100 ml. of 10% ether in petroleum ether (b. p. < 40° C), and finally with 100 ml. of petroleum ether (b. p. < 40° C). The column was sucked dry and the chlorophyll *a* zone dug out free of other zones and the pigment eluted with ether. The ether solution was diluted to 100 ml. and used for magnesium and spectroscopic determinations.

III. Physical Properties of Chlorophylls *a* and *b*.

Chlorophyll *a*.

Solubility. Chlorophyll *a* is soluble in alcohols, ether, benzene, and acetone. When it is pure it is only slightly soluble in petroleum ether. It is insoluble in water.

Spectral Absorption Properties of Chlorophyll a. The qualitative absorption curves of chlorophyll *a* in several solvents have been obtained by HARRIS and ZSCHEILE (1943). The quantitative absorption curves of chlorophyll *a* have been measured by WINTERSTEIN and STEIN (1933) in benzene, by MACKINNEY (1940) in acetone, by ZSCHEILE and COMAR (1941), by COMAR and ZSCHEILE (1942), and by the writers in ether. MACKINNEY's values were measured by use of pure dried chlorophyll *a*. ZSCHEILE and COMAR's and the writers' measurements were made with chlorophyll that had never been dried. Since the specific absorption coefficients of chlorophyll which has been isolated and dried may differ from samples

10*

which have not been so treated (Koski and Smith, 1948; Zscheile and Comar, 1941; Zscheile, Comar and Harris, 1944) it is preferable for analytical purposes to use a standard which has not been isolated in the solid state.

The quantitative absorption curve of chlorophyll a in ether as determined by the writers (cf. p. 186) is shown in Fig. 1. The standard absorption curve is perhaps the most useful analytical property of a chlorophyll. Its characteristic shape is valuable for qualitative identification and its absorption coefficients for quantitative estimation (cf. pp. 159, 188).

The absorption maxima and the corresponding absorption coefficients of chlorophyll a obtained by various workers are listed below. Small differences exist both in the wave-length positions of the maxima and in the absorption coefficients obtained by different workers using the same solvents. This emphasizes the fact that the spectroscopic constants should be determined in each laboratory under the conditions used for analysis.

Fluorescence Spectra of Chlorophyll a. Fluorescence spectrum curves of chlorophyll a have been published by Zscheile and Harris (1943) and by French (1954). Tables of fluorescence maxima in a number of organic solvents have been given by Zscheile and Harris (1943) and by Biermacher (1936). The positions of the fluorescence maxima of chlorophyll a in ether solution as reported by

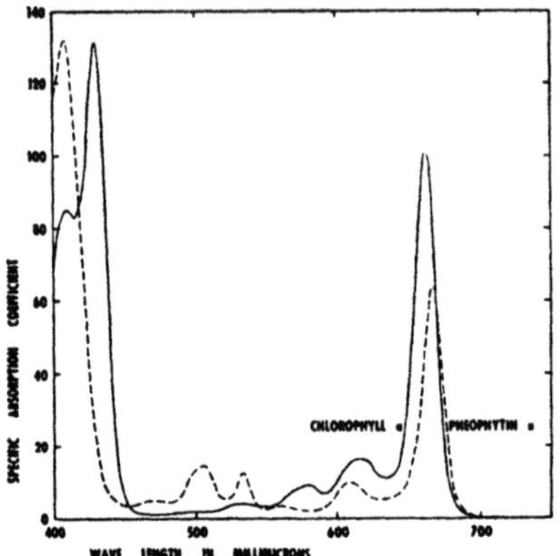

Fig. 1. Absorption spectra of chlorophyll a and pheophytin a in ether. The specific absorption coefficient, α, wherever used in this chapter is defined as

$$\alpha = \frac{1}{dC} \log \frac{I_0}{I} = \frac{D}{dC}$$

where d is the inside length of the absorption cell in centimeters, C is the concentration of the pigment in grams per liter, I_0 and I are the light intensities transmitted by pure solvent and solution, log is $\log_{10}$, and D is the optical density.

various authors are as follows: 664.5 and 720 mμ (Zscheile and Harris, 1943); 666 and 730 mμ (Livingston, Watson and McArdle, 1949); and 668 and 723 mμ (French, 1954).

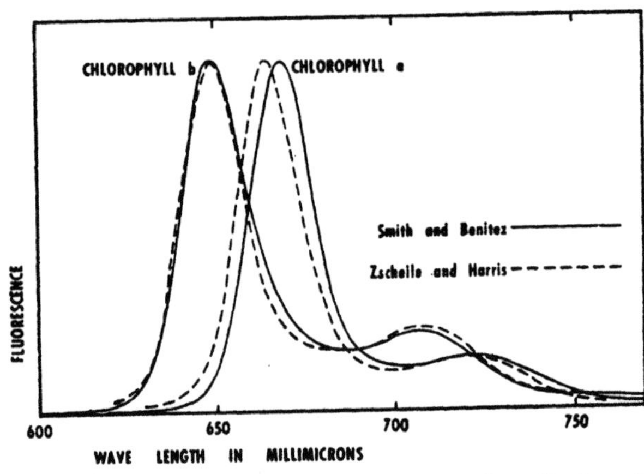

Fig. 2. Fluorescence spectra of chlorophylls a and b in ether. Comparison of values obtained by Zscheile and Harris (1943) and those obtained by French (1954) with material supplied by Smith and Benitez.

A comparison of the fluorescence curves obtained by Zscheile and Harris and in our own laboratory has been made by French and is reproduced in Fig. 2.

Absorption Maxima and Specific Absorption Coefficients: Chlorophyll a.

	Solvent: Ether					
Zscheile and Comar (1941)	660 102.1	614 15.5	576 8.51	532 4.14	429 135	410 mμ 85.2 l./gm. cm.
Smith and Benitez	662 100.9	615 16.3	578 9.27	533.5 4.22	430 131.5	410 mμ 85.2 l./gm. cm.
	Solvent: Acetone					
Mackinney (1940)	663 84.0	615 15.7	580 8.67	535 3.88	430 106.0	410 mμ 77.1 l./gm. cm.

These curves show the form of fluorescence-spectrum curves usually obtained with chlorophylls and closely related pigments.

Chlorophyll b.

Solubility. Chlorophyll *b* is soluble in alcohols, ether, acetone, and benzene. When pure, it is almost insoluble in petroleum ether. It is insoluble in water.

Spectral Absorption Properties of Chlorophyll b. The qualitative absorption spectrum curves of chlorophyll *b* in various solvents have been measured by Harris and Zscheile (1943). The quantitative absorption curves of chlorophyll *b* in benzene have been published by Winterstein and Stein (1933), in pure acetone by Mackinney (1940), and in ether by Zscheile and Comar (1941). The writers have determined the quantitative absorption spectrum curve of chlorophyll *b* in ether by the method described on p. 186 and plotted the results in Fig. 3.

The absorption maxima and the corresponding specific absorption coefficients of chlorophyll *b* in different solvents obtained by different workers are tabulated below:

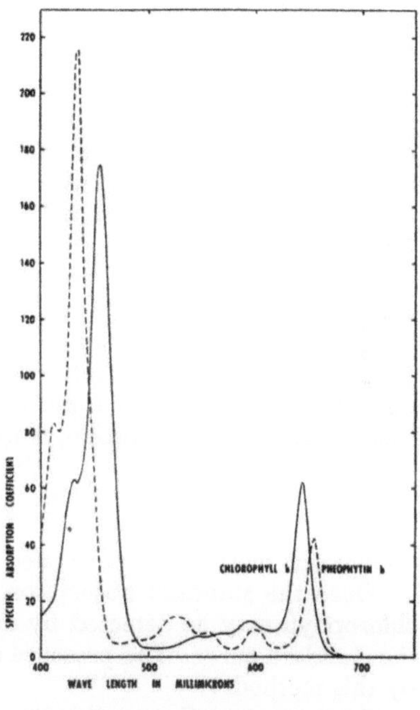

Fig. 3. The absorption spectra of chlorophyll *b* and pheophytin *b* in ether.

Absorption Maxima and Specific Absorption Coefficients: Chlorophyll b.

	Solvent: Ether				
Zscheile and Comar (1941)	642.5 56.8	594 10.9	550 6.20	453 171	430 mμ 58.5 l./gm. cm.
Smith and Benitez	644 62.0	595 12.7	549 7.07	455 174.8	430 mμ 62.7 l./gm. cm.
	Solvent: Acetone				
Mackinney (1940)	645 51.8	595 11.3	———— ————	455 146.9	———— mμ ———— l./gm. cm.

Fluorescence Spectra of Chlorophyll b. Fluorescence-spectrum curves of chlorophyll b have been published by Zscheile and Harris (1943) and by French (1954). A table of fluorescence maxima for chlorophyll b in various solvents has been published by Biermacher (1936). The positions of the fluorescence maxima in ether obtained from spectrophotometric fluorescence curves as reported by different investigators are 648.5 and 708 mμ (Zscheile and Harris, 1943); 649 and 708 mμ (French, 1954).

A comparison of the fluorescence curves obtained by Zscheile and Harris and by French is given in Fig. 2.

IV. Qualitative Test for Chlorophylls *a* and *b*.

Treatment of the chlorophylls a and b in ether solution with hydrochloric acid forms the pheophytins a and b which have characteristic absorption spectra (Fig. 1 and 3) and hydrochloric acid numbers (see pp. 152 and 154).

One of the most characteristic tests of the chlorophylls is the phase test described on p. 192. A mixture of chlorophylls a and b forms a brown ring at the interface between the ether and methanolic potassium hydroxide layers. On shaking the mixture, the color goes into the alkaline layer and leaves the ether layer colorless. After the solution has stood for a short time, the brown color gives way to green.

Chlorophyll a gives a transient yellow color which changes to green and chlorophyll b gives first a red coloration, then brown and finally green.

The fluorescence spectra of the individual chlorophylls are characteristic and can be used for their qualitative identification. This results from the fact that the fluorescence spectrum of a pure compound is independent of the exciting wave length and, in dilute solution, is independent of other solutes.

V. Tests for Purity.

Various methods may be used for testing the purity of chlorophylls a and b:

Once the standard absorption curve has been established, impurities in the chlorophyll may be detected by deviations of a measured absorption curve from the standard curve. The presence of one chlorophyll in the other may be revealed by this method.

Zscheile and Comar (1941) judged pheophytin contamination by the ratios, R_a and R_b. These are the ratios of the absorption coefficients at the long wavelength absorption maxima of chlorophylls a and b to their corresponding absorption coefficients at 505 and 520 mμ, where pheophytins a and b, respectively, have conspicuous absorption maxima. The value given for R_a is 52.4, and for R_b, 18.9. A decrease in R indicates contamination by pheophytin.

It is possible, however, that isomeric forms of chlorophyll possess slightly different absorption spectra, and the small divergences from a given value of R may be due to isomerization rather than to contamination by pheophytin. This supposition has not been tested experimentally.

Spectroscopic examination of the fluorescence emitted by chlorophylls may provide a very sensitive test for the presence of fluorescent contaminants. It is especially useful for discovering the presence of traces of one chlorophyll in another.

Carotenoid contamination may be detected qualitatively by an increase in the ratio of absorption in the blue region of the spectrum to that in the red (Strain, 1938, p. 130). Carotenoids may also be detected by treating an ether solution of chlorophyll with alkali which saponifies and extracts the chlorophylls leaving the carotenoids in the ether to color it yellow.

Contamination by derivatives of chlorophyll lacking the phytyl group may be detected by partitioning the pigment between ether and 22% hydrochloric acid. Owing to the lower hydrochloric acid numbers (cf. p. 193) of such derivatives they pass into the acid phase and color it, whereas the chlorophylls do not. These acidic contaminants may also be demonstrated by extracting the ether solution of the chlorophyll with 0.01 N alkali, sodium bicarbonate, disodium phosphate, or dilute ammonium hydroxide, in which case they are extracted and the alkaline solution becomes colored (cf. FISCHER and STERN, 1943, p. 326).

Allomerization of the chlorophylls may be detected by means of the phase test (p. 192). Allomerized chlorophyll gives no transient color, or gives a dull transient color as contrasted to the vivid colors of the nonallomerized chlorophyll.

Chromatography offers one of the most sensitive tests for discovering impurities. Under certain conditions, these impurities form distinct zones. The conditions may be such, however, that the impurities do not separate and for this reason a variety of conditions may have to be tried to demonstrate contamination. (For sensitivity of this test cf. p. 190.)

VI. Pheophytins *a* and *b*.

1. Pheophytin *a*.

The structural and empirical formulas of pheophytin *a* are given here (FISCHER and STERN, 1943, p. 55). The calculated molecular weight is 871.18. The calculated elementary composition is

C, 75.82; H, 8.56; O, 9.18; N, 6.43.

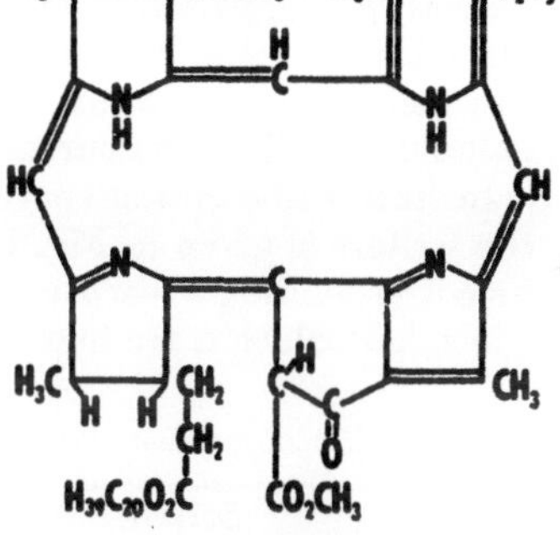

$C_{55}H_{74}O_5N_4$

PHEOPHYTIN *a*

Whether pheophytin *a* occurs naturally is not known. It almost always accompanies chlorophyll *a* in its preparation but its appearance is probably due to decomposition of chlorophyll *a*. It is easily prepared by treating chlorophyll *a* with acid. The method used by the writers to obtain an ether solution suitable for spectroscopic measurements is given in the next section.

Preparation of Pheophytin *a*. The individual pheophytins have been prepared in three ways: (a) from leaf powder directly without isolating the chlorophylls after which the pheophytin mixture was separated into its components (STOLL and WIEDEMANN, 1933a; WILLSTÄTTER and STOLL, 1913); (b) from the isolated chlorophyll mixture with subsequent separation of the pheophytin components by chromatography (Winterstein and Stein, 1933); or (c) by conversion of the chromatographically separated chlorophylls into their respective pheophytins (Zscheile and Comar, 1941).

In order to measure quantitatively the absorption spectra of the pheophytins, ether solutions of the separated chlorophylls of known concentrations and spectral properties were transformed quantitatively to ether solutions of the pheophytins.

An ether solution of chlorophyll a (50.0 ml. containing 0.3834 gm./liter as determined from magnesium analysis, p. 161) contained in a glass-stoppered vessel was cooled to about 5° C and mixed with 1 ml. of concentrated hydrochloric acid. After standing in the refrigerator over night the ether solution was washed with four 200 ml. portions of distilled water. The washings were extracted with fresh ether which was added to the main ether solution. The combined ether solutions were washed until the wash water had a pH > 5. This is essential because the absorption spectrum of pheophytin a varies with the pH (cf. Livingston et al., 1953).

If the ether was washed with sodium bicarbonate solution, it may become slightly colored by partially hydrolyzed pheophytin. In this case, the pigment was driven into fresh ether with a small excess of hydrochloric acid and the ether combined with the main body of ether and washed with water until the pH ≈ 5.

The ether solution was drawn into a 100 ml. glass-stoppered graduated cylinder and diluted to 100 ml. with ether. The specific absorption coefficients were calculated on the basis of the original chlorophyll concentration corrected for loss of magnesium.

Physical Properties of Pheophytin a.

Solubility. Pheophytin a is soluble in ether, acetone, benzene, and chloroform; slightly soluble in alcohol and petroleum ether; insoluble in water.

Spectral Absorption Properties of Pheophytin a. The absolute absorption curve of pheophytin a has been determined in ether by Zscheile and Comar (1941) and by Livingston et al. (1953) in methanol. Livingston et al. have determined the curves in acid, neutral, and alkaline media. The curves differ greatly under these different conditions. The quantitative absorption curve for pheophytin a in ether solution as determined by the writers is given in Fig. 1.

The wave lengths and the corresponding absorption coefficients of the absorption maxima of pheophytin in ether solution are here tabulated.

Absorption Maxima and Specific Absorption Coefficients: Pheophytin a.

	Solvent: Ether						
Zscheile and Comar (1941)	665	608	559	532	505	470	410 mμ
	59.0	8.8	3.2	11.3	13.5	4.5	126 l./gm. cm.
Smith and Benitez	667	609.5	560	534	505	471	408.5 mμ
	63.7	9.8	3.6	12.6	14.6	5.1	132 l./gm. cm.

Fluorescence Maxima of Pheophytin a. The fluorescence maxima of pheophytin a are in ether 672.5 and ca. 715 mμ, and in methanol containing 4% ether 672 and 720 mμ.

Qualitative Tests. The hydrochloric acid number of pheophytin a is ≈ 29 (Willstätter, 1915).

The phase test is positive (Fischer and Stern, 1943, p. 55; Willstätter and Stoll, 1913, p. 260).

When adsorbed on a powdered sugar adsorption column from petroleum ether, it forms a gray zone below the chlorophyll *a*. Dissolved in ether, pheophytin *a* produces a solution of almost neutral tint.

Pheophytin *a* is formed from chlorophyll *a* from seven to nine times faster than pheophytin *b* is formed from chlorophyll *b* (MACKINNEY and JOSLYN, 1940).

Methyl pheophorbide *a* was prepared by STOLL and WIEDEMANN (1933a) by treating a methyl alcoholic solution of pheophytin *a* with gaseous hydrochloric acid in the cold for about an hour. The pigments were transferred to ether and the alcohol and acid removed by washing with water. The methyl pheophorbide *a* was extracted completely with 22% hydrochloric acid. After retransferring to ether, it was crystallized from ether at low temperature. The phase test is positive. The hydrochloric acid number is ca. 16 (WILLSTÄTTER, 1915).

2. Pheophytin *b*.

The structural and empirical formulas of pheophytin *b* are given below (FISCHER and STERN, 1943, p. 244). The calculated molecular weight is 885.16. The calculated elementary composition is

C, 74.62; H, 8.20; O, 10.85; N, 6.33.

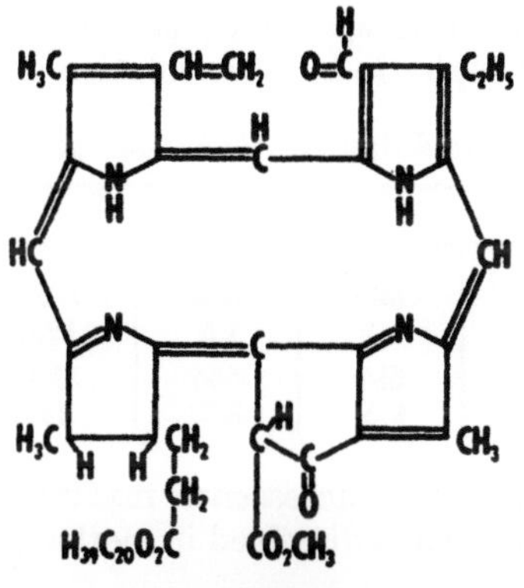

Pheophytin *b* is not known to occur naturally. Although it appears in extracts of leaves, it is assumed to be formed as a decomposition product of chlorophyll *b* during the extraction. The method used to prepare an ether solution of pheophytin *b* suitable for spectrophotometric examination is described in the next section.

Preparation of Pheophytin *b*. Various procedures for the preparation of pheophytin *b* have been cited already on p. 151. The procedure used by the writers for obtaining ether solutions suitable for spectroscopic measurements is given here in detail. An ether solution of chlorophyll *b* (20.0 ml.), which contained 0.1455 gm./l., was cooled to about 6° C and mixed with 1 ml. of concentrated hydrochloric acid solution in a glass-stoppered cylinder. This mixture, after being in the refrigerator overnight, was washed with 100 ml. portions of distilled water until the wash water had a pH $\approx$ 5. Actually four washings were made. Each washing was extracted with 10 ml. of ether which was added to the main volume of ether solution. After removal of the free water the ether extract was run into a 50 ml. graduated glass-stoppered cylinder and diluted to 40 ml. with ether. This solution was used for measuring the absorption spectrum. After the original weight of chlorophyll b had been reduced to the weight of pheophytin b the specific absorption coefficients were calculated. These are plotted against wave length in Fig. 3. The wave lengths of the absorption maxima and the corresponding specific absorption coefficients are given below. Although the pheophytin b prepared in this manner showed more than one zone when examined chromatographically, the main adsorption zone possessed essentially the same spectral absorption characteristics as the original solution. The minor zones were probably pheophorbides with absorption spectra nearly identical to pheophytin b. The absorption curve of the pheophytin obtained by the transformation of the chlorophyll is probably more reliable for analytical purposes than the absorption curve determined on the chromatographically purified pheophytin.

Physical Properties of Pheophytin b.

Solubility. Pheophytin b is soluble in ether, acetone, chloroform, and benzene; slightly soluble in methanol and petroleum ether; and insoluble in water.

Spectral Absorption Properties of Pheophytin b. The quantitative absorption curve of pheophytin b in ether has been measured by Zscheile and Comar (1941). Their values are in general about 10% lower than the values obtained by the writers which are recorded in Fig. 3.

The wave lengths and corresponding specific absorption coefficients at the absorption maxima of pheophytin in ether solution are here tabulated.

Absorption Maxima and Specific Absorption Coefficients: Pheophytin b.

	Solvent: Ether					
Zscheile and Comar (1941)	653	599	558	523	433	413 mμ
	37.0	8.1	7.5	12.6	197	77.0 l./gm. cm.
Smith and Benitez	655	599	555	525.5	434	412.5 mμ
	42.1	9.5	8.7	14.2	216	83.0 l./gm. cm.

Fluorescence Maxima. The fluorescence maxima of pheophytin b dissolved in ether lie at 657, 707 mμ; when dissolved in methanol containing 4% ether they are at 657.5, 707 mμ.

Qualitative Tests. The hydrochloric acid number of pheophytin b is ca. 35 (Willstätter, 1915).

The phase test is positive. At first a red coloration is produced, then brown. and finally green. The ether becomes colorless.

Pheophytin b when adsorbed on a column of confectioner's sugar from petroleum ether gives a brownish yellow zone. It is less strongly adsorbed than chlorophyll b and more strongly adsorbed than pheophytin a.

Dissolved in ether, it produces a brown solution.

Methylpheophorbide b was prepared by Stoll and Wiedemann (1933b) according to the method already described on p. 153. The pheophytin b required about three times more methyl alcoholic hydrochloric acid to dissolve it than did pheophytin a. The methyl pheophorbide gives a red phase test and possesses a hydrochloric acid number of ca. 21.

VII. Quantitative Determination of Chlorophylls a and b.

The estimation of chlorophyll in plant material depends fundamentally on two operations: the complete extraction of the pigment in known form from the plant material; and a reliable measurement of the pigment content in the extract.

There have been numerous methods proposed for the determination of chlorophyll each adapted to a particular purpose and to the equipment available. The methods, however, are based chiefly on four procedures — colorimetry, spectrophotometry, fluorometry, and the estimation of magnesium under such conditions that chlorophyll is the only magnesium-containing compound present. In some methods the chlorophylls are not determined as such but are converted quantitatively to derivatives which are estimated by one of the above-mentioned techniques.

If the content or ratio of the individual components in a mixture, e. g., chlorophylls *a* and *b*, is desired, either the components or their derivatives must be separated and determined independently, or each pigment must be measured by some distinctive property which is independent of the other components in the mixture.

Examples of the first type of method are (a) the separation of the chlorophylls *a* and *b* by chromatographic techniques followed by their estimation either colorimetrically or spectrophotometrically; and (b) the conversion of a mixture of chlorophylls *a* and *b* to phytochlorin *e* and phytorhodin *g* which are then separated by distribution between ether and hydrochloric acid of different concentrations and subsequently determined either colorimetrically or spectrophotometrically.

Examples of the second type are spectrophotometric or spectrofluorometrie determinations of the individual components of a mixture. Illustrations of each of these procedures will be given.

The extraction of the chlorophylls is usually much easier from fresh leaves than from dried leaves. If the analysis is to be put on a dry-weight basis, it is more desirable to determine the pigment content and the dry weight on separate comparable samples of fresh leaves and compute the desired relation.

Such diverse organisms as higher plants and green, brown, and red algae may require different methods of extraction.

1. Colorimetric Methods.

Determination of Chlorophyll (*a* + *b*).

As Chlorophylls. The method here described is that of PETERING, WOLMAN, and HIBBARD (1940). A 1 or 2 gram sample of fresh tissue was ground for about 30 seconds with quartz sand and a little sodium carbonate. The sample was moistened with acetone and ground further; then 25 ml. of pure acetone was added and the grinding continued. The extract was decanted and filtered by suction. The residue was extracted again with 25 or 30 ml. of 85% acetone and the mixture transferred to the filter. The extract was filtered and the residue washed with acetone until free of pigment.

The filtrates were combined in a 100 ml. volumetric flask and diluted to the mark. The transmission of a portion of this solution was measured in a colorimeter and corrected for the transmission of pure solvent. By use of this corrected transmission the chlorophyll concentration of the extract was read directly from a curve relating per cent transmission to chlorophyll content. This curve was constructed from measurements made with known concentrations of pure chlorophyll.

The colorimeter cells used for transmission measurements had 1 cm. light path. In order to eliminate the effect of absorption by yellow pigments the colorimeter (Cenco Photelometer) was fitted with a Corning No. 243 H. R. signal red filter 4.35 mm. thick and a No. 396 H. R. Aklo polished glass filter, light shade, 2.50 mm. thick. The precision of measurement was about 3 per cent.

As Derivatives of Chlorophylls. The method of Milner, French, Koenig, and Lawrence (1950) was developed only for isolated chloroplasts, but it appears to have possibilities for suspensions of other chlorophyll-containing material. A sample of from 0.1 to 2.0 ml. of chloroplast suspension was pipetted into a 6 × 125 mm. Pyrex test tube and the volume brought to 5 ml. with 5% methanolic potassium hydroxide. The suspension was thoroughly mixed and heated for three minutes in a water bath maintained at 63° C. After centrifugation, the clear green supernatant was decanted into a colorimeter tube and the transmission of the solution measured by means of a Klett-Summerson photoelectric colorimeter. So as to minimize interference by yellow pigments, red filter No. 66 was inserted in the light path. The chlorophyll content was determined from a standard curve of transmission versus chlorophyll content. This curve was prepared by treating known quantities of chlorophyll by the same procedure. The known quantities of chlorophyll were determined spectrophotometrically, using Comar and Zscheile's absorption constants.

The classical method of Willstätter and Stoll (1913, 1918) whereby the total chlorophyll was converted to potassium chlorophyllin and measured colorimetrically has been adequately described and will not be detailed here (cf. also Schertz, 1928a, b).

Determination of Chlorophylls *a* and *b*.

As Chlorophyll a and Chlorophyll b. The separation of chlorophyll components by chromatography and determination of each component either with a colorimeter or spectrophotometer has been used by Winterstein and Stein (1933), by Strott (1938), and by Seybold and Egle (1938/39a). The method of Winterstein and Stein is given here.

One or two leaves were wrapped in a wire and plunged into liquid nitrogen. The frozen leaves were ground in a 12 cm. mortar and extracted with 25 ml. of a mixture of benzine and benzene (9:1) with addition of 8 ml. of methanol. The solvent was filtered by suction and the residue washed with an equal volume of solvent. If properly done, the leaf residue was white. By careful washing with water in a separatory funnel the alcohol was removed. The pigment solutions were passed through a folded filter to remove water and then through a powdered sugar adsorption column 1 × 10 cm. The chromatogram was developed with benzine-benzene mixtures (from 4:1 to 19:1) of the proper concentration to separate the pigments. The yellow pigments were washed through the column and could be collected and quantitatively determined if desired. The two chlorophyll bands remained on the column. The column was washed with low-boiling petroleum ether and sucked dry in an atmosphere of CO_2. The chlorophyll bands were eluted separately with ether containing methanol. The methanol was washed out of the ether which was then diluted to 15 ml. with ether and the pigment content determined photometrically. A correction for the possible contamination of chlorophyll *b* by chlorophyll *a* was made from spectroscopic observations.

As Pheophytin a and Pheophytin b. Inasmuch as pheophytins are common contaminants of chlorophylls in crude extracts, especially in the extracts of acid leaves, Seybold and Egle (1938/39a) have proposed a method whereby the chlorophylls are converted quantitatively to the pheophytins by addition of oxalic acid, and the pheophytins *a* and *b* separated by chromatography on a sugar column. The two zones are removed from the column and the pigments eluted separately and determined photometrically. The values so obtained are then converted to chlorophyll equivalents.

Chlorophyll has also been determined as pheophytin by JACOBSON (1912) and by TAKASHIMA (1952). TAKASHIMA transferred the pheophytin to benzene and determined it spectrophotometrically.

Conversion of the chlorophyll to pheophytin for estimation assumes that a differential analysis of pheophytin and chlorophyll is not desired.

As Methylpheophorbide a and Methylpheophorbide b. In view of the instability of colorimetric standards made from chlorophyll or its derivatives, GUTHRIE (1928, 1929) proposed the use of substitute standards prepared from inorganic salts. Although analyses made with these substitute standards may not be as accurate as with authentic standards, they may adequately serve for many purposes. GUTHRIE's method follows: 30 gm. of leaves were placed in an 800 ml. KJELDAHL flask with 200 ml. of 0.6 N hydrochloric acid in methanol. An internal reflux condenser was hung in the neck of the flask and the mixture refluxed on the steam bath for one hour. The solution was decanted into a beaker and 100 ml. more of alcoholic acid added and refluxed for ten minutes. This was decanted as before and the procedure was repeated twice more. The collected extracts were cooled and filtered through a rapid filter. The extract was mixed with 350 ml. of ether in a 4 liter separatory funnel. Water, 1.5 liters, was added slowly. The ether containing the pigment was separated and the aqueous layer extracted once with 75 ml. of ether. The brownish water layer was discarded.

The ether solutions were combined in a 500 ml. separatory funnel and extracted three times with 25 ml. portions of hydrochloric acid (sp. g. 1.125) and twice with 15 ml. portions of hydrochloric acid (sp. g. 1.19). The extracts were run into a 500 ml. separatory funnel, 150 ml. of ether added, and water added with occasional vigorous shaking until the pigments had been transferred to ether.

The pigments were then fractionated. The ether was extracted with 50 ml. of 5.05 N hydrochloric acid. This extract was run into another separatory funnel and extracted with 20 ml. of ether. The acid was run into a 500 ml. volumetric flask. The ether was added to the main bulk of ether solution and the process repeated until the acid solution nearly reached the mark. The volume was brought to 500 ml. with the same acid and thoroughly mixed. This acid solution contained the methyl pheophorbide *a*. This solution was filtered through a rapid filter and compared with the standard for the *a* fraction.

Now the ether was extracted with 100 ml. portions of 6.75 N hydrochloric acid. These extracts were collected in a 500 ml. volumetric flask diluted to the mark with the same acid, mixed thoroughly, filtered, and compared with the standard for the *b* derivative.

The substitute standards were prepared in the following way and standardized by comparing against known concentrations of methyl pheophorbides *a* and *b* prepared from the respective chlorophylls by the method just described.

The stock solutions from which the standards were prepared consisted of the following solutions:

$CuSO_4 \cdot 5\,H_2O$	10 gm./l.
$K_2Cr_2O_7$	2 gm./l.
NH_4OH	2 N.

A standard was made by measuring the indicated volumes of copper sulfate and of potassium dichromate into a 100 ml. volumetric flask, adding 10 ml. of 2 N ammonium hydroxide and diluting to the mark. The substitutes when prepared as shown in the table corresponded to the designated equivalents of chlorophylls *a* or *b* when the two solutions were set at the same depth, whether 20, 30, or 40 mm. The color of the unknown solution was then matched with the substitute standard nearest to its own hue by means of a visual colorimeter (Bausch and Lomb was used) and the concentration of the unknown determined by the usual methods of colorimetry.

As Phytochlorin e and Phytorhodin g. The method of Willstätter and Stoll (1913, 1918) whereby chlorophylls *a* and *b*, or their pheophytins, are degraded by treatment with alkali and the resulting products separated as phytochlorin *e* and phytorhodin *g* by distribution between ether and hydrochloric acid of different concentrations has been adequately described previously (Treibs, 1932; Willstätter and Stoll, 1913, 1918).

| | Milliliters of | | Equivalent to mg in 500 ml | Chlorophyll |
	Copper Sulfate	Potassium Dichromate		
No. 1	40	23	54.1	*a*
	19	38	18.6	*b*
No. 2	25.5	20	41.6	*a*
	14.5	30	13.3	*b*
No. 3	15.5	14	27.0	*a*
	10.5	20	8.8	*b*

2. Spectrophotometric Methods.

Undoubtedly the most convenient, and probably the most accurate, method for the determination of the two chlorophylls in the presence of each other is the spectrophotometric method. Comar and Zscheile (1942) and Comar (1942) have used the following method for the analysis of chlorophylls *a* and *b*.

Fresh leaf tissue, from 2 to 10 gm., was accurately weighed and placed in a Waring Blendor containing acetone and a small amount of calcium or magnesium carbonate. The blendor was run about 15 minutes, which was usually long enough to effect complete extraction of the pigment. The solution was filtered through paper by suction and the residue washed with a little ether to collect the last trace of pigment. The filtrate was transferred to a 500 ml. volumetric flask and diluted to the mark. From this solution a 25 or 50 ml. portion was pipetted into a separatory funnel containing 50 ml. of ether. By cautiously adding distilled water the pigments were transferred to the ether and the ether solution freed of acetone by exhaustive washing with distilled water according to the method of Mackinney (p. 145). By repeated washings (from 15 to 20) the ether was completely freed of acetone. The chlorophyll solution was diluted to 100 ml with ether and dried with anhydrous sodium sulfate. When the solution was optically clear, its absorption was measured spectrophotometrically.

Comar and Zscheile used optical density measurements at wave lengths 6600 and 6425 Å to estimate the individual concentrations of chlorophylls *a* and *b*. From this measurement the total chlorophyll content as well as the ratio of the two chlorophylls could be determined. According to Comar and Zscheile (1942) the specific absorption coefficients are for chlorophyll *a* 102 at 6600 Å, and 16.3 at 6425 Å, and for chlorophyll *b* 4.50 at 6600 Å, and 57.5 at 6425 Å. At 6000 Å both chlorophylls have the same absorption coefficient, 9.95. If total chlorophyll is desired, the optical density can be measured at 6000 Å and the concentration of chlorophyll computed from the equation

$$C = \frac{1}{\alpha\, d} \log \frac{I_0}{I}$$

If *V* is the volume of solution in litres from which the sample was withdrawn for measurement, then the weight of chlorophyll, *W*, is

$$W = VC.$$

For calculation of concentrations of chlorophylls *a* and *b* and total chlorophyll the equations derived from these constants are:

$$\text{Chlorophyll } a \text{ (mg./l.)} = 9.93 \log\left[\frac{I_0}{I}\right]_{6600} - 0.777 \log\left[\frac{I_0}{I}\right]_{6425}$$

$$\text{Chlorophyll } b \text{ (mg./l.)} = 17.6 \log\left[\frac{I_0}{I}\right]_{6425} - 2.81 \log\left[\frac{I_0}{I}\right]_{6600}$$

$$\text{Total chlorophyll} \quad \text{(mg./l.)} = 7.12 \log\left[\frac{I_0}{I}\right]_{6600} + 16.8 \log\left[\frac{I_0}{I}\right]_{6425}$$

As stated on p. 142, it is essential for the accurate determination of the chlorophylls by spectrophotometric methods that the wave-length positions and the absorption coefficients used are checked under the conditions of analysis employed and the appropriate coefficients derived for the equations (cf. p. 186).

The writers have found from their measurements of the specific absorption coefficients in ether that the following equations are applicable to the determination of chlorophylls *a* and *b*.

$$D_{662} = 100.9\, C_a + 6.2\, C_b$$

$$D_{644} = 15.8\, C_a + 62.0\, C_b$$

From which the concentrations of chlorophyll *a* (C_a) and chlorophyll *b* (C_b) can be obtained from the equations:

$$C_a \text{ (gm./l.)} = 0.0101\, D_{662} - 0.00101\, D_{664}$$

$$C_b \text{ (gm./l.)} = 0.0164\, D_{644} - 0.00257\, D_{662}$$

If decomposition products of the chlorophylls are present in the solutions to be analyzed, these products may vitiate the use of spectrophotometry to obtain accurate values for the concentration of the chlorophylls. In this case methods must be resorted to for the determination of the isolated chlorophylls such as the chromatographic separation described on p. 156. Alternatively, additional spectroscopic constants must be determined and used.

MACKINNEY (1941) has developed a method whereby the extracted chlorophyll is not transferred to ether but is measured directly in aqueous acetone. Fresh leaves (5 to 10 grams) were weighed out and crushed in a solvent mixture composed of 20 volumes of water and 80 volumes of acetone. After decanting the first extract the leaves were extracted a second time with the same solvent. The extracts were brought to a known volume and portions diluted 1 to 5 and 1 to 25 for spectroscopic examination. In this way the solvent composition is made to differ very little from that used in the determination for the standard solutions. From the spectroscopic constants obtained with pure samples of the chlorophylls the following equations were set up for measuring chlorophyll concentrations:

$$\log\left[\frac{I_0}{I}\right]_{6630} = 82.04\, C_a + 9.27\, C_b$$

$$\log\left[\frac{I_0}{I}\right]_{6450} = 16.75\, C_a + 45.6\, C_b$$

These simultaneous equations can be solved for the concentrations, C_a and C_b, of chlorophylls *a* and *b*, expressed in gm./l.

MACKINNEY has given values for the specific absorption coefficients of chlorophylls *a* and *b* in the 80% acetone at a number of wave lengths.

3. Fluorometric Methods.

In dilute solution, the fluorescence intensity can be used for the quantitative determination of a fluorescing substance. This is based on the fact that the fluorescence intensity is proportional to the quantity of light absorbed which in turn is nearly proportional to the concentration of the absorbing substance.

In contrast to absorption spectrophotometry, fluorometry can be used even in cloudy solutions (GOODWIN, 1947).

Fluorometry provides a very sensitive test for the presence of these pigments: as little as 1 microgram per liter of chlorophyll can be determined (GOODWIN, 1947). Because of its great sensitivity, fluorescence can be used to detect small quantities of impurity in so-called pure substances. A published example exists in ZSCHEILE's (1935) first publication of the fluorescence curves of chlorophylls a and b. The fluorescence of chlorophyll b showed the typical fluorescence band of chlorophyll a, and BIERMACHER (1936) pointed out that this was due to contamination by chlorophyll a.

Quantitative Fluorometry of Chlorophylls a and b. The quantitative estimation of chlorophyll by fluorometry has been carried out by KAVANAGH (1941) and by GOODWIN (1947). The fluorescence analysis was made with a photoelectric KLETT fluorometer-colorimeter (cf. KAVANAGH. 1941). This instrument was arranged so that light from a single H-4 mercury lamp filtered by appropriate filters to isolate the desired mercury lines illuminated simultaneously a fluorescing standard (example, quinine in sulfuric acid) and the fluorescent compound whose concentration was to be measured. The fluorescent light from these substances was taken at right angles to the incident light and, after passing through suitable filters, was measured by barrier-layer photocells. The circuit for the two photocells was arranged so as to compensate for light fluctuations. The ratio of light intensities was measured by means of a potentiometer. A high sensitivity galvanometer was used as null-point indicator.

The quantitative relation between fluorescence, F, and concentration, c, of fluorescing substance is shown in the equation

$$F = KI\,(1{-}10^{-\alpha\,dc}) \approx KI\,\alpha\,dc$$

where K is a proportionality factor, I is the incident intensity of the exciting light, α is the absorption coefficient, d is the length of light path. At very low concentrations the expression in parenthesis approaches $\alpha\,dc$ and the fluorescence becomes proportional to c. This has been verified for chlorophyll (GOODWIN, 1947; KAVANAGH, 1941).

GOODWIN gives essentially the following method for fluorometric determination of chlorophyll in crude extracts: fresh tissue was accurately weighed, dipped for 30 seconds in boiling water and disintegrated in acetone by means of a special homogenizer. The solid plant material was centrifuged out and the clear supernatant decanted and diluted to 20 ml. The fluorescence of the extract, diluted to the proper concentration, was measured with exciting wave length 4047 Å and the concentration of chlorophyll determined from a calibration curve prepared with known chlorophyll concentrations. From the concentration and total dilution, the quantity of chlorophyll was calculated.

In dilute solutions of mixtures of chlorophylls a and b the two components fluoresce independently. The ratio of the fluorescences varies with the wave length of the exciting light (cf. p. 179). By measuring the ratio of the fluorescence with two exciting wave lengths at constant relative intensities and with constant weight of chlorophyll, it is possible to calculate the ratio of the two chlorophylls provided the fluorescence of each is known at the two wave lengths.

Another method for estimating individual components in a mixture is to use a fixed exciting wave length and determine the spectral fluorescence for chlorophyll mixtures (cf. FRENCH and YOUNG, 1952, and FRENCH, 1954). Since each chlorophyll has its own characteristic fluorescence curve and these curves are additive for mixtures, it is possible after proper calibration to estimate each chlorophyll independently.

It may be possible to determine chlorophyll directly in plant extracts by measuring the chlorophyll fluorescence initially present and after addition of known quantities of chlorophyll. From these measurements the relation between chlorophyll concentration and fluorescence intensity can be determined and the initial chlorophyll concentration computed (GOODWIN, 1947).

4. Magnesium Determination.

Gravimetric Magnesium Determination. DELEANO and DICK (1934) proposed the determination of chlorophyll under suitable conditions by means of magnesium analysis. They extracted the chlorophyll into chloroform in which inorganic magnesium salts were not soluble, then removed the chloroform and ashed the residue. The magnesium in the residue was determined as magnesium ammonium arsenate hexahydrate. From the magnesium determined the quantity of chlorophyll was calculated.

Colorimetric Magnesium Determination. A much more rapid and sensitive method for determining magnesium is the titan yellow colorimetric method. KOSKI and SMITH (1948) determined the chlorophyll concentration of ether solutions by estimating their magnesium content with titan yellow.

A known volume of the ether solution of the chromatographically purified pigments, enough to give 0.07 to 0.1 mg. of magnesium was transferred to a platinum dish and the ether carefully evaporated on a porcelain-covered water bath. To the residue was added 5 ml. of 0.3 N sulfuric acid. The acid was removed by heating with the flame of a Bunsen burner. The residue was completely ashed, 5 ml. of 0.3 N sulfuric acid added to the ash, and the sulfuric acid driven off by heating with a free flame.

The acid-treated ash was dissolved in 0.25 ml. of 1 N sulfuric acid and 3 ml. of water by heating on a water bath. The solution was quantitatively transferred to a 25 ml. glass-stoppered graduated cylinder and the platinum dish washed three times with 3 ml. portions of distilled water. The platinum dish was warmed after each addition of water and the transfers made with the aid of a rubber policeman. To the solution contained in the graduated cylinder was added successively 2 ml. of 1% starch solution (Merck's reagent soluble starch), 5 ml. of saturated calcium sulfate solution, and 1 ml. of 0.125% titan yellow solution. The volume was made up to 23 ml. by the addition of water and then 2 ml. of 2.5 N sodium hydroxide added. The solution was shaken vigorously for 5 minutes, and after standing for a further 5 minutes, its optical density was measured against a blank containing no magnesium. The optical density measurements were made by means of a Beckman spectrophotometer set at 540 mμ. The readings thus obtained were compared with the optical density readings determined on solutions of known magnesium content (cf. also LUDWIG and JOHNSON, 1942).

The blank and reference solutions of magnesium were prepared by pipetting 0, 2, 5, 10, and 12 ml. of a standard solution of magnesium sulfate containing 0.01 mg. of magnesium per ml. into 25 ml. glass-stoppered graduated cylinders. Then 0.25 ml. of 1 N sulfuric acid was added and the other reagents introduced in the order listed for the unknown. The solutions were then treated in the same way as the unknown and their optical densities determined. The blank was relatively stable and could be used as reference for some hours.

A calibration curve relating optical density to milligrams of magnesium was drawn (cf. Fig. 4) and from this curve the magnesium content of the unknown determined. Alternatively, the average ratio of magnesium content to optical density may be calculated and the magnesium content of the unknown reckoned from its optical density.

This method for the determination of magnesium has been refined so that quantities of the order of 5 micrograms of magnesium can be determined with an accuracy of $\pm$ 4 per cent (Granick, 1950).

5. Estimation of Chlorophyll in Dried Material.

It may be necessary, on occasion, to determine chlorophyll in dried material. The extraction of the material then may present certain difficulties. Willstätter and Stoll (1913, p. 73ff.; also Willstätter, 1915) showed that better extraction was obtained with 85 % aqueous acetone and 90 % ethanol than with the water-free solvents. However, pure methanol was better than aqueous methanol. They found that 85% acetone was the most suitable solvent. Milner (1948) has observed that lipids are slowly and usually incompletely extracted from dried *Chlorella* cells by anhydrous fat solvents. When the dried material is first treated with water and then dehydrated with methanol, the further extraction with fat solvents becomes quantitative and fairly rapid. Eyster (1950) reported that "Sodium nitrate decisively increases the amount of chloroplast pigments extracted from dry nettle

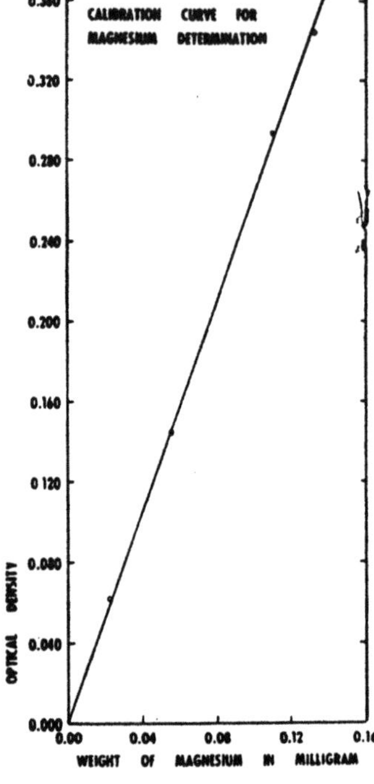

Fig. 4. Calibration curve for magnesium determination.

leaf powder in one hour by 80% acetone". Once the chlorophyll is extracted it can be estimated by any one of the methods already described.

B. Chlorophyll *c*.

The structural and empirical formulas of chlorophyll *c* are not known.

Chlorophyll *c* occurs naturally in the brown algae, diatoms, and dinoflagellates, and in symbiotic algae of sea-anemones (Strain, Manning and Hardin, 1943).

I. Preparation of Chlorophyll *c*.

Chlorophyll *c* has been prepared by Strain and Manning (1942), by Strain, Manning and Hardin (1943), and by Wassink and Kersten (1946). (For previous references, these articles should be consulted.)

The preparation of pure chlorophyll *c* is in a very unsatisfactory state. Strain and Manning (1942, p. 634) found that chlorophyll *c* obtained by partition and by adsorption possessed different absorption spectra. In the preparation of chlorophyll *c* from the brown alga, *Nereocystis luetkeana*, by chromatography, the writers have obtained several fractions which have absorption maxima at nearly the same wave lengths. The question arises as to which of these fractions is the true chlorophyll *c*.

The authors have prepared solutions of chlorophyll *c* for spectrophotometric measurements by the following chromatographic method:

The pigment was extracted from the blades of the brown alga *Nereocystis luetkeana* which was collected near Moss Beach, California. The blades were worked up in 1 kg. lots. Each lot was washed with fresh water, drained, chopped into pieces about 1 sq. cm. in area, and extracted successively with a 2 and a 1 liter portion of methanol each for 30 minutes. The two methanol extracts were then combined and treated with 200 ml. of petroleum ether which removed most of the chlorophyll *a*.

The alcoholic solutions from three such extractions were combined and 2 liter portions of the solution were each mixed with one liter of ether and then shaken with 3 liters of filtered saturated salt solution. The ether which separated was removed and the aqueous alcohol layer extracted again with 800 ml. of ether.

All of the first ether extracts were combined as were also the second extracts and each washed successively with two 1 liter portions of saturated salt solution and one liter of distilled water. The emulsion formed during each washing contained an interfacial precipitate which was rich in chlorophyll *c*. The emulsion was removed separately and dissolved in methanol. The methanol solution was mixed with ether and then shaken with saturated salt solution. This treatment transferred the chlorophyll *c* to ether.

All the ether extracts were combined and concentrated under reduced pressure. The ether solution was separated from a small quantity of water which deposited and added to four times its volume of petroleum ether. This solution was freed of water droplets with anhydrous sodium sulfate and poured over an adsorption column of confectioner's sugar 100 × 100 mm. The column was washed with petroleum ether to remove the bulk of the yellow pigment.

The filtrate was passed over another column of powdered sugar 60 × 340 mm. The column was washed with petroleum ether containing 20% ether. Most of the chlorophyll *c* remained on the column. The greenish pigment remaining on the column was eluted with ether containing 10% methanol.

The pigment on the first column was eluted with ether containing 10% methanol. The eluate was washed with water, dried with sodium sulfate, concentrated under reduced pressure to about 30 ml. The concentrate was diluted with 300 ml. of petroleum ether (b. p. < 40° C) and the pigment adsorbed an a column 60 × 330 mm. The column was washed with petroleum ether containing 30% ether. The greenish band was dug out and the pigment eluted with 10% methanol in ether.

The two eluates were combined and washed with four 150 ml. portions of distilled water. After being partially dried with sodium sulfate, the ether solution, about 125 ml., was combined with 300 ml. of petroleum ether and the pigment adsorbed on a sugar adsorption column 60 × 340 mm. The column was washed with 100 ml. of petroleum ether containing 30% ether, then with 200 ml. of benzene, and finally with 200 ml. of petroleum ether.

The greenish zone was dug out and the pigment eluted with 200 ml. of ether containing 10% methanol. The ether eluate was washed with three 100 ml. portions of water, dried with sodium sulfate, added to 100 ml. of petroleum ether and the pigment adsorbed on a sugar column 60 × 340 mm. The chromatogram was developed with 300 ml. of petroleum ether containing 2% dimethylanaline and 7% n-propanol. From top to bottom of the column there was a greenish band, a brownish band, and a wide diffused yellow-green band. The two upper bands and the lower diffuse band were dug out separately and treated with 10% methanol in ether.

The eluate of the lower greenish band was washed with three 100 ml. portions of water, treated with a small quantity of sodium sulfate to remove cloudiness and its absorption spectrum measured. This band showed the following absorption bands: maxima 628.5, 580, 446.5 mμ; minimum 507 mμ; relative absorptions 1.00, 0.89, 9.7 and the minimum 0.8 (this solution probably contained little alcohol, but undoubtedly contained some dimethyl aniline. The preparation of pigment solution for spectrophotometric measurements was continued with the two upper bands.

The two upper bands were dug out and the pigment eluted with 10% methanol in ether. The ether was washed with water, dried, and treated with 100 ml. of petroleum ether. This solution was put onto a sugar adsorption column 60 × 340 mm. and a chromatogram developed with 20% ether in petroleum ether. Two bands remained at the top of the column and a yellow-green band washed into the column.

The lower greenish band was eluted with 10% methanol in ether, washed with three 100 ml. portions of water, concentrated to about 80 ml., combined with 320 ml. of petroleum ether and the pigment adsorbed on a sugar adsorption column 60 × 300 mm. This pigment which was adsorbed at the very top of the column was eluted with 10% methanol in ether, the eluate washed with three 100 ml. portions of water, and the ether collected in a 50 ml. glass-stoppered graduated cylinder and diluted to 50.0 ml. with ether. This solution of chlorophyll c was diluted with ether to the proper concentration and its absorption measured (see Fig. 5). In ether the absorption maxima occurred at 626, 577.5, 443.5 mμ, the minimum at 500 mμ; relative absorption 1.00, 0.94, 10.1, minimum 0.14; in methanol, maxima 633, 584.5, 447 mμ, minimum, 515 mμ; relative absorption, 1.00, 0.87, 8.96, minimum 0.214. This was the chlorophyll c fraction used.

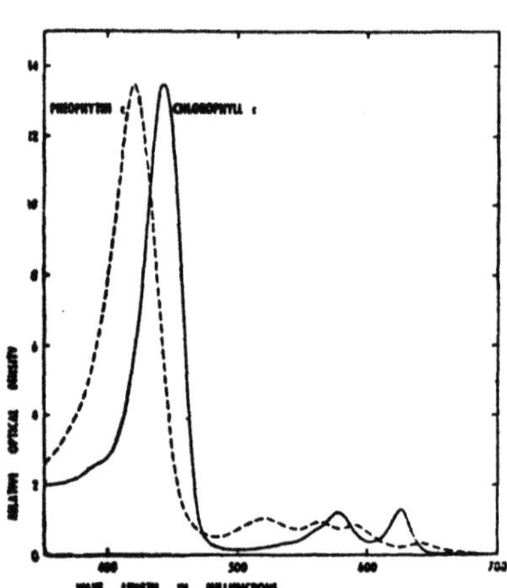

Fig. 5. The absorption spectra of chlorophyll c and pheophytin c in ether.

The pigments in the two upper bands were eluted separately with 10% methanol in ether. There was only an insignificant quantity of pigment in the uppermost zone. The eluate of the second zone contained a mixture of pigments as was demonstrated by distribution experiments between ether and 50% methanol saturated with salt solution.

II. Physical Properties of Chlorophyll c.

Solubility. Chlorophyll c is soluble in ether, acetone, ethyl acetate and methanol, but insoluble in petroleum ether (cf. STRAIN and MANNING, 1942).

Spectral Absorption Properties of Chlorophyll c. Qualitative absorption curves of chlorophyll c in methanol have been published by STRAIN and MANNING (1942) and in methanol and pyridine by WASSINK and KERSTEN (1946). A qualitative absorption curve of chlorophyll c in ether prepared according to the method described in the preceding section is shown in Fig. 5.

The wave lengths of the absorption maxima and of the absorption minimum of chlorophyll c in ether, acetone, and methanol are listed below:

Wave Lengths in mμ of the Absorption Maxima and Minimum for Chlorophyll c[1].

Solvent	Maxima			Minimum	
Ether	626	577.5	443.6	500	SMITH and BENITEZ
Ether	627	579.5	446		STRAIN and MANNING (1942)
Methanol	633	584.5	447	515	SMITH and BENITEZ
Methanol	635	585	450	516	STRAIN and MANNING (1942)
Acetone (80%)	631	581	446		STRAIN and MANNING (1942)

Fluorescence Maxima. STRAIN and MANNING (1942) reported that chlorophyll c in ether possessed two fluorescence bands, the principal band with maximum at 635 mμ, and a secondary maximum at about 690 mμ. From fluorescence spectrum curves of this pigment in ether, obtained by Dr. VIRGIN, the maxima were found to lie at 629 and ca. 690 mμ.

III. Qualitative Tests.

STRAIN, MANNING and HARDIN (1943) have suggested as a qualitative test for chlorophyll c in living organisms, that the organisms be extracted with methanol and the ratio of the optical densities of the extract at 665 and 635 mμ be determined. A methanolic solution of pure chlorophyll a gives a ratio of 4.55, whereas chlorophyll c gives a ratio of 0.111. A ratio considerably less than 4.55 indicates the presence of chlorophyll c.

The corresponding qualitative test can be applied to an ether extract in which the ratio of optical densities at 663 and 626 mμ is determined. For pure chlorophyll a the ratio is 7.48 and for chlorophyll c, 0.024.

GRANICK (1949) found the hydrochloric acid number to be about 12, which figure we have confirmed. In this property, chlorophyll c differs greatly from chlorophylls a, b, and d.

When chlorophyll c in methanol or acetone is treated with hydrochloric acid the solution changes color from green to yellow green. This treatment produces a pheophytin, the characteristic absorption of which in ether is given in Fig. 5.

The low hydrochloric acid number of chlorophyll c indicates the absence of a phytyl group. It is unlikely, however, that the carboxyl is free because the product is insoluble in aqueous solutions of sodium carbonate and ammonium hydroxide (STRAIN and MANNING, 1942).

Chlorophyll c gives a positive phase test (GRANICK, 1949; STRAIN and MANNING, 1942). The writers found that a red ring is formed at the interface between the ether and the methanolic potassium hydroxide. When the mixture is shaken, the alcoholic layer becomes red, then brown, and finally green.

KYLIN (1927) reported the separation of chlorophyll c by paper chromatography. In the adsorbed state, it undergoes characteristic color reactions; when dry it appears as a yellow-brown zone which turns green again when wet with water or alcohol. This is in contrast to chlorophyll a which is blue-green also when dry. (Cf. also p. 190).

---

[1] The writers have calculated from magnesium determinations provisional values for the specific absorption coefficients of chlorophyll c in ether solution by assuming it to have a molecular weight of 893.48

Wave lengths	628	579.5	447 mμ
Sp. Abs. Coeff.	22.0	20.6	227 l./gm. cm.

IV. Pheophytin c.

The empirical and structural formulas of pheophytin *c* are not known.

Pheophytin *c* is not known to occur in nature. It is formed by treating chlorophyll *c* with acid.

Preparation of Pheophytin c. In order to obtain an ether solution of pheophytin *c* for spectrophotometric examination it was prepared in the following manner:

An ether solution of chlorophyll *c* (25.0 ml.) of known optical density was mixed with 1 ml. of concentrated hydrochloric acid and stored in a darkened refrigerator for about two hours. It was then poured into a separatory funnel containing 100 ml. of distilled water. The water layer was drawn off and extracted successively with 20 and 10 ml. of ether which were combined with the bulk of the ether solution. The aqueous solution was discarded. If a quantitative transfer to ether is to be made, care must be taken that pigment is not adsorbed on the vessel walls or thrown away in the wash water. For this reason a rather elaborate washing procedure was followed.

The total ether extract was washed with 50 ml. of water. The water after being extracted with 10 ml. of ether was thrown away. The ether extracts were combined and washed twice with water, the two washings being extracted with another 10 ml. of ether. The final wash water had a pH value of 4. The main ether solution was drawn into a 100 ml. glass-stoppered graduated cylinder. To the separatory funnel containing the last 10 ml. of ether extract was added 5 ml. of methanol to desorb the pigment. This mixture was transferred to the separatory funnel from which the main ether extract had been drawn and the adsorbed pigment dissolved. The methanol was removed by three washings with water, 50, 25, and 10 ml. each. The first wash water was extracted with 10 ml. of ether. No significant amount of pigment was extracted, so the last washings were discarded. To the 10 ml. ether extract was added 5 ml. of methanol and the adsorbed pigment on the separatory funnel walls dissolved. After the methanol had been washed from the ether the ether was run into the volumetric cylinder. The volume was adjusted to 50 ml. with ether. The solution was reddish brown. The absorption spectrum of the solution was measured after the solution had stood in the refrigerator until clear. (Absorption spectrum, see Fig. 5.)

Physical Properties of Pheophytin c.

Solubility. Pheophytin *c* is soluble in ether, acetone, methanol, chloroform, and benzene. It is insoluble in petroleum ether.

Spectral Absorption Properties of Pheophytin c. No quantitative absorption spectrum has been determined for pheophytin *c*.

The qualitative absorption spectrum curve for pheophytin *c* in ether is shown in Fig. 5. It is very difficult to work quantitatively with pheophytin *c* in ether solution because of its great adsorbability on the walls of containing vessels. For this reason it is impossible to make quantitative dilutions of dilute solutions in this solvent. Also, it was impossible to get a significant ratio between the optical density measurements of chlorophyll *c* and pheophytin *c*.

The absorption peaks in ether solution were found at 641, 689.5, 564, 521.5, and 421.5 mμ.

Fluorescence Maxima. The fluorescence maxima of an ether solution of pheophytin *c*, as determined from a fluorescence spectrum curve obtained by Dr. Virgin, lie at 649 and 719 mμ. In contrast to many other pigments of the chlorophyll group, the fluorescence bands are very distinct, widely separated, and of nearly equal height.

Qualitative Tests. The pheophytin *c* prepared by this method was extracted from ether solution with dilute solutions of di-sodium hydrogen phosphate and ammonium hydroxide. For this reason it is possible that hydrolysis of ester groups occurred because chlorophyll *c* was not extracted under the same conditions.

The pigment goes into 22% hydrochloric acid with a blue color.

In the phase test, the pigment gives a yellow-brown color to the alcoholic potassium hydroxide which slowly changes to green and finally to a brownish yellow.

V. Quantitative Determination of Chlorophyll *c*.

No methods have been worked out for the quantitative determination of chlorophyll *c*. When the specific absorption coefficients are known with certainty its quantitative determination will become possible.

C. Chlorophyll *d*.

Neither the empirical nor structural formulas of chlorophyll *d* are known.

So far as is known, the occurrence of chlorophyll *d* is restricted to the red algae. Whether it is universally present in this group of algae has not been determined. The amount of this pigment varies considerably among the various species.

I. Preparation of Chlorophyll *d*.

Chlorophyll *d* has been prepared previously by MANNING and STRAIN (1943) from various red algae, chiefly *Gigartina Agardhii*. The writers have prepared this pigment for spectroscopic examination from *Gigartina Agardhii* by the method here described.

One kilogram of the red alga collected at Moss Beach, California, was washed with fresh water to free it of extraneous material. The plant material was drained, placed in a wooden bowl and chopped into small pieces with a knife. The chopped material was extracted with 2 liters of methanol for 30 minutes. The brei was poured into a 20 cm. BÜCHNER funnel and the extract filtered by suction. The residue was again extracted for 15 minutes with 1 liter of methanol. This extract was filtered and combined with the first filtrate. The total extract was combined with 300 ml. of petroleum ether (b. p. 40—50° C) in a 6 liter separatory funnel and treated with 500 ml. of a 10% sodium chloride solution saturated with magnesium carbonate. After the extraction, the petroleum ether layer containing the chlorophyll was separated and the alcoholic layer again extracted with 200 ml. of petroleum ether and the two petroleum ether extracts united.

The petroleum ether extracts from three 1 kg. lots of alga were combined and washed with three 1 liter portions of distilled water. The petroleum ether was dried with sodium sulfate and poured onto an adsorption column of confectioner's sugar, 100×100 mm. The filtrate contained a considerable amount of yellow pigment. The column retained the green pigment. After the column had been washed with 300 ml. of petroleum ether it was drained by suction. The green pigment adsorbate was dug out and eluted with ether. The ether eluate was stored in the refrigerator in the dark until it was used for further purification.

The ether eluates, obtained from extraction of 7.725 kg. fresh weight of alga, were combined and concentrated to about 50 ml. To the concentrate was added 20 ml. of benzene (Reagent grade thiophene-free) and 150 ml. of petroleum ether. The solution was poured onto an adsorption column of confectioner's sugar 60×300 mm. and was washed with 250 ml. of petroleum ether. From the top downward the column contained the following pigment zones; 0—98 mm. a faint blue-green zone which gave a spectroscopic test for chlorophyll d but which was probably allomerized pigment and was discarded; 98—167 mm. three zones containing chlorophyll d, namely, a green zone, an olive-green zone and a green-olive-green zone; 167—290 mm. a blue-green zone principally chlorophyll a; 290—295 mm. a yellow zone.

The three green zones were dug out, the pigment eluted with ether, the ether eluate concentrated to approximately 25 ml., and petroleum ether, 100 ml., added to the concentrated eluate. The solution was dried with sodium sulfate, and the pigment readsorbed on a powdered sugar column, 60 × 300 mm. The chromatogram was developed with 450 ml. of petroleum ether. Above the blue-green zone of chlorophyll a were three green zones which contained chlorophyll d mixed with yellow pigments. The pigments from these three zones were eluted with ether and transferred to 20 ml. of benzene, by the evaporation procedure. Petroleum ether (100 ml.) was added and the pigments readsorbed on a column 60 × 320 mm. The yellow pigments were removed by washing with 500 ml. of pure benzene. The column was washed with 200 ml. of petroleum ether to remove the benzene. Although there appeared to be three pigment zones on the column, these zones were indistinguishable inside the column, so all the green pigment was eluted with ether and readsorbed from petroleum ether on a sugar column

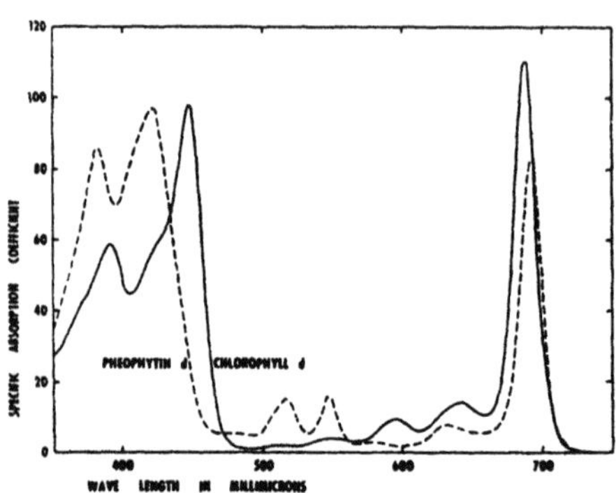

Fig. 6. The absorption spectra of chlorophyll d and pheophytin d in ether. (Molecular weight assumed for chlorophyll d, 893.48.)

60 × 310 mm. The chromatogram was developed with 350 ml. of petroleum ether containing 0.5% each of n-propanol and dimethylaniline. Five zones developed — an upper blue zone, a second brown zone (probably chlorophyll b), a narrow brown zone (probably chlorophyll b'), a grass green zone (chlorophyll d), and a narrow blue zone which passed to the lower part of the column.

The pigments from the grass green zone and part of the narrow brown zone were eluted with ether and readsorbed on a sugar column 60 × 210 mm. from petroleum ether. The chromatogram was developed with 500 ml. of the petroleum ether-n-propanol-dimethylaniline mixture. There resulted a narrow brown zone, a grass green zone, and a blue zone.

The pigment from the grass green zone was eluted with ether and readsorbed on a powdered sugar column 30 × 300 mm. from petroleum ether. The pigment was readsorbed in a zone 43 mm. deep. The chromatogram was again developed with the petroleum ether-n-propanol-dimethylaniline mixture until the narrow brown band and grass green zone were separated.

The pigment from the grass green band was eluted with ether and readsorbed from petroleum ether on a column 45 × 60 mm. It was washed free of propanol and dimethylaniline with petroleum ether (b. p. < 40° C). The pigment was eluted with ether, and the ether solution diluted to 50.0 ml. (This solution contained approximately 15 mg. of chlorophyll d.)

Aliquot portions of this solution were removed for magnesium analysis and for quantitative absorption spectrum measurements, from which determinations, specific absorption coefficients were calculated (cf. Fig. 6).

II. Physical Properties of Chlorophyll *d*.

Solubility. Chlorophyll *d* is soluble in ether, acetone, alcohol, and benzene. It appears to be very slightly soluble in petroleum ether.

Spectral Absorption Properties of Chlorophyll d. In 1943, MANNING and STRAIN published a qualitative absorption curve of chlorophyll *d* in methanol. They remarked that the absorption bands were not as pronounced in methanol as in ether. Their curve shows well defined maxima at three wave lengths only: 696, 456, and 400 mμ. The writers have determined the specific absorption coefficients of chlorophyll *d* in ether solution and plotted these values against wave length in Fig. 6. Actually, it was the molar absorption coefficients which were determined by magnesium analysis, but these values have been arbitrarily transformed to specific absorption coefficients by assuming the molecular weight of chlorophyll *d* to be equal to that of chlorophyll *a*. This transformation was made in order to make possible a more meaningful comparison between the various chlorophylls.

The values of the specific absorption coefficients at the various maxima are tabulated:

Absorption Maxima and Specific Absorption Cofficients: Chlorophyll d.

	Solvent: Ether						
SMITH and	688	643	595	548.5	512	447	392 mμ
BENITEZ	110.4	14.3	9.47	4.03	1.98	97.8	58.9 l./gm. cm.

Fluorescence Maxima. The values reported for the fluorescence maxima of chlorophyll *d* in ether by MANNING and STRAIN (1943) are 693 and ca. 750 mμ. The values taken from fluorescence spectrum curves measures for ether solution by Dr. VIRGIN are 696 and ca. 752 mμ.

III. Qualitative Tests.

Inasmuch as red algae contain chlorophyll *a* as their principal green pigment and chlorophyll *d* as their auxiliary green pigment, MANNING and STRAIN (1943) suggested that the ratio of the optical densities of a methanol extract of red algae at 665 and 700 mμ be used to test for the presence of chlorophyll *d*. For pure chlorophyll *a* the ratio is about 90, and for pure chlorophyll *d*, about 0.25. The rate of extraction of chlorophyll *d* is greater than for chlorophyll *a*, therefore a short extraction period is recommended to increase the sensitivity of the test

The hydrochloric acid number of chlorophyll *d* is greater than 22. The ether layer when shaken with the acid becomes brown.

In the phase test, chlorophyll *d* forms a red interface between the ether and alcoholic layers. When the mixture is shaken, the red color goes into the alcoholic phase which becomes brown and finally green. The upper ether layer becomes colorless.

MANNING and STRAIN (1943) pointed out that chlorophyll *d* readily forms an equilibrium of isomeric forms. This cannot be discussed here.

IV. Pheophytin *d*.

The empirical and structural formulas of pheophytin *d* are not known.

Pheophytin *d* is not known to occur naturally. It is formed from chlorophyll *d* by treatment with acid as is described in the next section.

Preparation of Pheophytin *d*. When chlorophyll *d* is acted on by hydrochloric acid, magnesium is removed and pheophytin is formed. According to MANNING and STRAIN (1943), this is a complicated reaction in that several pheophytins are formed. For a discussion of this complex subject, the reader is referred to the original article.

For analytical purposes, it seemed sufficient to transform chlorophyll d to its pheophytin in the usual manner and determine the absorption spectrum of the resulting pigment in the same solvent as was used for the chlorophyll. If this was done quantitatively, a relation between the absorption coefficients of chlorophyll d and pheophytin d would be obtained which would be useful in the spectrophotometric analysis of these pigments. Accordingly, the procedure used for this purpose is described here.

To 20.0 ml. of an ether solution of chlorophyll d of known optical density, which was contained in a glass-stoppered cylinder, was added 1 ml. of concentrated hydrochloric acid (sp. g. 1.19). The mixture was thoroughly mixed and set in the refrigerator at about 5° C for approximately four hours. The aqueous layer became blue, the ether layer a brownish color.

The mixture was transferred quantitatively to a separatory funnel containing 50 ml. of water and was washed four times with 50 ml. portions of water. Sufficient ether was added to keep the volume from becoming depleted. The last wash water was only faintly acid, pH $\approx$ 5, which demonstrated that the acid had been removed. The ether was collected in a 25 ml. glass-stoppered graduated cylinder.

The wash water was collected and extracted with 20 ml. of ether. This ether was washed with three 10 ml. portions of water to remove acid and was then united with the first portions of ether. The volume was brought to 20 ml. with ether and the absorption spectrum of this solution measured (cf. Fig. 6).

Since all operations were carried out quantitatively and the optical densities of the initial chlorophyll d solution and the final pheophytin d solution were known at their long-wave-length absorption maxima, the ratio of the absorption coefficients for the two pigments was calculated.

Physical Properties of Pheophytin d.

Solubility. Pheophytin d is soluble in ether, acetone, chloroform, benzene, and petroleum ether; it is only slightly soluble in methanol.

Spectral Absorption Properties of Pheophytin d. MANNING and STRAIN (1943) have published qualitative spectral absorption curves for pheophytin d and isopheophytin d in methanol. The absorption maxima, taken from the published curves, for pheophytin d lie at 693, 660, 547, 515, and 410 mμ; those for isopheophytin d lie at 663, 605, 534, 470, and 405 mμ. The curves for isopheophytin d and pheophytin a are nearly superposable.

The quantitative spectral absorption curve for pheophytin d in ether obtained in the manner described is plotted in Fig. 6.

The wave length maxima and the corresponding specific absorption coefficients are as follows:

Absorption Maxima and Specific Absorption Cofficients: Pheophytin d.

	Solvent: Ether						
SMITH and BENITEZ	692 82.7	631 7.75	576 2.88	547 15.9	516.5 15.1	476 5.52	421 97.2	382 mμ 85.8 l./gm. cm.

Fluorescence maximum. The fluorescence spectrum curve of pheophytin d in ether obtained by Dr. VIRGIN showed a maximum at 701 mμ and a secondary band to the long-wave-length side of the principal band, but without maximum.

Qualitative Tests. Pheophytin *d* is not extracted from ether solution with 22% hydrochloric acid.

The distribution of pheophytin *d* between equal volumes of ether saturated with concentrated hydrochloric acid ($\approx 35\%$) and concentrated hydrochloric acid saturated with ether was measured spectrophotometrically. The ether layer was separated and washed free of acid with water. The acid layer was removed, the pigment driven into ether, and the ether washed free of acid with water. The pheophytin in each of the ether solutions was measured spectrophotometrically and the distribution ratio calculated. It was computed to be 0.37.

Pheophytin *d* when subjected to the phase test gives first a brown color, then a grass-green color, and finally, after standing, a brown color again.

Under certain circumstances isopheophytin *d* and pheophytin *a* form a single zone when co-chromatographed. These compounds can be distinguished by treatment with acid and alkali (MANNING and STRAIN, 1943).

V. Quantitative Determination of Chlorophyll *d*.

The absolute absorption spectrum of chlorophyll *d* is now available (Fig. 6) consequently the quantitative estimation of this pigment in solution is possible. Also, procedures have been reported for completely extracting the chlorophyll from the red algae in which chlorophyll *d* appears. But the optical properties and extraction procedures have never been combined and adequately tested for determining chlorophyll *d* in plants.

Extraction procedures for red algae have been recommended by WURMSER (1921); MANNING and STRAIN (1943); and HAXO and BLINKS (1950). WURMSER and also HAXO and BLINKS both emphasized the necessity for extracting these organisms, first with either dilute acetone or methanol, and afterwards with increased concentrations of the organic solvent (cf. p. 185). MANNING and STRAIN found that after immersion of the red algae in boiling water for one minute the chlorophylls were extracted completely with methanol in from 10 to 30 minutes and that methanol was preferable to pyridine, or 80% acetone. If these extractions are complete and without damage to the chlorophylls, they may be combined with further procedures to determine chlorophyll *d*.

For example, the chlorophylls can be transferred quantitatively from these extracts to ether, the optical densities determined at appropriate wave lengths and the concentration of chlorophyll *d* computed from simultaneous equations. Since only chlorophylls *a* and *d* occur in red algae their spectral properties alone need to be considered. The preferred wave lengths and specific absorption coefficients in ether (cf. Figs. 1, 6) for setting up the simultaneous equations are here listed:

Wave lengths	663	688 mμ
Specific absorption coefficients		
Chlorophyll *d*	100.9	1.5
Chlorophyll *d*	11.3	110.4

The equations derived from these constants are

$$C_a = 0.00993\ D_{663} - 0.00102\ D_{688}$$
$$C_d = 0.00907\ D_{688} - 0.000135\ D_{663}$$

where C_a and C_d are grams per liter of chlorophylls *a* and *d* respectively, and D_{663} and D_{688} are the measured optical densities at the designated wave lengths when the length of the absorption cell is 1.00 cm.

Sometimes the red algae are contaminated with green epiphytes which give chlorophyll b on extraction. In this case, a correction for chlorophyll b may be made. This correction would undoubtedly be small because of the small absorption of chlorophyll b at such long wave lengths.

D. Chlorophyll e.

Strain (1943) reported the presence of a chlorophyll-like pigment in *Tribonema bombycinum* which he named chlorophyll e. This pigment resembles chlorophyll c in its adsorption properties. Its absorption maxima in methanol are given as 654 and 415 mμ with minima at 550 and 510 mμ (Strain, 1951, pp. 247, 248). No other data are available on this pigment at the present time.

E. Protochlorophyll.

The structural and empirical formulas of protochlorophyll are given below (Fischer, 1940; Fischer and Oestreicher, 1939/40). The calculated molecular weight is 891.46. The calculated elementary composition is

C, 74.10; H, 7.92; O, 8.97; N, 6.29; Mg, 2.73.

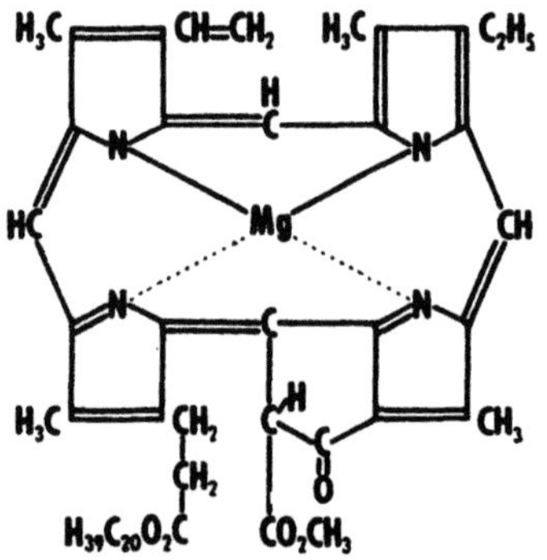

$$C_{55}H_{70}O_5N_4Mg$$

PROTOCHLOROPHYLL

Protochlorophyll occurs in small quantity in leaves of seedlings grown in the dark. It is present in the inner seed coats of the *Cucurbitaceae*. A very rich source is the inner seed coats of the Hubbard squash. The identity of the protochlorophylls from etiolated leaves and from inner seed coats is not yet conclusively proved.

I. Preparation of Protochlorophyll.

Noack and Kiessling (1929, 1930) have prepared protochlorophyll from cucurbits and Seybold (1936/37) has reported the separation of protochlorophyll from this source into two components which he called protochlorophylls a and b. Koski and Smith (1948) prepared protochlorophyll from etiolated barley leaves and because this protochlorophyll is known to be the precursor of chlorophyll a, a description of their method is reported here.

About 5 kg. of etiolated leaves from barley which had been grown from seed in the dark for nine days were harvested and divided into 300 gm. lots. Each lot was immersed in water at 90° C for five minutes. All operations before the hot-water immersion were performed in dim green light, in order to avoid the transformation of protochlorophyll to chlorophyll *a*. After the hot-water treatment the transformation no longer occurred and ordinary light could be used.

The killed leaves were pressed (35 kg. per sq. cm.) to remove water, and the pressed leaves (2.4 kg.) were chopped in a bowl and extracted in 80 gram lots in a Waring Blendor with 400 ml. portions of acetone.

The acetone extract was filtered. Liter portions of the extract were combined with 750 ml. of ether and the acetone removed by washing with two liters of water (cf. p. 145). The ether solutions were combined and filtered through a layer of powdered sugar 25.4 cm. diameter and 7.6 cm. deep. This filtration removed undesirable contaminants. The filtrate was concentrated by evaporation of the ether at reduced pressure. Petroleum ether was added and the solvent again evaporated at reduced pressure. This procedure was repeated once more to remove the last of the ether. To the concentrated pigment solution, about 1 liter, was added three liters of petroleum ether and the solution passed over an adsorption column of powdered sugar. This may be done in one column or several smaller columns with a volume of packed sugar equal to about 7,550 cc. The protochlorophyll was adsorbed as a blue-green zone which was washed with petroleum ether to remove fats and yellow pigments. The protochlorophyll was eluted with about 9 liters of ether.

The protochlorophyll was again transferred to petroleum ether by the evaporation procedure, adsorbed on sugar, and washed thoroughly with petroleum ether. (After each adsorption the protochlorophyll zone was dug out of the column, the pigment eluted with ether, and transferred to the solvent used for the next adsorption by the evaporation procedure.)

The pigment was adsorbed and washed successively as follows: Adsorbed from petroleum ether and washed with petroleum ether; adsorbed from benzene, washed with benzene; adsorbed from benzene, washed with 10% ether in benzene; adsorbed from benzene, washed with 20% ether in benzene; adsorbed from benzene, washed with petroleum ether. The pigment was concentrated in a zone 40 mm. in diameter and 15 mm. deep. The pigment was eluted with ether and diluted to 82 ml. Portions of this solution were used for spectral absorption measurements and magnesium determinations from which the specific absorption coefficients were determined.

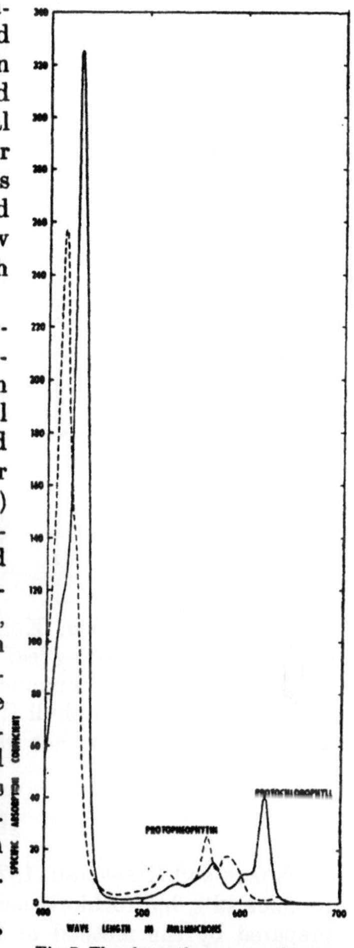

Fig. 7. The absorption spectra of protochlorophyll and protopheophytin in ether. (Cf. KOSKI, 1949; SMITH and YOUNG, 1954.)

II. Physical Properties of Protochlorophyll.

Solubility. Protochlorophyll is soluble in ether, acetone, alcohol, and benzene; almost insoluble in petroleum ether.

Spectral Absorption Properties of Protochlorophyll. Qualitative absorption spectrum curves of protochlorophyll from inner seed coats of certain *Cucurbitaceae* have been determined by RUDOLPH (1933/34), SEYBOLD (1948), and by KRASNOVSKII and VOINOVSKAYA (1949).

Quantitative absorption spectrum curves for precipitated protochlorophyll from etiolated barley leaves in ether, acetone, and methanol have been published by KOSKI and SMITH (1948). The absorption spectrum of unprecipitated protochlorophyll in ether solution determined by these workers is given in Fig. 7.

The absorption maxima and the corresponding specific absorption coefficients for unprecipitated protochlorophyll in ether, and for precipitated protochlorophyll in ether, acetone, and methanol are here tabulated.

Absorption Maxima and Specific Absorption Coefficients: Protochlorophyll.

Unprecipitated Pigment.

	Solvent: Ether			
KOSKI and SMITH (1948)	623 39.9	571 14.9	535 7.2	432 mμ 325.5 l./gm. cm.

Precipitated Pigment

	Solvent: Ether			
KOSKI and SMITH (1948)	623 36.9	571 14.0	535 6.8	432 mμ 305.9 l./gm. cm.
	Solvent: Acetone			
KOSKI and SMITH (1948)	623 34.9	571 13.5	535 6.9	432 mμ 270.5 l./gm. cm.
	Solvent: Methanol			
KOSKI and SMITH (1948)	629 27.2	578 10.5	— —	434 mμ 177.3 l./gm. cm.

Fluorescence Maxima. KOSKI (cf. SMITH and YOUNG, 1954) found the fluorescence maximum of barley leaf protochlorophyll in ether solution to lie at 629 mμ. Dr. VIRGIN (private communication) found for acetone extract of etiolated barley leaves the protochlorophyll fluorescence maxima to be at 627 and ca. 685 mμ, and for leaves killed by hot water immersion at 638 and ca. 695 mμ.

III. Analytical Tests.

NOACK and KIESSLING (1929, p. 16) found pumpkin seed coat protochlorophyll to color 12% hydrochloric acid whereas chlorophyll a did not. Protochlorophyll prepared by the method of KOSKI and SMITH (1948) was found by GRANICK (1950, p. 724) to have a hydrochloric acid number of $\approx$ 25.

SEYBOLD and EGLE (1938/39b) reported the phase test for both protochlorophylls from seed coats to be positive: protochlorophyll a becomes brown, then light yellow, dark yellow, and a red brown; the protochlorophyll b becomes lemon yellow and after a few seconds yellow green.

Chlorophyllase has not been demonstrated to act on protochlorophyll.

IV. Protopheophytin.

The empirical and structural formulas of protopheophytin are given here (FISCHER, 1940; FISCHER and OESTREICHER, 1939/40). The calculated molecular weight is 869.16. The calculated elementary composition is

C, 76.00; H, 8.35; O, 9.20; N, 6.45.

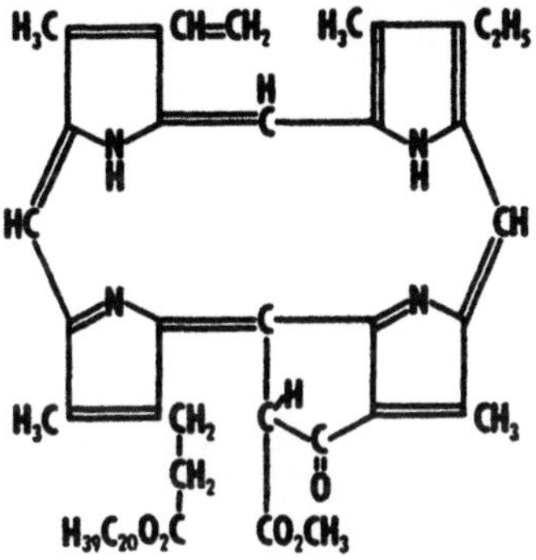

$C_{55}H_{72}O_5N_4$

PROTOPHEOPHYTIN

Protopheophytin has been prepared from protochlorophyll by treatment with acid and also by a series of transformations from chlorophyll a. It may occur naturally in the inner seed coats of certain cucurbits. (FISCHER and OESTREICHER, 1939/40, and perhaps in its phytol-free form in *Chlorella* mutant 31 examined by GRANICK, 1950.)

Preparation of Protopheophytin. NOACK and KIESSLING (1929, 1930) have prepared protopheophytin by treating an ether solution of seed-coat protochlorophyll with acid and precipitating out the protopheophytin with petroleum ether. There is some question whether this material contained the phytyl group (NOACK and KIESSLING, 1930, p. 118). By addition of oxalic acid to an ether solution of protochlorophyll from seed coats, SEYBOLD and EGLE (1938/39 b) have formed protopheophytin which they reported to be a mixture of protopheophytins a and b. This compound has been prepared synthetically from chlorophyll a (FISCHER, 1940; FISCHER and OESTREICHER, 1939/40). KOSKI (1949) prepared this pigment by addition of oxalic acid to an ether solution of pure protochlorophyll obtained from etiolated barley leaves. Her procedure is given here:

Pure solid protochlorophyll, 1.08 mg., was dissolved in 500 ml. of freshly distilled ether. Oxalic acid crystals, 0.3 gm., were added and the solution allowed to stand in the dark for 1.5 hours. The oxalic acid was removed by washing the ether with distilled water. The ether was brought to 50 ml. and the absorption spectrum determined. From the weight of protochlorophyll used, the concentration of protopheophytin was determined and the specific absorption coefficients calculated. The specific absorption coefficients are plotted against wave length in Fig. 7, and the optical constants of the absorption maxima are tabulated on p. 176.

Physical Properties of Protopheophytin.

Solubility. Protopheophytin is soluble in ether, dioxane, and pyridine. It is insoluble in petroleum ether.

Spectral Absorption Properties of Protopheophytin. Qualitative absorption curves of the reputed pheophytins *a* and *b* from cucurbit inner seed coats in ether were published by SEYBOLD (1948). A quantitative molar absorption spectrum curve of synthetic vinyl pheoporphyrin a_5 monomethyl ester in dioxane was published by GRANICK (1950). This curve agrees very well with the absorption curve of protopheophytin in ether obtained by KOSKI (1949) which is reproduced in Fig. 7 (cf. SMITH and YOUNG, 1954).

The wave-length maxima and the corresponding specific absorption coefficients of protopheophytin in ether and of vinyl pheoporphyrin a_5 in dioxane are tabulated below.

Absorption Maxima and Specific Absorption Coefficients: Protopheophytin.

	Solvent: Ether				
KOSKI (1949)	638 2.5	585 17.3	565 25.1	524 11.7	417 mμ 256.6 l./gm. cm.
	Solvent: Dioxane[1]				
GRANICK (1950)	638 2.5	588 16.1	567 20.4	525 9.5	419 mμ 222.0 l./gm. cm.

Fluorescence. The fluorescence maxima of protopheophytin have not been reported. SEYBOLD and EGLE (1938/39 b, p. 122) remarked its red color.

V. Quantitative Determination of Protochlorophyll.

Estimation of Protochlorophyll. The method of KOSKI, FRENCH and SMITH (1951), described here, was developed for the determination of protochlorophyll and chlorophyll *a* in etiolated leaves during the transformation of protochlorophyll to chlorophyll.

Etiolated leaves, 1 gm., were placed in a mortar with 12 ml. of acetone and washed sea sand. The leaf material was ground until disintegrated. The solution was decanted through a suction filter and the residue ground twice more with 7 ml. portions of acetone. These extracts were filtered by suction. The residue was transferred to the filter with the last extract. The debris was washed with 5 ml. of acetone. The filtrate and washings, which had been collected in a test tube, were poured into a beaker and the test tube rinsed with 20 ml. of ether which was added to the acetone extract.

The total extract was poured through a long-stemmed funnel which reached nearly to the bottom of a 180 ml. separatory funnel containing 100 ml. of water (cf. method of MACKINNEY, p. 145). After gently shaking the mixture the ether was allowed to separate completely and the water layer was drawn off and discarded. The ether layer was collected in the beaker from which it had been poured originally and was washed again with water in the same manner. This was repeated twice more. The ether was run into a glass-stoppered graduated cylinder to which was added the rinsings (5 ml.) of the beaker and funnel and the volume brought to 15.0 ml. by adding or evaporating ether. The solution was set in a refrigerator for from 30 to 60 minutes to clear and the optical density of the solution measured at the protochlorophyll absorption maximum, 624 mμ, by means of a Beckman Model DU Photoelectric Spectrophotometer. A 50 mm.

---

[1] Values of molar extinction $\times$ 10^{-4} in dioxane for vinyl pheoporphyrin a_5 monomethyl ester transformed to specific absorption coefficients of protopheophytin.

absorption cell was used. At the maximum (ca. 624 mμ) the specific absorption coefficient is 39.9 l./gm. cm. (KOSKI and SMITH, 1948). By substitution of the appropriate values into the equation for the LAMBERT-BEER's Law the concentration of the protochlorophyll was calculated.

If chlorophylls a and b were present the optical densities at three wave lengths were determined and the concentrations calculated by solving simultaneous equations. In ether the three wave lengths best suited for the determination and the specific absorption coefficients required are here tabulated.

KOSKI (1950) analyzed for these pigments but used slightly different values for the absorption coefficients of chlorophylls a and b.

Wave length in mμ.	662	644	624
	Specific Absorption Coefficients		
Chlorophyll a	100.9	15.8	14.2
Chlorophyll b	6.2	62.0	11.0
Protochlorophyll	0.2	1.0	39.9

For many purposes the ratio of protochlorophyll to chlorophyll a or of protochlorophyll to the sum of protochlorophyll plus chlorophyll a are the values desired. These ratios can be determined more accurately than the individual concentrations.

F. Bacteriochlorophyll.

The structural and empirical formulas of bacteriochlorophyll are as follows (FISCHER and STERN, 1943, p. 13): The calculated molecular weight is 911.50. The calculated elementary composition is

C, 72.47; H, 8.18; O, 10.53; N, 6.15; Mg, 2.67.

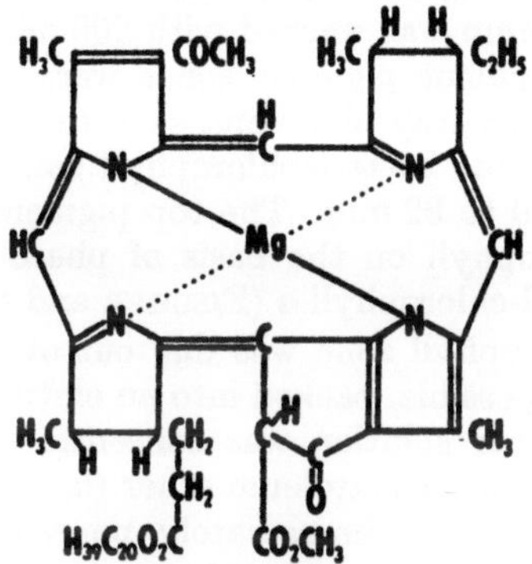

Bacteriochlorophyll occurs in the cells of purple and brown bacteria. VAN NIEL and ARNOLD (1938) found that the bacteriochlorophyll from six different strains of purple and brown bacteria was the same as that from the sulfur purple bacterium, *Thiocystis*, studied by H. FISCHER and his co-workers.

I. Preparation of Bacteriochlorophyll.

Bacteriochlorophyll has been prepared from purple bacteria by SCHNEIDER (1934a, b; cf. also FISCHER and STERN, p. 314). It has been separated chromatographically by MANTEN (1948). In order to obtain an ether solution of this pigment for measuring the absolute absorption spectrum, the writers have followed the procedure here described.

Rhodospirillum rubrum cells[1] were centrifuged from their culture medium. About 25 grams of very wet cells were treated with 250 ml. of redistilled methanol in a Waring Blendor. This operation was carried out under weak illumination and in a cold room at from 6 to 8° C. After standing for about 45 minutes in the dark and cold, the methanol was freed of cell material by centrifuging. The solid cell material was again extracted in a Waring Blendor with 250 ml. of methanol and afterwards collected by centrifuging. The sediment was finally washed with 60 ml. of methanol. The three methanol solutions were combined and extracted with 100 ml. of petroleum ether (b. p. 50 to 70° C). A small amount of water was added to the methanol which was then extracted successively with 100, 100, and 50 ml. of petroleum ether. More water was added to the methanol which was then extracted with 200 ml. of ether.

The petroleum ether and ether extracts were combined. The combined solution was washed with three 200 ml. portions of saturated magnesium carbonate solution and was concentrated under reduced pressure to 100 ml. The concentrate was poured onto a powdered sugar adsorption column 65 × 250 mm., care being taken that no water was poured onto the sugar column. The column was washed with 200 ml. of petroleum ether (b. p. 50 to 70° C).

The blue zone was dug out of the column and eluted with 100 ml. of ether. The ether solution was concentrated under reduced pressure and the concentrate, about 10 ml., treated with 100 ml. of 50 per cent mixture of benzene (thiophene-free) and petroleum ether (b. p. 50 to 70° C). This solution was poured onto a powdered sugar adsorption column (65 × 210 mm.) and a chromatogram developed with 300 ml. of 50 per cent benzene-petroleum ether mixture, and then 200 ml. of pure benzene. The column was washed with 200 ml. of petroleum ether (b. p. 20 to 40° C). Three prominent pigment zones were observed in the following order from the top: a brown-gray-blue band extending from 10 to 17 mm. below the top of the column; a blue bacteriochlorophyll band from 18 to 90 mm.; and a light green zone from 90 to 92 mm. The top pigment zone appeared to be an allomerized bacteriochlorophyll on the basis of phase test. The bottom green zone was possibly 2-acetyl-chlorophyll *a* (FISCHER and STERN, 1943, p. 311).

The blue bacteriochlorophyll zone was dug out of the adsorption column as free of other pigments as possible, packed into an eluting column and eluted with 130 ml. of ether. The ether solution was concentrated to about 10 ml. under reduced pressure and 100 ml. of petroleum ether (b. p. 50 to 70° C) added to the concentrate. This solution was immediately poured onto a powdered sugar adsorption column, 65 × 250 mm., and a chromatogram developed with 300 ml. of a solution containing 10% ether in petroleum ether, and then 200 ml. of a 15% ether in petroleum ether mixture. (Care must be exercised not to use too great a concentration of ether at first until it is certain that the pigments will be adsorbed. No general directions can be given. The percentage here prescribed was found satisfactory under the conditions used.)

The blue bacteriochlorophyll zone was dug out as free as possible from the brown zone above and the green zone below and eluted with 100 ml. of ether. The ether solution, after being concentrated under reduced pressure, was treated with 100 ml. of petroleum ether (b. p. 50 to 70° C) and rechromatographed on a column of powdered sugar, 65 × 200 mm., with 400 ml. of 20% ether in petroleum ether. The principal blue zone was again dug out and eluted with ether. This

---

[1] The writers wish to thank Dr. L. N. M. DUYSENS for supplying the bacteria used.

ether solution, about 55 ml., was used for magnesium determinations and for absorption-spectra measurements. In order to get an absorption spectrum in methanol, 1 ml. of the ether solution of bacteriochlorophyll was pipetted into a 25 ml. volumetric flask, the ether quickly removed by applying suction, and the residue dissolved in 25 ml. of methanol.

II. Physical Properties of Bacteriochlorophyll.

Solubility. Bacteriochlorophyll is soluble in ether, acetone, methanol, benzene, and pyridine. It is insoluble in petroleum ether.

Spectral Absorption Properties of Bacteriochlorophyll. The absorption curve of a crude methanolic extract of the pigments from purple bacteria has been published by FRENCH (1937). MANTEN (1948) has determined a qualitative absorption curve of "Pure Bacteriochlorophyll" in methanol saturated with hydrogen sulfide. A quantitative absorption curve of bacteriochlorophyll in ether has been determined by the writers according to the method described on p. 186 and is plotted in Fig. 8.

WEIGL (1953) published the wave lengths of the absorption maxima and their relative absorption coefficients for bacteriochlorophyll in several solvents. He also gave sufficient data for calculating the specific absorption coefficients at the maxima for an ether solution (cf. table below). Corresponding values have been determined by the writers and are listed in the following table:

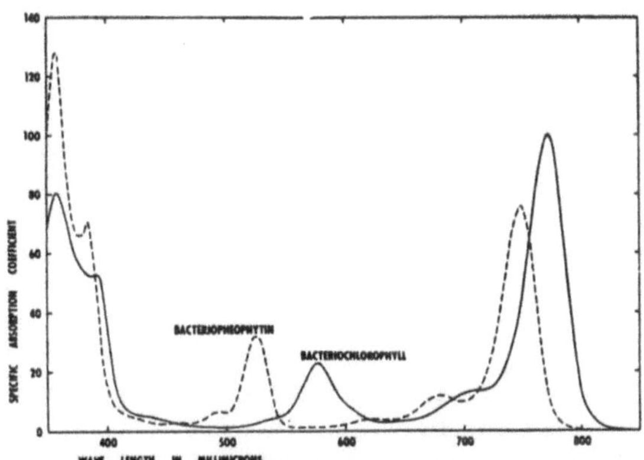

Fig. 8. The absorption spectra of bacteriochlorophyll and bacteriopheophytin in ether.

Absorption Maxima and Specific Absorption Coefficients: Bacteriochlorophyll.

	Solvent: Ether[1]					
SMITH and BENITEZ	773	(697)	577	(530)	391.5	358.5 mμ
	100	(10.0)	22.9	(3.0)	52.8	80.5 l./gm. cm.
WEIGL (1953)	772	(697)	575	526	391	358 mμ
	105	(12.6)	24.2	13.7	57.9	93.7 l./gm. cm.
	Solvent: Methanol					
SMITH and BENITEZ	772	(685)	608	—	––	365 mμ
	46.1	(9.5)	16.9	—	—	59.2 l./gm. cm.

Fluorescence Spectrum. A red fluorescence of bacteriochlorophyll in ether was reported by SCHNEIDER (1934a, b). A fluorescence curve for this pigment in alcohol, with maxima at 695 and 805 mμ, was published by VERMEULEN, WASSINK, and REMAN (1937). The red fluorescence observed was undoubtedly due to the 695 mμ band.

FRENCH (1954) found the fluorescence spectrum of chromatographically separated bacteriochlorophyll dissolved in ether to vary when excited by different

---

[1] The values in parentheses are estimated values because no well-defined maxima were present.

wave lengths. When excited with 436 mμ the red fluorescence at 687 mμ was more prominent than the infrared fluorescence, whereas when excited with 405 mμ the latter was more prominent. In a separate experiment he demonstrated that the fluorescence band at 687 mμ was due to the contaminating green pigment, already mentioned, presumably 2-acetyl-chlorophyll a (FISCHER and STERN, 1943, p. 311). It is the infrared fluorescence, therefore, which is characteristic of bacteriochlorophyll.

The method used by FRENCH is widely applicable. Change in the shape of a fluorescence curve with change in wave length of exciting light indicates that the substance being examined is not pure. This derives from the generalization that the fluorescence spectrum of a single substance is independent of the exciting wave length.

III. Qualitative Tests.

SCHNEIDER (1934 a, b) found that by shaking an ether solution of bacteriochlorophyll with 22.5% hydrochloric acid a trace of color was extracted whereas with 25% hydrochloric acid the pigment was extracted in large amounts. Treatment with hydrochloric acid turns the purplish ether solution to a brown-olive-green color.

Bacteriochlorophyll gives a positive phase test. It produces a transient yellow color in the alcoholic layer, then brown, and finally a permanent green.

The purity of bacteriochlorophyll can be judged from adsorption on confectioner's sugar. It is adsorbed as a blue band, but it is difficult to get it completely free of a more strongly adsorbed purplish component and a less strongly adsorbed green component.

Bacteriochlorophyll is acted on by chlorophyllase as chlorophylls a and b are (FISCHER, LAMBRECHT and MITTENZWEI, 1938).

IV. Bacteriopheophytin.

The structural and emperical formulas of bacteriopheophytin are presented here (FISCHER and STERN, 1943, p. 315). The calculated molecular weight is 889.20. The calculated elementary composition is

C, 74.29; H, 8.62; O, 10.80; N, 6.30.

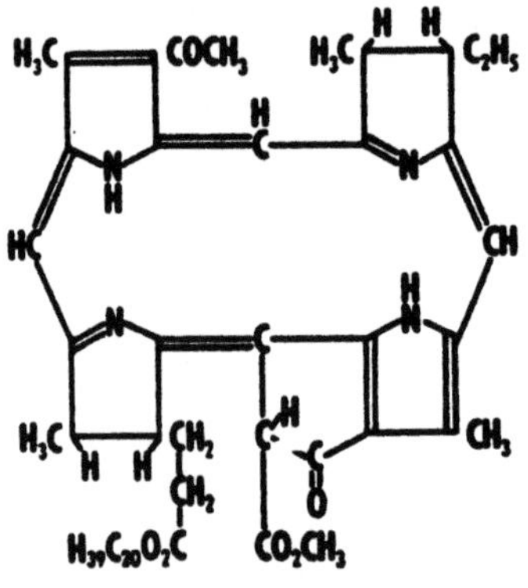

$$C_{55}H_{76}O_6N_4$$

BACTERIOPHEOPHYTIN

The bacteriopheophytin so far as is known does not occur naturally. It is prepared from bacteriochlorophyll by treatment with acid.

Preparation of Bacteriopheophytin. Bacteriopheophytin has been prepared by SCHNEIDER (1934 b), FISCHER, LAMBRECHT, und MITTENZWEI (1938), and FISCHER and HASENKAMP (1935). Solutions of bacteriopheophytin have been prepared for spectroscopic examination by VAN NIEL and ARNOLD (1938) and by FRENCH (1940).

In the following paragraph is described the method used by the writers for obtaining a solution of bacteriopheophytin from which absolute absorption coefficients were derived.

Five milliliters of an ether solution containing 0.1975 gm./l. of chromatographically purified bacteriochlorophyll was diluted to 50 ml. with ether. This solution, contained in a separatory funnel, was mixed with 10 ml. of 20 per cent hydrochloric acid. Both solutions had been previously cooled to about 6° C. The mixture was allowed to react at about 6° C for an hour. The ether solution was washed with five 100 ml. portions of water. The ether was drained into a 50 ml. volumetric flask and diluted to the mark with ether. This solution was used for the absorption-spectrum measurements shown in Fig. 8. The specific absorption coefficients were calculated on the basis of the bacteriochlorophyll initially present corrected for loss of weight of magnesium.

Physical Properties of Bacteriopheophytin.

Solubility. Bacteriopheophytin is soluble in ether, acetone, chloroform, and pyridine. It is little soluble in alcohol and is insoluble in petroleum ether.

Spectral Absorption Properties of Bacteriopheophytin. Absorption coefficients in various solvents have been determined by van Niel and ARNOLD (1938) with monochromatic light: the mercury line at 5461 Å and the helium line at 6678 Å. A qualitative absorption curve in carbon tetrachloride has been published by LARSEN (1953). A quantitative absorption curve of bacteriopheophytin in chloroform has been published by FRENCH (1940). An absorption curve in ether, obtained by the writers, is given in Figure 8.

The wave lengths of the maxima and the corresponding specific absorption coefficients of bacteriopheophytin in ether and chloroform are given in the following table. There is fair agreement between these various determinations except at the long wave-length maximum.

Absorption Maxima and Specific Absorption Coefficients: Bacteriopheophytin.

	Solvent: Ether					
WEIGL (1953)	750	680	620	525	384.5	357 mμ
	70.9	10.6	3.54	30.5	68.7	121.9 l./gm. cm.
SMITH and BENITEZ	749	680	625	525.5	384	357.5 mμ
	76.0	12.0	4.10	31.9	70.6	127.9 l./gm. cm.

	Solvent: Chloroform					
FRENCH (1940)	750	685	(625)	530	—	— mμ
	57.5	12.8	(4.0)	28.3	—	— l./gm. cm.
WEIGL (1953)	758	686	625	531.5	389	362 mμ
	65.2	13.0	4.6	28.0	58.1	108.3 l./gm. cm.
SMITH and BENITEZ	757.5	687	630	533	390	363 mμ
	71.4	13.0	4.4	29.5	59.4	111.9 l./gm. cm.

Fluorescence Maxima. Bacteriopheophytin in ether solution exhibits a fluorescence maximum at 760 mμ.

Qualitative Tests. The hydrochloric acid number is probably greater than 25. No bacteriopheophytin is removed from ether by 22% hydrochloric acid, but a considerable amount is removed by 25% (SCHNEIDER, 1934 b). The pigment cannot be extracted completely from ether with concentrated hydrochloric acid (Fischer and Hasenkamp, 1935). Bacteriopheophorbide gives a positive, red-brown, phase test (Schneider, 1934b).

Chlorophyllase acts on the bacteriopheophytin to form the bacterio-methyl-pheophorbide in methanol solution or the free pheophorbide in aqueous acetone (Fischer, Lambrecht and Mittenzwei, 1938).

The bacterio-methylpheophorbide can be formed by boiling bacteriopheophytin with methanolic hydrochloric acid under reflux.

V. Quantitative Determination of Bacteriochlorophyll.

Because bacteriochlorophyll is unstable in light, van Niel and Arnold (1938) developed a method for its determination as the pheophytin. The bacterial suspension, 1 vol., was mixed with absolute methanol, 10 vols., and centrifuged immediately. The supernatant was then added to 1.5 volumes of carbon tetra-chloride plus 0.1 volume of 25% hydrochloric acid. The optical density was measured immediately at wave length 6678 Å using a Keuffel & Esser Color Analyzer with "hot cathode" helium arc as the source of monochromatic light. From the optical density measurement the length of the absorption cell and the specific absorption coefficient for bacteriopheophytin, namely, 11.13 l./gm. cm., the concentration of the bacteriopheophytin was computed by the equation for Lambert-Beer's Law.

Exposure of a pheophytin solution to light for too long a time caused some decomposition. Small variations in the composition of the solvent had a negligible effect on the absorption coefficient. This method was applicable to the bacterio-chlorophyll from various species of organisms, however, waxy material from some strains of brown bacteria caused difficulty. In such cases the final solution was diluted and a longer absorption cell used.

French (1940) also determined bacteriochlorophyll as bacteriopheophytin. The samples he used were suspensions of *Spirillum rubrum* cells which had been disintegrated by supersonic vibration. These suspensions originally contained 30 cu. mm. of cells per ml.

The suspension, 2.55 ml., was treated in the dark with methanol, 22.5 ml., and, when extraction was complete, the cellular debris centrifuged out. The supernatant was removed and the sediment washed with 90% methanol. The extracts and washings were combined with cold ether and the ether washed four times with water. The ether solution was shaken with 20 ml. of 10% hydrochloric acid until the bacteriochlorophyll had been completely converted to bacterio-pheophytin. The ether solution was then washed with water, dried with two lots of calcium chloride and taken to dryness on the steam bath. The pheophytin residue was dissolved in 4.1 ml. of chloroform and the optical density measured at 530 mμ in a 1 cm. absorption cell.

At 530 mμ bacteriopheophytin in chloroform has a specific absorption coefficient of 28.2 l./gm. cm.

G. Bacterioviridin.

The empirical and structural formulas of bacterioviridin are not known with certainty. Schneider (1934b) and Fischer, Lambrecht, and Mittenzwei (1938) commented on the similarity of this pigment to the green pigment obtained by the oxidation of bacteriochlorophyll which Fischer *et al.* identified as 2-acetyl-chlorophyll *a* (Fischer and Stern, 1943, p. 311). Larsen (1953), however, states that "No experimental evidence is . . . presented to support this . . . "

Bacterioviridin occurs in the photosynthetic green sulfur bacteria.

LARSEN (1953) proposes that the name of this pigment be changed from bacterioviridin, so-named by METZNER (1922), to chlorobium-chlorophyll.

I. Preparation of Bacterioviridin.

Bacterioviridin has not been obtained as a pure pigment in the solid state. It is known only in solution. BUDER (1913) extracted the green pigment from the green bacterium, *Chloronium mirabile*, with alcohol. METZNER (1922) also used

alcohol to extract this chlorophyll from dried green sulfur bacteria of unknown species. KATZ and WASSINK (1939) obtained an ethyl alcoholic solution of the pigment from *Chlorobium limicola* from which they partially removed the yellow pigments by partition with petroleum ether. Bacterioviridin was not extracted by the petroleum ether. LARSEN (1953) extracted centrifuged cells of *Chlorobium thiosulfatophilum* with acetone. The absorption spectrum of this extract is shown in Fig. 9.

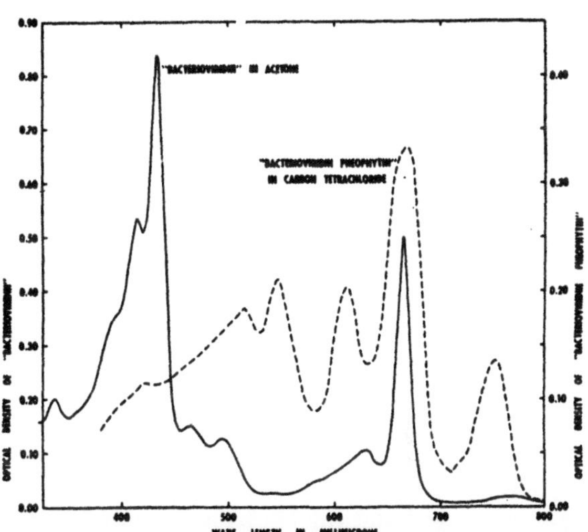

Fig. 9. The absorption spectra of bacterioviridin in acetone and bacterioviridin pheophytin in carbon tetrachloride redrawn from the curves of LARSEN (1953).

II. Physical Properties of Bacterioviridin.

Solubility. Bacterioviridin is soluble in alcohol and acetone. When partitioned between alcohol and petroleum ether it remains mostly in the alcohol solution (KATZ and WASSINK, 1939). EWART (1897) extracted what was reputed to be this pigment from *Bacterium photometricum* (cf. METZNER, 1922). The pigment he obtained could be transferred to ether and benzene from dilute alcoholic solution.

Spectral Absorption Properties of Bacterioviridin. Absorption spectrum curves have been published by KATZ and WASSINK (1939) for crude alcoholic extracts of bacterioviridin, and by LARSEN (1953) for crude acetone extracts. The writers are grateful to Dr. LARSEN for permission to publish his absorption curve, redrawn in Fig. 9.

The absorption maxima obtained by various workers are reviewed here. METZNER (1922) observed in alcoholic extracts of green bacteria a sharp, narrow, intense band at about 670 mμ and a weak, diffuse band almost beyond visibility in the red beginning at about 730 mμ. KATZ and WASSINK's (1939) measurements for alcoholic solutions of this pigment show an intense, sharp, narrow band with maximum at 670 mμ, and a very broad, weak band with maximum at about 770 mμ. The wave lengths of the maxima taken from LARSEN's curve (Fig. 9) for an acetone extract of *C. thiosulfatophilum* are ca. 770, 665, 630, 495, 465, 434, 415, and 337 mμ.

It should be remarked that the weak absorption band at ca. 770 mμ, obtained by all three investigators, corresponds to an absorption band of bacteriochlorophyll and might be due to contamination by this pigment. The maxima in Larsen's curve at 495 and 465 mμ are very probably due to carotenoid absorption (cf. Katz and Wassink, 1939).

Fluorescence. No wave lengths for the fluorescence emitted by this pigment have been published. Buder (1913) noted the fluorescence of the green bacteria when illuminated with near ultraviolet light. Metzner (1922) observed a red fluorescence when the alcoholic solution was irradiated with blue light.

III. Qualitative Tests.

When treated with acid, the green bacterioviridin solutions change to a muddy brown color (Larsen, 1953; Metzner, 1922).

Bacterioviridin gives a positive phase test, according to Buder (1913). The transient color changes were not reported.

IV. Bacterioviridin Pheophytin.

The structural and empirical formulas of bacterioviridin pheophytin [named chlorobium-pheophytin by Larsen (1953)] are not known.

This pigment does not occur naturally so far as is known. It is prepared from bacterioviridin by treatment with acid.

Preparation of Bacterioviridin Pheophytin. Larsen (1953, p. 92) has prepared bacterioviridin pheophytin by the procedure which van Niel and Arnold (1938) used for the transformation of bacteriochlorophyll to bacteriopheophytin (cf. p. 182). Larsen's procedure is quoted here: "A culture of *C. thiosulfatophilum* was centrifuged and the bacteria extracted with methanol. To the green extract was added some CCl_4, a few drops of 25% HCl, and a large volume of water. The solution immediately turned brown, and the pigment was taken up by the CCl_4 phase. After repeated washing with water and drying over anhydrous Na_2SO_4, the brown solution of the pigment was analyzed in the Coleman spectrophotometer, Type 10 S."

Physical Properties of Bacterioviridin Pheophytin. Little is known of the physical properties of bacterioviridin pheophytin.

Solubility. From Metzner's (1922) and Larsen's (1953) observations it may be deduced that this pigment is soluble in alcohol and in carbon tetrachloride, and insoluble in aqueous alcohol.

Spectral Absorption Properties of Bacterioviridin Pheophytin. A qualitative spectral absorption curve of bacterioviridin pheophytin in carbon tetrachloride has been determined by Larsen (1953). We are grateful to Dr. Larsen for permission to use his curve which is redrawn in Fig. 9.

Metzner (1922) published a diagram of the absorption bands of bacterioviridin pheophytin dissolved in acidified alcohol. From this diagram we have estimated the following approximate positions of the absorption maxima: 750, 667, 607, 553, and 519 mμ. Larsen (1953) gives the positions of the absorption maxima in carbon tetrachloride as 752, 668, 612, 546, and 515 mμ.

Fluorescence. Metzner (1922) states that this pigment shows weak red fluorescence.

Qualitative Tests. No data are available on characteristic qualitative tests for this pigment.

No methods have as yet been developed for the quantitative determination of either bacterioviridin or its pheophytin.

H. Special Topics.

Examples of a few types of chlorophyll analyses have been given in detail, but there are a multitude of variations in techniques which may be employed, some of which are mentioned below.

I. Preparatory Procedures.

Collection Procedures. Many materials, such as leaves or other plant organs and various algae, can be collected by hand. If the plant material cannot be collected by hand then special methods must be employed. For example, from cultures of unicellular algae the organisms are collected in centrifuge cups. The size of sample is determined either as the initial volume of suspension, as packed volume, or as weight (EMERSON and LEWIS, 1942; HAXO and CLENDENNING, 1953; ROHDE, 1948). If the average chlorophyll content of individual cells is desired, the number of cells per unit volume of medium is counted, a portion of the medium centrifuged, and the chlorophyll content of the sediment determined (WOLKEN and SCHWERTZ, 1953).

HAXO and CLENDENNING (1953) used a unique procedure to accumulate gametes of *Ulva lactuca*. They illuminated one side of a vessel containing the gametes and the gametes all migrated to that side. They collected the gametes with a medicine dropper and centrifuged them. After determining the packed volume of the sediment they extracted the chlorophyll and estimated it.

In order to assay the chlorophyll in plankton, KOZMINSKI (1938) collected samples of lake water at various depths by means of a hand-operated vacuum pump. He concentrated the organisms in the bowl of a continuous-flow Sharples centrifuge. MANNING and JUDAY (1941) collected lake water in special collecting bottles of 3 liter capacity and accumulated the plankton in the 3 in. bowl of a Foerst continuous centrifuge.

ATKINS and PARKE (1951) filtered sea water through a Whatman No. 50 filter or through a "Gradocol" collodian membrane, pore-diameter 1.09 microns, in order to collect plankton. They also entrained the organisms in a precipitate of aluminum hydroxide.

Extraction Procedures. Concentrated unicellular algae, phytoplankton, and chloroplast samples are frequently extracted directly in the cold with 80 to 100% acetone (ARNON, 1949; ATKINS and KENKINS, 1953; ATKINS and PARKE, 1951; KOZMINSKI, 1938; MANNING and JUDAY, 1941). Extraction has also been made by boiling the plant material with methanol for one or two minutes (FLEISCHER, 1935; JORGENSEN, 1953; ROHDE, 1948), or by heating previous to the addition of methanol (HAXO and CLENDENNING, 1953). Somctimes leaf material has been ground for a short time previous to addition of organic solvents, then ground in the presence of organic solvents (PETERING, WOLMAN and HIBBARD, 1940; SCHERTZ, 1928a; STRAIN, 1938). At other times it has been ground with solvent directly (DELEANO and DICK, 1934; JACOBSON, 1912; KOZMINSKI, 1938; MACKINNEY, 1941; SCHROTT, 1938; WILLSTÄTTER and STOLL, 1918; WURMSER, 1921). The leaves may be ground without abrasive (JACOBSON, 1912; MACKINNEY, 1941; WURMSER, 1921), or with various abrasives, such as glass powder (DELEANO and DICK, 1934) or quartz sand (PETERING, WOLMANN and HIBBARD, 1940; SCHERTZ, 1928a, b; STRAIN, 1938; STROTT, 1938; WILLSTÄTTER and STOLL, 1913, 1918).

Material has often been added to neutralize the plant acids; example, calcium carbonate (KOZMINSKI, 1938; WILLSTÄTTER and STOLL, 1918), sodium carbonate (PETERING, *et. al.* 1940), and dimethylaniline (STRAIN and MANNING, 1942).

Winterstein and Stein (1933) froze their leaves before extraction. Comar and Zscheile (1942) and Comar (1942) used a Waring Blendor to disintegrate leaves in acetone containing suspensions of calcium carbonate.

Solvents besides acetone and methanol have been used for extraction: 92% ethanol (Jacobson, 1912); benzine-benzene-methanol in volume ratios of 22.5 to 1 to 8 (Winterstein and Stein, 1933); and benzine containing a trace of methanol (Strott, 1938).

Filtration Procedures. After the plant material has been extracted it is necessary to separate extract from debris. This has been accomplished by centrifuging and by suction filtration through filter paper. Special methods have also been used: filtration through talc (Willstätter and Stoll, 1918); through Hyflo Super Cel (Strain, 1938); and through a porcelain filter crucible (Rodhe, 1948).

Photometric Analysis. After the extracts have been prepared, the absorption may be measured in 80% acetone (Mackinney, 1941) or the pigment transferred to ether and measured in ether (Comar, 1942; Comar and Zscheile, 1942), or it may be transferred to ether and diluted with ethanol for measurement (Emerson and Lewis, 1942; Strain, 1938). If the magnesium-free derivatives are used, the pigment may be transferred to chloroform for measurement (French, 1940; Guthrie, 1929; van Niel and Arnold, 1938).

The absorption may be measured with polychromatic light by the usual methods of colorimetry or, more reliably, with monochromatic light by the procedures of spectrophotometry.

Instruments Used. A great variety of instruments have been employed. Among the visual colorimeters used are the Duboscq, Zeiss-Pulfrich photometer, and Kober; photoelectric colorimeters, Cenco Photelometer, Klett-Summerson, Evelyn, and "EEL" portable colorimeter. Also a thermo-electric colorimeter was especially constructed by Johnston and Weintraub (1939). For colorimetric measurement either the yellow pigments must be removed (Schertz, 1928 b; Willstätter and Stoll, 1913, 1918) or the light absorbed by them must be taken out by the proper filters so that the light transmitted is absorbed only by chlorophyll. Spectrophotometers: Non-photoelectric spectrophotometers which have been used are Koenig-Martens, Bausch & Lomb, and Hilger ultraviolet spectrophotometer. Photoelectric spectrophotometers: The instruments which have been used are the following: one constructed by Smith (1936), one used by Comar and Zscheile (1942) employing a large Müller-Hilger Universal double monochromator, Cenco-Scheard Spectrophotelometer, Unicam spectrophotometer, and Beckman spectrophotometer, and the spectrophotometer developed by French (French and Young, 1952). With well-constructed spectrophotometers no filters are necessary. The proper wave lengths can be isolated which give maximum sensitivity for the estimation of the chlorophylls.

II. Spectrophotometry.

The determination of a chlorophyll spectrophotometrically depends upon knowing the quantitative relation between its light absorption and its concentration. This relation is expressed mathematically by the Lambert-Beer's law, in the equation

$$D = \log (I_0/I) = \alpha \, dC.$$

For analytical purposes, the specific absorption coefficients, α, at various wave lengths are most useful constants. A number of curves relating these values for certain pigments have been prepared by the writers and are given in the forepart of this chapter.

The determination of the correct values of the specific absorption coefficients depends upon the accurate determination of both the concentrations of the solutes and the optical densities of the solutions at the various wave lengths.

Determination of Solute Concentrations. The simplest method of determining the solute concentration is to dissolve a weighed quantity of solute in a known volume of solution. Whenever possible this method should be used. It is not possible to use this method for certain chlorophyllous substances, however, because of the change produced in the absorption coefficients of some pigments by drying (p. 147).

ZSCHEILE and COMAR (1941) determined the concentration of chlorophylls a and b by taking to dryness a known volume of the pure chlorophyll solution the absorption spectrum of which had been determined and weighing the thoroughly dried residue. This is satisfactory if the chlorophyll is the only solute present in the solution, if it does not form a solvated residue of unknown composition, and if it does not decompose when thoroughly dried.

KOSKI and SMITH (1948) and the writers have determined the chlorophyll concentrations of solutions by magnesium analysis (p. 161). The chlorophyll content of the solution in grams per liter was computed from the following equation

$$\text{Chlorophyll (gm./l.)} = \frac{\text{gm. Mg} \times 1000}{\text{Mg}_{chl} \times \text{V}}$$

Where the magnesium content of a known volume of solution (gm. Mg/V) was determined analytically and the magnesium content of the chlorophyll (Mg_{chl}) is known.

The regular precautions of quantitative analysis should be used in the preparation and dilution of pigment solutions for absorption spectra measurements. Strong light should be avoided during the preparation and storage of solutions. Changes sometimes occur during the storage of solutions so that storage should be avoided if possible. When not possible the solutions should be kept cold and dark.

Measurement of Optical Density. Optical density measurements to be of analytical significance must be accurately made at definite wave lengths. Such measurements may be made with visual, photographic, or photoelectric instruments. Because of their many advantages, photoelectric instruments have been used almost exclusively in recent years for quantitative measurements. The linearity of response to transmitted intensity can be established by showing that LAMBERT-BEER's law is obeyed at different concentrations or depths of solution when a solution known to obey this law is used; or by use of neutral filters of known transmission. The accuracy of the wave length readings may be checked against known emission lines of various elements, e. g., mercury, hydrogen, or helium. And the over-all behavior of the instrument can be determined by measuring the absorption curve of a calibrated colored glass or of a known solution at a series of wave lengths (SMITH, 1936; ZSCHEILE, COMAR, and MACKINNEY, 1942). The optical densities measured should ordinarily lie between 0.1 or 0.2 and 0.8. These limits depend somewhat on the precision of the instrument being used.

The absorption cells should either be exactly matched for transmissivity or the proper corrections made. For use with volatile solvents, ground-glass or ground-quartz stoppered cells should be used.

Because strong light oftentimes bleaches chlorophyll solutions, strong light should not be used in absorption-spectrum measurements.

Quantitative Determination of the Chlorophylls by Spectrophotometry. The quantitative determination of the chlorophylls and chlorophyll derivatives by spectrophotometry depends upon knowing the specific absorption coefficients of the pigments being determined and also on the absorption due to other pigments at the wave lengths employed. If a given chlorophyll is the only absorbing substance in solution its concentration can be calculated from the Lambert-Beer's law (pp. 158, 186).

If, in a mixture, there is no wave length at which the pigment to be estimated is the sole light absorber, then the optical densities at more than one wave length must be measured. Wave lengths are chosen at which the absorption coefficients of the absorbing substances differ as much as possible. From the measurements, the concentrations of the unknown pigments are calculated by means of simultaneous equations. For two substances the following equations are used:

$$\alpha_a' dC_a + \alpha_b' dC_b = D'$$
$$\alpha_a'' dC_a + \alpha_b'' dC_b = D''$$

where C_a and C_b are the concentrations of the two absorbing substances, d is the length of the light path, α_a and α_b are the specific absorption coefficients of the two substances, and D is the optical density. The prime (') and double prime (") refer to the two wave lengths employed. These simultaneous equations may be solved conveniently for C_a and C_b by determinants. An extension of the equations and their solutions to three or more substances can easily be made.

Qualitative Use of Absorption Spectra. Because the absorption curves of the chlorophylls and their derivatives have characteristic shapes, they can be used either for qualitative identification of the pigments or for testing their purity.

The absorption spectra of the porphyrins, chlorins, and pheophytins are affected by the pH (Granick and Gilder, 1947, p. 348; Livingston, Pariser, Thompson, and Weller, 1953).

The absorption spectra of the various chlorophylls differ in various solvents. Extensive qualitative data are available for chlorophylls a and b (Harris and Zscheile, 1943). The wave lengths of the absorption maxima vary with the solvent and attempts have been made to relate this variation to the refractive index of the solvent according to Kundt's rule. This rule states that the greater the refractive index the longer the wave length of the absorption maxima. This rule is not strictly valid, as the discussion by Rabinowitch (1951, pp. 637ff.) shows.

Infrared Spectra. Infrared spectra of various chlorophylls and their derivatives have been used very little for analytical purposes. Although infrared spectra are not suitable at present for quantitative analysis of the chlorophyll pigments, they appear to offer splendid possibilities for comparative and identification purposes. Granick, Bogorad, and Jaffe (1953) were able to identify synthetic and natural hematoporphyrins by such measurements and to demonstrate the presence of a secondary hydroxyl group.

Weigl and Livingston (1953) have analyzed the infrared spectra of chlorophylls a and b, pheophytin a, allomerized chlorophyll a, and bacteriochlorophyll. They have identified the bands that belong to the phytyl group and those that do not, the bands that are common to the various pigments and those that are unique. From this analysis, they have tentatively assigned frequencies to the various structural groups in the molecules.

Infrared absorption and fluorescence spectra have been used to advantage in demonstrating variations in the natural state of bacteriochlorophyll (Duysens, 1952; French, 1940; Wassink, Katz, and Dorrestein, 1939).

Further advance in the infrared techniques may furnish a powerful tool for chlorophyll analysis.

III. Chromatography.

Columnar Chromatography. The usefulness of the columnar chromatographic method for the separation of the various chlorophylls has been amply demonstrated in this chapter.

Adsorbents. Various adsorbents have been used for separating the chlorophylls from other pigments and from each other: calcium carbonate, inulin, or powdered sugar (TSWETT, 1906), powdered sugar mixed with talc, 6:1 (WINTERSTEIN and STEIN, 1933), starch (STROTT, 1938), inulin and magnesium citrate hexahydrate (MACKINNEY, 1940), and paper pulp (WEIGL and LIVINGSTON, 1953).

Powdered sugar is the most useful of the adsorbents because of its cheapness, availability, and simplicity of preparation (WINTERSTEIN and STEIN, 1933). In our own laboratory, commercial confectioner's sugar, containing 3% starch, can be used without further treatment. Preliminary drying and sifting may be advantageous under some conditions. Powdered sugar has been used for the separation of various chlorophylls by a large number of workers (HARRIS and ZSCHEILE, 1943; KOSKI and SMITH, 1948; MANNING and STRAIN, 1943; SEYBOLD and EGLE, 1938/39a, b; STRAIN, MANNING and HARDIN, 1943; WINTERSTEIN and STEIN, 1933; ZSCHEILE and COMAR, 1941).

Many adsorbents, particularly inorganic adsorbents, cause decomposition of chlorophyll in the adsorbed state. WINTERSTEIN and STEIN (1933) list as such aluminum oxide, calcium carbonate, and sodium sulfate. STRAIN (1942b, p. 123) observed that chlorophyll was rapidly decomposed when adsorbed on magnesium citrate which was acidic but that the destructive action was prevented when dimethylaniline was added to the solvents used.

Solvents. Petroleum ether has been almost universally used for the initial adsorption of chlorophyll from plant extracts. For the development of a chromatogram or in subsequent adsorptions for purposes of purification, other solvents have been added to the petroleum ether or used alone. Favorite combinations are mixtures of petroleum ether with benzene, ether, acetone, methanol, propanol, or dimethylaniline. Benzene, ether, or acetone may be used at times in the pure state. Benzene is especially useful for removing the yellow pigments from the chlorophylls. The alcohols and dimethylaniline are active eluants; consequently, they are usually added to petroleum ether in small concentrations, from 0.2 to 5.0 per cent. Toluene, xylene, and carbon bisulfide have also been used as developers (TSWETT, 1906).

Development. No predetermined recipe can be given for the chromatographic separation of pigments from plant extracts. The most suitable combination of solvents must be determined for each separation. This is done by developing the column with solutions containing increasing concentrations of the component added to move the bands down the column at the desired rate.

Inasmuch as each adsorbed component leaves a trail of adsorbed material behind it, the most adsorbed components are contaminated by traces of the pigments which preceded them on the column. To eliminate certain contaminants, reversal of the order of zones on the column is advantageous.

The reversal may be achieved by change of solvent as shown by STRAIN (1942a, p. 1155). When chlorophylls *a* and *b* and fucoxanthin are adsorbed on sucrose from petroleum ether the fucoxanthin is adsorbed below chlorophylls *a* and *b*; when adsorbed from petroleum ether containing several per cent of alcohol it is adsorbed above chlorophylls *a* and *b*. Lutein is adsorbed on cane sugar from petroleum ether solution below the two chlorophylls, but when a small percentage of alcohol is added to the petroleum ether it is adsorbed between the two chlorophylls.

Extremely small quantities of pigment are detectable by chromatography. STRAIN and MANNING (1942) reported that when chromatograms of chlorophyll b were developed on sugar columns with petroleum ether containing 1% methanol, 13 γ formed a band on a column 2 cm. in diameter, 1.3 γ a detectable band on a 1.4 cm. column, and 0.03 γ, a barely perceptible band on a 0.2 cm. column. Even from extracts of barley leaves in which the chlorophyll a to b ratio had been increased to 2000, 1.3 γ of chlorophyll b was detectable on a column 1.4 × 12 cm. One can detect on a sugar adsorption column 1 mm. in diameter 0.01 γ of chlorophyll a (STRAIN, 1942a). This great sensitivity enables the detection of pigment impurities in minute quantities and serves to check the purity of pigment preparations.

One of the most convincing tests for the identity or nonidentity of two chlorophyll samples is the homogeneity or nonhomogeneity of a mixed chromatogram. For this test the two pigments (only very small samples of which are needed) are chromatographed separately under identical conditions to insure their purity. If, when solutions of these pigments in the same solvent are mixed and chromatographed, two adsorption zones are formed, the pigments are not identical. If, however, only one zone is formed under a variety of conditions for developing the chromatogram it is highly probable that the two samples are the same pigment.

Elution. For elution from the sugar column the most used solvents are ether, acetone, and methanol. Mixtures of petroleum ether with these solvents are also widely used. Sometimes it is necessary to add methanol to ether or acetone to effect complete removal of a chlorophyll from the adsorbent.

The procedure for elution may be varied. The column may be developed with a combination of solvents so that the pigments are carried into the receiver and collected in series. This, however, usually requires more time and gives less pure fractions than the mechanical removal of the zones followed by elution of the individual pigments.

The pigment zones after being removed from the adsorption column may be put into the eluting solvent and the sugar removed by filtration, but a more convenient and a more economical procedure is to pack the adsorbate into a glass column of the proper size and pass the eluting mixture through the adsorbate. This method has the advantage that the adsorbent is being washed with progressively purer solvent.

Chromatographic separations combined with spectrophotometric measurements have been used to estimate the individual chlorophylls (cf. p. 156).

Paper Chromatography. For many years paper chromatography has been used to separate plant pigments. TSWETT (1906) separated the pigments of the green leaf by capillary analysis from aqueous alcohol extracts into the following zones: at the top a colorless zone; then a yellow border; a green zone; a yellow slightly greenish zone; and a bright green zone near the bottom where the strip of paper entered the solution.

KYLIN (1927) demonstrated the presence of chlorophyll c (chlorofucine) by capillary analysis: an alcoholic extract of a brown alga was treated with petroleum ether and water added to drive the chlorophyll a and most of the fucoxanthin into the petroleum ether. The alcoholic layer was then extracted with ether and the ether layer after addition of alcohol subjected to capillary analysis. Two bands were formed: a yellow-brown fucoxanthin band and above this a green band of chlorophyll *c*. The chlorophyll *c* band on being dried became yellow-brown and turned green again by wetting with water or alcohol. The band on treatment with 20% hydrochloric acid changed to yellow color and finally bleached.

BROWN (1939) separated the leaf pigments from carbon bisulfide solution on a blotting paper sandwiched between two six-inch square glass plates. The solution to be analyzed and the developing liquid were introduced onto the blotting paper through a quarter-inch hole at the center of the upper glass plate. "The brilliant colours of the carotenoid pigments together with the two chlorophyll pigments present a very striking display when resolved in this way." In some instances the paper was impregnated with other substances particularly with alumina.

The separation of chlorophylls *a* and *b* from leaf extracts of *Tradescantia albiflora* by paper chromatography has been described by BAUER (1952). In his method for one-dimensional ascending development of the chromatogram, a drop of the crude chlorophyll extract was placed 1.5 cm. from the lower edge of a paper strip (Schleicher & Schüll 2043 B) and dried. The paper was 10 to 15 cm. long. The lower end of the strip was dipped into the developing liquid placed in the bottom of a jar. The jar was tightly closed and the chromatogram allowed to develop. The upward movement of the solvent in the paper separated the pigment into the various zones. The paper should be dried at 105° C before use.

In the procedure for two-dimensional development a drop of the crude extract was placed 1.5 cm. in from the left-hand corner of a paper, 10 × 10 cm. or larger, and dried. The edge of the paper was immersed in developing fluid I and the chromatogram developed. The paper was then turned so the original spot was at the right-hand corner and developed with solvent II. After development, the pigment spots were distributed more or less diagonally across the paper in the following order beginning at the origin: chlorophyllins *a* and *b*, unidentified pigment, chlorophyll *b*, xanthophyll epoxide, chlorophyll *a*, pheophytin *b*, xanthophyll, pheophytin *a*, and carotene. It is recommended that the container be flushed with nitrogen and kept in the dark to obtain the best results.

The solvents used were: I) 10 parts special benzine, 2.5 parts petroleum ether, and 2 parts acetone; II) 10 parts special benzine, 2.5 parts petroleum ether, 1 part acetone, and 0.25 parts methanol.

Chlorophylls *a* and *b* have been separated by means of paper chromatography from extracts of soy-bean leaves by LIND, LANE, and GLEASON (1953). The leaves were extracted with acetone and the pigments transferred to petroleum ether (Skelly-b). The petroleum ether solution was washed with water and concentrated under reduced pressure. The solution (0.1 ml.) was placed at the lower left-hand corner of a filter paper 23 × 23 cm. (Whatman No. 1) which had previously been washed with petroleum ether and dried in the air. After the filter paper had been stapled together in the form of a cylinder it was set in about 100 ml. of the developing solvent contained in a tightly covered gallon jar. After each of the successive treatments the filter paper was air dried.

For development in the first dimension, acetone was used as solvent which carried all the pigments with it. The development with acetone was stopped when the pigments reached the top of the original spot. Development then proceeded with petroleum ether until the solvent front carrying the carotene reached 20 cm. from the bottom of the paper. Next, 1% n-propanol in petroleum ether was used until the front reached 20 cm. from the bottom of the paper which moved all the remaining pigments and separated the chlorophylls. The filter paper was rolled in the opposite direction, stapled, and a chromatogram developed in the second dimension with 25% chloroform, which had been washed and dried with calcium chloride, in petroleum ether (Skelly-b). With this solvent the chlorophylls moved relatively slowly and were freed of the faster moving xanthophylls. When the solvent front had moved up about 16 cm. the cylinder of paper was removed and dried. The chromatogram showed substances in the relative positions pictured in Fig. 10. The identification of the pigments was made spectroscopically.

STRAIN (1953) has separated the leaf pigments by ascending paper chromatography either on strips (3 × 20 cm.) of cellulose paper (Eaton-Dikeman Grade 301) or glass paper. The papers were used with or without impregnation by a second liquid phase such as methanol, glycerine, or vaseline. When the paper was unimpregnated or else impregnated with methanol or glycerine, petroleum ether or petroleum ether containing 0.5% propanol was used as developer. When the paper was impregnated with vaseline, 80% methanol was used as wash liquid.

In practice, the pigments were dissolved in petroleum ether and placed near the bottom of the paper in a spot 7 mm. in diameter. The bottom of the strip was placed in the developing liquid contained in a 2 liter beaker and the beaker covered. After development for from 20 to 40 minutes the separation of pigments was complete.

When the paper was impregnated with a little glycerine and developed with petroleum ether containing 0.5% propanol, the sequence of pigments from the bottom of the strip

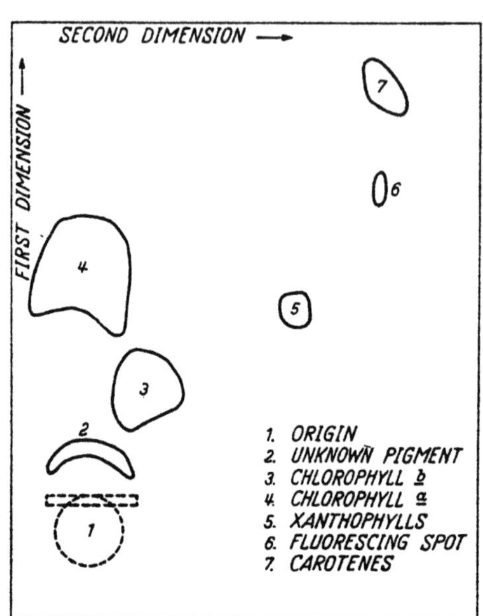

Fig. 10. The distribution of plastid pigments from soybean leaves in a paper chromatogram obtained by two-dimensional ascending development. [LIND, LANE and GLEASON (1953).]

upwards was neoxanthin, chlorophyll *b*, violaxanthin, chlorophyll *a*, lutein plus zeaxanthin, and carotenes. When the paper was impregnated with vaseline and developed with 80% methanol the sequence was nearly reversed, namely, carotenes, chlorophyll *a*, chlorophyll *b*, lutein plus zeaxanthin, violaxanthin, and neoxanthin.

By this change in procedure the sequence of pigments was altered, and the positions of the chlorophylls reversed. Such a change of sequence is very useful for the purification of pigments as has been explained on p. 189.

Impregnation of the filter paper with sucrose is advantageous for the separation of the chlorophylls by paper chromatographic procedures according to SPORER, FREED, and SANCIER (1954).

IV. Miscellaneous Procedures.

Phase Test. When an ether solution of a chlorophyll is underlaid with an equal volume of a 30% solution of potassium hydroxide in methanol, a colored ring is formed at the interface of the two phases which is characteristic of the chlorophyll being tested. When the solutions are mixed, this color permeates the alcoholic layer. The characteristic color is transient, lasting only from a few seconds to a minute or two. The alcoholic layer finally becomes about the same color as the original ether solution. At the end of the test, if the chlorophyll was pure, the ether layer is colorless.

The mechanism of the phase test is represented by STOLL and WIEDEMANN (1952) as enolization of the hydrogen atom on C-10 to the carbonyl group at C-9. This forms a new double bond between C-10 and C-9 which enters the conjugation of the phorbin system and causes the color change. At this stage the reaction is reversible, but further action of the alkali hydrolyzes the linkage between C-9 and C-10 forming the green compounds. These reactions are represented for pheophorbide a by the following structures:

Since chlorophyll is oxidized by air in the presence of alcohol to form a hydroperoxide on C-10 (FISCHER and STERN, p. 25) there is no longer an enolizable hydrogen on C-10 and the phase test fails. But the phase test fails only when a considerable fraction of the chlorophyll has been allomerized. Its failure is no guarantee for the absence of allomerization.

The Hydrochloric Acid Number. The distribution of porphyrin and phorbin compounds between organic solvents and aqueous solutions of various hydrogen ion concentrations has been very useful for separating mixtures of these compounds and for identifying individual members of the groups. The two-phase system most widely used is ether and hydrochloric acid. This method of separation was first introduced by WILLSTÄTTER and MIEG in 1906 for the separation of a number of different porphyrin-like substances.

WILLSTÄTTER assigned to each compound a characteristic number, the hydrochloric acid number ("Salzsäurezahl") which he defined as the percentage composition of the acid which extracts two thirds of the dissolved substance when thoroughly shaken with a like volume of the ether solution (WILLSTÄTTER and STOLL, 1913, p. 269).

For this test it is convenient to use a concentration of 0.02 gm. of substance in 100 ml. of ether, or, if the solubility is less, a saturated ether solution. The hydrochloric acid number can be determined sufficiently accurately in a test tube, since the distribution of the pigment between like volumes of ether and aqueous acid changes very rapidly with the acid concentration. A few per cent change in acid content may be enough to alter the percentage distribution of the pigment from zero to 100 per cent. Usually it is unnecessary to measure accurately the distribution but only to determine whether the substance in question goes mostly into one phase or the other.

A qualitative test of this kind is very useful in determining whether the chlorophyll molecule still contains the phytyl group. For example, if hydrochloric acid of 22% is used pheophytins a and b remain entirely in the ether phase, whereas the phytyl-free pheophorbides enter the acid phase.

There have been several procedures proposed for determining the hydrochloric acid number (FISCHER and KIRSTAHLER, 1931; GRANICK and BOGORAD, 1953; WILLSTÄTTER and MIEG, 1906; WILLSTÄTTER and STOLL, 1913, p. 269). There have also been new distribution constants proposed which are amenable to more exact measurement: a slightly different definition of the hydrochloric acid number by Granick and Bogorad (1953), the "Verteilungszahl" of Willstätter and Stoll (1913, p. 271), the "Halbierungszahl" of Zeile and Rau (1937), and the countercurrent distribution properties of chlorophyll-related compounds determined by Granick and Bogorad (1953) by use of the apparatus developed by Craig and Craig (1950).

Other distribution numbers are the phosphoric acid number and the pH number (Treibs, 1929; cf. Zeile and Rau, 1937).

Useful tables of hydrochloric acid numbers have been prepared (Oppenheimer and Pincussen, 1931; Treibs, 1929; Willstätter, 1915).

Estimation of Phytol Number. The method of phytol determination used by Willstätter and his co-workers is described by Willstätter and Stoll (1913, pp. 308—311). Treibs (1932) has summarized the method, and Fischer, Lambrecht, and Mittenzwei (1938) have refined it. According to the last-named authors it is applicable to samples of 0.1 to 0.2 gm.

Phytol either in the free state or esterified has been detected by polarography (van Rysselberghe, McGee, Gropp and Lane, 1947) and by infrared spectroscopy (Weigl and Livingston, 1953). It is possible these methods may be made quantitative [cf. especially van Rysselberghe et al. (1947)].

Optical Activity. Optical activity measurements have limited use for the analysis of chlorophylls and their derivatives. In case occasion should arise for the use of such measurements, the reader is referred to the works of Stoll and Wiedemann (1933c) and Fischer and Stern (1943, pp. 353—359) for methods and data.

V. Chlorophyllase.

Willstätter and Stoll (1910) discovered an enzyme in leaf material which catalyzes the replacement of the phytyl group in chlorophyll either with hydrogen or alkyl groups. They called this enzyme chlorophyllase. In their Chlorophyll Book (Willstätter and Stoll, 1913, pp. 172—209) they review the early work done in Willstätter's laboratory. This enzyme is widely distributed among photosynthetic organisms and unless care is exercised its action adversely affects the preparation of chlorophyll.

Various methods have been used for the preparation of chlorophyllase (Fischer and Lambrecht, 1938; Fischer, Lambrecht and Mittenzwei, 1938; Mayer, 1930) and for determining the specificity of its action (Fischer and Lambrecht, 1938). Its hydrolyzing (Mayer, 1930), esterifying (Fischer, Lambrecht and Mittenzwei, 1938), and transesterifying (Willstätter and Stoll, 1910), properties have been the subject of several investigations. The temperature optimum of its activity varies with the water content of the medium in which it is suspended (Weast and Mackinney, 1940). Its pH optimum lies near 6.2 (Mayer, 1930).

Note added in proof: After the manuscript of this chapter had been submitted for publication, two important articles by A. S. Holt and E. E. Jacobs have appeared in the American Journal of Botany, Vol. 41 (1954). These articles are entitled Spectroscopy of plant pigments. I. Ethyl chlorophyllides *a* and *b* (pp. 710—717) and II. Methyl bacteriochlorophyllide and bacteriochlorophyll (pp. 718—722).

Acknowledgements.

The writers wish to gratefully acknowledge the help received from the following: Professor Gilbert M. Smith of Stanford University, California, for his cooperation in the obtaining and naming of the algae used for the preparation of chlorophylls *c* and *d*, Professor ARTHUR STOLL, Sandoz, Ltd., Basel, Switzerland, for furnishing manuscript on the preparation of chlorophylls *a* and *b*, Professor C. S. FRENCH and Dr. H. I. VIRGIN of our own laboratory for obtaining numerous fluorescence curves of various pigment preparations, Dr. L. N. M. DUYSENS formerly of our laboratory for giving us the purple bacteria used in the preparation of bacteriochlorophyll, Dr. J. D. GUTHRIE for a personal communication on his method of chlorophyll analysis, Dr. HELGE LARSEN for permission to use material from his publication on the green sulfur bacteria, and to Dr. E. F. LIND and the Editor of *Plant Physiology* for permission to copy Fig. 10.

References.

ARNON, D. I.: Plant Physiol. **24**, 1 (1949). — ATKINS, W. R. G., and P. G. KENKINS: J. Marine Biol. Assoc. of U. K. **31**, 495 (1953). — ATKINS, W. R. G., and M. PARKE: J. Marine Biol. Assoc. of U. K. **29**, 609 (1951).

BAUER, L.: Naturwiss. **39**, 88 (1952). — BIERMACHER, O.: Thesis. University of Fribourg, Switzerland 1936. — BROWN, W. G.: Nature (London) **143**, 377 (1939). — BUDER, J.: Ber. dtsch. bot. Ges. **31**, (80) (1913).

COMAR, C. L.: Ind. Eng. Chem. Anal. Ed. **14**, 877 (1942). — COMAR, C. L., and F. P. ZSCHEILE: Plant Physiol. **17**, 198 (1942). — CRAIG, L. C., and D. CRAIG: "Technique of Organic Chemistry" Vol. III, p. 171 (Editor: A. WEISSBERGER). New York: Interscience Publishers Inc. 1950.

DELEANO, N. T., and J. DICK: Biochem. Z. **268**, 315 (1934). — DUYSENS, L. N. M.: Transfer of excitation energy in photosynthesis. Utrecht 1952.

EMERSON, R., and C. M. LEWIS: J. Gen. Physiol. **25**, 579 (1942). — EYSTER, H. C.: Ohio J. Sci. **50**, 79 (1950). — EWART, A. J.: Ann. Bot. **11**, 468 (1897).

FISCHER, H.: Naturwiss. **28**, 401 (1940). — FISCHER, H., and J. HASENKAMP: Liebig's Ann. **515**, 148 (1935). — FISCHER, H., and A. KIRSTAHLER: Hoppe-Seylers Z. **198**, 43 (1931). — FISCHER, H., and R. LAMBRECHT: Hoppe-Seylers Z. **253**, 253 (1938). — FISCHER, H., R. LAMBRECHT and H. MITTENZWEI: Hoppe-Seylers Z. **253**, 1 (1938). — FISCHER, H., and A. OESTREICHER: Hoppe-Seylers Z. **262**, 243 (1939/40). — FISCHER, H., and A. STERN: Die Chemie des Pyrrols. Bd. II/2. Pyrrolfarbstoffe. Edwards Brothers Inc. Ann. Arbor, Michigan 1943. — FLEISCHER, W. E.: J. Gen. Physiol. 18, 573 (1935). — FRENCH, C. S.: J. Gen. Physiol. 21, 71 (1937); **23**, 483 (1940); in press 1954. — FRENCH, C. S., and V. K. YOUNG: J. Gen. Physiol. **35**, 873 (1952).

GOODWIN, R. H.: Analyt. Chem. **19**, 789 (1947). — GRANICK, S.: J. Biol. Chem. **179**, 505 (1949); **183**, 713 (1950). — GRANICK, S., and L. BOGORAD: J. Biol. Chem. **202**, 781 (1953). — GRANICK, S., L. BOGORAD and H. JAFFE: J. Biol. Chem. **202**, 801 (1953). — GRANICK, S., and H. GILDER: Adv. in Enzymology 7, 305 (1947). — GUTHRIE, J. D.: Amer. J. Bot. 15, 86 (1928); 16, 716 (1929).

HARRIS, D. G., and F. P. ZSCHEILE: Botan. Gaz. **104**, 515 (1943). — HAXO, F. T., and L. R. BLINKS: J. Gen. Physiol. **33**, 389 (1950). — HAXO, F. T., and K. A. CLENDENNING: Biol. Bull. **105**, 103 (1953).

JACOBSON, C. A.: J. Amer. Chem. Soc. **34**, 1266 (1912). — JOHNSTON, E. S., and R. L. WEINTRAUB: Smithsonian Misc. Coll. **98**, No. 19 (1939). — JORGENSEN, E. G.: Physiologia Plantarum **6**, 301 (1953).

KATZ, E., and E. C. WASSINK: Enzymologia 7, 97 (1939). — KAVANAGH, F.: Ind. Eng. Chem. Anal. Ed. **13**, 108 (1941). — KOSKI, V. M.: Thesis. University of Minnesota 1949; Arch. Biochem. **29**, 339 (1950). — KOSKI, V. M., C. S. FRENCH and J. H. C. SMITH: Arch. Biochem. Biophys. **31**, 1 (1951). — KOSKI, V. M., and J. H. C. SMITH: J. Amer. Chem. Soc. 70, 3558 (1948). — KOZMINSKI, Z.: Trans. Wisconsin Acad. **31**, 411 (1938). — KRASNOVSKII, A. A., and K. K. VOINOVSKAYA: Doklady Akad. Nauk. S. S. S. R. **66**, 663 (1949). — KYLIN, H.: Hoppe-Seylers Z. **166**, 39 (1927).

LARSEN, H.: "On the microbiology and biochemistry of the photosynthetic green sulfur bacteria." Trondheim 1953. — LIND, E. F., H. C. LANE and L. S. GLEASON: Plant Physiol. **28**, 325 (1953). — LIVINGSTON, R., R. PARISER, L. THOMPSON and A. WELLER: J. Amer. Chem. Soc. **75**, 3025 (1953). — LIVINGSTON, R., W. F. WATSON and J. McARDLE: J. Amer. Chem. Soc. **71**, 1543 (1949). — LUDWIG, E. E., and C. R. JOHNSON: Ind. Eng. Chem. Anal. Ed 14, 895 (1942).

MACKINNEY, G.: J. Biol. Chem. **132**, 91 (1940); **140**, 315 (1941). — MACKINNEY, G., and M. A. JOSLYN: J. Amer. Chem. Soc. **62**, 231 (1940). — MANNING, W. M., and R. E. JUDAY:

Trans. Wisconsin Acad. **33**, 363 (1941). — Manning, W. M., and H. H. Strain: J. Biol. Chem. **151**, 1 (1943). — Manten, A.: "Phototaxis, phototropism, and photosynthesis in purple bacteria and blue-green algae." Utrecht 1948. — Mayer, H.: Planta **11**, 294 (1930). — Metzner, P.: Ber. dtsch. bot. Ges. **40**, 125 (1922). — Milner, H. W.: J. Biol. Chem. **176**, 813 (1948). — Milner, H. W., C. S. French, M. L. G. Koenig and N. S. Lawrence: Arch. Biochem. **28**, 193 (1950).

Noack, K., and W. Kießling: Hoppe-Seylers Z. **182**, 13 (1929); **193**, 97 (1930).

Oppenheimer, C., and L. Pincussen: "Tabulae Biologicae Periodicae." Bd. I, 1931.

Petering, H. G., W. Wolman and R. P. Hibbard: Ind. Eng. Chem. Anal. Ed. **12**, 148 (1940).

Rabinowitch, E. I.: "Photosynthesis" Vol. 2 Pt. 1. New York: Interscience Publishers Inc. 1951. — Rodhe, W.: Symbolae Botanicae Upsaliensis **10**, 1 (1948). — Rudolph, H.: Planta **21**, 104 (1933/34).

Schertz, F. M.: (a) Plant Physiol. **3**, 211 (1928); (b) Plant Physiol. **3**, 323 (1928). — Schneider, E.: (a) Ber. dtsch. bot. Ges. **52**, 96 (1934); (b) Hoppe-Seylers Z. **226**, 221 (1934). — Seybold, A.: Planta **26**, 712 (1936/37); **36**, 371 (1948). — Seybold, A., and K. Egle: (a) Planta **29**, 114 (1938/39); (b) Planta **29**, 119 (1938/39). — Smith, J. H. C.: J. Amer. Chem. Soc. **58**, 247 (1936). — Smith, J. H. C., and V. M. K. Young: Chlorophyll formation and accumulation in plants. "Radiation Biology" (Editor: A. Hollaender). In press 1954. — Sporer, A. H., S. Freed and K. M. Sancier: Science (Lancaster, Pa.) **119**, 68 (1954). — Stoll, A., and E. Wiedemann: (a) Helv. Chim. Acta **16**, 183 (1933); (b) Helv. Chim. Acta **16**, 739 (1933); (c) Helv. Chim. Acta **16**, 307 (1933); Fortschr. chem. Forschung **2**, 538 (1952). — Strain, H. H.: "Leaf Xanthophylls." Carnegie Institution of Washington Publication No. 490, 1938; (a) J. Phys. Chem. **46**, 1151 (1942); (b) "Chromatographic adsorption analysis." New York: Interscience Publishers Inc. 1942; Year Book No. 42, p. 79., Washington, D. C.: Carnegie Institution of Washington 1943; "Manual of Phycology", p. 243 (Editor: G. M. Smith). Waltham, Mass.: The Chronica Botanica Co. 1951; J. Phys. Chem. **57**, 638 (1953). — Strain, H., and W. M. Manning: J. Biol. Chem. **144**, 625 (1942). — Strain, H. H., W. M. Manning and G. Hardin: J. Biol. Chem. **148**, 655 (1943). — Strott, A.: Jahrb. wiss. Bot. **86**, 1 (1938).

Takashima, S.: Nature (London) **169**, 182 (1952). — Treibs, A.: Liebigs Ann. **476**, 1 (1929); Handbuch der Pflanzenanalyse. Bd. III/2, p. 1351. Wien: Julius Springer 1932. — Tswett, M.: Ber. dtsch. bot. Ges. **24**, 385 (1906).

Van Niel, C. B., and W. Arnold: Enzymologia **5**, 244 (1938). — Van Rysselberghe, P., J. M. McGee, A. H. Gropp and R. W. Lane: J. Amer. Chem. Soc. **69**, 809 (1947). — Vermeulen, D., E. C. Wassink and G. H. Reman: Enzymologia **4**, 254 (1937).

Wassink, E. C., E. Katz and R. Dorrestein: Enzymologia **7**, 113 (1939). — Wassink, E. C., and J. A. Kersten: Enzymologia **12**, 3 (1946). — Weast, C. A., and G. Mackinney: J. Biol. Chem. **133**, 551 (1940). — Weigl, J. W.: J. Amer. Chem. Soc. **75**, 999 (1953). — Weigl, J. W., and R. Livingston: J. Amer. Chem. Soc. **75**, 2173 (1953). — Willstätter, R.: J. Amer. Chem. Soc. **37**, 323 (1915). — Willstätter, R., and W. Mieg: Liebigs Ann. **350**, 1 (1906). — Willstätter, R., and A. Stoll: Liebigs Ann. **378**, 18 (1910); Untersuchungen über Chlorophyll. Berlin: Julius Springer 1913; „Untersuchungen über die Assimilation der Kohlensäure". Berlin: Julius Springer 1918. — Winterstein, A., and G. Stein: Hoppe-Seylers Z. **220**, 263 (1933). — Wolken, J. J., and F. A. Schwertz: J. Gen. Physiol. **37**, 111 (1953). — Wurmser, R.: «Recherches sur l'assimilation chlorophylliene». Paris: J. Hermann 1921.

Zeile, K., and B. Rau: Hoppe-Seylers Z. **250**, 197 (1937). — Zscheile, F. P.: Protoplasma **22**, 513 (1935). — Zscheile, F. P., and C. L. Comar: Bot. Gaz. **102**, 463 (1941). — Zscheile, F. P., C. L. Comar and D. G. Harris: Plant Physiol. **19**, 627 (1944). — Zscheile, F. P., C. L. Comar and G. Mackinney: Plant Physiol. **17**, 666 (1942). — Zscheile, F. P., and D. G. Harris: J. Phys. Chem. **47**, 623 (1943).

Haematin Compounds.

By

E. F. Hartree.

With 12 Figures.

A. Spectroscopic Methods.
I. Introduction.—

Although derivatives of haematin are very widely, if not universally, distributed in plants their concentrations are always low; much lower in fact than in animal tissues. The available information suggests that haematin occurs in combination with specific proteins while some of these "haemoproteins" function as intracellular catalysts. Their occurrence may be extremely wide (e. g. cytochromes) or narrowly restricted to certain specialized types of cell (e. g. haemoglobins in root nodules of leguminous plants).

The field of haematin derivatives in plants has been reviewed by SCARISBRICK (1947) and by HILL and HARTREE (1953). Valuable reviews of the properties and functions of haemoprotein enzymes have been contributed by ZEILE (1950), ZEILE and SIEDEL (1951) and PAUL (1951) while the treatise of LEMBERG and LEGGE (1949) deals comprehensively with all aspects of haematin and its derivatives.

The properties of haematin derivatives of greatest value to the analyst are a) their high and very specific catalytic activities and b) their characteristic absorption spectra in the visible region. It must be emphasized that each haemoprotein catalyst is a stoicheiometric compound of a specific haematin with a specific protein and that the slightest modification of either part will reduce, and often abolish, the specific catalytic function (although not necessarily all catalytic power) and will usually modify the absorption spectrum. It is the denaturation of the protein which is in practice the usual cause of loss of catalytic activity although a destructive oxidation of the haematin group may also take place under more drastic conditions. This lability of haemoproteins must be given consideration in the designing of analytical methods.

The disintegration of many plant tissues, unlike that of animal tissues, often initiates reactions in which endogenous phenolic bodies are rapidly oxidized through the agency of enzymes (catecholoxidases, tyrosinase) to quinones which exert a strong inhibitory action upon haemoprotein catalysts. It is for this reason that methods based upon spectroscopic examination of intact plant tissues are of great importance. The value of such observations on animal tissues by means of a small dispersion (direct-vision) spectroscope was first appreciated by MACMUNN (see below) but the full scope of this technique only became apparent with the rediscovery of MACMUNN's intracellular haematin derivatives by KEILIN (1925) who named them "cytochromes".

The commercial development of efficient and reliable photoelectric spectrophotometers has led to a general belief that a simple direct-vision spectroscope is now outmoded in the study of haematin compounds. However, no quantitative results can be obtained with a spectrophotometer except when clear solutions are used and the isolation of haematin pigments from plant sources in the form of clear solutions inevitably involves considerable losses and is sometimes impossible. Even the detection of absorption bands in plant tissues and homogenates is impossible with a conventional spectrophotometer although with a specialized type of instrument LUNDEGARDH (1953) has been able to study in great detail the reactions of the cytochromes in bundles of wheat roots. A direct-vision spectroscope, however, enables the absorption bands of haematin pigments to be detected in colloidal solutions, in particulate suspensions, and even in intact tissues. By using a spectroscope of this type fitted with a comparison prism accurate quantitative comparisons with standard solutions of pure pigments can be made.

The direct observation of cytochrome absorption bands in living plant tissues is possible in a limited number of cases. However, by means of simple fractionation procedures, which isolate fragmented structural elements, the characteristic bands have been observed in all plant tissues that have been examined (BHAGVAT and HILL, 1951). The enzymes peroxidase and catalase exhibit more diffuse absorption bands but the former has been detected spectroscopically in horseradish root slices at ordinary temperatures (KEILIN and MANN, 1937) and more readily at liquid air temperature (KEILIN and HARTREE, 1949a). The possibility of carrying out reactions on haemoproteins in intact tissues by means of easily-diffusible reagents is helpful in distinguishing between compounds with similar spectra, and is the basis of a valuable method for the determination of total haematin in a plant tissue (ELLIOTT and KEILIN, 1934). When a tissue section soaks up pyridine it becomes more translucent and, if a reducing agent is present, the entire haematin of the various haemoproteins is converted to pyridine haemochromogen, a compound of haem (reduced haematin) with two molecules of pyridine. Of all haematin derivatives haemochromogens exhibit the most intense absorption bands. If the spectrum is matched with that of a standard haemochromogen solution the haematin content of the tissue can be estimated with an error of < 10%. Application of this method to the distribution of haematin in plants (MANN, 1937, 1938) and to the estimation of haemoglobin in root nodules (SMITH, 1949) will be considered below.

For catalase, peroxidase and cytochrome oxidase numerous methods of estimation based upon direct measurements of catalytic activity are available. In the cases of catalase and cytochrome oxidase the methods have been worked out in the course of studies on the enzymes from animal sources but with certain limitations they are readily applicable to plant material.

Progress in biochemistry brings continual improvements and refinements in analytical technique and at first sight it would appear desirable that only the more recent methods should be fully described in the present work. However, it has frequently happened that one particular early method found wide acceptance and has since come to be regarded as the classical method for a particular haematin catalyst. It is, therefore, only possible to make quantitative comparisons with many earlier results if analyses are carried out by the old procedure or if a factor for conversion of a new unit to the old unit is determined. Under these circumstances it becomes essential that certain classical procedures should be fully described.

While it is inevitable that new or modified procedures should continue to appear it is unfortunate that few investigators attempt to relate their new units to one that is widely recognized. This lack of co-ordination is particularly apparent in the case of peroxidase for which many methods of estimation have been published. It is thus seldom possible to collect quantitative data of a type both suitable for tabulation and at the same time covering a comprehensive range of plant materials.

A summary of different schemes of nomenclature of haematin derivatives is given in Table 1.

Table 1. *Nomenclature of Haematin Derivatives.*

Four of the co-ordination valencies are directed to the corners of the square formed by the 4 porphyrin nitrogen atoms. The groups occupying the 5th and 6th positions (on each side of the plane of the porphyrin molecule) are indicated in the table. The nomenclature used in the present article is System I.

On the basis of spectroscopic behaviour peroxidase and catalase can be classified with acid methaemoglobin, cytochromes with haemochromogen $\rightleftharpoons$ parahaematin.

System I (ANSON and MIRSKY, KEILIN)	Fe valency	Co-ordination positions		System II (PAULING, BARRON)	System III (LEMBERG and LEGGE)
		5	6		
haem	2	H_2O	H_2O	ferroheme	heme
haematin	3	OH	H_2O	ferriheme hydroxide	hematin
haemin	3	Cl		ferriheme chloride	hemin
haemochromogen	2	N-cpd.	N-cpd.	ferrohemochromogen	hemochrome[1]
parahaematin	3	N-cpd.	N-cpd.	ferrihemochromogen	hemichrome[1]
haemoglobin	2	H_2O	protein	hemoglobin	hemoglobin
oxyhaemoglobin	2	O_2	protein	oxyhemoglobin	oxyhemoglobin
acid methaemo-globin	3	H_2O	protein	ferrihemoglobin	hemiglobin
alkaline methaemo-globin	3	OH	protein	ferrihemoglobin hydroxide	hemiglobin hydroxide

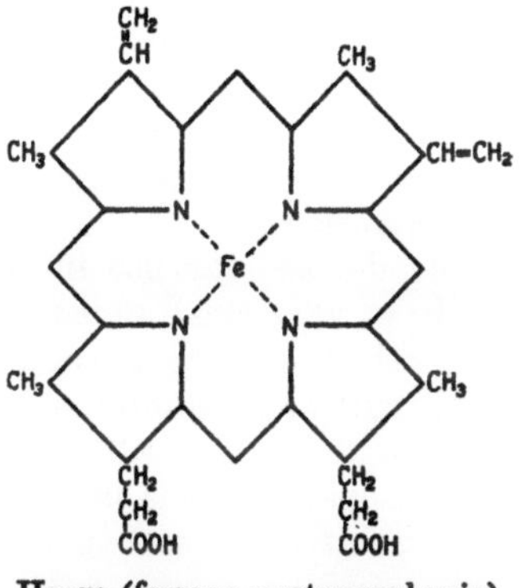

Haem (ferrous protoporphyrin).

Full lines represent coplanar bonds which, on account of resonance, are intermediate between single and double bonds. Water molecules attached to Fe are omitted (see Table 1).

II. Spectroscopes for Analytical Purposes.

The sensitivity of spectroscopic methods for the detection and characterization of pigments showing selective absorption varies inversely with the dispersion. The traditional type of spectroscope incorporating collimator and telescope has such a high dispersion that all but the strongest absorption bands are lost while opalescent or turbid materials transmit insufficient light to form a spectrum of

---

[1] The word hemochrome (no italicised o) is used as a collective term for hemochromes and hemichromes. Similarly hemoglobin includes hemoglobin and hemiglobin. *N*-compounds may be organic bases (e. g. pyridine) or denatured proteins.

any practical value. The direct-vision (hand) spectroscope, however, concentrates the transmitted light over a very short spectrum and suffers from neither of these drawbacks. Furthermore, many bands show a characteristic multiple structure when viewed at small dispersion but this diagnostic aid is lost in a high dispersion spectroscope and to a great extent in a spectrophotometer. It must, however, be realized that proficiency in the use of the small dispersion spectroscope can only come with experience.

Attention to the value of small dispersion instruments was first drawn by SORBY (1867, 1869, 1871) who, in developing a spectroscope-ocular for a microscope, was a pioneer in the application of spectroscopy to biology. His instrument, manufactured by Browning, consists essentially of a hand spectroscope equipped with a means of comparing the spectrum of material under examination with that of a pigment solution. The rays emerging from the two materials enter the one slit and the resulting spectra appear in the eyepiece as a single spectrum with a longitudinal dividing line. A description of the instrument as well as an account of the wide uses to which it was put by early investigators were published by MacMunn (1880). The Browning instrument was superseded by the Zeiss microspectroscope ocular which incorporates an improved optical system. Furthermore, it is possible with this instrument to swivel the upper part (containing the Amici prism) to one side and, by using the lower half as an ordinary eyepiece, to bring microscopical objects into focus before examining their spectra. This instrument and its uses were described in a second contribution by MacMunn (1914). A full description will be given below.

Although the microspectroscope is no longer manufactured, a new instrument supplied by ZEISS-WINKEL (Göttingen) under the name "Pupillen-Spektroskop" can be used in its place since the necessary features, a comparison prism and wavelength scale, are incorporated. Since the writer's main experience has been with the microspectroscope the methods that follow will be based upon this instrument. However, such modifications and accessories that are necessary for the adaptation of the microspectroscope to the techniques to be described can also be applied to the Pupillen-Spektroskop.

The microspectroscope was devised for use in conjunction with objectives of varying power for the study of very small objects. However, if the objective is removed from the microscope the brightness of the spectrum is greatly increased and, with sufficiently high sub-stage illumination it becomes possible to view the spectra of fairly opaque materials. Although the microscope is now no longer an essential part of the apparatus it is nevertheless the most convenient foundation for supporting the spectroscope and the necessary accessories.

III. Apparatus for Comparison Spectroscopy.

The apparatus developed at the Molteno Institute, Cambridge, for the estimation of haematin derivatives is shown in fig. 1 A and represented diagrammatically in fig. 2. The Zeiss ocular *a b d* fits into the draw-tube of a microscope and when the microspectroscope is used in conjunction with a low-power objective a clear rectangular spectrum image is obtained. In absence of an objective (as in fig. 2) the image becomes unevenly illuminated and elliptical. The introduction of a Ramsden ocular *i* restores the spectrum to normal. An alternative method of obtaining even illumination is as follows: The lens below the spectroscope slit (near *b*, fig. 2) is removed from the ocular and, by means of a suitable adhesive, is fitted into the orifice, at the base of the body-tube, into which the objective is normally screwed. The ocular is then replaced. While the focussing screw of the microscope does not affect the focus of the spectrum it will be found to affect the evenness of its illumination and should be adjusted accordingly.

For illuminating the material on the microscope stage a point-source lamp, or one with a very compact filament, is essential. Good results are obtained with

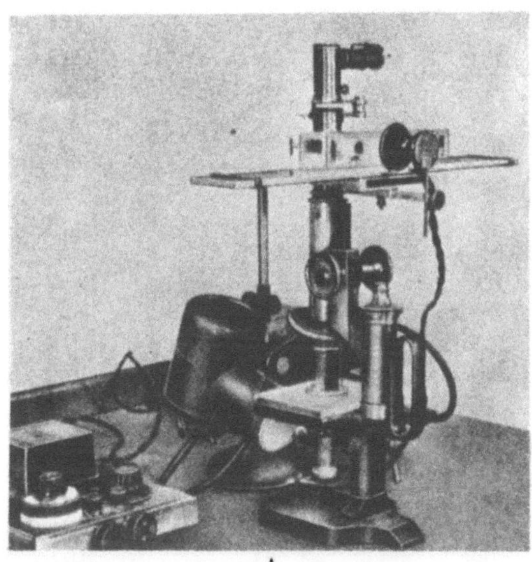

A

B

Fig. 1 A u. B. A. Apparatus for quantitative comparison of spectra incorporating a Zeiss microspectroscope and a double-wedge trough (see Fig. 2). The cell on the microscope stage has a screw plunger giving continuous variation of optical depth up to 2 cm. (Zeiss' "Absorptionsgefäß mit veränderlicher Schichtdicke"). B. Apparatus for maximal illumination of non-transparent specimens. Light from a 500 C. P. "Point-o-lite" lamp (Edison Swan Electric Co., Ltd., 155 Charing Cross Road, London, W. C. 2, England) passes through a CuSO₄ heat filter and is focussed on the absorption cell. Reproduced from the Ph. D. dissertation of C. H. CHIN (University of Cambridge, 1952) by kind permission of the author.

a low voltage (6—8) microscope lamp of at least 24 watts but preferably at least 36 watts. With opaque materials a 100 candle-power "Point-o-lite" (Ediswan) lamp gives better results while for maximum illumination a 500 c. p. "Point-o-lite" lamp can be used, as shown in fig. 1 B. The lamp should be equipped with a lens to focus an image of the filament upon the sub-stage mirror (except in the apparatus shown in fig. 1B where the image is focussed upon the absorption cell).

By means of the comparison prism situated behind the orifice d the spectrum of a standard or known solution ($e_1 e_2$) can be formed alongside that of material

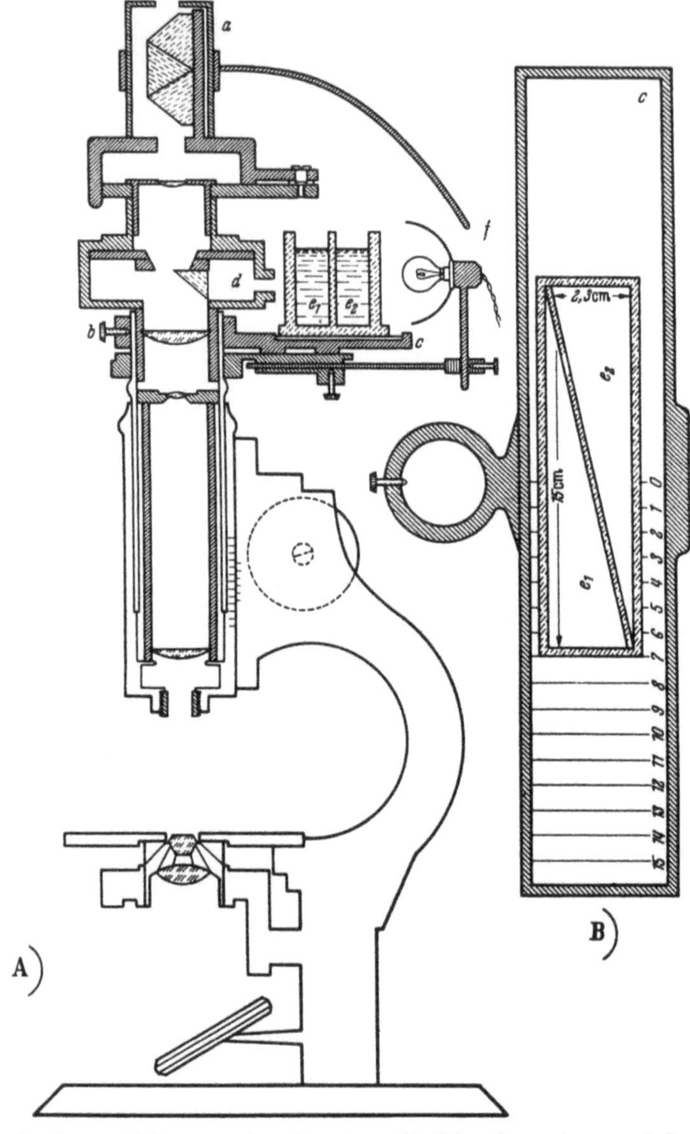

Fig. 2. Apparatus for quantitative comparison of spectra: $a\,d\,b$ Zeiss microspectroscope; i Ramsden ocular; $e_1 e_2$ Double-wedge trough. Reproduced from Keilin and Wang (1946) by permission of the Biochemical Journal.

placed over the sub-stage condenser (as described above for the Browning instrument). An attachment to the side of the instrument, consisting of two clips to hold a small test-tube against the orifice d and a swivel mirror, enables a comparison spectrum to be formed. The uses to which this attachment can be put are limited and it should be removed from the spectroscope. It is replaced by one of the platforms described below and a torch-lamp bulb f (2—3 watts). As a rule this type of lamp gives sufficient illumination since clear solutions are normally used as reference standards. It is adjustable, by means of two milled screws, both in a vertical direction and horizontally (i. e. in the plane of the diagram).

The two lamps are operated from independent transformers and are controlled by rheostats.

Cells containing reference solutions are placed upon a platform attached to a ring (fig. 2B) which fits closely over the draw-tube of the microscope. It is placed in position, and secured by the set-screw b, before the ocular is inserted into the draw-tube. Two types of platform have been used. The one illustrated is 36×5 cm. and is made of light aluminium sheet. The edges of the platform are turned up so as to form a 2 mm. wall around the platform. A piece of millimetre squared paper is cut to fit into the resulting shallow trough and graduations are marked on the paper as described below. A piece of glass, also cut to fit, covers the paper and is held in place by adhesive tape at the two ends of the platform. The second type of platform is smaller (5×5 cm.) and requires no millimetre scale. A device to hold the lamp f is attached to the under side of both platforms.

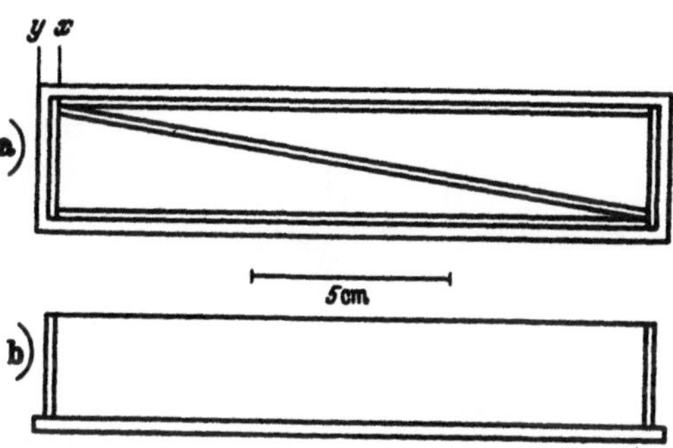

Fig. 3. Plan (*a*) and side-view (*b*) of fused glass double-wedge trough. (Made by C. J. Whilems Ltd., Forest Road, Ilford, Essex, England.)

The long platform serves to carry a fused double-wedge glass trough (fig. 3) which is 15 cm. long (internal) and has an optical depth of 2.3 cm. A centimetre scale is marked out on the squared paper of the platform so that the position of the trough with respect to the optical axis through f and d can be read off. The zero graduation of the scale is shown in fig. 2B to be in the exact centre of the length of the platform. Thus when the trough is situated so that the zero graduation is exactly beneath the *inner* edge of the shorter glass wall (position x, fig. 3) the light from f will pass entirely through the solution in compartment e_1, i. e. the spectrum observed *via* the comparison prism will contain no contribution from the solution in e_2. However, for ease of reading the scale it is better to arrange that when the trough is in such a position the zero on the scale coincides with the edge of the base-plate of the trough, i. e. to position y in fig. 3. The distance xy is about 5 mm., hence the scale should be offset by this distance; in other words the whole scale should be displaced from the centre (in a downward direction in fig. 2B) by a distance equal to xy. If the reading of the edge of the trough (i. e. at y) is r cm. on the offset scale then the observed spectrum will be equivalent to a mixture of the solutions e_1 and e_2 in the ratio $e_2/e_1 = r/(15 - r)$. For analytical purposes the solutions e_1 and e_2 will be either water and a standard solution of a haematin derivative respectively or, if it is required to determine the molar ratio of two haematin derivatives in a mixture, the two components of this mixture as solutions which are equimolar in terms of haematin. In the latter case the changes in spectrum that occur as the trough is moved will represent a changing ratio of two components whose total concentration is constant. The small 5×5 cm. platform is useful as a support for small optical cells containing solutions for purposes of comparison.

The appearance of striations running the whole length of the spectrum indicates that the slit of the spectroscope requires cleaning. For this purpose a piece of soft wood (e. g. matchstick), sharpened to a point, should be moved to and fro along the length of the opened slit. If striations are seen only in the comparison spectrum they are due to the fine structure of the filament of the lamp f and can be eliminated by placing a piece of ground glass immediately before f.

It is essential to protect the sub-stage condenser from damage either by scratching or by contact with tissues or spilled solutions. To this end a piece of plate glass to cover the stage completely should be attached to the stage by means of clips. For rapid examination, rigid or semi-solid material can now be placed directly on the stage while solutions can conveniently be examined in test-tubes. If the test-tube is held in an inclined position the depth of the solution can be varied by a slight movement of the tube across the stage. When a greater optical depth is necessary, or if the volume of solution is small, narrow flat-bottomed tubes can be used. These can be up to 6 cm. long if the microscope is racked up to the full extent. (Solutions in tubes up to 12 cm. in length can be examined if the glass stage cover-plate and the sub-stage condenser are removed and the tube is inserted through the stage so that its base is within 0.5 cm. of the sub-stage mirror. The decreased illumination under these circumstances is unimportant if the solutions are reasonably clear.)

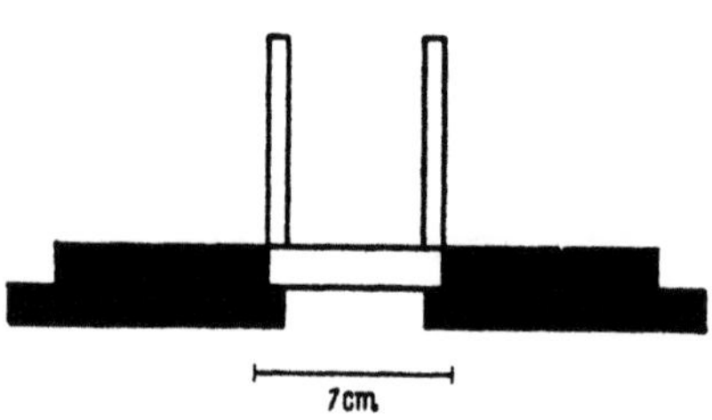

Fig. 4. Example of cylindrical absorption cell with fused optical glass base. The opaque mounting block prevents light passing through the length of the cylindrical wall. (Cells made by The Tintometer Ltd., The Colour Laboratory, Salisbury, England.)

For approximate quantitative work, a measured volume of fluid can be added to a flat-bottomed tube of known diameter and the depth calculated. The addition of further reagents to the tube causes the optical depth to increase in direct proportion to the dilution. Hence the spectrum band intensities will be independent of the degree of dilution (if there is no deviation from the Beer-Lambert Law) and for making spectroscopic comparisons by the methods to be described only the original depth need be known. In short, dilution of a solution in a vertical tube will not cause any "spectral dilution". For more accurate measurements, in which the meniscus error would be serious, tubes of the type shown in fig. 4 are convenient. These consist of cylindrical lengths of glass tubing (1, 2 and 3 cm. long) with fused-on optical glass end-plates. A convenient internal diameter is 9 mm. but smaller diameters (7 and 5 mm.) have advantages when the volumes are limited (5 mm. diameter tubes should not greatly exceed 1 cm. in length or interference from internal reflections may be experienced). The open ends of the tubes are ground flat and polished. In order to prevent light passing through the length of the tube wall a mounting block should be used (fig. 4) which can conveniently be made from two pieces of black polyvinyl chloride sheet (a little of the same material melted in a flame can act as adhesive). Alternatively the tubes can be made with a black glass wall. For quantitative work these tubes are filled completely and closed with a cover slip so as to exclude air bubbles. The examination of materials under anaerobic conditions is described below.

1. Adjusting the Microspectroscope.

The slit of the spectroscope is situated immediately above the comparison prism. The slit adjusting screw can be seen in fig.1 A immediately to the left of the double-wedge trough. Adjacent to this screw is a similar control to limit the width of the spectrum and a lever which allows the comparison prism to be swung out of the field so that only the spectrum arising from samples on the stage can be seen. Above the slit is an eye-piece which is movable vertically for focussing purposes. The cylindrical holder of the Amici prism can be swung aside about a swivel-joint shown in fig. 2 immediately above the letter e_1. When this is done the eye-piece can be removed and the slit then becomes accessible for cleaning. At the top of the spectroscope a horizontal tube (fig. 1 A) carries an engraved wavelength scale. Light from f strikes a mirror fixed to the extremity of this tube and an image of the scale is reflected upwards from the upper surface of the Amici prism. The wavelength scale adjusting screw can be seen in fig. 1 A below the horizontal tube. The scale can conveniently be set with mammalian oxy-haemoglobin (α-band: 577 mμ) or with a solution of didymium nitrate (sharp band at 521 mμ).

Before observations are made the spectrum should be focussed by means of the sliding eye-piece. A fairly sharp-banded spectrum (e. g. didymium nitrate) should be used. Since the spectrum will go out of focus as the slit is opened its aperture must be kept very low and since with materials that are fairly opaque little light is transmitted, observations should not be made in a brightly lit room. In particular, the emission bands of fluorescent lighting may cause considerable interference even if the light only reaches the spectroscope indirectly (e. g. by diffuse reflection from walls).

When a very small object is to be studied or when it is required to examine a very restricted area of a piece of tissue it is essential to ensure that only light passing through the specified tissue reaches the spectroscope. For such work the Ramsden ocular is not used and the lens beneath the spectroscope slit is not removed. A low-power objective is fitted to the microscope, the Amici prism is swung to one side and the slit opened to its full extent. (Alternatively, in the later models of the Zeiss microspectroscope the entire slit mounting can be shifted sideways and so out of the optical path.) The apparatus can now be used as a microscope and the specimen centred. The slit is then closed down and the spectrum width reduced until the image fills the resulting small rectangular aperture. Finally the Amici prism is returned to its normal position for spectroscopic observation.

2. Quantitative Comparison of Spectra.

It is self-evident that if the absorption spectra of two clear solutions are observed in the microspectroscope to match completely throughout their lengths then, if the Beer-Lambert relationship holds, the concentrations of absorbing pigment in the two solutions are inversely proportional to the optical depths. The term "complete match" indicates that not only are the corresponding absorption bands in each spectrum of the same intensity but also the intensity of the background illumination, i. e. the illumination of the spectral regions showing no selective absorption. There is no doubt that the Beer-Lambert Law is valid for haemoproteins and for those derivatives of haematin (e. g. pyridine haemochromogen) that are of analytical value. What is less widely realized is that results of comparable accuracy can be obtained if a clear standard solution is compared with material which contains the same pigment but which is very far from transparent. This point can be decisively illustrated as follows:—

A solution of, for example, oxyhaemoglobin is placed in a flat-bottomed tube on the microscope stage and matched against the same solution in one compartment of a double-wedge trough. The standard procedure is first to match the background intensities by means of the lamp rheostats and then the band intensities by sliding the double-wedge trough. The addition of water to the tube of pigment solution will cause no change in its spectrum and the identity of the spectra will be preserved. If, however, in place of water a diffusing material is added, such as a few drops of milk or a little kieselguhr, the spectra will no longer match. In fact the spectrum of the cloudy fluid will show a less bright background and the absorption bands will appear more intense. If, however, the rheostats are adjusted so that equality of background intensities is restored it will be seen that the band intensities also match; i. e. the spectra become identical again without the need for moving the double-wedge trough. It is in fact more strictly true to say that the two spectra can now be made to match absolutely *either* in the red-yellow-green region *or* in the blue region since the increase of light scattering by small particles at lower wavelengths will give the effect of a general absorption in the blue which is absent from the clear solution. However, if the lack of agreement between the spectra is limited to the blue region there will be no hindrance to the matching of absorption bands and their immediate backgrounds in the longer wave parts of the spectrum. In any case all biological materials contain pigments (flavines, carotenoids) which absorb in the blue region and it is thus never feasible to attempt to match the blue region of the spectra of tissue preparations when haematin pigments are being examined. Fortunately the essential bands of haematin pigments are in the green-red region of the spectrum.

With practice the errors in visual matching of spectra can be kept as low as 3—5%.

IV. Pupillen-Spektroskop.

This instrument (fig. 5) consists of a hand spectroscope *a b* incorporating a comparison prism, which is mounted in an adaptor *c* carrying two swivel-mirrors *d e*. The adaptor fits over a tubular carrier *f* for an eye-piece *g*. The carrier is fitted with an iris diaphragm which is in the focal plane when the instrument is used according to method 2 (below). The mirror *d* serves to illuminate the wave-length scale, an image of which can be viewed superimposed upon the spectrum. Mirror *e*, which is fitted to illuminate the comparison prism, should be removed together with its mounting. Platforms similar to those already described can be made for this instrument although the increased vertical distance between the top of the microscope draw-tube and the comparison orifice will necessitate some small changes in design. Focussing of the spectra is effected by sliding tube *a* within tube *b* while the controls for opening and closing the comparison orifice and for adjusting the slit width take the form of milled rings in the vicinity of the comparison orifice. The types of lamp recommended for the microspectroscope will also be suitable with the present instrument.

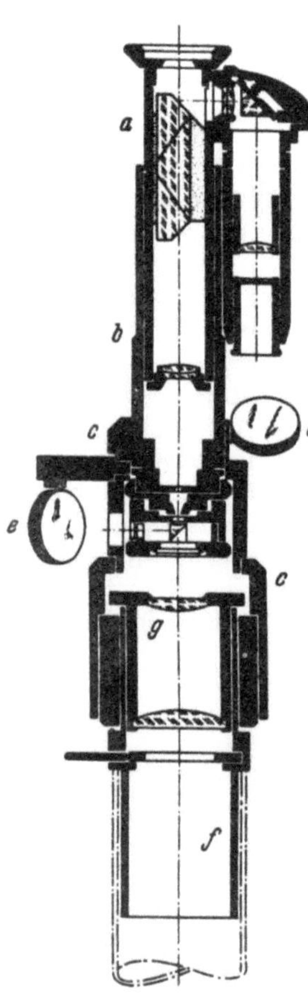

Fig. 5. Pupillen - Spektroskop. *a b* Direct - vision spectroscope; *c* Adaptor to hold spectroscope; *f* eye-piece carrier; *g* eye-piece; *d* wavelength scale mirror; *e* comparison prism mirror (this should be removed). Instrument made by R. Winkel G. m. b. H., Göttingen, Germany.

Like the microspectroscope the Pupillen-Spektroskop can be used 1) without an objective for the study of solutions and of larger tissue specimens or 2) with an objective for observing the spectra of small objects.

Spectroscopy without the Use of an Objective. Considerable distortion of the spectrum is observed unless the eye-piece is removed. In absence of the eye-piece a normal spectrum of uniform illumination is obtained. The intensity of this spectrum can, however, be increased in two ways: either by dropping a Ramsden ocular into the microscope draw-tube (as in fig. 2) or by removing the eye-piece supplied with the instrument and attaching the larger of its two lenses to the lower end of the microscope body-tube where an objective would be attached. The increased illumination obtained in this way is of value in the study of more opaque materials.

Spectroscopy with the Use of an Objective. As with the microspectroscope a Ramsden ocular or equivalent is not used in conjunction with an objective. The spectroscope and adaptor are lifted off and either the object or a certain restricted area of a larger specimen is brought to a focus in the centre of the field. The iris diaphragm is closed until the material in question fills the field, *the eye-piece is removed* and the adaptor and spectroscope are replaced.

V. A Comparison of the Microspectroscope and Pupillen-Spektroskop.

The fitting of a clamping screw to the Pupillen-Spektroskop to prevent its rotating within the draw-tube would be a distinct advantage. A drawback to this instrument is a small gap between the two spectra which makes matching more difficult. Hence, in ordering an instrument, it is advisable to specify that there should be no prominent dividing line. The various adjustments are easier to carry out with a microspectroscope and, with its longer slit, this instrument gives a rather better illumination, particularly when an objective is not used. On the other hand, only the Pupillen-Spektroskop is at present being manufactured[1].

As a general description for both instruments the term "spectroscope ocular" will be used.

VI. The Spectrocolorimeter.

An ingenious alternative to the double-wedge trough method of spectroscopic analysis was devised by HILL (1936). In this apparatus the unknown

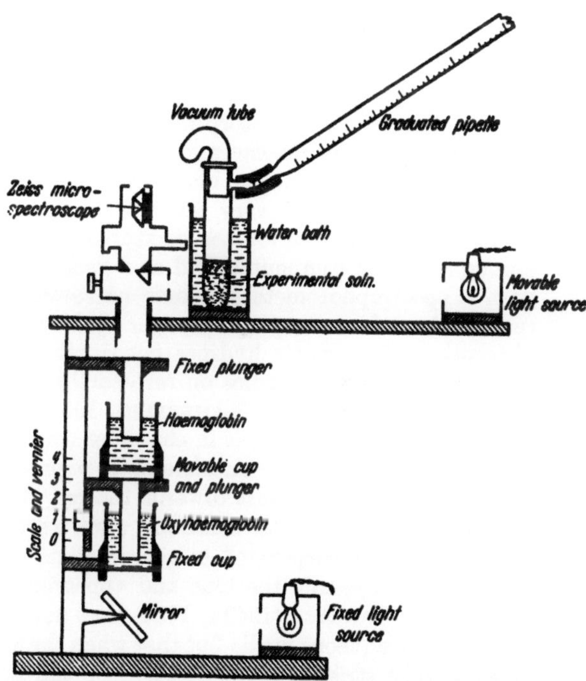

Fig. 6. Spectrocolorimeter. The diagram illustrates the estimation of haemoglobin and oxyhaemoglobin in a mixture at low O$_2$-tension. Replacement of the vacuum tube by absorption cells enables the apparatus to replace that shown in Fig. 2. Reproduced from HILL (1936) by permission of the Royal Society.

[1] After this Section was submitted it was learned that R. & J. Beck Ltd. (address on p. 208 were developing a microspectroscope. This instrument incorporating modifications suggested by the author, will be adaptable to all techniques described in this Section. It will be catalogued as "No. 3000 microspectroscope" and will become available in 1955.

solution is placed before the comparison prism while the light emerging from the standard solutions enters the spectroscope ocular from below. The diagrammatic representation of the device (fig. 6) is self-explanatory. The ocular is mounted above a single two-stage colorimeter attachment which permits optical mixing of the two components present in the unknown solution. Fig. 6 in fact illustrates a method for determining mixtures of haemoglobin and oxyhaemoglobin at various oxygen tensions; a problem that will arise in the section dealing with nodule haemoglobin (p. 239). The vacuum tube can be replaced by any other type of absorption cell and the intensity of illumination can be varied more conveniently by means of rheostats than by moving the light source.

VII. Hartridge Reversion Spectroscope.

The reversion spectroscope was devised by Hartridge (1912, 1923) for precise measurements of the wavelengths of bands in the visible region of the spectrum. With experience an accuracy of $0.1\,m\mu$ or better can be attained. The lamp which is supplied with the instrument is adequate only for the examination of clear solutions. For studying cloudy fluids or plant tissues it is necessary to use one of the more powerful lamps already described.

The instrument is manufactured by R. & J. Beck Ltd., 69, Mortimer Street, London, W. 1, England.

VIII. Intensification of Absorption Spectra at Low Temperatures.

1. The Scope of the Method.

The sharpening of absorption bands at very low temperatures, which may be accompanied by displacement and splitting of the bands, has until recently found little application to biological haematin pigments since solvents which remain transparent at very low temperatures $(ca.-180°)$ cannot be used for proteins. The effect of low temperatures upon absorption spectra is most marked among aromatic compounds and as a general rule the absorption bands, plotted with a spectrophotometer, become narrower and higher as the temperature is reduced [see, for example, the paper by Mayneord and Roe (1937) on the U.V. absorption of aromatic hydrocarbons]. Similar effects have been observed in the visible region in studies on rare earths (Spedding and Bear, 1933) and on porphyrins (Conant and Kamerling, 1931; Hausser, Kuhn and Seitz, 1935) which are strongly aromatic in character (see p. 199).

The first observations on a haematin pigment were made by Hartridge (1921) who detected a slight shift towards the red of the absorption bands of oxy- and CO-haemoglobin as their solutions were warmed from 8° to 40°. When these pigments were incorporated in gelatin films and cooled in liquid air the bands shifted $4\,m\mu$ towards the blue and appeared much stronger. It was shown by Keilin and Hartree (1949a, 1950) that not only could low temperature effects be studied in aqueous media but that when the solvent freezes to a microcrystalline mass a very striking intensification of the absorption bands occurs which is superimposed upon the sharpening already mentioned. Whereas the sharpening phenomenon leads to a change in the character of a band so that it can no longer be precisely matched against an increased depth of the same solution at room temperature, the intensification is a change that can be simulated by an increased depth at room temperature. The only difference in practice between absorption bands of an aqueous solution at room temperature and a thinner layer of the same solution frozen at $-180°$ in such a way as to exhibit the two phenomena of sharpening and intensification is that a band-shift accompanies the sharpening and the greatly intensified band is, as a rule, slightly nearer to the blue end of the spectrum. Naturally these effects can only be observed visually with a small dispersion spectroscope: the freezing of the solvent precludes any spectrophotometric measurements.

Among haematin derivatives the shift of an absorption band that accompanies cooling the solution in liquid air is not characteristic for each derivative. Thus at room temperature absorption bands will shift towards the red as a pigment is caused to aggregate; the maximum shift being attained when the pigment precipitates. KEILIN (1926) found that turacin (copper uroporphyrin), as it occurs in turaco feathers, shows absorption bands ca. 20 mμ nearer the red than does the pigment solution obtained by extracting the feather with ammonia. Various haemochromogens show similar variations as water is added to their solutions to produce turbidity (cf. Table 2). It appears that a band-shift of 6—10 mμ towards the blue is general for haematin derivatives when their solutions are cooled in liquid air but the freezing of·the solvent may lead to aggregation and to a superimposed shift, of the same order of magnitude, in the opposite direction. Thus an alkaline turacin solution shows a marked shift of bands towards the red when it is cooled in liquid air although as a general rule the net shift with haematin compounds is towards the blue.

The positions of absorption bands also vary with the refractive index of the solvent (KAYSER, 1905) but this variable will seldom be significant in the study of haemoproteins.

2. Experimental Procedure.

For spectroscopic examination at low temperatures the material is placed in a "Pyrex" glass tube 20—30 cm. long × 12 mm. diameter, the closed end of which is blown and flattened to form a disk of about 25 mm. diameter with an optical depth of about 2 mm. The material, which should occupy only the flattened portion of the tube, may be maintained at −180° by clamping the tube so that its lower end is immediately above the surface of liquid air in an unsilvered Dewar vessel. A simpler procedure is to cool the tube in liquid air and transfer it rapidly to the stage of the microscope assembly. In the former method a powerful point-source lamp is focussed on the frozen sample and the spectroscope ocular, detached from the microscope, is placed as close as possible to the Dewar vessel (KEILIN and HARTREE, 1950).

The loss of light by scattering naturally increases very considerably when the solvent freezes and when frost subsequently condenses on the cooled specimen. It is, therefore, desirable to incorporate means of varying the intensity of illumination over a wide range (e. g. neutral filters, iris diaphragms). By means of the double-wedge trough a fairly accurate measure of the degree of intensification can be obtained. If the same pigment solution is present in the trough and in the flattened tube a match point can usually be obtained after cooling although a more concentrated standard will be required when, for example, the degree of intensification is more than 10-fold (see below). The band-shift observed at low temperatures does not seriously affect the accuracy of the comparative method and the error should not exceed 10%.

The sensitivity of this method as a means of detecting absorption bands depends upon the degree of intensification which in turn depends upon the fine structure of the frozen solvent. Thus a haemoprotein in aqueous solution (e. g. cytochrome c, cf. KEILIN and HARTREE, 1949a), when cooled in liquid air, shows a spectral intensification of about 6-fold as measured by the double-wedge trough method. In aqueous glycerol, however, the intensification may be as high as 20-fold. To achieve this result the proportion of glycerol (on a volume basis) must be 45—50%. If such a mixture is used as solvent it will, if rapidly cooled, freeze to a clear glass and will then show sharpening (and shifting) of the solute bands but no intensification. If it is now allowed to warm up devitrification will occur; i. e. a mass of minute ice crystals will be formed. This process may be hastened by warming the disk at one point between the fingers until a turbidity

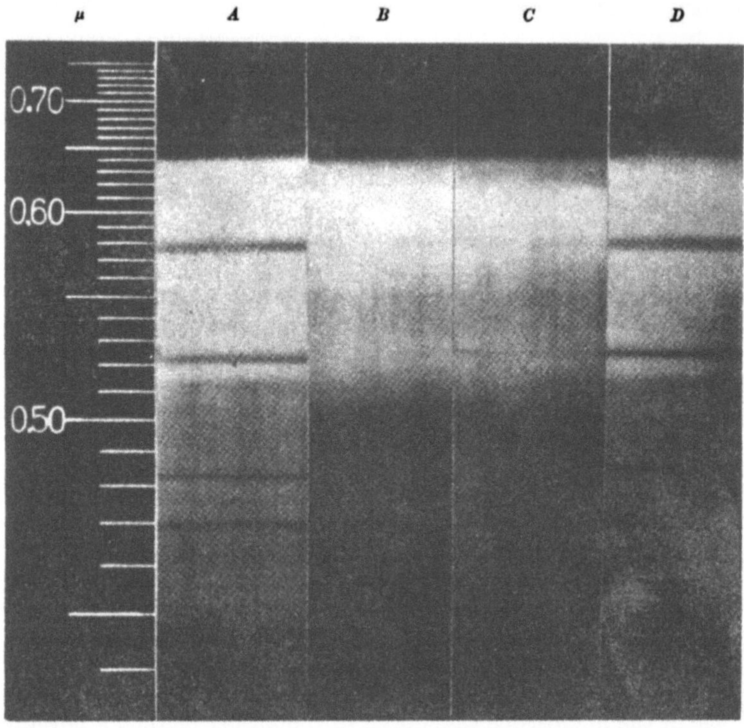

Fig. 7. Intensification of absorption spectra at low temperatures accompanying devitrification of the solvent. Solutions made up in 50% (v/v) aqueous glycerol. *A* 25% didymium nitrate (2 mm. depth). *B* 1.25% didymium nitrate (2 mm. depth). *C* As *B*, cooled to a glass in liquid air. *D* As B, cooled to a glass in liquid air, allowed to warm up until devitrification occurred and re-cooled in liquid air.

is produced. The turbidity will then spread spontaneously throughout the mass. A bluish zone will be seen at the devitrification front as it advances across the disk and the white devitrified mass appears strongly yellow by transmitted light. (These colours are independent of the solute and arise from scattering.) When devitrification is complete the material is again cooled in liquid air. A comparison with an unfrozen sample now reveals an intensification of about 20-fold in the case of certain haematin pigments (reduced cytochrome c, oxyhaemoglobin). Results obtained with didymium nitrate solutions are shown in fig. 7. If the concentration of glycerol is too high devitrification will not occur and no intensification will be observed. On the other hand if the concentration is too low the glass stage cannot be obtained and only the lower degree of intensification characteristic of aqueous solutions will be observed.

IX. Detection of Haematin in Plant Tissues.

A qualitative test for haematin in plants is scarcely necessary since it has been detected in all tissues so far examined and indeed is present in every living cell with the possible exception of some strictly anaerobic microorganisms. Two tests may, however, be considered.

a) Benzidine Test. Haematin in all forms, whether free or in combination with proteins or organic bases, gives a positive reaction. The enzyme peroxidase however is, on the basis of its haematin content, at least 10,000 times more active than haematin in any other form. The activity of haematin generally is thermostable since any denaturation of a haemoprotein has no appreciable effect upon the activity of the haematin moiety. The much higher activity of peroxidase is dependent upon the integrity of the entire molecule and its activity after heating will be insignificant compared with that of the native enzyme. Application of this test before and after heating will, therefore, give an indication of the presence of peroxidase. Since this enzyme is fairly resistant to heat in the crude state it is necessary to heat samples at 100° for at least 15 mins. Best results are obtained with a very dilute solution of benzidine which can be obtained by boiling the solid with molar acetate buffer, pH 5, cooling and filtering. Extracts of plant material are treated with one fifth of their volume of benzidine solution followed by one fifth of their volume of 1% H_2O_2. The presence of haematin is indicated by the progressive formation of a blue colour. Tissue specimens may be dipped successively in benzidine solution and H_2O_2 when the distribution of blue colour gives some indication of the distribution of haematin. Replacing buffer by ethanol, in which benzidine is more soluble, increases the sensitivity of the test but the colour is formed so rapidly that it is impossible to make use of the rate of colour formation as a measure of haematin content. Sensitivity is also increased by the use of 2 N acetic acid as solvent but peroxidase will be partly inhibited at the low pH.

In the case of nodule haemoglobin, details of its distribution have been obtained by microscopic examination of nodule sections after treatment with benzidine and H_2O_2 (SMITH, 1949).

b) Luminol Test. Luminol (o-aminophthalic cyclic hydrazide) exhibits a blue luminescence as it is oxidized by H_2O_2; a reaction which is catalysed by haematin and its derivatives. The test can be carried out as above using 1% luminol in 0.5 M Na_2CO_3 in place of benzidine.

X. Estimation of Total Haematin.

1. Rationale of the Method.

The standard for comparison is crystalline haemin (ferriheme chloride, $C_{34}H_{32}O_4N_4FeCl$, as formula on p. 199 with Cl attached to Fe). In haematin the 5th and 6th co-ordination positions are probably occupied by an OH group and a water molecule. If this structure is accepted the molecular weight of haematin is virtually equal to that of haemin. In haemoproteins the protein satisfies one Fe co-ordination valency and the prosthetic group, as it occurs in such derivatives, will have a lower molecular weight corresponding to the loss of either the OH group or the water molecule (KEILIN and HARTREE, 1951). The molecular weight of a haematin prosthetic group will thus be about 3% less than that of haemin. It is customary, however, to express haematin contents in terms of haemin, or alternatively to avoid confusion by expressing results on a molarity basis.

Since the original method of preparation of haemin described by SCHALFEJEFF (1885) a number of modifications have been introduced. The following method is simple and yields a product which requires no further purification.

14*

Preparation of Haemin. Blood of horse, ox or sheep is defibrinated by shaking with glass beads and straining off the coagulated fibrin. The blood is left standing 12—24 hrs. in glass cylinders at 5° and the upper layer of serum is sucked off. A mixture of 360 ml. glacial acetic acid, 20 ml. saturated KCl solution and 20 ml. water in a 1 l. "Pyrex" conical flask is brought to a vigorous boil on the direct heat of a bunsen burner (i. e. without a gauze). It is removed from the flame and the contents swirled vigorously while 80 ml. red cell suspension is added in a thin stream from a dropping funnel. Care should be taken that the blood falls directly into the hot acid and does not flow down the inside wall of the flask. This addition should be complete in 30—40 secs. The mixture is heated as rapidly as possible to boiling point and boiled briskly for 1—2 mins. The mixture is set aside to cool and, if necessary, further batches are worked up in the same way. It is inadvisable to work with larger volumes. When the heated mixtures have cooled somewhat they are poured into glass jars and allowed to stand 2—3 hrs. Most methods of preparation specify a longer period but this often leads to precipitation of amorphous material without an appreciable gain of haemin. The still warm mother liquor is now sucked off by means of a water-pump to leave a sediment of bluish-black crystalline haemin. The washing procedure for the combined product from 8 batches of 80 ml. red cells is as follows: The crystals are suspended in 400 ml. 10 N acetic acid and transferred to centrifuge tubes. The crystals are allowed to settle under gravity when a little flocculated material may remain at the surface. This material is sucked off together with the wash liquid. This washing is repeated twice with distilled water but in the final washing the mixture is centrifuged to pack the crystals. The haemin is transferred by means of absolute alcohol to a sintered glass funnel and filtered under reduced pressure. The product is finally washed with pure ether and left on the pump to dry. The yield is 0.5—0.7 g. per batch.

Solutions of haemin in sodium carbonate are unstable (Keilin, 1943). Standard haematin solutions (*ca.* 1 mM) should be prepared by dissolving haemin in N NaOH, diluting 10 times with distilled water and preserving the solution in a well-stoppered flask at $\not> 5°$. Such solutions remain unchanged for at least 4 weeks. Haemochromogen solutions prepared from haematin by addition of pyridine and sodium dithionite ($Na_2S_2O_4$) should be protected from air since their autoxidation leads to complete destruction of the porphyrin nucleus.

The methods for haematin estimation can be divided into 2 groups: a) Estimation in intact tissues and turbid suspensions using the double-wedge trough and spectroscope ocular. b) Estimation in clear solutions by spectrophotometry or by colorimetric comparison. Naturally the former method is also applicable to clear solutions.

a) Spectroscopic Comparison as Pyridine Haemochromogen with Standards.

The method was developed by Elliott and Keilin (1934) in the course of investigations on the haematin derivatives in horse-radish root. The haematin, whether present in the free state or as haemoprotein, is converted by treatment with pyridine and $Na_2S_2O_4$ into pyridine haemochromogen. In the visible region the haemochromogen derived from protohaematin (Table 1) shows a strong α-band at 556 mμ and a more diffuse β-band at 524 mμ. The intracellular haematin compounds which are derivatives of protohaematin are catalase, peroxidase and cytochrome b while the uncharacterized haematin compounds in plant tissues, which account for at least 50% of the total haematin, are also derivatives of protohaematin. Two haemoproteins are known, however, which have other haematins as prosthetic groups (see review by Paul, 1951). These are cytochromes a and c. The α-bands of their pyridine haemochromogens are at 587 and 550 mμ respectively. The concentrations of these pigments in plant tissues are low compared with that of protohaematin and the 550 band will thus fuse with the main 556 band and not be detectable. The 587 band will as a rule be invisible in treated plant material unless the cytochromes have been concentrated by the methods to be described (p. 224).

The minimum concentration of protohaematin that permits the α-band of haemochromogen to be observed in 1 cm. depth of fluid with a spectroscope ocular or hand spectroscope is *ca.* 2 μg./ml. (3 μM) at ordinary temperatures. Turbidity does not affect the sensitivity if spectral illumination can be maintained by increasing the intensity of the light source rather than by opening the slit. The following examples will illustrate the method.

(i) **Total Haematin in a Section of Horse-radish Root.** A 1 cm. cube of root is pricked on all sides with a needle and placed in a Thunberg tube with 1 ml. pyridine and 0.1—0.2 g. $Na_2S_2O_4$. The tube is evacuated and filled with air several times, to drive the pyridine into the root, and finally evacuated. (A more permanent specimen can be obtained by filling the tube with O_2-free nitrogen.) After the final evacuation there must be no free liquid pyridine in the tube. On standing for several hours the opaque root becomes translucent.

A standard haemochromogen is prepared by pipetting a known volume of haematin solution (*ca.* 0.5 mg. haemin) into a 25 ml. volumetric flask, adding 5 ml. pyridine, 100 mg. $Na_2S_2O_4$ and diluting to the mark. The comparison of the root cube in the Thunberg tube with the standard is made in the manner already described. Since the root will absorb (and scatter) light very strongly in the blue region a complete match cannot be expected but it is possible to match the backgrounds to the α-bands and then to match the α-bands themselves. For the haematin content of this root ELLIOTT and KEILIN (1934) give 0.035 mg./ml., KEILIN and MANN (1937) 0.03 mg./ml.

(ii) **Variations in Haematin Concentration in a Cross-section of Plant Tissue.** The method of ELLIOTT and KEILIN was extended by MANN (1937, 1938) to a study of the distribution of haematin in tissue sections. A slice of tissue, 1 cm. thick, is pricked with a needle to facilitate the entry of pyridine, placed in a Petri dish, sprinkled with $Na_2S_2O_4$ and treated with as much pyridine as it will absorb during 2—3 hours. Any free liquid is then poured off. The specimen must be kept covered an account of the susceptibility of haemochromogens to oxidation. A low power (16 mm.) objective is fitted to the microscope and the haematin concentration at various points is determined by comparison with suitable standards

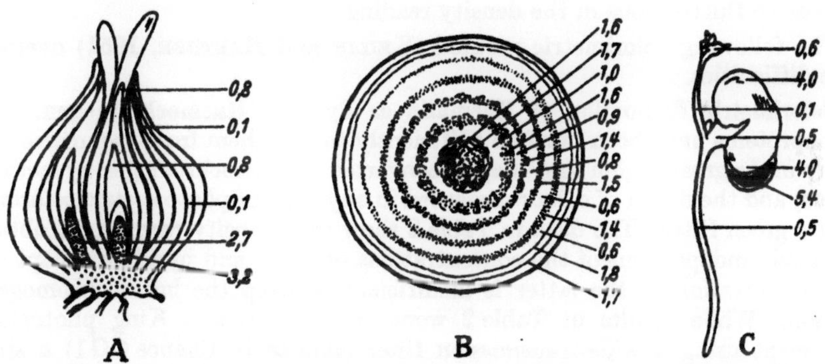

Fig. 8. Distribution of haematin in *A* onion bulb, *B* beet root, *C Phaseolus* seedling. Concentrations expressed as mg. haematin per 100 g. dry weight (after MANN, 1937, 1938). Reproduced from HILL and HARTREE (1953) by permission of Annual Reviews Inc.

in the double-wedge trough. Three typical examples of results that can be obtained by this method are given in fig. 8. Preparations should be examined within 2—3 hours; thereafter the irreversible oxidation of the haemochromogen will become appreciable.

(iii) **Estimation of Haematin in Aqueous Extracts of Plant Tissues and Homogenates.** A measured volume (5—10 ml.) is treated with NaOH to give a concentration of 0.1 N. After standing $^1/_2$—1 hr. sufficient pyridine is added to make a 20% (v/v) solution. If any precipitate is present it can be centrifuged off and discarded if it shows no haemochromogen bands when mixed with $Na_2S_2O_4$. The solution is reduced with $Na_2S_2O_4$ (final concentration *ca.* 1%) and a measured depth is compared with a standard which is made up to contain identical concentrations of NaOH, pyridine and $Na_2S_2O_4$. More purified and concentrated haemoproteins (e. g. catalase, peroxidase, haemoglobin) may require dilution but the concentrations of the various reagents should always be as present in the standard.

b) Direct Spectrophotometric and Colorimetric Estimation of Haematin.

The very pronounced α-band of pyridine protohaemochromogen forms the basis of a sensitive method for estimating haematin in clear solutions (or in liquids which are clarified by addition of NaOH and pyridine). The extinction coefficient at the α-maximum has, according to several authors, the value $\varepsilon_{mM} = 31 - 35.3$ where $\varepsilon_{mM} =$ optical density/mM concentration $\times$ optical depth (see LEMBERG and LEGGE, 1949, p. 177). This method can only be used in absence of other pigments which absorb in the green and is, as a rule, only applicable to haemoproteins from plant sources after considerable purification. It has, for example, been used by THEORELL (1942) to determine haematin in horse-radish peroxidase. The method is liable to the following errors:

(i) The peak of the haemochromogen α-band is very sharp and a slight error in the wavelength setting introduces a serious error. For this reason, sharp absorption maxima are generally less suitable for quantitative determinations than are spectral regions where there are no rapid changes of density with wavelength. Both THEORELL and the present writer (Table 2) find the α-maximum at 556 mμ while LEMBERG and LEGGE give 558 mμ.

(ii) The height of the α-band of pyridine haemochromogen can be made to vary over wider limits than those quoted by LEMBERG and LEGGE through variation in the concentrations of NaOH and pyridine (Table 2). Thus the buffering action of proteins and other ingredients of plant extracts may understandably give rise to fluctuations in the density reading.

The following colorimetric method (KEILIN and HARTREE, 1951) overcomes these difficulties.

Colorimetric Estimation of Haematin as Pyridine Haemochromogen. When haemoproteins have been purified sufficiently to free them from impurities which absorb in the green region they may be converted to haemochromogen by method iii (above) and the depth of colour measured in any type of photoelectric colorimeter using a green filter. The data in Table 2 show that density readings so obtained are largely independent of the concentrations of alkali and pyridine except when the concentration of the latter is insufficient to keep the haemochromogen in solution. While results in Table 2 were obtained with a King photoelectric instrument using a wide-transmission filter (similar to Chance OG 1) a similar independence of density reading may be obtained with a Hilger Absorptiometer and a narrow-transmission filter (Ilford 605). A calibration curve is prepared from standard solutions of haemin in presence of 0.1 N NaOH, 20% v/v pyridine and 1% $Na_2S_2O_4$. The solution to be analysed is converted to a haemochromogen containing the same concentrations of these reagents. Measurements are made against a blank solution containing everything but haematin. After addition of $Na_2S_2O_4$ it is essential that the solutions be left standing for 30 mins. before readings are taken since minor changes of density occur for a short time although the haemochromogen appears spectroscopically to be completely formed as soon as the reducing agent has dissolved.

Table 2. *Light absorption in the green region of the spectrum by solutions of pyridine haemochromogen. Effect of variations in the concentrations of pyridine and sodium hydroxide.*

Haematin Solution: 25.0 mg. haemin was dissolved in 2.5 ml. N NaOH and diluted with distilled water to 25 ml.

1 ml. haematin was placed in each of ten 25 ml. volumetric flasks, treated with pyridine and NaOH as shown, diluted to 25 ml. and reduced with 0.25 g. $Na_2S_2O_4$. The optical density of each solution was measured (against blanks containing all but the haematin) a) in a King colorimeter using a wide-transmission green filter (max. transmission at 535 mμ) and b) in a Beckman quartz spectrophotometer (1 cm. cells) at the α-band maximum.

Ingredients in 25 ml. additional to haematin and $Na_2S_2O_4$		a) Reading on King colorimeter (arbitrary units $\propto$ density)	b) Optical density at α-band maximum	
Pyridine (ml.)	N NaOH (ml.)		mμ	Density
4	0	56	556	1.59
4	0.25	55.5	556	1.67
4	1	55.5	556	1.74
4	5	56	556	1.82
1	0	74[1]		
2	0	72.2[1]	561.6	1.78
3	0	56[2]		
5	0	55	556	1.64
7	0	57.1	556	1.70
10	0	57.9	556	1.86

B. Cytochromes.

I. Spectral Characteristics.

The work of KEILIN (1925, 1933) established that cytochrome is very widely distributed in cells of animals, plants, yeasts and bacteria: in fact it is only in certain strictly anaerobic unicellular organisms that its absence has been adequately demonstrated. These pigments have been detected in every plant tissue which has been examined by suitable means and "it would be justifiable to conclude that the cytochrome system is a characteristic of the plant cell in the same sense as it is of the animal cell" (BHAGVAT and HILL, 1951).

It was shown by KEILIN that the bands of the cytochrome spectrum, which are very prominent in tissues, tissue extracts or cell suspensions of high respiratory activity, are those of a mixture of haemochromogen-like pigments. The most widely distributed type of cytochrome spectrum, which is also the one seen in suitable concentrates of plant material, is that typified by bakers' yeast (fig. 9, I). The bands a, b, c are the strong α-bands of components a, b, c of cytochrome while the band d represents the fused weaker β-bands of b and c. The position of the β-band of component a is still uncertain: in fact there is considerable evidence that this component shows only the one band (a) in the visible region (KEILIN and HARTREE, 1953). In addition to these more easily seen absorption bands there are the Soret- or γ-bands of the three components a — 448, b — 432, c — 416 mμ (KEILIN and HARTREE, 1939). These bands have not been detected in plant tissues because of the very marked general absorption and scattering in this region and because of the low cytochrome concentration: not more than 5% of that in bakers' yeast (BHAGVAT and HILL, 1951).

The bands of cytochrome only become visible when the oxygen supply to the tissue or preparation is limited. Under such conditions the dehydrogenase activities of the cells maintain the cytochromes in the ferrous (haemochromogen)

---

[1] Cloudy solution obtained.
[2] Solution becomes cloudy on standing.

state where they exhibit characteristic sharp bands. When oxygen is present the cytochromes become oxidized to the ferric (parahaematin) forms which, showing only diffuse bands, are invisible in plant preparations. The spectrum observed in intact tissues thus represents an equilibrium between the two forms which is based upon the relative activities of the two processes of oxidation and reduction. In tissue extracts, a deficiency of dehydrogenase substrates may prevent complete or even partial reduction of cytochromes and the maximal intensity of their bands may only be visible after adding $Na_2S_2O_4$ (KEILIN and HARTREE, 1940, 1949b).

Cytochromes a, b and c are not autoxidizable but their aerobic oxidation is catalysed by cytochrome oxidase (p. 223). This enzyme has been characterized as a fourth cytochrome component a_3 (KEILIN and HARTREE, 1939) the main absorption band of which, in the visible region, is superimposed upon the band a. This component is autoxidizable and shows many of the properties of cytochrome oxidase which is now considered as identical with Warburg's Atmungsferment (KEILIN and HARTREE, 1938). It is not possible to separate a and a_3 and there is evidence in favour of their being part of the one enzyme molecule or complex. It has been estimated that their molecular ratio, at least in a heart-muscle preparation, is $a:a_3 = 3:1$ and it is suggested that 4 molecules of a-haematin (which is chemically distinct from protohaematin) are linked to a single protein to form the a—a_3 complex. The linkage of one molecule to the protein, being of different type to the others, is considered as conferring upon that molecule the characteristic reactivities of cytochrome oxidase (BALL, STRITTMATTER and COOPER, 1951). Thus visual detection of cytochrome a in plant material is good evidence for the presence of cytochrome oxidase while proof of cytochrome oxidase activity means that cytochrome a, and by inference a complete cytochrome system, is present. The latter may be at too low a concentration for visual detection of bands even though independent tests for the oxidase are positive.

II. Types of Cytochrome Spectra in Plant Materials.

The absorption bands of cytochrome components are usually faint in plant material and it is essential that an intense illumination of the specimen be achieved. With insufficient light it becomes necessary to work with a wide slit and faint bands will then be lost. The low temperature techniques (p. 208) can be applied here to great advantage.

It is convenient to consider separately the cytochromes in chlorophyll-free portions of plants and in leaves.

1. Cytochromes in Tissues Devoid of Chlorophyll.

Fig. 9 (II) shows the spectrum observed on examining a 3—6 mm. section of the base of an eschallot bulb. On adding $Na_2S_2O_4$ the classical a b c d spectrum is obtained (III). The band d is only recorded in spectra I, II, III, XVIII, XIX, XX of fig. 9. It is of course present wherever b and c are detectable but, being comparatively weak, it is readily masked by other pigments and by light scattering. Since the β-bands of cytochromes b and c do not coincide exactly the position of band d will appear to vary according to the relative concentration of the two pigments. If c predominates it is nearer the blue (519 mμ) than if b predominates (523 mμ). In ungerminated soya-bean seeds, the sides of which had been flattened with sand-paper, YAKUSHIJI (1935) described 4 bands (IV), i. e. a, b, c and a fourth band at 635 mμ[1]. Soaking in water led to a fusion of b and c but if the soaked

---

[1] HILL and SCARISBRICK (1951), however, reported only bands a and c in dry *Phaseolus* seeds (XIV).

beans were powdered and reduced ($Na_2S_2O_4$) distinct a b c bands appeared while
the fourth band disappeared. The latter cannot be identified as a cytochrome
component and its presence in plant tissues has not been confirmed by later
workers. OKUNUKI (1939) found that a number of pollens show the "classical",
or bakers' yeast, type of spectrum (Table 3a; fig. 9, VII). Pollen is best examined
as a thick paste with aqueous glycerol. BHAGVAT and HILL (1951) examined the
chlorophyll-free parts of a number of plants (Table 3b) and detected cytochromes a,

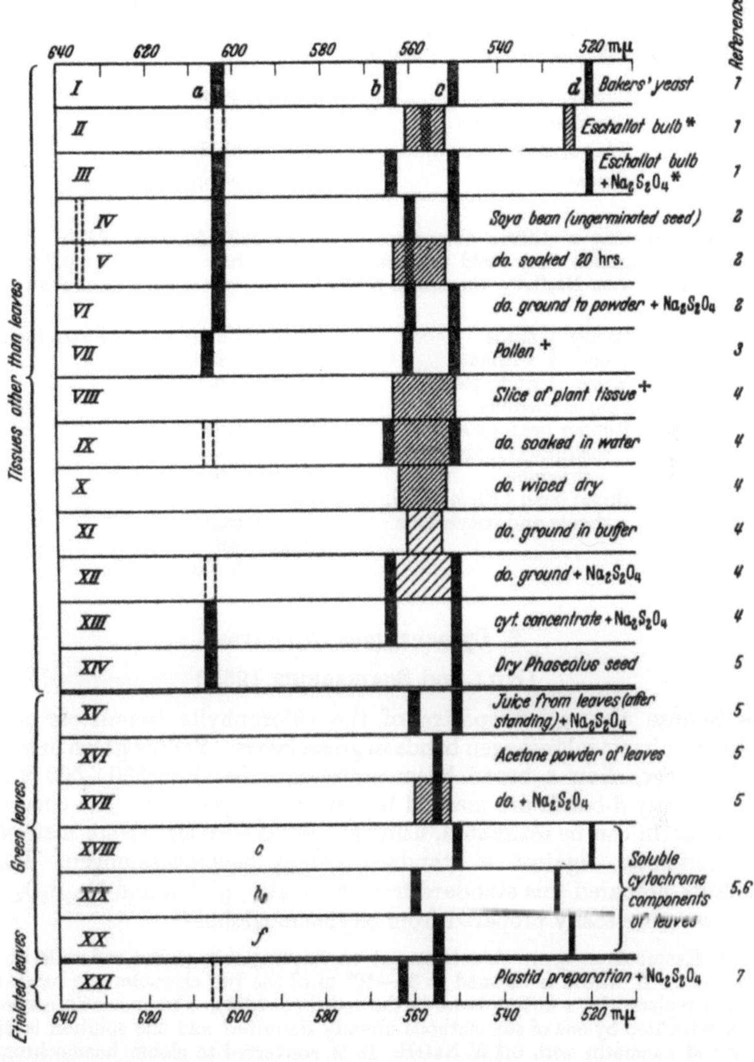

Fig. 9. Positions of absorption bands of reduced cytochromes in plant tissues and extracts. The β-bands (520–530 mμ) are only given in spectra I, II, III, XVIII, XIX and XX. As a rule the β-bands are obscured by other pigments and by light scattering. The band at 635 mμ. in IV and V is not due to a cytochrome component. No attempt has been made to represent relative band intensities within a spectrum except where bands are scarcely discernible (dashed lines) or where a broad area of general absorption occurs (shaded). Similarly no quantitative comparisons between spectra are justifiable: thus the spectrum of bakers' yeast (I) is about 20 times more intense than that of an equal thickness of eschallot bulb + reducer (III). The band positions correspond to wavelengths as measured with a spectroscope ocular or reversion spectroscope and, in XVIII, XIX and XX, do not necessarily correspond with maxima determined spectrophotometrically. Letters in italics refer to cytochrome components; roman letters to spectrum bands. (This convention is also observed in the text.) * Similar bands were observed in other monocotyledons, in leaves and in pollen. + See Table 3. — References: 1. KEILIN, 1925; 2. YAKUSHIJI, 1935; 3. OKUNUKI, 1939; 4. BHAGVAT and HILL, 1951; 5. HILL and SCARISBRICK, 1951; 6. DAVENPORT and HILL, 1952; 7. DAVENPORT, 1952.

b and *c* in all cases. The a b c spectrum can often be seen faintly after soaking the tissue in water but can invariably be seen distinctly after a suitable fractionation (fig. 9, VIII—XIII) such as that used by BHAGVAT and HILL for extraction of cytochrome oxidase (see p. 224). When treated with $Na_2S_2O_4$ such preparations show clear a b c bands.

Table 3. *Plant tissues in which cytochromes have been detected spectroscopically (cf. fig. 9).*

a) Pollens containing cytochromes *a, b* and *c;* also cytochrome oxidase as shown by the photoreversible inhibition by CO of their respiration in presence of glucose (OKUNUKI, 1939). Band positions: a — 605, b — 561, c — 550 mμ

Camellia japonica	*Lilium auratum*
Camellia sasanqua	*Lilium longiflorum*
Thea sinensis	*Lilium philippinense*
Paeonia albiflora	*Pinus densiflora*

b) Tissues showing a diffuse absorption band at 550—560 mμ. After grinding and differential centrifugation they yield preparations which show cytochrome oxidase activity and, on treatment with $Na_2S_2O_4$, the bands a, b and c of cytochrome (BHAGVAT and HILL, 1951).

Barley ⎫ seeds	Golden butter bean *(Phaseolus vulgaris),*
Wheat ⎬ without	shoots and cotyledons
Maize ⎭ endosperm	*Dolichos,* cotyledons
Pea, cotyledons	Leek, bulb
Runner bean *(Phaseolus*	Onion, bulb
multiflorus), shoots and	Dandelion, root
cotyledons	Seakale, root
Broad bean *(Vicia faba),*	Parsnip, root
shoots and cotyledons	Celery, root
	Sugar beet, root
	Rhubarb, flower buds.

2. Cytochromes in Leaves.

(HILL and SCARISBRICK 1951.)

The intense absorption spectra of the chlorophylls completely mask any cytochrome or haemochromogen bands in green leaves. Yellow-green or variegated leaves, however, show a broad haemochromogen band at 560—565 mμ (fig. 9, as VIII) but any β-bands are masked by carotenoid pigments. The concentration of this haematin can be estimated, using the double-wedge trough method or the spectrocolorimeter, against a standard globin haemochromogen. HILL and SCARISBRICK prepared this standard from haematin, globin and $Na_2S_2O_4$. It can, however, be more easily prepared from oxyhaemoglobin.

Globin Haemochromogen. A solution of oxyhaemoglobin (laked red cells or crystals, see p. 243) in 0.1 *N* NaOH is warmed at 35—40° until the two characteristic bands (577 and 542 mμ) are replaced by a diffuse band in the red (haematin). The haematin content of the solution is estimated by one of the methods already described and the solution is diluted to about 10^{-5} *M* haematin with 0.1 *N* NaOH. It is converted to globin haemochromogen by addition of $Na_2S_2O_4$ (ca. 5 mg./ml.).

Table 4,*a* gives analytical data for 3 species of chlorophyll-deficient leaves.

When green leaves are ground with acetone, chlorophyll and carotenoid pigments pass into solution. The insoluble residues from both yellow and green leaves show, when suspended in water, a band at 555 mμ (cytochrome *f*) which is sharper than the haemochromogen band in fresh leaves. On treatment with $Na_2S_2O_4$ the f band becomes more intense and spreads slightly towards the red (fig. 9, XVI, XVII). Although this band cannot be observed in fresh green leaves on account of other plastid pigments it can be seen, together with cytochromes a and b, in intact plastids prepared from leaves of etiolated barley seedlings (DAVENPORT, 1952; fig. 9, XXI). A plastid preparation was obtained by a method essentially the same as that given below for cytochrome oxidase. Cytochrome f has not been detected in any tissues apart from leaves.

Methods of estimation of haematin in green leaves are dependent upon the extent to which haematin is lost during the acetone extraction of chlorophyll. Although no appreciable amount of haematin will pass from a leaf mince into neutral acetone, the pigment will dissolve in acid acetone. While some tissues (e. g. horse-radish root) are sufficiently acid to give rise to a marked loss of haematin, the loss does not appear to be serious in the case of leaves (see discussion by HILL and SCARISBRICK, 1951). It is thus possible to estimate haematin in leaves by comparing the spectrum of a paste of the acetone powder with that of standard globin haemochromogen (Table 4, b). Addition to the acetone powder of $Na_2S_2O_4$ and NaOH gives rise to increases in band intensities which shows that more haematin is present than is immediately visible in the leaf or in the acetone powder. A comparison with the haematin content of potato tuber (which is representative of the level of haematin concentration in non-leafy parts of plants) shows that leaves contain at least 10 times as much haematin (Table 4, c). Even so, the haematin content of leaves is very much less than that of bakers' yeast (Table 4, d).

Table 4. *Concentrations of haematin compounds in leaves compared with potatoes and bakers' yeast.*
(HILL and SCARISBRICK, 1951.)

Concentrations expressed in col. 1 as m. moles haematin per litre of fresh tissue and in cols. 2, 3 and 4 as m. moles haematin per kg. fresh weight.

	Natural haemochromogen (untreated tissue)	Acetone powder	Acetone powder + $Na_2S_2O_4$	Acetone powder + $Na_2S_2O_4$ + NaOH
a) Chlorophyll-deficient leaves				
Euonymus japonica	0.2			
Sambucus racemosa	0.014		0.021	
(var. *aurea*)				
Beta vulgaris	0.024			
b) Green leaves				
Lamium album		0.016	0.046	0.046[1]
Stellaria media		0.020	0.030	0.041[1]
Triticum vulgare		0.026	0.052	0.052[1]
c) *Solanum* spp., tubers (average)	0.0015	(Total haematin: 0.006)		
d) Bakers' yeast		0	0.14	0.29[2]

The accuracy of the comparisons just described is less than is normally attained by methods of spectrum comparison owing to small differences between the unknown and standard spectra which are described in the footnotes below. However, the errors will not as a rule exceed 20 %.

---

[1] α-band is mainly that of cytochrome f (555 mμ) but is fused with a 557 mμ band of haemochromogen derived from other haematin compounds.

[2] α-band consists of fused bands of cytochrome c (550 mμ) and haemochromogen (557 mμ). The latter is derived in part from cytochrome b.

3. Cytochrome in Marine Algae.

Yakushiji (1935) found that a number of green, brown and red marine algae, when ground with acetone, yield a powder showing a broad absorption band at 550—560 mμ which is more intense on the long wave side (fig. 9, as VIII). When soaked in distilled water, fresh tissues of several species *(Phaeophyceae, Rhodophyceae, Chlorophyceae)* yielded extracts in which cytochrome c could be detected (α-band 550 mμ). *Porphyra tenera* thalli gave rise to a deep red solution containing phycoerythrin, phycocyanin and cytochrome c. Progressive addition of ammonium sulphate led to precipitation of the first two pigments while cytochrome c was precipitated by saturating the solution with ammonium sulphate at 50°.

4. Soluble Cytochrome Components.

The quantitative extraction of cytochromes in the form of clear solutions is not possible but estimates of their concentration in plant tissues and homogenates may be made by visual spectroscopic comparison with standardized solutions. Cytochrome c has been obtained in the completely pure state from horse- and ox-heart muscle and can be used as a standard of comparison for estimating the pigment in plant material. Hill and Scarisbrick (1951) were able to isolate c, b_3 and f from leaves (fig. 9, XVIII, XIX, XX). However, the preparation of b_3 in sufficient quantity for spectroscopic comparisons would be extremely tedious and since it has not been completely purified quantitative data would not be obtainable. If the green press-juice of leaf mince is allowed to stand at room temperature for 12 hrs. the chlorophyll is carried down on a copious precipitate. After treating the supernatant fluid with $Na_2S_2O_4$ the faint α-bands of cytochromes b_3 and c can be observed (fig. 9, XV). The former disappears on aeration since b_3 is autoxidizable (Hill and Scarisbrick, 1951). An improved preparation of cytochrome f from parsley (see below) has been reported by Davenport and Hill (1952). Although the pigment has not been purified sufficiently to establish its extinction coefficients, Davenport and Hill show that there is good reason to believe that the extinction at the α-maximum (555 mμ) is equal, within 5%, to that of cytochrome c at the corresponding maximum (550 mμ) and on this basis the concentration of f can be estimated by spectroscopic comparison with standard c solutions or by spectrophotometric measurement of the α-band.

Preparation of Cytochrome c from Heart Muscle. (Keilin and Hartree, 1945a.) One horse-heart (or 2—3 ox-hearts) is cut up, freed from fat and ligaments and finely minced. The mince is mixed with 0.154 N trichloracetic acid (1 l. per kg. of mince) and stirred occasionally during 3—4 hrs. The mass is pressed out through cheese-cloth and finally in a press. The cloudy fluid is neutralized to pH 7 with 10% NaOH and treated with solid ammonium sulphate (500 g. per l.). The mixture is placed on fluted filter papers and left to drain overnight. Next day the clear pink filtrate is again treated with ammonium sulphate (50 g. per l.) and left overnight in a refrigerator. Any precipitate is filtered off and the clear, cold solution is treated with 20% trichloracetic acid (25 ml. per l.). The precipitated cytochrome c is centrifuged down and resuspended in saturated ammonium sulphate solution (about 200 ml. per kg. original mince). The precipitate is again centrifuged down and is suspended in a few ml. of water. The suspension is dialysed against 0.5% NaCl until free from sulphate. The resulting red solution of oxidized c is saturated with chloroform and filtered. It will keep for several months at 5° or indefinitely at —15°. Alternatively it may be freeze-dried. Further purification on an ion-exchange column has been described by Margoliash (1952, 1954) but this is superfluous for most purposes.

The concentration of the pigment can be determined from the iron content (1 atom Fe per molecule of cytochrome), for which the methods of Keilin and Hartree (1945a) and Galston, Bonnichsen and Arnon (1951) are suitable. The most convenient method, however, is from the height of the α-band of the reduced form at 550 mμ, for which $\varepsilon_{550} = 27{,}700$ (Margoliash, 1954) where

ε = optical density of 1 cm. solution/molar concentration. For this determination the oxidized c should be suitably diluted in 0.1 M Na$_2$HPO$_4$ and reduced with a very small quantity of Na$_2$S$_2$O$_4$.

Cytochrome c has also been isolated from Delft yeast (KEILIN, 1933), pollen (OKUNUKI, 1939), wheat germ (GODDARD, 1944), the fungus *Ustilago sphaerogena* (NEILANDS, 1952) and from marine algae (see above).

Preparation of Cytochrome *f* from Parsley (DAVENPORT and HILL, 1952.) Leaves are picked from parsley *(Petroselinum sativum)* in its first year of growth. After removal of the petioles the leaves are weighed and soaked in tap water for about 30 mins. They are then drained, roughly dried by swinging in a cloth, and reweighed. Freshly picked parsley leaves were found to contain 84% water, so that the volume of water in the roughly dried leaves can be calculated. Immediately before grinding, that volume of cold (—10° C) 97% ethanol containing 1.5% v/v ammonia solution (sp. gr. 0.880) which will give 50% ethanol in the ground material is poured over the leaves. Grinding is then carried out in an electrically driven mill adapted from one originally designed for the manual grinding of small quantities of corn. The essential mechanism of the mill is a worm feed attached to a fluted male cone rotating within a fluted female cone. When feeding the mill care is taken to maintain as constant a proportion ethanol: leaves as possible.

The ground leaves are rapidly squeezed through a cloth, and the green extract centrifuged in a refrigerated centrifuge for 10 mins. The green precipitate is rejected and the yellow extract, which shows the α-band of ferrocytochrome f in a 2 cm. layer, is cooled further. When large amounts of leaves are to be treated better yields are obtained if the grinding and centrifuging operations are carried out successively upon small batches of leaves. The ethanol extract is accumulated at —15° C. Thus 3000 g. fresh leaves after soaking weighed 4250 g. Treatment in batches of 600 g. gave 5090 ml. ethanol extract containing 0.8×10^{-6} M cytochrome f haematin and having a dry weight of 22.5 mg./ml. To the cooled ethanol extract is added 1.1 vol. cold (—10° C) acetone with continuous stirring. The precipitate is filtered off on a Büchner funnel with 15 g. kieselguhr as filter aid. When nearly dry the filter pad is washed with a cold mixture of acetone/ethanol/water in the proportions 2:1:1 containing 0.5% 0.880 ammonia solution (all v/v). Washing is continued until the drainings are colourless. Without allowing the pad to dry it is washed with a 66% saturated ammonium sulphate solution, pH 8. This displaces the acetone:ethanol mixture and extracts some non-cytochrome protein together with brown pigments. The filter pad is now scraped from the paper and suspended in 50% saturated ammonium sulphate at pH 8. It is re-formed into a filter pad on the smallest possible Büchner funnel and washed with the ammonium sulphate solution. After sucking nearly dry the cytochrome f is eluted from the pad with the minimum volume of 0.06 M Na$_2$HPO$_4$. The yield at this stage is 220 ml. solution containing 1.6×10^{-5} M cytochrome f haematin representing a recovery of 86% from the ethanol extract. The solution is filtered with kieselguhr and dialysed against 3 l. 0.002 M ammonium carbonate solution.

The cytochrome is now adsorbed from solution on tricalcium phosphate gel[1]. Successive small volumes of the gel are added and centrifuged down. The proportion of cytochrome remaining in solution after each addition is estimated spectroscopically. The progress of adsorption is linear to 90%, and the adsorbed fractions between 10 and 80% are selected for further treatment. These fractions are eluted from the gel by three additions of 0.06 M Na$_2$HPO$_4$. The eluted cytochrome is brought to 20% saturation with ammonium sulphate by dialysis against a solution of that concentration (pH 8). The precipitate is centrifuged off and rejected and the supernatant taken to 45% saturation. The pink precipitate is dissolved in the minimum volume of 0.06 M Na$_2$HPO$_4$, dialysed against 3 l. 0.002 M ammonium carbonate and the sequence of adsorption on tricalcium phosphate gel followed by fractionation between 20 and 45% saturated ammonium sulphate repeated twice more.

The final precipitate is dialysed against 0.01 M Na$_2$HPO$_4$ to remove ammonium sulphate. The yield is 20 ml. of a solution 2.5×10^{-5} M with respect to cytochrome f haematin and representing 13% of that present in the original ethanol extract. The nitrogen concentration is 0.36 mg./ml., and the dry weight, corrected for phosphate, 2.25 mg./ml. This solution, which is 300 times as pure as the original extract, is freeze-dried and kept over phosphorus pentoxide, where it is stable for at least 2 years.

---

[1] *Calcium phosphate gel.* In a 10 l. jar are mixed 1 l. tap water, 250 ml. 0.6 M CaCl$_2$ and 250 ml. 0.4 M Na$_3$PO$_4$. The mixture is brought to pH 7.2 with N acetic acid and the jar is filled with tap water. The precipitate is washed six times by decantation, collected in a centrifuge and resuspended in a little water to give 400—500 ml. of suspension.

Notes on the Preparation. Manual grinding of small quantities of leaves (e. g. 50 g.) in a mortar gives the most efficient extraction. The yield decreases as the scale of the operation is increased. This is true to a smaller extent when mechanical methods of grinding are used. The ordinary meat mincer or the Waring blendor are very inefficient. A grinding or shearing action in which the leaf is rubbed between two hard surfaces appears to be essential. Moreover, the broken turgid cells must come into immediate contact with alkaline ethanol to a final concentration around 50%. Very little cytochrome is extracted when alkaline ethanol is added to previously ground leaves. It appears that rupture of the leaf tissue initiates a rapid and irreversible change in the protein complex of the cells which renders the cytochrome component insoluble.

The grinding operation is carried out at about 20° but the ethanol is cooled to compensate for heat arising from friction and from the heat of solution of ethanol. Removal of the leaf debris after grinding should be rapid since the yield diminishes rapidly on standing. Subsequent operations, until ethanol and acetone are removed by ammonium sulphate, are carried out below 0° C.

III. Quantitative Estimation of Cytochromes.

Cytochrome a cannot be determined absolutely in tissue preparations although the relative intensities of the a band in various preparations can be estimated by visual matching as in the case of fractions from animal tissues (SLATER, 1949a). In plant tissues the concentration of a is generally too low to permit anything but very approximate comparisons.

Cytochrome b cannot be estimated in plant tissues owing to the heavy background absorption (haemochromogen) between the band c bands (fig. 9, VIII, XI). However the cytochrome concentrates obtained from plant tissues other than leaves, when treated with $Na_2S_2O_4$ yield a clear a b c spectrum in which the b band can be compared with a standard globin haemochromogen (see above). SLATER (1949a) considers that the extinction coefficients of the α-bands of cytochromes b and c are approximately equal and he therefore determines the concentration of b approximately from an estimate of the relative intensities of the b and c bands in a tissue preparation. A rather more accurate method would be to compare the intensity of the α-band of cytochrome b with that of standard cytochrome c in a double-wedge trough.

Cytochrome b_3 can only be determined by the methods already described for cytochrome b. Comparisons with globin haemochromogen or with cytochrome c again depend upon the assumption that the absorption bands of the compounds will appear equal when concentrations are identical. This uncertainty may introduce errors of 20—30%.

Cytochrome c can be spectroscopically estimated by comparison with a standard solution and, if the background absorption is not too great, the error should not exceed 5%. When the pigment is in solution it can be estimated spectrophotometrically (above) or manometrically by its activating effect on the cytochrome oxidase activity of a heart muscle preparation (see p. 227).

Cytochrome f may be determined by direct spectroscopic comparison with a standard solution of the reduced pigment. A comparison with a standard cytochrome c has the advantage that the latter pigment is much easier to prepare. It appears from the work of DAVENPORT and HILL (1952) that the use of c as a standard would not greatly increase the experimental error. The standardization of a cytochrome f solution in the reduced state can be carried out with fair precision by comparison with cytochrome c or, if it is purified to the extent described above, by assuming that the extinction coefficient at the α-maximum is equal to that of cytochrome c ($\varepsilon_{\alpha\,max} = 27,700$).

Detection of Denaturation with Carbon Monoxide.

Although reduced cytochrome a_3 combines with CO and so gives rise to a shading on the short-wave side of the a band, none of the other cytochromes will combine with CO if they are in the native state. Furthermore at the low concentration of a which is characteristic of plant tissues it is generally impossible to see the a_3CO compound. Any change in the appearance of an absorption band of cytochrome (other than a) that may occur on treating the material with CO is a good indication that denaturation of the pigment has occurred with concomitant loss of catalytic activity.

IV. Cytochrome Oxidase.

Cytochrome oxidase may be defined as the intracellular catalyst required to bring about the oxidation of cytochrome c by oxygen. The enzyme may be estimated either spectrophotometrically by observing the rate of disappearance of the α-band of reduced cytochrome c in presence of the oxidase, or manometrically in terms of O_2 uptake when the oxidase preparation is shaken in air with reduced cytochrome c. It is not in practice possible to use cytochrome c as substrate in the manometric test since, on account of its high molecular weight (12,000), the quantity that can be added to a manometric vessel would have too low an oxygen equivalent (0.5 μl. per mg. cytochrome c). The standard manometric test system therefore consists of cytochrome oxidase + cytochrome c + a reducing agent for cytochrome c (H- or electron-donor). Examples of the latter are p-phenylene diamine, adrenaline, hydroquinone and ascorbic acid. Many plant tissues contain polyphenoloxidases (BHAGVAT and HILL, 1951) which, in an impure form, can oxidize these four donors while the widely distributed ascorbic acid oxidase (DAWSON and TARPLEY, 1951) may interfere if ascorbic acid is used. However, cytochrome oxidase activity can be distinguished from the activities of the other two oxidases in the following ways:— a) Cytochrome oxidase remains attached to insoluble particles when tissues are disintegrated while the other oxidases largely go into solution and can be washed away. b) The oxidations of the various donors are accelerated by addition of cytochrome c only if cytochrome oxidase is the catalyst. c) Whereas both cytochrome oxidase and polyphenoloxidase are inhibited by CO, only in the case of cytochrome oxidase is the inhibition abolished by light. (See WEBSTER, 1953; also HILL and HARTREE, 1953, for a discussion of the role of these enzymes in plant respiration.)

The most sensitive test for cytochrome oxidase is its ability to bring about the aerobic oxidation of reduced cytochrome c: a reaction which can be observed with any small-dispersion spectroscope.

Preparation of Reduced Cytochrome c. A solution of oxidized c (0.1—0.4 mM) is mixed with a little platinum black and treated with a stream of hydrogen until the bands of the reduced form attain an intensity equal to that of a sample of the same solution treated with $Na_2S_2O_4$. The platinum black is then centrifuged off. The pigment slowly oxidizes on storage. A more stable preparation has been described by MARGOLIASH (1952, 1954) in which the pigment is reduced with ascorbic acid, adsorbed on to "Amberlite IRC-50" resin, eluted with KCl solution, dialysed and freeze-dried.

A very simple method of obtaining a solution of reduced c is the following (MARGOLIASH, private communication). A solution of the oxidized pigment is dialysed until almost salt-free. It is then reduced with the minimum of $Na_2S_2O_4$ and shaken with a little of the above resin when the pigment is adsorbed. The now deep pink resin is washed thoroughly by decantation (tap-water) and collected on a small Büchner funnel. The reduced c can be eluted by addition of M NaCl or M phosphate buffer pH 7.3. If necessary, the concentration of salt can be reduced by a brief dialysis against distilled water.

Since cytochrome c may also be oxidized by the coloured oxidation products (quinones) of phenolic substances which occur in plants, this test is completely specific only in absence of such phenols or in absence of phenoloxidases which catalyse their oxidation. The formation of such oxidation products can be prevented by dialysing the preparation or, if it tends to darken rapidly in air, by addition of thiourea (Goddard and Holden, 1950) or phenylthiourea (Maxwell, 1950) which inhibit phenoloxidases. Alternatively, since phenoloxidases and ascorbic acid oxidase are soluble in water it is possible to obtain a cytochrome oxidase free from these enzymes by the methods described below.

1. Manometric Determination of Cytochrome Oxidase[1].

Slater's (1949 b) studies have shown that ascorbic acid is the best reducing agent for cytochrome c in manometric determinations of the oxidase although most work on plant cytochrome oxidase has been carried out with either hydroquinone (Goddard, 1944; Webster, 1952) or p-phenylene diamine (Rosenberg and Ducet, 1949; Bhagvat and Hill, 1951). The following methods of extraction of cytochrome oxidase from plant tissues are those of a) Bhagvat and Hili (1951) and b) Webster (1952).

Cytochrome Oxidase. a) Seeds of maize and beans are germinated for 24 hrs. Stems and roots are cut into thin slices and soaked in water for 2 hrs. The material is ground with 2 parts of 0.07 M Na$_2$HPO$_4$ and strained through glass wool. Coarse particles are removed by centrifuging at 3,000—4,000 RPM for $\geqslant$ 5 mins. and the cloudy supernatant is then centrifuged at 12,000 RPM for 10 mins. The resulting gelatinous residue is suspended in a little water.

b) The material is disintegrated in a Waring Blendor for $1^1/_2$ mins. at 3° in an equal weight of 0.1 M K$_2$HPO$_4$. The brei is centrifuged at 1000 g. for 10 mins. and the solid discarded The cloudy supernatant is then centrifuged at 16,000 g. for 30 mins. at 3° and the residue resuspended in 0.05 M K$_2$HPO$_4$. This mixture is again centrifuged cold at 16,000 g. and the gelatinous residue suspended in 3 times its volume of 0.01 M phosphate buffer pH 7.3. The percentage extraction of cytochrome oxidase by this method is at least 80%.

Although some workers have used higher temperatures 25° is recommended for the test as being more physiological. Preliminary tests should be carried out to determine the quantity of oxidase preparation required to give a convenient rate of O$_2$ uptake (say 30—60 μl. O$_2$/10 mins. in Warburg manometers). For this purpose manometers should be filled as described for manometer 4 of Table 5 with a constant phosphate concentration but with varying oxidase concentrations. A second test should be performed to determine the optimal phosphate concentration. A final test is then made to ascertain whether the system is saturated in respect to cytochrome c, i. e. whether an increase in the concentration of c causes no increase in the rate of O$_2$ uptake. According to Webster (1952) it is possible to reach saturation of plant cytochrome oxidase preparations with 50 μM cytochrome c in the manometer flask. If these conditions of saturation can be achieved it is possible to determine the activity of the enzyme preparation (Q_{O_2}) with one manometer and from this to calculate, if necessary, the Q_{O_2} of the original tissue. Q_{O_2} is defined as μl. O$_2$ uptake per hr. per mg. dry weight of the preparation and should be measured at optimal phosphate concentration. Since the oxidase may become slowly inactivated during the test the calculation should be based upon the initial steady rate of O$_2$ uptake; any lag during the first few minutes being ignored. The inclusion of versene (ethylenediamine tetraacetic acid) in the test system (Table 5) makes any control experiments unnecessary since autoxidation of ascorbic acid is then reduced to a negligible level.

It has become customary to use the fat-free dry weight in calculating the Q_{O_2} of cytochrome oxidase. This is determined as follows (Slater, 1949a).

---

[1] See Vol. 1 for a survey of manometric methods.

Table 5. *Manometer fillings for determination of cytochrome oxidase activity based upon the recommendations of* SLATER *(1949b). Volumes given in ml.*

	Manometer No.			
	1	2	3	4
Oxidase preparation in 0.1 M phosphate buffer, pH 7.3[1] . .	0.2	0.2	0.2	0.2
Phosphate buffer, 0.1 M[1]	2.0	2.0	2.0	2.0
Cytochrome c, 6.7×10^{-4} M in 0.5% NaCl	0.1	0.15	0.2	0.3
NaCl, 0.5%	0.2	0.15	0.1	—
Water	0.5	0.5	0.5	0.5
Ascorbic acid, 0.25 M, containing 0.01 M versene and neutralized to pH 7.3 with NaOH (in side bulb)	0.3	0.3	0.3	0.3
Total volume . . . . :	3.3	3.3	3.3	3.3

Fat-free Dry Weight. The oxidase preparation (1 ml.) is diluted in a weighed centrifuge tube with 5 ml. water and 1 ml. 20% trichloracetic acid is added. The flocculent precipitate is centrifuged down and the supernatant sucked off. The residue is washed by centrifugation, once with 5 ml. 50% ethanol and once with 5 ml. 96% ethanol. The residue is dried to constant weight at 100°.

If it is not possible to saturate the system in respect to cytochrome c it will also be found that the Q_{O_2} at a fixed c concentration will vary with the volume of oxidase taken. Under these circumstances it is necessary to determine the activity of the system at several cytochrome c concentrations and to extrapolate

Table 6. *Cytochrome oxidase activities of washed particulate preparations of various plant tissues measured i) spectrophotometrically and ii) manometrically. Examples taken from* WEBSTER *(1952) who gives data for 54 different species representing 23 families of dicotyledonous plants.*

(a) For details of spectrophotometric test see p. 228.

(b) System in manometers: 3 mg. hydroquinone as 0.1 ml. aqueous solution in side-bulb, 0.1 M phosphate buffer pH 7.1, total volume 2.4 ml. $T = 25°$. Alternatively, the system of Table 5 can be used.

			Cytochrome oxidase activity		
				Manometric test (Q_{O_2})	
Species	Common name	Tissue studied	Spectrophoto-metric test	$+ 5 \times 10^{-5}$ M-cyt. c	Without cyt. c
			(a)	(b)	(b)
Castanea dentata .	chestnut	leaf	1.33	23	6
Rheum rhaponticum	rhubarb	leaf	1.50	12	4
Brassica oleracea .	cabbage	leaf	1.62	45	10
Pyrus malus . . .	apple	fruit	1.09	36	32
Beta vulgaris . . .	beet	root	1.93	69	15
Phaseolus vulgaris	bean	root	2.40	50	7
Phaseolus vulgaris	bean	leaf	2.40	51	6
Pisum sativum . .	pea	root	2.22	56	9
Pisum sativum . .	pea	leaf	2.14	53	10
Daucus carota .	carrot	root	1.93	43	7
Daucus carota . .	carrot	young leaf	2.07	33	10
Daucus carota . .	carrot	mature leaf	1.62	28	9
Petroselinum crispum	parsley	root	1.39	22	14
Solanum tuberosum	potato	tuber	1.88	55	10

[1] The volume of oxidase preparation and the concentration of buffer will be varied, as described in the text, in order to provide conditions for optimal activity.

to infinite c concentration. The value so obtained will be independent of oxidase concentration in the manometer flask (Lineweaver and Burk, 1943; Slater, 1949b). The test should then be carried out, with appropriate standardized concentrations of oxidase and phosphate, in 4 manometers as shown in Table 5. (In Slater's original procedure versene was not used and, in consequence, it was necessary to incorporate 4 controls, without oxidase.) If the reciprocals of the Q_{O_2}'s are plotted against the reciprocal of the cytochrome c concentration, the intercept on the $1/Q_{O_2}$ axis gives the reciprocal of the Q_{O_2} at infinite cytochrome c concentration.

A selection of the results of Webster on cytochrome oxidase activities of various plant tissues is given in Table 6. The fact that the activities are stimulated by addition of cytochrome c indicates that cytochrome oxidase is present in the extracts and the findings that both young and mature carrot leaves yield preparations which are stimulated to the same extent by cytochrome c has an important bearing on the problem of the nature of the terminal oxidase in leaves of different ages (Hill and Hartree, 1953).

2. Other Manometric Methods.

Both hydroquinone and p-phenylene diamine can be used, as alternatives to ascorbic acid, in the manometric test. However, the oxidation products, particularly quinone formed from hydroquinone, are toxic for the oxidase and a constant rate of O_2 uptake may not be obtained. These substrates are useful if the material under test contains ascorbic acid oxidase, and the inhibition that occurs during the oxidation of 0.02 M hydroquinone can be prevented by the presence of 0.02 M 1-phenylsemicarbazide (Quinlan-Watson and Dewey, 1948).

Proof that the oxidation is mediated by cytochrome oxidase can be obtained from the effect of cytochrome c (see above) and also from the activating effect of light on the system inhibited by CO. It should, however, be remembered that while a positive result is clear proof of cytochrome oxidase activity, a negative result does not necessarily disprove it (Hill and Hartree, 1953).

For such studies the only necessary modification to a standard rectangular manometer bath is a window either in the bottom of the bath or in the side of the bath opposite to the one to which the manometers are attached. Perspex has significant advantages over glass since it is less liable to crack when water-proof gaskets are tightened or if the bath becomes distorted under the weight of its contents. In the writer's experience the most convenient light source is a projection lantern fitted with a 500-watt lamp. The lantern is focussed to give the narrowest possible beam and is placed close to a window in the wall of the bath. By means of an inclined mirror the light is deflected vertically upwards to illuminate the manometer flask containing CO. The light is adjusted so that the flask, while moving, remains in the light beam.

Gas mixtures in which the ratios CO/O_2 and N_2/O_2 are equal and of the order of 10 are made up in graduated Mariotte bottles and used for gassing pairs of manometers filled according to the last column of Table 5. The manometer bath is covered with a black cloth during equilibration and subsequently during dark periods. The manometer containing the CO/O_2 mixture is darkened and illuminated during alternate periods of 10—15 mins. after the ascorbic acid (or other donor) has been added. The control manometer (N_2/O_2) need not be illuminated. Some typical results with oxidase preparations of various sources and hydroquinone as donor are given in Table 7.

Table 7. *The influence of carbon monoxide in light and in darkness on the cytochrome oxidase activity of washed particulate preparations of various plant tissues.* (WEBSTER, 1952.)

$$K_A = \frac{n}{l-n} \cdot \frac{CO}{O_2}$$

where l = rate of O_2 uptake in N_2/O_2 and n = rate of O_2 uptake in CO/O_2 in the dark. For further details see Table 6.

Plant and tissue	O_2 consumption/ml. oxidase preparation per hour			Relative Affinity Constant K_A
	$N_2/O_2 = 14.4$	$CO/O_2 = 14.4$		
		Dark	Light	
Solanum melongena, fruit . . .	120	24	109	4.3
Daucus carota, leaf	60	15	50	4.8
Cucumis sativus, fruit	54	14	49	5.0
Medicago sativa, foot	56	12	52	3.9
Beta vulgaris, root	147	35	131	4.5

3. Application of the Manometric Test to the Determination of Cytochrome c.

GODDARD (1944) has described a method of estimation of cytochrome c based upon its accelerating effect on the oxidation of hydroquinone by cytochrome oxidase. By this method he was able to show that at very low concentrations of cytochrome c ($\not> 2\ \mu M$) its catalytic activity was the same whether it was obtained from ox heart or wheat germ (fig. 10). Thus at such concentrations a calibration curve prepared with heart cytochrome c (p. 220) may be used for estimation of c from wheat germ and probably also of c from other plant sources. This method has the advantage that cytochrome c can be estimated in concentrations that are too small for accurate determination by spectroscopic methods. (Thus the α-band in a 1 cm. depth of $2\ \mu M\ c$ would be barely visible.) On the other hand it will only estimate cytochrome c in solution: if it is attached to particles it will have no accelerating effect. In practice this difficulty is unlikely to arise since particles containing cytochrome c normally contain the oxidase as well and the test would then be invalid.

The test may be performed with the system shown in Table 5 or with hydroquinone in place of ascorbic acid and versene. It is advisable to choose the donor which gives rise to the slowest O_2 uptake in absence of added c and to use a highly active cytochrome oxidase preparation from heart-muscle.

Cytochrome Oxidase from Heart-Muscle. (KEILIN and HARTREE, 1947.) About 600 g. heart-muscle (horse, ox or pig) is cleaned from fat and ligaments, finely minced and washed 5 times with about 10 l. tap water and mechanical stirring, the mince being pressed out by hand each time in cheese-cloth. Of the resulting pulp 300 g. is ground in a mechanical mortar for 2 hrs. with 400 ml. 0.02 M phosphate buffer pH 7.3 and 150 g. coarse, acid-washed sand. The mixture is then diluted with a further 050 ml. buffer, centrifuged at 4.000 g. for 20 mins. and the upper cloudy layer collected and cooled in ice. About 250 ml. 0.02 M KH$_2$PO$_4$ is also cooled to 0°. To the cold supernatant fluid is added 100 g. crushed ice followed by N acetic acid, 2—3 ml. at a time, until the pH falls to 5.5 (methyl red). The mixture is immediately centrifuged for 10 mins. and the upper clear fluid discarded. The residue is suspended in the cold KH$_2$PO$_4$ solution plus a little ice and centrifuged again. The residue is now suspended in its own volume of borate-phosphate buffer, pH 7.4 (a mixture of 3 vols. 0.25 M boric acid and 1 vol. 0.25 M Na$_2$HPO$_4$). The preparation should be kept cold but must not be frozen. The presence of borate lengthens the useful life of the preparation from 4—5 days to several weeks (BONNER, 1954)[1].

This method is reliable when the washing of the mince is carried out at 15—20° with Cambridge tap-water containing 0.0013 M calcium as bicarbonate. Both the temperature and calcium content of the water appear to be important: low temperature washing with distilled water, for example, gives unsatisfactory results (the influence of these factors is

Note added in proof. If the suspending medium is a solution of 100 g. sucrose in 50 ml. 0.25 M Na$_2$HPO$_4$, the preparation can be stored for longer periods at —15°.

15*

228 E. F. HARTREE: Haematin Compounds.

discussed by CLELAND and SLATER, 1953). Replacement of the mechanical mortar by a Waring blendor also gives less satisfactory results at the stage of isoelectric precipitation. However, a blendor homogenate can be freed from coarse particles at 4,000 g and the active material subsequently collected by centrifuging in the cold at 10,000 g, washing the residue with KH_2PO_4 solution and resuspension in borate-phosphate buffer.

Fig. 10 shows the relationship between the activity of the system: oxidase, hydroquinone and varying concentrations of cytochrome c from two sources.

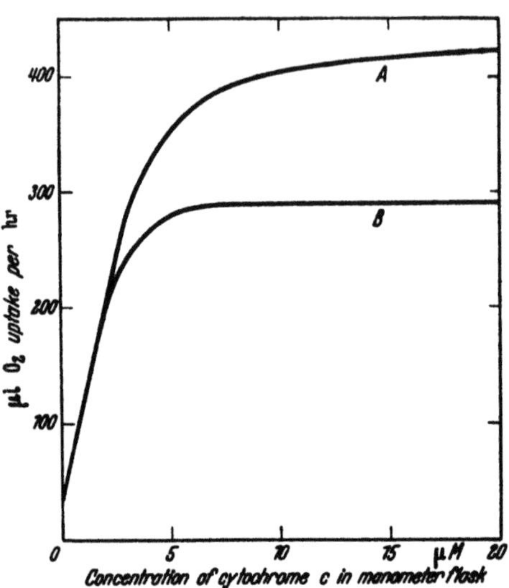

Fig. 10. Manometric estimation of cytochrome c isolated from ox heart (A) and wheat germ (B) based upon its acceleration of the cytochrome oxidase reaction in a heart-muscle preparation (GODDARD, 1944). System in manometers: 3.0 mg. hydroquinone, suitably diluted heart preparation, cytochrome c, phosphate buffer (0.05 M, pH 7.1). Final vol.: 2.4 ml. $T = 25°$. O_2 uptakes measured over 30 mins. Alternatively the system of Table 5 can be used.

The method will be slightly more sensitive if ascorbic acid is used as donor according to the scheme set out in Table 5. When solutions of cytochrome c are being analysed care must be taken that their addition to the manometer vessel does not disturb the standard conditions of pH and buffer concentration. This method will naturally only estimate native cytochrome c and it is essential to determine whether the solution used for standardization is devoid of denatured c. The presence of the latter will be revealed by a decrease in intensity of the characteristic bands of the reduced pigment when the solution is treated with CO at neutral pH. If the decrease, estimated by the double-wedge trough method, is slight (0 — 20 %) a correction can be made for it. If the percentage denaturation is larger the preparation should be purified by the method of MARGOLIASH (1954).

4. Spectrophotometric Estimation of Cytochrome Oxidase.

(COOPERSTEIN and LAZAROW, 1951.)

A solution of reduced cytochrome c is prepared as follows from the pigment isolated from heart-muscle. 30 ml. solution ($1.7 \times 10^{-5}\ M$) in 0.03 M phosphate buffer pH 7.4 is reduced by addition of 0.1 ml. freshly prepared 1.2 M $Na_2S_2O_4$. The solution is shaken in air for 2 mins. to oxidize excess of reducing agent. Alternative methods of obtaining reduced c have already been described (p. 223).

Measurements are carried out in 1 cm. spectrophotometer cells at 25°. 3 ml. reduced c is treated with an accurately measured volume (0.02—0.04 ml.) of oxidase preparation and readings of optical density are taken every 30 secs. for a period of 3—6 mins. A trace of potassium ferricyanide (*ca.* 1 mg.) is now added to oxidize c completely and a final reading is taken.

The density of fully oxidized c at 550 mμ is subtracted from the series of measured densities and the logarithm of the difference is plotted against time. WEBSTER (1952) expresses enzyme activity (A) as the decrease in log (molar concentration of reduced c) per min. per ml. of oxidase.

$$A = \frac{\triangle \log [\text{reduced } c]}{t \cdot v} = \frac{\log (D_1 - D_{ox}) - \log (D_2 - D_{ox})}{(t_2 - t_1) \cdot v}$$

where v = ml. oxidase taken for the test.

As such activities are not based upon dry weights there is no direct relationship between the activity obtained by this method and the manometric Q_{O_2} (Table 6). A standard enzyme activity (S) based upon fresh weight is defined by COOPER-STEIN and LAZAROW as the activity that would result if 0.03 g. fresh weight of tissue were dispersed in a final volume of 3.00 ml. cytochrome c solution (i. e. a 1:100 final dilution of the tissue). The standard activity can thus be calculated from

$$S = \frac{0.03\,A}{x}$$

where x = g. weight of fresh tissue that yields 1 ml. oxidase preparation.

These calculations neglect the dilution of the cytochrome c solution by the oxidase, which was in fact of the order of 1 % in the case of the very active animal tissue preparations used by COOPERSTEIN and LAZAROW. From WEBSTER's experiments it appears that the cytochrome oxidase concentrates from plant material are sufficiently active to make this dilution error negligible. The use of markedly larger volumes of oxidase would in any case make the reaction mixture too opaque for accurate spectrophotometry.

It was found that A is not affected by variations in cytochrome c concentration between 1.2 and 2.4 $\times$ 10^{-5} M. On the other hand the phosphate concentration quoted was worked out for animal tissue preparations and may require adjustment to give optimal activities with plant extracts.

The use of micro absorption cells (LOWRY and BESSEY, 1946) enables all volumes to be reduced to one tenth.

5. Colorimetric Methods.

The classical test for indophenol oxidase (= cytochrome oxidase + c) can serve as a rapid method for obtaining approximate comparisons of the cytochrome oxidase activities of different tissue preparations. The liquid under test, suitably diluted with phosphate buffer (0.1 M, pH 7.3), is treated with one third of its volume of Nadi reagent when the rate of formation of blue colour is a measure of cytochrome oxidase activity if cytochrome c is also present. From the work of WEBSTER (1952) it appears that 5 $\times$ 10^{-5} M cytochrome c would be adequate for optimal activity. The reagent is a freshly prepared mixture of equal volumes of 0.01 M dimethyl-p-phenylene diamine and 0.01 M α-naphthol, both in 50 % ethanol.

A more accurately quantitative colorimetric method is that of SMITH and STOTZ (1949) in which the leuco form of 2:6-dichlorbenzeneoneindo-3'-chloro-phenol is the donor. In presence of added c the rate of colour formation was found to be proportional to concentration in the case of cytochrome oxidases of animal origin.

C. Haematin Enzymes: Peroxidase and Catalase.

Both enzymes are compounds of protohaematin with specific proteins. The identity of their prosthetic groups with that of haemoglobin has been established by the isolation of crystalline haemin indistinguishable from the haemin obtained from blood. In the case of haemoglobin and peroxidase it is possible to dissociate haematin from protein, a process which leads to complete loss of the characteristic catalytic properties. On recombination, however, the original pigments are obtained in their native states, i. e. they show their normal catalytic activities (HILL and HOLDEN, 1926; THEORELL, 1940).

The two enzymes bear a very close resemblance, both spectroscopically and in terms of their reactions with peroxides and inhibitors, to methaemoglobin, the ferric form of haemoglobin (KEILIN and HARTREE, 1951). These relationships are summarized in fig. 11 and Table 8. The absorption data of fig. 11 enable the concentrations of the purified pigments to be determined rapidly. Although such methods are more sensitive in the Soret region (400—420 mμ) where the absorption is maximal, interference from other pigmented material will be least at the red end of the spectrum where each pigment shows a characteristic band. Less detailed provisional data on leaf catalase are given by GALSTON, BONNICHSEN and ARNON (1951). The reactions of peroxidase and catalase with respiratory inhibitors (Table 8), the study of which played an important part in the establishment of the haematin nature of these enzymes, are also of value in their spectroscopic identification. The chief characteristics of these reactions are a) the 1:1 molar ratio of inhibitor and haematin and b) their reversibility. The spectrum changes observed on addition of inhibitor can be reversed by dialysis and the return of the original spectrum accompanies a complete recovery of enzymic activity.

Peroxidase is essentially a plant enzyme although similar enzymes (in which the prosthetic groups are not protohaematin) can be isolated from milk and

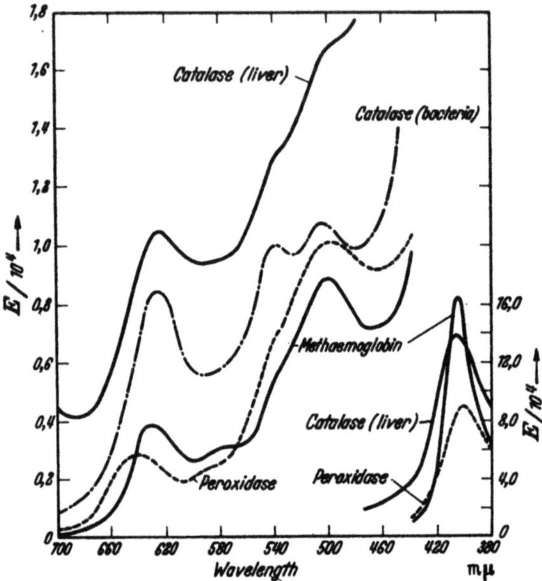

Fig. 11. Absorption curves of catalase (horse liver and *Micrococcus lysodeikticus*), peroxidase (horse-radish) and acid methaemoglobin (horse blood). Reproduced from KEILIN and HARTREE (1951) by permission of the Biochemical Journal.

leucocytes. In general, plant tissues contain either a strong peroxidase or a strong polyphenol oxidase. Tissues of the former type do not darken when bruised or minced (e. g. horse-radish root, turnip root) but give a powerful peroxidase reaction. The latter type (e. g. apple fruit, potato tuber) darken rapidly when damaged since the substrate for the oxidase is then brought into contact with the oxidase and with oxygen. Catalase, on the other hand, appears to be present in all plant cells.

Although papers on the distribution of these enzymes are very numerous there are no reviews which give comprehensive comparisons between different plant tissues. The methods used for the estimation of catalase and peroxidase have varied so extensively that quantitative comparisons between figures published by different authors are impossible. The earlier literature is surveyed very completely by EULER (1934). Later reviews reflect an increasing interest in reaction mechanisms and kinetics, and less attention has been given in recent times to the question of distribution. In the case of catalase the work of GRAČANIN (1926), who recorded the changes in catalase content of plants during growth, and of DELEANO, POPOVICI and IONESCU (1937), who studied the development of catalase in seedlings of maize, wheat and oats, can be mentioned.

General reviews on these enzymes have been contributed by ZEILE (1950), by LEMBERG and LEGGE (1949) and by various authors in the book of SUMNER and MYRBÄCK (1951).

I. Peroxidase.

Although the function of this enzyme has not yet been clearly established (HILL and HARTREE, 1953) it can be recognized by its ability to catalyse the oxidation of various "hydrogen donors" by H_2O_2. Examples of such donors are pyrogallol, p-cresol, benzidine, guaiacol, ascorbic acid, nitrite and cytochrome c. It is important to remember that the majority of such donors can reduce cytochrome c and hence may be oxidized, to similar coloured products, *via* the cytochrome oxidase system. However, peroxidase, unlike cytochrome oxidase, is a soluble enzyme.

The blueing of tincture of guaiacum by H_2O_2 in presence of plant and animal tissues has for long been used as a test for peroxidase but, as in the analogous oxidation of benzidine (see p. 211), the reaction is not specific for peroxidase but is given by all haematin derivatives. However, the activity of peroxidase is far

Table 8. *Comparison of the properties of Peroxidase, Catalase and Oxyhaemoglobin.*

	Peroxidase	Catalase		Oxyhaemoglobin	
	Horse-radish root	Liver and blood of mammals	Spinach leaf	Mammalian blood	Soya-bean root nodules
Properties of Purified Pigment					
Mol. wt.	40,000	225,000[1]		68,000	17,000[2]
Catalytic activity	P.N. = 1,220	Kat.f=60,000 — 100,000	Kat.f=23,600	—	—
% Haematin	1.61	1.05[1]		3.89	
% Fe	0.138	0.096[1]	0.049	0.334	0.35[2]
Wavelengths of absorption bands as seen with a microspectroscope.					
Haemoprotein	642 (583) (548) 497	623 (580) (544) 507	620 (545) (510)	577 542	575 540
do. + Na₂S₂O₄	(578) 556	no reaction		555	554
do. + Na₂S₂O₄+CO	572 542	no reaction		572 539	564 538 (presence of Na₂S₂O₄ unnecessary)
do. + CN′	(581) 538	(596) 557			
do. + F′	612 (561) (530)	(622) 598			
do. + N₃′	635 534[3]	621 (578) (530) 500			
				Methaemoglobin and deriv.	
do. + Fe(CN)₆‴ (pH < 7)	no reaction	no reaction		630 (578) (536) 500	626 (565) (534) 496
do. do. + CN′				(575) 542	540[4]
do. do. + F′				606 489	605 (532) 480
do. do. + N₃′				(572) 542	no data

Bands given in *heavy type* are those most readily detected *in vivo* or in crude extracts. Those in brackets would only be detectable in concentrated solutions.

[1] Average values.
[2] As usually prepared the material has mol. wt. 20,000 and contains 0.26—0.27% Fe.
[3] Azide-peroxidase is only stable between pH's 4.0 and 4.5.
[4] Main band — no data available on any weaker bands.

higher than that of other haematin compounds and any loss of activity that follows heating the plant material at 100° for 15 mins. may be considered as an indication of the presence of peroxidase. The enzyme is readily extracted by water from minced or homogenized tissues.

The most convenient source of the enzyme is horse-radish root *(Cochlearia armoracia)*. The enzyme was obtained in crystalline form by Theorell (1942) but a non-crystalline product of equivalent activity can be more readily obtained by a method worked out by Keilin and Hartree (1951)[1]. The properties of the pure enzyme are given in Table 8. Peroxidase in a high state of purity has also been isolated from fig-sap (Sumner and Howell, 1936). The isolation of peroxidase as a solution showing the characteristic band in the red (Table 8) has been reported a) by Yakushiji (1939) from roots of *Raphanus acanthiformis* (P.N. = 210) and from lac of *Rhus succedanea* (P.N. = 27) and b) by Hill and Scarisbrick (1951) as a by-product in the isolation of cytochrome b_3 from bean leaves (P.N. = 64). Purified horse-radish peroxidase has P.N. = 1,000—1,200. As a routine method of extracting peroxidase for quantitative estimation Purr (1950) recommends the following procedure:

Extraction of Peroxidase from Plant Tissues. 5 g. fresh tissue is rubbed in a mortar with 5 g. acid-washed sand for 3—5 mins. The mass is shaken with 25 ml. water and the suspension decanted from the sand. The sand is shaken twice with 25 ml. water and the combined liquids are diluted to 100 ml. Grains and grain products are finely ground, 5 g. of the material is rubbed with 10 ml. water and diluted to 100 ml. The mixture is shaken occasionally during 1 hr. and centrifuged. The solid layer is discarded.

The "classical" method for estimation of peroxidase, which found wide application, is the purpurogallin method of Willstätter and Stoll (1918). The oxidation of pyrogallol to purpurogallin, in presence of peroxidase and H_2O_2, is a complex series of reactions which can be summarized by the equation:

$$2\,C_6H_6O_3 + 3\,H_2O_2 \rightarrow C_{11}H_8O_5 + 5\,H_2O + CO_2 \qquad (1)$$

In the original method the plant extract was allowed to react with pyrogallol and H_2O_2 in a large volume (2 l.) of distilled water in order to minimize inactivation by H_2O_2. Willstätter introduced the term Purpurogallinzahl (P.Z.), or purpurogallin number (P.N.), to represent the purity of a preparation, i. e. the enzymic activity on a dry weight basis. P.N. represents the weight (in mg.) of purpurogallin formed under defined conditions from 1 mg. enzyme preparation. Thus P.N. = w/vd where w = mg. purpurogallin, v = volume of preparation taken for the test and d = mg. dry weight of 1 ml. preparation. The number of enzyme units (i. e. the quantity of enzyme present) in a peroxidase preparation is defined as P.U. (purpurogallin units) = $Vw/1000\,v$ where V = total volume of the preparation.

Preparation of Purpurogallin. (Evans and Dehn, 1930.) A solution of 8 g. sodium iodate in 100 ml. water is added slowly with stirring to a solution of 10 g. pyrogallol in 20 ml. water. After standing for 1 hr. the purpurogallin is filtered off, washed with water and dried. It is purified by two recrystallizations from aqueous ethanol. (A hot saturated solution of purpurogallin in ethanol is mixed with an equal volume of hot water and allowed to cool.) The product, well washed with water, is dried in a vacuum desiccator.

1. Estimation of Peroxidase by the Method of Elliott and Keilin (1934).

These authors reduced the volume of water to 500 ml. in the P.N. determination but maintained the concentrations specified in the original description. This change has made the method less tedious and has been generally accepted.

---

[1] *Note added in proof.* A simplified method of extraction has now been published by Kanten and Mann (1954).

Suitably diluted enzyme solution (0.1—1.0 ml.) is added to 500 ml. glass distilled water at 20° containing 1.25 g. pyrogallol. Commercial samples of pyrogallol are often unsatisfactory unless resublimed under reduced pressure (KEILIN and HARTREE, 1951). H_2O_2 (0.4 ml. 6% w/v) is added with stirring and exactly 5 mins. later the reaction is stopped by addition of 5 ml. 5 N H_2SO_4. The yellow purpurogallin is extracted with 3 lots (80, 30, 30 ml.) of peroxide-free ether[1]. The volume of the combined ethereal extracts is noted and the concentration of purpurogallin determined in a photoelectric colorimeter using a blue filter. The instrument is calibrated with ethereal solutions of the pure compound. The quantity of enzyme taken should be such as to yield ≯ 10 mg. purpurogallin.

Many investigators have found that the WILLSTÄTTER method shows unaccountable fluctuations. This matter has been studied at length by ETTORI (1949) who was, however, unable to trace their cause. A modification of the WILLSTÄTTER method was introduced by SUMNER and GJESSING (1943) with a view to increasing the reliability. In this method the estimation is carried out in a total volume of 21 ml. and is therefore more adaptable to routine analysis.

2. Estimation of Peroxidase by the Method of SUMNER and GJESSING.

To 15 ml. water in a 125 ml. conical flask are added 2 ml. 5% pyrogallol, 2 ml. 0.5 M phosphate buffer pH 6.0 and 1 ml. 1% H_2O_2. The mixture is brought to 20° in a water bath and treated with 1 ml. suitably diluted enzyme. After 5 mins. the reaction is stopped by addition of 1 ml. 2 N H_2SO_4. The purpurogallin is extracted with ether and estimated as above.

This procedure gives larger values of P.N. than does the WILLSTÄTTER method but again the results are not free from fluctuations when the same sample is tested repeatedly. At the same time there is a marked lack of agreement between the two methods (Table 9). ETTORI developed a manometric method for peroxidase estimation based upon the CO_2 produced when purpurogallin is formed (equation 1). The reaction mixture in the manometric flasks contained the concentrations of reagents specified by SUMNER and GJESSING but the volume was reduced to 2.3 ml. Peroxidase activities calculated from CO_2 evolution agreed well with those based upon the purpurogallin produced. ALTSCHUL and KARON (1947) developed a different type of manometric method in which calculations were based upon the H_2O_2 used up in the reaction. The residual H_2O_2 was determined by adding catalase and measuring the O_2 produced.

Table 9. *A comparison of the methods of* WILLSTÄTTER *and* STOLL *and of* SUMNER *and* GJESSING *for the determination of the P. N. of peroxidase based on results with various fractions of horse-radish enzyme.* (ETTORI, 1949.)

	WILLSTÄTTER and STOLL P. N.	SUMNER and GJESSING P. N.	P. N. ratio
Purified peroxidase (1)	735	1080	0.68
Purified peroxidase (2)	508	1460	0.35
	mg. purpurogallin formed/ml. of peroxidase preparation		
Horse-radish press juice	68	100	0.68
do. filtered	107	203	0.53
Two fractions isolated {	280	810	0.35
from press-juice {	83	228	0.36

The uncertainties of methods based upon purpurogallin (Table 9) have stimulated the search for more reliable methods based upon less complex chemical reactions. The following method appears to have overcome these uncertainties.

---

[1] Pure analytical grade ether should be left standing for at least 24 hrs. over ferrous sulphate crystals.

3. Estimation of Peroxidase by the Method of Purr (1950).

A special phosphate-citrate-oxalate buffer is used in this method. Purr recommends making dry phosphate-citrate and phosphate-oxalate buffer powders and dissolving them as required. A simpler method is to make up the buffer from stock reagents and to preserve it in the frozen state to prevent the growth of moulds.

The following reagents are required:

Buffer. 15.65 g. $Na_2HPO_4 \cdot 2 H_2O$, 7.70 g. citric acid $+ 1 H_2O$ and 0.686 g. oxalic acid $+ 2 H_2O$ are dissolved in glass distilled water and diluted to 250 ml.

0.05 *N ascorbic acid* is made up in 1 vol. buffer and 9 vols. H_2O. The concentration is periodically checked by titration with standard I_2 and starch. The solution is used as long as the titre does not fall below 90% of the original value.

0.1 *N* H_2O_2 is prepared from Perhydrol by dilution with the 1:10 diluted buffer.

0.5% *o-tolidine*. A 1% ethanolic solution is diluted with an equal vol. of water.

0.01 *N iodine* (freshly standardized).

To determine peroxidase 4 ml. ascorbic acid, 5 ml. H_2O_2 and 10 ml. buffer are brought to 25° in a thermostat. 1 ml. *o*-tolidine and a measured volume (> 1 ml.) of enzyme solution are added followed by water (at 25°) to a final volume of 50 ml. The mixture is incubated for 15 mins. (30—45 mins. if the enzyme is very dilute) and 10 or 20 ml. is pipetted rapidly into a second flask containing 2.5 or 5 ml. respectively of 2 *N* H_2SO_4. The residual ascorbic acid is estimated by titration (x ml.). The quantity of enzyme taken should be such that $> 20\%$ of the ascorbic acid is oxidized and the duration of the reaction should not exceed 45 mins. A blank titration is carried out by adding 12 ml. 2 *N* H_2SO_4 before the enzyme (a ml.). The first order velocity constant can then be calculated from

$$k = \frac{1}{t} \log \frac{a}{x}$$

and the Peroxydase Einheit (P.E.asc), or peroxidase unit (P.U.asc), is defined as 1000 k and measures the quantity of enzyme present. The Peroxydase Wert (P.W.asc), or peroxidase number (P.N.asc), is defined as the number of units in 1 g. dry weight of tissue.

Table 10. *Peroxidase activities of ground plant tissues, referred to the dry weight of the original tissue, and of purified horse-radish peroxidase determined by the ascorbic acid $+$ o-tolidine method* (Purr, 1950) *and by the classical purpurogallin method* (Elliott *and* Keilin, 1934).

P. N.asc = enzyme units (P. U.asc) in 1 g. dry weight of tissue and P. U.asc = 1000 × 1st. order velocity constant for oxidation of ascorbic acid under Purr's conditions.

P. N.pur = mg. purpurogallin formed from pyrogallol in presence of 1 mg. dry weight of tissue (or enzyme) under Elliott *and* Keilin's conditions.

a: data from Purr; *b:* author's results.

	P. N. asc	P. N. pur	$\dfrac{\text{P. N. asc}}{\text{P. N. pur}}$
a) Cabbage (leaf)	380		
Kohlrabi (stem)	4000		
Carrot (root)	15		
Potato (tuber)	60		
Oats (seed)	250		
Soya (whole seed)	118		
Soya (seed without testa)	81		
Soya testa	917		
b) Horse-radish (root)	4300—11000[1]	0.6—1.5[1]	7250
Purified peroxidase from horse-radish	5.2×10^6	1000	5200

---

[1] This range of activities is based upon data from Elliott and Keilin (1934), Keilin and Mann (1937), and the author's observations.

In Table 10 are given some of Purr's results with plant tissues together with comparative data, obtained by the author, on horse-radish peroxidase using Purr's procedure and the classical purpurogallin method. It will be seen that the unit conversion factors for very crude and for highly purified horse-radish peroxidase are of similar magnitude.

A possible improvement in the above method would be the incorporation of versene (0.01 M) in the ascorbic acid solution in order to abolish autoxidation.

4. Other Methods for Determination of Peroxidase.

Smith, Robinson and Stotz (1949) have used leuco-2,6-dichlorbenzene-oneindo-3'-chlorophenol as H-donor while Boiarkin (1951) recommends benzidine. A method similar to that of Smith *et al.* was devised by Diemair and Häuser (1941). They reduced 2,6-dichlorphenolindophenol to the leuco form with an exact equivalent of ascorbic acid. The addition of peroxidase and H_2O_2 then brought about a reoxidation to indophenol which was extracted with ether. Chapman and Saunders (1941) showed that mesidine (2,4,6-trimethylaniline) is oxidized by peroxidase and H_2O_2 to give a highly coloured anil in almost theoretical yield. A preliminary note from Avi-Dor and Paul (1953) indicates that this reaction may form the basis of a reliable method for estimation of peroxidase[1].

5. Possible Interference from Other Enzymes.

The possibility that cytochrome oxidase might interfere in the estimation of peroxidase has already been referred to. Fortunately it is possible to remove the particles carrying cytochrome oxidase from a solution by centrifuging at 20,000 g. Furthermore any reaction due to cytochrome oxidase will occur in absence of H_2O_2 and will be accelerated by addition of cytochrome c.

It is also possible that phenoloxidases may interfere especially since they are water soluble. But here again the test will be positive in absence of H_2O_2 and, unlike the peroxidase reaction, will be strongly inhibited by thiourea and phenylthiourea. Phenoloxidases may give positive reactions with pyrogallol or with ascorbic acid plus o-tolidine but will not oxidize pure mesidine to the purple anil. In fact, mesidine is not oxidised in presence of cytochrome oxidase (heart muscle), catechol oxidase (mushrooms) or laccase (lac)[2].

(Spectroscopic properties of peroxidase are discussed on p. 239.)

II. Catalase.

The classical catalase reaction is the decomposition of H_2O_2 into water and oxygen and for many years this was considered to be the reaction that it catalysed *in vivo*. Since a number of oxidases were known to reduce oxygen to H_2O_2 *in vitro*, catalase came to be regarded as a means of protecting living cells from the toxic action of H_2O_2. The work of Keilin and Hartree (1936, 1945b) showed, however, that catalase could promote peroxidatic reactions and reasons were adduced for supposing that such reactions were more closely related to the physiological function of the enzyme than was the "catalatic" decomposition into water and oxygen. However, the catalatic reaction provides the basis for detection and estimation of catalase since the peroxidatic power of this enzyme is comparatively feeble.

A description of a test for catalase in plants is scarcely called for since it is apparently present in all plant cells. The full intensity of O_2 evolution can only

---

[1] *Note added in proof.* Fuller details are now available (Paul and Avi-Dor, 1954).

[2] Unpublished observation by Dr. B. C. Saunders.

be attained after cellular disintegration (e. g. after grinding with sand according to the method of Purr on p. 232). In order to avoid rapid destruction of the enzyme the concentration of H_2O_2 should not exceed 0.5%.

Our knowledge of catalase is based largely upon studies of the enzyme isolated from animal tissues. (For references to methods of preparation see Bonnichsen, 1947, 1948.) This has arisen because the enzyme is present at much higher concentrations in animal tissues and is more stable than the plant enzyme after extraction. However Galston, Bonnichsen and Arnon (1951) were able to isolate a highly purified catalase from spinach leaves. The leaves were disintegrated under acetone to yield a powder from which the enzyme was extracted with Na_2HPO_4 solution. After 4 ammonium sulphate fractionations the enzyme was further purified by electrophoresis. Although a final ammonium sulphate fractionation resulted in the appearance of needle crystals it was not definitely established that these were catalase because they were always accompanied by amorphous material.

Like peroxidase, catalase is an intracellular enzyme but is readily extracted by water when the cells are broken. It is more readily damaged by organic solvents than horse-radish peroxidase and is more thermolabile, being rapidly destroyed at 60°. It is also more susceptible to variations in pH; the plant enzyme being rapidly destroyed below pH 5.3 and slowly above pH 8.9.

The most widely used method of determining catalase activity is based upon a determination of the 1st order reaction velocity constant of the decomposition of H_2O_2. The activity is then expressed as

$$\text{Kat.f. (Katalasefähigkeit)} = \frac{K}{w} = \frac{1}{w \cdot t} \log_{10} \frac{x_0}{x_t}$$

where w is the dry weight in *grams* of the sample taken for the estimation, x_0 and x_t are the peroxide concentrations at zero time and time t (mins.) respectively. An agreed standard procedure was never generally accepted but the following method of Euler (1934) has been widely used. The method of Purr (see p. 232) serves well for extracting catalase from plant material.

1. Determination of Kat.f. (Euler.)

Reagents required:— 0.01 N H_2O_2; 0.03 M phosphate buffer, pH 6.8; 0.01 N $KMnO_4$; N H_2SO_4.

All solutions are made up in glass-distilled water and cooled to 0°. A mixture of 7 ml. H_2O_2, 2 ml. buffer and 10 ml. H_2SO_4 is titrated with permanganate (x_0). A mixture of 35 ml. H_2O_2 and 10 ml. buffer is placed in an ice bath. A measured volume ($\not> 5$ ml.) of cold enzyme solution is added followed by cold water to make the volume up to 50 ml. Periodically 10 ml. portions are pipetted into equal volumes of H_2SO_4 and the residual peroxide is determined by titration (x_t). The quantity of enzyme taken should be such that 50—70% of the peroxide is decomposed in 10 mins. To this end Sumner and Somers (1947) recommend that K should lie between 0.025 and 0.04. The reaction is not strictly 1st order and the calculated reaction constant falls with time. It is therefore customary to transfer samples to acid at 3, 6 and 9 mins., to plot K against time and to extrapolate to zero time. The resulting value of K is used to calculate Kat.f.

In a careful analysis of the above procedure Bonnichsen, Chance and Theorell (1947) found that the extrapolation tended to give low values of K. This is on account of an initial inactivation which is complete well before the removal of the first sample, and which cannot be avoided at the high dilution of catalase (10^{-11} M) in the test. They therefore increased the catalase concentration to about $10^{-9} M$ and shortened the duration of the experiment. Under these conditions K becomes almost independent of t and the initial inactivation is avoided.

2. Determination of Kat.f. (BONNICHSEN, CHANCE and THEORELL.)

Reagents required:—$2.5 N H_2O_2$; $0.01 M$ phosphate buffer pH 7; $0.4 N H_2SO_4$; $0.01 N KMnO_4$.

The authors do not specify a temperature but it would be advisable to carry out the reaction at $0°$.

2 ml. H_2O_2 is added to 50 ml. buffer in a 500 ml. conical flask. A 2 ml. sample is withdrawn, blown into 5 ml. H_2SO_4 and titrated with permanganate (x_0). The catalase is suitably diluted with buffer and 0.03 ml. is placed on a small watchglass. This is dropped into the swirling contents of the flask and 2 ml. samples of the mixture are blown into acid at 13, 28 and 43 secs. (x_t). An alternative method which avoids mixing errors is to withdraw one sample after 5 secs. and a second about 15 secs. later. The constant is then obtained from

$$K = \frac{1}{t_2 - t_1} \log_{10} \frac{x_1}{x_2}$$

The quantity of enzyme should be such that most of the H_2O_2 has disappeared at 43 secs.

Values of Kat.f. of pure enzymes, obtained by this method, are given in Table 8.

3. Estimation of Catalase with Perborate.

FEINSTEIN (1949) found that H_2O_2 could be replaced by perborate and that the rate of disappearance of this substrate can be followed by permanganate titration as in the previous methods. While there is no doubt that the actual substrate is H_2O_2 arising from decomposition of the perborate, the maintenance of a very low peroxide concentration throughout the reaction greatly reduces inactivation of the enzyme. Consequently the determined enzyme activity is proportional to the amount of enzyme taken.

Reagents required:— 1.5% $NaBO_3 \cdot 4 H_2O$, filtered if necessary and adjusted to pH 6.8 with HCl; $0.067 M$ phosphate buffer, pH 6.8; $2 N H_2SO_4$; Standard $KMnO_4$ ($0.05 N$).

Four 125 ml. flasks containing 8 ml. perborate and 1.5 ml. buffer are warmed to $37°$ in a thermostat. Two flasks serve as controls and receive 0.5 ml. H_2O. The other flasks receive 0.5 ml. (at two concentrations) of ground tissue (or enzyme). After exactly 5 mins. 10 ml. H_2SO_4 is added to each flask and residual perborate is titrated with permanganate. The activity (A) is expressed as milli-equivalents perborate destroyed per mg. *wet weight* of tissue. Except where a tissue contains very little catalase the amount of homogenate required has no appreciable reducing value in terms of permanganate. However, a fifth flask without perborate can be included and a correction applied.

Comparative figures for catalase activities of various tissue homogenates determined by this method and by EULER's (1934) method are given by FEINSTEIN. If both calculations are made on the same weight basis (e. g. wet weight), the ratio Kat.f./$A = 110 \pm 30$ (S. D.). However, wide deviations from this figure are occasionally obtained.

4. Spectrophotometric Method for Estimation of Catalase.

BEERS and SIZER (1952) have shown that the decomposition of H_2O_2 can be followed by the decrease in its absorption in the ultra-violet. The molar extinction coefficients of H_2O_2 (optical density of 1 cm. solution/molar concentration) are

64.0 at 230 mμ, 39.8 at 240 mμ and 23.5 at 250 mμ[1]. For measurements at 240 mμ approximately 0.05 M H$_2$O$_2$ is prepared in 0.05 M phosphate buffer pH 7.0, and the catalase is diluted with the same buffer. Measurements are made at 20°. Two 1 cm. silica cells each receive 2 ml. catalase solution. The reference cell now receives 1 ml. buffer and the second cell 1 ml. H$_2$O$_2$. Mixing should be complete in 2—3 secs. Readings of density are taken every 10 secs. until at least 50% of the peroxide has been decomposed. Since, under the specified conditions the reaction is first order a plot of log density against time is a straight line. From this plot $t_{1/2}$, the time required for the density to be halved, can be determined. The velocity constant is then calculated as $\ln 2/t_{1/2}$ (where $\ln 2 = 0.693$). A measure of the purity of the enzyme can be obtained if the constant is divided by the dry weight. The standard deviation in experiments on ox-liver catalase by Beers and Sizer was 1.6%. Care must be taken that oxygen bubbles do not appear in the absorption cells during the test. This is likely to occur if the reaction is run for longer than 3 mins. or if the cells are not scrupulously clean.

5. Limitations of the above Methods of Determining Catalase Activity.

There is no doubt that the methods just described are dependable when catalases of a high degree of purity are being assessed. With cruder tissue extracts, however, their reliability becomes doubtful on account of 1) utilization of H$_2$O$_2$ in side reactions, 2) extra reduction of permanganate by impurities and 3) lack of transparency in the ultra-violet region of the spectrum. A manometric method should avoid these errors and although no such method has yet been generally accepted the following procedure is well adapted to the estimation of catalase in plant extracts and homogenates.

6. Manometric Estimation of Catalase According to Altschul, Kabon and Kyame (1948)[2].

The reaction is carried out at 0° in a Warburg manometer. The flask receives 1 ml. 0.07 M phosphate buffer, pH 6.8, together with enzyme and distilled water to a total volume of 3.15 ml. The side-bulb receives 0.15 ml. "1 volume" (0.18 N) H$_2$O$_2$. The enzyme concentration is adjusted so that between 25 and 55% of the peroxide is decomposed in 10 mins., under which conditions the reaction is first order and the rate constant is proportional to catalase concentration. The rate constant is calculated from

$$K = \frac{1}{t} \ln \frac{V}{v}$$

where v = volume of oxygen evolved in t mins. and V = total O$_2$ evolved when an excess of catalase is used. A unit of purity parallel to Kat. f is obtained by dividing the rate constant by the dry weight in g. of the catalase added to the manometer flask. It should be pointed out that natural logarithms are used here in the calculation of K although it has been customary to use logarithms to the base 10 in determinations of Kat.f. by the two methods described above.

The relationship between Kat.f. (Euler, 1934) and Q_{O_2} (μl. O$_2$ evolved/mg. dry weight of enzyme/hr., based upon initial rate) has been calculated by Herbert and Pinsent (1948). If the two reactions are carried out in the same buffer at 0° and if the H$_2$O$_2$ concentration in the manometric determination of Q_{O_2} is not appreciably greater than that in the Kat.f. measurement it can be shown that

$$Q_{O_2} = \text{Kat.f.} \times 77{,}500 \, S_0$$

where S_0 = initial molar H$_2$O$_2$ concentration.

---

[1] When a spectrophotometer is used at these wavelengths it is essential to guard against stray-light errors (see Glover, Vol. 1).

[2] *Note added in proof.* Another manometric method is described by Greenfield and Price (1954).

III. Spectroscopy of Peroxidase and Catalase.

No plant tissues are known in which the absorption spectrum of catalase or, after appropriate treatment, of any of its derivatives can be detected. The available data for purified spinach leaf catalase are given in Table 8 together with mcre complete data on liver and blood catalases. It can be expected that purified plant catalases will display the same types of spectral changes as animal catalases when treated with the listed reagents.

Since peroxidase is present at much higher concentrations in plants than is catalase, its spectral characteristics are of value in its detection. In fact, as KEILIN and MANN (1937) have shown, the enzyme can be detected spectroscopically in fresh horse-radish root. These authors found that a 1 cm. slice showed a fairly strong band at 556 mμ and a feeble band at 642 mμ. The former is the haemochromogen band that can be seen in many intact plant tissues (cf. Fig. 9, VIII) while the latter is the band of peroxidase. The opacity of the root renders this observation difficult but the light transmission can be increased if a number of very thin (1—2 mm.) slices of root are first soaked in glycerol and then built up to the equivalent of a 1 cm. slice. KEILIN and MANN estimated the content of haemochromogen in the root by comparison with a standard pyridine haemochromogen and obtained figures ranging from 0.00031 to 0.00046% of the wet weight or about 0.0017% of the dry weight. Since the absorption band of peroxidase-fluoride at 612 mμ is more intense than that of the free enzyme at 642 mμ it is easier to estimate peroxidase in tissues after treatment with fluoride. To this end KEILIN and MANN impregnated a 1 cm. cube of root with strong NaF solution by placing root and solution in a THUNBERG tube, evacuating the tube to remove all air from the root and then readmitting air. After standing 4 hrs. the 612 band reached maximal intensity. This band was matched against that of a standard solution of peroxidase (also treated with fluoride) using a microspectroscope. The concentration of peroxidase haematin was found to be 0.0018% of the dry weight. The sum of this figure and that of the haematin associated with the natural haemochromogen (0.0017%) is in satisfactory agreement with the figure for total haematin, estimated after treatment of the root with pyridine and $Na_2S_2O_4$ (p. 213), of 0.003—0.0035%.

The characteristic band of peroxidase at 642 mμ, and even the complete spectrum, has been detected in comparatively crude plant extracts. Thus KEILIN and MANN detected the complete spectrum in a solution of enzyme of P. N. = 100, while the detection of the band in the red in plant extracts of P.N. between 27 and 210 has been reported earlier (p. 232).

The intensification of absorption bands at very low temperatures (p. 208) can with advantage be applied to the study of peroxidase in plant tissues. For such experiments the sections should be soaked in glycerol.

D. Root Nodule Haemoglobin.

The identification of the red pigment in root nodules of soya-beans and of peas as a haemoglobin (Hb) was made by KUBO (1939) on the basis of the reactions of the pigment as extracted from minced nodules. In presence of air the extract showed 2 absorption bands at 575 and 540 mμ which, on addition of $Na_2S_2O_4$, were replaced by a single diffuse band centred at 555 mμ. Treatment of the extract with pyridine and $Na_2S_2O_4$ yielded a haemochromogen with bands at wavelengths identical with those similarly obtained from blood Hb or from crystalline haemin (556 and 524 mμ). KUBO found this nodule pigment to be very widely distributed among leguminosae (Table 11). Although later workers disputed Kubo's conclusion that the pigment was a haemoglobin showing a characteristic oxygen-binding property, Kubo's results were eventually fully confirmed and extended by Keilin and Wang (1945). These authors obtained the pigment in clear solution and demonstrated a strict analogy with blood haemoglobin in terms of reactions with O_2 and CO.

Table 11. *Occurrence of haemoglobin among Leguminosae.* (Kubo, 1939.) In all cases the pigment is found only in the root nodules.

Albizzia	*Glycine*	*Phaseolus*
Amorpha	*Indigofera*	*Pisum*
Arachis	*Kummerowia*	*Pueraria*
Astragalus	*Lathyrus*	*Sophora*
Canavalia	*Lespedeza*	*Trifolium*
Cassia	*Lotus*	*Vicia*
Crotalaria	*Lupinus*	*Vigna*
Desmodium	*Medicago*	*Wistaria*

It has been reported by many investigators that the pigment, when extracted, readily autoxidizes to brown methaemoglobin (MetHb) which is characterized by an absorption band in the red at 626 mμ and by its reactions with respiratory inhibitors which lead to marked spectroscopic changes. These reactions are summarized in Table 8 in which is also shown the close analogy between the nodule pigment and blood Hb. The band in the red of MetHb is feeble compared with those of oxyhaemoglobin (O_2Hb) and is of less value as a means of detecting the pigment. Thus if a solution of pH < 7 containing the two pigments in the ratio O_2Hb:MetHb $= 1:8$ is examined with a low-dispersion spectroscope the 626 band of MetHb and the 575 band of O_2Hb will appear to be of approximately equal intensity. At pH > 7 MetHb is transformed into a red alkaline form which is rather less readily differentiated spectroscopically from O_2Hb.

The nodule pigment can be estimated in the nodule or in extracts either by spectroscopic comparison with standard O_2Hb solutions or by a similar comparison, after conversion to pyridine haemochromogen, with a standard haemochromogen (p. 212). Analysis of clear solutions can be made by the standard methods worked out for blood pigments.

The spectroscopic detection of O_2Hb in root nodules of soya-bean and *Phaseolus vulgaris* has been described by Smith (1949). Fresh nodules are often too opaque to show bands in a spectroscope ocular but if they are sliced or crushed the two bands can readily be seen. Moistening the slices with glycerol renders them more translucent. On standing, the bands in moistened slices are slowly replaced by the broad band of the deoxygenated pigment on account of the respiration of the tissue. It must be remembered that Hb is only produced when nodules contain "effective" N-fixing strains of *Rhizobium* and that when plant growth ceases the pigment is replaced by a green bile-pigment type of chromoprotein (Virtanen, 1947). In addition, effective strains fail to induce Hb formation a) if plants are grown in the dark (unless glucose is supplied) and b) with certain varieties of host plant. Nodules formed on boron-deficient plants fail to fix nitrogen but no data on their Hb content are available. (The conditions under which Hb is formed are discussed by Smith, 1949.)

It has been claimed that there exists in nodules an equilibrium between O_2Hb and MetHb, which is largely controlled by the general level of illumination of the plant (Virtanen, 1945; Virtanen and Laine, 1946).

$$O_2Hb \underset{\text{Light}}{\overset{\text{Dark}}{\rightleftharpoons}} \text{Met Hb}$$

Investigations by Keilin and Smith (1947) failed to confirm the presence of MetHb. Although the 626 band of MetHb is comparatively feeble, the addition of NaF yields the fluoride derivative with a strong band at 605 mμ (cf. the reaction of peroxidase, p. 239). However, the 605 band could not be developed in nodules.

Smith's methods for estimation of Hb in nodules are based upon spectroscopic estimates of total haematin a) in nodule slices or b) in extracts. In the former method manipulative losses are avoided but there is no distinction between Hb and non-Hb haematin. Aqueous extraction, however, removes all Hb while the greater part of the other haematin derivatives (e. g. cytochromes) remains undissolved. The probable errors of the two methods will be discussed below.

1. Estimation of Hb as Pyridine Haemochromogen.

a) In Intact Nodules. A nodule is cut into slices 1—2 mm. thick. A slice is placed on a microscope slide, covered with $Na_2S_2O_4$ and a few drops of pyridine, and covered with a second slide. The two slides, separated by small pieces of plasticine at their corners, are pressed gently together so that they remain parallel as they come in contact with the slice. From the overall thickness and the thicknesses of the two slides the thickness of the slice can be calculated. The slice is placed on the microscope stage and a low power objective is fitted. The slice is carefully centred by observation through the spectroscope eye-piece while the slit is wide open. The comparison with a standard haemochromogen is carried out in the manner already described. The use of an objective enables estimates of haematin content to be made at different points in the slice. In the case of nodules from a single crop of soya-beans the concentration of haematin was found to be 0.234 mM ($\pm$15%).

b) In Nodule Extracts. Nodules freshly removed from the plant are washed, dried and weighed. Their total volume is then determined by adding them to water in a graduated cylinder and noting the displacement. The nodules are ground thoroughly in water and centrifuged at a speed high enough to yield a clear or almost clear supernatant. A second extraction of the solid is likely to contain $< 2\%$ of the total Hb and can, as a rule, be omitted. A measured volume (v) of the extract is placed in a cylindrical upright tube (cross-sectional area a), such as the type represented in fig. 4, and placed on the microscope stage. The Hb is converted to haemochromogen by addition of NaOH (final concentration 0.1 N) followed by pyridine (20% v/v) and $Na_2S_2O_4$. The experimental solution (optical depth v/a) is compared with a standard in the usual way (see Table 12).

Table 12. *Haemoglobin content of nodules determined in aqueous extracts as pyridine haemochromogen.* (Smith, 1949.)

Plant	Fresh wt. of sample (g.)	Mean fresh wt. of single nodule (mg.)	Volume of nodules (ml.)	Haematin in extract (mg.)	Mean value of the ratio $\frac{\text{Vol. of int. tiss.}}{\text{Total volume}}$	Conc. of Hb in internal tissue (mM)	$\frac{Hb_{17000}[1]}{\text{Fresh weight of nodules}}$ (mg./g.)
Phaseolus vulgaris .	1.7	16.8	1.7	0.072	0.59	0.109	1.09
Soya (medium sized nodules)	0.866	8.7	0.85	0.088	0.38[2]	0.42	2.68
do. (small nodules)	0.404	4.9	0.40	0.050	0.38[2]	0.50	3.25

[1] Mol. wt. 17,000 with one haematin group per molecule was assumed (see Ellfolk and Virtanen, 1952).

[2] S. D. 9.7%.

2. Relationship of Hb to Total Haematin in Nodules.

No pigment other than Hb was detected by Smith in the clear aqueous extract while the insoluble residue contained no Hb but showed distinct a b c bands of cytochrome on reduction. Since haemoglobin is limited to the bacteria-containing cells in the central part of the nodule the determination of Hb in these cells requires a correction for the proportion of the total volume that is occupied by this internal nodular tissue. If the nodules are assumed to be spherical this calculation can readily be made from observations with a micrometer eye-piece.

From Table 12 it is seen that the internal tissue of soya nodules represents 38% of the total nodule volume and that it contains *ca.* 0.46 mM Hb (in terms of extractable haematin). On a whole nodule basis the concentration would be 0.175 mM which can be compared with the figure 0.234 for total haematin (Table 12). Smith found that of the total haematin in nodules containing an effective strain of *Rhizobium* 92% was extractable (Table 13) and all the haematin in the extract appeared as Hb. On this basis the Hb concentration in the nodule is $0.92 \times 0.234 = 0.215$ mM. In view of the experimental difficulties and wide variations in Hb content the figures 0.175 and 0.215 can be considered to agree satisfactorily. The true value may reasonably be assumed to be within these limits.

Table 13. *Haematin contents of nodules inoculated with effective and ineffective strains of Rhizobium.* (From Smith, 1949.)

	Effective nodules (strain No. 505[1])		Ineffective nodules (strain No. 507[1])	
	mg. haematin per g. fresh weight	% of total	mg. haematin per g. fresh weight	% of total
Haematin in aqueous extract . .	0.071	92.2	0.0068	78.2
Haematin in residue	0.006	7.8	0.0019	21.8
Total haematin . . . . ·. . . .	0.077	—	0.0087	—

3. Estimation of Hb by Comparison with Hb Standards.

While it is in principle possible to estimate Hb in nodules or extracts by spectroscopic comparison with standard solutions the method does not appear to have been tested. Strictly speaking the comparison should be made with a solution of the nodule pigment but the latter is too readily autoxidizable in solution to make a satisfactory standard. Although the MetHb so formed can be readily reduced back to Hb, attempts to oxygenate it lead to a mixture of O_2Hb and MetHb (Sternberg and Virtanen, 1952). The CO derivative of nodule Hb is, however, more resistant to oxidation and a spectroscopic comparison as COHb would be feasible since, in a CO atmosphere, the pigment within the nodule would be converted quite rapidly to COHb. An alternative procedure would be to use mammalian O_2Hb as standard since the absorption curves of O_2Hb's from soya-bean and ox-blood are very similar (fig. 12).

The intensity of an absorption band as viewed in a direct-vision spectroscope is a measure of its height with respect to the minima of absorption on each side and not to its absolute height as depicted in fig. 12. Hence if spectra of equal quantities of nodule and blood Hb's were compared by means of a spectroscope ocular it would be found that the α-band of the former (575 mμ) would be slightly weaker than the α-band of the latter (577 mμ) while the β-bands would show a reverse

---

[1] Strains used by Wilson (1940).

relationship. Thus if the nodule pigment were being matched with a standard blood O_2Hb solution in a double-wedge trough the required relationship between the two spectra would be attained at a quite definite position of the trough on its platform.

Thus both nodule and blood haemoglobin could be used as standards. The latter can be crystallized to yield stable material or a solution can be prepared directly from red blood cells.

Preparation of Nodule Haemoglobin. (STERNBERG and VIRTANEN, 1952.) Soya-bean nodules are crushed in an equal weight of saturated ammonium sulphate and the mixture is left overnight at 0°. Next day the precipitate and nodule residue are centrifuged down and the clear liquid treated with ammonium sulphate to bring the saturation to 75% (S. G. 1.19). The mixture is again left overnight at 0° and next day the precipitate is collected on a filter and washed with 75% saturated ammonium sulphate. The precipitate is dialysed against running tapwater and finally against distilled water. The pigment so obtained is in the form of MetHb and can be preserved for long periods in the freeze-dried state. If a solution is saturated with CO and treated with the minimum of $Na_2S_2O_4$, COHb will be formed which will remain unchanged for short periods.

Preparation of Crystalline O_2Hb. (KEILIN and HARTREE, 1935.) Fresh horse blood is defibrinated by shaking with glass beads and straining off the coagulated fibrin. The red cells are collected by centrifuging and washed three times with 0.9% NaCl. Of the packed cell suspension 300 ml. is vigorously shaken in stoppered centrifuge tubes with 90 ml. ether and 90 g. crushed ice. 15 g. NaCl is dissolved in the mixture which is then centrifuged for 10 mins. The lower concentrated Hb solution is sucked off from beneath the ether-stromatin layer and dialysed overnight in cellophane tubing against running tap water. The dialysed solution is cooled in ice and one quarter of its volume of cold 90% ethanol is run in very slowly with constant and thorough shaking. Crystallization, which begins almost at once,

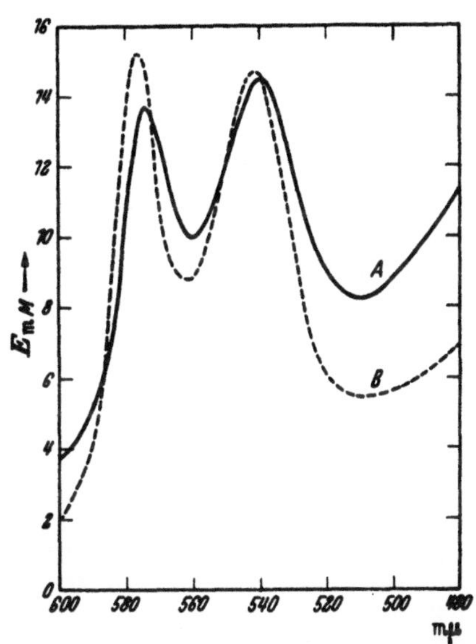

Fig. 12. Absorption curves of oxyhaemoglobin from soya-bean root nodules (A) and from ox blood (B) according to STERNBERG and VIRTANEN (1952).

$$\varepsilon_{mM} = \frac{\text{optical density}}{\text{m}M \text{ concentration of haematin} \times \text{optical depth.}}$$

is complete after 24 hrs. at 5°. The crystals are centrifuged down, resuspended in ice-water and centrifuged again. The wash liquor is drained away as completely as possible and the crystals are stored at 0°. A solution of O_2Hb can readily be made by adding to the crystal mass a few drops of 5% Na_2CO_3 and diluting the resulting solution with water.

Haemoglobin Solution from Red Blood Cells. Packed washed cells are lysed by addition of 10 volumes of water and the mixture is centrifuged to yield a perfectly clear solution.

The haematin content of the above O_2Hb solutions can be estimated as pyridine haemochromogen either by spectroscopic or by colorimetric comparison (p. 211). Alternatively, from the optical density (D) at the peak of the α-band (577 mμ) the concentration of mammalian O_2Hb can be obtained as 0.066 D mM haematin. In the case of nodule O_2Hb the factor would be 0.060 (see fig. 12).

An alternative method of estimating Hb and MetHb either alone or as mixtures is that of CHILCOTE and O'DEA (1953) in which the pigments are converted to MetHb cyanide and estimated colorimetrically.

4. Estimation of Hb and/or MetHb as MetHb Cyanide.

An aqueous solution containing 0.2% $K_3Fe(CN)_6$ and 0.1% KCN is adjusted to p_H 7.4 by addition of KH_2PO_4. It must be kept tightly stoppered to prevent loss of HCN. Clear solutions containing Hb or MetHb are treated with one tenth of their volume of this reagent and estimated either by measuring their optical density at 540 mμ in a spectrophotometer or by means of a colorimeter fitted with a green filter. Solutions for calibration purposes are standardized by one of the methods already mentioned. The molar ratio $K_3Fe(CN)_6$: haematin should be at least 20 in order that the cyanide derivative be formed completely. If the calculated result shows that the haematin content was too high the solution should be appropriately diluted and re-estimated.

Since virtually all the iron present in the Hb solutions described above is present as Hb iron, their concentrations can be derived from iron estimations. Suitable colorimetric methods are based upon TiCl$_3$ (King and Gilchrist, 1947), α,α'-dipyridyl (Keilin and Hartree, 1945a) and sulphosalicylic acid (Galston, Bonnichsen and Arnon, 1951).

5. Oxygen Dissociation Curves of Nodule Haemoglobin.

Owing to the rapid rate at which solutions of nodule Hb undergo oxidation to MetHb it has so far not been possible to make accurate determinations of its oxygen affinity. Should means of preventing this oxidation be found the method of Hill (1936) would be suitable. Fig. 6 illustrates Hill's apparatus. Keilin and Wang (1946) describe certain modifications of Hill's method and also correct an error in the formula for calculating the O$_2$ pressure in equilibrium with the Hb solution.

References.

Altschul, A. M., and M. L. Karon: Arch. Biochem. 13, 161 (1947). — Altschul, A. M., M. L. Karon and L. Kyame: Arch. Biochem. 18, 161 (1948). — Avi-Dor, Y., and K. G. Paul: Acta chem. scand. 7, 444 (1953).

Ball, E. G., C. F. Strittmatter and O. Cooper: J. Biol. Chem. 193, 635 (1951). — Beers, R. F. jr., and I. W. Sizer: J. Biol. Chem. 195, 133 (1952). — Bhagvat, K., and R. Hill: New Phytol. 50, 112 (1951). — Boiarkin, A. N.: Biochimia 16, 352 (1951). — Bonner, W. D. jr.: Biochem. J. 56, 274 1954). — Bonnichsen, R. K.: Acta chem. scand. 1, 114 (1947); 2, 561 (1948). — Bonnichsen, R. K., B. Chance and H. Theorell: Acta chem. scand. 1, 685 (1947).

Chapman, N. B., and B. C. Saunders: J. Chem. Soc. 496 (1941). — Chilcote, M. E., and A. E. O'Dea: J. Biol. Chem. 200, 117 (1953). — Cleland, K. W., and E. C. Slater: Biochem. J. 53, 547 (1953). — Conant, J. B., and S. E. Kamerling: J. Amer. Chem. Soc. 53, 3522 (1931). — Cooperstein, S. J., and A. Lazarow: J. Biol. Chem. 189, 665 (1951).

Davenport, H. E.: Nature (Lond.) 170, 1112 (1952). — Davenport, H. E., and R. Hill: Proc. Roy. Soc. (B) 139, 327 (1952). — Dawson, C. R., and W. B. Tarpley: in The Enzymes, by J. B. Sumner and K. Myrbäck, Vol. II, part 1, p. 454. New York: Academic Press. 1951.— Deleano, T., N. Popovici and L. Ionescu: Bull. Soc. Chim. biol. (Paris) 19, 898 (1937). — Diemair, W., and H. Häuser: Z. anal. Chem. 122, 12 (1941).

Ellfolk, N., and A. I. Virtanen: Acta chem. scand. 6, 411 (1952). — Elliott, K. A. C., and D. Keilin: Proc. Roy. Soc. (B) 114, 210 (1934). — Ettori, J.: Biochem. J. 44, 35 (1949). — von Euler, H.: Chemie der Enzyme. München: J. F. Bergmann 1934. — Evans, T. W., and W. M. Dehn: J. Amer. Chem. Soc. 52, 3649 (1930).

Feinstein, R. N.: J. Biol. Chem. 180, 1197 (1949).

Galston, A. W., R. K. Bonnichsen and D. I. Arnon: Acta chem. scand. 5, 781 (1951). — Goddard, D. R.: Amer. J. Bot. 31, 270 (1944). — Goddard, D. R., and C. Holden: Arch. Biochem. 27, 41 (1950). — Gračanin, M.: Biochem. Z. 168, 429 (1926). Greenfield, R. E., and V. E. Price: J. Biol. Chem. 209, 355 (1954)

Hartridge, H.: J. Physiol. 44, 1 (1912); 57, 47 (1923); Proc. Physiol. Soc. 54, 128 (1921).— Hausser, K. W., R. Kuhn and G. Seitz: Z. phys. Chem. 29 B, 391 (1935). — Herbert, D., and J. Pinsent: Biochem. J. 43, 193 (1948). — Hill, R.: Proc. Roy. Soc. (B) 120, 472 (1936). — Hill, R., and E. F. Hartree: Annu. Rev. Pl. Physiol 4, 115 (1953). — Hill, R., and H. F. Holden: Biochem. J. 20, 1326 (1926). — Hill, R., and R. Scarisbrick: New Phytol. 50, 98 (1951).

Kayser, H.: Handbuch der Spectroscopie, Vol. III. Leipzig: Hirzel 1905. — Keilin, D.: Proc. Roy. Soc. (B) 98, 312 (1925); 100, 131 (1926); Erg. Enzymforschg. 2, 239 (1933). —

KEILIN, D., and E. F. HARTREE: Proc. Roy. Soc. (B) 117, 1 (1935); 119, 141 (1936); 125, 171 (1938); 127, 167 (1939); 129, 277 (1940); (a) Biochem. J. 39, 289 (1945); (b) Biochem. J. 39, 293 (1945); 41, 500 (1947); (a) Nature (Lond.) 164, 254 (1949); (b) Biochem. J. 44, 205 (1949); Nature (Lond.) 165, 504 (1950); Biochem. J. 49, 88 (1951); Nature (Lond.) 171, 413 (1953). — KEILIN, D., and T. MANN: Proc. Roy. Soc. (B) 122, 119 (1937). — KEILIN, D., and J. D. SMITH: Nature (Lond.) 159, 692 (1947). — KEILIN, D., and Y. L. WANG: Nature (Lond.) 155, 227 (1945); Biochem. J. 40, 855 (1946). — KEILIN, J.: Biochem. J. 37, 281 (1943). — KENTEN, R. H., and P. J. G. MANN: Biochem. J. 57 (1954). KING, E. J., and M. GILCHRIST: Lancet 253, 201 (1947). — KUBO, H.: Acta Phytochim. (Tokyo) 11, 193 (1939).

LEMBERG, R., and J. W. LEGGE: Hematin Compounds and Bile Pigments. New York: Interscience 1949. — LINEWEAVER, H., and D. BURK: J. Amer. Chem. Soc. 56, 658 (1934). — LOWRY, O. H., and O. A. BESSEY: J. Biol. Chem. 163, 633 (1946). — LUNDEGÅRDH, H.: Arkiv Kemi 5, 97 (1953).

MacMUNN, C. A.: The Spectroscope in Medicine. London: Churchill 1880; Spectrum Analysis applied to Biology and Medicine. London: Longmans Green 1914. — MANN, T.: Haematin Compounds in Plants and their Relation to Peroxidase. Doctoral Thesis. Cambridge University 1937; Archiwum Towarzystwa Naukowego we Lwowie 9, 421 (1938) [These contributions by Mann are discussed by HILL and HARTREE (1953)]. — MARGOLIASH, E.: Nature (Lond.) 170, 1014 (1952); Biochem. J. 56, 529 and 535 (1954). — MAXWELL, R. E.: Plant Physiol. 25, 521 (1950). — MAYNEORD, W. V., and E. M. F. ROY: Proc. Roy. Soc. (A) 158, 634 (1937).

NEILANDS, J. B.: J. Biol. Chem. 197, 701 (1952).

OKUNUKI, K.: Acta Phytochim. (Tokyo) 11, 27 (1939).

PAUL, K. G.: in The Enzymes by J. B. SUMNER and K. MYRBÄCK, Vol. II, Part 1, p. 357. New York: Academic Press 1951. — PAUL K. G., and Y. AVI-DOR: Acta Chem. Scand. 8, 649 (1954). — PURR, A.: Biochem. Z. 321, 1 (1950).

QUINLAN-WATSON, T. A. F., and D. W. DEWEY: Aust. J. Sci. Res. 1, 139 (1948).

ROSENBERG, A. J., and A. DUCET: C. r. Acad. Sci. (Paris) 229, 391 (1949).

SCARISBRICK, R.: Rep. Progr. Chem. 44, 226 (1947). — SCHALFEJEFF, M.: Abstract in Ber. dtsch. chem. Ges. (Referat-Band) 18 232 (1885). — SLATER, E. C.: (a) Biochem. J. 45, 1 (1949); (b) Biochem. J. 44, 305 (1949). — SMITH, F. G., W. B. ROBINSON and E. STOTZ: J. biol. Chem. 179, 881 (1949). — SMITH, F. G., and E. STOTZ: J. Biol. Chem. 179, 890 (1949). — SMITH, J. D.: Biochem. J. 44, 585 (1949). — SORBY, H. C.: Proc. Roy. Soc. 15, 433 (1867); Quart. J. Micr. Sci. (new series) 9 358 (1869); Month. Microscop. J. 6, 124 (1871). — SPEDDING, F. H., and R. S. BEAR: Phys. Rev. 44, 287 (1933). — STERNBERG, H., and A. I. VIRTANEN: Acta chem. scand. 6, 1342 (1952). — SUMNER, J. B., and E. C. GJESSING: Arch. Biochem. 2, 291 (1943). — SUMNER, J. B., and S. F. HOWELL: Enzymologia 1, 133 (1936). — SUMNER, J. B., and K. MYRBÄCK: The Enzymes, Vol. II, Part 1. New York: Academic Press 1951. — SUMNER, J. B., and G. F. SOMERS: Chemistry and Methods of Enzymes, 2 nd. edition, p. 209. New York: Academic Press 1947.

THEORELL, H.: Ark. Kemi, Min. Geol. 14 B, No. 20 (1940); 16 A, No. 2 (1942).

VIRTANEN, A. I.: Nature (Lond.) 155, 747 (1945); Biol. Rev. 22, 239 (1947). — VIRTANEN, A. I., and T. LAINE: Nature (Lond.) 157, 25 (1946).

WEBSTER, G. C.: Amer. J. Bot. 39, 739 (1952); Science 117, 530 (1953). — WILLSTÄTTER, R., and A. STOLL: Liebigs Ann. 416, 221 (1918). — WILSON, P. W.: The Biochemistry of Symbiotic Nitrogen Fixation. Madison: University of Wisconsin Press 1940.

YAKUSHIJI, E.: Acta Phytochim. (Tokyo) 8, 325 (1935); 11, 187 (1939).

ZEILE, K.: Z. Vitamin-, Hormon- u. Fermentforsch. 3, 540 (1950). — ZEILE, K., and W. SIEDEL: In Physiologische Chemie, by B. FLASCHENTRÄGER and E. LEHNARTZ, Vol. I, p. 845. Berlin: Springer-Verlag 1951.

Nucleic Acids, Their Components and Related Compounds.

By

R. Markham.

With 39 Figures.

The nucleic acids have been known as major constituents of cell nuclei since their discovery in animal tissues by MIESCHER (1871). The first nucleic acid to be obtained from plants was that from yeast (ALTMANN, 1889), followed by the discovery of a similar substance, called "triticonucleic acid", by OSBORNE and HARRIS in 1902. As it happened, the two latter nucleic acids were of a different type from that in animal tissues, containing a pentose (ribose) as part of their structure, while the animal or "thymus nucleic" acid contained a sugar which eventually proved to be 2-deoxyribose. This led to the belief, which was to persist for many years that there were two kinds of nucleic acid: that from plants or "yeast nucleic acid", and that from animals or "thymus nucleic acid". Eventually, however, in 1924, JONES and PERKINS found the former in pancreas, and FEULGEN and ROSSENBECK found evidence of the latter in the nuclei of wheat germ.

It is now known that the two types of nucleic acid are both to be found in most cells, the "yeast" type or ribonucleic acid as it is now known, being mainly a cytoplasmic constituent, probably concerned with protein synthesis, while the "thymus" type is generally confined to the nuclei, where it is almost certainly the prime factor in deciding the heredity of the cell, and it probably plays a leading part in the regulation of the cells' activities.

In the last few years it has also become evident that many nucleic acids exist, characteristic of the various sources from which they have been isolated, and in some cases having important differences in their composition. The division into the two types, however, still holds, and the two differ greatly in their chemical behaviour, although their structure is very similar. The main reason for these differences appears to reside in the presence or absence of an —OH group in the sugar residues in position 2. This has such an outstanding influence on the resistance of the nucleic acids to enzymes and to chemical reagents that the two types might be regarded as belonging to different classes of substances.

A. Occurrence in Plants.

Plant tissues are relatively rich in nucleic acids, the order of magnitude being some 100 mg./kilo of wet leaf. In virus infected plants up to 1 g./kilo of virus nucleic acid may also be present. In dry tissues the quantities may be much greater, wheat germ containing some 3.6% total nucleic acid (OSBORNE and HARRIS, 1902). As far as the smaller constituents are concerned, it is reasonably safe to assume that many of them will be present in small amounts, but it is much more difficult to assess the value of many of the early records of the occurrence of these substances. In most cases the plant tissues or partially fractionated extracts derived from them have been treated in such a way that either the deoxyribonucleic acid or the ribonucleic acid present will have been degraded sufficiently to account for the results. Again, the methods used for the characterisation have often been extremely crude and equivocal. Plants also contain a sufficient variety of enzymes to ensure that, unless care is taken during the initial stages of an isolation, the substance eventually isolated might not have occurred free in the intact tissues.

Nevertheless, a number of interesting substances related to nucleic acids have been isolated from plant tissues, apparently occurring in them in the free state. Of these adenosinetriphosphoric acid (ATP) is discussed elsewhere. Other substances for which reasonably satisfactory evidence is available include guanosine (SCHULTZE and BOSSHARD, 1886) which has been found in a number of plant tissues, reaching the extraordinary level of 0.1% by weight in some samples of ergot, purine riboside (nebularine) which has recently been identified in *Agaricus (Clitocibe) nebularia* (LÖFGREN and LÜNING, 1953), and which will not be mentioned elsewhere in this article [$\varepsilon = 7.2 \times 10^3$ at 263 mμ (peak), 2.0×10^3 at 223 mμ (trough) in 0.1 N HCl], uridine-5'-pyrophosphate and its compounds, which occur in yeast (CAPUTTO, LELOIR, CARDINI and PALADINI, 1950) and various micro-organisms, and, of course, the methylated xanthine derivatives. An unusual observation concerns the existence of 1:3:7:9-tetramethyl, 2:6:8-trihydroxypurine in tea (JOHNSON, 1937). This substance, however, was isolated by hand-picking crystals found among the residues from one million pounds of tea, and this might well be the reason for the fact that this observation remains unconfirmed.

That the subject is by no means worked out is quite evident, and it only remains to observe that some time ago we examined in our laboratory a specimen of a dinucleotide isolated from pea plants (K. ROWAN, unpublished), and that recent work of some of my colleagues with the 8-azapurines suggests that nucleic acid metabolism in plants may involve quite complex transformations which take place among free bases.

B. Nucleic Acids: General Structure and Composition.

The nucleic acids, as we have seen, belong to one of two kinds: those containing D-ribose or RNAs, and those containing 2-deoxy D-ribose or DNAs. Both DNA and RNA are basically poly-sugar phosphates in which most or all of the sugar rings are joined by phosphate residues as diesters from C 3' to C 5' in adjacent rings. The sugars are themselves in the five-membered or furanose configuration so that the general structure is as in:

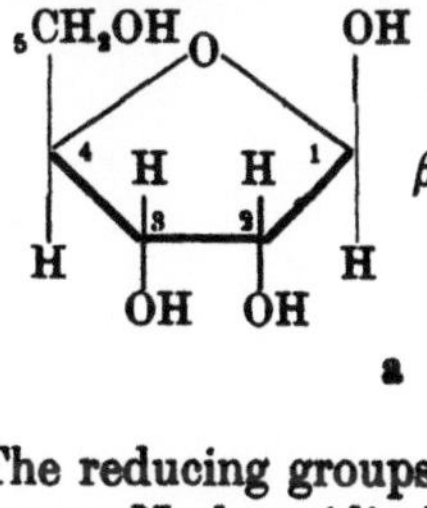

β-D-ribofuranose (a)

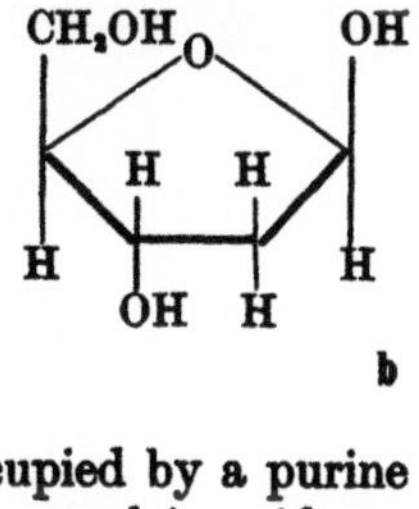

β-2-deoxy D-ribofuranose (b)

The reducing groups (C 1) are each occupied by a purine or a pyrimidine linked by an N-glycosidic linkage, so that the nucleic acids are non-reducing in their native state.

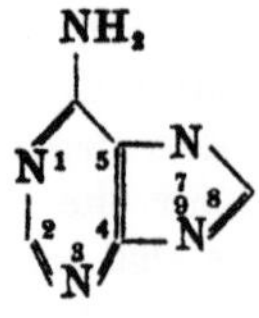
c

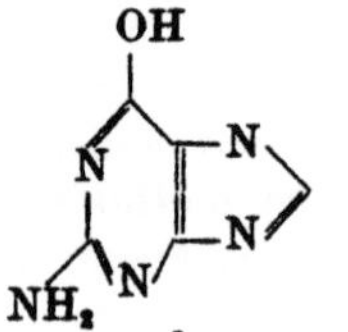
d

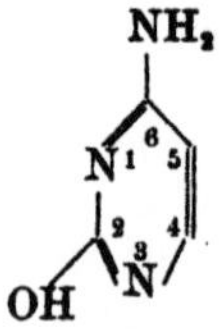
e

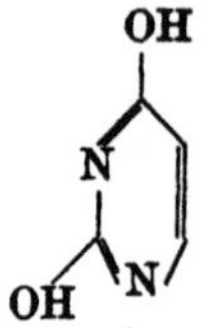
f

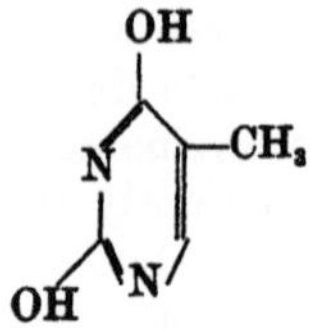
g

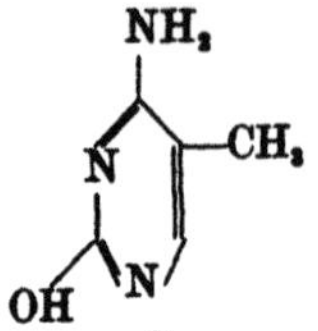
h

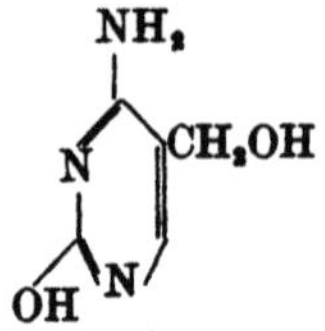
i

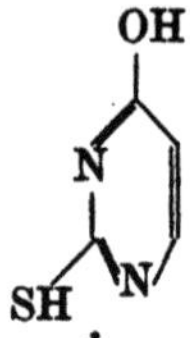
j

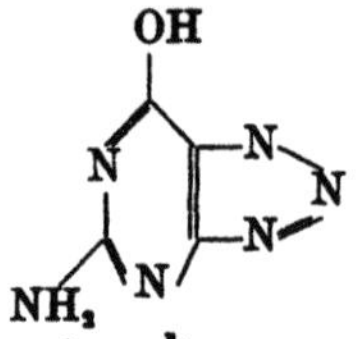
k

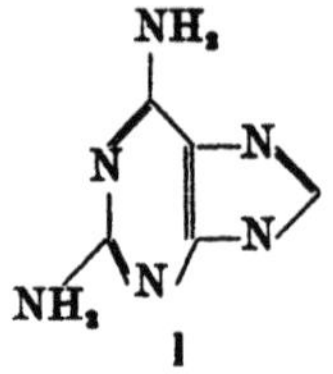
l

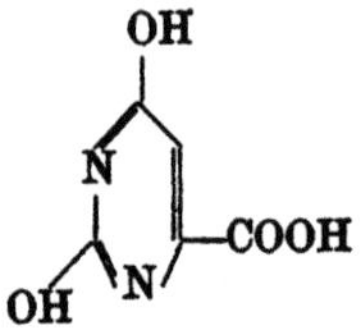
m

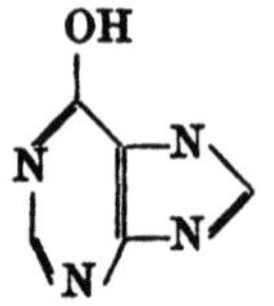
n

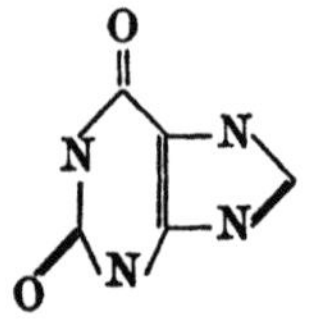
o

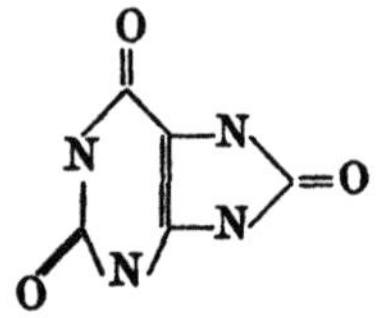
p

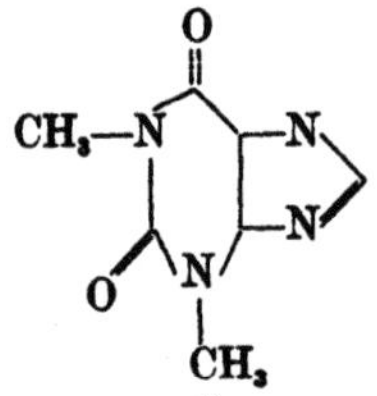
q

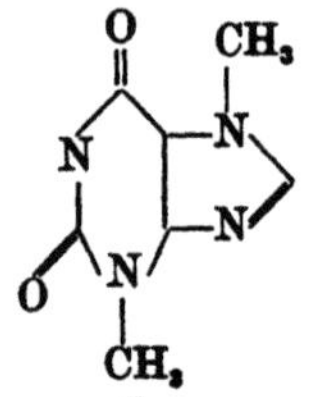
r

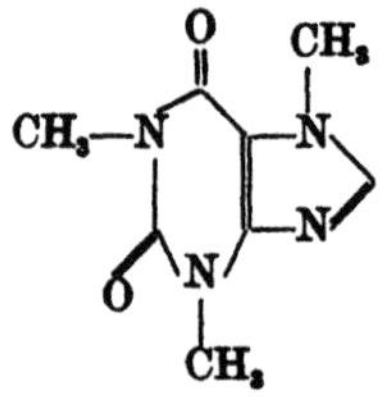
s

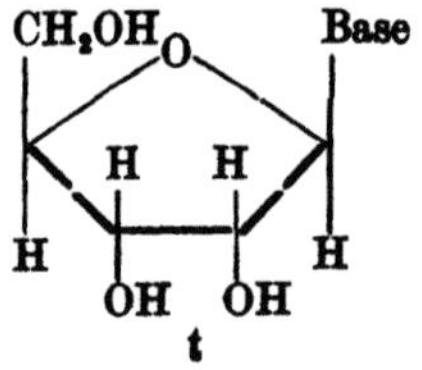
t

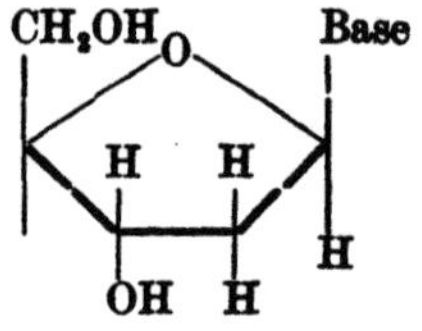
u

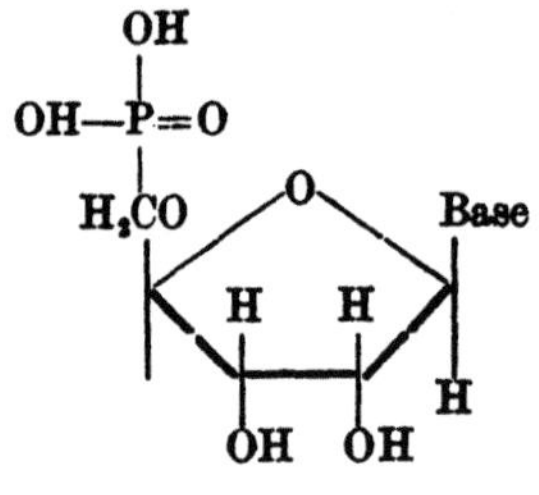
v

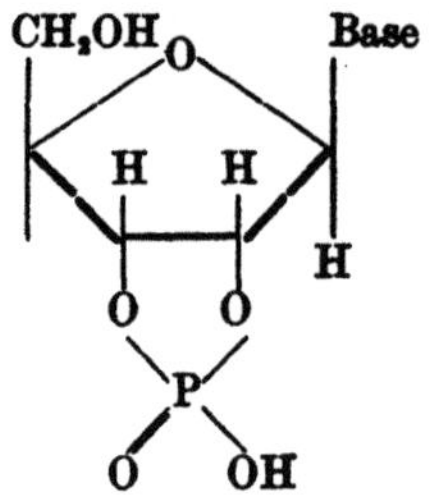
w

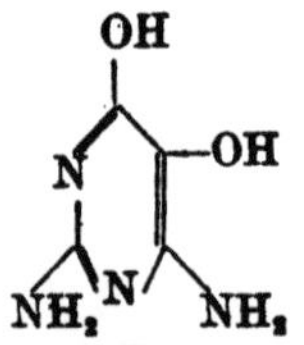
x

Explanation of formulae see opposite.

The purines found in nucleic acids are adenine (c) and guanine (d), and the pyrimidines are cytosine (e), uracil (f), thymine (g), 5-methyl cytosine (h) and 5-hydroxymethyl cytosine (i). In addition there are a number of related substances, which may participate in the structure or in the synthesis of nucleic acids, such as 2-thiouracil (j), 8-azaguanine (k), 2:6-diamino purine (l) and orotic acid (uracil-4-carboxylic acid) (m).

There are also a number of derivatives which can exist as compounds similar to those obtained from nucleic acids, such as hypoxanthine (n) and xanthine (o), which are formed from adenine and guanine by deamination, uric acid (p), which is an oxidation product of the two former substances and which has been reported to occur in fungal spores (SUMI, 1928), and also the methylated xanthine derivatives, caffein (q), theobromine (r) and theophylline (s).

Many of these purines and pyrimidines are also to be found in the forms of *nucleosides*, that is to say as the 2-deoxyribosides or the ribosides, and also as the nucleoside phosphates or *nucleotides* of which there are, of course, two possibilities in the former series: the nucleoside 3'-phosphates and the nucleoside 5'-phosphates (v), and four in the ribose series: these being the 2', 3' and 5'-phosphates and the 2':3' (cyclic) monohydrogen phosphates (w).

More complex derivatives are the nucleoside diphosphates, the dinucleoside monophosphates and then the large range of polynucleotides culminating in the "native" nucleic acids which are large molecules of the same order of molecular complexity as the proteins.

All or most of these substances may be obtained in plant or animal tissues, and a large number may be isolated or identified by the methods of chromatographic analysis on paper or on ion-exchange columns or by electrophoresis on paper strips. The detection and estimation of many of the compounds is made easy by the large and characteristic ultraviolet light absorption shown by the purines and pyrimidines and their compounds.

The nucleic acids are colourless substances (or should be) which dissolve in water to give colourless solutions, which may be extremely viscous in the case of the deoxyribonucleic acids. The latter are difficult to dissolve because the particles tend to form gelatinous masses. If difficulty is experienced in dissolving a nucleic acid it is safe to make the solution slightly alkaline (less than pH 9) if it is kept at 20° C. In acid solutions nucleic acids are precipitated as a characteristic white flocculent mass. 50% (v/v) ethanol will also precipitate deoxyribonucleic acids from neutral solution and 75% (v/v) ethanol will precipitate good specimens of ribonucleic acid. It may however be necessary to add a small quantity of 1 N acetic acid to ribonucleic acids in order to get complete precipitation. It is *essential* when precipitating nucleic acids by ethanol to have an adequate concentration of ions present in the solution. One drop of sat. $(NH_4)_2SO_4$ in 100 ml. is adequate. Magnesium salts are also good for this purpose.

The elementary composition of the nucleic acids is, of course, variable, but only over fairly small limits, and the nitrogen content is usually about 15%. They normally are to be found as salts, usually of potassium, sodium or magnesium, which substances are linked to the primary phosphate groupings present in the internucleotide bonds. The phosphorus content is one of the best indications of

---

c Adenine; d guanine; e cytosine; f uracil; g thymine; h 5-methyl cytosine; i 5-hydroxymethyl cytosine; j 2-thiouracil; k 8-azaguanine; l 2:6 diamino purine; m orotic acid; n hypoxanthine; o xanthine; p uric acid; q caffein; r theobromine; s theophylline; t N-riboside; u N-deoxyriboside; v nucleoside-5'-phosphate; w nucleoside-2':3'-(cyclic) phosphate; x divicine.

purity and should in general be between 9.5 and 10.0%. Usually, however, the nucleic acids are contaminated with protein which binds to them strongly, and commercial specimens of nucleic acids very often have P contents of 6% or less. Most of the protein may be removed by shaking the nucleic acid solutions in water with a mixture of chloroform and octyl alcohol (8:1 v/v) (Sevag, Lackmann and Smolens, 1938). If the shaking is violent (stirring in a Waring Blendor is excellent for this purpose) the protein is denatured and forms a gel with the chloroform. This gel may be centrifuged off and the aqueous layer decanted, and treated again until no further gel is formed. The nucleic acid is then precipitated by ethanol (50% v/v), acidified slightly if this should prove necessary.

Protein may be detected in specimens of nucleic acid by a very sensitive modification of the biuret reaction due to T. B. Osborne (unpublished, see Coghill 1931). In this method, one drop of 1% $CuSO_4$ is added to 2 ml. of solution + 1 ml. ethanol and an excess of KOH pellets is added. The latter dissolve and the saturated KOH salts the ethanol out of solution, and this takes with it the biuret colour. Less than 50 μg. of protein give a distinct pink tinge to the ethanol layer.

It now seems certain that nucleic acids are extremely fragile substances, and that for certain purposes connected with the testing of their biological activity they should not be subjected to boiling in aqueous solution or to extremes of pH. It is also undesirable to dry or to freeze the solutions. For general analytical work, however, such precautions may be neglected because they usually result in subtle changes which cannot be detected by analysis. It is as well however to treat nucleic acids as if they were enzymes rather than inorganic salts.

C. Absorption Spectra.

I. Nucleic Acids.

Nucleic acids absorb ultraviolet light by virtue of their individual purines and pyrimidines and they show a pronounced absorption maximum at about 263 mμ with a trough at about 227 mμ. The ratio of peak to trough is about 1.8:1 (Fig. 1). The exact form of the spectrum is, of course, dependent to some extent upon the composition of the nucleic acid, and for a relatively native nucleic acid the extinction per mole of phosphorus is about 8,000. On partial hydrolysis by acids, alkalies or by enzymes, this value rises to about 12,000, which is more like the extinction due to the nucleotides themselves (Table 1). The exact reason for this great increase is not known, but it is probably due to the resonance of individual rings being reduced by hydrogen bonding between various rings in intact nucleic acids.

Because of the variation in extinction of nucleic acids with their state of disaggregation, the measurement of density is not an accurate guide to the quantity of a nucleic acid in a solution unless it is either known to be native, or to have been disaggregated completely.

It is also desirable, although less essential, to have actual data on the particular type of nucleic acid which is being investigated. It should be noted that preparations of nucleic acid should have negligible absorption at 310 mμ. A shallow trough in the

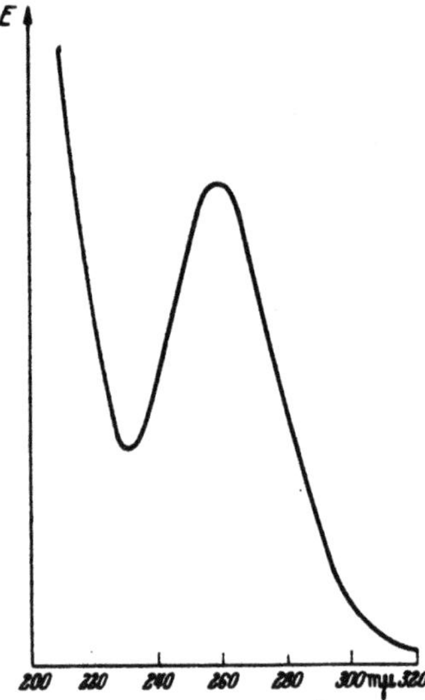

Fig. 1. Ultraviolet absorption spectrum of the ribonucleic acid from *Stellaria media*.

230 mμ region is also indicative of contamination, as is a shifting of the maximum towards the longer wavelengths.

II. The Smaller Constituents.

The absorption spectra of the more important purines and pyrimidines are given in Figs. 2—8.

These spectra are given for solutions in 0.1 N NaOH, water at pH 7 and 0.1 N HCl and it will be noted that the spectra differ with pH. This is associated with the presence of one or more ionising groupings in the molecule, and a study of the spectra at controlled pH values can in fact be used to determine the pK values of the various ionising groups, and can thus be useful for the identification of unknown substances.

For general analytical purposes it is most convenient to read absorption spectra in 0.1 N HCl. This is because a controlled pH is easily obtained, and the change of spectrum is small with large variation in the strength of the acid, and particularly because almost all the substances derived from nucleic acids are readily soluble in this solution. Moreover it will be noted that the extinction coefficients of the bases have a tendency to be highest in acid solution.

The extinction coefficients of a number of substances are tabulated in Table 1, which also gives simple conversion factors which will be of use. Some of these have been determined specially while others have been obtained from the literature from the sources indicated below the table. It is important to note that differences of 5% or more in the values are reported by various authors, this being partly due to spectroscopic errors, but also due to the fact that commercial specimens of most of these substances are usually impure, although it is now possible to obtain "chromatographically pure" specimens which probably contain less than 2% of impurity. The extinction coefficients of the various isomeric nucleoside phosphates also differ slightly, as do their spectra, but this is unlikely to cause serious errors.

For identification purposes it is often sufficient to know the chromatographic behaviour of the substance in question, but in some cases it is advantageous to check on the actual spectrum. For this it is often only necessary to read the absorption at 250, 260 and 280 mμ, because the ratios of the absorptions at these wavelengths are fairly characteristic. This procedure has been developed very largely by Cohn and his collaborators.

Effect of Substituents on Absorption Spectra. In general there is a tendency for the absorption maxima of the bases to shift towards longer wavelengths when the solvent is made alkaline. This effect is much less obvious among the nucleosides and nucleotides. The spectra of these latter are generally rather similar.

The substitution of a methyl or hydroxymethyl group in the pyrimidines at C 5 also shifts the spectra towards the red, about 50 Å units for a methyl group, and rather less for the hydroxymethyl group. This simple relationship played a large part in the identification of the methylated cytosines by Wyatt (Wyatt, 1950; Wyatt and Cohen, 1952).

The ribosides and nucleotides also have a tendency to have their peaks some 50 Å towards the red as compared with the substituted bases.

Characteristic Spectra. The characteristic spectrum of a substance is given by a plot of log D against wavelength, where D is the density (log I_0/I) of a solution of any suitable concentration in a cell of any path length. By plotting in this way one finds that, if the Lambert-Beer law is obeyed, as it usually is when workable concentrations are used, curves are obtained the shapes of which are independent of concentration, being merely shifted with respect to the ordinate. Such plots can therefore be used to determine the identity of samples of pure material. In the case of mixtures the problem is better served by a linear plot of D against wavelength, because an analysis of the spectra of a solution containing more than one component may be made by subtraction.

Table 1. *Absorption Data and Molecular Weights of Nucleic Acid Constituents and Related Substances.*

Substance	M. Wt.	pH	$E^{0.001\%}_{1\,cm.}$	$\epsilon \times 10^{-3}$	λ (mμ)
Adenine	135	1	0.96	13.0	260
Guanine	151	1	0.73	11.0	250
Cytosine	111	1	0.95	10.5	275
Uracil	112	1	0.705	7.9	260
Thymine	126	1	0.63	7.95	265
5-Methyl cytosine .	127	1	0.785	9.8	283
5-Hydroxymethyl cytosine	143	1	0.69	9.7	279
Xanthine	152	7.2	0.72	11.0	268
Hypoxanthine . .	136	7.2	0.77	10.5	257
Uric acid	168	7	0.73	12.2	290
2-Thiouracil . . .	128	2—6	1.10	14.0	274
8-Azaguanine . . .	152	2	0.87	13.2	247
Orotic acid . . .	156	1	0.48	7.52	280
Caffein	194	1	0.51	9.9	272
Theobromine . . .	180	1.3	0.54	9.8	273
Theophylline . . .	180	1	0.53	9.57	270
Adenosine	267	1	0.53	14.2	260
Guanosine	283	1	0.37	11.5	255
Uridine	244	1	0.41	10.1	262
Cytidine	243	1	0.55	13.4	280
Inosine	268	7.2	0.45	11.8	250
Thymidine	242	1	0.36	9.65	265
5-Methyl deoxy-cytidine	243	2	0.40	11.6	286
Adenylic acid . .	347	1	0.395	13.7	255
Guanylic acid . .	363	1	0.34	12.2	255
Cytidylic acid. . .	323	1	0.40	12.95	280
Uridylic acid . . .	324	1	0.305	9.89	260

Cavalieri, L. F., A. Bendich, J. F. Tucker and G. B. Brown: J. Amer. Chem. Soc. 70, 3875 (1948).

Dekker, C. A., and D. T. Elmore: J. Chem. Soc. 1951, 2864.

Fox, J. J., and D. Shugar: Biochim. Biophys. Acta 9, 199 (1952).

Hotchkiss, R. D.: J. Biol. Chem. 175, 315 (1948).

Kogan, L., F. J. DiCarlo and W. E. Maynard: Anal. Chem. 25, 1118 (1953).

Lockhart, E. E., and M. C. Merritt: J. Amer. Chem. Soc. 72, 5328 (1950).

Markham, R., and J. D. Smith: Biochem. J. 49, 401 (1951).

Shugar, D., and J. J. Fox: Biochim. Biophys. Acta 9, 199 (1952).

Wyatt, G. R.: Biochem. J. 48, 581 (1951); 48, 584 (1951).

Wyatt, G. R., and S. S. Cohen: Nature (Lond.) 170, 1072 (1952).

The simplest way to use characteristic spectra is to plot the spectrum of an authentic specimen on semilogarithmic paper and then to make a tracing of it on tracing paper. The spectrum of the unknown compound is then plotted on the semilogarithmic paper and the tracing placed over it. Care must be taken to fit the wavelength scales so that they match exactly.

D. The Reactions and the Identification of the Purines and Pyrimidines, Their Ribosides and Nucleotides.

In general, any mixture of purines, pyrimidines and their derivatives may be found occurring either naturally or in fractions prepared by artificial methods. In this section attention is drawn to the various characteristics of the most important of these substances which are of use in their identification, and reference is made to the content of other sections of this chapter which are relevant to the particular identification required.

I. The Purines: General Reactions.

The purines are heterocyclic compounds having substituent groups which may be acidic or basic in character. Because of their imidazole ring structure they are precipitated from neutral or slightly acid (pH 5) solutions by cuprous oxide suspensions as greyish brown complexes. Several methods of analysis have been published based upon this procedure. The cuprous oxide may be produced *in situ* by boiling in the presence of $CuSO_4$ and $NaHSO_3$, or preferably by adding freshly precipitated CuO made by reducing FEHLING's solution with glucose, taking care not to have excess glucose present. The precipitate is washed before use. It is then added to the solution of purine in a citrate buffer of pH 5. The cuprous purine may be decomposed by Na_2S solution.

Another method of precipitation which may be useful is by means of AgO in acid solution. This will be discussed in detail later. Silver purines precipitate from solution as a white mass which may be decomposed by adding chloride to the washed precipitate. As many such precipitates are photo-sensitive this reaction is best carried out in the absence of bright light.

All purines are decomposed by heating with strong acids at temperatures above 100°. The products of this reaction are glycine, formic acid, CO_2 and ammonia. The methylated purines give rise to methylamine instead of ammonia. The rate of decomposition is such that ordinary methods of hydrolysis used for protein analysis will destroy the bulk of the purine. Before complete destruction the amino purines are deaminated as well.

As a general rule all the purines and pyrimidines may be identified by chromatography on paper.

Adenine: 6-amino purine (c). Adenine is readily precipitated as the picrate by the addition of picric acid to a solution. This is best done after the removal of guanine (q. v.). The picrate may be decomposed in the usual way with hydrochloric or sulphuric acid, the picric acid being extracted with benzene.

Adenine may be deaminated to hypoxanthine by sodium nitrite and acetic acid (1 drop sat. $NaNO_2$ + 1 drop glacial HCOOH per 0.5 ml. of solution acting for 8 or more hours).

The amino group is very weak and is undissociated in neutral solutions. It may be detected by electrophoresis on paper (q. v.) at pH 3.5, by which adenine may be recognized.

The absorption spectrum is shown in Fig. 2.

Guanine: 2-amino:6-oxypurine (d). Guanine is extremely insoluble in water and particularly so in ammonia solution (1 N or thereabouts) and may be precipitated from solution by the latter almost quantitatively. It is best dissolved in 1 N HCl.

Its other reactions are similar to those of adenine but the amino group is even weaker. Guanine and its compounds fluoresce with a violet blue light when excited with light of wavelength 265 mμ. This effect is only shown by strongly acid solutions (pH $<$ 2) or on paper chromatograms which have been exposed to HCl fumes. Nitrous acid deaminates guanine to xanthine, as does an enzyme guanase, which is widely distributed in plants and animals. The spectrum is shown in Fig. 3.

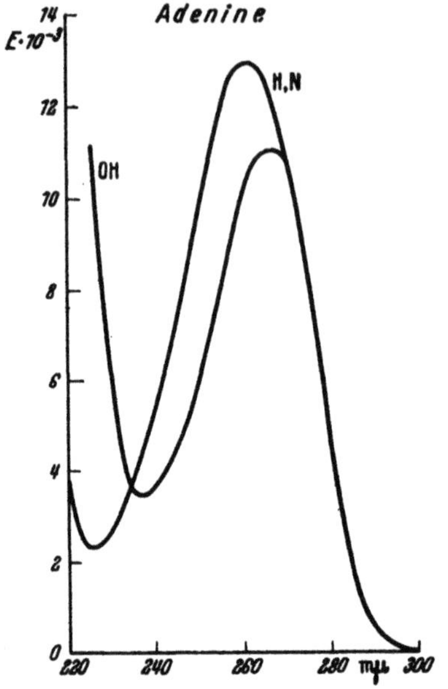

Fig. 2. Ultraviolet absorption spectra of adenine. H, N and OH refer to the acid (pH 1), neutral (pH 7) and alkaline (pH 13) spectra.

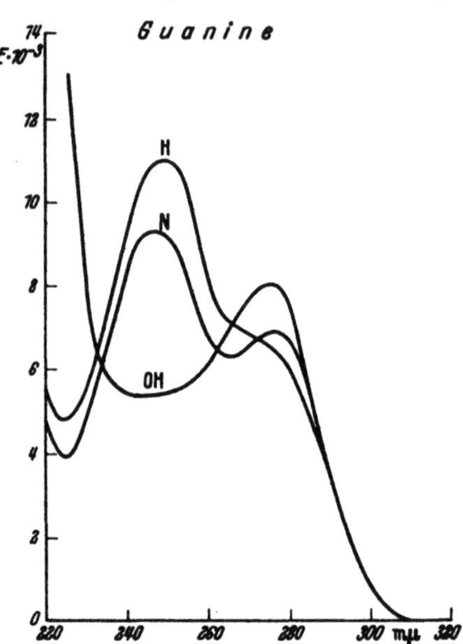

Fig. 3. Ultraviolet absorption spectra of guanine (see Fig. 2).

Hypoxanthine: 6-oxypurine (n). This purine is not found in nucleic acids as such, but may result by the deamination of adenine by enzymes or by chemical agents.

It may be differentiated from adenine by its lack of an amino group, and therefore its inability to form a picrate. It is also uncharged at pHs below 4, where adenine has a positive charge.

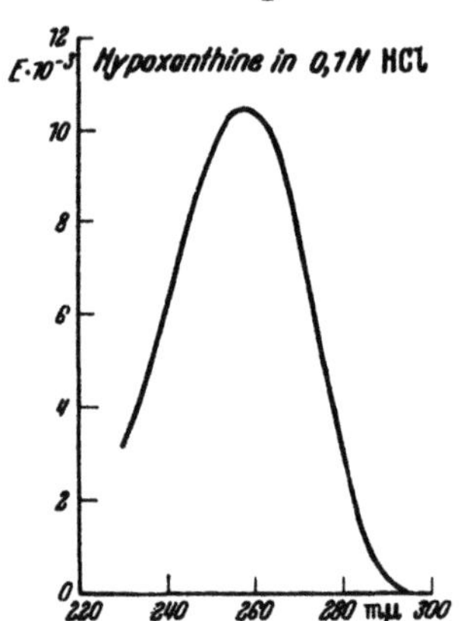

Fig. 4. Ultraviolet absorption spectrum of hypoxanthine in 0.1 N HCl.

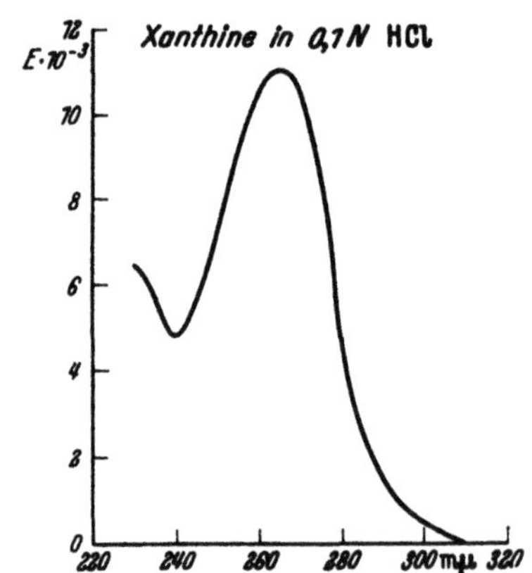

Fig. 5. Ultraviolet absorption spectrum of xanthine in 0.1 N HCl.

Hypoxanthine is reasonably soluble in aqueous solutions even at neutrality. The absorption maximum in 0,1 N HCl is at 250 mμ, with a trough below 220 mμ (Fig. 4).

Xanthine: 2:6-dioxy purine (o). Xanthine is to be found in many plants in the free state, presumably being the end product of purine oxidation. It is not very easily soluble in water, but dissolves in ammonia solutions or in HCl solutions. It is a weak acid and has no charge in neutral or acid solution. The absorption maximum in 0.1 N HCl is at 265 mμ, with a trough at 240 mμ (265/240 = 2.3) (Fig. 5).

II. The Pyrimidines.

The pyrimidines are much more stable to hydrolysis than are the purines and heating with mineral acids has little effect except to deaminate the amino-

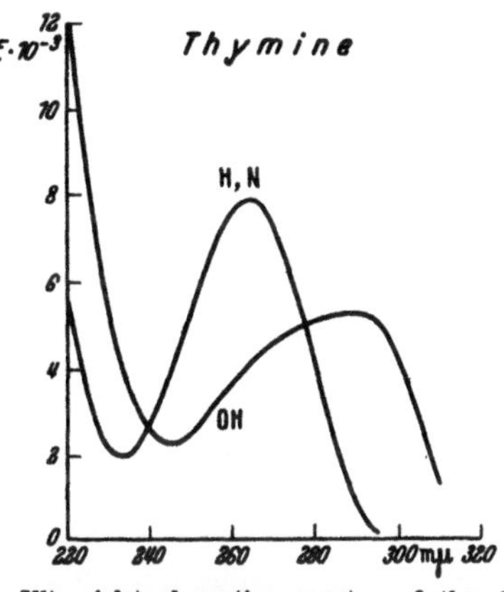

Fig. 6. Ultraviolet absorption spectra of thymine (see fig. 2).

Fig. 7. Ultraviolet absorption spectra of uracil (see fig. 2).

pyrimidines. An exception to this generalisation is 5-hydroxymethyl cytosine (i) which is destroyed by heating in strong acids. This pyrimidine is, however, only known to be present in certain bacteriophages.

The silver salts of the pyrimidines differ from those of the purines in being soluble in acid solution, but they may be precipitated by the addition of Ba(OH)$_2$ to a solution from which the purine silver salts have been removed.

Thymine: 2:6-dioxy-5-methyl pyrimidine (g). Thymine is the most insoluble pyrimidine, and may be precipitated from aqueous solutions by taking them to neutrality. It is readily identified by its chromatographic behaviour in organic solvents because of its methyl group, which increases the R_F value considerably. The absorption spectrum is shown in Fig. 6. Solutions are best made up in 0.1 N HCl or 0.1 N NH$_4$OH.

Uracil: 2:6-dioxypurine (f). Uracil resembles thymine in many respects but has no methyl group. This similarity is reflected in the absorption spectra, which are superficially similar, but the absence of the methyl group causes a shift towards the shorter wavelengths (Fig. 7). Uracil and thymine both have

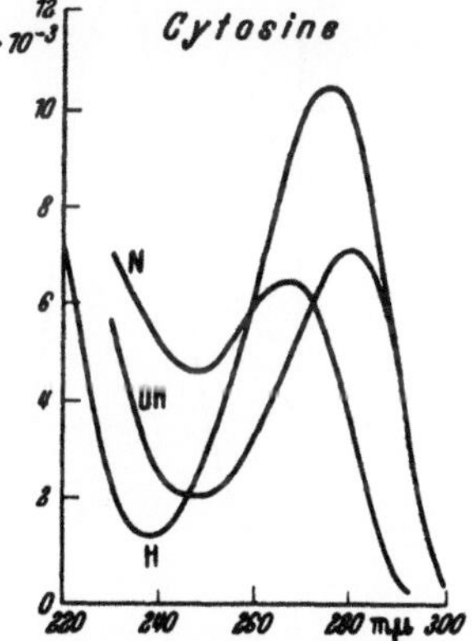

Fig. 8. Ultraviolet absorption spectra of cytosine (see fig. 2).

an acid dissociation at about pH 9.5, which differentiates them from the other pyrimidines. Having no amino group they do not form picrates.

Cytosine: 2-amino:6-oxypyrimidine (e). Cytosine is a weak base with a pK of at about pH 4 and so may be identified by paper electrophoresis and by the ability to form a picrate. It is deaminated by nitrous acid and by some bacteria to form uracil. The absorption spectra are shown in Fig. 8.

5-Methyl cytosine (h). (WYATT, 1950.) This pyrimidine which is most abundant in some plant deoxyribonucleic acids resembles cytosine closely, both in its properties and in its absorption spectra, which are similar in form but displaced some 5—8 mμ towards the longer wavelengths. The only simple method of separating cytosine and 5-methyl cytosine is by chromatography. It is deaminated by nitrous acid and by some bacteria to thymine.

5-Hydroxymethyl cytosine (i). (WYATT and COHEN, 1952.) This rare pyrimidine differs from the others in being destroyed by heating in HClO$_4$ (72% w/v). It otherwise resembles cytosine and 5-methyl cytosine from which it may be differentiated by chromatography. Its spectrum resembles that of cytosine but the peaks are shifted some 4 mμ towards the longer wavelengths.

III. The Nucleosides.

The nucleosides are the N-ribosides (t), or N-deoxyribosides (u), of the bases. The latter are attached at position 3 in the pyrimidine ring and position 9 in the purine ring. In general they resemble the bases in their chromatographic behaviour, but with reduced R_F values in most of the simpler solvent systems based upon n-butanol and similar alcohols. On electrophoresis they also have similar but somewhat retarded mobilities, because of their increased molecular size. However, because of the cis-glycol at C 2' and 3' in the nucleosides of the ribose series, these nucleosides have a negative charge at pH 9 in borate buffers. The ribose nucleosides also react with NaIO$_4$ to give a dialdehyde, which may be detected by SCHIFF's reagent.

The purine deoxyribonucleosides are extremely unstable to acid, and lose their purine base easily. They react with DISCHE's diphenylamine reagent, which has little effect on the pyrimidine analogues. The purine ribosides are easily hydrolysed by N H$_2$SO$_4$ or similar acids at 100° to give ribose and the free purine, and they also react with BIAL's orcinol-HCl-FeCl$_3$ reagent for pentoses.

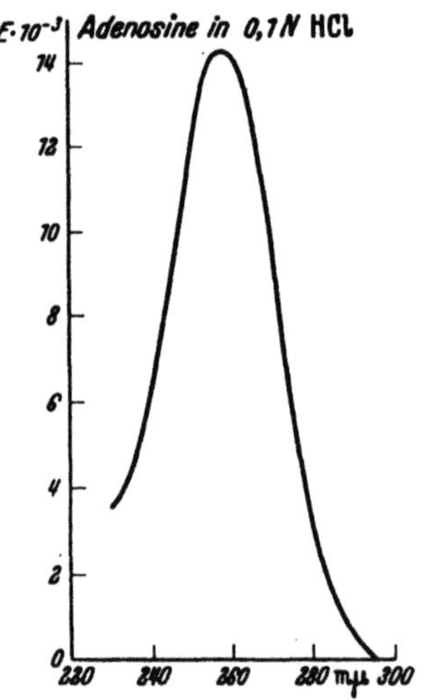

Fig. 9. Absorption spectrum of adenosine in 0.1 N NCl.

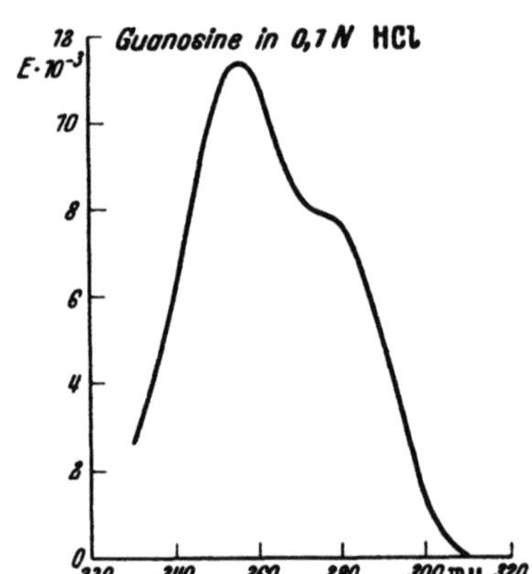

Fig. 10. Absorption spectrum of guanosine in 0.1 N HCl.

The pyrimidine nucleosides are much more stable and unreactive. All however may be decomposed by heating with 70% (w/v) HClO$_4$ at 100° for 1—2 hr., when the bases are liberated.

The various reactions of the bases with nitrous acid, picrate, etc. may be performed on the nucleosides.

The more common nucleosides are listed below: —

Adenosine: Adenine-9 riboside (Fig. 9).

Guanosine: Guanine-9 riboside (Fig. 10).

Cytidine: Cytosine-3 riboside (Fig. 11).

Uridine: Uracil-3 riboside (Fig. 12).

Deoxyadenosine: Adenine-9:(2-deoxy)riboside.

Deoxyguanosine: Guanine-9:(2-deoxy)riboside.

Deoxycytidine: Cytosine-3:(2-deoxy)riboside.

Deoxy 5-methyl cytidine: 5-methyl cytosine-3:(2-deoxy)riboside.

Thymidine: Thymine-3:(2-deoxy)riboside.

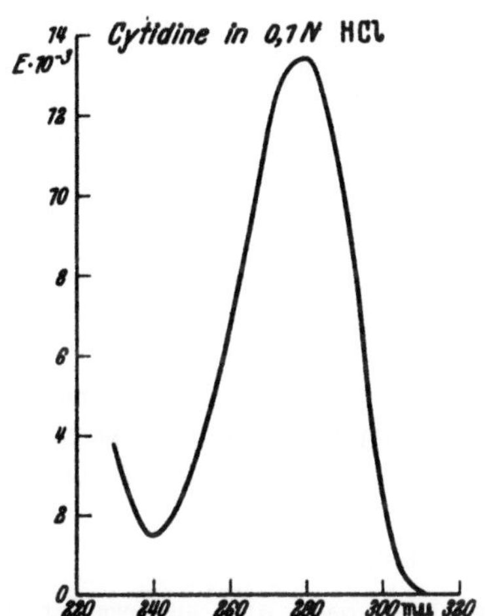

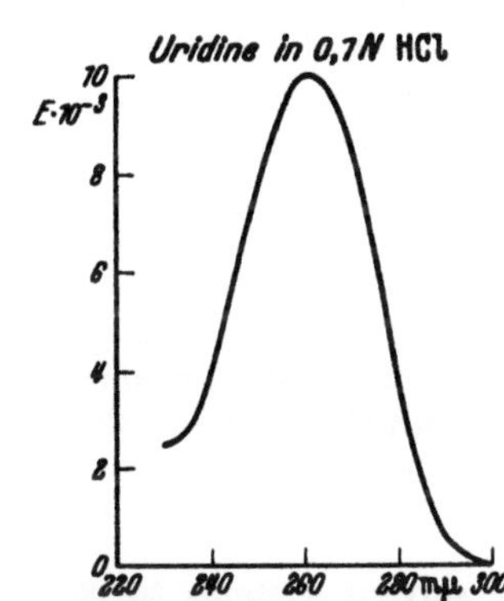

Fig. 11. Absorption spectrum of cytidine in 0.1 *N* HCl.   Fig. 12. Absorption spectrum of uridine in 0.1 *N* HCl.

IV. The Nucleotides.

The nucleotides represent a degree of complexity far above that of the previous groups of substances. Most of them occur in nature as such, although some of the isomers have not yet been reported from natural sources.

Ribose in the furanose form has —OH groups at C 2, C 3 and C 5 and is known to form four monophosphate derivatives or nucleotides, namely the nucleoside-2'-phosphates, the nucleoside-3'-phosphates, the nucleoside-5'-phosphates (v), and the nucleoside-2':3'-(cyclic) phosphates (w). Similarly 2-deoxyribose having two —OH groups available forms two series of nucleotides, the nucleoside-3'- and -5'-phosphates. Because of this, the identification of the various compounds is rather difficult. It may, however, be accomplished in a number of ways.

Firstly, all nucleotides (excepting the 2':3' phosphates) are dephosphorylated by the prostate enzyme (enzyme 1) to give the corresponding nucleosides. Secondly, those nucleotides containing 5'-phosphate groups may be dephosphorylated by snake venom (enzyme 5). Of the latter, the ribose series may be differentiated because they form a borate complex at pH 9 which is detectable by electrophoresis, and they are oxidised by periodate, a procedure which results in their subsequent dephosphorylation on subjecting the products to a pH of 10 or more.

The ribose 2':3'-phosphate derivatives, or the cyclic nucleotides as they are called, are easily differentiated, because they are not attacked by the prostate enzyme (Markham and Smith, 1952a). They also have no secondary phosphate group, and hence have a lower electrophoretic mobility at pH 7 than do the other nucleotides. The cyclic grouping may be split by alkali (1 N NaOH) or by acid (0.1 N HCl) at room temperature to give a mixture of the 2'- and 3'-phosphates.

The ribose (and probably the deoxyribose) 3'-phosphates are readily dephosphorylated by the 3'-nucleotidase from barley (enzyme 4) and the former will then react with $NaIO_4$ and H_3BO_3.

The ribonucleoside-2'-phosphates may be recognised by resistance to $NaIO_4$ and to H_3BO_3 complexing, and resistance to the barley 3'-nucleotidase.

A number of these compounds, especially the ribonucleoside 2' and 3' phosphates are also easily recognised by their chromatographic behaviour on paper and on ion exchange columns. Certain minor differences in the absorption spectra also exist.

E. Paper Chromatography.

In this section the chromatographic separation of a number of compounds will be described. All the possible compounds have, of course, not yet been examined, but a careful study of those which have been examined so far will act as a guide, enabling one to predict the behaviour of a certain compound or to deduce the possible structure of an unknown compound. This technique has in fact been applied with great success by Wyatt (1950, 1953) in the identification of the methylated cytosines.

Only paper chromatography will be discussed because chromatography on starch columns is now obsolete, although satisfactory results have been obtained by this method, which has the disadvantages of slowness and of requiring a great deal of labour.

I. General Considerations.

The solvent systems to be described have been used with success for some years in our laboratory and have been adapted for use in relatively small, all glass tanks (an ideal size for this type of work is about 20 ins. high $\times$ 14 ins. $\times$ 8 ins.). It must not be expected that the solvent systems given here will work satisfactorily with the large tanks used in two-dimensional amino acid chromatography, and a great deal of fruitless labour may be saved by using equipment suited to the technique. In addition, it is frequently necessary to run chromatograms further than the end of the paper. Consequently it is useless in many cases to employ the ascending techniques, which are, in any case, much slower.

As far as possible smelly or otherwise noxious solvents are avoided. Practically all of these are unnecessary and have no particular virtues. Most of the solvent systems are slightly undersaturated with respect to water so that temperature changes have no adverse effects, and no equilibration of the papers is necessary. As to the papers, Whatman No. 1 and No. 3 MM should prove adequate for any purpose. Acid-washed papers have been found very disappointing, and have given large diffuse spots.

The size of spots to be applied is not critical. In general we use 14—20 μl. of liquid, which forms spots of about 1.5 cm. diameter on the papers (Fig. 13). Putting on much smaller spots is generally a waste of time, because during the run they will usually expand to find their "natural" size, which is decided by the quantity of material applied, rather than by the volume in which it is put on the paper.

It is advisable to avoid the presence of salts in the samples. If this is impossible, then NH_3 and K salts are preferable, and the acids should be organic if possible. A very useful trick is to remove potassium by means of perchlorate and vice versa. Acids may be removed by extracting the solutions with 5% trioctylamine in chloroform (HUGHES and WILLIAMSON, 1951), without a serious loss of the phosphorylated derivatives. Most precipitation methods for the removal of undesirable substances are to be avoided in quantitative work. Thus silver precipitation of chloride, barium precipitation of sulphate, etc. may result in considerable losses. Excepting in the case of chromatograms run in the solvents containing strong hydrochloric acid, it is necessary to neutralise spots on the paper before chromatographing, and for this N NH_4OH or N acetic acid may be used according to whether the papers are acid or alkaline.

Fig. 13. A type of micropipette suitable for use in nucleic acid chromatography. The delivered volume should be 10—20 μl.

Many of the substances to be examined, for example guanine, may be rather insoluble, and, when solutions of these are made up as marker substances, 0.1 N HCl or 1 N NH_4OH may be used as solvents. Certain substances such as uric acid may have to be dissolved in 0.1 N KOH. In the case of the nucleotides, particularly the cyclic ones and the purine deoxyribonucleotides, it is essential to avoid the use of strongly acid solvents. It is important to apply the material to the papers as a solution rather than as a suspension, in order to avoid streaking, and it is better to concentrate a solution before applying it than to apply a number of spots in succession.

For a great deal of work it will be found to be more useful to use as marker spots mixtures of known composition rather than the individual pure substances. This is because the relative movement of a number of substances is a much better guide than is their absolute movement, which may vary from batch to batch of solvent, or with the temperature of the laboratory.

II. The Detection of Nucleic Acid Derivatives on Chromatograms.

The earlier workers in the field of nucleic acid chromatography employed indirect methods for the detection of the spots: for example HOTCHKISS (1948) cut his whole chromatogram into strips, extracted the latter with water and estimated all the samples individually. CHARGAFF and his colleagues (VISCHER and CHARGAFF, 1948) evolved several methods for forming insoluble mercury compounds of their various bases, nucleosides, etc. and then decomposed these with H_2S, the spots then becoming coloured grey to black with mercuric sulphide. Methods such as these are inconvenient, and except for certain specific purposes it is now customary to detect the spots by virtue of their ultraviolet light absorbing power. Almost all nucleic acid derivatives have characteristic absorption maxima in the region of 250—280 $m\mu$, and of these most have quite considerable absorption at 265 $m\mu$ and adequate absorption at 254 $m\mu$, the two wavelengths most suited to our purpose.

Two methods of exploiting this characteristic are in use, both having merits and demerits. The first is that of HOLIDAY and JOHNSON (1949) and of CARTER, which takes advantage of the fact that filter paper contains fluorescent impurities, and so, when exposed to ultraviolet light, purines and similar absorbing substances stand out against the background as *dark* spots. The other method, due to

Markham and Smith (1949a and b), takes advantage of the fact that filter paper is reasonably transparent to ultraviolet light, so that on making a contact photographic print of the filter paper, an image is produced having *white* spots on a black background.

Of the two, the first method is probably the simpler, but for serious work, the second, while involving more work, is more sensitive and has the very important quality of providing an *exact* picture of the whole chromatogram in a permanent form, which may be examined at leisure, and referred to at any time after the original filter paper has been destroyed. (It will be seen later that it is usually necessary to destroy the papers in order to complete the analyses.) As a matter of routine in our laboratory we employ both methods, and that is what I should recommend.

1. Ultraviolet Light Sources.

The most suitable light source for our purpose would be a monochromator with a high intensity source, which could be adjusted to the optimum wavelength, but for general convenience a mercury lamp is adequate. It should be pointed out here that the ordinary "Ultraviolet" lamp used for exciting fluorescence is quite useless, because this lamp, employing a nickel glass filter, transmits light of wavelength 365—366 mμ, a region unsuitable for our purpose. It is necessary to use a quartz bulb lamp with suitable filters.

There are three main kinds of mercury lamp and these are as follows:

(1) The Low Pressure Lamp Containing Argon. This is usually rather long and tubular, but may be spiral in form. The light emitted is mainly (about 90%) at the resonance Hg line — 254 mμ. These lamps, made for sterilisation, can give a useful photographic print if enough material is present in the spots. When fitted with a special glass filter, these lamps are very good, and they are available commercially under various names. They are rather expensive to buy.

(2) The Medium Pressure Arc. This lamp is of the "artificial sunlight" type, and emits a number of wavelengths and so *must* be filtered. It has the defect of a relatively low intensity per unit area and so is not to be recommended.

(3) The High Pressure Mercury Lamp. This type of lamp is made mainly for street lighting and general illumination, and consequently is readily available and cheap. Two types are made which are very suitable, and these are the 80 watt and the 120 watt MB/V lamps respectively. Either may be used, and it should be noted that the more powerful lamp is no better for our purpose because it is no brighter, the extra power being consumed because the source is larger. This does not interest us, because only a small area of the lamp can be used. This type of lamp emits much more light at 265 mμ than at 254 mμ, and when used with a suitable filter is more or less ideal.

2. Construction of a Light Source.
(Markham and Smith, 1951.)

Take an 80 watt mercury lamp (e. g. Mazda MB/V, 80 watt). This resembles a 100 watt pearl incandescent lamp, but may have a special base and holder. It consists (Fig. 14) of a small tubular quartz lamp enclosed in a glass bulb containing an inert gas. The glass bulb has to be removed, and this may be accomplished quite easily. Heat the glass bulb in a bunsen burner and allow a few drops of water to fall on to it. The bulb will crack, and then may be removed completely by nibbling with a pair of pliers. The quartz lamp so exposed should not be fingered, but if dirtied may be wiped with a piece of cotton wool damp with ethanol.

The lamp is now ready for use and its holder is connected to the A. C. mains through a special choke (M. R. G. 508) supplied by the lamp manufacturers. (It is possible to use these lamps on D. C. if a series resistance is used. The voltage drop across the lamp is about 40, and the resistance should be chosen to dissipate the rest of the voltage.)

The filter system which we now use is quite simple and is constructed from two 25 ml. round-bottomed flasks (Vitreosil T/A 4/407). One has a ground-in stopper and is filled with dry chlorine gas. The gas is generated by dropping HCl on $KMnO_4$, and passing the Cl_2 over $CaCl_2$, into the flask. The stopper is

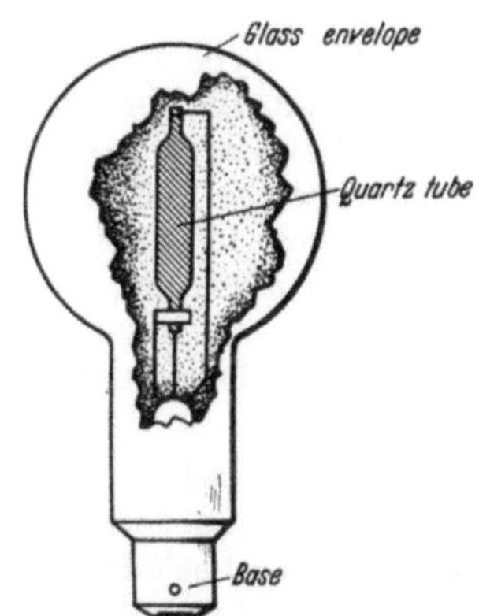

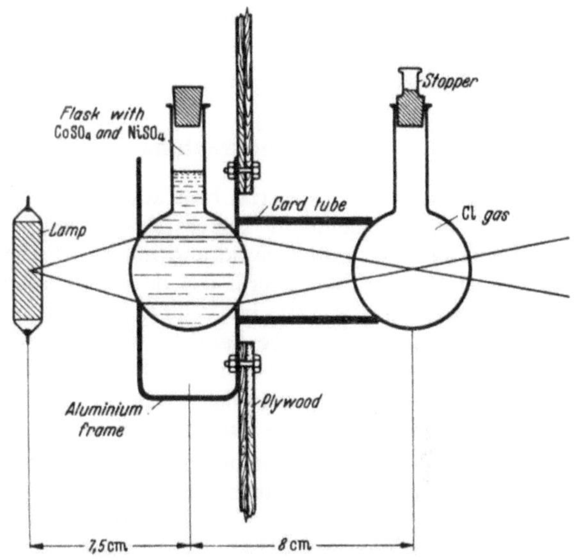

Fig. 14. Details of a commercial 80 watt mercury lamp.

Fig. 15. Details of the ultraviolet lamp with its filter system.

greased with petroleum jelly (silicone grease is useless) and fastened down with cellulose tape. This filter will last without deterioration for several years as long as the grease is not exposed to the ultraviolet light. The other flask, which is fitted with an ordinary cork, contains a solution of cobalt and nickel sulphates ($CoSO_4 \cdot 7H_2O$, 100 g., $NiSO_4 \cdot 7H_2O$, 350 g./l.) in water. It is essential that these salts should be pure and free from iron and chloride. This filter is stable and is placed nearest the lamp. The latter is best mounted inside a light-tight box having a hole of about 5 cm. diameter cut in the side about 45 cms. from the bottom (Fig. 15). Over this hole a U shaped piece of aluminium is fastened by means of screws. This piece of aluminium has two holes cut in it opposite each other, and 25 mm. in diameter. These serve the dual purpose of gripping the Co-Ni filter and of acting as diaphragm for the lens which the latter constitutes. On the outside of the box, a spring clip holds the neck of the Cl_2 filter, and the bulb of the latter is spaced away from the Co-Ni filter by a short length of cardboard tubing. The lamp position is adjusted so as to throw an image of the former in the centre of the Cl_2 filter. A suitable fluorescent screen is easily made by coating a glass or celluloid plate with gum or a gelatin solution and sprinkling on anthracene.

Caution. It is dangerous to expose the naked eyes to the ultraviolet light. When making adjustments wear glasses or look through a sheet of celluloid or glass.

When set up as described, the positions of the lamp and filter are sufficiently well fixed to enable one to remove the latter for a periodic cleaning. Occasionally it may be found necessary to rotate the Co-Ni filter because an opaque deposit may form on the side nearest the lamp. If serious it may be removed from the inside of the bulb by means of strong HNO_3.

This light source provides a strong cone of light predominantly of wavelength 265 mμ with a smaller amount of light of 254 mμ, and *no other* wavelengths until one gets to the mercury green and yellow lines, which are reduced to a small fraction of their initial intensity. These latter are of no importance for the purpose of photography because the photographic material used is completely insensitive to such long wavelengths, while for visual observation, their effect may be neutralised by viewing the chromatograms through a suitable light filter (e.g. Wratten 47) held before the eyes.

3. Photographing the Chromatograms.

The photographic material best adapted for this work is the type known as Reflex Contact Document Paper (this is made by most manufacturers, who usually stock two grades: rapid and normal. Either has about the same speed to ultraviolet light). For general use it is convenient to purchase the material as rolls (a useful size is $13^1/_2$ in. $\times$ 60 ft.). This paper is relatively intensitive to ordinary light, and may be handled with a bright yellow safelight, and, while the room should be darkened, it will be found that an open door will have little adverse effect unless direct sunlight falls on the paper. (For visual observation the room must, of course, be as dark as possible.)

An orthodox printing frame cannot be used because it would need to have a quartz window, so as a compromise the paper is kept flat and in contact with the chromatogram by pinning it with artist's drawing pins on to a sheet of plywood. A flat wood surface is relatively useless, but it will be found that if the wood is bent to form a plano-convex surface (Fig. 16), it is relatively easy to establish adequate contact between the two sheets of paper. Naturally the photographic paper is put underneath the filter paper. Recently an ingenious solution of the problem of holding the papers flat has been advanced by Smith and Allen (1953) who use nylon netting for this purpose.

For exposing, the chromatogram is held about 1 metre from the lamp, an average exposure being about 8—12 seconds for Whatman No. 1 paper and 3 minutes for Whatman No. 3 or 3 MM paper. The exposures are not critical, and it may be found advantageous to make several different exposures to resolve thick spots or bands. In the case of preparative chromatograms, when the applications are really thick, it is often found useful to wipe the filter paper before and during the exposure (of about 40 seconds for No. 3 paper) with carbon tetrachloride or ethyl acetate in order to make it less opaque.

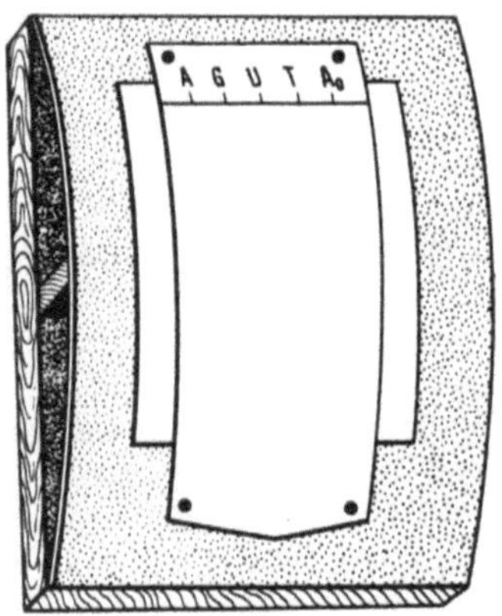

Fig. 16.
Mounting the chromatogram on a convex board for printing by ultraviolet light.

The exposed photographic paper is next developed for about 30 seconds to 1 minute in a suitable developer. A good formula is: Metol, 3 g., Na_2SO_3 (anhydrous), 50 g., hydroquinone, 13 g., Na_2CO_3 (anhydrous), 70 g., NaBr, 1 g., water to 1 l.

The print is then dipped in water and then into an acid fixing bath ($Na_2S_2O_3$, $5 H_2O$, 250 g., $K_2S_2O_5$, 25 g., water to 1 l.) for 30 seconds or longer. It is then washed for a few minutes in running water, blotted dry and hung up to dry. A short wash in acetone may be employed for rapid drying.

4. Finding the Position of the Spots.

The photographic print is placed on an illuminated box (an X-ray viewing box is ideal) and the chromatogram is registered by means of marks previously made on it and recorded on the print. A good system for this is to cut small V-shaped notches at intervals down the edges of the filter paper. These are easy to register and also provide a method of identifying the photograph. The spots are then ringed round with pencil. If the exposure has been adjusted to give a print in which the background is not quite devoid of detail, it is safe to take the line to within about 2 mm. of the photographic image, but naturally with increasing exposure the area of the spot diminishes. The limit of detection is of the order of 1 μg. of base.

Because of the fact that paper stretches and shrinks irregularly it should be noted that it might be impossible to register the whole of a large chromatogram at once. In such circumstances the registration marks nearest the appropriate spots are used.

5. The Recording of Fluorescent Spots.

(SMITH and MARKHAM, 1950.)

Guanine and its compounds when acid fluoresce markedly, while 8-azaguanine and some substituted xanthines fluoresce even at neutrality. It is possible to take advantage of this for detecting or identifying certain of these compounds. We shall take as an example guanine. If this has been run in an acid solvent (containing HCl), the spot will appear bluish violet when examined in the beam of the ultraviolet lamp. This fluorescence may be recorded photographically, and, because photography is an integrating process, the sensitivity of such a method exceeds that of any other. The printing is carried out just as before, but the paper used is Grade 0 or 1 bromide paper, and over it, between it and the chromatogram, is placed a sheet of thin nitrocellulose film (0.08 mm.). The latter is completely opaque to ultraviolet light, so that only fluorescence is recorded. The exposure time is nearly the same for No. 1 and No. 3 MM papers, and is of the same order as that used for No. 1 paper for an absorption picture, i. e. about 10 seconds. The spots appear as *black* on a greyish background. The fluorescence of a compound may be masked if the spot is sitting on another heavily absorbing spot. *Note:* Bromide paper needs $1^1/_2$—2 minutes development and is very sensitive. It needs to be processed in a real dark-room.

III. The Separation and Identification of Purines, Pyrimidines and Nucleosides.

(MARKHAM and SMITH, 1949.)

The separation of the simpler constituents of nucleic acids and related substances may be effected by simple solvents based upon n-butanol-water mixtures and amyl alcohol-water mixtures.

The composition of some of these solvents and the R_F values of a number of compounds are shown in Table 2. It will be noted that in these solvents the R_F

values are small and in some cases it will be desirable to cut the papers into notches at the bottom, and run the solvent front well off the end. It should be noted that uracil and cytosine, and their ribosides, are very sensitive to the ammonia in the vapour phase, and consequently the appropriate markers should be used when these compounds are being investigated.

Table 2. *R_F Values of Purines, Pyrimidines and Ribosides in Various Solvents.*

Solvent system[1]	n Butanol 86%: water 14%:	n Butanol 86%: water 14%: + NH₃[2]	n Butanol 77%: water 13%: formic acid 10%	n Butanol 50%: ethanol 15%: water 35%	Amyl alcohol saturated with water	Amyl alcohol saturated with water + NH₃[2]	Amyl alcohol saturated with water 90%: formic acid 10%
Purines:							
Hypoxanthine	0.26	0.12	0.30	—	0.15	0.02	0.19
Xanthine	0.18	0.05	0.24	—	0.10	0	0.16
6:8-Dihydroxypurine .	0.18	0.04	0.24	—	0.07	0	0.11
Uric acid	0.01	0	0.14	—	0	0	0.06
3-Methylxanthine	0.29	0.13	0.32	—	0.19	—	—
1:3-Dimethylxanthine (theophyllin)	0.52	0.22	0.64	—	0.54	0.12	0.55
3:7-Dimethylxanthine (theobromine)	0.42	0.27	0.47	—	0.28	0.15	0.36
3:8-Dimethylxanthine . .	0.42	0.18	0.48	—	0.34	—	—
1:3:7-Trimethylxanthine (caffeine)	0.63	0.65	0.71	—	0.56	0.50	0.67
Adenine	0.38	0.28	0.33	0.55	0.28	0.16	0.12
Guanine	0.15	0.11	0.13	0.37	0.05	—	0.04
Purine ribosides:							
Adenosine	0.20	0.22	0.12	0.50	0.11	0.09	0.04
Guanosine	0.15	0.03	0.17	0.40	0.02	0	0.04
Pyrimidines:							
Cytosine	0.22	0.24	0.26	0.53	0.09	0.09	0.07
Uracil	0.31	0.19	0.39	0.46	0.22	0.08	0.23
Thymine	0.52	0.35	0.56	—	0.40	—	—
Pyrimidine ribosides:							
Cytidine	0.12	0.11	0.18	0.42	0.02	0.03	0.03
Uridine	0.17	0.08	0.25	0.49	0.05	0.03	0.07

With these particular solvent systems one can make certain generalisations which may be of use for identifying unknown substances. These are:

(1) Pyrimidines move faster than purines having similar substituent groups. For example, uracil runs faster than xanthine and cytosine faster than guanine.

(2) In the presence of ammonia, the amino derivatives move much faster than the corresponding hydroxy compounds, but on changing to an acidic solvent this difference is decreased and may be reversed. This is particularly true of the pyrimidine.compounds, where this change takes place at neutrality.

(3) Increase in the number of hydroxyl groups in a compound decreases its R_F in all these solvents. This is illustrated in Fig. 17 where 6-oxypurine (hypoxanthine), 2:6-dioxypurine (xanthine) and 2:6:8-trioxypurine (uric acid) are shown after running in butanol-formic acid.

(4) Ribosides move more slowly than do the bases, but maintain the same relative positions as do the latter.

---

[1] Composition of solvents is given in terms of volume percentages.

[2] 5% by vol. of NH₃ solution (sp. gr. 0.880) was added to the solvent mixture in the bottom of the container.

(5) Increasing methylation tends to increase the R_F values in all solvents (Fig. 18).

It should be noted that, in solvents containing ammonia, guanine and uric acid are exceedingly insoluble, and they may not move or may streak.

F. The Quantitative Analysis of Nucleic Acids for their Bases.

I. Ribonucleic Acids.

The analysis of substances such as ribonucleic acid and polynucleotides derived from it depends to a great extent upon the hydrolysis procedure to be adopted. Many methods of hydrolysis have been proposed and almost all have their deficiencies, but a few are satisfactory for general use and will be described here.

1. Hydrolysis with N HCl at 100° for 1 Hour.
(SMITH and MARKHAM, 1950.)

This procedure converts RNA into a mixture of free purines plus pyrimidine nucleotides. Under these conditions about 5% of the latter are further broken down to nucleosides.

Procedure. The optimum quantity of RNA per spot is about 1—1.5 mg., although the method is sensitive enough to work accurately with one tenth this amount. The volume applied to the paper is about $10-20$ μl., so that a final concentration of some 50 mg./ml. is desirable. The minimum volume of N HCl recommended for the hydrolysis is 50 μl., if a tube of about 3—4 ml. internal volume is used for the hydrolysis vessel, because below this volume, the amount of water in the gas phase becomes an important factor in assessing the concentration of the hydrolysate. Nucleic acid has a fairly small specific volume and for all practical purposes 1 mg. may be taken to occupy 600 μl. Therefore if we take 2.5 mg. RNA and add 50 μl. of N HCl and dissolve, the final solution will have a concentration of RNA of $2.5/50 + 2.5 \times 0.6 \equiv 48.5$ mg./ml. of the starting material.

For accurate work, it is advisable to seal off the tube used for the hydrolysis and to immerse it completely in the water-bath at 100°. This is to minimise the local concentration of HCl, which may otherwise rise to 6 N because of the distillation of water to the top of the tube where it will condense. This may cause excessive breakdown of the pyrimidine nucleotides. When larger quantities of RNA are used, the volume of HCl becomes so large that these precautions are unnecessary, and a rubber stopper wired on may be used instead.

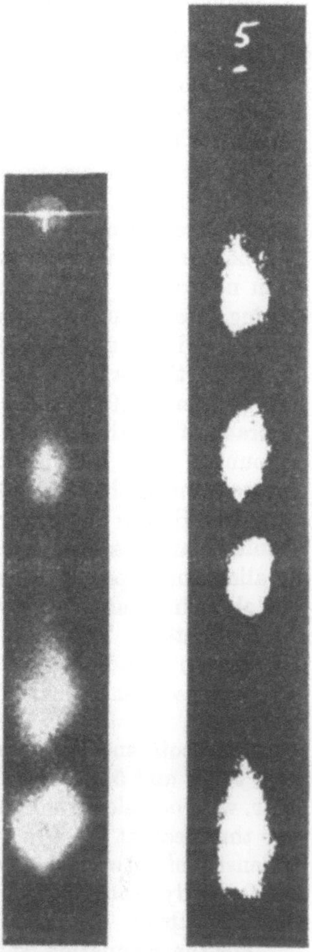

Fig. 17.   Fig. 18.

Fig. 17. Chromatogram showing in order of increasing R_F value hypoxanthine, xanthine and uric acid in n butanol-formic acid, to illustrate the effect of increasing numbers of hydroxyl groups (1, 2 and 3 OH groups respectively).

Fig. 18. Chromatogram illustrating the effect of methylation. Xanthine, 3-methyl xanthine, 3:7-dimethyl xanthine (theobromine) and 1:3:7-trimethyl xanthine (caffeine) running in n butanol-formic acid.

When the hydrolysis has reached completion the tube is cooled and shaken to mix thoroughly. The liquid is then deposited at the bottom of the tube before opening by centrifuging the tube for a few seconds at 1000 r. p. m. or more. Samples are taken for chromatography, and, if required (and this is recommended), an aliquot for a P determination.

Chromatography. The spots are applied to a Whatman No. 1 paper, and there is no need to neutralise, dry or otherwise treat them. A similar hydrolysate of yeast RNA may be used as a marker. The solvent recommended is *iso*propanol, 170 ml., conc. HCl, 44 ml., and water to 250 ml. (WYATT, 1951). This solvent need not be made up with any precautions other than to use commercial pure reagents. When the solvent has run about 25—30 cm., which takes some 16 to 18 hr., the sheets are removed and hung up in a fume cupboard with the fan running, some 40 cm. or more above a 1000 watt non-radiant heater. It is essential to use some warmth, but the paper must not get at all hot, otherwise it will scorch and be ruined. When all the smell of HCl has vanished (or if the paper is dry and then neutralized with ammonia vapour), it is ready for printing. *It is essential* during these manipulations to ensure that no volatile absorbing basic substances are allowed to come in contact with the paper, particularly pyridine or "collidine".

When the paper is printed the spots seen and their approximate R_F values will be guanine (0.27), adenine (0.40), cytidylic acid (0.61) and uridylic acid (0.79). A trace of cytidine (0.50) may also be seen in chromatograms having a large amount of RNA hydrolysate. When using HCl-containing solvents it is essential, before printing the chromatograms, to be sure that the excess HCl has evaporated. Any free HCl in the paper desensitises the photographic paper locally and may result in large patches of white being left all over the paper. If it is inconvenient to allow the paper to dry slowly, the excess HCl may be neutralized by hanging the sheet in a tank full of ammonia vapour.

Estimation. The spots containing the substances are located and a pencil line drawn round allowing 2—3 mm. margin. Equal areas at the same levels are also marked out in blank lanes which have been left between the spots for this purpose.

The spots and their appropriate blanks are then cut out and put into clean test tubes and 5 ml. of 0.1 N HCl is added. The tubes are allowed to stand for 6 hr. with occasional shaking, or overnight if possible. The density of the extracts are then read at the appropriate wavelengths. As a guide, the spots should have a density of between 0.5 and 1.5 at their peaks while the blanks should not exceed 0.1 for uridylic acid and 0.025 for adenine. (In practice they should be considerably less than this.)

For the calculation, the molar extinction coefficient is taken as:

$$
\begin{array}{lll}
\text{Guanine} & 11.0 & \times 10^3 \text{ at } 250 \text{ m}\mu \\
\text{Adenine} & 13.0 & \times 10^3 \text{ at } 260 \text{ m}\mu \\
\text{Cytidylic acid} & 12.3 & \times 10^3 \text{ at } 280 \text{ m}\mu \\
\text{Uridylic acid} & 9.45 & \times 10^3 \text{ at } 260 \text{ m}\mu.
\end{array}
$$

These factors include a correction for the slight loss of pyrimidine nucleotide in the hydrolysis. It happens that, owing to their differences in absorption spectrum, the nucleoside spots cannot cause any serious error (MARKHAM and SMITH, 1951).

This method may be used to determine the ratio of bases in RNA in the presence of protein, which appears to cause little error, because, little if any of the aromatic amino acids appear on the subsequent chromatograms. It is, of course, always preferable to carry out the analyses on more or less pure RNA.

2. Hydrolysis with 72% (w/w) HClO₄.

(MARSHAK and VOGEL, 1950.)

Similar quantities of RNA are used to those used in method I, and they must be in a dry state in a dry tube. The appropriate quantity of $HClO_4$ is then added, a final concentration of RNA of 50—100 mg./ml. being desirable. (*Note:* 60% (w/w) $HClO_4$ is quite useless and freedom from moisture is essential.) The tube is then stoppered (a wired-down rubber stopper is adequate) and heated on a water bath at 100° for 2 hours (1 hour is insufficient). After cooling, an equal volume of water is added to the $HClO_4$ — which may be as little in amount as 20—30 μl. The solution is mixed, and then is applied to the sheet of No. 1 Whatman paper directly or after centrifuging off any carbon deposit, if this is thought to be desirable (this may be the case if the RNA is contaminated with any carbohydrate).

The papers are then run without any further treatment in the same *iso*propanol HCl solvent as in method I. The method of hydrolysis converts the RNA to the bases, which run in the order (with R_F values): guanine (0.27), adenine (0.40), cytosine (0.47) and uracil (0.68). These are located and extracted as in the previous method, and the following extinction values used:

$$\begin{array}{lll} \text{Guanine} & 11.0 \times 10^3 & \text{at } 250 \text{ m}\mu \\ \text{Adenine} & 13.0 \times 10^3 & \text{at } 260 \text{ m}\mu \\ \text{Cytosine} & 10.5 \times 10^3 & \text{at } 275 \text{ m}\mu \\ \text{Uracil} & 7.9 \times 10^3 & \text{at } 260 \text{ m}\mu. \end{array}$$

Comment. This method is very useful for approximate quantitative work, especially in the analysis of dinucleotides and similar substances. For more exacting work, the other methods are rather more accurate. If thiouracil is present in the solution it is converted very largely to uracil.

II. The Quantitative Analysis of DNA.

(WYATT, 1951.)

For the hydrolysis of DNA only a few methods are available, and only one is really satisfactory for general use, and that is hydrolysis in formic acid at 175°.

In order to carry out the hydrolysis 1—30 mg. of DNA are put into a small Pyrex tube 13 mm. in diameter, 0.5 ml. 80—98% formic acid are added and the tube sealed leaving about 2 ml. of gas phase. It is preferable to seal the tube when it is full of formic acid vapour or to fill it with N_2, because an excessive volume of air appears to cause losses, particularly if smaller amounts of formic acid are used. The tube is then placed in a small bomb oven at 175° for 30 mins. The tube is removed, cooled, and opened with great care by heating the sealed tip cautiously with a flame until it softens and blows out. (If the tube is left for 1 hr. or more by mistake, the tube may explode on opening.)

The neck of the tube is then cut off neatly and the formic acid evaporated off completely. N HCl is added to make the desired concentration (about 50 mg./ml.) and the bases dissolved (the correction for the volume occupied by the hydrolysed DNA is the same as for RNA).

Spots are applied to Whatman No. 1 paper and the chromatogram is run in the solvent *iso*propanol, 170 ml., conc. HCl, 44 ml. water to 250 ml. Development is continued until the front has run 30 cm., by which time the spots revealed will be (with their R_F values): guanine (0.27), adenine (0.36), cytosine (0.47), 5-methylcytosine (0.55) and thymine (0.77) (Fig. 19). Occasionally uracil (0.68) may also be found.

The spots are eluted and read as in the methods for RNA. The molar extinction coefficient for thymine is 7.95×10^3 at 265 mμ.

The quantity of 5-methyl cytosine found may be exceedingly small. It may be determined with reasonable accuracy as follows:

Instead of running spots, a line some 25 cm. long containing the hydrolysate of 6—8 mg. of DNA is put on Whatman No. 1 paper and run in the usual solvent system. The positions of the cytosine *and* 5-methyl cytosine are located and the two bands are cut out together. The piece of filter paper is cut to a point and the

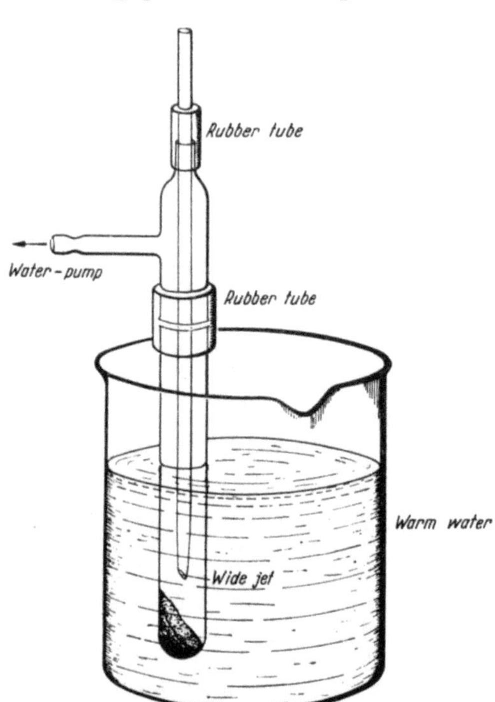

Fig. 19. A chromatogram of deoxyribonucleic acid from wheat germ. In order, from the top, are guanine, adenine, cytosine, 5-methyl cytosine and thymine.

Fig. 20. Apparatus for drying down small quantities of liquids, particularly suited to the concentration of eluates from filterpaper strips. A jet of air impinges on the surface of the liquid and a large bath of warm water prevents the liquid from becoming too cold to evaporate readily.

substances are eluted chromatographically with water, which is allowed to drop off the tip of the paper into a small tube. Some 0.5 ml. of water carry over all the cytosine and 5-methyl cytosine. This is then evaporated to dryness (Fig. 20) and 0.04 ml. of 0.1 N HCl added and allowed to dissolve the two bases. Aliquots of this are then applied to fresh chromatograms, which are developed in n-butanol NH$_3$ [n-butanol 86%, water 14%, v/v, plus about 5% of ammonia (s. g. 0.880) added to the solvent mixture in the bottom of the container]. The spots of 5-methyl cytosine and cytosine (the latter has the lower R_F value) are cut out, extracted and estimated (5-methyl cytosine: molar extinction = 9.8×10^3 at 283 mμ). Knowing from another analysis the proportion of cytosine in the whole nucleic acid, the proportion of 5-methyl cytosine is easily calculated.

A similar procedure could be used for identifying 5-hydroxymethyl cytosine. In the only nucleic acid in which it was found, no cytosine was present. The adenine band could, however, be used as a standard instead of cytosine.

An Alternative Method for the Analysis of DNA. DNA is hydrolysed by 72% (w/w) $HClO_4$ to the bases in 1 hour at 100°. This hydrolysis has already been described for RNA, but the time required is less. The rest of the analysis is carried out in the same way as for RNA. This method has the disadvantage of destroying some of the bases, particularly 5-hydroxymethyl cytosine, which may be nearly completely lost. On the other hand, for routine work, where great precision is not required, or for the analysis of dinucleotides and similar simple compounds, this is an excellent and rapid method.

G. Separation and Identification of Nucleosides and Nucleotides.

I. Paper Chromatography.

1. Detection of Ribonucleosides and Ribonucleoside 5′-Phosphates.

Substances having a *cis*-glycol structure as do the ribonucleosides and the ribonucleoside 5′-phosphates, react with periodates to give a dialdehyde structure. This reacts with SCHIFF's reagent to give a violet colour (BUCHANAN, DEKKER and LONG, 1950).

Procedure. Dry the chromatogram and spray with 1% sodium metaperiodate in water and leave for 10 mins. at room temperature. This oxidises the *cis*-glycols, and in order to remove excess periodate the chromatogram is treated with SO_2 gas in a jar. Iodine is liberated and then disappears from the paper. When this has happened, the papers are sprayed with SCHIFF's reagent and heated 10 mins. at 85—90°. The nucleosides, ets. show up as violet spots. This reaction is not an easy one to repeat every time, because SCHIFF's reagent is inclined to be temperamental. The reagent is made up as follows: dissolve 1 gm. basic fuchsin in 50 ml. water. Decolorise by bubbling SO_2 through the solution and filter after shaking with activated charcoal. Dilute to 1 l.

This method detects 20—50 μg. of nucleoside.

2. Detection of Deoxyribonucleotides.

The purine deoxyribonucleotides and deoxyribonucleosides react with SCHIFF's reagent to give violet colours. This reaction may be applied on paper chromatograms (BUCHANAN, DEKKER and LONG, 1950).

Procedure. Spray with SCHIFF's reagent (see previous section) diluted with $N\ H_2SO_4$ to give a final concentration of 0.1 $N\ H_2SO_4$. Heat at 60°. The spots should appear in about 6 minutes.

The function of the acid is to hydrolyse off the purine bases and to prevent the SCHIFF's reagent from losing its SO_2 under which circumstances it becomes coloured spontaneously. (The colour developed under these conditions has not the bluish tinge characteristic of the reaction with deoxyribose compounds.) This reaction is sensitive to 10 μg. of nucleotide.

An alternative method is to apply the DISCHE reagent. The papers are sprayed with 1% diphenylamine in glacial acetic acid containing 2.75% conc. H_2SO_4 (v/v). They are then placed between two glass plates and heated at 90°. Purine nucleotides (and nucleosides) react in 5—10 min. to give purple-blue spots. The pyrimidine derivatives need 30 mins. The sensitivity is 10 μg. for the former and 100 or more μg. for the latter.

3. Detection of Phosphorus Compounds.

Phosphorus compounds may be detected by a variety of methods. All of these are relatively difficult to apply with certainty on small amounts of material.

(a) The Method of Hanes and Isherwood (1949). The papers are sprayed with the reagent:

60% (w/w) $HClO_4$, 5 ml.; N HCl, 10 ml.; 4% ammonium molybdate, 25 ml.; water to 100 ml.

They are heated at 85° for 7 min. This hydrolyses the phosphate to give inorganic P which reacts with the molybdate. The complex is developed to a blue colour by dipping the paper into a jar containing H_2S in a moist atmosphere.

This method is more or less successful for purine bound P but not for pyrimidine bound P (or that esterifying C_5 of the sugars). An alternative is to spray the paper before applying the molybdate spray with a phosphatase solution (this may not react well with cyclic nucleotides or with nucleoside diphosphates) and then with the Hanes-Isherwood reagent, followed by H_2S treatment.

The H_2S development causes the background to become buff coloured. An alternative method, used by Boulanger and Montreuil (1951), is to expose the sprayed paper to bright daylight (ultraviolet light is also useful) after drying it. The phosphate compounds develop up as blue spots on a white background. This method seems to cause the hydrolysis of those phosphorus compounds which are not easily hydrolysed by warm acids.

(b) Method of Wade and Morgan. A method for the detection of phosphoric acid esters, applicable to nucleotides etc., has been described recently by Wade and Morgan (1953). This is dependent upon the formation of complexes between these substances and ferric ions, which may then be visualised by testing for free ferric ions.

The chromatogram, which must be at an acid pH (pH 1.5—2.5), is sprayed with a solution containing 0.1% $FeCl_3 \cdot 6H_2O$ in 80% ethanol. It is then allowed to dry at room temperature. The excess ferric ions are detected by spraying the chromatogram with 1% sulphosalicylic acid in 80% ethanol. The areas in which the iron is fixed stay colourless, while the rest of the paper becomes purple.

The advantages of this method are obvious. No hydrolysis is needed and therefore such substances as the pyrimidine nucleotides react readily. It is, however, evident that the critical factor, which is essential to success, is that the quantity of $FeCl_3$ per unit area should not exceed greatly the equivalent amount of the ester. The authors claim that the method can be made to detect 1—2 μg. P. In my experience the technique is best adapted for use with thin papers such as Whatman No. 1.

Owing to the spraying reagents used the paper cannot be used further, but the spots may be cut out and digested for phosphate determination. In the presence of large quantities of ester the spots may be detected after the first spraying as white areas on a yellow background.

4. Solvents for the Identification of Mononucleotides and Polynucleotides.

Two solvent systems are required and electrophoresis (q. v.) may also have to be employed, because so many substances may be encountered. The solvent systems are:

Solvent 1. 70% (v/v) *iso*propanol-water with 0.35 ml. NH_3 solution for each 1 l. tank volume added to the bottom of the tank. The separations possible with this solvent are shown in Fig. 21 and it will be noted that it has a tendency to separate compounds into groups having similar properties, and that it discriminates against guanine compounds, which always have lower R_F values than compounds of the other bases (this discrimination is also shown against xanthine compounds, which are not shown in the figure). The band numbers (MARKHAM and SMITH, 1952) are referred to in Table 2, which is in the section on electrophoresis.

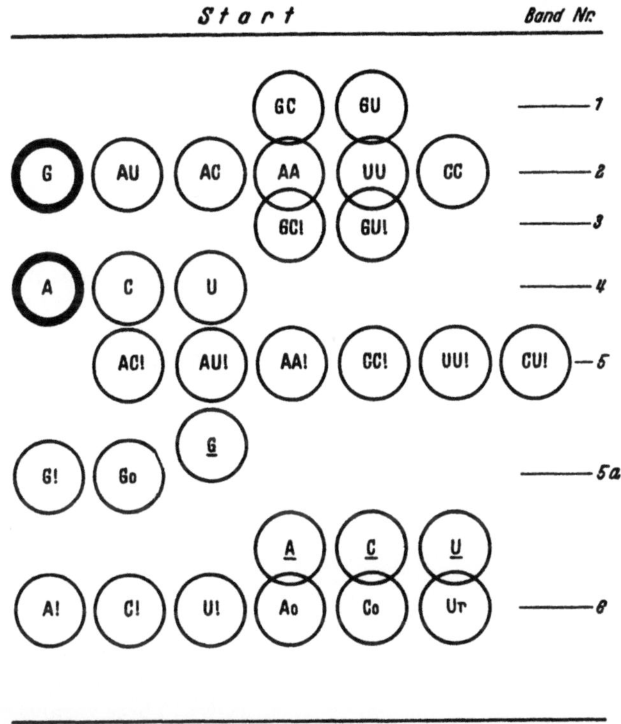

Fig. 21. The relative movement in 70% *isopropanol*—NH₃ of a number of compounds. A. G, C and U are adenylic, guanylic, cytidylic and uridylic acids respectively. The exclamation marks denote cyclic 2′:3′ phosphate terminations. Go, Ao, Co and Ur are guanosine, adenosine, cytidine and uridine, and *A*, *C*, *K* and *G* are adenine, cytosine, uracil and guanine. The xanthine and hypoxanthine derivatives move like the guanine and adenine analogues, while 8-azaguanine derivatives move rather more slowly than the corresponding guanine compounds. The two heavily outlined spots correspond with the alkaline hydrolysate of yeast RNA used as a marker. The positions of other substances in this solvent are indicated in the text. Compounds having a 5-phosphate termination have smaller and deoxyribose analogues rather larger R_F values than indicated here.

Solvent 2. This solvent (MARKHAM and SMITH, 1951) separates the isomeric nucleotides having amino groups. It consists of: saturated $(NH_4)_2SO_4$ in water, 80 parts, M sodium acetate, 20 parts, *isopropanol*, 2 parts (v/v/v).

The separations possible with this solvent are shown in Fig. 22. Although it contains so much salt, it is not safe to assume that salt-containing solutions will not interfere with the operation of this solvent. This solvent being aqueous runs extremely rapidly. It is sensitive to undue acidity or alkalinity of the spots, but is largely self-buffering owing to its composition.

II. Methods for the Identification of the Isomeric Nucleotides.

Until quite recently very little was known about the various nucleotides present in nature. LEVENE and HARRIS (1932) had shown by chemical degradation methods that the nucleotides obtained by the hydrolysis of yeast nucleic acid were the nucleoside 3′-phosphates, the structure of the deoxyribonucleotides was also unknown and "muscle adenylic acid" was known to be adenosine 5'-phosphate. The discovery of Carter (1950) and of Cohn (1950) that the ribonucleotides, thought to be the nucleoside 3'-phosphates by Levene and Harris, were in fact pairs of substances, changed the picture completely. Brown and Todd (1952) showed that these pairs of nucleotides were probably the nucleoside 2' and 3' phosphates, originating during the hydrolysis procedure from the 2':3' cyclic phosphates, so that in the ribose series alone, we now had 12 nucleotides derived from yeast nucleic acid alone. This number was soon increased to 16 by Cohn and Volkin (1952) who prepared the nucleoside 5'-phosphates from RNA by an enzymic degradation. The deoxyribonucleotides originating from the degradation of DNA by snake venom diesterase were shown by Hurst and Butler (1951) to be the nucleoside 5'-phosphates. The deoxyribonucleoside 3'-phosphates have not yet been found naturally but probably do exist somewhere. The deoxyribonucleoside 3':5'-diphosphates have been found in acid digests by Levene and Jacobs (1912).

These various substances can and do exist free and in combination, and it is necessary to have methods for identifying them. Fortunately a great many of the tedious procedures for accomplishing this have been superseded by simpler ones. The following few paragraphs indicate the approach to be made, assuming that the substance in question has already been isolated by one of the methods described in this chapter.

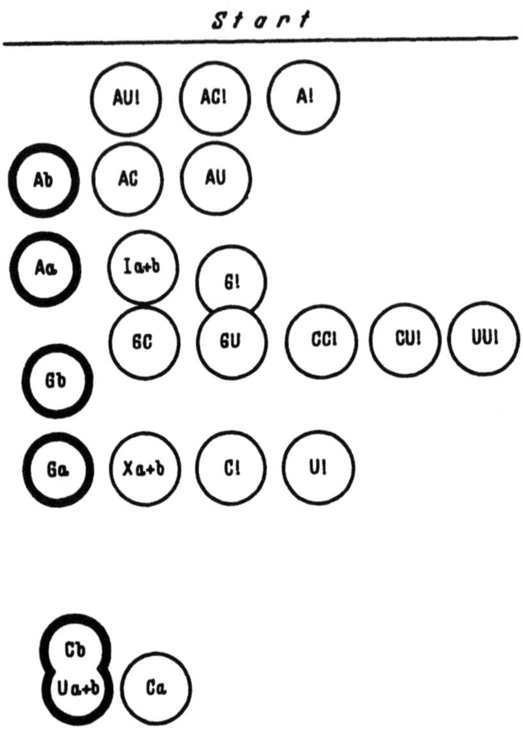

Fig. 22. *Explanation.* The positions of a number of mononucleotides and dinucleotides in the solvent: 80% sat. aqueous (NH₄)₂SO₄, 20% M NaCOOH (v/v) + 2% (v/v) *iso*propanol. The letters A, G, C and U denote adenylic, guanylic, cytidylic and uridylic acids respectively. The exclamation mark (!) denotes a cyclic 2':3'-phosphate termination. The a and b subscripts refer to the 2' and 3' phosphate terminations. All the non-cyclic dinucleotides shown are of the 3' form. The heavily outlined circles are those substances to be found in a suitable KOH hydrolysed yeast ribonucleic acid marker.

The Nucleoside 5'-Phosphates. These substances are readily identified because alone of all the nucleotides, they are dephosphorylated rapidly and completely by snake venom. To a solution (0.1—0.2 ml.) adjusted to pH 10 (thymol-blue as an internal indicator) a few grains of dry Russel viper venom are added, and the mixture incubated at 37° for a few hours. It is then concentrated if necessary, and chromatographed in 70% *iso*propanol-NH₃ (solvent 1). The nucleoside liberated is easily identified by its higher R_F value. Alternatively, analysis for inorganic P will show if the enzyme has acted.

The identification of the nucleotide as deoxyribose or ribose containing is less easy. Purine nucleotides may be identified by the sugar reactions, but the easiest method is to use electrophoresis in borate at pH 9, using a marker if possible, when the ribose nucleotides will form a borate complex, and the deoxyribose compounds do not.

An alternative method of identification is that the ribose derivatives react with $NaIO_3$ to give aldehyde derivatives, and after such treatment are hydrolysed by alkali to liberate inorganic P (BROWN, FRIED and TODD, 1953).

The Nucleoside 2′- and 3′-Phosphates. The purine ribonucleoside 2′- and 3′-phosphates are best identified by running in 80% sat. $(NH_4)_2SO_4$ in water, 18% M NaCOOH, 2% isopropanol (v/v/v) (solvent 2) when the 3′-phosphates have a lower R_F than do the 2′-phosphates (Figs. 22, 23). The deaminated derivatives, namely xanthosine and inosine 2′- and 3′-phosphates do not resolve, and run with the corresponding aminopurine riboside 2′-phosphates.

Cytidine 2′- and 3′-phosphates may be resolved in this solvent with a long run. Uridine 2′- and 3′-phosphates run with cytidine 2′-phosphate.

One of the easiest ways of separating the cytidylic acids (and the adenylic acids) is by ion exchange chromatography on Dowex 1-formate. This is described elsewhere.

Recently, SCHUSTER and KAPLAN (1953) have isolated an enzyme system from certain plant seeds, and this dephosphorylates nucleoside 3′-phosphates, and not the 2′- and 5′-phosphates, and so it may be used as an analytical tool. Although this enzyme has not been tested on deoxyribonucleoside 3′-phosphates, its action on the corresponding diphosphates suggests that it will act on these compounds as well.

The Nucleoside Cyclic 2′:3′-Phosphates. These substances are easily recognised by their R_F in *iso*propanol-NH_3 (solvent 1), their stability to prostate phosphatase (enzyme 1) and their conversion to a mixture of the 2′- and 3′-phosphates by N NaOH in the cold. They may be confused with the methyl or ethyl esters of the 2′- or 3′-nucleotides, but the latter are unaffected by $O·1$ N HCl at 20°, while the cyclic nucleotides are split to form the 2′- and 3′-phosphates.

The cyclic nucleotides are also readily identified by electrophoresis at pH 7.4 (q. v.).

III. Ion Exchange Chromatography.

Owing largely to the work of COHN and his colleagues in America methods for the separation of many constituents of nucleic acids have been developed using ion-exchange resin columns for the chromatography. These techniques are particularly well suited to the study of nucleic acid derivatives because of the number and variety of charged groupings which the latter possess.

Ion exchange chromatography has several great advantages and these will be enumerated here for convenience.

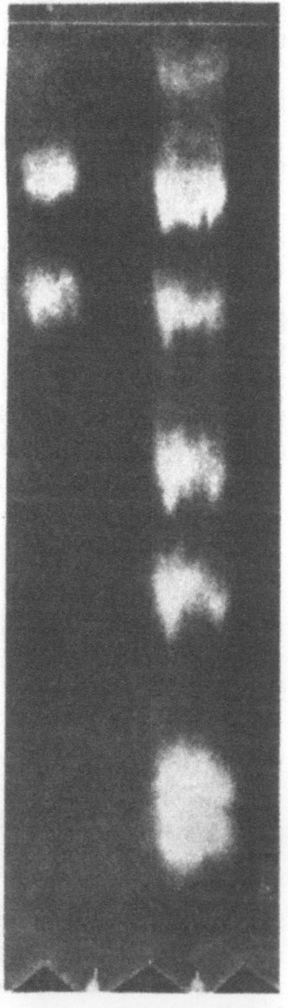

Fig. 23. Chromatographic separation of the purine nucleoside 2′ and 3′ phosphates in ammonium sulphate solution. In order from the top are adenosine 3′-phosphate, adenosine 2′-phosphate, guanosine 3′-phosphate, guanosine 2′-phosphate. The bottom spot contains uridylic and cytidylic acids only partially resolved. On the left side is a specimen of commercial "yeast adenylic acid". Showing the two constituents.

(1) Great resolving power may usually be attained when necessary.

(2) Large or small quantities of material may be handled, and the method is suitable for preparative work.

(3) Contaminants are of little importance because they are usually eluted in fractions which may be well removed from those in which the required substances are eluted. (This is in contrast both to paper electrophoresis and chromatography, where all the materials put on to the paper generally stay on the paper somewhere and thus may cause trouble.)

Against these advantages, there are several disadvantages, which may not be found serious. These are:

(1) Automatic fraction collectors are almost essential.

(2) The time taken for individual analyses may be several days.

(3) Large numbers of spectrophotometric readings have to be taken.

(4) Considerable volumes of liquid may have to be handled.

In practice it will usually be found that for much routine work paper chromatographic methods are equally accurate and much more rapid, but that for certain problems the ion-exchange method has definite superiority. As an example, the separation of the various isomeric pyrimidine nucleotides is much more conveniently carried out on ion-exchange columns, and in describing the technique of preparing the columns and using them, the separation of these substances will be given in some detail.

The Separation of Nucleotides by Ion-exchange Chromatography.

A suitable column consists of a tube having an internal diameter of 1 cm. and being some 25—30 cm. in length. The bottom of the tube is constricted and plugged at the constriction with a cotton plug. Alternatively the tube may be fitted with a sintered filter membrane, though this is far from essential. The tube contains Dowex-1 (200—400 mesh) resin in a column some 12—15 cm. in depth.

To prepare the resin it whould be cycled a few times with 1 N HCl, water, 1 N NH$_4$OH, water and so on. This process may be carried out in centrifuge tubes in bulk. The contact of the resin with the solutions need not be prolonged, but the intermediate water wash should be carried out with care in order to remove ions. The resin is now suspended in water, poured into the tube and allowed to settle under gravity. When this has completed, the layer of water is removed carefully, and the process repeated until the column is of the desired height.

The eluting solutions are admitted on to the column through a glass tube arranged in a rubber stopper in the top of the chromatogram tube, and as a considerable pressure is needed to allow the column to run at a reasonable rate (about 30—40 cm. water pressure) the eluting fluid is admitted through a sufficiently long polyethylene tube from a reservoir placed at an appropriate height above the column. Polyethylene is specified instead of rubber, because some acids, particularly formic acid, dissolve in the rubber tubing, and so cause difficulties if the apparatus is reused. Thin-walled polyethylene may be clipped in the same way as a rubber tube. Thick tube will have to have a glass tap inserted somewhere, but this should be avoided if possible.

If the flexible tubing is allowed to fall below the bottom of the column there will be no chance of the column running dry if left unattended. A column once allowed to run dry will have to be remade, whereas one kept moist can be used indefinitely.

To prepare the column in the formate form it is washed through with a little water and then with 1 N NH_4OH, which will displace any acid present. When the effluent liquid is alkaline, the wash liquid is changed to water, which is run through until it emerges substantially neutral (pH 8—8.5). At each change the stagnant liquid (about 6 cm. or so) above the resin is removed with great care and replaced by the next solution.

Having prepared the column in its basic form, the conversion to the formate is effected by running through a solution of 0.2 N HCOOH:0.2 M NaCOOH, until the effluent comes out with the same composition as the inflowing solution (take pH with methyl orange) and has a density not exceeding 0.05, when tested against the latter at 260 mμ in a 1 cm. cell.

Note: If the laboratory supply of distilled water contains much CO_2, this may convert the basic resin to the bicarbonate form. On subsequent addition of an acid solution, the CO_2 evolved may break up the column. The bicarbonate may be removed by displacing it with 0.01 M NaCOOH before adding the acid washing liquid.

Having converted the column to the formate form, it is washed with water until the effluent is neutral. It is then ready for use, and, as will be seen, the subsequent regeneration processes take place during the analytical runs, the column remaining in a usable condition for an indefinite period.

The specimen to be analysed is put on the column in a very dilute solution which has a low ionic strength (0.01 M or less) and at pH of about 8—9, using ammonia to make this adjustment. Having run the solution through the column, development may begin, after a wash with water or 0.01 M NaCOOH if it is desired to eliminate nucleosides and bases before the collection of fractions begins. A convenient volume for taking these fractions is 15—20 ml. and they must be of equal volume, or else the volume will have to be measured in each tube.

As a guide to the anticipated positions of the various ribonucleotides in a column of the dimensions given the following indication is given:

Table 3.

Solvent	Vol. to peak	Substance
0.01 M HCOOH	260 ml.	Cytidylic acid "a" (2′)
0.01 M HCOOH	340 ml.	Cytidylic acid "b" (3′)
0.15 M HCOOH	110 ml.	Adenylic acid "a" (2′)
0.15 M HCOOH	240 ml.	Adenylic acid "b" (3′)
0.05 M NaCOOH	430 ml.	Uridylic acid "a" (2′)
0.01 M HCOOH	550 ml.	Uridylic acid "b" (3′)
0.10 M NaCOOH 0.10 M HCOOH	550 ml.	Guanylic acid "a" (2′)
0.15 M NaCOOH 0.10 M HCOOH	300 ml.	Guanylic acid "b" (3′)

In practice a known substance or mixture is applied to the column and elution is started, using the eluting solvents in increasing order of ionic strength and/or pH, and the effluents are read on a BECKMAN spectrophotometer at 260 mμ and possibly also at 280 or 250 mμ. The curves are then plotted as in Fig. 24, the substances being eluted having been identified by their order in emerging, and by the absorption spectra in the neighbourhood of their peaks. The column has now been calibrated and will maintain the same characteristics for a number of runs provided that the eluting fluids are accurately made up. At the end of a run it is washed through with water, and is immediately ready for a new analysis.

The capacity of such a column is very large. Fig. 25 shows the analysis of a commercial adenylic acid specimen in which 20 mg. of material were used.

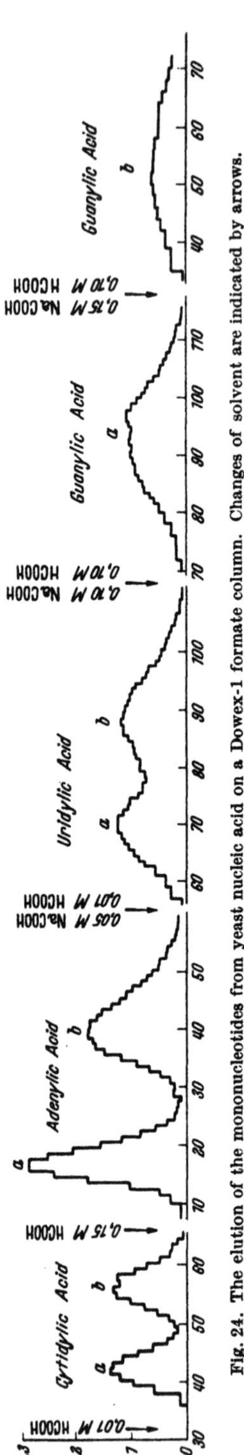

Fig. 24. The elution of the mononucleotides from yeast nucleic acid on a Dowex-1 formate column. Changes of solvent are indicated by arrows.

It will be noted that the bulk of this material was eluted in less than 80 ml. If it should be desired to use larger quantities of material the column may be made wider. The volume of eluting solution required to remove any peak is proportional to the amount of resin used.

The system described is particularly useful for detecting and estimating cytidylic acids "a" (2') and "b" (3') which are difficult to resolve on paper.

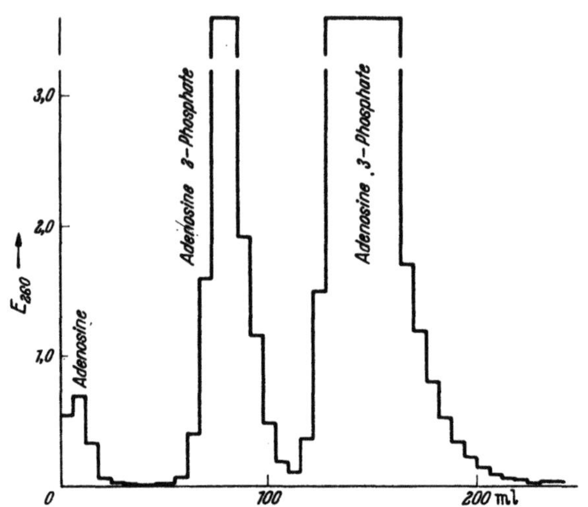

Fig. 25. The separation of the isomeric adenylic acids from a commercial specimen of "yeast" adenylic acid (20 mg. in all). Eluant is 0.15 M HCOOH.

The purine nucleotides may be readily and adequately resolved on paper. The uridylic acid isomers are rather more difficult to resolve on the column, but this separation cannot be effected on paper (without the use of enzymes).

The flow rates used with the column described were about 20—40 ml./hour, which seems quite a satisfactory rate. The actual rates used do not appear to be at all critical, and they may be reduced in order to enable a fraction collector to run unattended overnight or longer. The runs may be interrupted for a day or so without any apparent loss of resolving power.

It should be noted that the column described does not separate the cyclic nucleotides from the "b" (3') isomers, and may in fact be used with advantage to prepare these substances from digests which have been previously treated with phosphomonoesterase to remove the ordinary nucleotides.

Other Systems. The system described above appears to be the most generally useful analytical tool, but several other systems may also be used. It does not appear to matter much whether Dowex 1 or Dowex 2 is used for the columns, and it is often advantageous to use the resins in the chloride cycle.

To prepare this type of column the procedure is as for the formate columns, but 0.1 N HCl is used as a solution for forming the salt of the resin. Using such a column for the separation of the mononucleotides, and eluting with HCl solutions the pattern is as shown in Fig. 26 (COHN, 1950).

Deoxyribonucleotides. (VOLKIN, KHYM and COHN, 1953.) The same system may be used for the elution of the deoxyribonucleotides prepared by enzymic degradation of deoxyribonucleic acid. In this case thymidylic acid occurs in the

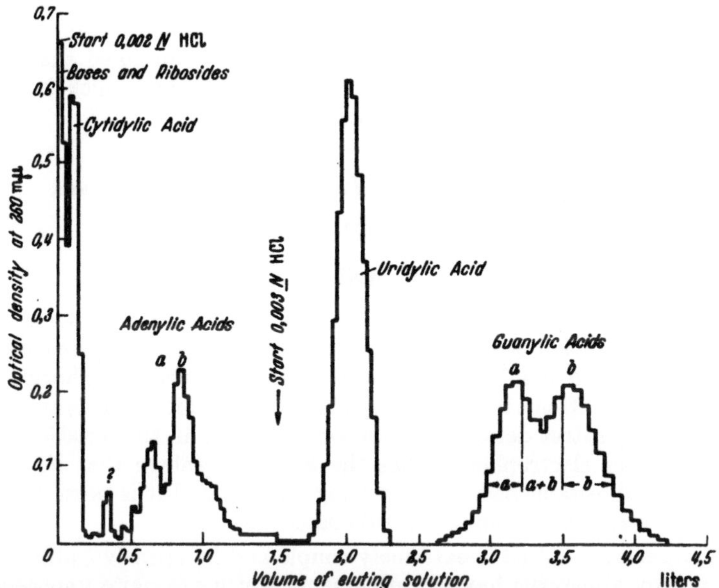

Fig. 26. The separation of ribomononucleotides on Dowex 1 chloride: column Dowex 1, 12.5 cm. × 0.74 sq. cm: eluant, dilute HCl at 0.5 ml./min., starting at 0.002 N, changing after 1.5 l. to 0.003 N. [COHN: J. Amer. Chem. Soc. 72, 1471 (1950).]

relative position occupied by uridylic acid among the ribonucleotides. The eluting volumes to the peaks are different because these nucleotides are the 5′ phosphates, and these as well as the ribonucleoside 5′ phosphates are more readily eluted than are the 2′ and 3′ phosphates.

Polynucleotides. The separation of polynucleotides by the use of ordinary ion exchange resins is very difficult, although MERRIFIELD and WOOLEY (1952) and SINSHEIMER and KOERNER (1952) have had considerable success in this field. COHN, DOHERTY and VOLKIN (1952) however, have managed to effect quite dramatic separations using 2% cross-linked Dowex 1 (the ordinary Dowex 1 is 10% cross-linked) in the chloride form. This resin is extremely soft and slow-running, and in the elution the use of strong acid is precluded because of the lability of the polynucleotides. It may be found that the pattern given by Dowex 1 (2% cross-linked) is not so reproducible as it is for the normal resin.

Nucleosides. An ingenious separation of the ribonucleosides has been reported by JAENICKE and VON DAHL (1952). These authors adsorb the nucleosides on a Dowex 1 borate column. The borate complexes of the nucleosides then behave as acids and may be eluted in the order: cytidine, adenosine, uridine and guanosine, by $N/100$ borax pH 9.2, $N/20$ borax pH 9.2, $N/20$ borax + $N/100$ HCl and $N/20$

borax $+$ $N/10$ HCl. The deoxyribonucleosides are, of course, not held on such a column because they do not complex with borate.

Bases. The bases may be separated on Dowex 1 chloride at a pH above 10 where they behave as acids and may be eluted by increasing the chloride concentration of ammonia solutions. Alternatively they may be eluted from Dowex 50 in its acid form, by 2 N HCl. There would, however, appear to be little advantage in the use of such a technique in the analysis of relatively pure specimens of nucleic acids or of bases over the normal paper methods, which are quite adequate for this purpose. In those cases where large amounts of interfering substances are be found, such as with crude extracts, the column analysis may well succeed where others fail.

In the separation of the bases on Dowex 1, it will be noted that the acidic character of these compounds is shared by their ribosides, which tend to elute at nearly the same position in the pattern.

IV. The Separation of Nucleic Acid Derivatives by Electrophoresis on Paper.

1. Method.

The method of electrophoresis on an inert base such as filter paper is extremely useful for the separation of practically all the derivatives of nucleic acid. So far, however, the method has been developed mainly for the phosphorylated compounds, which are difficult to separate by chromatographic methods and these will be discussed in greatest detail. In addition to its use in the isolation of known compounds, paper electrophoresis has the great advantage that its theoretical basis is quite well established, and by using this method it is possible to deduce what kinds of structure an unknown substance may have.

The equipment required is extremely simple and inexpensive, although many workers employ apparatus having refinements which are quite unnecessary and which do not result in appreciably better operation.

Apparatus. (MARKHAM and SMITH, 1953a.) In our laboratory it is customary to work with rather high voltages, because the separations may then be carried out in a relatively short time. A period of from one to three hours is usually adequate for making any required separation. In consequence of this high voltage precautions have to be taken about the heating of the papers, and the apparatus used (Fig. 27) is designed to avoid this. Vessels A and B contain the buffer solution, which is also used to soak the paper strip. The latter passes from A to B through C, which is filled with an inert, dense liquid, for which purpose carbon tetrachloride is nearly ideal. This liquid serves to prevent the evaporation of the water from the paper and also acts as a cooling bath. The filter paper is saturated with buffer solution, and the two halves in C are prevented from touching by means of a spacer of thin glass or of celluloid. The whole apparatus is covered by a thick glass plate, which serves the purpose of preventing evaporation of the carbon tetrachloride. The electrodes are conveniently made of two carbon sticks. Platinum can, of course, also be used, but any of the common metals should be avoided.

The power supply may be derived from a voltage doubler circuit or a whole-wave rectifier circuit, and should be designed to give an output of at least 10 ma. at 1000 V. Two suitable supplies are shown in Fig. 28 and Fig. 29. It is useful and almost essential to be able to vary the voltage on the electrodes from 500 to 1000 V., so as to be able to use buffers of differing ionic strengths. As a guide to the choice of voltage, it should be noted that 10—15 watts is a normal amount of energy to dissipate in this apparatus (watts = volts × milliamps × 1000).

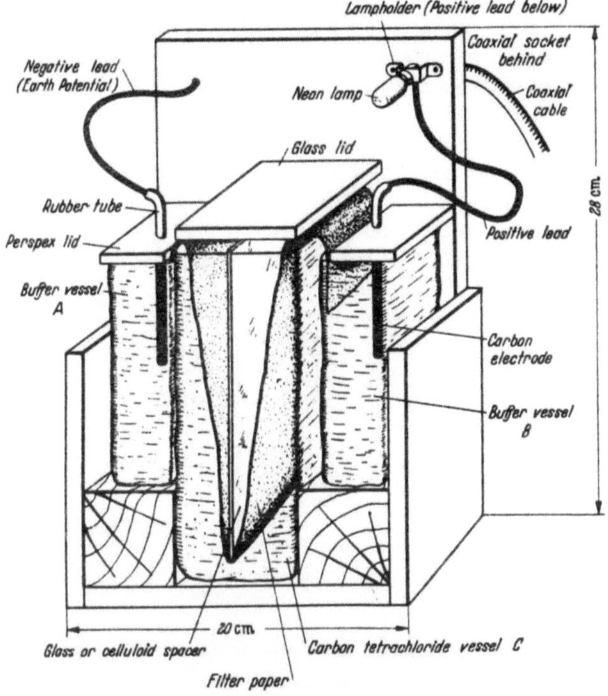

Fig. 27. *The electrophoresis apparatus.* The three small glass vessels (2 of 6 × 4 × 2 in., and 1 of 8.25 × 4.63 × 2.5 in.) are mounted in a wooden frame. The high tension lead (1000 V) is led to the back of the frame through a coaxial cable. The coaxial socket is fitted so that the lead passes through the wooden frame without touching it. The live end is protected by mounting a small lampholder directly over it. This carries a 0.2 ma. neon lamp which indicates whether the apparatus is live. The lamp is connected to the positive lead by two 2.2 Mω resistances (0.25 watt) in series. The electrodes are held in perspex lids for the buffer vessels and the tops are protected by short lengths of rubber tube. In the centre vessel the wet filterpaper is immersed in carbon tetrachloride, a spacer of glass or celluloid preventing contact between the two halves.

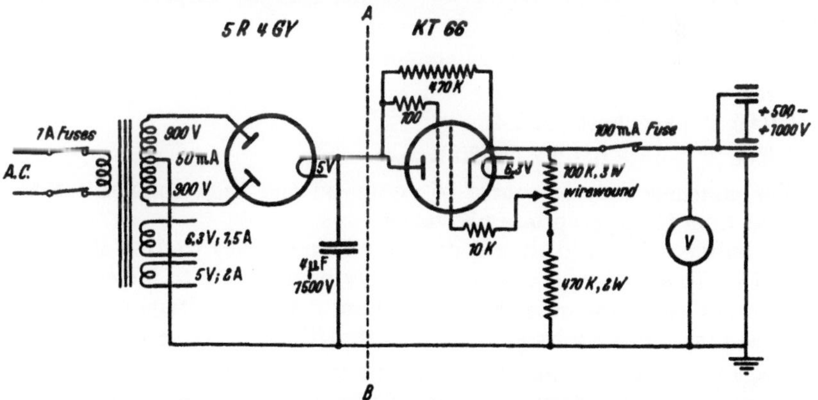

Fig. 28. A power supply suitable for running two electrophoresis apparatuses of the type shown in fig. 27. If more than two apparatuses are to be run, the transformer should be rated at 100 ma., and the cathode of the KT 66 valve should be connected to earth by a 4 μF. (1500 V. working) condenser. The circuit to the right of the dotted line AB can be applied to any D. C. supply to regulate its voltage over some 500 V., and may if desired be duplicated so as to allow two outlets to be run at different voltages. One KT 66 valve will permit up to 50 m/a. to flow for an output voltage of 500 V., and at higher output voltages more current may be taken up to 100 m/a. at 1000 V. For long life these values should only be approached occasionally.

Higher wattages can be tolerated however, and, if required, the carbon tetrachloride may be taken up to its boiling point. Immersing the vessel C in cold running water would also enable larger wattages to be used.

A number of methods may be used to regulate the voltage. An excellent but expensive method is to use a 0—250 V variable autotransformer in the primary circuit. An alternative to this is to use a variable resistance in the same place, but this is rather sensitive to changes in the conductivity of the buffer solutions. An excellent and inexpensive method is to use a series power triode regulating valve in the H. T. side (Fig. 28) and this will be found completely satisfactory. It should be pointed out here that with the voltage doubler-circuits *both* leads are live, whereas with the full wave rectifier circuit only the positive lead is live, and this may with advantage be led out through a shielded coaxial connector for safety (the voltage used is, of course, extremely dangerous).

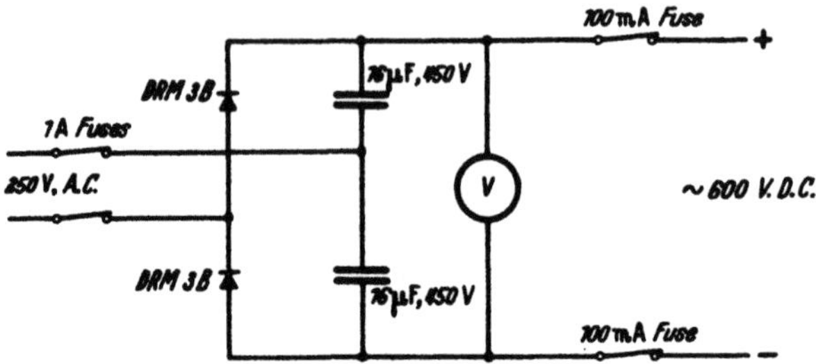

Fig. 29. A simple and inexpensive power supply giving some 600 V. at up to 100 m/a. It is not provided with a means of adjusting the voltage because this should be unnecessary. Note that both output leads are live.

Buffer Solutions. For general use uni-univalent buffers are preferable and they should also if possible be made from volatile constituents, and they *must* be transparent to ultraviolet light. It is usually advisable to avoid phosphate-containing buffers whenever possible, not only because of their relatively high conductivity, but also because it may be necessary or desirable to make tests for phosphorus on material eluted from the strip.

For most purposes the following buffer solutions are satisfactory:

(a) 0.05 M formate pH 3.5 (M ammonium formate 20 parts, N formic acid 30 parts, water to 1 litre).

(b) 0.05 M acetate pH 4.6 (M ammonium acetate 25 parts, N acetic acid 25 parts, water to 1 litre).

(c) 0.05 M phosphate pH 7.4.

(d) 0:05 M borate pH 9.2 ($M/20$ borax).

Other mixtures may of course be employed and their composition may be calculated from the appropriate pK values.

With buffers (c) and (d) it will probably be necessary to drop the voltage to 700—800 V. In addition, it may occasionally be desirable to increase the buffer strength when using the other buffer solutions. As the voltage is related to the wattage as $x:x^2$, when doubling the buffer strength it is only necessary to decrease the voltage to $\sqrt{2}$, and similarly for other buffer strengths.

Procedure. A strip of Whatman No. 3 MM paper 56 × 8 cms. is used, and is soaked in the appropriate buffer solution (this should be fresh, but that used in the electrode vessels may be used repeatedly until badly discoloured by the carbon from the positive electrode). It is then pinned up by one end, 8 cms. or so of which have been left dry for this purpose, and blotted lightly with filter paper to remove surface moisture. The sample to be analysed is put on in a thin line about 12 cm. from one end at right angles to the long axis of the paper. The quantity to be used depends upon the inclination of the worker and the concentration of the solution. With care 0.25—0.5 ml. can be loaded onto the paper, but 0.1 ml. is usually adequate. If required two half lines 4 cm. long, may be put on, each with a different substance on them, but it is not much use putting on spots. Lines, because of an interference effect, give a much better resolution than do spots.

When the paper has been prepared it is immersed in the centre vessel C, (Fig. 27) using the spacer to help in the process. The ends are then immersed in the buffer vessels. the electrodes and the lid fitted and the current switched on. With the dimensions given, it will be found that the voltage gradient is about 20 V./cm. if 1000 V. are applied to the electrodes.

With the usual pH values, it will be found that the amino bases and their ribosides will move towards the negative electrode, while phosphorylated derivatives move towards the positive. If it is not certain what substances are going to be found, the best place to put the starting material is in the middle of the strip, but, of course, the potential resolving power is thereby reduced.

When the run is completed, the papers are hung up to dry and printed in ultraviolet light in the usual way and a roll of filter paper put into vessel C to absorb the mist of water droplets which appear when the apparatus cools.

In general we use this method for preparing material for subsequent analysis by chromatography or for identification in comparison with known substances, rather than for quantitative analysis.

When it is desired to make a separation of two substances on a large scale it is, of course, possible, if they have greatly differing mobilities, to effect the separation in a distance of a few cm. In this case one may put on four or more bands at regular intervals along the strip of filter. In doing this it is as well to bear in mind that, owing to the difficulty of saturating the paper exactly, liquid flows in from the electrodes and moves the boundaries about 2—3 cm. towards the middle of the paper strip, if they are put on 12 cm. from the ends, but, of course, does not move material put on in the centre. This movement all takes place in the first half hour, so that for accurate mobility measurements, it is advisable (a) to put the material on at the centre of the paper and (b) wait half an hour before switching on the current.

It should also be noted that, because the buffer solutions used are necessarily weak, the presence of much conducting material in the sample may cause a localised reduction in the voltage gradient in the paper. This will cause the movement to "hang fire" for a time, and then the bands will move off fairly normally. It is therefore desirable to have as little salt as possible in the material to be analysed. The paper is able to support at least 2 mg./cm. width of individual substances, but this quantity is by no means the greatest which it can support. The main effect of adding more material is to make the bands wider. In the case of substances with greatly differing mobilities this is of little moment, but when two substances with nearly identical mobilities have to be separated, then the quantity applied to the paper should be reduced to a minimum.

2. The Separation of Mononucleotides.

The ionising groupings present in the mononucleotides are the primary and secondary phosphate groups, the amino groups of guanine, cytosine, 5-methyl cytosine and 5-hydroxymethyl cytosine, and the enolic, groups of guanine, thymine and uracil. Table 4 gives these values for a number of substances. The values for some of the various isomers are not given, because many have not been determined, and the pK's are so close to each other in these substances that for most purposes they may be taken as equal. Thus the values given for the 2' and 3'-phosphates may be used for the 5'-phosphates, and also for the corresponding deoxyribonucleotides. In the case of thymidylic acid, this may be

Table 4. *The pK' values of the dissociating groupings in nucleotides and related substances. Spaces with dashes indicate that the compound in question does not have this grouping. Blanks indicate that the data are not available. Primary phosphate pK's are about pH 1.*

Substance	pK'—NH$_3^+$	pK'—PO$_4^{--}$	pK'—OH$^-$
Adenosine 2'-phosphate	3.5	6.15	—
Adenosine 3'-phosphate	3.65	5.9	—
Adenosine 5'-phosphate	3.7	6.1	—
Guanylic acid	2.3	5.9	9.4
Inosine 5'-phosphate	—	6.0	
Cytidylic acid	4.2	6.0	>13
Uridylic acid	—	5.9	9.4
Deoxy-5-methyl cytidylic acid . .	4.5		
Adenosine	3.5	—	—
Adenine	4.15	—	—
Guanine	3.3	—	11
Cytosine	4.6	—	12
Uracil	—	—	9.45
Hypoxanthine	—	—	8.7
Xanthine	—	—	9.9

regarded as equivalent to uridylic acid (and similarly with the other methylated derivatives). It will be noted that the cyclic ribonucleotides have no secondary phosphate grouping.

The enolic groupings on most of the bases dissociate at pH values too high to be used advantageously, but in the case of 8-azaguanylic acid this group dissociates at about 2 pH units lower than does that of guanylic acid, so that at pH 9 or thereabouts 8-azaguanylic acid has an extra negative charge. This has enabled Matthews (1954) to isolate this nucleotide from certain nucleic acids into which the unnatural base 8-azaguanine has been incorporated.

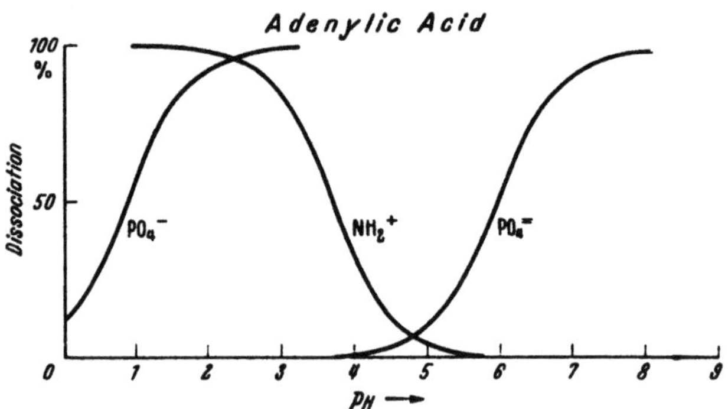

Fig. 30. Dissociation curves of groupings in adenylic acid.

The primary phosphate groups are completely dissociated at any pH value which is usable, and the main separation is made by taking advantage of the weak NH$_2$ groups, which have pK' values ranging from pH 2—5, while thymidylic and uridylic acids have no amino groups at all. Consequently it is possible by selecting a suitable pH value to obtain the maximum separation in mobility between any two mononucleotides. In doing this, it may be assumed that all the mononucleotides have very similar resistance to motion through liquids

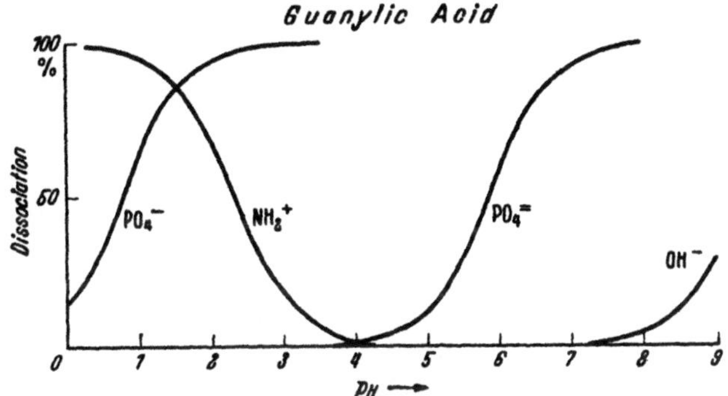

Fig. 31. Dissociation curves of groupings in guanylic acid.

(or diffusion constants, which are proportional to this force). In the case of a mixture of all the mononucleotides, it is found that the pH value which is most suitable for the separation is pH 3.5. Figs. 30, 31, 32, 33, give the dissociation

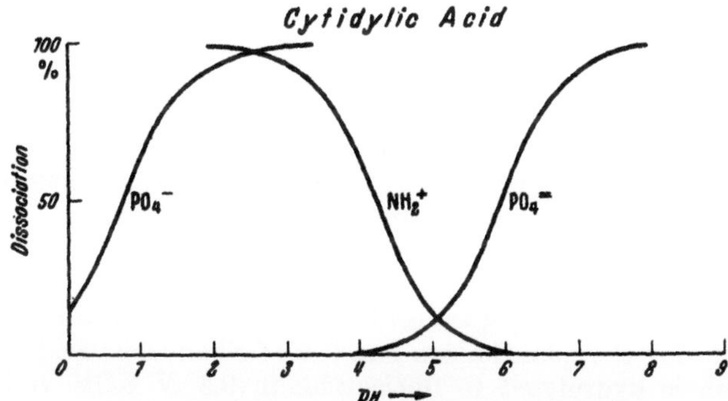

Fig. 32. Dissociation curves of groupings in cytidylic acid.

curves for the ribomononucleotides, and these may be used for the selection of suitable pH values for effecting difficult separations.

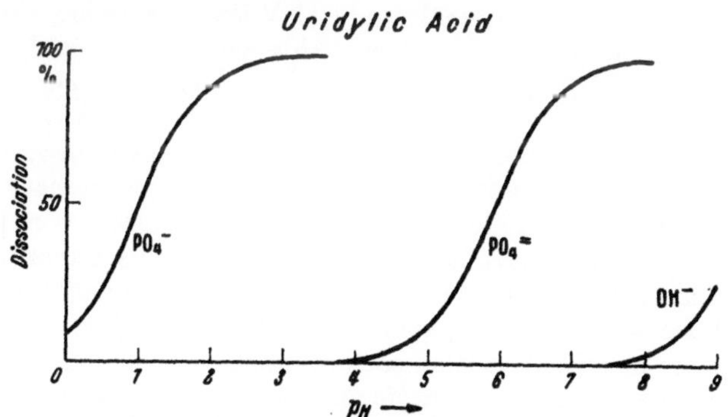

Fig. 33. Dissociation curves of groupings in uridylic acid.

In order to make use of the dissociation curves (either those shown here or those calculated from pK′ values) it is necessary to proceed as follows:

The rate of movement of a substance in a voltage gradient, dV/dx, is proportional to the latter, to the algebraic sum of the charges on the molecules, C, which is given by the fraction of dissociation of the various charged groupings, and inversely proportional to the frictional resistance to movement of the molecules through the buffer solution. The latter is, as we have noted, a function of the diffusion constant, and it is also dependent upon the buffer viscosity (which is a function of the composition and temperature of the solution).

Fortunately the frictional resistances to movement of all the nucleotides are approximately equal, so that the *relative* velocities of the various nucleotides may be calculated by summing the charges on each nucleotide at the particular pH value used. As an example, let us take adenylic acid and uridylic acid at pH 3.5. Adenylic acid has one primary phosphate group 100% dissociated, and one amino group 54% dissociated. Thus it has 1.00 negative charges and 0.54 positive charges, or a net 0.46 negative charge. Uridylic acid on the other hand has only one charged group, namely the primary phosphate group. Consequently it has a net charge of —1.0, and so it may be expected to electrophorese 1/0.46 times as fast as does adenylic acid. At pH 7.0 it will be seen that both molecules carry identical charges (—2.0), but, of course, the cyclic 2′:3′ phosphates will only have a charge of —1 at this pH, so that they can easily be separated from the former. (Similarly a methyl or ethyl ester of a nucleotide will only have one charged group at this pH value.)

It will be noted that some pairs of nucleotides have charges which are virtually equal at all pH values, for example thymidylic and uridylic acid (and xanthylic and inosinic acids) which have no amino groupings, and cytidylic and 5-methyl cytidylic acids which differ, as do the first two, only by one having a methyl group. Electrophoresis is, of course, useless for making separations of such substances, but this may be effected readily by other methods.

3. The Quantitative Analysis of Nucleic Acid.

Electrophoresis on filter paper has been used successfully by Davidson and Smellie (1952) for the quantitative analysis of ribonucleic acids. The RNA at about 5 mg/ml. is hydrolysed to nucleotides in 0.3 N KOH at 37° for 18 hr. The solution is brought to pH 1 with 60% (w/w) $HClO_4$, insoluble $KClO_4$ spun off and washed with cold water, and the combined supernatants brought to pH 4. Aliquots containing about 800—1000 μg. of ribonucleotides are run in pH 3.5 citrate buffer, 0.02 M (formate buffers would be better). These authors use paper suspended in a moist chamber, and apply 11 V./cm. for 18 hr., using a strip of Whatman No. 3 mm. paper 72 × 7 cm. This separation could, of course, be effected in 60 mins. in the apparatus already described (Fig. 27). The P recovery is about 97%.

Deoxyribonucleic acid could be analysed in a similar way, but using an enzymatic method for the hydrolysis. This involves a preliminary digestion by deoxyribonuclease at pH 7 followed by snake venom diesterase at pH 10, which liberates the nucleoside 5-phosphates quantitatively. It is important to ensure that the 5-nucleotidase of the venom is removed completely, otherwise some of the nucleotides will be dephosphorylated. Deoxycytidylic and deoxymethylcytidylic acids are not resolved from each other.

A deficiency in the electrophoretic method for the analysis of ribonucleic acids is that it assumes that the latter are converted quantitatively to nucleotides by the hydrolysis. Recent experiments in our laboratory suggest that this may not be so in the case of some ribonucleic acids. The discrepancy is due to the termination of the chains, and does not involve a serious error.

4. Cyclic Mononucleotides.

The cyclic 2′:3′ monohydrogen phosphates of the nucleotides have essentially the same charge as the equivalent 2′ or 3′ phosphates at any pH value where the secondary phosphate groups are unionised (actually the cyclic nucleotides move very slightly more slowly than do the 2′ or 3′ phosphates, possibly owing to their shape). At pH values above 5 however, the effect of the secondary phosphate group begins to become evident, and the group is more or less fully ionised at pH 7.0. Consequently, it is possible to resolve a mixture of cyclic and other nucleotides by running it at pH 7 or above, when the cyclic nucleotides have one positive charge and the non-cyclic nucleotides have two. (Fig. 34.)

5. The Ribonucleoside 5′-Phosphates.

The ribonucleoside 5′-phosphates are unusual in that they contain a *cis*-glycol structure not possessed by their deoxyribose analogues nor by the other ribonucleotides (2′, 3′ and 2′:3′ cyclic). This structure is able to form complexes with borate under suitable conditions of pH and concentration. This type of complex probably has the structure:

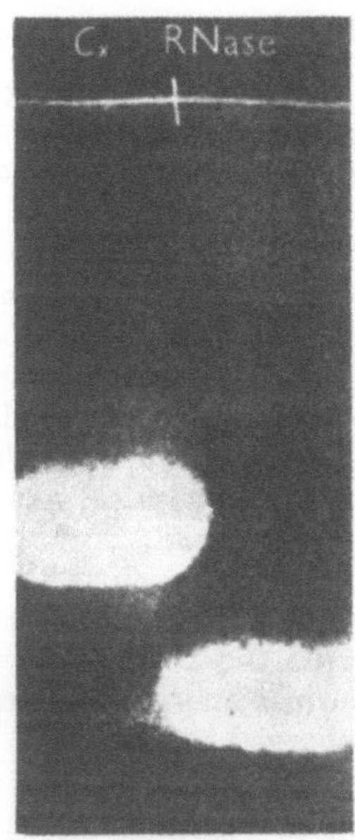

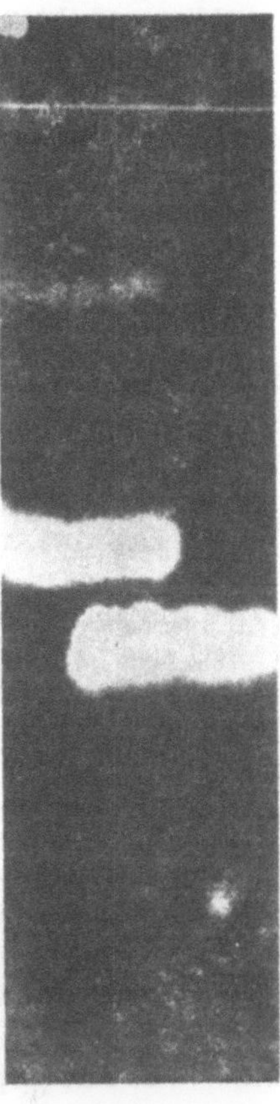

Fig. 34.

Fig. 35.

Fig. 34. The separation of cyclic cytidylic acid (left) from cytidine-3′-phosphate (right) by electrophoresis at pH 7.4. The former has only one phosphoric acid —OH⁻, compared with the two of the latter. [MARKHAM and SMITH: Biochem. J. 52, 552 (1952)]

Fig. 35. The separation of adenosine-5-phosphate (right), from adenosine-2′-phosphate (left), by electrophoresis in 0.05 M borate (pH 9.0). The former contains a *cis*-glycol and therefore forms a boric acid complex which is partly dissociated and thus adds an extra negative charge to the nucleotide, which does not happen with the corresponding 2′ or 3′ phosphates.

It will be noted that the effect of such complex formation is to add an extra acidic grouping to the nucleotide. At the pH at which this is normally done, pH 9.2, this extra grouping is about 50% dissociated, so that the ribonucleoside 5-phosphates carry 2.5 negative charges, compared with 2 negative charges for the 2′ and 3′-phosphates, and 1 negtive charge for the cyclic nucleotides. In addition, of course, certain of the nucleotides will carry a small extra charge from the partly dissociated ring —OH groups. An example of this separation is given in Fig. 35 which shows the separation in borate buffer of adenosine 5-phosphate from adenosine 3′-phosphate (JAENICKE and VOLLBRECHTSHAUSEN, 1952.)

6. The Separation of Nucleosides.

The separation of nucleosides by electrophoresis is similar to that of the nucleotides, but as they have no phosphate groups, only the amino groups need be considered. The amino group of guanosine is, however, so weak that it is almost impossible to separate this nucleoside from uridine in ordinary buffers excepting at pH values below pH 2. The optimum pH for separating adenosine, cytidine and this pair is about pH 3.3, where the mobility is about -1 for cytidine -0.5 for adenosine and 0.0 for guanosine and uridine.

The nucleosides, having a *cis* glycol on C2' and C3', form borate complexes, and then bear a negative charge, due to the borate, and also a small charge due to the dissociation of the enolic-OH groups of the bases. This separation in borate would seem to be of more use for differentiating between nucleosides and free bases than as an analytical method for resolving mixture of the former.

7. The Electrophoresis of the Bases.

The bases bear only two kinds of ionising substituents: the amino groups and the ring hydroxyl groups. Excepting in the case of the 8-azapurines, the latter groups are of little use in their separation and the amino groups are the greatest importance.

The only useful analytical separation which may be made by this method is the analysis of the mixture obtained by the acid hydrolysis of ribonucleic acid. This contains adenine, guanine, uridylic acid and cytidylic acid. This mixture when electrophoresed at pH 3.5 has the guanine more or less stationary, the adenine moving towards the negative electrode, and the cytidylic and uridylic acids moving towards the positive electrode. This separation may be made the basis of an extremely sensitive analytical method (EDSTRÖM, 1953) which may be used on less than 1 μg. of material.

In the case of the 8-azapurines, such as 8-azaguanine, the enolic $-OH$ group is fully dissociated at pH 9, so that on electrophoresis in borate buffer, these bases move rapidly towards the positive electrode. (MATTHEWS, 1954.)

The mobility of the purines and pyrimidines is about 15 cm./hr. at 20 V./cm. per one fully dissociated charged group

8. The Separation of Polynucleotides and Related Substances.

(MARKHAM and SMITH, 1952 b, c.)

The polynucleotides and their derivatives have ionising groupings very similar to those of the mononucleotides, but, because the structures are held together by secondary phosphate bridges, the proportion of secondary phosphate groups ionising may be equivalent only to a fraction of the available primary phosphate groups ionised, and may even be zero. In any case, the ionisation of the secondary phosphate groupings is only of consequence at pH values above 5, so that for many separations they may be disregarded. Consequently in a mononucleotide, a dinucleotide and a trinucleotide at pH 5, one has 1, 2 and 3 negative charges respectively per molecule. The resistance of these substances to motion through a liquid, however, is not as 1:2:3, but approximately 1:1.43:2, so that the net effect is for them to separate in increasing order of size as 1:1.4:1.5. This series converges, so that with higher polynucleotides, the mobility reaches a maximum just over 1.5 times that of the mononucleotide.

Naturally at pH 3.5, which is again the most useful general pH, this mobility is reduced by the presence of the amino groupings, and the general method of calculation is, as before, to sum the charges algebraically for the whole molecule, and for di- and trinucleotides to divide this charge by 1.43 and 2 respectively to allow for the increased frictional resistance. This then gives the ratio of the mobility to that of a substance having unit charge and unit frictional resistance — for which we have taken as a standard a mononucleotide, namely uridylic acid.

As an example, we may take the tri-nucleotide triadenylic acid. This has 3×0.54 positive charges and 3 negative charges, or a net negative charge of 1.38. The mobility is therefore 1.38/2 times (0.69 times) that of uridylic acid.

If we now raise the pH to 7.4, the triadenylic acid has 3 primary and 1 secondary phosphate group fully ionised, and uridylic acid 1 primary and 1 secondary group. The mobilities are therefore $4/2 = 2$ and 2 respectively, so that no separation is achieved. At pH 5.0 on the other hand, triadenylic acid has a charge of 3 and uridylic acid only of 1 and the mobilities then become as 1.5 : 1.

In these mobilities is included the flow of liquid in the filter paper, which has the effect of increasing the mobilities by a constant factor. This factor is also taken into account in Table 5 which tabulates the expected movement in 2 hours at pH 3.5 in a voltage gradient of 20 V./cm. for a number o

Table 5.

The *calculated mobilities* relative to uridylic acid of some nucleotides and polynucleotides at pH 3.5 and their *observed movement* in 0.05 M ammonium formate buffer pH 3.5 at 20 V./cm. The relative position of these substances on paper chromatograms in 70% *iso*propanol-NH$_3$ are also given (the bands are numbered 1—6 in order of increasing R_F value).

Substance	Calculated mobility relative to that of uridylic acid	Observed movement in 0.05 M ammonium formate buffer pH 3.5 (cm./ 2 H. at 20 V./cm.)	Position on the chromatogram in 70% *iso*propanol-NH$_3$(Bands numbered in order of increasing R_F value) band
A	0.46	8	4
G	0.95	14	2
C	0.16	6.5	4
U	1.00	16	4
A!	0.46	8	6
G!	0.95	14	5a (between 5 and 6)
C!	0.16	7	6
U!	1.00	16	6
AC!	0.43	9	5
AU!	1.02	16	5
UU!	1.4	22	5
AC	0.43	8	2
AU	1.02	16	2
AG	0.99	15	1
GC!	0.78	15	3
GU!	1.36	20	3
GC	0.78	13.5	1
GU	1.36	19.5	1
UUU!	1.50	24	5
ACC!	0.39	6	2
AAC	0.54	13.5	1
AAU	0.96	17	1
AGU	1.21	19	1

polyribonucleotides. The analogous deoxyribose compounds will, of course, have approximately the same mobilities. This table also indicates the position of these substances in *iso*propanol-NH$_3$ chromatograms, because the isolation of any one compound is best accomplished by a combination of both techniques.

The cyclic dinucleotides have a terminal 2':3'-phosphate residue, and consequently have no secondary phosphate group capable of ionising. They can be separated from non-cyclic dinucleotides by electrophoresis at pH 7.4.

The dinucleoside monophosphates may be treated in a similar way, but owing to the absence of a terminal phosphate residue, the resistance to movement is rather less than that of a dinucleotide, nor are there any available secondary phosphate groups. The *cis* glycol structure may be taken advantage of for electrophoresis in borate at pH 9.

The mononucleotide diphosphates have very high mobilities, their frictional constants being nearly those of mononucleotides. They are very characteristic in that they tend to more as extremely sharp, thin bands. Cyclic diphosphates may also be found, and may be identified as can other cyclic compounds by running at pH 7.4, if necessary after leaving in 0.1 N HCl overnight to open up the cyclic terminations.

9. Difficult Mixtures.

Some mixtures are exceedingly difficult to separate. One such mixture is that of the dinucleotide of guanine and cytosine (GC) and the trinucleotide of cytosine plus two adenine residues(AAC). This mixture runs at pH 3.5 as a single compound. By running at pH 5.0, however, one can take advantage of the difference of mobility between dinucleotide and a trinucleotide (in theory 1:1.07). Examination of the dissociation curves will also show that by dropping the pH to 2.5, the amino group of the guanylic acid becomes about half dissociated, and this holds the dinucleotide back but not as much as the trinucleotide which has little charge at this pH.

Similar difficult mixtures may be resolved by applying this type of reasoning before attempting to make a separation. In fact half an hour spent in considering the problem is often worth hours of random experiment.

H. Some Enzymes Used in Nucleic Acid Chemistry.

Many useful degradations in nucleic acid chemistry can only be accomplished by the use of specific phosphatases. Although little is known about phosphatases, it is fortunate that most of the enzymes useful in nucleic acid work are fairly well characterised and reasonably easy to make or obtain.

In the following section details of the use and preparation of a number of these enzymes is described.

1. Prostate Phosphomonoesterase. (Enzyme no. 1.)

This extremely useful enzyme is, unfortunately, only obtainable pure from the prostate glands of human beings or of the higher apes, although, of course, this and other phosphomonoesterases are obtainable from other sources, but accompanied by diesterases, which are undesirable for most purposes.

Hypertrophic human prostate glands are washed, freed of connective tissue and fat, and chopped into fine pieces with scissors. The tissue is then ground with sand in a mortar, 0.9% NaCl solution being added to make a smooth paste. The whole is then diluted with the salt solution (a volume of 50—100 ml. is adequate for one gland), chloroform added as a preservative (1—2 ml.), and the suspension incubated 24 hrs. at 37°. The liquid is then centrifuged to remove debris and an equal volume of $(NH_4)_2SO_4$ solution (sat.) added. The precipitate is discarded. The supernatant liquid is then taken to 75% saturation with $(NH_4)_2SO_4$, and the precipitate, which contains the enzyme, collected. This is redissolved and put through the $(NH_4)_2SO_4$ fractionation again. The enzyme is then dissolved (about 10 ml./gland) and dialysed. After centrifuging, it may be frozen dried. Solutions are adjusted to pH 5 (0.1 M acetate final concentration) and kept at 5° with a drop of chloroform as a preservative.

This enzyme is very active, and should be tested against known substances in order to assess the time and quantity to use. Its pH optimum is pH 5 and it is used at 37°. Suitable test materials are adenylic acid for the phosphomonoesterase activity, and the "core" from ribonuclease action as a control of diesterase activity.

Note: Prostatic tissue is very tough. It is quite useless to use a Waring Blendor for the preparation.

2. Deoxyribonuclease. (Enzyme no. 2.)
(McCarty, 1946.)

This enzyme isolated from pancreas is obtainable commercially. It is used at pH 7 in a solution containing about 0.01 M MgSO$_4$ at a concentration of 0.1 mg./ml. Very dilute solutions may be stabilised by the addition of 1% of gelatin.

3. Ribonuclease. (Enzyme no. 3.)

This enzyme, observed in pancreas by Jones (1920), was first purified by Kunitz (1940). It may be isolated by Kunitz's method but is now available commercially.

A suitable concentration for use is 0.1 mg./ml. in water. Its optimum pH lies around neutrality, but this is not critical. The solution may be heated to 80—90° in order to destroy proteolytic enzymes. The optimum temperature for ribonuclease is 50—55°.

Because this enzyme is so stable, it is difficult to stop its action, but by dropping acetone round spots applied to chromatograms, the action may be made to stop. This is probably due more to the removal of water than to the action of the acetone on the enzyme itself.

4. Barley 3'-Nucleotidase. (Enzyme no. 4.)

This enzyme was isolated from germinating barley by Shuster and Kaplan (1953) and is almost specific for nucleoside 3'-phosphates which it dephosphorylates rapidly. It is used at a pH of about 7.

500 g. barley is soaked 30 hrs. at 20° in water. It is homogenised in a Waring Blendor in 2 l. of water, and extracted for a short time (the authors used up to 3 weeks but say this is not essential). The fraction precipitating between 65—90% saturation with (NH$_4$)$_2$SO$_4$ contains the enzyme.

A further purification may be made by adsorbing on alumina C γ gel and then eluting in 15 ml. of M NaHCO$_3$. The enzyme is stable in a frozen dried condition.

Germinating rye grass is a better source of the enzyme.

5. Snake Venom. (Enzyme no. 5.)

Snake venom contains a number of enzymes including: 5'-nucleotidase, which is specific for dephosphorylating 5' nucleotides — but *not* nucleoside 3':5' nucleoside diphosphates (or the 2':5' diphosphates), a phosphodiesterase, which will split most phosphodiesters, but *not* 2':5' or 3':5' linked dinucleotides terminating with a 2' or a 3'-phosphate residue, and a pyrophosphatase.

For many purposes the snake venom may be used unpurified. Dried venom of the rattlesnake or Russel's viper is available commercially. It is used at pH 9.5—10 (grey to thymol blue), preferably adjusted with NH$_4$OH.

Note: Venom may interfere with electrophoretic separations if present in any quantity.

6. Snake Venom Phosphodiesterase. (Enzyme no. 6.)

This enzyme may be separated from the 5'-nucleotidase of the venom by adsorption of the latter on filter paper. Several methods have been published but these have proved capricious, possibly because snake venom is a poorly characterised material. A method for effecting the separation is as follows:

Take 5—10 mgs. of venom and dissolve in 0.5 ml. of water. This is placed in a line on a piece of Whatman No. 3 paper, and run as a chromatogram using pure water as the eluting fluid. The diesterase is found within 2 cm. of the solvent front, and is eluted in water.

The preparation must be tested against adenosine 5'-phosphate to ensure freedom from 5'-nucleotidase. The activity is best tested against DNA digested exhaustively with deoxyribonuclease. Nucleotides should be liberated without any nucleosides being produced. This is, of course, checked by chromatography on paper using 70% *iso*propanol-ammonia.

I. Determination of Phosphate and Carbohydrate.

I. The Analysis of Phosphorus.

The most generally useful and reliable method for analysing nucleic acid derivatives is that of ALLEN (1940). Like other colorimetric techniques it depends upon the formation of a blue complex on the reduction of phosphomolybdates in acid solution. In this method the acid used for hydrolysis is $HClO_4$ and the phosphomolybdate formed is reduced by amidol (2:4-diaminophenol hydrochloride).

Reagents.
a) $HClO_4$ 60% (w/w).
b) Ammonium molybdate 8.3% (w/v).
c) Amidol 2 gm., $NaHSO_3$ 40 gm., water to 200 ml. Filter and keep at 5°. This solution keeps for about 10 days.
d) Phosphorus standard (0.1 mg. P/ml.). KH_2PO_4, 439 gm. in 1 l. water. Add 1 ml. of chloroform as a preservative.

Method. For inorganic P and standards take 2 ml. *a*, 10 ml. of water and the P solution containing 0.01—0.1 mg. P in a 25 ml. cylinder. Make up to 20 ml. and add 1 ml. of *b*, mix and add 2 ml. of *c*. Make up to 25 ml. and read in a photoelectric colorimeter after 5 minutes and before 30 mins. against standards, using a deep red filter. The calibration curve is linear.

For purine bound P such as in adenosine 2'-phosphate, hydrolysis for 1 hr. at 100° in 2 ml. of *a* in 10—20 ml. final volume of water will be adequate. The rest of the procedure is as for inorganic P.

For total P and difficulty hydrolysed P the material is added to 2.2 ml. of *a* in a small KJELDAHL flask and incinerated for 30 mins. or more. Caution: These incinerations explode occasionally. This is infrequent but may be dangerous. After cooling the tubes proceed as for inorganic P without adding any more of *a*.

Modifications. These methods may be modified so as to use smaller volumes and micro cells in colorimeters. In this use may be made of the fact that 72% (w/w) $HClO_4$ (which is about 1.25 times as strong as 60%) hydrolyses all the nucleotides and their derivatives in 2 hr. at 100°. A small amount of carbon may form in the solution but this can generally be ignored. Volumes may be made constant by adding the various reagents with micropipettes, and the reagents may be changed in strength to permit this. Quantities of P of 0.2—1.0 μg. may be estimated in a total volume of 1 ml. using 1 cm. microcells.

II. The Diphenylamine Reaction for Deoxypentose.

This reaction is a qualitative test for deoxypentose and may be made quantitative under very special conditions. It is due to DISCHE (1930).
Reagents. 1 g. diphenylamine; 2 ml. conc. H_2SO_4; glacial acetic acid to 100 ml.

Procedure. To 1 volume of a solution containing deoxypentose or DNA or a purine containing derivative from DNA add 2 vols. of reagent and heat in a water bath at 100° for 3 mins. Cool at once. The purple colour is unstable, and the amount produced is interfered with by a number of substances. The reaction is only really reliable for quantitative purposes if pure DNAs or other pure derivatives are used. Only deoxyribose which is free or bound to purine is estimated. The pyrimidine nucleotides etc. do not give the reaction.

The reaction is not very sensitive and quantities of 0.25 mg. or thereabouts are needed.

III. The Estimation of Pentoses.

The quantitative estimation of pentoses may be carried out by a number of methods. Many of these, including one of the most popular, namely that due to MEJBAUM (1939), are unreliable. A method due to MILITZER (1946) has given us satisfaction for a considerable time, and this method shall be described in detail.

The principle of the method, which is common to many methods, is that pentoses react with HCl in the hot to give furfural. The latter reacts with orcinol in the presence of ferric ions to give a brilliant green colour, which is a function of the pentose concentration.

Solutions.

(a) Orcinol 1% in water. This solution will keep for several weeks in the refrigerator.

(b) Conc. HCl.

(c) 10% $FeCl_3 \cdot 6 H_2O$.

The reagent consists of 10 ml. (a), 40 ml. (b) and 1 ml. (c) and should be prepared fresh.

The pentose standard may be either a solution of "yeast adenylic acid" or 50 mg. xylose in 100 ml.

Method. 0.2 ml. of solution containing 5—50 μg. of pentose is mixed with 2 ml. of the reagent in a tube which is then stoppered and heated 8 mins. on a water bath at 100°. It is then cooled. Suitable standards are run at the same time.

The solutions may be read on a photoelectric colorimeter without dilution, using a red filter. The calibration is nearly linear. If the solutions are too strong they may be diluted to 10 ml. with *n*-butanol. This has the effect of suppressing the ionisation of the ferric ions and makes the colour bluish-green. When so diluted the colours may also be matched on a visual colorimeter.

Note: In the original paper MILITZER recommends 3 mins. heating. This is inadequate for combined ribose. Under the conditions described above ribose bound to pyrimidines does not react to any extent.

J. Other Methods for the Analysis and Determination of Nucleic Acids.

I. Analysis of Ribonucleic Acids by Means of Silver Precipitation.

(KERR, SERAIDARIAN and WARGON, 1949.)

As has been mentioned earlier purines form silver complexes in acid solution, whereas pyrimidines do not nor do their nucleotides. It is possible, therefore, to effect a separation of the two groups of substances by means of silver. As in the case of the precipitation by CuO, the silver precipitation is dependent upon the imidazole ring structure present in the purines.

Procedure. The ribonucleic acid is hydrolysed for 30 min. at 100° in 2 N H_2SO_4 in a volume of 5 ml. (about 50 mg. RNA). It is diluted to 25 ml. with water. 18 ml. are taken and 1 ml. of a 1 M suspension of AgO in water is added. The precipitate contains the purines and it is washed twice with 3 ml. volumes of water. The washings and the supernatant fluid contain the pyrimidine nucleotides. These are precipitated as silver salts on the addition of 3 volumes of *iso*propanol after adjusting the aqueous solution to an alkaline pH (pink to phenolphthalein). The two lots of silver salts are decomposed by adding a suitable volume of HCl.

The calculations are made as follows:—

Adenine and guanine are read in N HCl in which they both have a molar extinction coefficient of 7,100 at 276 mμ. Cytidylic acid and uridylic acid are read in 0.1 N HCl in which $\varepsilon = 9,300$ at 265 mμ.

A division of the pairs of substances may be accomplished by reading the combined spectra and applying suitable equations. This is to be avoided if possible, because the actual separations may be made with ease by chromatographic methods. The technique is only given as an illustration of a precipitation method. The application of the silver precipitation procedure to other problems is obvious. The main interference will come from chlorides and possibly phosphates present in extracts.

II. The Recognition of the Nucleic Acids in Tissues by Microscopical Methods.

The nucleic acids occur in tissues in two major locations which are to a large extent specific to the particular type of nucleic acid present. In general, deoxyribonucleic acid is confined to the nucleus and ribonucleic acid is mainly to be found in the cytoplasm and the nucleolus. This division is, of course, not absolute and there is evidence of the presence of ribonucleic acid in nuclei, and there is at least one authentic instance of the occurrence of deoxyribonucleic acid in the cytoplasm. In addition there may be cases where parasites having either or both types of nucleic acid occur in cells. These may be viruses or other small parasites, so it is possible that they may influence the distribution of these substances inside cells without any microscopically visible objects being noticed.

Staining techniques differentiate the two types of nucleic acid to some extent. Both ribonucleic acid and deoxyribonucleic acids are strong polyelectrolytes, having a large number of acidic groupings. In consequence they bind basic dyes very readily. In general deoxyribonucleic acids are more highly polymeric and so they have an increased affinity for such dyes.

One of the most generally used dye preparations is methyl green and pyronin. This mixture is supposed to differentiate deoxyribonucleic acid (which stains green) from ribonucleic acid (which stains orange) but this differentiation was based largely on observations of reasonably good deoxyribonucleic acids and extremely bad ribonucleic acids. In general a simple distinction cannot be made by these methods unless enzymes are used to remove one or other of the nucleic acids. Useful for this purpose are deoxyribonuclease and ribonuclease. The latter is best used after it has been heated to 80—90° for one or two minutes to destroy proteolytic enzymes.

A much more reliable method, but one which has also been criticised adversely by some authors is the Feulgen staining technique, which was used in the original demonstration of deoxyribonucleic acid in plant tissues. Feulgen (1914), in a systematic study of staining reactions, found that deoxyribonucleic acids hydrolysed at 60° in 0.1 N HCl had the ability to restore the colour of fuchsin which had been decolorised by sulphur dioxide, and Feulgen and Rossenbeck (1924) developed this technique into a staining method and by means of it demonstrated the presence of a nucleic acid other than ribonucleic acid in wheat germ [the ribonucleic acid from wheat germ, or triticonucleic acid as it was once called, was already well known from the work of OSBORNE and HARRIS (1902)].

The mechanism of the FEULGEN reaction is now well known, and if suitably used, the method would appear to be foolproof and certain. Under the conditions of the hydrolysis, deoxyribonucleic acid loses most or all of its purines leaving a polynucleotide chain interspersed with deoxyribose phosphate residues. The latter are predominantly in the aldehydic form and thus react with the SCHIFF reagent to give a coloured complex. If the reaction is properly carried out the nucleic acid is not disaggregated sufficiently to allow the complex to diffuse freely, and so the colour tends to remain in the site of the original unhydrolysed deoxyribonucleic acid. Naturally any aldehydic compounds will interfere with the reaction, but these are usually removed during the fixation and dehydration of the tissue.

Most plant cells unfortunately contain relatively little cytoplasm and a large vacuole, and it has been observed that the disorganisation which takes place during the hydrolysis is such that it makes the interpretation of the results obtained by this method extremely difficult if not impossible (CHAYEN and NORRIS, 1953).

The FEULGEN Staining Method. The material to be stained is fixed in one of the accepted ways, preferably not involving the use of formaldehyde. It is then hydrolysed in $1\ N$ HCl at 60° for 4 to 15 minutes. This breaks off purines and liberates the aldehyde groups. The material is stained by putting into a solution containing basic fuchsin decolorised by SO_2. This reagent is available commercially and may also be prepared in one of a number of ways. One recipe is as follows:

Boil 200 ml. of water, add 1 gm. basic fuchsin, cool to 50° and filter. Add 2 ml. $1\ N$ HCl. Cool to room temperature and add 1 g. anhydrous $NaHSO_3$ ($Na_2S_2O_5$). Allow to stand 24 hrs. before use. Some workers use activated charcoal to remove traces of colour.

After staining the solution is washed off in water containing some SO_2. The nuclei and substances containing deoxyribonucleic acid are stained reddish purple.

Ultraviolet Light Microscopy. As the nucleic acids absorb ultraviolet light strongly, ultraviolet light microscopy has been employed to detect and even to estimate the nucleic acid content of cells. This technique has reached its greatest refinement in the hands of CASPERSSON (1936). Sections or smears are photographed in light of 254 mμ wavelength, using microscopes having quartz optics or mirrors, and the plates are examined photometrically. The quantitative potentialities of such methods have been grossly overemphasised, but using less complex techniques ultraviolet microscopy may be used with success in observing cells, and if combined with suitable enzyme or chemical hydrolysis methods may give very useful results. It cannot, however, be too strongly emphasised that nucleic acids are not by any means the only substances which absorb ultraviolet light. The success of the method is due largely to the fact that in cells much of the nucleic acid is concentrated in relatively regular and easily recognised structures.

III. The Estimation of Nucleic Acids in Tissues.

All tissues contain nucleic acids and substances related to them. The estimation of such substances in plant tissues has however not been studied in much detail in recent years, and in consequence it is only possible to give an indication of the way in which one should proceed — and, in particular, what procedures to avoid.

The first matter to bear in mind is that the nucleic acids are extremely unstable substances, and are readily attacked by the enzyme systems which are distributed with prodigality throughout the vegetable and animal kingdoms. In consequence it is most unwise to subject tissues to any disintegration procedure, even at 0° C, in conditions in which enzymes may be active. In considering possible ways to avoid enzyme action, it may be remarked that the use of "specific" inhibitors is not of much general use. Examples of such procedures are to be found in the literature dealing with analogous problems in animal tissues. For example, the deoxyribonuclease of pancreas is an enzyme requiring Mg^{++} ions for its activity. Because of this, its activity may be inhibited by the addition of substances, such as citrate, which fix Mg^{++} in an unionised condition. Tissues other than pancreas also contain enzymes which will break down deoxyribonucleic acid, and a number of these have no requirement for Mg^{++}. The addition of citrate is however regarded as a palliative in such circumstances — erroneously, of course.

Probably the best procedure to follow is to drop the tissue into boiling ethanol. This is likely to destroy most enzymes encountered in plants, and is not a solvent for nucleic acids, although most of the small fragments such as mononucleotides will be dissolved in the final solution, which should be about 80% ethanol (v/v). An alternative procedure, which may be of service is to freeze the tissue at —10° C or less, grind in a cold mortar and to suspend the powder in about 10 times its weight of ethanol. This should effectively prevent any enzyme action.

It is customary when dealing with animal tissues to attempt to extract first "ribonucleoproteins" and then "deoxyribonucleoproteins", by extracting suspensions of ground tissues firstly with dilute NaCl solution (0.1 M) and then with stronger NaCl solutions (1 M). The first extraction removes ribonucleic acid and a protein fraction, and the second extraction deoxyribonucleic acid and more protein, which will precipitate on subsequent dilution of the extract with water. The rationale of such procedures is quite different from that visualised by their exponents. The nucleic acids are large polymers which have a structure resembling a strongly acidic (and also weakly basic) ion exchange resin, and these substances can (and do) combine readily with proteins when cells are disintegrated. The ribonucleic acids, being considerably smaller polymers than the deoxyribonucleic acids, are more readily displaced by the addition of extraneous ions, in this case derived from the NaCl (probably both Na^+ and Cl^- ions are involved in this displacement) and, in consequence, they are liberated from the relatively insoluble complexes which have formed during the disintegration process. Similarly the deoxyribonucleic acid-protein complexes are broken by the stronger salt solutions. The formation of such complexes does not, of course, imply any particular relationship between their components in the intact cells, although this is often claimed. The use of this type of procedure is to be avoided, if possible, because of the probability of enzyme action and consequent loss of material in the nucleotide and nucleoside fractions.

1. The Methods of Schneider and of Schmidt and Thannhauser.

The methods to be described in the next few paragraphs have been used very widely in work on animal tissues, where they have both proved most useful. As far as plant tissues are concerned they are to be avoided, and reasons for this will be evident. The procedures adopted can, however, be modified to give useful results in certain cases.

The SCHNEIDER (1945) method is as follows: The tissues to be analysed are treated with 10% cold trichloroacetic acid to remove acid soluble phosphorus compounds (such as mononucleotides, inorganic P, adenosine triphosphate, etc.). The phospholipids are then extracted by 80% ethanol, followed by 100% ethanol.

The defatted tissues are then boiled with 5% trichloroacetic acid, which extracts the nucleic acids (in a degraded state) more or less quantitatively. The relative amounts of each are then determined by sugar analyses using the diphenylamine reaction of DISCHE for deoxyribonucleic acids, and the Bial orcinol reaction for ribonucleic acids.

For animal tissues such a method is moderately successful if one bears in mind that only the purine combined sugars of nucleic acids give any reaction in these analyses, and so the composition of the nucleic acid has to be known (and usually is not). The diphenylamine reaction is also rather liable to interference. In plant tissues, however, the whole method breaks down because plants are full of substances which react with the Bial reagent, and they also contain substances reacting with diphenylamine, which, after all, is hardly a specific reagent, being used for the detection of substances as different from deoxyribose as ferrous ions and nitrites.

The SCHMIDT and THANNHAUSER (1945) method is based upon the differential stability of the two kinds of nucleic acid to alkali, and is part of a scheme for the analysis of animal tissues for the various P-containing materials. In this method the tissue is first minced and extracted with 7% trichloroacetic acid. It is then subjected to washing, and to a defatting procedure to remove phospholipids (ethanol-ether and methanol-chloroform in the hot), and dried.

The tissue is then extracted with 1 N KOH for 15 hr. at 37°. (*Note:* This extraction causes the deamination of part of the cytosine to uracil. This is unimportant in the SCHMIDT-THANNHAUSER procedure, but may be disadvantageous in methods adapted from it. The deamination may be avoided by the use of 0.33 N KOH at 37° or by making the extraction at 20° for 18 h.) This alkali treatment has the effect of making soluble almost all the tissue components. These include phosphate from phosphoproteins, nucleotides from ribonucleic acid, and partly degraded deoxyribonucleic acid. The latter is precipitated largely (but sometimes incompletely) by the addition of HCl to a final concentration of 0.6 N and trichloroacetic acid to a final concentration of 2.5%. This is done on a sample, and the precipitate contains the deoxyribonucleic acid P (= about 10% of the weight of the deoxyribonucleic acid).

Any inorganic phosphate from phosphoprotein is removed by adding $CaCl_2$ and $MgCO_3$ (DELORY, 1938) and estimated in the precipitate. The supernatant liquid contains all the ribonucleotides, which are also estimated from their P content.

This method has been criticised of recent years, because of its inherent assumption that the P containing constituents of tissues can be separated into this arbitrary grouping. There is now ample evidence that there are P containing substances which do not belong to these various categories. In the case of most animal tissues this is not a serious criticism, but in applications involving the use of radioactive isotopes, minor constituents having high specific activities may interfere greatly. The application of a method designed for use with animal tissues cannot be expected to be successful when applied without modification to plant and bacterial tissues. Certain aspects of the procedure are, however, extremely valuable, and the weakness of the method, which is largely due to the use of P analysis, may be eliminated by the use of more specific techniques for the estimation of the nucleic acid constituents.

2. The Method of Ogur and Rosen (1950).

This method, which has become popular recently, is an adaptation of the Schneider method designed particularly for application to plant tissues. As much of the estimation is carried out by spectrophotometric analysis of the fractions obtained, trichloroacetic acid, which absorbs heavily in the ultraviolet, is replaced by perchloric acid.

The tissues are first subjected to a washing procedure to remove small substances like mononucleotides, bases, etc. and fats (and incidentally nitrites and some other substances which react with diphenylamine), and they are then subjected to a differential extraction procedure with acid which is supposed to differentiate ribonucleic acid from deoxyribonucleic acid.

The procedure, with various comments including some as to possible sources of error, is summarised briefly as follows: — The tissue is first homogenised, during which process a certain amount of enzyme action may well take place. It is then extracted with 70% (v/v) ethanol and 70% ethanol containing 1% (v/v) of $HClO_4$. This will suffice to suppress further enzyme action and will remove purines, nucleotides and other absorbing material, sugars and nitrites. The tissue is then defatted by boiling with ethanol:ether (3:1, v/v) for 3 minutes. This extraction is repeated and the tissue is dried. The residue is then extracted with cold 0.2 N $HClO_4$ as rapidly as possible. This may have the effect of removing traces of material not already eliminated. The tissue is then extracted with 1 N $HClO_4$ at 4° for 3—4 hr. This treatment is said to hydrolyse ribonucleic acid quantitatively and to extract it without affecting the deoxyribonucleic acids. This has been questioned. The residue is extracted with 0.5 N $HClO_4$ for 20 mins. at 70°, and then with 2 N HCl. These two extracts are combined. The absorbing material in this extract is supposed to be exclusively deoxyribonucleic acid in the form of its hydrolysis products.

It is quite obvious that this method has its deficiencies. The differentiation between the extraction of (hydrolysed) ribonucleic acid and deoxyribonucleic acid is based upon a specificity of the hydrolytic method which the latter probably does not have (Di Stefano, 1952). The extracts also may not only contain absorbing materials derived from the nucleic acids, and it would appear likely that each particular plant tissue would have to be investigated thoroughly before it was safe to apply this method.

It is rather surprising that these authors have not used the hydrolysis method of Schmidt and Thannhauser for releasing the ribonucleotides, because this hydrolysis is extremely specific and relatively simple to carry out.

3. Tentative Suggestions for Approaching the Problem.

In view of the fact that little has been done in the way of investigating the analysis of the total nucleic acids in plant tissues, it might be worth while to summarise a few of the procedures which may prove useful in tackling this problem.

(1) The tissue should be treated in such a way that enzyme action is stopped at once. This can be effected by dropping the tissue into some three to four times its weight of boiling ethanol. This procedure will also serve to extract small components such as mononucleotides. The heating should not be prolonged beyond one or two minutes.

(2) Further extractions with aqueous solutions of pH 4 or with 50% ethanol with a trace of acetic acid to take it to a pH of below 5 may remove more material which would interfere, and should not hydrolyse the nucleic acid or extract it.

(3) Defatting may be advantageous, especially in such tissues as wheat embryo, and can be carried out with the use of ethanol-ether mixtures.

(4) The nucleic acids may be extracted by 1 N KOH for 18 hr. This splits the ribonucleic acid to mononucleotides. The deoxyribonucleic acid is extracted in a partially degraded state. The KOH is eliminated by neutralising exactly with $HClO_4$ and the solution is made slightly alkaline by adding NH_4OH. This is to avoid the loss of deoxyribonucleic acid when the $KClO_4$ is spun off. The deoxyribonucleic acid is then precipitated by the addition of one volume of ethanol and the adjustment of the solution to about pH 4 with acetic acid. The pellet which is centrifuged down contains the deoxyribonucleic acid (it may be necessary to add magnesium ions to cause the latter to precipitate).

(5) The ribonucleotides are best purified further. Solvent 1 is suitable for this purpose. Alternatively they may be resolved by paper electrophoresis, or by paper chromatography after hydrolysis, or by ion-exchange chromatography.

(6) The deoxyribonucleic acid may be hydrolysed to bases and chromatographed, or it may be converted to nucleosides by deoxyribonuclease followed by snake venom and then chromatographed.

(7) Alternative methods for the removal of ribonucleic acids could involve the use of enzymes such as ribonuclease.

(8) Intact nucleic acids may be obtained from some tissues by heating them with 1 M NaCl at pH 7—7.5 or by cold extraction at a more alkaline pH. The ribonucleic acids are reasonably stable to a pH of 9—10 at 20°.

(9) In some cases the ratio of thymine to uracil is a reliable indication of the proportions of the two nucleic acids.

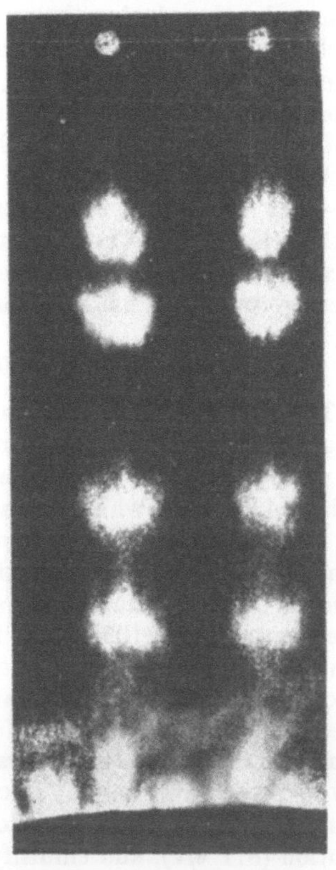

Fig. 36. The ribonucleic acid of Antirrhinum pollen separated into its constituents after the defatted pollen had been hydrolysed whole in N HCl for 1 hr. Solvent: *isopropanol*-HCl. From the top the spots are: Guanine, adenine, cytidylic acid, and uridylic acid.

(10) N HCl (or $HClO_4$) at room temperature for 24 hrs. will release the ribonucleotides, and also the purines from deoxyribonucleic acid. These may then be estimated by any of the methods described.

In the case of some plant tissues the quantity of ribonucleic acid is so great that a fractionation is hardly necessary. As an example, take certain of the pollens. Fig. 36 shows a chromatographic analysis of the pollen from *Antirrhinum*. The pollen was heated with 80% ethanol, dried and hydrolysed whole in N HCl at 100° for 1 hour. The solution was then chromatographed without any further treatment. The success of this particular experiment is due to the fact that pollens contain very little deoxyribonucleic acid. The sensitivity may be assessed by the fact that the pollen sample was obtained from six flowers.

K. The Isolation of Nucleic Acids from Plant Tissues.

The nucleic acids probably occur in cells largely in the free state, but they may often be extracted as "nucleoproteins", salt complexes of nucleic acid with proteins. It is unlikely that the proteins in such complexes are the same as those with which the nucleic acid is associated in the intact cell.

The ease of extraction of nucleoproteins from tissues is dependent upon the strength of combination of the two components which have to be dissociated from each other during the extraction procedure. As deoxyribonucleic acids are much larger substances than are the ribonucleic acids, the affinity which they have for large basic molecules, especially proteins is much greater, and in order to break up the salt complex, it is necessary to add a high concentration of ions to effect this dissociation. In the case of deoxyribonucleic acid-protein complexes 1—2 M NaCl is necessary for this dissociation. On diluting such an extract to 0.1—0.15 M, the complex forms again and precipitates from solution. This method (MIRSKY and POLLISTER, 1946) is the basis of many techniques for isolating deoxyribonucleic acids and has been applied to the extraction of the latter from wheat germ, which has a reasonable content of deoxyribonucleic acid.

I. The Isolation of Deoxyribonucleic Acid from Wheat Germ.

The material to be used must not have been baked or otherwise treated to denature the proteins and it must be defatted before any other manipulations are performed. This is best done with light petroleum. The germ is next washed with 0.14 M NaCl which extracts some of the soluble content including most of the ribonucleic acid. It is then suspended at 0° for 18 hr. in 1.0 M NaCl with occasional stirring. The solution rapidly becomes viscous and this makes the removal of the residue of the germ very difficult. Straining, followed by high-speed centrifuging (10,000 r. p. m. for 20 min.) is one possible method, but decantation is the simplest way if quantitative recoveries are not wanted. The extract is poured into 10 volumes of water, when the nucleoprotein precipitates in threads. These are collected by centrifuging or by collecting on a rod, and may be decomposed in a number of ways. Possibly the most convenient method is by the SEVAG procedure (SEVAG, LACKMANN and SMOLENS, 1938). To do this take a solution in 1 M NaCl and add an equal volume of chloroform:octyl alcohol solution (8:1 v/v), and emulsify in a Waring Blendor for 20 mins. Centrifuge off the chloroform emulsion and take the aqueous (upper) layer. Repeat this procedure until little or no emulsion forms. The aqueous layer then contains the nucleic acid more or less pure. It is diluted to 0.5 M NaCl and the nucleic acid is precipitated by the addition of 1—2 vols. ethanol. It may then be washed with alcohol and ether and stored in a dry state. Alternatively the salt may be removed by dialysis, but it is to be noted that nucleic acid is not precipitated from salt-free solutions by ethanol. The addition of a few drops of 0.1 M MgSO$_4$, however, will cause the nucleic acid to precipitate from such solutions. The main deficiency of this method is that enzyme action is not suppressed. Consequently the product may be partially degraded unless care is taken to keep the solutions cold.

It is possible that this method as described above would enable deoxyribonucleic acid to be isolated from tissues other than those of wheat germ and similar materials, but most plant tissues contain very little deoxyribonucleic acid, presumably because the nuclei of plant cells constitute such a small proportion of the whole. Deoxyribonucleic acid is probably mainly concerned with nuclear activity and is relatively static in non-dividing cells, but on the other hand, ribonucleic acid would appear to be connected with a number of metabolic activities. It is not surprising therefore that the predominant nucleic acid of plant tissues is ribonucleic acid. Reasonable quantities of partially purified ribonucleic acid may be isolated from many plants by a simple procedure.

II. Isolation of Ribonucleic Acid from Plant Tissues.

About 200 gm. of whole *Stellaria media* plants (or tobacco leaves, pea embryos, or *Brassica oleracea* leaves) are frozen (—10°) and minced in a chilled mincer, preferably with the addition of some solid CO_2 to keep the mixture cold. The powder has 500 ml. of ethanol added. The suspension is then raised in temperature until the alcohol boils. This removes the possibility of enzyme action during the later steps. The suspension is filtered off on a sintered funnel, and the solid has some 200 ml. of acetone layered above it in the funnel. Using very slow suction the acetone is allowed to percolate through the layer of tissue thus removing most of the leaf pigments. A second washing with acetone ensures complete removal of the latter. The tissue is then dried at room temperature. The acetone-free powder is suspended in 500—700 ml. of 0.5 M NaCl and brought to boiling, care being taken that the suspension is neutral to bromothymol blue. After boiling for 5—10 mins. the solution is filtered. This may present difficulties unless thick pleated filters are used. The filtrate has 50% (v/v) ethanol added, and enough acetic acid to make it acid to methyl red. Nucleic acid precipitates and is centrifuged down. This precipitate also contains polysaccharide, most of which is soluble in aqueous solutions when they are acid, but which is insoluble in dilute NaOH or KOH (0.01—1.0 N). Most of the polysaccharide may be removed by dissolving the precipitate in water plus sufficient 1 N NH_4OH to make the solution neutral to bromothymol blue. It is centrifuged to remove any solid and 1 N HCl is added slowly until a precipitate forms. (The quantity used should not exceed 0.1 vol.) A few drops of 0.1 M $MgSO_4$ will help in the precipitation. The precipitate is fairly pure ribonucleic acid. Any contaminants present will probably be insoluble in 1 N KOH, in which the nucleic acid may be hydrolysed for analysis.

This method produces a ribonucleic acid which can be taken as fairly representative in its composition to the material in the plant, but it has certain undesirable features. Nucleic acids should not be subjected to such high temperatures if that can be avoided and the pH used for the precipitation is possibly rather low, although at 20° it will not in fact break any of the links known to be present in ribonucleic acids. These preparations of ribonucleic acid do not appear to contain any deoxyribonucleic acid. An analysis of a specimen of the ribonucleic acid of *Stellaria media* gave the following figures:

Adenylic acid,	21.8%	
Guanylic acid,	29.6%	of the nucleotide residues.
Cytidylic acid,	23.7%	
Uridylic acid,	24.8%	

L. Other Purines and Pyrimidines in Plants.
I. Vicine and Divicine.

The seeds of some leguminous plants, notably *Vicia sativa*, the vetch, contain an unusual pyrimidine derivative which has also been reported to occur in the sugar beet. This substance is the β-D-glucopyranoside of the pyrimidine divicine, which for many years has been regarded (LEVENE and SENIOR, 1916) as 4:6 dioxy, 2:5-diaminopyrimidine. Recently BENDICH and CLEMENTS (1953) have given considerable evidence that the structure of divicine is in fact 2:4 diamino, 5:6-di-oxypyrimidine (x), and that it is linked to the sugar by an O-glycosydic linkage to C 5 of the pyrimidine.

Vicine is extracted from the ground seeds of vetch by 2—5% H_2SO_4 (v/v) in water. The extract is neutralised with $Ba(OH)_2$ and filtered. An excess of a solution containing 10% $HgSO_4$ in 5% H_2SO_4 is then added. The solution is neutralised with $Ba(OH)_2$ and the precipitate is filtered off and suspended in water, excess $BaCO_3$ being added to the suspension. This is treated with H_2S to decompose the mercury compound of vicine. The filtrate is concentrated *in vacuo* to a small volume and ethanol added as long as a precipitate is formed. The ethanol suspension is brought to the boil and filtered. The solution containing the vicine is concentrated *in vacuo*, when the compound crystallises out. Vetch seeds contain some 0.25% by weight of vicine.

Analysis. Vicine is easily recognised by its various properties as described above. In *iso*propanol-NH_3 (solvent 1) it runs just ahead of adenylic acid, and may be detected by its absorption of ultraviolet light, and by giving a brown spot when sprayed with 0.1 M aniline hydrogen phthalate (Partridge, 1949) followed by heating at 105° for a few minutes.

On electrophoresis vicine moves towards the negative electrode at pH 3.5, the mobility being 3 cm./hr. at 20 V./cm. Again the spray with aniline hydrogen phthalate will distinguish it readily from other compounds having similar ultra-violet absorption.

For quantitative analysis chromatography is to be recommended. $\varepsilon = 16.4 \times \times 10^3$ in 0.1 N HCl, the absorption peak being at 274 mμ.

Vicine is readily hydrolysed by 1 N HCl at 100° to give divicine and glucose. The base itself is very unstable.

II. The Methylated Xanthine Derivatives.

Three methylated derivatives of xanthine are found naturally in plants, namely 1:3 dimethyl xanthine (q) or theophylline; 3:7 dimethyl xanthine (r) or theobromine; and 1:3:7 trimethyl xanthine or caffein (s). These occur in the leaves and fruits of certain plants either singly or together, and in part at least would appear to form compounds with other leaf constituents. The recognition of these purines presents little difficulty because they are easily separated chromatographically, the best general solvent system being (Table 2, p. 264) butanol-formic acid.

The main difficulty in quantitative analysis of these compounds is presented by the extraction procedure and by the presence of foreign substances, notably tannins, in the extracts. The latter are relatively efficiently removed by MgO. Of the three compounds, theobromine, which occurs in the cacao bean to the extent of some 3% by weight, is perhaps the most difficult to estimate, largely because of its limited solubility in most solvents.

The usual methods of estimation which have been applied until recently have been based upon selective extraction procedures, followed generally by a gravimetric determination, although in certain cases titrimetry has been used. As the methylated purines have characteristic and well defined spectra (Figs. 37, 38, 39), however, it would appear that chromatographic separations could be employed to advantage, and a procedure for estimating the caffein content of coffee has been worked out in detail by Kogan, Di Carlo and Maynard (1953).

The total content of xanthine bases in plants of commercial importance varies from about 0.5% to 3% of the dry weight.

1. The Determination of Caffein in Tea and Coffee.

The specimen to be analysed is ground to pass a 30 mesh sieve. 10 g. are taken, moistened with ethanol and transferred to a SOXHLET apparatus and extracted 8 hr. with ethanol. Transfer the extract with the aid of hot water to a porcelain dish containing 10 g. heavy MgO suspended in 100 ml. of water. Evaporate slowly to dryness, stirring frequently. Pour boiling water on the residue and stir into a paste. Transfer to a filter with boiling water, and filter, collecting 250 ml. filtrate in all. Add 20 ml. 10% (v/v) H_2SO_4 and boil gently for 30 min. Cool and filter into a separating funnel and wash in with small portions of 0.5% H_2SO_4. Extract with 6 successive 25 ml. portions of chloroform and combine the extracts. Wash the

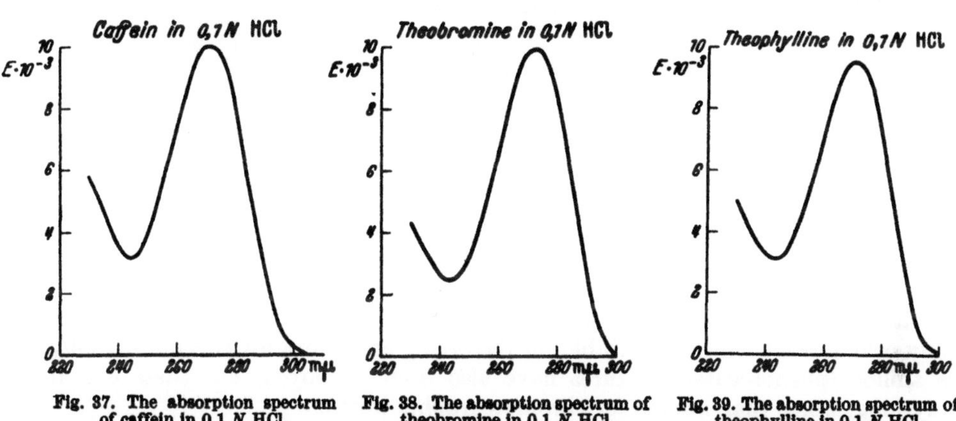

Fig. 37. The absorption spectrum of caffein in 0.1 N HCl.   Fig. 38. The absorption spectrum of theobromine in 0.1 N HCl.   Fig. 39. The absorption spectrum of theophylline in 0.1 N HCl.

extracts with 5 ml. 1% KOH solution. Filter the chloroform solution into an Erlenmeyer flask and add two 10 ml. chloroform extracts made from the 1% KOH solution. Evaporate extracts down to a small volume and transfer quantitatively to a weighed beaker. Dry at 100° for 30 min. and weigh. Check the purity by determining total N (factor = 3.464) (Official Methods of Analysis, Assn. of Official Agricultural Chemists, 7th edn., Washington, 1950).

This method may be modified to allow chromatographic determination of the bases. The alcohol extract is made as above, and this is shaken up with heavy MgO, which removes some of the colour. The filtrate from this extraction contains the caffein (and the theophylline if present). Measured volumes of the alcohol are chromatographed in *iso*-propanol-HCl or butanol-formic acid where the caffein runs ahead and the theophylline behind.

2. The Determination of Caffein in Coffee.

(KOGAN, DI CARLO and MAYNARD, 1953.)

In this method the caffein is extracted by dilute acid, the solution clarified, and the quantity of the base determined chromatographically.

Procedure. 10 g. of coffee beans are ground finely. 200 ml. of water and 50 ml. 0.05 N H_2SO_4 are added and the mixture boiled for 20 min. 25 g. MgO (heavy) are added and the mixture boiled for a further 20 min. The extract is cooled and made up to 285 g. weight (approx. 250 ml. of aqueous phase). The solution is filtered and 0.06 ml. are taken for each chromatographic spot ($3 \times 20 \mu l.$). The chromatograms are developed in the *iso*propanol-HCl solvent of WYATT (conc. HCl, 44 ml., *iso*propanol, 170 ml., water to 250 ml.). Caffein has an R_F value of 0.78. An additional spot is also given by trigonelline ($R_F = 0.59$, $E = 297$ for 10 mg./ml. at 297 mμ). The caffein absorption maximum in 0.1 N HCl is at 272 mμ. $\varepsilon = 9.9 \times 10^3$.

The solvent system used does not give a very good separation of caffein from other methylated xanthines, and might with advantage be replaced by butanol-formic acid [n-butanol, 86, water, 14, HCOOH 10 (v/v/v)] (Table 2, p. 264).

3. The Determination of Theobromine.

Theobromine is found in cocoa, which contains up to 3% by weight and is found in the leaves and other tissues of *Theobroma cacao*. It is also to be found in small quantities in other plants notably cola.

As has already been indicated, this purine has solubility properties which make its quantitative estimation difficult. The solubility at room temperature and in suitable solvents at their boiling points are:

Solvent	Cold (15.5°) %	Boiling %
Trichloroethylene . .	0.09	0.87
Water	0.06	0.70
Aniline	0.65	8

It will be seen that the choice of solvent is somewhat limited, and that the only cold solvent which is suitable for dissolving specimens for chromatography is aniline (phenol-water mixtures have also been employed, but these are, if anything, worse than aniline for chromatographic work).

WADSWORTH's (1921) Method for Estimating Theobromine in Cocoa. The material is ground very fine, and if fatty is extracted with light petroleum (B. pt. less than 80° C) which does not remove appreciable quantities of theobromine (but some of the caffein, which is present to the extent of about one third the weight of the theobromine). Cocoa itself does not need to be defatted.

Mix 10 g. of the powder with 2—3 g. of freshly calcined MgO (heated to 900° C and kept in a stoppered bottle). To the mixture add 14 ml. water and mix to a uniform damp paste. Heat on a water-bath with occasional mixing for 30 min. This treatment serves to release bound theobromine.

The paste, which should not have been allowed to dry completely, is transferred to a flask, 150 ml. of trichloroethylene added and the mixture boiled under reflux for 30 min. The extract is filtered off and the residue is extracted for another 30 min. with 120 ml. boiling trichloroethylene. This serves to reduce the effect of the slight adsorption of the theobromine by the MgO. The filtrates are combined and the trichloroethylene is distilled off down to 3—5 ml. On cooling 60—70 ml. of ethyl ether are added, and the precipitate formed contains the theobromine. This is filtered off after standing overnight and weighed. A minor correction is made to allow for the solubility of theobromine in ethyl ether (4 mgs. in 70 ml. at room temperature).

Chromatographic Procedure. WADSWORTH's method lends itself to a chromatographic estimation procedure in which both the theobromine and the caffein are estimated independently.

The extraction with trichloroethylene is carried out as before, but the distillation is taken to completion *in vacuo*. A measured quantity of aniline (freshly distilled) is then added to the dry residue to give a solution containing some 50—100 μg. of base per 20 μl. The flask is warmed to facilitate solution.

20 μl. portions are placed on Whatman No. 1 paper, and the paper is allowed to dry in a current of warm air until the smell of aniline disappears. Most of the aniline can be removed by this treatment. The chromatograms are best run in *iso*propanol-NH$_3$ (solvent 1), in which theobromine has an R_F of 0.6 and caffein an R_F of 0.8. During the chromatographic process the remainder of the aniline disappears from the paper and does not interfere with the estimation of the bases.

Chromatograms may also be run in *iso*propanol-HCl (p. 266), but in this case the aniline (hydrochloride) runs in the position of caffein. The theobromine spot (R_F 0.6), however, is well resolved and runs quite clear of the aniline spot ($R_F = 0.78$).

References.

ALLEN, R. J. L.: Biochem. J. **34**, 858 (1940). — ALTMANN, R.: Arch. Anat. u. Physiol. 526 (1889).

BENDICH, A., and G. C. CLEMENTS: Biochim. Biophys. Acta **12**, 462 (1953). — BROWN, D. M., M. FRIED and A. R. TODD: Chem. and Ind. **1953**, 352. — BROWN, D. M., and A. R. TODD: J. Chem. Soc. **1952**, 44. — BUCHANAN, J. G., C. A. DEKKER and A. G. LONG: J. Chem. Soc. **1950**, 3162.

CARTER, C. E.: J. Amer. Chem. Soc. **72**, 1466 (1950). — CAPUTTO, R., L. F. LELOIR, C. E. CARDINI and A. C. PALADINI: J. Biol. Chem. **184**, 333 (1950). — CASPERSSON, T.: Skand. Arch. Physiol. **73**, Suppl. 8 (1936). — CHAYEN, J., and J. P. NORRIS: Nature (Lond.) **171**, 472 (1953). — COGHILL, R. D.: J. Biol. Chem. **90**, 57 (1931). — COHN, W. E.: J. Amer. Chem. Soc. **72**, 1471 (1950); **72**, 2811 (1950); J. Cell. Comp. Physiol. **38**, Suppl. 1, 21 (1951). — COHN, W. E., D. G. DOHERTY and E. VOLKIN: Phosphorus Metabolism. **2**, 339 (1952). — COHN, W. E., and E. VOLKIN: Nature (Lond.) **167**, 483 (1951); Arch. Biochem. **35**, 465 (1952).

DI STEFANO, H. S.: Science (Lancaster, Pa.) **115**, 316 (1952). — DISCHE, Z.: Mikrochemie **8**, 4 (1930). — DELORY, G. E.: Biochem. J. **32**, 1161 (1938). — DAVIDSON, J. N., and R. M. SMELLIE: Biochem. J. **52**, 599 (1952).

EDSTRÖM, J. E.: Nature (Lond.) **172**, 809 (1953).

FEULGEN, R.: Hoppe-Seylers Z. **92**, 154 (1914). — FEULGEN, R., and H. ROSSENBECK: Hoppe-Seylers Z. **135**, 203 (1924).

HANES, C. A., and F. A. ISHERWOOD: Nature (Lond.) **164**, 1107 (1949). — HOLIDAY, E. R., and E. A. JOHNSON: Nature (Lond.) **163**, 250 (1949). — HOTCHKISS, R. D.: J. Biol. Chem. **175**, 315 (1948). — HUGHES, D. E., and D. H. WILLIAMSON: Biochem. J. **48**, 487 (1951). — HURST, R. O., and G. C. BUTLER: J. Biol. Chem. **193**, 91 (1951).

JAENICKE, L., and I. VOLLBRECHTSHAUSEN: Naturwiss. **39**, 87 (1952). — JAENICKE, L., and K. VON DAHL: Naturwiss. **39**, 87 (1952). — JOHNSON, T. B.: J. Amer. Chem. Soc. **59**, 1261 (1937). — JONES, W. E.: Amer. J. Physiol. **52**, 203 (1920). — JONES, W. E., and M. E. PERKINS: J. Biol. Chem. **62**, 291 (1924/25).

KERR, S. E., K. SERAIDARIAN and M. WARGON: J. Biol. Chem. **181**, 761 (1949). — KOGAN, L., F. J. DI CARLO and W. E. MAYNARD: Anal. Chem. **25**, 1118 (1953). — KUNITZ, M.: J. Gen. Physiol. **25**, 15 (1940).

LEVENE, P. A., and S. A. HARRIS: J. Biol. Chem. **98**, 9 (1932). — LEVENE, P. A., and W. A. JACOBS: J. Biol. Chem. **12**, 411 (1912). — LEVENE, P. A., and J. K. SENIOR: J. Biol. Chem. **15**, 60 (1916). — LÖFGREN, N., and B. LÜNING: Acta Chem. Scand. **7**, 225 (1953).

McCARTY, M.: J. Gen. Physiol. **29**, 123 (1946). — MARKHAM, R., and J. D. SMITH: (a) Nature, (Lond.) **163**, 250 (1949); (b) Biochem. J. **45**, 294 (1949); **49**, 401 (1951); (a) Biochem. J. **52**, 552 (1952); (b) Biochem. J. **52**, 558 (1952); (c) Biochem. J. **52**, 565 (1952). — MARSHAK, A., and H. J. VOGEL: Fed. Proc. **9**, 85 (1950). — MATTHEWS, R. E. F.: J. Gen. Microbiol. **10**, 521 (1954). — MEJBAUM, W.: Hoppe-Seylers Z. **258**, 117 (1939). — MERRIFIELD, R. B., and D. W. WOOLEY: J. Biol. Chem. **197**, 521 (1952). — MIESCHER, F.: Hoppe-Seylers Med. Chem. Unters. **4**, 441 (1871). — MILITZER, W. E.: Arch. Biochem. **9**, 85 (1946). — MIRSKY, A. E., and A. W. POLLISTER: J. Gen. Physiol. **30**, 117 (1946). —

OGUR, M., and G. ROSEN: Arch. Biochem. **25**, 262 (1950). — OSBORNE, T. B., and I. F. HARRIS: Hoppe-Seylers Z. **36**, 85 (1902).

PARTRIDGE, S. M.: Nature (Lond.) **164**, 443 (1949).

Schmidt, G., and S. J. Thannhauser: J. Biol. Chem. 161, 83 (1945). — Schneider, W.: J. Biol. Chem. 161, 293 (1945). — Schultze, E., and E. Bosshard: Hoppe-Seylers Z. 10, 80 (1886). — Shuster, L., and N. O. Kaplan: J. Biol. Chem. 201, 535 (1953). — Sevag, M. G., D. B. Lackmann and J. Smolens: J. Biol. Chem. 124, 425 (1938). — Sinsheimer, R. L., and J. F. Koerner: J. Amer. Chem. Soc. 74, 283 (1952). — Smith, J. D., and R. Markham: Biochem. J. 46, 509 (1950). — Smith, K. C., and F. W. Allen: J. Amer. Chem. Soc. 75, 2131 (1953). — Sumi, M.: Biochem. Z. 195, 161 (1928).

Vischer, E., and E. Chargaff: J. Biol. Chem. 176, 715 (1948). — Volkin, E., J. X. Khym and W. E. Cohn: J. Amer. Chem. Soc. 73, 1533 (1953).

Wade, H. E., and D. M. Morgan: Nature (Lond.) 171, 529 (1953). — Wadsworth, R. V.: Analyst 46, 32 (1921). — Wyatt, G. R.: Nature (Lond.) 166, 237 (1950); (a) Biochem. J. 48, 581 (1951); (b) Biochem. J. 48, 584 (1951). — Wyatt, G. R., and S. S. Cohen: Nature (Lond.) 170. 1072 (1952).

Also:
Official Methods of Analysis, Assn. of Official Agricultural Chemists, 7th. edn. Washington 1950.

Adenosine Diphosphate, Adenosine Triphosphate.

By

Harry G. Albaum.

With 1 Figure.

The increasing importance for animals and bacteria of adenosine diphosphate (ADP) and adenosine triphosphate (ATP) in energy-transfer mechanisms involving such widely diverse processes as muscular contraction, nerve conduction, fire fly luminescence, coenzyme syntheses, tumor growth, polysaccharide synthesis and many other key reactions (McELROY and GLASS, 1951, 1952) has focused attention, in recent years, on the possible role of similar compounds in higher plants. The main difficulties encountered in demonstrating these effects in plants have centered around the problem of isolating and characterizing these compounds from plants. Methods developed primarily for the isolation of these compounds from animal tissues, as will become apparent later, have proven unsatisfactory for the isolation of these compounds from plants.

However, in spite of these difficulties, it has been possible to show that animal ATP can participate in certain enzyme reactions in plants and, conversely, that partially purified preparations of plant ATP can function in much the same way in animal tissues. So, for example, ADP and ATP from animal tissues can act as phosphate acceptors during the oxidation of KREBS cycle intermediates by mitochondria prepared from the hypocotyls of mung beans (BONNER and MILLERD, 1953), and cauliflower [LATIES, 1953 (1) and LATIES, 1953 (2)]. Mung bean mitochondria, supplemented with ATP from animal sources in the presence of spinach grana and light are able to carry out photosynthetic phosphorylation (VISHNIAC and OCHOA, 1952). ATP from animal sources can satisfy one of the co-factor requirements in a flavin-adenine dinucleotide, ascorbic acid system from pea leaf lyophilisates which decomposes fructose diphosphate (TEWFIK and STUMPF, 1951). ATP, again from animal sources, can be utilized in an isolated phospho-hexokinase system from pea meal (AXELROD, SALTMAN, BANDURSKI and BAKER, 1052). In this reaction fructose 6-phosphate is phosphorylated to fructose 1—6 diphosphate. The ability of ATP from animal sources to function in this way has also been demonstrated for the hexokinase system from higher plants (SALTMAN, 1953).

On the other hand, the adenine nucleotide requirement of the "Cyclophorase" system can be satisfied by an adenosine polyphosphate isolated from mung beans (CROSS, TAGGART, COVO and GREEN, 1949). This preparation can also be substituted for animal ATP in the hexokinase system isolated from yeast (ALBAUM, OGUR and HIRSHFELD, 1950). A closely related nucleotide from oats is able to function in the phosphorylation of pyridoxal to pyridoxal phosphate in the tyrosine decarboxylase system of bacteria (ALBAUM and OGUR, 1947).

It is clear from the experiments which have been cited above that plant and animal adenosine polyphosphates can be substituted for each other in a wide variety of reactions.

A. Occurrence and Distribution.

In the "Cyclophorase" system referred to earlier, in fresh preparations which possessed activity, ADP and ATP appeared to be firmly bound to the mitochondria. Exhaustive washing with isotonic KCl did not remove them. When the preparation, however, was treated with agents like trichloracetic acid, the adenine nucleotides along with other co-factors were split away (Albaum, 1949). The loss of adenine nucleotide from such preparations also occurs normally on aging, even at $0°$ C, after several hours. Maximal activity can be restored to such aged preparations by the addition of AMP, ADP or ATP and Mg ions (Green, 1951). Fresh preparations of plant mitochondria like those of animals are able to oxidize pyruvate completely to CO_2 and H_2O without the addition of adenine nucleotides (Millerd, Bonner, Axelrod and Bandurski, 1951). However, the rate of oxidation can be increased by the addition of both Mg ions and adenine nucleotides (Millerd, 1953). One may conclude from these observations that at least part of the ATP normally found in both plant and animal tissues is "intimately" bound to mitochondria. This nucleotide can apparently be split away from such formed elements quite readily with protein precipitants like trichloracetic acid.

There are no figures available at the present time on the relative distribution and concentration of the adenine nucleotides in the various portions of the plant, e. g. root, stem, or leaf. Some data collected in our laboratory show that during germination of the mung bean there is a progressive increase in the ATP level of the seedling. Between the time of soaking and the fourth day of growth at $28°$ C, the ATP content (expressed as adenine) increases from about four to twelve micrograms per seedling (Albaum, 1952) (1).

B. Isolation of Adenosine Triphosphate.

I. Difficulties in Isolation.

Several methods are available for the isolation of ATP from vertebrate tissues (Needham, 1942; Kerr, 1941; Szent-Györgi, 1947; Dounce, Rothstein, Thannhauser, Beyer, Meier and Freer, 1948). Some of these have been used with varying degrees of success on tissues of invertebrates, a few bacteria and yeast. In the technique which has heretofore been most widely used for the isolation of ATP from rabbit muscle (Needham, 1942), the tissue is extracted with trichloracetic acid; the extracts are neutralized and treated vith barium. Under these conditions, ADP and ATP are precipitated as the barium salts along with several other phosphorus-containing compounds. The barium precipitate is now redissolved in dilute nitric acid and treated with mercuric nitrate. The mercury salts of the nucleotides which are precipitated are collected, freed of mercury with hydrogen sulfide and once more precipitated as the barium salts. Most of the other procedures are essentially modifications of this one in one form or another. This original Needham procedure has been used extensively for the preparation of most of the ATP isolated for laboratory work.

Our first attempts to apply this method and its various modifications to the isolation and characterization of ATP from plant sources met with little success. A brief discussion of some of the difficulties encountered is of some import for other investigators. The tissue used was the oat seedling. This was selected because of the author's initial interest in mechanisms of auxin action, for which the oat had been extensively used. When oat grains which had been soaked in water for several hours or one or two day-old seedlings were used, it was almost impossible to obtain clear trichloracetic acid extracts. This, as it turned out, was due largely to the presence of starch which appeared to carried along in all filtrates and extracts in colloidal form. This difficulty became less troublesome as the seedling grew older and the starch stores became depleted. However, even in the oldest seedlings used (7 days) the presence of starch interfered with the isolation procedure. Where the extracts were completely clarified, the adenine nucleotides were found to be present in very low concentration. We came to the conclusion that the adenine nucleotides had become adsorbed on the starch particles to such an extent that when we were able to remove the starch we also removed most of the nucleotide.

In those cases, where we were able to get rid of the starch and obtained clear extracts, a new kind of difficulty was encountered. Oats and grasses in general contain large amcunts of phytic acid. This unique compound, which probably has a role in the storage of inorganic orthophosphate in the grain is precipitated from trichloracetic acid extracts as the barium salts under the same conditions as is ATP. This similarity extends to the precipitation of both compounds as the mercury salts. Since the phytic acid is not normally encountered in animal tissues, one need not be concerned with it. However, in plant tissues the concentration of phytic acid, especially in the younger seedlings, is many times in excess of the adenine nucleotides. We were able to get around this problem, in part, by taking cognizance of the fact that barium phytate is more insoluble than barium ATP. Barium phytate is completely insoluble at pH 4. Barium ATP, on the other hand, is still partly in solution up to about pH 6.8. The removal of barium phytate at pH 4.0 made it possible to demonstrate in oat seedlings an ADP-like compound with some of the properties of ATP (ALBAUM and OGUR, 1947).

II. Criteria for Purity.

The identification of ATP is based on the ratio of adenine; pentose; labile phosphorus; stable phosphorus. In ATP the ratio of labile phosphorus: stable phosphorus is 2 to 1. Labile phosphorus is that phosphorus which is split from ATP in seven minutes at 100° C in N hydrochloric acid. The lability of this phosphorus is due to the presence of pyrophosphate linkages in contrast to the normal ester linkage which is involved in the attachment of the stable phosphorus to the 5-carbon of pentose. In the early literature, the presence of readily-hydrolizable (labile) phosphorus was taken as an index of the presence of ATP. This interpretation, especially when applied to crude plant extracts must be treated with extreme caution. Trichloracetic extracts of plant tissues contain many compounds whose phosphorus can be split away at least in part in seven minutes at 100° C in HCl. These include fructose-1:6-diphosphate which loses approximately 25% of its phosphorus under these conditions, glucose-1-phosphate which gives up all of its phosphorus, arginine phosphate, triphosphopyridine nucleotide, and many others. In phytic acid, the phophorus is extremely stable. However, when large amounts of this compound are present, some phosphorus is split away even after a 7 minute heating period. One might conclude that labile phosphorus was present. Actually, nothing could be further from the truth. The criterion of phosphorus lability can be applied only where one is dealing with a compound which is relatively pure and where data are available on stable phosphorus as well. The ratio of labile to stable phosphorus must be 2 to 1. In ADP, the ratio is 1 to 1. Where there is a mixture of both compounds and where other substances have been ruled out, the ratio of labile to stable phosphorus is somewhere between 2 to 1 and 1 to 1.

Phosphorus Analysis. In our laboratory, phosphorus analyses are carried out according to the method of Fiske and Subarrow (1929) with slight modification. Inorganic orthophosphate is determined by making the aliquot (usually under 1.0 ml.) to 8.6 ml. with H_2O, followed by the addition of 1.0 ml. 2.5% ammonium molybdate in 5 N H_2SO_4 and 0.4 ml. of reducer. (The reducer is prepared by grinding 0.5 g. 1-amino 2-naphthol-4-sulfonic acid in 15% $NaHSO_3$, making to a volume of 195 ml. with 15% $NaHSO_3$, adding 5 ml. 20% Na_2SO_3 and warming until all the reagents are in solution.) Ten minutes after the reducer is added, the color is measured in a photoelectric colorimeter using a 6600 Å filter. We have employed the Evelyn as well as the Klett-Summerson Colorimeters for our work. A standard is run using inorganic orthophosphate in the range of 5—50 micrograms.

Labile phosphorus is measured by making the sample to 1.0 ml. with water, adding 1.0 ml. 2 N HCl and heating in a boiling water bath for 8 minutes. It takes approximately 1 minute for the tube contents to come to bath equilibrium so that the actual heating time at 100° C is 7 minutes. After cooling, water to 8.6 ml. and 1.0 ml. of 2.5% ammonium molybdate in 3 N HSO_4 are added, followed by 0.4 ml. of reducer. The color is read as before. If there is any inorganic orthophosphate in the preparation, the reading due to it must be substracted from the reading obtained.

Stable phophorus is determined by adding 0.2 ml. 10 N H_2SO_4 to the sample and heating on a hot plate just to fuming. The tube is removed and cooled after which one or more drops of superoxol (30% hydrogen peroxide) are added. The tube is replaced on the hot plate and heated once more to fuming. At the end of this second fuming the digest should be colorless. The tube is once more cooled, made to 8.6 ml. with water: 1.0 ml. 2.5% ammonium molybdate in 3 N H_2SO_4 and 0.4 ml. of reducer are added and the color once more read. The stable phosphorus is calculated by substracting the readings due to inorganic and labile phosphate.

The second criterion used in the identification of ATP is the presence of ribose. This can be estimated quantitatively using the method of Mejbaum (1939). In this procedure, ribose, either free or bound, is heated in HCl in the presence of orcinol. During the first part of the heating procedure, the purine is split from the 1-carbon of the pentose. The labile phosphorus, is, of course, also split away. The remainder of the molecule is converted into a furfural which reacts with the orcinol to produce a green color. Color is developed more rapidly when the sugar is phosphorylated in the 5 position than when it is in the free form or phosphorylated in the 2 or 3 position. In the original Mejbaum procedure, a heating time of 20 minutes is recommended. For complete color development, in many cases, a longer heating time is required (Albaum and Umbreit, 1947). This longer heating time must be used when one is quantitatively estimating the pentose content of ATP using a free pentose standard. Since free pentose develops its color more slowly and is incomplete at the end of twenty minutes, while the 5-phosphorylated pentose has developed its color completely in twenty minutes, one obtains erroneous results in estimating pentose. It is in this connection, too, that the presence of starch, even in small quantities, may give confusing results. The presence of starch in partially purified plant ATP leads to the development of a brownish color in the orcinol-pentose reaction due to the formation of a related furfuralorcinol complex. While this complex has an absorption maximum which does not correspond to the maximum of the pentose-orcinol complex, it provides an added density which tends to give high values for pentose. Pentose estimations on crude plant extracts have little value in demonstrating ATP, since plants, in general, have large quantities of pentoses, other than ribose, as well as high concentrations of pentosans.

Ribose Estimation. The estimation of ribose in our laboratory is carried out in the following way: The sample, usually containing 10—30 micrograms of ribose, in a volume of less than 3.0 ml. is made to 3.0 ml. with water, followed by the addition 3.0 ml. of Fe'''—HCl reagent (0.990 g. of $FeNH_4(SO_4)_2 \cdot 12\,H_2O$ dissolved in 1 liter of 12 N HCl). To this is now added 0.3 ml. of freshly prepared orcinol reagent (1 g. orcinol dissolved in 10 ml. 95% alcohol). The contents of the tube are mixed and the latter heated in a boiling water bath for 40 minutes. The tubes are covered with large glass beads to prevent loss of fluid. The color produced is read in the photoelectric colorimeter with a 6600 Å filter. Ribose is used as a standard. If pure adenine nucleotides are available as standards, the heating time may be reduced to 20 minutes.

The third criterion used in identifying ATP is its adenine content. Adenine, 6-amino purine, has a sharp absorption maximum at 2600 Å. The pure compound has virtually no absorption at 2300 Å and 2800 Å. Pure ATP preparations give this typical spectrum, since the absorption observed is due solely to the adenine. The density at 2600 Å of the adenine in ATP and ADP and AMP, however, is greater than it is for free adenine or for adenosine. This is presumably due to esterification. In calculating adenine content from absorption spectra, the appropriate compound should be used as a standard. Our determinations are carried out in the BECKMANN Spectrophotometer.

In crude preparations made from oat seedlings, one rarely obtains a clean adenine spectrum. This appears to be due to the fact that crude extracts contain molecules which absorb light non-specifically in the ultra-violet regions of the spectrum. This absorption becomes less and less marked as one approaches the visible end of the spectrum. However, it is still considerable in the significant 2600 Å absorbing region for adenine nucleotides.

In conclusion, it should be emphasized that the usual methods used for identifying and characterizing ATP from animal sources cannot be applied to crude plant preparations. They can be applied only to the preparations of relatively high purity. This is true for ATP preparations made from oat seedlings and is probably true of monocotyldonous seedlings in general.

III. The Isolation of Adenosine Triphosphate from Mung Beans (Phaseolus aureus).

A method which has consistently yielded ATP preparations with a purity of 70% or better when applied to mung been seedlings is outlined in detail below (ALBAUM, OGUR and HIRSHFELD, 1950). For convenience the procedure is divided into four purification steps.

The first purification step is outlined in Scheme 1. In this isolation, a 10 pound batch of seedlings was used. These had been germinated in the dark at room temperature for five days and had hypocotyls approximately 50 mm. in length. Mung beans seedlings were selected for this isolation procedure for several reasons: First, they have relatively little starch; second, even in crude extracts, one obtains a relatively clean adenine absorption spectrum. This is important in following the ATP through the different purification stages so that some quantitative estimate of recovery can be made; third, as indicated earlier, the ATP content per seedling increases during germination. Mung bean seedlings, in germinated form are available in large quantity in this country as part of the food industry (Chinese Restaurants) so that one does not have to germinate them in the laboratory.

The first purification step yields a fraction which we refer to as residue I Scheme 1). Parallel studies of samples of residue I, prepared from mung beans and from rabbit muscle by an identical procedure, help to point up some additional difficulties encountered in applying existing isolation procedures devised for

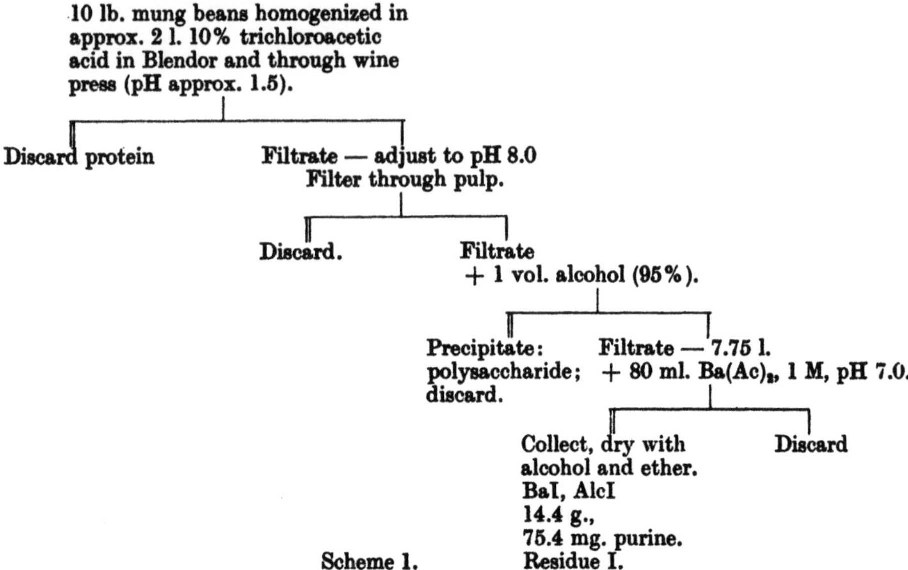

animal tissues to plant tissues. 97% of the adenine from residue I of rabbit muscle was extracted at pH 2.6 as compared with approximately 25% from plant material. Even when the plant residue was suspended in water and adjusted to pH 1, it was found that most of the adenine remained with the insoluble residue after centrifugation. Since this residue is normally discarded in existing procedure (e. g. that of NEEDHAM for animal tissue), a part of the background leading to the low yields obtained in earlier studies on plant tissues was illuminated.

Since more than 90% of the inorganic phosphate of residue I was also extractable at pH 2 under conditions, where the bulk of the adenine nucleotide was still insoluble, a fractionation at pH 2 leading to the elimination of the bulk of the inorganic phosphorus was carried out, resulting in residue II (Scheme 2). At this

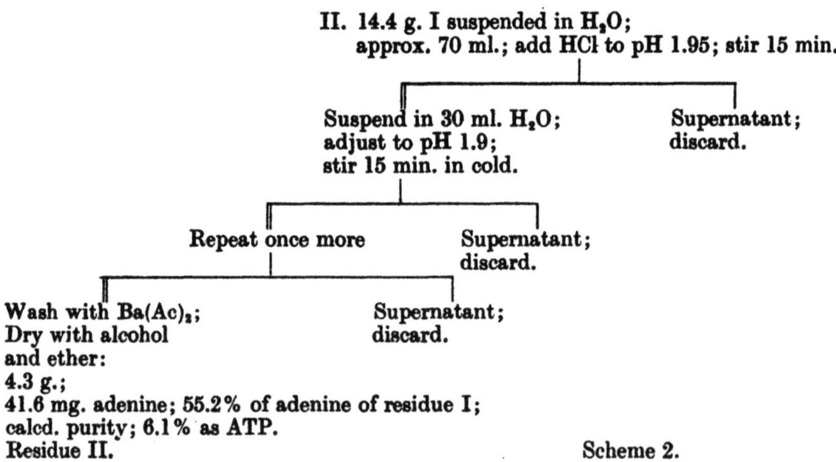

point, another difficulty encountered in isolating ATP from plant sources as compared to animal sources should be emphasized. Animal tissue, when rapidly removed from the freshly killed organism, has relatively small quantities of inorganic orthophosphate. Plant material, on the other hand, has large quantities, especially as germination continues. Where small amounts are present, no difficulty is encountered in the subsequent mercury precipitation step. However, where there are large amounts present, these are precipitated along with the adenine nucleotides, as mercury salts, so that the mercury step effects poor separation.

Residue II, as indicated above, still contained 55% of the adenine originally present in residue I. The purity of ATP at this stage, calculated from the adenine, was approximately 7%. The ratio of adenine: labile phosphorus: stable phosphorus in this preparation was found to be 1.00:1.85:1.29. The very low pH required to bring residue II into solution with subsequent splitting of the ATP, made it impractical to dissolve the nucleotide by this method. In further purification, another method had to be devised to bring the barium salt into solution without lowering the pH too drastically. By stirring residue II in the cold in an excess of sodium sulphate at pH 2, the greater part of the adenine was extracted as the sodium salt; barium sulphate produced as a result of the exchange was centrifugated off. To obtain complete exchange, it was found necessary to use an excess of sodium sulphate, some of which was still present after the removal of the barium sulfate. The presence of this excess sulfate made it difficult to

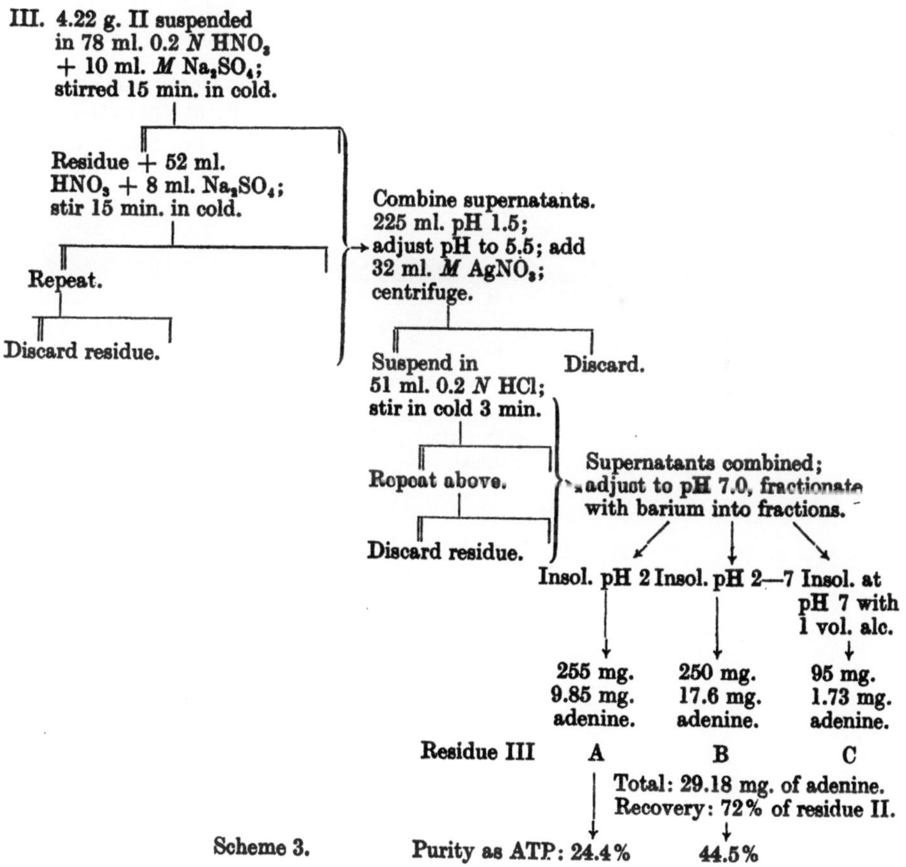

precipitate the nucleotide as the mercury salt. Silver, however, could be used. The addition of silver as the nitrate followed by decomposition with hydrochloric acid proved to be satisfactory. Precipitation with barium, at this point, yielded from a fraction initially insoluble at pH 2, three fractions; one still precipitable as the barium salt at pH 2, a second insoluble between pH 2—7, and a third precipitable with one volume of alcohol at pH 7. The preparation of these fractions (residue III) is outlined in Scheme 3. Table 1 shows the analytical data for these fraction III A and III B analyzed as ATP whereas fraction III C appears to be ADP. The purity of fraction III A and III B has increased markedly over that of residue II.

Table 1. *Fractionation after Sodium-Sulfate Exchange and Silver Treatment Analytical Data of Residues III A, B, and C.*

	A BaI, p_H 2.0	B BaI, pH 7.0 (between 2—7)	C BaS (pptd. 1 vol. alc.)
	Average 4 runs	Average 4 runs	Average 4 runs
Total wt. of ppt., mg.	242	179	51
Total adenine, mg.	10.15	11.65	1.59
Ratio: adenine	1.00	1.00	1.00
$\Delta 7'$ P	2.06	1.80	0.91
Stable P	1.11	1.09	1.12
Pentose	1.03	1.00	1.04
Purity as ATP, per cent	27.1	39.6	
Recovery of adenine in fractions A, B, C	64.6% of adenine of residue II.		

Further purification has been carried out thus far only on fraction III B, since it was soluble and could be handled readily. Fraction III B, at this stage, was yellow in color and showed some absorption at 3200 Å. The procedure used in further purification is shown in Scheme 4 (residue 4). Table 2 shows the analytical

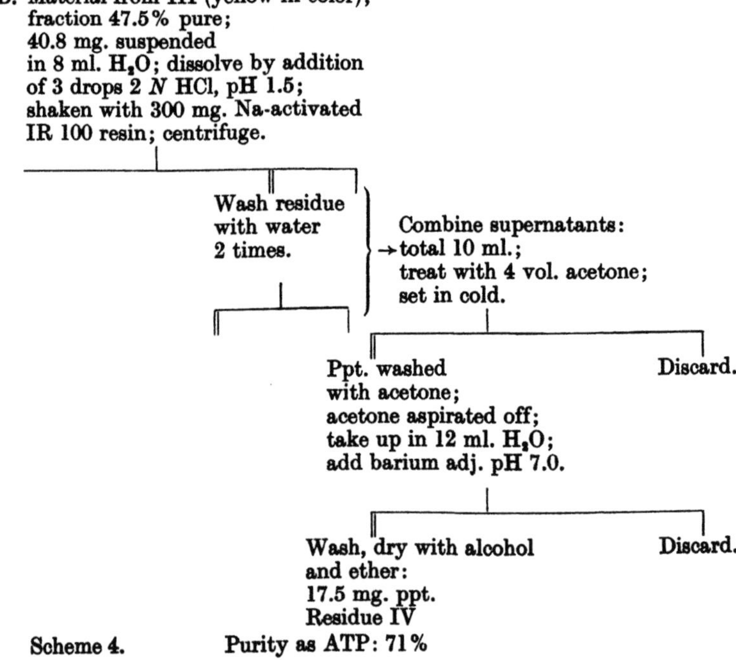

Scheme 4.   Purity as ATP: 71%

data on a sample of residue IV. As a result of the step just outlined, residue IV was now white in color, showed no absorption at 3200 Å and assayed 64% pure as ATP with the requisite molar ratios of adenine: labile phosphorus; stable phosphorus: pentose.

The procedure outlined above, therefore, yields a preparation of about 70% purity, which satisfies all of the analytical criteria for ATP. It has the proper ratio of adenine: labile phosphorus: total phosphorus; pentose. It is free of inorganic phosphorus. The adenine absorption spectrum is clean. There is no excess phosphorus. In the orcinol pentose reaction, it behaves just as does ATP from animal source. As indicated earlier, it can be substituted for the nucleotide from animal sources in the "cyclophorase" and tyrosine decarboxylase systems. Yet, it differs in some important respects from the nucleotide obtained from rabbit muscle, and animal tissue in general. These differences are associated largely with the behavior of this compound in certain enzymatic reactions and on a paper chromatogram. Since one of these sets of enzymatic reaction has been employed by us for the quantitative estimation of AA, ADP and ATP both in plant and in animal tissues, it is described below in some detail.

Table 2. *Analytical Data for Residue IV*. Combined B fractions (BaI, pH 7.0) from 3 different batches; 500 mg. purified with acetone and amberlite; final yield: 118 mg. barium salt.

Total salt, mg.	118
Adenine, mg.	11.1
Purity as ATP, per cent	64.0
Ratio: adenine	1.00
7' P	2.16
total P	0.92
pentose	0.98

C. Determination of Adenosine Triphosphate.

I. Enzymatic Determination.

The method (ALBAUM and LIPSHITZ, 1950) is an extension of a series of techniques developed by KALCKAR [1947 (1), (2), (3).] When combined with KALCKAR's procedures, the method can be used for the simultaneous determination of adenylic acid (AA), adenosine diphosphate (ADP), and adenosine triphosphate (ATP) at the micro level. In principle the method consists of converting adenosine triphosphate enzymatically into adenylic acid with a combined hexokinase-myokinase system and measuring the rate of inosinic acid formation with a muscle deaminase in the BECKMANN spectrophotometer.

Adenylic acid deaminase (SCHMIDT's deaminase) was prepared according to the method of KALCKAR [1947, (3)]; myokinase, according to COLOWICK and KALCKAR (1943). Hexokinase was prepared in partially pure form according to the method of BERGER, SLEIN, COLOWICK and CORI (1946) through step 3 (fractionation with ethanol at 0° C) of the isolation procedure. The enzymes were kept in deep freeze and were found to be stable over long periods of time.

Assay Method. The assays were carried out in the following way: Varying amounts of ATP were used, 0.5 ml., 0.1 M malonate buffer pH 5.9, 0.1 ml., 0.5 M glucose, 0.1 ml. $MgCl_2$ (10 mg./ml.), 0.01 ml. myokinase (containing 4 μg. protein N). 0.01 ml. hexokinase (containing 7 μg. protein N water to 4.9 ml., and 0.03 ml. of the deaminase added last (containing 15 μg. protein N). Time recording was begun as soon as the deaminase was added. The cuvettes were transferred to the spectrophotometer, set at 2650 Å, and readings were taken at intervals.

In early experiments the change in density was recorded every 2 minutes. The course of the reaction was linear for at least the first 10 minutes. Plotting

ATP (expressed as micrograms of adenine) against 10 min. change in density at 2650 Å gives a straight line (Fig. 1). In assay procedure, 10 min. readings are used and levels of ATP are read directly from the curve. In the above series of reactions it is necessary to scale the concentrations of myokinase and hexokinase in such a way that the adenylic acid is formed more rapidly than it is deaminated by the muscle enzyme. If this is done, the method can also be used for the assay of adenylic acid, and adenosine diphosphate as well as ATP. In such a case the concentration curve for all three compounds plotted on an adenine basis should be identical (Fig. 1). In applying these methods, therefore, preliminary experiments should be carried out to determine nonlimiting levels of hexokinase. In our experience some preparations of hexokinase gave low activity. To insure complete conversion of ATP and ADP to AA all the ingredients were mixed together with the exception of the deaminase which was added after a time interval of about 30 minutes. In this connection, the same result could have been achieved in a shorter time interval by raising the level of hexokinase. This was not done because it was desirable to keep the density due to the enzymes themselves at a minimum. Since all three compounds give identical results when concentration, expressed as micrograms of adenine, is plotted against density, it is possible by using different combinations of enzymes to determine all three constituents in an unknown solution. This has been done in assaying plant ATP preparations.

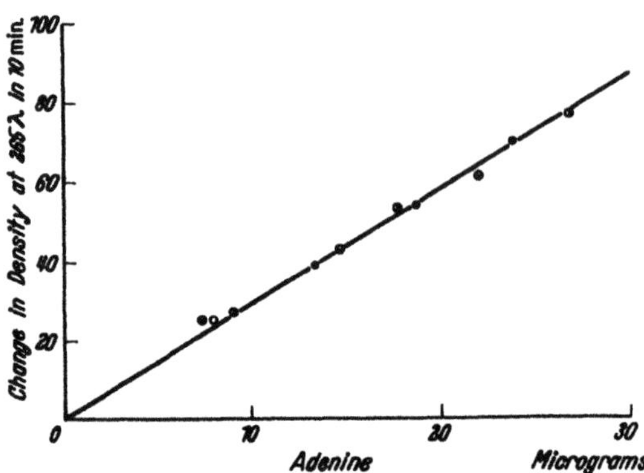

Fig. 1. Relationship between 10 min. density change at 265 mμ. and concentration of adenine (micrograms) in various adenine nucleotides treated with different enzyme combinations: ◐, adenylic acid; ○, adenosine diphosphate; ◑, adenosine triphosphate; ●, coenzyme I.

Applying the enzymatic procedure outlined above to preparations of plant ATP from mung beans at the residue IV stage of purification yields the result shown in Table 3. These results are interesting from several points of view. First, in spite of the fact that all of the preparations are indistinguishable chemically from animal ATP, in no case does one obtain as complete or as rapid deamination. The data suggest as one alternative the presence of an inhibitor which primarily affect the enzyme system.

Table 3. *Course of Deamination of Plant and Animal Adenosine Triphosphate with a Combined Hexokinase-Myokinase-Deaminase System.*

	Decrease in density at 265 mμ.			
Time, min.	4	8	12	16
Plant ATP (Residue IV)	.007	.016	.027	.0 37
Animal ATP	.014	.029	.045	.063
Total	.021	.045	.072	.100
Plant and animal ATP run together in same tube	.021	.049	.074	.100

The aliquots used in this experiment contained identical amounts of adenine, labile P and total P.

However, in all cases where inhibitors were looked for by doing recoveries of animal ATP in the presence of plant ATP, no evidence for inhibitors was obtained (Table 3). From these data we have come to the tentative conclusion that the nucleotide as it occurs in the intact plant is perhaps polymerized in the form of a larger molecule which cannot be attacked completely by the enzyme systems outlined above. The anamolous behaviour of the plant ATP disappears when the preparation is treated with mercury, a procedure used in isolating animal ATP (Table 4). The rate of deamination in the system outlined above is now similar to that obtained with animal ATP. We believe that what has happened is that the mercury has split the compound from its "native state". The enzymatic procedure outlined above may then be used to quantitatively assay preparations of plant ATP provided they have gone through the mercury procedure in which case there is excellent correlation between the chemical and the enzymatic data. The method can be used effectively for samples containing between 3—30 micrograms of ATP adenine.

Table 4. *Attempt at Further Purification of Residue IV.* Pure material (64%) fractionated once with Hg(Ac)₂ in acetic acid, decomposed with H_2S; barium salt formed.

Total salt: fractionated, mg.	30.0
recovered, mg.	10.0
Recovered: total adenine, mg.	1.08
initial adenine, mg.	2.84
Purity as ATP, per cent	68.2
Ratio: Adenine	1.02
7′ P	2.00
stable P	1.03
pentose	1.03
N	5.08
Rate of deamination, per cent	64.9/69.0
	94.5

Luminescence Method. Another enzymatic procedure which has recently appeared in the literature can normally be used to quantitatively assay as little as 1 microgram of ATP. The method utilized the luminescence produced by extracts of fireflies in the presence of ATP as a quantitative index of the nucleotide (Strehler and Totter, 1952). The luminescence is measured in the Farrand photofluorometer. By the use of finer recording equipment, the sensitivity of the method can be increased up to 1000-fold. Strehler and Totter have applied this procedure to ATP changes in *Chlorella* during photosynthesis. The assays are carried out on hot water extracts of the alga.

The luminescent enzyme system used in the assays with the Farrand photofluorometer is prepared in the following way: 50 mg. (10—12 luminous organs) of vacuum-dried firefly *(Photinus pyralis)* lanterns are extracted with 5 ml. of ice-cold 0.1 M sodium arsenate buffer at pH 7.4 for 5—10 min. with grinding. The material is filtered into a test tube and kept in an ice bath. Fifty mg. of $MgSO_4 \cdot 7 H_2O$ is dissolved in the solution. This preparation remains active for several days if kept at 4° C.

For routine analysis, 0.2 ml. of enzyme plus ATP or unknown and water are made to a volume of 0.8 ml. The photofluorometer is set at maximum sensitivity and galvanometer readings are made at a standard time after mixing (usually 30 sec. to 1 minute).

II. Identification of ATP by Paper Chromatography.

Evidence of another sort has been used in our laboratory to differentiate between plant and animal ATP. This has been done thus far, only with purified preparations. There is no reason, however, why this same technique cannot be applied to crude preparations. The procedure is a chromatographic one involving the use of paper.

Method. The extracts to be tested or the nucleotide in partially purified form are spotted on Schleicher and Schüll, No. 597 paper, cut to 37×42 cm. Anywhere between 5—25 micrograms of purine should be spotted on the paper. The quantity of solution used will depend, of course, upon the concentration of the nucleotide. With dilute solutions, we have spotted .01 ml. at a time and have built up concentrations of 0.2—0.3 ml. After spotting, the filter papers are rolled into cylinders and stapled on top and on bottom. Chromatographic runs are carried out in 15×50 cm. pyrex jars. Migration is upward in a developer consisting of 1% malonate buffer at pH 6 mixed with approximately $1/_3$ volume of isoamyl alcohol. Migration is allowed to take place for a period of 16 hours at room temperature. After drying at room temperature, the paper is examined in the dark with a Mineralite No. 44 Lamp (Ultra-Violet Products, Incorporated, Los Angeles, California). The purine containing material produces a quenching of fluorescence resulting in a dark spot which can be readily observed and marked for subsequent analysis. The R_f values for the plant ATP as compared to the animal preparation are shown in Table 5. The difference between the mean R_f values has a high degree of statistical significance.

Table 5. *Chromatographic Data, R_f Values*[1].

	mean	n	σ
Plant ATP . . .	.846	32	.019
Animal ATP . .	.873	37	.014
	d/S.E.[2] 7.2		

Two Dimensional Method. Another chromatographic procedure which has been utilized to detect adenosine triphosphate in tissues of plant extracts has recently been described by BANDURSKI and AXELROD (1951), using a two dimensional chromatogram. In their procedure, Schleicher and Schüll, No. 589, filter paper 28 centimeters square was used. They found that the chromatography of phosphorus esters in general is best accomplished with the free acids. The salts of the esters are readily converted to the corresponding acids and the metallic ions removed from the extracts by shaking the solutions or suspensions of the salts with a cation exchanger such as Dowex No. 50. The chromatographic runs are done in cylindrical jars similar to those employed in our own laboratory. Development is carried out first in the acid solvent for 6—6.5 hours at 2° at which time the solvent reaches the upper edge of the paper. After removal from the solvent the paper is air-dried in front of a fan at room temperature and developed again, at right angles to the first development, in the alkaline solvent. The second development in the alkaline solvent requires from 12—15 hours. The R_f value for adenosine triphosphate in the acid solvent is .12, in the alkaline solvent, .15. This procedure was used to identify ATP in orange flavedo extracts and in pea meal (AXELROD, BANDURSKI and SALTMAN, 1951).

The acid solvent consists of 80 volumes of methanol, 15 volumes of formic acid (88% by weight) and 5 volumes of water. The alkaline consists of 60 volumes of methanol, 10 volumes of ammonium hydroxide (specific gravity 0.9015) and 30 volumes water.

Our procedure gives high R_f values for the adenine nucleotides; that of BANDURSKI and AXELROD tends to give low values. However, the latter authors were interested in the identification of other phosphorus esters as well. Their method is more useful from that point of view. We have described another developer for detecting phosphorus esters other than adenine nucleotides in another publication [ALBAUM, 1952 (2)].

---

[1] Residue IV was used in these experiments. See text for details.
[2] Ratio of deviation to standard error.

Successive Solvent Method. The most recent modification of paper chromatographic procedures to the separation and identification of adenosine phosphates is that of EGGLESTON and HEMS (1952). In this procedure 2 different solvents (acid and alkaline) are used successively as developers in the *same* direction, the first ascending and the second descending on the reversed strip. The method, according to the authors, has the advantage over the two dimensional chromatogram (of which it is a modification) that smaller pieces of paper are required and therefore larger numbers of determinations may be run simultaneously; in addition the developed spots have a greater compactness.

Whatman No. 1 paper (45×19 cm.) is first treated with 0.2% ethylene-diaminetetraacetic acid (neutralized top H 8.5 with NaOH) for 30 minutes, and washed thoroughly with water and dried in a current of hot air. This preliminary treatment of the paper results in greater compactness of spots and cuts down on "trailing", presumably because metal impurities are removed.

The paper is spotted 8 cm. from one end with 3—10 micrograms of adenosine phosphate phosphorus. Development is carried out on 8 papers simultaneously in glass tanks 40 cm. high with a 30×22 cm. base area, first in the acid solvent upward for 3—4 hours at 22° C. The papers are then removed, air dried, reversed and developed downward (4 papers at a time) for 27—30 hours at 28° C in the alkaline solvent. The spots are detected with a U. V. lamp. The solvents are made up as follows: 1) acid solvent — 90 ml. isopropyl ether, 60 ml. 90% (v/v formic acid). 2) alkaline solvent — 240 ml. *n*-propanol, 120 ml. ammonia (sp. gr. 0.880), 40 ml Methylenediaminetetraacetic acid.

Tissue extracts are prepared by deproteinizing with 0.1 volume 30% trichloracetic acid (v/w) at 2—4° C. It is not necessary to neutralize these extracts for chromatography. They may be stored for several weeks at —14° C without breakdown of ATP.

III. The Quantitative Estimation of ATP by Column Chromatography.

An excellent procedure available for quantitatively estimating adenosine triphosphate in crude plant extracts at the microgram level is based on the methods of COHN and CARTER (1950) utilizing column chromatography. We have applied this procedure to mung beans. One is able to take up nucleotides from crude extracts on a Dowex resin and successfully elute them one at a time. The procedure employed in our work was the following: a barium alcohol precipitate is prepared from approximately 3 gm. of tissue (about 50 seeds to 8—10 fourday seedlings). The barium alcohol precipitate is freed of barium with sodium sulfate and made 1 M with respect to ammonium hydroxide. It is then poured through a column containing an anion exchange resin, Dowex No. 1. In our

Table 6. *Adenine and Organic Phosphorus Content of Three Successive Eluates Collected From Dowex Column Treated With Extract From Four-Day Mung Bean Seedling.*

Eluate	μM adenine	μM organic P
1	0.67	1.95
2	0.128	0.352
3	0.091	0.219
	0.889	2.515
Molar ratio	0.96	3.00

work, the column was approximately 25 mm in height and 8 mm in diameter. The eluate is poured through the resin several times to make sure that all the nucleotides are taken up. Under these conditions, adenosine, adenine, adenylic acid, ATP, a portion of the inorganic orthophosphate and ATP are taken up on the column. The column is then rinsed with water after which the different compounds can be eluted successively with the following solutions: solution 1 for adenosine 0.01 M NH$_4$Cl in 0.1 M NH$_4$OH; 2 for adenine 0.01 M NH$_4$Cl in H$_2$O; 3 for AMP .003 M HCl; 4 for ADP and inorganic PO$_4$, 0.02 M NaCl in 0.01 M HCl; and 5 for ATP, 0.2 M NaCl in 0.01 M HCl. In the elution procedure if solution 4, for example, is used first, all the compounds removed by solutions 1, 2, and 3 as well as solution 4, are transferred into the eluate. We were primarily interested in ATP so that we treated the column first with solution 4 and then several times with solution 5. Identity with ATP was established by the correspondence between phosphorus and adenine. A typical result from four-day seedlings is shown in Table 6.

This technique may also be employed with extracts from plants soaked in radioactive phosphorus in which case it becomes possible to measure the specific activity of the phosphorus in ATP [ALBAUM, 1952 (1)]. The entire column procedure is summarized in Scheme 5.

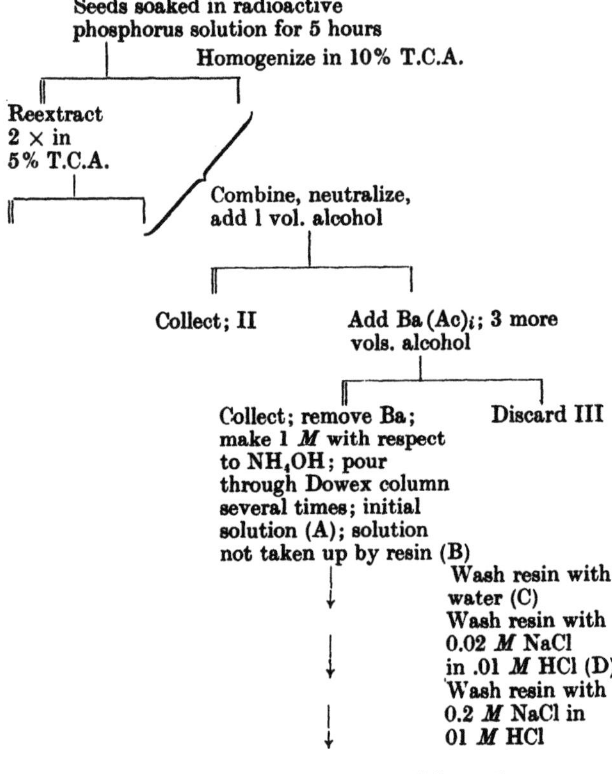

Scheme 5.

D. The Status of Adenosine Diphosphate.

Adenosine diphosphate thus far has not been isolated from plant tissues. The preparation from oats (ALBAUM and OGUR, 1947) resembles ADP chemically; however, sufficient material was not available to test its activity in enzyme systems which are known to require it. Residue IIIC in the mung bean isolation procedure (Table 1) is also identical chamically with ADP. However, this is admittedly a breakdown product produced during the purification of ATP.

Indirect evidence for the presence of ADP in plants comes from studies on mung bean mitochondria (BONNER and MILLERD, 1953) where the enzyme myokinase has been demonstrated. This enzyme catalyzes the reaction in which 2 moles of ADP are converted to 1 mole of adenylic acid and 1 mole of ATP.

References.

ALBAUM, H. G., and M. OGUR: Arch. of Biochem. Biophys. 15, 158 (1947). — ALBAUM, H. G., and W. W. UMBREIT: J. Biol. Chem. 167, 369 (1947). — ALBAUM, H. G., M. OGUR and A. HIRSHFELD: Arch. of Biochem. Biophys. 27, 130 (1950). — ALBAUM, H. G., and R. LIPSHITZ: Arch. of Biochem. Biophys. 27, 102 (1950). — ALBAUM, H. G.: Arch. of Biochem. Biophys. 24, 375 (1949); (1), in "The Biology of Phosphorus" edited by L. F. WOLTERINK. Lansing, Michigan: Michigan State College Press 1952; (2) Ann. Rev. of Plant Physiol. 3, 35 (1952). — AXELROD, B., R. S. BANDURSKI and P. SALTMAN: Fed. Proc. 10, 158 (1951). — AXELROD, B., P. SALTMAN, R. S. BANDURSKI and R. S. BAKER: J. Biol. Chem. 197, 89 (1952).

BANDURSKI, R. S., and B. AXELROD: J. Biol. Chem. 193, 405 (1951). — BERGER, L., M. W. SLEIN, S. P. COLOWICK and C. F. CORI: J. Gen. Physiol. 29, 379 (1946). — BONNER, J., and A. MILLERD: Arch. of Biochem. Biophys. 42, 135 (1953).

COHN, W. E., and C. E. CARTER: Fed. Proc. 2, 161 (1950). — COLOWICK, S. P., and H. M. KALCKAR: J. Biol. Chem. 148, 117 (1943). — CROSS, R. J., J. V. TAGGART, G. A. Covo and D. E. GREEN: J. Biol. Chem. 177, 655 (1949).

DOUNCE, A. L., A. ROTHSTEIN, G. THANNHAUSER-BEYER, R. MEIER and R. M. FREER: J. of Biol. Chem. 174, 361 (1948).

EGGLESTON, L. V., and R. HEMS: Biochem. J. 52, 156 (1952).

FISKE, C. H., and Y. SUBBAROW: J. Biol. Chem. 81, 629 (1929).

GREEN, D. E.: in "Enzymes and Enzyme Systems" edited by J. T. EDSALL. Cambridge, Mass.: Harvard University Press 1951.

KALCKAR, H. M.: (1) J. Biol. Chem. 167, 429 (1947); (2) J. Biol. Chem. 167, 445 (1947); (3) J. Biol. Chem. 167, 461 (1947). — KERR, S. E.: J. Biol. Chem. 139, 121 (1941).

LATIES, G.: (1) Physiologia Plantarum 6, 199 (1953); (2) Physiologia Plantarum 6, 215 (1953).

McELROY, W. D., and B. GLASS: Phosphorus Symposium V. 1. Baltimore: Johns Hopkins Press 1951; Phosphorus Symposium V. 2. Baltimore: Johns Hopkins Press 1952. — MEJBAUM, W.: Hoppe-Seylers Z. 258, 117 (1939). — MILLERD, A., J. BONNER, B. AXELROD and R. S. BANDURSKI: Proc. Nat. Acad. Sci. USA 37, 855 (1951). — MILLERD, A.: Arch. of Biochem. Biophys. 42, 149 (1953).

NEEDHAM, D.: Biochem. J. 36, 114 (1942).

SALTMAN, P.: J. Biol. Chem. 200, 145 (1953). — STREHLER, B. L., and J. R. TOTTER: Arch. of Biochem. Biophys. 40, 28 (1952). — SZENT-GYÖRGI, A.: Chemistry of Muscular Contraction. New York: Academic Press 1947.

TEWFIK, S., and P. K. STUMPF: J. Biol. Chem. 192, 527 (1951).

VISHNIAC, W., and S. OCHOA: J. Biol. Chem. 198, 501 (1952).

Codehydrasen I und II

(Diphospho-pyridin-nucleotid und Triphospho-pyridin-nucleotid).

Von

K. Hasse.

Mit 3 Abbildungen.

A. Einleitung.

Diphospho-pyridin-nucleotid und Triphospho-pyridin-nucleotid beanspruchen unter den Nucleotiden ein besonderes Interesse, da sie die Coenzyme einer großen Zahl von Dehydrasen sind. Ihre quantitative Bestimmung durch enzymatische Analysen hat zu ihrer Reindarstellung geführt. Die Ermittlung ihrer Konstitution und ihrer Eigenschaften hat den Einblick in den Chemismus dieser Klasse der Pyridin-dehydrasen ermöglicht. In diesen Verbindungen hat das Vitamin Nicotin-säureamid seine Wirkungsform. Die Bestimmung des Gehaltes in verschiedenen Gewebearten hat deswegen eine Bedeutung.

Ihre Analyse gestattet es, die Kinetik der Dehydrierungsvorgänge, an denen sie beteiligt sind, zu messen. Auf dieser analytischen Grundlage konnten auch die Eiweißkörper der Dehydrasen und solcher Enzyme, deren Reaktionen mit den Dehydrierungen gekoppelt sind, auf ihre Aktivität getestet und in hohem Reinheitszustand erhalten werden. Verschiedene organische Verbindungen, die durch Codehydrase I oder II-abhängige Enzyme hydriert oder dehydriert werden, lassen sich durch die Analyse der Cofermente quantitativ bestimmen.

Biologische Funktion. Diphospho- und Triphospho-pyridin-nucleotid sind obligate Komponenten zahlreicher Dehydrasen. Diese sogenannten „Pyridin-dehydrasen" sind Proteide, in denen Eiweißanteil (Apodehydrase) und Codehydrase dissoziierbar miteinander verbunden sind. Es sind zu unterscheiden spezifisch auf DPN (Co I) eingestellte Diphospho-pyridin-proteide und auf TPN (Co II) eingestellte Triphospho-pyridin-proteide. In einzelnen Fällen kann die Apodehydrase sowohl mit Co I als auch mit Co II zu wirksamen Enzymen ergänzt werden. Die Codehydrase enthält die Wirkungsgruppe. Durch reversible Änderung des Oxydationszustandes (Pyridin $\rightleftarrows$ Dihydropyridin) überträgt sie Wasserstoff vom Wasserstoffdonator auf den Wasserstoffacceptor. Die primär gebildete Dihydro-codehydrase ist nicht autoxydabel, ihre Reoxydation unter gleichzeitiger Hydrierung eines Wasserstoffacceptors geschieht unter Mitwirkung von spezifischen Dehydrasen. Die 1. Phase der Hydrierung des Coenzyms läßt sich realisieren, wenn solche Enzyme fehlen.

Nomenklatur:

Diphospho-pyridin-nucleotid (DPN)	Triphospho-pyridin-nucleotid (TPN)
Cozymase	
Codehydrase I (Co I)	Codehydrase II (Co II)
Codehydrogenase I	Codehydrogenase II
Coenzym I	Coenzym II
Dihydro-DPN, DPN-H$_2$, DPNH, Co I-H$_2$	Dihydro-TPN, TPN-H$_2$, TPNH, Co H-H$_2$

Konstitution. Diphospho-pyridin-nucleotid und Triphospho-pyridin-nucleotid sind Dinucleotide. Im DPN sind Nicotinsäureamid-ribonucleotid und Adenylsäure (5') durch eine Pyrophosphatbindung miteinander verknüpft. Die Struktur von TPN unterscheidet sich von DPN nur dadurch, daß zusätzlich eine Hydroxylgruppe (2') der Adenylsäure mit Phosphorsäure verestert ist.

Diphospho-pyridin-nucleotid (R = H)
Triphospho-pyridin-nucleotid (R = — PO$_3$H$_2$)

Einige Eigenschaften:

	Mol. Gewicht	Summenformel
DPN	663	$C_{21}H_{27}O_{14}N_7P_2$
TPN	743	$C_{21}H_{28}O_{17}N_7P_3$

Diese Pyridin-nucleotide sind amorphe, weiße hygroskopische Pulver. Sie sind einander chemisch sehr ähnlich. Exsikkatortrocken enthalten sie noch etwa 5—7% Wasser, das sie im Hochvakuum bei 80—100° weitgehend abgeben. DPN ist eine einbasische, TPN eine dreibasische Säure. Sie sind beide leicht löslich in Wasser, unlöslich in organischen Lösungsmitteln, wie Aceton, Alkohol und Essigester, mäßig löslich in Phenol und angesäuertem Methanol. Beide Nucleotide bilden unlösliche Salze mit Schwermetallen, wie Blei, Quecksilber und Silber, aber in Wasser lösliche Alkali- und Bariumsalze. Das Bariumsalz vom TPN ist in verdünntem Alkohol schwerlöslich, das Bariumsalz von DPN ist leicht löslich in verdünntem Alkohol.

Leichte Spaltbarkeit der glykosidischen Bindung der Moleküle und der Pyrophosphatbindung bedingt schon unter milden Bedingungen eine Instabilität der Nucleotide. Die festen Präparate sind begrenzt haltbar und zeigen in Monaten eine sehr langsame Abnahme ihrer Aktivität. Neutrale Lösungen sind im Eisschrank einige Tage ohne Wirkungsverlust aufzubewahren. In der Hitze tritt selbst in neutralem Medium starke Schädigung ein. Verschiedene Anionen, wie Citrat- oder Phosphat-Ionen, beschleunigen die Hitzeinaktivierung (DPN p$_H$ 7, 5 min bei 100°: ungepuffert 25%, mit Phosphatpuffer 60%ige Spaltung). Verdünntes Alkali spaltet die Nicotinsäureamid-Ribose-Bindung und die Pyrophosphat-Bindung (DPN und TPN in n/10 NaOH, 15 min bei 20°: 50%ige

Spaltung). Hingegen sind die Coenzyme in saurer Lösung (n/10 HCl) bei Zimmertemperatur beständig. In der Hitze tritt jedoch auch unter diesen Bedingungen schnell Hydrolyse ein (DPN und TPN in n/10 HCl, 8 min bei 100°: 50% Spaltung).

In Gewebeextrakten können Pyridinnucleotide enzymatisch inaktiviert werden. In verschiedenen Preßsäften ist hinzugefügte Codehydrase nach wenigen Minuten nicht mehr nachweisbar (DPN, 10 min bei 20°: in Extrakten aus Spinat, Kartoffeln, Erbsen: 30, 100, 0% Spaltung). Verschiedene Mechanismen des Abbaues der Dinucleotide sind bekannt.

Die Reaktivität der Pyridiniumgruppierung in den Dinucleotiden ist in chemischer wie auch in biologischer Hinsicht die bemerkenswerteste. Chemisch und enzymatisch lassen sich DPN und TPN leicht partiell zu den Dihydroverbindungen reduzieren. Diese Aufnahme von Wasserstoff ist reversibel.

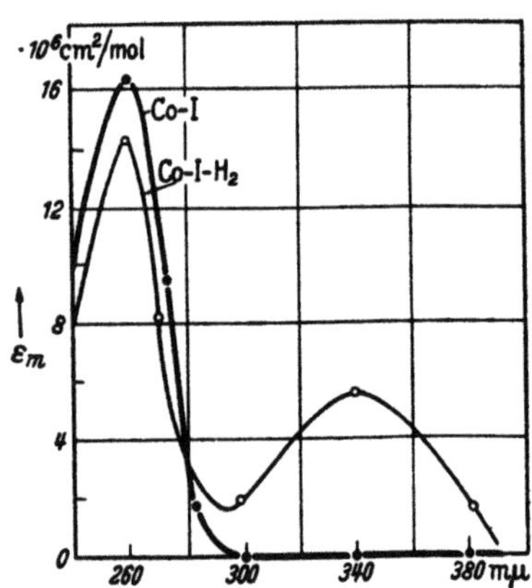

Im Gegensatz zu den oxydierten Formen sind die Dihydro-pyridin-nucleotide gegenüber Alkalien recht beständig. 10 min langes Erhitzen in n/10 NaOH auf 100° übersteht DPN-H_2 ohne nennenswerten Wirkungsverlust. Dagegen sind sie gegen Säure höchst empfindlich und werden durch Mineralsäuren n/10 bei 20° unmittelbar zerstört. In alkalischer Lösung sind sie gegen Luftsauerstoff beständig. Bei p_H 7 erfolgt langsam bei Zimmertemperatur, rascher bei erhöhter Temperatur, spontane Oxydation.

Von großer Bedeutung für die quantitative Bestimmung der Codehydrasen und ihre Charakterisierung sind die Unterschiede in den Absorptionsspektren der oxydierten und hydrierten Formen im UV-Bereich. Die Spektren von DPN und TPN unterscheiden sich nicht. Auch DPN-H_2 und TPN-H_2 haben praktisch die gleiche Absorption. Hingegen zeigen die reduzierten Cofermente ein Absorptionsmaximum bei 340 mμ, das in den Spektren der oxydierten Formen fehlt (Abb. 1).

Abb. 1. Absorptionsspektren der oxydierten und der reduzierten Codehydrase I.

B. Analytische Verfahren zur Bestimmung der Codehydrasen.

I. Allgemeines.

Die Bestimmung der beiden Pyridin-codehydrasen in pflanzlichem Material bereitet erhebliche Schwierigkeiten. Diese sind bedingt durch ihre Ähnlichkeit in chemischen und physikalischen Eigenschaften zueinander und zu anderen Nucleotiden. Die Schwierigkeiten einer quantitativen Erfassung werden weiter hervorgerufen durch die chemische Labilität der beiden Dinucleotide und ihre

Angreifbarkeit durch Enzyme. Deshalb sind die in der Literatur angegebenen Werte für die Codehydrasen in pflanzlichen und auch in tierischen Gewebearten unsicher. Ergebnisse über den Coenzymgehalt in höheren Pflanzen liegen nur spärlich vor.

Exakt bestimmbar ist jedoch jedes Coenzym in den Extrakten und den verschiedenen Fraktionen im Verlaufe von Reinigungsverfahren. Auch der Gehalt von Präparaten ist genau zu ermitteln, sofern nicht die analytische Bestimmung störende Verbindungen vorliegen.

Die chemische Charakterisierung der Dinucleotide ist schwierig. Die Daten der Elementaranalyse allein sind nicht ausreichend, um quantitative Aussagen über die Reinheit eines Präparates zu machen, da inaktive Substanzen der gleichen Zusammensetzung als Begleiter auftreten, und da sich die beiden Dinucleotide selbst nur wenig unterscheiden. Zur Orientierung über die Reinheit von Coenzympräparaten kann der Phosphorsäuregehalt oder der Quotient Phosphorsäure/Pentose u. a. dienen.

Alle chemischen Agentien erfassen beide Verbindungen gleichzeitig, so daß durch die Charakterisierung der Abwandlungsprodukte stets die Summe der Codehydrasen ermittelt wird. Doch sind chemische Methoden anwendbar zur Gehaltsbestimmung von Konzentraten, in denen jeweils nur eines der beiden Dinucleotide enthalten ist.

Durch *enzymatische Analyse* ist eine eindeutige Bestimmung der beiden Codehydrasen nebeneinander möglich. Diese Methoden haben den Vorteil einer Spezifität, wie sie mit chemischen und physikalischen Verfahren nicht erreichbar ist. Die Empfindlichkeit des enzymatischen Testes ist sehr hoch, so daß quantitative Bestimmungen im Bereiche von wenigen γ im ml möglich sind. Die Nachteile der enzymatischen Analyse bestehen in der notwendigen Bereitung der z. T. nur begrenzt haltbaren Komponenten des Enzymsystems.

Die 1. Gruppe der enzymatischen Methoden umfaßt solche, bei denen das zu bestimmende Coenzym vollständig enzymatisch zur Dihydroverbindung hydriert und diese dann auf spektrophotometrischem Wege analysiert wird (WARBURG). Zur enzymatischen Hydrierung werden Testsysteme verwendet, die sich aus einer DPN- bzw. TPN-spezifischen Apodehydrase und dem dazugehörigen Substrat zusammensetzen. Die Dihydroverbindungen haben im Vergleich zu den oxydierten Formen eine charakteristische Absorptionsbande mit einem Maximum bei 340 mμ. In Kenntnis des molekularen Extinktionskoeffizienten für eine Wellenlänge in diesem Bereich ist die Extinktion ein absolutes Maß für die Coenzymkonzentration. Führt die enzymatische Hydrierung durch das verwendete Enzymsystem jedoch zu einem meßbaren Gleichgewicht, so ergibt die Absorptionsmessung nur relative Werte der Konzentration. Durch Vergleich mit einem Standardpräparat ist dann eine Umrechnung in γ/ml möglich. Der Meßbereich liegt etwa zwischen 10 und 100 γ Coenzym im ml.

Die 2. Gruppe der enzymatischen Methoden bestimmt die Coenzyme auf Grund ihrer katalytischen Funktion in einem enzymatischen Dehydrierungsvorgang oder in komplexen Enzymsystemen.

Ist das Coenzym der geschwindigkeitsbestimmende Faktor der Reaktion, so ist die Geschwindigkeit in einem beschränkten Bereich ein relatives Maß für die Coenzymkonzentration. Die absolute Bestimmung der vorliegenden Menge ist durch einen Vergleich mit einem Standardpräparat möglich. Die hier verwendeten Testsysteme enthalten neben der DPN- oder TPN-spezifischen Apodehydrase und dem Substrat ein oder mehrere weitere Enzyme, um den Wasserstoff von den nicht autoxydablen Dihydro-coenzymen auf einen Wasserstoffacceptor weiter zu leiten. Die Geschwindigkeit der Reaktion wird gemessen durch die Konzentrationsänderung eines geeigneten Partners. Dabei dominieren manometrische und optische Methoden. Mit diesem Typ von enzymatischen Analysen lassen sich Mengen bis und unter 1 γ im ml erfassen.

Bei der Anwendung *chemischer Methoden* wird ein Abwandlungsprodukt der Coenzyme analytisch erfaßt. Auch hier spielt die spektrophotometrische Bestimmung der Dihydroverbindungen eine große Rolle.

Eine Reihe anderer Verbindungen, in die sich die Coenzyme leicht überführen lassen, können ebenfalls für die Analyse, insbesondere unter Verwendung optischer Methoden, dienen. In allen Fällen wird die Summe der beiden Coenzyme ermittelt.

Der *qualitative Nachweis* der Codehydrasen besitzt bei weitem nicht die Bedeutung der quantitativen Analyse. Grundsätzlich kommen die für die Bestimmung geeigneten enzymatischen Testsysteme in Frage. Der Nachweis geschieht durch die Auslösung eines Dehydrierungsvorganges. Apparativ besonders einfach ist die Anwendung der Thunberg-Technik (s. unten).

Die Auswahl unter den sehr zahlreichen Analysenmethoden ist z. T. willkürlich, andererseits aber auch weitgehend abhängig von der Art des zu untersuchenden Materials und der Begleitstoffe.

Ohne Zweifel ist das eleganteste Verfahren die spezifische, quantitative enzymatische Hydrierung mit anschließender spektrophotometrischer Bestimmung der Dihydroverbindung. Doch sind die Voraussetzungen dieser Meßmethoden häufig nicht erfüllt. Die Gehaltsbestimmung in rohen Extrakten und trüben Lösungen ist nur mit den manometrischen Methoden und der Thunberg-Technik möglich. Die hierfür erforderlichen Standardpräparate müssen wiederum mit der optischen Methode getestet werden. Es ergibt sich, daß oft mehrere Methoden nebeneinander benötigt werden.

1. Der Zustand der Codehydrasen im pflanzlichen Material und ihre Extraktion.

Während der Extraktion kann eine chemische und enzymatische Spaltung eintreten, selbst wenn die Extraktion auf wenige Minuten beschränkt bleibt. Die oxydierten und reduzierten Formen der Pyridin-nucleotide liegen in den verschiedenen Gewebearten nebeneinander vor (ruhende Hefe: $DPN/DPN\text{-}H_2 = 12/I$; gärende Hefe: $DPN/DPN\text{-}H_2 = 2/I$). Die Berücksichtigung ihrer unterschiedlichen Stabilität gegenüber Säuren und Alkalien ist für ihre Erfassung von Bedeutung.

Die Wirkung der schädigenden Enzyme läßt sich durch Hitze- und Säure-Inaktivierung beseitigen. Die häufig angewendete Heißwasser-Extraktion bei 80—90° kann in neutralem Medium selbst in 5 min einen Verlust von 10—20% bewirken. Die reduzierten Coenzyme werden unter diesen Bedingungen spontan durch Sauerstoff in die oxydierten Formen verwandelt. Die saure Extraktion in der Kälte $(0,5\,\mathrm{n}\,H_2SO_4, 2\%$ Trichloressigsäure) schädigt die oxydierten Formen der Coenzyme nicht. Jedoch werden die vorliegenden hydrierten Anteile momentan zerstört.

Die oxydierten und reduzierten Anteile der Coenzyme lassen sich durch geeignete Extraktion bestimmen. Durch Extraktion mit verdünntem Alkali bleiben die reduzierten, durch Extraktion mit verdünnter Säure die oxydierten Formen erhalten (Adler, 1936, 1940; v. Euler, 1939; Holzer, 1954).

Thermostabile Hemmungskörper machen häufig den qualitativen Nachweis der Codehydrasen unmöglich (v. Euler, 1924; Hasse, 1952). So lassen sich in den Extrakten von manchen Blättern und höheren Pilzen oft keine oder nur spärliche Andeutungen für das Vorhandensein der Codehydrasen finden. Solche Extrakte setzen die Wirkung einer bekannten zugesetzten Menge an Codehydrase herab, so daß es nicht angängig ist, aus den mit Rohsäften erhaltenen Werten auf den Coenzymgehalt des Materials zu schließen. Spuren von Schwermetallen

— auch in den Coenzym-Präparaten hohen Reinheitsgrades — setzen die Aktivität der enzymatischen Testsysteme herab. Die chemische Natur organischer, niedermolekularer störender Begleitstoffe ist weitgehend unbekannt. In manchem pflanzlichen Material können es Gerbstoffe sein, aber auch gerbstofffreie Extrakte vermögen die Aktivität der Enzymsysteme vollständig zu hemmen. In solchen Fällen ist es bisweilen gelungen, durch Abtrennung der Hemmungskörper bzw. durch Anreicherung der Coenzyme zu einer Gehaltsbestimmung zu kommen. Eine Konzentrierung kann bewirkt werden durch Ausschütteln der Coenzyme aus Extrakten mit Phenol (WHATLEY, 1951) oder durch Adsorption an Kohle (PFLEIDERER, 1953; ANDERSON, 1954).

Verbreitung. Bei der Bedeutung, die den Pyridin-Codehydrasen zukommt, fehlen sie wahrscheinlich in keiner Zelle mit aktivem Stoffwechsel. Ihr Gehalt in den verschiedenen Gewebearten ist jedoch außerordentlich unterschiedlich. In tierischen Zellen und in den Hefen liegen die Werte für DPN wesentlich höher als für TPN. WHATLEY (1951) findet in Blättern den DPN-Gehalt wesentlich niedriger als den des TPN. Hingegen sind die beiden Coenzyme nach ANDERSON (1954) in Blättern in etwa gleichen Konzentrationen vorhanden. Jedoch ist die enzymatische Bestimmung in vielen Extrakten und Konzentraten durch die Gegenwart starker Hemmstoffe erschwert.

In Phenol-Extrakten fand WHATLEY für TPN bei einer Reihe von Pflanzenarten (Blätter, Knollen, Samen) 0,4—8,6 γ/g Frischgewicht des Ausgangsmaterials. Hingegen ließ sich der DPN-Gehalt sogar in Konzentraten aus Blättern nicht enzymatisch bestimmen, da Enzymgifte als hartnäckige Begleiter auftreten.

Die erhaltenen Werte bedeuten deshalb vielfach nur Annäherungen.

Tabelle 1. *DPN-Gehalt verschiedener Pflanzengewebe.*

	γ DPN/g Frischgewicht (oxyd. u. red. Form)	Referent
Bierhefe	1000—1500	SCHLENK (1942)
Bäckerhefe	1000—1500	SCHLENK (1942)
Pollen (Salix, Populus)	700—1000	v. EULER (1944)
Weizenkeime[1]	8—10	PFLEIDERER (1953)
Erbsensprosse	10—20	HASSE
Wickensprosse	10	HASSE
Bohne, Blatt	13,7	ANDERSON (1954)
Tomate, Blatt	2,9	ANDERSON (1954)
Spinat, Blatt	13,6	ANDERSON (1954)

Tabelle 2. *TPN-Gehalt verschiedener Pflanzengewebe.*

	γ TPN/g Frischgewicht (oxyd. u. red. Form)	Referent
Bäckerhefe	8,6	WHATLEY (1951)
Klee, Blatt	1,7—4,6	WHATLEY (1951)
Klee, Blattstiel	0,7	WHATLEY (1951)
Klee, Blütenstand	0,4	WHATLEY (1951)
Hafer, Blatt	4,4	WHATLEY (1951)
Kartoffel, Blättchen	8,1	WHATLEY (1951)
Kartoffel, junge Knollen	1,0	WHATLEY (1951)
Schwarzer Holunder, Blättchen	3,1	WHATLEY (1951)
Stangenbohnen, Blättchen	2,0	WHATLEY (1951)
Weizenkeime[1]	3—4	PFLEIDERER (1953)
Bohne, Blatt	9,7	ANDERSON (1954)
Tomate, Blatt	2,6	ANDERSON (1954)
Spinat, Blatt	8,2	ANDERSON (1954)

[1] Lufttrocken, saurer Extrakt.

2. Bestimmung von Diphospho-pyridin-nucleotid.

Optische Methoden. Verschiedene Dehydrierungsysteme, insbesondere unter Verwendung von Methylenblau als Wasserstoffacceptor, sind geeignet: Milchsäure-dehydrase mit Diaphorase (Whatley, 1951; Clark, 1949 und Sumner, 1947). Zymasesystem (Sumner, 1947).

Die spektrophotometrische Bestimmung als Dihydro-cozymase erfolgt durch Messung der Extinktion um 340 mμ. Die Reduktion geschieht chemisch (Warburg, 1936) oder enzymatisch mit Alkohol-dehydrase (Racker, 1949), Milchsäuredehydrase (Clark, 1949), Phosphoglycerinaldehyd-dehydrase (Adler, 1938). Wird Ferri-Cytochrom c als Wasserstoffacceptor verwendet (Alkohol-dehydrase und DPN-Cytochrom c-reduktase), läßt sich dessen Reduktion durch Messung der Extinktion bei 550 mμ bestimmen (Hogeboom, 1948).

Blausäure reagiert mit N-substituierten Nicotinsäureamid-Verbindungen unter Bildung dissoziabler Komplexe, die fluoreszieren und im Bereich von 325 mμ Absorptionsbanden zeigen (Colowick, 1951). Bei der Behandlung von DPN mit starkem Alkali entsteht ein fluoreszierendes stabiles Produkt, das zur Bestimmung dienen kann (Colowick, 1951).

Manometrische Methoden. Die Geschwindigkeit der alkoholischen Gärung mit bestimmten Präparaten aus Trockenhefe kann proportional der hinzugefügten Menge des Coenzyms sein. Es wird die CO_2-Bildung gemessen (Myrbäck, 1933; v. Euler, 1936). Mit dem Zymasesystem kann in Gegenwart von Methylenblau die Geschwindigkeit der Dehydrierung des Triosephosphats auch durch die Bestimmung der Sauerstoffabsorption ermittelt werden (Krishnan, 1948). Warburg (1936) verwendet einen Test mit Enzymfraktionen aus Lebedew-Saft. Jandorf (1941) bedient sich der Glykolyse. Die Reaktion besteht in der Umwandlung des Hexose-diphosphats in Phosphoglycerin und Phosphoglycerinsäure. Die Dehydrierung der Milchsäure kann in Gegenwart von Methylenblau unter aeroben Bedingungen zur Bestimmung herangezogen werden (Green, 1936).

Die quantitative Hydrierung mit Natriumhydrosulfit läßt sich manometrisch auswerten (Haas, 1936; Drabkin, 1945).

Bakteriologische Methode. Gewisse Mikroorganismen (z. B. *Haemophilus parainfluenzae*) wachsen auf synthetischer Basis nur, wenn ein Faktor V zugegen ist. Dieser Faktor läßt sich durch DPN oder TPN ersetzen, ein Befund, der zur Bestimmung von Coenzymen dienen könnte (Lwoff, 1937).

Bestimmung der Dihydro-cozymase erfolgt durch enzymatische Dehydrierung mit DPN-abhängigen Dehydrasen und gelben Fermenten und Messung des Extinktionsabfalls um 340 mμ (Racker, 1950; Kubowitz, 1943 u. a.), chemisch durch direkte Titration mit Jod (Drabkin, 1945), durch manometrische Titration mit Ferricyanid (Haas, 1937) oder durch Dehydrierung mit Dichlorphenolindophenol (Haas, 1944).

3. Bestimmung von Triphospho-pyridin-nucleotid.

Optische Methoden. Für die Thunberg-Methodik sind geeignet der Glucose-6-phosphorsäure-dehydrase-Test (Adler, 1940 u. a.) und der Isocitronensäure-dehydrase-Test (Whatley, 1951). Die spektrophotometrische Bestimmung als Dihydroverbindung wird um 340 mμ ausgeführt. Die quantitative Reduktion kann chemisch mit Natriumhydrosulfit erfolgen (Warburg, 1936; LePage, 1947; Pfleiderer, 1953). Die Reduktion wird spezifisch mittels TPN-abhängiger Apodehydrasen bewirkt: Glucose-6-phosphorsäure-dehydrase (Warburg, 1936), Isocitronensäure-dehydrase (Ochoa, 1948), Phospho-gluconsäure-dehydrase (Horecker, 1951; Kaplan, 1952). Messung der Geschwindigkeit, mit der Ferri-

Cytochrom c durch Glucose-6-phosphat reduziert wird (Glucose-6-phosphorsäure-dehydrase, TPN-Cytochrom c-reduktase) (HAAS, 1942).

Manometrische Methoden. Die Bestimmung von TPN kann auf Grund der katalytischen Wirkung des Coenzyms durch Messung des Sauerstoffverbrauchs bei der Oxydation von Glucose-6-phosphat erfolgen (mit Glucose-6-phosphorsäure-dehydrase und „altem gelbem Ferment") (WARBURG, 1935). Die quantitative Hydrierung mit Hydrosulfit kann manometrisch ausgewertet werden durch die Messung des CO_2, das durch gebildetes saures Sulfit aus Bicarbonat ausgetrieben wird (WARBURG, 1936; DRABKIN, 1946).

Die Bestimmung der Dihydroverbindung erfolgt durch enzymatische oder chemische Dehydrierung unter spektrophotometrischer Messung des Extinktions-abfalls um 340 mμ.

II. Die „THUNBERG-Technik".

Diese Acceptormethodik beruht auf der Beobachtung der Reduktion der Wasserstoff-Acceptorsubstanz durch den Wasserstoff, den die Dehydrase von ihrem spezifischen Substrat mobilisiert. Besonders geeignet sind Versuchs-anordnungen, bei denen die Reduktionswirkung durch Farbveränderungen verfolgt werden kann. In den meisten Fällen handelt es sich um reversible Oxydo-Reduktions-Indikatoren, die aus einer stark gefärbten Form in eine farblose Leukoform übergehen.

Auf der Verwendung von Methylenblau als Acceptor beruht die THUNBERG-Technik. Diese Methode führte zur Entdeckung einer großen Anzahl von Enzymen, die die Dehydrierung verschiedener Stoffwechselprodukte bewirken. Die sekundäre Übertragung des Wasserstoffs von dem hydrierten Pyridin-nucleotid auf das Methylenblau erfolgt unter Mitwirkung von Überträgersystemen (Flavin-proteide, die zwischen das Co-ferment-System und den Farb-stoff eingeschaltet sind). Zum Nachweis einer Codehydrase ist ein System erforderlich, das eine auf DPN oder TPN spezifische Apodehydrase, das dazu gehörige Substrat, ein auf Co-H$_2$ eingestell-tes Flavoprotein und Methylen-blau enthält. Dieses Enzymsystem wird durch das Coenzym kom-plettiert. Die Entfärbung des Farbstoffs muß unter anaeroben Bedingungen gemessen werden, da Leuko-Methylenblau durch Sauer-stoff reoxydiert wird. Bei der hohen Farbintensität des Methy-lenblaus ist die Empfindlichkeit der Methode eine außerordentlich große. Werden die Versuchs-bedingungen so gewählt, daß alle Komponenten des Systems mit

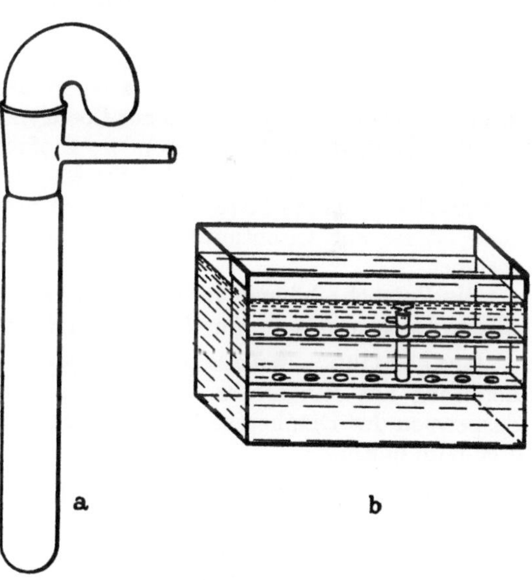

Abb. 2a u. b. Einrichtungen für die THUNBERG-Methode.
a THUNBERG-Röhrchen; b Thermostat.

Ausnahme der Codehydrase im Überschuß sind, die Codehydrase den begrenzenden Faktor der Reaktion bildet, so kann die Entfärbungsgeschwindigkeit ein Maß für die relative Konzentration des Coenzyms sein. Die Reduktionsintensitäten sind bei gleicher Methylenblaumenge umgekehrt proportional der Entfärbungszeit. Für eine schnelle Orientierung über den Coenzymgehalt in Rohextrakten erweist sich die Thunberg-Methodik als besonders geeignet, vorausgesetzt, daß keine das Enzymsystem störenden Begleitstoffe vorhanden sind.

Ausführung. (Ahlgreen, 1925.) Die Bestimmung der Methylenblau-Entfärbung wird in evakuierten Vakuum-Röhren (Thunberg-Röhren) durchgeführt. Diese haben ein Volumen von etwa 10 ml. Der Glasstöpsel ist mit einem Loch von 2—3 mm Durchmesser in der Höhe des Seitenrohres versehen (Abb. 2). Ist es von Interesse, erst nach der Evakuierung die Reaktion durch Zugabe einer Komponente auszulösen, so verwendet man vorteilhaft Vakuumröhren mit seitlichem Ansatz oder ausgehöhltem Stöpsel. Für die quantitative Messung des Entfärbungsgrades in einem Photometer sind verschiedene modifizierte Gefäße mit planparallelen Glasplatten beschrieben worden. Um luftdichten Verschluß zu garantieren, werden die Hähne vor jedem Versuch eingefettet.

Nach der Beschickung mit der Reaktionsmischung werden die Röhrchen mit einer Wasserstrahlpumpe evakuiert. Durch Drehen des Hahnes kann die Verbindung zur Pumpe bzw. der Abschluß von der Außenluft bewirkt werden. Das Evakuieren wird beendet, wenn die Flüssigkeit beim Erwärmen mit der Hand zu sieden beginnt. Die Röhrchen werden in einen Wasserthermostaten mit Thermoregulator und Rührwerk bei 30° oder 35° gebracht, und die Zeit der Entfärbung wird beobachtet. Die Thermostaten (etwa ein Aquarium mit den Ausmaßen $40 \times 30 \times 30$ cm) sind so eingerichtet, daß man die Reduktion des Methylenblaus leicht verfolgen kann. Im allgemeinen verfährt man so, daß man die Zeitdauer bis zur vollständigen Entfärbung ermittelt. Doch ist es häufig zweckmäßig, statt dessen den Zeitpunkt 90%iger Entfärbung durch Vergleich mit einem Kontrollversuch (aérob, kein Coenzym) zu bestimmen, der nur 10% der Methylenblaumenge des Hauptversuchs enthält.

Die Versuche lassen sich auch in Reagenzgläsern durchführen, durch die ein kontinuierlicher Strom sauerstofffreien Stickstoffs geleitet wird. Die Gläser werden mit Korkstopfen verschlossen, die einen Schlitz haben, damit das Gas entweichen kann (Sumner, 1948).

1. Bestimmung von Diphospho-pyridin-nucleotid nach der Thunberg-Technik.

Milchsäure-dehydrase-Test. (Sumner, 1948.) Das Enzymsystem besteht aus Natriumlactat, Milchsäure-apodehydrase, Diaphorase, Blausäure, Methylenblau und DPN.

Werden die Versuchsbedingungen so gewählt, daß DPN der die Reaktion begrenzende Faktor ist, so gibt es einen Bereich der Proportionalität der Entfärbungszeit des Methylenblaus zur Coenzymkonzentration.

Diese Methode ist durch besondere Einfachheit ausgezeichnet. Die benötigten Enzympräparate sind in der hierfür erforderlichen Reinheit leicht zu erhalten und haltbar. Die Methode erhebt keinen besonderen Anspruch auf Genauigkeit, doch ist sie zur Orientierung über den DPN-Gehalt von Rohextrakten sehr geeignet.

Die Versuche können in Thunberg-Röhren im Vakuum ausgeführt werden oder in Reagenzgläsern, durch die sauerstofffreier Stickstoff geleitet wird.

Reagentien: *Milchsäure-apodehydrase und Diaphorase* (Sumner, 1948). Bereitung aus Rinderherz. Die Lösung der Milchsäure-apodehydrase und die Suspension des Diaphorase-Präparats sind unter Toluol in der Kälte mindestens 6 Monate haltbar.

DPN-Lösung. Die angewendete Menge ist so zu wählen, daß Entfärbungszeiten zwischen 1—5 min gemessen werden (etwa 1—6 γ DPN).

Ansatz: 1,0 ml Diaphorase-Suspension,
       3 Tropfen Milchsäure-apodehydrase-Lösung,
       1 ml 2%ige Kaliumcyanid-Lösung (neutral),
       2 Tropfen 10%ig D,L-Natriumlactat,
       2—3 Tropfen Caprylalkohol,
       0,1 ml Methylenblau 1:5000,
       0,1 ml DPN-Lösung.

Kontrolle enthält statt 0,1 ml DPN-Lösung 0,1 ml Wasser. Badtemperatur 25°.

Beispiel.

γ DPN	1	2	3	6.5
Entfärbungszeit in min	4	2,5	1,5	1

Bereitung eines Milchsäure-apodehydrase-Präparats (SUMNER, 1948).

Frisches Rinderherz wird von Fett befreit und unter Zusatz von 3 l Wasser und etwas Toluol im „Waring Blendor" fein gemahlen. Die Lösung wird in der Kälte filtriert, der Rückstand zur Bereitung des Diaphorase-Präparats aufgehoben. Unter Rühren gibt man auf je 100 ml des Filtrats 46 g Ammonsulfat. Es wird in der Kälte filtriert und die Lösung verworfen. Der Niederschlag wird in 500 ml Wasser gelöst und erneut durch Zugabe von 46 g Ammonsulfat auf je 100 ml der Enzymlösung gefällt. Der Niederschlag wird mit etwa 40 ml Wasser und etwas Toluol vermischt und in der Kälte über Nacht gegen dest. Wasser dialysiert. Der Dialysiersack darf nur $^1/_4$ in das Wasser eintauchen, um eine zu starke Verdünnung zu vermeiden.

Bereitung einer Diaphorase-Suspension.

Der fein gemahlene unlösliche Rückstand aus Rinderherz (s. oben), der bei der Bereitung der Milchsäure-apodehydrase abfiltriert wurde, wird in 2 l 1%igem neutralem Phosphatpuffer suspendiert und im „Waring-Blendor" unter Zusatz von etwas Toluol zerkleinert. Die Suspension wird in der Kälte filtriert. Der Rückstand wird erneut in verdünntem Puffer suspendiert, zerkleinert und filtriert. Der Niederschlag wird in etwa 100 ml 5%igem neutralem Phosphat suspendiert, noch einmal gemahlen und dann unter Toluol im Eisschrank aufbewahrt.

2. Bestimmung von Triphospho-pyridin-nucleotid nach der THUNBERG-Technik.

Glucose-6-phosphorsäure-dehydrase-Test. (ADLER, 1940.) In dem System Glucose-6-phosphat, Glucose-6-phosphat-apodehydrase, Flavinenzym, Methylenblau und TPN wird die Entfärbung von Methylenblau gemessen. Werden die Bedingungen so gewählt, daß TPN der die Reaktion begrenzende Faktor ist, so gibt es einen Proportionalitätsbereich zwischen Coenzymkonzentration und Entfärbungsgeschwindigkeit.

Reagentien: *Glucose-6-phosphorsäure-apodehydrase* (Vorschrift S. 333). Bereitung nach LE PAGE (1951). Die Konzentration des Enzyms wird so gewählt, daß TPN-Mengen zwischen 2 und 30 γ gut meßbare Entfärbungszeiten geben.

„*Altes gelbes Ferment*". Bereitung nach WARBURG (1933) (Vorschrift S. 333).

Glucose-6-phosphorsaures Kalium. Bereitung nach WARBURG (1932) (Vorschrift S. 333).

Die Versuche werden in THUNBERG-Gefäßen ausgeführt.

Ansatz: 0,25 ml m/10 Kaliumsalz von Glucose-6-phosphat,
      0,25 ml Hexose-6-Phosphorsäure-apodehydrase,
      0,25 ml „altes gelbes Ferment", entsprechend etwa 0,8 γ Lactoflavin,
      0,25 ml m/2 Phosphatpuffer p_H 7,6,
      0,5 ml 0,02%ig Methylenblau-Lösung (aus Ansatzbirne zugekippt),
      x ml TPN-Lösung,
      0,5—x ml Wasser.

Beispiel.

γ TPN	5	10	15	20
Entfärbungszeit in min	20	10	6	4

III. Manometrische Methoden.

Manometrische Verfahren haben bei der Isolierung der Codehydrasen eine große Rolle gespielt. Die Testmethode bei der fortschreitenden Reinigung der Cozymase in v. Eulers Institut war die Gärprobe mit gewaschener Unterhefe, in Warburgs Institut ein Test mit homogenen Enzymfraktionen. Der Leitfaden bei der Isolierung von Triphospho-pyridin-nucleotid war die Sauerstoff-Messung bei der enzymatischen Oxydation von Glucose-6-phosphat. Von entscheidender Bedeutung für die Durchführung der Teste ist die von Warburg geschaffene Apparatur (Warburg-Technik). Vgl. dazu: R. H. Kenten: Gasometric Analysis in Plant Investigation. Bd. I dieses Handbuchs.

1. Manometrische Bestimmung von Diphospho-pyridin-nucleotid im Hefe-Gärtest.
(Schlenk, 1947; Axelrod, 1939.)

Werden zu einem gewaschenen Trockenhefepräparat (Apozymase) variable Mengen an DPN hinzugefügt, so ist die Geschwindigkeit der alkoholischen Gärung in bestimmten Grenzen proportional der Menge des angewendeten Coenzyms.

Das gebildete CO_2 wird in der Warburg-Apparatur gemessen. Die Gärung wird mit Luft im Gasraum ausgeführt. Das p_H-Optimum liegt bei 6,0—6,4. Nach einer Induktionszeit von 10—20 min erreicht die Gärungsgeschwindigkeit ein Maximum über eine Zeitspanne von 30 min und länger. Diese Induktionsperiode wird durch Fructose-1,6-diphosphat verkürzt. Mg- und Mn-Ionen haben einen bemerkenswerten Einfluß auf die Gärungsgeschwindigkeit. Die maximale Geschwindigkeit der CO_2-Bildung ist proportional der vorhandenen Coenzymmenge im Bereich von etwa 10—30 γ DPN/Ansatz. Es gelingt nicht, die Trockenhefe unter Erhaltung der enzymatischen Aktivität vollständig von Cozymase zu befreien. Der Leerwert des Ansatzes ohne Coenzym-Zusatz ist in Abzug zu bringen. Die Fehlergrenze in Parallelversuchen beträgt ± 10%. Die relativen Werte dieser Methode müssen mit dem eines DPN-Präparats von bekanntem Gehalt (spektrophotometrische Bestimmung) verglichen werden.

Gewisse Schwierigkeiten bereitet die Unbeständigkeit der Apozymasesuspension. Sie verändert in Stunden schon wesentlich ihre Aktivität, so daß die Vergleichslösung des Standardpräparates bei jeder Versuchsreihe mitgemessen werden muß. Die Stabilisierung der Apozymase ist möglich durch Aufbewahren der Suspension in eingefrorenem Zustand. Unter diesen Bedingungen ist die Aktivität etwa 24 Std. ausreichend. Eine größere Stabilität über 1—2 Wochen wird durch Trocknen erhalten. Die etwas umständliche getrennte Einwaage der festen Apozymase für jeden Versuch bringt den Vorteil, daß für zahlreiche Versuchsserien ein Versuch mit dem Standardpräparat ausreichend ist.

Reagentien:

Fructose-1,6-diphosphorsaures Kalium. Bereitung durch biologische Phosphorylierung von Glucose mit Bierhefe (Neuberg, 1928). Salze des Hexosediphosphats im Handel erhältlich. Reinigung nach Bioch. Preps. 2, 52.

Apozymase. Bereitung durch Auswaschen von Trockenbierhefe (Schlenk, 1947).

Ansatz: 200 mg trockene Apozymase,
   0,6 ml m/10 Phosphatpuffer p_H 6,2,
   0,2 ml 40%ige Glucoselösung in m/10 Phosphatpuffer p_H 6,2,
   0,4 ml 5,5%ige Lösung von Fructose-1,6-diphosphorsaurem Kalium
    p_H 6,2,
   0,1 ml Mg-chlorid + Mn-chlorid je 0,1%ig,
   x ml DPN-Lösung (entsprechend 10—40 γ) aus der Birne zugekippt,
   0,7—x ml Wasser.

Bedingungen: Temperatur 30°, Gasraum: Luft.

Beispiel.

Min. nach Einkippen der DPN-Lösung	DPN-Präparat in γ/Ansatz			
	0	10	20	40
	mm³CO_2	mm³CO_2	mm³CO_2	mm³CO_2
10	15	23	30	48
20	25	45	66	104
30	38	69	99	158

Darstellung von Fructose-1,6-diphosphorsaurem Kalium (NEUBERG, 1928).

Darstellung des Calciumsalzes. 200 g Rohrzucker und 36 g KH_2PO_4 werden in einem 3 l-Kolben in 1 l Wasser von 40° gelöst. Dann erfolgt Zugabe von 11 g $NaHCO_3$ (p_H-Optimum 6,2—6,6). In dieser Lösung werden 300 g gut ausgewaschene Unterhefe (Bierhefe) suspendiert und 50 ml Toluol zugesetzt. Man bewahrt das Reaktionsgemisch unter häufigem Umschwenken bei 37° auf. CO_2-Bildung und Veresterung beginnen bald. Es werden Proben entnommen, um die Abnahme des anorganischen Phosphats durch eine kolorimetrische Methode zu verfolgen. Die Reaktion ist nach 2—5 Std. beendet. Das Reaktionsgemisch wird nun mit 2 n-Natronlauge gegen Lackmus neutralisiert und etwa 20 min auf dem siedenden Wasserbad erhitzt. Dabei wird alles Eiweiß ausgefällt. Das Filtrat enthält Hexosediphosphat neben etwas Monophosphat. In einem Rundkolben wird die Lösung auf 95° erwärmt und mit einer siedenden Lösung von kristallisiertem Calciumchlorid in 100 ml Wasser versetzt. Das in der Hitze schwerlösliche Calciumsalz fällt sofort aus und wird heiß abgesaugt. Es wird mit wenig siedendem Wasser nachgewaschen. Die Ausbeute beträgt etwa 30 g.

Reinigung. Der noch feuchte Niederschlag wird in 50 ml 2 n-Essigsäure unter Zusatz von 50 ml Wasser gelöst. Das Filtrat wird mit 2 n-Natronlauge gegen Lackmus neutralisiert. Es wird 15 min auf dem siedenden Wasserbad erhitzt, dann möglichst heiß abgesaugt und mit wenig siedendem Wasser nachgewaschen.

Lösliches Calciumsalz. Das rohe unlösliche Calciumsalz wird in Milchsäure gelöst. Nach Abfiltrieren des Rückstandes und Neutralisieren mit Ammoniak wird die lösliche Modifikation des Calciumsalzes mit Alkohol ausgefällt.

Lösung des Kaliumsalzes. Das lösliche Calciumsalz wird entsprechend seinem Gehalt mit der äquivalenten Menge einer Lösung von Kaliumoxalat umgesetzt und vom Calciumoxalat durch Zentrifugieren abgetrennt.

Der Gehalt der Hexose-diphosphat-Präparate an organisch gebundenem Phosphat wird aus der Zunahme des Orthophosphats nach 3 stündiger Hydrolyse in 1 n-HCl bei 100° ermittelt (LOHMANN, 1926).

Apozymase aus Trockenhefe. (SCHLENK, 1947.) Als Ausgangsmaterial dient getrocknete Bierhefe. Bierhefe wird mit viel Wasser einige Stunden ausgewaschen und die abgepreßte Hefe in dünner Schicht 24 Std. bei 20° getrocknet. Diese Trockenhefe ist zur Bereitung von Apozymase mindestens 1 Jahr geeignet.

Apozymase wird aus dieser getrockneten Bierhefe durch Waschen mit Wasser erhalten. Bäckerhefe ist ungeeignet. Die optimale Dauer des Auswaschens muß empirisch ermittelt werden. Langes Waschen setzt zwar den Leerwert des Gärversuchs herab, führt aber zu einer geringeren Aktivität der Apozymase.

Vorschrift: 2 g Trockenhefe werden in einem Zentrifugenglas in 100 ml kaltem Wasser suspendiert und nach 7 min langem Schütteln wird 7—10 min bei 2000 Touren zentrifugiert. 3—4 maliges Auswaschen im Verlaufe einer Stunde ist ausreichend.

Anmerkung: Besonders einfach zugänglich ist Apozymase aus käuflicher technischer Trockenbierhefe. Halbstündiges, einmaliges Auswaschen von 2 g mit 800 ml Wasser unter Rühren ist nach eigener Erfahrung ausreichend.

2. Manometrische Bestimmung von Triphospho-pyridin-nucleotid im Glucose-6-phosphorsäure-Dehydrase-Test.
(WARBURG, 1935.)

In dem System Glucose-6-phosphat, Glucose-6-phosphorsäure-apodehydrase, „altes gelbes Ferment", Sauerstoff und TPN wird die Geschwindigkeit der Oxydation manometrisch gemessen unter Bedingungen, unter denen die Geschwindigkeit proportional der Konzentration des Coenzyms ist.

Ein Zusatz von Kaliumcyanid dient der Vergiftung der Katalase des Enzympräparats. Ist W die Wirkung, die m mg des zu prüfenden Präparats erzeugen, und W_0 die Wirkung, die m_0 mg des Standardpräparats erzeugen, so ist der Reinheitsgrad R des zu prüfenden Präparats $R = \dfrac{W_o}{W} \times \dfrac{m_0}{m}$.

Reagentien:

Glucose-6-phosphorsaures Kalium. Präparat im Handel erhältlich.. Enzymatische Phosphorylierung von Glucose mit Hefe führt zu einem Gemisch von Hexosemonophosphaten, aus denen reines Glucose-6-phosphat erhalten werden kann (Warburg, 1932 u. a.). Bereitung durch enzymatische Isomerisierung von d-Fructose-6-phosphat (Bioch. Preps. 3, 71). Chemische Synthese über Tetraacetyl-β-d-Glucose (Lardy, 1946). Chemische Phosphorylierung von Glucose mit Polyphosphat (Seegmiller, 1951), Vorschrift S. 332. 1 ml der Lösung soll etwa 8 mg Phosphor enthalten.

Glucose-6-phosphorsäure-apodehydrase (nach Negelein, 1936; Le Page, 1947). 2 mg der Apodehydrase nach Le Page in 10 ml Wasser. Haltbarkeit der Lösung etwa 2 Tage.

„*Altes gelbes Ferment*" (Warburg, 1933). 1 g des mit Chloroform gereinigten Präparates wird in 10 ml Wasser gelöst und hochtourig zentrifugiert. Die Fermentlösung wird mit Wasser so verdünnt, daß die Konzentration 4×10^{-8} Mol/ml beträgt. Die Konzentrationsbestimmung erfolgt durch lichtelektrische Messung bei 465 mμ.

Ansatz: 0,2 ml Hexose-monophosphorsaures Kalium,
0,2 ml Glucose-6-phosphorsäure-apodehydrase,
0,25 ml „altes gelbes Ferment" (etwa 10^{-8} Mol), aus der Birne zugekippt,
0,2 ml n/10 Kaliumcyanid-Lösung,
x ml TPN-Lösung,
$3-x$ ml Wasser.

Bedingungen: 38°, Gasraum: Sauerstoff, Einsatz: 0,2 ml 5%ige Kalilauge.

Beispiel.

	TPN-Präparat in γ/Ansatz		
	0	2,5	5
Vor Einkippen des gelben Ferments in 10 min	0 nl O_2	0 nl O_2	0 nl O_2
Nach Einkippen des gelben Ferments in 10 min	0 nl O_2	—25 nl O_2	—50 nl O_2

Hexose-monophosphorsaures Kalium (Gleichgewichtsester). (Nach Warburg, 1932.)

Darstellung des Bariumsalzes. 1200 ml frischer Lebedew-Saft und 120 g Glucose werden in einen 10 l-Kolben gebracht, auf 29° erwärmt und mechanisch gerührt. In Abständen von 6 min fügt man jedesmal 100 ml einer Glucose/Phosphatlösung zu (300 g Glucose + 300 g $Na_2HPO_4 \cdot 10 H_2O$ + 20 g KH_2PO_4 mit Wasser auf 1500 ml), bis in 90 min 1500 ml zugegeben sind. Nach weiteren 15 min fällt man mit 80 g Trichloressigsäure (gelöst in 100 l Wasser) und läßt nach Zusatz einiger Tropfen Oktylalkohol etwa 12 Std. bei 0° stehen. Der Niederschlag wird abzentrifugiert, die überstehende Lösung zur Klärung durch ein Filter gesaugt. Zu dem Filtrat gibt man 320 g festes Bariumacetat, neutralisiert mit heißgesättigter Barytlösung (Phenolphthalein rosa) fügt $^1/_{10}$ des Volumens Alkohol hinzu, schüttelt um und saugt ab. Aus dem Filtrat wird die Monosäure mit Bleisubacetat gefällt, je nach dem Gehalt an Monosäure sind 0,8—2,7 kg Liquor plumbi subacetici (DAB 6) nötig. Das abzentrifugierte Bleisalz wird mit Wasser gewaschen (Volumen des Waschwassers etwa gleich der Hälfte des Volumens der Mutterlauge), dann mit Wasser in einer Reibschale zerrieben und auf der Schüttelmaschine 12 Std. mit Schwefelwasserstoff geschüttelt. Das Filtrat wird durch Auspumpen von Schwefelwasserstoff befreit, mit Baryt neutralisiert (je nach Ausbeute 40—130 g Baryt) (Phenolphthalein rosa) und nach Klärung durch ein Faltenfilter in das doppelte Volumen Alkohol gegossen. Das Hexose-monophosphorsaure Barium wird abgesaugt, mit abs. Alkohol in einer Reibschale zerrieben, wieder abgesaugt und im Vakuum-exsikkator über Schwefelsäure getrocknet. Die Ausbeute an rohem Bariumsalz beträgt 100—300 g.

Darstellung des Calciumsalzes. Man löst 100 g des Bariumsalzes in 500 ml Wasser und gibt bei schwach alkalischer Reaktion eine 15%ige Lösung von Mercuri-acetat zu, bis nichts mehr fällt (etwa 75 ml). Das Filtrat, das das Bariumsalz der Monosäure enthält, wird mit Schwefelwasserstoff entquecksilbert und nach dem Abfiltrieren des Quecksilbersulfids 12 Std. durchlüftet. Die gelbe Farbe und den Geruch der Lösung entfernt man durch drei-maliges Schütteln mit je 3 g Tierkohle. Aus der farblosen Lösung wird das Barium mit einem kleinen Überschuß von Schwefelsäure gefällt. Das Bariumsulfat wird abzentrifugiert. Zu der überstehenden Lösung gibt man etwas mehr als die berechnete Menge Calciumacetat (35 g), gelöst in 170 ml warmen Wassers, wobei kein Niederschlag fallen darf und dann unter kräftigem Umschütteln Alkohol (etwa 200 ml), bis der an der Eingußstelle sich bildende Niederschlag beim Umschütteln gerade noch gelöst wird. Erwärmt man diese Lösung auf dem siedenden Wasserbad, so fällt das Calciumsalz der Hexose-monophosphorsäure als schwerer Niederschlag, der sich schnell absetzt. Er besteht aus regelmäßig ausgebildeten Kugeln mit radialer Zeichnung, oft aus sternförmig gruppierten Nadeln. Nach dem Röntgen-bild ist der Niederschlag kristallinisch. Da sich der Niederschlag beim Abkühlen löst, so muß er heiß abgesaugt werden. Man wäscht ihn auf der Nutsche zuerst mit heißem 30%igem Alkohol, dann mit kaltem abs. Alkohol (Fraktion I).

Die Mutterlauge von Fraktion I kühlt man mit Eiswasser ab, setzt in der Kälte wieder Alkohol zu (etwa 300 ml), bis gerade kein bleibender Niederschlag entsteht und erwärmt nun und behandelt weiter wie bei der Gewinnung der Fraktion I. Die Ausbeuten an den Fraktionen I und II betragen zusammen etwa 32 g.

Man erhält weitere Fraktionen, wenn man Zusatz des Alkohols und Erwärmen wiederholt, doch sind die späteren Fraktionen unreiner. Fraktion I und II haben gleichen Wasser- und Calciumgehalt, etwa gleiche spezifische Drehung und Aldosezahl.

Das Calciumsalz löst sich in kaltem Wasser verhältnismäßig leicht. Die Lösungen trüben sich beim Erwärmen und werden beim Abkühlen wieder klar. Das im Vakuum über Schwefel-säure getrocknete Salz verliert im Hochvakuum bei 60° 1 Mol Wasser.

Kaliumsalz. 2 g Calciumsalz werden in einer Reibschale mit 15 ml Wasser verrieben und mit 5 ml Kaliumoxalat-Lösung (23 g in 100 ml) versetzt. Nach einer halben Stunde wird zentrifugiert. Die überstehende Lösung prüft man auf Calcium und Oxalat und fällt nun je nach der Komponente, die im Überschuß ist, tropfenweise mit Oxalat oder mit hexose-monophosphorsaurem Calcium, indem man nach jedem Zusatz zentrifugiert und prüft. Die Lösung wird als Stammlösung im Eisschrank aufbewahrt.

Darstellung von Glucose-6-phosphorsäure-apodehydrase (LE PAGE, 1947; WARBURG, 1932). Ausgangsmaterial: Bierhefe wird dreimal durch Suspendieren in dest. Wasser gewaschen und jedesmal in einer Filterpresse abgepreßt. Sie wird in dünner Schicht ausgebreitet und mit einem Luftstrom bei 30° C getrocknet.

Vorschrift: 400 g dieser Trockenhefe werden in 1200 ml dest. Wasser suspendiert und zur Autolyse 10 Std. bei 37° C sich selbst überlassen. Diese Suspension wird zentrifugiert. 150 ml der überstehenden Lösung werden auf 1500 ml mit dest. Wasser verdünnt und auf 0° abgekühlt. Diese Lösung wird mit CO_2 gesättigt (30 min). Der gebildete Niederschlag wird bei 0° durch Zentrifugieren abgetrennt und verworfen. Die überstehende Lösung wird mit CO_2-gesättigtem dest. Wasser von 30° C auf 15 l verdünnt. Diese Lösung bleibt 30 min bei 30° stehen, dann über Nacht bei 0°. Die klare Lösung wird dekantiert, der Rest zentrifugiert. Der Niederschlag enthält die gewünschte Dehydrase. Er wird über Phosphorpentoxyd im Vakuum getrocknet. Ausbeute 1,0—1,5 g. Das Präparat zusammen mit Glucose-6-phosphat hydriert DPN nicht.

„Altes gelbes Ferment" aus Hefe (WARBURG, 1933). Vorschrift: Bierhefe wird gewaschen, getrocknet und nach v. LEBEDEW mit Wasser extrahiert. Der frische Lebedew-Saft wird mit Bleisubacetat (pro Liter Saft 400 ml Liquor plumbi subacetici DAB 6) versetzt und unter Zusatz von etwas Oktanol kräftig geschüttelt. Man läßt zur Vervollständigung der Fällung 12 Std. im Kältezimmer stehen, zentrifugiert und fällt aus der überstehenden Lösung, die das Ferment enthält, das überschüssige Blei mit Phosphat (pro Liter Lebedew-Saft 200 ml einer Phosphatmischung aus 85 Vol. m/2 Na_2HPO_4 und 15 Vol. m/2 KH_2PO_4). Das Blei-phosphat wird abzentrifugiert, die überstehende Lösung, die das Ferment enthält, wird mit dem halben Volumen Aceton versetzt und bleibt 24 Std. bei 0° stehen. Der abgeschiedene weiße Niederschlag wird abzentrifugiert, die überstehende Lösung, die das Ferment enthält, wird bei 0° mit Kohlensäure gesättigt und mit so viel Aceton (knapp dem halbem Volumen) versetzt, daß der gesamte gelbe Farbstoff als zähes gelbes Öl ausfällt. Man gießt das Aceton ab, löst den Bodensatz in Wasser, zentrifugiert einen weißen Bodensatz ab, sättigt wieder bei 0° mit Kohlensäure und wiederholt die Fällung mit Aceton aus kohlensaurer Lösung noch zweimal. Bleibt dabei das Öl in der gelben Flüssigkeit emulgiert, so muß es durch Zentrifugieren abgetrennt werden. Nach der dritten Acetonfällung löst man das Öl in Wasser,

fällt bei 0° mit dem gleichen Volumen eisgekühltem Methanol, wobei alles Ferment ausfällt, zentrifugiert das gelbe Öl ab, löst es in Wasser, fällt es bei 0° mit dem dreifachen Volumen Methanol unter kräftigem Schütteln, saugt den körnigen Niederschlag ab, wäscht ihn auf der Nutsche mit abs. Methanol und trocknet ihn im Vakuum-Exsikkator über Schwefelsäure. Aus 118 kg Trockenhefe wurden so 1700 g Rohprodukt als gelbes Pulver erhalten. Bei dem Verfahren ist gute Kühlung der Flüssigkeiten bei Zugabe des Acetons oder Methanols wesentlich. Wird nicht hinreichend gekühlt, so hat man Verluste durch Farbstoffabspaltung (grünlich-gelbe Färbung der Mutterlauge).

Reinigungsverfahren: 2 g rohes Fermentpulver werden mit 40 ml 1%iger Kochsalzlösung verrieben und nach Zusatz von 0,5 ml Chloroform und 0,2 ml Oktanol bei 38° 24 Std. auf der Schüttelmaschine geschüttelt. Der weiße Niederschlag wird abzentrifugiert, die überstehende Lösung wird bei 0° in Cellophanschläuchen gegen fließendes dest. Wasser 1—2 Tage dialysiert und dann im Vakuum-Exsikkator zur Trockne verdampft. Der Rückstand (etwa 0,5 g) besteht aus durchsichtigen gelben Lamellen, die sich in Wasser zu einer klaren gelben Flüssigkeit lösen.

IV. Spektrophotometrische Methoden.

Lichtelektrische Methoden haben bei der Aufklärung der Wirkungen der Pyridin-dehydrasen eine entscheidende Rolle gespielt, da die Coenzyme charakteristische Absorptionsspektren besitzen, die sich beim Wechsel ihres Oxydationszustandes ändern. So ist es leicht möglich, die Reaktionsgeschwindigkeiten der Dehydrierungsvorgänge durch Veränderung der Absorption zu verfolgen oder durch nur eine Messung die Konzentration der Codehydrasen zu bestimmen.

Die spektrophotometrische Bestimmung der beiden Coenzyme gründet sich darauf, daß ihre Dihydroverbindungen im UV gleichstarke Absorptionsbanden mit einem Maximum bei 340 mμ besitzen, die den oxydierten Formen fehlen. Die Bestimmung von DPN bzw. TPN erfolgt nach ihrer Hydrierung als Dihydroverbindung durch Messung der Absorption in dem für sie charakteristischen Spektralbereich. Ist der molekulare Absorptionskoeffizient der reinen Substanz für eine bestimmte Wellenlänge um 340 mμ bekannt, so ist direkt die Bestimmung des Coenzymgehalts der Lösung möglich. Die Bestimmung erfolgt mit einem Spektrophotometer unter Verwendung von Quarzküvetten bei meist 340 mμ. Apparativ einfacher ist die Messung bei 366 mμ, der stärksten und sehr einfach isolierbaren Spektrallinie des Quecksilberdampfspektrums.

Für die Lichtschwächung beim Durchgang durch Materie gilt:

$$E = \lg \frac{I_0}{I} = \varepsilon \cdot c \cdot d$$

E gemessene Extinktion,
d Schichtdicke in [cm],
c Konzentration des absorbierenden Stoffes,
ε Extinktionskoeffizient.

Den molaren Extinktionskoeffizienten ε_m erhält man, wenn man die Konzentration in [Mol/Volumen] ausdrückt. Im folgenden wird stets mit dem molaren Extinktionskoeffizienten gerechnet, seine Dimension ist [cm^2/Mol], die Konzentration wird entsprechend in [Mol/ml] angegeben.

Die in der Literatur für ε_m angeführten Werte weichen z. T. beträchtlich voneinander ab. Schuld daran tragen die Schwierigkeiten, die bei der Herstellung reinster Coenzym-Präparate auftreten. Die Zahlenwerte schwanken zwischen $4,3 \times 10^6$ und $6,3 \times 10^6$ [cm^2/Mol]. Die heute gebräuchlichsten Werte sind: $5,6 \times 10^6$ [cm^2/Mol] (Warburg, 1934; LePage, 1938) und $6,22 \times 10^6$ [cm^2/Mol] (Horecker, 1948). Der zuletzt genannte ist der wahrscheinlichste, er fällt auch sehr nahe mit dem von Ohlmeyer (1938) an isoliertem DPN-H$_2$ gemessenen Wert $6,27 \times 10^6$ [cm^2/Mol] zusammen.

Die Konzentration der Coenzyme erhält man sehr einfach, wenn man die bei einer Schichtdicke von 1 cm erhaltenen Extinktionswerte mit einem der folgenden Faktoren multipliziert. Für DPN: 106,7, TPN: 120. Man erhält so direkt die Konzentration des Coenzyms in γ/ml.

Der für lichtelektrische Messungen günstigste Bereich für E liegt zwischen 0,1 und 1,0. Damit empfiehlt sich bei der Schichtdicke 1 cm für die Coenzyme ein Konzentrationsbereich zwischen 10 und 100 γ/ml.

Die Reduktion der Coenzyme kann chemisch oder enzymatisch erfolgen. Die chemische Reduktion mit Dithionit erfaßt unspezifisch beide Coenzyme, ebenso auch ihre Spaltprodukte mit quartärem Stickstoffatom am Pyridinkern.

Die spezifische Reduktion ist unter Verwendung DPN- bzw. TPN-abhängiger Apodehydrasen mit den entsprechenden Substraten möglich. Die dafür benötigten Verbindungen zeigen in dem auszumessenden Wellenlängenbereich häufig nur geringe Absorption, auch kann man diese leicht bei der Messung kompensieren. Die enzymatische Hydrierung führt zu Gleichgewichten. Für die direkte quantitative Analyse der Codehydrasen ist eine praktisch vollständige Überführung in eine Dihydroverbindung erforderlich. Dies läßt sich erreichen durch geeignete Auswahl unter den Dehydrasen und durch geeignete Bedingungen. Insbesondere durch die Anwendung von Hilfssubstanzen ist das Gleichgewicht weitgehend im Sinne der Hydrierung der Coenzyme zu verschieben. Die Verwendung der spektrophotometrischen Methoden ist an eine Reihe von Voraussetzungen gebunden: Nur klare Lösungen sind der Messung zugänglich. Die Eigenabsorption soll nicht zu hoch sein. Die verwendeten Apodehydrasen müssen mindestens partiell gereinigt sein, damit nicht durch die Wirkung begleitender Enzyme die Messung gestört wird. So muß vermieden werden, daß das Coenzym in der Zeit des Versuchs gespalten wird, und Co-H_2-spezifische Flavinenzyme dürfen nur in suboptimalen Konzentrationen vorliegen, um eine Reoxydation der zu bestimmenden Verbindung zu vermeiden. Die Apodehydrase-Präparate brauchen in bezug auf ihre dehydrierende Wirkung nicht katalytisch einheitlich zu sein, wenn — wie z. B. bei der Gehaltsbestimmung von Coenzym-Präparaten — nur ein spezifisches Substrat vorliegt. Rohextrakte und Konzentrate sind zur Coenzym-Bestimmung nach dem spektrophotometrischen Verfahren häufig nicht zu verwenden. Auch Enzymgifte, wie Schwermetallspuren u. a., können wie bei anderen enzymatischen Analysenmethoden die Hydrierung stören. Abhilfe schafft ein Zusatz von Leim, Glykokoll, Pyrophosphat u. a.

Im Verlauf der Messung können Trübungen der Analysenlösung auftreten. Durch spezifische enzymatische Reoxydation der gebildeten Dihydro-coenzyme (Abfall der Extinktion) ist eine Kontrolle möglich.

1. Spektrophotometrische Bestimmung der beiden Codehydrasen als Dihydroverbindungen.

Chemische Reduktion.

Diphospho-pyridin-nucleotid und Triphospho-pyridin-nucleotid werden chemisch quantitativ hydriert zu den Dihydroverbindungen und diese spektrophotometrisch um 340 mμ bestimmt.

Die Reduktion von Co I- und Co II-Präparaten kann chemisch mit Dithionit (Hydrosulfit) in Gegenwart von Natriumbicarbonat ausgeführt werden, wobei die Arbeitsbedingungen recht unterschiedlich gewählt werden (WARBURG, 1936; LePAGE, 1947; GUTCHO, 1948). Natriumborhydrid läßt sich für die Hydrierung ebenfalls verwenden (MATHEWS, 1948).

Methode von GUTCHO und STEWART (1948). Bei dieser Schnellmethode ist das Coenzym ohne Luftausschluß bei 100° nach einer Minute quantitativ reduziert, nach etwa 10 min liegt bereits das Analysenergebnis vor. Die Methode eignet sich besonders für Reihenbestimmung fester Coenzym-Präparate mit nicht zu geringem Gehalt.

Reagentien:

Dithionitlösung: 200 mg Dithionit und 1 g Natriumbicarbonat in 100 ml Wasser. Die Lösung wird für jede Meßreihe frisch angesetzt.

Bicarbonat-Carbonat-Puffer p_H 9,7: 1 g Natriumbicarbonat und 1 g Natriumcarbonat (wasserfrei) in 100 ml Wasser.

Vorschrift: In ein Reagenzglas 13 × 100 mm wägt man etwa 10 mg des Coenzym-Präparates genau ein, setzt 2 ml der Dithionit-Lösung zu und taucht das Glas sofort für genau 1 min in siedendes Wasser. Die anfänglich auftretende Gelbfärbung soll nach etwa 30 sec verschwunden sein. Man schreckt in Eiswasser ab und füllt dabei das Gläschen mit der Pufferlösung fast voll. Nach einer halben Minute spült man in einen 50 ml-Meßkolben über und füllt mit Puffer zur Marke auf. Dithionit zeigt bei 330 mμ eine starke Absorption. Zur Beseitigung überschüssigen Reduktionsmittels leitet man durch die Lösung 5 min lang einen kräftigen Luftstrom. Dann mißt man die Extinktion der Lösung bei 340 mμ gegen den Bicarbonat-Carbonat-Puffer. In einer gesonderten Einwaage ohne Dithionit-Zusatz bestimmt man die Eigenabsorption des nicht reduzierten Coenzyms.

Beispiel: Gehaltsbestimmung eines DPN-Präparats.

Nr.	DPN-Präparat in der Analysen-lösung γ/ml	Extinktion $E = \lg \dfrac{I_0}{I}$	gefunden % Co I
1	298	1,042	37,3
2	159	0,568	38,1
3	148	0,507	36,5

Messungen mit UV-Spektrophotometer (Unicam SP 500), Schichtdicke 1 cm, Quarzküvetten, molarer Extinktionskoeffizient 6,22 × 10⁶ ml/Mol. Die Werte sind bereits für die Eigenabsorption des oxydierten Coenzym-Präparats korrigiert.

Mittelwert (aus 5 Messungen): Mittlerer Fehler der Einzelbestimmung:
37,5 ± 0,4% Co I. ± 0,7% Co I.

2. Spektrophotometrische Bestimmung von DPN als Dihydro-Verbindung.

Enzymatische Reduktion: Alkohol-Dehydrase-Test.

DPN wird in Gegenwart von Alkohol-apodehydrase durch Alkohol hydriert, und das gebildete DPN-H₂ wird spektrophotometrisch bei 340 mμ bestimmt. Die Vollständigkeit der Reduktion kann durch aldehydbindende Stoffe (Semicarbazid u. a.) erzwungen werden (NEGELEIN, 1937). Bei geeigneter Wahl der Bedingungen (großer Alkohol-Überschuß und alkalisches Medium) verläuft die Reaktion auch ohne solche Zusätze quantitativ (RACKER, 1949; BONNICHSEN, 1950). Das Enzympräparat ist aus Leber (BONNICHSEN, 1950) und besonders leicht aus Hefe zugänglich (RACKER, 1949).

Reagentien:

DPN-Lösung: Einwaage des festen Präparats bzw. Verdünnung der zu analysierenden Lösung ist so zu bemessen, daß 1 ml 100—1000 γ an reinem DPN enthält.

Alkohol-Apodehydrase-Lösung (RACKER, 1949): Kristallisiertes Präparat im Handel erhältlich. Bereitung aus Hefe. Reinheit der kristallisierten Dehydrase nicht erforderlich. Enzymkonzentration ist so zu wählen, daß die Reduktion des Coenzyms in 1—2 min beendet ist.

Puffer-Lösung: 0,01 mol Natriumpyrophosphat p_H 8,8.

Ausführung: In 2 Reagenzgläsern 13 × 100 mm bereitet man die folgenden Ansätze:

	I	II
	0,5 ml DPN-Lösung	0,5 ml DPN-Lösung
	3,5 ml Puffer	3,5 ml Puffer
	0,1 ml Enzymlösung	0,1 ml Enzymlösung
	0,5 ml Äthylalkohol	

In II fehlt lediglich das Substrat. Man mißt die Extinktion von I gegen II bei 340 mμ, Schichtdicke 1 cm. Aus der gemessenen Extinktion berechnet man die DPN-Konzentration, wobei man als Gesamtvolumen der Lösung 4,6 ml einsetzt.

Beispiel (HASSE).

Nr.	DPN-Präparat in der Analysen-lösung γ/ml	Extinktion $E = \lg \frac{I_0}{I}$	gefunden % Co I
1	91,7	0,323	37,6
2	114,0	0,416	38,7
3	206,0	0,736	38,1

Mittelwert (aus 10 Messungen):   Mittlerer Fehler der Einzelmessung:
38,1 ± 0,2% Co I.   ± 0,7% Co I.

Messung mit Unicam-Spektrophotometer SP 500, bei 340 mμ, Schichtdicke 1 cm, Quarzküvetten, molarer Extinktionskoeffizient 6,22 × 10^6 cm^2/Mol.

Mit dieser Methode lassen sich auch unreine Co-Präparate untersuchen mit Gehalten bis zu 1% und darunter. Man muß dazu die Einwaagen in einem geeigneten Puffer aufnehmen und durch Zentrifugieren klare Lösungen gewinnen.

Darstellung kristallisierter Alkohol-Apodehydrase aus Bäckerhefe. (RACKER, 1949.) Bäckerhefe wird in dünner Schicht 4—5 Tage bei Zimmertemperatur an der Luft getrocknet. Die Trockenhefe wird in einer Kugelmühle bei 0° 16 Std. gemahlen und kühl im gut verschlossenen Gefäß aufbewahrt.

200 g dieser Trockenhefe werden mit 600 ml 0,066 mol Dinatriumphosphatlösung unter Rühren (2 Std.) bei 37° extrahiert, anschließend weitere 3 Std. bei Zimmertemperatur. Der Heferückstand wird durch Zentrifugieren (15min bei 13000 U/min) entfernt, er kann zur Erhöhung der Ausbeuten noch mehrmals unter gleichen Bedingungen extrahiert werden. Die klare Lösung wird schnell

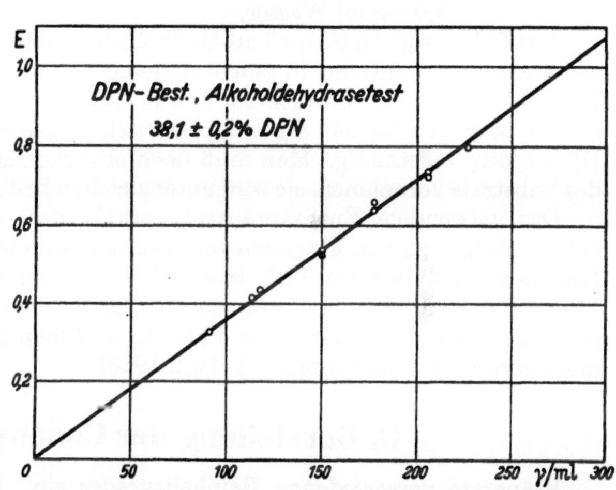

Abb. 3. Eichkurve für die photometrische Bestimmung von DPN-H$_2$.

auf 55° erwärmt und 15 min bei dieser Temperatur gehalten. Nach dem Abkühlen wird zentrifugiert. Die klare Lösung läßt sich bei 0° über Nacht aufbewahren. Zu je 100 ml des Hefeextraktes fügt man bei —2° langsam 50 ml eiskaltes Aceton hinzu. Der gebildete Niederschlag wird bei 0° zentrifugiert und verworfen. Zu je 100 ml der Lösung gibt man (wieder bei —2°) 55 ml gekühltes Aceton. Die entstandene Fällung, die die Alkoholapodehydrase enthält, wird in ungefähr 50 ml kalten Wassers gelöst und 3 Std. gegen fließendes Wasser dialysiert. Nach erneutem Zentrifugieren setzt man zu der klaren überstehenden Lösung für je 100 ml 36 g Ammonsulfat zu. Nach halbstündigem Stehen bei 0° wird, ebenfalls in der Kälte, zentrifugiert (20 min 15000 U/min). Den Niederschlag löst man in etwa 20 ml Wasser und gibt weitere 4 g festes Ammonsulfat zu. Der Niederschlag wird abzentrifugiert und verworfen. Aus der überstehenden Lösung beginnt, wenige Minuten nach Zugabe von weiterem Ammonsulfat, die Alkohol-Apodehydrase zu kristallisieren. Man setzt in kleinen Mengen im Verlauf mehrerer Stunden soviel Ammonsulfat zu, daß eine 60%ige Sättigung erreicht wird. Die Kristalle werden abzentrifugiert. In wenig Wasser gelöst, beginnen sie auf Zusatz kleiner Mengen gesättigter Ammonsulfatlösung erneut zu kristallisieren. Das umkristallisierte Material wird unter Rühren 2 Std. gegen dest. Wasser dialysiert und dann lyophilisiert. Dieses Trockenpräparat läßt sich mit nur geringem Aktivitätsverlust mehrere Wochen im Vakuumexsikkator aufbewahren.

3. Spektrophotometrische Bestimmung von TPN als Dihydroverbindung.

Enzymatische Reduktion: Glucose-6-phosphat-dehydrase-Test (Warburg, 1935; Adler, 1940; Pfleiderer, 1953).

TPN wird in Gegenwart von Glucose-6-phosphat-Apodehydrase durch Glucose-6-phosphat quantitativ in die Dihydro-Verbindung übergeführt. Aus der lichtelektrisch bei 340 mμ gemessenen Extinktion wird seine Konzentration bestimmt.

Reagentien:

Glucose-6-phosphorsaures Kalium. Methoden der Darstellung s. S. 332. Vorschrift S. 332.

Glucose-6-phosphorsäure-Apodehydrase. Bereitung aus Hefe (Warburg, 1932; LePage, 1951). Vorschrift S. 333. 60 mg des Rohpräparats werden in 10 ml 0,1%iger Natriumbicarbonatlösung suspendiert, bei Raumtemperatur 20 min stehengelassen. Die zentrifugierte Lösung wird verwendet. Sie behält ihre Aktivität unvermindert mindestens 1 Woche, sofern sie bei 0° aufbewahrt wird (Pfleiderer, 1953).

Ansatz: 0,4 ml Glucose-6-phosphorsaures Kalium 3%ig,
0,2 ml Glucose-6-phosphat-Apodehydrase-Lösung,
x ml TPN-Lösung, entsprechend 40—400 γ,
3,0—x ml Wasser.

Ausführung: In einem 3 ml-Meßkölbchen werden TPN, Substrat und Enzym wie angegeben gemischt, in einem anderen kommt zu der TPN-Lösung nur das Enzym. Nach dem Auffüllen zur Marke wird die Differenz der Extinktionen der beiden Lösungen bei 340 mμ lichtelektrisch gemessen. Nach etwa 15 min ist die Hydrierung vollständig. Man muß noch eine Korrektur für die Eigenabsorption des Substrats vornehmen, sie wird unter gleichen Bedingungen gesondert bestimmt.

Das verwendete Enzympräparat enthält stets noch Alkohol-apodehydrase, evtl. auch Spuren von Flavinenzym. Befindet sich in der zu analysierenden TPN-Lösung auch hydriertes TPN bzw. DPN, so können Störungen auftreten. Die Absorption von DPN-H$_2$ kann durch Zugabe eines Tropfens Acetaldehyd m/2 beseitigt werden, den Gehalt an TPN-H$_2$ muß man gesondert durch Reoxydation mit Flavinenzym bestimmen (Adler, 1946).

C. Darstellung der Codehydrasen.

Präparate verschiedenen Reinheitsgrades sind im Handel erhältlich (1953: DPN 75%ig 100 mg 12 DM; DPN 100%ig 100 mg 22 DM; TPN 80%ig 10 mg 100 DM).

Konzentrate mit einem Reinheitsgrad von 0,2—0,5 lassen sich aus verschiedenen Rohstoffen ohne besondere Schwierigkeiten gewinnen, für die Reindarstellung — insbesondere der Cozymase — ist Erfahrung erforderlich. In den meisten Fällen wird jedoch die Verwendung der reinen Substanzen nicht notwendig sein.

Die Schwierigkeiten der Reindarstellung bestehen in der Trennung der beiden Coenzyme und in ihrer Abtrennung von der Adenylsäure, die stets als hartnäckiger Begleiter auftritt. In den älteren Verfahren wird die Trennung durch eine lange Folge von Fraktionierungen erreicht mit Hilfe von Schwermetall- und Bariumsalzen. Durch selektive Adsorption an Kohle unter geeigneten Bedingungen, durch die Anwendung von Ionen-Austauschern und die der Gegenstromverteilung läßt sich die Reinigung jedoch wesentlich vereinfachen. Da das Adsorptionsvermögen der verwendeten Kohlen sehr unterschiedlich sein kann, müssen optimale Bedingungen für eine Reinigung und Trennung erst empirisch ermittelt werden. Die für ein Ausgangsmaterial ausgearbeiteten Methoden sind nur bedingt auf ein anderes zu übertragen. Für eine einmalige Bereitung eines Konzentrates haben deshalb die älteren Verfahren immer noch ihre Berechtigung.

1. Darstellung von Diphospho-pyridin-nucleotid.

Bäcker- und Bierhefen werden am häufigsten für die Darstellung von DPN verwendet. Frischer Kaninchenmuskel (OCHOA, 1937), Schafleber (KORNBERG 1953) und Pferdeblutzellen (WARBURG, 1935) sind ebenfalls geeignet. Aus höheren Pflanzen ist DPN bisher nicht in reiner Form erhalten worden.

Nach älteren Verfahren extrahiert man die frische Hefe mit Wasser bei 80—90°. Ein Teil der Begleitstoffe wird mit Bleiacetat gefällt. Das Coenzym wird mit Bleiacetat, Quecksilbernitrat, Phosphorwolframsäure, Silbernitrat, Kupfer -I-chlorid und Alkohol u. ä. fraktioniert (v. EULER, 1936; OHLMEYER, 1938; WILLIAMSON, 1947 u. a.). Hefe läßt sich vorteilhaft auch in der Kälte mit organischen Lösungsmitteln in Gegenwart von Säure extrahieren (SUMNER, 1947). JANDORF (1941) führte die Technik der Kohleadsorption für die Reinigung von DPN ein, LEPAGE (1947) und SUMNER (1948) arbeiteten Methoden aus, die sich der Kohleadsorption bedienen. NEILANDS (1951) beschrieb die Reinigung von DPN mit Hilfe von Ionenaustauschern. Durch die Methode der Gegenstromverteilung nach CRAIG wurde der Weg zu reinsten Präparaten wesentlich erleichtert (HOGEBOOM, 1948; 97%iges DPN). Durch kombinierte Anwendung von Schwermetallsalzen und Ionenaustauschern lassen sich 25 g 100%iges DPN aus 100 kg Hefe bereiten. (DRP 913406 (1954); C. F. Boehringer und Söhne; KORNBERG, 1953).

Begleiter der Cozymase sind Adenosin-5'-monophosphat und -diphosphat, Adenin-ribose-pyrophosphat-ribose. Diese Verbindungen folgen der Cozymase über viele Reinigungsstufen. Sie werden auch beim Aufbewahren in festem Zustand, vor allem aber in der wäßrigen Lösung oder durch Alkalibehandlung aus dem Coenzym gebildet.

Methode nach CLARK (1949). Diese Methode basiert auf dem Verfahren von SUMNER (1948). Sie liefert Präparate mit einer Reinheit bis zu 85%. Die einzelnen Stufen lassen sich rasch durchführen. Sie bestehen in der Extraktion frischer Hefe mit einem Gemisch aus Aceton, Äthylalkohol und Schwefelsäure, fraktionierter Fällung der Cozymase mit mehr Alkohol, Umfällen des Rohprodukts aus wäßriger Lösung mit Alkohol und Reinigung durch selektive Adsorption an Kohle. Das Endprodukt enthält höchstens 1% TPN. 4,5 kg Bäckerhefe ergeben 600—800 mg DPN vom Reinheitsgrad 65—78% bzw. 450—500 mg DPN vom Reinheitsgrad 85%.

Extraktion und Fällung der Cozymase. 4,5 kg frische Bäckerhefe werden zerkleinert in 3500 ml Aceton eingetragen. Nach 5 min langem Rühren gibt man ein kaltes Gemisch von 550 ml konz. Schwefelsäure und 3500 ml Äthylalkohol 95%ig unter Rühren vorsichtig hinzu. Nach 15 min lebhaften Rührens wird filtriert oder zentrifugiert. Die klare braune Lösung (etwa 6,8 l) wird mit dem gleichen Volumen Alkohol vermischt und die Mischung filtriert. Der Niederschlag, der hauptsächlich aus KHSO₄ besteht, wird verworfen. Das Filtrat mischt man mit einem weiteren Volumen Alkohol und läßt über Nacht bei —15° C oder bei 0—4° C 24—48 Std.stehen. Der Niederschlag wird abfiltriert.

Umfällen mit Äthylalkohol. Das Rohprodukt wird in 200 ml Wasser gelöst und das gleiche Volumen Alkohol unter Rühren hinzugefügt. Die Mischung wird filtriert und dann mit weiteren 400 ml Alkohol versetzt. Dann läßt man 12—24 Std: bei 0° C stehen. Das gefällte

DPN wird in der Kälte zentrifugiert und die überstehende Lösung verworfen. Den Niederschlag suspendiert man in etwa 150 ml abs. Alkohol recht fein. Nach dem Zentrifugieren wird der Niederschlag erneut in 40 ml abs. Alkohol suspendiert. In gleicher Weise wäscht man den abgeschleuderten Niederschlag noch mit 40 ml Äther. Das Rohprodukt wird zuerst an der Luft, dann im Vakuum über Phosphorpentoxyd getrocknet. Ausbeute: 4—4,5 g; Reinheit 15—20%. Abweichungen im Reinheitsgrad in dieser Stufe des Verfahrens hängen vor allem vom Frischezustand der Hefe ab.

Reinigung des Rohprodukts durch Kohleadsorption. Eine Menge des Rohproduktes, die etwa 400 mg reiner Cozymase entspricht, wird in 800 ml Wasser gelöst und 5 min mit vorbehandeltem Norit[1] verrührt. Der DPN-Norit-Komplex wird abzentrifugiert, zweimal mit je 800 ml Wasser und je zweimal mit dem gleichen Volumen verdünnten Ammoniaks (0,0028prozentig) ausgewaschen. Nach dem Waschen mit Ammoniak ist längeres Zentrifugieren erforderlich, kleine Verluste an Norit lassen sich kaum vermeiden. Man wäscht dann noch zweimal mit destilliertem Wasser nach.

Von der Kohle trennt man die Cozymase durch einstündiges Schütteln einer Aufschlämmung in einem Gemisch von 100 ml Wasser und 10 ml Iso-amylalkohol. Die Kohle wird abzentrifugiert und die Elution wiederholt. Die vereinigten Eluate filtriert man zur Entfernung von Kohleresten durch ein feuchtes Filter, sie lassen sich 24 Std. bei 0—4° C aufbewahren. Die Lösung des eluierten DPN wird mit 1—2 Tropfen 7 n Ammoniak-Lösung schwach alkalisch gemacht (Phenolrot), dann bei vermindertem Druck bei einer Badtemperatur von 35—40° auf 2—3 ml eingeengt. Die Lösung wird mit etwa der gleichen Menge Wasser quantitativ in ein Zentrifugenglas gespült, diese konz. DPN-Lösung mit dem gleichen Volumen Alkohol vermischt und zur Entfernung von Kohleresten und ausgefallenen Begleitstoffen zentrifugiert. Aus der klaren Lösung fällt man mit 25—30 ml Alkohol (wasserfrei) die Cozymase. Zusatz von 2 Tropfen 7,5 n H_2SO_4 erleichtert das Ausflocken. Nach dem Zentrifugieren wird der Niederschlag mit 20—30 ml abs. Alkohol und danach mit Äther gewaschen und dann getrocknet. Ausbeute 300—400 mg, Reinheit 65—78%. Durch eine anschließende zweite Kohleadsorption läßt sich bei Ausbeuten bis 70% die Reinheit auf 85% steigern.

2. Darstellung von Triphospho-pyridin-nucleotid.

Als Ausgangsmaterial sind Blutzellen (Warburg, 1935) und Leber (LePage, 1949. Kornberg, 1953) geeignet. Konzentrate sind auch aus Hefe (LePage, 1949) und aus Maiskeimen (Pfleiderer, 1953) erhalten worden.

Warburg extrahierte Pferdeblutzellen mit Aceton. Er erzielte nach Fraktionierung mit Quecksilber-II-acetat eine Trennung der beiden Codehydrasen über die Bariumsalze. Die weitere Reinigung gelingt durch Fällen mit Essigester aus angesäuertem Methanol und durch Fraktionierung mit Bleiacetat. Aus 100 l Blutzellen werden erhalten 1150 mg TPN (Reinheitsgrad 0,52) oder 340 mg (Reinheitsgrad 1,0). Gesamtausbeute, bezogen auf den TPN-Gehalt des ersten Extraktes, 25%.

LePage adsorbiert den Heißwasser-Extrakt aus Leber an Aktivkohle und erhält durch Chromatographie an granulierter Kohle und anschließende Fällung mit Essigester aus angesäuertem Methanol ein 100%iges Produkt, frei von DPN. Aus 15 kg Schweineleber werden erhalten 500 mg TPN (Reinheitsgrad 1,0). Gesamtausbeute 80%, bezogen auf den TPN-Gehalt des ersten Extraktes.

Methode nach Warburg (1935).

Acetonextrakt. Pferdeblut wird zentrifugiert, die Zellen werden dreimal mit 0,9%iger Kochsalzlösung gewaschen und dicht zusammenzentrifugiert. 100 l Zellen werden mit 200 l Wasser cytolysiert und sofort mit 300 l Aceton gefällt. Der Niederschlag wird abzentrifugiert, die überstehende Flüssigkeit, die das Coferment enthält, wird mit einem ihr gleichen Volumen Aceton versetzt, der Niederschlag wird durch Faltenfilter filtriert. Das Filtrat, das das Coenzym enthält, wird unter vermindertem Druck auf $^1/_3$ seines Volumens eingeengt, wobei das Aceton bis auf einige Prozent entfernt wird. Das Volumen des Acetonextraktes beträgt dann etwa 220 l.

Fraktionierung mit Quecksilber-II-acetat. Gibt man zu der schwach essigsauren Lösung des Acetonextraktes Quecksilber-II-acetat, so fällt zunächst ein Niederschlag, der viel Substanz, aber kaum Coenzym enthält. Mit mehr Quecksilber-II-acetat fällt aus dem Filtrat

---

[1] Norit wird mit kalter konz. HCl verrührt, der Reihe nach mit Wasser („ad nauseam"), Ammoniak und wieder mit Wasser gewaschen und getrocknet (Sumner, 1947).

alles Coenzym. Die zur Fraktionierung geeigneten Mengen an Mercuriacetat werden durch eine Vorprobe ermittelt. Zu 220 l Acetonextrakt gibt man 0,44 l 2 n Essigsäure und 4,4 l 10%ige Quecksilberacetatlösung, entfernt den Niederschlag, der sehr wenig Coenzym enthält mit der SHARPLESS-Zentrifuge und fällt die Lösung mit 12 l 10%igem Quecksilberacetat. Nach 12 Std., wenn sich der Niederschlag gesenkt hat, wird die überstehende Flüssigkeit abgehebert. Der Niederschlag, der das Coenzym enthält, wird dicht zusammenzentrifugiert, in 2 l Wasser zerteilt und mit Schwefelwasserstoff auf der Schüttelmaschine geschüttelt. Das Sulfid wird abzentrifugiert und dreimal mit Wasser gewaschen. Das Filtrat und das erste Waschwasser vom Quecksilbersulfid fällt man mit dem $3^1/_2$fachen Volumen Aceton, läßt einige Stunden stehen, zentrifugiert und wäscht den Niederschlag, der das Coenzym enthält, auf der Zentrifuge mit reinem Aceton. Fraktion A: 4740 mg, Reinheit 17%. Die Mutterlauge der Acetonfällung und das zweite und dritte Waschwasser des Sulfids dampft man unter vermindertem Druck auf 300 ml ein, fällt mit dem $3^1/_2$fachen Volumen Aceton, löst den Niederschlag, der viel Coenzym enthält, in 5 l Wasser, fällt mit 0,27 l 10%igem Quecksilberacetat, zerlegt den Niederschlag wie oben mit Schwefelwasserstoff, engt das Filtrat und die Waschwässer unter vermindertem Druck auf 20 ml ein, fällt mit dem $3^1/_2$fachen Volumen Aceton das Coenzym und wäscht den Niederschlag auf der Zentrifuge mit reinem Aceton. Fraktion B: 1250 mg, Reinheit 11%.

Fraktionierung der Bariumsalze. 5990 mg Coenzym vom mittlerem Reinheitsgrad 15,7% (aus 100 l Pferdeblutzellen) werden in 500 ml Wasser gelöst und mit 53 ml 22%igem Bariumacetat gefällt. Der Niederschlag wird abzentrifugiert und mit 50 ml einer 2%igen Bariumacetat-Lösung gewaschen. Die überstehende Flüssigkeit, die das Coenzym enthält, und die Waschflüssigkeit werden vereinigt und mit kaltgesättigter Barytlösung versetzt, bis die Farbe der Flüssigkeit in gelb umschlägt (p_H etwa 9), wozu etwa 50 ml Barytlösung notwendig sind. Es entsteht ein zweiter Niederschlag von coenzymfreien Bariumsalzen, der abzentrifugiert und mit 50 ml 2%igem Bariumacetat gewaschen wird. Die überstehende Flüssigkeit, die das Coenzym enthält, wird mit der Waschflüssigkeit vereinigt und mit dem dreifachen Volumen Alkohol versetzt, wodurch das Bariumsalz des Coenzyms (mit anderen Bariumsalzen) niedergeschlagen wird. Man löst den Niederschlag in 1 l Wasser, fällt das Coenzym mit 7 g Mercuriacetat, zentrifugiert, wäscht mit 0,5%igem Quecksilberacetat, suspendiert den Niederschlag in Wasser, zerlegt ihn mit Schwefelwasserstoff, zentrifugiert das Quecksilbersulfid ab, wäscht es mit Wasser, dampft das Filtrat vom Quecksilbersulfid und die Waschwässer unter vermindertem Druck auf 20 ml ein, fällt das Coenzym mit dem dreifachen Volumen Aceton, und wäscht es auf der Zentrifuge mit reinem Aceton. Ausbeute 2000 mg, Reinheit 32%.

Fällung mit Essigester aus salzsaurem Methanol. 500 mg Coferment vom Reinheitsgrad 0,32 werden bei 0° mit 250 ml Methanol, das bezogen auf Salzsäure n/10 ist, verrieben, wobei sich, wenn das Methanol hinreichend wasserfrei ist, fast alles löst. Ein geringer Rest wird schnell abzentrifugiert. Die überstehende Lösung wird sofort mit dem dreifachen Volumen eiskalten Essigesters gefällt, der flockige Niederschlag wird abzentrifugiert und auf der Zentrifuge dreimal mit Essigester gewaschen, wobei darauf zu achten ist, daß der Niederschlag beim Aufwirbeln mit Essigester jedesmal fein verteilt wird. Man trocknet im Exsikkator über Schwefelsäure und Kaliumhydroxyd. Ausbeute 1150 mg, Reinheit 52%.

Fraktionierung mit Bleiacetat. 1000 mg Coencym vom Reinheitsgrad 52% werden in 200 ml wäßriger normaler Essigsäure bei Zimmertemperatur gelöst. Durch 32 ml mol. Bleiacetat-Lösung entsteht ein Niederschlag. Man stellt 1 Std. in Eiswasser, zentrifugiert bei 0° ab und erhält Fraktion I der Bleisalze. Die Mutterlauge der Fraktion I erwärmt man auf 30°. Dann gibt man unter Schütteln im feinen Strahl 60 ml Alkohol hinzu, wobei ein Niederschlag fällt, der abzentrifugiert wird (die Temperatur fällt während des Zentrifugierens auf etwa 20°). Der Niederschlag ist Fraktion II der Bleisalze. Fraktion III erhält man, indem man die Mutterlauge der Fraktion II 17 Std. in den Eisschrank stellt und das ausgeschiedene Pulver bei 0° abzentrifugiert. Die Mutterlauge der Fraktion III erwärmt man auf 30°, gibt unter Schütteln in feinem Strahl 120 ml Alkohol hinzu, wobei die Flüssigkeit klar bleiben soll oder sich nur eben trüben darf. Man stellt in den Eisschrank und zentrifugiert Fraktion IV der Bleisalze nach 17 Std. bei 0° ab. Die Bleisalze werden nicht gewaschen. Man vereinigt die Fraktionen I und II (da sich eine getrennte Bearbeitung nicht lohnt) und gewinnt folgendermaßen aus den nunmehr 3 Teilen das freie Coenzym: Jeder Teil wird mit 10 ml Wasser fein zerschüttelt, dann wird mit Schwefelwasserstoff zerlegt, zentrifugiert und das Bleisulfid einmal mit 5 ml Wasser und zweimal mit 5 ml n/10 Salpetersäure gewaschen. Die Zersetzungs- und Waschflüssigkeiten filtriert man in je 125 ml eiskaltem Alkohol, mit etwa 2 ml 2 n Salpetersäure hinzu, damit die Alkoholfällung flockig wird, dann noch je 50 ml Äther, zentrifugiert die ausgefallenen Niederschläge und wäscht sie auf der Zentrifuge gründlich (Entfernung der Salpetersäure) mit absolutem Alkohol. Fraktion I und II: 131 mg, Reinheit 62%; Fraktion III: 261 mg, Reinheit 100%; Fraktion IV: 76 mg, Reinheit 100%. Anschließend kann Behandlung mit Bromwasser zur Entfernung fluoreszierender Verunreinigungen erfolgen.

3. Darstellung der Dihydro-codehydrasen.

Beide Verbindungen sind sowohl durch partielle chemische Reduktion mit Hydrosulfit als auch durch enzymatische Hydrierung zugänglich. Für spektrophotometrische Untersuchungen ist es meist ausreichend, Lösungen der Verbindungen zu bereiten. Nach der Reduktion wird überschüssiges Hydrosulfit durch Einleiten von Luft zerstört. Bei enzymatischer Hydrierung wird das Protein durch Hitze- oder Alkalidenaturierung entfernt. Im alkalischen Medium sind die Lösungen der Dihydroverbindungen im eingefrorenen Zustand längere Zeit haltbar.

Dihydro-coenzym I ist im Handel erhältlich. Feste Präparate aus DPN werden mit Hydrosulfit in Bicarbonat erhalten (OHLMEYER, 1938; Bioch. Preps. 2, 92). Enzymatische Reduktion kann erfolgen mittels Alkohol-Apodehydrase (BONNICHSEN, 1950; PULLMAN, 1952) oder Ameisensäure-Apodehydrase (MATHEWS, 1950).

Festes *Dihydro-coenzym II* ist in gleicher Weise aus TPN durch Reduktion mit Hydrosulfit gewonnen worden (CONN, 1952).

Darstellung von Dihydro-coenzym I durch Reduktion von Coenzym I mit Hydrosulfit (OHLMEYER, 1938). In der Birne eines Manometer-Gefäßes werden 80 mg Cozymase mit 8,5% Wassergehalt, vom Reinheitsgrad 0,9 (oder von niedrigerem Reinheitsgrad) und 40 mg Natriumhydrosulfit eingewogen. Im Hauptraum befinden sich 6,6 ml 1,3%iges Natriumbicarbonat, die im Thermostaten bei 20° mit 5% CO_2 in N_2 gesättigt werden. Hydrierung und CO_2-Entwicklung beginnen beim Herüberspülen der trockenen Substanzen und sind nach 2-stündigem Schütteln beendet. Man gießt die Lösung in 20 ml Alkohol und kühlt auf —10° ab. Nach 20 min wird zur Entfernung der ausgefallenen Salze zentrifugiert, aus der Lösung fallen mit 80 ml Alkohol etwa 65 mg Na-dihydrocozymase vom Reinheitsgrad 0,95. Sie wird mit Äther gewaschen und getrocknet. Löst man sie in 8 ml Methylalkohol, zentrifugiert und fällt mit 50 ml Äthylalkohol und 40 ml Äther, so gewinnt man etwa 40 mg der reinen Substanz mit Na- und Wassergehalt, entsprechend 42 mg DPN-H_2. Die Ausbeute beträgt also 60%.

Anhang.

Verfahren zum Nachweis und zur Bestimmung der Pyridin-Apodehydrasen.

Nachweis und Bestimmung der Pyridin-Apodehydrasen erfolgt — wie bei allen Enzymen — auf Grund ihrer spezifisch-katalytischen Aktivität. Grundsätzlich sind diejenigen enzymatischen Testsysteme, die zur Analyse der Codehydrasen dienen können, auch für diese spezifischen Proteine geeignet. Ausreichende Charakteristika zur Ermittlung des absoluten Reinheitsgrades fehlen. Alle Analysenverfahren beruhen auf der Ermittlung einer relativen Aktivität, die bezogen wird auf das reinste jeweils zugängliche Präparat, oder deren Einheit willkürlich festgelegt wird. Werden die Bedingungen des Testes so gewählt, daß die zu bestimmende Apodehydrase den begrenzenden Faktor der Reaktion darstellt, so ist in einem beschränkten Bereich die Konzentration des Proteins proportional der Anfangsgeschwindigkeit des ausgelösten Dehydrierungs- oder Hydrierungsvorgangs.

Für den *Nachweis* der Pyridin-Apodehydrasen in Rohextrakten ist die THUNBERG-Methodik von unschätzbarem Wert. Fehlen in den zu untersuchenden Extrakten die Diaphorasen, so muß das Enzymsystem durch sie ergänzt werden.— Die Verwendung von Triphenyl-tetrazoliumsalzen anstelle von Methylenblau kann in einzelnen Fällen vorteilhaft sein (KUHN, 1941, 1952). Die farblose Triphenyltetrazolium-Verbindung geht bei der Reduktion in eine rote „Leuko-Form"(Formazan) über, die gegen Sauerstoff beständig ist. Die Menge des gebildeten unlöslichen Farbstoffs kann nach der Extraktion auf kolorimetrischem Wege bestimmt werden. Die Reduktion der Triphenyl-tetrazolium-Verbindung durch keimende Samen ist die Grundlage eines topographischen Verfahrens zur

Bestimmung der Keimfähigkeit (LAKON, 1942). Die Reduktion der Tetrazolium-Verbindungen beruht — wie die des Methylenblaus — auf der Reaktion mit den beiden Dihydro-coenzymen.

Spektrophotometrisch kann der Nachweis der Apodehydrasen in Extrakten erbracht werden, die den Voraussetzungen für diese Methodik genügen. Diese Extrakte müssen frei sein von Substraten und dürfen Co-H_2-dehydrierende Fermente (gelbe Fermente) nur in sehr geringer Menge enthalten.

In Gegenwart des spezifischen Substrats und Coenzyms für das gesuchte Ferment wird der Extinktionsanstieg bei 340 mμ (Co-H_2) ermittelt (v. EULER, 1937).

Grundsätzlich sind die manometrischen Analysenverfahren für den Nachweis anwendbar.

Die *quantitative Bestimmung* der Apodehydrasen in Gewebsextrakten ist nicht befriedigend gelöst. Ist vor der Bestimmung des Ferments eine partielle Reinigung erforderlich, so ist die Ausbeute an Enzym bei derartigen Operationen unsicher (v. EULER, 1937, 1938). Doch ist eine relative quantitative Bestimmung der Apodehydrasen in geeigneten Extrakten sowohl nach manometrischen als auch spektrophotometrischen Verfahren möglich. Sie bilden die Grundlage für die Reinigung dieser Proteine. Man ermittelt den Quotienten Q

$$Q = \frac{\text{enzymatische Wirkung}}{\text{g Protein}}$$

und setzt die Reinigung der Apodehydrase fort, bis der Quotient nicht mehr ansteigt. So erhielt NEGELEIN (1936, 1937) kristallisierte Alkohol-apodehydrase aus Hefe mit einem manometrischen Test: Der Sauerstoffverbrauch bei der Oxydation von Äthylalkohol in Gegenwart von DPN, der Alkohol-dehydrase und „alten gelben Ferments" ist hier das Maß für die enzymatische Wirkung. Im Bereich von 0—1 γ Protein/ml des Ansatzes ist die Reaktionsgeschwindigkeit der Enzymkonzentration proportional.

Die spektrophotometrische Methode hat sich für die Isolierung der Apo-dehydrasen — insbesondere in den letzten Stufen ihrer Reinigung — als überaus wertvoll erwiesen. Die Geschwindigkeit der Dehydrierung (oder der Hydrierung) des Substrats läßt sich durch die Messung des gebildeten Dihydro-coenzyms (bzw. des verschwundenen Co-H_2) bei 340 mμ ermitteln.

Da auch die in dem Ansatz vorhandene Proteinkonzentration bei 260 oder 280 mμ bestimmt werden kann, ist die Aktivitätsbestimmung eines Präparats durch 2 Messungen möglich. Auf dieser Grundlage isolierte WARBURG (1939) Phospho-glycerinaldehyd-Dehydrase aus Hefe.

Literatur.

ADLER, E., u. H. v. EULER: Svensk Kem. Tidskr. 48, 221 (1936). — ADLER, E., u. F. CALVET: Arch. Kemi Mineral. Geol. 12 B, 32 (1936). — ADLER, E., u. G. GÜNTHER: Hoppe-Seylers Z. 253, 143 (1938). — ADLER, E., H. v. EULER, G. GÜNTHER and M. PLASS: Biochem. J. 33, 1028 (1939). — ADLER, E., ST. ELLIOT and L. ELLIOT: Enzymologia 8, 80 (1940). — AHLGREN, G.: Habilitationsschrift, Lund 1925. — ANDERSON, D. G., and B. VENNESLAND: J. Biol. Chem. 207, 613 (1954). — AXELROD, A. E., and C. A. ELVEHJEM: J. Biol. Chem. 131, 77 (1939).

BONNICHSEN, R. K.: Acta Chem. scand. 4, 714 (1950); 4, 715 (1950).

CLARK, H. W., A. L. DOUNCE and E. STOTZ: J. Biol. Chem. 181, 459 (1949). — COLOWICK, S. P., N. O. KAPLAN and M. M. CIOTTI: J. Biol. Chem. 191, 447, 461 (1951). — CONN, E. E., L. M. KRAEMER, L. PEI-NAN and B. VENNESLAND: J. Biol. Chem. 194, 143 (1952).

DRABKIN, D. L.: J. Biol. Chem. 157, 563 (1945). — DRABKIN, D. L., and O. MEYERHOF: J. Biol. Chem. 157, 571 (1945).

v. EULER, H., u. S. STEFFENBURG: Hoppe-Seylers Z. 175, 38 (1928). — v. EULER, H.: Ergebn. Physiol. 38, 1 (1936). — v. EULER, H., H. ALBERS u. F. SCHLENK: Hoppe-Seylers Z. 240, 113 (1936). — v. EULER, H., L. AHLSTRÖM, B. HÖGBERG u. J. PETTERSON: Arkiv

Kemi, Mineral. Geol. 19 A, 4 (1944). — v. Euler, H., E. Adler, G. Günther u. H. Hell-ström: Hoppe-Seylers Z. 245, 217 (1937). — v. Euler, H., E. Adler u. G. Günther: Hoppe-Seylers Z. 247, 65 (1937).

Gingrich, W., and F. Schlenk: J. Bacteriol. 47, 535 (1944). — Green, D. E., and O. Brosteaux: Biochem. J. 30, 1489 (1936). — Green, D. E., D. M. Needham and J. G. Dewan: Biochem. J. 31, 2327 (1937). — Gutcho, S., and E. D. Stewart: Analyt. Chem. 20, 1185 (1948).

Haas, E.: Biochem. Z. 285, 368 (1936); 291, 79 (1937); 298, 378 (1938); J. Biol. Chem. 155, 333 (1944). — Haas, E., Carter J. Harrer and T. R. Hogness: J. Biol. Chem. 142, 835 (1942). — Hasse, K., u. M. Specht: Biochem. Z. 322, 502 (1952). — Hogeboom, G. H., and G. T. Barry: J. Biol. Chem. 176, 935 (1948). — Hogness, T. R., and van R. Potter: Ann. Rev. Biochem. 10, 509 (1947). — Horecker, B. L., and A. Kornberg: J. Biol. Chem. 175, 385 (1948). — Horecker, B. L., and H. A. Lardy: In H. E. Carter, Biochem. Preparations, Vol. 3, 71, New York 1953. — Horecker, B. L., and P. Z. Smyrniotis: J. Biol. Chem. 193, 371 (1951). — Holzer, H., St. Goldschmidt, W. Lamprecht, E. Helmreich: Hoppe-Seylers Z. 297, 1 (1954).

Jandorf, B. J., F. W. Klemperer and A. B. Hastings: J. Biol. Chem. 138, 311 (1941).

Kaplan, N. O., S. P. Colowick and E. F. Neufeld: J. Biol. Chem. 195, 107 (1952). — Kornberg, A., and W. E. Pricer, jr.: In H. E. Carter, Biochem. Preparations, Vol. 3, 20, New York 1953. — Kornberg, A., and B. L. Horecker: In H. E. Carter, Biochem. Preparations, Vol. 3, 24, New York 1953. — Krishnan, P. S.: Arch. Biochem. 16, 291 (1948). — Kubowitz, F., and P. Ott: Biochem. Z. 314, 94 (1943). — Kuhn, R., u. D. Jerchel: Ber. dtsch. chem. Ges. 74, 949 (1941). — Kuhn, R., G. Pfleiderer u. W. Schulz: Liebigs Annalen 578, 159 (1952).

Lakon, G.: Ber. dtsch. bot. Ges. 60, 299, 434 (1942). — LePage, G. A.: J. Biol. Chem. 168, 623 (1947). — LePage, G. A., and G. C. Mueller: J. Biol. Chem. 180, 775, 975 (1949). — Lohmann, K.: Biochem. Z. 194, 326 (1928). — Lwoff, A., and M. Lwoff: Proc. Roy. Soc. London B 122, 352, 360 (1937). — Lardy, H. A., and H. O. L. Fischer: J. Biol. Chem. 164, 513 (1946); Bioch. Preps. 2, 39 (1952).

Mathews, M. B.: J. Biol. Chem. 176, 229 (1948). — Mathews, M. B., and B. Vennes-land: J. Biol. Chem. 186, 667 (1950). — Myrbäck, K.: Hoppe-Seylers Z. 177, 158 (1928); Ergebn. Enzymforschung 2, 139 (1933).

Negelein, E., and W. Gerischer: Biochem. Z. 284, 289 (1936). — Negelein, E., u. H. J. Wulff: Biochem. Z. 293, 351 (1937). — Neilands, J. B., and Å. Åkeson: J. Biol. Chem. 188, 307 (1951). — Neuberg, C., u. M. Kobel: In Oppenauer u. Pincussen, „Metho-den der Fermente", S. 406, 1929.

Ochoa, S.: Biochem. Z. 292, 68 (1937); J. Biol. Chem. 174, 123 (1948). — Ohlmeyer, P.: Biochem. Z. 297, 66 (1938).

Pfleiderer, G., u. W. Schulz: Liebigs Annalen 580, 237 (1953). — Pullman, M. E., S. P. Colowick and N. O. Kaplan: J. Biol. Chem. 194, 593 (1952). — Potter, van R.: J. Biol. Chem. 165, 311 (1946).

Racker, E.: J. Biol. Chem. 184, 313 (1949).

Seegmiller, J. E., and B. L. Horecker: J. Biol. Chem. 192, 175 (1951). — Sumner, J. B., P. S. Krishnan and E. B. Sisler: Archiv of Biochem. 12, 19 (1947). — Sumner, J. B., and P. S. Krishnan: Enzymologia 12, 232 (1948). — Schlenk, F.: In Symposium on Respi-ratory Enzymes. Madison, Wis.: Univ. of Wisconsins Press 1942; In Bamann und Myrbäck: „Methoden der Fermentforschung", S. 2295, 1941. — Schlenk, F., and T. Schlenk: Arch. Biochem. 14, 131 (1947). — Schlenk, F.: In Sumner u. Myrbäck: The Enzymes, Bd. II, 1, S. 250, 1951.

Warburg, O., W. Christian: Biochem. Z. 254, 438 (1932); 266, 377 (1933). — Warburg, O., W. Christian u. A. Griese: Biochem. Z. 282, 157 (1935). — Warburg, O., u. W. Christian: Biochem. Z. 287, 291 (1936); 303, 40 (1939). — Whatley, F. R.: New Phytologist 50, 244 (1951). — Williamson, S., and D. E. Green: J. Biol. Chem. 135, 345 (1940).

Thiamine and Its Derivatives.

By

Sir R. A. Peters and J. R. P. O'Brien.

Thiamine (I) and co-carboxylase (II) are the names given to the vitamin known as vitamin B_1 and its pyrophosphate ester. It has now been agreed internationally that the name aneurin should be discontinued; this was given to vitamin B_1 by JANSEN (1935) who, together with DONATH, was the first to isolate the vitamin in crystalline form from rice polishings in 1926. This was actually the first isolation of any vitamin; though nicotinic acid was discovered earlier in yeast extracts, it was not recognised to be a vitamin until much later.

The enzyme carboxylase was discovered by NEUBERG and co-workers and proved to decarboxylate α-keto-acids (NEUBERG and KARCZAG, 1911). Later, a thermostable co-factor was found to be part of carboxylase, and the name co-carboxylase given to this by AUHAGEN (1932, 1933). In 1932, SIMOLA showed that the concentration of co-carboxylase was lowered in the brain of animals deficient in vitamin B. In 1928 KINNERSLEY and PETERS demonstrated that there was a form of thiamine precipitable by neutral lead acetate. It was left for LOHMANN and SCHUSTER (1937a, b) to make the important discovery that co-carboxylase was the pyrophosphate of thiamine. Since it had been found in Oxford (PETERS and co-workers) that thiamine was concerned in pyruvate oxidation in brain tissue, the discovery of LOHMANN and SCHUSTER formed the connection between the animal and yeast work.

That it was concerned in the oxidation of carbohydrates suggested that thiamine was alternatively oxidised and reduced. Reduction of the vitamin with dithionite gives a biologically inert product, but with catalytic hydrogenation, dihydrothiamine (III) which is as active as its parent substance (LIPMANN, 1937; LIPMANN and PERLMAN, 1938; BARRON and LYMAN, 1940). But neither of these products is autoxidisable. When oxidised by hydrogen peroxide at pH 7.5 or by iodine in an alkaline medium, thiamine and its pyrophosphate yield the corresponding disulphide derivatives (IV) (ZIMA and WILLIAMS, 1940; ZIMA, RITSERT and MOLL 1941) which can be reduced by glutathione or cysteine. In the catatorulin test, the disulphide is as active as thiamine but only if it is first reduced by cysteine is it active in the decarboxylation of pyruvate by washed yeast (PETERS, 1946a; KARRER and VISCONTI, 1946). However from an analytical point of view it is conceivable that a disulphide or monosulphide of thiamine may arise in tissues or in extracts therefrom and constitute a source of error in some types of determination although not in the thiochrome method.

Lipothiamide. Thiamine may exist in a new intricate form, lipothiamide (V), a condensation product of thiamine and lipoic acid (VI). The discovery of this compound originated in observations that natural materials contained a group of chemically related substances which could replace acetate in its growth promoting action upon certain lactic acid bacteria and were needed for the oxidative decarboxylation of pyruvate. These factors were designated by the names "acetate replacing factor"; and "pyruvate oxidation factor" (Guirard, Snell and Williams, 1946a, b; O'Kane and Gunsalus, 1947, 1948). In biological activity these factors were similar to concentrates of "protogen" a growth factor for *Tetrahymena geleii* (Dewey, 1944; Stokstad, Hoffmann, Regan, Fordham and Jukes, 1949). They may be freed from natural materials by acid or enzyme hydrolysis. Two of these factors obtained by acid hydrolysis are acidic in nature;

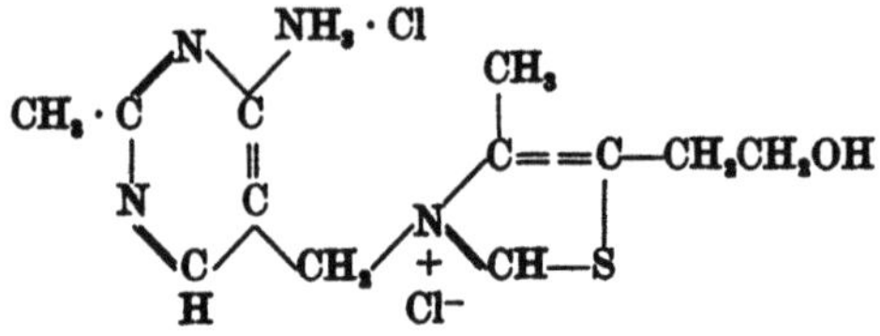

I

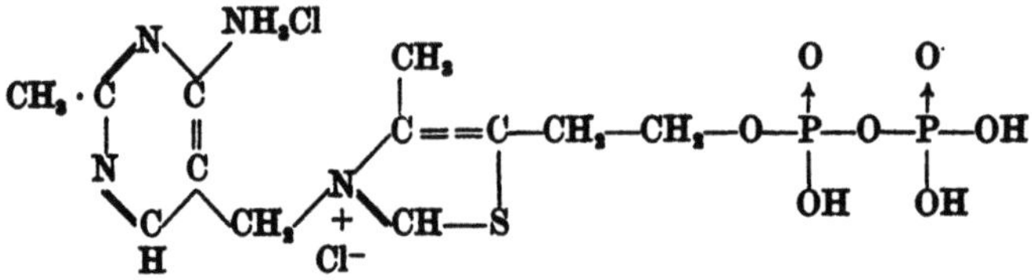

II

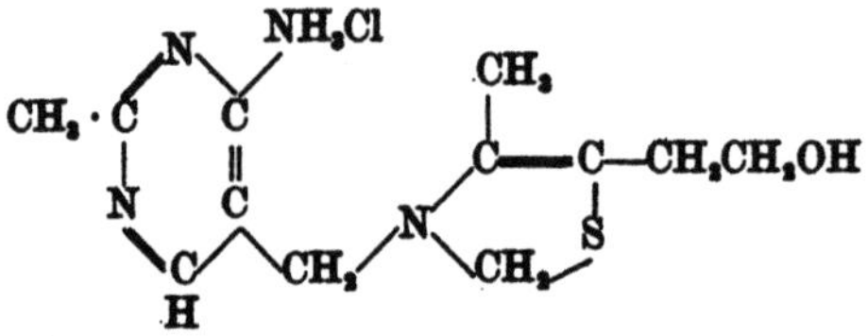

III

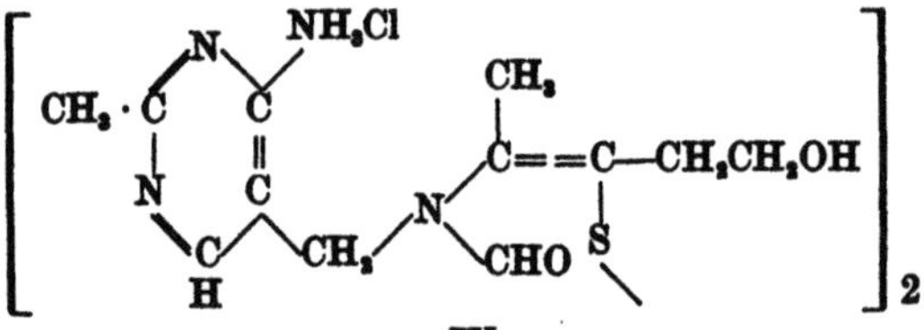

IV

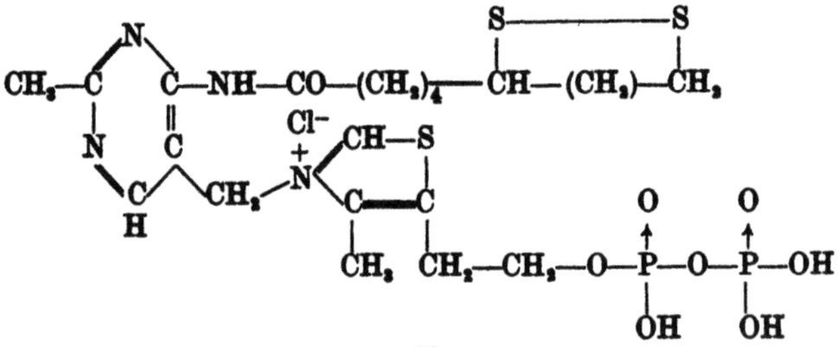

V

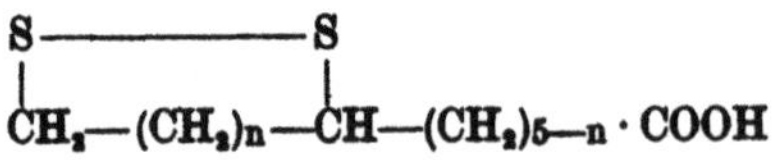

VI

the less polar of the two, α-lipoic acid has been isolated in crystalline form from liver and the more polar β-lipoic acid resembles in many of its properties Protogen B which has been isolated as its crystalline S-benzyl thiuronium salt from liver.

The structure of α-lipoic acid is that of a cyclic disulphide, namely 6-8 dimercapto-octanoic acid, a view supported by chemical synthesis and biological tests. α-Lipoic and β-lipoic acid may be changed one into the other by chemical procedures whose mode of action would suggest that β-lipoic acid is an oxidised form of α-lipoic acid perhaps a sulphoxide (BULLOCK, BROCKMAN, PATTERSON, PIERCE, STOKSTAD, 1952; REED, GUNSALUS, SCHNAKENBERG, SOPER, BOAZ, KERN and PARKE, 1953; REED, DE BUSK, HORNBERGER and GUNSALUS, 1953; HORNBERGER, HEIT-MILLER, GUNSALUS, SCHNAKENBERG and REED, 1953; REED, 1953).

That lipoic acid like thiamine pyrophosphate is involved in the oxidation of pyruvate points to a relationship between them. Microbiological and chemical evidence has been produced to show that there exist conjugates in which lipoic acid is combined with thiamine through an amide link (REED and DE BUSK, 1952a, b). Indeed, a wild type of *Escherichia coli* contains a conjugase capable of forming lipothiamide pyrophosphate from thiamine pyrophosphate acid and bound lipoic acid (REED and DE BUSK, 1952c). Such a conjugate is lipothiamide which give a positive azo reaction but a negative thiochrome. Without lipo-thiamide or its phosphorylated forms, a mutant strain of *Escherichia coli* will not oxidise pyruvate or α-ketoglutarate. A mechanism whereby lipothiamide brings about oxidative decarboxylation of α-keto acids has been proposed by REED and DE BUSK (1952d, 1953) and the role of thioctoic acids in photosynthesis discussed by CALVIN (1953). Confirmation of these views is actively awaited.

The very probable existence of thiamine in plant tissue as amide of lipoic acid or even as more complex forms may ultimately simplify methods for the determination of the vitamin. For once the modes of combination of thiamine have been recognised it should not be difficult to devise suitable modes of degrada-tion to free it from substances with which it is combined. At present it must be realised that lipothiamide has not been isolated from plant tissue and that acid or basic hydrolysis releases predominantly lipoic acid from tissues and enzymatic hydrolysis liberates complex forms of lipoic acid. Future analysis of tissues for thiamine may be expected to be concerned with simultaneous determinations, chemical or biological, of thiamine and lipoic acid and/or determination of the conjugate, lipothiamide.

This outline of present knowledge upon lipoic acid has been presented to show that thiamine may exist in its active biological forms as a highly complex co-enzyme molecule, yet to be defined. Once these forms are established a revision in the methods of determination of this vitamin may be expected. Such analysis would include not only the free vitamin but also both its simple and complex derivatives.

A. Properties of Thiamine Relevant to Its Determination.

I. Physical and Chemical Properties.

Thiamine forms salts with mineral acids of which the hydrochloride chloride is the most familiar. Usually in the form of white monoclinic plates of m. p. 248 to 250° (for the hemihydrate), it is easily soluble in water (1 g./1 ml.), less soluble in ethyl alcohol but insoluble in chloroform, ether acetone and benzene. Its spectral absorption varies with hydrogen ion concentration; in 0.005 N HCl it shows a maximum at 247 mμ, $E_{1cm}^{1\%}$ 425—450, and at pH 7, two maxima at 235 mμ

and 267 mμ (Wintersteiner, Williams and Ruehle, 1935; Smakula, 1934; Peters and Philpot, 1933; Holiday, 1935). It may be estimated spectroscopically when in the pure state (Holiday and Irwin; Gillam, Morton, Moore and Wang, 1946).

The stability of thiamine under different conditions is of importance in the preparation and storage of standard solutions and in procedures used in its extraction from plant and animal materials. The vitamin can be safely stored without loss of activity in a dry state or in acidic solutions. It is, however, particularly unstable in neutral or alkaline solutions, the more so when such solutions are heated or exposed to air. Its destruction in solution is not solely dependent upon hydrogen ion concentration but is related to the nature and concentration of salts that are present. Of buffers, phosphate is the one of several least harmful to the vitamin (Farrer, 1941, 1945, 1947, 1948; Beadle, Greenwood and Kraybill, 1943; Booth, 1943).

Thiamine hydrochloride is particularly susceptible towards the sulphite in weakly acidic conditions — a fact used in some methods of assay, for example, the yeast fermentation method to produce control solutions. At a pH 4.8—5 sodium sulphite splits thiamine into 2-methyl-4-amino-pyrimidine-5-methyl-sulphonic acid (VII) and 4-methyl-5-hydroxy-ethyl thiazole (VIII).

Thiamine

$\downarrow$ NaHSO$_3$

VII   VIII

Thiochrome. In 1935, Peters made an observation upon the reaction of thiamine towards oxidising agents which is the basis of the most extensively used method of determining the vitamin. Thiamine when oxidised gives a substance with an intense blue fluorescence (Kinnersley, O'Brien and Peters, 1935 a, b). This substance was identified as thiochrome (IX), a yellow compound isolated by Kuhn and his colleagues (1935) from yeast. It was later prepared in crystalline form from thiamine and its constitution established by synthesis (Barger, Bergel and Todd, 1935 a, b; Bergel and Todd, 1936). A number of oxidising agents, potassium permanganate, manganese dioxide, potassium ferricyanide and mercuric oxide under alkaline conditions convert thiamine into thiochrome.

IX

Thiochrome has an absorption spectrum characterised by maxima at 358 and 375 mμ. Its formation from thiamine was once thought to be through the intermediary formation of a pseudo-base. Another explanation has come from the work of Sykes and Todd (1951) combined with that of Zima and Williams (1940). Probably the first product in the oxidation of thiamine is thiamine disulphide which may in the absence of adequate oxidation be accompanied by a water insoluble colourless substance thiamine thiazolone (XIV). The second stage may be the cyclisation of the formyl group with the 4 amino group of the pyrimidine group and splitting of the disulphide to form two thiol radicles (X, XI). Within the radicle takes place an intramolecular addition across the C=N whereby arises a dihydro thiochrome radical (XII), whence develops dihydro thiochrome (XIII). The latter substance is readily oxidised to thiochrome even spontaneously by air.

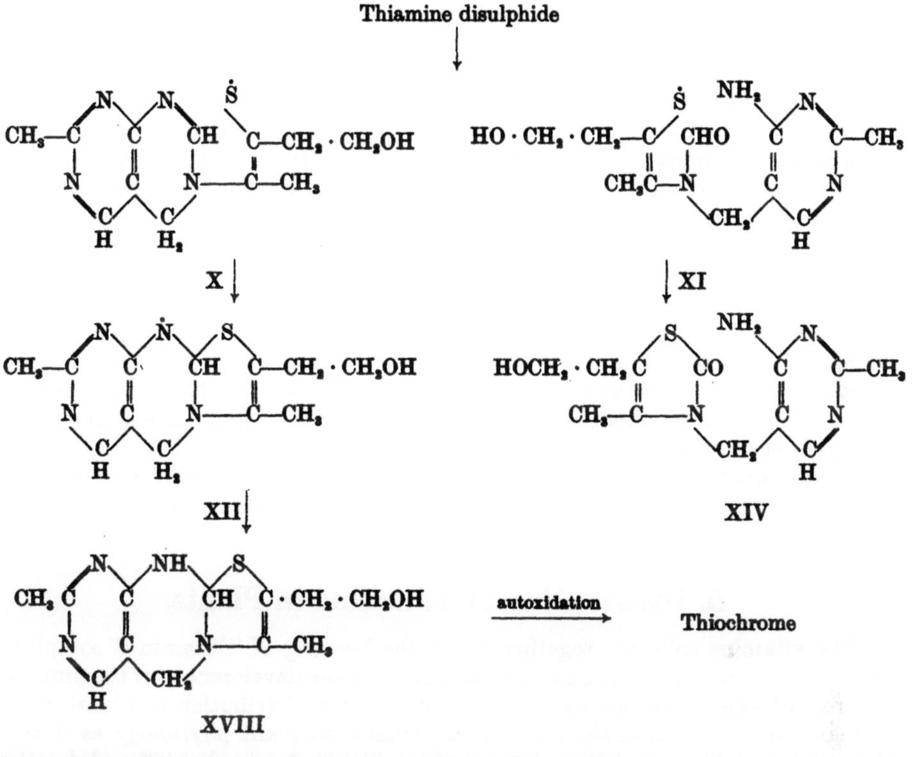

Thiamine disulphide

X   XI

XII   XIV

XVIII   autoxidation ⟶ Thiochrome

Two points of interest to analysts come from this explanation; adequate amounts of oxidant must be present to ensure conversion of thiamine to thiochrome and not to thiamine thiazolone and the virtue of ferricyanide and mercuric oxide (Holman, 1944) as agents in converting thiamine to thiochrome lies in their being one electron oxidants which oxidise the sodium salt of the cyclised thiol form of thiamine to the free radicle.

With diazonium compounds thiamine reacts in alkaline conditions to form coloured products. Kinnersley and Peters (1934) were the first to investigate the reaction and to make use of it as a means of estimating thiamine. No satisfactory mechanism of the reaction has so far been proposed.

II. Biological Effects of Thiamine Related to Its Determination.

The first clue to the relationship of thiamine to carbohydrate metabolism came from the work of Peters and his colleagues (for review see Peters, 1946b). It was found by a manometric technique that minced brain of an avitaminous pigeon in a lactate-phosphate medium did not respire at the same level as a normal brain. But the addition of minute amounts of thiamine to such brain preparation caused a specific absorption of O_2. This action of thiamine, called the catatorulin effect, has been used as a basis for the estimation of thiamine. The vitamin was not active in the presence of succinate or other carbohydrate intermediary compounds that were not transformable to pyruvate. The defect in the brain of the thiamine deficient pigeon was a failure to oxydise pyruvate, a defect repaired by the addition of thiamine *in vivo* and *in vitro*. In yeast, pyruvate is decomposed into acetaldehyde and carbon dioxide by an enzyme carboxylase which until Auhagen's work (1932, 1933) was considered not to have a coenzyme. Auhagen was able to inactivate carboxylase by washing yeast with alkaline phosphate solutions and to restore its decarboxylating activity by adding yeast "Kochsaft" and Mg ions at pH 6.5. The dissociation of the holoenzyme into apoenzyme and co-carboxylase and its synthesis from these components has been extensively investigated by Westenbrink and Goudsmit (1938, 1941); this work is of particular interest in manometric methods of assay of thiamine and thiamine pyrophosphate. By making cocarboxylase determinations Simola (1932) proved that this coenzyme of yeast fermentation was present in animal tissues and was less in amount than normal in organs of thiamine deficient pigeons. The isolation and crystallization of co-carboxylase by Lohmann and Schuster (1937a, b) and its identification as thiamine pyrophosphate provided a connection between the animal and microorganism. In animal tissues thiamine exists mainly as pyrophosphate but definite proof that this is the active form of thiamine was only obtained when Banga, Ochoa and Peters (1939) showed that with a labile brain dispersion in presence of fumarate thiamine pyrophosphate and not thiamine is active in the oxidation of pyruvate.

Upon these established metabolic effects of thiamine have been based methods of assaying of the thiamine and its ester, which although not widely used, are worthy of note because of their sensitivity and specificity.

B. Distribution of Thiamine in Plants.

The vitamins collected together under the heading of "Vitamin B complex" are important as plant hormones and as factors in the development of the embryo, roots and leaves. For this reason knowledge of the distribution and localization of these vitamins within the plant is as valuable to plant physiology as it is in the assessment of the nutritional value of the plant. A co-ordinated and detailed account of the infinite number of determinations of thiamine upon plant tissue is impossible within the limits of this section. Only certain features can therefore be mentioned.

Like all the B vitamins thiamine has been found in many species and almost all organs of the higher plants (see Boas-Fixsen and Roscoe, 1938, 1939—1940; Cheldelin and Williams, 1942; Cravioto et al., 1945). Perhaps the most extensive investigations have been made upon cereals because of their nutritional importance. Whole grains of wheat, barley, maize, oats, rye and rice contain thiamine in amounts ranging 0.35—1.0 mg./100 g. for wheats to 0.3—0.4 mg./100 g. for rice; the highest concentration of the vitamin is localised in the germ, particularly the scutellum. Vegetables have a variable content of thiamine; cauliflower, peas and beans contain 0.12—1.2 mg./100 g. but most vegetables, such as lettuce, spinach, potato, contain about 0.1—0.2 mg./100 g. Fruits do not have much thiamine: about 0.1—0.15 mg./100 g. is present in apples, pears, plums etc. In nuts a somewhat higher content of thiamine, 0.24—1.0 mg./100 g. than in fruits is found. As far as can be seen the amounts of thiamine in different parts of a plant are not in accordance with any general rule.

More detailed analysis of the thiamine content of plant organs brings forth interesting facts. The highest concentration of thiamine is apparently in the youngest leaves of the vegetative plant with a general gradient of falling concentration towards the older parts. In the tomato the concentration of thiamine is highest (19.88 μg./g.) in the shoot apex and younger leaves and falls with the age of the leaf (BONNER, 1942; BONNER and DORLAND, 1943a, b). A similar distribution has been noted in the pea plant (RYTZ, 1939) and in maize (BURCHHOLDER and MCVEIGH, 1940). Flowers contain appreciable amounts of thiamine (CRAVIOTO et al., 1945); RYTZ (1939) found 14—23 μg./g. of dried tissue in the young buds, calyx, corolla and stamens of the pea and tulip. In the seeds of cereals, thiamine is present in relatively low amounts in the endosperm, in higher amounts in the germ and extremely high in the scutellum (HINTON, 1942, 1944; WARD, 1943). The significance of this localization and distribution of thiamine is as yet not fully explained.

C. Methods for Determination of Thiamine.

Methods for the assay or determination of thiamine are of three particular types (a) biological (b) chemical physical and (c) microbiological. The purposes behind estimation of th'amine are several. One may be purely nutritional in nature; a collection of in.ormation upon the distribution of thiamine in the plant and animal kingdom, the sources of the food supplies of man. It may be extended to a detailed examination of the thiamine content of species of a plant to decide how far environmental and genetical influences determine vitamin content and thus pave the way for the selection and breeding of vitamin enriched plants. A second important objective is the distribution of vitamin among the tissues of the plant. Not only is such information valuable in understanding the physiological function of thiamine in plants but also in commercial preparation of plants for food as in the milling of grain.

The biological methods of assaying thiamine have been the chief means of checking the reliability of other types of methods of thiamine determination. Numerous comparisons of one form or another have been made of biological and non-biological methods. They often show that the colorimetric, thiochrome, yeast fermentation and microbiological tests give estimates of the thiochrome content of material which agree satisfactorily with those of bioassay (MELNICK and FIELD, 1939c; HENNESSEY and CERECEDO, 1939; SCHULTZ, ATKINS and FREY, 1937b; FITZGERALD and HUGHES, 1949). Often they are confined to the analysis of easily accessible substances of high vitamin content or, as when the choice of a routine method is being made, to a particular type of vitamin preparation. For instance determination of thiamine in pharmaceutical products of relatively simple composition by the thiochrome fermentation and rat curative gave results that showed no great difference among the methods (HENNESSEY, WAPNER and TRUHLAR, 1944). To judge the general applicability of methods, comparisons of assays on a wide variety of materials of low and high vitamin content is exceedingly useful. An example of such a comparison is that of thiochrome and biological methods, particularly the bradycardia one, by HARRIS and WANG (1941), on 50 foodstuffs including vegetables, fruits and cereals of vitamin content from 0.1—800 μg./g. In 75% of the instances the biological and chemical values differed by less than 15% or in 85% of the instances by 30%; the larger percentage errors occurred only with foods of low vitamin content or in the dessicated products. The results of this and other comparisons (Eppright and Williams, 1944; Brown, Hartzler, Peacock and Emmett, 1943) suggests that, with proper preparation of material and care in experimental technique, the thiochrome method gives estimates of thiamine content às reliable as those of bioassay, and probably less liable to error than other types of determination. Although most comparisons bring out the degree of agreement among different methods they do not always allow an easy assessment of the relative accuracy of a method owing to inadequate statistical analysis. An example of a comparison of methods which shows their relative accuracies is that by the vitamin B_1 Sub-Committee of Accessory Food Factors Committee of the Medical Research Council and the Lister Institute (1943). The thiamine content of national flour and bread was determined by the rat growth, rat bradycardia and rat convulsions methods, by the fermentation method of Schultz, Atkin and Frey (1942) by the thiochrome and azo methods. Good agreement was obtained by these methods for a flour and bread made from the flour (Table 1). By the chemical tests the vitamin

Table 1. *Comparison of thiamine content (µg./g.) of flour C and breadcrumb C, and accuracy of methods.* (Adapted from Vitamin B_1 Sub-Committee 1943.)

	Biological tests[1]		Chemical tests[1]		Biological tests, %		Chemical tests, %	
	As received	Calculated on dry wt.	As received	Calculated on dry wt.	P = 0.95	P = 0.99	P = 0.95	P = 0.99
Flour C . . .	3.9	4.4	3.4	3.9	76.4—130.9	70.2—142.5	±1.71	±2.72
Breadcrumb C	3.6	3.9	2.8	3.0	70.8—141.2	63.5—157.5	±4.18	±5.49

content of the flour was significantly higher than the bread but by the biological test there was no such difference because it was covered by the error of the test. The results of the chemical tests apparently indicate a loss of vitamin; a conclusion which, owing to the limits of error, the biological tests cannot confirm. The relative accuracy of the thiochrome method and azo methods in the determination of pure thiamine and thiamine in flour and bread can be seen from Table 2. It is

Table 2. *Limits of error obtainable in chemical determinations of pure thiamine and of thiaminel in extracts obtained from national flours and bread.* (Adapted from Vitamin B_1 Sub-Committee 1943.)

	Thiochrome method, % Photoelectric comparison		Azo method, %	
	P = 0.95	P = 0.99	P = 0.95	P = 0.99
Pure vitamin B_1[2]:				
In duplicate	1.98	2.60	2.8	3.7
In triplicate	1.62	2.12	2.3	3.0
In quadruplicate	1.40	1.84	2.0	2.6
Flour or bread[3]:				
In duplicate	13.2	17.4	8.9	11.6
In triplicate	10.8	14.2	7.2	9.5
In quadruplicate	9.3	12.2	6.3	8.3

[1] The average figures given were obtained by weighting the various results according to their accuracy.

[2] The two sets of results obtained for the comparison of the thiochrome and the azo methods, are each the results of work done in only one laboratory. The limits of error by the azo method are for an amount of 30 µg. Similar values have been obtained for amounts of 20—40 µg.

[3] The errors for flours and bread include errors due to differences between laboratories.

doubtful whether the azo method is more accurate than the thiochrome since the estimation of its accuracy is based on fewer results than the latter.

I. Extraction of Thiamine from Biological Materials.

Although it is very soluble in water, thiamine is sometimes not easily extracted from plant and animal tissues by aqueous solutions. This difficulty is in the main due to insufficient knowledge of the state in which thiamine exists in plant tissue. In certain yeasts it seems clear that the greater part of thiamine is present as its pyrophosphate ester combined together with magnesium with the apoenzyme of carboxylase (GREEN, HERBERT and SUBRAHMANYAN, 1940). Even in yeasts thiamine pyrophosphate may be combined with proteins other than carboxylase (PARVÉ and WESTENBRINK, 1944). The lack of extensive investigations prevents any generalisation upon the distribution in plant materials of thiamine pyrophosphate or carboxylase or of their combination as the holoenzyme. Determinations upon easily accessible materials such as rice polishings, cereal grains and wheat germ show that in these materials, in contrast to yeast and some animal tissues, thiamine pyrophosphate is present in small amounts compared with the free vitamin. Nor is carboxylase always associated with thiamine: the jack bean and soya bean have about the same thiamine content, but the first has a very active carboxylase system and the second does not (COHEN, 1946). It is possible that thiamine pyrophosphate exists in different combinations in yeast and wheat germs, for the thiamine pyrophosphate of wheat germ is far more easily hydrolysed by commercial phosphatases at low pH values than that of brewers yeast (SURE, 1946). With such limited knowledge about the nature of thiamine in most plants, methods of extraction are to an extent empirical and may need investigation before use in determination of thiamine in plants.

The numerous methods that have been used in the extraction of thiamine from animal and plant materials can be classified under extraction with dilute acid, digestion with enzymes or a combination of both. Acid seems at present to be the most effective agent in extracting thiamine and its pyrophosphate from animal and plant tissue. Cereals are finely ground before extraction but with plant tissues (potatoes, cauliflower, spinach, etc.) a weighed amount (10 g.) is ground with a few drops of concentrated HCl to a fine pulp, treated with 30—40 ml. water and heated in boiling water for 10 min. Acid extraction preserves thiamine against destruction and is usually carried out at the temperature of a boiling water bath. Although BOOTH (1942) considers acid extraction as effective as enzyme digestion, hydrolysis of thiamine pyrophosphate with N acid for 15 mins. at 100° C produces thiamine monophosphate (LOHMANN and SCHUSTER, 1951). To ensure complete hydrolysis of the pyrophosphate in the acid extracts phosphatases from different sources have been used; takadiastase (KINNERSLEY and PETERS, 1938; HARRIS and WANG, 1941), kidney phosphatase (HENNESSEY and CERECEDO, 1939), yeast phosphatase (MELNICK and FIELD, 1939c; WESTENBRINK and WILLEBRANDS, 1944). Together with takadiastase, papain may be used; a useful combination, for papain destroys protein in the acid extracts and also factors that may inhibit hydrolysis of the pyrophosphate by takadiastase (HARRIS and WANG, 1941). Generally digestion of extracts with phosphatase preparations is carried out at a pH value 4—5. From a survey of the literature of the various phosphatase preparations recommended, takadiastase would seem to be a most effective enzyme preparation. With some products the appropriate phosphatase preparations must be chosen since they vary in their efficiency in hydrolysis of thiamine pyrophosphate as has been noted in yeast and wheat germ (SURE, 1946).

A point of some importance is the occurrence in abundance of phosphatases in certain yeasts (see Westenbrink, van Dorp, Gruber and Veldman, 1940), higher plants, seeds, potatoes (Roche, 1950). Like the phosphatases generally used in thiamine determination, these enzymes are most active in acidic conditions at pH values 5.2—5.4. In determinations of thiamine pyrophosphate, they deserve attention because given appropriate conditions they might destroy what little pyrophosphate the plant tissue contains. For instance, during the autolysis of yeast complete disappearance of thiamine pyrophosphate activity has been noted (Auhagen, 1933). Possibly the activity of these phosphatases might be reduced by inhibitors such as fluoride or by thiamine (Ochoa, 1938; Ochoa and Peters, 1938; Westenbrink et al., 1940; Westenbrink and van Dorp, 1942).

Another factor, the importance of which in the determination of thiamine is yet to be assessed, is the existence in animals and plants of substances capable of destroying thiamine. In the tissues of fish, especially carp, exists an enzyme, thiaminase which splits thiamine into its pyrimidine and thiazole components (see review by Harris, 1951). Similar substances are found in plants (Bhagvat and Devi, 1944; Weswig, Freed and Haag, 1946; Evans and Evans, 1949; Evans, Jones and Evans, 1950; Evans and Evans, 1952) in bracken (*Pteris aquilina*), horse tail (*Equisetum arvense*) male fern (*Aspidium filix-mas*) and certain cereals and oil seeds, but how general is their distribution is unknown. It is probable that such substances, if like bracken thiaminase which is more easily extracted at an alkaline than acid pH, will not be present in acid extracts of plant tissue to be of serious consequence.

II. Biological Methods for the Determination of Thiamine.

Most of the early methods of assay of thiamine were based upon its biological effect: its promotion of growth and cure of signs and symptoms in the deficient animals. These methods are still regarded as the ultimate standard against which others of non-biological character must be checked. Their adaption as a standard is largely due to emphasis on the nutritional importance of thiamine. They are the methods which assay the amount of thiamine available in any particular foodstuff to the whole animal and, in this respect, they provide information given by no others. On the other hand from the analytical viewpoint the validity of a method which determines total thiamine and the proportions of it and its derivatives in foodstuffs may be gauged by its conformity with analytical criteria or by comparison with a less tedious and yet specific biological method such as the method of Ochoa and Peters (1938).

Methods of biological assay that use living animals have unavoidable limitations from which the non-biological ones are free. They are laborious and time consuming, and to reduce individual variation, demand large number of animals which need special care and attention. In most, an added burden is the preparation of special diets. Furthermore though based on a few principles, the numerous methods of bioassay have taken forms due to differences in technique that their relative merits are difficult to assess. For this reason comparison of methods of bioassay by groups of workers are most valuable (Coward, Burn, Ling and Morgan, 1933; Birch and Harris, 1934), but equally important is the proper statistical analysis of the results.

Methods of biological assay of thiamine on live animals are of three main types: (a) Curative or preventive, (b) growth response, (c) physiological results. Examples of each will be described briefly.

Curative or Preventive. Pigeons, chicks and rats are used as test animals. Carried out on thiamine deficient pigeons the test consists in comparing the efficiency of an amount of test substance with that of a standard in curing the classical symptoms of head retraction. A relation has been shown to exist between the dose of thiamine and the percentage of birds that are cured. The conditions and accuracy have been thoroughly discussed (KINNERSLEY, PETERS and READER, 1928; KINNERSLEY and PETERS, 1936; COWARD et al., 1933; COWARD and MORGAN, 1939; COWARD, 1947). With a supply of deficient birds the test is quick and simple; it is specific for thiamine and requires less work and attention than many tests. A similar curative test has developed on the same principle (SMITH, 1930; BIRCH and HARRIS, 1934) for the rat and is the method recommended in the United States Pharmacopoeia (1947).

Growth Response. This test has been and is used as much as the curative (SHERMAN and SPOHN, 1923; GUHA and DRUMMOND, 1929; COWARD et al., 1933; COWARD, 1936) and is the one recommended in the British Pharmacopoeia (1932). It makes use of the fact that young rats fed a thiamine deficient diet cease to put on weight in 1—2 weeks and unless they are given thiamine lose weight and die. But given a daily dose of a substance containing thiamine, they gain weight at a rate proportional to the log. of the dose (COWARD, 1947). The standard and the unknown are compared on litter mates of the same weight and sex of preferably an inbred strain of rats. Although test periods of 4—8 weeks, have been recommended, there is little gain in accuracy by extending it beyond two weeks when the error of the determination may be about ± 12.5% (COWARD, 1936).

Physiological Response. A severe bradycardia is a feature in the rat of thiamine deprivation (DRURY, HARRIS and MAUDSLEY, 1930) and is rapidly cured by administration of the vitamin. Upon the cure of the abnormality BIRCH and HARRIS (1934) developed a simple and rapid method of estimating thiamine which gives results comparable with those of the curative and growth response methods. The relationship between dose of thiamine and increase in heart rate in 24 hrs. is apparently linear.

III. Enzymatic Methods.

The metabolic effects of thiamine have allowed the design of methods of determination enzymatic in nature. Although, as yet, they have found no general use in plant analysis, they are worthy of a brief description because of their specificity and of their adaptation to small amounts of material.

Catatorulin Test. The method is the earliest example of the use of an enzyme system in the assay of thiamine. It is based on the stimulatory effect of thiamine upon the oxygen uptake of minces of avitaminous pigeon's brain (PASSMORE, PETERS and SINCLAIR, 1933; PETERS, RYDIN and THOMPSON, 1935). Its usefulness lies in its specificity for small amounts of thiamine, about 0.2 μg., low sensitivity to thiamine pyrophosphoric esters and lack of response to thiochrome (PETERS, 1937, 1938; OGSTON and PETERS, 1936). Thiamine disulphide is as active as thiamine in the test (PETERS, 1946a).

Decarboxylation of Pyruvate by Yeast Carboxylase. In 1938, OCHOA and PETERS developed a method which allowed the separate quantitative estimation of thiamine and thiamine pyrophosphate. It depends upon the stimulatory action of the two compounds upon the decarboxylation of pyruvate by yeast carboxylase. The enzyme is prepared by washing dried baker's yeast with 0.1 M Na_2HPO_4 to free it from thiamine pyrophosphate. As a suspension in phosphate buffer pH 6.2 the enzyme is placed in the bottle of a WARBURG apparatus (see Vol. 1 of this handbook) together with magnesium chloride and the test solution. The reaction is started by adding pyruvate from the side arm of the bottle or from a dangling cup and the production of carbon dioxide followed manometrically.

Whereas thiamine cannot replace co-carboxylase it greatly stimulates the decarboxylation of pyruvate by its pyrophosphate (OCHOA, 1938); thiamine monophosphate, and pyrimidines similar structurally to the pyrimidine part of thiamine act in the same way (OCHOA and PETERS, 1938). This effect is apparent with 0.1 μg. thiamine and increases with increasing concentrations; with 1.0 μg. of thiamine pyrophosphate it is maximal with 15 μg. of thiamine hydrochloride. By using constant amounts of one of the two substances, the other can be estimated. The method is remarkably sensitive: in the presence of excess thiamine 0.01 to 0.02 μg. of thiamine pyrophosphate can be estimated and with excess thiamine pyrophosphate 0.05 μg. thiamine.

Particularly for the determination of thiamine pyrophosphate the method has been investigated, refined and extensively used by WESTENBRINK and his colleagues whose papers should be consulted for details (WESTENBRINK, GOUDSMIT and VELDMAN, 1941). As mentioned elsewhere WESTENBRINK showed that the stimulatory effect of thiamine and other substances upon thiamine pyrophosphate activity was due to their inhibition of phosphatase that hydrolysed the pyrophosphate (WESTENBRINK, VAN DORP, GRUBER and VELDMAN, 1940; WESTENBRINK and VAN DORP, 1942). WESTENBRINK (1940a, b) made a neat refinement of the method by adapting it to the Cartesian diver increasing its sensitivity to the measurement of 0.05 mμg. co-carboxylase and 0.5 mμg. thiamine. The sensitivity of the method depends upon the kind of yeast used as a source of carboxylase and the manner in which the yeast is free of thiamine pyrophosphate. The method has been of considerable value in studies upon the distribution of thiamine pyrophosphate in the blood and tissues of animals (WESTENBRINK, PARVE and THOMASSEN, 1943; WESTENBRINK, PARVE, VAN DEN LINDEN and VAN DEN BROCK, 1943; SMITS and FLORIJN, 1949) but has, as yet, found no great use in plant analysis.

IV. Microbiological Methods.

Some twenty odd years ago a search was made for a suitable microbiological test to replace the laborious bioassays for thiamine. The stimulatory effect of thiamine on yeast multiplication (WILLIAMS, 1919) and on the growth of *Streptothrix corallinus* (ORR-EWING and READER, 1928; PETERS, KINNERSLEY, ORR-EWING and READER, 1928) was the basis of tests advocated as substitutes for the laborious bioassays. That little came of these and other efforts was in part due to the unspecific nature of the effects and to the necessity at that time of working with crude preparations and to sparseness of knowledge on the growth requirements of micro organisms. In more recent years, several microbiological methods for the assay of thiamine have been recommended: the yeast fermentation (SCHULTZ, ATKINS and FREY, 1937a, b, 1941, 1942), yeast growth (WILLIAMS, MCMAHAN and EAKIN, 1941), growth of *Lactobacillus fermenti* (SARETT and CHELDELIN, 1944; FITZGERALD and HUGHES, 1949). These methods, often of extreme sensitivity, are subject to errors due to the effect of non-specific substances upon the growth or metabolic activity of the organism.

Of the microbiological methods the yeast fermentation has gained preference. In comparative tests it has given estimates of the thiamine content of flour (Vitamin B$_1$ Sub-Committee, 1943) pharmaceuticals (HENNESSEY et al., 1944), a number of natural products (EPPRIGHT and WILLIAMS, 1944) and yeast preparations (SCHULTZ et al., 1937b) comparable with those of biological and chemical methods. The method depends on the fact that thiamine has as powerful a stimulatory effect upon the rate of yeast fermentation as upon yeast growth (WILLIAMS and ROEHM, 1930; SCHULTZ et al., 1937a). As little as 1.0 μg. thiamine/ml. of culture medium will produce a substantial fermentation by baker's yeast, measurable gasometrically in terms of carbon dioxide output. The increase in fermentation bears a definite relation to the amount of thiamine; the carbon dioxide output increases with time but the relative percentage increase produced by thiamine concentrations is the same for each period. It is thus possible to estimate the thiamine content in an extract by comparing its effect upon fermentation with that of standard thiamine solution. Usually two standards of 2 μg. and 4 μg. are used because the difference in gas production in these two tests, equivalent to 2 μg., is reasonably constant in day to day determinations. By use of the WARBURG apparatus in place of a fermentometer, 0.005—0.025 μg. thiamine can be assayed with an accuracy of $\pm$ 5% (ATKINS, SCHULTZ and FREY, 1939; see also LASER, 1941).

The method is not without difficulties. Yeasts vary in their response: *Torula utilis* does not respond at all to thiamine and some samples of baker's yeast show an increase in fermentation only after several hours of incubation (LASER, 1941). As, in many microbiological tests, there is the problem of non-specific effects of substances other than thiamine. Of these, the most troublesome are the pyrimidine and thiazole compounds of thiamine for it is not easy to decide from present facts how far their presence in extracts affects the fermentative response. The pyrimidine component does stimulate fermentation (SCHULTZ et al., 1937b) and is more effective at low concentrations than is thiamine (DEUTSCH, 1944). The evidence for the effect of the thiazole component and of the products of cleavage of thiamine and thiamine pyrophosphate is not decisive; whereas SCHULTZ et al. (1941) state they are inactive, DEUTSCH (1944) claims that they are slightly active. Moreover, OBERMEYER and MEYER (1945) has found that yeast can synthesise thiamine from its components. Besides these substances others may affect fermentation and to assess the residual fermentative activity of extracts, SCHULTZ et al. (1942) treat an aliquot of the thiamine extract with sodium sulphite which at appropriate conditions of pH and temperature quantitatively splits thiamine and its pyrophosphate into pyrimidine and thiazole components.

The description which follows of the yeast fermentation method contains its essential features. A more detailed account is given in "Methods of Vitamin Assay", Ed. the Association of Vitamin Chemists (1947).

Yeast Fermentation Method. Preparation of samples follows procedures common to most methods. The solution or suspension of the sample is made acid to congo red with sulphuric acid and heated for 20 mins. at 100°. When cool, the solution is made up to known volume; a portion is taken for sulphite treatment and the remainder, neutralised with NaOH to litmus paper, used for determination of total fermentation activity.

Treatment with sulphite. An aliquot of the sample solution (not greater than 20 ml.) is treated with 0.7 ml. N H_2SO_4 and 5 ml. of 4% (w/v) sodium sulphite, (Na_2SO_3 7 H_2O). The reaction is adjusted to pH 5.2—5.6. The mixture is heated at 100° C for 30 min., and then cooled. The excess sulphite is destroyed by the cautious addition of 3% H_2O_2 solution; the end point is detected by starch iodide solution. The mixture is neutralised to litmus with NaOH and an aliquot taken for the determination of the residual fermentation.

Solution A. 180 g. ammonium dihydrogen phosphate, 72 g. diammonium hydrogen phosphate and 0.2 g. nicotinic acid are dissolved in 1 litre distilled water. It is sterilised by heating at 100° C for 30 min.

Solution B. 200 g. anhydrous glucose, 7 g. $MgSO_4 \cdot 7 H_2O$, 2.2 g. KH_2PO_4, 1.7 g KCl, 0.5 g. $CaCl_2 \cdot 2 H_2O$, 0.01 g. $FeCl_3 \cdot 6 H_2O$ and 0.01 g. $MnSO_4 \cdot 4 H_2O$

are dissolved in 1 litre distilled water. This solution should be distributed in 500 ml. Erlenmeyer flasks and sterilised at 100° for 3 successive days.

Yeast Suspension. 10 g. of baker's yeast is suspended in 200 ml. of H_2O; yeast of high thiamine content should not be used.

Apparatus. A fermentometer which consists of a thermostat, a shaking device and a number of gasometers. (For description of fermentometer see SCHULTZ et al., 1942.)

In each of four reaction bottles of the gasometers 2.5 ml. solution A and 7.5 ml. of solution B are placed, followed by known volume test solutions: in the first bottle the prepared sample, in the second, the sulphite treated solution together with 2 ml. of a solution containing 2 μg. thiamine; the third, 2 μg. thiamine and to the fourth 4 μg. of thiamine. The volume of solution in each bottle is then made up to 40 ml. and then 10 ml. of the yeast suspension is added as rapidly as possible. The bottles are placed in the thermostat and connected to the gasometers. After shaking for 3 mins. the first reading of the gas volume is made and after 3 hours the final one.

Calculation.

Let W Weight of sample.

T = Volume of gas produced in bottle containing test solution.

B = Volume of gas produced in bottle containing sulphite treated sample and 2 μg. thiamine.

S_2 = Volume of gas produced in bottle containing 2 μg. thiamine.

S_4 = Volume of gas produced in bottle containing 4 μg. thiamine.

F = Factor to correct for use of aliquot of sample.

Then μg./g. of sample $= 2 + 2 \times \dfrac{T-B}{S_4-S_2} \times \dfrac{1}{W} \times F.$

A microbiological assay which is as sensitive as the yeast fermentation method but which requires no complicated apparatus other than a photoelectric colorimeter is that introduced by SARETT and CHELDELIN (1944) and subsequently improved by FITZGERALD and HUGHES (1949). It is based on the effect of thiamine upon the growth of *Lactobacillus fermenti 36*. Over the range 0.02 to 0.04 μg. the response of the micro-organism is linear and the thiamine content of the sample under test can be obtained by interpolation. The inhibitory or stimulating effects of substances other than thiamine is counteracted by adding to the standard tubes portions of the extract of the sample which have been treated with sulphite. The method has given estimates of the thiamine content of food materials which agree reasonably well with those of chemical assay.

A brief mention must be made of the use of the mold *Phycomyces Blakesleeanus* in the assay of thiamine (SCHOPFER, 1935). The growth of the fungus is sensitive to amounts of thiamine of the order of 0.1 μg. The growth response of the mold has been used to assay thiamine in blood (SCHOPFER, 1937; SCHOPFER and JUNG, 1937; MEIKLEJOHN, 1937; SINCLAIR, 1938, 1939a, b) and plant materials (BURKHOLDER and McVEIGH, 1940; ROBBINS, 1939; BONNER and ERICKSON, 1938; MEIKLEJOHN, 1943).

V. Chemical Methods for the Determination of Thiamine.

There are two chemical methods that are generally used in the determination of thiamine: the azo and the thiochrome methods. About equal in accuracy for amounts of thiamine of the order of 20—40 μg. they differ in sensitivity. Whereas the azo method is usually applicable to materials of relatively high thiamine content, the thiochrome method with greater sensitivity can be used in the analysis of low and high concentration of thiamine. Both methods are described because in particular instances one may be more appropriate than the other.

1. The Azo Method.

PETERS and KINNERSLEY (1934) introduced a chemical reaction for the detection and estimation of thiamine which they had found useful in the isolation of the vitamin from yeast. With diazotised sulphanilic acid in a defined alkaline medium, thiamine reacts to form a red substance which is stabilised by the presence of formaldehyde. Later they improved their method by extracting the compound with butyl alcohol from which it was later recovered by shaking with acid and applied it to plant materials (KINNERSLEY and PETERS, 1938). A point of procedure worthy of note because of its possible use with other plant tissue is the mode of extraction of thiamine: the finely ground plant material is extracted with 50% acid ethyl alcohol and the thiamine subsequently precipitated as its phosphotungstate at pH 6.0. In this way extracts with a reduced content of interfering substances were obtained.

Subsequent modifications were aimed at increasing the sensitivity and specificity of the method. Of amines proposed in place of sulphanilic acid, p-amino-acetophenone, introduced by PREBLUDA and McCOLLUM (1936, 1939), is satisfactory; it does not react with a large number of biological substances and the coloured products it forms with others (e. g. inositol) is not extracted by xylene. The trichloracetate of p-aminobenzoic acid, so far not thoroughly investigated, is claimed to be sufficiently specific to avoid preliminary adsorption of thiamine (KIRCH and BERGEIM, 1942). The mechanism of the reaction between thiamine and aromatic amines is obscure but apparently involves the thiazole component rather than the pyrimidine (PREBLUDA and McCOLLUM, 1939; KIRCH and BERGEIM, 1942). MELNICK and FIELD (1939a, b, c) developed the method of PREBLUDA and McCOLLUM and applied it to the estimation of thiamine in rice polishings, wheat germ, yeast and liver. Their essential modifications were the use of a mixture of ethyl alcohol and phenol to increase the sensitivity of the reaction and reduction of interference from substances in the initial extract by adsorbing thiamine upon permutit (CERECEDO and HENNESSEY, 1937) and hydrolysis of the phosphorylated forms of thiamine by yeast phosphatase. The method is as follows:

Method. A sample, finely ground or in solution, containing 150 μg. of thiamine is submitted to a procedure which in its early stages of extraction, enzymic digestion, and adsorption on and elution from decalso is practically identical with those of the thiochrome method (see p. 300). An aliquot of the eluate from the decalso is pipetted into a 50 ml. centrifuge tube, followed by an equal volume of 95% ethyl alcohol containing 5 mg. phenol/ml. One drop thymol blue indicator is added and while a fine stream of N_2 is bubbled through the solution, 1 N NaOH is added dropwise until a faint but positive blue colour appears. 6 ml. of the PREBLUDA-McCOLLUM reagent is added, the tube stoppered, and the mixture allowed to stand overnight. 2 ml. xylene are added and the mixture shaken vigorously for 1.5 min. After centrifugation the colour in the xylene is compared with a standard similarly treated. The xylene contains the coloured vitamin derivative; the aqueous layer, the indicator and coloured derivatives of phenol and other substances reacting with the reagent.

The standard solution is 0.5 ml. of a solution containing 50 μg. thiamine chloride added to 2.5 ml. of a blank KCl eluate (10 ml.). The blank is essential because the amount of pigment recovered from the decalso eluates may vary from 85—95% of the theoretical.

Prebluda and McCollum *Reagent*. Solution A.   3.18 g. p-aminoaceto-phenone are dissolved in 45 ml. conc. HCl and the solution diluted to 500 ml. distilled water. It is stored in the cold in a stoppered bottle. Solution B. 22.5 g. $NaNO_2$ are dissolved in 500 ml. distilled water. It is stored in the cold in a stoppered bottle.   Solution C.   20 g. NaOH are dissolved in 600 ml. distilled water; 28.8 g. $NaHCO_3$ are then added and the solution made up to 1 litre with distilled water.

Diazotisation. In an ice bath 1 part of solution A is mixed with 1 part of solution B and the mixture stirred for 10 min. Then 4 parts of solution B is added, the mixture stirred and kept at $0-5°$ C for 20 min. Fresh solutions should be made daily.

The Reagent. 20 ml. of freshly diazotised p-amino acetophenone is added to a flask containing 275 ml. of Solution C. When these solutions are mixed a purple colour appears but, on stirring for 5—10 min., it disappears.

Discussion. The method is claimed to estimate amounts of thiamine of the order of 5 μg. with an error of $\pm 2\%$; it is not so sensitive as the thiochrome or manometric or microbiological methods. It does not directly estimate the phosphorylated forms of thiamine because these compounds yield coloured products with the diazonium salts which are not extracted by solvents from alkaline solution. Estimation of thiamine in extracts before and after incubation with phosphatase give results whose difference allows a calculation of thiamine pyrophosphate concentration (Melnick and Field, 1939c). Treatment of extracts with other enzymes such as takadiastase would be equally effective. Some materials such as liver may contain substances that seriously affect the adsorption of thiamine upon permutit; in such cases Melnick and Field recommend an additional step in their procedure: extraction of thiamine with benzyl alcohol. Among definite substances that seriously interfere with the reaction are ascorbic acid and uric acid (Emmett, Peacock and Brown, 1940; Alexander and Levi, 1942) but only at high concentrations. In some plant tissues e. g. walnut where high concentrations of ascorbic acid are apt to occur this interference might be significant. Emmett et al. (1940) recommend adsorption of thiamine on superfiltrol and others have used oxidants to eliminate ascorbic acid.

Checked against biological assay Melnick and Field's method gives satisfactory estimates of the thiamine content of yeast, wheat germ and rice polishings. It is a convenient method for the assay upon materials relatively rich in thiamine and is readily adaptable for use in the detection of the thiamine (see Kinnersley and Peters, 1934; Prebluda and McCollum, 1939).

2. Thiochrome Method.

Use of the oxidation of thiamine to thiochrome as a means of estimating the vitamin was made by Jansen (1936). His method contained many essential features which are retained in later procedures: oxidation with alkaline ferricyanide, extraction of thiochrome from the reaction mixture with isobutanol and comparison of fluorescence against quinine standards. Since 1936 it has become a general method for the determination of thiamine in biological fluids, animal and plant tissues. Several modifications of the original Jansen method have been devised mainly to circumvent interference from substances that might affect thiochrome formation or its fluorescence. Adsorbents such as frankonite were used to separate thiamine from extracts, but the most significant improvement was the introduction of an artificial zeolite decalso by Hennessey and Cerecedo (1937). The adsorption of thiamine upon decalso is a recognised essential stage in determination of the vitamin in natural materials but it does not eliminate odd analytical difficulties that some biological materials present, owing to peculiarities in their composition. The adsorption of thiamine by decalso may be reduced by the inorganic salts present in high concentration or other substances from plant material. In the case of urine the difficulty has been avoided by MELNICK and FIELD (1937c) by a preliminary extraction of thiamine by benzyl-alcohol, a technique that might be of use with plant materials. Some fluorescent materials in foodstuffs may enhance the fluorescence of thiochrome; their contribution may be estimated by their destruction (WANG and HARRIS, 1942) or that of thiamine. Other substances in cereal extracts may quench the fluorescence of thiochrome, an effect which ORGAN and WOKES (1944) overcame by adding known amounts of thiamine to the test sample and then subtracting the amount added from the results. These examples suffice to show that the thiochrome method does, on occasions, demand of the analyst ingenuity to surmount difficulties created by the properties of his materials. However its sensitivity, its general agreement with other kinds of thiamine assay and its accuracy (p. 351) and its wide applicability recommends its use in plant analysis.

Method. In its recent forms the method consists of the following stages.

(a) *Extraction.* Thiamine and its phosphate esters are extracted from the material by acid and the acid extract digested with an enzyme preparation to convert the phosphate esters into free thiamine. In the determination of total thiamine content, enzymic hydrolysis of thiamine phosphates is necessary because their oxidised forms are not extracted by isobutanol.

(b) *Purification.* Thiamine is adsorbed on decalso and eluted with potassium chloride.

(c) *Oxidation.* Thiamine is oxidised to thiochrome which is extracted from the reaction mixture by isobutanol.

(d) *Measurement of fluorescence.*

The following procedure is in its essentials that of HENNESSEY and CERECEDO (1937) but is in detail an abridged version of those recommended by the Association of Vitamin Chemists Inc. (1947) and the Sub-Committee on Vitamin Estimations of the Analytical Methods Committee of the Society of Public Analysts (1951).

Reagents. All chemicals should be of analytical reagent standard or of the purest quality that is obtainable. Reagents and solutions should be stored in clean vessels and not allowed contact with materials such as corks, rubber stoppers or lubricants likely to yield fluorescent materials. Glycerine or silicone may be used as tap or stopper grease.

Ethanol. Redistilled from an all-glass apparatus.

Sodium Hydroxide. 15% (w/v).

Potassium Ferricyanide. 1% (w/v). This reagent is stable if it is kept cool and in the dark preferably in a brown bottle.

Alkaline Potassium Ferricyanide Solution. The reagent is prepared immediately before use. Dilute 3 ml. 1% potassium ferricyanide solution to 100 ml. with cool 15% NaOH.

Hydrochloric Acid Solution. 0.2 N (approx.).

Sodium Acetate Solution. 2.5 M. Dissolve 205 g. anhydrous sodium acetate (CH_3COONa) or 340 g $CH_3COONa \cdot 3 H_2O$ in water and dilute to 1 litre.

Isobutanol (water saturated). Commercial isobutyl alcohol is distilled in an all glass apparatus. The fluorescence of the distillate should not exceed that of a solution containing 0.01 μg./1 ml. quinine sulphate.

Enzyme Solution. A fresh solution should be prepared daily from taka-diastase (Parke Davis & Co., London),Clarase (Takamine Laboratories, Clifton, N. J., USA.) or other suitable source of phosphatase. The enzyme (6 g.) is suspend-ed in 2.5 M sodium acetate and the mixture is thoroughly shaken and diluted to 100 ml. with more sodium acetate. Each sample of enzyme should be tested for thiamine by the fluorescence procedure, and if not almost thiamine-free should be replaced by a fresh batch of enzyme.

Potassium Chloride Solution. 25% (w/v).

Acid Potassium Chloride. Dilute 8.5 ml. conc. HCl to 1 litre with 25% KCl solution.

Base-exchange Silicate. A suitable base exchange substance is Decalso F (Permutit Co., London), an artificial zeolite. Different methods are used in its activation. The following is satisfactory. A quantity of 100—500 g. of the silicate is placed in a beaker and covered with hot 3% acetic acid. The mixture is kept at 100° C for 10—15 min. and stirred frequently. The decalso is allowed to settle and the supernatant fluid is decanted. The adsorbent is washed three times with hot 25% KCl and with boiling water until the last washings give no reaction for chloride. It is dried in an oven at a temperature not above 100°.

Stock Thiamine Solution. Thiamine hydrochloride is dried over P_2O_5 in a desiccator for 24 hr. A stock solution may be prepared by dissolving 100 mg. in 1 litre of 25% ethanol or of 0.01 N HCl or of 0.2 N HCl. These solutions are stable for several months provided they are stored at 5° C or below.

Standard Thiamine Solution. The stock solution (5 ml.) warmed to room temperature is diluted to 100 ml. Of this dilution 10 ml. are transferred to a flask containing 200 ml. of approximately 0.1 N H_2SO_4 and 12.5 ml. sodium acetate solution and then diluted to 250 ml. The final concentration is 0.2 μg./ml. thiamine. This solution is prepared anew daily.

Stock Quinine Standard. Dissolve 25 mg. quinine sulphate (B. P.) in 250 ml. 0.1 N H_2SO_4. Stored in a dark brown bottle and at a temperature below 5° C the solution is stable indefinitely.

Standard Quinine Solution. Dilute 10 ml. of stock quinine solution to 1 litre with 0.1 N H_2SO_4. Stored in a brown bottle and at a temperature below 5° C the solution is stable for a few months.

Procedure. A sample finely ground or in solution containing less than 50 μg. of thiamine is weighed or pipetted into a large boiling tube. To it are added 65 ml. 0.1 N HCl. It is digested for 30 min. in a boiling water-bath and is often mixed. During digestion, the mixture must remain at a pH below 4.5. Cooled to below 50° C, the mixture is neutralised to pH 4.5 with 2.5 M sodium acetate and, to it, is added 5 ml. freshly prepared enzyme suspension. The mixture is incubated at 45—50° C for 3 hr. or at 37° C overnight. It is cooled and then diluted to 100 ml. and filtered.

The extract is now purified. A base-exchange tube (15 × 0.8—1.0 cm.) is filled with 5 g. of activated decalso, suspended in water. When almost all the water has drained away the column is treated with 5 ml. 3% acetic acid which is also allowed to drain away. 25 ml. extract is passed through the column which is then washed with three successive 10 ml. portions of boiling water. The thiamine is now eluted by pouring 10 ml. of almost boiling acid potassium chloride through the column. The eluate is collected in a graduated cylinder. When the first 10 ml. of acid potassium chloride has entered the column, a second 10 ml. is added. The total eluate is cooled and diluted to 25 ml. with acid potassium chloride. The volume of acid potassium chloride used for elution may, of course, be adjusted to the expected amounts of thiamine. The decalso column may be repeatedly used if, after a determination, it is washed with hot water to remove potassium chloride and its efficiency during use may be gauged by a recovery experiment with pure thiamine.

Into each of two stoppered measuring cylinders or separatory funnels 5 ml. potassium chloride eluate is pipetted. While air is bubbling through the eluate in one vessel, 5 ml. alkaline ferricyanide are added and then 25 ml. water-saturated isobutanol; the mixture is shaken for 1—5 min. The eluate in the other vessel is similarly treated except that 5 ml. 15% NaOH is added instead of 5 ml. alkaline ferricyanide; it constitutes the unknown blank. In the same way, two 5 ml. thiamine standard solution are treated; they provide the standard and the standard blank. After the four solutions have stood for a few minutes, 1 ml. ethanol is added to the isobutanol layer in each vessel. The isobutanol layer is gently stirred until it is clear. If the isobutanol solution is separated from the lower aqueous layer, it may be cleared with anhydrous sodium sulphate. The fluorescence of the isobutanol extracts is measured in a fluorimeter which is checked between readings by means of the quinine standard.

The calculation of the thiamine content of the sample in μg./g. is:

$$\frac{U - U_B}{S - S_B} \times \frac{100}{25} \times \frac{5}{\text{wt. of sample}}$$

U = unknown reading $\qquad$ U_B = Unknown blank reading
S = standard reading $\qquad$ S_B = standard blank reading

Clearly the calculation must be modified if the initial volume of the extract or the volume of extract used for adsorption differs from those of the above description.

Discussion. An indication of the effect of interfering factors in an assay may be obtained by carrying out recovery experiments. Although the results may show whether an assay is satisfactory or not, they may not prove suitable in making allowance for interfering factors in a complex system of analysis such as the thiochrome method [see RIDYARD (1950) for a valuable analysis of the thiochrome method and recovery experiments.]

With a suitable fluorimeter 0.1 μg. thiamine may be estimated (HENNESSEY and CERECEDO, 1937). Besides the total thiamine of a sample the method can be used to give estimates of free and phosphorylated thiamine. This may be done by omission of the enzymic hydrolysis of the initial acid extract. The difference between the total free thiamine is a measure of amounts of phosphorylated thiamine.

The accuracy of the method has been commented upon (p. 351). In addition, it may be noted that a collaborative trial undertaken by the Sub-Committee on Vitamin Estimations of the Analytical Methods Committee of the Society of Public Analysts (1951) showed that the thiochrome method, essentially in the form described above, gave results on malt extract, yeast products and wheat products substantially in agreement with those of biological and microbiological methods. The precision of the method may, however, vary when applied to different classes of substance. It was found that agreement among collaborators was better for some substances than other, thus for results upon two wheat products, the coefficient of variation was 4.9%, for four yeast products 13.5%.

3. Micromethods for Estimation of Thiamine by Thiochrome Method.

HINTON (1943) has described a procedure based on the thiochrome technique which he had successfully applied to small amounts (1—50 mg.) of cereal products The test solution is contained in a capillary which, supported in a metal holder, allows measurements of fluorescence intensity with the Spekker fluorimeter. With solutions of pure thiamine or with cereal extracts of thiamine, the method is accurate to $\pm 20\%$ at concentrations of 0.001 μg. and $\pm 3\%$ at 0.05 μg. The value of technique is well illustrated by Hinton's beautiful analysis of the distribution of thiamine in grains of wheat, rye, barley, oats, maize and rice (Hinton, 1944, 1947, 1948, 1949). By careful dissection he has separated, the pericarp and testa, aleurone layer, embryo, scutellum and endosperm and on 10—20 mg. samples determined their thiamine content. High concentrations of thiamine were found localised in the scutellum amounting to as much as 60% of the total thiamine content (see Pace p. 80). Such localisation of thiamine is of the utmost importance to the practical problems of milling of grain but of uncertain biological significance although it may be related to the growing tips of the seedlings.

Histochemical Detection of Thiamine. A technique which may be of value in studying qualitatively the distribution of thiamine in plant tissues has been developed by Somers and Coolidge (1945a, b). Wheat grains embedded in paraffin are sectioned to obtain the desired region with a smooth surface. A piece of cellophane soaked in alkaline ferricyanide is pressed against the grain which is then illuminated with ultraviolet light. From the relative degree of fluorescence of different parts of the grain Somers and Coolidge concluded that thiamine is located principally in the aleurone layer, in the endosperm cells adjoining the aleurone layer at the base of the crease, in the scutellum and in the endosperm adjoining the scutellum. These results agreeing substantially with those of Hinton support the validity of the method. Nevertheless, other substances may contribute to the fluorescence and treatment with alkaline ferricyanide may change the permeability of cells so that thiamine diffuses from one part to another in the tissue.

VI. Spectrophotometric Estimation of Thiamine.

Spectrophotometric methods based upon the absorption of thiamine in the ultraviolet have not been greatly used in the determination of thiamine in plant and animal tissues, probably because they required extract or preparations of thiamine in a highly purified form. Their suitability for the assay of crystalline thiamine preparations is well illustrated by work described by Holiday and Irwin (1946). Consisting of collaborative tests by four laboratories this work is noteworthy because of the careful analysis of the errors affecting the accuracy of the spectrophotometric method and of the influence upon the spectral absorption of the vitamin of such variables as solvent, drying and irradiation of the vitamin. It includes a notable feature in a mathematical treatment of the data of the spectral absorption of thiamine which gives greater and more accurate information than the specific extinction coefficient at maximum absorption. From this treatment arises a standard extinction coefficient which is defined in terms of several points in the absorption curve and which was found to be more accurate than the extinction coefficient measured at any other wavelength.

Measuring the concentration of thiamine under defined conditions Holiday and Irwin report that the overall coefficient of variation is 2%. They also give a specification, together with a tolerance, for the ultraviolet absorption spectrum of thiamine. Their work should prove valuable, not only for the accurate estimation of thiamine in carefully purified preparations, but also for the definition of those crystalline preparations used as standards in the various methods which have been described.

References.

ALEXANDER, B., and J. E. LEVI: J. Biol. Chem. **146**, 399 (1942). — Association of Vitamin Chemists Inc.: Methods of Vitamin Assay 1947. — ATKINS, L., A. S. SCHULTZ and C. N. FREY: J. Biol. Chem. **129**, 471 (1939). — AUHAGEN, E.: Hoppe-Seylers Z. **204**, 149 (1932); Biochem. Z. **258**, 330 (1933).

BANGA, I., S. OCHOA and R. A. PETERS: Biochem. J. **33**, 1109 (1930). — BARGER, G., F. BERGEL and A. R. TODD: (a) Nature (London) **136**, 259 (1935); (b) Ber. dtsch. chem. Ges. **68**, 2257 (1935). — BARRON, E. S. G., and C. M. LYMAN: Science (Lancaster, Pa.) **92**, 337 (1940). — BEADLE, B. W., D. A. GREENWOOD and H. R. KRAYBILL: J. Biol. Chem. **149**, 349 (1943). — BERGEL, F., and A. R. TODD: Nature (London) **138**, 406 (1936). — BHAGUAT, K., and P. DEVI: Indian J. Med. Res. **32**, 123, 131, 139 (1944). — BIRCH, T. W., and L. J. HARRIS: Biochem. J. 28, 602 (1934). — BOAS-FIXSEN, M. A., and M. H. ROSCOE: Nutr. Abs. 7, 837 (1938); 9, 815 (1939/40). — BONNER, J.: Amer. J. Bot. **29**, 136 (1942). — BONNER, J., and R. DORLAND: (a) Amer. J. Bot. **30**, 414 (1943); (b) Archiv. Biochem. **2**, 451 (1943). — BONNER, J., and J. ERICKSON: Amer. J. Bot. **25**, 685 (1938). — BOOTH, R. G.: Analyst 67, 162 (1942); Biochem. J. **37**, 518 (1943). — British Pharmacopoeia (1932); 7 th Addendum. Appendix xxiii, 1936. — BROWN, R. A., E. HARTZLER, G. PEACOCK and A. D. EMMETT: Ind. Eng. Chem. Anal. Ed. 15, 494 (1943). — BULLOCK, M. W., J. A. BROCKMAN, E. L. PATTERSON, J. V. PIERCE and E. L. R. STOKSTAD: J. Amer. Chem. Soc. 74, 3455 (1952). — BURCHHOLDER, P. R., and I. MCVEIGH: Amer. J. Bot. **27**, 853 (1940).

CALVIN, M.: Chem. and Engin. News **31**, 1735 (1953). — CHELDELIN, V. H., and R. J. WILLIAMS: Univ. Texas Pub. No. 4237, 105 (1942). — COHEN, P. P.: J. Biol. Chem. **164**, 685 (1946). — COWARD, K. H.: Biochem. J. **30**, 2012 (1936); "The Biological Standardisation of the Vitamins" 2nd. Ed. London: Baillière, Tindall & Cox. 1947. — COWARD, K. H., J. H. BURN, H. W. LING and B. G. E. MORGAN: Biochem. J. 27, 1719 (1933). — COWARD, K. H., and B. G. E. MORGAN: Biochem. J. **33**, 658 (1939). — CRAVIOTO, R. B., E. E. LOCKHART, R. K. ANDERSON, F. DE P. MIRANDA, R. S. HARRIS, E. AGUILAR, E. W. TAPIA, H. S. LOCKHART, M. K. NUTTER and L. P. GUILD: J. Nutrition **29**, 317 (1945).

DEUTSCH, H. F.: J. Biol. Chem. **152**, 431 (1944). — DEWEY, V.: Biol. Bull. 87, 107 (1944).— DRURY, A., L. J. HARRIS and C. MAUDSLEY: Biochem. J. 24, 1632 (1930).

EMMETT, A. D., G. PEACOCK and R. A. BROWN: J. Biol. Chem. **135**, 131 (1940). — EPPRIGHT, M. A., and R. J. WILLIAMS: Ind. Eng. Chem. Anal. Ed. 16, 576 (1944). — EVANS, E. T. R., and W. C. EVANS: Biochem. J. **44**, ix (1949). — EVANS, W. C., N. R. JONES and R. A. EVANS: Biochem. J. **46**, xxxVIII (1950). — EVANS, W. C., and N. R. JONES: Biochem. J. **50**, xxVIII (1952).

FARRER, K. T. H.: Proc. Austral. Chem. Inst. 8, 113 (1941); Biochem. J. **39**, 128 (1945); 41, 162 (1947); Brit. J. Nutrition **2**, 242 (1948). — FITZGERALD, E. E., and E. B. HUGHES: Analyst 74, 340 (1949).

GUIRARD, B. M., E. E. SNELL and R. J. WILLIAMS: (a) Archiv. Biochem. **9**, 361 (1946); (b) Archiv. Biochem. **9**, 381 (1946). — GREEN, D. E., D. HERBERT and V. SUBRAHMANYAN: J. Biol. Chem. **135**, 795 (1940). — GUHA, B. C., and J. C. DRUMMOND: Biochem. J. **23**, 880 (1929).

HARRIS, L. J., and Y. L. WANG: Biochem. J. **35**, 1050 (1941). — HARRIS, R. S.: In "The Enzymes I. Part 2 Ed. J. B. SUMNER and K. MYRBACH". New York N. Y.: Academic Press Inc. 1951. — HENNESSEY, D. J., and L. R. CERECEDO: J. Amer. Chem. Soc. 61, 179 (1939).— HENNESSEY, D. J., S. WAPNER and J. TRUHLAR: Ind. Eng. Chem. Anal. Ed. 16, 479 (1944). — HINTON, J. J. C.: Biochem. J. **37**, 585 (1943); **38**, 214 (1944); Proc. Roy. Soc. B. **134**, 418 (1947); Brit. J. Nutr. **2**, 237 (1948/49). — HOLIDAY, E. R.: Biochem. J. **29**, 718 (1935). — HOLIDAY, E. R., J. O. IRWIN, WITH A. E. GILLAM, R. A. MORTON, T. MOORE and Y. L. WANG: Quart. J. Pharm. and Pharmacol. **19**, 155 (1946). — HOLMAN, B.: Biochem. J. 38, 388 (1944). HOMBERGER, C. S., K. F. HEITMILLER, I. C. GUNSALUS, G. H. F. SCHNAKENBERG and L. J. REED: J. Amer. Chem. Soc. **75**, 1273 (1953).

JANSEN, B. C. P.: Rec. trav. chim. **55**, 1046 (1936); Nature (London) **135**, 267 (1935). — JANSEN, B. C. P., and W. F. DONATH: Proc. K. Akad. Wetensch Amsterdam **29**, 1390 (1926).

KARRER, P., and M. VISCONTI: Helv. Chim. Acta **29**, 711 (1946). — KINNERSLEY, H. W., J. R. P. O'BRIEN and R. A. PETERS: Biochem. J. **29**, 701 (1935); **29**, 2369 (1935). — KINNERSLEY, H. W., and R. A. PETERS: Biochem. J. **22**, 419 (1928); **28**, 667 (1934); **30**, 985 (1936); **32**, 1516 (1938). — KINNERSLEY, H. W., R. A. PETERS and V. READER: Biochem. J. **32**, 276 (1928). — KIRCH, E. R., and O. BERGEIM: J. Biol. Chem. **143**, 575 (1942). — KUHN, R., T. WAGNER-JAUREGG, F. W. VAN KLAVEREN and H. VETTER: Hoppel-Seylers Z. **234**, 196 (1935).

LASER, H.: Biochem. J. **35**, 488 (1941). — LIPMANN, F.: Nature (London) **138**, 1097 (1937). — LIPMANN, F., and G. PERLMAN: J. Amer. Chem. Soc. **60**, 2574 (1938). — LOHMANN, K., and P. SCHUSTER: (a) Naturwissenschaften **25**, 26 (1937); (b) Biochem. Z. **294**, 188 (1937).

Meiklejohn, A. P.: Biochem. J. **31**, 1441 (1937). — Meiklejohn, J.: Biochem. J. **37**, 49 (1943). — Melnick, D., and H. Field: (a) J. Biol. Chem. **127**, 505 (1939); (b) J. Biol. Chem. **127**, 515 (1939); (c) J. Biol. Chem. **127**, 531 (1939).

Neuberg, C., and L. Karczag: Biochem. Z. **37**, 70 (1911).

Obermeyer, H. G., and L. Chen: J. Biol. Chem. **159**, 117 (1945). — Ochoa, S.: Nature (London) **141**, 831 (1938). — Ochoa, S., and R. A. Peters: Biochem. J. **32**, 1501 (1938). — Ogston, A. G., and R. A. Peters: Biochem. J. **30**, 736 (1936). — O'Kane, D. J., and I. C. Gunsalus: J. Bact. **54**, 20 (1947); **56**, 499 (1948). — Organ, J. G., and F. Wokes: J. Soc. Chem. Ind. Trans. **63**, 165 (1944). — Orr-Ewing, J., and V. Reader: Biochem. J. **22**, 440 (1928).

Parvé, E. P. S., and H. G. K. Westenbrink: Z. Vitaminforschg. **15**, 1 (1944). — Passmore, R., R. A. Peters and H. M. Sinclair: Biochem. J. **27**, 842 (1933). — Pharmacopeia of USA.: 13 th Ed. Easton, Pa.: Mack Printing Co., 1947. — Peters, R. A.: Nature (London) **135**, 107 (1935); Biochem. J. **30**, 2206 (1937); **32**, 2031 (1938); (a) Nature (London) **158**, 707 (1946); (b) Bull. Soc. Chim. Biol. (Paris) **28**, 700 (1946). — Peters, R. A., H. W. Kinnersley, J. Orr-Ewing and V. Reader: Biochem. J. **22**, 445 (1928). — Peters, R. A., and I. St. Philpot: Proc. Roy. Soc. B **113**, 48 (1933). — Peters, R. A., H. Rydin and R. H. S. Thompson: Biochem. J. **29**, 37 (1935). — Prebluda, H., J., and E. V. McCollum: Science (Lancaster, Pa.) **84**, 488 (1936); J. Biol. Chem. **127**, 495 (1939). — Reed, L. J.: Physiol. Rev. **33**, 544 (1953); Reed, L. J., and B. G. de Busk: (a) J. Amer. Chem. Soc. **74**, 3457 (1952); b) J. Biol. Chem. **199**, 872 (1952); (c) J. Amer. Chem. Soc. **74**, 4727 (1952); (d) J. Amer. Chem. Soc. **74**, 3457 (1952); J. Amer. Chem. Soc. **75**, 1261 (1953). — Reed, L. J., B. G. de Busk, C. S. Hornberger and I. C. Gunsalus: J. Amer. Chem. Soc. **75**, 1271 (1953). — Reed, L. J., I. C. Gunsalus, G. H. F. Schnakenberg, Q. F. Soper, H. E. Boaz, S. F. Kern and T. V. Parke: J. Amer. Chem. Soc. **75**, 1267 (1953). — Ridyard, H. N.: Analyst. **75**, 634 (1950). — Robbins, W. J.: Bot. Gaz. **101**, 428 (1939). — Roche, J.: In "The Enzymes" I. Ed. J. B. Sumner, and K. Myrbäck. New York: Academic Press Inc. 1950. — von Rytz, W.: Ber. schweiz. botan. Ges. **49**, 339 (1939).

Sarett, H. P., and V. H. Cheldelin: J. Biol. Chem. **155**, 153 (1941). — Schopfer, W. H.: Z. Vitaminforschg. **4**, 67 (1935). — Schopfer, W. H., and A. Jung: C. r. Acad. Sci. Paris **204**, 1500 (1937). — Schultz, A. S., L. Atkins and C. N. Frey: (a) J. Amer. Chem. Soc. **59**, 948 (1937); (b) J. Amer. Chem. Soc. **59**, 2457 (1937); Ind. Eng. Chem. Anal. Ed. **14**, 35 (1942). — Schultz, A. S., L. Atkins, C. N. Frey and R. Williams: J. Amer. Chem. Soc. **63**, 632 (1941). — Sherman, H. C., and A. Spohn: J. Amer. Chem. Soc. **45**, 583 (1923). — Simola, P. E.: Biochem. Z. **254**, 229 (1932). — Sinclair, H. M.: Biochem. J. **32**, 2185 (1938); (a) Biochem. J. **33**, 1816 (1939); (b) Biochem. J. **33**, 2027 (1939). — Smakula, A.: Hoppe-Seylers Z. **230**, 231 (1934). — Smits, G., and E. Florijn: Biochim. et Biophys. Acta **3**, 44 (1949). — Smith, M. I.: U.S. Publ. Health Rep. **45**, 116 (1930). — Somers, C. F., and M. H. Coolidge: (a) Science **101**, 88 (1945); (b) Cereal Chemistry **22**, 333 (1946). — Stockstad, E. L. R., C. E. Hoffmann, M. A. Regan, D. Fordham and T. H. Jukes: Arch. Biochem. **20**, 75 (1949). — Sub-Committee on Vitamin Estimations of the Analytical Methods Committee of the Society of Public Analysts and other Analytical Chemists.: Analyst **76**, 177 (1951). — Sure, B.: J. Biol. Chem. **162**, 139 (1946). — Sykes, P., and A. R. Todd: J. Chem. Soc. 1951, 534.

Vitamin B$_1$ Sub-Committee of the Accessory Factors Committee of The Medical Research Council and the Lister Institute: Biochem. J. **37**, 433 (1943).

Wang, Y. L., and L. J. Harris: Chem. and Ind. **20**, 27 (1942). — Ward, A. H.: Chem. and Ind. **21**, 11 (1943). — Westenbrink, H. G. K.: (a) C. r. Lab. Carlsberg **23**, 195 (1940); (b) Enzymologia **8**, 97 (1940). — Westenbrink, H. G. K., and J. Goudsmit: Enzymologia **5**, 307 (1938); **10**, 146 (1941). — Westenbrink, H. G. K., J. Goudsmit and H. Veldman: Vitamine u. Hormone **1**, 15 (1941). — Westenbrink, H. G. K., E. P. S. Parvé and A. C. van der Linden and W. A. van den Brock: Z. Vitaminforschg. **13**, 218 (1943). — Westenbrink, H. G. K., E. P. S. Parvé and H. J. Thomassen: Z. Vitaminforschg. **13**, 101 (1943). — Westenbrink, H. G. K., and D. A. van Dorp: Enzymologia **10**, 212 (1942). — Westenbrink, H. G. K., D. A. van Dorp, M. Gruber and H. Veldman: Enzymologia **9**, 73 (1940). — Westenbrink, H. G. K., and A. F. Willebrands: Z. Vitaminforschg. **14**, 291 (1944). — Weswig, P. H., A. M. Freed and J. R. Haag: J. Biol. Chem. **165**, 737 (1946). — Williams, R. J.: J. Biol. Chem. **38**, 440 (1919). — Williams, R. J., J. R. McMahan and R. E. Eakin: Univ. Texas Publ. No. 4137, 31 (1941). — Williams, R. J., and R. R. Roehm: J. Biol. Chem. **87**, 581 (1930). — Wintersteiner, C., R. R. Williams and A. E. Ruehle: J. Amer. Chem. Soc. **57**, 517 (1935).

Zima, O., and R. R. Williams: Ber. dtsch. chem. Ges. **73**, 941 (1940). — Zima, O., K. Ritsert and T. Moll: Hoppe-Seylers Z. **267**, 210 (1941).

The Alkaloids.

By

B. T. Cromwell.

With 2 Figures.

A. General Introduction.

The presence of alkaloids has been established in about 40 families of flowering plants representing approximately one seventh of the total number of families. The majority of these families belong to the Dicotyledones and the number of Monocotyledonous families showing the presence of alkaloids is relatively small. The Gymnosperms and Pteridophytes are poorly represented and so far as is known, alkaloids are virtually absent from the lower groups of plants with the exception of one or two families of the Fungi. The distribution of alkaloids throughout the families of plants is shown in Table 1.

Table 1. *The Orders and Families of Plants Containing Alkaloids.*

Order	*Family*	*Principal genera containing alkaloids*
	(1) Dicotyledones.	
Piperales	*Piperaceae*	*Piper*
Garryales	*Garryaceae*	*Garrya*
Aristolochiales	*Aristolochiaceae*	*Aristolochia*
Centrospermae	*Chenopodiaceae*	*Anabasis, Salsola*
	Aizoaceae	*Mesembryanthemum*
Ranales	*Nymphaeaceae*	*Nymphaea*
	Ranunculaceae	*Aconitum, Delphinium, Hydrastis, Coptis*
	Berberidaceae	*Berberis, Mahonia*
	Menispermaceae	*Chondrodendron, Jateorhiza, Sinomenium, Arcangelisia*
	Magnoliaceae	*Magnolia*
	Calycanthaceae	*Calycanthus*
	Anonaceae	*Anona, Xylopia*
	Monimiaceae	*Laurelia, Atherosperma, Daphnandra Peumus (Boldea)*
	Lauraceae	*Actinodaphne, Litsea, Nectandra*
	Hernandiaceae	*Hernandia, Illigera*
Rhoeadales	*Papaveraceae*	*Papaver, Chelidonium, Dicentra, Corydalis, Bocconia*
Rosales	*Leguminosae*	*Lupinus, Cytisus, Ulex, Physostigma, Erythrophloeum, Anagyris, Sophora, Erythrina*
Geraniales	*Erythroxylaceae*	*Erythroxylum*
	Rutaceae	*Pilocarpus, Skimmia, Galipea, Casimiroa, Evodia, Peganum, Acronychia etc.*

Table 1. (Continued.)

Order	Family	Principal genera containing alkaloids.
Sapindales	Buxaceae	Buxus
	Celastraceae	Celastrus, Catha
Rhamnales	Rhamnaceae	Rhamnus
Opuntiales	Cactaceae	Anhalonium
Myrtiflorae	Punicaceae	Punica
Umbelliflorae	Umbelliferae	Conium
Contortae	Loganiaceae	Strychnos, Gelsemium
	Apocynaceae	Holarrhena, Tabernanthe, Aspidosperma, Geissospermum
	Asclepiadaceae	Asclepias
Tubiflorae	Convolvulaceae	Convolvulus
	Boraginaceae	Echium, Anchusa
	Solanaceae	Atropa, Datura, Solanum, Hyoscyamus, Nicotiana, Duboisia
Rubiales	Rubiaceae	Cinchona, Psychotria, Pausinystalia (Corynanthe)
	Campanulaceae	Lobelia
	Compositae	Senecio, Artemisia

(2) Monocotyledones.

Order	Family	Principal genera containing alkaloids.
Glumiflorae	Gramineae	Lolium
Principes	Palmae	Areca
Liliflorae	Stemonaceae	Stemona
	Liliaceae	Fritillaria, Colchicum, Schoenocaulon, Veratrum
	Amaryllidaceae	Narcissus, Lycoris, Clivia, Amaryllis, Buphane
	Dioscoraceae	Dioscorea
Microspermae	Orchidaceae	Orchis

(3) Gymnospermae.

Order	Family	Principal genera containing alkaloids.
Coniferae	Taxaceae	Taxus
Gnetales	Gnetaceae	Ephedra

(4) Pteridophyta.

Order	Family	Principal genera containing alkaloids.
Equisetales	Equisetaceae	Equisetum
Lycopodiales	Lycopodiaceae	Lycopodium

(5) Fungi.

Order	Family	Principal genera containing alkaloids.
Ascomycetes	Hypocreaceae	Claviceps

I. Distribution in the Plant.

Alkaloids occur in all organs of plants though not necessarily in all organs of individual plants. In certain species they are synthesized in young metabolically active tissues of roots and stems and from these tissues are translocated to other parts of the plant. Synthesis takes place in seedlings which have developed from seeds lacking alkaloids *(Nicotiana tabacum, Papaver somniferum)* a few days after germination, and during the subsequent growth period continued synthesis resulting in gradual accumulation of alkaloids takes place. Seeds of the *Leguminosae* which are rich in alkaloids, germinate to give seedlings which begin to synthesize alkaloids shortly after germination has taken place. There is however an initial loss of alkaloids from the cotyledons during the early stages of germination. The ultimate distribution in the mature plant varies considerably with the type of plant. In annuals there is a gradual accumulation in the organs of the shoot system during the vegetative period with a maximum at or near the time of flowering. After flowering the alkaloids may move from stems and leaves into the developing fruits (e. g. *Papaver somniferum*). In trees and woody perennial plants considerable accumulation of alkaloids may take place from year to year in the bark of stems and roots.

The histological distribution of alkaloids in individual organs varies to some extent with the type of plant. In addition to the undifferentiated meristematic tissues, alkaloids are normally found in high concentration in epidermal tissues (including the piliferous layer of the root) and hypodermal tissues. Guard cells however are usually free from alkaloids. In *Atropa belladonna* the epidermal cells of leaves, petioles, calyx and corolla accumulate alkaloids. Carpels and ovules also show a high concentration. The epidermal cells of the leaves of *Cinchona* are free from alkaloids but accumulation takes place in the hypodermal cells. The cortical parenchyma of stems and roots of many plants contains an abundance of alkaloids (*Cinchona* and *Berberis*). Traumatic tissue produced as the result of wounding is often rich in alkaloids and the regenerated tissue which develops after the removal of bark from *Cinchona* trees contains a high concentration of alkaloids. Cutting or wounding of the potato leads to the production of a wound tissue rich in solanine. The tissues of the endodermis, pericycle and bundle sheaths very frequently accumulate alkaloids (*Solanum, Colchicum*) but sieve tubes and vessels are normally free. Companion cells of the phloem in *Narcissus* and other members of the *Amaryllidaceae* are rich in alkaloids, and in many plants (*Atropa belladonna*) the cells of the pith contain alkaloids. The laticiferous system of the opium poppy is a notably rich source of morphine and its associated alkaloids. The distribution of alkaloids in the tissues of *Berberis* spp. provides an interesting example of their occurrence in dead cell elements. In addition to an abundance of alkaloid in the cortical cells, the walls of the xylem elements become impregnated. In etiolated shoots of *Berberis* the inner cortical tissues immediately surrounding the vascular bundles and interior to the phellogen are engorged with berberine. JAMES (1950) has given a full account of the distribution of alkaloids in plants which produce them and for further details the reader is referred to this work.

Alkaloids usually occur in plants in the form of salts of malic, oxalic, succinic, tannic or other plant acid. In some cases the alkaloids occur in combination with special acids. Aconitine occurs in combination with aconitic acid, the opium alkaloids with meconic acid and the cinchona alkaloids with quinic or cinchotannic acid.

II. The Isolation of Alkaloids.

1. Extraction.

Special methods for the isolation of individual alkaloids are given in the following sections and general principles only will be dealt with here.

Plant material is dried at a temperature of 60° in an oven provided with forced draught. For tissues which contain unstable or volatile alkaloids the temperature should not exceed 40°. The dried material is finely powdered and stored if necessary in a desiccator in the dark.

For details of extraction apparatus the reader is referred to the section "General Methods for Separation", Vol. 1, of this work. The SOXHLET type of continuous extractor is generally suitable for the extraction of alkaloids with organic solvents. For large scale extractions special equipment is necessary but quantities of dried material of 1 kg. in weight can be handled satisfactorily in large SOXHLET extractors constructed of sheet copper. A useful accessory to the SOXHLET is a small stopcock fused into the base of the extractor for withdrawing samples from time to time in order to test for complete extraction. Unstable ester alkaloids (e. g. aconitines) are usually extracted by percolation with solvents at room temperature and for this purpose a percolator of the Toogood pattern is suitable. The extraction of the total alkaloids is a comparatively simple procedure and is based on the principle of partition between immiscible solvents. The alkaloids as a rule are basic substances (colchicine and rutaecarpine are non-basic) more or less soluble in organic solvents and in many cases little soluble in water. On the contrary, the salts of the alkaloids are usually little soluble in organic solvents but readily soluble in water (the hydrochlorides of some alkaloids e. g. lobeline, are soluble in chloroform and are extracted by this solvent from aqueous solution). The conventional procedure for the extraction of alkaloids therefore consists of the following steps: (1) liberation of the free base by treatment of the dried material with alkali $(Ca(OH)_2$, Na_2CO_3 or $NH_4OH)$; (2) extraction of the free base with an organic solvent (chloroform, ether, benzene, ethanol, methanol, methylene chloride); (3) removal of the base from the organic solvent by shaking with dilute acid $(H_2SO_4$, HCl). Alternatively as most of the alkaloids and their salts are soluble in ethanol, this solvent can be used for the extraction of the plant material without previous treatment with alkali. After removal of the ethanol the residue is treated with dilute aqueous acid and the solution filtered to remove resins and fatty material which separates out. The alkaloids are then extracted from the acid solution by making alkaline and shaking out with ether or chloroform (or a mixture of both). The addition of a solution of gum tragacanth to the aqueous phase often prevents the formation of troublesome emulsions. If emulsions continue to form, the phases may be separated by centrifugation. Alkaloids which are volatile in steam (sparteine, nicotine, nornicotine) are usually steam distilled from an aqueous alkaline solution. For the extraction, separation and purification of alkaloids it is essential to use pure organic solvents only. In particular, ether should be free from ethanol and peroxide, and ethanol should be free from acetone. For the extraction of seeds containing much fat or oil, it is often advantageous to remove the fatty material by a preliminary extraction with light petroleum in which most alkaloids are little soluble. However it is always advisable to shake out the light petroleum solution with dilute acid and to test the acid solution for the presence of alkaloids in case they are removed by this procedure.

2. Fractionation.

As practically all the alkaloid-producing plants elaborate a mixture of closely related alkaloids, separation of the individual alkaloids from the solution of total alkaloids obtained as above becomes necessary. The process of separation is often difficult and tedious and until comparatively recent times the methods of separation relied upon were fractional crystallization, fractional distillation and the preparation of derivatives which were little soluble in certain solvents. Modern biochemical methods including partition chromatography on columns and countercurrent distribution have now been employed for the separation of alkaloid mixtures with considerable success. The general principles of these methods are discussed in Vol. 1 of this handbook and references to the use of the methods for the separation of individual alkaloids will be found in the sections following. A few remarks on *fractional crystallization* as a means of separating alkaloids from mixtures may be helpful. It sometimes happens that a single solvent can be used for the separation of a mixture of two or even more alkaloids, but often it is necessary to use a mixture of miscible solvents. The majority of alkaloids are easily soluble in chloroform and less so in the other organic solvents, the general order being chloroform > acetone > ethanol > methanol > ethyl acetate > ether > benzene > > hexane. Thus if all the alkaloids of a mixture are easily soluble in chloroform and one of the alkaloids is much less soluble in ethanol than chloroform, fractional crystallization of this alkaloid is possible if the mixture in chloroform is concentrated to a suitable degree, and hot ethanol added in small portions. On cooling crystals of the one alkaloid less soluble in ethanol separate out. The remaining alkaloids of the mixture are then subjected to treatment with other solvents in the hope that further separation may be effected. If fractional crystallization of the alkaloids fails, fractional crystallization of the salts should be tried. The salts most frequently used for this purpose are the hydrochlorides, hydrobromides, hydriodides, nitrates, perchlorates, oxalates and picrates. The acids may either be used in aqueous or methanolic solution. From methanolic solution the salts can be precipitated by the addition of ether and hydrochlorides can often be crystallized from hot acetone containing a small proportion of methanol. Perchlorates, oxalates and picrates may in some cases be precipitated from solutions in acetone by the addition of ethyl acetate. The picrates, aurichlorides and reineckates are useful for purposes of identification.

Precipitation. Quarternary alkaloids are very soluble in water and cannot as a rule be extracted from aqueous solution with organic solvents. It is necessary therefore to precipitate these bases with mercuric chloride, potassium mercuric iodide or a solution of iodine in KI. The precipitates given with mercuric chloride and potassium mercuric iodide are decomposed with H_2S and the iodine precipitate with thiosulphate. The alkaloids remain in solution as the hydrochlorides or hydriodides. For the separation of mixtures of tertiary and secondary bases, the latter are acetylated. On treatment of the products of acetylation with dilute HCl, the tertiary base is removed as the soluble hydrochloride and the insoluble acetyl derivative removed and subjected to mild acid hydrolysis. The methiodides of the tertiary bases and the benzoyl derivatives of the secondary amines are useful for characterization. Phenolic alkaloids are soluble in NaOH or KOH solutions and are not as a rule extracted from these solutions by ether. They may be removed either by saturating the aqueous alkaline solution with CO_2 or neutralizing and making faintly alkaline with $NaHCO_3$ or dilute ammonia when the alkaloid is precipitated (e. g. morphine). Phenolic alkaloids are usually characterized by conversion to their methyl ethers. The steam-volatile alkaloids often may be separated by fractional distillation and converted to salts for purification (e. g. the tobacco and hemlock alkaloids). The melting points, optical rotation and other physical constants are used as criteria of purity.

Qualitative Tests. The precipitation reactions of alkaloids are not sufficiently specific for purposes of identification but they are useful as presumptive tests for the presence of alkaloids in a solution. If no precipitate or turbidity is produced in a solution on the addition of a range of alkaloid reagents, alkaloids are absent from the solution. As indicated above for the quaternary bases, the alkaloid reagents have been used for the isolation of alkaloids by precipitation and also have been used for the quantitative determination of total alkaloids.

Most of the colour reactions of alkaloids are not sufficiently specific for identification purposes but, used in conjunction with other tests, serve as useful supplementary or confirmatory tests. Certain of the more specific colour reactions however have been adapted for use in quantitative estimations. In carrying out colour tests it is essential to use pure reagents and pure alkaloids. Colour reactions given by impure alkaloids very often are not given by the alkaloids in a state of purity. These tests are best carried out in small porcelain evaporating dishes or, where heat is not required, on spot plates.

24*

III. The Quantitative Estimation of Alkaloids.

Methods for the quantitative estimation of alkaloids exist for comparatively few of the large number of alkaloids which have been isolated from plants. The small number for which quantitative methods have been developed includes alkaloids of economic importance and more especially those which have found use in medicine. In the following sections only those alkaloids for which quantitative methods of estimation are available have been included and some important groups, e. g. the alkaloids of *Senecio*, *Corydalis* and *Delphinium* and the acridine alkaloids of the Australian representatives of the *Rutaceae*, have therefore been omitted. In many plants the estimation of total alkaloids only has been attempted and the methods used are based on the principles of extraction and separation previously outlined. Total alkaloids are determined after purification by partition in immiscible solvents by (a) titrimetric methods, (b) gravimetric methods, (c) colorimetric methods and (d) precipitation methods. Examples of these methods will be found in subsequent sections. The Soxhlet apparatus is generally suitable for quantitative extraction except when dealing with an unstable alkaloid, which is best extracted at room temperature. In quantitative work tests for complete extraction should be made on samples of the extract withdrawn during the extraction, by means of Mayer's reagent and iodine solution. The solvents most often used for extraction are methanol, ethanol, methylene chloride, ether, chloroform and occasionally benzene. For purification by partition, ether, chloroform or benzene are the solvents most frequently used. *Titration* is best carried out by dissolving the residue obtained after removal of the solvent in excess of standard acid and back titrating with standard alkali. Methyl red is the indicator best suited to the majority of alkaloid titrations. *Gravimetric methods* are either carried out on the alkaloid residue obtained as above or the alkaloid is weighed as a salt or derivative. If the residue of alkaloids is weighed, the amount of alkaloids present can be checked by titration. *Colorimetric methods* for the determination of the total alkaloids of plant extracts are useful if all components of the mixture contain the same reactive groups, e. g. the indole nucleus of the ergot alkaloids, in which case the colour is developed by the use of a reagent specific for the group. Colorimetric determination may follow *precipitation* of the alkaloids, e. g. reineckate precipitation followed by determination of the coloured reineckate ion. Other precipitation methods involving the use of picrolonic acid, iodine in potassium iodide, Mayer's reagent and Dragendorff's reagent have been employed for the determination of total alkaloids. Picrolonates are weighed as such and the precipitates formed by Mayer's reagent, iodine in potassium iodide and Dragendorff's reagent are decomposed and the mercuric iodide or iodine liberated is estimated volumetrically (cf. the estimation of the alkaloids of *Lolium*). For the estimation of nicotine and other steam-volatile alkaloids, precipitation with silicotungstic acid has been used with success. The estimation of individual alkaloids in the total alkaloid fraction depends largely on quantitative separation of the constituents of the mixture unless a colorimetric or spectrophotometric method is available which will selectively determine one or more of the component alkaloids. Methods of chromatographic adsorption and ion exchange when applied to mixtures of alkaloids have in many instances resulted in satisfactory quantitative separations of individual alkaloids. These methods possess the virtue of rapidity and comparative simplicity in operation and will doubtless find extensive use in alkaloid analysis. Colorimetric methods can be applied for the determination of individual alkaloids in a mixture if one of the alkaloids possesses a reactive group, not present in the remaining alkaloids, which can be exploited for colour development.

Thus morphine, a phenolic alkaloid, can be determined in the presence of codeine, narcotine and papaverine by use either of the RADULESCU reagent or the FOLIN-CIOCALTEU phenol reagent. Similarly, spectrophotometric methods may be used for the determination of individual alkaloids in a mixture if the absorption maxima of the component alkaloids occur at wavelengths of sufficient difference to eliminate undue interference.

KIRKPATRICK (1945, 1946, 1947) has made a polarographic study of the alkaloids and has shown that satisfactory results are obtained by use of the polarographic method if the alkaloids are isolated in a state of purity. Up to the present time however this method has not proved satisfactory for the estimation of alkaloids in plant tissues.

Fluorometric methods have been used for the determination of alkaloids in mixtures and have been found to give satisfactory results (cf. hydrastine).

Biological methods, making use of specific physiological properties of alkaloids, have been used for assay of certain alkaloids, but for accurate estimations these methods generally have not found favour. The principles underlying the above methods of analysis are discussed fully in Vol. 1 of this handbook.

For a full account of their structure and chemistry, and for references to the extensive literature of the alkaloids, the reader should consult the standard works on the subject, a list of which is given below. These works have been used by the writer as the principal sources of information on the physical constants etc. of the alkaloids dealt with in the chapters following.

IV. Reagents.

1. General Alkaloid Reagents.

In this group are included those reagents which when added to solutions of salts of the alkaloids give insoluble or slightly soluble amorphous or crystalline precipitates. In very dilute solutions of the alkaloids a turbidity is produced. Some of these reagents give precipitates with proteins, purines, ammonium salts, methylated amines, betaines etc. and are therefore not specific for alkaloids. However if no precipitate is produced in a solution to which certain of these reagents are added, the absence of alkaloids in the solution can be assumed. In this connexion the general alkaloid reagents are useful for testing for the complete extraction of alkaloids from plant material. As a presumptive test for the presence of alkaloids in plants the ether or chloroform extract of the aqueous alkaline solution in the standard method of isolation is evaporated to dryness, the residue dissolved in dilute H_2SO_4, filtered, and portions of the filtrate placed in several test tubes. A few drops of the more sensitive reagents are added and in the presence of an alkaloid (or other basic substance extractable with ether or chloroform) a distinct precipitate or turbidity is produced.

Potassium Mercuric Iodide (MAYERS's *Reagent*). 1.36 g. of $HgCl_2$ are dissolved in 60 ml. of distilled water and 5 g. of KI in 10 ml. of water. The two solutions are mixed and diluted to 100 ml. with distilled water. This reagent is perhaps the most generally useful of the alkaloid reagents and gives precipitates with the hydrochlorides of most alkaloids in very dilute solution.

The reagent should be added to solutions rendered distinctly acid with dilute HCl or H_2SO_4. The solution should not contain acetic acid and more than a small amount of ethanol in which the alkaloid precipitates are soluble. A few drops only of the reagent should be added as the precipitates with some alkaloids are soluble in excess of the reagent.

Iodine-Potassium Iodide Solution (WAGNER's *Reagent*). 1.27 g. of iodine and 2 g. of potassium iodide are dissolved in 5 ml. of water and the solution diluted to 100 ml. This reagent gives brown flocculent precipitates with most of the alkaloids and is the reagent of choice for testing for the complete extraction of colchicine.

Potassium Bismuth Iodide (DRAGENDORFF's *Reagent* — KRAUT's *Reagent*). 8 g. of $Bi(NO_3)_3 \cdot 5 H_2O$ are dissolved in 20 ml. of HNO_3 (sp. gr. 1.18) and 27.2 g. of KI in 50 ml. of water. The two solutions are mixed and allowed to stand. When KNO_3 crystallizes out the supernatant is decanted off and made up to 100 ml. with distilled water. The alkaloids are regenerated from precipitates by treatment with sodium carbonate followed by extraction of the liberated base with ether.

Potassium cadmium iodide (Marmé's *Reagent*). 20 g. of KI are dissolved in about 20 ml. of warm water and the solution added to a solution of 10 g. of CdI_2 in 50 ml. of water. The mixed solutions are diluted to 100 ml. with water. This reagent gives white or yellowish precipitates with alkaloids in dilute solutions of H_2SO_4. The precipitates are amorphous at first but become crystalline on standing. The precipitates are soluble in excess of the reagent and in ethanol.

Phosphomolybdic Acid (Sonnenschein's *Reagent*). A strong solution of ammonium molybdate is completely precipitated at approx. 40° with a solution of Na_2HPO_4. The yellow precipitate is well washed, suspended in water and warmed with a concentrated solution of Na_2CO_3 until completely dissolved. The solution is evaporated to dryness and the residue ignited until all the ammonia is driven off. If reduction (blue colour) takes place the residue is moistened with nitric acid and again ignited. The residue is dissolved in hot water with the addition of nitric acid so that the solution is strongly acid. 10 parts of solution are prepared from 1 part of residue. The yellow solution should be stored in a well-stoppered bottle. This reagent gives precipitates with most of the alkaloids. The precipitates are yellow and amorphous. The alkaloids may be recovered from the precipitate by treating with ammonia and shaking out the liberated alkaloid with a suitable solvent. If the alkaloid is easily oxidizable, e. g. morphine, berberine, the molybdic acid is reduced to molybdic oxide which has a blue colour. When this happens the alkaloid is best recovered by moistening the precipitate with Na_2CO_3 and extracting with ethanol. Ammonium salts give a precipitate with the reagent.

Phosphotungstic Acid (Scheibler's *Reagent*). 100 g. of sodium tungstate and 70 g. of Na_2HPO_4 are dissolved in 500 ml. of water and the solution is acidified with nitric acid. Amorphous precipitates are produced when this reagent is added to solutions of alkaloids in dilute H_2SO_4. The reagent is a very delicate precipitant for many alkaloids especially strychnine and quinine. The alkaloids are recovered by treating the precipitates with Na_2CO_3 or baryta solutions.

Silicotungstic Acid. 5 g. of $4\,H_2O \cdot SiO_2 \cdot 12\,WO_3 \cdot 22\,H_2O$ are dissolved in 100 ml. of $6\,N\,H_2SO_4$.

Picric Acid (Hager's *Reagent*). A saturated aqueous solution is usually employed.

Picrolonic Acid. 26.4 g. of picrolonic acid is dissolved in 1 litre of ethanol. A saturated solution of picrolonic acid in ether is also used.

Tannic Acid. A freshly prepared 5% aqueous solution of tannic acid gives a precipitate with most alkaloids. The precipitates are usually soluble in dilute acid or ammonia solutions.

Ammonium Reineckate. A saturated aqueous solution (approx. 4%) slightly acidified with HCl gives with most alkaloids a pink flocculent precipitate. The precipitate is soluble in 50% acetone solution from which the alkaloid reineckate can be crystallized. The reineckates are useful for characterization purposes.

Gold Chloride Solution. 1 g. of $HAuCl_4 \cdot 4\,H_2O$ is dissolved in 20 ml. of water.

Platinic Chloride Solution. 5 g of $H_2PtCl_6 \cdot 6\,H_2O$ are dissolved in 100 ml. of water.

2. Special Reagents.

Fröhde's *Reagent (Sulphomolybdic Acid)*. 5 mg. of molybdic acid or sodium molybdate are dissolved in pure conc. H_2SO_4. The reagent should be freshly prepared before use.

Erdmann's *Reagent*. 10 drops of a mixture of 10 drops of HNO_3 and 100 ml. of water are added to 20 ml. conc. H_2SO_4 (pure).

Marquis *Reagent*. 2—3 drops of 40% formaldehyde solution are mixed with 3 ml. of conc. H_2SO_4 (pure).

Mandelin's *Reagent (Sulphovanadic Acid)*. 1 g. of finely powdered ammonium vanadate is dissolved in 200 g. of pure conc. H_2SO_4.

Mecke's *Reagent*. 1 g. of selenious acid is dissolved in 200 g. pure conc. H_2SO_4.

Rosenthaler's *Reagent* (for opium bases). 1 g. of potassium arsenate is dissolved in 100 g. of pure conc. H_2SO_4.

Schaer's *Reagent*. 1 volume of pure 30% H_2O_2 is carefully mixed with 10 volumes of pure conc. H_2SO_4. The reagent is freshly prepared before use.

Folin-Ciocalteu *Phenol Reagent*. 100 g. of $Na_2WO_4 \cdot 2\,H_2O$ and 25 g. of $Na_2MoO_4 \cdot 2\,H_2O$ are dissolved in 700 ml. of water in a 1500 ml. flask connected by a ground-glass joint to a reflux condenser. 50 ml. of syrupy phosphoric acid and 100 ml. of HCl are added and the mixture refluxed gently for 10 hours. After cooling, 150 g. of Li_2SO_4, 50 ml. of water and 4—6 drops of bromine are added and the mixture boiled for 15 minutes without a condenser, to expel excess bromine. After cooling, the solution is transferred to a 1 litre volumetric flask, diluted to volume with water and filtered. The reagent should have a golden-yellow tint and should be stored in a refrigerator. For use, 1 volume of this stock solution is diluted with 2 volumes of water.

B. The Alkaloids of *Aconitum* spp.

Two types of alkaloid are present in members of the genus *Aconitum (Ranunculaceae)*, (a) the aconitines, which are diacyl esters of polyhydric amino-alcohols the aconines and (b) atisines, which are amino-alcohols. The aconitines generally are highly poisonous but the atisines are of low toxicity. A large number of species of *Aconitum* have been examined and the principal alkaloids with their main sources are given in Table 2.

Table 2. *The Alkaloids of Aconitum spp.*

(a) Aconitines	
Aconitine	*A. napellus* L. (Europe) and other spp. belonging to Japan.
Pseudaconitine . .	*A. balfourii* STAPF. and *A. deinorrhizum* STAPF. (India)
Indaconitine	*A. chasmanthum* STAPF. (India)
Bikhaconitine . . .	*A. spicatum* STAPF. (India)
Mesaconitine	*A. manshuricum* NAKAI and other Japanese spp.
Hypaconitine . . .	*A. calliantum* KOIDZ. and other Japanese spp.
Jesaconitine	*A. sachalinense* FR. SCHMIDT and other Japanese spp.
Lappaconitine . . .	*A. septentrionale* KOELLE (Europe)
Septentrionaline . .	*A. septentrionale*
Lycaconitine . . .	*A. lycoctonum* LINN. (Europe)
Myoctonine	*A. lycoctonum*
(b) Atisines.	
Atisine. Hetisine . .	*A. heterophyllum* WALL. (India)
Palmatisine	*A. palmatum* DON. (India)
Talatisine	*A. talassicum* POPOV (Central Asia)
Lucidusculine . . .	*A. lucidusculum* NAKAI (Japan)
Anthorine	*A. anthora* LINN. (Europe)

The isomeric form of aconitine known as japaconitine occurs in some Japanese species.

I. Properties.

Aconitine, $C_{34}H_{47}O_{11}N$, crystallizes in rhombic prisms, m. p. 202—203, $[\alpha]_D + 14.61°$ or $+ 18.7°$ (CHCl$_3$). The alkaloid is soluble in chloroform or benzene, less soluble in ether or dry ethanol and almost insoluble in water or light petroleum. It is precipitated on addition of alkali to solutions of its salts in water as an amorphous white precipitate which becomes crystalline on the addition of ether.

Aconitine (acetyl benzoylaconine) is readily hydrolysed by boiling with dilute acids or alkalies or with water under pressure, acetic and benzoic acids being split off. The salts of aconitine crystallize well and are laevorotatory. The hydrobromide, $B \cdot HBr \cdot 2.5 H_2O$, crystallizes in hexagonal tablets from water and sinters at 160° and melts indefinitely between 200° and 207°. If dry, it melts at 115—120°. From a mixture of ethanol and ether it crystallizes with $0.5 H_2O$ and melts at 206—207°, $[\alpha]_D$ —30.8° (H_2O). The hydrochloride, $B \cdot HCl \cdot 3.5 H_2O$, has m. p. 149—153° or 194—195° (dry), $[\alpha]_D$ —30.9° (H_2O). The hydriodide, $B \cdot HI \cdot 3.5 H_2O$, has m. p. 226° and is sparingly soluble in water. The aurichloride, $B \cdot HAuCl_4 \cdot 3 H_2O$, is formed when a solution of gold chloride is added to a solution of an aconitine salt to which HCl has been added. It is very sparingly soluble in dilute HCl and crystallizes in golden needles in three forms. The α-form is obtained by crystallizing from acetone, aqueous ethanol or chloroform and has m. p. 135° (dec.). The β-form, m. p. 152° is obtained by crystallization from absolute ethanol and the γ-form, m. p. 176°, by recrystallizing the β-form from a mixture of chloroform and ether. The perchlorate has m. p. 215—222°, $[\alpha]_D$ —18.9°. The diacetyl derivative, $C_{34}H_{45}(Ac_2)O_{11}N$, has m. p. 158°.

Pseudaconitine (acetylveratroylpseudaconine), $C_{36}H_{51}O_{12}N$, crystallizes from hot ethanol in colourless rhombs, m. p. 212—213°, $[\alpha]_D^{20°} +17.06°$ (ethanol) or 22.75° (chloroform). This base resembles aconitine in its solubility in common solvents but is rather less soluble in ether than aconitine. It is readily hydrolysed with the formation of acetic acid and *veratroylpseudaconine*. On alkaline hydrolysis the latter yields *veratric acid* and *pseudaconine*. Pseudaconitine is intensely poisonous, and is about twice as toxic as aconitine. The salts are crystalline and laevorotatory in solution. The hydrobromide, $B \cdot HBr \cdot 3 H_2O$, forms rosettes of triangular prisms, m. p. 199°, $[\alpha]_D^{20°}$ —18.5° (H_2O). The hydriodide $B \cdot HI \cdot H_2O$, forms prisms, m. p.

230°, and the nitrate $B \cdot HNO_3 \cdot H_2O$, prisms, m. p. 198° $[\alpha]_D^{18°}$ —17.95° (H_2O). The auri-chloride, $B \cdot HAuCl_4$, forms golden-yellow needles, m. p. 233°, and the picrate orange-yellow prisms, m. p. 196°.

Indaconitine (acetylbenzoylpseudaconine), $C_{34}H_{47}O_{10}N$, crystallizes in rosettes of needles or hexagonal prisms, m. p. 202—203°, $[\alpha]_D$ +18.3 (ethanol). It is soluble in ethanol, chloro-form, acetone and ether but almost insoluble in water and light petroleum. Hydrolysis takes place in two stages, acetic acid and *benzoylpseudaconine* being first formed. On further hydrolysis with alkali benzoic acid and pseudaconine are the resultant products. The hydro-bromide, $B \cdot HBr$, forms hexagonal prisms from water, m. p. 183—187°, $[\alpha]_D$ —17.3°, or, when crystallized from ethanol and ether, m. p. 217—218°. The aurichloride, $B \cdot HAuCl_4$, forms rosettes of yellow needles, m. p. 147—152°.

Bikhaconitine (acetylveratroylbikhaconine), $C_{36}H_{51}O_{11}N \cdot H_2O$, crystallizes from ethanol on addition of water m. p. 113—116° (dry), $[\alpha]_D^{20°}$ +12.21 (ethanol). From ether it separates as white, semi-spherical or button-shaped masses m. p. 118—123°. Bikhaconitine is easily soluble in ethanol, ether or chloroform, but is practically insoluble in water or light petroleum. Hydrolysis takes place in two stages, acetic acid and *veratroylbikhaconine* being produced in the first stage and veratric acid and *bikhaconine* in the second stage. The salts of bikhaconitine crystallize well, the hydrobromide, $B \cdot HBr \cdot 5 H_2O$, melts at 173—175° (dry), $[\alpha]_D^{20°}$ —12.42°. The hydrochloride, $B \cdot HCl \cdot 5 H_2O$, has m. p. 159—161° (dry) and the hydriodide has m. p. 193—194° (dry). The aurichloride $B \cdot HAuCl_4$ crystallizes by adding ethanol, ether or light petroleum to its solution in chloroform and is obtained in yellow needles, m. p. 232—233°.

Mesaconitine (acetylbenzmesaconine), $C_{33}H_{45}O_{11}N$, crystallizes from boiling methanol in prisms, m. p. 208—209°, $[\alpha]_D^{17°}$ +25.7° (chloroform). The hydrobromide $B \cdot HBr \cdot 3.5 H_2O$ has m. p. 172—173° (dry, dec.), $[\alpha]_D$ —24.8°. The perchlorate crystallizes in prisms, m. p. 217—225° (dec.) and the aurichloride $B \cdot HAuCl_4$ forms yellow prisms, m. p. 224—226° (dec.). On hydrolysis acetic acid and *benzmesaconine* are formed. Further hydrolysis of benzmes-aconine yields benzoic acid and *mesaconine*.

Hypaconitine (acetylbenzhypaconine), $C_{33}H_{45}O_{10}N$, crystallizes from methanol in prisms, m. p. 197—198° (dec.), $[\alpha]_D^{17°}$ +22.7° (chloroform). The hydrobromide, $B \cdot HBr \cdot 2.5 H_2O$, has m. p. 178—179° (dry), $[\alpha]_D$ —20.5° (H_2O). The perchlorate has m. p. 178—180° (dec.) and the aurichloride, $B \cdot HAuCl_4$, forms prisms m. p. 243—245° (dec.). On hydrolysis with boiling dilute H_2SO_4, acetic acid and benzhypaconine are formed. On hydrolysis by water in a sealed tube at 160—170°, hypaconitine yields acetic acid, benzoic acid and *hypaconine*.

Jesaconitine (acetyl-p-anisoylaconine), $C_{35}H_{49}O_{12}N$, is an amorphous base, m. p. 128—131°. It yields a crystalline perchlorate, m. p. 230—232° (dec.), $[\alpha]_D$ —16.7° (methanol), an auri-chloride crystallizing in prisms, m. p. 208—209° (dec.) and a perbromide, m. p. 181—182° (dec.). On boiling with dilute H_2SO_4, acetic acid and *p-anisoylaconine* are formed. On alkaline hydrolysis, jesaconitine yields acetic acid, *p-anisic acid* and *aconine*.

Lappaconitine (acetyl-anthranoyl-lappaconine), $C_{32}H_{44}O_8N_2$ (or $C_{32}H_{42}O_9N_2$) crystallizes in hexagonal tablets, m. p. 213.5—214.5°, $[\alpha]_D$ +22.3° (benzene). Crystalline salts have not been prepared. On complete hydrolysis, acetic acid, *anthranilic acid* and *lappaconine* are produced. Lappaconitine is soluble in ethanol and ether but sparingly soluble in water. Its solutions have a reddish-violet fluorescence.

Septentrionaline, $C_{33}H_{46}O_9N_2$, occurs as a white amorphous powder, m. p. 131°, $[\alpha]_D^{19.5°}$ +32.7° (ethanol). It is soluble in ethanol, ether and water and does not give fluorescent solutions.

Lycaconitine (succinyl-anthranoyl-lycoctonine), $C_{36}H_{46}O_{10}N_2$, is an amorphous powder, $[\alpha]_D^{20°}$ +42.5° (ethanol). It melts at 112—115° with decomposition and is sparingly soluble in water but readily soluble in absolute ethanol, chloroform and benzene. It is less soluble in ether and almost insoluble in light petroleum. The salts have not been crystallized. On hydrolysis with water or dilute acid lycaconitine yields succinic acid and *anthranoyl-lycoctonine* and alkaline hydrolysis leads to the production of *lycoctonic acid* (succinyl-anthranilic acid) and *lycoctonine* which separates in crystalline form and can be extracted with ether. Lycoc-tonine has m. p. 139° and produces a hydrochloride, m. p. 145° (dec.), a picrate, m. p. 154°, and a methiodide m. p. 188°. Some of the delphinium alkaloids on hydrolysis produce lycoctonine.

Myoctonine $(C_{36}H_{42}O_{10}N_2)_2$ is an amorphous base, $[\alpha]_D^{20°}$ + 44.7 (ethanol). It is sparingly soluble in water but very soluble in ethanol, chloroform and benzene. In ether and light petroleum it is very slightly soluble. On hydrolysis it yields the same products as lycaconitine.

Atisine $C_{22}H_{33}O_2N$, occurs as a colourless varnish readily soluble in water but insoluble in light petroleum. It melts at about 85° and is unstable in solution more especially in presence of alkali. The salts crystallize well; the hydrochloride, $B \cdot HCl$, has m. p. 311—312° (dec.). $[\alpha]_D^{25°}$ +28° (water), the hydrobromide, $B \cdot HBr$, melts at 273°.

Palmatisine can be crystallized from a mixture of ethanol, ether and light petroleum. It melts at 285° and forms a hydrobromide melting at 245°. **Talatisine,** $C_{20}H_{29}O_3N$ melts at 246°, $[\alpha]_D$ +37.7° (ethanol). The hydrochloride B · HCl, has m. p. 256—257°, the hydriodide, B · HI, m. p. 265° (dec.) the perchlorate, B · HClO$_4$, m. p. 222° (dec.) and the picrate has m. p. 247—250° (dec.).

Anthorine is an amorphous dextrorotatory alkaloid soluble in ether which has not been characterized.

II. Isolation.

1. The Isolation of Aconitine.

The finely powdered root is extracted by percolation with a mixture of 5 parts of methanol with 1 part of amyl alcohol. The extract is concentrated by distillation *in vacuo* until the temperature of the extract rises to about 60—65°. The extract is transferred to a separating funnel and the alkaloids shaken out with 0.5% hydrochloric acid until extraction is complete. The acid solution is washed well with ether to remove the residue of amyl alcohol, resins and fatty material, made alkaline with a small excess of ammonia or sodium carbonate and the alkaloids extracted with ether. The ethereal solution is extracted with successive small quantities of 1% HCl until extraction is complete and the combined acid solutions are made alkaline with a slight excess of ammonia. The alkaline solution is gently agitated with a small volume of ether from which aconitine crystallizes out. Aconitine remaining in the ether solution is removed by agitating with dilute hydrobromic acid solution and allowing the aconitine hydrobromide to crystallize. Aconitine is best purified by recrystallization of its salts with final liberation of the base and recrystallization from ethanol or methanol. Aconitine is found in all parts of *A. napellus* but the root is the richest source. However there is great variation in the percentage of total alkaloids in roots, values from 0.2% to 1.5% having been recorded. The daughter tubers are usually a better source of aconitine than the old parent tubers and good quality material yields from 0.4% to 0.5% of total alkaloids.

Qualitative Tests for Aconitine. When a trace of a very dilute solution of aconitine is placed upon the tongue a characteristic tingling sensation is produced followed by numbness, which may spread to the lips. This test is given by the other aconitines in varying degree. Very great care should be exercised in carrying out the test in view of the fact that these alkaloids are intensely poisonous.

The alkaloid reagents give precipitates with slightly acidified solutions of aconitine or its salts. Iodine in potassium iodide gives a reddish-brown precipitate in a dilution of 1:20,000 and MAYER's reagent gives a yellowish-white precipitate in a dilution of 1:10,000. DUNSTAN (1896) has described a characteristic reaction with potassium permanganate which is useful for the identification of aconitine in dilutions up to 1:4,000. The solution of aconitine or one of its salts is slightly acidified with acetic acid and a few drops of 0.2 N KMnO$_4$ solution added. Rosettes of red prismatic crystals are formed (or in very dilute solution isolated prisms). Cocaine, hydrastine and papaverine give similar precipitates in concentrated solution but not at dilutions above 1:500. The sensitivity of the test is decreased by the presence of other aconite alkaloids. The crystalline form of aconitine itself may be used for identification. If one or two drops of 5% Na$_2$CO$_3$ solution are added to 1 ml. of a solution of an aconitine salt and the solution is heated to 50° and cooled, irregular hexagonal plates are produced. DENIGÈS (1944) has suggested the use of the hydriodide, hydrobromide and hydrochloride of aconitine as a means of identification. A drop of a 2% solution of aconitine nitrate in 10% acetic acid is mixed on a slide with one drop of a saturated solution of KI, KBr or NaCl. The characteristic micro-crystalline salt of the alkaloid is obtained with as little as 0.02 mg. of aconitine. Ammonium reineckate in slightly acid solution gives with aconitine in high dilution a precipitate which on microscopic examination consists of rosettes of fine crystals. Colour tests are not reliable for the identification of aconitine but the following test with resorcinol may be used as a supplementary test. When a few crystals of aconitine are warmed on a water bath with 0.5 ml. of sulphuric acid (sp. gr. 1.75) benzoic acid is produced. If a few crystals of resorcinol are added after an interval of 5 minutes and heating continued, a reddish-yellow colour changing to red develops. The production of acetic acid and benzoic acid on hydrolysis has also been used for the identification of aconitine.

2. The Isolation of Pseudaconitine.

Henry and Sharp (1928) have used the following procedure for the extraction of pseudaconitine from the air-dried roots of *A. balfourii* and *A. deinorrhizum*. The finely ground root material is percolated to exhaustion with cold 95% ethanol, the percolate concentrated to low bulk under reduced pressure and the alkaloids extracted from the residue with 0.5% HCl. The acid solution is agitated with ether to remove acids resin and oil and made alkaline with Na_2CO_3 solution. The alkaline solution is extracted with ether, several portions being used to ensure complete extraction. The combined ether extracts are evaporated when crude pseudaconitine crystallizes out. Purification is achieved by recrystallization of pseudaconitine as the hydrobromide which separates from ethanol in rosettes of triangular prisms. The yield of pseudaconitine is about 0.4% expressed on the air-dried roots. Pseudaconitine resembles aconitine in its precipitation reactions with the alkaloid reagents. It is distinguished from aconitine by the high melting point of its aurichloride and by the fact that on hydrolysis with ethanolic potash it yields veratric acid, m. p. 178°.

3. The Isolation of Indaconitine and Bikhaconitine.

Dunstan and Andrews (1905a, b) describe the following methods for the extraction of these alkaloids. Indaconitine is prepared from the finely powdered root of *A. chasmanthum* by percolation at room temperature with a mixture of 5 parts of methanol with 1 part of amyl alcohol. The methanol is removed from the percolate by distillation under reduced pressure, the amyl alcohol containing the alkaloids is transferred to a separating funnel and repeatedly agitated with dilute HCl which is then well washed with ether to remove amyl alcohol and other extractable matter. The indaconitine is fractionally precipitated from the aqueous liquid by the addition of dilute ammonia, extracted as completely as possible with ether and finally with chloroform. The ethereal solution is washed with water to remove alkali, dried with calcium chloride and evaporated to small bulk, when crystals of the base are deposited in large irregular rosettes. The alkaloid is purified through the hydrobromide which can be crystallized by the addition of dry ether to a solution in ethanol. When the pure alkaloid is regenerated it crystallizes from ether in masses of fine white needles. Indaconitine crystallizes in several forms from the same solvent, a property which distinguishes it from aconitine. In the permanganate test of Dunstan, the crystals are smaller and decompose more readily than those of aconitine. The salts show a lower rotatory power than those of aconitine. Bikhaconitine is prepared from the root of *A. spicatum* by a method similar to that used for the preparation of indaconitine. The roots contain about 0.4% of bikhaconitine. Purification of the alkaloid is carried out by conversion to the hydrobromide with regeneration of the base from this salt. The form of the crystals separating from ether (semi-spherical or button-shaped masses) distinguishes this base from aconitine. Addition of water to an ethanolic solution of the alkaloid is the best means of crystallization. On alkaline hydrolysis veratric acid, m. p. 178° is produced.

4. Isolation of other Aconitum Alkaloids.

Mesaconitine and other alkaloids of Japanese species of *Aconitum* are isolated by a procedure similar to that for aconitine (MAJIMA, SUGINOME and MORIO, 1924).

Lycaconitine is extracted from the roots of *A. lycoctonum* by exhausting with ethanol acidified with tartaric acid. The extract is freed from ethanol, mixed with water, filtered and repeatedly shaken with ether while still acid. The aqueous solution is made alkaline with $NaHCO_3$ and lycaconitine extracted with ether. The ether layer is run off and the aqueous phase is shaken with chloroform to remove lycaconitine remaining in aqueous solution together with myoctonine. The content of crude lycaconitine amounts to about 1.13%, and of myoctonine to 0.8%.

On alkaline hydrolysis, lycaconitine yields lycoctonic acid (succinyl-anthranilic acid) and lycoctonine, m. p. 139° (sinters at 120°). The following *colour tests* for lycacomtine have been described. With syrupy phosphoric acid a violet colouration is produced on warming and when treated with a mixture of water (8 ml.), conc. H_2SO_4 (6 ml.) and sodium selenate (0.3 g.) lycaconitine gives a rose or pale reddish-violet colour. Both lycaconitine and lycoctonine give precipitates with the alkaloid reagents.

The alkaloid fractions of *A. heterophyllum* have been isolated and studied by JACOBS and CRAIG (1942) and by JACOBS and HUEBNER (1947).

III. The Estimation of the Total Alkaloids of Aconite Root.

Up to the present time no entirely satisfactory method has been devised for accurate estimation of the alkaloids of aconite. Existing methods depend on separation of the alkaloids and subsequent titration with acid, the results being expressed in terms of aconitine. Alkaline hydrolysis with the production of acetic and benzoic acids which are steam distilled, also has been used for purposes of estimation. The following titrimetric method has been recommended by LYNCH and EVERS (1942). 10 g. of the root in No. 60 powder are introduced into a stoppered percolator of suitable capacity, 100 ml. of a mixture of 3 vols. of ether and 1 vol. of chloroform are added, and the mixture shaken well and set aside for 15 minutes. 5 ml. of a dilute solution of ammonia (10%) are added and the mixture shaken for one minute at ten-minute intervals for one hour. The solution is allowed to percolate into a separating funnel. When the solution ceases to flow the drug is packed firmly and extraction continued with further quantities of the mixed solvent until the drug is exhausted. The percolate is tested for complete extraction by collecting about 2 ml. in a dish, evaporating the solvent, dissolving the residue in a few drops of 0.1 N H_2SO_4 and adding a drop of 0.1 N iodine solution. In absence of alkaloids no precipitate or turbidity should appear. To the percolate, 30 ml. of N H_2SO_4, or sufficient to render the mixture faintly acid, is added, the mixture shaken well and the two layers allowed to separate. The lower layer is run off into a separating funnel A. The extraction is continued with 10 ml. portions of 0.1 N H_2SO_4 until extraction of the alkaloids is complete as shown by the iodine test. The mixed solutions in A are washed with about 10 ml. of chloroform which is run off into another separating funnel B, containing 20 ml. of 0.1 N H_2SO_4. This mixture is shaken, allowed to separate and the chloroform rejected. The washing of the liquid in separating funnel A is continued with two further portions of 5 ml. of chloroform which in turn are transferred to separating funnel B, washed with the same aqueous acid liquid, allowed to separate and the chloroform rejected. The acid liquid from funnel B is transferred to funnel A, the solution made just alkaline with dilute ammonia and 2 ml. added in excess. The alkaline solution is shaken with successive portions of chloroform until extraction of the alkaloids is complete. Each chloroform extract is washed with the same 20 ml. of water contained in another separating funnel. The combined chloroform solutions are evaporated and 2 ml. of absolute ethanol are added to the residue. The ethanol is evaporated at a temperature not exceeding 60° and the residue is dried at a temperature below 60° for 30 minutes. The residue is now dissolved with warming in 2 ml. of 95% ethanol (neutral), 20 ml. of 0.02 N H_2SO_4 and 10 ml. of water added, the solution cooled and titrated with 0.02 N NaOH using methyl red as indicator. 1 ml. of 0.02 N acid is equivalent to 0.01291 g. of alkaloids calculated as aconitine.

Mühlemann and Weil (1949) have described a similar titrimetric method for the total alkaloids of aconite in which ether is used throughout for extraction, and methyl-red — methylene blue used as indicator. These authors have also determined aconitine and benzoylaconine by hydrolysis of the titrated total alkaloid solution, followed by steam distillation and titration of the acetic and benzoic acids formed.

C. The Alkaloids of Areca Nut.

The areca or betel nut palm, *Areca catechu L. (Palmaceae)*, is cultivated in India and the islands of the Eastern Archipelago where the seeds are used as a masticatory. The seeds contain six alkaloids, arecoline, arecaidine, arecolidine, guvacine, guvacoline and isoguvacine. Of these alkaloids, arecoline and arecaidine are present in the highest concentrations, the remaining bases occurring in very small amounts. Arecoline is the most important alkaloid of the group and occurs to the extent of from 0.1—0.5%. Arecaidine is present to the extent of about 0.1%.

I. Properties.

Arecoline, $C_8H_{13}O_2N$, is a colourless, odourless strongly alkaline oil, b. p. 209°, which is volatile in steam and miscible with most organic solvents and water. It can only be extracted by ether from aqueous solution in the presence of dissolved salts. Arecoline is the methyl ester of *arecaidine* and on hydrolysis yields the latter base and methanol. The salts are crystalline but usually deliquescent. The hydrochloride, B · HCl, forms slender needles, m. p. 157—158°, and is readily soluble in water and ethanol. The hydrobromide, B · HBr, forms slender prisms, m. p. 170°, and is easily soluble in water and in hot ethanol, much less soluble in cold ethanol and sparingly soluble in ether and chloroform. The platinichloride, $B_2 · H_2PtCl_6$, m. p. 176°, crystallizes from water in orange-red rhombs. The methiodide forms prisms, m. p. 173—174°.

Arecaidine (Arecaine), $C_7H_{11}O_2N · H_2O$, forms colourless four or six-sided tablets, m. p. 222—224° (Jahns, 1890), 232° (Hess and Leibbrandt, 1919). It is readily soluble in water, slightly soluble in absolute ethanol but almost insoluble in ether, chloroform and benzene. Arecaidine forms salts which are crystalline. The hydrochloride, B · HCl, forms colourless needles, m. p. 261° (dec.) and the hydrobromide melts at 242—243°. The platinichloride, $B_2 · H_2PtCl_6$ crystallizes in octahedra, m. p. 225—226° and the aurichloride, B · HAuCl_4, in prisms, m. p. 197—198°, from hot dilute HCl. Arecaidine is readily esterified with methanol yielding arecoline.

Guvacine, $C_6H_9O_2N$, the simplest of the areca alkaloids, forms small lustrous prisms, m. p. 271—272°, 293—295° (dec.). This base is neutral to litmus and is soluble in water or dilute ethanol but practically insoluble in absolute ethanol, ether, chloroform and benzene. The salts are crystalline and acid in reaction. The hydrochloride, B · HCl, forms prisms, m. p. 312—316°, soluble in water and sparingly soluble in dilute HCl.

The aurichloride, B · HAuCl_4, crystallizes in broad flat prisms, m. p. 197—199°, and the platinichloride, $B_2 · H_2PtCl_6 · 4 H_2O$, forms hexagonal prisms from water, m. p. 220—221° (Freudenberg, 1918).

Guvacoline, $C_7H_{11}O_2N$, is an oil, b. p. 114°/13—14 mm., m. p. 27°. It is the methyl ester of *guvacine* and forms the latter on hydrolysis with barium hydroxide. The hydrochloride, B · HCl, forms leaflets, m. p. 121—122°, and the hydrobromide occurs in prisms, m. p. 144 to

145°. The platinichloride crystallizes in golden leaflets, m. p. 211°. **Arecolidine,** $C_8H_{13}O_2N$, is isomeric with arecoline and crystallizes from ether in colourless needles, m. p. 105° (after sublimation m. p. 110°). It is hygroscopic and forms stable salts. The hydrochloride, $B \cdot HCl \cdot H_2O$, forms hygroscopic prisms from absolute ethanol which when dry, decompose at 250°. The hydrobromide has m. p. 268—271° (dec.) and the platinichloride crystallizes in dark orange needles, m. p. 222—223° (dec.). The methiodide crystallizes in prisms, m. p. 264° (dec.). The base is soluble in water, ether and acetone and is the methyl ester of N-methylguvacine.

Isoguvacine is probably impure arecaidine.

II. The Isolation of the Areca Alkaloids.

In a series of papers, JAHNS (1888, 1890, 1891) describes the isolation of the alkaloids by precipitation with bismuth potassium iodide. The ground seeds are digested three times with water containing 2 g. of H_2SO_4 per kg. of seeds used. The extract is concentrated and precipitated with bismuth potassium iodide, excess of the reagent in which the alkaloid precipitate is soluble being avoided. The solution is set aside for one day to allow the precipitate to settle and then filtered. The precipitate is washed with water and the alkaloids regenerated by boiling the precipitate with water containing barium carbonate. The solution is concentrated and, after the addition of barium hydroxide, is extracted with ether. Arecoline passes into the ether and is purified by recrystallization of the hydrobromide. The mother liquors are neutralized with H_2SO_4 and treated with silver sulphate, barium hydroxide and carbon dioxide in succession to remove barium iodide etc. and the final f ltrate is evaporated to dryness and extracted with dry ethanol. Arecaidine and guvacine are slightly soluble in absolute ethanol and the two bases can therefore be extracted from the residue by continued treatment with this solvent. Fractional precipitation from dilute ethanol, in which guvacine is less soluble than arecaidine, is used for the separation of the bases. CHEMNITIUS (1927) has described a method for the large scale extraction the alkaloids.

III. The Detection and Estimation of Arecoline.

1. Qualitative Tests. When bismuth potassium iodide is added to arecoline in very dilute solution, a red microcrystalline precipitate is formed. Excess of the reagent should be avoided. This test is very delicate. MAYER's reagent also gives a precipitate soluble in excess of the reagent. Phosphomolybdic acid produces a white precipitate but neither mercuric chloride nor tannic acid give a precipitate. There are no specific colour tests for arecoline.

2. Quantitative Estimation. The following method is based on a method for the assay of arecoline hydrobromide described by BOND (1942). 20 g. of the ground Areca nut is moistened with 10% ammonia solution and exhaustively extracted with ether in a SOXHLET. An equal volume of 10% hydrobromic acid solution is added to the ether extract, the mixture shaken and the ether carefully removed by distillation. The cooled acid solution is filtered, shaken with ether to remove impurities and the ether discarded. The solution is saturated with solid $NaHCO_3$ until a small quantity of bicarbonate remains undissolved and thoroughly extracted with chloroform, using successive portions of 30, 25, 20, 15, 10, 10 and 10 ml. A further extraction is made with 10 ml. of chloroform and the extract shaken with 0.02 N acid. The chloroform is evaporated off and the acid solution tested for arecoline with bismuth potassium iodide. Each chloroform extract is drawn off into a 500 ml. glass stoppered Erlenmeyer flask through a funnel, the stem of which holds a fairly tight cotton wool pledget moistened with chloroform. The funnel stems are rinsed with chloroform which is added to the filtrate in the Erlenmeyer flask, 35 ml. of 0.02 N H_2SO_4 are added to the flask, the mixture shaken thoroughly and, using glass beads to prevent superheating, the chloroform completely evaporated on the steam bath. The acid solution is cooled and titrated with 0.02 N NaOH using methyl red as indicator. 1 ml. of 0.02 N acid is equivalent to 0.00472 g. of arecoline hydrobromide.

D. The Berberis and Hydrastis Alkaloids.

Berberine and its associated alkaloids are isoquinoline derivatives found in a large number of plants belonging to the families Berberidaceae, Ranunculaceae, Anonaceae, Menispermaceae, Papaveraceae and Rutaceae. The more important alkaloids of this group with their principal sources will be found in Table 3.

Table 3. *The Occurrence and Distribution of Berberine and Associated Alkaloids.*

Family	Genera and Species	Alkaloids
Berberidaceae	Berberis darwinii HOOK.	Berberine
	B. buxifolia LAM.	
	B. glauca DC.	
	B. aetnensis PRESL.	
	B. nervosa PURSH.	
	B. vulgaris L.	Berberine, berbamine, oxyacanthine, jatrorrhizine, columbamine
	B. heteropoda SCHRENK.	Berberine, palmatine, berbamine, oxyacanthine, jatrorrhizine, columbamine
	B. thunbergii DC.	Berberine, oxyacanthine, jatrorrhizine, columbamine
	B. insignis HOOK.	Umbellatine
	Mahonia aquifolium NUTT.	Berberine, berbamine, oxyacanthine
	M. nepalensis DC.	Umbellatine, neprotine
	M. swaseyi FEDDE.	Berberine
	M. trifoliolata FEDDE.	Berberine
	Nandina domestica THUNB.	Berberine, nandinine, domesticine
Ranunculaceae	Hydrastis canadensis L.	Hydrastine, canadine, berberine
	Coptis japonica MAK.	Berberine, palmatine, coptisine, columbamine
	C. occidentalis SALIS.	Berberine
	C. teeta WALL.	
	C. trifolia SALIS.	
	Xanthorrhiza apiifolia L'HERIT.	Berberine
	Thalictrum foliolosum DC.	Berberine, thalictrine
Anonaceae	Xylopia polycarpa OLW.	Berberine
Menispermaceae	Arcangelisia flava L.	Berberine, jatrorrhizine, columbamine
	Coscinium blumeanum MIERS.	Berberine, jatrorrhizine, palmatine
	C. fenestratum COLEBR.	Berberine
Papaveraceae	Argemone mexicana L.	Berberine
	Chelidonium majus L.	Berberine
Rutaceae	Phellodendron amurense RUPR.	Berberine, palmatine
	Evodia meliifolia BENTH.	Berberine
	Toddalia aculeata PERS.	Berberine
	Xanthoxylum cariboeum LAM.	Berberine

Many other species of *Berberis* in addition to those above-listed contain berberine. Palmatine, jatrorrhizine and columbamine are the principal alkaloids of calumba root *Jateorhiza palmata* LAM., MIERS., and belong to the protoberberine section of the isoquinoline group of alkaloids. The tertiary tetra-hydro derivatives of these alkaloids occur frequently in the *Rhoeadales*, especially in the genus *Corydalis*. These alkaloids will not be further treated in this work and for a full account of their distribution and chemistry, the reader should consult the standard work by HENRY (1949). Berbamine and oxyacanthine contain two isoquinoline nuclei and are therefore included in the bisbenzylisoquinoline sub-group. Domesticine belongs to the aporphine sub-group of isoquinoline alkaloids.

I. Properties.

Berberine, $C_{20}H_{19}O_5N$, crystallizes from water or dilute ethanol in bright yellow needles with 5.5 H_2O. On heating to 100° the crystals lose 3 H_2O and on further heating, decompose at 160°. Berberine shows a tendency to combine with neutral solvents and from chloroform it forms triclinic tablets containing 1 $CHCl_3$, m. p. 145°, and from acetone reddish-yellow tablets having the composition $B \cdot C_3H_6O$. Berberine dissolves in cold water (1 in 4.5 at 21°) or ethanol (1 in 100) and is easily soluble in hot water or hot ethanol. It is slightly soluble in benzene or chloroform, insoluble in ether or light petroleum. The aqueous solution is bitter to the taste, neutral to litmus and is optically inactive. Berberine forms well defined crystalline salts with acids. The hydrochloride, $C_{20}H_{17}O_4N \cdot HCl \cdot 2 H_2O$, crystallizes in small needles on adding HCl to a warm aqueous solution of the alkaloid. It is soluble in about 500 parts of cold water but almost insoluble in ethanol or dilute HCl. The nitrate separates in greenish yellow needles when a warm aqueous or ethanolic solution of berberine is acidified with nitric acid. It is soluble in about 500 parts of cold water, more readily soluble in hot water and almost insoluble in ethanol or dilute HNO_3. The hydriodide, $C_{20}H_{17}O_4N \cdot HI$, forms very small needles sparingly soluble in cold water (1 in 2130) and the acid sulphate forms slender yellow needles soluble in about 100 parts of cold water. This compound which has the formula $C_{20}H_{17}O_4N \cdot HS_2O_4$ is formed when a solution of berberine (or one of its salts) in absolute ethanol is treted with etahanolic H_2SO_4. The aurichloride $B \cdot HAuCl_4$, crystallizes from ethanolic HCl in brown needles, the platinichloride, $B_2H_2PtCl_6$, forms a yellowish precipitate almost insoluble in ordinary solvents and the picrate forms a yellow precipitate insoluble in cold water. The normal or crystalline (ammonium) form of berberine is soluble in water and insoluble in ether but the so-called aldehyde form obtained by treating a solution of the normal form with excess of NaOH is very little soluble in water but soluble in ether. This behaviour of berberine on treatment with excess of NaOH has been utilid for the determination of the alkaloid. The melting points of berberine and its salts are indefinite and consequently are of little value for characterization.

Canadine (l-tetrahydroberberine), $C_{20}H_{21}O_4N$, forms silky needles, m. p. 133°, $[\alpha]_D$ —299° (chloroform), insoluble in water but readily soluble in ethanol, ether, chloroform and benzene. The base is difficultly soluble in light petroleum. The salts are crystalline and very little soluble in water and dilute solutions of the corresponding acids. The hydrochloride, $B \cdot HCl$, is a white crystalline powder slightly soluble in cold water, more soluble in hot water, giving a neutral solution which is laevo-rotatory. The nitrate, $B \cdot HNO_3$, forms white leaflets less soluble in water than the hydrochloride. The sulphate, $B \cdot H_2SO_4$, crystallizes in large colourless plates and is easily soluble in water. The aurichloride is a reddish-brown flocculent precipitate and the platinichloride gives an amorphous yellow precipitate. The α-methiodide, $C_{21}H_{24}O_4NI$, crystallizes from hot water in prisms and melts at 220°. At 230° the α-methiodide is converted to the β-methiodide which melts at 264° (dec.).

1-α-Canadine methochloride, $C_{21}H_{24}O_4NCl \cdot H_2O$, was isolated by JOWETT and PYMAN (1913) from *Zanthoxylum brachyacanthum* F. MUELL. and crystallizes from dry ethanol in colourless prismatic needles, m. p. 262° (corr., dec.), $[\alpha]_D$ —137° (H_2O).

Oxyacanthine, $C_{37}H_{40}O_6N_2$, crystallizes from ethanol in needles, m. p. 208—209°, $[\alpha]_D$ +279° (chloroform). The base is sparingly soluble in water, more soluble in ethanol, readily soluble in chloroform and benzene but sparingly soluble in light petroleum. The hydrochloride, $B \cdot 2 HCl$, forms small needles, m. p. 270—271°, $[\alpha]_D^{20°}$ +188.5° (H_2O) and is sparingly soluble in dilute HCl. The hydrobromide, $B \cdot 2 HBr$, crystallizes in white needles, m. p. 273—275° (vac.) and the nitrate, $B \cdot 2 HNO_3$, forms needles or wart-like crystals, m. p. 195—200°, sparingly soluble in cold water.

Berbamine, $C_{37}H_{40}O_6N_2$, is isomeric with oxyacanthine and crystallizes from ethanol in leaflets with 4 H_2O, m. p. 156° (or 172° dry) or from light petroleum in warty masses, m. p. 197—210°, $[\alpha]_D$ +108.6° (chloroform). The sulphate, $B_2 \cdot H_2SO_4 \cdot 4 H_2O$, crystallizes in leaflets or needles and the hydrochloride, $B \cdot HCl \cdot 2 H_2O$, forms wart-like crystals. The platinichloride, $(B \cdot HCl)_2PtCl_4 \cdot 5 H_2O$, forms a yellow precipitate practically insoluble in water.

Hydrastine, $C_{21}H_{21}O_6N$, forms colourless rhombic prisms from ethanol. The base reacts alkaline to litmus and is practically insoluble in water but easily soluble in chloroform (1 in 14 at 25°) or benzene, less soluble in ethanol (1 in 170 at 25°) or ether (1 in 175 at 25°). It is insoluble in light petroleum. Hydrastine melts at 132° and has $[\alpha]_D$ —67.8° (chloroform). It is a weak base and can be extracted from acid solutions by chloroform. The salts are unstable in water and do not as a rule crystallize well. The hydrochloride, $B \cdot HCl$, has m. p. 116°, $[\alpha]_D$ +127.3° (dil. HCl) and may be obtained by passing HCl gas into a solution of the alkaloid in ether. The acid tartrate, $B \cdot H_2C_4H_4O_6 \cdot 4 H_2O$, crystallizes from hot water and the picrate, $B \cdot C_6H_2(NO_2)_3OH \cdot H_2O$, crystallizes in yellow needles, m. p. 184°. Hydrastine behaves as a tertiary base and when treated with dilute HNO_3 it undergoes hydrolytic oxidation with the formation of opianic acid and a derivative base hydrastinine, $C_{11}H_{13}O_3N$, which crystallizes in colourless needles from light petroleum and has m. p. 116 to 117°. In aqueous solution hydrastinine shows a blue fluorescence.

Umbellatine, $C_{21}H_{21}O_8N$, crystallizes from water in yellow silky needles with 5.5 H_2O, m. p. 205—207° (dec.). The base is optically inactive and forms crystalline salts. The sulphate has m. p. 271—273° (dec.) and the nitrate has m. p. 265—267° (dec.). The picrate melts at 231° (dec.).

Neprotine, $C_{19}H_{21}O_6N$, forms garnet-red crystals which melt with decomposition above 200°. The base is optically inactive and forms a hydrochloride crystallizing in garnet-red elongated prisms and a platinichloride which decomposes without melting.

Nandinine, $C_{19}H_{19}O_4N$, crystallizes from ethanol in leaflets, m. p. 145—146°, $[\alpha]_D$ +63.2° (ethanol). On exposure to air it develops a reddish colour.

Domesticine, $C_{19}H_{19}O_4N$, crystallizes from methanol or ethanol and has m. p. 115—117° $[\alpha]_D$ +60.51°.

Thalictrine, $C_{20}H_{27}O_4N \cdot 3\,H_2O$, has m. p. 208°, $[\alpha]_D^{24°}$ +308° (H_2O). The anhydrous base melts at 224—225° (dec.) and has $[\alpha]_D^{24°}$ +370° (H_2O). It is a quaternary base and the chloride, $C_{20}H_{26}O_3NCl$, has m. p. 163—165° (dec.); the iodide, m. p. 265° (dec.); the picrate, m. p. 207—208°, and the platinichloride $(C_{20}H_{26}O_3NCl)_2 \cdot PtCl_4$, m. p. 233—234° (dec.).

II. The Distribution of the Alkaloids within the Plant.

The alkaloids of *Berberis* and *Hydrastis* are located chiefly in the cortical tissues of the roots and stems. The bark of old roots contains the highest concentration of alkaloids. In the upper parts of the shoot system the concentration is low and in young leaves alkaloids cannot be detected. The histological distribution of berberine in *B. darwinii* has been studied by CROMWELL (1933) and in *M. swaseyi* and *M. trifoliolata* by GREATHOUSE and WATKINS (1938). CHATTERJEE (1943, 1944) has examined a number of Indian species of *Berberis* and has studied the histological distribution of berberine, umbellatine and neprotine in these plants. The average concentration of berberine in various tissues of 3, 5, 7 and 30-year old plants of *B. darwinii* is given in Table 4.

Table 4. *The Average Concentration of Berberine g./100 g. Dry Tissue in the Tissues of B. darwinii.*

Age of Plant (years)	Roots (whole)	Roots (subsidiary)	Main Stem	Branches (subsidiary)
3	1.27	—	0.29	0.12
5	1.44	—	0.40	0.15
7	—	1.84	0.48	0.11
30 bark	11.0	—	2.67—4.93	0.33—0.76
wood	2.8	2.8	—	—

The mean value for young actively growing shoots is 0.04% and for young parenchymatous roots 1.41%. There is a progressive increase in the berberine content of the plants with increase in age. Similar results have been obtained by GREATHOUSE and WATKINS (1938) and by CHATTERJEE (1943, 1944) for *Mahonia* spp. and other species of *Berberis*. For the determination of the histological distribution of berberine, the tissue sections are treated with 2% HCl or 2% HNO_3 and the development of crystals of berberine hydrochloride (needles) or nitrate (needles) within the cells noted. The colour reaction with chlorine water and HCl is useful for the study of the occurrence of berberine in individual cells and precipitation of the alkaloid as berberine-acetone has also proved of value. In the very young roots of *B. darwinii* the alkaloid is most abundant in the ring of parenchymatous tissue surrounding the vascular bundle and the walls of the xylem elements are impregnated with berberine. Fibrous roots show a similar distribution with the ring of tissue surrounding the vascular bundles richest in berberine. Further penetration of the xylem has taken place and many cells of the cortex now show the presence of alkaloid. In the main and subsidiary woody roots berberine is distributed throughout the section with the exception of the phloem which remains free. The distribution in the main stem is similar to that of the main root but a fall in concentration is apparent, more especially in the bundle sheaths and the pith. In young shoots berberine cannot be detected in the walls of the xylem elements and in the bundle sheaths. Etiolated shoots show a high concentration of berberine and in sections the inner cortex is engorged, the wood fibres show impregnation, while the vessels and phloem are free from alkaloid. Berberine accumulates from year to year in the tissues which take no part in metabolic activity. Ringing experiments on plants grown in light show an accumulation of berberine in the tissues above the ring. The histological distribution of umbellatine and neprotine in the tissues of *M. nepalensis* (CHATTER-JEE, 1944) follows closely the pattern above-described for *B. darwinii*, with the highest concentration of umbellatine (0.48%) and neprotine (0.02%) occuring in the roots of old plants during the winter months. The rhizome of *Hydrastis canadensis* yields from 1.5 to 4% of hydrastine and from 1 to 5% of berberine and the root of *Mahonia aquifolium* 2.3% of berberine and 2.8% of oxyacanthine. *Coptis teeta* may contain as much as 8.5% of berberine in the root and *Coptis trifolia* contains 4% of berberine. In the wood of *Archangelisia flava* the berberine content may reach a value of 4.8%.

III. The Isolation of the Alkaloids.

1. Berberine.

From the tissues of plants in which berberine is the principal alkaloid the following procedure is recommended. The powdered plant material is exhaustively extracted with ethanol in a continuous extractor of the SOXHLET type. The ethanol is removed by evaporation and the residue dissolved in hot water. At this stage resins separate out and are removed by filtration of the hot solution. The solution is now treated with excess HCl and berberine hydrochloride allowed to crystallize. The hydrochloride after separation is purified by dissolving in ethanol and precipitating with ether. In order to obtain pure berberine, the hydrochloride is dissolved in hot water and the solution made alkaline with a few drops of 10% NaOH. Acetone is now added and the solution diluted with an equal volume of water. The precipitate of berberine-acetone is allowed to settle overnight at low temperature. After separation and washing with ice-cold water the dry berberine-acetone is dissolved in a mixture of absolute ethanol and chloroform (10:1), the solution heated to boiling and allowed to cool. Crystals of pure berberine separate out.

From the tissues of *Mahonia aquifolium* which contain oxyacanthine and berbamine in addition to berberine the method adopted for the isolation and separation of the alkaloids is as follows. The dried and finely powdered plant material is exhaustively extracted with ethanol in a continuous extractor. The ethanol is removed by evaporation and the residue taken up with hot dilute HCl. Resins which separate are removed by filtration and the filtrate when cool is made alkaline with dilute ammonia or Na_2CO_3 solution. Oxyacanthine and berbamine are precipitated if the concentration is high and are removed by centrifuging the solution. If a precipitate is not formed the ammoniacal solution is thoroughly shaken out with several portions of ether in which oxyacanthine and berbamine are appreciably soluble whereas berberine is insoluble and remains in the aqueous phase. The precipitate (if formed) containing oxyacanthine and berbamine is likewise exhaustively extracted with ether, the solvent removed and oxyacanthine separated from berbamine by conversion to the sulphate in the following manner. The residue is dissolved in dilute H_2SO_4 and the solution saturated with Na_2SO_4 at 50°. On cooling, oxyacanthine sulphate separates as a resinous mass which is removed, washed with a little water and dissolved in the minimum amount of the same solvent. From this solution oxyacanthine base is precipitated on addition of dilute Na_2CO_3 solution. The precipitate is removed, dried, and extracted with ether in a SOXHLET. The ether is removed by evaporation and the resinous residue dissolved in a mixture of equal parts of ethanol and dilute H_2SO_4. On concentration, the solution deposits white crystals of oxyacanthine sulphate. The free base is recovered by treating the solution of the sulphate with ammonia when a white flocculent precipitate is formed which is separated and dried. The mother-liquors remaining after the first separation of oxyacanthine as the sulphate contain berbamine. This base is precipitated as the nitrate on adding sodium nitrate to the mother-liquors. After some time in the cold, the nitrate is removed and the free base liberated by treatment of the nitrate (in solution) with ammonia. Berbamine is purified by re-crystallization from ethanol. Berberine, which remains in the aqueous phase after removal of oxyacanthine and berbamine with ether, is best recovered by making the solution strongly alkaline with 10% NaOH and shaking out with ether. After removal of the ether, the residue is dissolved in hot dilute HCl to convert the berberine to the hydrochloride which crystallizes out on cooling the solution.

2. Hydrastine.

The rhizome of *Hydrastis canadensis* contains berberine and canadine in addition to hydrastine. Berberine being a stronger base than hydrastine or canadine is not removed by ether from an ammoniacal solution of the mixed alkaloids. Separation of berberine can therefore be accomplished by taking advantage of this property. Finely powdered *Hydrastis* rhizome is uniformly moistened with 10% ammonia solution and allowed to stand for 30 minutes. The mass is transferred to the thimble of a Soxhlet and exhaustively extracted with ether or benzene. The ethereal or benzene solution is transferred to a separating funnel and shaken out with several small portions of dilute HCl. The combined acid solutions are filtered, made alkaline with ammonia and shaken out with ether. On removal of the solvent a residue is obtained which consists largely of hydrastine with a small proportion of canadine. The latter base has been separated from hydrastine by dissolving the crude mixture in dilute H_2SO_4, adding HNO_3, and allowing the solution to stand for a few days. The precipitate which forms is removed, dissolved in hot water and re-precipitated with ammonia. The treatment with dilute H_2SO_4 followed by the addition of HNO_3 is repeated until white crystals of canadine nitrate are obtained. The liberated base is recrystallized from boiling light petroleum and finally from ethanol. Crude hydrastine is purified by repeated recrystallization from hot ethanol.

If isolation of berberine is desired, this base can be removed from the original rhizome material from which the remaining alkaloids have been extracted by extraction with ethanol and subsequent precipitation as the hydrochloride.

3. Umbellatine.

For the isolation of this alkaloid from *Mahonia nepalensis*, CHATTERJEE (1944) extracted 1 kg. of the powdered air-dried root with ethanol in a SOXHLET extractor. The solvent was removed, the brown residue treated with 500 ml. of water and filtered. To the filtrate sufficient HCl was added to give a 1% solution and the solution was allowed to stand overnight. Crystals of the yellow hydrochloride of umbellatine separated out and were well washed. The brown-coloured filtrate was concentrated on a water bath to 400 ml., filtered and allowed to stand overnight when a second crop of crystals appeared. The yellow hydrochloride was recrystallized from hot water containing a few drops of HCl and then further purified by repeated crystallization from hot ethanol. The free base was obtained by dissolving the hydrochloride in hot water and making alkaline with a dilute solution of Na_2CO_3. On cooling yellow silky needles of the free alkaloid were obtained.

IV. Qualitative Tests for the Alkaloids.

Berberine gives precipitates with most of the alkaloid reagents including picrolonic acid and ammonium reineckate in acid solution. Concentrated HNO_3 gives with berberine a reddish-brown colour. Concentrated H_2SO_4 dissolves berberine with an orange-yellow colour changing to olive-green on warming. On adding a crystal of potassium dichromate to a solution of the alkaloid in H_2SO_4 a black colour changing to brownish-violet is produced. With FRÖHDE's reagent berberine gives a brown or green colour and with MANDELIN's reagent a violet colour. The reaction of berberine with chlorine or bromine water in the presence of HCl provides a sensitive test which is highly specific. To a solution of berberine or one of its salts a drop of chlorine or bromine water is added followed by a few drops of HCl. A bright red colour is produced with very small amounts of berberine. Excess of chlorine or bromine water must be avoided and one drop of a light amber coloured solution of bromine in water is quite sufficient to give the test. The colour is stable for at least 12 hours. LABAT's test for hydrastine is positive for berberine if gallic acid is used. Many of the salts of berberine, e. g. the chromate, hydriodide, picrate, aurichloride and platinichloride are very little soluble in water and may be used for characterization.

Hydrastine gives precipitates with a solution of iodine in KI, with MAYER's reagent, bismuth potassium iodide, picric acid, picrolonic acid and ammonium reineckate. Concentrated H_2SO_4 dissolves hydrastine with a faint yellow colour changing to violet on warming. In the presence of traces of an oxidizing agent, e. g. $K_2Cr_2O_7$, H_2SO_4 gives with hydrastine a green to brown colour. With H_2SO_4 and MnO_2 the colour is at first orange which changes to a deep carmine-red. FRÖHDE's reagent gives with hydrastine a sage-green colour which changes to brown and then fades. MANDELIN's reagent dissolves hydrastine with a rose-red colour changing to orange-red and then fading. A solution of hydrastine in dilute H_2SO_4 does not show fluorescence but on addition of a few drops of dilute $KMnO_4$ solution an intense blue fluorescence develops due to the formation of hydrastinine. LABAT (1909) has described the following sensitive test for hydrastine. A crystal of hydrastine is dissolved in 0.1 ml. of 10% H_2SO_4 in an evaporating dish and 2 ml. of pure H_2SO_4 are added with mixing. 0.1 ml. of a 5% solution of gallic acid in ethanol is added and the mixture warmed on a water bath. An intense emerald-green colour is produced which slowly turns blue. On addition of an excess of glacial acetic acid the colour changes to violet. If catechol or guaiacol is substituted for gallic acid the colour is red. Berberine gives the test when gallic acid is used but not with catechol or guaiacol.

Canadine dissolves in concentrated H_2SO_4 yielding a yellow solution which slowly turns red. With FRÖHDE's reagent canadine gives a transient olive-green colour changing to reddish-brown. Canadine liberates iodine from an iodic acid solution and reduces a mixed solution of potassium ferricyanide and ferric chloride to Prussian blue.

Oxyacanthine dissolves in concentrated H_2SO_4 giving a colourless solution. On addition of a drop of nitric acid a yellow colour develops changing to reddish-brown and reddish-yellow. Fröhde's reagent gives a violet colour which turns dirty green and finally yellow. With Mandelin's reagent oxyacanthine gives a faint violet colour changing to reddish-violet. Bromine water gives a yellow precipitate with oxyacanthine. A solution of potassium ferricyanide and ferric chloride is reduced by oxyacanthine which also liberates iodine from a solution of iodic acid. The colour reactions of *berbamine* are similar to those of oxyacanthine.

V. The Quantitative Estimation of Berberine and Hydrastine.

For the *gravimetric estimation of berberine* in plant tissues, the method of Richter (1914) is suitable. This method is based upon the fact that berberine in the crystalline form (ammonium form) is soluble in water but insoluble in ether whereas the presumed aldehyde form, produced by the action of strong alkali, is soluble in ether and almost insoluble in water. 5 g. of powdered material is extracted with ethanol in a Soxhlet apparatus until the extract siphoning over is colourless. The ethanolic extract is evaporated to dryness on a water bath and the residue taken up with hot water, which precipitates much resinous matter. The resin is removed by filtration and the filtrate reduced in volume by evaporation to about 20 ml. 10 ml. of 15% w/v NaOH solution are added and the alkaline solution shaken with 110 ml. of ether for 10—15 minutes, 5 ml. of a 2% solution of gum tragacanth being added prior to shaking. 55 ml., representing one half of the original volume of ether added, is removed for subsequent estimation. This portion of the ether extract is treated with a saturated solution of picrolonic acid in ether until precipitation is complete. The precipitate is washed with ether to remove excess picrolonic acid, dried at 105° and weighed. The weight multiplied by the factor 0.561 gives the equivalent amount of berberine. The accuracy of the method is within $\pm$ 5%. The above method is best suited for the estimation of berberine in those species of *Berberis* which contain berberine as the major alkaloid, associated alkaloids either being absent or present in small amounts only. The method can however be used for the estimation of berberine in *Mahonia aquifolium, Berberis vulgaris* and *Hydrastis canadensis* provided that the associated alkaloids are first removed.

Quantitative separation of berberine from the associated alkaloids can best be achieved by the adoption of the adsorption procedure described below. Extraction of an ammoniacal solution of the mixed alkaloids with ether removes a small amount of berberine in addition to the associated alkaloids and although this method is suitable for purposes of isolation it cannot be relied upon to give entirely satisfactory results in quantitative work. A method for the *estimation of hydrastine* in the rhizomes of *Hydrastis canadensis* has been described by Mendez, Kirch and Voigt (1949). This method is based on the separation of hydrastine from berberine by adsorption on a Florisil column with selective elution of the hydrastine with ammoniacal ethanol. The separated hydrastine is determined by spectro-photometric estimation of the colour produced by oxidation with potassium dichromate and H_2SO_4 and berberine is estimated spectrophotometrically by measuring its absorption at a wavelength of 430 mμ. The powdered rhizome material is extracted with ether after treatment with dilute ammonia and the extract purified according to the official methods for assay of hydrastine or by the method of van Arkel and Meijst (1952) described below. Pyrex adsorption tubes (19 mm. diam. and 20 cm. in height) with standard tapered glass joint are used for the column. With the aid of a two-hole rubber stopper, the tube is inserted at the top of a small percolator provided with a glass stopcock which permits the removal of quantities of collected liquid. Through the second hole of the stopper is attached a controlled vacuum line. A 250 ml. separating funnel is placed in the upper end of the adsorption tube and a rubber tube attached to a glass tube inserted in the rubber stopper and provided with a pinch clamp allows the adjustment of the level of the solution from which the alkaloids are adsorbed. A pad of glass wool is placed at the bottom of the adsorption tube and Florisil (60—100 mesh, Floridin Co., Warren, Pa.) is then packed to a height of approx. 10—13 cm. with sufficient pressure to ensure a uniform column. A layer of glass wool is placed at the top of the column and 100 ml. of 10% ammonia solution is drawn through the column with the aid of suction in order to activate the Florisil. Excess of ammonia is removed by the use of 200 ml. of distilled water. The column is now ready to receive the alkaloid solution. A measured quantity (25 ml.) of the ethereal extract of *Hydrastis* rhizome is transferred to a 250 ml. beaker and 10 ml. of approx. 0.5 N H_2SO_4 added. The ether is completely removed by evaporation on a water bath, the resulting acid solution filtered through glass wool and passed through the column. The beaker is rinsed repeatedly with several portions of distilled water totalling 300 ml. and each rinsing is passed through the column. 50 ml. of ether is now passed through the column in order to displace most of the water. Suction is continued during this procedure until the water-ether interface has descended to within 2 cm. from the bottom of the column. The ether is drained off and retained for further use. 100 ml. of ammoniacal ethanol are then passed through the column and this eluate is combined with the ether fraction in a 250 ml. beaker. The solution in the beaker is now evaporated to dryness on a water bath and the residue dissolved in 10 ml. of approx. 0.1 N H_2SO_4 by heating on a water bath. The solution is transferred to a 50 ml. volumetric flask, washing the beaker with a further 10 ml. of 0.1 N H_2SO_4, and the volume made up to 50 ml. with distilled water. Aliquots of this solution (1—2 ml.) representing from 1 to 3 mg. of hydrastine are transferred to a 25 ml. volumetric flask, 4 ml. of 1% potassium dichromate solution are added, followed by 5 ml. of conc. H_2SO_4. The mixture is well shaken and allowed to stand for 5 minutes to develop the maximum colour. Distilled water is carefully added to within a few ml. of the mark and the final volume is adjusted at 20°. The transmittance of the solution is read in a spectrophotometer at 600 mμ against a reagent blank. The corresponding values for hydrastine are obtained from a standard curve prepared by the use of solutions of pure hydrastine. Any berberine which may be present in the eluate aliquot is estimated directly by means of spectrophotometric readings at 430 mμ with reference to a standard curve. Results obtained by the above procedure are lower than those obtained by use of the official methods for hydrastine. The difference represents berberine and other ether-soluble extractives included in the official assay. The average value for hydrastine in samples of rhizome was found to be 2.51% as compared with 2.67% by the official method.

A fluorometric method for the quantitative estimation of hydrastine has been described by BROCHMANN-HANSSEN and EVERS (1951). This method is based on the oxidation of hydrastine to hydrastinine which shows a very strong fluorescence. The method is claimed to be specific and to give accurate results. A standard reference curve for hydrastine is first of all prepared by dissolving 50 mg. of the pure alkaloid in 0.1 N H_2SO_4 and adjusting the volume to 250 ml. Standard solutions containing from 0 to 6 mg./100 ml. are prepared by diluting the stock solution with 0.1 N H_2SO_4. 5 ml. of the hydrastine solution and 10 ml. of nitric acid, sp. gr. 1.30, are mixed in a 100 ml. volumetric flask and placed in a constant temperature water bath at 50° for 30 minutes. After cooling under running water, the flask is filled up to the mark with distilled water at 20°. 10 ml. of this solution is diluted in another flask to 100 ml. with distilled water and the intensity of the fluorescence measured. The authors used a Coleman Photofluorometer (Model 12A) equipped with a B-1 primary filter and a PC-2 secondary filter. The sensitivity of the instrument was adjusted to a reading of 50 with a solution of quinine sulphate in 0.1 N H_2SO_4 (3.6 $\times$ 10^{-4} mg./ml.) and all readings were taken at 25°. Hydrastinine is slowly decomposed when exposed to ultraviolet light and therefore only the first few readings are used. The method is applied for the determination of hydrastine in the rhizome of *Hydrastis* as follows:— 0.4 g. of finely powdered rhizome is placed in a 100 ml. Erlenmeyer flask and 50 ml. of peroxide-free ether added. The flask is stoppered and shaken for 2 minutes. 1 ml. of dilute (10%) ammonia is added and the mixture shaken mechanically for 30 minutes. At the end of this time 25 ml. of the ether extract are transferred to a 100 ml. volumetric flask and the ether removed almost completely on a steam bath. 10 ml. of fresh ether are added and evaporated to approx. 2 ml. Finally 10—15 ml. of 0.1 N H_2SO_4 are added and the ether removed completely. After cooling, the volume is adjusted to 100 ml. with 0.1 N H_2SO_4. 5 ml. of this solution is oxidized with nitric acid as previously described and the intensity of the fluorescence measured. From the standard curve the content of hydrastine can be read ($Y \times 10^{-4}$ mg./ml. corresponds to Y per cent). The presence of small amounts of berberine and canadine do not interfere in the determination.

A simple gravimetric method for the determination of hydrastine in the rhizome of *Hydrastis* has been described by van Arkel and Meijst (1952). The powdered rhizome (3 g.) is shaken with 30 g. of ether and 2.5 ml. of 10% ammonia solution for 30 minutes. 30 g. of light petroleum (b. p. 65—80°) and 3 ml. of water are added and the mixture shaken vigorously. After allowing to settle, the ether solution is removed and shaken with successive portions of 10, 5, 5 and 5 ml. of 1% HCl, the combined acid solutions filtered, made alkaline with 5 ml. of 10% ammonia and shaken out once more with ether for 3 minutes. 3 g. of gum tragacanth are added and the mixture shaken vigorously. After separation of the phases, 20 g. of the ether solution are transferred to a tared flask and the solvent distilled off to approx. 5 ml. The evaporation is completed at room temperature and the residue dried at 100° and weighed as hydrastine.

Ether extraction of *Hydrastis* rhizome in the presence of ammonia removes canadine in addition to hydrastine and a little berberine. The gravimetric method described above estimates *canadine* as well as hydrastine. If the amount of hydrastine present is determined by the fluorometric method the amount of canadine present can be determined by difference provided that the extract contains negligible amounts of berberine. The fluorometric method gives values for hydrastine lower than those obtained by the colorimetric method of Mendez, Kirch and Voigt (1949). It is probable that canadine is included in the values for hydrastine given by the latter method as this alkaloid is a very weak base and would probably be eluted from a Florisil column along with hydrastine. Moreover canadine gives a colour with H_2SO_4 and dichromate somewhat similar to that given by hydrastine. For the estimation of berberine in the rhizome of *Hydrastis* the extraction procedure of Richter (1914) is carried out followed by spectrophotometric determination at 430 mμ.

The alkaloids of *Hydrastis canadensis* have been separated by Erbring and Wulf (1952) using the method of two-dimensional *paper chromatography*. The solvents employed were firstly 2:4:6-collidine saturated with water and secondly 94% ethanol. Berberine and hydrastinine were detected by the fluorescence of the spots. Hydrastine oxidizes on the paper to hydrastinine after a period of two days. Canadine cannot be detected at low concentration.

E. The Alkaloids of the Calabar Bean.

Calabar beans are the seeds of *Physostigma venenosum* BALF. *(Leguminosae)*, a woody climbing plant indigenous to West Africa. The seeds contain the alkaloids physostigmine (eserine), eseramine, geneserine and physovenine. Physostigmine is the major alkaloid and is present in the cotyledons to the extent of 0.04—0.3%, the average content being about 0.15%. The seeds of *Physostigma cylindrospermum* HOLMES, *Mucuna urens* DE CANDOLLE *(Leguminosae)* and *Entada scandens* BENTH. *(Leguminosae)* are also said to contain physostigmine.

I. The Isolation of Physostigmine from Calabar Beans.

SALWAY (1911) extracted finely powdered material (122.7 kg.) by continuous percolation with hot ethanol. The ethanol was evaporated off and the residue mixed with water, when resins and fatty material separated out and were removed. The aqueous solution was made alkaline with an excess of sodium carbonate and repeatedly extracted with ether until all the alkaloid was removed. The ethereal extracts were combined, concentrated to small bulk, and shaken out with successive portions of 5% sulphuric acid until the extract became just acid in reaction. The neutral solution of the alkaloid sulphate was treated with an excess of a saturated solution of sodium salicylate. Physostigmine salicylate was precipitated as an almost colourless crystalline powder which was washed with water and dried over sulphuric acid. A further quantity of the salicylate was obtained from the filtrate by making alkaline with sodium carbonate, extracting with ether and repeating the above process. The free base was recovered from the salicylate by treatment with aqueous sodium carbonate followed by. immediate extraction with ether.

Physostigmine ($C_{15}H_{21}O_2N_3$) crystallizes in two forms, an unstable form, m. p. 86—87° and a stable form, m. p. 105—106° (SALWAY, 1911). The colourless crystals are slightly soluble in water, giving a solution which reacts alkaline, easily soluble in ethanol, ether, chloroform and benzene, but insoluble in light petroleum. Solutions are laevorotatory, the values for $[\alpha]_D$ being —75.8° in chloroform and —120° in benzene. Physostigmine behaves as a mono acidic tertiary base and the following salts have been used for characterization. The sulphate, $B_2 \cdot H_2SO_4$, is a white microcrystalline deliquescent powder, m. p. 145°, very soluble in water and ethanol. The salicylate, $B \cdot C_7H_6O_3$, occurs in colourless crystals, m. p. 186—187°, and is soluble in 150 parts of cold water and in 12 parts of cold ethanol. The aurichloride, $B \cdot 2\,HAuCl_4$, occurs in yellow leaflets, m. p. 163—165°, and the picrate, $B \cdot C_6H_3O_7N_3$, in yellow feathery needles, m. p. 114°. The picrate is sparingly soluble in hot water but readily soluble in ethanol. Physostigmine is very poisonous and is a powerful myotic.

Eseramine ($C_{16}H_{25}O_3N_4$) and physovenine ($C_{14}H_{18}O_3N_2$) were isolated in very small quantities from Calabar beans by SALWAY (1911). Eseramine forms colourless needles m. p. 245° (dec.) and is easily soluble in hot ethanol but sparingly soluble in ether, chloroform and benzene.

Physovenine crystallizes in small colourless prisms from a mixture of benzene and light petroleum, and melts at 123°. It is a weak base insoluble in water and light petroleum, moderately soluble in ether and easily soluble in ethanol, benzene and chloroform.

Geneserine was isolated from Calabar beans by POLONOVSKI and NITZBERG (1915) and forms orthogonal crystals, m. p. 128—129°. It is a weak base and yields a salicylate, m. p. 89—90° and a picrate, m. p. 175°.

II. Methods for the Detection and Estimation of Physostigmine.

1. Qualitative Tests.

Aqueous solutions of physostigmine are strongly alkaline and on exposure to air, oxidation takes place with the formation of rubreserine $C_{13}H_{16}O_2N_2$. According to SALWAY (1912a) rubreserine crystallizes from water in deep red needles and is readily soluble in water, ethanol and chloroform, imparting a blood red colour to the solution. The aqueous solution of rubreserine is decolorized by reducing agents such as sodium thiosulphate and H_2S. Traces of alkali facilitate the oxidation of physostigmine and if a drop of dilute NaOH solution is added to a solution of a salt of physostigmine the red colour of rubreserine begins to develop immediately. Numerous other colour tests have been proposed for the detection of physostigmine and the following may be given as useful for purposes of identification. A yellowish-red colour is produced when a small quantity of physostigmine or one of its salts is warmed with 1 ml. of strong ammonia solution. On evaporation to dryness on a steam bath a bluish residue (eserine blue) remains which is soluble in ethanol forming a blue solution. Addition of acetic acid to this solution gives a strong red fluorescence which is intensified on dilution with water. The residue of eserine blue is also soluble in H_2SO_4 forming a green solution changing to red on the addition of ethanol. If a solution of physostigmine or one of its salts is heated to 100° and a few drops of strong nitric acid are added, an orange coloured solution results. On addition of a slight excess of NaOH to the cooled solution an intense violet colour develops (Saul, 1887). A modification of this test has been proposed by Ferreira da Silva (1893). A small quantity of physostigmine is heated in a porcelain basin on a steam bath with a drop or two of fuming nitric acid, when a yellow solution is formed which on evaporation to dryness forms a green residue (chloreserine), readily soluble in ethanol to give a green solution. When added to a solution of potassium ferricyanide and ferric chloride, physostigmine gives a blue colour. Ellis (1943) has made a chemical study of the coloured breakdown products of physostigmine and references to other colour reactions are given by Manske and Holmes (1952).

2. The Estimation of Physostigmine.

A volumetric method for the assay of Calabar beans has been described by Salway (1912b). 20 g. of finely powdered Calabar beans are mixed with 200 ml. of ether and 10 ml. of 10% sodium carbonate solution and the mixture is shaken at intervals during a period of 4 hours. 100 ml. of ether (representing 10 g. of the drug) are withdrawn into a separating funnel and 20 ml. of 0.1 N H_2SO_4 are added and the mixture shaken. The acid layer is run off into another separating funnel and the operation repeated with two further portions of 10 ml. of 0.1 N H_2SO_4. The combined acid liquids are made alkaline with sodium carbonate solution and shaken out with 200 ml. of ether in 20 ml. portions. The combined ether extracts are washed once with water and the water removed. The ether is evaporated off and the residue dissolved in 5 ml. of 0.1 N H_2SO_4 and titrated with 0.1 N NaOH (or borax) using methyl red as indicator. 1 ml. of 0.1 N H_2SO_4 is equivalent to 0.02752 g. alkaloid calculated as physostigmine. During recent years, more accurate methods for the estimation of physostigmine have been described by Ellis, Plachte and Straus (1943); Hellberg (1947, 1948) and Ehrlés (1948). These methods are based on the conversion of physostigmine to rubreserine which can be determined colorimetrically. In addition, Ellis et al. (1943) have devised an enzymic method for the estimation of very small amounts of physostigmine which depends on the specific inhibition by physostigmine of the hydrolysis of acetyl choline, and the dissociation constant of the cholinesterase — physostigmine complex. Physostigmine is easily hydrolysed by the action of alkalies with the formation of eseroline ($C_{13}H_{18}O_2N_2$), methylamine and carbon dioxide. Eseroline undergoes oxidation with the formation of the red compound rubreserine. The colorimetric method of Ellis et al. (1943) is suitable for the estimation of physostigmine in the Calabar bean. This method will determine 0.05—0.5 mg. of physostigmine contained in 1 ml. of solution, but larger amounts may be determined if they are diluted so that the concentration falls within the stated range. The lower limit of the method is 0.01 mg. but accuracy in this range is not good. Finely powdered Calabar bean (1—5 g.) is extracted with ether according to the method of SALWAY (1912b) described above. For the smaller weight of material extracted, proportionately smaller amounts of solvent are employed and the final alkaloid is taken up with 2—5 ml. of 0.1 N H_2SO_4 or such volume as will give a concentration of physostigmine of 0.05—0.5 mg./ml. 1 ml. of the acid extract is pipetted into a tube, neutralized with 10% NaOH and one drop is added in excess.

The tube is shaken frequently during a period of 30 minutes to permit full development of colour. At the end of this period one drop of HCl solution (conc. HCl diluted with an equal volume of distilled water) is added and the mixture again neutralized by the addition of small amounts of solid sodium bicarbonate. The volume is made up exactly to 5 ml. with $M/15$ phosphate buffer pH 7.0 to prevent changes in pH, which may affect the colour of the rubreserine. The colour is measured in a suitable colorimeter or absorptiometer and the concentration of physostigmine read from a calibration curve. Rubreserine in water shows an absorption maximum at about 300 mμ. HELLBERG (1949) describes a method for the estimation of physostigmine based on the hydrolytic production of methylamine which is distilled off and determined colorimetrically with ninhydrin. The rubreserine method however is simpler and more rapid and gives results with an accuracy of $\pm$ 2%.

F. The Cinchona Alkaloids.

The barks of various species of *Cinchona* and *Remijia (Rubiaceae)* yield a number of closely related alkaloids belonging to the quinoline group. The numerous species of *Cinchona* are indigenous to the Andes of S. America where they grow at an elevation of 4,000—7,000 feet. Old trees may reach a height of 100 feet. The great importance of quinine as an anti-malarial drug led to the extensive cultivation of *Cinchona* spp. in most tropical regions including Java, India, Ceylon and Tanganyika. The chief species yielding commercially important barks are: *C. officinalis* L. (pale or crown bark), *C. calisaya* WEDD. (yellow bark), *C. ledgeriana* MOENS, *C. succirubra* PAVON, *C. lancifolia* MUTIS, *C. nitida* RUIZ et PAVON, *C. micrantha* RUIZ et PAVON and *C. peruviana* HOWARD. Cuprea bark is obtained from *Remijia pedunculata* FLUECK and false cuprea bark from *R. purdieana* WEDD. The alkaloids are present in the cortical tissues of the stems and roots of plants of all ages and have also been isolated in small quantity from the seeds of some species.

For the cultivation of plants and collection of bark the following systems have been adopted. (a) The Uprooting System: Plants raised from seed are planted a few feet apart and after a period of six or seven years of growth the trees are thinned out and the bark collected from roots and stems. Each subsequent year the process of thinning is continued until all the plants have been removed. The ground is then re-planted. This system is the most economical and most widely employed. (b) The Coppicing System: After seven or eight years of growth the trees are cut down and the bark removed. Adventitious shoots are allowed to grow from the stool and a bushy growth encouraged. Fine quills of bark are removed from the shoots and eventually the tree is dug out and the root bark, which is rich in alkaloids, removed. (c) The Mossing System: This system was introduced in 1863 to obviate the necessity for felling old trees. Narrow strips of bark are removed and the wound covered with moss or other protective material. Cambial activity leads to the production of new bark which is rich in alkaloids. This system proved too expensive to maintain and has now been discontinued.

The major alkaloids of *Cinchona* spp. are quinine, quinidine, cinchonine and cinchonidine, but over twenty other alkaloids have been isolated and characterized. A list of the alkaloids of *Cinchona* and *Remijia* species with their chief sources is given in Table 5.

Table 5. *The Alkaloids of Cinchona and Remijia spp.*

Alkaloid	Formula	Chief Source	$[\alpha]_D$	m. p.
Paricine	$C_{16}H_{18}ON_2$	*C. succirubra* (Darjeeling)	± 0	136°
Cinchonine	$C_{19}H_{22}ON_2$	esp. *C. micrantha* R. and P.	$+263.7°$ at 15° c = $M/40$; $N/10$ H_2SO_4	264°
Cinchonidine		esp. *C. succirubra*	$-178°$ at 15° c = $M/40$; $N/10$ H_2SO_4	204.5°
Cinchotine	$C_{19}H_{24}ON_2$	In commercial cinchonine	$+225.8°$ at 15° c = $M/40$; $N/10$ H_2SO_4	267°
Cinchamidine		In commercial cinchonidine	$-95.8°$ c=0.377 (ethanol)	232°
Cinchonamine		*Remijia purdieana*	$+121.1°$ (ethanol)	194°
Cupreine	$C_{19}H_{22}O_2N_2$	*Remijia pedunculata*	$-175.5°$ at 17° (dry: ethanol)	198°
Quinamine	$C_{19}H_{24}O_2N_2$	esp. *C. ledgeriana* and *C. succirubra*	$+104.5°$ (ethanol)	185—186°
Conquinamine		esp. *C. ledgeriana* and *C. succirubra*	$+204.6°$ (ethanol)	121°
Quinine	$C_{20}H_{24}O_2N_2$	*C. calisaya* etc.	$-284.5°$ at 15° (dry base, c = $M/40$ in $N/10$ H_2SO_4)	173—175°
Quinidine		esp. *C. calisaya* and *C. pitayensis*	$+334.2°$ at 15° dry base, c = $M/40$ in $N/10$ H_2SO_4	173.5° (anhydr.
Quinicine		By heating quinine acid sulphate	$+38.6°$ ($CHCl_3$)	—
epi-Quinine		In residues after separation of the major alkaloids	$+43.3°$ at 22° c = 0.95	—
epi-Quinidine			$+102.4°$ at 19° c = 0.8648, ethanol	113°
Hydroquinine	$C_{20}H_{26}O_2N_2$	In commercial quinine	$-235.7°$ at 15° = $M/40$, $N/10$ H_2SO_4	173.5° (dry)
Hydroquinidine		In commercial quinidine	$+299°$ at 15°, dry base c = $M/40$, $N/10$ H_2SO_4	169.5°
Chairamine	$C_{22}H_{26}O_4N_2$	*Remijia purdieana*	$+100°$ (ethanol)	233° (dry)
Conchairamine			$+68.4$ at 15° (ethanol)	120° (dry)
Chairamidine			$+7.3°$ at 15° (ethanol)	126—128°
Conchairamidine			$-60°$ at 15° (ethanol)	114° (dry)
Concusconine	$C_{23}H_{26}O_4N_2$	*Remijia purdieana*	$+40.8°$ at 15°	144°
Cusconine		*Cinchona pelletieriana*, WEDD.	$-54.3°$ (ethanol)	110° (dry)
Aricine		*C. pelletieriana* WEDD.	$-58.3°$ (ethanol)	188°
Dicinchonine	$C_{38}H_{44}O_2N_4$	*C. succirubra*	$+65.6°$ at 15° (ethanol)	40°
Diconquinine	$C_{40}H_{46}O_3N_4$	*C. calisaja* etc.	—	—

I. Properties.

Quinine, the most important of the cinchona alkaloids, crystallizes from boiling benzene in needles containing solvent. On exposure to air, the solvent is lost and a micro-crystalline powder remains. On evaporation of an ethereal solution, the base separates as a gelatinous or resinoid mass. On precipitation of one of its salts by alkali a white precipitate of quinine is formed which is at first amorphous but gradually passes into the crystalline efflorescent trihydrate, B · 3 H_2O, m. p. 57°. The trihydrate, on exposure to air, changes to B · 2 H_2O, and when dried over H_2SO_4, another molecule of water is lost. At 125° it becomes anhydrous. The anhydrous base is sparingly soluble in water (1 in 1960 at 15°) but easily soluble in ethanol (1 in 0.6 at 25°), ether (1 in 4.5 at 25°), chloroform (1 in 1.9 at 25°) and boiling benzene. It is sparingly soluble in light petroleum. Quinine is a strong diacidic base and forms both neutral and acid salts, the most important of which will now be described. Commercial quinine sulphate, B_2 · H_2SO_4 · 8 H_2O (or 7.5 H_2O) is prepared by neutralizing the base with dilute H_2SO_4 and recrystallizing from boiling water. Colourless needles are formed which effloresce in dry air to yield the stable dihydrate, m. p. 205°. The heptahydrate is sparingly soluble in water (1 in 720 at 25°, 1 in 30 at 100°), soluble in ethanol (1 in 86 at 25°, 1 in 9 at 60°), easily soluble in a mixture (2:1) of chloroform and dry ethanol. It is insoluble in ether, chloroform and light petroleum, but very soluble in dilute acids. The heptahydrate has $[\alpha]_D^{15°}$ $-166.36°$ (ethanol). The acid sulphate, B · H_2SO_4 · 7 H_2O, forms transparent ortho-rhombic crystals, m. p. 160° (dec.), $[\alpha]_D^{15°}$ $-216.1°$ (dry salt, c = $M/40$, H_2O). The acid salt is soluble in water (1 in 8.5 at 25°), in ethanol (1 in 18 at 25°) and sparingly soluble in ether (1 in 1770 at 25°) or chloroform (1 in 920 at 25°). The aqueous solution is acid to litmus and shows a strong fluorescence.

Quinine hydrochloride, $B \cdot HCl \cdot 2 H_2O$, forms prisms which effloresce in dry air. When heated at 100° it becomes anhydrous and melts at 159—160°, $[\alpha]_D^{17°}$ —133.7°(H_2O). This salt is soluble in water (1 in 18 at 25°), in ethanol (1 in 0.6 at 25°) and in chloroform (1 in 0.8 at 25°). It is sparingly soluble in ether (1 in 240 at 25°). The aqueous solution is neutral and is not fluorescent. The acid hydrochloride, $B \cdot 2 HCl$, is obtained by treating a solution of the acid sulphate with $BaCl_2$ and crystallizes in needles. It is soluble in water (1 in 0.75), in ethanol (1 in 5), in chloroform (1 in 7) and insoluble in ether. Quinine salicylate, $2 (B \cdot C_6H_4(OH)COOH) \cdot H_2O$, forms colourless needles, m. p. 187° (dec.) which slowly turn pink on exposure to air. It is soluble in water (1 in 77 at 25°) in ethanol (1 in 11 at 25°) and in chloroform (1 in 37 at 25°).

Quinine oxalate, $B_2 \cdot H_2C_2O_4 \cdot 6 H_2O$, forms needles which are very sparingly soluble in water (1 in 1030 at 10°) in contradistinction to the oxalates of the other major alkaloids which are much more soluble.

Quinidine is isomeric with quinine and crystallizes in prisms from boiling ethanol with 2.5 mols. of solvent or from dry ethanol with 1 mol. of solvent. From ether it crystallizes with $^1/_2$ mol. of solvent in trimetric tablets and from boiling water in leaflets with 1.5 H_2O. When free from solvents of crystallization it melts at 173.5° and is sparingly soluble in water (1 in 2000 at 15°) and in chloroform, benzene, and light petroleum. It is easily soluble in ethanol (1 in 26 of 80% ethanol at 20°) and ether (1 in 35 at 10°). Quinidine is a diacidic base and forms two series of salts the most important of which are the sulphates, the hydrochlorides and the neutral hydriodide. The neutral sulphate, $B_2 \cdot H_2SO_4 \cdot 2 H_2O$, crystallizes from hot water in needles or prisms and is soluble in water (1 in 98 at 15°) and in hot water (1 in 7 at 100°). It is soluble in chloroform (1 in 20) but almost insoluble in ether. $[\alpha]_D +184 \cdot 17°(CHCl_3)$. The acid sulphate, $B \cdot H_2SO_4 \cdot 4 H_2O$, forms fine needles easily soluble in water (1 in 8.7 at 10°): $[\alpha]_D^{15°} +247.8°$ (c $= M/10, H_2O$). The neutral hydrochloride, $B \cdot HCl \cdot H_2O$, forms prisms easily soluble in hot water or ethanol, less soluble in cold water (1 in 62.5 at 10°). This salt has m. p. 258—259° (dry, dec.) and $[\alpha]_D^{20°} +200°$ (H_2O). The acid hydrochloride, $B \cdot 2 HCl \cdot H_2O$, forms prisms, easily soluble in ethanol and sparingly soluble in water or chloroform. The neutral hydriodide, $B \cdot HI$, is formed when KI is added to a neutral aqueous solution of a salt of quinidine and is sparingly soluble in water (1 in 1250 at 15°). Quinidine is readily isolated in the form of its hydriodide.

Cinchonine crystallizes from ethanol in rhombic prisms and is sparingly soluble in water (1 in 3810 at 10°, 1 in 3670 at 20°), more soluble in ethanol (sp. gr. 0.852, 1 in 140 at 10°) and in boiling ethanol (1 in 28), in ether (sp. gr. 0.7305, 1 in 371 at 10°), or amyl alcohol (1 in 109 at 15°). Cinchonine is a diacidic base and gives two series of salts. The sulphate, $B \cdot H_2SO_4 \cdot 2 H_2O$, forms rhombic crystals, m. p. 200° (dry, dec.) and is fairly soluble in water (1 in 65.5 at 13°) and readily soluble in 80% ethanol (1 in 5.8 at 11°), $[\alpha]_D +133°$ ($CHCl_3$). The acid sulphate, $B \cdot H_2SO_4 \cdot 4 H_2O$, forms octahedral crystals easily soluble in water and ethanol. The hydrochloride, $B \cdot HCl \cdot 2 H_2O$, has m. p. 217—218° (dry) and forms monoclinic crystals easily soluble in water and ethanol.

Cinchonidine crystallizes in short prisms and is sparingly soluble in water (1 in 5263 at 11.5°), soluble in ethanol (1 in 16.3 of 97% ethanol at 13°), in ether (1 in 188 of ether sp. gr. 0.72 at 15°). In dry ether it is much less soluble (1 in 1053 at 11.5°). Cinchonidine is not fluorescent in H_2SO_4 solution. It is a diacidic base and yields two series of salts. The neutral sulphate, $B_2 \cdot H_2SO_4$ m. p. 205° (dry, dec.) forms a number of hydrates. From cold water it forms monoclinic prisms with 6 H_2O and from hot water it crystallizes with 3 H_2O. It is soluble in water (1 in 63 at 25°) and in ethanol (1 in 72 at 25°). The acid sulphate, $B \cdot H_2SO_4 \cdot 5 H_2O$, has $[\alpha]_D^{15°}$ —133.6° (dry salt, c $= M/40, H_2O$). The neutral hydrochloride, $B \cdot HCl \cdot H_2O$, has m. p. 242° (dry), $[\alpha]_D$ —117.6° (dry salt, c $= 1.214, H_2O$) and forms monoclinic double pyramids, or prisms with 2 H_2O from saturated aqueous solution. The dry salt is soluble in water (1 in 38.5 at 10°), in ether (1 in 325 at 10°) and is easily soluble in chloroform. The acid hydrochloride, $B \cdot 2 HCl \cdot H_2O$, forms monoclinic prisms easily soluble in water or ethanol. The dihydrobromide, $B \cdot 2 HBr \cdot 2 H_2O$, has $[\alpha]_D^{15°}$ —114.3° (dry salt, c $= M/40, H_2O$). The tartrate, $B_2 \cdot H_2C_4H_4O_6 \cdot 2 H_2O$, gives a crystalline precipitate sparingly soluble in water (1 in 1265 at 10°) and almost insoluble in a strong solution of Rochelle salt. This salt is useful in the separation and estimation of cinchonidine.

Quinicine occurs as a yellow alkaline varnish. It forms crystalline salts, the most important of which are the oxalate, $B_2 \cdot H_2C_2O_4 \cdot 9 H_2O$, which crystallizes in small prisms, m. p. 166 to 167°, $[\alpha]_D^{20°} + 24°$. This salt is sparingly soluble in water (1 in 120) and can be purified by recrystallization from ethanol or chloroform. The hydrochloride, $B \cdot HCl$, crystallizes in leaflets, m. p. 180—182°, $[\alpha]_D +16.26°$. The benzoyl derivative crystallizes from benzene or methanol and melts at 113—114°, $[\alpha]_D^{22°} +36°$ (c $= 2$, ethanol).

epi-Quinine is an oil, $[\alpha]_D^{20°} + 43.3°$ (c $= 0.95$ ethanol) and forms a dihydrochloride which melts at 196° (dec.) and has $[\alpha]_D +33.3°$ (c $= 0.796$).

epi-Quinidine crystallizes from ether in leaflets, m. p. 113°, $[\alpha]_D^{19°}$ +102.4° (c = 0.8648, ethanol). The dihydrochloride melts at 195—196° (dec.) and has $[\alpha]_D^{20°}$ +45.5° (c = 0.8012, ethanol). epi-Quinidine is fluorescent in dilute H_2SO_4 solution.

Hydroquinine crystallizes from ether or benzene in needles which are easily soluble in organic solvents. The solution in dilute H_2SO_4 is fluorescent. The neutral sulphate, $B_2 \cdot H_2SO_4 \cdot 6\,H_2O$, crystallizes in prisms and is moderately soluble in water (1 in 348 at 15°), but sparingly soluble in chloroform, $[\alpha]_D^{15°}$ —204.6° (dry salt, c = $M/40$, $N/10\ H_2SO_4$). The acid sulphate, $B \cdot H_2SO_4 \cdot 3\,H_2O$, forms long needles, soluble in water or ethanol and the hydrochloride, $B \cdot HCl \cdot 2\,H_2O$, melts at 206—208° (dry) and has $[\alpha]_D^{21°}$ —123.9° (H_2O). A crystalline benzoyl derivative is formed which melts at 102—107°. Hydroquinine when precipitated from a salt by alkali crystallizes with $2\,H_2O$.

Hydroquinidine occurs in the form of prismatic needles containing $2.5\,H_2O$. In physical and chemical properties it closely resembles quinidine. The sulphate, $B_2 \cdot H_2SO_4 \cdot 2\,H_2O$, forms white needles, soluble in 92.3 parts of water at 16°. The hydrochloride, $B \cdot HCl$, occurs in prismatic plates m. p. 273—274° (dec.), $[\alpha]_D^{26°}$ +183.9°. This salt is easily soluble in water, methanol or chloroform.

Cinchotine crystallizes from ethanol in prisms and is soluble in ethanol (1 in 136 parts of 90% ethanol at 15°) and in 534 parts of ether at 20°. It is practically insoluble in cold water but dissolves in hot water. The neutral sulphate, $B_2 \cdot H_2SO_4 \cdot 11\,H_2O$, forms fine needles, m. p. 194.8°—195° (dry) and is soluble in water (1 in 37.6 at 12°). The hydrochloride, $B \cdot HCl \cdot 2\,H_2O$, melts at 220—221° (dry, dec.) and has $[\alpha]_D$ +155—159° (H_2O). The platinichloride, $B_2 \cdot 2\,HCl \cdot PtCl_4$ occurs in orange-red needles sparingly soluble in water.

Cinchamidine (hydrocinchonidine) crystallizes from dilute ethanol in six-sided leaflets and is practically insoluble in water, ether or chloroform The neutral sulphate, $B_2 \cdot H_2SO_4 \cdot 7\,H_2O$, forms needles soluble in water (1 in 57 at 10°) and the acid sulphate, $B \cdot H_2SO_4 \cdot 5\,H_2O$, forms leaflets which are slightly soluble in water, $[\alpha]_D^{15°}$ —106.8° (dry salt, c = $M/40$, H_2O). The hydrochloride, $B \cdot HCl \cdot 2\,H_2O$, forms short prisms readily soluble in water or ethanol, and melts at 202—203° (dry), $[\alpha]_D$ —89.4° (c = 1.19, H_2O).

Cinchonamine crystallizes in needles and is soluble in ethanol, ether, chloroform or benzene but only sparingly soluble in water and light petroleum. It is a diacidic base and forms two series of salts. The sulphate, $B_2 \cdot H_2SO_4$, forms colourless prisms very soluble in water but less soluble in ethanol. The nitrate, $B \cdot HNO_3$, forms short prisms, m. p. 196°, practically insoluble in cold water. The hydrochloride, $B \cdot HCl$, forms laminae or, with $1H_2O$, cubical crystals. Cinchonamine forms a methiodide which melts at 208°.

Cupreine crystallizes from ethanol in the anhydrous form but from ether in concentric prisms with $2\,H_2O$. It is sparingly soluble in ether or chloroform but dissolves readily in ethanol or solutions of caustic alkalies but not in ammonia. Cupreine is a diacidic base: the neutral sulphate, $B_2 \cdot H_2SO_4$, forms small needles, m. p. 257°, and is slightly soluble in water (1 in 813 at 17°). The acid sulphate, $B \cdot H_2SO_4 \cdot H_2O$ crystallizes in prisms and is soluble in 73.4 parts of water at 17°. Cupreine on methylation is converted into quinine.

Quinamine crystallizes in delicate anhydrous needles and is practically insoluble in cold water, more readily soluble in boiling water. It is easily soluble in hot ethanol and other boiling organic solvents. Solutions of quinamine in dilute H_2SO_4 are not fluorescent. The hydrochloride, $B \cdot HCl \cdot H_2O$, has m. p. 166—167°, $[\alpha]_D$ +102.8° (H_2O) and the hydriodide has m. p. 224°, $[\alpha]_D$ +84.8° (ethanol). The nitrate melts at 186—188° and has $[\alpha]_D$ +94.9° (H_2O). The picrate crystallizes from boiling water in needles, m. p. 175—176°.

Conquinamine forms triclinic crystals, m. p. 121°. It is easily soluble in ether, chloroform and benzene, $[\alpha]_D^{15°}$ +204.6° (ethanol). The sulphate, $B_2 \cdot H_2SO_4 \cdot x\,H_2O$, is soluble in water and the aurichloride occurs as a yellow precipitate turning to purple on standing.

Paricine crystallizes with $0.5\,H_2O$ and has m. p. 136°, $[\alpha]_D$ ±0.

Aricine forms prisms, m. p. 188°, $[\alpha]_D$ —58.3° (ethanol). It forms a sparingly soluble oxalate and gives a green colour with nitric acid.

Cusconine, $B \cdot 2\,H_2O$, forms leaflets, m. p. 110° (dry), $[\alpha]_D$ —54.3° (ethanol).

Concusconine, $B \cdot H_2O$ crystallizes in monoclinic needles, m. p. 144° (re-melts at 206 to 208°), $[\alpha]_D^{15°}$ +40.8°. It is insoluble in water, very sparingly soluble in cold ethanol and easily soluble in ether or chloroform. This base dissolves in conc. H_2SO_4 giving a bluish-green colour which changes to olive-green on warming.

Chairamine, $B \cdot H_2O$, forms needles or prisms, m. p. 233° (dry), $[\alpha]_D$ +100° (ethanol). A solution of the base in acetic acid gives with nitric acid a dark green colour.

Conchairamine, $B \cdot H_2O$, forms prisms, m. p. 120° (dry), $[\alpha]_D^{15°}$ +68.4° ethanol). This base gives the same colour reaction as described above for chairamine. Conc. H_2SO_4 gives a brown colour turning to green.

Chairamidine, $B \cdot H_2O$, is amorphous and has m. p. 126—128°, $[\alpha]_D^{15°}$ +7.3 (ethanol). With conc. H_2SO_4 it gives a green colour. A solution of the base in HCl is coloured green on addition of HNO_3.

Conchairamidine, $B \cdot H_2O$, is crystalline and melts at 114° (dry), $[\alpha]_D^{15°}$ —60° (ethanol). Conc. H_2SO_4 gives a deep green colour.

II. The Distribution of Alkaloids in Cinchona Bark.

The alkaloids exist chiefly in combination with quinic acid, $C_7H_{12}O_6$, and to a lesser extent with cinchotannic acid, $C_{14}H_{16}O_9$. The content of quinic acid varies from 5% to 8% and cinchotannic acid from 3% to 4%. The distribution of the crystallizable and the amorphous alkaloids in the bark of *Cinchona* species grown in India is given in Table 6.

Table 6. *The Distribution of the Major Alkaloids of Cinchona spp.*

Source of bark		Quinine	Cinchon-idine	Quinidine	Cinchonine	Amorphous alkaloids	Total alkaloids
Species	Type of bark						
C. officinalis	Natural	2.77	0.39	0.16	1.57	0.5	5.39
C. officinalis	Mossed	3.40	0.45	0.20	1.50	0.62	6.17
C. officinalis	Renewed	4.21	0.65	0.22	0.85	0.70	6.63
C. ledgeriana	Natural	5.49	0.82	—	1.33	0.88	8.52
C. ledgeriana	Branch	2.21	1.07	—	0.49	0.50	4.27
C. succirubra	Natural	1.91	1.14	—	2.11	0.88	6.04
C. succirubra	Mossed	1.69	1.68	—	2.03	0.98	6.34
C. succirubra	Renewed	1.84	1.25	—	1.48	0.71	5.28
C. succirubra	Branch	1.38	1.59	—	2.28	1.16	6.41
C. succirubra	Root	1.24	1.43	0.41	0.77	1.27	5.12
C. calisaya	Natural	1.21	2.13	—	2.32	0.29	5.95
C. calisaya	Branch	0.59	1.93	—	0.73	0.48	3.73
C. robusta	Natural	1.43	1.58	—	2.08	0.31	5.40
C. robusta	Mossed	1.92	0.77	—	3.16	0.35	6.20
C. robusta	Renewed	4.40	0.51	—	2.54	1.65	9.10
C. robusta	Branch	1.64	1.17	—	2.17	0.50	6.02
C. micrantha	Natural	—	1.92	—	—	0.40	2.32
C. micrantha	Renewed	Trace	1.12	—	2.45	1.02	4.54
C. micrantha	Branch	—	1.60	—	—	0.45	2.05
C. pitayensis	Natural	2.34	1.93	1.10	0.56	0.39	6.32
C. pitayensis	Mossed	3.81	1.91	0.63	0.95	0.37	7.67
C. pitayensis	Renewed	2.50	2.33	0.78	0.52	0.55	6.68

Note: *C. robusta* = *C. succiruba* × *C. officinalis*. The figures represent per cent dry weight.

The absolute and relative proportions of the alkaloids in *Cinchona* bark varies with cultural and climatic conditions and also in relation to the position on the tree from which the bark is removed. In general, the content of total alkaloids is greater in the root bark and at the base of the tree than in the upper parts of the trunk. BIRCH and DOUGHTY (1948) have made a study of the distribution and interrelationships of the alkaloids in the bark of *C. ledgeriana*. These workers found that increases in bark weight and total alkaloids are associated with significant increases in quinine, cinchonidine and cinchonine at the expense of the amorphous alkaloids. Local thickening of bark at the forks of trees and thickening due to injury leads to a high alkaloid content with low ratios for quinine: cinchonidine, and quinine: cinchonine. Cambial activity is an important factor in the determination of the alkaloid content of the bark, the degree of conversion of the amorphous to the crystallizable alkaloids, and the proportions in which the crystallizable alkaloids occur.

III. The Isolation of the Crystallizable Alkaloids.

The general principles of the method for the large-scale isolation of the alkaloids are as follows. The finely powdered bark is intimately mixed with about 30% of its weight of $Ca(OH)_2$ and the mixture made into a stiff paste with a sufficient quantity of a 5% solution of NaOH. The mass is transferred to a continuous extractor and exhausted with light petroleum (b. p. 80—100°) or crude benzene. The alkaloids are extracted from the organic solvent by agitating with successive quantities of dilute H_2SO_4 at 60° until extraction is complete. The combined acid extracts while still warm are adjusted to pH 6.5 with dilute NaOH and the solution allowed to cool. Crystals of the neutral sulphate of quinine separate out. The sulphate is freed from cinchonidine and cinchonine by repeated re-crystallization from hot water using activated charcoal for removal of colouring matter. The base is prepared from the sulphate by dissolving the latter in dilute H_2SO_4 (heating if necessary to effect solution) and with continual stirring adding dilute ammonia or Na_2CO_3 solution until the solution is slightly alkaline. The amorphous precipitate gradually becomes micro-crystalline and is washed free from sodium or ammonium salts and dried at low temperature. The mother-liquors from which the quinine sulphate has been separated contain the remaining alkaloids which are precipitated by making the solution slightly alkaline and removed by filtration. The precipitated alkaloids are dried at low temperature and extracted repeatedly with small quantities of ether which removes the more soluble cinchonidine and quinidine leaving behind the more sparingly soluble cinchonine. The residue containing cinchonine is purified by repeated re-crystallization from ethanol, using charcoal to decolourize the solution. The ethereal solution containing cinchonidine and quinidine is agitated with dilute HCl and the alkaloids converted to their hydrochlorides. The aqueous acid solution is treated with a strong solution of sodium tartrate or sodium potassium tartrate and allowed to stand. Cinchonidine is precipitated as the tartrate which is separated, washed, and treated with aqueous alkali to liberated the base. The base is filtered off, dried *in vacuo* at 40—45° and washed with absolute ether to remove traces of quinine. The cinchonidine is finally purified by re-crystallization from ethanol.

Quinidine remains dissolved in the liquors from which the cinchonidine tartrate has been separated and is precipitated as the sparingly soluble hydriodide by the addition of potassium iodide solution. After standing for a few hours the precipitate is removed and treated with a solution of dilute ammonia to liberate the base. The base is dissolved in acetic acid, the solution decolourized with charcoal and the quinidine re-precipitated with ammonia and re-crystallized from boiling ethanol. The amorphous alkaloids remain in the mother-liquors from the separation of quinine sulphate, and, after removal of the other crystallizable alkaloids, can be obtained in crude mixture by extraction with ether. Henry (1949) gives a list of references dealing with the separation of the amorphous alkaloids and of the alkaloids of *Cinchona pelletieriana* and *Remijia purdieana*.

Cupreine is isolated from the bark of *Remijia pedunculata* by a process of extraction identical with that for quinine. The total alkaloids are removed from the organic solvent by shaking with dilute H_2SO_4 and the acid solution is exactly neutralized with dilute NaOH. As quinine is present in the bark, the sulphate of a compound of quinine and cupreine (homoquinine) crystallizes out. This is dissolved in dilute H_2SO_4, and excess of dilute NaOH added to make the solution alkaline and the quinine removed by shaking with ether. Cupreine which remains in the alkaline solution is obtained by neutralizing the solution with dilute H_2SO_4

when cupreine sulphate crystallizes out. The base is obtained by dissolving the salt in dilute H_2SO_4, making slightly alkaline with ammonia and crystallizing the precipitate from ethanol.

IV. The Qualitative Reactions of the Cinchona Alkaloids.

The Cinchona bases give precipitates with most of the alkaloid reagents, MAYER's reagent, bismuth potassium iodide and picric acid being especially useful precipitants. Ammonium reineckate in slightly acid solution gives a bulky precipitate with very small quantities of the alkaloids.

Quinine in solution in dilute H_2SO_4 or acetic acid shows a strong blue fluorescence which is very marked in extremely dilute solutions. The hydrochloride and other halogen compounds do not give fluorescence in solution.

1. The Thalleioquin Test.

This test is best carried out by dissolving a small amount of quinine in dilute acetic acid and adding 1—2 ml. of water. Two or three drops of bromine water are added and the mixture shaken. On the addition of a drop of strong ammonia an emerald green colour is produced. Careful neutralization with dilute H_2SO_4 results in the formation of a bluish colour which changes to red on the addition of more acid.

2. The Erythroquinine Test.

To a solution of quinine in dilute acetic acid, one or two drops of bromine water are added followed by a drop of a 10% solution of potassium ferrocyanide. On the addition of a drop of strong ammonia the solution turns red. If shaken immediately with 1 ml. of chloroform, the colour is taken up by the chloroform. Modifications of these tests to distinguish between quinine and quinidine are described below. The thalleioquin test is positive for quinidine, quinicine, hydroquinine and cupreine and negative for the remaining alkaloids. Quinidine resembles quinine in its solubility in ether and in the fluorescence of its solution in dilute H_2SO_4. It differs from quinine in being strongly dextro-rotatory and in forming a sparingly soluble hydriodide. DAVID (1949) distinguishes quinine from quinidine by the following tests:— 0.01 g. of the alkaloid salt is well mixed with 0.25 ml. of bromine water in a basin and the solution transferred to a test tube with the aid of 1 ml. of water. 1 ml. of chloroform is added and the mixture allowed to stand for 3 minutes. The solution is now treated with 1 drop of 10% potassium ferrocyanide solution, shaken and 3 ml. of 5 N NaOH added with shaking. In the presence of quinidine the chloroform assumes a red colour. With quinine the chloroform remains colourless. The two alkaloids are also distinguished by a solubility test. 0.01 g. of the salt and 2 ml. of water are heated to boiling and treated with 1 drop of 10% potassium ferrocyanide solution. In the presence of quinidine a yellowish precipitate of the ferrocyanide separates out but with quinine the solution remains clear at first but later becomes milky and minute spherical particles appear on the walls of the tube. Cinchonine is easily distinguished from quinine by its limited solubility in ether and by its dextro-rotatory power. Its solution in dilute H_2SO_4 does not show fluorescence and it does not give the thalleioquin reaction. The neutral tartrate of cinchonine is soluble in water but the neutral tartrate of quinine is very little soluble. Cinchonidine resembles quinine in the sign of its optical activity and in the sparing solubility of the tartrate in water. It is distinguished from quinine by its lower specific rotation, by its non-fluorescence and a negative thalleioquin test. SUMERFORD (1947) uses picric acid as a means for distinguishing between the four alkaloids. 1 ml. quantities of 1% solutions of the respective alkaloids in glacial acetic acid are each treated with 1 ml. of a solution of picric acid (1 in 70) in glacial acetic acid and allowed to stand for approx. 5 minutes. A precipitate is formed only with cinchonidine, but on scratching the sides of the remaining tubes cinchonine picrate begins to crystallize and after a period of about 1 hour quinine picrate is precipitated. Quinidine picrate does not crystallize. Cinchonine and cinchonidine are further differentiated by the solubility of the benzoates. 1 ml. of a 1% aqueous solution of sodium benzoate gives an immediate crystalline precipitate with 1ml. of a solution of cinchonidine sulphate but no precipitate is given with a 1% solution of cinchonine sulphate.

Hydroquinine like quinine is laevo-rotatory, its solution in dilute H_2SO_4 is fluorescent and it gives the thalleioquin reaction. It differs from quinine in decolourizing a solution of $KMnO_4$ in 10% H_2SO_4 at 0° very slowly. Hydroquinidine also differs from quinidine in its stability towards $KMnO_4$. Cupreine in ethanolic solution gives a reddish-brown colour with ferric chloride and gives the thalleioquin reaction. It may be distinguished (and separated) from quinine by its solubility in NaOH solution.

Hamet (1941, 1945a) has shown that cinchonamine gives the typical colour reactions of yohimbine and also has an ultraviolet absorption spectrum characteristic of indoles. Quinamine and aricine also possess an indole structure and give colour reactions typical of this nucleus (Hamet, 1945b).

V. The Estimation of the Total Alkaloids of Cinchona Bark.

The estimation of the total alkaloids may be carried out either by gravimetric or titrimetric methods. The gravimetric method gives a greater degree of accuracy than the titrimetric method in view of the fact that the molecular weights of the cinchona alkaloids are high and small errors in titration lead to large errors in the calculated amount of total alkaloids. Birch and Doughty (1948) have adopted the following method for the gravimetric estimation of the total alkaloids of *C. ledgeriana*. Air-dried bark (20 g.) is mixed with CaO (12 g.), water (15 ml.) and 5 N NaOH (2 ml.). After allowing to stand overnight the mixture is transferred to a Soxhlet apparatus and extracted with benzene. The benzene extract is shaken out with successive portions of 5% w/v H_2SO_4. The combined acid extracts are made alkaline and the liberated alkaloids extracted with chloroform. The bulk of the chloroform is distilled off and the residual extract washed with chloroform into a small tared beaker. After the addition of a small quantity of ethanol the solution is evaporated and the residue dried to constant weight at 100°. For the estimation of the total alkaloids by the titrimetric method Loustalot and Pagan (1947) use the following procedure. 2 g. of finely powdered bark are weighed into a 100 ml. beaker and approx. 0.5 g. of finely powdered CaO is added. Sufficient water (7—10 ml.) is now added to make a smooth homogeneous paste. After allowing to stand for 10 minutes, the paste is transferred to a 200 ml. volumetric flask with 100—150 ml. of ethanol. The ethanolic bark suspension is shaken vigorously and allowed to stand for 1 hour or longer with occasional shaking. After this period it is made to volume with ethanol, shaken and filtered through a Whatman No. 5 paper (24 cm.). A watch glass is placed over the funnel and a plug of cotton wool in the mouth of the flask receiving the filtrate to minimize loss of ethanol by evaporation. 100 ml. of the ethanolic bark extract (representing 1 g. of bark) is pipetted into a 400 ml. beaker, 10 ml. of 0.1 N HCl and 100 ml. of water are added and the excess acid is titrated to pH 6.2 with 0.05 N NaOH using a pH meter. A blank determination is made by titrating the ethanol and acid. 1 ml. of 0.05 N NaOH is equivalent to 0.0155 g. of total alkaloids.

VI. The Determination of the Individual Crystallizable Alkaloids.

1. B. P. Method (1932).

The method of the British Pharmacopoeia (1932) for totaquina is suitable for the estimation of quinine, quinidine, cinchonine and cinchonidine in the dried residue from the total alkaloid extraction as carried out by BIRCH and DOUGHTY (1948). 2 g. of the dried residue is dissolved in a mixture of 20 ml. N H_2SO_4, 40 ml. of water and 40 ml. of 95% ethanol. The solution is heated to boiling and 0.1 N NaOH added (keeping the solution hot during the addition) until the solution is faintly alkaline to litmus. The solution is cooled and 0.1 N H_2SO_4 added dropwise until slightly acid to litmus. The acid solution is boiled for 1—2 minutes, cooled and if necessary again rendered slightly acid to litmus. The solution is boiled once more and filtered into a tared flask, the original flask and the filter being washed with boiling water. The combined acid filtrate and washings are evaporated down to about 120 g. 30 g. of powdered sodium potassium tartrate are added, the solution shaken until the tartrate has dissolved, and the flask set aside for 24 hours. After this period the precipitate is filtered off through a hardened filter and the flask and filter washed with 80 ml. of a 25% w/v solution of sodium potassium tartrate in water, added in portions. The filtrate and washings are reserved for the determination of cinchonine and quinidine. The filter and precipitate are returned to the flask, and 40 ml. of dilute NaOH solution added followed by 80 ml. of chloroform. The mixture is shaken to effect complete solution of the precipitate and extraction of the alkaloids. The chloroform solution is separated and the flask and aqueous solution washed with further portions of chloroform until complete extraction of the alkaloids is effected. The combined chloroform solutions are washed with a little water and the water rejected. The chloroform is removed by evaporation, 5 ml. of 95% ethanol added, evaporated and the residue dried at 100° and weighed. The weight represents the amount of quinine and cinchonidine present in the sample. The quinine in the mixture of the two alkaloids is determined by estimation of the methoxyl group by the method of ZEISEL. 0.2 g. of the mixed alkaloid is taken for estimation. 1% of methoxyl is equivalent to 10.45% of anhydrous quinine. Alternatively the quinine may be determined by the spectrophotometric methods described below. The filtrate and washings from the precipitated tartrates are run into a separating funnel containing 80 ml. of ether and 20 ml. of a dilute (5%) solution of NaOH and the funnel shaken. The aqueous phase is run off into a second separator and shaken with two further portions of 80 ml. of ether, each portion of ether being returned to the first separator. The combined ethereal solutions are washed with a little water and the alkaloids extracted by shaking with successive portions of 10, 10 and 5 ml. of N H_2SO_4 and finally with 10 ml. of water. The combined acid and aqueous solutions are run into a separating funnel containing 25 ml. of ether and 30 ml. of N NaOH, the mixture shaken and set aside for 1 hour. The precipitated cinchonine is collected on a tared filter using a little water to effect complete transference of the precipitate to the filter. The ether is separated from the filtrate and run through the precipitate on the filter. The aqueous solution is shaken once more with two portions of 25 ml. of ether and the ethereal washings used to wash the precipitate. The precipitate is dried at 100° and weighed. To the weight obtained, 0.08 g. is added to correct for loss of cinchonine due to its solubility in ether. The ethereal filtrate from the cinchonine is run into a separator and the filter flask washed out with a little water and ether, these washings being added to the liquid in the separator. The aqueous layer is separated and the quinidine extracted from the ethereal solution by shaking with successive quantities of 10, 10, 5 and 5 ml. of a 10% w/w aqueous solution of glacial acetic acid, which have previously been used to wash out any alkaloid remaining in the filter flask or the stem of the funnel. The mixed acetic acid solutions are heated to boiling, neutralized with dilute ammonia and 5 g. of potassium iodide added. The solution is allowed to stand overnight and the clear supernatant decanted through a filter. The precipitate is warmed with 5 ml. of 50% ethanol the liquid filtered off and the crystalline residue washed on the filter with 5 ml. of 50% ethanol. The residue is dried at 100° and weighed as quinidine hydriodide, 0.008 g. being added to the weight to correct for the solubility of quinidine hydriodide. 1 g. of quinidine hydriodide is equivalent to 0.717 g. of quinidine. The sum of the percentages of quinine, cinchonine, cinchonidine and quinidine gives the percentage of crystallizable alkaloids. The amount of amorphous alkaloids present is obtained by calculating the difference between total and crystallizable alkaloids.

2. Spectrophotometric Methods.

LOUSTALOT and PAGAN (1947) have applied the spectrophotometric method to the determination of quinine in *Cinchona* bark. These workers found that at a wavelength of 380 mμ cinchonine and cinchonidine had no effect on the absorption of light, which was proportional only to the amount of quinine and quinidine present. Since quinidine does not occur in all *Cinchona* barks and is, when present, seldom met with in large amount, the method to be described gives a reasonably accurate determination of the quinine present. The instrument used for this work was a Coleman double monochromator spectrophotometer, Model 105, utilising the 30 mμ slit in conjunction with a Coleman electrometer, Model 310. For the determination of quinine 25 ml. of the ethanolic extract, prepared as above for the estimation of total alkaloids by the titrimetric method, is pipetted into a 50 ml. Erlenmeyer flask and 25 mg. of Norite A added. The flask is shaken for 15—30 seconds and the solution filtered. 8 ml. of the clarified extract (representing 80 mg. of bark) is pipetted into a 100 ml. volumetric flask, 5 ml. of 0.1 N HCl added and the solution made to volume. The percentage transmittance is determined at 380 mμ. The percentage of quinine as sulphate is obtained from a standard curve prepared as follows. Five graded standard stock solutions of pure quinine sulphate in 0.1 N HCl are prepared so as to contain in 5 ml. aliquots 1.6, 3.2, 4.8, 6.4 and 8 mg. of anhydrous quinine sulphate. This series is equivalent to 2, 4, 6, 8 and 10% respectively of quinine sulphate, using an 80 mg. sample of bark. 5 ml. of each quinine sulphate stock solution are pipetted into 100 ml. volumetric flasks, 8 ml. of 95% ethanol added and the solutions made to volume with water. The percentage transmittance at 380 mμ is determined and a curve of log of percentage transmittance against concentration is plotted. A blank solution containing the acid and ethanol is used as a reference. For each set of determinations a new curve is prepared. GRANT and JONES (1950) have described a simple and rapid spectrophotometric method for the determination of the cinchona alkaloids.

These authors have shown that the composition of a mixture of quinine (or quinidine) and cinchonine (or cinchonidine) can be accurately calculated from the optical density at 316 mμ and 348 mμ of a solution of the alkaloids in 0.1 N HCl by the use of simultaneous equations. The mixture of the cinchona alkaloids from *Cinchona* bark behaves spectrophotometrically as a simple two-component system and the amount of quinine-type and cinchona-type alkaloids in such mixtures can be determined by the same procedure. This method has been used for the determination of the total alkaloid content of *Cinchona* bark (in addition to the amount of quinine-type alkaloids) in view of the fact that the amorphous alkaloids have absorption spectra very similar to that of either quinine or cinchonine. The spectrophotometric determination can also be applied for the accurate estimation of quinine and cinchonidine in the tartrate precipitate obtained in the above described method for totaquina. Also the quinidine content of the iodide and the cinchonine content of the crude cinchonine precipitate obtained in the same method can be determined spectrophotometrically. For this purpose the precipitates are dissolved in a measured volume of 0.1 N HCl and the optical density of the solutions at the appropriate wavelength (or wavelengths) determined. The method for the estimation of the total alkaloids of *Cinchona* bark is as follows: 1 g. of powdered bark is mixed with a few ml. of 10% NaOH solution in a 200 ml. flask, 100 ml. of benzene are added and the flask and contents weighed. The flask is connected to a reflux condenser and the benzene boiled gently until the alkaloids are completely extracted (3—6 hours). After cooling to room temperature any benzene lost is replaced. A 50 ml. aliquot of the benzene is transferred to a separating funnel and the alkaloids are extracted with several small portions of 0.1 N HCl. The combined acid extracts are boiled for 2—3 minutes to expel benzene, cooled and diluted to exactly 1000 ml. The optical density of the acid solution of the alkaloids is determined at 316 mμ and 348 mμ. From these data and the density of standard solutions of quinine and cinchonine (30 mg./litre in 0.1 N HCl) the alkaloid content is calculated from the simultaneous equations,

$$D_{316} = X D_q^{316} + Y D_c^{316}$$
$$D_{348} = X D_q^{348} + Y D_c^{348}$$

where X and Y are the amounts of quinine-type and cinchonine-type alkaloids; D_{316} and D_{348} are the optical densities of the sample solutions at 316 mμ and 348 mμ; and D_q^{316}, D_q^{348}, D_c^{316} and D_c^{348} are the optical densities per unit weight of the two standards at the respective wavelengths. In most cases the results by the spectrophotometric and gravimetric methods do not differ by more than 2% of the alkaloid content of the sample. The above data were obtained with a Beckmann Model DU spectrophotometer. Slit width approx. 2 mμ 1 cm. Corex cells. The spectrophotometric methods can usefully be applied to the determination of the alkaloid content of *Cinchona* seedlings, saplings and the bark of young trees.

3. The Colorimetric Estimation of the Cinchona Alkaloids.

PRUDHOMME (1940) showed that quinine, ephedrine, pilocarpine, physostigmine and atropine combine with eosin to form coloured compounds which can be extracted with chloroform. MARSHALL and ROGERS (1945) found the eosin colours too weak for practical use in the determination of quinine and suggested the use of bromothymol blue, which forms a similar coloured compound with quinine, quinidine, cinchonine and cinchonidine. The coloured compounds given by the other alkaloids with bromothymol blue were less intense than those produced by equal amounts of the cinchona alkaloids. MARSHALL and ROGERS (1945) have used bromothymol blue for the estimation of small amounts of cinchona alkaloids in blood, but the method could doubtless be adapted for use with plant material. An outline of the method using an acid solution of the alkaloids will now be given. To 5 ml. of a solution of the mixed alkaloids in 0.05 N HCl, 1 ml. of a 0.04% w/v bromothymol blue solution is added and the solution neutralized to approx. pH 7.0 (green colour) with N NaOH followed by 0.1 N NaOH as the end point is approached. The neutralized solution is transferred to a narrow-bore tube, 1 ml. of SORENSEN buffer (pH 7.0) added, the solution mixed and allowed to stand for 60 minutes. Exactly 1 ml. of benzene is added and the tube closed with a tightly fitting cork covered with cellophan. The tube is placed on a shaker for 40 minutes. Shaking should take place along the line of the long axis of the tube at 200 oscillations per minute. The solution is now centrifuged at moderate speed for 10 minutes and the benzene layer pipetted off into a 0.5 ml. colorimeter cell. The depth of the colour is measured in an Absorptiometer (Spekker Absorptiometer, Ilford Spectrum violet filter No. 601). A standard curve is prepared with known amounts of alkaloid in 0.05 N HCl. If so desired, 10 ml. of the acid solution of the alkaloids may be used with 5 ml. of bromothymol blue solution and 5 ml. of buffer. The final colour is then extracted with 5—15 ml. of benzene. Good recoveries were obtained when 1.5—18 mμ of cinchonidine were added to blood samples. Brodie and Udenfriend (1945) have used methyl orange in a similar manner for the colorimetric estimation of cinchona alkaloids in blood plasma and urine.

4. The Separation of the Cinchona Alkaloids by Paper Chromatography.

Lussman, Kirch and Webster (1951) have described a method for the separation of the alkaloids using cyclohexanol or cyclohexanone and dilute HCl as the solvent. The apparatus for the ascending method consists of a glass chamber 24 inches high and 6 inches in diameter covered with a glass plate. The paper (Whatman No. 1) is rolled into the form of a cylinder with the lower edge dipping into the solvent which is placed in the bottom of the chamber. The descending chromatograms are run in apparatus of the conventional pattern. Determinations are carried out under two sets of conditions. 1. The organic solvent (cyclohexanol or cyclohexanone) is saturated with water and dilute HCl placed in a separate container within the chamber. 2. The organic solvent is saturated with water-acid mixture. After separation of the phases, the solvent layer is placed in the bottom of the container and the aqueous-acid layer in a separate container within the chamber. The solutions are prepared from equal volumes of solvent and acid-water mixture. The latter solvent system gives the best results and the most efficient separations are obtained by use of HCl in dilution from N to 4 N. The fluorescence of quinine and quinidine provides a ready means of identification of these alkaloids on the paper. For cinchonine and cinchonidine the paper is sprayed with a solution of potassium iodoplatinate prepared as follows. 5 ml. of 5% platinum chloride in N HCl are mixed with 45 ml. of 10% KI solution and 100 ml. of distilled water. The solution is stored in a dark glass-stoppered bottle and diluted with 4 parts of distilled water before use. The cyclohexanol chromatogram requires 36—48 hours for development and the cyclohexanone 6—8 hours. The cyclohexanol system gives the better separation. The R_F values of the crystallizable alkaloids using these solvent system are given in Table 6a.

Table 6a. *R_F values of cinchona alkaloids using cyclohexanol saturated with water and dilute acid in separate container.*

HCl concentration	Quinine	Quinidine	Cinchonine	Cinchonidine
Approx. 2.5N	0.25±0.05	0.75±0.05	0.20±0.05	0.32±0.05
Approx. 3.5N	0.34±0.05	0.55±0.05	0.27±0.05	0.48±0.05

R_F values using cyclohexanol saturated with dilute acid.   Excess dilute acid in separate container.

HCl concentration	Quinine	Quinidine	Cinchonine	Cinchonidine
Approx. 2.5N	0.45±0.03	0.51±0.03	0.35±0.05	0.27±0.05
Approx. 3.5N	0.51±0.03	0.58±0.03	0.42±0.05	0.34±0.05
Approx. 2.5N	0.32±0.05	0.43±0.05	0.09±0.03	0.06±0.03
Approx. 3.5N	0.21±0.04	0.31±0.04	0.03±0.03	0.05±0.03

The colour reaction with potassium iodoplatinate is more sensitive with cinchonidine than with cinchonine. Approx. 25 μg. of cinchonidine are easily identified by use of the reagent but approx. 100 μg. of cinchonine is required before identification is possible.

5. Other Methods for the Determination of Total Alkaloids.

JINDRA and POHORSKY (1951) have used the method of ion exchange for the determination of the total alkaloids of *Cinchona*. A solution of the alkaloids extracted from 80 mg. of bark is passed through a column of Amberlite IR-4B resin and titrated with 0.01 N HCl using the antimony electrode. Good agreement with other methods of analysis can be obtained by this rapid and convenient means of estimation. KYKER, WEBB and ANDREWS (1941) have described a method for the estimation of total alkaloids by precipitation with silicotungstic acid.

G. The Curare Alkaloids.

Curare is the name given to plant extracts used as arrow poisons by tribes of Indians in the valleys of the Orinoco and Amazon. Three types of curare have appeared in commerce distinguished by the containers in which they are packed. The tube or bamboo curare was packed in bamboo tubes but is now exported in sealed tins. The calabash curare was packed in gourds and the pot curare in small earthenware pots. Curare is obtained as a thick black or dark brown syrup, as a resin or as a powder, depending on the water content of the drug. It contains tertiary and quaternary alkaloids, the distribution of which varies with the type of curare. The preparation known as *Radix pareirae bravae* contains alkaloids related to the curare alkaloids. During recent years the botanical sources and the chemistry of curare have been subjects of detailed investigations. It has now been established that the botanical source of tube curare is *Chondrodendron tomentosum* RUIZ and PAVON *(Menispermaceae)* while that of pot curare is other plants belonging to the *Menispermaceae*. The calabash or gourd curare appears to be derived from *Strychnos toxifera* SCHOMB. (Loganiaceae) and *Radix pareirae bravae* from *Chondrodendron platyphyllum* (St. Hil.) MIERS and *Ch. microphyllum* (Eichl.) MOLD. (KING, 1940). KING (1935, 1937, 1940, 1947, 1948, 1949a, 1949b) has investigated the sources and the chemistry of the curare alkaloids and the reader is referred to these papers for a full account of the work. DUTCHER (1946) also has examined the alkaloid fractions of *Ch. tomentosum*, and has isolated D-tubocurarine chloride, L-curine, D-isochondrodendrine, D-isochondrodendrine dimethyl ether and chondrocurine. KING (1947) however found L-curine and L-tubocurarine chloride in a specimen of *Ch. tomentosum*, which raises the question of the identity of the material of the two workers. KING has suggested that two closely allied species of *Chondrodendron* have been given the same specific name. The tertiary alkaloids of curare are of little pharmacological interest but D-tubocurarine is now extensively used to promote muscular relaxation during surgical operations.

I. Properties.

1. The Alkaloids of Tube Curare and Ch. tomentosum.

D-Tubocurarine Chloride, $C_{38}H_{44}O_6N_2Cl_2 \cdot 5\,H_2O$, is the quaternary active alkaloid of tube curare, $[\alpha]_{5461}^{20°}$ $+235°$ (c = 0.97) or for the anhydrous salt $+264.8°$, and for the ion $+295° \cdot (H_2O)$. On crystallization from water D-tubocurarine forms microscopic leaflets with a characteristic sheen, m. p. 274—275°. On addition of solid $NaHCO_3$ to a saturated solution of D-tubocurarine chloride, a granular amorphous powder consisting of the phenolic betaine separates out. On treatment with methanolic potash and methyl iodide, 0-methyl-tubocurarine iodide, $C_{40}H_{48}O_6N_2I_2 \cdot 3\,H_2O$, is formed. This derivative has m. p. 267° (dec.) and $[\alpha]_{5461}$ $+188.7$ for the anhydrous salt.

l-Curine (l-Bebeerine), $C_{36}H_{38}O_6N_2$, crystallizes with 1 mol. of benzene, m.p. 213° (solvent-free) or from methanol, m. p. 214°. The base free from methanol of crystallization gives in 0.1 N HCl $[\alpha]_{5461}$ —340°. The hydrochloride has m. p. 271—273°. l-Curine is a tertiary alkaloid and from the tertiary fraction the alkaloids described below also have been isolated.

d-iso-Chondrodendrine, $C_{36}H_{38}O_6N_2$, crystallizes from methanol in fine needles m. p. 316° (dec.), $[\alpha]_D^{22} +120°$ (0.1 N HCl). The sulphate, $B \cdot H_2SO_4 \cdot 15\,H_2O$ or $B \cdot H_2SO_4 \cdot 7\,H_2O$, has m. p. 291—292° (dry, dec.). The anhydrous sulphate has $[\alpha]_{5461}^{19} +158.9°$ (H_2O). The hydrochloride crystallizes in plates, m. p. 333°, the methiodide, $B \cdot 2\,CH_3I \cdot 8\,H_2O$, crystallizes from methanol in prisms, m. p. 287° (dec.) and the benzoyl derivative has m. p. 215°.

d-Chondrocurine, $C_{36}H_{38}O_6N_2$, has m. p. 232—234°, $\alpha]_D^{24} +200°$ (0.1 N HCl) or $+105°$ (pyridine). The hydrochloride, $B \cdot HCl$, has m. p. 280—282°, and the sulphate, $B \cdot H_2SO_4 \cdot 4H_2O$ has m. p. 263—265° (dec.), $[\alpha]_D^{24} +193°$ (dry salt; H_2O).

2. The Alkaloids of Pot Curare.

King (1937) has shown that pot curare derived from Menispermaceous plants of the Amazon contains in the "non-quaternary" fraction many phenolic alkaloids. Two of these alkaloids, protocuridine and neo-protocuridine have been characterized.

Protocuridine, $C_{36}H_{38}O_6N_2$, crystallizes in plates with the constitution $B \cdot O \cdot 5\,C_5H_5N$, when water is added dropwise to a solution of the base in boiling pyridine. The base crystallizing with pyridine melts at 295°. The hydrochloride, $B \cdot 2\,HCl \cdot 6\,H_2O$, crystallizes in octahedra, m. p. 295° (efferv.) $[\alpha]_{5461}^{20} +7.6°$ (dry salt, H_2O).

Neo-Protocuridine, $C_{36}H_{38}O_6N_2 \cdot 8\,H_2O$, crystallizes from hot water in diamond-shaped leaflets, m. p. 232° (dec.). The hydrochloride, $B \cdot 2\,HCl \cdot 6\,H_2O$, crystallizes from water in octahedra or prisms.

3. The Alkaloids of Calabash Curare.

Karrer and Schmid (1946, 1947) have studied the alkaloid fractions of calabash curare and have found C-curarine I to be the chief component. King (1949 b) has made an examination of the bark of *Strychnos toxifera* Schomb., the reputed source of calabash curare and has isolated twelve crystalline quaternary salts, toxiferines I—XII. These salts gave colour reactions with conc. H_2SO_4, HNO_3 and potassium dichromate in 50% H_2SO_4 suggesting a biogenetic relationship with tryptophane. Recently, Schmid and Karrer (1950) have used the method of paper chromatography for the separation and detection of the alkaloids of calabash curare. The chromatograms show that at least thirty quaternary alkaloids of similar constitution are present.

As limitations of space preclude the possibility of further attention being given to the alkaloids of pot and calabash curare, the isolation and estimation of the two major alkaloids of tube curare alone will be described.

II. The Isolation of l-Curine and d-Tubocurarine Chloride.

King (1935) describes the following procedure for the isolation of these alkaloids from tube curare. 25 g. of the drug is dissolved in 625 ml. of a warm aqueous solution of tartaric acid (1%), the solution cooled, filtered and made alkaline with 250 ml. of a saturated solution of sodium hydrogen carbonate. The alkaline solution is extracted four times with ether and crude curine (3 g.) recovered from the combined ether solutions. Further extractions with chloroform remove an additional amount of still cruder curine (1 g.). The aqueous solution is made acid to Congo-red paper with 2 N H_2SO_4 (100 ml.) and 800 ml. of 0.5 N basic lead acetate solution added. The precipitate and filtrate are separately decomposed with H_2S, the PbS removed and the filtrates concentrated to yield solutions A and B respectively. Solution B (420 ml.) is treated with H_2SO_4 (21 g.) followed by 70 ml. of 25% phosphotungstic acid in 5% (by weight) H_2SO_4. The collected precipitate is decomposed with hot saturated baryta solution in excess, and the alkaline filtrate from the $BaSO_4$ and Ba phosphotungstate exactly neutralized to

Congo-red paper with 50% H_2SO_4. The $BaSO_4$ is again removed and the filtrate subjected to exact double decomposition with $BaCl_2$ solution. The filtrate (250 ml.) which now contains the alkaloid as chloride, is saturated with finely powdered $HgCl_2$ (25 g.) the mixture being stirred mechanically during the addition. Crude tubocurarine mercurichloride is precipitated as a granular creamy solid. The precipitate is collected, decomposed with H_2S, mercuric sulphide removed and the filtrate evaporated to dryness. An amorphous gum (1.96 g.) remains. Solution A is treated likewise and yields 1.44 g. of an amorphous gum. The gummy residue from solution B is dissolved in a little water and kept at 0° when a white microcrystalline powder is deposited (1.38 g.). This serves as the source of an inoculum for the residue from solution A which is dissolved in a little water and cooled likewise. By recrystallization from water D-tubocurarine chloride (1.18 g.) is obtained in a pure state. The crude L-curine obtained from the ether and chloroform extracts is recrystallized from methanol.

KING (1948) has described a procedure for the isolation of the alkaloids of *Ch. tomentosum* using ammonium reineckate as the precipitant. The finely powdered stem material (1.08 kg.) is extracted with 15 litres of 1% tartaric acid solution and the extract concentrated to 3.6 litres. A portion of this solution (250 ml.) is treated with 100 ml. of 0.5 N basic lead acetate solution and the filtrate freed from lead with H_2S. The filtrate (500 ml.) is freed from H_2S and made alkaline with $NaHCO_3$ (25 g.). After standing for a few days the solid A (1.93 g.) is removed and the aqueous solution extracted repeatedly with chloroform which removes non-quaternary alkaloids B (0.18 g.). The extracted aqueous solution is made weakly acid with conc. HCl (27.5 ml.) and treated with 100 ml. of saturated ammonium reineckate solution. The precipitated reineckate is converted into chloride and the solution on concentration gives 0.37 g. of crystalline D-tubocurarine chloride. A small second crop is obtained from the mother-liquor. Solid A is treated with chloroform which removes non-quaternary bases. This chloroform solution is mixed with fraction B, the chloroform removed and the residue dissolved in methanol. On standing, crystals of D-isochondrodendrine separate out. The chloroform insoluble solid (1.01 g.) is neutralized with N HCl (5 ml.), treated with 40 ml. of saturated ammonium reineckate solution and the precipitate collected and converted to the chloride. The solution on concentration yields D-tubocurarine chloride (0.32 g.). The remainder (3350 ml.) of the original tartaric acid extract may be used for the isolation of L-bebeerine, D-chondrocurine, D-isochondrodendrine and tomentocurine.

III. Qualitative Reactions of D-Tubocurarine Chloride.

A saturated aqueous solution of D-tubocurarine chloride gives a weak green colour on the addition of ferric chloride solution. The colour is intensified by warming and cooling. Addition of Na_2CO_3 solution to a solution of D-tubocurarine chloride causes the formation of a yellow-brown precipitate. The alkaloid reduces ammoniacal silver on warming and gives amorphous precipitates with gold and platinum reagents, potassium iodide, potassium bromide, potassium thiocyanate, mercuric chloride and perchloric acid. The addition of solid $NaHCO_3$ to a saturated solution of D-tubocurarine chloride leads to the separation of the phenolic betaine in the form of a granular amorphous powder. It is readily soluble in the solution on warming and separates on cooling. It is soluble in caustic alkalies. The MILLON reaction is given by D-tubocurarine chloride, bebeerine and by protocuridine but not by neoprotocuridine (KING, 1937). A biological test which is useful for the rough determination of curare activity depends upon the paralytic effect of some of the curare alkaloids on the voluntary muscles of the frog.

IV. The Quantitative Estimation of D-Tubocurarine Chloride.

Foster and Turner (1947) have described tentative polarimetric and colorimetric methods for the estimation of the D-tubocurarine content and the total quaternary alkaloidal content respectively, of curare. For the assay of D-tubocurarine by the polarimetric method the following procedure has been used. 2 g. of curare are weighed into a 100 ml. Pyrex centrifuge tube and 50 ml. of 1% w/v H_2SO_4 added. The mixture is stirred until a uniform suspension results and allowed to stand for 30 minutes. The mixture is centrifuged at 2,000 r. p. m. for 2 minutes and the supernatant decanted into a 250 ml. beaker. The residue is re-suspended in 20 ml. of 1% w/v H_2SO_4, heated on a boiling water bath for 10 minutes, cooled and centrifuged. The remaining solid is finally washed with a little water. 2 g. of $NaHCO_3$ are added to the combined acid extracts and washings and the mixture centrifuged. The supernatant is decanted into a 250 ml. separating funnel and the solid washed with a little 1% w/v $NaHCO_3$ solution. The combined alkaline solutions and washings are extracted with three portions of 50 ml. of ether and then with three portions of 50 ml. of chloroform. Emulsions are broken by the addition of ethanol in the case of the ether extraction and by centrifuging in the case of the chloroform. Any solid matter which may have separated during centrifuging is returned to the alkaline extract. The tertiary alkaloids are removed from aqueous solution by the organic solvents but the quaternary alkaloids (D-tubocurarine fraction) are too soluble in water to be removed by organic solvents from aqueous solution. The alkaline extract is now acidified with 50% w/v H_2SO_4 and 10 ml. of 25% w/v phosphotungstic acid in 5% w/v H_2SO_4 added. The precipitate is centrifuged down, the supernatant rejected and the precipitate washed with water and the washings rejected. The phosphotungstate precipitate is decomposed with a hot solution of baryta containing 10 g. of $Ba(OH)_2 \cdot 8H_2O$ in 10 ml. of water. The mixture is stirred well, centrifuged, and the supernatant is transferred to a 250 ml. beaker. The solid residue is washed with water and the washings added to the alkaline extract. Excess barium is removed by the addition of 50% w/v H_2SO_4 to the extract until acid to Congo-red, and the precipitated $BaSO_4$ centrifuged off. The supernatant is transferred to a 200 ml. graduated glass-stoppered cylinder and the $BaSO_4$ precipitate washed well with water, centrifuged, and the washings added to the solution in the cylinder. The volume of the solution in the cylinder is adjusted to 200 ml. and the solution mixed thoroughly. The optical rotation, $[\alpha]_D^{20°}$ of the solution is determined using a 2 dcm. or 4 dcm. tube. The D-tubocurarine content of the sample calculated as $C_{38}H_{44}O_6N_2Cl_2 \cdot 5H_2O$, is given by

$$\frac{(\alpha)_D^{20°} \times 100}{190} \times 100 \text{ per cent.}$$

D-tubocurarine chloride has a specific rotation of $+190°$ in aqueous solution in sodium light. The accuracy of the estimation may be improved by determining the optical rotation in mercury green light under which conditions the hydrated salt has $[\alpha]_{5461}^{20°} +235°$. Although a high degree of accuracy is not claimed for the method (about $\pm 5\%$) it is useful in giving a good estimate of the activity of a sample of curare.

The colorimetric assay of the total quaternary alkaloidal content of curare is carried out by using the Folin-Ciocalteu phenol reagent to produce a blue colour with the alkaloids. This reagent is prepared as follows. 100 g. of sodium tungstate, $Na_2WO_4 \cdot 2H_2O$, and 25 g. of sodium molybdate, $Na_2MoO_4 \cdot 2H_2O$, are dissolved in 700 ml. of water in a 1500 ml. flask connected by a ground-glass joint to a reflux condenser. 50 ml. of syrupy H_3PO_4 and 100 ml. of conc. HCl are added and the solution refluxed gently for 10 hours. After cooling, 150 g. of lithium sulphate, Li_2SO_4, 50 ml. of water and 4—6 drops of bromine are added and the solution boiled without condenser for 15 minutes to remove excess bromine. The solution is cooled, transferred to a 1 litre flask, diluted to volume with water and filtered. The reagent should have a golden-yellow colour and is stored in the refrigerator. For use, 1 vol. of this stock solution is diluted with 2 vols. of water. A standard solution of D-tubocurarine chloride is prepared by dissolving 10 mg. of $C_{38}H_{44}O_6N_2Cl_2 \cdot 5H_2O$ in water and making up to 100 ml. The test solution is prepared by diluting the solution used for the polarimetric assay to contain approx. 10 mg. tubocurarine in 100 ml. of solution. The colour reaction is carried out by measuring 2 ml. of the standard or test solution into a 25 ml. stoppered measuring cylinder, adding 3 ml. of the reagent and making to 25 ml. with water. 2 ml. of 20% w/v sodium carbonate solution are added, the solution well mixed and a 5 ml. sample removed in a test-tube and heated in a boiling water-bath for 3 minutes. The solution is cooled and the colour read in a suitable photometer at 650 mμ. Reaction mixtures for the standard and test solutions are prepared under identical conditions. The colorimetric method is of limited application in view of the fact that many substances will give a colour reaction with the FOLIN reagent. PRYDE and SMITH (1949) have proposed a method for the estimation of D-tubocurarine chloride which is based on the cherry-red colour given with MILLON's reagent. The reagent is prepared by dissolving pure mercury in twice its weight of conc. HNO_3, warming if necessary to complete solution. The solution is diluted with twice its volume of water, allowed to stand overnight and filtered if necessary. A standard 0.1% solution of D-tubocurarine chloride in water is prepared and 1, 2, 3, 4 and 5 ml. portions of this solution are diluted to 10 ml. with water and 5 ml. of MILLON's reagent added to each. The contents of each tube are well mixed and allowed to stand for 2 hours at room temperature. Maximum colour development is reached in this time, after which, precipitation may take place in the more concentrated solutions. The colour is read in an absorptiometer using a 2 cm. cell and a filter with max. transmission at 430 mμ. A straight line graph is obtained by plotting the logarithmic (density) readings against concentration. This method can be applied to the extract of curare prepared for the determination of the optical rotation in the previous method. Other substances (including bebeerine and protocuridine) possessing a phenolic structure interfere and must be removed prior to estimation of D-tubocurarine. An accuracy of $\pm$ 2% is claimed by the authors for this method, but it cannot be said that chemical assays of D-tubocurarine are entirely satisfactory.

The spectrophotometric determination of dimethyltubocurarine and D-tubocurarine has been described by SWANN (1951) who found that D-tubocurarine in 0.1 N NaOH has an absorption maximum at 292 mμ. Provided that sources of interference could be eliminated it would seem that a spectrophotometric method would possess advantages over chemical methods for the determination of D-tubocurarine in plant extracts.

H. The Alkaloids of Ergot.

Ergot of rye is the sclerotium of the fungus *Claviceps purpurea* TULASNE *(Hypocreaceae)* and occurs in the ovary of *Secale cereale* L. *(Gramineae)*. Other species of *Claviceps* (*C. microcephala* WALLR., *C. nigricans* TULASNE) produce ergots in the ovaries of other members of the Gramineae *(Triticum, Avena, Festuca, Poa, Lolium, Molinia, Nardus)* and *Cyperaceae* (*Scirpus* and *Ampelodesma*). Up to the present time, the following six pairs of interconvertible alkaloids have been isolated from ergot and each pair consists of an L-base (named first) and its D-isomer. The L-components of the pairs are physiologically active but the D-isomers are inert.

1. Ergotamine — ergotaminine. 2. ergosine — ergosinine. 3. ergometrine — ergometrinine. 4. ergocristine — ergocristinine. 5. ergocryptine — ergocryptinine. 6. ergocornine — ergocorninine. On alkaline hydrolysis, all the alkaloids give rise to lysergic acid. Complexes of the L-bases with D-bases are formed. A number of amino-acids and simple bases accompany the alkaloids in ergot. These substances include tyramine, leucine, tyrosine, histidine, tryptophane, aspartic acid, histamine, betaine, ergothioneine, choline, acetylcholine and the diamines, putrescine, cadaverine and agmatine. The average content of total alkaloids in the scerotia varies between 0.025 and 0.4% but the alkaloid content of individual sclerotia is also variable some sclerotia containing very little alkaloid, while others may contain as much as 1%. The relative proportions of the various alkaloids in the sclerotia also appear to vary widely. Ergotamine and ergotaminine appear to be absent from ergot of rye but occur in the ergot of the tall fescue (*Festuca elatior* L.). Of the total alkaloids, the ergometrine pair are water soluble and assays now include the determination of the total alkaloids and the water soluble alkaloids, giving the water insoluble alkaloids by difference.

The ergot alkaloids are more or less soluble in organic solvents with the exception of light petroleum in which they are practically insoluble. The latter solvent is therefore used for the removal of fats and oils from the powdered sclerotia prior to extraction with ether, acetone or benzene. The solubility in ether is much less than in benzene, chloroform, acetone or the alcohols. In view of the number of synonyms which have been applied to the ergot alkaloids it will be an advantage to consider the nomenclature and properties before giving methods of isolation.

I. Properties.

1. The Water Soluble Alkaloids.

Ergometrine (ergonovine), $C_{19}H_{23}O_2N_3$, crystallizes with solvent from benzene in needles and from methyl ethyl ketone with solvent in prisms. The m. p. of both forms is 162—163° (dec.). Both these forms are somewhat unstable, but a more stable form, m. p. 212° (dec.), has been obtained by crystallization from acetone. Ergometrine has $[\alpha]_D$ —44° (chloroform). The hydrochloride forms needles, m. p. 245—246° and the oxalate, needles which melt at 193° (dec.). The maleate has $[\alpha]_D^{25°}$ +48—57° (c = 1; H_2O). The picrate exists in two forms, the hydrated form crystallizing in yellow needles, m. p. 148° (dec.) and the anhydrous form in ruby-red prismatic columns which decompose at 188—189°.

Ergometrinine crystallizes from acetone in prisms, m. p. 195—197° (dec.), $[\alpha]_D^{20°}$ +414° (c = 1, chloroform). The hydrochloride, $B \cdot HCl \cdot H_2O$, forms small needles, m. p. 175—180° (dec.), and the hydrobromide, $B \cdot HBr \cdot H_2O$, crystallizes from aqueous acetone on addition of ether in needles, m. p. indefinite (130—190°). The nitrate, $B \cdot HNO_3$, forms prisms, m. p. 235° (dec.) and crystallizes from aqueous methanol on addition of ether $[\alpha]_D^{20°}$ +282° (c = 0.98, H_2O).

2. The Water Insoluble Alkaloids.

Ergotamine, $C_{33}H_{35}O_5N_5$, crystallizes from aqueous acetone (with both acetone and water of crystallization) in rectangular plates, $B \cdot 2 H_2O \cdot 2 C_3H_6O$, and decomposes at 180°. $[\alpha]_D^{20°}$ —160°. Ergotamine is a weak monoacidic base and forms crystalline salts. The hydrochloride, $B \cdot HCl$, melts at 212° (dec.), and the sulphate, $B_2 \cdot H_2SO_4$, has m. p. 205° (dec.). The ethane sulphonate, $B \cdot C_2H_5SO_3H$, melts at 207° (dec.) and the tartrate crystallizes from methanol in rhombic plates, $B_2 \cdot C_4H_6O_6 \cdot 2 CH_3OH$, m. p. 203° (dec.). The di-(p-toluyl)-L-tartrate crystallizes in rectangular plates and melts at 190° Ergotamine is converted into ergotaminine by refluxing in an atmosphere of nitrogen, a methanolic solution of the base containing 0.1 ml. of glacial acetic acid.

Ergotaminine is sparingly soluble in most organic solvents and separates free from solvent. From boiling methanol, it crystallizes in rhombic plates, m. p. 241—243° (dec.), $[\alpha]_D^{20°}$ +369° (c = 0.5, chloroform). Ergotaminine does not form crystalline salts and has weaker basic properties than ergotamine. It is converted into ergotamine by allowing to stand in darkness in the presence of glacial acetic acid, methanol and sulphuric acid.

Ergocornine (ergotoxine), $C_{31}H_{39}O_5N_5$, was isolated from the mixture formerly known as ergotoxine by fractional crystallization of the di-(p-toluyl)-L-tartrates (STOLL and HOFMANN, 1943). It crystallizes from methanol (1 in 20) in polyhedra and melts at 182—184° (dec.). $[\alpha]_D^{20°}$ —188°. The pure base forms crystalline salts. The hydrochloride, B · HCl, melts at 223° (dec.), the phosphate, B · H₃PO₄, has m. p. 190—195°, the ethanesulphonate, B · C₂H₅SO₃H, melts at 209° (dec.) and the di-(p-toluyl)-L-tartrate, B · C₂₀H₁₈O₆, has m. p. 180—181° (dec.).

Ergocorninine is formed when ergocornine is allowed to stand in aqueous ethanolic potash solution. It crystallizes from boiling ethanol (1 in 15) in prisms, m. p. 228° (dec.). $[\alpha]_D^{20°}$ +409°.

Ergocristine, $C_{35}H_{39}O_5N_5$, was isolated from the ergotoxine fraction by precipitation as the di-(p-toluyl)-L-tartrate, B₂ · C₂₀H₁₈O₆, m. p. 191° (dec.) $[\alpha]_D^{20°}$ +58° (c = 0.2, Ethanol). The base crystallizes from acetone in prisms with 1 mol. of solvent, and has m. p. 160—175° (dec.) $[\alpha]_D^{20°}$ —183°. The phosphate, B · H₃PO₄, has m. p. 195° (dec.), the ethanesulphonate, B · C₂H₅SO₃H, m. p. 207° (dec.) and the D-tartrate, B · C₄H₆O₆, m. p. 185—190° (dec.). On boiling in methanol, ergocristine forms the D-isomer ergocristinine (ergotinine) m. p. 226° (dec.), $[\alpha]_D^{20°}$ +366°, which is reconverted to ergocristine by boiling with a 1% ethanolic solution of phosphoric acid.

Ergocryptine, $C_{32}H_{41}O_5N_5$, was separated as the di-(p-toluyl)-L-tartrate from the ergotoxine fraction by STOLL and HOFMAN (1943). This compound is sparingly soluble in absolute methanol (or ethanol) and crystallizes from this solvent. The free base on liberation, may be crystallized from ethanol, acetone or hot benzene. From the last named solvent, ergocryptine crystallizes in prisms, m. p. 212° (dec.), $[\alpha]_D^{20°}$ —187°, and forms crystalline salts. The di-(p-toluyl)-L-tartrate, B · C₂₀H₁₈O₆, forms needles, m. p. 186° (dec.), the hydrochloride, B · HCl, needles, m. p. 208° (dec.), the phosphate, B · H₃PO₄, hexagonal plates, m. p. 198 to 200° (dec.), the D-tartrate, B · C₄H₆O₆, rectangular plates, m. p. 209° (dec.) and the ethanesulphonate, B · C₂H₅SO₃H, prisms, m. p. 204° (dec.). Ergocryptine is converted into ergocryptinine by boiling in methanol from which the D-isomer crystallizes in needles, m. p. 240 to 242° (dec.), $[\alpha]_D^{20°}$ +408°.

Ergosine, $C_{30}H_{37}O_5N_5$, and ergosinine were isolated by SMITH and TIMMIS (1937). Ergosine crystallizes readily from ethyl acetate in prisms, m. p. 228° (dec.); $[\alpha]_D^{20°}$ —161° (c = 1, chloroform). The hydrochloride, B · HCl · COMe₂, crystallizes from acetone in diamond shaped plates, m. p. 235° (dec.), and the hydrobromide, B · HBr · COMe₂, crystallizes in needles from acetone, m. p. 230° (dec.). The nitrate, B · HNO₃ · COMe₂, crystallizes in needles from acetone m. p. 215° (dec.). *Ergosinine* crystallizes in solvent free prisms from aqueous actone, benzene or ethyl acetate, m. p. 228° (dec.), or from methanol in needles, B · O · 5 CH₃OH, m. p. 220° (dec.). The solvent free base has $[\alpha]_D^{20°}$ +420° (c = 1, chloroform). The hydrochloride is amorphous and decomposes at 206°. Ergosinine is converted to ergosine by boiling with a mixture of acetone, phosphoric acid, water and ethanol for 9 hours in an atmosphere of nitrogen or by treatment with a mixture of N ethanolic potash and water in an atmosphere of nitrogen for 45 minutes.

The ergot alkaloids on alkaline hydrolysis yield lysergic acid or isolysergic acid. On acid hydrolysis the peptide portion of the molecule is broken down to the constituent amino acids. The hydrolytic products of the ergot alkaloids in addition to lysergic acid are given in Table 7.

Table 7. *Hydrolytic Products of Ergot Alkaloids.*

Alkaloids	Hydrolytic products
Ergocornine — ergocorninine	Dimethylpyruvic acid, D-proline, L-valine
Ergocristine — ergocristinine	Dimethylpyruvic acid, D-proline, L-phenylalanine
Ergocryptine — ergocryptinine	Dimethylpyruvic acid, D-proline, L-leucine
Ergosine — ergosinine	Pyruvic acid, D-proline, L-leucine
Ergotamine — ergotaminine	Pyruvic acid, D-proline L-phenylalanine
Ergometrine — ergometrinine	l(+)-β-aminopropanol (combined in amide linkage with lysergic acid)

II. The Isolation of the Ergot Alkaloids.

The ergotoxine and ergotinine groups of alkaloids are isolated by the method of BARGER and CARR (1907). The finely ground ergot is, extracted with 96%

ethanol, the solvent removed by evaporation, and the residue extracted with light petroleum to remove oil. The residue is dissolved in ethyl acetate and shaken out with successive portions of citric acid solution (1—2% w/v) until all the alkaloids have been removed. Sodium bromide is added to the acid solution to precipitate the hydrobromides of the alkaloids. The precipitate is collected, treated with dilute NaOH solution and shaken with ether. The ergotinine is removed by the ether and may be obtained in a fairly pure state by recrystallization from ethanol. The ergotoxine fraction remains in the NaOH solution which is neutralized, made alkaline with sodium carbonate solution and extracted with ether. The residue left on distilling off the ether is dissolved in 80% ethanol and a slight excess of a solution of phosphoric acid in ethanol added and the mixture set aside for a few days. Ergotoxine phosphate crystallizes out and may be purified by recrystallization from 90% ethanol. 50 ml. of boiling 90% ethanol is used for each gram of the phosphate taken. Ergocristine can be obtained from the ergotinine, isolated as above, by dissolving the ergotinine in ethanol and precipitating with a solution of di-(p-toluyl)-L-tartaric acid, in absolute ethanol (two equivalents of the acid to one of the base). The ergocristine salt easily crystallizes out. Ergocryptine and ergocornine are obtained from the ergotoxine prepared above by fractional crystallization of the di-(p-toluyl)-L-tartrates in methanolic solution. After the di-(p-toluyl)-L-tartrate of ergocryptine has crystallized out, the methanolic mother liquor is diluted with water until the methanol content is about 80% when the salt of ergocornine begins to crystallize. The isolation of ergosinine has been described by Smith and Timmis (1937). The total alkaloids are extracted with boiling benzene, the cooled solution is separated from sparingly soluble substances and extracted with 1% NaOH solution. The extract is made just acid to Congo-red by the addition of dilute H_2SO_4 and the precipitate of sparingly soluble sulphates is removed by filtration. The filtrate is made alkaline with sodium bicarbonate when a precipitate forms. This is filtered off, dried in a vacuum desiccator over H_2SO_4, and dissolved in methanol. The crystalline precipitate of crude ergosinine is purified by recrystallization from hot aqueous acetone. A further quantity may be obtained by extracting the precipitate of sparingly soluble sulphates with water, making the extract alkaline with sodium bicarbonate and treating the resulting precipitate with methanol as described above. Ergosinine may be converted to ergosine by acid or alkaline treatment as described above.

Ergotamine has been isolated by Stoll (1945) as follows. Coarsely ground ergot is mixed with aluminium sulphate and 300 ml. water and blended into a finely divided state with cooling. The mixture is continuously extracted with hot benzene (1.5 l.), the solvent removed, and the residue stirred with 4 litres of benzene and made slightly alkaline with ammonia gas. After several hours of continuous stirring the solution is filtered and the residue washed with benzene until free from alkaloids. The filtrate and washings are concentrated under reduced pressure to 75 ml. when ergotamine crystallizes out. A further crop of crystals may be obtained by treating the mother liquor with light petroleum. The base is purified by recrystallization from aqueous acetone. Ergometrine may be isolated from ergot by extracting the defatted powder with hot dilute H_2SO_4 (Dudley and Moir, 1935). The extract is treated with excess of baryta and excess Ba removed with CO_2. The filtrate is concentrated in vacuo and treated with ethanol. The ethanolic extract is concentrated and the concentrate extracted with chloroform. The chloroform solution is shaken out with dilute H_2SO_4. The base may be separated by neutralizing the acid extract with ammonia, adding a slight excess, saturating with NaCl and extracting with ether.

III. Detection and Estimation of the Ergot Alkaloids.

1. Qualitative Tests.

The ergot alkaloids are precipitated by the alkaloidal reagents, the most sensitive being MAYER's reagent which gives an opalescence in dilutions of 1 part per million of the alkaloid. Iodine in potassium iodide also gives a precipitate with very dilute solutions. A number of colour tests, which apparently depend on the presence of an indole nucleus in the lysergic acid component of the alkaloid molecule, have been described and the most important of these tests are given as follows.

KELLER's Test. A trace of ferric chloride (anhydrous) is added to a solution of the alkaloid in glacial acetic acid and concentrated H_2SO_4 carefully added. An intense blue coloration is produced at the boundary of the two layers. A deep blue colour appears when a solution containing ergot alkaloids is mixed with a reagent consisting of 0.125 g. p-dimethylaminobenzaldehyde + 0.1 ml. of 5% $FeCl_3$ + 65% v/v H_2SO_4 to make 100 ml. Glyoxylic acid reagent in the presence of concentrated H_2SO_4 gives a blue colour with ergot alkaloids. The amino acids produced by acid hydrolysis of the water insoluble alkaloids may be used for the identification of members of this group. Valine occurs only in the ergocornine pair, phenylalanine occurs in the molecules of ergotamine and ergocristine or their isomers, and leucine results from the hydrolysis of the ergosine and ergocryptine pairs. Subsequent identi:ication of the amino acids by the method of paper chromatography therefore provides a useful means of identification of the respective alkaloids. Acid hydrolysis is effected by heating the alkaloid with conc. HCl in a sealed tube at 100° for 16 hours, evaporating the solution to dryness on a steam bath, dissolving the residue in water and applying the aqueous solution to the paper.

2. Quantitative Estimation of the Total and Water Soluble Alkaloids.

SMITH (1947) has described the results of a collaborative study of the methods available for the assay of ergot and has selected a method based on that of POWELL, REAGAN, STEVENS and SWANSON (1941) as meeting the requirements of accuracy and speed. This method gives satisfactory results for total alkaloids and ergometrine, using a colorimetric procedure for the final determination. Finely ground ergot powder is defatted by extraction with light petroleum and allowed to dry in the air. 15 g. of defatted powder are placed in a 250 ml. flask, 147 ml. of reagent acetone and 3 ml. of 10% ammonia solution added, the flask stoppered and shaken on a mechanical shaker for 1 hour. The solution is filtered and 100 ml. of the filtrate (representing 10 g. of the powdered ergot) are transferred to an evaporating dish. The solution is evaporated under an air current to approx. 20 ml. and transferred to a 125 ml. separating funnel. Approx. 3 vols. of pure ether are added (some of which is used to wash out the evaporating dish) and the solution is made acid with 0.2 ml. of 20% tartaric acid solution. The acetone-ether solution is now extracted with four 10 ml. portions of 1% tartaric acid solution, each portion being drawn off into a small round bottomed flask. Traces of ether and acetone are removed from the combined portions by evacuation at approx. 40°. The solution is transferred to a 50 ml. volumetric flask, washing with small volumes of 1% tartaric acid solution to make to volume. A 5 ml. aliquot is removed (representing 1 g. of ergot), diluted to 50 ml. (or other suitable volume) and estimated colorimetrically for total alkaloids using ergotoxine ethanesulphonate as standard. For the colorimetric estimation, the following procedure is recommended. 4 ml. of the ergot solution are mixed with 8 ml. of the ergot reagent (0.125 g. p-dimethylaminobenzaldehyde added to 0.1 ml. of 5% solution of $FeCl_3$ + 65% v/v H_2SO_4 to make 100 ml.) in a 25 or 50 ml. volumetric flask.

The mixture is allowed to stand 30 minutes and the colour read at 550 mμ. A standard curve is prepared using pure ergotoxine ethanesulphonate. The above procedure is applied for the estimation of the total alkaloids and the water soluble ergometrine is estimated as follows. The pipette used for the removal of the 5 ml. aliquot of total alkaloids is rinsed with about 2 ml. of water and the rinsing returned to the 45 ml. of 1% tartaric acid solution which represents 9 g. of ergot. The solution is made alkaline to pH 8 with sodium bicarbonate and the volume made up to 50 ml. with distilled water. The water insoluble ergotoxine alkaloids are filtered off and a 45 ml. aliquot of the filtrate (representing 8.1 g. of ergot) is pipetted into a separating funnel and shaken successively with two 50 ml. portions of carbon tetrachloride (allowing a period of 10 minutes for each shaking) to remove the last traces of the water-insoluble alkaloids. The carbon tetrachloride extracts are discarded and all traces of carbon tetrachloride are removed from the aqueous fraction by evaporation at reduced pressure at 40° in a round bottomed flask. The aqueous solution is now transferred to a 125 ml. separating funnel, saturated with NaCl, and shaken with five successive portions of 50 ml. of ether. The combined ether extracts are evaporated to dryness under an air current and the residue dissolved in water adjusted to a pH of 4 with tartaric acid. The final volume is made up to 50 ml. (representing 8.1 g. of ergot) for colorimetric estimation as above described. A standard curve is prepared using pure ergometrine maleate. McGillivray and Metcalf (1943) have developed a rapid and accurate method for the estimation of the total alkaloids in bulk samples and in individual sclerotia based on the colour given by the p-dimethyl-aminobenzaldehyde reagent and also have described a micro-method for the estimation of ergometrine in a mixture of ergometrine and ergotoxine applicable to individual sclerotia and bulk samples.

Fuchs and Himmelbauer (1950) have described a similar method for the determination of the total and water soluble alkaloids of ergot. These workers use ether for the extraction of the mixed alkaloids. A micromethod for the estimation of the two alkaloidal groups of ergot has been developed by Fischer and Hecht (1951) in which small samples (0.03—0.3 g.) of powder are used for extraction with benzene. This method has been successfully employed for the analysis of single sclerotia. Both the above methods use the p-dimethylamino-benzaldehyde reagent for colorimetric determination. The colorimetric method lacks specificity and any other alkaloid or degradation product containing an indole nucleus gives a colour with the reagent. Moreover, the total alkaloid content is expressed in terms of ergotoxine and estimation of the individual alkaloids of the water insoluble group is not possible.

Paper Chromatography. During recent years much attention has been given to the development of methods involving paper chromatography for the separation and estimation of the alkaloids. Foster, Macdonald and Jones (1949) using the technique of partition paper chromatography have shown that the water soluble ergot alkaloids can be separated from the water insoluble alkaloids with the solvent system prepared by shaking together 4 vols. n-butanol, 1 vol. glacial acetic acid and 5 vols. of water. The water insoluble alkaloids pass down the paper with the solvent front but ergometrine and ergometrinine become separated on the paper, the R_F values being 0.59 and 0.68 respectively. The chromatogram is prepared by placing 0.05 ml. of a solution containing 5—10 μg. of the alkaloids as tartrates, lactates or maleates on the paper and running for 12—18 hours.

When the alkaloids are present as sulphates little movement on the paper takes place. After drying in air the paper is examined in ultra-violet light when the alkaloid spots become visible. The above procedure is applied by these workers to the determination of ergometrine in ergot. 5 g. of ergot in No. 60 powder is defatted with light petroleum, allowed to dry in air and mixed thoroughly with 0.3 g. of $NaHCO_3$. Water is added drop by drop to the mixture with stirring until the mass is well damped. The mixture is transferred to a percolator and extracted with peroxide free ether containing 5% ethanol. Extraction of the alkaloids is slow and is best carried out by drawing off 10 ml. of percolate at intervals of 1 hour until 70 ml. is collected. The powder is left in contact with solvent overnight and the percolation continued as above until another 100 ml. has been withdrawn. The percolation is once more stopped and the powder allowed to remain in contact with solvent overnight. The extraction is completed by drawing off portions of percolate at 30 minute intervals until the total volume of extract amounts to 200—250 ml. The percolate is collected in an amber glass bottle (the entire operation should be conducted in a dark room). The ethereal extract is transferred to a separating funnel and the alkaloids removed by shaking with six 10 ml. portions of 5% lactic acid. The combined extracts are collected in a graduated cylinder and the volume adjusted to 100 ml. with distilled water. Portions of this extract are suitably diluted with 1% lactic acid until 0.05 ml. placed on Whatman No. 1 paper and developed with the n-butanol-acetic-acid-water mixture, gives a fluorescent spot approx. equal in intensity to that obtained with an ergometrine standard containing 0.2—0.5 μg. of ergometrine in 0.05 ml. By running a series of standards on the same paper the ergometrine content of ergot is determined. The precision of this method is not high, due to errors which may occur in matching the intensities of the fluorescent spots, but the method gives reasonable agreement with the above described colorimetric method of SMITH (1947). However, in comparison with the method of the British Pharmacopoeia, the results are somewhat low, but the authors have shown that the use in the B. P. method of boiling ether for extraction results in partial hydrolysis of the alkaloids with the production of lysergic acid which is removed together with the water soluble alkaloids and estimated as ergometrine. BRINDLE, CARLESS and WOODHEAD (1951) have successfully used buffered filter paper for the separation and estimation of the water insoluble alkaloids of ergot. In this method, filter paper (Whatman No. 1) is soaked in the buffer solution, excess buffer removed with blotting paper, and the paper allowed to dry in air for 10 hours. Before a chromatogram is run, the paper is allowed to equilibrate with the vapour phase of the solvent system for 24 hours. The buffer solutions used are 0.1 M citric acid — 0.2 M Na_2HPO_4, 0.1 M tartaric acid — 0.2 M Na_2HPO_4 and 0.1 M maleic acid — 0.2 M Na_2HPO_4. The solvent is pure anaesthetic ether and the alkaloids used are ergocristinine, ergotoxine (from ergotoxine ethanesulphonate), ergotamine (from ergotamine tartrate), lysergic acid and isolysergic acid. The bases are prepared by dissolving the salts in aqueous acetone and precipitating the base by the addition of freshly prepared $NaHCO_3$ solution. The base is filtered off, washed with water to remove alkali and dried over P_2O_5. Solutions are prepared in chloroform or in 90% ethanol, 0.01 ml. containing 10.30 μg. of alkaloid applied to the buffered paper, and the chromatogram run for 3—5 hours. The alkaloid spots are located by means of filtered ultra-violet light. The optimum pH for maximum separation of the alkaloids lies between 3 and 4. For quantitative estimation, the ascending method gives more satisfactory results than the descending method. Pyrex test tubes 19 cm. × 2.9 cm., each fitted with a rubber bung slit on the under surface to hold a filter paper strip 2.6 cm. wide, are used. The lower end of the strip is cut away leaving a thin strip 2—5 mm. wide dipping into the solvent at the bottom of the tube. The width of the strip and the depth to which it is immersed in the solvent are important factors in the determination of the rate of upward flow of the solvent. For equilibration of the buffered strips with ether saturated water, an identical set of tubes is used containing about 5 ml. of ether saturated water. The strips are folded and mounted in the tube so that the paper does not come into contact with the water. After 2—6 hours the bungs and damp paper are transferred to the tubes containing the solvent for the ascending run. Addition of 2—5% ethanol to the ether before it is shaken with water reduces the tendency for the formation of ghost spots. The solvent front travels about 10 cm. in a period of 2—3 hours. The alkaloids are eluted from the paper by treatment with 50% ethanol. A strip of paper containing the alkaloid spot is cut out and one end cut to a point. The strip is suspended with the straight edge dipping into a small glass trough containing 50% ethanol. The solvent travels down the strip and is collected in a small dish. The whole is enclosed in a glass container to prevent evaporation of solvent from the paper. After 1—2 ml. eluate is collected (usually after a period of 2—3 hours), the strips are given a final washing with 50% ethanol and examined under ultra-violet light to check completeness of extraction. The eluate and washings are evaporated to dryness using gentle heat and the dimethylaminobenzaldehyde colour reaction applied to the residue. 5—40 μg. can be estimated by this method. Before measuring the intensity of the colour in the photometer the coloured solution is filtered through a No. 3 sintered glass microfilter. For the preparation of the calibration curve, 0.1025% solution of ergotoxine in 50% ethanol is prepared. 1 ml. of this solution is diluted to 50 ml. with 95% ethanol. Each ml. of the diluted solution therefore contains 20.5 μg./ml. This solution is placed in a 2 ml. burette and volumes from 0.25 to 1.5 ml. are run into small Pyrex evaporating dishes. The solvent is removed by evaporation, 0.5 ml. of 1% tartaric acid solution added, followed by 1 ml. of dimethylaminobenzaldehyde reagent. The contents of the dish are mixed and the colour intensity read after a period of 10—20 minutes. For the blank cell, 2 vols. of reagent added to 1 vol. 1% tartaric acid solution are used. Buffered cellulose columns have also been used by these workers for the separation and estimation of the ergot alkaloids. In addition to the above methods, Tyler and Schwarting (1952) have described a method for the separation of the ergot alkaloids by paper partition chromatography and Fuchs and Pöhm (1951) have determined the composition of the ergotoxine group by paper chromatographic separation of the amino-acids resulting from acid hydrolysis of the alkaloids.

I. The Alkaloids of Gelsemium spp.

The rhizome and root of *Gelsemium sempervirens* Ait. *(Loganiaceae)*, a twining shrubby plant indigenous to N. America, contain the alkaloids gelsemine, gelsemicine and sempervirine. In addition, the presence of two amorphous bases has been reported by Forsyth, Marrian and Stevens (1945). *Gelsemium elegans* Benth., a native of China, has been found to contain in addition to gelsemine, the alkaloids koumine, kouminine and koumidine about which little is known. The concentration of total alkaloids in the root and rhizome varies from 0.15 to 0.25%.

I. Properties.

Gelsemine, $C_{20}H_{22}O_2N_2$, is a tertiary base possessing a very bitter taste and is poisonous. It crystallizes from acetone in prisms with one molecule of solvent which is lost at 120°. Gelsemine melts at 178° and the acetone free base has $[\alpha]_D^{20°}$ +17.8° in chloroform (c=2.026).

It is sparingly soluble in water (1 in 644), more readily soluble in ethanol and easily soluble in ether and chloroform. With acids it forms well defined crystalline salts; the hydrochloride, $B \cdot HCl$, forms sparingly soluble prisms, m. p. 333° (dec.), $[\alpha]_D +2.6°$ (H_2O), the hydrobromide has m. p. 325° (dec.) and the nitrate forms prisms, m. p. >280°. The methiodide, $B \cdot CH_3I$, crystallizes in prisms, m. p. 284° (indef.).

Gelsemicine, $C_{20}H_{24}O_4N_2$, crystallizes from acetone in prisms, m. p. 170—171^d, $[\alpha]_D^{24°}$ —140° (ethanol). It forms salts with one equivalent of acid. The hydrochloride, $B \cdot HCl$, crystallizes in micro-prisms, m. p. 140° and the picrate in yellow plates, m. p. 203°. The benzoyl derivative crystallizes from ethanol in colourless needles, m. p. 232°.

Sempervirine, $C_{19}H_{16}N_2 \cdot H_2O$, forms orange yellow needles from chloroform, m. p. 228°, or yellow to reddish brown crystals from ethanol, m. p. 258—260°. The salts crystallize well; the hydrochloride, $B \cdot HCl \cdot 2 H_2O$, forms yellow needles, m. p. 352°, the nitrate, $B \cdot HNO_3 \cdot 2 H_2O$, has m. p. 282° (dec.) and is sparingly soluble in water. The picrate melts at 268° (dec.). A dilute solution of the alkaloid in ethanol shows an intense fluorescence.

II. Isolation of Alkaloids.

FORSYTH, MARRIAN and STEVENS (1945) have described a method for the isolation of the Gelsemium alkaloids. The finely powdered root or rhizome is shaken with four parts of rectified spirit for 15—20 hours on a mechanical shaker and the extract evaporated in a vacuum. After making the residue alkaline with ammonia, the alkaloids are extracted with chloroform. The chloroform is evaporated and the residue taken up with 0.5% HCl. The resinous matter which remains is removed and the solution of alkaloid hydrochlorides is treated with 1/20 of its volume of saturated sodium nitrate solution when sempervirine nitrate crystallizes out and is removed by filtration. The bases in the filtrate are liberated by ammonia, extracted with chloroform and the concentrated extract shaken with 2 N HCl until no more bases are extracted. The acid solution is made alkaline with ammonia and shaken with three portions of ether (A) and subsequently with three portions of chloroform (B). The residue from extract (A) is dissolved in acetone and from the acetone solution gelsemine can be crystallized. The amorphous residue not dissolved by acetone gives no crystalline derivatives. An ethanolic solution of the residue from extract (B) is saturated with hydrogen chloride and the light brown amorphous hydrochlorides precipitated with ether. These hydrochlorides are mainly soluble in chloroform and from this solution amorphous picrates are obtained. The amorphous picrates are converted into the free bases and these are treated in cold chloroform with excess of benzoyl chloride and NaOH solution. An ethereal solution of the non-basic portion deposits benzoylgelsemicine.

III. The Detection and Estimation of the *Gelsemium* Alkaloids.

1. Qualitative Tests.

Pure gelsemine dissolves in strong H_2SO_4 without colour. On the addition of a small crystal of potassium dichromate to the solution in H_2SO_4, a red colour is produced which changes through violet to green (cf. strychnine). The solution of alkaloid in H_2SO_4 should not be heated before addition of the potassium dichromate. Other oxidizing agents (manganese dioxide) also give the test. A solution of gelsemine in ethanol gives a pink colouration with a solution of p-dimethylaminobenzaldehyde in HCl.

Gelsemine gives precipitates with most of the alkaloid reagents. Iodine in KI gives a brown precipitate in a solution of 1 in 10,000, and ammonium reineckate in slightly acid solution gives a precipitate of gelsemine reineckate in very dilute solution. Apart from showing an intense fluorescence in dilute ethanolic solution, sempervirine does not appear to give any characteristic reactions.

2. Quantitative Estimation.

Methods for the estimation of the total alkaloids of *Gelsemium* (Sayre, 1911) follow the conventional practice of percolating the dried and powdered drug with ether or chloroform in the presence of alkali and extracting the percolate with acid. The acid solution is made alkaline, the alkaloids shaken out with ether or chloroform, the solvent is removed and the residue titrated with acid. Lynch and Evers (1941) have made a study of the available methods and recommend the following procedure. 10 g. of the root powder and 5 g. of acid-washed sand, both in No. 60 powder, are introduced into a pear shaped separating funnel of 300 ml. capacity. 100 ml. of a mixture of 3 vols. of ether and 1 vol. of chloroform are added, the funnel shaken well and set aside for 10 minutes. 5 ml. of a dilute solution of ammonia are added and the funnel shaken for one minute at ten-minute intervals during one hour. A plug of cotton wool is inserted into the stem of the funnel and the solution is allowed to percolate into another separating funnel. When the liquid ceases to flow the drug is packed firmly and percolation continued with further quantities of solvent until extraction is complete. The percolate is tested for alkaloids by evaporating a little of the solution in a dish, dissolving the residue in a few drops of 0.1 N H_2SO_4 and adding a drop of 0.1 N iodine. No precipitate (or turbidity) is formed in absence of alkaloids. 30 ml. of N H_2SO_4 are added to the percolate and the mixture shaken well, the phases allowed to separate, and the lower layer run off. The extraction is continued with 10 ml. portions of 0.1 N H_2SO_4 until extraction of the alkaloids is complete. The combined acid solutions are washed with 10 ml. of chloroform which is run off into a second separating funnel containing 20 ml. of 0.1 N H_2SO_4. The second funnel is shaken, the solutions allowed to separate, and the chloroform rejected. The washing of the solution in the first separating funnel is repeated with two further quantities of 5 ml. each of chloroform and each in turn is transferred to the second separating funnel and washed with the same aqueous acid liquid. The phases are allowed to separate and the chloroform discarded. The acid solution from the second separating funnel is transferred to the first funnel, the combined solution made just alkaline with dilute ammonia, and 2 ml. added in excess. The solution is now shaken with successive portions of chloroform until extraction of the alkaloids is complete, each chloroform extract being washed with the same 20 ml. of distilled water in another separating funnel. The chloroform is removed from the combined extracts by distillation, 2 ml. of absolute ethanol added to the residue, evaporated, and the residue dried at 60° for 30 minutes. The residue is now dissolved in 2 ml. of 95% ethanol, warming if necessary, 2 ml. of 0.1 N H_2SO_4 and 10 ml. of water added, the solution cooled and titrated with 0.1 N NaOH from a micro-burette, using methyl red as indicator. 1 ml. of 0.1 N acid is equivalent to 0.0366 g. of alkaloids calculated as gelsemine.

J. Alkaloids of Gymnosperms.

I. The Alkaloids of *Ephedra* spp.

Several species of *Ephedra (Gnetaceae)* contain the alkaloidal amine ephedrine. The Chinese species *E. sinica* Stapf. (Tsaopen Ma-Huang) and *E. equisetina* Bunge (Mupen Ma-Huang) are the principal commercial sources of the alkaloid, but the Indian species *E. intermedia* Schrenk and Mayer, *E. gerardiana* Wall., *E. nebrodensis* Tineo, *E. pachycladia* Boiss. and *E. foliata* Boiss. also have been used as sources of the alkaloid. The European species *E. distachya* L. (*E. vulgaris* Rich.) also contains ephedrine. The leaves of *Taxus baccata* L. have

been found to contain small amounts of ephedrine (GULLAND and VIRDEN, 1931)
and the base formerly known as cathine isolated from *Catha edulis* FORSK. has
been shown to be D-nor-pseudoephedrine. In addition to ephedrine, the closely
related alkaloids pseudoephedrine, L-nor-ephedrine, nor-D-pseudoephedrine,
N-methylephedrine and D-N-methylpseudoephedrine have been isolated from
the various species of *Ephedra*. Ephedrine is the major alkaloid and is present in
E. sinica in proportions up to 1.3%. The percentage of total alkaloids in *Ephedra*
species varies from about 1% to 2.5% of which approx. 80% is ephedrine. The
ephedrine content of the plant appears to be greatest in the autumn. The slender
green stems are the most abundant source of the alkaloids while the woody parts
of the plant are considered valueless. The roots contain little alkaloid.

II. Properties.

Ephedrine, $C_{10}H_{15}ON$, occurs in the anhydrous form, m. p. 38.1°, or as a hemihydrate,
m. p. 40°, b. p. 225° or 152—153°/25 mm. The base is laevorotatory, $[\alpha]_D^{20°}$ —6.3° (ethanol)
or dextrorotatory in water, $[\alpha]_D$ +11.2°. It is deliquescent and is readily soluble in water,
ethanol or ether. When heated ephedrine volatilises and it can be distilled from a solution
of 50% NaOH. The hydrochloride, B · HCl, forms colourless needles, m. p. 217—218°,
$[\alpha]_D^{20°}$ —34° (H_2O). This salt is formed when a solution of the base in chloroform is evaporated
or allowed to stand. The hydrobromide, B · HBr, has m. p. 205° and the sulphate has m. p.
243°, $[\alpha]_D^{22°}$ —30°, and is readily soluble in water (1 in 2) and in ethanol (1 in 60). The oxalate,
$B_2 · H_2C_2O_4$, forms prismatic needles, m. p. 249° (dec.) and is sparingly soluble in cold water.
The aurichloride forms yellow needles, m. p. 128—131° and the platinichloride also crystallizes
in needles, m. p. 186° (dec.). The dibenzoyl derivative has m. p. 134° and the nitrosamine
m. p. 92°. On boiling with 25% HCl ephedrine is partially converted into pseudoephedrine.
The change is reversible and an equilibrium mixture of the bases is formed.

Pseudoephedrine, $C_{10}H_{15}ON$, crystallizes from ether in rhombic tablets, m. p. 118—119°
$[\alpha]_D^{20°}$ +51.2° (ethanol). In contradistinction to ephedrine it is sparingly soluble in water.
The hydrochloride, B · HCl, crystallizes in slender needles m.p. 181—182°, $[\alpha]_D^{20°}$ +62.05 (H_2O)
and, in contrast to the hydrochloride of ephedrine, is soluble in chloroform. The sulphate
forms prisms easily soluble in water or ethanol, $[\alpha]_D$ +52.5° and the oxalate, m. p. 218°
(dec.), unlike that of ephedrine is easily soluble in water. The aurichloride melts at 126.5 to
127.5°, nitroso-pseudoephedrine has m. p. 86° and the dibenzoyl derivative melts at 119—120°.

L-nor-Ephedrine, $C_9H_{13}ON$, forms a crystalline mass, m. p. 51°, b. p. 167—168°/22 mm.
The hydrochloride has m. p. 173° (corr.), $[\alpha]_D^{20°}$ —33.14° (H_2O), the platinichloride m. p. 221°
(dec.), the aurichloride m. p. 188° and the p-nitrobenzoyl derivative melts at 175—176°.

nor-D-Pseudoephedrine, $C_9H_{13}ON$, crystallizes in plates, m. p. 77—78° (corr.),
$[\alpha]_D^{20°}$+37.9° (methanol). The sulphate crystallizes in hexagonal plates, m. p. 295° (corr.,
dec.) and has $[\alpha]_{D_{26.1}}^{20°}$ +48.7° (H_2O). The hydrochloride crystallizes in prisms, m. p. 178—179°,
$[\alpha]_D^{24°}$ +42.1° (H_2O). The oxalate crystallizes in needles, m. p. 235° and the dibenzoyl derivative has m. p. 156—157° (corr.).

N-Methylephedrine, $C_{11}H_{17}ON$, crystallizes in stout needles, m. p. 87—88° (corr.), b. p.
137—139°/14 mm., $[\alpha]_D$ —29.2° (methanol). The hydrochloride, B · HCl, has m. p. 188—189°,
$[\alpha]_D^{20°}$ —29.8° (H_2O). The picrate has m. p. 144°, the methiodide, m. p. 212—213°,
$[\alpha]_D$ —21.8° (H_2O), the oxalate m. p. 187° and the aurichloride, m. p. 128—120°.

D-N-Methylpseudoephedrine, $C_{11}H_{17}ON$, crystallizes in needles, m. p. 30°, b. p. 145°/24 mm.
$[\alpha]_D^{21°}$ +48.1° (methanol). The hydrochloride has $[\alpha]_D^{22°}$ +58.11° (H_2O). The aurichloride
forms yellow plates, m. p. 126—127°, and the picrate melts at 152—153° (corr.). The methiodide has m. p. 216—217°, $[\alpha]_D$ +42.3° (H_2O).

III. The Isolation of Ephedrine and Pseudoephedrine from *Ephedra* spp.

CHOU (1926) has described the following procedure for the isolation of the
alkaloids. 1 kg. of the finely powdered material is moistened with a solution of
Na_2CO_3 to give a uniformly wetted friable mass from which the alkaloids are
completely extracted by shaking with cold benzene. The benzene solution is
filtered off from the residual drug and extracted with successive portions of dilute
HCl. The combined acid solutions are made alkaline by the addition of solid
K_2CO_3 and completely extracted with successive portions of chloroform. The

combined chloroform solutions are dried with anhydrous Na_2SO_4 and the solvent distilled off. The crude residue containing ephedrine and pseudoephedrine is treated with a solution of oxalic acid, the mixture warmed and filtered. Ephedrine oxalate crystallizes out on cooling and the soluble oxalate of pseudoephedrine remains in solution. The bases are regenerated, shaken out from alkaline solution with a mixture of chloroform and ether (1:3) and converted to their hydrochlorides.

IV. Qualitative Tests for Ephedrine.

Of the alkaloid reagents, iodine in potassium iodide and phosphotungstic acid give immediate precipitates with ephedrine, but with most of the other reagents precipitates are formed slowly or not at all. A colour test has been described by CHEN and KAO (1926) in which a purple colour extractable with ether is produced when 0.1 ml. of a 10% solution of $CuSO_4$ and 1 ml. of 20% NaOH solution are added to a solution of ephedrine in water. PESEZ (1938) has shown that a small amount of ephedrine dissolved in 2 ml. of conc. H_2SO_4 develops a pink to red colour when 3 to 4 drops of 40% formaldehyde are added. On warming the solution in a hot water bath the colour changes to wine-red. WACHSMUTH (1949) has described a colour reaction with ninhydrin. A slightly alkaline solution of ephedrine is warmed with a drop of a 1% solution of ninhydrin. A violet colour is produced which is soluble in amyl alcohol. The colour shows a maximum absorption at 550 mμ and the sensitivity is about 1 in 5000. This reaction is considered to be fairly specific for ephedrine. WICKSTRØM and SALVESON (1952) have shown that acetaldehyde is produced by the action of periodic acid on ephedrine in slightly alkaline solution. This reaction has been used as a basis for the quantitative determination of ephedrine. By the action of potassium ferricyanide and NaOH, ephedrine yields benzaldehyde.

V. The Quantitative Estimation of Ephedrine.

For the estimation of the total alkaloids of *Ephedra* a suitable method is described in the British Pharmaceutical Codex (1934). 20 g. of powdered material (No. 40 powder) is shaken frequently with 200 ml. of a mixture of chloroform and ether (1:3) and 10 ml. of 10% ammonia solution and 1 g. of anhydrous Na_2CO_3 are added. The mixture is shaken at frequent intervals for 4 hours and allowed to stand overnight. The mixture is now transferred to a percolator and percolation continued with 100 ml. of the ether-chloroform mixture and finally with ether until the alkaloids are completely extracted. The combined percolates are shaken with successive portions of 40, 30, 20 and 20 ml. of $N/3$ HCl and to the combined and filtered acid extracts is added sufficient $N/1$ NaOH solution to render the solution slightly acid. 10 g. of anhydrous Na_2CO_3 are now added and sufficient NaCl to saturate the solution. The clear alkaline solution is extracted with 4 successive portions of 60, 50, 50 and 30 ml. of ether and then with 25 ml. portions of ether until extraction of the alkaloids is complete. Five shakings are usually sufficient. The combined ether extracts are allowed to stand until clear and decanted through a filter into a beaker. The ether solution is warmed and poured off from any crystals which may separate. The ether is now evaporated to a volume of approx. 10 ml. and the residual solvent allowed to evaporate in air. The residue is dissolved in excess of $0.1 N H_2SO_4$, 20 ml. of distilled water added and the excess of acid back titrated with $0.1 N$ NaOH using methyl red as indicator. 1 ml. of $0.1 N H_2SO_4$ is equivalent to 0.01651 g. of total alkaloids calculated as ephedrine. A method for the determination of ephedrine by steam distillation in the presence of an excess of strong alkali has been described by SCHOEN (1944). An aliquot of the solution containing from 5—500 mg. of ephedrine is pipetted into the digestion flask of a Micro-KJELDAHL apparatus. The flask is connected and 10—15 ml. of 50% NaOH solution is added. The mixture is distilled with a rapid current of steam and the distillate collected in an excess of standard H_2SO_4. From 50 to 150 ml. of distillate is collected according to the amount of ephedrine present. The neutral reaction of the distillate to litmus gives an indication of the completeness of distillation. The excess acid is back titrated with standard NaOH using methyl red as indicator and titrating to a salmon pink colour. It is convenient to use a slight excess of acid only and to titrate back with 0.02 N NaOH using a micro-burette. 1 ml. of 0.1 N H_2SO_4 is equivalent to 0.01652 g. of anhydrous ephedrine. Under the conditions of the experiment there appears to be no decomposition of ephedrine and with care in manipulation the average error should not exceed 1% with values for ephedrine within the range 8—400 mg. WICKSTRØM and SALVESON (1952) have used a method based on the colorimetric determination of acetaldehyde formed by the action of periodic acid on ephedrine. A solution containing from 40 to 150 μg. of ephedrine in 1 ml. of water is oxidized by 0.5 ml. of a solution 0.005 M with respect to KIO_4 and 0.1 M with respect to $NaHCO_3$. After allowing the solution to stand at room temperature for 10 minutes, 0.5 ml. of a 0.05 M solution of stannous chloride in 0.25 N-HCl is added, followed by 0.1 ml. of a 5% solution of p-hydroxydiphenyl in glacial acetic acid. Finally 18 ml. of conc. H_2SO_4 is added slowly, the mixture being kept at 0° during the addition. After a period of 4 hours at room temperature the colour reaches its maximum intensity and is stable for a period of at least 20 hours. The intensity of the colour is read in an absorptiometer or spectrophotometer and a calibration curve prepared. CHATTEN and PUGSLEY (1952) have described a spectrophotometric method for the determination of ephedrine in which the intensity of the colour produced by the reaction of ephedrine with picryl chloride in benzene solution is measured. Amounts of ephedrine between 0.1 mg. and 0.6 mg. show a linear relationship when concentration is plotted against absorbancy. A standard curve is prepared by taking 1, 2, 3, 4, 5 and 6 ml. of a 0.01% solution of pure ephedrine in benzene and transferring each volume to each of six colorimeter tubes. The volume of solution in each tube is adjusted to 9 ml. with benzene and 1 ml. of a 0.3% solution of picryl chloride in benzene is added to each tube and to a blank containing 9 ml. of benzene. Beginning with the addition to the first tube and until all the tubes have received picryl chloride and have been stoppered, a period of exactly 2 minutes is allowed. The stoppered tubes are immediately placed in a water bath at 75—77° for exactly 20 minutes. At the end of this time the tubes are removed from the water bath and after exactly 3 minutes at room temperature the intensity of the colour is read at 400 mμ. The reaction with picryl chloride is not specific for ephedrine and the other secondary amines would probably give a similar colour complex. Both this method and the preceding method can be adapted for the estimation of ephedrine in *Ephedra* spp. For the oxidation method the final residue left after evaporation of the solvent in the method of the British Pharmaceutical Codex (1934) is dissolved in water, made to volume and an aliquot taken, or the solution after titration is diluted to volume and an aliquot taken for the estimation of ephedrine. For the application of the picryl chloride method the final residue of alkaloids is dissolved in benzene, made to volume with benzene and an aliquot taken. The colour reaction of CHEN and KAO (1926) described above has also been used for the colorimetric determination of ephedrine but the methods based on this reaction are not sufficiently sensitive to give reliable quantitative results.

VI. The Alkaloid of Yew *(Taxus baccata L.)*.

Taxine occurs in the leaves, shoots and fruits of the yew tree and is present in highest concentration in the leaves during the winter months. The alkaloid is poisonous and has been responsible for the death of farm animals which have eaten the leaves or shoots. The fleshy parts of the fruit do not appear to contain taxine but the seeds are poisonous.

1. Properties.

Taxine, $C_{37}H_{51}O_{10}N$, forms a colourles granular powder and the purest material which has yet been obtained melts at 121—124° after softening at 115°; $[\alpha]_D^{17°}$ +95.7° (ethanol). The base is insoluble in water or light petroleum but soluble in ethanol, ether, chloroform or benzene. The salts are amorphous, including the aurichloride which melts at 132—134°. The methiodide has m. p. 123—124° and produces trimethylamine when treated with alkali. *Cinnamic acid* is formed by the action of acids or alkalies on taxine and on oxidation with permanganate, benzamide, benzoic acid, acetic acid, oxalic acid and benzonitrile are produced. The alkaloid reduces Tollen's reagent but not Fehling's solution.

2. The Isolation of Taxine.

Callow, Gulland and Virden (1931) have used the following method for the isolation of taxine. 2.15 kg. of leaves, dried at 37°, are coarsely powdered and extracted with 8 litres of 1% H_2SO_4 for a period of 4 days. The extract is filtered and the filtrate mixed with "norite" charcoal and stirred mechanically for 12 hours at a low temperature. After filtration, the clear yellow solution is made alkaline with ammonia and thoroughly extracted with ether. The ethereal solution is shaken out with 1% H_2SO_4, the acid solution basified with ammonia and the alkaloid once more shaken out with ether. The extract is dried and evaporated, first at 45°, and then in a vacuum desiccator over H_2SO_4. About 20 g. of crude taxine is obtained by this procedure. For purification the crude taxine is dissolved in ether, shaken out with normal H_2SO_4, liberated with ammonia, and again shaken out with ether (the alkaloid should not remain in acid solution for a longer period than is necessary and acid solutions are kept ice-cold). The ethereal solution is dried, concentrated to small volume, and the residue slowly mixed with light petroleum. The viscous precipitate which separates at first is discarded and more light petroleum is gradually added to the solution cooled to 0°. The powdery precipitate obtained is dissolved in 5% H_2SO_4, the acid solution made alkaline with ammonia and the alkaloid extracted with ether. The ethereal solution is dried and the solvent evaporated. The colourless residue is dissolved in acid, the solution made alkaline with ammonia and once more the base is shaken out with ether. The product is subjected to four successive precipitations from ether by means of light petroleum. The material is now dissolved in 1% H_2SO_4, the alkaloid precipitated with ice-cold ammonia, collected, washed with water and dried in a vacuum desiccator. The product melts at 121—124° after softening at 115°.

3. Reactions and Estimation.

Taxine thus purified gives with conc. H_2SO_4 a deep red colour and with conc. H_2SO_4 and $K_2Cr_2O_7$ a yellow colour becoming dull olive-brown. The Salkowski test for sterols gives a reddish-orange colour in the acid layer and pink in the chloroform layer. With the Lifschütz sterol reagent taxine gives a red colour changing rapidly to dark blue and slowly to dull green with a green fluorescence. Taxine gives amorphous precipitates with the alkaloid reagents. Methods for the determination of taxine in dry yew leaves have been described by Winterstein and Iatrides (1921). 100 g. of powdered yew leaves are covered with 1% H_2SO_4

solution and the mixture allowed to stand for 3 days with frequent stirring. The extract is separated by pressing out the exhausted plant material which is well washed with water. The acid extract and washings are filtered, the filtrate made alkaline with NaOH solution and the alkaloid shaken out with benzene. The benzene solution is washed with water and the taxine extracted by shaking with dilute H_2SO_4. The acid solution is made alkaline with NaOH extracted with ether and the ethereal solution dried with Na_2SO_4. The ether is evaporated off to small volume, the solution transferred to a weighing dish and the remainder of the solvent removed. The residue is dried in a vacuum desiccator and weighed as taxine. By this process 0.7 to 1.4% of alkaloid was found in the dried leaves. Kuhn and Schäfer (1937) use a titrimetric method for the determination of the alkaloid residue.

K. The Alkaloids of Hemlock.

The common hemlock, *Conium maculatum* L. *(Umbelliferae)* is a biennial plant widely distributed throughout Europe and N. America. The whole plant possesses an acrid taste and when bruised, an unpleasant smell. Treatment of the crushed tissues with a dilute solution of caustic soda liberates the alkaloids from their salts and an odour resembling that of mice is produced. Five alkaloids have been isolated from the plant, namely, coniine, N-methylconiine, γ-coniceine, conhydrine and pseudoconhydrine. According to Farr and Wright (1895), stems contain 0.01—0.06% of total alkaloids; leaves 0.03—0.18%; flowers 0.09—0.24% and green unripe fruits, 0.73—0.98%. Madaus and Schindler (1938) have studied the changes in alkaloidal content during the growth phases of the plant and have shown that there is an increase from 0.096% to 1.49% in leaves before flowering and a decline during the flowering period to 0.35—0.47%. Green unripe seeds yielded 1.62% and ripe seeds 1.00%.

I. Properties.

Coniine, $C_8H_{17}N$, the major alkaloid of *Conium maculatum*, occurs in both the D- and L-forms. In the pure state, D-coniine is a colourless strongly alkaline liquid with an odour similar to piperidine; it boils at 166—168° has the refractive index $n_D^{23°}$ 1.4505, and is dextrorotatory; $[\alpha]_D$ + 15.7. At —2° it solidifies to a soft crystalline mass. Coniine is slightly soluble in cold water (1 in 90) but less soluble in hot water and therefore a clear cold solution becomes turbid on warming. The base dissolves about 25% of water at ordinary temperatures. It is easily soluble in ethanol and in most organic solvents. It is removed from aqueous solution by shaking with organic solvents not miscible with water: ether removes small quantities of coniine from a faintly acid solution. Coniine is slowly oxidized in contact with air and gradually develops a yellow colour. The salts crystallize well and are soluble in water and ethanol. The hydrochloride, B · HCl, obtained as a crystalline mass by dissolving coniine in anhydrous ether and passing dry HCl gas through the solution, has m. p. 220° and is easily soluble in chloroform, less so in ether and ethyl acetate, and insoluble in light petroleum. The hydrobromide, m. p. 211°, crystallizes in needles and the hydrofluoride in long fine needles or leaflets. The latter salt is insoluble in acetone and chloroform and has been used as a microchemical test for coniine. The D-acid tartrate, B · $C_4H_6O_6$ · 2 H_2O, has m. p. 54° and the aurichloride, B · $HAuCl_4$, crystallizes on standing and melts at 77°. The picrate forms yellow needles m. p. 75° and crystallizes from hot water. The 2:4-dinitrobenzoyl derivative melts at 139—140° and the 3:5-dinitrobenzoyl derivative at 108—109°. Coniine combines with carbon disulphide to form the coniine salt of coniylthiocarbamic acid, $C_8H_{16}N$ · CSSH · NC_8H_{17}, which crystallizes in colourless needles, m. p. 71—72°.

L-Coniine has $[\alpha]_D^{21°}$ —15° and resembles D-coniine in properties. It occurs in small quantities in *Conium maculatum* and forms salts which have slightly different melting points from those of the D-isomeride. Both forms of coniine are volatile in steam and in the vapour of ethanol and are powerful poisons. N-Methyl-D-coniine, $C_8H_{16}N$ · CH_3, is a colourless oily liquid, b. p. 173—174°, $_D^{24·3°}$ 0.8318 and $[\alpha]_D^{24·3}$ +81.33°. The hydrochloride, B · HCl, forms needles, m. p. 188° and the platinichloride, B_2 · H_2PtCl_6, melts at 158°. N-Methyl-L-coniine has properties similar to the D-form and forms a hydrochloride crystallizing in leaflets,

m. p. 191—192°. The hydrobromide melts at 189—190°; the platinichloride forms orange crystals, m. p. 153—154°, the aurichloride, leaflets, m. p. 77—78°, and the picrate, needles, m. p. 121—122°.

γ-Conicëine, $C_8H_{15}N$, is a secondary base with a strongly alkaline reaction. It is an oily liquid, b. p. 171—172°/746 mm., $\frac{15\cdot4}{D}$ 0.8740, is volatile in steam and optically inactive. The base is slightly soluble in water and forms crystalline salts. The hydrochoride, m. p. 143° is hygroscopic; the hydrobromide, m. p. 139° is soluble in acetone; the aurichloride, m. p. 69° and the picrate, m. p. 62° crystallize on standing. The platinichloride melts at 192° and the cadmium iodide double salt, $B \cdot HI \cdot CdI_2$, m. p. 146—147° crystallizes from water in needles. γ-Conicëine is a very powerful poison.

Conhydrine, $C_8H_{17}ON$, is a solid and crystallizes in colourless leaflets m. p. 121°; it may be sublimed and distils without decomposition at 226°. It is optically active $[\alpha]_D +10°$, moderately soluble in water (1 pt. in 25.6 pts. at 15°) in ether (1 in 66) and easily soluble in ethanol and chloroform. The salts are crystalline, the aurichloride forming small prisms, m. p. 133°. The benzoyl derivative has m. p. 132°. **Pseudoconhydrine**, $C_8H_{17}ON$, crystallizes from dry ether in needles, m. p. 105—106°, b. p. 236—236.5°, $[\alpha]_D +10.98°$. On crystallization from moist ether, plates of the monohydrate are formed which melt at 60°.

The base is strongly alkaline and forms a hydrochloride, m. p. 212—213°, an aurichloride, m. p. 133—134° and a platinichloride forming golden yellow needles, m. p. 185—186°. The benzoyl derivative melts at 132—133°.

II. Isolation of the Alkaloids.

For the large scale isolation of coniine Chemnitius (1928) has devised the following process. 300 kg. of ground seed moistened with 10% of its weight of 15% NaOH solution is packed in layers separated by wood shavings into an extraction cylinder fitted with a heating coil and a condenser. Ether is forced in under pressure (from a cylinder of nitrogen) in sufficient quantity to cover the material and heat is applied until the ether is evaporated and condenses back into the extractor. After a period of 3—4 hours, the extract is drawn into a receiver and the ether distilled and returned to the extractor. This procedure is repeated three times. The concentrated extract is poured into several 15 litre flasks and 50% acetic acid is added to each until a turbidity results, after which the ether is removed by distillation. The residues are cooled after combination in an earthenware container. The floating layer of fat is removed and heated twice with 10% acetic acid, cooled, and the aqueous layers added to the main solution. The aqueous solutions is then extracted with ether, rendered alkaline with an excess of 35% NaOH solution and again extracted with ether. The ether extracts containing the alkaloids are dried over K_2CO_3, filtered, and the ether removed by distillation. Fractional distillation of the crude alkaloid in a stream of hydrogen yields a fraction, b. p. 165°, which consists substantially of coniine and can be used for the preparation of salts. Wolffenstein (1895) separated the remaining bases as follows. γ-Conicëine together with a little coniine appears in the fraction which boils at 169—173° and the two bases were separated by utilising the different solubilities of their hydrochlorides in acetone. The fraction boiling at 173—174° was found to contain mostly N-methyl-coniine with a small amount of coniine. These two bases were separated by fractional crystallization of the hydrobromides. Von Braun (1905) has described a method for the quantitative separation of the bases involving fractional distillation to remove the lower boiling coniine and benzoylation of the minor alkaloids in the next fraction. Conhydrine and pseudoconhydrine remain in the residue.

III. The Detection and Estimation of the Hemlock Alkaloids.

1. Qualitative Tests.

Melzer (1898) has shown that coniine being a secondary base reacts with carbon disulphide with the formation of the coniine salt of coniyl thiocarbamic acid. DILLING (1909) used and elaborated this reaction as a test for coniine and associated alkaloids in the following manner. To 0.5 ml. of an ethanolic solution of the free alkaloid, add a few drops of carbon disulphide, boil and add excess of water. On the addition of a few drops of copper sulphate, nickel chloride or uranium nitrate solution, coloured substances are formed which in some cases dissolve in ether or toluene. Evaporation of the ethereal solutions gives crystals of varying form. These reactions are summarised in Table 8.

The tests may be carried out with salts of the alkaloids by making 0.5 ml. of an aqueous solution alkaline with sodium carbonate, adding a few drops of ethanol and carbon disulphide, and proceeding as above. Nicotine, which in common with the Hemlock alkaloids is volatile in steam, does not give any of the above reactions and may be further distinguished from coniine by the fact that aqueous solutions of both alkaloids give a red colour with phenolphthalein which in the case of nicotine is discharged on shaking with chloroform but is permanent with coniine. GABUTTI (1906) has proposed the following test for coniine*. A dilute solution of sodium nitroprusside when added to a dilute aqueous solution of coniine produces a strong red colour which disappears on heating and reappears on cooling the solution. A trace of an aldehyde added to the red solution causes the colour to change to blue-violet. If small amounts of coniine or related alkaloids are warmed with dilute H_2SO_4 and a crystal of potassium dichromate is added, butyric acid is formed which may be detected by its odour.

Table 8. *Colour Reactions of Hemlock Alkaloids.*

Alkaloid	Copper sulphate	Nickel chloride	Uranium nitrate
Coniine	Brown colour sol. in ether. Long flat rhomboid plates	Green colour sol. in ether. Thin green rhomboid plates	Orange-red colour sol. in ether or toluene. Brown needles in clumps.
Conhydrine	Brown colour sol. in ether. Long flat brown plates-rhomboid	Green colour sol. in ether. Flat green rhomboid crystals	Orange colour slightly sol. in ether, insol. toluene. Amorphous residue
Pseudoconhydrine	Brown colour sol. in ether. Flat brown rhombic plates	Green colour sol. in ether. Green rhombic crystals	Orange colour almost insol. in ether. Amorphous residue
γ-Coniceïne	Brown colour sol. in ether	Red colour sol. in ether	Orange colour partly sol. in ether, insol. in toluene

Note: Piperidine gives colour reactions similar to coniine.

2. Quantitative Determination.

For quantitative determinations, the plant material should be dried in an oven with forced draught at a temperature not exceeding 60°.

Two processes for the estimation of total alkaloids have been described by FARR and WRIGHT (1904).

Gravimetric Estimation. 10 g. of finely powdered material is exhausted by percolation with 70% ethanol, and, after the addition of 25 ml. of water acidified with 1—2 ml. of N H_2SO_4, the ethanol is removed by evaporation. The acid solution is transferred to a separating funnel and shaken twice with chloroform, any alkaloid removed by the chloroform being recovered by agitation with a little acidified water. The chloroform is discarded and the aqueous solution made alkaline with potassium hydroxide and the alkaloids removed by shaking with

---

* Recent work by the author has shown that γ-Coniceïne (base) but *not* pure coniine (base) gives a red colour with sodium nitroprusside.

three successive portions of 5 ml. of chloroform. The alkaloids are further purified by shaking the combined chloroform solutions with dilute acid, making the acid solution alkaline and once again extracting with chloroform. The final combined chloroform solutions are shaken with three drops of strong HCl, evaporated and the resulting hydrochlorides dried at a temperature not above 90° and weighed.

Volumetric Estimation. To 10 g. of the finely powdered tissue, 5 ml. of 5% HCl and 50 ml. of light petroleum are added, the mixture shaken well for 5 minutes and allowed to settle. The light petroleum is removed by decantation and the operation repeated with a further 50 ml. to remove fats completely. The light petroleum is completely removed by decantation and evaporation in a current of air. 80 ml. of light petroleum and 1 g. of potassium carbonate are added to the moist material, the mixture shaken well for several minutes and allowed to stand overnight. 40 ml. of the light petroleum extract (representing 5 g. of material) is filtered through a plug of cotton wool into another flask, 10 ml. of 0.1 N H_2SO_4 are added and the light petroleum evaporated in a current of warm air. Excess acid is titrated with 0.02 N alkali using methyl red as indicator. 1 ml. of 0.1 N H_2SO_4 is equivalent to 0.0127 g. of alkaloids calculated as coniine.

L. The Alkaloids of *Holarrhena* spp.

The bark and seeds of *Holarrhena antidysenterica* Wall. *(Apocynaceae)*, known in India as Kurchi, yield a number of alkaloids, the best known of which is conessine. The African species *H. africana* DC and *H. congolensis* Stapf. also contain alkaloids and from the latter species Pyman (1919) isolated conessine and a second alkaloid holarrhenine. Other alkaloids isolated from *H. antidysenterica* by Bertho (1939) are conessidine, conkurchine, kurchine, holarrhimine, kurchicine and conkurchinine. The bark of *H. antidysenterica* contains about 2% of total alkaloids.

1. Properties.

Conessine, $C_{24}H_{40}N_2$, crystallizes from boiling acetone in colourless plates, m. p. 125°, $[\alpha]_D$ —1.9° (chloroform) or +21.6° (ethanol). It is little soluble in water but soluble in ethanol, chloroform or light petroleum. The hydrochloride B · 2 HCl · H_2O, forms silky needles which darken at 235° but do not melt up to 340°, $[\alpha]_D^{20°}$ +9.3° (water). The hydrobromide has $[\alpha]_D$ +7.4° and the acid oxalate B · 2 $H_2C_2O_4$, crystallizes in prisms, m. p. 280° (dec.). This salt is sparingly soluble in cold water but readily soluble in hot water and is used for the isolation of the base. The picrate has m. p. 222—224° (dec.) and the dimethiodide, B · 2 CH_3I · 3 H_2O has m. p. 303—304° (dec.) and $[\alpha]_D^{23°}$ +11.5° (water). *Conessidine,* $C_{21}H_{32}N_2$, has m. p. 123°, $[\alpha]_D^{21°}$ —63.5° (chloroform) and forms a dihydriodide, m. p. 259° (dec.), a diperchlorate, B · 2 $HClO_4$ · H_2O, m. p. 243° (dec.), and a dimethiodide, m. p. 269° (dec.).

Conkurchine, $C_{21}H_{32}N_2$, has m. p. 152—153°, $[\alpha]_D^{20°}$ —51.9° (ethanol) and forms a carbonate, B_2 · H_2CO_3, m. p. 149—150°. The dihydriodide, B · 2 HI, melts at 278° (dec.) and the diacetyl monohydrate has m. p. 182—183° (dec.).

Kurchine, $C_{23}H_{38}N_2$, has m. p. 75°, $[\alpha]_D^{22°}$ —7.6° (chloroform) or +6.4° (ethanol). The hydrogen oxalate has m. p. 221° and the aurichloride m. p. 160—166°.

Holarrhenine, $C_{24}H_{38}ON_2$, crystallizes from ethyl acetate in silky needles, m. p. 197—198°, $[\alpha]_D$ —7.1° (chloroform). It is insoluble in water, easily soluble in ethanol or chloroform, sparingly soluble in cold ethyl acetate, acetone or ether. The hydrobromide, B · 2 HBr · 3 H_2O, crystallizes from water in needles, m. p. 265—268° (dry), $[\alpha]_D$ +11.0° (water). The acetyl derivative melts at 180° (corr.).

Holarrhimine, $C_{21}H_{31}(NH_2)_2(OH)$, has m. p. 183°, $[\alpha]_D^{25°}$ —14.2° (chloroform). It forms a dihydrochloride, m. p. 345°, $[\alpha]_D^{25°}$ —22.8° (methanol) and a picrate, m. p. 198—200° (dec.).

Kurchicine, $C_{20}H_{36}ON_2$, has m. p. 175°, $[\alpha]_D^{32°}$ —11.4° (chloroform). The dihydrobromide has m. p. 260° (dec.) and $[\alpha]_D^{32°}$ —27.2° (water).

Conkurchinine, $C_{25}H_{36}N_2$, has m. p. 161°, $[\alpha]_D^{22°}$ —47.0° (ethanol). It forms a dimethiodide which melts at 255—256° (dec.).

2. The Isolation of Conessine.

PYMAN (1919) describes the following method for the isolation of conessine and holarrhenine from the bark of *H. congolensis* STAPF. 29 kg. of bark are percolated with very dilute HCl, the extract made alkaline with ammonia and extracted with chloroform. The solvent is distilled off and the dark viscous residue extracted first with light petroleum and then with ether. The light petroleum extract is shaken with dilute HCl, the acid extract made alkaline with Na_2CO_3 and extracted with light petroleum. The light petroleum is removed by distillation and the residue dissolved in a solution of 0.7 parts of hydrated oxalic acid in 4 parts by weight of ethanol. On allowing to stand, the colourless crystalline hydrogen oxalate of conessine separates out, giving a yield of approx. 0.9% of the weight of the bark. The oxalate is dissolved in water, the base regenerated by Na_2CO_3 and extracted with light petroleum. After removal of the solvent, the residue is crystallized from acetone when conessine in a pure state is obtained. The ethereal extract of the total alkaloids is extracted with dilute HCl, the acid solution made alkaline with ammonia, and extracted first with light petroleum and then with ether. From the light petroleum extract a further amount of conessine is obtained. The ethereal extract is concentrated and allowed to stand when a small quantity of holarrhenine separates out. The crude base is purified by crystallization from ethyl acetate and then converted to the hydrobromide. This salt is crystallized from water and washed with acetone. The base is liberated from the hydrobromide and recrystallized from ethyl acetate. For the isolation of conessine from the seeds of *H. antidysenterica*, KANGA, AYYAR and SIMONSEN (1926) extract with 88% ethanol. 500 g. of the finely-crushed seeds are extracted in a copper Soxhlet apparatus with light petroleum (b. p. 40—60°) to remove oil. The residue of material is freed from solvent and mixed with milk of lime (CaO, 100 g.). The mixture is exposed to air overnight and then extracted with 88% ethanol for approx. 18 hours. The deep brown extract is freed from ethanol by distillation, finally under reduced pressure, and the residue shaken with dilute HCl. The acid solution is filtered and shaken with ether to remove non-basic impurities. The crude alkaloid is precipitated from the acid solution by the addition of ammonia and extracted with ether. From the dried extract the alkaloid is obtained as a viscid brown oil which on standing, partly crystallizes. The oil dissolved in the minimum quantity of ethanol is warmed with a concentrated ethanolic solution of sufficient oxalic acid for the formation of the hydrogen oxalate. The salt crystallizes in colourless prisms on cooling and is recrystallized from ethanol. The following colour reactions of conessine have been described. When warmed on a water bath with a few drops of conc. H_2SO_4 a green colour is produced which on the addition of a drop of water changes to blue. The solution of conessine in MARQUIS reagent is coloured brownish-yellow but on standing a crimson colour with a green fluorescence develops.

MANDELIN's reagent dissolves conessine with the production of a green colour on standing. On the addition of dilute bromine water to a solution of the alkaloid in H_2SO_4, a green ring is produced at the phase boundary of the two liquids after a few hours.

3. The Estimation of the Total Alkaloids of *H. antidysenterica*.

DUTTA and GHOSH (1949) have described the following method for estimation of the alkaloids. 10 g. of the powdered material (60 mesh) is shaken with a mixture of ether and chloroform (3:1) and after period of 10 minutes, 10 ml. of concentrated ammonia solution is added. The mixture is shaken for 1 hour and allowed to stand overnight. The solvent is filtered off and the material percolated with a quantity of fresh solvent until the alkaloids are completely removed. The combined ether-chloroform solutions are shaken with N HCl, the acid solution made alkaline with ammonia and the alkaloids extracted with chloroform. After removal of the solvent the dried residue is weighed. The residue may also be dissolved in an excess of standard acid and back titrated with standard alkali with methyl red as indicator. The total alkaloids are calculated in terms of conessine.

M. The Ipecacuanha Alkaloids.

I. Properties.

Cephaelis ipecacuanha (Brot.) A. RICH. (= *Psychotria ipecacuanha* STOKES), a small plant about 30 cm. in height belonging to the *Rubiaceae* and found in most parts of Brazil, is the principal source of the alkaloids emetine, cephaeline, psychotrine, O-methylpsychotrine and emetamine. *C. acuminata* KARSTEN, an allied species from Colombia, also contains the above-named alkaloids but in different proportions from *C. ipecacuanha*. The alkaloids are located chiefly in the roots but also occur in smaller amount in the stems of the plants. The roots of average Brazilian plants contain up to 2.5% of total alkaloids and CARR and PYMAN (1914) isolated from roots yielding 2.7% of total alkaloids, 1.35% of emetine and 0.25% of cephaeline. The amount of psychotrine present appears to be very small, an average yield of 0.04 to 0.06% having been obtained. According to PYMAN (1917) the content of emetamine varies from 0.002 to 0.006% and of methylpsychotrine from 0.015 to 0.033%. The roots of *C. acuminata* contain a higher proportion of cephaeline than emetine, the value for cephaeline ranging from 50 to 55% of the total alkaloids, and for emetine 40—50% of the total alkaloids. The alkaloids occur chiefly in the cortical tissues of the root and stem. Woody tissues contain about 1% of total alkaloids.

Emetine, $C_{29}H_{40}O_4N_2$, occurs as a white amorphous powder, m. p. 74° $[\alpha]_D$ —25.8° to 32.7° (50% ethanol, $c = 1.8$—4.1) or —50° (chloroform). The base is sparingly soluble in water but readily soluble in methanol, ethanol, ether or chloroform, less soluble in benzene or light petroleum. The hydrochloride, $B \cdot 2 HCl \cdot 7 H_2O$, crystallizes from hot water in woolly needles or from a cold saturated solution in water in transparent prisms, m. p. 235 to 255° (dry, dec.), $[\alpha]_D$ +53° (chloroform). The hydrobromide, $B \cdot 2 HBr \cdot 4 H_2O$, forms slender needles from water, m. p. 250—260°, $[\alpha]_D$ +12° to 15.2° (H_2O, $c = 1.4$—3.9). The hydriodide, $B \cdot 2 HI \cdot 3 H_2O$, separates from ethanol in colourless needles, m. p. 235—238°, and is sparingly soluble in water. The nitrate, $B \cdot 2 HNO_3 \cdot 3 H_2O$, crystallizes from water or ethanol in rosettes of fine needles and sinters at 188° and gradually melts as the temperature is raised to 245°. The nitrate is sparingly soluble in water. The sulphate, $B \cdot H_2SO_4 \cdot 7 H_2O$, forms white wooly needles, m. p. 205—245°, and is readily soluble in water. The platinichloride separates as an amorphous buff-coloured precipitate, m. p. 253—265°.

Cephaeline, $C_{28}H_{38}O_4N_2$, forms colourless needles, m. p. 115—116°. When dried at 100° it melts between 120 and 130°, $[\alpha]_D$ —43.4° (chloroform). Cephaeline is insoluble in water, readily soluble in ethanol, acetone or chloroform, sparingly soluble in ether or light petroleum. It is soluble in NaOH solution in virtue of its phenolic structure, a property which makes possible its separation from emetine which is insoluble in alkali. The hydrochloride, $B \cdot 2 HCl \cdot 7 H_2O$, crystallizes from dilute HCl in stout prisms, m. p. 245—270°, $[\alpha]_D$ +25° to 29.5° (H_2O, $c = 1.7$—6.7). From strongly acid solutions an acid hydrochloride, $B \cdot 5 HCl$, separates in fine needles, m. p. 84—86°. The hydrobromide, $B \cdot 2 HBr \cdot 7 H_2O$, forms colourless prisms, m. p. 266—293°, from dilute HBr. The nitrate, sulphate and hydriodide are amorphous.

Psychotrine, $C_{28}H_{36}O_4N_2 \cdot 4 H_2O$, crystallizes from moist acetone or ethanol in yellow prisms showing a blue fluorescence. After drying at 100°, it sinters at 120° and melts at 138°. It is sparingly soluble in water, ether, benzene or light petroleum and more readily soluble in ethanol, chloroform or acetone. The sulphate, $B \cdot H_2SO_4 \cdot 3 H_2O$, is very soluble in water but can be crystallized from this solvent in faintly yellow shining scales, m. p. 214—217° (when dried at 100°), $[\alpha]_D$ +39.2° (dry salt, H_2O). The hydriodide, $B \cdot 2 HI$, crystallizes from dilute HI in microscopic yellow needles, m. p. 200—222°, and the nitrate, $B \cdot 2 HNO_3 \cdot H_2O$, crystallizes from water in colourless needles, m. p. 184—187° (dried at 100°). The acid oxalate $B \cdot 2 H_2C_2O_4 \cdot 4.5 H_2O$, forms almost colourless needles which soften at 130° and decompose at 145°.

0-Methylpsychotrine, $C_{29}H_{38}O_4N_2$, crystallizes from hot dry ether in prisms, m. p. 123 to 124°, $[\alpha]_D +43.2°$ (ethanol). The base is readily soluble in chloroform or in dry ether but sparingly soluble in moist ether or water. The sulphate, $B \cdot H_2SO_4 \cdot 7 H_2O$, crystallizes from water in large hard colourless prisms, m. p. 247° (dec., after drying at 160—170°), $[\alpha]_D +54.1°$ (dry salt, H_2O). Dilute solutions of the sulphate show a beautiful blue fluorescence. The hydrobromide, $B \cdot 2 HBr \cdot 4 H_2O$, forms pale yellow silky needles, m. p. 190—200° (dried *in vacuo*), $[\alpha]_D +48°$ (dry salt, H_2O). The hydrogen oxalate, $B \cdot 2 H_2C_2O_4 \cdot 3.5 H_2O$, crystallizes in rosettes of needles, m. p. 150—162° (dec.) and the picrate produces octagonal plates from acetone and melts slowly from 142—170°.

Emetamine, $C_{29}H_{36}O_4N_2$, crystallizes from ethyl acetate in colourless needles, m. p. 153 to 154°, $[\alpha]_D +13.6°$ (ethanol). The base is insoluble in water or alkalies, sparingly soluble in ether, readily soluble in ethanol, benzene or chloroform, almost insoluble in light petroleum. Neither the base nor its salts give fluorescent solutions. The hydrochloride, $B \cdot 2 HCl \cdot 8.5 H_2O$, crystallizes from HCl (d = 1.16) in glistening needles, m. p. 77—80° or 218—223° (dry), $[\alpha]_D -17.5°$ (air-dry salt, H_2O). The hydrobromide, $B \cdot 2 HBr \cdot 7 H_2O$, crystallizes in glistening prismatic needles from water, m. p. 210—225°, $[\alpha]_D -24.3°$ to $-22°$ (H_2O, c = 8.04 to 4.15). The hydrogen oxalate, $B \cdot H_2C_2O_4 \cdot 3 H_2O$, forms rosettes of colourless needles m. p. 172° (dec.), $[\alpha]_D -6.1°$ (H_2O). The nitrate, $B \cdot 2 HNO_3 \cdot 2 H_2O$, forms prismatic needles, m. p. 165—166°, and the picrate crystallizes from acetone in long needles softening at 147° and gradually melting up to 173°.

II. The Isolation of the Alkaloids.

PAUL and COWNLEY (1894) used the following procedure for the isolation of emetine, cephaeline and psychotrine from root material of Brazilian ipecacuanha. The dried finely powdered drug is exhaustively extracted with ethanol and basic lead acetate added to the extract until no further precipitate is formed. The liquor is filtered and the filtrate carefully evaporated to dryness and the residue dissolved in dilute acid. The acid solution is filtered, made alkaline with ammonia and thoroughly shaken out with ether. The ethereal solution is separated and the alkaloids shaken out with dilute H_2SO_4. The combined acid extracts are treated with an excess of NaOH solution which precipitates the emetine. Cephaeline which possesses a phenolic structure remains dissolved in the alkaline solution. The precipitate of emetine is removed and purified by dissolving in dilute HCl, making the solution alkaline with NaOH and shaking out with ether. This operation is repeated until all traces of cephaeline have been removed. The insoluble emetine is converted into the hydrochloride and the salt recrystallized from water. The base is finally liberated by treatment of a solution of the hydrochloride with ammonia. The alkaline solution from which the emetine has been removed is neutralized, made alkaline with ammonia, and the cephaeline extracted with ether. The ethereal solution on evaporation leaves a residue of crude cephaeline which is converted to the hydrochloride. The salt is recrystallized several times from water and the base finally liberated with ammonia. The ammoniacal mother liquors from which emetine and cephaeline have been separated contain small amounts of the third base, psychotrine. This is extracted from the liquors by shaking out with chloroform. On removal of the solvent a residue containing crude psychotrine remains. The pure base is obtained by repeated recrystallization from acetone. PYMAN (1917) has described the following method for the isolation of O-methylpsychotrine and emetamine from crude emetine. The procedure for the separation of cephaeline from emetine is followed and the non-phenolic alkaloids remaining in the ethereal solution are converted to their hydrobromides and crystallized from water, when emetine hydrobromide separates out. The mother liquors gradually deposit further crops of this salt after being concentrated and allowed to stand.

When the separation of the crystals is complete, the uncrystallizable residues
are made alkaline and extracted with ether or chloroform and the solvent removed.
The remaining syrup is dissolved in twice its volume of ethanol and mixed with
a hot solution of hydrated oxalic acid (0.5 part) in ethanol (5 parts) when a
hydrogen oxalate separates in small rosettes of fine needles which consist of a
mixture of the hydrogen oxalates of methylpsychotrine and emetamine. For the
separation of the two bases Pyman (1917) dissolves the crude hydrogen oxalate
mixture (600 g.) in water and makes the solution alkaline with NaOH. The bases
are extracted with chloroform and the chloroform solution extracted with an acid
solution containing 61.5 g. of H_2SO_4 in 1200 ml. of water. After separation of
the phases the chloroform solution (A) is washed with water and the acid solution
washed with chloroform. The base is liberated from the acid solution by treatment
with NaOH and extracted with chloroform. The chloroform solution is again
extracted with dilute H_2SO_4 as above, leaving a chloroform solution (B). The
aqueous acid solutions are evaporated to dryness *in vacuo* on a water bath, the
residue boiled with a little ethanol and again evaporated to dryness, the last two
operations being repeated until the residue appears wholly crystalline. It is
then boiled with a large volume of ethanol and the insoluble monohydrate of
methylpsychotrine sulphate collected (yield 306 g.). The chloroform solution (A)
is extracted three times with 5% H_2SO_4, using 100 to 110 ml. each time and the
extracts evaporated to give further amounts of the monohydrate in the manner
described above. The remaining chloroform is then distilled and the residue
converted into the hydrogen oxalate (17.6 g.) having $[\alpha]_D$ —1.4°. Further
fractionation of this material gives finally a number of fractions having $[\alpha]_D$ —5°.
This partly purified emetamine hydrogen oxalate is dissolved in water, the
solution made alkaline with NaOH and extracted with ether. After drying the
ethereal extract with anhydrous K_2CO_3 and distilling the solvent, the residue is
converted into the hydrobromide and recrystallized from water. The base is
liberated from the hydrobromide by shaking with chloroform and aqueous NaOH.
After distilling the chloroform the base soon becomes crystalline and is purified
by recrystallization from ethyl acetate.

III. Qualitative Tests for the Ipecacuanha Alkaloids.

The ipecacuanha alkaloids give precipitates with most of the alkaloid reagents.
Iodine in potassium iodide, Mayer's reagent and bismuth potassium iodide are
especially sensitive. Characteristic colour reactions are not given with any
reagents other than Fröhde's reagent which gives with emetine and cephaeline a
dirty greenish yellow coloration which in the case of emetine fades on the addition
of a drop of HCl while that of cephaeline changes to a dull greenish-blue. Emetine
hydrochloride however gives a bright green colour with Fröhde's reagent.
Cephaeline and psychotrine couple with p-nitrodiazobenzene to give a dye which
is soluble in aqueous NaOH giving a purple solution. Psychotrine gives a sherry-
red colour when dissolved in conc. H_2SO_4 to which a trace of HNO_3 has been
added. With Fröhde's reagent psychotrine gives a pale green colour. Emetamine
and O-methylpsychotrine dissolve in conc. H_2SO_4 to form a pale yellow solution
with a green tinge and give with Fröhde's reagent an emerald-green colour.
A solution of the ipecacuanha alkaloids in dilute HCl gives with a solution of
calcium hypochlorite an orange colour.

IV. The Quantitative Estimation of the Total alkaloids.

Lowdell (1952) has described a method for the rapid determination of the
total alkaloids in ipecacuanha root. 10 g. of finely powdered root material is

weighed into a flask and 100 ml. of a mixture of ether and chloroform (3:1) are added, the flask well shaken and set aside for 10 minutes. After this period 7.5 ml. of a 10% solution of ammonia are added and the flask shaken frequently during a period of 1 hour. The mixture is transferred to a small percolator plugged with cotton wool and after the liquid has ceased to flow, the drug is packed firmly and the percolation continued with the mixture of ether and chloroform until the alkaloids are completely extracted as shown by absence of precipitate or turbidity with MAYER's reagent. The solution of the alkaloids in the ether-chloroform mixture is now shaken with successive quantities each of about 15 ml. of water, shaking each water extract after separation with the same 50 ml. of ether contained in a second separator. This procedure is continued until the ether-washed water is no longer alkaline to litmus. To the combined ethereal solutions 25 ml. of 0.1 N H_2SO_4 are added, the mixture shaken well and the acid solution run into a beaker. The ether solution is washed with further quantities of water until the washings are no longer acid to litmus and each washing is added to the acid solution in the beaker. The excess acid is now titrated with 0.1 N NaOH using methyl red as indicator. 1 ml. of 0.1 N H_2SO_4 is equivalent to 0.0240 g. of total alkaloids calculated as emetine. For the determination of the non-phenolic alkaloids in the above titration liquid, the method of the British Pharmacopoeia (1932) may be followed. The titration liquid is transferred to a separator and 5 ml. of 5% NaOH solution are added, followed by 50 ml. of ether. The mixture is shaken, the ether separated and extracted with further portions of 10 ml. and 5 ml. of NaOH solution. The alkaline solutions are mixed and shaken with two 15 ml. portions of ether. The ethereal solutions are mixed and washed with successive quantities of 5 ml. of water until free from alkali. Each portion of water is washed with the same 10 ml. of ether contained in a second separator. The combined ethereal solutions are evaporated and the residue dissolved in 10 ml. of 0.1 N H_2SO_4. The excess of acid is titrated with 0.1 N NaOH using methyl red as indicator. 1 ml. of 0.1 N H_2SO_4 is equivalent to 0.0240 g. of non-phenolic alkaloids calculated as emetine.

Methods for assay of the total alkaloids of ipecacuanha using the principles of ion-exchange have been described by LASSLO and WEBSTER (1950) and JINDRA and POHORSKY (1951). In each case the classical methods of partition between miscible solvents for purification of the alkaloids have been substituted by use of an ion-exchange process. For the estimation of the alkaloids LASSLO and WEBSTER (1950) have adopted the following procedure. 10 g. of powdered material are placed in a dry 250 ml. Erlenmeyer flask together with 100 ml. of peroxide-free ether. The flask is stoppered tightly with a rubber stopper, the mixture shaken thoroughly and allowed to stand for 5 minutes. 10 ml. of a 10% solution of ammonia are added and the mixture shaken for 1 hour on a mechanical shaker. The mixture is now allowed to stand for 4 hours and then shaken again for 1 hour an a mechanical shaker. After the drug has settled, 50 ml. of the clear supernatant ethereal solution are pipetted into a 125 ml. Erlenmeyer flask and the pipette rinsed with ether, the rinsings being added to the contents of the flask. The ether solution is evaporated on a steam bath to approx. 20 ml. and 10 ml. of 0.5 N H_2SO_4 added. The ether is now completely removed on the steam bath, shaking the flask occasionally. The adsorption apparatus is similar to that used by MENDEZ, KIRCH and VOIGT (1949) for the purification of hydrastine (q. v.). The Florisil column is 19 mm. in diameter and 135 mm. in height and is activated by allowing 100 ml. of a 10% solution of ammonia to pass through. Excess ammonia is now removed by drawing 200 ml. of distilled water in two portions through the column. When all but the last 20 ml. of water has run from the separating funnel at the top of the column, a small funnel with a coarse sintered-glass filtering plate is fitted into the top of the separating funnel and the solution of the extracted alkaloids in dilute H_2SO_4 is passed through the filter. The flask which contained the acid solution, and the filter are washed 3 times with 30 ml. portions of distilled water. The filter is removed and the separating funnel and the column are washed by drawing through an additional 200 ml. of distilled water. When only 5 ml. of water remains in the funnel, 50 ml. of peroxide-free ether are added and allowed to run through the column until the interface of ether and water is about 2 cm. from the bottom of the column. All the liquid which has passed through the column up to this point is discarded and all of the subsequent eluate collected and combined for the titration of the alkaloids. 100 ml. of an approx. 10% solution of ammonia in 95% ethanol is added to the separating funnel before the last 2 ml. of ether has run out. The ethanolic solution is allowed to run through without the application of suction and when it almost reaches the bottom of the column, the rate of elution is adjusted to about 4 ml. per minute. After approx. 50 ml. of the ethanolic solution has flowed from the separating funnel the elution is stopped for 15 minutes by closing the tap on the funnel. The eluate collected up to this point is removed from the receiver into a 250 ml. beaker and placed on a steam bath to remove the solvent. The flow of solvent through the column is now resumed at the same rate until all the ammoniacal ethanol has passed through the system. The eluate is removed from the receiver, combined with that previously collected, and evaporated almost to dryness. The same column can be used for three consecutive assays and to this end sufficient ammoniacal ethanol is used to provide 100 ml. of eluate and to keep a small layer of solvent above the adsorbent column at all times. The column is prepared for the next experiment by drawing through 100 ml. of a 10% aqueous solution of ammonia. For the titration of the alkaloids, 30 ml. of peroxide-free ether are added to the alkaloid residue in the beaker, the walls of the beaker being rinsed during the addition. Exactly 10 ml. of 0.1 N H_2SO_4 solution are added and the ether completely evaporated on the steam bath. 50 ml. of CO_2-free distilled water and two drops of methyl red indicator are added and the excess acid titrated with 0.1 N NaOH to a faint pink end-point (pH 5.2).

Jindra and Pohorsky (1951) use a column of Amberlite IR-4B resin for the purification of the alkaloids, and for the micro-method 0.3 g. of powdered root is extracted as follows. To this amount of material accurately weighed, 10 g. of ether and 1 ml. of 5% solution of ammonia are added and the mixture shaken vigorously for 15 minutes. After separation, 7 g. of the ethereal layer is carefully weighed into a 20 ml. conical flask and evaporated to dryness. 1 ml. of 0.2% H_2SO_4 and 5 ml. of 90% ethanol are added and the resulting solution passed through the ion-exchange column. The eluate is diluted with 20 ml. of water and titrated with 0.01 N HCl using the antimony electrode. 1 ml. of 0.01 N HCl is equivalent to 0.002402 g. of alkaloids. The ion-exchange methods permit of the rapid estimation of the alkaloids and give results which agree well with the older conventional methods.

N. The Alkaloids of Liliaceae.

I. The Alkaloids of *Colchicum* spp.

The meadow saffron or autumn crocus (*Colchicum autumnale* L., *Liliaceae*) a perennial plant widely distributed over Europe contains colchicine, a poisonous substance which was formerly used in medicine for the treatment of gout. Recent work has shown that colchicine possesses a tropolone structure and being non-basic in character, it cannot strictly be considered as an alkaloid. However, following convention, it is included in this section as an alkaloid and will be referred to as such. In addition to colchicine other minor alkaloids chemically related to colchicine have been isolated from *C. autumnale* (ŠANTAVY and REICHSTEIN, 1950). In addition to *C. autumnale*, colchicine occurs in other species of Colchicum and also in *Merendera sobolifera* FISCH. and MEY, and *Merendera caucasia* BIEB. (ALBO, 1901). KLEIN and POLLAUF (1929) isolated colchicine from plants of the following genera of Liliaceae: *Hemerocallis*, *Veratrum*,*Tulipa*, *Bulbocodium*, *Tofieldia* and *Ornithogalum*. CLEWER, GREEN and TUTIN (1915) found 0.3% of colchicine in the dried tubers of *Gloriosa superba* L. The distribution of colchicine in *C. autumnale* has been studied by a number of workers and it has been shown that all parts of the plant contain the alkaloid. Seeds contain an average of 0.72—0.75% and corms 0.38—0.4%: leaves yield smaller amounts, young leaves containing more than old leaves. The histochemical distribution of colchicine in the tissues of *C. autumnale* has been studied by ALBO (1901), LIPTÁK (1928) and KLEIN and POLLAUF (1929). According to LIPTAK the alkaloid is concentrated chiefly in the endosperm and third layer of the seed coat. NIEMANN (1933) has stated that colchicum flowers may contain as much as 0.8% of alkaloid.

1. Properties.

Pure Crystalline Colchicine, $C_{22}H_{27}O_6N$, was obtained by CLEWER, GREEN and TUTIN (1915) in pale yellow needles from ethyl acetate, m. p. 155—157°. With water it forms a sesquihydrate $(B_2 \cdot 3 H_2O)$ and with chloroform, two crystalline compounds are obtained $B \cdot CHCl_3$ and $B_2 \cdot CHCl_3$. From benzene it crystallizes with one molecule of solvent and has m. p. 140°. In chloroform $[\alpha]_D^{16.5°} = -120.8°$. Colchicine is soluble in water, and in aqueous ethanol, and is easily soluble in chloroform. In absolute ethanol it is less soluble than in aqueous ethanol. Hot benzene and amyl alcohol dissolve colchicine but it is insoluble in absolute ether and in light petroleum.

The basic characters of colchicine are very feeble indeed. It is neutral to litmus and it is extracted from both acid and alkaline solutions by shaking with chloroform. During recent years colchicine has assumed great importance in cytological investigations in virtue of its property of arresting mitosis at the stage of early metaphase.

2. The Isolation of Colchicine.

COOK and LOUDON in their monograph on colchicine (MANSKE and HOLMES, 1952) describe the following process for the isolation of colchicine based on the method of ASHLEY and HARRIS (1944). Finely ground *Colchicum* seeds are exhausted with 90% ethanol and the ethanol removed by distillation. The dark brown gummy residue which contains resinous matter is treated with water and the resinous material is removed by heating with paraffin wax until the wax becomes molten (the isolation described utilizes 330 g. of the gummy residue which is treated with 450 ml. of water and 75 g. of paraffin wax). The mixture is stirred and allowed to cool and the solid crust of wax is removed from the surface and the process repeated twice with fresh wax. In this manner most of the resinous material passes into the wax. The combined wax layers are extracted three times with 100 ml. of boiling water and the aqueous extracts added to the solution of the alkaloid. A paste of filter paper pulp (50 g.) is then added to the colchicine solution (the pulp is prepared by boiling filter paper with conc. HCl to disintegrate and washing the mass with water until neutral). The mixture is filtered on a filter bed to which some paper pulp has already been added. A clear brown filtrate should result. The filter bed is boiled with a little water and the solution filtered. The combined filtrates are extracted with twelve portions of 200 ml. of chloroform, care being taken to ensure that the chloroform is free from HCl. Addition of potassium carbonate to the yellow extract causes the precipitation of some brown flocculent material which is filtered off. The chloroform is evaporated off and a golden brown syrupy residue remains. The residue is dissolved in 150 ml. of chloroform and the solution passed through a column of alumina (B. D. H. chromatographic alumina), 25 cm. long and 3.5 cm. in diameter, which has previously been saturated with benzene. Three bands are formed, an upper reddish-brown band, a larger bright yellow band and a lower almost colourless band which contains the colchicine. The column is washed with chloroform until the yellowish eluate becomes colourless and yields no residue on evaporation. Distillation of the chloroform from the total eluate gives a golden yellow syrup which is distilled three times with an equal volume of absolute ethanol to remove the residual chloroform. The residue is finally crystallized from ethyl acetate and yields 10 g. of colchicine as fine colourless needles, m. p. 148—150°. A second chromatographic purification followed by crystallization from ethyl acetate raises the m. p. to 155°.

3. Tests for Colchicine.

A number of useful colour and precipitation tests for colchicine have been described by Zeisel (1886). When a few drops of mineral acid are added to solutions of colchicine an intense yellow colour develops. Conc. nitric acid gives with colchicine a dirty violet colour which changes through brown to yellow. If the violet solution is diluted with water the colour changes to yellow and on addition of excess NaOH, an orange-red colour is formed. If 1 ml. of conc. H_2SO_4 is added to a particle of colchicine, a yellow colour is produced which on the addition of one or two drops of conc. nitric acid turns a fine green changing through violet and red to yellow. On dilution with water and addition of excess NaOH, a red colour is produced. Acid hydrolysis of colchicine yields colchiceine which gives an olive green colour with ferric chloride, therefore if a solution of colchicine acidified with dilute HCl is boiled after addition of a few drops of diluted ferric chloride solution, the green colour develops almost immediately. An ethanolic solution of colchicine gives with ferric chloride a garnet-red colour. Kippenberger (1897) has shown that an orange colour is produced when a solution of colchicine is warmed with an equal volume of a solution of hydroxylamine hydrochloride containing a few drops of dilute NaOH. Freshly prepared bromine water precipitates colchicine from aqueous solution and Wagner's reagent (iodine in KI) gives an immediate brown precipitate in the presence of a little dilute acid. The latter reagent should be used when testing for the complete extraction of colchicine from plant material. Colchicine is precipitated from dilute solution by bismuth potassium iodide (Dragendorff's reagent) and by tannic, phosphotungstic and phosphomolybdic acids. Mayer's reagent does not precipitate colchicine unless dilute mineral acid is present when a lemon-yellow precipitate appears. Cadmium potassium iodide (Marme's reagent) behaves similarly. The aurichloride B · HAuCl$_4$, m. p. 209°, is precipitated on addition of auric chloride and crystallizes on standing. It is easily soluble in ethanol.

4. The Estimation of Colchicine.

Older methods for the determination of colchicine involved a lengthy procedure for the isolation of the alkaloid in a pure state and subsequent gravimetric estimation of the purified colchicine. During recent years, attention has been directed to more rapid methods of purification using the principles of chromatographic adsorption followed by colorimetric determination.

Gravimetric Methods. Davies (1921) in the following method utilizes phosphotungstate precipitation as a means for the determination of colchicine. 5 g. of finely powdered seeds of *Colchicum autumnale* are exhausted by percolation with 100 ml. of 70% ethanol. The ethanol is removed on the water bath and the residue is taken up with three successive portions of 20 ml. of water. The combined aqueous solutions are quantitatively transferred to a separating funnel and shaken out with four successive portions of chloroform (20, 15, 15 and 10 ml.). Saturation of the aqueous phase with sodium chloride or sodium sulphate facilitates extraction of the colchicine with chloroform. The combined chloroform solutions are evaporated, the residue dissolved in 30 ml. of hot water and filtered into a beaker. To the cooled filtrate 3 ml. of phosphotungstic acid solution (10 g. sodium tungstate, 6 g. sodium phosphate, 50 ml. of water acidified with nitric acid) and 7 ml. of dilute H_2SO_4 are added. The precipitate is allowed to settle and centrifuged down. The supernatant is poured off, the precipitate suspended in water, transferred to a separating funnel and decomposed by shaking with 30 ml. of chloroform. The chloroform is run off and the operation repeated with a further portion of 10 ml. of chloroform. The combined chloroform solutions are evaporated and the residue treated with 5 ml. of ethanol, again evaporated and dried at 100° to constant weight. Further treatment with ethanol may be necessary to drive off the last traces of chloroform from combination with colchicine which is weighed as such.

Colorimetric Methods. SEIFERT (1943) and KING (1951) have described colorimetric methods for the estimation of colchicine based on the green colour formed when colchiceine reacts with ferric chloride. MACK and FINN (1950) have used the colour reaction with hydroxylamine and NaOH for the determination of colchicine in pharmaceutical preparations. Using colorimetric methods satisfactory results can only be obtained if the colchicine is present in a reasonably pure state and to achieve this object it is necessary to effect a preliminary purification of plant extracts before the application of a colorimetric method. The chromatographic adsorption method of ASHLEY and HARRIS (1944) gives an almost pure product when applied to chloroform solutions of crude colchicine. The present writer finds the following method suitable for the estimation of colchicine. 5 g. of colchicum seed or corm (finely powdered) are exhausted by extraction with 100 ml. of 70% ethanol. The ethanol is removed by evaporation on the water bath. 10 ml. of water is added to the aqueous solution which remains and resins and oil allowed to separate out. The aqueous solution is filtered through glass wool and the resinous mass extracted further with two or more portions of 10 ml. of water until extraction is complete. The combined aqueous solutions are shaken three times with light petroleum to remove colouring matter and oil. The aqueous solution is now saturated with sodium chloride and the alkaloid shaken out with chloroform until extraction is complete. The combined chloroform solutions are dried with anhydrous sodium sulphate and passed through a column of alumina according to the method of ASHLEY and HARRIS (1944). A 15 cm. × 1 cm. column of alumina (B. D. H. chromatographic alumina) is saturated with benzene and the chloroform solution of the alkaloid (approx. 50 ml.) is passed through slowly. The alkaloid is eluted with approx. 100 ml. of chloroform which is removed by evaporation leaving a yellow gummy residue. This residue is taken up in a known volume of water and the following colorimetric methods applied to an aliquot. (a) The method of KING (1951) is suitable for the estimation of amounts of colchicine from 0.05—0.5 mg. A solution of pure colchicine containing 2.5 mg. is made up to 25 ml. with N HCl in a 125 ml. Erlenmeyer flask. A funnel is inserted in the neck of the flask and the flask warmed on a steam bath for one hour. The solution is cooled to room temperature and the volume made up to 25 ml. exactly with N-HCl. Aliquots containing up to 0.5 mg. of colchicine are taken, made up to 5 ml. with N HCl and 0.1 ml. of 5% ferric chloride solution is added to each. The colour is read at 470 mμ and a calibration curve drawn. Samples of the aqueous solution of colchicine after chromatographic purification are treated likewise and the amount of colchicine present determined by reference to the calibration curve. The colorimetric method of MACK and FINN (1950) is applied as follows. 40.5 mg. of pure colchicine salicylate is weighed out, dissolved in 75 ml. of warm water in a 100 ml. volumetric flask and, after cooling, the volume is adjusted to the mark. 1, 2, 3 and 4 mg. aliquots are taken and diluted with water to exactly 15 ml. in 25 ml. test tubes marked at 15 ml. To each tube is added 2 ml. of 2% aqueous solution of hydroxylamine hydrochloride and 1.5 ml. of 0.5 N NaOH The contents of the tubes are mixed and all tubes are placed in a water bath at 75° for exactly 5 minutes. The tubes are removed, and chilled in an ice bath to room temperature and the colours read in 1 cm. cells in a spectrophotometer at 500 mμ. The colour appears to be light sensitive and all analyses are conducted in subdued light. A calibration curve is drawn and samples containing unknown amounts of colchicine are treated with 1 mg. of sodium salicylate, diluted to 15 ml. and treated in the same manner as for the standard tubes. A chromatographic purification followed by a titrimetric estimation of colchicine has been described by MÜHLEMANN and TOBLER (1946). These workers precipitate colchicine with iodine solution in the presence of a known amount of standard acid. The precipitate of the alkaloid hydriodide is removed and the residual acid titrated with alkali.

II. The Veratrum Alkaloids.

The alkaloids belonging to this group are present in the tissues of *Veratrum album* L., *Veratrum viride* AIT. and *Schoenocaulon officinale* A. GRAY, all members of the *Liliaceae*. *V. album* (European White Hellebore) is a herbaceous plant widely distributed over Central and Southern Europe and *V. viride* (American Green Hellebore) is common in the woods of the eastern United States. These plants are not to be confused with the true Hellebores, *Helleborus* spp., belonging to the *Ranunculaceae*. *Schoenocaulon officinale* is a tall herbaceous plant growing in Mexico and S. America, the seeds of which (cevadilla or sabadilla seeds) are rich in alkaloids.

1. Properties.

The Alkaloids of *Veratrum album*.

Jervine, $C_{27}H_{39}O_3N$, is a secondary base and is the major alkaloid of *V. album*. It crystallizes in prisms, m. p. 244—246°, $[\alpha]_D^{25°}$ —147° (ethanol), and is almost insoluble in water and benzene, insoluble in light petroleum, soluble in ethanol, ether and chloroform. Jervine is alkaline to litmus and forms well crystallized salts. The hydrochloride, B · HCl · 2 H$_2$O, m. p. 330—334°, is sparingly soluble in water, the hydriodide has m. p. 302—305° and the picrate m. p. 274—284°. The sulphate, $B_2 \cdot \cdot$ H$_2$SO$_4$ and the nitrate, B · HNO$_3$, are both very sparingly soluble in water and are easily prepared from the acetate by treatment with dilute H$_2$SO$_4$ or HNO$_3$.

Rubijervine, $C_{27}H_{43}O_2N$, crystallizes from ethanol in needles, m. p. 240—242°, $[\alpha]_D^{25°}$ +19.0° (ethanol). It is sparingly soluble in ethanol and ether but easily soluble in chloroform. The hydrobromide has m. p. 265—270° and the hydriodide, m. p. 293—296°. It forms a diacetyl derivative which melts at 160—163°.

Pseudojervine, $C_{33}H_{49}O_8N$, crystallizes from ethanol in hexagonal tablets, m. p. 300—307° (dec.), $[\alpha]_D^{20°}$ —139° (chloroform). It is sparingly soluble in ethanol, almost insoluble in ether and easily soluble in chloroform. The hydrochloride, B · HCl · 2 H$_2$O, has m. p. 254—256° (dec.).

Isorubijervine, $C_{27}H_{43}O_2N$, crystallizes with solvent from hot ethanol on rapid cooling in needles, m. p. 218° or on cooling slowly, in prisms, m. p. 241—244°, $[\alpha]_D^{25°}$ +9.4 (ethanol).

Protoveratrine, $C_{36}H_{61}O_{12}N$, crystallizes from ethanol in platelets and decomposes at 275—283°, $[\alpha]_D^{27°}$ —40° (pyridine). The base is practically insoluble in water, benzene and light petroleum but is sparingly soluble in chloroform and hot ethanol and very little soluble in ether. The hydrochloride, $B \cdot HCl \cdot H_2O$, forms small tablets, m. p. 234—236° (dec.); the hydrobromide, $B \cdot HBr \cdot 3 H_2O$, has m. p. 230—232°; the picrate m. p. 216—220° (dec.) and the aurichloride m. p. 199° (dec.). Protoveratrine is very poisonous and causes violent sneezing if applied to the nose. On alkaline hydrolysis it yields acetic acid, L-methylethyl-acetic acid and methylethylglycollic acid, together with the alkamine protoverine.

Germerine, $C_{37}H_{59}O_{11}N$, crystallizes from benzene in leaflets, m. p. 193—195° (dec.), $[\alpha]_D^{20°}$ +10.8° (chloroform). It forms well crystallized salts; the hydrochloride, $B \cdot HCl \cdot 2 H_2O$, crystallizes in needles, m. p. 215° (dec.); the hydrobromide, $B \cdot HBr$, forms needles, m. p. 212 to 213° (dec.) and the picrate has m. p. 186—187° (dec.). *Protoveratridine*, $C_{32}H_{51}O_9N$, does not occur naturally but is formed from germerine in the process of extraction.

Germine, $C_{27}H_{43}O_8N$, is produced by hydrolysis of *germerine* or *protoveratridine* but occurs naturally in *V. album*. It crystallizes with solvents; from methanol in prisms, $B \cdot 2 CH_3OH$, m. p. 220°, effervescing at 163—173°, $[\alpha]_D^{25°}$ +5° (ethanol) or +21.1° (dilute acetic acid). The aurichloride, $B \cdot HAuCl_4$, crystallizes in golden-yellow leaflets and the picrate has m. p. 190—205° (dec.).

Veralbidine. STOLL and SEEBECK (1952) have recently isolated a new alkaloid, veralbidine, from *V. album*. This base crystallizes from dilute acetone in pentagonal plates, from dilute methanol in prisms and from ether in fine needles, m. p. 181—183°, $[\alpha]_D^{20°}$ —11.7° (pyridine) or +5.4° (chloroform). It has the empirical formula $C_{37}H_{61}O_{12}N$ and is insoluble in water, sparingly soluble in ethanol, ether and acetone but easily soluble in chloroform. The hydrochloride, $B \cdot HCl$, has m.p. 250—251° (dec.) and is soluble in ethanol or water. The thiocyanate, $B \cdot HNCS$, has m. p. 235—236° (dec.) and is sparingly soluble in water and easily soluble in methanol or acetone. GLEN, MYERS, BARBER, MOROZOVITCH and GRANT (1952) have also isolated two new ester alkaloids from *V. album*, which have been named germitetrine and veratetrine. These alkaloids were isolated from the drug extract by countercurrent distribution using benzene and $2 M$ acetate buffer at p_H 5.5.

Germitetrine, $C_{41}H_{63}O_{14}N$, has m. p. 229—230°, $[\alpha]_D^{25°}$ —74° (1% in pyridine) and —12° (1% in chloroform).

Veratetrine, $C_{43}H_{64}O_{16}N$, has m. p. 269—270° (dec.), $[\alpha]_D^{26°}$ —32° (1% in pyridine) or $[\alpha]_D^{27°}$ —6.8 (2% in chloroform).

The Alkaloids of *Veratrum viride*.

In addition to *jervine, rubijervine, pseudojervine,* and *germine* this plant contains *veratrosine, veratramine, veratridine* and *cevadine*.

Veratrosine, $C_{33}H_{49}O_7N$, has m. p. 242—243°, $[\alpha]_D^{25°}$ —53° (ethanol + chloroform) and is a glucoside giving on hydrolysis D-glucose and *veratramine*. *Veratramine*, $C_{27}H_{39}O_2N$, has m. p. 209—210.5° $[\alpha]_D^{27°}$ —68° (CH_3OH). It yields a dihydroderivative, m. p. 198—200° and a triacetyl-derivative, m. p. 204—206°. *Veratridine*, $C_{36}H_{51}O_{11}N$, is a colourless amorphous powder, m. p. 160—180°, $[\alpha]_D^{22°}$ +8.0 (ethanol). The nitrate is very little soluble in water and the sulphate, $B \cdot H_2SO_4 \cdot 9 H_2O$, forms fine needles. On alkaline hydrolysis it yields *veratric acid* and *cevine*. FRIED, WHITE and WINTERSTEINER (1950) in a study of the hypotensive principles of *V. viride* have carried out large scale extractions of dried roots and rhizomes and, using the method of countercurrent distribution in conjunction with chromatography on columns of alumina, have isolated and separated two powerfully hypotensive ester alkaloids germitrine and germidine.

Germitrine, $C_{39}H_{61}O_{12}N$, melts at 216—219° (dec.) and after drying at 110° has $[\alpha]_D^{25°}$ —69° (c = 0.85 in pyridine). The thiocyanate, $B \cdot HNCS$, has m. p. 231—232° (dec.).

Germidine, $C_{34}H_{53}O_{10}N$, melts at 230° and has $[\alpha]_D^{25°}$ +13° (c = 1.67 in chloroform) or —11° (c = 1.84 in pyridine). The thiocyanate, $B \cdot HNCS$, has m. p. 242—244° (dec.).

Neogermitrine. FRIED, NUMEROF and COY (1952) have isolated another new alkaloid neogermitrine from the amorphous fraction remaining after all previously known crystalline alkaloids had been removed. It is an ester alkaloid and has the empirical formula $C_{36}H_{55}O_{11}N$. From the roots and rhizomes of *V. viride*, KLOHS, ARONS, DRAPER, KELLER, KOSTER, MALESH and PETRACEK (1952) have isolated neoprotoveratrine, a new hypotensively active tetra-ester of protoverine and have also shown that protoveratrine is present in *V. viride*.

Neoprotoveratrine, $C_{41}H_{63}O_{15}N$, has m. p. 255.4—255.8° and $[\alpha]_D^{24°}$ —39° ± 2° (c = 1.0 in pyridine) or —3.8° ± 2° (c = 1.0 in $CHCl_3$). The picrate has m. p. 233° (dec.).

The Alkaloids of *Schoenocaulon officinale*.

The seeds of this plant contain cevadine, cevine, cevadilline (sabadilline), sabadine and veratridine. Cevadine is present in greater amount than the other alkaloids.

Cevadine, $C_{32}H_{49}O_9N$, formerly known as "crystallized veratrine", crystallizes in rhombic prisms with 2 mols. of ethanol which are lost at 130—140°. The dry alkaloid melts at 205°, $[\alpha]_D^{17°}$ +12.5° (ethanol). It is insoluble in water, sparingly soluble in ether, more soluble in chloroform and easily soluble in hot ethanol. It is poisonous and sternutatory. The hydrochloride, $B \cdot HCl$, forms needles and the aurichloride, $B \cdot HAuCl_4$, crystallizes in yellow needles, m. p. 182° (dec.). The mercurichloride, $B \cdot HCl \cdot HgCl_2$, forms scales, m. p. 172° (dec.). When warmed with ethanolic potash cevadine is hydrolysed with the formation of cevine, tiglic acid and angelic acid. It forms a benzoyl derivative, m. p. 255° and an o-nitrobenzoyl derivative, m. p. 236°.

Cevine, $C_{27}H_{43}O_8N$ (with 3.5 H_2O), forms triclinic prisms which sinter at 155—160° and melt at 195—200°, $[\alpha]_D$ —17.52° (ethanol). It reduces Fehling's solution. The hydrochloride, $B \cdot HCl$, crystallizes in needles, m. p. 247°, the aurichloride melts at 162° (dec.) and the methiodide, $B \cdot CH_3I \cdot 2 H_2O$, has m. p. 257° (dec.). Dibenzoylcevine has m. p. 195 to 196°, diacetylcevine melts at 190° and di-(o-nitrobenzoyl)-cevine has m. p. 175°, $[\alpha]_D$ —54.7° (ethanol).

Cevadilline, $C_{34}H_{53}O_8N$, is an amorphous base insoluble in ether, which on hydrolysis with ethanolic soda yields *tiglic acid* and *cevilline*, $C_{29}H_{47}O_7N$.

Sabadine, $C_{29}H_{51}O_8N$, crystallizes from ether in needles, m. p. 238—240° (dec.). The hydrochloride, $B \cdot HCl \cdot 2 H_2O$, melts at 282—284° (dec.). The nitrate, $B \cdot HNO_3$, is sparingly soluble in water, and the aurichloride, $B \cdot HAuCl_4$, crystallizes in golden-yellow needles. The base is difficultly soluble in water and ether and insoluble in light petroleum.

The veratrum alkaloids are classified in three groups according to their chemical constitution. Group I includes cevadine, germerine, protoveratridine, protoveratrine, veratridine, germitrine, neogermitrine, germidine, germitetrine, veratetrine, veralbidine and neoprotoveratrine which are ester alkaloids. On hydrolysis these alkaloids yield an alkamine and one or more acids. Group II includes pseudojervine and veratrosine which are glucosides of the veratrum alkamines. Group III includes the alkamines, cevine, jervine, germine, protoverine, rubijervine, isorubijervine and veratramine which are all 27 C compounds. The alkaloid content of *V. album* and *V. viride* is subject to wide variation according to season and climatic conditions. Manceau, Montant and Netién (1946) in an examination of the tissues of *V. album* found an average yield from 0.6% to 0.7% of total alkaloids for the best years. The maximum yield obtained was 1%. The distribution of the total alkaloids in different parts of the plant was as follows: roots contained 75% of the total alkaloids, rhizomes 15% and stems 10%. These workers advocate slow drying of the plants in the shade as exposure to sunlight results in destruction of appreciable amounts of the alkaloids. The seeds of *Schoenocaulon officinale* are stated to contain up to 4.5% of total alkaloids.

2. The Isolation of the Major Alkaloids of *Veratrum* spp.

The general principles underlying the methods of extraction of the alkaloids from *Veratrum* spp. and *Schoenocaulon* are similar. Extraction is carried out in alkaline solution (ammonia) with benzene, chloroform or trichloroethylene. Craig and Jacobs (1942) have used the following procedure for the isolation of *protoveratrine*. Ground commercial roots of *V. album* in 2 kg. portions are thoroughly mixed with 7 litres of benzene and stirred with a mixture of 100 ml. of ammonia (sp. gr. 0.9) and 1 litre of water. The next day the solvent is filtered and the moist solid pressed dry and the latter re-extracted with an additional 5 litres of benzene. This is followed by a third extraction with 5 litres of benzene which is used directly for the first extraction of a second portion of root. The first two extracts obtained from a total of 8 kg. of roots are concentrated *in vacuo* to 3 litres. The clear, dark coloured solution is extracted six times with 1 litre portions of 5% acetic acid. The acid extracts are shaken with 2 litres of benzene and made alkaline with excess of 25% NaOH. The extraction with repeated portions of benzene must be carried out as quickly as possible to avoid crystallization of the alkaloid. The benzene extract is washed with water several times, during which a small amount of sparingly soluble alkaloid material (proto-veratridine) remains suspended in the aqueous phase. The benzene layer is dried and evaporated to dryness. The resinous residue is dissolved in about 500 ml. of dry ether when crystallization occurs. After 18 hours the solid is collected. From 8 kg. of root, 9.2 g. of crude protoveratrine is obtained. For recrystallization this is suspended in 10 parts of hot 95% ethanol and an excess of acetic acid added. On addition of a slight excess of ammonia the alkaloid rapidly separates as a crystalline powder. On concentration of the mother liquor *in vacuo* a second crop of crystals can be obtained.

For the isolation of *jervine* JACOBS and CRAIG (1945) extract the powdered roots and rhizomes of *V. viride* with benzene and dilute ammonia as described above. The combined benzene extracts from 6 kg. of material are concentrated under reduced pressure and brought to 4 litres in benzene. This solution is extracted repeatedly with portions of a total of 5 litres of 5% aqueous acetic acid. The free bases are re-precipitated from the aqueous extract with excess of 25% NaOH and then re-extracted with benzene. A small persistent precipitate which gradually separates at the interface is discarded and the benzene solution after washing with water is dried over anhydrous sodium sulphate. The remaining alkaline aqueous phase by further repeated extractions with chloroform yields *germine*. The benzene solution is concentrated to dryness and gives a partly crystalline residue (approx. 65 g.) of a crude alkaloid mixture. A solution of this material in 950 ml. of 5% acetic acid is treated with 135 ml. of a saturated solution of $(NH_4)_2SO_4$. The precipitate is re-suspended in fresh wash solution and re-centrifuged several times to ensure complete washing of the crude sulphate. The sulphate is re-suspended in water and treated with excess of NaOH solution and the mixture shaken thoroughly with chloroform to re-dissolve the liberated bases. The extract is separated by centrifugation and after washing with water is cleared with Na_2SO_4 and concentrated to small bulk. After addition of absolute ethanol, all residual chloroform is boiled off. To the mixture (250 ml.) which partly crystallizes, HCl (sp. gr. 1.19) is carefully added in slight excess. Crystals of *jervine hydrochloride* are deposited. After cooling and collection with absolute ethanol, the yield is approx. 32 g. To obtain the base, the hydrochloride is dissolved in hot, somewhat diluted ethanol and decomposed with ammonia when the base separates as needles.

Veratramine is obtained from the mother liquor from crude jervine hydrochloride by treating with excess of ammonia and carefully diluting to incipient turbidity. *Veratramine* gradually separates and may be purified by re-crystallization from 95% ethanol. The combined mother liquor and washings of the crude insoluble sulphate fraction yield a mixture of bases including *rubijervine* and *isorubijervine*. From the above-mentioned alkaline aqueous phase remaining after the re-extraction of the crude alkaloid mixture with benzene, germine is obtained by extraction with twelve portions of chloroform. The combined chloroform extracts after drying and concentration yield a residue which when dissolved in methanol and seeded with germine give approx. 0.23 g. of the alkaloid.

The isolation of the glycosidic alkaloids *pseudojervine* and *veratrosine* is described by JACOBS and CRAIG (1944). Roots and rhizomes of *V. viride* are dried and powdered. The material in amounts of 2 kg. is moistened with dilute ammonia and first extracted with benzene as described above. The remaining material is covered with 6 litres of 95% ethanol and, after standing at room temperature for a day or so, is filtered and the residue pressed. The solid is again suspended in 4 litres of solvent to which 50 ml. of ammonia (sp. gr. 0.9) have been added. The extraction is repeated a third time with 4 litres of ethanol.

The combined extracts are concentrated *in vacuo* to remove all the ethanol, and towards the end of the operation a little octyl alcohol is added to control foaming. The dark-coloured resinous aqueous mixture is treated with an excess of 20% Na_2CO_3 solution and shaken with an equal volume of chloroform. The mixture is allowed to stand overnight for separation. The large fraction of undissolved resinous material is macerated with about 250 ml. of water, added to the original aqueous phase and extracted with fresh chloroform. A resin again forms at the interface and its extraction with chloroform, after suspension in water, is repeated several times. In this manner, from a total of 6 kg. of root, about 5 litres of combined chloroform extract are obtained. The combined extracts are washed with water and repeatedly extracted with 2.5% tartaric acid solution. The resinous suspension which collects at the interface is separated by centrifugation and re-extracted with acid after re-suspension. The combined tartaric acid extracts are made alkaline with excess of NaOH and re-extracted with chloroform. The chloroform extract is cleared with Na_2SO_4 and concentrated to a thin syrup. Addition of ether precipitates about 20 g. of crude alkaloid. 15 g. of this material are warmed with 75 ml. of 95% ethanol. Solution is followed by rapid crystallization of a fine coloured powder. This powder is dissolved in a mixture of 50 ml. of methanol and 3 ml. of acetic acid, cleared with charcoal and the filtrate treated with an excess of ammonia. On dilution, lustrous leaflets of *pseudojervine* are obtained. The ethanolic mother liquor remaining after the separation of the crude pseudojervine contains veratrosine. On careful dilution of the mother liquor, a small amount of a suspension of very fine crystals of pseudojervine appears which cannot easily be separated by filtration. Dilution is continued to incipient turbidity and after several days a coloured crystalline deposit is obtained. This material is dissolved in 100 ml. of methanol with the aid of 4 ml. of 50% acetic acid, the solution treated with 5 ml. of ammonia and allowed to stand. Apart from flocculent material no crystallization occurs and the mixture is filtered and washed with methanol. To the filtrate (150 ml.) 50 ml. of water are added when crystallization gradually takes place. After standing, the crystals are collected with a mixture of methanol and water (3:1) in which a certain amount of solid dissolves. About 3 g. of crude *veratrosine* are obtained which can be purified by dissolving in 25 ml. of hot methanol and filtering rapidly. From the filtrate after dilution, masses of delicate needles separate which melt with effervescence at 242—243°.

3. Qualitative Tests for the Veratrum Alkaloids.

a) **Precipitation Reactions.** Most of the alkaloid reagents give precipitates with the veratrum alkaloids. Mayer's reagent is sensitive and is the most useful reagent in testing for completeness of extraction in isolation processes.

b) **Colour Tests.** When treated with conc. H_2SO_4 the veratrum alkaloids give a series of colours ranging from yellow through green to violet or from orange through red to crimson. When mixed with about 5 times their weight of sucrose and moistened with conc. H_2SO_4, some of the alkaloids give a green, blue or violet coloration. Cons. HCl gives on warming with protoveratrine a cherry-red colour. Fröhde's reagent gives colours resembling those given by H_2SO_4. Commerical "veratrine" gives a red colour when evaporated to dryness with a little fuming nitric acid and the residue treated with ethanolic potash (Vitali test). The addition of a drop of bromine water to a solution of the alkaloids in cons. H_3SO_4 accelerates the colour changes given by H_2SO_4 alone. A summary of the colour tests is given in Table 9.

Table 9. *Colour Reactions of the Veratrum Alkaloids.*

Alkaloid	Reagents		
	H_2SO_4	H_2SO_4 + sucrose	HCl (on warming)
Cevadine	yellow-green fluorescence-orange — crimson	deep green-blue	—
Cevine	crimson	brown	—
Sabadine	yellow-green fluorescence-blood-red — violet	—	—
Jervine	yellow-bright green	violet-blue	—
Rubijervine	yellow — reddish brown	—	reddish violet
Pseudojervine	yellow — bright green	—	—
Protoveratrine	green-blue-violet	green-dark brown	cherry-red
Neoprotoveratrine	slow greenish blue	—	—
Veratridine	orange-red-crimson	—	—
Germerine	Colourless at first, carmine-red on standing or warming	—	—

A modification of the H_2SO_4 — sucrose test using furfural in place of sucrose gives a characteristic play of colours. When the alkaloid is treated with a few drops of a solution containing 1 ml. of 1% furfural in 10 ml. of conc. H_2SO_4, a blue coloration is formed which changes to olive-green, dark green, blue and finally violet. On warming gently, the colour changes to cherry-red, violet-red and dark red.

4. The Quantitative Estimation of the Veratrum Alkaloids.

Of the few available chemical methods for the determination of the total alkaloids those depending on extraction of the alkaloids from ammoniacal solution by organic solvents with subsequent titration of the liberated bases give the most satisfactory results. The method of GSTIRNER (1932) estimates the total alkaloids in the seeds of *Schoenocaulon officinale*. 5 g. of the finely powdered seed material are placed in a 150 ml. bottle; 50 g. of ether and 5 g. of 10% aqueous ammonia are added and the mixture shaken at intervals for 30 minutes. The ether solution is passed through a 15 cm. folded filter into a 200 ml. flask. 30 g. of the filtrate are evaporated on a steam bath and the weight of the flask is determined when cold. The residue is dissolved in 10 ml. of ether and 25 ml. of 0.5% HCl are added. The flask is shaken vigorously and the ether distilled off. After cooling, more 0.5% HCl is added until the liquid in the flask amounts to 30 g. The solution is filtered through a 10 cm. folded filter and 25 g. of the filtrate made alkaline with 10% aqueous ammonia. The solution is extracted with three portions of 15, 10 and 10 ml. of chloroform by shaking for a period of 2 minutes. Each portion is passed through a 5.5 cm. filter into a 150 ml. flask, the chloroform distilled off and 10 ml. ethanol added to the residue. After the residue is dissolved, 10 ml. of water are added and the solution titrated with 0.1 N HCl using methyl red as indicator. 1 ml. of 0.1 N HCl is equivalent to 0.0625 g. of alkaloids calculated as "veratrine" (crystallized veratrine = cevadine). A biological method based on the lowering of blood pressure in the dog has been described by MAISON, GOTZ and STUTZMANN (1951). Biological methods however are time-consuming and give an approximate estimate of the hypotensive alkaloids alone. WALASZEK and PIRCIO (1952) have made a polarographic study of some of the alkaloids of *V. viride* and have shown that polarographic analysis may be used satisfactorily for the determination of the alkaloids provided they are present in pure solution.

For the determination of the hypotensive ester alkaloids PAPINEAU-COUTURE and BURLEY (1952) have used an infra-red spectrophotometric method. They found that the intensity of the infra-red band due to the carbonyl double bond of the ester grouping provided an index of the concentration of these alkaloids. The intensity of the ester band was found to be parallel to the biological activity.

O. The Lobelia Alkaloids.

The genus *Lobelia* contains alkaloids which have been classified by WIELAND into three groups (MANSKE and HOLMES, 1950). The lobeline group contains the major alkaloids and will be considered in some detail.

I. Properties.

1. Lobeline Group.

L-Lobeline ($C_{22}H_{27}O_2N$) is a monoacidic tertiary base, m. p. 130—131°, $[\alpha]_D^{15°}$ —42.85° (ethanol), easily soluble in hot ethanol or benzene, sparingly soluble in water and light petroleum. It crystallizes from ethanol, ether or benzene in colourless needles and forms crystalline salts; the hydrochloride, m. p. 182°, is soluble in chloroform. With benzoyl chloride it forms benzoyl lobeline hydrochloride, m. p. 155—157° (dec.). On warming with water lobeline decomposes yielding acetophenone and on reduction with sodium amalgam and acetic acid it is converted in *lobelanidine*.

DL-Lobeline melts at 110° and forms a hydrochloride, m. p. 170°.

Lobelanine ($C_{22}H_{25}O_2N$) crystallizes in needles from ether or light petroleum, m. p. 99°. It is optically inactive and is easily soluble in ethanol, benzene and chloroform but almost insoluble in water, and forms salts which crystallize well. The hydrochloride, B · HCl, melts at 188° with decomposition, the hydrobromide also melts at 188°, the nitrate has m. p. 153 to 154°, the hydriodide m. p. 169—172° and the perchlorate, B · HClO$_4$, m. p. 173—174°.

Norlobelanine ($C_{21}H_{23}O_2N$) melts at 120—121° and forms a nitrate, B · HNO$_3$, m. p. 193° (dec.) and hydrochloride, m. p. 201—202°.

Lobelanidine ($C_{22}H_{29}O_2N$) forms colourless scales, m. p. 155°, and is insoluble in water but easily soluble in acetone, benzene and pyridine. It is much less soluble in cold ethanol and sparingly soluble in ether and light petroleum. The principal salts are the hydrochloride, m. p. 138°, and the hydrobromide, m. p. 188—190°. The dibenzoyl derivative ($C_{36}H_{37}O_4N$) has m. p. 109—110°.

Norlobelanidine ($C_{21}H_{27}O_2N$) forms needles, m. p. 120°. The hydrochloride also forms needles which melt at 244° with decomposition.

2. Lelobine Group.

The minor alkaloids included in this group are DL-*lelobanidine*, L-*lelobanidine*-I, L-*lelobanidine*-II and D-*Norlelobanidine*.

3. Lobinine Group.

This group contains the minor alkaloids *lobinine*, *isolobinine*, *lobinanidine* and *isolobinanidine*. Groups 2 and 3 will not be further considered as the alkaloids in these groups form but a small proportion of the total alkaloids present in the plant.

The chief source of the Lobelia alkaloids is the annual plant *Lobelia inflata* L. widely distributed over the eastern parts of N. America. Lobeline is the major alkaloid of this plant and the content increases rapidly with growth of the plant and reaches a first maximum shortly before blooming and a second maximum at the time flowering ceases. As the plant dies off, a rapid decrease takes place. The lobeline values for different parts of the plant are given by ESDORN (1940) as follows. The flowering apex contains 0.9—1.1%, the unripe capsule 0.88 to 1.05% the leaves 0.42—0.43% the stems 0.35—0.38% and the roots 0.54—0.56%. According to KALASHNIKOV (1939) the seeds contain 0.21—0.23% and this author states that the quantity of alkaloid in the plant varies with conditions of growth. Other species of *Lobelia* which have been examined are *L. cardinalis* L. (0.445% total alkaloids), *L. urens* L. (0.752%) and *L. syphilitica* L. (0.535%). *L. erinus* L. contains traces only.

II. Estimation of Alkaloids in Lobelia Tissues.

Titrimetric Method. As lobeline is unstable when warmed in aqueous solution, it is recommended that tissues should be dried at a temperature not exceeding 40°. For the estimation of total alkaloids of *Lobelia*, LYNCH and EVERS (1939) have subjected to critical examination the method of MARKWELL (1936) which was found to give satisfactory results. The method is as follows:

10 g. of finely powdered material and 10 g. of ignited sand are introduced into a pear-shaped separating funnel provided with a plug of cotton wool in the tube below the stopcock. 75 ml. of a mixture of 4 vols. ether and 1 vol. 95% ethanol are added and the funnel shaken and set aside for 15 mins. 5 ml. of 10% solution of ammonia are added and the funnel shaken in a mechanical shaker for 1 hour or by hand for 1 minute at 10 minute intervals during 1 hour. The liquid is allowed to percolate into another separating funnel and when percolation is complete, the drug is packed firmly in the separating funnel by means of a button-ended glass rod. The percolation is continued, first with 25 ml. of the ether-ethanol mixture and then with ether until the alkaloid is completely extracted as shown in the usual manner (see "General Introduction", p. 373). 30 ml. of N H_2SO_4 is added to the percolate (HCl should not be used) and the funnel is shaken well and the phases allowed to separate. The lower layer is run off into another separator and the extraction repeated with a mixture of 25 ml. of 0.5 N H_2SO_4 with 5 ml. of ethanol (95%). The lower layer is run off and the extraction repeated with three further quantities of 20 ml. of the acid-ethanol mixture or until the alkaloids are completely extracted. The mixed acid solutions are washed, first with 10 ml. and then with successive quantities of 5 ml. of chloroform, each chloroform solution being washed with the same 20 ml. of 0.5 N H_2SO_4 contained in another separator. The chloroform is rejected and the acid solution transferred from the second to the first separator and neutralized to litmus with dilute ammonia solution. A further 5 ml. of ammonia solution is added in excess. The alkaloids are now extracted by shaking with successive quantities of 10 ml. of chloroform. The chloroform solutions are combined, washed with 3 ml. of distilled water, and filtered through a 7 cm. filter paper into a flask. The filter is washed thoroughly with more chloroform and the washings collected in a flask. The chloroform is next removed on the water bath until about 2 ml. remains. 2 ml. of absolute ethanol are added and the evaporation continued on the water bath using a gentle air blast to complete the process. To ensure dehydration of the residue the process is repeated with two further quantities of absolute ethanol. The residue is now heated for 1 hour at 80° and 2 ml. of 95% ethanol are added, warming if necessary, to dissolve the residue. 10 ml. of 0.02 N H_2SO_4 are added and the solution is titrated with 0.02 N NaOH or sodium borate using methyl red as indicator. 1 ml. of 0.02 N H_2SO_4 is equivalent to 0.00674 g. of the alkaloids of *Lobelia* calculated as lobeline.

Acetophenone Method. UFFELIE (1946) has developed a specific method for the determination of lobeline in *Lobelia* which depends on the fact that when lobeline is distilled with steam from a solution of pH 6—10, one molecule of acetophenone is quantitatively split off. The acetophenone can be estimated by conversion to iodoform. The author claims that the results obtained by the application of this method are more constant than those obtained by the titration method. For the determination of lobeline in *Lobelia* 10 g. of the finely powdered drug is shaken in a 200 ml. flask for 30 minutes with 100 g. ether and 7 g. of ammonia water. The ethereal solution is filtered through a plug of cotton wool and 70 g. representing 7 g. of the drug is then filtered through filter paper into a separating funnel. This solution is shaken with successive quantities of 10,5 and 5 ml. of 0.1 N HCl. The acidified aqueous solution is filtered into a 100 ml. round bottomed flask and the filter washed with water. Two drops of aqueous bromthymol blue (1 in 500) are added followed by a saturated solution of sodium phosphate (Na_2HPO_4) in sufficient quantity to turn the solution green. The solution is distilled with steam and the distillate condensed in an upright condenser and received in an iodine addition flask under a small quantity of water. Distillation is continued until 50 ml. have passed over. 5 ml. of 50% NaOH and 50 ml. of 0.01 N iodine are added to the distillate, the flask stoppered, and the solution allowed to stand for 30 minutes. 10 ml. of 38% HCl are then added and the flask again stoppered. After cooling, the free iodine is back titrated with 0.01 N sodium thiosulphate. A blank is run containing 50 ml. of 0.01 N iodine. The difference in ml. of 0.01 N sodium thiosulphate multiplied by 0.562 is equivalent to the concentration in mg. of lobeline in 7 g. of the original drug.

Silicotungstic Method. Another method has been described by Mascré and Caron (1930) who determine total alkaloids by precipitation with silicotungstic acid. 12.5 g. of powdered drug is treated with a mixture of 40 g. ethanol (95%) and 4 g. NH_4OH. After allowing to stand for some hours, 360 g. ether are added and the mixture shaken periodically during 4 hours. The solution is filtered and 360 ml. are taken and evaporated down to 10 ml. 150 ml. ether are added and the alkaloids are shaken out with 15 ml. N HCl and then four times with 5 ml. N HCl. The ether is evaporated from the combined acid extracts at low temperature and 10 ml. of silicotungstic acid (5%) are added. After a period of 12 hours the precipitate of alkaloid silicotungstate is filtered off, washed well with N HCl and incinerated. The weight of residue × 0.414 gives the weight of total alkaloids.

This method tends to give slightly higher results than the titrimetric method.

P. The Alkaloids of *Lolium perenne* L. and Other Grasses.

The fluorescent alkaloid perloline was isolated from rye grass by the New Zealand workers Grimmett and Melville (1943). In addition to perloline other alkaloids including perlolidine, an oxidation product of perloline, are also present in small quantity in the tissues of rye grass. Perloline also occurs in *Lolium temulentum* L., *Festuca arundinacea* Schreb., *Setaria lutescens* (Weigel) F. T. Hubb and in traces in *Lolium multiflorum* Lam. At the time of flowering of *L. temulentum* and *S. lutescens* the highest concentration of perloline was found in the stems (65—120 mg. per cent), with less in the leaves (a trace to 30 mg. per cent) and heads (19—28 mg. per cent). Traces were also found in the roots but none in the seeds. Many other species of *Gramineae*, *Cyperaceae* and *Juncaceae* were examined for perloline but none was found. This alkaloid produces mild toxic effects when introduced parenterally into animals of various species but the amounts ingested by grazing animals appear not to have harmful effects.

1. Properties.

Perloline, $C_{40}H_{24}O_7N_4 \cdot H_2O$ (White and Reifer, 1945), occurs as pale yellow needles, m. p. 181°. The alkaloid is somewhat soluble in ethanol, chloroform, or acetone and very slightly soluble in water, ether or benzene. Solutions of the base in ethanol or chloroform are yellow and show a marked green fluorescence in reflected light which can be detected at a concentration of 1 part in 5 million in ordinary light. The solution of the base in chloroform is unstable in bright sunlight and loses its fluorescense quickly. The dihydrochloride, $B \cdot 2\,HCl \cdot 7\,H_2O$, crystallizes in golden-yellow needles, m. p. 220—265° and the diperchlorate crystallizes from acetone-water in yellow needles, m. p. 284° (dec.) with slight sintering from 235°. The mercurichloride exists in two forms. When neutral solutions of perloline are treated with an excess of mercuric chloride yellow crystals are formed which when recrystallized from hot water melt at 265°. From acid solutions the precipitate is orange-red and when recrystallized from hot water melts at 201°. The dipicrate on recrystallization from acetone-water forms yellow needles, m. p. 242°. The reineckate forms yellow prisms from acetone-water which decompose at 195—205°. Perloline is optically inactive.

2. The Isolation of Perloline.

GRIMMETT and WATERS (1942) have described a method for the large-scale extraction of perloline from fresh rye grass but dried material is to be preferred and has been used by REIFER and BATHURST (1943) for the isolation of the alkaloid. Dried and ground rye grass in 500 g. lots are boiled under reflux with 4 litres of ethanol and 80 ml. of acetic acid for 30 minutes. The extract is separated by pressing and the press-cake boiled once more with the same volume of ethanol containing 40 ml. of acetic acid. The ethanol is removed by distillation which is allowed to proceed until the temperature reaches 91—92°. The residue is diluted with 3 volumes of water, neutralized with NaOH, slightly acidified with HCl and allowed to cool. The greater part of the aqueous solution can be removed by decantation from the slurge of waxes and pigments which separate out. The sludge is washed with warm water and the extract and washings filtered, 20 ml. of a saturated solution of potassium dichromate added and the solution allowed to stand overnight. It is now made alkaline with Na_2CO_3 and exhaustively extracted with chloroform until no further colour appears in the solvent layer. The chloroform extracts are shaken out with dilute HCl as soon as possible after separation. The combined chloroform residues from which no further alkaloid can be extracted with HCl are distilled from a flask containing dilute acid until the solvent is completely removed. The acid layer contains considerable amounts of fluorescent material which can be recovered. For purification, the acid solutions of perloline are made alkaline with Na_2CO_3 and shaken out with chloroform until the extracted layer is practically colourless. This operation is repeated several times. Emulsions are likely to occur and use of the centrifuge may be necessary. The free base is then precipitated by the addition of Na_2CO_3 solutions to an approx. 1% solution of the hydrochloride until no further precipitate forms. The precipitate is centrifuged off, washed once with cold water and dissolved in dilute acid. The free base is once more liberated, extracted with chloroform and then shaken out into a small volume of 2% HCl. This solution is evaporated *in vacuo* to a small bulk and left overnight in the refrigerator when the hydrochloride of perloline separates out in golden-yellow needles. Further purification of the hydrochloride is effected by crystallization from hot water.

SHORLAND (1943) has shown that the absorption curves for the free base in chloroform and the hydrochloride in 0.1 N HCl have maxima at 470 mμ and 398 mμ respectively. The corresponding values for $E_{1\,cm}^{1\%}$ are 325 and 530 respectively.

3. Precipitation and Colour Reactions of Perloline.

REIFER and BATHURST (1943) give the following details of precipitation reactions, with numbers in parentheses representing the greatest dilution at which the reaction can be detected. Picric acid gives a crystalline precipitate (1:5,000), silver nitrate (1:5,000), mercuric chloride (1:2,000), phosphomolybdic acid (1:20,000), phosphotungstic acid (1:100,000), gold chloride (1:5,000, crystalline), platinic chloride (1:1,000 crystalline), potassium periodide (1:100,000, crystalline), potassium bismuth iodide (1:10,000, crystalline), potassium mercuric iodide (1:100,000), ammonium reineckate (1:10,000).

Perloline gives very few characteristic colours with the usual reagents. With Mandelin's reagent it gives a brown colour, and with titanous oxide in H_2SO_4, a brick-red colour.

4. The Quantitative Estimation of the Total Alkaloids.

Reifer and Bathurst (1942) have described a method for the determination of the total alkaloids of rye grass in which the alkaloids are precipitated with potassium mercuric iodide followed by titrimetric estimation with ceric sulphate. The basis of the method is the oxidation of the potassium mercuric iodide precipitate to iodine and undissociated mercuric iodide.

5 g. (accurately weighed) of dried rye grass are placed in the thimble of a Soxhlet and the material covered with a cotton wool plug. The extraction is carried out with a solution of ethanol containing phosphoric acid (100 ml. ethanol containing 1 ml. of phosphoric acid). The grass powder is moistened with a little of this solution and the remainder placed in the extraction flask. Extraction proceeds for 1 hour using an electric hot plate. After cooling, the thimble is removed and the bulk of the ethanol distilled over into the empty extraction compartment. After the addition of about 25 ml. of water into the flask, the remainder of the ethanol is carefully evaporated on the hotplate. The cooled solution is filtered into a 100 ml. separating funnel, the filter washed with water, and the solution made alkaline with Na_2CO_3. The liquid is then shaken with 20 ml. of chloroform. After separation the chloroform solution is transferred to a 50 ml. separating funnel containing about 10 ml. of dilute HCl (2 ml. conc. HCl in 100 ml.) and the alkaloid shaken from the chloroform into the acid. The chloroform is then removed. The extraction of the alkaline solution is then repeated once or twice using 20 ml. of chloroform as above and shaken into the 10 ml. of dilute acid already containing the greater portion of the extracted alkaloids. This acid solution is again made alkaline with Na_2CO_3, extracted twice with 10 ml. of chloroform and the alkaloids removed by shaking with 8 ml. of 0.1 N HCl in a 25 ml. separating funnel. The dilute solution containing the alkaloids is now partly neutralized by making the volume to 10 ml. with 0.25 N NaOH. For determination of total alkaloids by precipitation with potassium mercuric iodide, 4 ml. of this solution, corresponding to 2 g. of dried grass, are transferred to a 15 ml. centrifuge tube and 1 ml. of alum solution (2 g. of potassium alum, $KAl(SO_4)_2 \cdot 12 H_2O$, made up to 100 ml. with water) added, followed by 1 ml. of phosphate solution (20 g. $KH_2PO_4/100$ ml.). The precipitated aluminium phosphate aids the precipitation of the alkaloid and prevents the precipitate from floating on the surface of the solution. 0.5 ml. of Mayer's reagent is added dropwise from a burette and the tube shaken after each addition (Mayer's reagent is prepared by dissolving 10 g. KI and 15 g. HgI_2 in a little water and making up to 100 ml., 10 ml. of this solution is diluted to 100 ml., left to stand for 24 hours and filtered). After allowing to stand for 5 minutes, the tubes are centrifuged for 10 minutes at 2000—2500 r. p. m., the supernatant sucked off and the precipitate stirred (using a glass rod) with 10 ml. of water. After centrifuging for 10 minutes the supernatant is sucked off and the centrifuge tube left standing upside down on a piece of filter paper for 2—3 minutes. The precipitate is suspended in 2 ml. of 5% Na_2CO_3 solution and with the aid of a little water transferred quantitatively into a 50 ml. separating funnel. The centrifuge tube is washed with 10 ml. of

chloroform, the tube stoppered, well shaken and the chloroform transferred to the separating funnel. The alkaloid is shaken into the chloroform and after standing for 2 minutes, the chloroform layer is removed. The shaking is repeated with 10 ml. of chloroform as above and the chloroform layer removed. The supernatant alkaline solution is transferred to a 50 ml. Erlenmeyer flask and the separating funnel washed first with 2 ml. of dilute H_2SO_4 solution (3% v/v) and then twice with 2—3 ml. of water, the washings being added to the solution in the Erlenmeyer flask. The combined solutions are boiled for 3 minutes after the addition of about 0.5 g. of glucose which makes the end point of the titration clearer. After cooling, 0.1 ml. of indicator solution (100 mg. diphenyl benzidine dissolved in 10 ml. of conc. H_2SO_4 and made up to 100 ml. with glacial acetic acid) is added, followed by 3 ml. of carbon tetrachloride and the solution titrated with 0.001 N ceric sulphate (standardized against arsenious oxide). As free iodine is produced during the titration it is necessary to add very small quantities (0.2 to 0.3 ml.) of ceric sulphate at a time with shaking after each addition. The end point (i. e. the first drop in excess of ceric sulphate) is reached when the aqueous layer turns bluish-violet. A blank is run, substituting water for the alkaloid solution. The factor for calculation of perloline hydrochloride is 0.423 and for perloline, 0.382. The difference between the titration values in ml. of the estimation and the blank multiplied by the factor gives the amount of alkaloid present. Precipitation with potassium mercuric iodide is not a specific method for the estimation of alkaloids but according to REIFER and BATHURST (1942) most of the non-alkaloidal compounds precipitated with MAYER's reagent are not soluble in chloroform. As different alkaloids combine with different quantities of MAYER's reagent the titration results are expressed in terms of perloline, using the factor. CLARE and MORICE (1945) have used DRAGENDORFF's reagent for precipitation of the total alkaloids of rye grass. The precipitate is decomposed with Na_2CO_3 and the alkaloids extracted with chloroform. The chloroform solutions are extracted with a little dilute HCl and the solution dried to constant weight under reduced pressure. The amount of alkaloid is found by weighing.

Photometric Method. BATHURST, REIFER and CLARE (1943) have described a photometric or spectrophotometric method for the estimation of perloline in small samples of rye grass. The high tinctorial power of a chloroform solution of the alkaloid is the basis of the method. According to the concentration of the alkaloid, 1 g. or 2 g. of the dried and finely ground tissue are weighed out from a well-mixed bulk sample and boiled for 2 minutes with 100 ml. of dilute HCl (2 ml. conc. HCl s.g. 1.18 in 100 ml. of water) for each gram of tissue taken for analysis. The hot suspension is filtered on a small BUCHNER funnel and the residue washed twice with approx. 20 ml. of water. The combined extract and washings are cooled, made distinctly alkaline with Na_2CO_3 solution and shaken with chloroform (three or more successive portions of 20 ml. of chloroform) until clear. The chloroform layers require separation by use of the centrifuge. The combined chloroform extracts are shaken with 5 ml. of dilute HCl and the chloroform discarded. The acid solution is again made alkaline and shaken with successive portions of chloroform until no further colour is extracted. The combined chloroform extracts are made to volume, dried with Na_2SO_4 and the colour measured either with a Zeiss Pulfrich photometer (S 47 filter and 1 cm. cell) or spectrophotometrically at a wavelength of 470 mμ using a 2 cm. cell. Calibration curves are prepared using purified samples of perloline. The validity of this method for the estimation of perloline depends on the absence of significant amounts of other pigmented substances in the chloroform extract used for the final estimation.

Q. The Alkaloids of Opium.

Opium, the dried latex of the unripe capsules of *Papaver somniferum* L. contains, so far as is known, twenty five alkaloids which have been divided into the following groups.

1. **Morphine Type.** Morphine, $C_{17}H_{19}O_3N$. Pseudomorphine, $(C_{17}H_{18}O_3N)_2$. Codeine, $C_{18}H_{21}O_3N$. Neopine, $C_{18}H_{21}O_3N$. Thebaine, $C_{19}H_{21}O_3N$. Porphyroxine, $C_{19}H_{23}O_4N$.

2. **Benzylisoquinoline Derivatives.** Papaverine, $C_{20}H_{21}O_4N$. Papaveraldine (Xanthaline), $C_{20}H_{19}O_5N$. DL-Laudanine, $C_{20}H_{25}O_4N$. L-Laudanine. Laudanosine, $C_{21}H_{27}O_4N$. Codamine, $C_{20}H_{25}O_4N$.

3. **Phthalideisoquinoline Derivatives.** Narcotine, $C_{22}H_{23}O_7N$. Narcotoline, $C_{21}H_{21}O_7N$. Narceine, $C_{22}H_{27}O_8N$. Oxynarcotine, $C_{22}H_{23}O_8N$.

4. **Cryptopine Type.** Protopine, $C_{20}H_{19}O_5N$. Cryptopine, $C_{21}H_{23}O_5N$.

5. **Tetrahydroisoquinoline Derivatives.** Hydrocotarnine, $C_{12}H_{15}O_3N$.

6. **Alkaloids of Unknown Constitution.** Aporeine, $C_{18}H_{16}O_2N$. Rhoeadine, $C_{21}H_{21}O_6N$. Meconidine, $C_{21}H_{23}O_4N$. Papaveramine, $C_{21}H_{25}O_6N$. Lanthopine, $C_{23}H_{25}O_4N$.

The major alkaloids are morphine, narcotine and codeine. Macedonian and Turkey opium yields 15—21% of morphine, and good quality Persian opium contains 10—12% of morphine. Indian opium yields from 3 to 15%, average 7.5—10%. Indian smoking opium contains 4—6.5% and Chinese opium contains from 4.3 to 11.2% of morphine. The alkaloid exists in combination with meconic acid and sulphuric acid. The content of narcotine varies from 2 to 8% on average, but may reach 12%. Indian and Persian opium are richer in narcotine than the Turkey variety. Narcotine, being a weak base, is said to exist in the free state in the plant, but probably exists to some extent in combination with meconic acid. Codeine occurs in combination with acids, to the extent of from 0.3—4% and is present in greatest amount in Persian and Indian opium. The minor alkaloids, including papaverine, thebaine and narceine occur in much smaller amounts. The content of thebaine varies from 0.2 to 0.5%, narceine from 0.1 to 0.4%, and the papaverine content is approx. 0.8%. The remaining bases are present in amounts less than 0.1%. Meconic acid, $C_7H_4O_7 \cdot 3H_2O$, is characteristic of opium and exists to the extent of approx. 5% in combination with morphine and other alkaloids. Although the concentration is greatest in opium, alkaloids are found in all organs of *P. somniferum*. Poppy straw contains on average from 0.07—0.19% of alkaloids, leaves, 0.06—0.07% and capsules 0.3—0.5%.

Poethke and Arnold (1951) and Wegner (1951b) have determined the amount of morphine present in the different organs of the poppy plant in relation to growth. Young roots of the experimental plants had a concentration of 0.34 to 0.45% whereas in old roots the percentage had fallen to 0.09—0.19%.

During the active vegetative period the content of leaves increased to 0.10 to 0.24% at the time of bud formation and then decreased to 0.03—0.05% at harvest time. At the time of bud formation the morphine content of stalks was 0.14 to 0.34% and increased to 0.34—0.47% at the time of capsule formation. By harvest time the content of stalks had decreased to 0.08—0.15%. It would appear that most of the morphine moves to the capsule during the development of that organ.

I. Properties.

1. Alkaloids of the Morphine Type.

Morphine, $C_{17}H_{19}O_3N$, is the most important of the opium alkaloids and crystallizes from dilute ethanol in colourless trimetric prisms containing one molecule of water. At 100° it loses the water of crystallization and melts with decomposition at 254°. Morphine is bitter to the taste and is sparingly soluble in most solvents. The solubility in water is 1 in 3533; in ether, 1 in 7632; in benzene, 1 in 1599; in chloroform, 1 in 1525; in cold ethanol, 1 in 250, and in boiling ethanol, 1 in 30 to 1 in 36. Morphine is soluble in lime-water (1 in 100 at 25°), in alkali hydroxide solutions and to a limited extent in ammonia solution (1 in 117, sp. gr. 0.97). Solutions of morphine are strongly laevorotatory, $[\alpha]_D^{23°}$ —130.9° (CH_3OH). It is a monoacidic base and forms salts which crystallize well and are neutral to litmus. Morphine can be titrated with accuracy with standard acid if methyl red is used as the indicator. The hydrochloride, $B \cdot HCl \cdot 3 H_2O$, forms white silky needles from water, $[\alpha]_D^{15°}$ —100.67°. It is soluble in water (1 in 17.2 at 25°) and in ethanol (1 in 42 at 25°). The sulphate, $B_2 \cdot H_2SO_4 \cdot 5 H_2O$, forms masses of silky needles and at 15° is soluble in 20 parts of water. It is soluble in ethanol (1 in 565 at 25°) but insoluble in ether and chloroform. It is laevorotatory, $[\alpha]_D^{15°}$ —100.47°. The acetate, $B \cdot CH_3 \cdot COOH \cdot 3 H_2O$, is a white crystalline powder, m. p. 200° (dec.), $[\alpha]_D$ —77° (H_2O). It is very soluble in water (1 in 2.25 at 25°), less so in ethanol (1 in 21.6 at 25°) and sparingly soluble in chloroform (1 in 480 at 25°). The tartrate, $B_2 \cdot C_4H_6O_6 \cdot 3 H_2O$, is easily soluble in water but the hydrogen tartrate, $B \cdot C_4H_6O_6$, is sparingly soluble in water. The picrate is soluble in 450 parts of water and has m. p. 163—165°.

Codeine, $C_{18}H_{21}O_3N$, crystallizes from water with 1 H_2O in large translucent orthorhombic prisms, m. p. 155° (dry), $[\alpha]_D$ —137.7° (ethanol) or —111.5° ($CHCl_3$). It has a slightly bitter taste and is moderately soluble in water (1 in 120 at 25°), more so in ether (1 in 75 at 15.5°) and radily soluble in ethanol (1 in 2 at 25°), or chloroform (1 in 0.5 at 25°). It differs from morphine in being fairly soluble in cold benzene (1 in 13) and in its sparing solubility in aqueous solutions of alkali hydroxides. Codeine is a strong monoacidic base and forms salts which are neutral to litmus. Methyl red is a suitable indicator for the titration of codeine. The hydrochloride, $B \cdot HCl \cdot 2 H_2O$, forms needles which are soluble in water (1 in 28.5 at 15.5°), $[\alpha]_D^{22.5°}$ —108.2° (H_2O), m. p. 264° (dry). The sulphate, $B_2 \cdot H_2SO_4 \cdot 5 H_2O$, forms rhombic prisms, m. p. 278° (dec.), $[\alpha]_D^{15°}$ —101.2° (H_2O). It is soluble in water (1 in 30 at 25°) sparingly soluble in ethanol (1 in 1280 at 25°) and insoluble in ether. The phosphate, $B \cdot H_3PO_4 \cdot 1, 1.5$ or $2 H_2O$, forms needle shaped crystals, m. p. 235° (dec.) and is soluble in water (1 in 2.5 at 25°) and in ethanol (1 in 325 at 25°). The picrate, m. p. 196—197°, crystallizes from 50% ethanol.

Thebaine, $C_{19}H_{21}O_3N$, crystallizes from dilute ethanol in leaflets and from dry ethanol in prisms, m. p. 193°, $[\alpha]_D^{15°}$ —218.6° (ethanol). It is almost insoluble in water and sparingly soluble in lime-water and ammonia. It is soluble in ethanol (1 in 10), in ether, (1 in 140 at 10°), and easily soluble in chloroform and benzene. Thebaine behaves as a monoacidic base. The hydrochloride, $B \cdot HCl \cdot H_2O$, forms rhombic prisms soluble in 15.8 parts of water at 10°. The picrate melts at 217° and is very little soluble in water and ethanol. The salicylate, which is sparingly soluble in water, may be used for the separation of thebaine from other opium bases.

2. Benzylisoquinoline Alkaloids.

Papaverine, $C_{20}H_{21}O_4N$, crystallizes in rhombic prisms or needles, m. p. 147°, and is almost inactive optically. It is insoluble in water, soluble in hot ethanol or chloroform and slightly soluble in cold ethanol or ether. It is a weak base and forms a hydrochloride, $B \cdot HCl$, m. p. 225—226°, sparingly soluble in water (1 in 37 at 18°). The picrate forms plates, m. p. 186°; the acid oxalate is sparingly soluble in water and has m. p. 196°.

Laudanosine, $C_{21}H_{27}O_4N$, crystallizes from hot benzene in needles, m. p. 89°, $[\alpha]_D^{15°}$ +103.23° (ethanol). It is soluble in ethanol, chloroform, hot benzene or ether, but insoluble in water and alkalies. The hydriodide, $B \cdot HI \cdot {}^1/_2H_2O$, forms prisms soluble in ethanol and sparingly soluble in water. The acid oxalate, $B \cdot H_2C_2O_4 \cdot 3 H_2O$, forms prisms easily soluble in water.

Laudanine, $C_{20}H_{25}O_4N$, crystallizes from dilute ethanol in rhombic prisms, m. p. 166°. It is soluble in solutions of alkali hydroxides but almost insoluble in ammonia solution. Laudanine is a strong base and forms salts which crystallize well. The hydrobromide, $B \cdot HBr \cdot 3 H_2O$, melts at 76—77° and the hydriodide, $B \cdot HI \cdot H_2O$, is sparingly soluble in water (1 in 500 at 15°). The picrate has m. p. 176—177°.

Laudanidine (L-laudanine) melts at 184—185°. In physical properties this compound resembles laudanine and gives the same colour reactions.

Codamine, $C_{20}H_{25}O_4N$, crystallizes in hexagonal prisms, m. p. 126°. The hydriodide $B \cdot HI \cdot 1.5 H_2O$, is crystalline and sparingly soluble in water. Codamine dissolves in water, giving an alkaline solution, and is soluble in ethanol. It is precipitated by dilute alkali hydroxides and ammonia but is soluble in excess of either reagent.

3. Phthalide-isoquinoline Derivatives.

Narcotine, $C_{22}H_{23}O_7N$, crystallizes from ethanol in needles, m. p. 176°, $[\alpha]_D$ —207.35° (ethanol), —198.0° (chloroform). This base is almost insoluble in water, sparingly soluble in cold 85% ethanol or ether, readily soluble in benzene, acetone or ethyl acetate. It is insoluble in cold alkali hydroxide solutions but soluble in hot alkali hydroxide solutions and in lime water. Narcotine is a very weak base and forms dextrorotatory salts which are unstable and readily dissociated in water. The hydrochloride, $B \cdot HCl$, crystallizes with 0.5 to 4 H_2O and is very soluble in water. The oxalate, $B \cdot H_2C_2O_4$, has m. p. 174° and $[\alpha]_D^{20°}$ +39.5° (H_2O), and the phthalate has m. p. 160° and $[\alpha]_D^{23°}$ +115° (chloroform). The picrate melts at 175° and is little soluble in absolute ethanol.

Narceine, $C_{23}H_{27}O_8N \cdot 3\ H_2O$, forms slender needles or prisms, m. p. 170° or 140—145° (dry). It is optically inactive and is sparingly soluble in cold water (1 in 1285 at 13°) or 80% ethanol (1 in 945 at 13°) but dissolves on heating. Narceine is slightly soluble in chloroform and insoluble in ether and benzene, and is a weak monoacidic base forming salts which are crystalline. The hydrochloride, $B \cdot HCl$, crystallizes from dilute HCl with 5.5 H_2O in the cold and with 3 H_2O from hot solutions. The hydrochloride crystallizes with 1 mol. of CH_3OH if a methanolic solution of hydrogen chloride is used to prepare the salt. This compound melts at 190—192°. The aurichloride forms reddish-yellow needles, m. p. 130° and the picrate melts at 195°.

4. The Cryptopine Type.

Cryptopine, $C_{21}H_{23}O_5N$, occurs in small quantity in opium but is widely distributed throughout the genera *Corydalis* and *Dicentra* of the *Fumariaceae* (HENRY, 1949). This alkaloid, which is optically inactive, crystallizes from ethanol in prisms, m. p. 220—221° (corr.). It is soluble in boiling ethanol (1 in 80) and sparingly soluble in ether or benzene. The salts often appear gelatinous but crystallization can be induced by warming the solution. The hydrochloride, $B \cdot HCl \cdot 5$ or 6 H_2O, forms gelatinous masses of feathery crystals soluble in water or chloroform. The oxalate, $B \cdot H_2C_2O_4 \cdot 4\ H_2O$, is often used for purification of the alkaloid. The picrate forms yellow needles, m. p. 161—163°, the aurichloride, brownishyellow needles, m. p. 205° (dec.) and the platinichloride concentrically arranged needles, m. p. 204° (dec.).

Protopine (Fumarine), $C_{20}H_{19}O_5N$, is present in small quantity in opium but occurs in many genera of the *Papaveraceae* and *Fumariaceae* (HENRY, 1949), DANCKWORTT (1912). The best source of protopine is *Dicentra spectabilis*. This base forms monoclinic crystals, m. p. 207°. It is optically inactive and is soluble in chloroform (1 in 15), less soluble in ethanol (1 in 1,000) acetone or ammonia solution. The hydrochloride, $B \cdot HCl$, forms prisms slightly soluble in water, and the nitrate $B \cdot HNO_3$ is sparingly soluble in water. These salts do not form gelatinous masses as in the case of cryptopine. The aurichloride occurs as a yellow crystalline powder, m. p. 198°, and the platinichloride, $B_2 \cdot H_2PtCl_6 \cdot 2$ or 4 H_2O, forms warty crystals.

II. Isolation of the Major Alkaloids from Opium.

For the extraction of the alkaloids, opium is made into a thin paste with calcium chloride solution and the mass extracted with warm water. The morphine and other bases are thus converted into their hydrochlorides and the acids with which they were combined are precipitated as insoluble salts. The insoluble matter is removed by filtration and the filtrate concentrated in vacuo to a thin syrup. The method of DOTT (1929) can be applied to a solution of the mixed hydrochlorides obtained by the above procedure. The solution is treated with a 10% solution of NaOH when narcotine, papaverine and thebaine are precipitated and removed. The alkaline solution contains the morphine, codeine and narceine. On shaking the solution with chloroform, the codeine is extracted and on separating the alkaline liquid, making acid, and rendering it faintly alkaline with ammonia, the morphine is precipitated. Narceine remaining in solution is recovered by evaporating the liquid to dryness and treating the residue with strong ethanol. Thebaine is separated from the bases precipitated by NaOH by crystallization as the acid tartrate or alternatively the narcotine and papaverine are removed from thebaine by dissolving the mixture of the bases in dilute ethanol, making the solution faintly acid with acetic acid, and adding 3 volumes of boiling water, when the narcotine and papaverine are precipitated. Narcotine and papaverine are separated by solution in boiling water containing 0.33% of oxalic acid when an acid papaverine oxalate crystallizes on standing. This process is repeated several times and the narcotine finally precipitated with ammonia and crystallized from boiling ethanol.

ADAMSON, HANDISYDE and HODGSON (1947) have described a procedure for
the analysis of papaveretum, a pharmaceutical preparation consisting of a mixture
of the hydrochlorides of the opium alkaloids. The method, which involves the
extraction and separation of narcotine, papaverine and codeine, is as follows.
1 g. of the sample of papaveretum is weighed out, dissolved in water, and washed
into a separating funnel containing 40 ml. of chloroform, using a total of 20 ml. of
water. One drop of dilute HCl is added and the mixture shaken vigorously for
30—40 seconds. This and the following extractions are carried out in subdued
light. After separation, the lower layer is run into a second separator containing
10 ml. of 1 N NaOH and the mixture shaken. The extraction is repeated with
four 20 ml. portions of chloroform, each extract being washed in turn with the
same 10 ml. portion of NaOH solution and with the same 10 ml. portion of water
contained in a third separator. The washed chloroform extracts are filtered into
a dry flask through a paper moistened with chloroform. The chloroform is
evaporated, 2 ml. of absolute ethanol added and evaporated in a current of air
on a water bath with rotation of the flask to prevent undue heating. The residue
contains narcotine and papaverine. The codeine remains in the aqueous layer,
the sodium hydroxide washings and the water washings. For the extraction and
determination of codeine, the aqueous layer remaining after the extraction of the
narcotine and papaverine is transferred to a 100 ml. beaker and washed in with
first, the aqueous washing in the third separator followed by 5 ml. of water passed
in turn through the separators. The alkaline washings in the second separator
are now added and washed in with 5 ml. of water. The contents of the beaker are
warmed to remove chloroform, cooled, 3 ml. of dilute NaOH solution added, the
solution stirred and set aside for 30 minutes. The solution is filtered into a 100 ml.
separator, washing the beaker and filter with two portions of 5 ml. of water. The
combined filtrate and washings are extracted with three portions of 20 ml. of
benzene contained in three separators, each extract being washed in turn with the
same portion of 10 ml. of water. The benzene extracts are filtered into a clean
separator, washed with two 5 ml. portions of benzene and extracted with 25 ml.
of dilute H_2SO_4, shaking vigorously for 5 minutes. The acid solution is transferred
to a 100 ml. beaker, the benzene layer washed with 20 ml. of distilled water and
the washings added to the acid extract. The combined extract and washings are
made alkaline with 3.5—4 ml. of ammonia (sp. gr. 0.880) and boiled until the
volume has been reduced to 20—30 ml. The pH of the solution, which should be
between 4.0 and 3.6, is checked with bromocresol green paper (yellowish green).
The solution is filtered into a 100 ml. separator and washed in with two 5 ml.
portions of distilled water. The combined filtrate and washings are extracted
with three portions of 20 ml. of chloroform and each extract washed in turn with
the same 10 ml. portion of water. The chloroform extracts are discarded and the
combined aqueous layers, after making alkaline with 2 ml. of dilute ammonia,
are extracted with three portions of 20 ml. of chloroform and each extract washed
with the same portion of 10 ml. of water. The chloroform extracts are filtered
into a tared flask, the chloroform removed and the residue dried with ethanol as
described above. The codeine is weighed and checked for purity. For the separation and determination of narcotine and papaverine, the crude preparation
obtained above is dissolved in 6 ml. of benzene, 3 ml. of a 10% w/v solution
of KOH in absolute ethanol added, and the solution placed in a water bath at
20° for 40 minutes. The solution is transferred to a separator and washed in with
three 5 ml. portions of benzene and then with 10 ml. of N NaOH. The mixture is
well shaken, allowed to separate and the aqueous layer transferred to a small
conical flask. The benzene layer is washed with two portions of 5 ml. N NaOH

and finally with two portions of 5 ml. of water. All the washings are added to the
first aqueous layer. The benzene layers contain papaverine. Narcotine, present
in the combined aqueous layers, is isolated by adding 6.5 ml. of conc. HCl and
placing the solution in a boiling water bath until the temperature reaches 95°.
The solution is cooled, transferred to a separator and washed in with 3 portions
of 3 ml. of water. 7 ml. of ammonia solution (sp. gr. 0.880) are added and the
alkaloid extracted with portions of 20, 10, 10 and 10 ml. of chloroform, each
extract being washed in turn with a portion of 10 ml. of water in a second separator.
The chloroform extracts are filtered through a filter moistened with chloroform
into a tared flask. The chloroform is evaporated off and the residue of narcotine
is treated with 2 ml. of ethanol and heated in a current of air with rotation of the
flask. The residue is finally dried at 100—105° for 10 minutes, cooled and weighed
as narcotine. For the isolation of papaverine, the benzene layer from the first
separator is filtered into a dry flask and the separator washed first with the
benzene washings in the second separator and finally with 5 ml. of benzene. The
benzene is evaporated and after the addition of 2 ml. of ethanol, the residue is
dried as above. The residue is next moistened with 1 ml. of conc. HCl, covered,
and allowed to stand for 15 minutes. The mixture is washed into a separator with
20 ml. of water in portions and the papaverine extracted with portions of 20, 10,
10 and 10 ml. of chloroform, each extract being washed in turn with the same
portion of 10 ml. of N NaOH, and then with the same portion of 10 ml. of water.
The chloroform extracts are filtered into a tared flask, the chloroform evaporated,
the residue dried as above, cooled and weighed.

A scheme for the separation of the minor alkaloids of opium devised by Hesse
is outlined by Henry (1924). The opium is extracted with water and calcium
chloride added to the extract to precipitate calcium meconate and convert the
alkaloids to the hydrochlorides. The calcium meconate is filtered off and the
filtrate concentrated gradually when the hydrochlorides of morphine, pseudomorphine and codeine are precipitated in this order. The mother liquor is diluted
with an equal volume of boiling water, excess of ammonia added, the precipitate
removed by filtration, dissolved in acetic acid, filtered, purified by shaking
with ether and then made alkaline with NaOH solution. Papaverine, narcotine,
thebaine, some cryptopine, protopine, laudanosine and hydrocotarnine are
precipitated while lanthopine, laudanine, codamine and some cryptopine remain
in solution. The precipitate is digested with dilute ethanol and acetic acid added
until the solution is slightly acid. On the addition of 3 volumes of boiling water,
a crystalline precipitate of papaverine and narcotine comes down. The filtrate
is evaporated to remove ethanol and tartaric acid added which precipitates
thebaine hydrogen tartrate. The mother liquor is neutralized with ammonia,
mixed with 3% by weight of sodium bicarbonate made into a paste with water,
and set aside for one week when a black mass separates. The solution is filtered
and ammonia added. The precipitate which forms is removed and extracted with
hot benzene; the filtrate also is agitated with benzene. The combined benzene
solutions are shaken with a saturated aqueous solution of sodium bicarbonate
when laudanosine crystallizes out. The benzene filtrate is treated with HCl
gas and hydrocotarnine hydrochloride is obtained. The bases not dissolved by
hot benzene are cryptopine and protopine and they are converted to their hydrochlorides by treatment with strong HCl. The mixed hydrochlorides are washed
with a little water in which the cryptopine salt is freely soluble, the protopine
salt only slightly so. The alkaloids which remained in solution in the caustic soda
solution are obtained by neutralizing the solution with dilute HCl, making alkaline
with ammonia, extracting with ether and shaking out the ethereal solution with

acetic acid. Lanthopine separates out during a period of 24 hours when the acetic acid is neutralized with ammonia. The lanthopine is removed by filtration and excess of ammonia is added to the filtrate when laudanine, cryptopine and codamine separate out. The precipitate is dissolved in a small quantity of boiling dilute ethanol which on cooling deposits laudanine and cryptopine, and from the mother liquor codamine is obtained by evaporation and treatment of the residue with ether.

III. Qualitative Tests for the Opium Alkaloids.

For purposes of identification, the following tests may be used to supplement data obtained from determinations of the physical constants of the bases. The opium alkaloids give precipitates with the alkaloid reagents. The most sensitive reagents are bismuth potassium iodide, iodine in KI, MAYER's reagent, gold chloride, phosphotungstic acid, phosphomolybdic acid, picrolonic acid and ammonium reineckate (in slightly acid solution). Many of the alkaloids give colour reactions when treated with strong acids, particularly in the presence of oxidizing reagents. For satisfactory results the reagents must be pure and the alkaloids also should be in a reasonably high state of purity. The tests are best carried out in a small evaporating dish either with a small crystal of the alkaloid or on the residue obtained by the evaporation of an ethanolic solution of the base. The

Table 10. *Colour Reactions of the Opium Bases.*

Reagent	Morphine	Codeine	Narcotine	Papaverine	Thebaine	Narceine
Pure Nitric acid	orange-red yellow on warming	yellow	red	yellow	yellow	yellow fading
Pure H_2SO_4 cold	no colour	no colour	pale yellow to orange	no colour if pure	blood-red to orange yellow	yellow to brown
MARQUIS reagent	purple to blue	violet	violet to olive-green to yellow	yellowish-green to brown	red	yellow
FRÖHDE's reagent	violet to blue to dirty green	dirty green to blue	pink to green to orange	no colour if pure	blood-red to orange to yellow	brown-green to yellow to red
MECKE's reagent	blue to green to olive-green	blue to emerald-green to olive-green	greenish to cherry-red	greenish to deep violet	deep orange fading	greenish to yellow to violet
MANDELIN's reagent	red to blue to violet	green to blue on warming	red to crimson	blue-green to blue	orange-red	violet to reddish yellow
ERDMANN's reagent	—	blue on warming	pink to orange to violet on warming	no colour if pure	orange-red	brown to yellow
ROSENTHALER's reagent (cold)	greenish-blue to green	light-blue	greenish-yellow	—	—	—
ROSENTHALER's reagent (warm)	blue to dark green	dark blue	cherry-red	—	—	—
ROSENTHALER's reagent warm + + HCl	red to violet	purple to red	yellow to red	—	—	—
Ferric chloride (neutral)	greenish-blue	no colour	no colour	no colour	no colour	—

composition and method of preparation of the reagents will be found in the appendix to the introductory chapter of this work. Colour reactions of the more important opium bases with a selection of the better known reagents are given in Table 10.

Additional Tests for the Opium Alkaloids.

Morphine. Two important colour tests which have been adapted for the quantitative estimation of morphine have been described by RADULESCU (1913) and GUARINO (1946). *The* RADULESCU *test* is carried out as follows: To a solution of morphine in dilute HCl, a few drops of a 1% solution of sodium nitrite are added. After allowing to stand for a few minutes the solution is made alkaline with dilute ammonia when a red colour develops. This reaction is given by phenolic substances other than morphine and is therefore not specific for morphine, but it sharply distinguishes morphine from codeine. *The* GUARINO *test* depends on the following series of reactions which are given by morphine only. To a solution of morphine in dilute HCl a few crystals of iodic acid are added when a yellowish-brown colour develops. On addition of a saturated solution of ammonium carbonate a golden-yellow colour is formed. When treated with a few drops of ferric chloride solution acidified with dilute HCl, the solution now turns reddish-violet. Half of the solution is treated with a few drops of dilute oxalic acid which gives a stable blue colour. To the other half, acetic acid is added with the development of an emerald-green colour. The green solution is decolourized by adding $N/1$ oxalic acid and dilute ammonia added dropwise. An almost black or reddish-violet ring is produced. The reaction is said to be sensitive to dilutions of more than 1 to 20,000. BROUARDEL-BOUTMY *Test*. A few drops of a mixture of dilute solutions of potassium ferricyanide and ferric chloride added to an aqueous solution of a salt of morphine gives a deep blue colour. This test depends on the reducing action of morphine on potassium ferricyanide which is converted to ferrocyanide. Codeine does not give the test. *Iodic Acid Test*. A solution of morphine in dilute H_2SO_4 reduces a solution of pure iodic acid or potassium iodate. If the mixture is shaken with chloroform a violet colour is imparted to the chloroform.

Codeine. Although codeine in its reactions behaves similarly to morphine, it is distinguished by its solubility in ether and chloroform and by the fact it is precipitated by excess of NaOH.

Narcotine. Being a very weak base, narcotine can be extracted with chloroform from an aqueous tartaric acid solution. This property serves to distinguish narcotine from other opium bases.

Narceine. If narceine is warmed in an evaporating dish with 1—2 ml. of dilute H_2SO_4, a violet colour develops as the acid becomes more concentrated. This colour changes to cherry-red on longer heating and on allowing to cool, blue-violet streaks appear. A crystal of narceine when stirred with a mixture of 0.01 g. of resorcinol in 10 drops of conc. H_2SO_4 develops a yellow colour. On warming on a water-bath the colour changes to crimson-red or cherry-red and on cooling passes through blood-red to orange-yellow.

Papaverine. WARREN's test distinguishes papaverine from most of the other alkaloids. A very small crystal of $KMnO_4$ is crushed with a glass rod and intimately mixed with about 0.5 mg. of papaverine. This mixture is treated with 0.2 ml. of MARQUIS' reagent with stirring. A green colour appears which rapidly changes to blue. The blue colour deepens to an intense violet-blue and after standing for some time becomes bluish-green, green and finally dirty brown.

Other Alkaloids. *Laudanine*, yields a green colour when treated with ferric chloride solution and with nitric acid an orange-red colour develops. The solution of laudanine in conc. H_2SO_4 shows a faint pink tint. On warming the solution a reddish-violet colour develops. *Laudanosine* gives a rose-red colouration with conc. H_2SO_4 which on warming to 150° changes to deep reddish-violet. No reaction is given with ferric chloride. *Codamine* gives with nitric acid a dark green colouration and with strong H_2SO_4 in presence of a trace of ferric chloride a greenish-blue colour develops. *Protopine* gives with conc. H_2SO_4 a yellow colour changing to red. If a trace of oxide of iron is added to the sulphuric acid a deep violet colour is produced. The base dissolved in acetic acid gives with conc. H_2SO_4 a blue-violet colour which on addition of water changes to red. MANDELIN's reagent gives a reddish-violet colour changing to deep blue. *Cryptopine.* This base gives with conc. H_2SO_4 a violet colour which changes to green on warming to 150°. No reaction is given with ferric chloride. *Neopine* resembles codeine in its colour reactions but gives with ferric chloride solution a bluish-green colour. Of the alkaloids of unknown constitution, *papaveramine* gives a bluish-violet colouration with H_2SO_4 and *meconidine* gives a green solution with the same reagent. Mineral acids give with *rhoeadine* solutions which are intensely purple-red in colour. The colour is destroyed by alkali but is restored on adding an excess of acid. This colour test is given by very small amounts of rhoeadine. *Aporeine* gives with a mixture of conc. H_2SO_4 and nitric acid a violet colour which changes to brown and yellow. MARQUIS reagent gives a green colour changing to blue and finally black.

IV. The Quantitative Estimation of the Opium Alkaloids.

For the estimation of the individual alkaloids of opium other than morphine it is necessary to effect complete separation of the bases from each other. With the use of conventional methods involving precipitation reactions the individual alkaloids have been isolated but the procedure is lengthy and time-consuming and is not generally suitable for quantitative estimations. However, ADAMSON, HANDISYDE and HODGSON (1947) have successfully separated and determined codeine, papaverine and narcotine in admixture in papaveretum (see under methods of isolation). The presence of a phenolic group in morphine serves to distinguish this alkaloid from the remaining major alkaloids of opium and a number of colorimetric methods are dependent on the phenolic structure. In the present work attention will mainly be directed to the estimation of morphine in the tissues of the opium poppy. For the extraction of morphine it is preferable to exhaust the powdered plant material, after liberation of the base with alkali, with pure methanol or methylene chloride in a Soxhlet-type extractor. Alternatively the alkaloid may be extracted by refluxing the powder after treatment with alkali with a mixture of n-butanol and benzene for about 1 hour.

1. Colorimetric Methods for the Determination of Morphine.

The method of ADAMSON and HANDISYDE (1946) is based on the RADULESCU *colour reaction.* An aliquot of a 0.1 N HCl solution of morphine containing not more than 1 mg. of the latter is diluted to 20 ml. with 0.1 N HCl in a 50 ml. Nessler cylinder and 8 ml. of a 1% solution of sodium nitrite added with mixing. After a period of exactly 15 minutes, 12 ml. of a 10% solution of ammonia are added and the solution diluted to exactly 50 ml. and mixed. The colour is read in a Spekker absorptiometer using 4 cm. cells and Ilford No. 601 spectrum violet filter. The colour tends to fade in the light and it is advisable to take readings in the dark. An accuracy of $\pm$ 2% is claimed for the method. The reaction is not specific for morphine and other phenols give the colour. This method has been used by WEGNER (1951) for the estimation of morphine in the capsules and other organs of *Papaver somniferum.*

5 g. of the plant material dried at 75° and ground to coarse powder are placed in a mortar with 5 ml. of a solution of Na_2CO_3 (7.5 g./100 g. solution). The powder is thoroughly moistened, the mixture lightly triturated and allowed to stand for 1 hour. The material is now transferred to the extraction thimble of a Soxhlet, the thimble closed with a plug of cotton wool, and extracted on a water bath with methylene chloride for 8—10 hours. After extraction is complete the solution is allowed to stand overnight. The bulk of the solvent is distilled off except for about 10 ml. According to the amount of morphine expected to be present, either the entire extract is used or an aliquot is taken for determination. The amount of morphine in the extract taken for analysis should lie between 1 and 6 mg. For the determination of morphine in ripe capsules it is convenient to transfer the extract to a 50 ml. volumetric flask and make to volume with the solvent. 10 ml. of the solution representing 1 g. of capsule material are transferred to a separating funnel and well shaken with 10 ml. of 0.1 N HCl for 2 minutes. After allowing to settle for 10 minutes, the layers are separated and the morphine solution in HCl is further purified by shaking with two 20 ml. portions of chloroform for 2 minutes. The chloroform washings are rejected and the acid solution is made alkaline with 2 ml. of N NaOH solution and, by the addition of 13 ml. of water, the volume of the solution is brought up to 25 ml. After mixing, the liquid is filtered and 10 ml. of the alkaline filtrate are shaken out in a separating funnel with two portions of 20 ml. of chloroform for 2 minutes. After settling the chloroform solutions are run off and discarded. To the aqueous solution in the separating funnel are added 1 ml. of N HCl and 12.5% ammonia. The solution is now shaken out with three portions (20 ml., 10 ml. and 10 ml.) of a mixture of chloroform and isopropanol (3:1) for 2 minutes on each occasion. The chloroform-isopropanol solutions are run off into an Erlenmeyer flask (100 ml.) with a ground-glass stopper and the solvent removed by placing the flask in a boiling water bath for 20 minutes. The residue is taken up with 5 ml. of chloroform and shaken out for 2 minutes in the closed flask with 20 ml. of 0.1 N HCl. After standing, the upper aqueous layer is decanted or pipetted off and filtered. The photometric method of Adamson and Handisyde (1947) is now employed for the determination of morphine in the colourless or slightly yellow filtrate. 5 ml. of the filtrate are placed in each of two test tubes (1) and (2) with ground glass stoppers. Into a third tube 5 ml. of 0.1 N HCl are placed to serve as a blank. 2 ml. of water are added to tube (2) and to tubes (1) and (3) are added 2 ml. of a 1% solution of sodium nitrite. The tubes are shaken to mix the solutions and exactly 15 minutes after the addition of sodium nitrite all three tubes receive 3 ml. of 12.5% ammonia and after mixing, 2.5 ml. of water. The colour is now read in an Absorptiometer and the morphine content of the solution read off from a calibration curve prepared from standard solutions of pure morphine hydrochloride.

The *colorimetric method of* Guarino (1946) has been used by a number of workers for the determination of morphine after removal of thebaine. Cramer and Voermann (1949) have described the following procedure for the estimation of morphine, using a modified Guarino reaction in which nickel sulphate is substituted for ferric chloride with good results. An aliquot of morphine solution containing not more than 8 mg. of morphine is run into a 50 ml. volumetric flask and diluted with water to 15 ml. To this solution 15 ml. of 0.1 N HCl are added followed by 2 ml. of a 5% solution of iodic acid. After a period of 2 minutes has elapsed, 5 ml. of a saturated solution of ammonium carbonate are added and the solution made to volume (50 ml.) with a 5% solution of ammonium carbonate. After 30 minutes 1 ml. of a 1% solution of nickel sulphate is added, the mixture shaken and set aside for 90 minutes. A blank is prepared with water substituted for iodic acid. For colorimetric determination a red filter is used. Neither meconic acid nor the other major alkaloids of opium interfere.

MOORHOFF (1952) has described a method for the estimation of morphine in poppy heads using ferric chloride in the GUARINO reaction. A weak acid extract containing from 15—40 mg. of morphine is made strongly alkaline with NaOH solution and shaken out with three 25 ml. portions of benzene. The combined benzene solutions are washed with 20 ml. of 0.1 N NaOH solution. This procedure removes thebaine which interferes. The mixed alkaline solutions are brought to pH 9 by the addition of acid and NH_4Cl and are shaken out with a mixture of chloroform and isopropanol (3:1). The combined chloroform-isopropanol extracts are filtered through anhydrous Na_2SO_4 and shaken out with dilute H_2SO_4. The combined acid extracts are diluted to 100 ml. and 15 ml. of the extract placed in each of two flasks (1) and (2). 15 ml. of 0.1 N HCl are added to (1) and 17 ml. to (2). To flask (1) 2 ml. of a 5% solution of iodic acid are added and after a period of exactly 2 minutes, 17 ml. of a 10% solution of ammonium carbonate are added to both flasks. Exactly 1 minute later, 1 ml. of a freshly prepared 0.3% solution of ferric chloride is added to both flasks. After 15—30 minutes the colour is measured against a blank. MARIANI, GUARINO and MARELLI (1951) state that the maximum absorption (for spectrophotometric determination) occurs at 510—520 mμ. Colorimetric methods based on the RADULESCU reaction have also be described by REITH and INDEMANNS (1950) and SVENDSEN (1948). STEPHENS (1951) substitutes M-phosphoric acid for HCl in the RADULESCU reaction and obtains a colour which is less sensitive to light than that obtained with HCl. ORAM (1950) has used the reaction of morphine with $CuSO_4$ and H_2O_2 in the presence of ammonia for the colorimetric determination of the alkaloid. WEGNER (1950) has determined codeine, narcotine and thebaine in pure solution by a colorimetric method involving reaction with diazobenzene-sulphonic acid.

2. The Gravimetric Estimation of Morphine.

MANNICH (1935, 1942) has described a gravimetric method for morphine depending on the conversion of the latter to its dinitrophenyl ether. In this method the hydrochloride of morphine is treated with a solution of 2:4-dinitrochlorobenzene in aqueous methanol and the derivative which crystallizes easily, weighed.

3. The Polarographic Estimation of Morphine.

BAGGESGAARD-RASMUSSEN and LANNG (1948) have determined morphine by extraction of poppy capsules (after treatment of the powdered material with Na_2CO_3 solution) with an n-butanol-benzene mixture (7:3). The powdered material is refluxed with the solvent for 1 hour and the extract shaken out with N-HCl. The acid extract is treated with excess of KOH and the morphine determined polarographically.

4. The Determination of Morphine by Paper Chromatography.

SVENDSEN (1951) has developed the following method for the determination of small amounts of morphine. Approx. 0.25 g. of morphine is placed in a small dish and rubbed down with 0.5 ml. of concentrated formic acid. The mixture is transferred with the aid of 2.5 ml. of water to a sintered glass filter of medium porosity and filtered. The residue is treated with a further quantity of 0.5 ml. of formic acid (5%) for 2 minutes and the solution transferred to the filter. This operation is repeated until a total of 5 ml. of extract has been collected. The solvent mixture used for the chromatogram consists of ethyl acetate 10 vols. formic acid 1 vol., water 3 vols. Six spots each of 0.02 ml. of the extract are placed

on the paper and the chromatogram run. A strip of paper including one spot is
cut off and developed by spraying with a 2% solution of $NaNO_2$ and after 2 minutes
the paper is placed in an atmosphere of ammonia. This strip serves for the determination of the position of the morphine on the paper. The remaining five spots
are cut out, each placed in a test tube and treated with 2 ml. of a 1% solution of
HCl and 2 ml. of a 0.5% solution of $NaNO_2$. After a period of exactly 10 minutes,
1 ml. of a 5% solution of ammonia is added and the solutions filtered on a sintered
glass filter if necessary. The intensity of colour is read at 450 mμ. The amount
of morphine present is calculated from a standard curve prepared with pure
morphine.

5. Adsorption and Spectrophotometric Methods.

A method for the determination of morphine in opium using both adsorption
and colorimetric techniques has been described by KLEE and KIRCH (1953). An
adsorption column is used to remove interfering substances and a colorimetric
determination of the morphine in the eluate is carried out by means of the FOLIN-
CIOCALTEU phenol reagent. The adsorption column is prepared as follows: A small
plug of cotton wool is fitted into the bottom of a glass tube 285 mm. in height
and 12 mm. in internal diameter. Florisil (60—100 mesh) is packed to a height
of 200 mm. in the tube and a plug of cotton wool placed on the top of the column.
Approx. 40 ml. of absolute methanol is passed through the column and air pockets

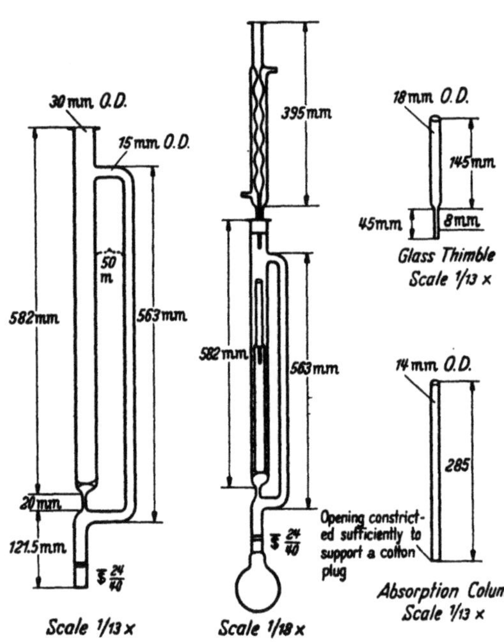

Fig. 1. Elution apparatus (STOLMAN and STEWART 1949).

which might form between the
cotton wool and Florisil pressed
out with a glass rod of diameter
slightly less than that of the tube.
A sample of opium (0.5 g.) previously dried to constant weight
at a temperature not exceeding
70° is weighed into a folded, cone-
shaped filter paper (9 cm.). The
opium is covered with cotton wool
and the sample placed in the base
of a glass thimble 145 mm. in
length and 18 mm. in outer diameter. The stem of the thimble,
approx. 45 mm. in length an 8 mm.
in external diameter extends into
the adsorption column when the
glass thimble is in place over the
column. The column and thimble
are placed in a modified form
of the elution apparatus of STOL-
MAN and STEWART (1949) as shown
in Fig. 1. The flask (500 ml.)
contains 100 ml. of absolute methanol and a few pieces of por-
celain (or glass beads) to prevent bumping. The solvent is boiled at such a
rate as to maintain a small layer of methanol above the Florisil column and
care is taken not to allow overflow of the methanol down the vapour tube into
the flask. The rate of percolation through the column may be slow at first but
the rate of flow increases after a short time. After a period of 3 hours extraction
the solution in the flask is transferred quantitatively (using absolute methanol

as a wash) to an evaporating dish and the solvent completely removed on a steam bath. Approx. 30 ml. of distilled water is added to the residue and the mixture swirled to loosen the residue and to prevent the formation of a colloidal suspension of the resinous matter extracted. Such a suspension makes filtration difficult. The solution is now quantitatively transferred and filtered into a 200 ml. volumetric flask, distilled water being used to rinse the dish and to make to volume. 1 ml. of this extract (representing approx. 250 μg. of morphine) is transferred to a 100 ml. volumetric flask and treated as described below for the pure morphine solutions. With the size of column used and an extraction period of 3 hours, the sample of opium must not exceed 0.5 g. The final volume of opium extracts which contain a high percentage of morphine is adjusted so that 250 μg. will be present in an aliquot of not less than 1 ml. For the colorimetric determination a standard morphine solution containing 100 μg./ml. is prepared by diluting 10 ml. of a stock solution (containing 100 mg. of morphine in 100 ml.) to 100 ml. A standard curve is prepared by taking aliquots of the morphine solution representing 100—400 μg. and transferring them to 100 ml. volumetric flasks. 2 ml. of the FOLIN-CIOCALTEU reagent (prepared as described in the introductory chapter) are added, followed by 3 ml. of a saturated solution of Na_2CO_3. Sufficient distilled water is added to bring the volume to 100 ml. After a period of 5 hours, the maximum blue colour has developed and remains stable for at least 18 hours. The colour is read at 765 mμ in a spectrophotometer. In running a blank the morphine solution is omitted.

Solutions of pure codeine, narcotine and papaverine do not give a colour with the FOLIN-CIOCALTEU reagent. Meconic acid, with which morphine is combined in opium, gives a blue colour with the reagent but as it is completely adsorbed on the Florisil column no interference takes place. The above method of estimation of morphine gives higher results than the official methods but the authors claim that this is due to the fact that there is little loss of morphine in manipulation and that other disadvantages of the official methods are not present. Desirable features of the above method are that a number of extractions can be made in one day with a sufficient number of extractors, the colour can be developed as soon as extraction is complete and the blue colour which is stable for 18 hours can be read on the following morning. GRANT and HILTY (1953) have shown that morphine can be separated from codeine by passing a mixture of the alkaloids through a column of a strongly basic quaternary amine-type of anion exchange resin. The phenolic morphine exchanges with the resin while codeine does not. The codeine base is quantitatively removed from the column by elution with methanol and is then determined by titration. The morphine is subsequently removed by elution with dilute acid and determined spectrophotometrically. The most satisfactory resin for the separation is Amberlite XE 75 which is converted to the hydroxyl form by allowing it to stand in a 10% solution of NaOH overnight. The resin is washed with successive portions of distilled water until the washings are neutral to phenolphthalein. The resin is now packed tightly in a column 25 cm. in length.

For the spectrophotometric determination of morphine a series of standard aqueous solutions of pure crystalline morphine sulphate are prepared in the range of 1—5 mg. per 25 ml. of solution containing 0.2 ml. of H_2SO_4. Acidified aqueous solutions of morphine sulphate have an absorption maximum at 285 mμ. The absorbancy is measured in a spectrophotometer using matched 1 cm. silica cells and a calibration curve obtained. For the determination of codeine 50 ml. of distilled water are added to a 20 ml. aliquot of the eluate and the solution is titrated with 0.01 N H_2SO_4 using methyl red-methylene blue mixed indicator. The separation and determination of codeine phosphate and morphine sulphate from a mixture is carried out as follows: The column of resin is prepared and washed with 250 ml. of 95% methanol and further washed with an additional 50 ml. of 95% methanol. The eluate is used as a blank in the codeine titration. A mixture of 0.1345 g. of morphine sulphate and 0.1635 g. of codeine phosphate is dissolved in 20 ml. of distilled water and passed through the column. The column is washed with 95% methanol until a combined eluate of 100 ml. is obtained. An aliquot of 20 ml. of this eluate and also of the methanol blank is titrated. Recovery of codeine phosphate equivalent to 0.1629 g. is obtained. The column is now washed by adding 100 ml. of distilled water to displace the methanol and an additional 50 ml. to serve as a blank in the determination of morphine. The morphine is now eluted with 125 ml. of 2% phosphoric acid followed by distilled water until the combined eluate reaches a volume of 250 ml. An aliquot of 10 ml. is transferred to a 25 ml. volumetric flask, 0.2 ml. of H_2SO_4 added and the solution made to volume with distilled water. The blank is treated in the same manner. Recovery of morphine sulphate equivalent to 0.1358 g. is obtained.

Foster and Macdonald (1951) have made a study of the ultra-violet absorption spectrum of papaverine hydrochloride which may prove of value in quantitative work. These workers found that under standardized conditions of pH, papaverine hydrochloride shows a maximum absorption in the region of 240 mμ.

It is probable that the future will see considerable development of adsorption and spectrophotometric methods for the separation and estimation of the opium bases. Meanwhile the most reliable methods available for the determination of the individual alkaloids involve a tedious process of separation before estimation is possible.

R. The Alkaloids of the *Papilionaceae* (Lupinane Group).

Members of the sub-order *Papilionaceae* of the *Leguminosae* contain alkaloids having the lupinane structure. The sub-orders Mimosaceae and Caesalpinaceae have also been examined and the occurrence of alkaloids has been established in some genera and species. Jaretzky and Axer (1934) have examined over one hundred species of the Papilionaceae growing in Europe and White (1943,

Table 11. *The Alkaloids of the Papilionaceae.*

Alkaloid	Principal Sources
Sparteine . . .	*Cytisus scoparius* Link. (up to 1.5%), *Lupinus arboreus* Sims. (seeds, 2%). Also present in *Chelidonium majus* (*Papaveraceae*)
Cytisine	*Cytisus laburnum* L. (*Laburnum vulgare* Presl.) seeds av. 1.5%. Widely distributed in the *Papilionaceae*
N-Methylcytisine	*Cytisus stenopetalus* Christ., *Cytisus canariensis* Steud., *Baptisia australis* R. Br.
Anagyrine . . .	*Anagyris foetida* L., *Cytisus linifolius* Lam.
Calycotomine . .	*Calycotome spinosa* Link., 0.8—1.2%
Monspessulanine	*Cytisus monspessulanus* L. (about 0.2% in tops)
Retamine . . .	*Retama sphaerocarpa* Boiss. (*Genista sphaerocarpa* Lam.)
Genisteine . . .	*Cytisus scoparius* Link.
Sarothamnine .	*Cytisus scoparius* Link.
Lupinine	*Lupinus luteus* L. (also in *Anabasis aphylla, Chenopodiaceae*)
Lupanine . . .	*Lupinus angustifolius* L. and *Cytisus* spp.
Hydroxylupanine	*Lupinus albus* L., *L. hilarianus* Benth.
D-Isolupinine . .	*Lupinus pilosus* L.
Virgiline	*Virgilia capensis* L.
Virgilidine . . .	*Virgilia capensis* L.
Ammodendrine	*Ammodendron Conollyi* Bge.
Matrine	*Sophora* spp.

1944, 1946, 1951) has made an extensive investigation of the alkaloids of a large number of New Zealand species. The reader is referred to these papers for a full account of the alkaloids and their distribution in the tissues examined. The most important of the alkaloids and their principal cources are given in Table 11.

I. Sparteine.

1. Properties.

Sparteine (Lupinidine), $C_{15}H_{26}N_2$, is a colourless alkaline oil volatile in steam, $D^{20°}$ 1.0196, $[\alpha]_D^{21°}$ —16.42° (ethanol, b. p. 188°/18 mm. or 325°/154 mm. in hydrogen. It has a bitter taste and a characteristic odour and is sparingly soluble in water (1 in 328 at 22°) but easily soluble in ethanol, chloroform or ether, insoluble in benzene or light petroleum. The base is monoacidic to phenolphthalein and diacidic to methyl orange. The salts crystallize well; the sulphate, $B \cdot H_2SO_4 \cdot 5\,H_2O$, forms rhombohedra m. p. 159—162° (dry), $[\alpha]_D^{15°}$—27° (H_2O), and is easily soluble in water or ethanol. The monohydriodide, $B \cdot HI$ has m. p. 226—228° and the dihydriodide, $B \cdot 2\,HI$, m. p. 257—258°. The platinichloride, $B \cdot H_2PtCl_6 \cdot 2\,H_2O$, forms rhombic prisms, m. p. 243.5° (dec.) and the picrate forms yellow needles from boiling ethanol, m. p. 208°. The perchlorate, $B \cdot HClO_4$ (mono) has m. p. 171—172°.

Sparteine is present in several species of the genera *Cytisus*, *Genista* and *Lupinus* and is the major alkaloid of *Cytisus scoparius* LINK., the common broom. The alkaloid is present in all parts of the plant but the amount present in the lower parts is small. The cortex of the stems contains nearly all the alkaloid with relatively little in the woody tissues. In New Zealand the leaves were found to contain only traces of alkaloid in contrast to the leaves of plants growing in Europe which contained 1% in May. The maximum content of sparteine in the tops (stems and leaves) of the European plants is about 1.5% in May followed by a decline (after flowering) in June to 0.66%. From June onwards a rise occurs steadily until the maximum is reached in the following May. In New Zealand plants the maximum of 1.45% occurs in young stems in September. On seeding, the alkaloid tends to move from the stems to the developing seeds. The floral parts contain 0.01% (petals), 0.05% (calyx), 0.06% (stamens) and 0.22% (style). The pedicels contain at the flowering stage 0.46%; with pods forming, 0.56%; at ripe pod stage, a trace. The green seeds at red pod stage contain 0.53% and the ripe seeds 0.02%.

2. The Isolation of Sparteine.

The alkaloid is isolated by extracting finely ground broom tops with a 2% solution of H_2SO_4. The acid extract is made alkaline with excess of NaOH solution and the liberated alkaloid steam distilled. The distillate is exactly neutralized with HCl and carefully evaporated to dryness. The residue is treated with solid KOH and the sparteine distilled off. The distillate is further purified by re-distillation in a current of hydrogen, the last traces of water being removed from the distillate by metallic sodium. Alternatively the powdered plant material is exhausted with 60% ethanol, the solvent distilled off, and the residue taken up with 5% tartaric acid solution. The acid solution is filtered, treated with excess of potassium carbonate and the alkaloid shaken out with ether. The alkaloid is removed from the ether by shaking with tartaric acid solution, potassium carbonate is added in excess to the separated acid solution and the alkaloid once more shaken out with ether. On evaporation of the ether, sparteine is obtained in fairly pure state.

3. Qualitative Tests for Sparteine.

Sparteine gives white precipitates with MAYER's reagent, silicotungstic acid and sodium phosphomolybdate, the precipitate with phosphomolybdate being soluble in hot water. Bismuth potassium iodide in ethanolic solution gives an orange-red precipitate with an ethanolic solution of sparteine slightly acidified with acetic acid. For microchemical slide tests, potassium cadmium iodide gives characteristic needle and plate forms with transverse outgrowths, potassium triiodide small clusters of curved blades and gold chloride plate clusters. A bulky red precipitate is given when H_2S is passed through a suspension of sulphur in an ethereal solution of the base. Couch (1925) has modified Grants' test as follows. A chloroform solution of the base is allowed to evaporate on a filter paper and the spot exposed to bromine vapour for a few seconds. In the presence of sparteine a yellow stain develops which disappears on holding the paper in ammonia fumes. If the spot is now warmed over a hot plate, a pink colour appears. This test is said to be specific for sparteine.

4. The Quantitative Estimation of Sparteine.

White (1943) in his studies on the alkaloids of the Papilionaceae has used the following modification of the method of Jaretzky and Axer (1934) for the estimation of sparteine in the broom. Air-dried powdered material is extracted in a Soxhlet with 2% acetic acid in 50% ethanol as solvent. After concentration, the extract is cleared with lead acetate and made alkaline. The alkaline extract is distilled into dilute acid until the distillate gives no turbidity with potassium triiodide. The acid solution is concentrated to small bulk (at most 5 ml.) and the alkaloid precipitated with an excess of potassium cadmium iodide solution. After standing for 30 minutes the precipitate is filtered through a sintered glass filter, washed twice with a little dilute potassium iodide solution, once with a little water, dried at 40° and weighed as $H_2CdI_4 \cdot C_{15}H_{26}N_2$.

The method of Hirt (1929) differs from the above method in that the alkaloid is extracted with 5% HCl, the solution made alkaline with ammonia and the alkaloid shaken out with successive portions of ether. The combined ether extracts are shaken with successive portions of 10% H_2SO_4 solution and the alkaloid precipitated with silicotungstic acid from the acid solution. The precipitate is ignited and weighed. The base may be titrated if extracted from alkaline solution with chloroform. The chloroform is evaporated off and the sparteine titrated with methyl orange as indicator.

5. Sarothamine and Genisteine.

Two minor alkaloids, sarothamnine and genisteine, have been isolated from the mother liquors remaining after the isolation of sparteine from *Cytisus scoparius*.

Sarothamnine, $C_{15}H_{24}N_2$, forms crystalline additive products with solvents. With chloroform, $C_{15}H_{24}N_2 \cdot {}^1/_2 CHCl_3$ is formed, m. p. 127°, $[\alpha]_D - 38.7°$, and with ethanol, $C_{15}H_{24}N_2 \cdot {}^1/_2 C_2H_5OH$ is formed, m. p. 99°, $[\alpha]_D -25.6°$.

Genisteine, $C_{16}H_{28}N_2$, is a volatile base, m. p. 60.5°, b. p. 139.5—140.5°/5 mm. It forms a hydrate $B \cdot H_2O$, m. p. 117°, $[\alpha]_D -52.3°$ (ethanol), a platinichloride, m. p. 215°, $B \cdot H_2PtCl_6 \cdot 2.5 H_2O$, and a picrate, $B \cdot 2 C_6H_2 (NO_2)_3 OH$. m. p. 215°.

Munier and Macheboeuf (1951) have described a method for the separation of sparteine and genisteine by paper chromatography.

II. Cytisine.

Cytisine is the major alkaloid of *Cytisus laburnum* L. (*Laburnum vulgare* Presl.), but is fairly widely distributed throughout the Papilionaceae, e. g. in *Baptisia* spp., *Sophora* spp., *Anagyris foetida* L., *Ulex europaeus* L., *Genista* spp. and species of *Cytisus* other than *C. laburnum*. The alkaloid is found in all parts of the plant but principally in the seeds. White (1943) in an examination of *Cytisus laburnum* grown in New Zealand found a maximum of 1.2% in the unripe seed and 1.96% in late seed. Values for European plants appear to vary from 1.5% to 3%. The distribution of cytisine within the plant and with season follows that given above for sparteine in *Cytisus scoparius* and follows a course common to most of the Leguminous plants examined. The amount of cytisine present in seeds of *Ulex europaea* varies with development from about 0.3% to 1.1%.

1. Properties.

Cytisine, $C_{11}H_{14}ON_2$, forms rhombic crystals, m. p. 153°, b. p. 218°/2 mm., $[\alpha]_D^{17°}$—119.6° (water). It is a strong base and is soluble in water, ethanol and chloroform but almost insoluble in light petroleum, ether or benzene. It forms crystalline deliquescent salts. The hydrochloride, $B \cdot HCl \cdot H_2O$ forms colourless prisms and the dihydrochloride, $B \cdot 2\,HCl \cdot 3\,H_2O$, yellowish needles. The nitrate, $B \cdot HNO_3 \cdot H_2O$, forms needles or leaflets and has $[\alpha]_D$—81.5°. The perchlorate melts at 296° (dec.) and the picrate has m. p. 277°. The picrolonate melts at 270° (dec.) and the aurichloride, $B \cdot HAuCl_4$, forms reddish-brown needles, m. p. 220° (dec.) sparingly soluble in warm water. Two platinichlorides are known, $B_2 \cdot H_2PtCl_6$ forming sparingly soluble pale yellow plates or needles and $B \cdot H_2PtCl_6$, golden yellow needles fairly soluble in water. The acetyl derivative has m. p. 208° and the benzoyl derivative melts at 116°. The p-toluenesulphonyl derivative crystallizes in prisms, m. p. 207—208° and is insoluble in cold water but can be recrystallized from hot water. It is soluble in dilute acid and warm ethanol.

2. The Isolation of Cytisine.

ING (1931) has described the following process for the extraction of cytisine from laburnum seeds. 1 kg. of seeds ground to pass a 20-mesh sieve are intimately mixed with 100 g. of slaked lime and 500 ml. of water added. The mixture is well stirred and extracted with chloroform for 20 hours. The chloroform extract is evaporated *in vacuo* to remove the last traces of solvent and the residue is stirred with 1000 ml. of light petroleum and allowed to stand overnight. Most of the alkaloid crystallizes and is separated, the mother-liquor being extracted with dilute acid. The crude alkaloid is purified by boiling its solution in dilute HCl with charcoal, filtering, making strongly alkaline, and extracting with chloroform. The chloroform solution is dried over anhydrous sodium sulphate and evaporated when the alkaloid crystallizes out (yield 20 g.). Cytisine is purified by distillation *in vacuo* followed by crystallization from dry acetone.

3. Qualitative Tests for Cytisine.

Cytisine gives precipitates with the alkaloid reagents. Phosphomolybdic acid and phosphotungstic acids give precipitates in a dilution of 1 in 30,000. Bromine water gives an orange-red precipitate in a dilution of 1 in 15,000. Ferric chloride gives with cytisine a red colouration which disappears on addition of a drop of H_2O_2 solution. If the solution is now warmed on a water bath a blue colour appears (VAN DER MOER's reaction). When a crystal of thymol is added to a solution of cytisine in conc. H_2SO_4 and the solution warmed a yellow colouration passing into red is produced (MAGALHAËs reaction).

4. Quantitative Estimation of Cytisine.

For the quantitative estimation of cytisine (WHITE, 1943) the plant material is dried in air, finely powdered, and extracted in a Soxhlet with a solvent consisting of 50% ethanol containing 2% of acetic acid. The extract after concentration is cleared with lead acetate, concentrated, made alkaline with NaOH and extracted with six portions of an equal volume of chloroform. The crude alkaloid residue after evaporation of the combined chloroform extracts is titrated with 0.01 N HCl using screened methyl red as indicator, to which cytisine is monobasic.

III. N-Methylcytisine.

N-Methylcytisine was first obtained by Power and Salway (1913) from the rhizomes and roots of *Caulophyllum thalictroides* Michx. *(Berberidaceae)* a herbaceous plant indigenous to N. America. This base has since been found to occur in association with cytisine and other alkaloids in a number of plants belonging to the Papilionaceae. The tops of *Cytisus canariensis* Steud., *C. stenopetalus* Christ. and *C. hillebrantii* Briq. contain methylcytisine as the main alkaloid. According to White (1946) the tops of these plants contain up to 0.4% of alkaloids, the greater portion of which consists of methylcytisine. The tops of *Baptisia australis* R. Br. also contain methylcytisine in association with cytisine, and tops of *Cytisus monspessulanus* L. contain up to 0.9% of alkaloids consisting of 50 to 60% of methylcytisine, 15% of cytisine and the remainder of monspessulanine (White, 1946).

1. Properties.

Methylcytisine, $C_{12}H_{16}ON_2$, forms colourless needles, m.p. 136—137°, $[\alpha]_D^{20°}$—221° (water). It is readily soluble in water, ethanol, chloroform and benzene, less soluble in ether. The hydrochloride, $B \cdot 2 HCl \cdot H_2O$, melts at 250—255° (dec.) and the perchlorate has m. p. 254° (dec.). The aurichloride, $B \cdot HAuCl_4$, crystallizes in golden yellow needles, m. p. 205—206° (dec.) and the picrate forms long yellow needles, m. p. 228° (sintering at 220°).

The isolation of methylcytisine from the seeds of *Anagyris foetida* is described below but the method may be applied to its isolation from the above mentioned species of *Cytisus* in which it is present in association with cytisine.

2. Qualitative Tests for Methylcytisine.

White (1946) gives the following microchemical reactions of methylcytisine. Neutral or slightly acid solutions are used. Bismuth potassium iodide gives a heavy amorphous precipitate and potassium mercuric iodide in acid solution gives an amorphous precipitate from which tiny spheroids separate (distinction from cytisine). Potassium cadmium iodide gives in acid solution an amorphous precipitate and potassium triiodide large black needle clusters. Gold chloride gives curved needle clusters developing to thin blades and plates with transverse outgrowths, and gold bromide forms characteristic blades with transverse outgrowths.

3. Quantitative Estimation of Methylcytisine.

Power and Salway (1913) give details of a titrimetric method for the determination of methylcytisine in *Caulophyllum thalictroides* where it appears to be the only alkaloid present. In mixtures with other alkaloids it is essential to effect a quantitative separation of the bases before proceeding to a titration. The titration procedure follows the method given above for cytisine.

IV. Anagyrine.

Anagyrine is present in the seeds of *Anagyris foetida* L. where it occurs together with cytisine, methylcytisine and D-sparteine (Ing, 1935). This base also occurs in the tops of *Cytisus linifolius* Lam. to the extent of about 1% and in young shoots of *Ulex europaeus* (0.02%).

The separation of the alkaloids of *Anagyris* is carried out as follows (Ing, 1933, 1935). The powdered seed is mixed with 10% of its weight of $Ca(OH)_2$, damped and percolated with 90% ethanol until exhausted. The ethanol is distilled off and the alkaloids shaken out with chloroform. The chloroform solution is shaken out with dilute HCl and the mixed alkaloids obtained in solution as their hydrochlorides. This solution is made alkaline with strong ammonia and extracted five times with benzene to remove anagyrine which is much more readily extracted by benzene from aqueous solution than cytisine and the other alkaloids. The benzene extract is evaporated, the residue taken up in water and neutralized with perchloric acid (d. 1.12). The anagyrine crystallizes as the perchlorate and a second crop of crystals may be obtained by evaporation of the mother-liquor. The crude perchlorate is purified by boiling in aqueous solution with charcoal, filtering while hot and allowing to cool, when the perchlorate crystallizes. Anagyrine is obtained from the perchlorate by treatment with ammonia and extraction with chloroform. The aqueous mother-liquor remaining after the benzene extraction is shaken with five portions of chloroform, the extract dried with Na_2SO_4 and evaporated. The semi-solid residue is extracted with hot ethyl acetate and on concentrating the extract, crude cytisine crystallizes out. The ethyl acetate mother-liquors from the crystallization of cytisine are evaporated and the gummy residue dissolved in water. The aqueous solution is treated with benzenesulphonyl chloride and alkali to precipitate residual cytisine as the benzenesulphonyl derivative which is sparingly soluble even in hot water but crystallizes from ethanol in diamond-shaped plates, m. p. 263—264°. The aqueous filtrate is made strongly alkaline and extracted first with benzene and then with chloroform. The benzene extract leaves on evaporation a further small quantity of anagyrine and the chloroform extract on evaporation leaves a residue of crude methyl-cytisine which is dissolved in benzene and light petroleum added. Crystals of methylcytisine m. p. 136—137°, are formed and removed by filtration.

1. Properties.

Anagyrine, $C_{15}H_{20}ON_2$, forms a pale yellow glass which darkens on exposure to light, b. p. 210—215°/4 mm., $[\alpha]_D^{20°}$ —165.3° (ING, 1933). It is less soluble in hot than in cold water and is readily soluble in ethanol, ether, chloroform and benzene but insoluble in light petroleum. The hydrochloride forms colourless needles, $B \cdot HCl \cdot 3\ H_2O$, and has m. p. 295—297° (dry), $[\alpha]_D^{15°}$ —142.5°. The methiodide, $B \cdot MeI$, forms colourless needles, m. p. 264° (dec.) and the platinichloride $B \cdot 2\ HCl \cdot PtCl_4 \cdot 1.5\ H_2O$, ruby-red needles, m. p. 250—251°. The per-chlorate, $B \cdot HClO_4$, forms colourless needles decomposing at 297—298°. The picrate crystal-lizes from ethanol in needles, m. p. 242°.

2. Reactions of Anagyrine.

WHITE (1944a) has recorded the following microchemical slide tests for anagyrine (as pure base or hydrochloride in water). Potassium cadmium iodide gives characteristic needle clusters, gold chloride fern-like crystals, mercuric chloride an amorphous precipitate changing to characteristic spheroids, and potassium mercuric iodide gives amorphous material crystallizing to clusters of needles. Gold bromide, picric acid, potassium triiodide and bismuth potassium iodide give amorphous precipitates crystallizing indefinitely. The presence of acid hinders crystallization. Anagyrine gives a red colour with ferric chloride, turned blue-green with H_2O_2 (VAN DER MOER's reaction) and COUCH (1939) has shown that anagyrine gives in the GRANT test, an orange-yellow with bromine, turned to blue with ammonia and finally to brown on heating.

V. Other Alkaloids.

1. Calycotomine.

Calycotomine, $C_{12}H_{17}O_3N$, is present in the seeds of *Calycotome spinosa* LINK. to the extent of about 1% (WHITE, 1944b). The base is also found in the stems of *Cytisus nigricans* var. *elongatus* WILLD. and has been isolated in the DL-form from seeds of *Cytisus proliferus* L.

Calycotomine, in the pure state melts at 141° and is readily soluble in ethanol and acetone, slightly soluble in ether and soluble to about 12% in water, $[\alpha]_D^{20°}$ $+21°(H_2O)$ The hydrochloride, B · HCl, has m. p. 193°, $[\alpha]_D^{20°}$ $+15°(H_2O)$, the picrate, B · $C_6H_4O_7N_3$ · H_2O,. has m. p. 163—166°, the perchlorate, m. p. 176—177°, and the mercurichloride, m. p. 118 to 119°.

On demethylation, calycotomine gives a phenolic product which gives with ferric chloride an intense green colour. The base is isolated from the ripe seeds of *Calycotoma spinosa* by Soxhlet extraction with 50% ethanol containing 2% of acetic acid. The extract is concentrated, clarified with lead acetate, made alkaline with NaOH and extracted with chloroform. The crude residue after evaporation of the chloroform forms needles which are dissolved in ethanol and precipitated by the addition of ether as rods, m. p. 139—141°. To the mother-liquor, HCl is added and the solution dried. Crude hydrochloride which remains, is dissolved in ethanol and precipitated with acetone, giving colourless rods, m. p. 190—191°.

Microchemical Reactions of Calycotomine. In slightly acid solution the following reagents give good reactions. Bismuth potassium iodide gives at first an amorphous precipitate which later forms large crystals. Potassium cadmium iodide gives an oil which later forms characteristic plate clusters, mercuric chloride sheaves of long needles, gold chloride an oil forming plates, gold bromide an oil forming characteristic large plate clusters and rods, picric acid an oil forming fine needles, and potassium triiodide an oil forming large curved plate-like clusters.

2. Monspessulanine.

Monspessulanine, $C_{15}H_{22}ON_2$, crystallizes in rods, m. p. 101°, $[\alpha]_D$ $+117°$ (ethanol). It is sparingly soluble in cold water, much less soluble in hot water, readily soluble in ether and most organic solvents including light petroleum. The base is faintly alkaline to litmus and the salts are dissociated. The hydrochloride, B · HCl, is extremely deliquescent and melts at 244°. The perchlorate has m. p. 214—215° on recrystallization from acetone-ethyl acetate, and the methiodide has m. p. 247°.

The extraction of monspessulanine and its separation from cytisine and methyl-cytisine is described by White (1946b). It does not give the van der Moer test but gives precipitates with most of the alkaloid reagents.

3. Retamine.

Retamine, $C_{15}H_{26}ON_2$, is isomeric with oxysparteine and occurs in the tops of *Genista sphaerocarpa* Lam. to the extent of about 0.4—0.5%. It is a strongly alkaline diacidic base and is slightly volatile in steam, m. p. 162°. Retamine is almost insoluble in cold water, slightly soluble in hot water and sparingly soluble in ethanol, but soluble in light petroleum. It has $[\alpha]_D$ $+43.15$ (ethanol) and yields crystalline salts. The dihydrochloride, B · 2 HCl, melts at 298° and the methiodide on recrystallization from hot acetone melts at 217°. The picrate, B · 2 $C_6H_3O_7N_3$ · 2 H_2O, melts at 165°. Retamine has been isolated from *G. sphaero-carpa* by White (1946c) who gives details of microchemical tests with the alkaloid reagents. Retamine does not give the van der Moer reaction.

4. Lupanine.

Lupanine, $C_{15}H_{24}ON_2$, occurs in D-, L- and DL-forms in the blue lupin, *Lupinus angustifolius*, and is distributed throughout the plant to the extent of from 0.2 to 0.5%. During development the alkaloid accumulates in the tops and in the cortical tissues and at flowering the alkaloid moves to the flower parts, then to pods and finally to the seeds. The cortex of stems producing flowers may contain as much as 0.6—0.8% of alkaloid. Petals may contain from 0.45—0.8% and the calyx 0.93%. Seeds show variations from 0.7—1.6%, and the pods give high values during the early stages of their growth falling to 0.05% when ripe.

l-Lupanine forms prismatic plates which melt at 98—99°, b. p. 185—195°/1 mm. It is strongly alkaline, behaves as a monoacidic base, and is soluble in water and most organic solvents. The monohydrochloride, B · HCl · 2 H₂O, is deliquescent and has m. p. 127—128° (hydrated) or 250—252° (dry). The hydriodide, B · HI · 2 H₂O has m. p. 184—185° (dry), the aurichloride, m. p. 177—178°, and the thiocyanate, m. p. 124°.

D-Lupanine is the major alkaloid of all strains of *L. angustifolius* growing in New Zealand (WHITE, 1943). It distils at 185—186°/0.08 mm. and crystallizes on standing, m. p. 40°, [α]$_D$ +61.4° (acetone). The hydrochloride, B · HCl · 2 H₂O, has m. p. 127° (dry), [α]$_D$ +62°, and the hydriodide, B · HI · 2 H₂O, forms colourless prisms, m. p. 189°, [α]$_D$ +45.5° (H₂O). The thiocyanate melts at 184° and the picrate has m. p. 180°.

Lupanine may be isolated and estimated by the method of WHITE (1943). The powdered material is extracted in a SOXHLET with 50% ethanol containing 2% of acetic acid as for cytisine. The alkaloid is shaken out from the alkaline solution with seven portions of chloroform, the combined chloroform solutions evaporated, the residue dissolved in water and titrated with screened methyl red as indicator (to which lupanines are monobasic). Lupanine gives characteristic precipitates with most of the alkaloid reagents.

5. Lupinine.

Lupinine C₁₀H₁₉ON, is the characteristic alkaloid of the seeds of the yellow lupin. The base crystallizes from light petroleum in rhombs, m. p. 68.5—69.2° b. p. 255—257° (in a current of hydrogen), [α]$_D^{17°}$ —19°. The solution in water is strongly alkaline to litmus and the base liberates ammonia from its salts. It is readily soluble in cold water, less soluble in hot water, and soluble in organic solvents including light petroleum. An ethereal solution of lupinine containing sulphur gives a light orange precipitate when H₂S is passed in (CLEMO and RAPER, 1929). The hydrochloride, B · HCl, forms rhombic prisms, m. p. 212—213°, [α]$_D$ —14° (H₂O); the D-tartrate has m. p. 171°; the methiodide m. p. 295—296° and the picrate, m. p. 136—137°. The aurichloride, B · HAuCl₄, forms needles, m. p. 196—197° and the platinichloride, B₂ · H₂PtCl₆, has m. p. 163—164°.

Lupinine is isolated from the ground seeds of *Lupinus luteus* by continuous extraction with 50% ethanol containing 2% of acetic acid or by percolation of the seed material with ethanol containing 1% HCl. The ethanol is removed by distillation and the residue boiled with water, filtered from fat if present, made alkaline with NaOH and extracted with ether. The ethereal solution is extracted with dilute HCl and sparteine precipitated as the mercurichloride by treatment with mercuric chloride. The filtrate is diluted with water, the mercury removed as sulphide and the lupinine removed by making alkaline and shaking with ether. It is purified by crystallization from light petroleum. Alternatively the alkaloids in HCl solution may be separated by making the solution alkaline and shaking out with light petroleum which extracts lupinine but not sparteine. Lupinine may be estimated in the light petroleum extract by removal of the solvent and titration of the residue dissolved in water with acid, using screened methyl red as indicator. Lupinine gives precipitates with most of the alkaloid reagents.

6. Hydroxylupanine.

Hydroxylupanine, C₁₅H₂₄O₂N₂, may be isolated by a similar process from *Lupinus albus*. It crystallizes in rhombic prisms with 2 H₂O, m. p. 76—77°, or 172—174° (dry), [α]$_D$ +64.12. The aurichloride, B · HAuCl₄, forms prisms from dry ethanol, m. p. 205—206°. On boiling for 24 hours with hydriodic acid and phosphorus, it is converted into D-lupanine.

D-Isolupinine is present to the extent of about 0.3% in the tops of *Lupinus pilosus*. It crystallizes from hot solution in light petroleum giving plates, m. p. 76°, [α]$_D^{17°}$ +32° (ethanol). It is less soluble in hot water than in cold water. The hydrochloride, B · HCl, on repeated crystallization from ethanol-acetone has m. p. 173°, [α]$_D^{17°}$ +20.3 (water). The picrate has m. p. 144°, and the methiodide m. p. 247—248° (WHITE, 1951).

7. Virgiline and Virgilidine.

Virgiline and *virgilidine* are present in the tops of *Virgilia capensis* which contain 0.5 to 1.1% total alkaloids of which about a third consists of virgiline and about a tenth, of virgilidine (WHITE, 1946d).

Virgiline, $C_{16}H_{26}O_2N_2$, crystallizes from hot acetone containing methanol in colourless shining prisms, m. p. 247—248°, $[\alpha]_D$ —46° (ethanol). The base is sparingly soluble in acetone and water, readily soluble in methanol and ethanol but very slightly soluble in ether and light petroleum. The hydrochloride $B \cdot HCl \cdot H_2O$, crystallizes in needles, m. p. 260—261° (dec.), the methiodide melts at 176° and the picrate hexahydrate has m. p. 188—189°. The monoacetyl derivative crystallizes from acetone-ether in prisms, m. p. 173—174°, soluble in most organic solvents with the exception of light petroleum.

Virgilidine, $C_{10}H_{19}ON$, is isomeric with lupinine and boils at 90°/0.01 mm. It forms a semi-crystalline syrup, $[\alpha]_D$ +12° (ethanol), $n_D^{17°} = 1.5128$. The base is volatile in steam and is soluble in cold water, less soluble in hot water and soluble in most organic solvents including light petroleum. The hydrochloride, $B \cdot HCl \cdot H_2O$, forms needles (when precipitated from ethanol by addition of acetone) m. p. 260—261° (dec.). The picrate crystallizes with $6 H_2O$, m. p. 188—189°. The dry picrate precipitated from acetone by the addition of water forms fine needles, m. p. 203° (dec.). Both virgiline and virgilidine in slightly acid solution give precipitates with the alkaloid reagents, and the crystal forms may be used in microchemical slide reactions as an aid to the identification of the bases (White, 1946d).

8. Ammodendrine and Matrine.

Ammodendrine, $C_{12}H_{20}ON_2 \cdot H_2O$, is acetyltetrahydroanabasine and is an optically inactive base, m. p. 73—74°. The salts are generally amorphous and deliquescent but the hydriodide, $B \cdot HI$, forms a crystalline precipitate from ethanol, m. p. 218—220°, and a perchlorate, m. p. 199—200°.

Matrine, $C_{15}H_{24}ON_2$, is an isomeride of lupanine and occurs in several species of *Sophora*. In the seeds of *S. tetraptera* Ait., Briggs and Taylor (1938) found 3% of crude alkaloid most of which is matrine. This base exists in four forms, the α-form occurs in flat prisms or needles m. p. 77°; the β-form in rhombs, m. p. 87°; the γ-form (freshly distilled base) a liquid, b. p. 223° at 6 mm. pressure, and the δ-form gives prisms m. p. 84°. The α-form is most usually found and is soluble in cold water, much less soluble in hot water from which it separates. It is soluble in most organic solvents and is optically active, $[\alpha]_D^{10°}$ +39.11°. The hydrobromide melts at 272—273° and the hydriodide has m. p. 267°. The aurichloride, $B \cdot HAuCl_4$, forms leaflets, m. p. 199° and the platinichloride $B \cdot H_2PtCl_6$, decomposes at 229—230°. Matrine gives precipitates with most of the alkaloid reagents.

S. The Alkaloids of *Peganum harmala* L.

The seeds and roots of *Peganum harmala (Rutaceae)*, a herbaceous plant widely distributed in S. Europe, contain harmaline, harmine and harmalol belonging to the indole group of alkaloids. Harmine also occurs in the S. American liane *Banisteria caapi* Spruce *(Malpighiaceae)* and other species of *Banisteria*. The seeds of *Peganum harmala* also contain a small proportion of vasicine (peganine) an alkaloid belonging to the quinazoline group and found in the leaves of *Adhatoda vasica* Nees *(Acanthaceae)*. Harman, a degradation product of harmine occurs in *Symplocos racemosa* Roxb. *(Styracaceae)* where it was formerly known as loturine. The root of *Peganum harmala* yields about 3% of harmine and the total alkaloid content of the seeds varies between 2.5 and 3% of which 50—75% is harmaline. It is stated that the alkaloids of the seed occur chiefly in the seed coat.

I. Properties.

Harmaline, $C_{13}H_{14}ON_2$, occurs in colourless or pale yellow plates, m. p. 239—240° (dec.). It is very little soluble in water but soluble in hot ethanol and chloroform. The base is optically inactive and forms crystalline salts with one equivalent of acid. The hydrochloride, $B \cdot HCl \cdot 2 H_2O$, occurs in yellow needles, m. p. 212°. The N-acetyl derivative forms colourless needles, m. p. 204—205°. Harmaline reacts with methyl iodide to give N-methylharmaline iodide from which N-methylharmaline (needles, m. p. 162°) is prepared by treatment with baryta. The benzylidene derivative is prepared by boiling finely powdered harmaline with freshly distilled benzaldehyde and methanol under reflux. Crystals of benzylidenediharmaline, m. p. 245° (dec.) appear on cooling. It is practically insoluble in organic solvents.

Harmine, $C_{13}H_{12}ON_2$, crystallizes from methanol in colourless prisms, m. p. 257-259°, and is sparingly soluble in water, ethanol and ether. The base is optically inactive and forms colourless salts which show a deep blue fluorescence in dilute solution. The hydrochloride, $B \cdot HCl$, m. p. 262°, and the platinichloride, m. p. 264—266°, crystallize well. Harmine

condenses with benzaldehyde to form benzylidene-harmine which separates from ethanol
in pale yellow prisms or needles, m. p. 191—192°. Its solution in neutral solvents shows
an intense violet-blue fluorescence. The p-nitrobenzylidene derivative crystallizes from
ethyl acetate in red needles, m. p. 266°.

Harmalol, $C_{12}H_{12}ON_2 \cdot 3 H_2O$, crystallizes in needles from water and melts at 212° (dec.).
It is soluble in chloroform, acetone and hot water and sparingly soluble in benzene. It is
unstable in air. Harman, $C_{12}H_{10}N_2$, crystallises in rhombic pyramids from dry ether or prisms
with 8 H_2O from wet ether, m. p. 229°. It is soluble in 7760 parts of water at 23° and soluble
in ethanol.

II. The Isolation of Harmine and Harmaline.

These alkaloids are isolated from the crushed seeds by extraction with acidified
water followed by precipitation as the hydrochlorides. Use is made of the fact
that on careful addition of ammonia to a solution of the mixed hydrochlorides
harmine is precipitated first and not until the decomposition of harmine hydrochloride is complete is harmaline precipitated. The powdered seeds are covered
with three times their weight of 3% v/v acetic acid and the mixture is shaken
occasionally during a period of 2—3 days. The seed residue is pressed free of
solvent and the press cake again subjected to extraction with dilute acetic acid
and once more pressed out. The combined extracts are treated with NaCl
(100 g./litre of extract) to convert the acetates of the alkaloids to the hydrochlorides which are insoluble in cold NaCl solution and are precipitated on cooling.
The bulk of the supernatant liquid is siphoned off and the crystalline mass filtered
at the pump. The mixture of crude hydrochlorides is dissolved in hot water and
NaCl again added to precipitate the hydrochlorides. The process is repeated
until the hydrochlorides have acquired a yellow tint. For the separation of
harmine and harmaline, the mixture of hydrochlorides is dissolved in warm water
(50—60°) and ammonia carefully added, when harmine (needles) begins to crystallize
out. The addition of ammonia is stopped as soon as crystals of harmaline (plates)
begin to appear. The harmine is filtered off and on further addition of ammonia
to the filtrate harmaline is precipitated. Purification of both alkaloids is effected
by recrystallization of the hydrochlorides (HASENFRATZ, 1927)..

For the isolation of harmalol, the acid extract is concentrated to one third
of its volume, filtered and made distinctly alkaline with KOH. The precipitated
harmine and harmaline are filtered off and the alkaline filtrate acidified by the
addition of acetic acid. The filtrate is next made alkaline with Na_2CO_3, the
harmalol extracted with chloroform, and purified by recrystallization from water
or ethanol (FISCHER, 1901).

III. The Detection and Estimation of the Harmala Alkaloids.

1. Qualitative Tests.

The alkaloids are precipitated by the general reagents and according to
BRUSTIER, BOURBON and VIGNES (1950) harmine and harmaline give a positive
reaction with the glyoxylic-sulphuric acid test for indole derivatives. Apart from
this colour test, no other characteristic tests have been described. PERKIN and
ROBINSON (1919) however have shown that harmaline reacts in dilute acetic acid
solution with p-nitrobenzenediazonium chloride to form a chocolate brown
precipitate which develops slowly and is soluble in concentrated H_2SO_4 to give an
intense royal-blue solution which changes to reddish-purple and brown.

2. Quantitative Estimation.

Methods for the estimation of harmine and harmaline are based on the separation of the two alkaloids as described above for the extraction of the alkaloids.

After separation the bases are titrated with 0.1 N H_2SO_4 or in chloroform solution with p-toluenesulphonic acid. Buntzelman (1931) has described the following method for harmine and harmaline in the seeds of *Peganum harmala*. The powdered material is treated with 15% acetic acid solution and allowed to stand overnight. The extract is filtered through cotton and then through a filter paper moistened with 15% acetic acid solution. An aliquot of the filtrate is made strongly alkaline with a 30% KOH solution and the alkaloids shaken out with six portions of ether. The combined ether solutions are evaporated and the residue dissolved in 5 ml. of 0.1 N H_2SO_4 (or such volume as will ensure an excess). 15—20 ml. of water are added and the excess acid titrated with 0.1 N NaOH using methyl red as indicator. 1 ml. of 0.1 N H_2SO_4 is equivalent to 0.0213 g. of alkaloids. Harmine is separated from harmaline by bringing the bases into solution in 2% acetic acid (keeping the concentration below 0.1%). An aliquot of the solution (20 ml.) is made alkaline to phenolphthalein first with 0.5 ml. of 30% NaOH solution and then with N NaOH solution until the pink colour appears. The solution is next made slightly acid with dilute acetic acid and 0.1 N NaOH solution is added until the pink colour again appears. 0.4 ml. excess of 0.1 N NaOH is then added and the solution heated on a water bath for 5—10 minutes until crystals appear. These are collected on a filter and the filtrate run into a separating funnel. The crystals are dissolved in a hot 2% solution of acetic acid and the volume made up to 20 ml. 0.1 N NaOH solution is added to make the solution just alkaline to phenolphthalein and 0.1 ml. excess run in. A second separation of harmine is effected as before and the crystals are removed on a filter and the filtrate collected in a separating funnel. The combined filtrates are made strongly alkaline with 30% NaOH solution and the harmaline extracted with four portions of ether. The combined ether solutions are evaporated and the residue dissolved in an excess of 0.1 N H_2SO_4. The excess acid is back titrated with 0.1 N NaOH. 1 ml. of 0.1 N H_2SO_4 is equivalent to 0.0214 g. of harmaline. The harmine remaining on the filter is dissolved in a hot 2% solution of acetic acid, the alkaloid liberated with NaOH solution, shaken out with ether and determined as above described. 1 ml. of 0.1 N H_2SO_4 is equivalent to 0.0212 g. of harmine. Ya Khait (1945) has determined harmine in the roots of *Peganum harmala* by titrating a chloroform solution of the alkaloid after extraction, with 0.05 N p-toluenesulphonic acid.

It cannot be said that the existing methods for the estimation of the alkaloids of *Peganum harmala* are entirely satisfactory, and further study of these alkaloids is necessary in order that more accurate methods may be evolved.

T. The Alkaloids of *Pilocarpus* spp.

The leaves of *Pilocarpus (Rutaceae)*, a South American genus of small trees and shrubs, contain the alkaloids pilocarpine, isopilocarpine, pilocarpidine and pilosine. The alkaloid content of the principal species which have been examined is given below. *Pilocarpus microphyllus* Stapf. contains pilocarpine, isopilocarpine and pilosine (total alkaloid content, 0.76—0.78%: pilocarpine nitrate, 0.45%). *P. Jaborandi* Holmes; pilocarpine, isopilocarpine and pilocarpidine (total alkaloids 0.72%: pilocarpine nitrate 0.67%). *P. pennatifolius* Lemaire; pilocarpine and isopilocarpine (total alkaloids 0.2—0.3%). *P. trachylophus* Holmes; total alkaloids 0.4%. *P. Spicatus* St. Hilaire contains the minor alkaloids ψ-pilocarpine and ψ-Jaborine (total alkaloids, 0.16%). The alkaloids are located chiefly in the upper epidermal cells of the leaves and in the cells of the mesophyll bordering upon the lower epidermis.

I. Isolation of the Alkaloids.

The finely powdered leaf material is extracted with ethanol containing 1% of HCl. The solvent is distilled off, the residue taken up with a little water and neutralized by the addition of dilute ammonia. Resins which separate out are filtered off and the filtrate is concentrated to small bulk. Excess of ammonia is added and the alkaloids are shaken out with successive portions of chloroform. The chloroform is distilled off from the combined extracts and the residue is dissolved in a small volume of water and neutralized with dilute nitric acid. After cooling, the nitrates of pilocarpine and isopilocarpine crystallize and are separated by fractional crystallization from ethanol.

II. Properties.

Pilocarpine ($C_{11}H_{16}O_2N_2$) is a colourless oil b. p. 260°/5 mm. but has been obtained in the crystalline state, m. p. 34°. It is dextro-rotatory, $[\alpha]_D = +100.5°$ in chloroform. The base is easily soluble in water, ethanol and chloroform but almost insoluble in ether and light petroleum. The nitrate, $B \cdot HNO_3$ crystallizes in prisms, m. p. 178° and is soluble in 6.4 parts of water at 20° and 269 parts of absolute ethanol at 20°: $[\alpha]_D = +82.9°$ (water). The hydrochloride, $B \cdot HCl$ crystallizes in prisms m. p. 204—205°: $[\alpha]_D = +91.74°$. The aurichloride, $B \cdot HAuCl_4 \cdot H_2O$ forms yellow needles, m. p. 124—125°, and the picrate, $B \cdot C_6H_3O_7N_3$, needles which melt at 161—162°.

Isopilocarpine ($C_{11}H_{16}O_2N_2$) is the isomeride of pilocarpine and like the latter, occurs as a colourless oil, b. p. 261°/10 mm. It is dextro-rotatory $[\alpha]_D = +42.8°$. The nitrate, m. p. 159°, crystallizes from water in prisms and is soluble in 8.4 parts of water at 19° and 357 parts of absolute ethanol at 20°. The aurichloride forms yellow needles, m. p. 158—159°.

Pilocarpidine ($C_{10}H_{14}O_2N_2$) is a colourless oil and forms salts which crystallize well. The nitrate, $B \cdot HNO_3$ melts at 137°: $[\alpha]_D = +73.2$. The aurichloride crystallizes from acetic acid and melts at 124—125°. The picrate occurs as an oil and does not crystallize well.

Pilosine ($C_{16}H_{18}O_3N_2$) crystallizes from ethanol in colourless plates, m. p. 187°: $[\alpha]_D +39.9°$ (ethanol). It is a weak monoacidic base and forms salts which do not readily crystallize. The sulphate, $B_2 \cdot H_2SO_4$, occurs in plates, m. p. 194—195°: $[\alpha]_D = +21°$. The hydrogen tartrate, $B \cdot H_2C_4H_4O_6$ crystallizes slowly and melts at 135—136°, while the aurichloride $B \cdot HAuCL_4$ crystallizes from acetic acid in golden yellow plates, m. p. 143—144°. The minor alkaloids *ψ-pilocarpine* and *ψ-Jaborine* are optically inactive. The former occurs as a syrup and forms a nitrate, m. p. 142°. The nitrate of ψ-Jaborine melts at 158°.

III. The Detection and Estimation of the Alkaloids of *Pilocarpus*.

1. Qualitative Tests.

HELCH (1902) showed that pilocarpine forms a violet coloured compound when a solution of the base or one of its salts is treated with hydrogen peroxide and potassium dichromate in the presence of dilute acid. The violet coloured compound is soluble in benzene and chloroform and was characterized by BIEDEBACH (1933) as pilocarpine perchromate. The test is applied to the leaves of *Pilocarpus* as follows. 0.5 g. of powdered leaf material is extracted with 5 ml. of ammoniacal chloroform. The extract is filtered, the solvent distilled off and the residue dissolved in a little water. One drop of dilute H_2SO_4 is added, followed by 1 ml. of 3% H_2O_2, 1 ml. chloroform and a particle of potassium dichromate. On shaking vigorously for 1 minute, the chloroform assumes a bluish violet colour. Strychnine gives a blue colour with this test but the colour soon fades and does not pass into the chloroform.

EKKERT (1925) has devised the following test for the alkaloids of *Pilocarpus*. To 1 ml. of a solution of pilocarpine (or salt) is added 1 ml. of a freshly prepared solution of sodium nitroprusside (2%) and 1 ml. of N NaOH. After allowing to stand for 5 minutes, dilute HCl is added, with the development of a wine red colour. Half of this solution is treated with a few drops of 0.1 N sodium thiosulphate solution whereupon a green colour develops. To the other half, a few drops of dilute H_2O_2 are added and a carmine colour is obtained. This test is sensitive to 0.2 mg. of pilocarpine in 1 ml. of solution.

2. Quantitative Estimation of the Total Alkaloids of Leaves of *Pilocarpus*.

(a) Volumetric Method. LYNCH and EVERS (1948) have described a method suitable for samples of dried leaf of 10 g. To 10 g. of leaf powder contained in a flask is added 50 ml. of chloroform. The flask is shaken well and allowed to stand for 10 minutes. 5 ml. of dilute ammonia ($10\% NH_3$) are added and the flask shaken continuously for 1 hour. The mixture is transferred quantitatively by means of further portions of chloroform to a percolator plugged with cotton wool and percolation is continued using a total volume of 160 ml. of chloroform. Complete extraction is tested for by evaporating 2 ml. of the percolate to dryness, dissolving the residue in 5 drops of 0.1 N H_2SO_4 and testing with MAYER's reagent. The chloroform solutions are mixed and concentrated to a volume of about 40 ml. The concentrate is transferred quantitatively to a separating funnel, 150 ml. ether are added followed by 20 ml. of 0.5 N H_2SO_4 and the funnel shaken well. After separation the lower aqueous layer is run off and the ether solution extracted with three further portions of 10 ml. 0.1 N H_2SO_4. The combined acid extracts are filtered into another separator, made alkaline with dilute ammonia and shaken with four successive 20 ml. portions of chloroform. Each of the chloroform extracts is washed with the same 10 ml. portion of water. The chloroform extracts are combined and the solvent removed by distillation: the residue is treated with 2 ml. of water free ethanol and evaporated to dryness. The alkaloidal residue is dried for 5 minutes on a boiling water bath and dissolved in 1—2 ml. neutral ethanol. 15 ml. of 0.05 N H_2SO_4 are added and the solution is titrated with 0.05 N NaOH, using methyl red as indicator. Each ml. of 0.05 N H_2SO_4 is equivalent to 0.0104 g. of alkaloid calculated as pilocarpine. As pilocarpine and its salts when present together in a solution constitute a buffer system, some difficulty may be experienced in determining the exact end point in the titration. Moreover, pilocarpine base is easily soluble in water and it is essential to make certain that extraction is complete from water solution to chloroform when the latter solvent is being used. Pilocarpine contains a lactone group which can be saponified with sodium hydroxide to give a water soluble salt which cannot be extracted with chloroform. Sodium hydroxide therefore must not be used to free the base from its salts prior to extraction with chloroform.

(b) Colorimetric Methods. ELVIDGE (1947) has described a method for the estimation of the total alkaloids of *Pilocarpus* leaves, based on the colour reaction of EKKERT (1925). Pilosine does not react. After extraction of leaf material with ammoniacal chloroform as described above, an aliquot of the sulphuric acid extract (containing approx. 1—5 mg. alkaloids) is placed in a 25 ml. volumetric flask and neutralized with NaOH. 1 ml. of 2% sodium nitroprusside solution (freshly prepared) is added, followed by 1 ml. N NaOH. After allowing the solution to stand for 5 minutes, 5 ml. of 0.01 N $KMnO_4$ is added, followed immediately by 3 ml. of 3 N H_2SO_4. The solution is made up to 25 ml. with distilled water and the colour read in a colorimeter. If the Spekker Absorptiometer is used, 4 cm. cells and the Ilford Spectrum green filter, 604 are suitable. At the same time, a blank is conducted and a calibration curve is prepared using pure pilocarpine nitrate (a 0.13% solution of pilocarpine nitrate contains 1 mg. pilocarpine in 1 ml.). As the colour is somewhat unstable, readings should be taken within 5—10 minutes after development.

Shupe (1941) has used the Helch reaction as the basis of colorimetric determination of pilocarpine with satisfactory results and Webb, Kelley and McBay (1952) have further modified the method of Shupe, for assay of pilocarpine and its salts.

U. The Alkaloids of Pomegranate.

From the bark of the pomegranate, *Punica granatum* L. *(Punicaceae)*, a shrub or small tree indigenous to N. W. India and generally cultivated in the Mediterranean countries, the alkaloids pelletierine (isopelletierine of Tanret), pseudopelletierine, isopelletierine and methylisopelletierine (methylpelletierine of Tanret; isomethylpelletierine of Piccinini) have been isolated. There appears to be considerable variation in the content of total alkaloids in the bark and Ewers (1899) states that the stem bark contains about 0.5% and the root bark from 0.6 to 0.7%. Goodson (1940) found variation from 0.074 to 0.58% in commercial samples of bark. On storage the alkaloid content of the bark tends to diminish and wide variations found by various workers may be attributable to different periods of storage before analyses were carried out. In addition to cortical tissues the wood of the root contains alkaloids. Hess (1919) obtained from 100 kg. of bark, 179 g. of pseudopelletierine, 52.5 g. of pelletierine, 23 g. of methylisopelletierine and approx. 1.5 g. of isopelletierine. Chilton and Partridge (1950) have applied the method of partition chromatography to the separation and distribution of the pomegranate alkaloids and their results for the major alkaloids are summarised in Table 12.

Table 12. *The Distribution of the Alkaloids of Punica granatum.*

Tissue sample	Total Alkaloids % dry weight. (calc. as pseudo-pelletierine)	Pseudo-pelletierine % of total	Methyliso-pelletierine % of total	Pelletierine % of total
1. Old bark	0.10	58	24	17
2. Old bark	0.10	27	41	32
3. Old bark	0.23	25	36	38
4. Old bark	0.19	38	33	28
5. Fresh bark of young roots . .	0.38	32	36	32
6. Fresh wood of young roots .	0.115	21	32	47
7. Fresh bark of older roots . .	0.44	54	20	25
8. Fresh whole roots	0.39	22	35	42

I. Properties.

Pelletierine. $C_8H_{15}ON$ is a colourless alkaline oil, b. p. 106°/21 mm. which rapidly takes up oxygen and resinifies. It is soluble in water and organic solvents and readily forms crystalline salts with acids. The hydrochloride, B · HCl, forms needles, m. p. 143—144°, the hydrobromide forms feathery needles m. p. 140° and the aurichloride crystallizes in orange leaflets and melts at 82—82.5°. The picrate crystallizes well and melts at 150—151°; the picrolonate melts at 172—173°. The reineckate forms clusters of needles from aqueous acetone, m. p. 254° (dec.). As isolated from the plant, pelletierine occurs as the DL-form but may occur in the plant in the L-form which is known to undergo racemization very easily. The DL-base can be resolved through the tartrates yielding L-pelletierine-L-tartrate, m. p. 129°, $[\alpha]_D^{20°}$—20.94° and D-pelletierine-D-tartrate, m. p. 129°, $[\alpha]_D^{20°}$ +20.93°. Pelletierine is a secondary base and forms an N-acetyl derivative, b. p. 173—174°/18 mm. $[\alpha]_D$ +32.6°, and a benzoyl derivative crystallizing in plates, m. p. 75°. The oxygen of pelletierine is present in an aldehyde group and an oxime and semicarbazone are produced. The hydrochloride of the semicarbazone crystallizes in prisms, m. p. 168—170°, and the oxime crystallizes in two forms, m. p. 96—97° (from light petroleum) and m. p. 80° (from ether).

Methylisopelletierine, $C_9H_{17}ON$, is an oily alkaline liquid, b. p. 114—117°/26 mm. which is miscible with water and optically inactive. It readily forms crystalline salts. The hydrochloride, B · HCl, has m. p. 156—158°, the hydrobromide melts at 151—153° and the picrate has m. p. 158°. The aurichloride crystallizes in orange-yellow rosettes, m. p. 115—117° and the reineckate crystallizes from aqueous acetone in leaflets, m. p. 228° (dec.). The base can be resolved into the D- and L-forms through the tartrates. Methylisopelletierine forms a semicarbazone, m. p. 169°, the hydrochloride of which melts at 208—209°; a liquid hydrazone, b. p. 154—155°/29 mm., and an oxime, b. p. 160°/12 mm. which yields a picrate, m. p. 106°. The methiodide crystallizes in cubes and has m. p. 156°.

Isopelletierine, $C_8H_{15}ON$, is an optically inactive oil, b. p. 102—107°/11 mm. It forms a hydrobromide, m. p. 135° and a hydrochloride, m. p. 143°. The picrate melts at 147—149°. It reacts with ethyl chloroformate to form an N-carbethoxy derivative which on hydrolysis yields the unchanged base and therefore can be used for the isolation of the alkaloid.

Pseudopelletierine (N-methylgranatonine), $C_9H_{15}ON$ is an optically inactive base homologous with *tropinone* and crystallizes from light petroleum in prismatic plates, m. p. 53—54°, b. p. 246°. It is easily soluble in water, ethanol, ether and chloroform but less soluble in light petroleum. Pseudopelletierine is a strong tertiary base and forms crystalline salts. The hydrochloride, B·HCl, forms rhombohedra, the platinichloride, $B_2H_2PtCl_6$, reddish needles, and the aurichloride occurs in yellow crystals melting at 162° (dec.). The picrate crystallizes from water in yellow needles, m. p. 252—253° (dec.) and the reineckate in plates from aqueous acetone, m. p. 216° (dec.). The oxime crystallizes in tablets m. p. 128°, and the dipiperonylidene derivative (which may be used for characterization) crystallizes from ethanol in yellow triangular microscopic plates, m. p. 226—227°.

II. The Isolation of the Alkaloids.

For the large scale isolation of the pomegranate alkaloids Hess (1919) has used the following method. The finely powdered bark is intimately mixed with slaked lime, moistened with water and allowed to stand for 4—6 hours. The mixture is thoroughly extracted with water and the alkaloids removed from the aqueous solution by shaking out with chloroform. The alkaloids are then extracted from the chloroform solution with dilute H_2SO_4, the acid solution made alkaline with NaOH and the alkaloids extracted with ether. The ethereal solution is dried over KOH or K_2CO_3, filtered, the ether distilled off and the residue subjected to fractional distillation, under a pressure of 15—20 mm. The fraction distilling over at 100—120° consists of a mixture of the pelletierines and methylpelletierine. At 145°, pseudopelletierine distils over and solidifies to a crystalline mass in the receiver. At this stage, distillation is stopped, the residue is dissolved in light petroleum and the solution cooled in ice water. Pure pseudopelletierine crystallizes out. For the separation of the pelletierines and methylisopelletierine, the fraction containing these alkaloids is treated with ethyl chloroformate to convert the secondary bases into the urethanes. The tertiary base, methylisopelletierine, does not react. The mixture is now fractionally distilled and methylisopelletierine passes over at 105—106°/15 mm. and the urethanes of pelletierine and isopelletierine pass over at 150—165°/13 mm. Pelletierine urethane heated with conc. HCl at 125—135° yields pure pelletierine with elimination of CO_2 and ethyl chloride. The isopelletierine urethane (present in small quantity) is saponified with ethanolic potash and the liberated base is recovered by distillation. Pelletierine urethane on treatment with ethanolic potash yields a non-volatile resin and therefore the two bases can be separated through their urethanes. An alternative means of isolating pelletierine is made possible by crystallization as the hydrobromide (at low temperature) from the mixture of the crude alkaloids.

III. The Estimation of the Pomegranate Alkaloids.

1. Qualitative Tests.

Pelletierine gives precipitates with Mayer's reagent, iodine in potassium iodide, phosphomolybdic acid, bromine water, potassium cadmium iodide and tannic acid. Picric acid and ammonium reineckate both give crystalline precipitates which are useful for the identification of the pomegranate alkaloids. Pseudopelletierine gives precipitates with MAYER's reagent, phosphomolybdic acid, tannic acid and bismuth potassium iodide. With sulphuric acid containing a trace of potassium dichromate, pseudopelletierine gives an intense green colour. The dipiperonylidene derivative gives with strong sulphuric acid an intense blue colour which changes to green and yellow on dilution with water. This derivative is prepared by adding a solution of 0.5 g. KOH in 0.5 ml. of water to a solution containing 1 g. of pseudopelletierine and 3 g. of piperonal in 25 ml. of ethanol. The mixture is gently boiled for 10 minutes and on cooling, yellow crystals of the dipiperonylidene derivative separate out (MENZIES and ROBINSON, 1924).

2. Quantitative Determination.

Methods for the estimation of the total alkaloids have been described by EWERS (1899) and STÖDER (1902). The method of STÖDER is as follows. 20 g. of the dry powdered bark (dried at a temperature not exceeding 60°) is shaken with 100 ml. of chloroform and 5 ml. of ammonia for 12 hours on a mechanical shaker. 20 ml. of water, or sufficient to cause the powder to agglutinate, are added and the mixture allowed to settle. 75 ml. of the chloroform solution are filtered off and the filter well washed with chloroform. About 50 ml. of chloroform are distilled off, the residue transferred to a separating funnel and washed in with two 5 ml. portions of chloroform. The chloroform solution is shaken with 10 ml. of 0.1 N HCl, the acid solution is filtered, the chloroform washed with water and the washings poured through the filter and added to the acid. Excess acid is titrated with 0.1 N NaOH using methyl red as indicator.

CHILTON and PARTRIDGE (1950) have separated the pomegranate alkaloids on buffered columns of kieselguhr using as the source of the alkaloids, commercial "pelletierine tannate" and samples of bark, wood and whole roots. The "pelletierine tannate" is first assayed for total alkaloids by the following method. The alkaloids are liberated with NaOH, extracted with chloroform and the chloroform solution shaken with excess of standard acid (H_2SO_4). After evaporation of the chloroform the excess acid is back titrated with bromocresol green as indicator. For chromatographic separation on the kieselguhr column, the "pelletierine tannate" is dissolved in N NaOH, the mixed bases extracted with chloroform and shaken out with a small excess of N H_2SO_4. The acid solution is concentrated under reduced pressure to a small volume, the solution saturated with sodium phosphate and 4 N NaOH equivalent to the N H_2SO_4 used in extracting the alkaloids is added. This solution is mixed with twice its weight of kieselguhr and the mixture packed on top of the chromatographic column. The preparation of the buffered column and the technique for the elution of the alkaloids with ether or chloroform as described by EVANS and PARTRIDGE (1948) is given in the section on the tropane alkaloids (q. v.). Separation of mixed bases equivalent to about 30—40 mg. of pseudopelletierine is possible on a column of 10 g. of kieselguhr with 5 ml. of 0.5 M phosphate buffer, pH 6.8. The alkaloids in the eluate fractions, after titration with 0.005 N H_2SO_4, are characterized by conversion to the picrates and reineckates. For the separation of alkaloids from "pelletierine tannate" equivalent to 5 g. of pseudopelletierine, a CLAESSON multiple column is used with 650 g. of kieselguhr and 325 ml. of 0.5 M phosphate buffer, pH 6.8. For the fractionation of samples of root bark, wood and roots, the powdered material, dried below 60°, is mixed with 30% of its weight of calcium hydroxide, moistened with water and set aside for 4 hours. After packing in a percolator the mixture is macerated with 70% ethanol for 24 hours and then percolated with 70% ethanol. The percolate is acidified with dilute H_2SO_4 and the precipitated $CaSO_4$ removed. The filtrate is concentrated to small bulk, filtered to remove resin, and washed with chloroform. The alkaloids are liberated with NaOH and collected in chloroform as quickly as possible. An aliquot of the chloroform solution is used for a titration and the percentage of alkaloids in the drug calculated. An equivalent of N H$_2$SO$_4$ is shaken with the chloroform and the aqueous solution of the sulphates treated in the manner above described for "pelletierine tannate". Separation of alkaloids from root bark equivalent to 600 mg. of pseudopelletierine is effected on a column of 160 g. of kieselguhr and 80 ml. of 0.5 M phosphate buffer, pH 7.0, with chloroform as the eluant. CHILTON and PARTRIDGE (1950) in the course of their experiments found that the kieselguhr used adsorbed some of the alkaloids. The alkaloids adsorbed by kieselguhr could however be separated by the use in the column of "Pyrex" glass in powder form (No. 60 powder). 7 g. of powdered glass supported 1 ml. of the buffer solution. An advantage in the use of powdered glass is that samples are standard and reproducible whereas samples of kieselguhr may vary widely in their efficiency in chromatographic separations.

V. The Alkaloids of *Solanum* spp.

A number of species of the genus *Solanum* contain alkaloidal glycosides which undergo acid hydrolysis to yield alkaloids possessing a steroidal structure, and a mixture of sugars. The best known of these alkaloidal glycosides is solanine which is present in the tissues of *S. tuberosum* L. (potato), *S. nigrum* L., *S. dulcamara* L., and *S. lycopersicum* L. (tomato). Alkaloidal glycosides closely related to solanine have been isolated from other species of *Solanum*. Solasonine occurs in the berries of *S. sodomaeum*, *S. aviculare* FORST., and *S. xanthocarpum* (BELL and BRIGGS, 1942; BRIGGS, NEWBOLD and STACE, 1942) and the berries of *S. auriculatum* AIT. contain Solauricine (BELL, BRIGGS and CARROLL, 1942). TUTIN and CLEWER (1914) isolated solangustine from *S. angustifolium* and BARGER and FRAENKEL-CONRAT (1936) obtained solanocapsine from the berries of *S. pseudocapsicum* L.

I. Properties.

Solanine, $C_{45}H_{73}O_{15}N$, in the pure state crystallizes from water in needles and melts at 285° after drying at 100°. At 235° shrinkage of the crystals from the walls of the melting point tube occurs. It has $[\alpha]_D^{20°}$ —42.16° in dilute HCl, is sparingly soluble in water (1 in 8,000), soluble in hot ethanol and amyl alcohol, but almost insoluble in ether, chloroform, benzene and light petroleum. It is a feeble base with a faintly alkaline reaction. Solanine does not reduce FEHLING's solution but reduces ammoniacal silver nitrate on warming. It possesses a very bitter taste and is poisonous. Solanine is not affected by alkalies but when warmed with acids it is hydrolysed to solanidine and a mixture of glucose, rhamnose and galactose. The salts are amorphous. However, a crystalline hydrochloride has been prepared which has m. p. 212° (dec.). The platinichloride B$_2$ · H$_2$PtCl$_6$ occurs as a yellow flocculent precipitate soluble in hot water, insoluble in ether. Solanine forms an acetyl derivative which melts at 204—205°.

Solanidine, $C_{27}H_{43}ON$, formed on hydrolysis of solanine, crystallizes from ethanol in needles, m. p. 219°: $[\alpha]_D^{21°}$ —28.5 (ethanol). The hydrochloride, B · HCl, forms rhombic prisms, m. p. 345° (dec.); the platinichloride, B$_2$ · H$_2$PtCl$_6$ is a yellow amorphous compound sparingly soluble in water. Solanidine is very slightly soluble in water, soluble in ethanol and ether. It does not reduce FEHLING's solution or ammoniacal silver nitrate.

Solasonine, $C_{45}H_{73}O_{16}N$, crystallizes from 60—80% ethanol or 80% dioxan-water and melts at 284—285° (dec.). $[\alpha]_D^{25°} = $ —68.7° (ethanol). It is slightly soluble in water and forms a hydrochloride melting at 265°. The picrate melts at 199—201°, the picrolonate at 230—231°, and the acetyl derivative at 135—138°.

Solasodine, $C_{27}H_{43}O_2N$ · H$_2$O, results from the acid hydrolysis of solasonine and crystallizes from dilute alcohol or dioxan in scales, m. p. 197—198°. $[\alpha]_D^{25°} = $ —97.1° (methanol). The picrate has m. p. 144°, the picrolonate m. p. 234°, the hydrochloride m. p. 314°, and the oxalate m. p. 248°. The acetyl derivative is prepared by refluxing solasodine with dry pyridine and acetic anhydride for 2 hours and after cooling, pouring the mixture into water and making alkaline with ammonia. After successive re-crystallizations from dioxan-water and ethyl acetate, narrow plates are formed which melt at 195°.

Solauricine, $C_{45}H_{73}O_{16}N$, is closely related to solasonine and has m. p. 272°. On acid hydrolysis *solauricidine*, $C_{27}H_{43}O_2N$, is formed.

Solangustine, $C_{33}H_{53}O_7N \cdot H_2O$, separates from hot amyl alcohol in small yellow crystals, m. p. 235° (dec.), which are soluble in pyridine. On acid hydrolysis, solangustine yields *solangustidine*, $C_{27}H_{43}O_3N$, and glucose. Solangustidine is amorphous but forms crystalline salts. The hydrochloride, $B \cdot HCl$, forms plates, m. p. 325°, the nitrate colourless leaflets, m. p. 290° (dec.), from ethanol. The acetyl derivative crystallizes from ethyl acetate in needles, m. p. 256°.

Solanocapsine, $C_{26}H_{44}O_2N_2 \cdot H_2O$, crystallizes from 50% ethanol in flat prisms, m. p. 222°. $[\alpha]_D = +25.5°$. The hydrochloride, $B \cdot 2\,HCl \cdot H_2O$, has m. p. 280° and separates as needles, the sulphate $B \cdot H_2SO_4$, has m. p. 324° and the picrate, m. p. 194°.

II. The Isolation of Solanine.

1. From *Solanum tuberosum*.

ROOKE, BUSHILL, LAMPITT and JACKSON (1943) have adopted the method of PFANKUCH (1937) for the isolation of solanine from sprouting potatoes. 4 kg. of sprouting potatoes are finely minced, treated with 4 litres of 96% ethanol followed by 80 ml of glacial acetic acid and the mixture shaken occasionally during a period of 18 hours. The mixture is filtered on a BUCHNER funnel (Whatman No. 4 paper) and two further extractions of the residue are made with portions of 4 litres of 64% ethanol. The ethanolic extracts are combined and concentrated to approx. 1 350 ml. under reduced pressure, 140 g. sodium sulphate added and the mixture warmed on a water bath for 30 minutes. A flocculent precipitate of protein appears and after cooling and the addition of 50—60 ml. of 20% H_2SO_4, the mixture is filtered on a BUCHNER funnel (Whatman No. 4 paper) and the residue washed with 250 ml. of water. The filtrate is made alkaline with strong ammonia and allowed to stand overnight at 4°. The precipitate is filtered off (Whatman No. 5 paper) and washed with 60 ml. of 2% ammonia. The crude solanine is re-dissolved in dilute acetic acid, filtered and re-precipitated with ammonia. This process is repeated four times and the product is dried at 100°, ground and extracted with hot amyl alcohol until the residue no longer gives a colour with H_2SO_4 and formaldehyde. The amyl alcohol extract is evaporated under reduced pressure until the solution of solanine forms a gel. The gel is dissolved in 0.2% acetic acid, the last traces of amyl alcohol distilled off, and the solution made alkaline with ammonia. On warming, a flocculent precipitate forms which, after the mixture has been kept at 5° overnight, is separated by centrifuging. Further purification is carried out by extracting the precipitate with acetone to remove any free solanidine which may be present and re crystallizing the residue from ethanol. The final product should, after drying at 100°, melt at about 280°.

Solanidine is prepared from solanine by acid hydrolysis. 0.1 g. of solanine is dissolved in 100 ml. of 2.5% HCl and heated on a boiling water bath for 1 hour. The solution is cooled, made slightly alkaline with ammonia and the solanidine flocculated by gently warming. The mixture is kept overnight at 4°, filtered, and the precipitate washed with 2% ammonia and re-crystallized from ethanol.

2. The Isolation of Solasonine from the Berries of *Solanum sodomeum*.

BELL and BRIGGS (1942) extract the minced and dried green berries with ethanol in an all-metal extractor of the Soxhlet type capable of treating up to 56 lb. of material. Most of the ethanol is removed from the extract by distillation and excess of 2% aqueous acetic acid is added and the residual ethanol removed by steam distillation. The aqueous solution is filtered, the filtrate boiled, and the crude solasonine precipitated in granular form by the addition of ammonia to the hot filtrate. The crude alkaloid is purified by dissolving in dilute acetic acid and re-precipitating with ammonia followed by repeated crystallization from 60—80% dioxan-water. The final product occurs in colourless plates, m. p. 284 to 285° (dec.). A second method has been described by Briggs, Newbold and Stace (1942). Green berries are minced and pressed and the press-juice allowed to stand for 4—8 hours. The clear supernatant liquid is drawn off by suction and filtered in a gravity filter with frequent changes of paper. The filtrate is boiled and filtered while hot. The clear filtrate is boiled and ammonia passed through it until it becomes alkaline; boiling is then continued for 5 minutes. After cooling, a granular precipitate is formed which is collected, dissolved in 2% acetic acid and re-precipitated with ammonia from the boiling solution. This treatment is repeated and further purification carried out by re-crystallization from 60—80% ethanol and 80% dioxan-water.

Solasodine is prepared from solasonine by heating the latter with an excess of 3% HCl for 3 hours at 100°. After cooling, the hydrochloride of solasodine crystallizes out and is collected, re-crystallized twice from 80% ethanol, suspended in hot water, made alkaline with ammonia and heated at 100° for 30 minutes. After cooling, the solasodine is collected and repeatedly crystallized from 80% ethanol, m. p. 197.5—198.5°.

3. The Distribution of Solanine in *Solanum tuberosum*.

Lampitt, Bushill, Rooke and Jackson (1943) have made a study of the distribution of solanine in the potato and a summary of this work is given in Table 12a.

Table 12a. *The Distribution of Solanine in Solanum tuberosum.*

	Original material (mg./100 g.)	Calc. on dry solids
Skin (2—3% of tubers)	30—64	106—270
Peel (10—12% of tubers)	15	66
Peel including eye ($^1/_8$ inch disc)	30	130
Peel excluding eye ($^1/_8$ inch disc)	19	83
Flesh	1.2—10	6—40
Whole potato	7.5	27
Sprouts (formed during irradiation of tuber)	420—730	565—4070
Flower	215—415	1580—3540
Leaf	55—60	506—620
Stem of haulm	2.3—3.3	25—55

In sprouted potatoes the amount of solanine varies with the variety and free solanidine occurs in the sprouts of some varieties but not in others. In the plant solanine (but not solanidine) is present in all parts, the content being highest in the flowers. As the plant matures, the solanine content increases in the flowers, stolons and tubers.

III. The Detection and Estimation of the Alkaloids of *Solanum*.

1. Qualitative Tests.

Colour reactions employing reagents which have been used for the detection of sterols have been described by various workers. Alberti (1932) has shown that solanine and solanidine give an intense purple colour with H_2SO_4 and formaldehyde. This reaction has been used as the basis for the colorimetric determination of solanine. A study of the reactions of solanine prepared from the potato has been made by ROOKE et al. (1943) and a scheme of tests is summarised in Table 13.

Table 13. *Colour and Precipitation Tests for Solanine and Solanidine.*

Reagent	Solanine (solid)	Solanidine (solid)
Conc. H_2SO_4 warmed	Yellow → red	Yellow → brown
ERDMANN Reagent	Yellow	Yellow
FRÖHDE's reagent warmed	Yellow → brown	Yellow → brown
Ethyl sulphuric acid	Cherry-red	Cherry-red
Resorcinol in glacial acetic acid + H_2SO_4 warmed	Wine-red → blood-red → brownish red	Yellow → orange → red → brown with green fluorescence
Vanillin in glacial acetic acid + H_2SO_4 cold	Brown ring at junction of liquid, purple in acetic acid layer	As for solanine
Vanillin in glacial acetic acid + H_2SO_4 warmed	Yellow → brownish yellow → reddish brown → blood-red → brown	Yellow → orange-red → red → brownish-red → dark brown → black
	Solanine (in dilute H_2SO_4)	*Solanidine (in dilute H_2SO_4)*
Iodine in KI solution	No precipitate	No precipitate
MAYER's reagent	No precipitate	No precipitate
Picric acid	No precipitate	No precipitate
Tannic acid	Turbidity	No precipitate
Phosphomolybdic acid	Precipitate	No precipitate

BRIGGS et al. (1942) have described a colour test which distinguishes solasonine and solasodine from solanine and solanidine. Conc. H_2SO_4 (1 ml.) is carefully added to a solution of solasonine or solasodine in hot ethanol (1 ml.). A characteristic intense greenish-yellow fluorescence is produced. This reaction is not given by solanine or solanidine. These authors also give details of colour reactions of solasodine and solanidine with aromatic aldehydes in acetic acid — sulphuric acid solution. A few mg. of solasodine or related alkaloid are dissolved, together with a small quantity of the aldehyde, in 1—2 ml. of acetic acid. An equal volume of conc. H_2SO_4 is carefully added. Characteristic colours develop at the interface and on mixing and heating, the full range of colours appears. The aldehydes used for these tests are vanillin, anisaldehyde and p-hydroxybenzaldehyde.

2. Quantitative Estimation.

PFANKUCH (1937) has developed a *colorimetric method* for the determination of solanine and solanidine based on the reaction with the MARQUIS reagent (formaldehyde and sulphuric acid), and CONNER (1937) has described a volumetric method which measures the concentration of reducing sugars present after acid hydrolysis of solanine. The latter method determines solanine only and by combination of the two methods, both solanine and solanidine may be determined when present together in tissues. ROOKE et al. (1943) have used both methods in their studies on the solanine contant of potato tissues.

The method of PFANKUCH (1937) is carried out in the following manner. Up to 150 g. of finely minced material (tubers, sprouts, leaves, flowers) is treated with 150 ml. of 96% ethanol followed by 3 ml. of glacial acetic acid and the mixture is shaken occasionally during a period of 18 hours. The mixture is now filtered in a BUCHNER funnel (Whatman No. 4 paper) and two further extractions of the residue are made with portions of 150 ml. of 64% ethanol. The combined ethanolic extracts are concentrated to 50 ml. under reduced pressure, 5 g. of sodium sulphate are added and the mixture warmed on a water bath for 30 min.

A flocculent precipitate of protein appears at this stage and after cooling and addition of 2 ml. of 20% H_2SO_4 the mixture is filtered in a Buchner funnel (Whatman No. 4 paper) and the residue washed with 10 ml. of water. The filtrate (plus washings) is made alkaline with strong ammonia and allowed to stand overnight at 4°. The precipitate is filtered off (Whatman No. 5 paper) and washed with 10 ml. of 2% ammonia. The volume of the filtrate (plus washings) is noted so that a correction may be applied for the solubility of solanine. The precipitate is dissolved in 15 ml. of 1% H_2SO_4 and the solution diluted with water to 25 ml. This solution is used for the determination of solanine as described below.

2.5 ml. of the solution are measured into a small dry flask cooled in ice, and 5 ml. of conc. H_2SO_4 are added dropwise from a burette. The solution is well shaken during the addition of acid. After 1 minute, 2.5 ml. of 1% formaldehyde are added drop by drop and the solution kept for 90 minutes at room temperature to allow the purplish-red colour to develop. The extinction coefficient of the solution is measured in a photometer (Zeiss, filter S. 53, 1 cm. cell) and the concentration of solanine plus solanidine expressed as solanine is obtained by reference to a calibration curve. Allowance is made for the solubility of solanine (0.2 mg./100 ml.) in the ammoniacal filtrate. The extinction coefficient is most readily determined at a wave-length of 530 mμ and a linear relationship exists between the concentration of solanine and the extinction coefficient at this wave-length. Using this method amounts greater than 3 mg. of solanine may be determined with an accuracy of $\pm 5\%$.

The *volumetric method* of Conner (1937) is carried out as follows. 5 ml. of the solution containing solanine are placed in each of two 25 ml. graduated flasks and 5 N HCl is added to each to a final concentration of 2.5% HCl. The solution in one flask is neutralized immediately with 5 N NaOH using methyl red as indicator and diluted to 25 ml. The second solution is heated for 1 hour in a boiling water bath, cooled, neutralized with 5 N NaOH and diluted to 25 ml. The reducing sugar is then determined in 10 ml. samples of both solutions using either the Pavy copper reduction method or the Hanes-Hagedorn-Jensen method as modified by Hulme and Narain (1931) (cf. Vol. 2 of this handbook). Solutions containing known weights of pure solanine are treated as above and a curve constructed for reference. The concentration of solanine in the original material is then calculated, allowance being made for the solubility of solanine in the ammoniacal solution. Not less than 15 mg. of solanine may be determined with an accuracy of $\pm 5\%$.

W. Alkaloids of *Strychnos* spp.

A number of species of *Strychnos (Loganiaceae)*, a genus of small trees or stout climbing shrubs widely distributed throughout India and the tropics contain Strychnine and Brucine as the major alkaloids. In addition, the minor alkaloids Vomicine, Pseudostrychnine, α-Colubrine and β-Colubrine have been found in the mother liquors remaining after the isolation of strychnine and brucine. Strychnicine has been isolated in small quantity from the leaves of *Strychnos nux-vomica* L. grown in Java. The alkaloids are located chiefly in the seeds, but in some species they are present also in the wood, bark, leaves and roots. The occurrence and distribution of the alkaloids in the principal species examined are given in Table 14.

In the seeds of *S. nux-vomica* the alkaloids occur in the large thick walled cells of the endosperm, strychnine being located in the cells near the centre of the seed and brucine in the cells near the epidermis.

Table 14. *The Occurrence and Distribution of Alkaloids in Strychnos spp.*

Species	Constituents	Habitat
S. nux-vomica L.	Seeds, strychnine, 0.25—2% brucine, 0.5 —2% Bark, brucine (anhyd.) 7.78% Wood, brucine (anhyd.) 2.26% Leaves, brucine (anhyd.) 0.33%	India
S. ignatii BERG.	Seeds, strychnine 0.52—1.5% brucine 1.43%	Phillipine Islands
S. colubrina L.	Wood, 0.96% total alkaloids Bark 5.54% total alkaloids (Strychnine and brucine present)	East Indies
S. rheedii CLARKE	Wood and bark, strychnine and brucine Seeds 0.06% brucine	Malabar
S. potatorum L.	Seeds, brucine (traces)	East Indies, Ceylon
S. guianensis MART.	Bark, strychnine and brucine	Guiana
S. tieute LESCH	Root bark, strychnine Leaves and seeds, strychnine 1.4% brucine (trace)	Java
S. lucida R. BR.	Seeds, strychnine 0.30 brucine 1.5%	W. Australia
S. ligustrina BL.	Seeds, brucine 0.6—1.5% Stem bark, strychnine 0.5% brucine 1.7%	East Indies
S. javanica	Bark, brucine 2.7%	Cochin-China
S. icaja BAILL.	Bark, leaves, roots, strychnine	Africa
S. aculeata SOL.	Seeds, brucine (trace)	W. Africa

I. The Isolation of Strychnine and Brucine.

Finely ground seeds of *S. nux-vomica* are mixed with slaked lime and made into a paste by the addition of a little water. The mass is dried at 100° and extracted with hot chloroform in a continuous extractor. The alkaloids are removed from the chloroform solution by shaking with successive portions of dilute H_2SO_4. The combined acid extracts are filtered and excess of ammonia added to precipitate the alkaloids. The precipitate is extracted with 25% ethanol which dissolves the brucine. The undissolved strychnine is filtered off and is purified by repeated recrystallization from ethanol. The filtrate is concentrated to small bulk and the brucine precipitated as oxalate by neutralization with oxalic acid. The precipitate is washed with cold ethanol, dissolved in hot water, decolourized with charcoal and the solution evaporated to dryness with magnesia on a water bath. The brucine is extracted from the residue with acetone, the acetone removed, and the crude brucine recrystallized from dilute ethanol. Separation of strychnine and brucine may also be effected by means of the relatively insoluble salts of strychnine such as the ferrocyanide, chromate and hydriodide, which are much less soluble in water than the corresponding salts of brucine.

II. Properties.

Strychnine, $C_{21}H_{22}O_2N_2$, occurs as a white crystalline powder (when precipitated) or as colourless prisms of the rhombic system. $[\alpha]_D = -109.9°$ (80% ethanol) or $-139.3°$ (chloroform). The pure alkaloid melts at 286—288°. Strychnine is slightly soluble in water (1 in 6400 at 25°, 1 in 2500 at 100°), soluble in cold 90% ethanol (1 in 160) and in boiling 90% ethanol (1 in 12), in cold absolute ethanol (1 in 350) and in boiling absolute ethanol (1 in 40). It is practically insoluble in absolute ether and is very little soluble in acetone and light petroleum. The base dissolves readily in chloroform (1 in 6) and in a mixture of equal parts of ether and chloroform. It is soluble in benzene (1 in 160) and in amyl alcohol (1 in 180). Strychnine behaves as a strong monoacidic base and forms salts which are crystalline. Aqueous solutions react alkaline and have an intensely bitter taste. The base and its salts are very powerful poisons. Strychnine sulphate, $B_2 \cdot H_2SO_4 \cdot 5 H_2O$, forms colourless prisms or prismatic needles, m. p. 200° (dry), and is soluble in water (1 in 31 at 25°), in ethanol (1 in 65 at 25°) but much less soluble in chloroform (1 in 325 at 25°). The nitrate, $B \cdot HNO_3$, occurs in colourless needles and is soluble in water (1 in 42 at 25°), ethanol (1 in 120 at 25°) and chloroform (1 in 156 at 25°). The hydrochloride, $B \cdot HCl \cdot 2 H_2O$, forms colourless prisms and is soluble in cold water (1 in 35) and ethanol (1 in 80). The hydriodide, $B \cdot HI \cdot H_2O$, and the periodide, $B \cdot HI \cdot I_2$, are sparingly soluble in water. The aurichloride, $B \cdot HAuCl_4$, crystallizes from ethanol in orange-yellow needles. The chromate, ferrocyanide, hydroferrocyanide and picrolonate are very little soluble in water and may be used for the isolation of the alkaloid.

    Brucine, $C_{23}H_{26}O_4N_2$, crystallizes from water or aqueous ethanol in monoclinic prisms containing $4 H_2O$. It melts at 178° (dry) and has $[\alpha]_D$ —119° to —127° (chloroform) or —80.1° (ethanol). Aqueous solutions, are intensely bitter and react alkaline. The alkaloid is sparingly soluble in cold water (1 in 320), more soluble in boiling water (1 in 150), and easily soluble in ethanol, chloroform and acetone. It is very little soluble in absolute ether. Brucine is practically insoluble in dilute solutions of NaOH and KOH and is sparingly soluble in ammonia solutions. It behaves as a monoacidic base and forms salts which crystallize well. It is much less poisonous than strychnine. The hydrochloride, $B \cdot HCl$, forms needles and is readily soluble in water. The sulphate, $B_2 \cdot H_2SO_4 \cdot 7 H_2O$, forms long needles soluble in water and the hydriodide, $B \cdot HI$, forms leaflets sparingly soluble in water.

    Pseudostrychnine, $C_{21}H_{22}O_3N_2$, forms a crystalline powder, m. p. 266—268°, and has $[\alpha]_D^{25°}$ —43.8° (ethanol). It is a weak base and is much less poisonous than strychnine. Pseudostrychnine forms crystalline salts, the best known of which are the hydrochloride, $B \cdot HCl \cdot 2 H_2O$, $[\alpha]_D^{19°}$ +3.9° (water); the nitrate $B \cdot HNO_3$, $[\alpha]_D$ +7.6 (ethanol), and the ferrichloride, $C_{21}H_{22}O_2N_2Cl_4Fe$, which crystallizes as orange-red plates, m. p. 234—235° (dec.) from acetic acid.

    α-Colubrine, $C_{22}H_{24}O_3N_2 \cdot 4 H_2O$, crystallizes from hot dilute ethanol and in the anhydrous form melts at 184°. $[\alpha]_D^{19°} = $ —76.5° (ethanol, 80%). The hydrochloride, $B \cdot HCl \cdot 3 H_2O$, forms long leaflets and the sulphate, $B_2 \cdot H_2SO_4 \cdot 10 H_2O$, glancing leaflets. *β-Colubrine,* $C_{22}H_{24}O_3N_2$, crystallizes from dilute ethanol and has m. p. 222°: $[\alpha]_D^{19°} = $ —107.7° (ethanol, 80%). The hydrochloride, $B \cdot HCl \cdot H_2O$, is crystalline.

    Vomicine, $C_{22}H_{24}O_4N_2$, is a weak monoacidic base and forms colourless needles from 80% ethanol or hexagonal prisms from acetone. It has m. p. 282°; $[\alpha]_D^{22°}$ +80.4° (ethanol). The hydrochloride, $B \cdot HCl \cdot 3 H_2O$, has m. p. 245° (dec.) and in common with the sulphate, nitrate, hydrobromide and hydriodide, is sparingly soluble in water. The acetyl derivative has m. p. 204—205°.

III. The Detection of Strychnine.

1. Precipitation Tests.

Strychnine is precipitated from solutions of its soluble salts on addition of alkali hydroxides, alkali carbonates and ammonia, but not by addition of bicarbonates. On standing, the precipitate becomes crystalline and the free base may be recovered by shaking out with a mixture of equal volumes of ether and chloroform and removing the solvent on a water bath. The alkaloid reagents give precipitates in slightly acid solution with very dilute solutions of strychnine and advantage may be taken of the delicacy of these reactions for the removal and subsequent recovery of small quantities of strychnine from plant extracts. The precipitate (e. g. phosphotungstate) is suspended in water, treated with excess of ammonia and extracted with the ether-chloroform mixture as above. A solution of potassium or ammonium thiocyanate added to the solution of a strychnine salt gives characteristic crystals (four sided columns) of strychnine thiocyanate. This test is best carried out on a microscope slide and the form of the crystals examined with the low power objective. Potassium chromate and potassium dichromate give with strychnine crystalline precipitates which are very little soluble in water. Strychnine chromate, $B_2 \cdot H_2CrO_4$, appears as a yellowish-brown precipitate when potassium chromate solution is added to a solution of strychnine or one of its soluble salts. It crystallizes from hot water in orange-yellow needles. The dichromate, $B_2 \cdot H_2Cr_2O_7$, crystallizes in orange-yellow needles and is soluble in 1815 parts of water at 18°. Strychnine ferrocyanide is precipitated from solutions of strychnine or its salts on addition of potassium ferrocyanide solution. According to ALLEN and ALLPORT (1940) the precipitation is greatly facilitated if the ferrocyanide solution is added to an acid solution of strychnine which is $N/14$ with respect to H_2SO_4 and $N/3$ to oxalic acid. The chromate and ferrocyanide are very useful in the separation of strychnine from brucine as the corresponding salts of the latter are much more soluble than those of strychnine. Strychnine is precipitated almost quantitatively by a solution of picrolonic acid in ethanol or ether. The picrolonate melts at 286° and the test is useful as a confirmatory test for strychnine.

2. Colour Tests.

Pure strychnine, free from brucine, dissolves without colour in concentrated H_2SO_4, ERDMANN's and FROEHDE's reagents. Concentrated nitric acid dissolves strychnine with a yellowish colour. *The sulphuric acid-dichromate test.* If a trace of strychnine is dissolved in conc. H_2SO_4 in a small basin and a few minute crystals of potassium dichromate are dusted on to the surface of the solution and stirred with a glass rod, a deep blue-violet colour is formed which changes through purple and crimson to cherry-red. On long standing the colour changes to orange or yellow. Other oxidizing agents, e. g. potassium permanganate and manganese dioxide, may be used in place of potassium dichromate. Strychnine dichromate gives the test directly on the addition of conc. H_2SO_4. This test is extremely delicate and sensitive but it should be borne in mind that other alkaloids containing the indole nucleus such as gelsemine and yohimbine give colour reactions with H_2SO_4 and an oxidizing agent. Brucine, which often accompanies strychnine in the plant, interferes with the test if present in substantial amounts. Most substances likely to interfere can however be removed by the procedure outlined below. The presence of nitric acid or nitrates prevents the development of colour in the test, consequently strychnine nitrate gives a negative result. MANDELIN's *reagent* (H_2SO_4 + ammonium vanadate) gives with strychnine a violet-blue colour which changes through purple to cherry-red and finally to a fairly stable orange colour. MALAQUIN (1909) has described the following test which is said to be characteristic for strychnine. 1 ml. of a very dilute solution of a salt of strychnine is treated with 2 ml. of HCl and 1 g. of granulated zinc. After standing for 2—3 minutes, the tube is placed in a water bath at 100° for 5—10 minutes and cooled. The unchanged zinc is filtered off and a small crystal of sodium nitrite is added to the filtrate. A fine red colour develops immediately. The test is very delicate and will detect strychnine in a dilution of 1 in 100,000. If a particle of strychnine is dissolved in one or two drops of nitric acid and the solution evaporated to dryness at 100°, the residue gives a reddish-purple colour on treatment with ammonia. An intense scarlet coloration is produced when a small crystal of strychnine is dissolved in 20% nitric acid, the solution gently heated, and a trace of potassium chromate added. In the application of the above tests for the detection of strychnine in plant tissues, the finely powdered tissue is moistened with ammonia and exhaustively extracted with chloroform or, as recommended by ALLEN and ALLPORT (1940), the free alkaloid is liberated by piperazine and extracted with tetrachloroethylene. The alkaloid is shaken out with dilute H_2SO_4 and the acid solution shaken with a mixture of equal volumes of ether and chloroform to remove substances such as salicin and piperine which give colours with conc. H_2SO_4 alone. The ether-chloroform solution is run off and the acid solution rendered alkaline with ammonia and again shaken out with ether-chloroform mixture. Strychnine is separated either as the chromate or ferrocyanide as described above. In this manner, strychnine is removed from most substances likely to interfere in the colour tests. The substance proto-curarine, isolated as a red powder from certain species of *Strychnos*, gives a colour reaction with H_2SO_4 and oxidizing agents almost identical with that of strychnine but in contradistinction to strychnine it is easily soluble in water and may be removed from a mixture of the two substances by making alkaline with ammonia and filtering. Injection of as little as 0.05 mg. of strychnine in the form of a soluble salt into the lymph sac of the frog will cause tetanic convulsions to appear in approx. 30 minutes.

IV. Estimation of Strychnine.

1. The Colorimetric Estimation of Strychnine.

ALLEN and ALLPORT (1940) have described a method for the estimation of strychnine in nux vomica seed which is based upon the colour test of MALAQUIN. 1 g. of finely powdered seed is accurately weighed and placed in a 100 ml. conical flask, 3 ml. of 95% ethanol are added and the mixture heated on a boiling water bath until most of the ethanol has evaporated. 10 ml. of tetrachloroethylene (tech.) and 0.6 g. of piperazine are added, the flask connected to a reflux condenser and heated so that the solvent boils vigorously for 15 minutes. The flask is allowed to cool slightly and as much as possible of the solvent and powdered drug is transferred to a small dry percolator (16 cm. × 8 mm. bore) previously plugged with a little cotton wool and suspended within a stoppered 100 ml. measuring cylinder. If necessary, the powder is compressed with a small glass rod so that the solvent percolates at the rate of about one drop a second. The transference of the powder to the percolator is completed by washing out the flask with hot tetrachloroethylene and extraction of the powder is continued with the heated solvent until the volume of the cooled percolate is 42 ml. Sufficient $N/8\ H_2SO_4$ is added to bring the level of the liquid to the 90 ml. graduation mark and the cylinder is stoppered and shaken for about 15 seconds. After the immiscible liquids have separated about 10 ml. of the upper layer are pipetted off and filtered through a dry filter paper. Exactly 5 ml. of the filtrate are transferred to a conical centrifuge tube of 15 ml. capacity and exactly 1.5 ml. of a 9% w/v aqueous solution of oxalic acid and 1 ml. of a 5% w/v solution of potassium ferrocyanide in 0.2% w/v aqueous solution of sodium carbonate are added. The tube is allowed to stand at room temperature for 15 minutes and then is immersed in a freezing mixture. The tube is agitated until the contents are frozen and then removed and gently warmed until melting takes place. Approx. 0.1 g. of acid washed kieselguhr is added and the precipitate centrifuged down. The clear supernatant is poured off and about 2 ml. of a 0.1% w/v aqueous solution of H_2SO_4 containing 1% of oxalic acid are added. The precipitate is washed by stirring with a glass rod which is rinsed free from precipitate by adding about 8 ml. more of the washing solution and again centrifuged. The washing is repeated once more and the clear supernatant is decanted as completely as possible. 1 ml. of conc. HCl is added to the precipitate and the mixture heated over a small luminous flame until the kieselguhr is dispersed throughout the acid solution of strychnine ferrocyanide. 10 ml. of 10% (w/v) HCl are added and the mixture is transferred to a 100 ml. volumetric flask, the centrifuge tube being rinsed well with dilute HCl and the volumetric flask filled to the mark with the washings. A portion of this solution is filtered and 5 ml. of the filtrate are trans-ferred to a test tube containing 0.2 g. of zinc amalgam (20 mesh: 40% Hg). The tube is immersed in a boiling water bath for 7 minutes, cooled under the tap and 0.05 ml. of a freshly prepared solution (0.1%) of sodium nitrite added. The colour is measured in an absorptiometer and the strychnine determined by reference to a calibration curve prepared by using solutions of pure strychnine ferrocyanide. The relationship between the colour produced and the quantity of strychnine is linear provided that not more than 0.10 mg. of the alkaloid is present. Precipitation of strychnine as ferrocyanide eliminates interference from brucine. ROLLAND-LECLERCQ (1947) states that maximum absorption of the colour produced occurs at the wave-lengths 440 mμ and 520 mμ. RASMUSSEN (1943) has used the colour reaction with a modification of MANDELIN's reagent for the estimation of strychnine and states that for mixtures containing equal parts of strychnine and brucine, the strychnine value appears only 2% high.

2. Spectrophotometric Estimation of Strychnine.

EL RIDI and KHALIFA (1952) have described a method for the chromatographic purification and ultra-violet spectrophotometric assay of strychnine in galenical preparations and have shown that in absolute ethanol strychnine shows a maximum absorption at 254 mμ and brucine a maximum absorption at 264 mμ and another less distinct band at 301 mμ. Destruction of brucine with nitric acid was not permissible in the spectrophotometric assay as the nitrate ion absorbs at 302 mμ. Brucine was therefore eliminated by treatment of the mixture containing strychnine and brucine with potassium persulphate at 60—70° in the presence of 3% H_2SO_4 for 1 hour. Strychnine remained unchanged by this treatment and was extracted from alkaline solution with chloroform, the chloroform removed by distillation, and the residue dissolved in absolute ethanol. The strychnine was determined by measuring the E 254 mμ value. The method can be used for the determination of strychnine in the Liquid Extract of Nux vomica and following the extraction procedure of ALLEN and ALLPORT (1940), for the determination of strychnine in the seeds of nux vomica. The assay method for the Liquid Extract is carried out as follows. Into a glass tube 25—30 cm. long, 1.3 cm. in diameter, with a constricted end, 15 g. of active alumina is packed dry in portions forming an adsorption column 14 cm. long. The column is connected to a suction apparatus and 2 ml. of the Liquid Extract is poured on and gentle suction applied. Before the liquid disappears from above the column, 86% ethanol is added a little at a time to wash down the alkaloids on the side of the tube. The column is then washed with larger amounts of 86% ethanol until the percolate is alkaloid free. About 50 ml. of 86% ethanol is usually sufficient for complete washing. The clear percolate is transferred quantitatively to a 100 ml. standard flask and made to volume with 96% ethanol. An aliquot (20—30 ml.) is distilled and the residue dissolved in 10 ml. of 3% H_2SO_4, 0.5 g. of potassium persulphate is added and dissolved by shaking and the solution kept in a water bath at 60—70° for 1 hour. After cooling, the solution is transferred quantitatively to a 100 ml. standard flask with distilled water. The solution is mixed by shaking, filtered through a dry filter paper into a dry flask. The strychnine is then determined by measuring the E 254 mμ value of the solution, and the amount calculated from the equation $C = E/E_1^{1\,cm.}\,\%$ where C = concentration in g. per cent. $E_1^{1\,cm.}\,\% = 390$ at 254 mμ. E = measured extinction at 254 mμ. The accuracy of the estimation of strychnine by the spectrophotometric method depends on the removal of interfering ultra-violet absorbing impurities from the extracts used. For a number of determinations the average error was found to be of the order of —0.07%.

3. Determination of Strychnine and Brucine Separately by Chromatographic Adsorption Followed by Titration.

Fischer and Buchegger (1950) estimate strychnine and brucine separately by adsorption of the two alkaloids on a column of alumina followed by elution and titration of the eluates with acid. 2 g. samples of the finely powdered material (nux vomica seeds or seeds of *Strychnos ignatii*) are exhausted with chloroform after treatment with alkali or are extracted with acid and water. The alkaloids in solution in trichloro-ethylene are adsorbed on 10 g. of alumina (Merck, no. 95, 581) in a column. Strychnine is eluted with 70 ml. of carbon tetrachloride containing 9% of acetone and brucine with 25 ml. of ethanol. The solvents are distilled off and the residue titrated with acid. The authors claim very satisfactory results by use of this method.

4. The Determination of Total Alkaloids of *Strychnos nux vomica* by Ion Exchange.

Jindra and Pohorski (1951) have applied the method of ion exchange to the determination of the total alkaloids of nux vomica seeds. Amberlite IR—4 B resin is used in the column and the alkaloids are electrometrically titrated against $0.01 N$ HCl. For nux vomica seeds approx. 0.3 g. of finely powdered material is accurately weighed and shaken vigorously for 15 minutes with 5 g. of chloroform, 15 g. of ether and 1 ml. of 10% ammonia. After separation the ether-chloroform layer is run off into 5 ml. of water and shaken well for 30 seconds. 14 g. of the ether chloroform solution is carefully weighed and evaporated almost to dryness. 1 ml. of 90% ethanol is then added and the solution completely evaporated. To the dry residue 3 drops of 2% H_2SO_4 are added followed by 10 ml. of 90% ethanol and the solution passed through the column. The eluate is diluted with 25 ml. of water and titrated against $0.01 N$ acid. 1 ml. of $0.01 N$ HCl is equivalent to 0.003642 g. alkaloids. Using the antimony electrode for the titration, the authors obtained satisfactory results which compared favourably with other methods of analysis.

5. The Gravimetric Determination of Total Alkaloids.

Picrolonic acid has been used for the estimation of total alkaloids in nux vomica (Matthes and Rammstedt, 1907). 15 g. of finely powdered material is extracted by shaking with a mixture of 100 g. of ether, 50 g. of chloroform and 10 ml. of 10% NaOH solution. 15 ml. of water is added and the mixture shaken to agglutinate the powder. After allowing the layers to separate, 50 g. of the chloroform-ether solution which has been filtered through a dry filter (and which represents 5 g. of the powder) are evaporated to about half the volume. 5 ml. of $N/10$ solution of picrolonic acid in ethanol are added and the solution set aside for 24 hours to complete precipitation. The precipitate of strychnine-brucine picrolonate is collected in a Gooch crucible, washed with 2 ml. of a mixture of ethanol and ether (1:3), dried for 30 minutes at 110° and weighed. On the assumption that strychnine and brucine are present in nux vomica in approximately equal proportions the mean of the molecular weights of the picrolonates of the two alkaloids is used in the calculation. The weight of the mixed alkaloids is thus obtained by multiplying the weight of the picrolonate precipitate by 0.5798. Bandelin (1950) has used ammonium reineckate for the precipitation and estimation of alkaloids including strychnine. The alkaloid reineckate is dissolved in acetone and the colour of the reineckate ion measured at 525 mμ.

6. The Colorimetric Determination of Brucine.

For the colorimetric determination of brucine in the presence of strychnine
WÖBER (1918) has utilized the colour test given by brucine with nitric acid.
Strychnine gives with nitric acid a yellow coloration which may interfere with
the colour development due to brucine if both alkaloids are present. A mixture
of equal volumes of strong nitric acid (sp. gr. 1.4) and 20% H_2SO_4 is used as the
reagent. The estimation is carried out as follows. 0.1 g. of the sample alkaloid
mixture is dissolved in 20 ml. of 1% H_2SO_4 with gentle heating. This solution is
transferred to a 100 ml. graduated cylinder, rinsed and made up to 30 ml. with
1% H_2SO_4. If the alkaloids are present as salts they must be converted to the
bases by treatment with alkali and extraction with chloroform. After evaporation
of the chloroform, 0.1 g. of the dry residue is treated as above. 20 ml. of a solution
of strychnine containing 0.1 g. in 1% H_2SO_4 is placed in a 100 ml. graduated
cylinder. A suitable amount of standard brucine solution (to be determined by
preliminary estimation) is added to the strychnine solution. To both sample
and standard are added 10 ml. of the mixture (1:1) of nitric acid and 20% H_2SO_4.
The solution is stirred and after 1 minute, 2 ml. of a saturated aqueous solution
of potassium chlorate is added to sample and standard. The solution is well
stirred and if necessary diluted to 50 or 100 ml. according to colour intensity.
The colour is then measured in a suitable absorptiometer and the amount of
brucine calculated from a calibration curve. The error of the method is 1—4%
for 5—50 mg. of brucine in a mixture of strychnine and brucine.

7. The Paper Partition Chromatography of Strychnine and Brucine.

The paper partition chromatography of strychnine and brucine has been
studied by GORE and ADSHEAD (1952). Good separation of the two alkaloids
is obtained on Whatman No. 1 paper with the solvent systems n-butanol 10,
formic acid 1, water 10, and n-butanol 10, propionic acid 1, water 10. In the
former solvent system average R_F values for brucine and brucine hydrochloride
are 0.44 and 0.39 and for strychnine and strychnine hydrochloride, 0.58 and 0.56.
For the second solvent system the R_F values are, brucine and brucine hydrochloride 0.58 and 0.53, and for strychnine and strychnine hydrochloride 0.70
and 0.69. The spots are developed by spraying the dry paper with aqueous
iodine solution.

Addendum.

In a recent study of the alkaloids of Australian species of *Strychnos*, ANET,
HUGHES and RITCHIE (1953) have isolated two new alkaloids, *strychnospermine*
and *spermostrychnine*, from the leaves of *S. psilosperma* F. MUELL. *Strychnospermine* $C_{22}H_{28}O_3N_2$, has m. p. 208—209° and $[\alpha]_D^{185}$ +60.3°. It forms a picrate,
m. p. 254°, and gives a positive OTTO test. *Spermostrychnine*, $C_{21}H_{26}O_2N_2$, has
m. p. 208—209° and $[\alpha]_D^{22°}$ +88° and is similar to strychnospermine in giving
the same colour reactions and forming similar derivatives. The picrate has
m. p. 173—175°.

X. The Tobacco Alkaloids.

The genus *Nicotiana (Solanaceae)*, represented by about 100 species and
sub-species, is widely distributed in many parts of the world. The cultivated
species, *N. tabacum* L., and *N. rustica* L., are the principal sources of tobacco,
the latter species furnishing East Indian and Turkish tobaccos. Many species
have been examined for their alkaloidal content and the bases listed in Table 1
have been isolated and characterized.

Table 15. *The Bases of Nicotiana spp.*

L-nicotine, $C_{10}H_{14}N_2$	anabasine, $C_{10}H_{14}N_2$	piperidine
nicotyrine, $C_{10}H_{10}N_2$	anatabine, $C_{10}H_{12}N_2$	pyrrolidine
L-nornicotine, $C_9H_{12}N_2$	N-methylanabasine, $C_{11}H_{16}N_2$	N-methylpyrrolidine
D-nornicotine, $C_9H_{12}N_2$	N-methylanatabine, $C_{11}H_{14}N_2$	3'.2-dipyridyl
nicotelline, $C_{10}H_8N_2$	nicotöine, $C_8H_{11}N$	

The species of *Nicotiana* have been classified into groups according to the predominance of one or other of the alkaloids present. Group A contains those species in which nicotine is the main alkaloid and a secondary alkaloid is either absent or present in small quantity. Representatives of this group are *N. rustica* L. *N. alata* Link and Otto and *N. gossei* Domin. Group B contains species in which nornicotine is the major alkaloid and a secondary alkaloid is either absent or present in small quantity. Representatives of this group are *N. glutinosa* L., *N. tomentosa* Ruiz and Pav. and *N. maritima* Wheeler. Group C includes species in which both nicotine and nornicotine are present in quantity. Examples of this group are *N. tabacum* L., *N. suaveolens* Lehm., and *N. repanda* Willd. Group D includes *N. debneyi* Domin. and *N. glauca* R. Grah. in which the major alkaloid is anabasine. The wild species of *Nicotiana* generally have a lower alkaloidal content than the cultivated species and hybridization of species may result in variation in the relative proportions of the constituent alkaloids. The reader is referred to the monograph by Marion (Manske and Holmes, 1950) for a complete list of the various species of *Nicotiana* which have been examined, and for an account of the effects of hybridization and grafting. Leaves of different species of *Nicotiana* show a wide variation in the total alkaloidal content. Low alkaloid types may have as little as 0.08% whereas normal and mixed alkaloid types show variation between 0.3% and 4% calculated on the basis of air-dry weight. The normal alkaloid types show a preponderance of nicotine, the content of nornicotine varying between 0.03% and 0.2%. In the mixed alkaloid types however, the content of nornicotine may vary from 0.25% to 1.4% and therefore exceed that of nicotine. The alkaloids are present in all parts of the plant, the leaves yielding the highest values. Dawson (1941) has shown that synthesis of nicotine takes place in the root tissues. The tobacco alkaloids occur in the plant as salts of citric, malic or oxalic acids, but glycosides yielding nicotine on hydrolysis also have been isolated from tobacco plants. Nicotine, nornicotine and anabasine have during recent years been found in a wide variety of plants. *Duboisia hopwoodii* F. Muell. contains nicotine and nornicotine (Bottomley, Nottle and White, 1945) and *D. myoporoides* R. Br. has recently been shown to contain in addition to hyoscine, both nicotine and nornicotine (Hills, Bottomley and Mortimer, 1953). Nicotine also occurs in *Asclepias syriaca* L. (Marion, 1939), in *Lycopodium flabelliforme* L. (Manske and Marion, 1942), *L. clavatum* L. (Marion and Manske, 1944), *L. lucidulum* Michx. (Manske and Marion, 1946) and *L. sabinaefolium* Willd. (Marion and Manske, 1946). *Equisetum arvense* L. also contains nicotine in small quantity (Manske and Marion, 1942) and Marion (1945) has recorded its occurrence in *Sedum acre* L. Anabasine was first recorded in *Anabasis aphylla* L. *(Chenopodiaceae)* by Orechoff (1929).

I. Properties.

L-Nicotine in the pure state is a colourless oil, b. p. 246.1°/730.5 mm., $D_{4}^{20°}$ 1.00925, $[\alpha]_D^{20°}$ —168.66°. On exposure to air nicotine turns yellow and eventually brown. It possesses a burning taste, but when pure little odour. It is hygroscopic and distils unchanged in steam. Nicotine is miscible in all proportions with water below 60° and above 210° but at intermediate temperatures soluble hydrates are not formed and miscibility becomes much reduced. An azeotrope is formed containing 2.45g. of nicotine in 100 ml. of water which boils at 99.6/750 mm. Aqueous solutions are strongly alkaline to litmus. Nicotine is easily soluble in ethanol, ether, light petroleum and benzene. The salts with mineral acids are soluble in water and are dextrorotatory. $B \cdot HCl$ has $[\alpha]_D +102.2$ and $B_2 \cdot H_2SO_4$ has $[\alpha]_D +84.8°$. The zincichloride, $B \cdot ZnCl_2 \cdot 2 HCl \cdot H_2O$, crystallizes well and can be used for the purification of nicotine. The acid D-tartrate, $B \cdot 2 H_2C_4H_4O_6 \cdot 2 H_2O$ has m. p. 88—89° (hydrated) and $[\alpha]_D^{27°} +26.6°$ (dry salt) and the neutral D-tartrate $B \cdot C_4H_6O_6 \cdot 2 H_2O$ has m. p. 68.5° (hydrated) and $[\alpha]_D^{27°} +29.5°$ (dry salt). The dipicrate $B \cdot 2 C_6H_2(NO_2)_3OH$ crystallizes in short yellow prisms, m. p. 224°, and the tetrachloriodide $B \cdot 2(HICl_4)$ in orange prisms, m. p. 150° (dec.). The dipicrolonate melts at 228° (corr.).

L-Nornicotine is a colourless iol, b.p. 120°/0.5 mm. with a faintly basic odour, $[\alpha]_D^{20°} -88.8°$. If forms a dipicrate, m. p. 190—191°, a dipicrolonate m. p. 253°, and a diperchlorate, m. p. 183—186°.

D-Nornicotine has $[\alpha]_D^{20°} +88.8°$ and forms a dipicrate and dipicrolonate which have the same m. p. as the L-form.

L-Anabasine occurs as an oil, b. p. 104—105°/2 mm. It freezes at 9° and has $D_{20°}^{20°}$ 1.0455 and $[\alpha]_D^{15°} -81.7°$. It is a strongly basic substance, miscible with water in all proportions and is only slightly volatile in steam. The hydrochloride is deliquescent and dextrorotatory $[\alpha]_D^{20°} +9.23°$. It forms a dipicrate, m. p. 205—207°, a dipicrolonate, m. p. 235—237° and a fluorosilicate $B \cdot H_2SiF_6 \cdot H_2O$, m. p. 239° (dec.).

Nicotyrine is a colourless oil, b.p. 280—281° or 150°/15 mm. and has $D^{13°}$ 1.124. It yields a dipicrate, m. p. 170—171°, a chloroplatinate, m. p. 160° (dec.) and a methiodide, m. p. 211—213°

L-Anatabine occurs in the lower boiling fractions of crude nicotine and is a colourless oil boiling at 145—146°/10 mm. $[\alpha]_D^{17°} -177.8°$ (no solvent). It forms a dipicrate, m. p. 191—193° a dipicrolonate m. p. 234—235° and a trinitro-m-cresolate, m. p. 191—192°. N-methyl-anabasine is a colourless oil, b. p. 127—128°/12 mm., $[\alpha]_D^{15°} -85.1°$. It forms a dipicrate, m. p. 237—238° (dec.) and a dipicrolonate, m. p. 234—236° (dec.). The trinitro-m-cresolate melts at 231—232°.

N-methylanatabine is an oil, b. p. 120°/1 mm., $[\alpha]_D^{18°} -171.4°$ (methanol), and forms a dipicrate, m. p. 207—208° and a trinitro-m-cresolate $B \cdot 2 C_7H_5O_7N_3$, m. p. 228—229°.

Nicotelline in aqueous solution is neutral to litmus and is not volatile with steam. It is a solid, melting at 147—148°.

Nicotöine is a colourless oil, b. p. 208°.

II. The Isolation of the Major Alkaloids.

For the isolation of nicotine from *N. tabacum*, or other species in which nicotine is the principal alkaloid, the finely powdered material is moistened with 20% NaOH solution and exhausted with ether or light petroleum in a continuous extractor. The solution of the alkaloid in the organic solvent is shaken out with successive portions of dilute acid until all the alkaloid has passed into the acid solution. The solution is made alkaline with NaOH and once again shaken out with ether. The ether is evaporated and the oily residue of nicotine purified by fractional distillation or through the picrate. A similar procedure is adopted for the extraction of nornicotine from *N. glutinosa* and anabasine from *N. glauca*. If species of *Nicotiana* are chosen which represent the mixed alkaloid types of group C above, it is necessary to use methods which will give adequate separation of the alkaloids present and such methods have been described by SCHMUK (1941), SMITH and SMITH (1942), and DAWSON (1945). Anabasine is isolated, by the above process, from *Anabasis aphylla* as a thick oil, in a yield of about 2.33% of the dried plant. It can be fractionated by distillation *in vacuo* into a fraction (85%), b. p. 136—138.5°/12 mm. and a fraction (15%) b. p. 200°/12 mm. The low boiling fraction is separated by benzoylation and fractional distillation into lupinine and benzoylanabasine, b. p. 222°/2 mm., m. p. 83—84°. Anabasine can also be separated from lupinine by the addition of sodium to a solution of the basic mixture in toluene. The sodium salt of lupinine is filtered off and the anabasine recovered from the filtrate (SADYKOV and SPASOKUKOTSKII, 1943). MARKWOOD and BARTHEL (1943) use the following method for the release of the alkaloids on a small scale. 2 g. of powdered material is placed in a 100 ml. beaker in a moderately cool water bath and mixed with 6 ml. of H_2SO_4 (90%) until the powder is completely disintegrated (2—3 minutes). When the reaction has ceased, the beaker is removed from the water bath and 50 ml. of water added without cooling. The beaker is covered with a watch glass and placed on the steam bath for 30 minutes. The mixture is cooled to room temperature and filtered. The filter and contents are washed with water until silicotungstic acid gives a negative test. The total volume should now be over 100 ml. The solution is made strongly alkaline with solid NaOH and then exhaustively extracted with benzene(5×30ml.). The benzene solution is dried if necessary with solid NaOH and is extracted with a slight excess (10 ml.) of $0.1\ N$ HCl. The benzene is washed twice with 5—10 ml. of water and the washings are added to the acid solution.

The acid solution is neutralized (phenolphthalein) and saturated aqueous picric acid solution (10.25 ml.) is added to precipitate the alkaloids. The solution is boiled to dissolve the precipitate and cooled slowly. The crystals are filtered off, washed first with dilute picric acid and then with water, dried and the m. p. taken.

III. The Detection of the Alkaloids of *Nicotiana* spp.

As adequate analytical data for the minor alkaloids are not available, consideration will be given only to the major alkaloids, nicotine, nornicotine, and anabasine. Sanchez (1922) described a colour test for nicotine in which a rose-red colour is produced when nicotine is added to a solution of vanillin in conc. HCl. Feinstein and McCabe (1951) have studied this test and have substituted 85% syrupy phosphoric acid for HCl. They found that pure nicotine and nornicotine failed to give a red colour with the test. Anabasine gave a faint brown-violet colour. Pyrrole and nicotyrine gave an immediate red colour but carbazole gave the colour only after some time. However, when nicotine, nornicotine and anabasine were dehydrogenated with platinum black, nicotine and nornicotine gave the red colour but anabasine failed to respond to the test.

The modified test is carried out as follows. A small quantity of the alkaloid is mixed with a little platinum black and heated in a vessel to 170° for 1.5 hours. Care is taken to prevent loss of the alkaloid by evaporation. After the sample is cooled, 5 ml. of distilled water are added and the whole is filtered. To 3 ml. of the filtrate are added 12 ml. of vanillin-phosphoric acid solution (0.5 g. of vanillin in 100 ml. of 85% orthophosphoric acid) and the colour change is noted. The platinum black is prepared according to the method of Linstead and Thomas (1940). Norit charcoal is heated on a steam bath for 24 hours with 10% nitric acid, washed free from acid, and dried at 100°. It is then heated at 340° in a flask which is slowly evacuated to 4 mm., left for 1 hour under these conditions, and allowed to cool. A solution of platinum chloride (free from nitrate) from 5 g. of platinum in 50 ml. of water and 5 ml. of conc. HCl is cooled in a freezing mixture and treated with 50 ml. of 40% formaldehyde and 11 g. of charcoal. The mixture is stirred mechanically during the slow addition of a solution of 50 g. of potassium hydroxide in 50 ml. of water. The temperature is kept below 5° during the addition and finally raised to 60° for 15 minutes. The catalyst is washed thoroughly by decantation with water and finally with dilute acetic acid, collected on a filter and washed with hot water until the washings give no reaction for chloride, alkali or reducing material. The catalyst is dried at 100° and stored in a desiccator.

Feinstein and McCabe (1951) describe tests for nornicotine and anabasine based on the reaction with quinhydrone. A solution of quinhydrone (0.5 g. in 100 ml. of ethanol) when added to a solution of nornicotine or anabasine in phosphate buffer at pH 7 gives a cherry-red colour. Nicotine does not react. When a solution of nornicotine (363 μg. lower limit.) in 5 ml. of acetone is added to a mixture of 15 ml. of di-isopropylketone and 2 ml. of 0.3% 1:3-diketohydrindene in di-isopropylketone, a violet colour is produced. Under the same conditions, nicotine and anabasine do not give a violet colour. The test described for coniine by MELZER (see Hemlock Alkaloids) is negative for nicotine but anabasine and nornicotine give brown-black colours.

MELZER (1898) showed that nicotine (but not coniine) on heating with epichlorhydrin gives a deep red colour. A few drops of a dilute solution of nicotine in ethanol are added to 2 ml. of epichlorhydrin and the solution heated. A red colour develops which varies in intensity with the amount of nicotine present. The colour is orange with very small amounts of nicotine and the lower limit of the test is stated to be 0.25 mg. In common with many other pyridine derivatives, nicotine gives with cyanogen bromide and β-naphthylamine (or aniline) an orange red colour. The alkaloid reagents give precipitates with the *Nicotiana* alkaloids in very dilute solution. Ammonium reineckate precipitates the alkaloids from dilute acid solutions as reineckates which are very little soluble. Mercuric chloride gives with nicotine a white crystalline precipitate soluble in dilute acetic acid or dilute HCl. When a solution of nicotine in dry ether is added to a solution of iodine in dry ether, red crystals of an iodo derivative are gradually formed. Anabasine does not give this test.

IV. The Quantitative Estimation of the Alkaloids.

For purposes of quantitative determination plant material should be dried at a temperature not exceeding 60° and preferably in an oven with forced draught. Three methods have been used for the extraction or release of the alkaloids from plant material. (a) The powdered material is treated with a solution of NaOH and the alkaline solution subjected to steam distillation into dilute acid. (b) The material is treated with alkali and the free bases extracted by shaking with light petroleum or a mixture of light petroleum and ether. (c) Treatment of the powder with 90% H_2SO_4 according to the method of MARKWOOD and BARTHEL (1943). The total alkaloids can be determined with a fair degree of accuracy by precipitation as silicotungstates, by ultraviolet absorption, and by titration. An entirely satisfactory method for the determination of nicotine, nornicotine and anabasine when present together has not yet been devised but HOUSTON (1952) has described a method for the separation of nicotine and nornicotine by partition on a column of starch, followed by titration of the separated alkaloids.

1. The Estimation of Total Alkaloids by Titration.

5 g. of the finely powdered material is placed in a small basin and intimately mixed with 1 g. of powdered barium hydroxide, $Ba(OH)_2 \cdot 8\ H_2O$. Sufficient baryta water is added with stirring to make a stiff paste, the mixture dried by the addition of starch and transferred quantitatively to the thimble of a SOXHLET apparatus. The alkaloids are extracted with light petroleum for 3 hours. The extract is transferred to a separating funnel, the flask rinsed with light petroleum and the rinsings added to the extract which is then shaken out with 50 ml. of 0.1 N HCl. An aliquot of the acid solution is titrated with 0.1 N NaOH using methyl red as indicator. 1 ml. of 0.1 N acid is equivalent to 0.01629 g. of alkaloids calculated as nicotine.

2. The Estimation of Alkaloids by Steam Distillation Followed by Precipitation with Silicotungstic Acid.

The method is based on that of AVENS and PEARCE (1939). The distillation apparatus consists of a 500 ml. Pyrex flask for the generation of steam fitted with a three-hole rubber stopper through which passes the steam outlet tube, a 90 cm. length of 6 mm. tube to serve as a pressure gauge and a tube fitted with a glass stopcock for relieving pressure if necessary. The distilling flask is a 50 ml. round bottomed Pyrex flask fitted with a two-hole stopper carrying a safety trap and the inlet tube for the steam which should reach to the bottom of the flask. The safety trap is connected to a small vertical condenser the delivery tube of which dips beneath the surface of the acid solution receiving the distillate. Rubber connections are used throughout. Changes in steam pressure are obtained by adjusting the opening of the stop-cock or by controlling the flame of the burner. A small micro-burner is used to keep the liquid in the distillation flask as low as desired. A sample containing from 5—10 mg. of nicotine is desirable, but as little as 2 mg. can be determined by this method. The powder is weighed directly into the distillation flask, covered with 2—3 ml. of water and 2 drops of phenolphthalein indicator solution added. 40% NaOH solution is added dropwise to give a slight excess as determined by the indicator. The distillation flask is connected to the apparatus and steam passed in. A steam pressure of 45—60 cm. on the gauge is maintained throughout the period of distillation. The beaker used to collect the distillate contains 3 ml. of HCl (1 to 4) and about 5 ml. of water. When distillation is proceeding at a smooth rate the micro-burner is used to reduce the volume of liquid in the distillation flask. Distillation is continued for 30 minutes at the end of which time the liquid in the distillation flask should be reduced almost to dryness and the volume of distillate should not exceed 100 ml. When distillation is complete, the condenser and delivery tube are washed out and the volume of distillate adjusted to about 100 ml. 1 ml. of silicotungstic acid solution ($12\%\ 4\,H_2O \cdot SiO_2 \cdot 12\,WO_3 \cdot 22\,H_2O$) is used for every 10 mg. or less of nicotine. After precipitation the solution is covered with a watch glass and heated on the steam bath for 15 minutes, cooled slowly to room temperature and placed in a refrigerator at 7° overnight. The precipitate is filtered off in a tared Selas crucible of medium porosity previously chilled. Dilute HCl (1 in 2000) is used for washing out the precipitate and subsequent washing on the filter to free the precipitate from silicotungstic acid. The precipitate is ignited at 650° in a muffle furnace for 1 hour, cooled to room temperature in a desiccator and weighed. The weight of the residue multiplied by 0.1141 gives the weight of nicotine present. Alternatively the precipitate may be dried at 100° for 3 hours, cooled and weighed. The weight multiplied by the factor 0.1012 gives the weight of nicotine. After drying at 100° the anhydrous salt, $2\,B \cdot 2\,H_2O \cdot SiO_2 \cdot 12\,WO_3$, is formed.

An improved steam distillation apparatus for use in the determination of nicotine has been described by Griffith and Jeffrey (1948). By use of this apparatus the nicotine in a 0.1—2 g. sample of tobacco can be distilled in 5 to 7 minutes from a volume of 5—15 ml., the final volume of distillate ranging from 75 to 250 ml. These authors determined nicotine in dried samples and in green tobacco leaves. For green tobacco leaves 25—100 g. are extracted with acetone in a Waring Blendor for 5—10 minutes. Any leaf particles on the side of the Blendor cup are washed down once during the extraction with pure acetone in a wash bottle. The volume of acetone originally added to the cup will vary with the size of the sample and should be chosen so that the acetone concentration is at least 80% by volume with the quantity of water in the tissue considered. The resulting solution is removed from the residue by filtration through a Büchner funnel. The residue in the funnel is washed with pure acetone until the washings are colourless. The volume of acetone extract is measured in a graduated cylinder or made to a convenient volume in a volumetric flask. A 5—25 ml. portion of the extract (depending on nicotine content of the tissue) is pipetted into a 25 ml.

separating funnel and one or two drops of conc. H_2SO_4 are added. The extract is run into the distillation chamber of the steam distillation apparatus and the acetone distilled off and discarded. It is essential to remove all the acetone as it interferes with the precipitation of nicotine by silicotungstic acid. After the acetone is completely removed, a receiver containing 3 ml. of dilute HCl $(1 + 4)$ is placed beneath the condenser, magnesium oxide added to the chamber and distillation of the nicotine completed. The nicotine in the distillate is then determined with silicotungstic acid. The advantages of the use of green samples are that drying, with possible loss of nicotine, is eliminated and much time is saved in the preparation of samples. The method gives good quantitative results if the nicotine is distilled immediately after preparation of the acetone extract. Loss of nicotine occurs if the acetone extract is allowed to stand.

In the adoption of the above methods, the assumption is made that the distillate contains nicotine only and in the analysis of *N. tabacum* the assumption is largely justified as the error due to the presence of nornicotine is not large. However, with the introduction of crosses between *N. tabacum* and wild species of *Nicotiana* for increased disease resistance, the resultant hybrids may contain considerable amounts of nornicotine which interferes in the determination of nicotine. Methods for the quantitative determination of nicotine and nornicotine when present together have been described by MARKWOOD (1943) and BOWEN and BARTHEL (1943).

3. Estimation of Nornicotine in Presence of Nicotine.

The method of MARKWOOD (1943) is based upon the fact that nornicotine, being a secondary amine, is converted to the non-volatile nitroso derivative when treated with nitrous acid whereas nicotine, a tertiary amine, does not react and can be steam distilled in the usual manner. As nornicotine is not so readily distilled as nicotine, the second part of the analysis involves the methylation of nornicotine, in a separate sample, to nicotine and determination of the total steam volatile alkaloids present. The content of nornicotine is then found by difference. The aqueous alkaloid solution containing nicotine and nornicotine is diluted to about 15 ml. in a 300 ml. KJELDAHL flask and treated with 2 ml. of 30% acetic acid and 10 ml. of freshly prepared 5% $NaNO_2$ solution. The solution is allowed to stand at room temperature for 15—20 minutes and then made slightly alkaline to phenolphthalein with NaOH (30% at first followed by 5%). The neck of the flask is washed down and 2% acetic acid is added dropwise until the pink colour of the phenolphthalein is just discharged. To remove vapours of acetic and nitrous acids a current of air is passed through the solution. 10 ml. of buffer solution, pH 10, is added and the solution steam distilled at low volume (15 ml.) until all the nicotine has passed over. Alkalinity is controlled at pH 10 since the nitrosamine in small quantity may distil over at higher pH values. The nicotine in the distillate is then determined by silicotungstic acid precipitation or by titration. To a second sample containing 5 ml. of aqueous alkaloid solution, 0.1 ml. of formic acid (87%) and 5 ml. of formaldehyde (37%) are added and the solution gently refluxed in a 300 ml. KJELDAHL flask for 15 minutes. Pumice chips are added to promote smooth boiling. The solution is cooled and 10 ml. of NaOH solution (300 g./litre) are added. After standing for 15 minutes, the solution is steam distilled and the distillate treated as above. The difference between the two silicotungstic precipitations or titration values represents the nornicotine equivalent. 1 ml. of 0.05 N HCl $= 0.0074$ g. of nornicotine or 0.0081 g. of nicotine.

The procedure of BOWEN and BARTHEL (1943) which does not involve methylation is carried out as follows. 2.5 g. of powdered tobacco sample together with 10 ml. of 30% w/v NaOH solution, 10 g. of sodium chloride and a small piece of paraffin are placed in the distillation flask of the steam distillation apparatus and distilled into 3 ml. of dilute HCl (1 + 4). Distillation is continued until a few drops of the fresh distillate fail to give an opalescence with silicotungstic acid solution. The distillate is concentrated by boiling (porcelain chips added) until the volume is less than 25 ml. and then transferred to a 25 ml. standard flask. Washings are added to the flask and the solution made to volume. A 10 ml. aliquot is taken for precipitation of the nicotine and nornicotine with silicotungstic acid. Another 10 ml. aliquot is neutralized to phenolphthalein, 2 ml. of 30% acetic acid added followed by 0.5 g. of solid $NaNO_2$. After standing at room temperature for 20 minutes, the solution is made slightly alkaline with NaOH and steam distilled into 3 ml. of dilute HCl (1 + 4). The distillate contains nicotine only, which is precipitated with silicotungstic acid, ignited and weighed. The weight of the residue from the second aliquot multiplied by 0.1140 gives the weight of nicotine present and the weight of the residue from the first aliquot minus the weight of the residue from the second aliquot, multiplied by 0.1042 gives the weight of nornicotine. BOWEN (1947) recommends the use of excess magnesium oxide in place of NaOH for making the solution alkaline prior to steam distillation in view of the risk of nitroso-nornicotine distilling over if any excess of NaOH is used. Excess of magnesium oxide always results in solutions of the same alkalinity.

JEFFREY (1951) has made a comparative study of the methods for estimating the tobacco alkaloids and has drawn attention to discrepancies between the methods when they are applied to tobaccos of the mixed alkaloid types. In these types, the use of strong NaOH solution and NaCl for making the solution alkaline before steam distillation results in an appreciable amount of nornicotine appearing in the distillate with consequent estimation as nicotine. Moreover the use of the milder alkali magnesium oxide does not entirely prevent the distillation of nornicotine, therefore in species of *Nicotiana* containing nicotine and much nornicotine, the methods of BOWEN and BARTHEL (1943) and MARKWOOD (1943) should be employed.

4. Column Partition.

The recent method of HOUSTON (1952) in which nicotine and nornicotine are separated by partition on a column of starch eliminates steam distillation and chemical means of separating the bases. This method gives promise of fairly accurate results, and has the advantage of speed and simplicity compared with other methods. The column consists of a borosilicate glass tube 75 cm. in length and 20 mm. in diameter constricted near one end to support a perforated aluminium disc which is covered with a small plug of cotton wool. The column is prepared by taking 5 g. of starch and suspending it in 35—40 ml. of n-butanol saturated with water at room temperature. The starch need not be dried for the preparation of the column. The suspension is poured into the glass column and excess of solvent forced out with gentle air pressure until the solvent level falls to the level of the top of the starch column. A tightly fitting cotton wool plug is then forced down the tube until it rests firmly on the surface of the starch. About 25 ml. of hexane or similar light hydrocarbon is forced through the column to wash out the excess n-butanol, 1 g. of the dried and finely powdered tobacco sample is mixed with 0.25 g. of powdered barium hydroxide octahydrate, the mixture moistened with 1 ml. of a saturated solution of barium hydroxide and thoroughly stirred. If the sample contains more than 0.1% ammonia it should be left in a desiccator over conc. H_2SO_4 for 15—24 hours, to remove the greater part of the ammonia. About 1 g. of dry starch is added and thoroughly mixed with the sample. It is now sufficiently dry to enable it to be transferred through a funnel to the top of the column and packed in place with a tightly fitting cotton wool plug. The sample and column are washed with 200 ml. of hexane under gentle air pressure. The effluent is collected in a 500 ml. Erlenmeyer flask. Nornicotine, which is retained on the column under these conditions, is eluted by washing the column with 200 ml. of chloroform. The fractions are analysed by titration. Exactly 100 ml. of distilled water are added to the effluent followed by two drops of 0.1% bromo-cresol green indicator and 10 ml. of 0.02 N HCl. The flask is stoppered and the mixture shaken vigorously. The phases are allowed to separate, a 50 ml. aliquot of the water phase (110 ml.) is removed by pipette and excess acid titrated with 0.02 N NaOH using 1 drop of 0.1% methyl red as indicator. Anabasine if present would interfere and therefore the method should not be used on material of those species of *Nicotiana* which contain much anabasine.

5. UV Spectrophotometry.

A rapid and accurate method for the determination of nicotine by ultra-violet spectrophotometry has been described by WILLITS, SWAIN, CONNELLY and BRICE (1950). The ultra-violet absorption spectrum of nicotine shows a sharp absorption maximum at 260 mμ. The exact position and intensity of the maximum is influenced by the nature of the solvent and is affected by the presence of acid. The sensitivity of detection however is greater in acid solution and therefore the method is well suited to the determination of nicotine in a steam distillate. Nornicotine has an absorption peak almost identical with that of nicotine and if present will be estimated as nicotine. The positive error due to nornicotine is usually unimportant if species of *Nicotiana* in group A are being examined.

6. Colorimetric Methods.

Colorimetric methods based on the intensity of colour produced when nicotine reacts with cyanogen bromide and β-naphthylamine have been proposed by MARKWOOD (1939) and TRIM (1948). Aniline has been used in place of β-naphthylamine by WERLE and BECKER (1942). These methods, while useful in certain circumstances, are not specific for nicotine and are subject to interference by a number of other pyridine derivatives.

7. The Estimation of Anabasine.

Anabasine in presence of nicotine and nornicotine is difficult to estimate accurately. SMITH and SMITH (1942) have described a method based on the observation that nicotine forms an azeotropic mixture with water at the boiling temperature in which the concentration of nicotine is 2.5% w/v. Nornicotine and anabasine do not form such a mixture. Total alkaloids are first determined followed by distillation and estimation of the nicotine. The non-volatile alkaloids are precipitated with picric acid, the picrates methylated and the bases recovered. Nornicotine is converted to nicotine and removed as the azeotrope. BURKAT (1937) found that anabasine could be distilled from a strong solution of NaOH saturated with NaCl. The anabasine is distilled into standard phosphomolybdic acid, the precipitate removed and excess acid titrated. Other methods for the estimation of anabasine have been reviewed by FUKS (1941).

8. Paper Chromatography of the Tobacco Alkaloids.

WERLE and KOCH (1951) describe a method for the separation of the tobacco alkaloids using paper chromatography.

Y. The Tropane Alkaloids.

The alkaloids included in this group occur in the families, *Solanaceae*, *Erythroxylaceae*, *Convolvulaceae* and *Dioscoreaceae*. They are esters, which on hydrolysis yield an organic acid and an amino-alcohol. The more important of the tropane alkaloids and the families of plants in which they occur are given in Table 16.

Table 16. *The Tropane Alkaloids.*

Family	Alkaloid	Formula	Products of hydrolysis
Solanaceae	Atropine	$C_{17}H_{23}O_3N$	tropine-DL-tropic acid
	Apoatropine	$C_{17}H_{21}O_2N$	tropine-atropic acid
	Hyoscyamine	$C_{17}H_{23}O_3N$	tropine-L-tropic acid
	Norhyoscyamine	$C_{16}H_{21}O_3N$	nortropine-L-tropic acid
	Hyoscine (Scopolamine)	$C_{17}H_{21}O_4N$	scopine-L-tropic acid
	Meteloidine	$C_{13}H_{21}O_4N$	teloidine-tiglic acid
	Tigloidine	$C_{13}H_{21}O_2N$	pseudotropine-tiglic acid
	Valeroidine	$C_{13}H_{23}O_3N$	dihydroxytropane-isovaleric acid
	Poroidine	$C_{12}H_{21}O_2N$	nortropine-isovaleric acid
	iso-Poroidine	$C_{12}H_{21}O_2N$	nortropine-D-α-methylbutyric acid
Erythroxylaceae	Cocaine	$C_{17}H_{21}O_4N$	L-ecgonine-benzoic acid-methanol
	Cinnamylcocaine	$C_{19}H_{23}O_4N$	L-ecgonine-cinnamic acid-methanol
	α-Truxilline	$C_{38}H_{46}O_8N_2$	L-ecgonine-α-truxillic acid-methanol
	β-Truxilline	$C_{38}H_{46}O_8N_2$	L-ecgonine-β-truxillic acid-methanol
	Tropacocaine	$C_{15}H_{19}O_2N$	pseudotropine-benzoic acid
	Hygrine	$C_8H_{15}ON$ ⎫	derivatives of
	Cuscohygrine	$C_{13}H_{24}ON_2$ ⎭	pyrrolidine
Dioscoraceae	Dioscorine	$C_{13}H_{19}O_2N$	lactone
Convolvulaceae	Convolvine	$C_{16}H_{21}O_4N$	nortropine-veratric acid
	Convolvamine	$C_{17}H_{23}O_4N$	tropine-veratric acid

The distribution and content of total alkaloids in the various genera and species of the above families will now be given. Percentages are based on dry weight.

Atropa belladonna L. Chiefly L-hyoscyamine with some hyoscine: leaves 0.4%; roots 0.5%; seeds 0.8%; entire plant 0.2—1%. Apoatropine is also said to occur. Atropine has been found but probably is the result of racemization during extraction.

Atropa lutescens JACQ. Leaves and roots 0.45—0.47%; whole plant 0.32—0.38%.

Atropa boetica WILLK. Leaves 0.82—1.06%; roots 0.94%; fruit 1.09, chiefly hyoscyamine. *Datura stramonium* L. Leaves 0.2—0.45%; seeds 0.2—0.5% (chiefly hyoscyamine); roots 0.21—0.25% (hyoscyamine and hyoscine); leaves and tops 0.6—0.7%. *Datura metel* L. Leaves 0.2—0.5%; roots 0.1—0.2%; seeds 0.2—0.5%; fruits 0.12%. Hyoscine with small quantities of hyoscyamine, and a trace of norhyoscyamine. *Datura meteloides* D. C. Entire plant 0.4% of which 0.13% is hyoscine. Meteloidine (0.07%) has also been isolated.

Datura fastuosa L. var. niger, fruits 0.2%; leaves and stems 0.12%; roots 0.1%, var flora coerulea plena, seeds 0.25%. Var. flora alba plena, seeds 0.22%. Hyoscine is the chief alkaloid with some hyoscyamine. *Datura quercifolia* H. B. and K. Leaves 0.42%; seeds, 0.29%; hyoscine and hyoscyamine. EVANS and PARTRIDGE (1948) give the total alkaloid content of *Datura ferox* L. as 0.61%, of which 0.40% is hyoscine and the remainder chiefly meteloidine. *Duboisia leichhardtii* MUELL. contains 0.8—3.7% of total alkaloids. Leaves have yielded L-hyoscyamine 1.97%; L-hyoscine 0.06%; DL-hyoscine 0.05%; norhyoscyamine 0.01%. According to HILLS, TRAUTNER and RODWELL (1945) hyoscine is the dominant alkaloid in some samples. *Duboisia myoporoides* R. BR. This Australian species has been extensively investigated by HILLS, TRAUTNER and RODWELL (1945) and HILLS and RODWELL (1951). The northern "hyoscine type" contains 0.9—4% total alkaloids and the southern "hyoscyamine type" 1.0—2.7% total alkaloids in the leaf. In the northern types, hyoscine may account for a high proportion of the total alkaloids present so that in these types hyoscine is the major alkaloid. In the southern types hyoscyamine is the major alkaloid. In addition, valeroidine, tigloidine, poroidine, isoporoidine and DL-hyoscine are present.

Hyoscyamus niger L. contains hyoscyamine as the major alkaloid with some hyoscine. Leaves 0.045—0.08%; roots 0.16%; seeds 0.06—0.1%. *Hyoscyamus muticus* L. Leaves 1.4%;

Leaves and stems 0.6%; seeds 0.9—1.3%; stems 0.6%. Hyoscyamine is the major alkaloid.
Hyoscyamus albus L. Leaves 0.2—0.56%; roots 0.1—0.14%; seeds 0.16%. Hyoscyamine
and hyoscine present, the former predominant.

 Hyoscyamus reticulatus L. Whole plant 0.12—0.24% total alkaloids.

 Scopolia carniolica Jacq. Rhizomes 0.43—0.51%; hyoscyamine and hyoscine.

 Scopolia lurida Dun. Roots 2—2.8%. Hyoscine, hyoscyamine and cuscohygrine.

 Solandra laevis Hook. 0.16%. Norhyoscyamine, hyoscyamine.

 Dioscorea hirsuta Blume and *D. hispida* Dennst. Dioscorine 0.21% in tubers.

 Convolvulus pseudocantabricus Schrenk. Total alkaloids in seed 0.5%. Convolvine
with small quantities of convolvamine, convolvidine and convolvicine. *Erythroxylon* spp.
Two cultivated species, *E. coca* Lamarck and *E. truxillense* Rusby form the principal sources
of coca leaves. Huanuco or Bolivian leaves are derived from *E. coca* and Peruvian (Truxillo)
and Java leaves from *E. truxillense.* Bolivian leaves usually contain 0.5—1.5% total alkaloids of
which 70—80% may be cocaine. Java leaves contain 1.0—2.5% total alkaloids of which about
50% is cocaine and the remainder cinnamylcocaine (mostly in young leaves) with small amounts
of tropacocaine and truxillines. The Peruvian leaves also yield small quantities of hygrine.

 Histological Distribution of the Alkaloids. In belladonna and stramonium the alkaloids
are present in the young undifferentiated cells of the stem apex, but when tissue differentiation
is complete, the alkaloids are found in greatest amount in the epidermal cells, the outer
cortical cells, the phloem parenchyma, the xylem parenchyma, medullary rays and outer
regions of the pith. Alkaloids accumulate in the epidermal cells of the petiole, leaf and berries
and are present in abundance in the epidermal cells of the floral parts. The carpels and
ovules contain alkaloids in all parts. In young roots, the root cap is rich in alkaloids and in
older roots the concentration is highest in the outer cortex and the phloem parenchyme.
The coca alkaloids occur in greatest amount in the upper epidermal cells of the leaves.

I. Properties.

 Atropine (DL-hyoscyamine), $C_{17}H_{23}O_3N$. It is doubtful whether atropine occurs as such
in the plant but is formed from hyoscyamine during the process of extraction. It occurs
in long prisms, m. p. 118°, and crystallizes from ethanol on addition of water or from chloroform on addition of light petroleum. When heated rapidly it sublimes unchanged. It is
easily soluble in ethanol (1 in 3), in chloroform (1 in 2) and less soluble in ether (1 in 60)
and benzene (1 in 60). It is soluble in boiling water (1 in 35), sparingly soluble in cold water
(1 in 450 at 25°) and very little soluble in light petroleum. The aqueous solution is distinctly
alkaline to litmus and phenolphthalein. Atropine is readily hydrolysed by treatment with
dilute acid or alkali yielding *tropic acid* and *tropine*. Even in very dilute solution atropine
causes dilation of the pupil of the eye. The salts crystallize well. The sulphate $B_2 \cdot H_2SO_4 \cdot H_2O$
is a colourless crystalline powder easily soluble in water (1 in 0.38) and ethanol (1 in 3.7),
sparingly soluble in chloroform (1 in 620 or ether (1 in 2140). It has m. p. 195—196°. The
hydrochloride, $B \cdot HCl$, forms needles, m. p. 162° and the hydrobromide, $B \cdot HBr$, has
m. p. 163—164°. The oxalate $B_2 \cdot H_2C_2O_4$, crystallizes readily in prisms and the aurichloride,
$B \cdot HAuCl_4$, separates as an oil which crystallizes on standing and can be recrystallized from
very dilute HCl. It has m. p. 137—139° and is useful for identifying the alkaloid. The picrate
forms rectangular plates m. p. 175—176°.

 apo-Atropine, $C_{17}H_{21}O_2N$, crystallizes from ether in prisms, m. p. 60°. It is slightly
soluble in water but readily soluble in organic solvents with the exception of light petroleum.
The aurichloride crystallizes from water in fine yellow needles, m. p. 110—112°, and the
picrate, m. p. 166—168°, also crystallizes from water in yellow needles.

 L-Hyoscyamine, $C_{17}H_{23}O_3N$, is the most widely distributed alkaloid of the group. It
crystallizes from dilute ethanol in needles which melt at 108.5°. It has $[\alpha]_D$ —22° in 50%
ethanol or, as the basic ion as salt in water —32.4°.

 L-Hyoscyamine resembles atropine in its solubility in organic solvents and water, and is
readily converted into atropine by treating an ethanolic solution with dilute ethanolic caustic
potash. When heated with dilute acid or alkali it yields *tropic* acid and *tropine*, the tropic
acid asuming the DL-form if alkali is used, but the L-form if hot water alone is used. The
salts crystallize well. The sulphate, $B_2 \cdot H_2SO_4 \cdot 2 H_2O$, crystallizes from ethanol in needles,
m. p. 206° (dry), $[\alpha]_D = -27.8°$ (H_2O). It is very soluble in water. The hydrobromide,
$B \cdot HBr \cdot 2 H_2O$, forms prisms and has m. p. 150—151°. The aurichloride, $B \cdot HAuCl_4$,
crystallizes from ethanol or dilute HCl in golden hexagonal plates, m. p. 165°, and the auribromide, $B \cdot HAuBr_4 \cdot H_2O$, as deep red lustrous needles, m. p. 160° (anhydrous). The
platinichloride, orange coloured prisms, melts at 206° and the picrate, which crystallizes in
plates has m. p. 165°. The oxalate forms prisms, m. p. 176°.

 Norhyoscyamine, $C_{16}H_{21}O_3N$, is usually separated from hyoscyamine by extraction with
ether and crystallization of the mixed oxalates from water. The oxalate of norhyoscyamine

separates first. It crystallizes in prisms, m. p. 140°, $[\alpha]_D$ —23.0° (50% ethanol) and is soluble in ethanol and chloroform, less soluble in ether and acetone and sparingly soluble in water (1 in 270° at 14°). It is strongly alkaline. The hydrochloride, $B \cdot HCl$, forms rosettes of needles from ethanol, m. p. 207°. The sulphate, $B_2 \cdot H_2SO_4 \cdot 3 H_2O$ crystallizes from acetone and water in slender needles, m. p. 249°, and the oxalate forms long prisms, m. p. 245—246°, soluble in water (1 in 20 at 15°). The aurichloride, $B \cdot HAuCl_4$, forms yellow scales, m. p. 178 to 179° and the platinichloride, $B_2 \cdot H_2PtCl_6 \cdot 3 H_2O$, forms reddish yellow prisms, m. p. between 132—141°. The picrate forms needles, m. p. 220°. Norhyoscyamine, being a secondary amine, forms a nitrosamine.

Hyoscine (Scopolamine), $C_{17}H_{21}O_4N$, occurs as a syrup and is readily soluble in water, ethanol, chloroform and ether but much less soluble in light petroleum and benzene. $[\alpha]_D^{20°} = $ —18° (ethanol) or —28°(H_2O). The hydrobromide, $B \cdot HBr \cdot 3 H_2O$, forms rhombic tablets, m. p. 193—194° (dry), and is readily soluble in water and ethanol but sparingly soluble in chloroform and insoluble in ether. $[\alpha]_D = $ —15.72° (ethanol) and —28°(H_2O). On hydrolysis, hyoscine yields L-*tropic* acid and *oscine* (scopoline) but *scopine* is the true amino-alcohol of hyoscine. The aurichloride, $B \cdot HAuCl_4$, m. p. 208—209° crystallizes in needles with serrated edges. The auribromide, $B \cdot HAuBr_4$, crystallizes in chocolate coloured leaflets from boiling 2.5% HCl, m. p. 191—192°. The picrate has m. p. 187—188° and crystallizes in pale yellow needles.

Meteloidine, $C_{13}H_{21}O_4N$, crystallizes from benzene in tabular needles, m. p. 141—142°. It is easily soluble in ethanol and chloroform but sparingly soluble in water, ether or benzene. The hydrobromide, $B \cdot HBr \cdot 2 H_2O$, crystallizes in chisel shaped needles and has m. p. 250° (dry); the aurichloride, $B \cdot HAuCl_4 \cdot ^1/_2H_2O$, forms yellow needles, m. p. 149—150°, and the picrate yellow hexagonal plates from ethanol, m. p. 177—180°. On hydrolysis, meteloidine yields *tiglic* acid and *teloidine*.

Tigloidine, $C_{13}H_{21}O_2N$, occurs as a colourless syrup. The hydrobromide forms tabular crystals which melt at 234—235° (corr.). The aurichloride, m. p. 213.5—214°, crystallizes from dilute acetone in golden-yellow plates and the picrate, m. p. 239°, in rectangular plates from ethanol. On hydrolysis by boiling with baryta water tigloidine yields *tiglic acid* and *pseudotropine*.

Valeroidine, $C_{13}H_{23}O_3N$, forms colourless laminar crystals, m. p. 85°, $[\alpha]_D^{20°}$ —9.0° (ethanol) or —4°(H_2O). The hydrobromide forms small needles and melts at 170—172°; the picrate, needles from water, m. p. 152—153°, and the hydrobromide of the acetyl derivative, m. p. 197° crystallizes in colourless needles from a mixture of ether and ethanol.

Poroidine, $C_{13}H_{21}O_2N$, forms a hydrobromide, m. p. 224—225°, which crystallizes in small plates from ether-ethanol, an oxalate, m. p. 301—302°, and a picrate, golden-yellow prisms, m. p. 172°.

Dioscorine, $C_{13}H_{19}O_2N$, forms greenish-yellow plates, m. p. 43.5°, and is soluble in water, ethanol and chloroform, but sparingly soluble in ether or benzene. The hydrochloride, $B \cdot HCl \cdot 2 H_2O$, forms needles or diamond shaped plates from ethanol and melts at 204°, $[\alpha]_D$ +4.66°. The picrate, m. p. 184°, forms yellow needles and the aurichloride yellow needles melting at 171° (dry).

Convolvine, $C_{16}H_{21}O_4N$, forms crystals from light petroleum melting at 115°. The hydrochloride, $B \cdot HCl$, has m. p. 260—261°; the oxalate crystallizes in glistening leaflets, m. p. 265—266° (dec.); the picrate in yellow needles, m. p. 261—263°, and the aurichloride in yellow leaflets, m. p. 217° (dec.).

Convolvamine, $C_{17}H_{23}O_4N$, melts at 114—115° and forms a picrate m. p. 263—264° (dec.), an aurichloride, m. p. 201—202°, and a hydrochloride, m. p. 237—239°.

Cocaine, $C_{17}H_{21}O_4N$, crystallizes from ethanol in prisms, m. p. 97—98°, b. p. 187 to 188°/0.1 mm., $[\alpha]_D = $ —15.8° (chloroform). It is slightly soluble in cold water and readily soluble in ethanol, ether, benzene and light petroleum. The aqueous solution is alkaline to litmus. In slightly alkaline solutions cocaine readily undergoes hydrolysis and even in hot water some decomposition takes place. Cocaine produces an intense local anaesthetic effect when a drop of a dilute solution is placed on the tongue or on a mucous surface. The salts crystallize well; the hydrochloride, $B \cdot HCl$, crystallizing from ethanol in prisms, m. p. 200 to 202° (dry), $[\alpha]_D = $ —71.95° (H_2O). It is easily soluble in water (1 in 0.4 at 25°) and ethanol (1 in 2.6 at 25°) but insoluble in ether or light petroleum. The nitrate $B \cdot HNO_3 \cdot 2 H_2O$. forms deliquescent crystals, m. p. 58—63° and the chromate, $B \cdot H_2CrO_4 \cdot H_2O$, is precipitated as orange-yellow leaflets, m. p. 127°, when a solution of potassium chromate is added to a solution of the hydrochloride acidified with HCl. Mercuric chloride in aqueous solution gives a white precipitate of the mercurichloride, $B \cdot HCl \cdot HgCl_2$, which can be crystallized from ethanol. The periodide, $B \cdot HI \cdot I_2$ has m. p. 161°.

Cinnamylcocaine, $C_{19}H_{23}O_4N$, is almost insoluble in water but soluble in organic solvents. It crystallizes from benzene or light petroleum in needles, m. p. 121°, $[\alpha]_D$ —4.7° (chloroform). The hydrochloride, $B \cdot HCl \cdot 2 H_2O$, crystallizes from water in shining flat needles, m. p. 176°

(dry). The platinichloride, m. p. 217°, crystallizes in fine needles on standing, and the auri-chloride forms yellow needles, m. p. 156°. On hydrolysis with HCl cinnamylcocaine yields methanol, L-*ecgonine* and *cinnamic acid*.

α-**Truxilline**, $C_{38}H_{46}O_8N_2$, occurs as an amorphous white powder m. p. 80°, which is soluble in organic solvents with the exception of light petroleum. It is practically insoluble in water. On hydrolysis, α-truxilline yields L-*ecgonine*, methanol and α-*truxillic acid*.

β-**Truxilline** resembles-α-truxilline in properties. It sinters at 45° and decomposes above 120°. On hydrolysis it yields L-*ecgonine*, methanol and β-*truxillic acid*.

Tropacocaine, $C_{15}H_{19}O_2N$, crystallizes in needles, m. p. 49°, insoluble in water but soluble in ethanol, ether, chloroform and benzene. The hydrochloride, B · HCl, forms needles and melts at 271° (dec.). The aurichloride separates in small yellow needles from hot aqueous solution, m. p. 208°, and the picrate melts at 238—239° and occurs as yellow needles. The oxime melts at 116—120°.

Hygrine, $C_8H_{15}ON$, is a liquid boiling at 92—94°/20 mm. or 193—195°/760 mm. d_4^{17} 0.940, $[\alpha]_D$ —1.3°. It is a strongly alkaline tertiary amine. On oxidation with chromic acid it yields hygric acid, m. p. 164°. The picrate forms yellow needles, m. p. 158°.

Cuscohygrine, $C_{13}H_{24}ON_2$, boils at 169—170°/23 mm. and is miscible with water forming a crystalline hydrate, B · $3^1/_2$ H_2O, m. p. 40°. It forms a hydrobromide, m. p. 234°, a nitrate, m. p. 209° (dec.) and an oxime, m. p. 53—54°.

II. Isolation of the Tropane Alkaloids.

For the large scale extraction of belladonna leaves, the finely powdered material is exhausted with 95% ethanol. The ethanol is removed by distillation and the syrupy residue treated with 1% HCl. Resinous matter is separated and the acid solution is further purified by shaking out with light petroleum. The solution is made alkaline with ammonia and shaken out with chloroform. The chloroform solution is run off, shaken with dilute acid, and the acid solution made alkaline with ammonia and again extracted with chloroform. The solvent is removed by distillation and the crude alkaloids neutralized with oxalic acid. The oxalates of atropine and hyoscyamine may be separated by fractional crystallization from acetone and ether in which hyoscyamine oxalate is the more soluble.

The aurichlorides and picrates of atropine and hyoscyamine are useful for characterization. Hyoscine is similarly isolated from *Datura metel, D. ferox* or *Duboisia myoporoides* by moistening the finely ground material with 10% $NaHCO_3$ solution and extracting with ether. After purification of the ether extract hyoscine is crystallized as the hydrobromide from a mixture of ethanol and acetone. Dioscorine is isolated from the dried tubers of *Dioscorea hirsuta* or *D. hispida* by extraction with 96% ethanol acidified with acetic acid. The syrupy residue after removal of the solvent is treated with aqueous sodium carbonate solution and the alkaloid extracted with chloroform. The chloroform is removed by distillation and the residue neutralized with hydrobromic acid. The hydrobromide is purified by crystallization from a mixture of ethanol and acetone. Cocaine may be obtained from coca leaves by mixing the powdered leaf with sufficient slaked lime to give a stiff paste. The mass is extracted with benzene at room temperature, the solvent distilled off and the residue shaken out with 3% H_2SO_4. Potassium permanganate is added in sufficient quantity to saturate the solution which is kept at a temperature below 5°. The cinnamylcocaine is oxidized by this treatment and the cocaine is extracted with ether after liberation with dilute ammonia.

III. The Detection of the Tropane Alkaloids.

Most of the alkaloidal reagents precipitate the tropane alkaloids from very dilute solution. Especially sensitive reagents are iodo-potassium iodide, potassium bismuth iodide, potassium mercuric iodide and phosphotungstic acid. Ammonium reineckate in slightly acid solution also precipitates atropine, cocaine, hyoscyamine and hyoscine from very dilute solution. The reineckates are useful for characterization and for the quantitative estimation of the alkaloids. Atropine reineckate melts at 157—158° (dec.); hyoscyamine reineckate has m. p. 156—157°, and hyoscine reineckate melts at 171—172° (dec.). Atropine, hyoscyamine and hyoscine are strongly mydriatic and when instilled into the eye of the cat cause dilation of the pupil in very dilute solution. The mydriatic action of cocaine is much weaker than that of the solanaceous alkaloids. Cocaine however possesses the distinctive property of producing a characteristic numbness when a drop of very dilute solution is placed upon the tongue. Atropine and hyoscyamine in aqueous solution react alkaline to phenolphthalein. Hyoscine and cocaine, being weaker bases, are alkaline to litmus but not to phenolphthalein.

Schaer's reagent (a freshly prepared solution of 1 vol. of 30% H_2O_2 mixed with 10 vols. conc. H_2SO_4) gives a green colour with the tropane alkaloids. A few drops of the reagent are added to a particle of the alkaloid in an evaporating dish. After 30 seconds a green colour develops.

The Vitali-Morin Reaction. This colour test is by far the most useful for the detection of the solanaceous alkaloids and has been the subject of a special study by James and Roberts (1945). These authors have shown that the reaction is a special case of the colour reaction of polynitrobenzene derivatives with alkalies. The colour is given by methyl and ethyl tropate and the truxillines but not by cocaine. To carry out the test, a little of the solution containing atropine, hyoscyamine or hyoscine is evaporated to dryness in a porcelain basin on a water bath, the residue treated with about 0.2 ml. of fuming nitric acid and again evaporated to dryness on the water bath. The residue is dissolved in 10 ml. of pure water-free acetone and a solution of 3% potassium hydroxide in pure methanol added dropwise. A characteristic deep purple colour is obtained. As little as 0.0001 mg. of any of the above alkaloids will give a positive reaction. "Veratrine" gives a reddish colour but according to Beckmann (1886) interference from this alkaloid is eliminated if nitrous acid or a nitrite instead of nitric acid and aqueous instead of methanolic potassium hydroxide are used. "Veratrine" gives a yellow colour under these conditions.

Rathenasinkam (1950) Reaction. A mixture of nitric and sulphuric acids brings about nitration of the benzene ring in cocaine whereas nitric acid alone is sufficient for the nitration of the ring in atropine, hyoscyamine and hyoscine. From this reaction the following test has been developed. To about 0.5 mg. of cocaine in a test tube about 100 mg. of potassium nitrate and 10 drops of conc. H_2SO_4 are added and the mixture heated in a boiling water bath for 10 minutes. The solution is cooled, diluted with water to about 30 ml. and extracted with chloroform. The chloroform is removed and discarded. The solution is made alkaline with ammonia and again extracted with chloroform. The chloroform is evaporated off and the residue dissolved in about 2 ml. of acetone and 1 or 2 drops of 10% NaOH solution added. A bluish-purple colour is formed, the intensity of which depends on the amount of cocaine present. The colour is distinct from that given by atropine under the same conditions. Dioscorine, the double bond of which is associated with a lactone ring, gives a reddish-violet colour with sodium nitroprusside in the presence of alkali, and with conc. H_2SO_4 and potassium iodate a yellow colour slowly changing to blue-violet.

IV. The Quantitative Estimation of Total Alkaloids of Belladonna.

1. Colorimetric Methods.

(a) Allport and Wilson (1939) have described a method for the determination of total alkaloids which is based on the Vitali-Morin reaction. 1 g. of the dry finely powdered sample is accurately weighed and transferred to a 50 ml. beaker.

1 ml. of ethanol (95%) and 0.1 ml. of ammonia solution (10% w/w NH₃) are added
and the mixture stirred and evenly wetted. 5 ml. of chloroform are added, the
mixture heated to boiling point and as much of the sample as possible is transferred
to a miniature percolator previously plugged with cotton wool wetted with
chloroform and suspended within a stoppered 100 ml. measuring cylinder. (The
percolator is made from glass tubing of 8 mm. bore and has an overall length of
17.5 cm. The tube is sharply narrowed about 3 cm. from one end and the other
end is spun out to form a flange.) If necessary, the sample is gently compressed
with a glass rod so that the solvent percolates at the rate of about 1 drop per
second. More chloroform is added to the beaker and the transference of the
sample to the percolator is completed. The extraction is continued until the
volume of the percolate is 31 ml. 6% acetic acid (made up with approx. 5%
ethanol) is added to the percolate to bring the level of the liquid up to the 80 ml.
graduation mark. The cylinder is stoppered and shaken gently for 15—20 seconds.
When the phases have separated out completely, 5 ml. of the upper layer is
pipetted off, filtered through a dry paper, and exactly 1 ml. of the filtrate transferred to an evaporating dish of 5 cm. diameter. The dish is placed on a boiling
water bath and the liquid evaporated just to dryness and 0.2 ml. of fuming nitric
acid (s. g. 1.5: Analar) immediately added from a pipette. The acid is allowed
to make contact with the whole of the alkaloidal residue after which it is evaporated
off on the water bath. 3 minutes are allowed for this operation and the residue
is dissolved in about 3 ml. of dry acetone. The solution is transferred with the
aid of a small glass rod to a standard stoppered 10 ml. measuring cylinder, the
dish washed with further small quantities of acetone and the washings transferred
to the measure until the latter contains exactly 10 ml. of solvent. After allowing
to cool, 0.1 ml. of a freshly prepared 3% solution of potassium hydroxide (pure)
in methanol (pure) is added, the stopper inserted, the cylinder inverted once and
the solution allowed to stand for exactly 5 minutes. The intensity of the colour is
measured immediately in an absorptiometer using a green filter. A calibration
curve is prepared by dissolving a known amount of pure hyoscyamine sulphate
in chloroform and applying the above procedure to aliquots. This method gives
reasonably accurate results provided careful attention is given to details of
manipulation and pure reagents are used. As the colour is unstable and sensitive
to light, all readings should be taken exactly 5 minutes after development of the
colour. Dilution of the acetone extract may be necessary if the alkaloid content
of the material is high. It should be remembered that if errors are made when
aliquots are being determined, these errors will be multiplied in the final calculation. The colour reaction is very sensitive and as little as 0.01 mg. of alkaloids
can be determined, thus rendering possible the estimation of the alkaloids in
individual seedling plants. ASHLEY (1952) has pointed out that this method may
give variable results if the water content of the solvent in which the colour is
developed is not controlled. This author uses pure pyridine (Analar) containing
less than 0.17% w/v of water in place of acetone for the extraction of the nitrated
base and reduces the concentration of the methanolic solution of potassium
hydroxide from 3% to 0.5%.

(b) In a combined precipitation and colorimetric method, COLBY and BEAL
(1952) use ammonium reineckate to precipitate the alkaloids as reineckates which
are then dissolved in acetone to form a coloured solution. 1 g. of the powdered
material is moistened with 1 ml. of 95% ethanol and 0.1 ml. of 10% ammonium
hydroxide. 5 ml. of chloroform are added and the mixture heated to boiling for
2 minutes. The material is transferred to a miniature percolator previously
plugged with a small wad of cotton wool, soaked in chloroform and suspended

in a 60 ml. separating funnel. The powder is percolated with warm chloroform at the rate of one drop per second until at least 30 ml. has run into the separating funnel. The chloroform solution is extracted with five successive 10 ml. portions of 0.5 N H_2SO_4. The combined acid solutions are made alkaline with 10% ammonia solution and completely extracted with chloroform. The chloroform solution is evaporated on the water bath until nearly dry and 5 ml. of 0.5 N H_2SO_4 added. The remainder of the chloroform is then evaporated off. 10 ml. of 1% ammonium reineckate solution are added to precipitate the alkaloid and the beaker allowed to stand at room temperature for 30 minutes. The solution is filtered through a sintered glass filter of medium porosity with suction and the beaker and precipitate are washed with three 2 ml. portions of cold water. A 25 ml. volumetric test tube is placed inside a suction flask with the stem of the sintered glass funnel adaptor within the neck of the test tube. The precipitate is dissolved in acetone by passing small portions through the filter. The beaker which contained the precipitate is also washed with a little acetone and the washings added to the filtrate. The volume of the filtrate is made up to 25 ml. with acetone and the solution well mixed. The colour is measured in an absorptiometer at 525 mμ. A calibration curve is prepared by dissolving 100 mg. of pure hyoscyamine sulphate in 15 ml. of 0.5 N H_2SO_4. The solution is made alkaline with 10% ammonia solution and extracted with chloroform. Aliquot portions are removed and transferred to 50 ml. beakers. The chloroform solution is evaporated nearly to dryness and 5 ml. 0.5 N H_2SO_4 added. The chloroform is completely removed and the acid solution precipitated with ammonium reineckate as described above. The reineckate method probably gives better accuracy than the method of Allport and Wilson (1939) but is not as sensitive. The colour of the reineckate ion however is quite stable in contrast to that of the purple colour of the Vitali-Morin reaction.

(c) Durick, King, Ware and Cronheim (1950) have described a procedure for the estimation of atropine and hyoscyamine based on the observation that number of alkaloids and other organic bases form with some acidic dyes, salt-like compounds which are soluble in certain organic solvents. After subsequent decomposition with alkali, the dye can be estimated colorimetrically. The authors use bromocresol purple as the acid component and benzene as the organic solvent. This method permits the quantitative estimation of the above mentioned alkaloids in amounts ranging from 0.1 to 2 mg. in 10 ml. of solution.

In applying the method to plant extracts, a preliminary purification to remove chlorophyll and other extraneous substances is necessary.

2. Titrimetric, Adsorption and Chromatographic Methods.

(a) Roberts and James (1947) in a method for the accurate estimation of the total alkaloids in small samples of belladonna and stramonium, have developed an adsorption technique for the removal of impurities followed by titration or colorimetric determination of the released alkaloids. About 1 g. of finely powdered material is accurately weighed and transferred to a small beaker. Five or six drops of 0.5 N H_2SO_4 are added and stirred in until the powder is uniformly moistened. 15 ml. of ether are added to the mixture which is stirred vigorously and allowed to settle. The liquid is decanted off, care being taken not to remove any solid. The operation is repeated with a further 15 ml. of ether. The solvent still remaining with the powder is driven off by warming on a water bath for a few minutes. If root material is being used the above procedure may be omitted. The dried powder is mixed with five or six drops of 10% w/w ammonium hydroxide and stirred until evenly moistened. Benzene (pure) is added up to 5—10 ml.

and brought to the boil on a water bath. The suspension is washed into a small percolator and the powder compressed tightly with glass rod with a flattened end. The powder is further extracted with benzene until 60—70 ml. have been collected. The time taken should be 2—4 hours. A further 2—3 ml. of benzene percolate when evaporated to dryness should give a negative VITALI-MORIN reaction. The adsorption column (Fig. 2) consists of two parts. The top section A, is furnished with a pad of cotton wool, C, dried at 100°. A column of activated alumina (Type H, 100/200 mesh: Spence, Widnes, England) 2.0—2.5 cm. long is tightly packed above by means of the glass rod, T. Cotton wool and a silica column (8—12 cm. granulated neosyl, 80/100 mesh; Spence) are introduced similarly into the lower section, S. The two sections, A and S, are then put together. The ground glass joint should not be greased and should fit well enough to prevent leakage. The columns should be packed so as to allow a slow drip when the liquids are poured into A. If necessary the receiving flask, R, can be attached to a suction pump. The benzene solution of the alkaloids is poured into the bulb of the apparatus and the container washed out with a few ml. of the solvent. When the benzene has passed through the column, 30 ml. of absolute ethanol is pipetted on to the alumina. As soon as the alumina becomes dry, the top section, A, is removed and the cone of the ground joint washed with 5 ml. of

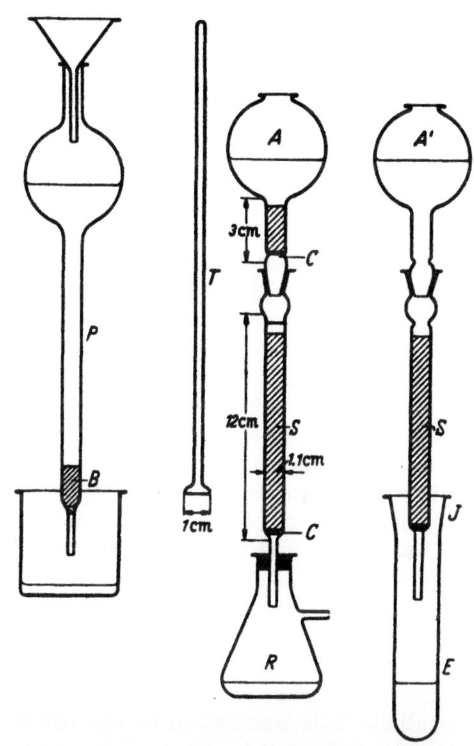

Fig. 2. Apparatus used in the ROBERTS and JAMES (1947) method for total alkaloids in belladonna and stramonium.

acetone, the washings being put through the silica. A is then replaced by a similar glass part containing no adsorbent. 4 ml. of 20% ammonium hydroxide (w/w) or a little less if the silica column is shorter than 12 cm., are added to deactivate the adsorbent. As soon as the aqueous solvent has disappeared from the surface of the silica, chloroform is placed in the top bulb and the flask, R, is replaced by the titration tube, E. Elution with chloroform is continued until 45—50 ml. have been collected. Throughout the separation, the top of the silica column must remain wet with solvent. The chloroform solution of the alkaloids is evaporated on a warm water bath under reduced pressure. This is conveniently carried out by providing the collecting and titration tube with a ground-in head with a side arm. The last traces of solvent are removed by a current of air at room temperature. The alkaloid residue is dissolved in a few drops of ethanol and 1 ml. of 0.02 N H_2SO_4 is added from an accurately calibrated pipette of narrow bore. The solution is made up to about 15 ml. with CO_2 free distilled water and methyl red indicator added. The residual acid is back titrated with 0.01 N NaOH in a CO_2 free atmosphere using a micro-burette of the REHBERG type. 1 ml. of 0.01 N acid or alkali is equivalent to 2.892 mg. of alkaloid calculated as L-hyoscyamine or atropine. The above method has proved very useful in studies on the biosynthesis of the belladonna alkaloids.

(b) In species of *Datura* and *Duboisia* hyoscyamine and hyoscine occur together and the latter may constitute a high proportion of the total alkaloids present. A method for the separation and assay of hyoscyamine and hyoscine in *Duboisia* spp. has been described by Trautner and Roberts (1948). The alkaloids are adsorbed from benzene solution on silica columns and hyoscine eluted with absolute ethanol, collected and titrated. Hyoscyamine was eluted with ammoniacal chloroform and estimated after removal of the solvent. The assay of *Duboisia myoporoides* is carried out as follows. About 0.5 g. of powdered leaf is accurately weighed, moistened with dilute H_2SO_4, and washed with ether to remove ether soluble matter. The alkaloids are extracted with equal volumes of chloroform previously saturated with ammonia gas and ether. The extract is evaporated to dryness, the residue dissolved in chloroform, the solvent removed, and the residue again dissolved in chloroform. The proportion of non-volatile bases is approximately determined by titration with p-toluene sulphonic acid in chloroform, using dimethyl yellow as indicator. (In an experiment, the volume of acid used corresponded to 17—20 mg. of total alkaloids calc. as hyoscine.) After a further purification of the total alkaloids by transferring to dilute acid and then to ether, the mixed ether extracts are dried, the solvent removed, and the residue dissolved in 25 ml. of pure benzene. The chromatographic column consists of a small column, 2.5 × 1 cm., of activated alumina arranged above a column of activated silica, 12 × 1 cm. Activated alumina is prepared by treating a sample of alumina (type H, Spence) with 50% acetic acid, washing with distilled water, drying and calcining for 24 hours. Activated silica is prepared by washing the sample (Granulated Neosyl, 80/100 mesh: Spence, Widnes) with a 50% aqueous solution of acetic acid followed by water, and, after drying, calcining for 12 hours. Water and moist solvents inactivate the silica rapidly and therefore all solvents should be pure and be dried before use. The purpose of the column of alumina is to effect a preliminary purification by adsorption of colouring matter and other material. 10 ml. of the benzene solution containing the alkaloids is transferred to the combined column and allowed to run slowly through. The combined column is eluted with one 15 ml. portion and two 5 ml. portions of absolute ethanol. The second and third fractions are virtually free from bases. The alumina column is removed and the silica eluted with 40 ml. of ammoniacal chloroform. The ethanol fraction is diluted to 25 ml. and the tropic esters determined on a 1 ml. aliquot by the colorimetric method of Allport and Wilson (1939). The remainder of the ethanol fraction is used for the characterization of the hyoscine. The solution is quickly evaporated to dryness under reduced pressure, the residue dissolved in 2 ml. of chloroform and slightly over-titrated with 0.05 N picric acid in chloroform. An equal volume of benzene is added, followed after a few hours by several drops of light petroleum. On standing overnight crystals separate which are washed with a mixture of equal volumes of chloroform and benzene containing several drops of light petroleum. The crystals of hyoscine picrate are dried at 105° and melting points taken. The ammoniacal chloroform eluate is made up to exactly 50 ml. with chloroform and an aliquot of 1 ml. taken for determination of the total alkaloids by the method of Allport and Wilson (1939). The remaining 49 ml. of eluate is divided into two portions of 25 and 24 ml. The 25 ml. portion is evaporated to dryness, the residue dissolved in a little ethanol, and the bases titrated as a check to the colorimetric determination. The 24 ml. portion is evaporated to dryness, the residue dissolved in chloroform, just over-titrated with picric acid as above, and, after addition of 1 volume of ether and 1 volume of light petroleum, the mixture is left overnight. The crystals are washed as above described for hyoscine, dried and melting points taken including a mixed melting point with hyoscyamine picrate.

The above method of separation and assay is suitable only when the proportion of hyoscine to hyoscyamine in the mixture of the two is greater than 10%. The separation fails when applied to extracts of *Atropa belladonna* in which the proportion of hyoscine is very low. The authors state that the efficiency of the separation of hyoscine and hyoscyamine by fractional elution from a silica column with absolute ethanol is to some extent dependant on the dimensions of the column. A column 12 × 1 cm. allows the removal of as much as 10—12 mg. of hyoscine in the first 20 ml. of ethanol, while retaining up to 4 mg. of hyoscyamine even if the volume of eluant is increased to 30 ml. If the conditions are suitable, the first 15 ml. of ethanol contains the bulk of the hyoscine and the fractions between 15 and 30 ml. are almost free from bases. Hyoscyamine begins to be eluted as the volume of the solvent exceeds 30 ml. but removal of hyoscyamine is better carried out with ammoniacal chloroform. With care, the total recovery of alkaloids by the elutions with absolute ethanol and ammoniacal chloroform combined, is within 3% of the correct amount. A small quantity of hyoscine is not eluted by absolute ethanol but is removed by ammoniacal chloroform. The amount of hyoscine so held depends on the quantity and quality of the silica used and in the presence of hyoscyamine this amount is considerably reduced. For the most efficient separation under the prescribed conditions the total alkaloids submitted to the separation should contain between 3 and 4 mg. of hyoscyamine and the best results are obtained by using mixtures of alkaloids in which the ratio of hyoscine to hyoscyamine lies between the limits 1:4 and 6:1.

(c) EVANS and PARTRIDGE (1948) have studied the separation of the solanaceous alkaloids on columns of kieselguhr buffered with 0.5 M phosphate. The buffer solution is intimately mixed with purified kieselguhr and the column packed. Before use, the column is washed with pure ether to remove traces of fatty material from the kieselguhr. For separation of the alkaloids, 100 g. quantities of *D. stramonium* and *A. belladonna* and 70 g. of *D. ferox* are extracted in the usual manner and the resinous residue treated with successive quantities of warm ether. The ethereal solution is chromatographically fractionated at pH 7.2 on a 30 g. column containing 50% of phosphate buffer. Ether saturated with buffer and chloroform saturated with buffer are used as the developing solvents. The eluate is collected in fractions by means of siphons delivering 2—8 ml. Each fraction is received in a hard glass test tube containing 1 ml. of water and a drop of a solution of bromocresol green adjusted to pH 4.4 by the addition of a trace of $0.005\ N\ H_2SO_4$. The presence of an alkaloid in the fraction becomes evident from the colour of the indicator, and the quantity is determined by titration with either $0.02\ N\ H_2SO_4$ or $0.005\ N\ H_2SO_4$, using a buffer, pH 4.4, containing bromocresol green for comparison. Development is continued with chloroform previously saturated with buffer. An equal volume of ether is added to the eluate before titration.

Reineckates were found very useful for characterization of the solanaceous alkaloids. Saturated aqueous ammonium reineckate is added to a solution of the alkaloids slightly acidified and the precipitated reineckate is recrystallized from 20—30% aqueous acetone. The reineckate radical is converted to thiocyanate as follows. About 40 mg. are accurately weighed into a boiling tube and boiled under reflux with 5 ml. of 10% NaOH solution. The resulting chromium hydroxide suspension is diluted with 10 ml. of water and filtered. The residue on the filter is well washed with water. To the combined filtrate and washings, 2 ml. of a 5% solution of ferric ammonium sulphate are added, the solution neutralized with dilute nitric acid (free from nitrous acid) and 10 ml. excess of dilute nitric acid added. After cooling to 15°, the solution is titrated with 0.05 N mercuric nitrate. The thiocyanate radical in the reineckate is thus determined. The load of total alkaloids on the kieselguhr column can be varied between 6 and 120 mg. for columns containing weights of kieselguhr from 2.5 to 50 g. and the proportions of buffer solution to kieselguhr varied from 15 to 50%. JINDRA and POHORSKY (1951) have described a method for the determination of the solanaceous alkaloids by exchange of ions. Amberlite resin IR—4 B is used for anion exchange and the eluate titrated against 0.01 N acid using the antimony electrode for the electrometric titration. This method can be used for the determination of the total alkaloids of *A. belladonna* and *Hyoscyamus* spp. and the results compare well with the pharmacopoeial methods of assay.

(d) A recent method for the adsorption analysis of the tropane group of alkaloids has been published by BJÖRLING and BERGGREN (1953). Decalso F. (Permutit Co., London) is used as the adsorbent and elution is effected with moderately strong solutions of acids or salts. The alkaloids are quantitatively taken up by Decalso from water and some organic solvents.

For their studies on the variation in the alkaloids of *Duboisia myoporoides* HILLS and RODWELL (1951) determined hyoscine by the picrate method and also separated the bases by partition chromatography with subsequent estimation of hyoscine by titration.

3. The Paper Chromatography of the Solanaceous Alkaloids.

MUNIER and MACHEBOEUF (1951) have applied the methods of paper chromatography for the separation of the solanaceous alkaloids. The solvents used for development are mixtures of n-butanol with varying amounts of acetic or hydrochloric acids and isobutanol with varying amounts of hydrochloric acid. The spots are developed with a modified DRAGENDORFF reagent. Stock solution A contains 850 mg. of bismuth subnitrate, 40 ml. of distilled water and 10 ml. of glacial acetic acid. Stock solution B contains 8 g. of potassium iodide in 20 ml. of distilled water. For use, solutions A and B are mixed and 10 ml. of the mixed solution are diluted with 20 ml. of glacial acetic acid and 100 ml. of distilled water. MUNIER, MACHEBOEUF and CHERRIER (1951) obtained improved separation of the bases by treating the paper with KH_2PO_4. BRINDLE, CARLESS and WOODHEAD (1951) used buffered filter paper for the separation of the alkaloids. Whatman No. 1 paper soaked in M/15 phosphate buffer (SÖRENSEN) of pH 7.4 and air dried before use was found to give good separation of atropine, apoatropine and hyoscine with n-butanol saturated with water as the solvent. The spots were developed by spraying with a solution of 0.5% iodine in aqueous KI or with the modified DRAGENDORFF reagent of MUNIER and MACHEBOEUF (1951). The method was applied to the separation of the alkaloids in extracts of *Atropa belladonna* and *Datura stramonium*. GORE and ADSHEAD (1952) in a study of the paper partition method as applied to the separation of alkaloids, advocate the use of the following solvent systems for general use. 1) n-Butanol, 40 vols., glacial acetic acid, 10 vols., water 50 vols., 2) n-Butanol, 10 vols., water, 10 vols., formic acid, 1 vol. 3) n-Butanol, 10 vols., water, 10 vols, propionic acid, 1 vol. After shaking, the saturated upper layers are used in the trough and the aqueous phase is placed in beakers at the bottom of the cabinet. The solvent is run for 16 hours and removed from the paper in an air draught at 80—85°. The position of the alkaloids is revealed by spraying with a 0.2% w/v aqueous solution of iodine. Whatman No. 1 paper gives the best separations. Average R_F values for the tropane alkaloids are given in Table 17.

The authors recommend the capillary ascent technique as possessing some advantages over the descending flow method. The former method gives more compact spots, and the reproducibility of the R_F values is superior to that of the descending flow method. In the writer's laboratory, a solution of 0.25% w/v bismuth potassium iodide in absolute ethanol has been used with success for the development of the spots, and has proved more satisfactory than the iodine reagent. With the tropane alkaloids, orange to brick-red spots are developed against a pale yellow background. The ethanolic solution is easily sprayed on to the paper and evaporates quickly.

Table 17. *Average R_F values of the Tropane Alkaloids.*

Alkaloid	Solvent 1 (above)	Solvent 2	Solvent 3
Atropine	0.73	0.58	0.68
Atropine sulphate	0.72	0.56	0.64
Hyoscyamine	0.75	0.57	0.66
Hyoscyamine sulphate . .	0.73	0.55	0.60
Hyoscine	0.63	0.42	0.51
Hyoscine hydrobromide . .	0.62	0.41	0.49
Cocaine	0.76	0.72	0.79
Cocaine hydrochloride . .	—	0.66	0.69

V. The Estimation of the Coca Alkaloids.

The total alkaloids of Coca leaves may be estimated by the following titrimetric method described by van Itallie (1938). 20 g. of finely powdered leaves are triturated with 20 ml. of 2 N Na$_2$CO$_3$ until the solution is homogeneous, and the mixture is allowed to stand for 30 minutes with occasional grinding. The mass is transferred quantitatively to a continuous extractor and exhausted with ether. The ether extract is shaken with three portions of 20, 15 and 10 ml. of 0.1 N HCl and the combined acid extracts filtered into a second separator. The acid solution is covered with 30 ml. of a mixture of 2 parts of ether and 1 part of light petroleum (b. p. 40—60°) and 1 g. of NaHCO$_3$ added in small portions. After shaking and separation, the extraction is repeated with two further portions of ether-light petroleum mixture. The combined ether-petroleum extracts are dried with anhydrous Na$_2$SO$_4$, filtered and the solvent removed by evaporation. The residue is dissolved in 5 ml. of ethanol, methyl red indicator added, and titrated with 0.1 N acid to pale orange. 50 ml. of boiled and cooled distilled water are added and the titration continued to the red end point. 1 ml. of 0.1 N acid is equivalent to 0.0303 g. of alkaloids calculated as cocaine. De Jong (1940) recommends treatment of the leaf sample (20 g.) with 5 ml. of 2 N ammonium carbonate solution and subsequent grinding with 20 ml. of benzene prior to extraction with ether. Treatment with benzene is said to facilitate the extraction of the alkaloids with ether. Young (1948) has estimated cocaine by hydrolysis with dilute alkali and determination of liberated methanol colorimetrically. A standard cocaine solution is prepared by weighing accurately, 50 mg. of pure cocaine hydrochloride and dissolving in 10 ml. of distilled water. A calibration curve is prepared by taking seven aliquots (from 0.4 to 1 ml.) of the standard solution and introducing them into 25 ml. round bottomed distilling flasks with long necks. 2 ml. of 2% NaOH solution are added to each flask, followed by 5 ml. of water. The flasks are connected to small condensers (ground glass joints to all connections) and the contents carefully distilled until 2 ml. of distillate is collected in a 6 inch Nessler tube having a 2 ml. marking. To the distillate is added 0.25 ml. of 24% v/v ethanol and 2.75 ml. of water. 2 ml. of a solution containing 3 g. $KMnO_4$ and 15 ml. of syrupy phosphoric acid (85%) in 100 ml. of distilled water, are added, and the tubes allowed to stand with occasional shaking for 10 minutes. After this period has elapsed, 2 ml. of oxalic acid solution (5 g. oxalic acid in 100 ml. of 50% v/v H_2SO_4) are added, followed by 5 ml. of Schiff's reagent prepared as follows. 0.2 g. of Kahlbaum rosaniline hydrochloride is dissolved in about 120 ml. of hot water and the solution cooled. A solution containing 2 g. of Na_2SO_3 in 20 ml. of water is added followed by 2 ml. of conc. HCl. The solution is diluted to 200 ml. and placed in a refrigerator for at least 24 hours before use. It should be stored in a cool place. After the tubes have been allowed to stand for two hours the colour is read in a photometer at 560 mμ. The intensity of the colour is a function of the amount of cocaine. Samples of cocaine solutions of unknown strength are subjected to the above procedure and the amount of cocaine present read off from the calibration curve. The oxidation of the samples and standards should be started at the same time and if the colour of the sample exceeds that of the 1 ml. standard, an appropriate dilution must be made.

The above method can be applied to the determination of cocaine in extracts of Coca leaves if other alkaloids, which on hydrolysis give methanol, are first removed. Cinnamylcocaine and the truxillines give methanol on hydrolysis but tropacocaine is not a methyl ester. Cinnamylcocaine can bo oxidized with permanganate in 3% H_2SO_4 at 5°. Cocaine is not oxidized and is removed by rendering the permanganate solution alkaline with ammonia and shaking out with light petroleum. The truxillines are insoluble in light petroleum.

Z. The Yohimbe and Quebracho Alkaloids.

From the bark of *Pausinystalia yohimba* Pierre (syn. *Corynanthe yohimbe* Schum.), a tree indigenous to the Cameroons and the French Congo and belonging to the *Rubiaceae*, the alkaloid yohimbine and a number of isomers are obtained. The bark of *Pseudocinchona africana* Chev. *(Rubiaceae)* contains the alkaloids corynanthidine (α-yohimbine), corynanthine (isomeric with yohimbine), corynanthëine and its isomer corynanthëidine, together with pseudoyohimbine. *Mitragyna stipulosa* Kuntze *(Rubiaceae)* contains mitraphylline isomeric with yohimbine. *Aspidosperma quebrachoblanco* Schlecht *(Apocynaceae)* contains quebrachine, which has been shown to be identical with yohimbine, and aspidospermine as the major alkaloids. *Vallesia glabra* Link and other species of *Vallesia (Apocynaceae)* contain vallesine and aspidospermine. A full account of these alkaloids, their occurrence and distribution is to be found in the standard works on the plant alkaloids. The total alkaloid content of yohimbe bark varies considerably, but average samples contain 1.6—3.2%. The upper limit is about 6.1% and the lower limit 0.5%. *Aspidosperma* spp. contain 2—4% of total alkaloids and the leaves of *Vallesia glabra* are said to contain 4% of aspidospermine and the stems 2%.

I. Properties.

Yohimbine, $C_{21}H_{26}O_3N_2$, is the major alkaloid of yohimbe bark and is the most important of the group. It crystallizes from dilute ethanol in colourless needles, m. p. 234° $[\alpha]_D^{20°} +107.9$ (pyridine). Yohimbine is soluble in ethanol, chloroform and hot benzene but sparingly soluble in water or ether. It is a monoacidic base and forms crystalline salts. The hydrochloride, B · HCl, forms platelets, m. p. 301° (dec.), $[\alpha]_D$ 101.9°. The thiocyanate crystallizes from hot water and melts at 233—234°. On acetylation, yohimbine yields an acetyl derivative m. p. 133°. Yohimbine is the methyl ester of *yohimbic acid*, $C_{20}H_{24}O_3N_2 \cdot H_2O$.

Yohimbene, $C_{21}H_{26}O_2N_2$, the methyl ester of *yohimbenic acid*, has been isolated from the mother liquors remaining after the isolation of yohimbine. It melts at 276—277° (dec.).

$[\alpha]_D$ +43.7° (pyridine). In the pure state it is sparingly soluble in ethanol. The hydrochloride, $B \cdot HCl \cdot 3 H_2O$, forms needles, m. p. 234°. Other isomers which have been isolated from the mother liquors in the preparation of yohimbine are *isoyohimbine*, needles m. p. 240 to 243°; *alloyohimbine*, leaflets m. p. ($B \cdot H_2O$) 135—140°; *β-yohimbine*, leaflets, m. p. 235 to 236°; *γ-yohimbine* leaflets with $3 H_2O$, m. p. 240° (dec.); *δ-yohimbine*, prisms, m. p. 254°, and *pseudoyohimbine*, rhombic plates, m. p. 264—265°.

Corynanthine, $C_{21}H_{26}O_3N_2$, also isomeric with yohimbine, is laevo-rotatory $[\alpha]_D$ —73° (pyridine). It crystallizes in platelets or needles with $2 H_2O$ and melts at 241—242° (dry).

Corynanthidine is an isomer of corynanthine and crystallizes in needles, m. p. 243—244°, $[\alpha]_d^{12°}$ —11.5° (methanol). The hydrochloride crystallizes in needles, m. p. 288°, and is almost insoluble in ethanol. The picrate is crystalline and melts at 231—232°.

Corynanthëine is an amorphous base which yields a crystalline hydrochloride, m. p. 205°, soluble in chloroform. The ultraviolet absorption of corynanthëine differs from that of corynanthine and yohimbine.

Corynanthëidine is isomeric with corynanthëine and crystallizes from acetone with one molecule of solvent. It forms crystalline salts, a hydrochloride, $B \cdot HCl \cdot 2 H_2O$, m. p. 213°, picrate, m. p. 252°, and styphnate, $B \cdot C_6H_3O_8N_3$, m. p. 246°.

Aspidospermine, $C_{22}H_{30}O_2N_2$, crystallizes from ethanol in needles, m. p. 208°, b. p. 220°/1—2 mm., $[\alpha]_D^{18°}$ —99° (ethanol). This base does not form well crystallized salts. When boiled with dilute HCl it loses one molecule of acetic acid to give *deacetylaspidospermine*.

Quebrachamine, $C_{19}H_{26}N_2$, is a feeble base which forms crystalline salts with dibasic acids only. From dry ethanol it forms rhombohedral leaflets, m. p. 147°, $[\alpha]_D$ —100.5° (acetone). It occurs in small amount in the bark of *Aspidosperma quebracho-blanco*. The sulphate $B \cdot H_2SO_4 \cdot 2 H_2O$ forms prisms sparingly soluble in cold water, the oxalate $B \cdot H_2C_2O_4$ also forms prisms, m. p. 217°. The picrate on recrystallization from absolute ethanol forms scarlet needles, m. p. 195—196°.

Vallesine, $C_{20}H_{26}O_2N_2$, crystallizes from acetone or ether in needles, m. p. 154—156°, $[\alpha]_D^{24°}$ —91° (ethanol). The hydrochloride, $B \cdot HCl$, crystallizes in needles, m. p. 247—251° and the oxalate, $B \cdot C_2H_2O_4$, in prisms, m. p. 233—234° (dec.). On hydrolysis with 2 N—HCl at room temperature, formic acid and *desformylvallesine* are produced.

II. The Isolation of Yohimbine.

The method of Chemnitius (1927) is as follows. The powdered bark (1.5 kg.) is moistened with sufficient 20% sodium carbonate solution to make a soft plastic mass. The mass is allowed to stand for 1 day with occasional stirring and extracted with ether in a Soxhlet, the flask of which contains, in addition to the solvent, 2 g. of oxalic acid and 50 ml. of water for each 100 g. of the drug. The ether is distilled off and the cooled aqueous solution filtered into a separating funnel, the filter being washed with a little water. The aqueous solution is covered with ether saturated with powdered sodium carbonate and shaken vigorously. After separation of the ether layer the aqueous solution is further extracted with several portions of ether until extraction is complete. The combined ether extracts are concentrated, dried over fused $CaCl_2$, filtered, and exactly neutralized with anhydrous ethanolic hydrogen chloride. The brown precipitate which is formed is treated with anhydrous acetone for 24 hours with occasional stirring to remove resinous matter. The yellow yohimbine hydrochloride which remains is washed with acetone and dried.

III. The Detection of the Yohimbe and Quebracho Alkaloids.

Yohimbine gives precipitates with the alkaloidal reagents and with ammonium reineckate in slightly acid solution. With Fröhde's reagent it gives a deep blue colour which gradually changes to green, and when a small crystal of potassium dichromate is added to a solution of the alkaloid in conc. H_2SO_4 violet streaks changing to blue and finally green are produced. A number of tests for yohimbine based on the colour formed by indole derivatives with aldehydes in the presence of H_2SO_4 have been described. With the glyoxylic reagent (Adamkiewicz-Hopkins-Cole) yohimbine gives a blue ring and when warmed with H_2SO_4

(2 vols. to 1 vol. water) containing a little vanillin or piperonal, a rose-red to violet colour is produced. A rose-violet colour results when conc. HCl is used instead of H_2SO_4. Malowan (1950) has described the following colour test for yohimbine. A small quantity of the alkaloid is dissolved in 1 ml. of conc. HCl, 4 drops of a 2% solution of p-dimethylamino-benzaldehyde in conc. HCl added, the solution gently warmed and 2 drops of a 0.05% solution of sodium nitrite added. In the presence of yohimbine a deep violet-blue ring appears after a short time and on shaking the solution a blue colour develops. If a drop of fuming nitric acid is added to a particle of yohimbine in an evaporating dish and the solution gently taken to dryness and the residue moistened with ethanolic potash, a cherry-red colour is produced. A reagent consisting of 1 g. of chloral hydrate dissolved in 5 ml. of ethanol to which 10 ml. of conc. H_2SO_4 is added with cooling, gives on warming with yohimbine a deep blue colour.

Aspidospermine dissolves unchanged in cold conc. H_2SO_4 and on addition of a crystal of potassium dichromate a brown colouration develops which finally changes to olive-green. In the presence of conc. H_2SO_4, or HCl, vanillin or piperonal give a violet colour with aspidospermine. When the base is warmed with perchloric acid, a rose-red colour develops. Quebrachamine dissolves unchanged in conc. H_2SO_4 but on the addition of a trace of an oxidizing agent such as potassium dichromate, a blue colour develops. With vanillin and conc. HCl it gives a violet colour on warming, with Ehrlich's reagent a purple colour and with the Adamkiewicz-Hopkins-Cole reagent a blue colour.

IV. The Quantitative Estimation of the Total Alkaloids of Yohimbe Bark.

Lynch and Evers (1948) describe the following method for the estimation of total alkaloids. 2 g. of bark in No. 60 powder are accurately weighed, transferred to a glass mortar and mixed with 1 g. of ignited sand also in No. 60 powder. 2 ml. of a solution of 10% ammonia are added, the mixture triturated thoroughly for 10 minutes, covered, and allowed to stand for 2 hours. The mixture is transferred quantitatively to an apparatus for continuous extraction and extracted for 1 hour with 100 ml. of boiling chloroform. After extraction has proceeded for 1 hour, the drug is packed down tightly with a flat ended glass rod and the extraction continued for a further hour. At the end of this period a test for complete extraction is made by evaporating 2 ml. of the extract to dryness dissolving the residue in 5 drops of N H_2SO_4 and adding Mayer's reagent. Extraction is continued until a negative test for alkaloids is obtained. The drug is removed from the percolator or thimble, transferred to the glass mortar and the chloroform allowed to evaporate in air. 0.5 ml. of 10% ammonia solution is added to the mass which is triturated thoroughly, repacked in the percolator or extraction thimble and extracted once more with chloroform (50 ml.). The combined chloroform extracts are transferred to a separating funnel, the flask of the extraction apparatus being well washed with small portions of chloroform which are added to the separating funnel. The alkaloids are extracted from the chloroform by shaking first with 30 ml. N H_2SO_4, followed by 10 ml. portions of acid until extraction is complete. The acid extracts are combined and washed by shaking with three successive 5 ml. portions of chloroform. Each of the chloroform solutions is washed in succession with the same 10 ml. portion of 0.1 N H_2SO_4 and the chloroform discarded. The acid washing is added to the combined acid extracts, the alkaloids liberated by making alkaline with ammonia and shaken out with successive 10 ml. portions of chloroform until extraction is complete.

The chloroform extracts are washed in succession with the same 20 ml. of water. The combined chloroform extracts are transferred to a flask, the chloroform distilled off and the residue treated with 2 ml. of neutral absolute ethanol or acetone. The solution is heated on the water bath until the solvent is evaporated and to the residue is added 0.5 ml. of chloroform and 20 ml. of 0.02 N H_2SO_4. The chloroform is boiled off by heating on the water bath and the acid solution is titrated with 0.02 N NaOH using 5 drops of methyl red (0.1% solution in ethanol) solution as indicator. 1 ml. of 0.02 N H_2SO_4 is equivalent to 0.00709 g. of alkaloids calculated as yohimbine.

PERROT and HAMET (1932) estimate yohimbine as the hydrochloride by precipitation with acetone. 40 g. of the powdered bark is moistened with a 10% solution of sodium carbonate (about 25 ml. is required) and the mass is exhaustively extracted in a continuous extractor with benzene. The alkaloids are removed from the benzene by shaking with successive portions of 60, 20 and 10 ml. of a 2% solution of formic acid. The combined acid solutions are filtered, made alkaline with $NaHCO_3$ and shaken out with successive portions of 60, 20, 10 and 10 ml. of chloroform. The combined chloroform solutions are dried with anhydrous Na_2SO_4, filtered, and the chloroform distilled off. The dried residue represents the total alkaloids. This residue is dissolved in a small quantity of ethanol and HCl is added until the reaction is just acid to Congo red. Pure acetone is now added and the solution set aside at 0° for 24 hours. The yohimbine hydrochloride crystallizes out and is washed with acetone and ether and weighed. MUNIER and MACHEBOEUF (1951) have effected a good separation of yohimbine, corynanthine, corynanthidine, corynanthëine and corynanthëidine by the method of paper chromatography using n-butanol-HCl as the solvent system.

General References.

Allen's Commercial Organic Analysis 5th ed. Vol. VII. London: Churchill 1929.
HENRY, T. A.: The Plant Alkaloids, 4th ed. London: Churchill 1949.
MANSKE, R. H. F., and H. L. HOLMES: The Alkaloids, Vol. I. New York: Academic Press Inc. 1950; Vol. II, New York: Academic Press Inc. 1952. Vol. III 1953.
WINTERSTEIN, E., and G. TRIER: Die Alkaloide, Pt. I, 2nd ed. Berlin: Borntraeger 1928: Pt. II, 2nd ed. Berlin: Borntraeger 1931.

References.

Section A.

JAMES, W.O.: The Alkaloids, Vol. I edited by R. H. F. MANSKE and H. L. HOLMES. New York: Academic Press Inc. 1950.
KIRKPATRICK, H. F. W.: Quart. J. Pharm. 18, 245, 338 (1945); 19, 8, 127, 526 (1946); 20, 87 (1947).

Section B.

DENIGÈS, G.: Bull. Soc. Pharm. Bordeaux 82, 85 (1944). — DUNSTAN, W. R.: Pharm. J. (iv) 2, 122 (1896). — DUNSTAN, W. R., and A. E. ANDREWS: (a) J. chem. Soc. 87, 1620 (1905); (b) J. chem. Soc. 87, 1636 (1905).
HENRY, T. A., and T. M. SHARP: J. chem. Soc. 1928, 1105.
JACOBS, W. A., and L. C. CRAIG: J. Biol. Chem. 143, 605 (1942). — JACOBS, W. A., and C. F. HUEBNER: J. Biol. Chem. 170, 189 (1947).
LYNCH, G. R., and N. EVERS: Analyst 67, 289 (1942).
MAJIMA, R., H. SUGINOME and S. MORIO: Ber. dtsch. chem. Ges. 57, 1461, 1472 (1924). —
MÜHLEMANN, H., and R. WEIL: Pharm. Acta. Helv. 24, 419 (1949).

Section C.

BOND, H. R.: J. Ass. Off. Agric. Chem. Wash. 25, 817 (1942).
CHEMNITIUS, F.: J. prakt. Chem. 117, 147 (1927).
FREUDENBERG, K.: Ber. dtsch. chem. Ges. 51, 976, 1669 (1918).
HESS, K., and F. LEIBBRANDT: Ber. dtsch. chem. Ges. 51, 806 (1918).
JAHNS, E.: Ber. dtsch. chem. Ges. 21, 3404 (1888); 23, 2972 (1890); 24, 2615 (1891).

Section D.

van Arkel, C. G., and M. Meijst: Pharm. Weekbl. 87, 853 (1952).
Brochmann-Hanssen, E., and J. A. Evers: J. Amer. Pharm. Ass. Sci. Ed. 40, 620 (1951).
Chatterjee, R.: J. Amer. Pharm. Ass. Sci. Ed. 32, 1 (1943); 33, 205, 210 (1944). — Cromwell, B. T.: Biochem. J. 27, 860 (1933).
Erbring, H., and W. Wulf: Kolloidztschr. 125, 99 (1952).
Greathouse, G. A., and G. M. Watkins: Amer. J. Bot. 25, 743 (1938).
Henry, T. A.: The Plant Alkaloids, 4th ed. London: Churchill 1949.
Jowett, H. A. D., and F. L. Pyman: J. Chem. Soc. 103, 291, 825 (1913).
Labat, A.: Bull. Soc. chim. France 5, 742 (1909).
Mendez, J., E. R. Kirch and R. F. Voigt: J. Amer. Pharm. Ass. Sci. Ed. 38, 538 (1949).
Richter, V. E.: Arch. Pharm. 253, 192 (1914).

Section E.

Ehrlés, I.: Farm. Revy. 47, 519 (1948). — Ellis, S.: J. Pharmacol. Exp. Ther. 79, 364 (1943). — Ellis, S., F. L. Plachte and O. H. Straus: J. Pharmacol. Exp. Ther. 79, 295 (1943).
Ferreira da Silva, M. A. J.: Bull. Soc. Chim. Biol. 9, 753 (1893).
Hellberg, H.: Svensk. Farm. Tidskr. 51, 560 (1947); 53, 638 (1949).
Manske, R. H. F., and H. L. Holmes: The Alkaloids, Vol. II. New York: Academic Press Inc. 1952.
Polonovski, M., and C. Nitzberg: Bull. Soc. Chim. France (4) 17, 244 (1915).
Salway, A. H.: J. Chem. Soc. 99, 2148 (1911); (a) J. Chem. Soc. 101, 978 (1912); (b) Amer. J. Pharm. 84, 49 (1912). — Saul, J. E.: Pharm. J. (III) 17, 642 (1887).

Section F.

Birch, H. F., and L. R. Doughty: Biochem. J. 43, 38 (1948). — British Pharmacopoeia p. 457, 1932. — Brodie, B. B., and S. Udenfriend: J. Biol. Chem. 158, 705 (1945).
David, L.: Pharm. Acta. Helv. 24, 427 (1949).
Grant, H. S., and J. H. Jones: Analyt. Chem. 22, 679 (1950).
Hamet, R.: C. r. Acad. Sci. Paris 212, 135 (1941); (a) C. r. Acad. Sci. Paris 220, 670 (1945); (b) C. r. Acad. Sci. Paris 221, 307 (1945). — Henry, T. A.: The Plant Alkaloids, 4th ed. London: Churchill 1949.
Jindra, A., and J. Pohorsky: J. Pharm. Pharmacol 3, 344 (1951).
Kyker, G. C., B. D. Webb and J. C. Andrews: J. Biol. Chem. 139, 551 (1941).
Loustalot, A. J., and C. Pagan: J. Ass. Off Agric. Chem. Wash. 30, 153 (1947). — Lussman, D. J., E. R. Kirch and G. L. Webster: J. Amer. Pharm. Ass. Sci. Ed. 40, 368 (1951).
Marshall, P. B., and E. W. Rogers: Biochem. J. 39, 258 (1945).
Prudhomme, R. O.: J. Pharm. Chim. (9), 1, 8 (1940).
Sumerford, W. T.: Bull. Nat. Form. 15, 90 (1947).

Section G.

Dutcher, J. D.: J. Amer. Chem. Soc. 68, 419 (1946).
Foster, G. E., and J. V. Turner: Quart. J. Pharm. 20, 228 (1947).
Karrer, P., and H. Schmid: Helv. chim. Acta 29, 1853 (1946); 30, 1162, 2081 (1947). — King, H.: J. Chem. Soc. 1935, 1381; 1937, 1472; 1940, 737; 1947, 936; 1948, 1945; (a) J. Chem. Soc. 1949, 955; (b) J. Chem. Soc. 1949, 3263.
Pryde, A. M., and F. R. Smith: J. Pharm. Pharmacol. 1, 192 (1949).
Schmid, H., and P. Karrer: Helv. chim. Acta 33, 512 (1950). — Swann, R. V.: J. Pharm. Pharmacol 3, 843 (1951).

Section H.

Barger, G., and F. H. Carr: J. Chem. Soc. 91, 337 (1907). — Berg, A. M.: Pharm. Weekbl. 87, 69, 282 (1952). — Brindle, H., J. E. Carless and H. B. Woodhead: J. Pharm. Pharmacol 3, 793 (1951).
Dudley, H. W., and J. C. Moir: Brit. Med. J. 1, 520 (1935).
Fischer, R., and M. Hecht: Mikrochemie 38, 538 (1951). — Foster, G. E., J. Macdonald and T. S. G. Jones: J. Pharm. Pharmacol 1, 802 (1949). — Fuchs, L., and W. Himmelbauer: Scientia Pharm. 18, 93 (1950). — Fuchs, L., and M. Pöhm: Scientia Pharm. 19, 232 (1951).
Hellberg, H.: Farm. Rev. 50, 17, 33 (1951); Chem. Abs. 1951, 4404.
McGillivray, W. A., and W. S. Metcalf: N. Z. J. Sci. Tech. B, 25, 115 (1943).
Powell, C. E., O. W. Reagan, A. Stevens and E. E. Swanson: J. Amer. Pharm. Ass. Sci. Ed. 30, 255 (1941).

SMITH, R. G.: J. Amer. Pharm. Ass. Sci. Ed. **36**, 321 (1947). — SMITH, S., and G. M. TIMMIS:
J. Chem. Soc. **1937**, 396. — STOLL, A.: Helv. chim. Acta. **28**, 1283 (1945).
TYLER, V. E. JR., and A. E. SCHWARTING: J. Amer. Pharm. Ass. Sci. Ed. **41**, 351 (1952).

Section I.

FORSYTH, W. G. C., S. F. MARRIAN and T. S. STEVENS: J. Chem. Soc. **1945**, 579.
LYNCH, G. R., and N. EVERS: Analyst **66**, 108 (1941).
SAYRE, L. E.: Drug. Circ. **1**, 55 (1911).

Section J.

British Pharmaceutical Codex, p. 410, 1934.
CALLOW, R. K., J. M. GULLAND and C. J. VIRDEN: J. Chem. Soc. **1931**, 2138. — CHATTEN,
L. G., and L. I. PUGSLEY: J. Amer. Pharm. Ass. Sci. Ed. **41**, 108 (1952). — CHEN, K. K.,
and C. H. KAO: J. Amer. Pharm. Ass. Sci. Ed. **15**, 625 (1926). — CHOU, T. Q.: J. Biol.
Chem. **70**, 109 (1926).
GULLAND, J. M., and C. J. VIRDEN: J. Chem. Soc. **1931**, 2148.
KUHN, A., and G. SCHÄFER: Dtsch. Apotheker-Ztg. **52**, 1265 (1937).
PESEZ, M.: J. Pharm. Chim. **27**, 120 (1938).
SCHOEN, K.: J. Amer. Pharm. Ass. Sci. Ed. **33**, 116 (1944).
WACHSMUTH, H.: J. Pharm. Belg. **4**, 186 (1949). — WICKSTROM, A., and B. SALVESON:
Medd. Norsk. Farm. Selskap. **14**, 97 (1952). — WINTERSTEIN, E., and D. IATRIDES: Hoppe-
Seylers Z. **117**, 240 (1921).

Section K.

v. BRAUN, J.: Ber. dtsch. chem. Ges. **38**, 3108 (1905).
CHEMNITIUS, F.: J. prakt. Chem. **118**, 25 (1928).
DILLING, W. J.: Pharm. J. London (IV), **29**, 34, 70 (1909).
FARR, E. H., and R. WRIGHT: Pharm. J. London (IV) **1**, 89 (1895); (IV) **18**, 185 (1904).
GABUTTI, E.: Boll. chim. farm. **45**, 289 (1906).
MADAUS, G., and H. SCHINDLER: Arch. Pharm. **276**, 280 (1938). — MELZER, H.: Arch. Pharm.
236, 701 (1898).
WOLFFENSTEIN, R.: Ber. dtsch. chem. Ges. **28**, 302 (1895).

Section L.

BERTHO, A.: Arch. Pharm. **277**, 237 (1939).
DUTTA, A., and B. K. GHOSH: Ind. J. Pharm. **11**, 74 (1949).
KANGA, D. D., P. R. AYYAR and J. L. SIMONSEN: J. Chem. Soc. **1926**, 2123.
PYMAN, F. L.: J. Chem. Soc. **115**, 163 (1919).

Section M.

British Pharmacopoeia, p. 238, 1932.
CARR, F. H., and F. L. PYMAN: J. Chem. Soc. **105**, 1591 (1914).
JINDRA, A., and J. POHORSKY: J. Pharm. Pharmacol **3**, 344 (1951).
LASSLO, A., and G. L. WEBSTER: J. Amer. Pharm. Ass. Sci. Ed. **39**, 193 (1950). — LOWDELL,
D. P.: J. Pharm. Pharmacol. **4**, 123 (1952).
MENDEZ, J., E. R. KIRCH and R. F. VOIGT: J. Amer. Pharm. Ass. Sci. Ed. **38**, 538 (1949).
PAUL, B. H., and A. J. COWNLEY: Pharm. J. (III) **25**, 111, 373, 690 (1894). — PYMAN, F. L.:
J. Chem. Soc. **111**, 419 (1917).

Section N.

ALBO, G.: Arch. sci. phys. nat. **12**, 227 (1901). — ASHLEY, J. N., and J. O. HARRIS: J. Chem.
Soc. **1944**, 677.
CLEWER, H. W. B., S. J. GREEN and F. TUTIN: J. Chem. Soc. **107**, 835 (1915). — CRAIG, L. C.,
and W. A. JACOBS: J. Biol. Chem. **143**, 427 (1942).
DAVIES, E. C.: Pharm. J. **106**, 480 (1921).
FRIED, J., H. L. WHITE and O. WINTERSTEINER: J. Amer. Chem. Soc. **72**, 4621 (1950).—
FRIED, J., P. NUMEROF and N. H. COY: J. Amer. Chem. Soc. **74**, 3041 (1952).
GLEN, W. L., G. S. MYERS, R. BARBER, P. MOROZOVITCH and G. A. GRANT: Nature (Lond.)
170, 932 (1952). — GSTIRNER, F.: Pharm. Ztg. (Berlin) **77**, 509 (1932).
JACOBS, W. A., and L. C. CRAIG: J. Biol. Chem. **155**, 565 (1944); **160**, 555 (1945).
KING, J. S.: J. Amer. Pharm. Ass. Sci. Ed. **40**, 424 (1951). — KIPPENBERGER, C.: Nachw.
Giftstoffen 1897, 104. — KLEIN, G., and G. POLLAUF: Österr. Bot. Z. **78**, 251 (1929). —
KLOHS, M. W., R. ARONS, M. D. DRAPER, F. KELLER, S. KOSTER, W. MALESH and F. J.
PETRACEK: J. Amer. Chem. Soc. **74**, 5107 (1952).
LIPTÁK, P.: Chem. Zentralal. **1928**, 1534.

Mack, H., and E. J. Finn: J. Amer. Pharm. Ass. Sci. Edn. **39**, 532 (1950). — Maison, G. L., E. Gotz and J. W. Stutzmann: J. Pharmacol. Exp. Ther. **103**, 74 (1951). — Manceau, P., F. Montant and G. Netien: Ann. pharm. franc. **8**, 11 (1946). — Manske, R. H. F., and H. L. Holmes: The Alkaloids, Vol. II. New York: Academic Press Inc. 1952. — Mühlemann, V. H., and R. Tobler: Pharm. Acta. Helv. **21**, 34 (1946).
Niemann, E.: Pharm. Acta. Helv. **8**, 92 (1933).
Papineau-Couture, G., and R. A. Burley: Analyt. Chem. **24**, 1918 (1952).
Santavy, F., and T. Reichstein: Helv. Chim. Acta **33**, 1606 (1950). — Seifert, R.: Dtsch. Apoth.-Ztg. **58**, 77 (1943). — Stoll, A., and E. Seebeck: Science (Lancaster, Pa.) **115**, 678 (1952).
Walaszek, E. J., and A. Pircio: J. Amer. Pharm. Ass. Sci. Ed. **41**, 270 (1952).
Zeisel, S.: Monatsh. Chem. **7**, 557 (1886).

Section O.

Esdorn, I.: Heil- u. Gewürz-Pflanzen **19**, 1 (1940).
Kalashnikov, V. P.: Farmatsiya 1, 20 (1939).
Lynch, G. R., and N. Evers: Analyst **64**, 581 (1939).
Manske, R. H. F., and H. L. Holmes: The Alkaloids, Vol. I. New York: Academic Press Inc. 1950. — Markwell, W. A. N.: Pharm. J. **136**, 617 (1936). — Mascré, M., and M. Caron: Bull. Sci. Pharmacol. **37**, 657 (1930).
Uffelie, O. F.: Pharm. Weekbl. **81**, 41 (1946).

Section P.

Bathurst, N. O., I. Reifer and E. M. Clare: N. Z. J. Sci. Tech. B **24**, 161 (1943).
Clare, E. M., and I. M. Morice: N. Z. J. Sci. Tech. B, **27**, 36 (1945).
Grimmett, R. E. R., and J. Melville: N. Z. J. Sci. Tech. B, **24**, 149 (1943). — Grimmett, R. E. R., and D. F. Waters: N. Z. J. Sci. Tech. B, **24**, 151 (1942).
Reifer, I., and N. O. Bathurst: N. Z. J. Sci. Tech. B, **24**, 17 (1942); **24**, 155 (1943).
Shorland, F. B.: N. Z. J. Sci. Tech. B, **24**, 159 (1943).
White, E. P., and I. Reifer: N. Z. J. Sci. Tech. B, **27**, 38 (1945).

Section Q.

Adamson, D. C. M., and F. P. Handisyde: Quart. J. Pharm. **19**, 350 (1946). — Adamson, D. C. M., F. P. Handisyde and H. W. Hodgson: Quart. J. Pharm. **20**, 218 (1947).
Baggesgaard-Rasmussen, H., and O. Lanng: Dansk. Tidsskr. Farm. **22**, 203 (1948).
Cramer, J. S. N., and J. G. Voermann: Pharm. Weekbl. **84**, 129 (1949).
Dott, D. B.: Allen's Commercial Organic Analysis, 5th ed. Vol. VII, p. 671. London: Churchill 1929.
Foster, G. E., and J. Macdonald: J. Pharm. Pharmacol. **3**, 127 (1951).
Grant, E. W., and W. W. Hilty: J. Amer. Pharm. Ass. Sci. Ed. **42**, 150 (1953). — Guarino, S.: Arch. Sci. biol. **31**, 115 (1946).
Henry, T. A.: The Plant Alkaloids, 2nd ed. London: Churchill 1924.
Klee, F. C., and E. R. Kirch: J. Amer. Pharm. Ass. Sci. Ed. **42**, 146 (1953).
Mannich, C.: Arch. Pharm. **273**, 97 (1935); **280**, 386 (1942). — Mariani, A., S. Guarino and O. M. Marelli: Annali Chim. **51**, 661 (1951). — Moorhoff, C. F.: Pharm. Weekbl. **87**, 593 (1952).
Oram, J. B.: Amer. J. Pharm. **122**, 60 (1950).
Poethke, W., and E. Arnold: Pharmazie **6**, 406 (1951).
Radulescu, D.: Boll. chim. Farm. **51**, 865 (1913). — Reith, J. F., and A. W. M. Indemans: Pharm. Weekbl. **85**, 309 (1950).
Stephens, R. L.: J. Pharm. Pharmacol. **3**, 815 (1951). — Stolman, A., and C. P. Stewart: Analyst. **74**, 536 (1949). — Svendsen, A. B.: Dansk Tidsskr. Farm. **22**, 131 (1948); Pharm. Acta Helv. **26**, 323 (1951).
Wegner, E.: Pharmazie **5**, 33 (1950); (a) Pharmazie **6**, 55 (1951); (b) Pharmazie **6**, 420 (1951).

Section R.

Briggs, L. H., and W. S. Taylor: J. Chem. Soc. **1938**, 1206.
Clemo, G. R., and R. Raper: J. Chem. Soc. **1929**, 1927. — Couch, J. F.: Amer. J. Pharm. **25**, 37 (1925); J. Amer. Chem. Soc. **61**, 3327 (1939).
Ing, H. R.: J. Chem. Soc. **1931**, 2195; **1933**, 504; **1935**, 1053.
Jaretzky, R., and B. Axer: Arch. Pharm. **272**, 152 (1934).
Munier, R., and M. Macheboeuf: Bull. Soc. Chim. biol. Paris **33**, 846 (1951).
Power, F. B., and A. H. Salway: J. Chem. Soc. **103**, 191 (1913).

WHITE, E. P.: N. Z. J. Sci. Tech. B, **25**, 93 (1943); (a) N. Z. J. Sci. Tech. B **25** 143 (1944); (b) N. Z. J. Sci. Tech. B, **25**, 152 (1944); (a) N. Z. J. Sci. Tech. B, **27**, 335 (1946); (b) N. Z. J. Sci. Tech. B, **27**, 339 (1946); (c) N. Z. J. Sci. Tech. B, **27**, 474 (1946); (d) N. Z. J. Sci. Tech. B, **27**, 478 (1946); N.Z.J. Sci. Tech. B, **33**, 50 (1951).

Section S.

BRUSTIER, V., P. BOURBON and R. VIGNES: Bull. Soc. chim. France 17, 113 (1950). — BUNTZEL-MAN, N. I.: Byull, Nauch. Isstedovatel Khim-Farm. Inst. **1931**, 71; Chem. Abstr. **27**, 5474 (1933).
FISCHER, O.: Chem. Zentralbl. **1901** I.
HASENFRATZ, V.: Ann. chim. (10) 7, 151 (1927).
PERKIN, W. H., and R. ROBINSON: J. Chem. Soc. 115, 962 (1919).
YA KHAIT, G.: Farmatsiya, 8, No. 6, 25 (1945); Chem. Abstr. 41, 2206 (1947).

Section T.

BIEDEBACH, F.: Apothek.-Ztg. 48, 723 (1933)
EKKERT, L.: Pharm. Zentralh. **66**, 36 (1925). — ELVIDGE, W. F.: Quart. J. Pharm. **20**, 234 (1947).
HELCH, M. H.: Pharm. Post **35**, 289 (1902).
LYNCH, G. R., and N. EVERS: Analyst **73**, 311 (1948).
SHUPE, I. S.: J. Ass. Off. Chem. Wash. 24, 757 (1941).
WEBB, J. W., R. S. KELLEY and A. J. McBAY: J. Amer. Pharm. Ass. Sci. Ed. 41, 278 (1952).

Section U.

CHILTON, J., and M. W. PARTRIDGE: J. Pharm. Pharmacol. **2**, 784 (1950);
EWERS, E.: Arch. Pharm. **237**, 49 (1899).
GOODSON, J. A.: Quart. J. Pharm. **13**, 57 (1940)
HESS, K.: Ber. dtsch. chem. Ges. **52**, 1005 (1919).
MENZIES, R. C., and R. ROBINSON: J. Chem. Soc. **125**, 2163 (1924).
STÖDER, W.: Pharm. Ztg. 46, 451 (1902).

Section V.

ALBERTI, B.: Z. Unters. Lebensm. 64, 260 (1932).
BARGER, G., and H. L. FRAENKEL-CONRAT: J. Chem. Soc. **1936**, 1537. — BELL, R. C., and L.H. BRIGGS: J.Chem. Soc. **1942**, 1. — BELL, R. C., L. H. BRIGGS and J. J. CARROLL: J. Chem. Soc. **1942**, 12. — BRIGGS, L. H., R. P. NEWBOLD and N. E. STACE: J. Chem. Soc. **1942**, 3.
CONNER, H. W.: Plant Physiol. 12, 79 (1937).
HULME, A. C., and R. NARAIN: Biochem. J. **25**, 1051 (1931).
LAMPITT, L. H., J. H. BUSHILL, H. S. ROOKE and E. M. JACKSON: J. Soc. Chem. Ind., Lond. 62, 48T (1943).
PFANKUCH, E.: Biochem. Z. **295**, 44 (1937).
ROOKE, H. S., J. H. BUSHILL, L. H. LAMPITT and E. M. JACKSON: J. Soc. Chem. Ind., Lond. 62, 20T (1943).
TUTIN, F., and H. W. B. CLEWER: J. Chem. Soc. 105, 559 (1914).

Section W.

ALLEN, J., and N. L. ALLPORT: Quart. J. Pharm. **13**, 252 (1940). — ANET, F. A. L., G. K. HUGHES and E. RITCHIE: Austr. J. Chem. 1, 58 (1953).
BANDELIN, F. J.: J. Amer. pharm. Ass. Sci. Ed. **39**, 493 (1950).
EL RIDI, M. S., and K. KHALIFA: J. Pharm. Pharmacol. 4, 190 (1952).
FISCHER, R., and E. BUCHEGGER: Pharm. Zentralh. 89, 146 (1950)
GORE, D. N., and J. M. ADSHEAD: J. Pharm. Pharmacol. 4, 803 (1952).
JINDRA, A., and J. POHORSKI: J. Pharm. Pharmacol. **3**, 344 (1951).
MALAQUIN, P.: J. Pharm. et Chim. (vi) **30**, 546 (1909). — MATTHES, H., and O. RAMMSTEDT: Arch. Pharm. **245**, 124 (1907).
RASSMUSSEN, S. K.: Dansk Tidsskr. Farm. **16**, 11 (1942). — ROLLAND-LECLERCQ, S.: J. Pharm. Belg. **2**, 283 (1947).
WÖBER, A.: Z. Angew. Chem. **31**, 124 (1918).

Section X.

AVENS, A. W., and G. W. PEARCE: Industr. Engng. Chem. (Anal. ed.) 11, 505 (1939).
BOTTOMLEY, W., R. A. NOTTLE and D. E. WHITE: Austr. J. Sci. 8, 18 (1945). — BOWEN, C. V.: J. Ass. Off. Agric. Chem. Wash. **30**, 315 (1947). — BOWEN, C. V., and W. F. BARTHEL; Industr. Engng. Chem. (Anal. ed.) 15, 740 (1943). — BURKAT, S.: Farm. Zhur. 41, 228 (1937).

Dawson, R. F.: Science (Lancaster, Pa.) **94**, 396 (1941); Amer. J. Bot. **32**, 416 (1945).
Feinstein, L., and E. T. McCabe: Analyt. Chem. **23**, 385 (1951). — Fuks, M. G.: Lab. Prakt. USSR., **16**, No. 9, 23 (1941).
Griffith, R. B., and R. N. Jeffrey: Analyt. Chem. **20**, 307 (1948).
Hills, K. L., W. Bottomley and P. I. Mortimer: Nature (Lond.) **171**, 435 (1953). — Houston, F. G.: Analyt. Chem. **24**, 1831 (1952).
Jeffrey, R. N.: J. Ass. Off. Agric. Chem. Wash. **34**, 843 (1951).
Linstead, R. P., and S. L. S. Thomas: J. Chem. Soc. **1940**, 1127.
Manske, R. H. F., and L. Marion: Canad. J. Res. B., **20**, 87 (1942); **24**, 57 (1946). — Manske, R. H. F., and H. L. Holmes: The Alkaloids, Vol. I. New York: Academic Press Inc. 1950.— Marion, L.: Canad. J. Res. B., **17**, 21 (1939); **23**, 165 (1945). — Marion, L., and R. H. F. Manske: Canad. J. Res. B., **22**, 137 (1944); **24**, 63 (1946). — Markwood, L. N.: J. Ass. Off. Agric. Chem. Wash. **22**, 427 (1939); **26**, 283 (1943). — Markwood, L. N., and W. F. Barthel: J. Ass. Off. Agric. Chem. Wash. **26**, 280 (1943). — Melzer, H.: Z. anal. Chem. **37**, 357 (1898).
Orechoff, A.: Compt. rend. Acad. Sci. Paris **189**, 945 (1929).
Sadykov, A., and N. Spasokukotskii: J. Gen. Chem. USSR. **13**, 830 (1943). — Sanchez, J. A.: Semana Méd. (Buenos Aires) **28**, 61 (1922). — Shmuk, A.: J. App. Chem. USSR. **14**, 864 (1941). — Smith, H. H., and C. R. Smith: J. Agric. Res. **65**, 347 (1942).
Trim, A. R.: Biochem. J. **43**, 57 (1948).
Werle, E., and H. W. Becker: Biochem. Z. **313**, 182 (1942). — Werle, E., and J. Koch: Naturwissenschaften **38**, 333 (1951). — Willits, C. O., M. L. Swain, J. A. Connelly and B. A. Brice: Analyt Chem. **22**, 430 (1950).

Section Y.

Allport, N. L., and E. S. Wilson: Quart. J. Pharm. **12**, 399 (1939). — Ashley, M. G.: J. Pharm. Pharmacol. **4**, 181 (1952).
Beckmann, E.: Arch. Pharm. (iii) **24**, 481 (1886). — Björling, C. O., and A. Berggren: J. Pharm. Pharmacol. **5**, 169 (1953).
Colby, A. B., and J. L. Beal: J. Amer. Pharm. Ass. Sci. Ed. **41**, 351 (1952).
Durick, F., J. S. King jr., P. A. Ware and G. Cronheim: J. Amer. Pharm. Ass. Sci. Ed. **39**, 680 (1950).
Evans, W. C., and M. W. Partridge: Quart. J. Pharm. **21**, 126 (1948).
Hills, K. L., E. M. Trautner and C. N. Rodwell: J. Council Sci. Ind. Res. (Austr.) **18**, 234 (1945). — Hills, K. L., and C. N. Rodwell: Austr. J. Sci. Res. B. **4**, 486 (1951).
van Itallie, L.: Pharm. Weekbl. **75**, 909 (1938).
James, W. O., and M. Roberts: Quart. J. Pharm. **18**, 29 (1945). — Jindra, A., and J. Pohorsky: J. Pharm. Pharmacol **3**, 344 (1951).
Munier, R., and M. Macheboeuf: Bull. Soc. Chim. biol. Paris **33**, 846 (1951). — Munier, R., M. Macheboeuf and N. Cherrier: Bull. Soc. Chim. biol. Paris **33**, 1919 (1951).
Rathenasinkam, E.: Analyst **75**, 169 (1950). — Roberts, M., and W. O. James: Quart. J. Pharm. **20**, 1 (1947).
Trautner, E. M., and M. Roberts: Analyst **73**, 140 (1948).
Young, J. L.: J. Ass. Off. Agric. Chem. Wash. **31**, 781 (1948).

Section Z.

Chemnitius, F.: Chemiker-Ztg. **51**, 469 (1927).
Lynch, G. R., and N. Evers: Analyst **73**, 309 (1948).
Malowan, L. S.: Analyst **75**, 338 (1950). — Munier, R., and M. Macheboeuf: Bull. Soc. Chim. biol. Paris **33**, 846 (1951).
Perrot, E., and R. Hamet: Bull. Sci. pharm. **39**, 593 (1932).

Amine und Betaine.

Von

E. Werle.

Mit 6 Abbildungen.

Die hier zu behandelnden Amine sind organische Basen mit einer oder zwei endständigen, freien oder substituierten Aminogruppen. Es gehören dazu aliphatische, aromatische und einfache heterocyclische Verbindungen, deren pflanzenphysiologische Bedeutung noch so gut wie unbekannt ist (s. GUGGENHEIM, 1951). Lediglich für Putrescin ist nachgewiesen, daß es einen Wuchsstoff für Bakterien [*Haemophilus parainfluencae* (HERBST und SNELL, 1948—49) und *Neisseria perflava* (MARTIN, PELCZAR und HANSEN, 1952) und für eine Mutante des Schimmelpilzes *Aspergillus nidulans* (SNEATH, 1955)] darstellt.

Die Amine sind gasförmige, flüssige oder feste Stoffe. Sie sind meist wasserlöslich und hygroskopisch; in mit Wasser nicht mischbaren organischen Lösungsmitteln sind sie nicht oder doch nur sehr wenig löslich. Sie unterscheiden sich so von den Alkaloiden, die umgekehrt in organischen Lösungsmitteln leicht, in Wasser nur wenig löslich sind. Die höheren Alkylamine (Isobutyl- und Isoamylamin) und die Phenylalkylamine ohne freies phenolisches Hydroxyl werden aus alkalischer Lösung durch Äther und Chloroform, Histamin und o-Oxyphenyläthylamin durch Amylalkohol aufgenommen. Aus stark alkalischer Lösung gehen geringe Mengen Histamin auch in Chloroform über (s. GUGGENHEIM, 1951).

A. Allgemeine Verfahren zur Anreicherung und zum Nachweis der Amine.

1. Anreicherung und Abtrennung.

Amine können in Pflanzen genuin enthalten sein, sie können nach der Probenahme, also während der Vorbereitung zur Analyse, entstehen vermehrt oder abgebaut werden. So entsteht Cholin aus Phosphatiden, z. B. aus Lecithin durch Einwirkung der Lecithinase, sobald Blätter zu welken beginnen.

Freie Amine können im zerkleinerten Pflanzenmaterial durch Aminoxydase abgebaut oder in unsterilem Material aus Aminosäuren durch Decarboxylasen von Bakterien erst gebildet werden. Es kann also die Vorgeschichte des zu analysierenden Materials (z. B. das Trocknungs- und Extraktionsverfahren) für die Analysenergebnisse entscheidend sein, vgl. auch Bd. 1, S. 1, dieses Handbuches). Zur Isolierung und Charakterisierung müssen die Amine von der Masse der Begleitsubstanzen befreit werden, was je nach der Beschaffenheit des Ausgangsmaterials und der Natur der Amine auf verschiedene Weise erfolgen kann. So lassen sich die niedrigen aliphatischen Monoalkylamine aus alkalischer Lösung mit Wasserdampf abdestillieren, wobei zweckmäßig flüchtige Säuren oder neutrale Substanzen durch vorhergehende Destillation bei saurer Reaktion entfernt werden. Aminoäthanol ist auch mit Dämpfen von Äther und Toluol flüchtig.

Zur Isolierung der flüchtigen Amine wird nach STEINER und STEIN v. KAMIENSKI (1953) das zu untersuchende Pflanzenmaterial im Homogenisator, z. B. mit

Wasser unter Zusatz von Zitronensäure (s. S. 537), homogenisiert. Anschließend wird das Homogenat in einem geeigneten Apparat lackmusalkalisch gemacht und dann der Wasserdampfdestillation unterworfen. Die übergegangenen Basen werden in 0,1 n HCl aufgefangen, das Destillat im Vakuum bei 70° eingeengt und im Vakuumexsiccator, der mit festem Ätznatron als Trockenmittel beschickt ist, zur Trockne gebracht. Nach Aufnahme des Rückstandes in wenig Wasser werden die Amine z. B. papierchromatographisch identifiziert (s. S. 537).

Zur Isolierung der nicht flüchtigen Amine müssen makromolekulare Stoffe, insbesondere Eiweiße aus den Pflanzenextrakten, entfernt werden. Eine Enteiweißung ist möglich durch Hitzekoagulation oder Säurehydrolyse, ferner durch Versetzen der Extrakte mit der 4—5fachen Menge Alkohol oder Aceton, wobei die Salze der Amine in Lösung bleiben (Guggenheim, 1951). Ohne Zweifel können auch Verfahren, die sich bei der Enteiweißung von Extrakten aus tierischen Geweben bewährt haben, Anwendung finden, wobei zu berücksichtigen ist, daß der Eiweißgehalt pflanzlicher Extrakte wesentlich niedriger sein kann als der vieler tierischer Extrakte.

Es kann also enteiweißt werden mit 10%iger Trichloressigsäure nach Chang und Gaddum (1939) — Bisset (1954) enteiweißt z. B. wäßrige Extrakte aus Bohnensamen durch Zugabe von 1 g Trichloressigsäure auf 20 ml Extrakt. Nach 45 min langem Stehen kann der Niederschlag abfiltriert werden — oder durch Aufkochen mit 5 Teilen eines Gemisches aus Isopropylalkohol und Aceton 1:1 bei Gegenwart von 0,5 Vol.-% Phosphorwolframsäure und schwach salzsaurer Reaktion (Lesure, 1937). Es kann auch der Pflanzenextrakt zur Sirupdicke eingeengt, mit Sand oder einem anderen feinkörnigen Material verrieben und die feste Masse mit Aceton ausgezogen werden. Dabei darf das Wasser nicht vollständig entfernt werden, weil die Salze der Amine in wasserfreiem Zustand in Alkohol oder Aceton meist nicht löslich sind (Guggenheim, 1951). Nach Eggleton und Eggleton (1932) ist eine schonende Enteiweißung und Konzentrierung in der Weise durchführbar, daß der Gewebsbrei bei 32° mit 40 Gewichtsprozent wasserfreiem Natriumsulfat gemischt und anschließend auf 0° abgekühlt wird. Dabei kristallisiert $Na_2SO_4 \cdot 10\,H_2O$ aus, und es bleibt ein proteinfreier konzentrierter Extrakt zurück.

Durch neutrales Bleiacetat werden neben Eiweiß bei saurer Reaktion Zucker und andere Kohlenhydrate sowie verschiedene Carbonsäuren als schwerlösliche Bleiverbindungen ausgefällt. Durch Zugabe von basischem Bleiacetat wird die Abscheidung vervollständigt. Durch Schwefelwasserstoff, Schwefel- oder Phosphorsäure wird das überschüssige Blei aus dem Filtrat entfernt. Die Enteiweißung kann auch durch 10%ige Natriumwolframatlösung erfolgen (Bisset, 1954). Eine Abtrennung der Amine von hochmolekularen wasserlöslichen Kolloiden ist auch durch Dialyse möglich. Dabei fallen größere Flüssigkeitsmengen an, die eingeengt werden müssen.

Ein Abbau oder eine Bildung von Aminen durch Fermente des Extraktes oder durch Bakterien kann dabei z. B. durch Einstellen von stärker saurer Reaktion verhindert werden. Gegen eine Entwicklung von Bakterien hat sich die Dialyse in Cellophanbeuteln gegen Chloroformwasser bei 0—2° bewährt (Bisset, 1954).

Eine Enteiweißung ist auch möglich mit Hilfe von Tannin bei schwach saurer Reaktion (Kossel, 1910). Eine Enteiweißung mit Hilfe von kolloidalen Lösungen von Eisenhydroxyd oder Aluminiumhydroxyd würde zu größeren Verlusten an Aminen durch meist irreversible Adsorption führen; ebenso können fein verteilte amorphe oder kristalline Niederschläge (Talkum, Kaolin, Tierkohle, Bariumsulfat usw.) Aminverluste durch Adsorption verursachen. Schließlich können makromolekulare Stoffe auch mit Hilfe starker Harzaustauscher entfernt werden

(BOARDMANN und PARTRIDGE, 1953). Durch Elektrodialyse unter Einschaltung einer „semipermeablen Membran" können die Amine aus wäßrigen Extrakten an der Kathode abgeschieden werden (FUELNEGG und KENDALL, 1931). MACGREGOR und THORPE (1933) verwendeten zu diesem Zweck eine 3teilige Apparatur, deren Kammern durch Pergament voneinander getrennt sind. Anode aus Kohle, Kathode aus Nickel in den äußeren Kammern. Durch einen Strom von 0,2 bis 0,5 Ampere wird z. B. Histamin aus einem in die mittlere Kammer gegebenen Gewebsbrei innerhalb 2 Std. in den Kathodenraum übergeführt. Auf Grund verschiedener Wanderungsgeschwindigkeiten der Amine, die durch den Aciditätsgrad der Elektrolytlösung variiert werden können, sind weitere Auftrennungen in die einzelnen Basen möglich (FOSTER und SCHMIDT, 1921; GUGGENHEIM, 1951). Nach Abtrennung makromolekularer Stoffe mit Hilfe eines der genannten Verfahren und nach der Entfernung eventuell vorhandener Polyphenole durch Adsorption an Kohle können anorganische Kationen und organische Basen einschließlich der Aminosäuren von einem sauren Austauscher aufgenommen werden. Es können dann die schwächeren organischen Basen mit 0,15 n Ammoniak und danach die stärkeren Basen mit 0,1 n Natronlauge eluiert werden. Im Filtrat von der sauren Säule befinden sich Säuren und neutrale Stoffe wie z. B. Zucker. Die erhaltenen Gruppen von Aminen können durch Verteilungsanalyse an geeigneten Austauschern weiter aufgetrennt werden (PARTRIDGE, 1952).

In besonderem Maße sind die modernen Verfahren der Chromatographie, nämlich der Verteilungschromatographie auf Papier (s. S. 537), der Säulenchromatographie, der Gas-Flüssigkeits-Chromatographie (JAMES und MARTIN, 1952), der Chromatographie an Ionenaustauschern (PARTRIDGE, 1952), der Papierelektrophorese sowie das Verfahren der Gegenstromverteilung zur Isolierung und Trennung der Amine geeignet (RAUEN, 1953, HECKER, 1955, vgl. auch Bd. 1.) Es sei an dieser Stelle lediglich auf das Verfahren der Gas-Flüssigkeits-Chromatographie etwas näher eingegangen. Es beruht im Prinzip auf den gleichen physikalischen Gegebenheiten wie die Verteilungschromatographie mit Flüssigkeiten. Es wird bei der Gas-Flüssigkeits-Chromatographie das zu trennende Stoffgemisch auf eine Flüssigkeitssäule aufgetragen, die durch eine indifferente Substanz, z. B. Kieselgur, stabilisiert ist. Durch diese Säule wird Stickstoff geleitet, der die flüchtigen Substanzen nach Maßgabe ihres Partialdruckes mitnimmt. Der Partialdruck kann durch Aufheizen der Säule beliebig erhöht werden. Während bei der Verteilungschromatographie in erster Linie der R_f-Wert zur Identifizierung verwendet wird, dient bei der Gas-Flüssigkeits-Chromatographie als Kriterium allgemein das „Retensionsvolumen", nämlich das Produkt, das aus der Zeit t, die verstreicht bis das Zentrum einer Zone aus der Kolonne austritt und dem in dieser Zeit aus der Kolonne ausgetretenen Gasvolumen V, also $V^R = V \cdot t$. Das Retensionsvolumen ermöglicht unter anderem auch eine Unterscheidung von primären, sekundären und tertiären Aminen, die sich auf die Tatsache gründet, daß die primären und die sekundären Amine nur mit polaren, nicht aber mit apolaren stationären Flüssigkeitsphasen Wasserstoffverbindungen eingehen können. Quantitativ machen sich diese Verhältnisse im Retensionsvolumen bemerkbar. Die Gas-Flüssigkeits-Verteilungschromatographie ist nur für die Trennung von Aminen in γ-Mengen geeignet; zur Aufarbeitung in größerem Maßstab würde die Apparatur zu umfangreich (JAMES und MARTIN, 1952).

2. Qualitativer Nachweis der Amine.

Die flüchtigen aliphatischen Amine wie z. B. Trimethylamin besitzen einen ausgeprägten Geruch. Anhaltspunkte über die Art des vorliegenden Materials lassen sich mit folgenden Reaktionen, zu denen allerdings z. T. größere Substanzmengen erforderlich sind, erhalten:

Reaktion mit Nesslers-Reagens: Nach Zusatz von Nesslers Reagens zur wäßrigen Lösung der Amine können folgende Erscheinungen beobachtet werden: Braunfärbung bei Gegenwart größerer, Gelbfärbung bei Gegenwart geringer Ammoniakmengen. Methylamin verursacht eine gelbe oder orangerote in Wasser schwer lösliche Fällung. Dimethylamin ergibt einen rotbraunen Niederschlag. Trimethylamin und Butylamin einen weißen in Wasser ziemlich löslichen Niederschlag.

Nitritreaktion. Primäre Amine bilden mit salpetriger Säure unter Stickstoffentwicklung primäre Alkohole. Die Reaktion kann nach van Slyke quantitativ gestaltet werden. Sekundäre Amine bilden mit salpetriger Säure, gelbe in Wasser schwer lösliche Nitrosamine. Die Nitrosamine der niederen Glieder sind ziemlich beständig und lassen sich mit Wasserdampf destillieren. Aus den Nitrosaminen können durch Erhitzen mit starker Säure die sekundären Amine regeneriert werden. Erwärmt man die Nitrosamine mit Phenol und konzentrierter Schwefelsäure und übersättigt nach Verdünnen mit Wasser die Probe mit Natronlauge, so tritt eine starke Blau- bis violettblaue Färbung auf (Liebermannsche Reaktion).

$$\begin{array}{c} H_3C \\ \\ H_3C \end{array}\!\!\!>\!NH + HONO \qquad\qquad \begin{array}{c} H_3C \\ \\ H_3C \end{array}\!\!\!>\!N-NO + H_2O$$

Nitrosodimethylamin

Zum Nachweis tertiärer Amine, die mit salpetriger Säure in der Kälte nicht reagieren, kann z. B. das Mayersche Reagens (45 g HgJ_2 + 35 g KJ auf 100 ml Wasser) verwendet werden. Trimethylamin wird in 0,01 n Lösung in Form gelber Nadeln der Zusammensetzung $(CH_3)_3N \cdot HJ \cdot HgO_2$ Fp. 136° ausgefällt. Sekundäre Amine werden erst aus 0,4%iger Lösung gefällt. Mit Ammoniak erhält man keinen Niederschlag.

Isonitrilreaktion: Beim Erwärmen der alkoholischen Lösung primärer Amine tritt in Gegenwart von Chloroform und Kalilauge oder in konzentrierter wäßriger Lösung mit Chloroform und Pottasche der sehr intensive, charakteristische Isonitrilgeruch auf. Der Isonitrilbildung liegt folgende Gleichung zugrunde:

$$CH_3NH_2 + CHCl_3 + 3\,KOH = CH_3{-}N{=}C + 3\,KCl + 3\,H_2O.$$

Senfölreaktion: Die primären Amine verbinden sich mit Schwefelkohlenstoff zu alkyldithiocarbaminsauren Salzen (s. S. 532). Versetzt man diese mit Schwermetallsalzen (z. B. $HgCl_2$; $AgNO_3$ u. a.), so entsteht das betreffende Schwermetallsalz der Alkyldithiocarbaminsäure, das beim Kochen in Metallsulfid und Senföl (Isorhodanwasserstoffsäureester) zerfällt. Die Senföle besitzen einen scharfen, an Senf erinnernden Geruch.

Reaktion mit Nitroprussidnatrium: Aliphatische Amine geben mit brenztraubensäurehaltiger Nitroprussidnatriumlösung eine violette, nach Zusatz von Essigsäure nach Blau umschlagende rasch vorübergehende Färbung; primäre Amine geben mit Nitroprussidnatrium und Aceton eine orangerote Färbung. Sekundäre Amine geben mit Natrium und Acetaldehyd eine Blaufärbung, tertiäre Amine rufen keine Färbung hervor.

3. Mikrochemischer Nachweis für primäre, sekundäre und tertiäre Amine nach Feigl und Feigl (1954).

Er gründet sich auf die Schwefelwasserstoffabspaltung beim Erhitzen von geschmolzenem Alkalithiocyanat mit Ammonsalzen oder Salzen (Chloride, Sulfide, Nitride) von primären, sekundären und tertiären, aromatischen und aliphatischen Aminen. Bei der „topochemischen" Reaktion von Aminsalzen wird die Bildung von NH_4SCN, Isothioharnstoff bzw. der entsprechenden substituierten Verbindungen und Zerfall der letzteren in Cyanamid und Schwefelwasserstoff angenommen. Hydroxylamin und Hydrazinsalze reagieren analog, wobei elementarer Schwefel gebildet wird. Die Reaktion wird besonders zum Nachweis aliphatischer und aromatischer Amine empfohlen, wobei freie Basen in die Chloride übergeführt und organische Verbindungen die beim Erhitzen Wasser abspalten, sowie Ammonsalze ausgeschaltet werden müssen.

Ausführung der Reaktion: Ein Tropfen Aminlösung in organischem Lösungsmittel wird mit einem Tropfen HCl bei 110° eingedampft. Der Rückstand wird mit überschüssigem trockenem KSCN gemischt, auf 200—250° erhitzt, und der abgespaltene Schwefelwasserstoff mit feuchtem Bleiacetatpapier nachgewiesen. Es können so z. B. noch 15 γ Äthylendiamin nachgewiesen werden.

B. Verfahren zur Isolierung der Amine.
(Guggenheim 1951.)

Nach der oben beschriebenen Vorreinigung können die Amine aus ihren wäßrigen Lösungen in Form schwerlöslicher Salze oder Doppelsalze ausgefällt werden.

1. Fällung mit Phosphorwolframsäure.

Mit Phosphorwolframsäure bilden die Amine schwer lösliche Verbindungen der allgemeinen Formel $Am_3PO_4 \cdot 12\,WO_3 \times 3\,H_2O$ (Drummond, 1918), wobei Am ein einsäuriges Amin bedeutet. Die Phosphorwolframsäure ist leicht löslich in Äther, Aceton und Wasser. Die Fällbarkeit der Amine ist p_H-abhängig (Peters, 1933). Die Ausfällung der Amine erfolgt am besten in Gegenwart von 5%iger Schwefelsäure. Die Phosphorwolframsäure wird bis zu einem geringen Überschuß in 5%iger Schwefelsäurelösung zugefügt. Ein größerer Überschuß ist zu vermeiden, da er auf den Phosphorwolframatniederschlag wieder lösend wirkt. Ammonium- und Kaliumsalze werden zweckmäßig vor der Ausfällung der Amine entfernt, da sie ebenfalls mit Phosphorwolframsäure ausfallen. Ammoniak wird mit den flüchtigen Alkylaminen aus schwach alkalischer Lösung abdestilliert. Das Kalium kann zum größten Teil dadurch entfernt werden, daß die Aminlösungen vor dem Zusatz der Phosphorwolframsäure weitgehend konzentriert und dann mit Alkohol oder Aceton extrahiert werden, wobei die Kaliumsalze größtenteils ungelöst zurückbleiben. Der Phosphorwolframatniederschlag wird nach 24 Std. abfiltriert und nach dem Auswaschen mit 5%iger Schwefelsäure durch gelöstes oder fein gepulvertes Barythydrat zerlegt. Es bildet sich unlösliches Bariumwolframat, während die Basen mit überschüssigem Baryt in die wäßrige Lösung gehen. Zur Erleichterung der Zerlegung kann man vor dem Zusatz des Baryts die Phosphorwolframate mit Aceton oder Aceton-Wasser behandeln, wobei 3 Teile Aceton mit 4 Teilen Wasser gemischt werden. Die meisten Phosphorwolframate sind in Aceton leicht löslich und setzen sich dann rascher und vollständiger um als beim Zerreiben mit Baryt. Gelegentlich kann die Zerlegung auch in saurer Lösung erfolgen, wobei die Phosphorwolframate mit Salz- oder Schwefelsäure versetzt werden und die freie Phosphorwolframsäure mit Äther oder einem Amylalkohol-Äther-Gemisch ausgeschüttelt wird, in welchem sie leicht löslich ist. Das Gemisch der mit Phosphorwolframsäure gefällten Amine kann durch das sog. Silberbarytverfahren in 3 Fraktionen aufgeteilt werden. 1. Zugabe von Silbernitrat oder Silbersulfat zur schwach salpeter- oder schwefelsauren Lösung fällt Purinbasen und Chlorionen, wenn sie in der Lösung vorhanden sind. 2. Aus dem mit Baryt oder Bariumhydrat schwach alkalisch gemachten Filtrat fällt nach Zusatz von Silbersalz und weiterer Zugabe von Alkali die sog. Histidinfraktion aus, die auch andere Imidazolbasen, z. B. Histamin, enthält. 3. Bei stark barytalkalischer Reaktion fällt Silber neben Arginin und anderen Guanidinbasen noch Imidazolbasen, die in der 2. Silberfällung nicht abgeschieden worden waren. Im Filtrat der 3. Silberfällung finden sich neben Lysin die Diamine Putrescin und Cadaverin. Ferner Betaine und ω-Aminosäuren, Cholin und andere quaternäre Ammoniumbasen.

2. Gewinnung der Chloroplatinate und Goldchloride der Amine.
(Guggenheim 1951).

Chloroplatinate: Zur Fällung der Basen wird Platinchlorid in wäßriger oder alkoholischer Lösung angewandt. Die Chloroplatinate der Amine entsprechen der allgemeinen Formel $Am_2 \cdot H_2PtCl_6$, wobei Am ein einwertiges Amin bedeutet. Die Chloroplatinate lösen sich leicht in Wasser, weniger leicht in Alkohol und können deshalb aus verdünnter alkoholischer Lösung umkristallisiert werden.

Chloroaurate: Die Chloroaurate entsprechen der allgemeinen Formel $Am \cdot HAuCl_4$. Zur Vermeidung einer Zersetzung und Abscheidung von metallischem Gold erfolgt die Fällung aus wäßriger Lösung in Gegenwart von 0,5 bis 1%iger Salzsäure und etwas überschüssigem Aurichlorid.

Spezielle Angaben über Chloroaurate und -Platinate finden sich bei den einzelnen Aminen.

3. Fällung mit Quecksilbersalzen.

Nach Engeland (1908) können basische Amine gefällt werden durch Zusatz von Sublimatlösung und Natriumacetat. Man fügt abwechselnd die gesättigten wäßrigen Lösungen der beiden Reagentien so lange zu, bis die Zugabe eines großen Überschusses kalt gesättigter wäßriger Quecksilberchlorid- und Natriumacetatlösung zu einer abfiltrierten Probe sich nach längerem Stehen nicht mehr trübt. Der körnige Niederschlag wird mit Sublimat-Natriumacetatlösung gewaschen, in heißer verdünnter Salzsäure gelöst und nach dem Abfiltrieren des Quecksilbersalzes und nach dem Einengen durch Aufnahme in Methanol von anorganischen Salzen befreit. Nach Verdampfen des Methylalkohols enthält der Rückstand neben Ammonchlorid fast die gesamten biogenen Amine, die durch Ausfällen mit alkoholischer Sublimat- und Cadmiumchloridlösung und durch Herstellung der Platin- und Goldchloriddoppelsalze weiter aufgetrennt werden können.

4. Fällung mit Kaliumwismutjodid.
(Dragendorff- oder Krautsches Reagens.)

Dieses Reagens, welches vornehmlich zur Isolierung von Cholin und Betain aus Pflanzenextrakten verwendet wird, eignet sich ganz allgemein zur Ausfällung der quaternären Ammoniumbasen, die nach dem Phosphorwolframsäureverfahren in die Lysinfraktion gehen, mit der es ziegelrote, meist amorphe Niederschläge bildet. Zur Herstellung des Reagens werden 80 g basisches Wismutnitrat in 200 ml Salpetersäure (D. 1,18) gelöst und langsam unter Umschütteln eine konzentrierte wäßrige Lösung von 277 g Jodkalium zugegeben. Der anfangs entstehende braune Niederschlag löst sich zu einer goldroten Flüssigkeit. Nach starker Abkühlung kristallisiert Salpeter aus, der abfiltriert wird, die verbleibende Flüssigkeit wird 1:1 verdünnt. Andere Vorschriften der Darstellung des Reagens wurden von Kossel und Weiss (1910) sowie von Neuberg (1910) angegeben. Die Wismutjodidverbindungen der allgemeinen Formel $AmHJ_2 \cdot BiJ_3$ (Neuberg, 1910) werden durch frisch gefälltes Bleihydroxyd zerlegt. Bleioxyjodid wird abfiltriert. Überschüssiges Blei wird mit Schwefelwasserstoff entfernt, dann werden die Basen in alkoholische Lösung übergeführt und mit Sublimat fraktioniert.

5. Fällung mit Reineckesalz.

Reineckesalz ist das Ammoniumsalz der Chromtetrarhodanwasserstoffsäure. $NH_4Cr(NH_3)_2 \cdot (SCN)_4 \cdot H_2O$. Darstellung des Reineckesalzes s. GUGGENHEIM (1951). Nach DUQUENOIS (1939) verwendet man zur Fällung der Amine am besten eine Reineckesalzlösung, die etwa 4%ig ist. Sie wird in der Weise gewonnen, daß man Reineckesalz in der Wärme löst und in der Kälte den wieder ausgefallenen Anteil abfiltriert. Es wird von dem Reagens so lange zugefügt, bis die überstehende Lösung schwach rosa gefärbt ist. Die Fällung ist bei allen untersuchten Basen bei 0° vollständiger als bei 15°; häufig ist sie vollständiger in schwach saurem Medium als in neutralem. (Zugabe von 1 Tropfen verdünnter Salzsäure.)

Die Löslichkeit der Reineckesalze wird auch durch die Zugabe von Kochsalz vermindert (HEIN und SEGITZ, 1927). In bezug auf ihre Fällbarkeit durch Reineckesäure kann man nach DUQUENOIS und FALLER (1939) 3 Gruppen von organischen Substanzen unterscheiden: Stoffe, die sofort, Stoffe, die unvollständig und erst nach einer gewissen Zeit, und Stoffe, die nicht gefällt werden. Die quaternären Ammoniumsalze, z. B. Cholin, bilden die unlöslichsten Reineckate, sie werden augenblicklich niedergeschlagen. Auch die tertiären Amine, z. B. Trimethylamin, werden sofort gefällt.

Die meisten sekundären Amine wie Dimethylamin werden langsam, die primären, aliphatischen Amine sehr langsam niedergeschlagen (Äthylamin, Propylamin, Allylamin). Trotzdem erhält man z. B. mit Cadaverin eine vollständige Fällung nach 24 Std. Die Aminoalkohole, deren Aminfunktion sekundär oder tertiär ist, geben eine Reaktion in neutralem oder schwach saurem Medium, z. B. Diäthylaminoäthanol. Alle untersuchten Alkaloide (40 verschiedene) wurden sofort niedergeschlagen. Proteide trüben sich sofort und fallen langsam mit Reineckesäure aus. Alle Reineckate haben rosarote bis weinrote Farbe. Sie zersetzen sich vor dem Schmelzen. Sie haben die allgemeine Formel $Cr(NH_3)_2 \cdot (SCN)_4 \cdot H \cdot Am$, wobei Am eine einsäurige Base bedeutet. Zur Analyse der erhaltenen, meist wasserfreien Salze wird deren Chromgehalt bestimmt. Zur Wiedergewinnung der Basen werden die Reineckate in warmem Wasser unter Zugabe von Aceton gelöst. Man zerlegt mit Silbersulfat, entfernt das Silber mit Schwefelwasserstoff und das Sulfat mit Baryt. (Alle Reineckate sind in kaltem Aceton löslich.)

6. Fällung mit Flaviansäure.

Neben anderen Basen werden vor allem Diamin-, Imidazol- und Guanidinbasen mit 1-Naphthol-2,4-dinitro-7-sulfosäure, $C_{10}H_6O_8N_2S$ der Flaviansäure, in schwerlösliche Salze übergeführt, die sich zur Charakterisierung verschiedener Amine eignen (KOSSEL und GROSS, 1924). Zur Fällung der Amine mit Flaviansäure (MÜLLER und SIEVERS, 1929) werden deren alkoholische Lösungen mit einer konzentrierten alkoholischen Lösung von Flaviansäure versetzt, wobei meist sofort ein reichlicher Niederschlag auftritt, der abgenutscht und mit Alkohol gewaschen wird. Die Flavianate haben keine Schmelzpunkte, sie zersetzen sich bei höherer Temperatur. Ihre Löslichkeit nimmt mit steigender Methylierung der Amine zu, während ihr Zersetzungspunkt fällt (s. Tab. 1).

Die Verwendung von Flaviansäure als Fällungsmittel hat den großen Vorteil, daß viele Flavianate aus gewöhnlichen Lösungsmitteln auskristallisieren. Die Flaviansäure kann von den Niederschlägen der Basen auf dreierlei Weise getrennt werden. Fällung der Flaviansäure als Bariumsalz und Adsorption der geringen

Tabelle 1. *Flaviansäurederivate einiger Amine.*

Flavianat von	Verhältnis Base:Säure	N-Gehalt in %	g löslich bei 18° in		Zersetzung
			100 g Alkohol	100 g Wasser	
Methylamin . . .	1:1	12,17	0,28	3,42	265—268°
Dimethylamin . .	1:1	11,70	0,51	7,54	230—235°
Trimethylamin . .	1:1	11,26	0,85	12,14	217—223°
Cholin	1:1	10,17	0,50	30,36	—

in der Lösung zurückbleibenden Menge an Tierkohle (MÜLLER und SIEVERS, 1929).
Ansäuern der Lösung und Extraktion der Flaviansäure durch organische Lösungs-
mittel (PRATT, 1926) und Adsorption der Flaviansäure an Wolle (s. S. 527).
Zur Identifizierung und quantitativen Bestimmung von Flavianaten in geringen
Mengen (einige mg) haben LANGLEY und ALBRECHT (1935) eine kolorimetrische
Bestimmungsmethode der Flaviansäure, die in wenigen Minuten mit 1 mg Flavia-
nat durchführbar ist, angegeben. Sie gründet sich auf folgende Tatsachen: Wird
Flaviansäure durch ein Metall in saurer Lösung in Abwesenheit von Sauerstoff
reduziert, so wird sie entfärbt. Wird durch die farblose Lösung Luft geleitet, so
wird sie orangerot, wobei wahrscheinlich ein Chinon gebildet wird. Die Farbintensität ist proportional der Menge ursprünglich vorhandener Flaviansäure.
Für die Bestimmungen muß reine umkristallisierte Flaviansäure verwendet

Tabelle 2. *Flaviansäure-Derivate der Amine* (Aus LANGLEY u. ALBRECHT 1955).

Flavianat von	Schmelzpunkt	Flavian-säure-gehalt	Kristallform	Lösungsmittel
Ammonium	291—292	94,8	Nadeln	H_2O
Flaviansäure	150	100,0	Nadeln	wäßrige Säure
Trimethylamin-oxyd	218	80,7	Prismen	H_2O
Putrescin (mono-)	285	74,7	Prismen	H_2O
Acetylcholin	222,5—225	68,2	Prismen	Äthanol
Betain	242—243	72,8	Prismen	H_2O oder Äthanol
Trigonellin	236,8	69,5	Nadeln	Äthanol 60%
Tyramin	235	67,4	Nadeln	Butanol
Monomethylamin	262,5	91,0	Prismen	Äthanol
Äthanolamin	211—212	83,7	Prismen	feuchtes Butanol
Histamin (mono-)	260	73,9	Nadeln	H_2O
Nikotinsäure	250—251	71,8	Prismen	H_2O
Trimethylamin	239	84,2	Platten	Äthanol
Guanidin (mono-)	279,5—280	84,2	Prismen	H_2O
Cholin	162	75,1	Prismen	feuchtes Butanol
Methylguanidin (mono-)	230	81,1	Platten	Butanol oder H_2O
Kalium	best. 300	88,9	Prismen	H_2O
Dimethylamin (mono-)	234—236	87,4	Platten Tabletten	Äthanol

werden. Eine mikroskopische Prüfung von Flavianaten mit polarisiertem Licht ergibt Daten, die eine optische Identifizierung erlauben (LANGLEY und AL-BRECHT 1955). In der Tab. 2 sind einige wichtige Charakteristika einer Reihe von Aminen zusammengestellt.

7. Fällung mit Pikrinsäure.

Amine werden aus neutraler wäßriger Lösung durch Zugabe einer alkoholischen oder wäßrigen Lösung von Pikrinsäure als Pikrate ausgefällt. Eventuell vorhandene Ammoniumsalze, die schwer lösliches Ammoniumpikrat geben, sollten vorher entfernt werden. Die Pikrate lassen sich aus Wasser, wäßrigem oder reinem Alkohol umkristallisieren. Die Pikrate werden durch Mineralsäure zerlegt, wobei die Amine als Hydrochloride oder Sulfate in Lösung gehen, und die Pikrinsäure z. T. sich kristallin abscheidet. Der Rest der Pikrinsäure wird der Lösung mit Äther oder Benzol entzogen. Die Pikrate können mit Hilfe von Nitron 1,4-Diphenyl-end-anilo-dihydrotriazol $C_{20}H_{16}N_4$, welches die Pikrinsäure aus wäßriger Lösung ausfällt und die Basen unverändert zurückläßt, quantitativ bestimmt werden.

Die Umsetzung von Aminen mit 2,4,5-Trinitrotoluol in 10%iger alkoholischer Lösung (BARGER und TUTIN, 1918) ist von Vorteil, sofern sie auf die niederen Glieder der aliphatischen Aminreihe beschränkt bleibt; Verwendung des Reagens gegenüber höheren Homologen der aliphatischen primären Amine führt nämlich zur Bildung öliger Produkte.

8. Fällung mit 3,5-Dinitrobenzoesäure.

Eine Charakterisierung von Aminen ist nach BUCHLER und CALFEE (1934) auch mit 2,4- und 3,5-Dinitrobenzoesäure möglich. Zum Nachweis wird die alkoholische Lösung von je 0,01 Mol der Aminlösung mit der alkoholischen Lösung der Dinitrobenzoesäure zusammen gegeben und der Kristallisation überlassen. (Bisher noch keine biogenen Amine untersucht.)

9. Fällung von Aminen mit Nitranilsäure nach MÜLLER (1941).

Die Nitranilsäure (3,6-Dioxy-2,5-dinitro-benzochinon(1,4)) bildet mit den folgenden Verbindungen in wäßrigen Lösungen schwer lösliche Salze: Agmatin, β-Alanin, Anserin, Arcain, Arginin, Cadaverin, Canavanin, Carnosin, Glycocyamin, Glykokoll, Guanidin, Harnstoff, Histamin, Histidin, Imidazol, Kreatinin, Lysin, Methylamin, Methylguanidin, Methylhistidin, Nicotinsäure, Nicotinsäureamid, Putrescin, Trigonellin und Tyramin. Geringe Fällung wird erst nach längerer Einwirkung der Säure, erhalten mit: Trimethylamin, Glykokoll-Betain, Stachydrin, Colamin, Asparaginsäure, Glutaminsäure, Ornithin, Kreatin, Methylhydantoin, Dijodtyrosin und Dibromtyrosin.

Keine Fällung geben: α-Alanin, Leucin, Valin, Prolin, Oxyprolin, Cystein, Tryptophan, Tyrosin, Cholin, Trimethylaminoxyd, Sarcosin und Glycosamin.

Wegen ihrer guten Löslichkeit in Wasser und Alkohol ist die Nitranilsäure der Pikrinsäure als Fällungsmittel für stickstoffhaltige organische Basen überlegen, ebenso wegen der Schwerlöslichkeit ihrer Salze in den genannten Solventien.

Die Zersetzung der Nitranilate zur freien Base bzw. zur Gewinnung von Chloriden, Chlorauraten oder Chloroplatinaten gelingt durch Schütteln mit Baryt oder Essigester aus schwefelsaurer Lösung. Im folgenden sind die Löslichkeitsverhältnisse der Nitranilide angegeben.

Agmatin: In Wasser bei 22° löslich. 1:1538. In Alkohol unlöslich.
$$C_5H_{14}N_4 \cdot C_5H_2O_8N_2$$

β-Alanin: In Wasser bei 20° löslich 1:284. In Alkohol fast unlöslich.
$$(C_3H_7O_2N) \cdot C_6H_2O_8N_2$$

Anserin: In Wasser bei 24° löslich 1:225. In Alkohol unlöslich.
$$C_{10}H_{16}O_2N_4 \cdot C_6H_2O_8N_2$$

Arcain: In Wasser bei 22° löslich 1:10000. In Alkohol unlöslich.
$$C_6H_{16}N_6 \cdot C_6H_2O_8N_2$$

Arginin: In Wasser bei 19° löslich 1:227. In Alkohol unlöslich.
$$C_6H_{14}O_2N_4 \cdot C_6H_2O_8N_2$$

Lysin: In Wasser bei 22° löslich 1:65,5. In Alkohol in Spuren löslich.
$$C_6H_{14}O_2N_2 \cdot C_6H_2O_8N_2$$

Methylamin: In Wasser bei 21° löslich 1:241. In Alkohol unlöslich.
$$(CH_3N)_2 \cdot C_6H_2O_8N_2$$

Methylguanidin: In Wasser bei 21° löslich 1:132. In Alkohol mit ganz hellgelber Farbe wenig löslich. Das Pikrat löst sich in Wasser bei 21° 1:385.
$$(C_2H_7N_3)_2 \cdot C_6H_2O_8N_2$$

Methylhistidin: In Wasser bei 20° löslich 1:396. In Alkohol unlöslich.
$$C_7H_{11}O_2N_3 \cdot C_6H_2O_8N_2$$

Nicotinsäure: In Wasser bei 21° löslich 1:294. In Alkohol mit schwach hellgelber Farbe löslich.
$$(C_6H_5O_2N)_2 \cdot C_6H_2O_8N_2$$

Nicotinsäureamid: In Wasser bei 21° löslich 1:294. In Alkohol in Spuren löslich.
$$(C_6H_6ON_2)_2 \cdot C_6H_2O_8N_2$$

Putrescin: In Wasser bei 21° löslich 1:2353. In Alkohol unlöslich. Das Pikrat löst sich bei 16° 1:2222.
$$C_4H_{12}N_2 \cdot C_6H_2O_8N_2$$

Trigonellin: In Wasser bei 19° löslich 1:44. In Alkohol mit schwach gelber Farbe löslich.
$$(C_7H_7O_2N)_2 \cdot C_6H_2O_8N_2$$

Tyramin: In Wasser bei 20° löslich 1:667. In Alkohol fast unlöslich. Das Pikrat ist bei 24° in Wasser 1:322 löslich.
$$(C_8H_{11}ON)_2 \cdot C_6H_2O_8N_2$$

Cadaverin: In Wasser bei 20° löslich 1:3636. In Alkohol unlöslich. Das Pikrat dagegen in Wasser bei 21° schon löslich 1:455.
$$C_5H_{14}N_2 \cdot C_6H_2O_8N_2$$

Canalin: In Wasser bei 20° löslich 1:256. In Alkohol nur in Spuren löslich.
$$C_4H_{10}O_3N_2 \cdot C_6H_2O_8N_2$$

Canavanin: In Wasser bei 20° löslich 1:82. In Alkohol in Spuren löslich.
$$C_5H_{12}O_3N_4 \cdot C_6H_2O_8N_2$$

Carnosin: In Wasser bei 21° löslich 1:37. In Alkohol unlöslich.
$$C_9H_{14}O_3N_4 \cdot C_6H_2O_8N_2$$

Guanidin: In Wasser bei 21° löslich 1:2500. In Alkohol unlöslich. Das Pikrat in Wasser bei 21° löslich 1:1290.
$$(CH_5N_3)_2 \cdot C_6H_2O_8N_2$$

Harnstoff: In Wasser bei 21° löslich 1:625. In Alkohol löslich 1:1482.
$$(CH_4ON_2)_2 \cdot C_6H_2O_8N_2$$

Histamin: In Wasser bei 20° löslich 1:2660. In Alkohol unlöslich.
$$C_5H_9N_3 \cdot C_6H_2O_8N_2$$

Histidin: In Wasser bei 20° löslich 1:533. In Alkohol in Spuren löslich.
$$C_6H_9O_2N_3 \cdot C_6H_2O_8N_2$$

Imidazol: In Wasser bei 20° löslich 1:889. In Alkohol unlöslich.
$$(C_3H_4N_2)_2 \cdot C_6H_2O_8N_2$$

10. Fällung mit Dinitro-α-naphthol und Rufiansäure.

Amine können auch in Form ihrer Dinitro-α-naphthole charakterisiert werden (s. S. 534). Durch Rufiansäure, 1,4-Dioxyanthrachinon-2-sulfosäure, Chinizarinsulfosäure, werden Amine in alkoholischer Lösung gefällt. Die Rufianate können in einigen Fällen unzersetzt aus Wasser umkristallisiert werden. Am besten wird eine 4%ige Lösung der auf dem Wasser getrockneten Rufiansäure in abs. Äthylalkohol verwendet.

11. Zerlegung der Flavianate, Pikrate, Pikrolonate und Rufianate.

Flavianate, Pikrate, Pikrolonate und Rufianate können nach Müller (1932) zerlegt werden durch Adsorption der freigesetzten Säuren aus wäßriger Lösung oder Suspension an Wolle, wobei die basische Komponente unter sehr milden Bedingungen, z. B. auch in neutraler Lösung, gewonnen wird. Es wird Lammwolle verwendet, die durch wiederholtes Auskochen mit Wasser von löslichen Verunreinigungen befreit wurde. Die Methode ist zur Aufarbeitung kleiner Mengen (weniger als 10 mg) sehr geeignet. Größere Niederschlagsmengen benötigen verhältnismäßig viel Wolle und man zersetzt sie zweckmäßiger mit Baryt oder basischem Kupfercarbonat, wenn die zu isolierenden Amine diese weniger schonende Aufarbeitung vertragen. Durch 1 g Wolle werden aus überschüssiger wäßriger Lösung eines Basenfällungsmittels bei 48stündigem Stehen bei 37° 0,196 g Rufiansäure, 0,190 g Flaviansäure, 0,140 g Pikrinsäure und 0,180 g Pikrolonsäure aufgenommen.

Beispiel: 0,238 g Betainrufianat werden in 6,5 ml 0,1 n HCl unter Erwärmung gelöst. In diese Lösung wird 1 g Wolle gegeben und über Nacht belassen. Die fast farblos gewordene Lösung wird mit Wasser gut ausgewaschen und das Filtrat zur Trockne verdampft. Betainausbeute: 94% der Theorie.

12. Charakterisierung aliphatischer Amine als Thioharnstoffderivate nach Brown und Campbell (1937).

Umsetzung von Aminen mit Phenyl-isothiocyanat oder besser mit 4-Diphenyl-isothiocyanat oder β-Naphthyl-isothiocyanat zu den entsprechenden Thioharnstoffen führt in glatter Reaktion zu gut kristallisierenden Produkten.

Tabelle 3. *Summenformeln und Schmelzpunkte der Reaktionsprodukte einiger Amine mit 4-Diphenyl- und β-Naphthylisothiocyanat.*

Amine	Formel	Schmelzpunkt	Formel	Schmelzpunkt
Methyl-	$C_{14}H_{14}N_2S$	142°	$C_{12}H_{12}N_2S$	127°
Äthyl-	$C_{15}H_{16}N_2S$	165°	$C_{13}H_{14}H_2S$	142°
n-Propyl-	$C_{16}H_{18}N_2S$	156°	$C_{14}H_{16}N_2S$	114°
n-Butyl-	$C_{17}H_{20}N_2S$	155°	$C_{15}H_{18}N_2S$	119°
iso-Butyl-		157°		137°
n-Amyl-	$C_{18}H_{22}N_2S$	147°	$C_{16}H_{20}N_2S$	114°
iso-Amyl-		130°		116°
n-Heptyl-	$C_{20}H_{26}N_2S$	149°	$C_{18}H_{24}N_2S$	115°
Dimethyl-	$C_{15}H_{16}N_2S$	225°	$C_{13}H_{14}N_2S$	173°

13. 2-Nitroindandion-(1,3) als Basenfällungsmittel.
(Müller 1941.)

Nach Wanag u. Lode (1937) bildet das durch Nitrierung von α, γ-Diketohydrinden oder Indandion- (1,3) gewonnene 2-Nitro-indandion- (1,3) mit Basen Salze, die sich durch sehr große Kristallisationsfähigkeit und zum Teil auch durch Schwerlöslichkeit auszeichnen Nach Müller (1941) sind Betain, Cholin, Colamin, Lysin, Harnstoff auf diesem Wege nicht niederzuschlagen. β-Alanin, Cadaverin, Putrescin, Histamin, Histidin fallen sofort, während Arginin, Carnosin, Kreatinin, Tyramin und Glykokoll gleichfalls gut ausgebildete Kristalle entstehen lassen, allerdings zögernd, so daß manchmal erst nach Stehen über Nacht die Fällung deutlich ist. Die Löslichkeiten dieser Salze sind fast alle größer als die der entsprechenden von Müller (1941) beschriebenen Verbindungen mit Nitranilsäure. Demgegenüber liegt ein großer Vorzug des

Nitroindandions und seiner Salze darin, daß die Zersetzlichkeit geringer ist, als im Falle der Nitranilsäure.

Das Fällungsmittel ist leicht nach der Vorschrift WANAGs (1936) aus Indandion darzustellen, das seinerseits im Handel ist. Als Vorteil ist noch anzusehen, daß das Nitroindandion nicht nur in Wasser, sondern auch in Alkohol ziemlich leicht löslich ist.

Salze des 2-Nitro-indandions-(1,3).

Cadaverin: $C_5H_{14}N_2 \cdot C_9H_5O_4N$, löslich in Wasser 1:606 bei 20,5°. In Alkohol unlöslich

Histamin: $C_5H_9N_3 \cdot C_{18}H_{10}O_8N_2$. In Wasser 1:714 löslich bei 20,5°.

Methylguanidin: $C_2H_7N_3 \cdot C_9H_5O_4N$. Löslich in Wasser 1:157 bei 20,5°. In siedendem Alkohol etwas löslich.

Tyramin: $C_8H_{11}ON \cdot C_9H_5O_4N$. Löslich in Wasser 1:175 bei 20,5°. In Alkohol geringe Löslichkeit in der Hitze. Die Millonsche Probe bleibt positiv. Die Zersetzungsprodukte dieser Salze sind so unscharf, daß auf deren Wiedergabe verzichtet wird.

Die 2-Nitroindandione von Methyl-, Äthyl-, Propyl-, Butyl- und Allylamin sind zu leicht löslich, als daß diese Verbindungen zu ihrer Isolierung oder Charakterisierung herangezogen werden könnten (WANAG u. DOMBROWSKI, 1942).

14. Farbreaktion mit Bindon nach WANAG (1940).

Bindon (Anhydro-bis-indandion) gibt nach WANAG (1940, 1938) in Eisessig eine Farbreaktion (blau bis grün bei aromatischen, violett bei aliphatischen Basen) Diese Farbreaktion wird auch mit primären aliphatischen und aromatischen Di- und Polyaminen erhalten. Aus Tab. 4 ist die Empfindlichkeit der Farbreaktion mit einigen Aminen zu ersehen. Sie dürfte in den meisten Fällen zu gering sein für den Nachweis von Aminen aus pflanzlichem Material.

Tabelle 4. *Empfindlichkeit der Farbreaktion einiger Amine mit Bindon.*

Base	Verdünnung, bei der die violette Farbe noch zu bemerken ist
Ammoniak	n/20
Methylamin	n/1500
Äthylamin	n/2000
Propylamin	n/1000
Isopropylamin	n/800
n-Butylamin	n/1000
Isobutylamin	n/800
n-Amylamin	n/1000
Isoamylamin	n/1000
Isohexylamin	n/1200
n-Heptylamin	n/1000
n-Heptadecylamin	n/600
Allylamin	n/2000
Benzylamin	n/1600
α-Phenyl-Äthylamin	n/800
β-Phenyl-Äthylamin	n/1000
Benzhydrylamin	n/1500
Camphylamin	n/1200
Bornylamin	n/2000

15. Charakterisierung von Aminen als „Diliturate".

(REDEMANN und NIERMANN 1940.)

Dilitursäure, 5-Nitrobarbitursäure, wird durch die direkte Nitrierung von Barbitursäure hergestellt. Die Löslichkeit der Säure ist (bei 25°) in Methanol 99, in 95%igem Äthanol 85, in Wasser 63, in absolutem Äthanol 35, in Aceton 25, in Äthyläther 0,9 und in Benzol 0,4 Millimol/Ltr. Die Säurestärke liegt zwischen der der Pikrinsäure und der Salzsäure. In Tab. 5 sind die mit 5-Nitrobarbitursäure gebildeten Salze einiger Basen angegeben. Die 5-Nitrobarbiturate der primären aliphatischen Amine sind überraschend unlöslich in Wasser bei 25°; innerhalb bestimmter Grenzen nimmt ihre Löslichkeit in Wasser mit zunehmender Kettenlänge zu. Der Übergang eines primären Amins in ein sekundäres Amin vermindert die Löslichkeit des Diliturates. Dagegen sind die Salze der tertiären

Amine stärker löslich als die der entsprechenden sekundären Amine. Die heterocyclischen Basen, z. B. Imidazol und Pyridin, bilden wenig lösliche Nitrobarbiturate. Die Beobachtung, daß Äthanolamindiliturat nur eine geringe Löslichkeit in Wasser besitzt, bietet eine Möglichkeit der Isolierung und quantitativen Bestimmung von Äthanolamin und anderen Aminoalkoholen. Im Gegensatz zu Dilitursäure ist ihr Trimethylaminsalz in Wasser leicht löslich. Die Amine können leicht regeneriert werden, indem diese entweder durch Äthylendiamin, Magnesium oder Ammonium, deren Dilitursäuren-Salze außerordentlich unlöslich in Wasser sind, verdrängt werden. Dilitursäure besitzt eine Reihe von Eigenschaften eines idealen sauren Fällungsmittels. Sie ist billig, in wäßriger Lösung verhält sie sich wie eine starke einbasische Säure, wodurch die Möglichkeit der Bildung gemischter Salze stark vermindert ist. Sie ist mäßig leicht löslich in Wasser und Alkoholen. Mehrere sehr lösliche Salze sind bekannt.

Alle Diliturate sind wohl kristallisiert, was ihre Isolierung erleichtert. Ihre Salze sind durch eine große Verschiedenheit ihrer Löslichkeit gekennzeichnet, wodurch eine Fraktionierung der Mischung ermöglicht wird. Weiterhin besitzen die Säuren und die Salze einen hohen Temperaturquotienten der Löslichkeit, wodurch Umkristallisierungen erleichtert werden, und endlich kann in den meisten Fällen die Base frei erhalten werden durch die erwähnte doppelte Umsetzung. Dilitursäure wird gewonnen nach Organic Synthesis 12, 58 (1932).

Gewinnung von Dilituraten. 5 Millimol/l Dilitursäure und 5 Millimol/l der Base (2,5 Millimol/l einer 2säurigen Base) werden in möglichst wenig kochendem Wasser aufgenommen. Dann wird auf 25° abgekühlt.

Tabelle 5. *Diliturate von Aminen.*

Amin	Mol.-Gew.	Löslich in Wasser mM/Liter bei 25°	Aussehen
Äthylamin	218,1	13	Nadeln
Dimethylamin	218,1	15	Puder
Phenylisopropylamin	326,2	15	Nadeln
n-Butylamin	246,1	17	Schuppen
n-Amylamin	260,2	24	Schuppen
Methylamin	204,1	37	Puder
Imidazol	242,1	36	Prismen
d,l-Ephedrin	338,2	87	Platten
β-Oxyphenyl-Äthylamin	319,2	13	Flache Nadeln
l-Ephedrin	338,2	30	Platten
Äthanolamin	234,1	53	Tafeln
Betain	290,1	34	Nadeln
Guanidin	232,1	7,0	Nadeln
Tyramin	310,2	8,7	Platten
Histamin	284,1	17	Nadeln
Phenyläthylamin	303,2	19	Nadeln
Kreatinin	286,1	23	Puder
Harnstoff	233,1	36	Puder

16. Charakterisierung primärer, aliphatischer Amine mit o-Acetoacetylphenol.

(BAKER, HARBORNE und OLLIS 1952.)

Primäre, aliphatische Amine setzen sich mit o-Acetoacetylphenol zu o-β-Alkylaminocrotonylphenolen um, Substanzen, die im UV-Licht eine charakteristische grünlich-gelbe Fluoreszenz zeigen. Die Amine werden dabei als solche oder in Äthanol gelöst zu einer äthylalkoholischen Lösung von o-Acetoacetylphenol bei

Zimmertemperatur gegeben. Der Beginn der Reaktion zeigt sich sofort in einer Verfärbung der Lösungen nach grün-gelb, wobei im UV-Licht starke Fluoreszenz beobachtet wird. Die Kristallisation der o-β-Alkylaminocrotonyl-phenole beginnt nach wenigen Stunden; nach 2tägigem Stehen werden die Kristalle gesammelt und aus Äthanol umkristallisiert. Blaßgelbe oder grüngelbe Prismen oder Nadeln von intensiv grüngelber Fluoreszenz im UV-Licht. Ausbeuten mit Methylamin 60%, Äthylamin 83%.

Viele primäre Amine von biologischer Wichtigkeit, wie Tyramin und Tryptamin, reagieren mit äthanolischem Acetoacetylphenol bei Zimmertemperatur. Aliphatische, primäre Diamine reagieren ebenfalls und geben die erwarteten Bis-Kondensationsprodukte (Tab. 6). Alle diese Derivate sind stabil, leicht durch Kristallisieren zu reinigen und geben Kupferkomplexe mit Cupriacetat und sind zur Charakterisierung primärer aliphatischer Amine

Tabelle 6. *Kondensationsprodukte einiger Mono- und Diamine mit o-Acetoacetylphenol.*

Amine	Umkrist. aus	Schmelzpunkt ° C
Methyl- . . .	e	101
Äthyl-	a	97,5
n-Propyl- . . .	b	77—78
iso-Propyl . .	a	89
n-Butyl	a	106
iso-Butyl . . .	a	111—112
+ Tyramin . .	c	147
+ Tryptamin .	a	167
Diamine		
Äthylen- . . .	d	183
Trimethylen .	d	171—172
Tetramethylen	d	210 (decomp.)
Pentamethylen	d	190—191

a = Äthanol, b = Methanol, c = wäßriges Äthanol, d = Benzol, e = Benzol-Leuchtpetroleum, f = wäßriges Methanol.

sehr geeignet. Acetoacetylphenol reagiert nicht mit sekundären Aminen, primären aromatischen Aminen oder Aminosäuren.

17. Identifizierung von Aminen als Amide der o-Nitrobenzolsulfensäure nach Billmann und Mahoni.

Die Umsetzung von o-Nitrobenzol-sulfenchlorid mit aliphatischen und aromatischen Aminen geht in ätherischer Lösung leicht vonstatten und führt zu den in 10%igem Alkali unlöslichen Amiden der o-Nitrobenzol-sulfensäure. [Synthese des Säurechlorids "Organic Syntheses" 13, 45 (1935).] Die Spaltung der erhaltenen Amide in Amin und das genannte Säurechlorid erfolgt in ätherischer Lösung durch 3 min langes Einleiten von trockenem Chlorwasserstoffgas. Nach Abfiltrieren des in Äther unlöslichen Aminhydrochlorids wird dasselbe mit trockenem Äther gewaschen. Ausbeuten 95—99%.

Tabelle 7. *Eigenschaften einiger Aminderivate der o-Nitrobenzolsulfensäure.*

Amin	Schmelzpunkt der Aminderivate° C	Schmelzpunkt der Hydrochloride ° C
n-Butylamin	27 —28	142—142,5
Dimethylamin	62,5—63	171
Äthylamin	32,5—33	108
Methylamin	35,5—36	225—226

$$\text{C}_6\text{H}_4(\text{NO}_2)\text{—S—Cl} + 2\,\text{NH}(\text{C}_2\text{H}_5)_2 \longrightarrow \text{C}_6\text{H}_4(\text{NO}_2)\text{—S—N—}(\text{C}_2\text{H}_5)_2 + (\text{C}_2\text{H}_5)_2\text{NH} \cdot \text{HCl}$$

$$\text{C}_6\text{H}_4(\text{NO}_2)\text{—S—N}(\text{C}_2\text{H}_5)_2 + \text{HCl} \longrightarrow \text{C}_6\text{H}_4(\text{NO}_2)\text{—S—Cl} + \text{HN}(\text{C}_2\text{H}_5)_2$$

18. Charakterisierung primärer Amine mit Ninhydrin und Analogen.

Primäre Amine der Typen $\text{R—CH}_2 \cdot \text{NH}_2$ und $\frac{\text{R}}{\text{R}}{>}\text{CH} \cdot \text{NH}_2$ reagieren außer mit Ninhydrin mit Perinaphthindantrion unter Bildung eines blauen Farbstoffes wie mit Ninhydrin. Beim Ninhydrin bilden sich bekanntlich Doppelsalze der überschüssigen Amine mit dem aus Ninhydrin erhaltenen Reduktionsprodukt 2-Oxydiketohydrinden. Werden die Amine mit Ninhydrin in Kohlensäureatmosphäre 30 min lang gekocht, so gehen die dem Amin entsprechenden Aldehyde, die als 2,4-Dinitrophenylhydrazone charakterisiert werden können, in das Destillat über. Die Reaktion wurde z. B. mit Butylamin, 1,3-Diaminopropan, Putrescin, Cadaverin und Histamin durchgeführt. Sie kann nach VAN SLYKE, DILLON, FAYDEN und HAMILTON (1941) quantitativ gestaltet werden. Die Methode wurde kürzlich von BISSET (1954) mit geringfügigen Abänderungen zur quantitativen Bestimmung des Nicht-Protein-Stickstoffgehaltes von Extrakten aus *Pisum sativum* angewandt.

Reaktion für die $> CH \cdot NH_2$-Gruppe (SCHÖNBERG und AWAD, 1949). Wird eine Verbindung oder ihr Salz welche die $>CH \cdot NH_2$-Gruppe enthält, in wäßrigem Alkohol in Gegenwart von Ammoniumchlorid mit Phenanthrachinonimin erwärmt, so scheidet sich Tetrabenzophenoxazin „Phenanthroxazin" ab. Phenanthroxazin ist praktisch unlöslich in niedrig siedenden organischen Lösungsmitteln und in Wasser und kann aus Nitrobenzol umkristallisiert werden. Die Verbindung bildet charakteristische Kristalle und gibt eine tief-indigoblaue beständige Farbe mit konzentrierter Schwefelsäure. Die Farbe kann bei 1 cm dicker Schicht in einer Konzentration von 1 mg in 300 ml konzentrierter Schwefelsäure noch festgestellt werden. Der Test kann also als Mikrotest ausgeführt werden.

Es besteht ein großer Unterschied zwischen der hier beschriebenen und der Ninhydrinreaktion. Ninhydrin gibt eine violette Farbe nicht nur mit Substanzen, die die $>CHNH_2$-Gruppe enthalten, sondern auch mit Ammoniak, Kaliumcyanid und alkoholischen Lösungen von Anilin selbst in der Kälte. Die folgenden Substanzen, die die $CHNH_2$-Gruppe enthalten, geben einen positiven Test:

Methyl-, Äthyl-, n-Propyl-, n-Butyl-, Isobutyl-, sek. Butyl-, Benzyl-, n-Octylamin (-hydrochlorid), 1,3-Diaminopropan und dessen Hydrochlorid, Putrescin und Cadaverinhydrochlorid, Spermintetrahydrochlorid, Glykokoll, d,l-Alanin, Cystein, Asparaginsäure, d,l-Threonin, d-Glutaminsäure, d,l-Valin, l-Histidindihydrochlorid, Leucin (synth.), d,l-Isoleucin, d,l-Norleucin, Glucosamin, l(+)-Lysin und l(+)-Argininmonohydrochlorid, -Phenylalanin, l(—)-Tyrosin, Glutathion, d,l-Tryptophan, Tyroxin u. a.

19. β-Resorcylsäure.

Nach WILSON, ANDERSON und DONOHOE (1951) können Amine durch Bildung von Salzen mit β-Resorcylsäure identifiziert und voneinander getrennt werden.

Die nach Organic Syntheses Coll. Vol. II, p. 757. New York, 1943 gewonnene β-Resorcylsäure wird in trockenem Äther gelöst und die ätherische Lösung der Amine tropfenweise unter Schütteln zugegeben. Es bildet sich entweder sofort oder erst nach Stehen über Nacht eine Kristallisation. Fällt ein Öl aus, so kann es durch kräftiges Zerdrücken mit einem Metallspatel zur Kristallisation gebracht werden. Die Salze sind fast unlöslich in Äther und werden in etwa 80%iger Ausbeute erhalten. Resorcylsäure gibt mit den meisten organischen Aminen keine Salze und kann daher zur Trennung von aliphatischen Aminen, z. B. von Pyridin oder Chinolin verwendet werden. Aus den β-Resorcylaten können die freien Amine durch Schütteln mit Natriumcarbonat oder NaOH-Lösung gewonnen werden. Es wurden z. B. die Resorcylate folgender Amine dargestellt:

Mono-n-propylamin $C_{10}H_{15}O_4N$, F. 123°,
Mono-n-butylamin $C_{11}H_{17}O_4$, F. 133°,
Monoisoamylamin $C_{12}H_{19}N$, F. 138°,
Diäthylamin $C_{11}H_{17}O_4N$, F. 128°.

Angaben über die Empfindlichkeit des Nachweises fehlen.

34*

C. Quantitative Bestimmung von Aminen.

1. Maßanalytische Bestimmungsmethode für primäre aliphatische Mono- und Diamine.
(Binder 1954.)

Die Methode basiert auf der Bildung von Komplexverbindungen aliphatischer Diamine und einiger primärer Monamine mit löslichen Silbersalzen, wobei 2 Aminogruppen auf ein Silbersalzmolekül entfallen. Ein Diaminmolekül tritt mit einem Silbernitratmolekül zur Komplexverbindung zusammen.

$$\left[Ag \left(\begin{array}{c} H_2N \\ H_2N \end{array} \hspace{-4pt} {>} R \right) \right]^+ NO_3^-$$

Der Titrationsendpunkt wird durch wenige Tropfen Kaliumchromat, Kaliumcarbonat oder Natronlauge erkennbar. — Sekundäre Amine und tertiäre Amine, aliphatischer, aromatischer oder heterocyclischer Natur geben mit Silbersalzen keine Komplexverbindungen. Die Bestimmung der Diamine und Monamine ergibt in Gegenwart von Lösungsmitteln wie Methanol oder Aceton zu niedrige Werte, sie soll daher nur in wäßrigen Medien ausgeführt werden. Säureamide, ebenso ringförmige Säureamide, wie Lactame, geben mit Silbersalzen in wäßriger Lösung keine Komplexverbindungen. Es ist daher möglich, aliphatische, primäre Diamine sowie tertiäre Monamine neben Säureamiden titrimetrisch zu erfassen. Die Methode von Binder wurde bisher noch nicht auf pflanzliches Material angewandt.

2. Mikrotitration von Aminen.
(Keen und Fritz 1952.)

Zur Bestimmung von 1—10 mg Amin (A) in 5—10 ml Lösungsmittel mit 0,01 n $HClO_4$ (0,85 ml 72%ige $HClO_4$ in 1 Liter Eisessig [Lösung I]) erfolgt der Zusatz von Lösung I kurz vor dem Endpunkt in Mengen von 0,01 ml. Der Endpunkt der Titration wird mit Methylviolett als Indikator (Umschlagspunkt blau) oder potentiometrisch mit einer Glas-Silberelektrode nach Fritz (1950) festgestellt. Aminmengen bis zu 0,2 mg (B) werden in 1 ml Lösungsmittel in einem Gefäß von 5 ml Inhalt unter Zusatz von $HClO_4$-Lösung in Mengen von 0,005 ml (Durchmischung mit einem Rührer) bestimmt. 15—20 γ (C) aromatischer und aliphatischer Amine werden in 1 ml Benzol gelöst und mit 0,001 n $HClO_4$ (gewonnen durch Verdünnen von Lösung I) in einem Zentrifugenröhrchen unter Zusatz eines Tropfens verdünnter Indikatorlösung titriert. Das Titrationsmittel wird in allen 3 Fällen mit Diphenylguanidin eingestellt. Als Lösungsmittel haben sich Eisessig, Chlorbenzol, Acetonitril, Benzol, Tetrachlorkohlenstoff, Chloroform, Diäthylcellosolve, Äthylacetat, Äthylenchlorid und Nitrobenzol bewährt. Die Genauigkeit beträgt für A 0,3, für B 0,5 und für C 1—2%.

3. Quantitative Bestimmung von Aminen über die Dithiocarbamate.

Wie bereits erwähnt (s. S. 520), bilden Amine mit CS_2 Dithiocarbamate. Die Reaktion wurde von Stanley und Savacool (1949) zur quantitativen Bestimmung von Methyl- und Dimethylamin ausgebaut.

Prinzip: Methylamin- und Dimethylamin, nicht aber Trimethylamin bilden mit CS_2 die entsprechenden Dithiocarbamate:

$$1.\ R{-}NH_2 + CS_2 \rightarrow S{=}C \hspace{-2pt} \begin{array}{l} {}^{\nearrow NH_2} \\ {}_{\searrow SR} \end{array}$$

$$2.\ R{-}NHR + CS_2 \rightarrow S{=}C \hspace{-2pt} \begin{array}{l} {}^{\nearrow NHR} \\ {}_{\searrow SR} \end{array}$$

Diese können entweder rasch durch Titration mit Silbernitratlösung oder durch eine länger dauernde, aber sehr genaue Methode bestimmt werden, bei der das Dithiocarbamat mit Säure zerstört wird. Der entstehende Schwefelkohlenstoff wird durch Überleiten über Bleiacetat gereinigt und durch Reaktion mit alkoholischem K_2CO_3 in Xanthat übergeführt, welches mit einer eingestellten Jodlösung titriert wird.

4. Bestimmung von tertiären, aliphatischen Aminen
in Gegenwart von primären und sekundären Aminen und Ammoniak.
(Wagner, Brown und Peters 1947.)

Die meisten für die Bestimmung tertiärer aliphatischer Amine in Gegenwart von primären und sekundären Aminen vorgeschlagenen Methoden gründen sich auf die Isolierung und Titration von tertiären Aminen nach Behandlung der Mischung mit salpetriger Säure. Mitchell, Hawkins und Smith (1944) bestimten primäre und sekundäre Amine in Gegenwart von tertiären durch Acetylierung der Amine mit einer definierten Menge von Essigsäureanhydrid, hydrolysieren den Überschuß mit einer bekannten Menge Wasser und bestimmen das unverwendete Wasser mit Carl-Fischer-Reagens (vgl. Bd. 1 dieses Handbuches). Der Gehalt der Probe an tertiären Aminen kann dann vom totalen Basenwert durch Differenzbestimmung berechnet werden. Blumrich und Bandel (1941) haben eine einfache Methode vorgeschlagen für die direkte Bestimmung tertiärer Amine nämlich die Acetylierung von primären und sekundären Aminen und Ammoniak mit Hilfe von Essigsäureanhydrid und folgende potentiometrische Titration der nicht reagierenden tertiären Amine, mit Hilfe von Perchlorsäure in Essigsäure. Die Methode wurde ursprünglich nur auf eine Mischung von Diäthylamin und Triäthylamin angewandt. Sie ist aber allgemein anwendbar.

Ausführung: Eine Probe, die nicht mehr als 2 g wiegt und nicht mehr als 1 g Wasser enthält, wird zu 20 ml Essigsäureanhydrid gegeben, enthaltend 2 ml Essigsäure. Die Mischung bleibt 3 Std. bei Zimmertemperatur stehen. 30 ml Essigsäure werden hinzugefügt und die Mischung wird potentiometrisch mit 0,1 n Perchlorsäure in Essigsäure titriert. Bei sterisch gehinderten sekundären Aminen werden 20 ml Essigsäureanhydrid und 2 ml Essigsäure hinzugegeben. Nach einstündigem Erhitzen am Rückflußkühler werden 30 ml Essigsäure hinzugegeben; dann wird potentiometrisch titriert.

5. Charakterisierung von Aminen als Dinitro-α-naphthole und
quantitative mikrochemische Bestimmung.
(Klein und Steiner 1928.)
a) Trennung der Alkylamine von Ammoniak.

In ein dickwandiges Reagenzglas von ungefähr 10 ml Inhalt kommt eine Löffelspitze (0,2—0,4 g) gelbes Quecksilberoxyd (pro Analysi, Merck), dazu 0,8 ml einer Lösung, bestehend aus 7 Teilen 30%iger Natronlauge und 10 Teilen 20%iger Natriumcarbonatlösung. Man füllt mit dest. Wasser soweit auf, daß die Analysenlösung (1 ml) noch bequem zugefügt werden kann und verschließt mit einem gut sitzenden Gummistopfen. Nach zweistündigem Schütteln auf der Schüttelmaschine wird zentrifugiert, die klare Lösung mit einer Pipette abgehoben und in Salzsäure eingegossen. Der Quecksilberniederschlag wird mit einer Waschflüssigkeit, welche in 1 l 20 ml 30%ige Natronlauge und 40 ml 20%iges Natriumcarbonat enthält, versetzt, mehrmals kurz geschüttelt, zentrifugiert, abgehoben und die Waschflüssigkeit mit der ersten Fraktion vereinigt, wobei zu beachten ist, daß die Reaktion sauer bleibt.

Die weitere Verarbeitung der Ammoniakfraktion erfolgt nach Franzen und Schneider (1921) durch Zersetzen der Quecksilberammoniakverbindung mit

Ameisensäure. Diese wird dabei unter gleichzeitiger Oxydation der Ameisensäure zu metallischem Quecksilber reduziert. Es genügt hierbei, die Quecksilberverbindung mit einem Überschuß von Ameisensäure auf dem Drahtnetz kurz aufzukochen und langsam erkalten zu lassen. Franzen und Schneider (1921) spalten durch 20 min langes Erhitzen auf dem Wasserbad. Das als Formiat vorliegende Ammoniak kann nun über Alkali in 0,1 n Schwefelsäure destilliert und durch Rücktitrieren mit Natronlauge quantitativ bestimmt werden, wenn man es nicht vorzieht, nach Destillation in verdünnte Salzsäure einzuengen und die unten beschriebene Methode zu verwenden, die zwar nur schätzungsweise quantitative Resultate liefert, aber die Vorzüge viel kürzerer Dauer und größerer Bequemlichkeit der Ausführung in sich vereinigt.

b) Bestimmung der Alkylamine und des Ammoniaks mit Dinitro-α-naphthol.

Prinzip: Ein Tropfen der zu untersuchenden Flüssigkeit wird zu einem Tropfen Natronlauge gegeben, der sich in einer kleinen Kammer befindet. Die Kammer wird mit einem Deckglas versehen, welches auf der Unterseite einen Tropfen einer wäßrigen Suspension von Dinitro-α-naphthol trägt. Das aus der alkalischen Lösung entweichende Amin reagiert mit dem Dinitronaphthol unter Bildung von Kristallen die für jedes einzelne Amin charakteristisch sind.

Ausführung: Ein oben und unten abgeschliffener Glasring, wie er für die Mikrosublimation verwendet wird (1 cm innerer Durchmesser, 6 mm Höhe) wird mit venezianischem Terpentin auf einen Objektträger aufgekittet. Auf den Boden dieser Kammer bringt man 1—2 Tropfen 30%ige Natronlauge, hierzu fügt man mit einer Kapillarpipette 0,04 ml der auf Amine zu prüfenden neutralen oder schwach sauren Flüssigkeit. Man bedeckt sofort mit einem Deckgläschen, welches auf der Unterseite einen Tropfen dest. Wasser trägt, in welchem mit einer Nadel einige Kriställchen von Dinitronaphthol aufgeschwemmt wurden. Es ist darauf zu achten, daß die Natronlauge nicht mit dem Reagens in Berührung kommt, der Reagenstropfen darf vor allem den oberen Rand des Glasringes nirgends berühren. Bei Anwesenheit von Ammoniak oder Aminen setzen sich die schwerlöslichen Dinitronaphtholverbindungen der Amine an den Kristallen des Reagens fest: Bei Anwesenheit genügender Basenmengen wird das ganze feste Reagens allmählich in die Aminverbindung verwandelt. Zur weiteren Untersuchung wird das Deckglas abgehoben mit der Kristallseite nach unten auf einen Objektträger gebracht und mit ganz wenig Vaseline an einer Ecke festgeklebt. Sind größere Basenmengen vorhanden, so muß das Deckgläschen mit dem Reagens öfter erneuert werden. In manchen Fällen gelingt dabei gleichzeitig eine Fraktionierung, da sich die Amine infolge ihrer verschiedenen Flüchtigkeit verschieden rasch anreichern. Finden sich beispielsweise in Pflanzendestillaten Trimethylamin und Isoamylamin nebeneinander, so findet man regelmäßig auf den zuerst aufgelegten Deckgläschen das Trimethylamin stark angereichert, während in späteren Fraktionen Isoamylamin vorherrscht. Die Verbindungen der Amine und des Ammoniaks mit Dinitro-α-naphthol sind durch scharf umschriebene, wohl definierte Eigenschaften ausgezeichnet und heben sich sowohl untereinander als auch gegen überschüssiges Reagens sehr gut ab. Dadurch wird die Abschätzung vorhandener Mengen ermöglicht, sie ist nur mit einem kleinen Fehler behaftet. Um einerseits die Erfassungsgrenze feststellen zu können, andererseits Vergleichswerte für die Abschätzung der bei den Pflanzenuntersuchungen gefundenen Werte zu erhalten, sind mit bekannten, reinen Aminlösungen Verdünnungsreihen herzustellen und mit je einem Tropfen die oben beschriebene Reaktion durchzuführen. Es empfiehlt sich in jedem Falle das Untersuchungsobjekt mit Präparaten aus reiner Substanz zu vergleichen. Unter dieser Voraussetzung läßt sich die Identifizierung von Aminen äußerst rasch, bequem und sicher durchführen. Nachstehend sind die Resultate eines solchen Vergleichsversuches mit Trimethylaminchlorhydratlösung wiedergegeben. Die Größe der zu untersuchenden Tropfen (0,04 ml, Kapillarpipette) ist in allen Fällen gleichzuhalten.

```
                                                    Trimethylamin
1 Tropfen einer 0,3%igen Lösung (120 γ)   ++++          = > 100 γ
1    „       „  0,09   „     „  ( 30 γ)    +++—++++  = ca. 50 γ
1    „       „  0,027  „     „  (10,8 γ)   +++           = „ 10 γ
1    „       „  0,0081 „     „  ( 3,2 γ)   ++            = „  3 γ
1    „       „  0,002  „     „  ( 1,0 γ)   +             = „  1 γ
1    „       „  0,0007 „     „  ( 0,3 γ)   0 — +         = „  0,5 γ
1    „       „  0,0002 „     „  ( 0,1 γ)   Spur          = „ < 0,5 γ
```

Wesentlich sind Form und Farbe der Kristalle und einige optische Konstanten bei der Untersuchung im Polarisationsmikroskop sowohl für das ursprüngliche Präparat als auch für die durch Erwärmen „umgelagerten" Kristalle. Wichtig sind ferner die Temperaturen der Umlagerung und die Schmelzpunkte. (Mikroapparatur nach G. KLEIN, 1928.) Zur Identifizierung von Ammoniak und einigen aliphatischen Aminen wird auf Grund der Eigenschaften ihrer Dinitro-α-naphtholverbindungen folgender Bestimmungsschlüssel verwendet:

Gekrümmte, peitschenartige Kristallnadeln	Methylamin
Wohlausgebildete Kristalle	1
1. Kristalle ± hellzitronengelb	2
Kristalle ± dunkelgelb bis braun	11
2. Auslöschung, gerade	3
Auslöschung, schief	8
3. Schmelzpunkt unter 100°	4
Schmelzpunkt über 100°	5
4. Lange Nadeln	Triäthylamin
Kurze ± isodiametrische Kristalle	Tripropylamin
5. Kristalle eng, untereinander zu Bündeln, diese zu sternartigen Büscheln vereinigt. Umlagerung in hellgelbe, rechtwinklige Tafeln	Allylamin
Kristalle einzeln oder zu Büscheln vereinigt	6
6. Kristalle, spitz	7
Kristalle, stumpf oder giebelartig endigend	Trimethylamin
7. Spieße, Umlagerung zu ± isodiametrischen Platten	Propylamin
Prismen mit spitzem Ende, Umlagerung zu langgestreckten Platten .	i-Propylamin
8. Umlagerung zu ebenen Platten, Rand glatt	9
9. Umlagerungsprodukt ± isodiametrische, hellgelbe Platten, Endigungen treppenartig[1]	i-Amylamin
Umlagerungsprodukt ± langgestreckte, hellgelbe Platten, nach dem Erkalten durch Querrisse septiert, Endigung oft treppenartig . . .	n-Heptylamin
10. Kristalle, hellbraun	11
Kristalle, orangegelb	15
11. Kristalle, spitze Nadeln	Ammoniak
Kristalle von anderem Habitus	12
12. Auslöschung, gerade	Diäthylamin
Auslöschung, schief	13
13. Schmelzpunkt, ohne oder fast ohne vorherige Umlagerung	15
14. Umlagerung in ± isodiametrische, braune Platten	Dimethylamin
Umlagerung in langgestreckte, dunkelgelbe Platten und Balken . . .	Di-i-Butylamin
15. Kristalle, nadelförmig, Endwinkel stumpf	Äthylamin
Kristalle, entweder Platten, Prismen oder Klumpen mit spitzen Endwinkeln oder ± isodiametrischen Platten	i-Butylamin

Die Umrechnung der Untersuchungsresultate auf direkt vergleichbare Werte ist nach KLEIN (1928) durch eine einfache Formel möglich. Die Angaben beziehen

---

[1] Falls keiner der beiden angegebenen Gegensätze zutrifft, vgl. auch i-Butylamin, 15.

sich auf 100 g frisches Pflanzenmaterial, dessen Exhalat resp. Destillat auf 2 ml eingeengt wurde, von denen wieder 1 Tropfen zur Mikroreaktion verwendet wird. Als Zeiteinheit (bei Exhalaten) wurden 24 Std. gewählt. Für die Berechnung ergeben sich als bekannte Größen:

p = Frischgewicht in Gramm des verarbeiteten Pflanzenmaterials.

v = Volumen in ml, zu welchem das Exhalat oder Destillat eingeengt wurde.

w = Volumen, welches davon zur weiteren Verarbeitung (Trennung) entnommen wurde (ml).

f = Volumen der nach der Trennung erhaltenen NH_3- bzw. Aminfraktion in ml.

n = Anzahl der Tropfen Probelösung, die für die Mikroreaktion verwendet wurden (1 Tropfen = 0,04 ml).

l = Die der erhaltenen abgeschätzten Menge des Reaktionsproduktes ungefähr entsprechende Basenmenge.

t = Dauer in Stunden bei Durchlüftungsversuchen.

Durch einfache Proportionen ergibt sich dann für

$$X = \frac{50 \cdot v \cdot l \cdot w}{n \cdot p \cdot f}$$

bzw. bei Exhalaten (wenn $t \geqq 24$) $= \dfrac{50 \cdot v \cdot l \cdot w \cdot 24}{n \cdot p \cdot f \cdot t}$.

Um eine möglichst einfache Rechnung zu erzielen, engt Klein die Destillate und Exhalate fast ausnahmslos auf 2 ml ein. 1 ml wird nach Francois getrennt und jede der dabei erhaltenen Teilfraktionen ebenfalls auf ein Volumen von 1 ml gebracht. Für die Reaktion wurden im allgemeinen 1—3 Tropfen, seltener mehr, je nach dem Gehalt an Basen, verwendet.

Die abgeschätzten Basenmengen werden durch folgende Zeichen angegeben:

++++	= 100 γ	+—++	=	ca. 2 γ
+++—++++	= ca. 50 γ	+	=	„ 1 γ
+++	= „ 10 γ	0—+	=	„ <0,5 γ
++—+++	= „ 6 γ	Spur	=	0,5 γ (nahe der
++	= „ 3 γ			Fehlergrenze).

Aus der obigen Erklärung folgt, daß die dem angegebenen Symbol entsprechende Ausbeute nur mit 50 multipliziert werden muß, um die in 100 g Pflanzenmaterial ungefähr enthaltene absolute Aminmenge zu ermitteln.

Beispiel. 60 g Sproßteile (p = 60) von *Chenopodium vulvaria* wurden 48 Std. (t = 48) durchlüftet, das Exhalat auf 3 ml (v = 3) eingeengt, davon 1 ml (w = 1) dem Trennungsgang nach Francois unterworfen. Die erhaltene Aminfraktion hatte ein Volumen von 0,9 ml (f = 0,9). 3 Tropfen davon (n = 3) ergaben das Resultat:

Trimethylamin +—++.

Daher ist $x = \dfrac{50 \cdot 3 \cdot 2 \cdot 24}{3 \cdot 60 \cdot 0,9 \cdot 48} = 9,4$ als umgerechnetes Resultat erscheint

Trimethylamin +++ = ca. 10 γ.

Nach Wacek und Löffler (1935) können außer Dinitro-1-naphthol im Verfahren von Klein und Steiner auch 4,8-Dinitro-1-naphthol und 2,4,5-Trinitro-1-naphthol zur Identifizierung von Aminen herangezogen werden. Bezüglich Farbe, Auslöschung, Dichroismus, Umlagerung und F. P. ihrer Reaktionsprodukte siehe Wacek und Löffler (1935). Durch stufenweise Steigerung der Basizität in der Kammer erreicht man eine rohe Fraktionierung des Amingemisches. Das 4,8-Dinitro-α-naphthol besitzt vor anderen Substanzen den Vorzug mit primären Aminen, besonders mit CH_3NH_2, gut kristallisierte Reaktionsprodukte zu bilden, während die Kristallformen der Reaktionsprodukte mit sekundären und tertiären Aminen und auch mit NH_3 wenig charakteristisch sind, da sie leicht verwittern. Die Kristallisate mit 2,4,5-Trinitro-1-naphthol sind zwar nicht so charakteristisch wie mit 2,4-Dinitro-1-naphthol, lassen sich aber zum Teil umlagern und sind für Vergleichszwecke oft recht geeignet. Zur Identifizierung eines durch die angegebene Methode nicht einwandfrei zu charakterisierenden primären Amins kann mit Vorteil die Isonitrilreaktion herangezogen werden, indem man die Deckglasprodukte dieser Reaktion unterwirft. Die Isonitrilreaktion hat neben dem Nachteil, eine Geruchsprobe zu sein, den Vorteil der größeren Empfindlichkeit; mit ihr lassen sich noch 0,5 γ primäres Amin/ml nachweisen, während bei dem vorgeschlagenen Nachweis im ungünstigen Falle die Erfassungsgrenze nur 10 γ beträgt [Jahrb. wiss. Botanik 68, 602 (1929)]. DIHLMANN (vgl. ZORN und DIHLMANN, 1953) sowie DIHLMANN (1954) hat das vorstehend geschilderte Verfahren mit der papierchromatographischen Bestimmung der Amine kombiniert (s. S.541).

D. Nachweis und Identifizierung von Aminen mit Hilfe der Papierchromatographie.

(STEINER und STEIN v. KAMIENSKI[1].)

I. Isolierung der Amine aus der Pflanze.

Etwa 20—50 g pflanzliches Material, z. B. Blüten, werden mit 10 ml einer 1%igen Zitronensäure und einer ausreichenden Menge Wasser im Mixapparat zerkleinert. Das zerkleinerte Material wird einer alkalischen Destillation unterworfen, wobei folgende Apparaturen verwendet werden: Rundkolben 1 oder 2 l, Dampfeinleitungsrohr, bis fast zum Boden reichend, ableitendes Rohr mit senkrecht absteigendem Liebig-Kühler verbunden, der in einer Peligotröhre endigt. In der Peligotröhre als Vorlage 10 ml 0,1 n HCl; bei sehr aminreichem Material (Pilze) entsprechend mehr. Das Destillationsgut wird durch Zusatz von 5% NaCl + 5% Na_2CO_3 deutlich lackmusalkalisch gemacht. In der Regel genügen 20 ml des Gemisches. Man destilliert im Wasserdampf so lange, bis etwa 100 ml übergegangen sind. Dazu sind in der Regel etwa 20 min erforderlich.

Einengen des Destillates. Das Destillat wird zunächst aus einem Kolben im Vakuum mit vorgeschaltetem Kühler bei höchstens 70° auf etwa 5—10 ml eingeengt. Die eingeengte Lösung wird quantitativ in ein Schälchen gebracht und über Calciumchlorid in einem Vakuum-Exsiccator getrocknet, in dem sich ein Schälchen mit etwas festem NaOH befindet. Der trockene Rückstand wird mit 2 ml Wasser aufgenommen und in einem Tablettengläschen unter Thymolzusatz für die weitere Bearbeitung bereitgestellt.

II. Trennung und Nachweis durch Papierchromatographie.

Aufsteigende Methode, Chromatographierzylinder nach HELLMANN (vgl. Bd. 1 dieses Handbuches).

Papier.
a) Schleicher und Schüll 2043 b M ohne Vorbehandlung.
b) Schleicher und Schüll 2043 b vorbehandelt mit 0,15 Mol Natriumacetat. Papier in die Lösung eintauchen und trocknen. Streifengröße 7 × 40 cm.

Lösungsmittel.
1. Butanol — Wasser — Eisessig 50:49:1. (B50/W49/E1) Die Epiphase wird verwendet.
2. Butanol — Wasser — Eisessig 4:1:1 (B4/W1/E1). Das homogene Gemisch wird verwendet.
3. Collidin mit H_2O gesättigt, ausschließlich für zyklische Amine und aliphatische Amino-Alkohole; empfehlenswert zur Differenzierung dieser Amine von anderen Aminen mit nahebeieinanderliegenden R_f-Werten.

Sämtliche Lösungsgemische dürfen nicht länger als höchstens 7 Tage aufbewahrt werden.

---

[1] Private Mitteilung der Autoren an den Verfasser.

1. Nachweis der Flecke.

Primäre Amine, weniger gut sekundäre Amine mit 0,1% Ninhydrin in Butanol. Besprühen der trockenen Chromatogramme, erhitzen im Thermostaten auf 90°, ca. 15 min.

Sekundäre und tertiäre Amine: die getrockneten Chromatogramme kommen über Nacht in einen Exsiccator, dessen Atmosphäre mit Joddampf gesättigt ist. Die Flecke treten in einer vergänglichen Braunfärbung auf. Tertiäre Amine zeigen stärkere Färbung als sekundäre. Das an sich wegen der R_f-Werte günstige Natriumacetatpapier kann für diese Nachweismethode nicht verwendet werden, da es sich zu stark mit anfärbt.

Besonders bei den mit Jod nachgewiesenen Aminen wird eine Überprüfung durch die mikrokristallographische Methode (s. S. 535) sehr empfohlen.

Die R_f-Werte für 35 Amine sind in Tab. 8 verzeichnet. Die Werte unterliegen gewissen Schwankungen. Es empfiehlt sich deshalb, stets ein oder mehrere bekannte Amine mitlaufen zu lassen.

Tabelle 8. R_f-Werte × 100. STEINER und STEIN v. KAMIENSKI (unveröffentlicht).

Nr. Amin	B4/W1/E1 Papier b	B4/W1/E1 Papier a	B 50/W 49/E1 Papier b	B 50/W 49/E1 Papier a	Entwicklung, Empfindlichkeit	Collidin H₂O gesättigt Papier b	Entwicklung Empfindlichkeit
1 Methyl . . .	43	30	15	10	N 0,6	—	—
2 Dimethyl .	44	35	20	11	N 61 Y 10	—	—
3 Trimethyl . .	40	32	19	10	N-Y 2 P5	—	—
4 Acetyl . . .	54	40	25	15	N 0,6	—	—
5 Diacetyl . . .	60	45	44	25	N 10 Y5	—	—
6 Triäthyl . .	64	60	45	30	N-Y 3 P5	—	—
7 n-Propyl . .	68	50	41	24	N 0,6	—	—
8 T-Propyl . .	66	48	39	25	N 0,6	—	—
9 D-n-Propyl .	82	62	50	42	N 10 Y5	—	—
10 D-i-Propyl .	80	64	49	43	N 10 Y5	—	—
11 T-Butyl . . .	75	60	55	40	N 0,6	—	—
12 n-Butyl . . .	77	64	57	41	N 0,6	—	—
13 D-n-Butyl . .	96	85	82	70	N 10—15 Y5	—	—
14 T-n-Butyl . .	97	90	89	80	N-Y 3 P5	—	—
15 n-Amyl . . .	82	73	70	54	N 0,6	—	—
16 i-Amyl . . .	80	71	69	53	N 0,6	—	—
17 D-n-Amyl . .	94	92	94	85	N 10 Y5	—	—
18 D-i-Amyl . .	96	88	93	78	N 10 Y5	—	—
19 T-n-Amyl . .	—	70	—	90	N-Y 4 P5	—	—
20 n-Hexyl . .	85	77	45	65	N1	—	—
21 n-Heptyl . .	86	80	77	66	N1	—	—
22 n-Octyl . . .	87	82	83	72	N1	—	—
23 n-Decyl . . .	89	85	85	74	N1	—	—
24 Benzyl . . .	77	62	60	43	N1	80	N 0,5
25 Tyramin . . .	75	62	56	40	N1	79	N1
26 Tyramin . . .	74	65	60	45	N1	80	N1
27 β-Phenylacetyl	80	62	66	45	N1	77	N 0,5
28 Cadaverin . .	30	10	2	01	N 0,5	09	N 0,5
29 Putrescin . .	27	10	01	01	N 0,5	12	N 0,5
30 Allyl	59	42	35	20	N 0,6	—	—
31 Histamin . .	30	42	05	01	N1	46	N1
32 Äthanol . . .	40	23	10	06	N 0,6	35	N1
33 Diäthanol . .	—	30	26	10	N 30 Y 6	40	N 5
34 Triäthanol .	—	29	—	09	N-Y 2 P5	70	Y 3
35 NH₃	40	—	12	—	N 100	40—50	N 4

N = Ninhydrin 0,1% in n-Butanol; Y = Joddampf; C = Chinon 1% in n-Butanol + Pyridin 1% zur Dimethyl; P = Phosphormolybdänsäure Empfindlichkeit in γ untere Erfassungsgrenze.

Die quantitative Bestimmung der mit Ninhydrin angefärbten Flecke durch Photometrie bereitet noch methodische Schwierigkeiten. Eine weitere Identifizierung der Amine ist dadurch möglich, daß man in einem Chromatogramm die Aminflecken ausschneidet, in ein oben abgeschliffenes Röhrchen bringt, mit Lauge die Amine freisetzt und mit aufgelegtem Deckglashängetropfen mit Dinitro-α-naphthol (nach KLEIN und STEINER, s. S. 534) die kristallographische Identifizierung durchführt. Flecke, die mit Ninhydrin angefärbt wurden, sind hierfür nicht mehr verwendbar. Hingegen stört der Nachweis mit Joddampf diese Methode nicht. Tertiäre Amine können auch mit Phosphormolybdänsäure (nach MUNIER u. MACHEBOEUF, 1951) und anschließende Reduktion durch $SnCl_2$ im Chromatogramm nachgewiesen werden.

2. Nachweis primärer, sekundärer und tertiärer Amine.

Bei der Anwendung der Methode durch STEINER und STEIN v. KAMIENSKI (1954) wurden erstmals aufgefunden: In Blüten von *Sambucus nigra* Äthyl- und i-Butylamin, in *Filipendula ulmaria* i-Butylamin, in *Arum italicum* (Appendix), Äthylamin, in *Heraclum sphondilium* Trimethylamin. In *Phallus impudicus* Trimethylamin und i-Amylamin; in *Russula aurata* Dimethylamin und in *Boletus sanguineus* i-Amylamin.

Ein Reagens, das außer komplizierter gebauten Alkaloiden nur die Flecke von tertiären Aminen und quaternären Ammoniumbasen sichtbar macht und daher zur Unterscheidung primärer und sekundärer Amine von tertiären und quaternären herangezogen werden kann, wurde von MUNIER und MACHEBOEUF (1949, 1950) angegeben. Es handelt sich um ein modifiziertes DRAGENDORFFsches Reagens, das neuerdings von THIES und REUTHER (1954) wesentlich verbessert worden ist. Durch das modifizierte Reagens werden das Fließen der Farbflecke, verschwommene Ränder und Höfe vermieden und scharfe begrenzte Farbflecke erzielt. Bei diesem Reagens wird an Stelle von Wasser der leicht flüchtige Essigester als Lösungsmittel und an Stelle von Kaliumjodid das in Essigester besser lösliche Natriumjodid verwendet. Die Vorteile sind rascheres Verdunsten des Essigesters, beschränkte Mischbarkeit der Reagenslösung mit Wasser, insbesondere auch mit dem Wasser der Hydrathülle der Cellulosefasern des Papiers. Die Farbflecke erscheinen erst gegen Ende der Trocknung. Es wird eine gleichmäßige Anfärbung erzielt, wie sie für quantitative photometrische Auswertung erforderlich ist.

Herstellung der Reagenslösung nach THIES und REUTHER (1954). 2,6 g basisches Wismutcarbonat (DAB 6) und 7,0 g Natriumjodid (etwa 24 Std. über konzentrierter Schwefelsäure im Exsiccator getrocknet) werden mit 25 ml Essigsäure einige Minuten gekocht. Beim Stehen über Nacht scheiden sich reichliche Mengen von Natriumacetat aus, die abgetrennt werden. 20 ml der verbleibenden klaren, tiefroten Lösung werden mit 80 ml Essigester versetzt und bilden die Stammlösung, die im gut verschlossenen Gefäß längere Zeit haltbar ist. Zur Herstellung der eigentlichen Färbelösung werden 20 ml Stammlösung mit 50 ml Essigsäure und 120 ml Essigester vermischt und weiterhin unter ständigem Umschütteln tropfenweise mit 10 ml Wasser versetzt. Diese Färbelösung ist vor allem zum Anfärben von Papierstreifen durch Eintauchen geeignet; 200 ml reichen zum Färben von etwa 30 Streifen (10 × 40 cm) aus. Zur Verwendung als Sprühreagens kann es zweckmäßig sein, die Konzentration der Färbelösung je nach Papiersorte durch Verringerung der zuzusetzenden Menge Stammlösung bis zur Hälfte zu erniedrigen.

Zur papierchromatographischen Trennung der Amine verwenden BAKER, HARBORNE und OLLIS (1952) Whatman-Papier Nr. 1 und ein Lösungsmittelgemisch von n-Butanol, Eisessig und Wasser (5:4:4 oder 2:1:1). Chromatographie bei Zimmertemperatur. Die primären Amine werden nach Besprühen mit einer 1%igen Lösung von o-Acetoacetylphenol in n-Butanol (s. S. 529) als grüngelb fluoreszierende Flecke identifiziert.

Die in Tab. 9 verzeichneten R_f-Werte einer Reihe von Aminen wurden von Bremner und Kenten (1952) in der Weise ermittelt, daß die Amine als Hydrochloride, Agmatin als Sulfat, verwendet wurden. Die nur als Basen vorhandenen Amine wurden auf neutrale Reaktion eingestellt. 1—3 ml der Lösung, enthaltend etwa 6—40 γ der Amine, wurden horizontal an der Spitze eines Whatman Nr. 4-Filtrierpapierstreifens aufgetragen und das Papier nach dem Trocknen über Nacht im abgeschlossenen Raum mit dem Dampf der wäßrigen Phase des verwendeten Lösungsmittelgemisches in Berührung gehalten, sodann erst die übrigen Lösungsmittelkomponenten zugegeben. Als Lösungsmittelgemisch dienten (sie beziehen sich sämtlich auf v/v): Phenol, Collidin (gleiche Volumina von 2,4-Lutidin und 2,4,6-Collidin mit Wasser gesättigt), n-Butanol, n-Butanol-Eisessig

Tabelle 9. *R_f-Werte von Aminen in verschiedenen Lösungsmitteln auf Whatman-Nr. 4-Papier bei Zimmertemperatur.*

Amine	... n-Butanol	n-Butanol (40%) Eisessig 10% Wasser (50%)	„Collidin"	m-Kresol (50%) Essigsäure (2%) Wasser (48%)	Phenol	Phenol	Phenol	Farbe mit Ninhydrin
unter Zufügung von ...	—	Diäthylamin	—	NH₃	HCN	Essigsäure		
Methylamin	0,10	0,37	· - -	0,52	0,96	0,91	0,72	P
Äthylamin	0,18	0,45	...)	0,67	0,95	0,94	0,80	P
n-Propylamin . . .	0,28	0,58	...)	0,76	0,97	0,98	0,86	P
n-Butylamin	0,31	0,70	...)	0,86	0,96	0,98	0,91	P
n-Amylamin	0,53	0,77	...):	0,98	0,96	0,98	0,92	P
n-Heptylamin . . .	0,58	0,85	...)	0,93	0,95	0,98	0,94	P
iso-Propylamin . .	0,27	0,57	...)	—	—	—	—	P
iso-Amylamin . . .	0,52	0,77	...	—	0,97	—	—	P
1:2-Diaminoäthan .	0,02	0,14	0,20	0,02	0,93	0,23	0,18	P
1:3-Diaminopropan	0,02	0,15	0,30	0,03	0,93	0,35	0,25	P
1:4-Diaminobutan (Putrescin). . . .	0,02	0,16	0,20	0,06 (0,38)[2]	0,94	0,89	0,45	P
1:5-Diaminopentan (Cadaverin) . . .	0,02	0,17	0,22	0,09 (0,50)[2]	0,94	0,94	0,59	P
1:6-Diaminohexan .	0,03	0,20	0,24	0,14 (0,64)[2]	0,94	0,97	0,67	P
Benzylamin	0,38	0,68	0,79	0,86	0,95	0,98	0,91	RP[3]
β-Phenyläthylamin .	0,51	0,72	0,80	0,88	0,95	0,97	0,91	B
ββ-Diphenyläthylamin	0,50	0,72	0,80	0,86	0,94	0,97	0,90	B
β-Phenyl-β-hydroxyäthylamin	0,35	0,65	0,79	0,84	0,94	0,96	0,86	P
Adrenalin	0,14	0,45	Streifen	0,58	Streifen	0,91	0,74	P =
Agmatin.	0,00	0,05	0,16	0,06 (0,64)[2]	0,94	0,60 (0,93)[1]	0,42 (0,75)[2]	P
Allylamin	0,03	0,50	... :	0,14	0,95	0,95	0,86	P
Äthanolamin . . .	0,09	0,33	0,44	0,47	0,80	0,85	0,65	P
Dimethylamin . . .	0,12	0,43	... :	—	—	0,97	0,95	P =
Ephedrin	0,53	0,75	0,86	0,95	0,94	0,96	0,94	P
Glucosamin	0,04	0,24	0,50	0,03	0,67	0,42	0,30	P
Histamin	0,03	0,19	0,44	0,09	0,96	0,94	0,68	BG
Spermin	0,00	0,07	0,10	0,02	0,93	0,33	0,24	P
Tryptamin	0,35	0,67	0,84	0,86	0,95	0,96	0,91	GBr
Tyramin.	0,30	0,62	0,84	0,70	0,95	0,96	0,85	GP

[1] Endgültige Farben auf den Chromatogrammen die in n-Butanol-Eisessig liefen. P = rötlich, R = rot, B = blau, G = grau, Br = braun, : Diese Amine waren durch Ninhydrin in diesem Lösungsmittel nicht nachweisbar. [2] Werte beziehen sich auf schwächere Flecke. [3] = Farbe anfänglich gelb. [4] Schwache Farbe mit Ninhydrin.

(40% n-Butanol, 10% Eisessig und 50% Wasser, die Lösung wird nach Durchschütteln 3 Tage lang vor Gebrauch aufbewahrt); 50% m-Kresol, 2% Eisessig und 48% Wasser. Bei Verwendung von Phenol-Wasser-Gemisch wird in Gegenwart von Ammoniak oder von Blausäure sowie in einer Atmosphäre, die mit einem Gemisch von 50% Eisessig und 50% Wasser gesättigt ist, chromatographiert. Bei Verwendung von Collidin wird unter Zusatz von Diäthylamin gearbeitet.

Bei dem papierchromatographischen Nachweis, der Trennung und der Bestimmung der aliphatischen Amine von Pflanzen verfährt SCHWYZER (1952) wie folgt: Die Amine werden mit 0,1—2% Eisessig enthaltendem abs. Alkohol extrahiert. Die Lösung wird auf das schnell laufende Munktell-OB-Papier aufgetragen. Durchmesser der Startflecke höchstens 6—8 mm. Während des Auftragens wird die Verdampfung des Lösungsmittels durch einen warmen Luftstrom beschleunigt. Verbringen des Papiers wie üblich in ein Lösungsmittelgemisch bestehend aus n-Butanol mit 25% Eisessig (das Gemisch blieb 3 Tage vor Verwendung zur teilweisen Veresterung bei Zimmertemperatur stehen). Nach einer Stunde Laufzeit, wobei das Gemisch in der Regel 13—14 cm hoch steigt, wird das Papier herausgenommen, 30—45 min lang bei Zimmertemperatur getrocknet und anschließend durch Besprühen mit 0,2% Bromphenolblau in abs. Alkohol entwickelt. Die Amine erscheinen als blaue Flecke auf gelbem Hintergrund. Eine halbquantitative Bestimmung wird durch Auftragen von Verdünnungsserien der unbekannten Lösung zusammen mit Verdünnungsserien bekannter Aminmengen auf demselben Papierstreifen ermöglicht. Die Identifizierung stark flüchtiger Amine geschieht nach dem Entwickeln mit Hilfe der Mikrodiffusionsmethode nach RICHTER oder nach OPFER-SCHAUM (1944) durch Schmelzpunktbestimmung. Im Folgenden sind die gefundenen R_f-Werte und die geringste nachweisbare Aminmenge (Werte in Klammern) einer Reihe von Aminen in γ angegeben: Methylamin 0,28 (3), Dimethylamin 0,30 (3), Trimethylamin 0,31 (4), Äthylamin 0,35 (3), n-Propylamin 0,46 (2), Isopropylamin 0,45 (2), Benzylamin 0,59 (2), Phenyläthylamin 0,64 (2), Cadaverin 0,13 (1) freie Base.

Weitere Angaben zum papierchromatographischen Nachweis und zur Trennung von Aminen mit Hilfe der papierchromatographischen Methode s. MUNIER (1952), ROCHE, FELIX und ROBIN-THOAI (1951).

Die Sichtbarmachung der Flecke kann anstatt mit Ninhydrin in besonderen Fällen, z. B. bei Agmatin, mit dem SAKAGUCHI-Reagens, bei Histamin und Tyramin durch diazotiertes Sulfanilamid (BLOCK, 1950) erfolgen. Die Farbstärke der Flecke, die mit einem photoelektrischen Instrument gemessen wird, ist der Konzentration der betreffenden Verbindung proportional. Über weitere Möglichkeiten zur Sichtbarmachung siehe bei den einzelnen Aminen. DIHLMANN (1954) vorfährt zum papierchromatographischen Nachweis von Aminen wie folgt: Freisetzen der Amine durch Vakuumdestillation des sodaalkalischen Materials bei 20—25° in eine eisgekühlte Vorlage mit 5 ml Essigsäure. Einengen der Vorlageflüssigkeit unter den gleichen Bedingungen auf etwa 0,5 ml. Auftragen der Lösung mit einer auf 4 mm³ geeichten Mikropipette auf die Startpunkte des Papiers. Papiersorte: W F l der Fa. Gessner und Kreuzig, Niederschlah i. Erzgebirge. Lösungsmittelgemisch: n-Butanol, Eisessig, Wasser 4:1:5 nach PARTRIDGE (1948). Aufsteigendes Chromatogramm. Entwicklung durch Behandlung mit 0,2%iger Ninhydrin-Butanollösung oder mit 1%iger Natrium-β-naphthochinonsulfonat-Lösung in 5%iger Sodalösung und nachfolgendes Trocknen, wodurch primäre und sekundäre Amine sichtbar werden. Die tertiären Amine werden durch Behandeln der Chromatogramme mit Joddampf bei Zimmertemperatur im Exsiccator identifiziert. R_f-Werte: Für Methylamin 0,34, Dimethylamin 0,28, Trimethylamin 0,33, Äthylamin 0,38, Diäthylamin 0,57,

Tabelle 10. *R_f-Werte ausgewählter Amine bei aufsteigender Papierchromatographie;
Lösungsmittel: Butanol—Wasser.*

Amine[1]	R_f	Farbe	R_f-Werte aus der Literatur
aliphatische:			
Methylamin	0,10	blaupurpur	0,10
Äthylamin	0,17	purpur	0,18
2-Mercaptoäthylamin	0,013	purpur	0,52
n-Butylamin	0,44	purpur	0,31
3-Methyl-butylamin	0,53	purpur	0,52
n-Hexylamin	0,66	purpur	
Cyclohexyl-amin	0,54	purpurrot	
2-Hydroxy-hexylamin	0,19	purpur	
Cystamin-2 HCl	0,11	purpur	
Glucosamin	0,052	purpur	0,04
Glutamin	0,045	purpur	
n-Heptylamin	0,71	purpur	0,58
1,3-Dimethyl-pentylamin	0,69	purpur	
1-Methyl-heptylamin	0,75	purpur	
n-Decylamin	0,75	purpur	
n-Dodecylamin	0,76	purpur	
n-Tetradecylamin	0,80	purpur	
n-Octadecylamin	0,94	purpur	
aromatisch substituierte:			
2-Phenäthylamin	0,48	purpur	0,51
2-(4-Oxyphenyl)-äthylamin	0,33	braunrot	0,30
2-Oxy, 2-(3,4-Dioxy-phenyl)-äthylamin	0,13	braun	
2 Oxy, 2-(3,4-dimethoxy-phenyl)-äthylamin	0,32	rotbraun	

Triäthylamin 0,58, n-Propylamin 0,53, i-Propylamin 0,55, Dipropylamin 0,78, n-Butylamin 0,67, Di-n-Butylamin 0,86, i-Butylamin 0,64, Di-i-Butylamin 0,87, i-Amylamin 0,76, i-Hexylamin 0,64, Allylamin 0,47. Unterscheidung gesättigter Amine von ungesättigten gelingt nach Anfärben mit Folins Reagens: Gesättigte Amine zeigen eine taubenblaue Farbe, Allylamin eine gelbgrüne. Wird nach Besprühen mit dem Reagens mit Essigsäure besprüht, so ergibt sich bei gesättigten und ungesättigten ein Farbumschlag nach Orange. Die Amine wurden (bisher aus tierischem Material) in eine eisgekühlte Vorlage in 5 ml Eisessig destilliert, dann wurde die Lösung im Vakuum eingeengt. Empfindlichkeit des Nachweises bei Verwendung von Folin-Reagens 5 γ, bei Ninhydrin etwas größer.

Nach Davies, Wolfe und Perry (1953) ist zur Trennung einer großen Reihe von Aminen die eindimensionale aufsteigende Papierchromatographie mit wassergesättigtem Butanol am vorteilhaftesten. Besprüht wird mit einer 0,2%igen Ninhydrinlösung in Isopropanol, das nach Fitzpatrick (1949) 20% Pyridin enthält. Tab. 10 verzeichnet die durchschnittlichen R_f-Werte und die Farben der Ninhydrinkomplexe einer Reihe von Aminen im Vergleich mit R_f-Werten aus der Literatur. Die Veränderung der R_f-Werte mit zunehmender Kettenlänge der Amine ist in Abb. 1 veranschaulicht. Bei Verbindungen mit weniger als 7 Kohlenstoff-Atomen nimmt der R_f-Wert bei Verlängerung der Kette um 1 C-Atom um 0,11 zu. Die R_f-Werte von Isoamylamin und anderen verzweigten Aminen entsprechen angenähert den aus der Gesamtzahl ihrer C-Atome sich ergebenden Zahlen. Sulfhydryl- und Hydroxyl-Substitutionen beim Äthylamin erniedrigen die R_f-Werte. Bei aromatisch substituierten Aminen vermindert eine Hydroxylgruppe in p-Stellung den R_f-Wert und verschiebt die Farbe des Ninhydrinkomplexes nach Rotbraun. Eine weitere Hydroxylgruppe im Ring bewirkt eine

---

[1] Alle Amine wurden in Form ihrer Hydrochloride verwendet.

weitere Herabsetzung der R_f-Werte. Die Substitution der Ringhydroxyle mit Methoxy-Gruppen erhöht die R_f-Werte. Ein Hydroxyl in α-Stellung zur Aminogruppe ändert die gewöhnlich purpurrote Farbe der aliphatischen Aminreihe nicht. Histamin und Tryptamin können von den einfachen und den phenylsubstituierten aliphatischen Aminen durch ihre braunrote Farbe und durch ihre R_f-Werte unterschieden werden. Unter den angegebenen Bedingungen geben sekundäre oder tertiäre Amine oder andere Verbindungen mit Ninhydrin keine Farbreaktion, ebensowenig primäre Aminogruppen, die direkt am Benzolring sitzen, Amide, sekundäre und tertiäre Amine.

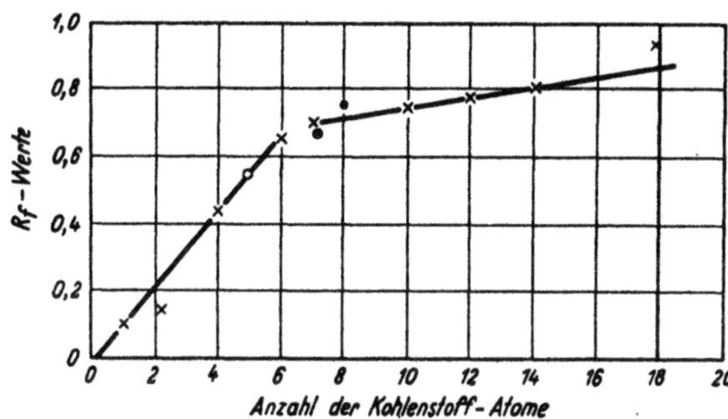

Abb. 1. Beziehung zwischen Länge der Kohlenstoff-Kette und Anzahl der Kohlenstoff-Atome und der R_f-Werte von aliphatischen primären Aminen in Butanol—Wasser. $\times$ = unverzweigte gesättigte primäre Amine; O = Isoamylamin; $\ominus$ = 1,3-Dimethylpentylamin; $\bullet$ = 1-Methylheptylamin.

Das Verhalten von N-haltigen Verbindungen gegen Ninhydrin in Gegenwart von Pyridin ist aus der Tab. 11 ersichtlich.

Tabelle 11. *Reaktion von Ninhydrin mit Stickstoff-Verbindungen in Gegenwart von Pyridin.*

Reaktions-geschwindigkeit	Art der Verbindung	Beispiele
rasch	H | R—C—NH₂ | H	Äthylamin, Butylamin, Isoamylamin, Hexylamin, Heptylamin, Decylamin, Dodecylamin, Tetradecylamin, Octadecylamin, Tyramin, Tryptamin, Histamin, Arterenol, Glucosamin, Glutamin, 2-Phenyläthylamin. Ausnahme: Methylamin.
	H | R—C—COOH | NH₂	Glycin, Alanin, Leuzin, Methionin, Phenylalanin, Tryptophan, Tyrosin, Histidin, Lysin, Arginin, Ornithin, Citrullin, Glutaminsäure, Asparaginsäure.
verzögert	H | R—C—NH₂ | R′	1,3-Dimethyl-pentylamin; 1-Methyl-heptylamin; 2-(4-oxyphenyl)-äthylamin; 1-Butyl,2-oxy-2-phenyläthylamin; 1-Phenyläthylamin, Cyclohexylamin. Ausnahme: Aminosäuren.
	H | R—C—C—NH₂ || | O H	α-Dimethyl-amino, 2,4-Dioxy-acetophenon; α-Amino 3,4,-Dioxy-acetophenon.
	andere	Methylamin, Prolin.
keine Reaktion	⟩—NH₂	Anilin, p-Bromanilin, p-Nitranilin, Adenin.
	R—C—NH₂ || O	Harnstoff Acetamid.
	H | R—N—R′ R″NH₂	Adrenalin, Xanthin. Ausnahme: Prolin. Ammonium, Hydroxylamin.

R und R′ bedeuten organische Substituenten; R″ OH oder H.

3. Quantitative Bestimmungen von Aminen mit der Ninhydrin-Reaktion.

Es werden Aminlösungen, die 2 und 20 $\times$ 10^{-4} mMol/ml enthalten, in saurem Isobutanol hergestellt. Die Lösungen mit unbekanntem Amingehalt werden mit saurem Isopropanol verdünnt, so daß ihre Konzentration etwa in diesen Bereich fällt. 1 ml jeder Standardlösung wird in ein Pyrexglas mit 30 ml Fassungsvermögen gegeben. Es werden 5 ml Pyridin, 2 ml Ninhydrinreagens und Säure-Isopropanolmischung hinzugefügt, so daß sich 10 ml ergeben. Man verwendet nicht mehr als 3 ml der zu analysierenden Lösung und behandelt sie in gleicher

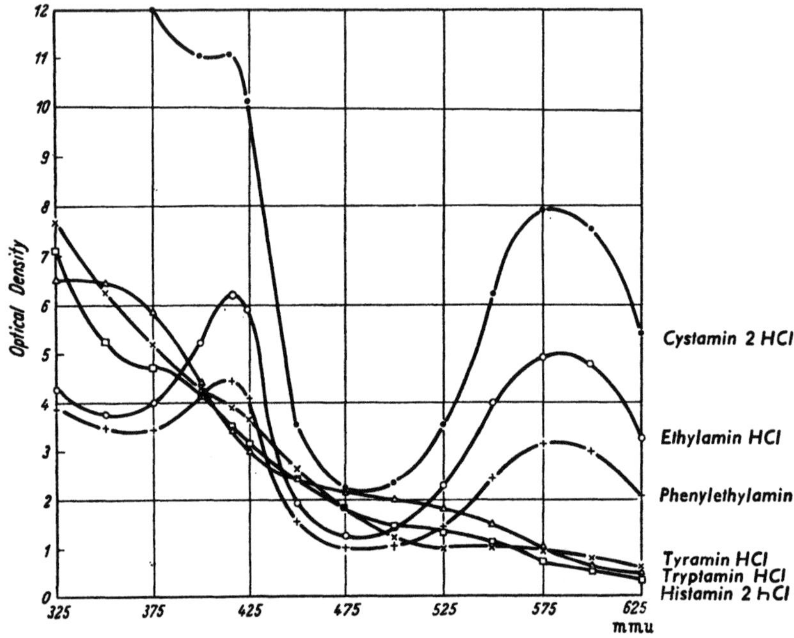

Abb. 2. Spektrale Absorptionskurven von Ninhydrin-Amin-Komplexen ausgewählter primärer Amine. Die Konzentration der Amine in den gefärbten Lösungen war jeweils 1,0 $\times$ 10^{-4} M. 2 Gipfel sind charakteristisch für einfache aliphatische Amine und α-Phenyl-Äthylamin. Ein einzelner Gipfel für andere aromatische und heterocyclische Verbindungen.

Weise. Ein Leeransatz wird hergestellt, bei welchem die Aminlösung durch Isopropanol ersetzt wird. Es wird dann gründlich gemischt, wobei die Berührung der Lösungen mit Stopfen oder Haut vermieden werden muß. Die Ansätze werden genau 7 min lang in ein Wasserbad von 85° gegeben, dann sofort in kaltem Wasser abgekühlt. Die Farbintensität kann innerhalb der ersten Stunde ohne merkliche Abblassung bei 575 mμ abgelesen und aus einer Eichkurve des zugehörigen Amins kann die Konzentration der zu bestimmenden Lösung entnommen werden. Bei einer Wellenlänge von 575 mμ und einer Küvettenweite von 1 mm ist im Beckman-Spektrophotometer der Spektralbereich 30 μ groß, innerhalb dessen alle Amine ihr Absorptionsmaximum haben. Aus den Absorptionswerten in diesem Bereich kann die Konzentration eines Amingemisches abgelesen werden, allerdings mit einem Fehler bis zu 50% belastet. Die Absorptionskurven einer Reihe biologisch wichtiger Amine sind in Abb. 2 wiedergegeben. Auch eine Elution der Amin-Ninhydrin-Komplexe aus den Chromatogrammen mit Wasser und quantitative Bestimmung der in ihnen enthaltenen Amine ist mit dem Beckman-Spektrophotometer möglich.

Tabelle 12. *Relative Absorption bei 400 und 570 mμ ausgewählter aliphatisch und aromatisch substituierter Ninhydrinamine, eluiert aus Papierchromatogrammen.*

Aminhydrochlorid	Konzentrat im Eluat μ M	Absorption[1] bei 400 mμ	Absorption[1] bei 570 mμ	Verhältnis 400/570 mμ
Einfache aliphatische				
Methylamin	500	0,134	0,142	0,944
Äthylamin	333	0,254	0,291	0,863
n-Propylamin	333	0,292	0,323	0,904
n-Butylamin	333	0,349	0,365	0,956
n-Amylamin	333	0,427	0,437	0,977
Isoamylamin	333	0,423	0,434	0,975
	333	0,266	0,280	0,950
	333	0,423	0,453	0,934
n-Hexylamin	500	0,490	0,498	0,984
1-Methyl-Heptylamin	250	0,106	0,084	1,262
Decylamin	400	0,208	0,207	1,005
S- und O-substituierte aliphatische Amine				
Glutamin	1000	0,423	0,468	0,904
β-Mercaptoäthylamin	500	0,237	0,172	1,378
Cystamin	333	0,430	0,460	0,935
Ringsubstituierte				
β-Phenyläthylamin	2500	0,261	0,174	1,500
Tyramin	1000	0,529	0,387	1,367
Nor-adrenalin	333	0,305	0,204	1,495
Arterenon	333	0,173	0,106	1,632
Tryptamin	400	0,256	0,162	1,580
Serotonin Kreatin SO$_4$ (5-OH Tryptamin)	166	0,228	0,099	2,303
Histamin.	500	0,220	0,147	1,497
d,l-p-Oxy-a-methylphenyl-äthylamin	333	0,116	0,096	1,208

Wird ein und dieselbe Aminkonzentration gemessen, nach Elution des Amin-Ninhydrin-Komplexes aus dem Papierchromatogramm oder nach direkter Reaktion im Reagenzglas, so ist das Verhältnis der gewonnenen Absorptionswerte, die bei 400 und 570 mμ abgelesen werden, bei aliphatischen Aminen etwa gleich 1, bei ringsubstituierten Aminen steigt das Verhältnis bis zu 2,3. Findet sich diese Verschiebung des Absorptionsverhältnisses bei einem Amin, so kann mit Sicherheit angenommen werden, daß es sich um ein ringsubstituiertes Amin handelt. In Tab. 12 sind diese Verhältnisse für eine größere Reihe von Aminen belegt.

4. Papierchromatographie quaternärer Ammoniumbasen und verwandter Verbindungen.

(BREGOFF, ROBERTS und DELWICHE 1953.)

Bereitung der Extrakte: Die Gewebe werden nach dem Zerkleinern in einem Homogenisator oder durch Zerreiben in einem Mörser mit Glas oder Sand mit 70%igem Alkohol extrahiert. Der alkoholische Extrakt wird bei 40—50° in einem warmen Luftstrom zur Trockne verdampft und der Rückstand in Wasser aufgenommen, so daß 1 ml Lösung 3 g frischem Gewebe entspricht. Der wäßrige Extrakt wird enteiweißt durch Behandeln mit $^1/_{10}$ bis $^1/_5$ seines Volumens mit Trichloressigsäure oder durch Dialyse gegen das 5fache Volumen Wasser. Das

---

[1] Das Maximum der Absorption beider Gipfel des Ninhydrin-Amin-Komplexes verschiebt sich gegen die ultraviolette Region hin, wenn vom Papier mit Wasser eluiert wurde. Deswegen werden besser 400 u. 570 mμ, denn 415 u. 575 mμ Ablesungen benützt.

Dialysat wird zum ursprünglichen Volumen konzentriert bei 40—50° in Gegenwart von 0,1—0,2 Vol.-% Benzyl- oder Octylalkohol.

Fraktionierung der Reineckesalze: Die Reineckate der Verbindungen, die keine freie Carboxylgruppe enthalten, werden bei alkalischer Reaktion zuerst niedergeschlagen. Die Reineckate der freien Betaine werden dann nach Ansäuern des Filtrates erhalten. 2—5 ml der enteiweißten Lösung, die wenigstens 1 mg methylierter Stickstoffverbindungen enthalten soll, werden in Zentrifugengläser gegeben und auf p_H 8,5—9 durch Zugabe von 1 n Ammoniak oder 1—2 n NH_3—NH_4Cl-Puffer von p_H 9—10 gebracht.

2 ml einer 5%igen Reineckesalzlösung in absolutem Methanol werden zugefügt und die Lösung wird 2 Std. im Eisschrank belassen. Die Suspension wird in der Kälte zentrifugiert und die überstehende Lösung so vollständig wie möglich in ein anderes Zentrifugenglas übergeführt und aufbewahrt. Der Niederschlag wird 3 mal mit 1—2,5 ml n-Propanol ausgewaschen und die Waschflüssigkeit so vollständig wie möglich entfernt. Hat man es mit einem Extrakt unbekannter Zusammensetzung zu tun, so sollte die Propanol-Waschflüssigkeit aufbewahrt werden, da einige der Alkaloid-Reineckate in n-Propanol löslich sind. Der gewaschene Niederschlag wird in 1—2 ml Aceton gelöst und eine etwaige Trübung durch Zentrifugieren entfernt. Die intensiv gefärbte Acetonlösung wird in ein graduiertes Gefäß abgegossen und der graue amorphe Rückstand mit Aceton gewaschen. Das Acetonwaschwasser ist gewöhnlich nahezu farblos, es wird zu der überstehenden Lösung hinzugefügt und die Lösung zu einem geeigneten Volumen mit Aceton aufgefüllt. Die Lösung kann nun einer quantitativen spektrophotometrischen Bestimmung nach Glick (1944) vor der Chromatographie unterworfen werden. Die alkalische wäßrige Mutterlauge des Reineckeniederschlags wird durch Zufügung 6 n HCl auf p_H 1,2 eingestellt und die Lösung wiederum 2 Std. in der Kälte belassen. Der saure Reineckeniederschlag wird dann genau so behandelt wie der alkalische.

Chromatographie: Der dialysierte wäßrige Extrakt oder die Acetonlösung des Reineckeniederschlages wird punktförmig auf Whatman Nr. 1-Papier in Anteilen, die 50—100 γ von jedem stickstoffhaltigen Bestandteil enthalten, aufgetragen. Flecke von alkalischen Reineckeniederschlägen, welche Cholin enthalten, müssen vor der Chromatographie mit Silbernitrat behandelt werden, um das Cholin aus dem Reineckat freizusetzen. Zu diesem Zweck wird der getrocknete Fleck mit einem Überschuß von 0,1 n $AgNO_3$ behandelt. Die Chromatographie wird entweder mit einem Äthanol-Ammoniak-Lösungsmittel oder in dem Butanol-Eisessig-Lösungsmittel von Munier und Macheboeuf durchgeführt. Das Äthanol-Ammoniak-Lösungsmittel besteht aus 95 ml 95%igem Äthanol + 5 ml konzentriertem NH_4OH und das Butanol-Eisessig-Lösungsmittel aus 100 ml n-Butanol, 30 ml Eisessig und so viel Wasser, daß die Lösung eben sich trübt (80—90 ml). Die obere Phase des Gemisches dient zur Chromatographie. Die untere Phase wird zur Sättigung der Atmosphäre des Chromatographiergefäßes verwendet. Das Äthanol-NH_3-Lösungsmittel ist ein einphasiges System und bedarf eines gut abgeschlossenen Gefäßes. Ab- oder aufsteigende Chromatographie ist möglich. Das Phosphormolybdänsäurereagens von Chargaff und Vischer wird zur Sichtbarmachung der Flecke verwendet. Es gibt dunkelblaue Flecke auf blaßblauem Grund. Ein allgemeiner verwendbares Reagens ist das modifizierte $KBiJ_4$ (Kaliumwismutjodid), welches orange- bis braunrötliche Flecke gibt auf gelbem Grund. Es wird in folgender Weise hergestellt: Eine Stammlösung wird nach Guggenheim bereitet. 8 g Wismutsubnitrat werden in 20—25 ml 30 % HNO_3 gelöst. Spezifisches Gewicht

1,18. Diese Lösung wird langsam unter Rühren zu einer Lösung, die 28 g Kaliumjodid und 1 ml 6 n HCl enthält in etwa 6 ml Wasser zugefügt. Der dunkle Niederschlag löst sich wieder auf, wobei eine orangerote Lösung resultiert, die im Eisschrank abgekühlt wird. Sie wird filtriert und auf 100 ml mit Wasser verdünnt. Die orangerote Stammlösung ist für einige Wochen stabil, wenn sie in einer dunklen Flasche im Kühlschrank aufbewahrt wird. Die Entwicklungslösung (das $KBiJ_4$-Reagens besteht aus 20 ml Wasser, 5 ml 6 n HCl und 2 ml der DRAGENDORFF-Stammlösung) und 5 ml 6 n NaOH werden in dieser Weise zugegeben. Wenn sich nicht alles $Bi(OH)_3$ löst, müssen unter Schütteln einige Tropfen 6 n HCl zugegeben werden. Das $KBiJ_4$-Reagens ist in der Kälte aufbewahrt 10 Tage lang stabil. Die Papiere werden mit $KBiJ_4$-Reagens entwickelt, indem man sie durch das Lösungsmittel zieht, wobei man die Papiere immer in derselben Richtung, nämlich vom Auftropfpunkt bis zur Front zieht, da die Löslichkeit einiger Verbindungen in Wasser eine leichte, doch reproduzierbare Verschiebung der Flecke hervorruft.

In der Tab. 13 sind die R_f-Werte in Butanol-Eisessig und in Alkohol-NH_3 einer Reihe bekannter, methylierter Stickstoffbasen und ihr Verhalten gegenüber dem Nachweisreagens und die Bedingungen, unter der sie durch Reineckesäure gefällt werden, angegeben. Das Nachweisreagens wurde, wie S. 539 erwähnt, von THIES und REUTHER (1954) verbessert.

Tabelle 13. *R_f-Werte und chemisches Verhalten von bekannten methylierten Stickstoffbasen.*

Verbindung	Bu-Eisessig	Äthanol NH_3	$KBiJ_4$	CHARGAFF, LEVINE	Ultraviolett 250 mμ	Fällung mit Reineckelösung p_H
Tetramethylammonium- bromid	0,49		I.	+		alkalisch
Betainhydrochlorid . . .	0,43	0,30	A.	—	—	sauer
Cholinchlorid	0,50	0,48	I., A.	+	—	alkalisch
Acetylcholinchlorid . . .	0,59	hydrol.	I.	+	—	alkalisch
γ-Butyrobetainbromid . .	0,52	0,13	A.	—	—	sauer
Trigonellin		0,20	A.	schwach	absorb. sehrstark	sauer
Hordeninsulfat		0,94	I.	schwach	absorb. mäßig	alkalisch

I. = Sofortiges Erscheinen des Fleckes.

A. = Fleck erscheint erst nach dem Trocknen.

I. A. = Sofortige Reaktion, nur wenn mehr als 50 γ der Substanz vorhanden. Die sofortige Farbreaktion ist dunkelbraun. Beim Trocknen geht sie in einen orangefarbenen Fleck über. Geringe Cholinmengen geben nur diese letztere Reaktion. Verblaßt beim Trocknen ganz oder teilweise, wenn der Farbfleck schwach war.

Die absoluten R_f-Werte der Tab. 13 können um 10% variieren in Abhängigkeit von der Gegenwart anderer in der Lösung enthaltener Substanzen.

Das Äthanol-NH_3-Gemisch wird für Routinezwecke bevorzugt. Die Lösungsmittelfront bewegt sich bei diesem Reagens rasch. Salze in der zu chromatographierenden Lösung stören nur wenig. Zur Identifizierung der Verbindungen werden noch Nachweisreaktionen herangezogen, besonders wenn Verschiebungen der R_f-Werte eintreten. So ist ein positiver LEVY-CHARGAFF-Test charakteristisch für Cholin. Das Auftreten eines intensiven orangeroten Fleckes mit $KBiJ_4$ weist auf eine Verbindung hin, die keine freie Carboxylgruppe besitzt. Verbindungen mit freier Carboxylgruppe geben erst eine Reaktion mit $KBiJ_4$, nachdem das Reagens getrocknet worden war. Diese Verbindungen sind allerdings relativ unempfindlich gegenüber dem Reagens, mindestens 50 γ der freien Betaine werden benötigt, um einen erkennbaren Fleck zu erzielen, während die sofort reagierenden bei etwa 15 γ schon sichtbar sind. Die Methode wird demonstriert an dem Ausfall der Reaktionen an Blättern von 2 Monate alten Zuckerrübenpflanzen. Die Dialysate der Blattextrakte geben sehr saubere Chromatogramme für Betain und Cholin, welche die einzigen quaternären Substanzen sind, die in dieser Pflanze bisher gefunden wurden. Es wurde dabei festgestellt, daß die Extrakte der Blätter von Zuckerrüben und von Keimlingen in verschiedenen Stadien der Entwicklung immer Betain und Cholin enthielten. Jedoch wechselten die relativen Mengen. Jüngere Pflanzen sind gewöhnlich reicher an Cholin, aber die Konzentration dieser Verbindung übersteigt nicht $^1/_{10}$ des Betains in 1 bis 2 Monate alten Pflanzen. Betain und Cholin sind auch in den Wurzeln und Speicherorganen der Zuckerrübe und der roten Rübe enthalten. Die Fraktionierung über die Reineckate wird notwendig, wenn die dialysierten Extrakte große Mengen von störenden Substanzen enthalten, wie die Speichergewebe der roten und der Zuckerrüben und wie bei Weizensamen. Die erfolgreiche Anwendung der Reineckefraktionierung hängt davon ab, daß während der Chromatographie die Reineckeverbindungen sauber in die freien methylierten Stickstoffbasen aufgetrennt werden, die zu ihren charakteristischen Stellen wandern und in das Reineckereagens, welches ein R_f von 0,6—0,7 in der Alkohol-Ammoniumlösung hat. Es erscheint ein rosa Fleck. Nur Cholinreineckat zersetzt sich nicht, und muß daher vor der Chromatographie mit Silbernitrat behandelt werden.

III. Mikrobiologische und pharmakologische Bestimmung von Aminen.

Außer mit Hilfe der geschilderten chemischen Methoden können Amine auch qualitativ nachgewiesen und quantitativ bestimmt werden mit biologischen Methoden, so Cholin (s. S. 569) und Putrescin (s. S. 575) auf Grund ihrer Wuchsstoffwirkung bei Pilzen bzw. Bakterien. Histamin, Tyramin, Tryptamin und Acetylcholin können auf Grund ihrer pharmakologischen Wirkungen mit Hilfe von Standardlösungen rasch quantitativ bestimmt werden, z. B. Histamin am Blutdruck der Katze oder am isolierten Meerschweinchendarm. Eine automatische Apparatur für pharmakologische Messungen an isolierten Präparaten stammt von Bourer, Mongar und Schild (1954). Die biologischen Meßverfahren sind zeitraubend, die pharmakologischen rasch durchführbar. Beide können äußerst empfindlich gestaltet werden und benötigen daher nur sehr geringe Substanzmengen.

E. Spezieller Teil.

I. Methylamine.

1. Methylamin: CH_3NH_2.
(Klein 1931.)

CH_5N: C 38,7%, H 16,1%, N 45,2%. Mol.-Gew. 31.

Eigenschaften. Farbloses, ammoniakartig riechendes Gas, Kp. bei 755 mm Hg. 1 Vol. Wasser löst bei 12,5°, 1150 Vol. bei 25° 959 Vol. Methylamin. Leicht löslich in Alkohol.

Derivate.

Chlorhydrat. $CH_3NH_2 \cdot HCl$. Zerfl. Blätter, Fp. 226—227°, löslich in Alkohol.

Platinsalz. $(CH_3NH_2 \cdot)_2 \cdot H_2PtCl_6$. *Goldsalz* $(CH_3NH_2 \cdot HAuCl_4 + H_2O)$, Tetraeder wasserfrei, monoklin, in Wasser leicht löslich. Goldgelbe hexagonale Prismen, unlöslich in abs. Alkohol, löslich in 50 Teilen kaltem Wasser.

Pikrat. $CH_3NH_2C_6H_3O_7N_3$, Prismen oder Tafeln, Fp. 215°, 1,3 Teile in 100 Teilen Wasser löslich. Blaßgelbe Kristalle vom Zersp. 244°. Schwer löslich in H_2O.

Pikrolonat. $H_2N \cdot CH_3C_{10}H_8O_5N_4$, blaßgelbe Nadeln, die sich bei 244° zersetzen. Löslich in 1073 Teilen kaltem und in 369 Teilen kochendem Wasser, in 4717 Teilen kaltem und in 135 Teilen siedendem Alkohol löslich.

α-Naphthylisocyanatverbindung. Fp. 196—197°.

Vorkommen. Methylamin ist in höheren Pflanzen in Pilzen und Bakterien nachgewiesen. Eine sekundäre Entstehung in Pflanzen ist nicht immer ausgeschlossen worden. CROM-WELL (1949) isolierte Methylamin als Pikrolonat aus Schößlingen von *Mercurialis perennis* L. und bestimmte die Konzentration der Base spektrophotometrisch (s. S. 550). Da die Infiltration einer Glycin- und Äthanolaminlösung in Blätter von *Mercurialis perennis* L. zur Methylaminvermehrung führt, nimmt CROMWELL an, daß Methylamin aus Glycin über die Zwischenstufen Äthanolamin und Methyläthanolamin vermutlich durch Bakterieneinwirkung gebildet wird.

<h3 align="center">Qualitativer Nachweis.</h3>

(s. auch S. 519 ff.)

Reaktion nach TSALAPANTI mit Chloranil. Wird eine methylaminhaltige Lösung mit Salzsäure neutralisiert, zur Trockne verdampft, der Rückstand in 95%igem Alkohol aufgelöst und die Lösung mit einer Spur Chloranil (Tetrachlorchinon) auf 70° erwärmt, so tritt Violettfärbung auf. Die Reaktion wird durch Ammoniak nicht hervorgerufen.

<h3 align="center">Isolierung von Methylamin aus Mercurialis perennis
und quantitative Bestimmung.
(CROMWELL 1949 a.)</h3>

Blühende Schößlinge (15 kg) wurden in 50 g-Portionen etwa 10 sec lang in Äther suspendiert. Nach dem Verdampfen des anhaftenden Äthers wurde das Material 5 min lang in 100 ml Wasser, das mit 0,01 n HCl auf p_H 4,5 angesäuert war, homogenisiert. Das Homogenat wurde durch Mull filtriert und ausgepreßt. Die vereinigten Filtrate wurden zur Ausfällung von Eiweiß rasch auf 80° erhitzt. Nach dem Zentrifugieren wurde das Überstehende mit einem Überschuß von $Mg(OH)_2$ behandelt und solange in 25 ml 0,1 n HCl dampfdestilliert, bis 100 ml Destillat übergegangen waren. Aus den vereinigten Destillaten (1875 ml) wurde Ammoniak bei p_H 7,4 nach FRANCOIS, PUGH und QUASTEL (1937) entfernt, das Filtrat der FRANCOIS-Trennung wurde mit einem Überschuß an $Mg(OH)_2$ versetzt und in 50 ml einer Lösung von Pikrolonsäure (70 mg in Äthanol) abdestilliert. Die Pikrolonatlösung wurde eingeengt. Das abgeschiedene Methylaminpikrolonat wurde aus wäßrigem Äthanol zweimal umkristallisiert.

Ausbeute: 35 mg Pikrolonat, Fp. 242°. Synthetisches Methylaminpikrolonat: Fp. 243°. Mischschmelzpunkt 242°.

<h3 align="center">Nachweis und Bestimmung von Methylamin.</h3>

1. Methylaminbestimmung nach ORMSBY und JOHNSON (1950). Die Methode gründet sich auf eine Rotfärbung mit Laktose, Absorptionsmaximum bei 540 mμ. Die Farbreaktion wird durch NH_3 verstärkt. Di- und Trimethylamin geben keine Farbreaktion, beeinflussen jedoch diejenige mit Methylamin im Sinne einer Abschwächung der Farbintensität wenn ihre Konzentration mehr als 20—40mal so hoch ist wie die des Methylamins.

Reagentien. 1. Standardmethylaminlösung: 20 g Methylaminhydrochlorid werden in einem Liter Wasser gelöst. Der Stickstoffgehalt (etwa 4 mg Stickstoff/ml) wird nach der Mikrokjeldahlmethode bestimmt. Durch geeignete Verdünnung der Stammlösung wird eine Arbeitsstandardlösung, die 0,1 mg Stickstoff/ml enthält, hergestellt. Beide Lösungen werden durch einige Tropfen Chloroform konserviert.

2. 3%ige wäßrige Laktoselösung mit Thymol konserviert.

3. Ammoniumsulfatlösung: 6,607 g reines trockenes $(NH_4)_2SO_4$ werden in Wasser gelöst, gegen Lackmus neutralisiert und auf 100 ml mit Wasser aufgefüllt. Die Lösung enthält 14 mg NH_3N/ml. Sie wird mit Chloroform konserviert.

4. 20%ige Natronlauge.

Ausführung der Bestimmung: Das Methylamin wird mit 5 ml 30%iger Natronlauge aus 10 ml der zu untersuchenden Lösung unter Zusatz einiger Tropfen Mineralöl zum Verhüten des Schäumens übergetrieben. Das Absorptionsgefäß enthält 10 ml 0,2 n HCl. Während des Destillierens muß das p_H der Absorptionsflüssigkeit mit Reagenspapier kontrolliert werden. Gegebenenfalls müssen einige Tropfen 2 n HCl nachgegeben werden, um die Reaktion sauer zu erhalten. Es wird nach dem Eintreten des Dampfes in den Kühler 10 min lang destilliert. Dann wird das Absorptionsgefäß gesenkt, so daß das Kühlerende sich über dem Flüssigkeitsspiegel befindet und noch einige Minuten weiter destilliert. Das Destillat wird auf p_H 6—7 gebracht. Ist der Methylamingehalt geringer als 1 mg Stickstoff/100 ml, so muß die zu untersuchende Lösung vor der Destillation eingeengt werden. Proben der zu untersuchenden Lösung (maximal 6 ml), die 0,01—0,15 mg Methylamin-N enthalten, werden in Reagenzröhrchen gegeben. Von der verdünnten Standardlösung werden 0,01, 0,05 und 0,1 mg Amin-N enthaltende Volumina in Reagenzgläser gegeben, ferner 6 ml Wasser als Leerversuch in ein weiteres Reagenzglas. Alle Gläser werden nun auf 6 ml mit Wasser aufgefüllt und mit 0,5 ml Ammoniumsulfatlösung, 0,5 ml Laktoselösung und schließlich mit 0,2 ml 20%iger Natronlauge versetzt. Der Inhalt der Gläser wird durch vorsichtiges Umdrehen gemischt. Kräftiges Schütteln muß vermieden werden, weil Belüftung der Lösungen die endgültige Farbintensität herabsetzt. Nach Verschluß der Reagenzgläser mit Glasstopfen oder Kugeln werden sie 30 min lang bei 56° (Wasserbad), dann 1 Std. lang bei Raumtemperatur belassen. Dann wird kolorimetriert, z. B. photoelektrisch unter Benutzung von Filter Nr. 54 in einem Klett-Summerson-Photometer. Auf der geraden Eichkurve (direkte Proportionalität zwischen Farbe und Konzentration) werden die Methylaminwerte abgelesen. In Kontrollversuchen wurden 100% der vorgelegten Methylaminmenge wiedergefunden.

Quantitative Bestimmung von Methylamin in Blättern nach Cromwell (1949).

20—50 g Blätter werden 10 sec lang in Äther suspendiert. Nach dem Verdampfen von anhaftendem Äther werden die Blätter in einem Homogenisator mit 50—75 ml dest. Wasser, das mit 0,01 n HCl auf p_H 4,5 angesäuert ist, homogenisiert. Das Homogenat wird durch feinen Mull filtriert und ausgepreßt. Die Zelltrümmer werden 2mal mit je 10 ml dest. Wasser gewaschen. Die vereinigten Filtrate und Waschwässer werden rasch auf 80° erhitzt, der entstehende Niederschlag wird abzentrifugiert, das klare Überstehende abgegossen und der Rückstand 1mal mit 10 ml dest. Wasser gewaschen. Der Extrakt wird in ein Destillationsgefäß übergeführt, mit einem Überschuß an $Mg(OH)_2$ behandelt und im Vakuum bei 40° in 5 ml 0,01 n HCl destilliert bis zu einem Gesamtvolumen von 15 ml.

Die Methylaminlösungen werden dann bei niedriger Temperatur mit Hilfe der Mikrodiffusionstechnik von Conway und Byrne (1933, vgl. auch Bd. 1 dieses Handbuches) konzentriert. Es werden breite Schalen von 9,5—4,5 cm Durchmesser verwendet. 10 ml des Destillates werden in das äußere und 1 ml 0,01 n HCl in das innere Gefäß gegeben. In das äußere Gefäß werden dann 5 ml gesättigte K_2CO_3-Lösung pipettiert und die Schalen bei 30° 16—18 Std. belassen. Der Inhalt des inneren Gefäßes wird quantitativ in einen Mikrodestillationsapparat übergeführt und das Methylamin in Anlehnung an das Verfahren zur Glycinbestimmung nach Alexander (1945) gemessen.

Die Methode vermag noch 3 γ Methylamin zu erfassen. Einem Homogenat zugefügtes Methylamin wurde zu 95% wiedergefunden.

Methylaminbestimmung nach ALEXANDER, LANDWEHR und SELIGMAN 1945.

Prinzip. Methylamin, Glycin und N-Methyläthanolamin reagieren in saurer Lösung mit Ninhydrin unter Bildung von Formaldehyd und Ammoniak. Die Reaktion kann für Methylamin spezifisch gestaltet werden, da es auf Grund seiner großen Flüchtigkeit bei niedriger Temperatur und unter reduziertem Druck vom Glycin und N-Methyläthanolamin abgetrennt werden kann. Unter diesen Bedingungen geht N-Methyläthanolamin, obwohl es auch etwas flüchtig ist, nicht über. Der bei der Reaktion des Methylamins mit Ninhydrin sich bildende Formaldehyd wird quantitativ bestimmt.

Reagentien. Ninhydrinlösung 1%ig, Phosphatpuffer p_H 5,5: 3,5 g Trikaliumphosphat werden zu 100 ml einer 20%igen Lösung von Mono-Kaliumphosphat gegeben. Schwefelsäure 0,66 n; Natriumwolframatlösung 10%ig. Konzentrierte Schwefelsäure, spez. Gewicht 1,84; 5%ige Chromotropsäurelösung (1,8-Dioxynaphthalin-3,6-disulfosäure) wird im Eisschrank aufbewahrt. Sie ist mindestens 2 Wochen haltbar. 5 ml der zu analysierenden Methylaminlösung werden in ein Destillationsgefäß gebracht, das mit 2 ml Phosphatpuffer, 1 ml Ninhydrinlösung und einem Glaskügelchen beschickt ist. Nach Verbindung des Destillationsgefäßes mit einem Kühler wird rasch in ein Reagenzglas, das 1 ml 0,66 n H_2SO_4 enthält und bei 10 ml kalibriert ist, destilliert. Nachdem etwa 7 ml Destillat übergegangen sind, wird der Destillationskolben auf Raumtemperatur abgekühlt, dann werden 2 ml Wasser einpipettiert und bis zur Trockne weiter destilliert. Die gesamte Destillationszeit soll 15 min nicht überschreiten. Das Absorptionsgefäß wird auf 10 ml mit Wasser aufgefüllt und durchgeschüttelt. 5 ml der Lösung werden in einem Eisbad abgekühlt und langsam mit 4 ml konzentrierter Schwefelsäure versetzt. Nach Abkühlen auf Zimmertemperatur werden zur Lösung 3 Tropfen Chromotropsäure gegeben. Der Ansatz wird durchgeschüttelt, mit einem Stopfen verschlossen und 30 min lang in ein kochendes Wasserbad gebracht. Nach Abkühlung der Lösung wird in einem Klett-Summerson-Photometer mit Filter 565 oder 540 photometriert. Anhand einer Eichkurve werden die Methylaminwerte ermittelt.

Eine Methode zur Bestimmung von Methylaminen unter besonderer Berücksichtigung der Reaktionsprodukte von Formaldehyd mit NH_3, Mono-, Di- und Trimethylaminen wurde von ATTWOOD, FORD, HOYLE und PARKES (1950) angegeben. Sie wurde bei biologischem Material bisher nicht angewandt.

Über Absorptionsspektren der dunkelvioletten Methyl- und Äthylaminsalze mit Violantin s. GAIND und DUTT (1933).

2. Dimethylamin $(CH_3)_2NH$.

C_2H_7N: C 53,3%. H 15,6%, N 31,1%, Mol.-Gew. 45

Eigenschaften. Farbloses Gas, das sich bei $+7,2°$ verflüssigt; von ammoniakähnlichem an Trimethylamin erinnerndem Geruch. Leicht löslich in Wasser und Alkohol.

Derivate.

Chlorhydrat. $(CH_3)_2NH \cdot HCl$. Nadeln vom Fp. 171°. Im Gegensatz zum Monomethylaminchlorhydrat in Chloroform löslich; unlöslich in Alkohol. Perjodid $(CH_3)_2NH \cdot HJ \cdot J_2$. Hexagonale, an trockener Luft beständige Tafeln vom Fp. 83—85°.

Platinsalz. $[(CH_3)_2NH]_2 \cdot H_2PtCl_6$. Die orangegelben Blättchen sind dimorph, rhombisch. Fp. 206°. Leicht löslich in heißem, weniger in kaltem Wasser. In Alkohol ist es löslicher als das Salz des Methylamins.

Pikrat. $(CH_3)_2NH \cdot C_6(NO_2)_3OH$. Hellgelbe, feine Nadeln vom Zersp. 222°. Löslich in 764 Teilen kaltem, 33 Teilen siedendem Wasser; in 853 Teilen kaltem, 38 Teilen heißem Alkohol.

Rufianat. 0,3 g Dimethylaminchlorhydrat werden in 5 ml Äthylalkohol gelöst und mit 4%iger alkoholischer Rufiansäurelösung versetzt. Roter, kristalliner Niederschlag, Waschen mit absolutem Alkohol, Ausbeute an Dimethylaminmonorufianat 0,6 g. Fp. 255—256°.

Vorkommen. Steiner und Stein v. Kamienski (1953). Im Fruchtkörper von *Russula aurata*. In grünen Pflanzen kann Dimethylamin als Spaltprodukt aus Gramin dem β-Indolylmethyl-dimethylamin entstehen (Wieland, Fischer und Moewus, 1948), ferner in Pilzen, die durch Fäulnisbakterien zersetzt sind. *Phallus impudicus* L., Stink-Morchel; im Fruchtkörper von *Courbonia virgata* (neben Ammoniak, Trimethylamin und Isoamylamin) (Henry und Grindlay, 1949). Zum qualitativen Nachweis s. S. 519 ff.

Quantitative polarographische Bestimmung.
(English 1951).

Zur Bestimmung von $(CH_3)_2 \cdot NH$ in Gegenwart von NH_3, Mono- und Trimethylamin behandelt man das Gemisch mit HNO_2, wodurch NH_3 und CH_3NH_2 zerstört und Dimethylamin in das Nitrosamin übergeführt wird. Nitrosamin kann bei p_H 0,8 (jedoch nicht in neutralem oder alkalischem Medium) polarographisch bestimmt werden.

Es sind folgende Reagentien zur Bestimmung notwendig: Sulfaminsäure (NH_2SO_3H). Man löst 15 g in Wasser auf 100 ml 30%ige Natriumhydroxydlösung. Natriumhydrosulfit, gewöhnlich als Hydrosulfit bezeichnet, $Na_2S_2O_4$. Es ist ein Handelspräparat von hohem Reinheitsgrad anwendbar.

Jod: etwa 0,2 n, bereitet durch Lösung von 50 g Kaliumjodid und 26 g Jod in 50 ml Wasser und Auffüllen auf 1 Liter. Es ist nicht notwendig, die Lösung zu standardisieren. Gelatine: 0,5 g Gelatine (Knox Nr. 1) gelöst in 50 ml warmem Wasser. Die Gelatinelösung muß täglich frisch bereitet werden.

Ausführung. Von der 15—25% Gesamtamine enthaltenden wäßrigen Lösung entnimmt man einen 0,05—0,15 g Dimethylamin enthaltenden aliquoten Teil, verdünnt mit 5 ml Wasser und gibt unter Eiskühlung 3,0 ± 0,05 g $NaNO_2$ und 5 ml Eisessig hinzu und läßt 10 min in einem Wasserbad von 25° stehen. Nach Zerstörung der überschüssigen HNO_2 durch etwa 17 ml 15%ige Sulfaminsäurelösung neutralisiert man mit 30%iger NaOH gegen Phenolphthalein, gibt noch 0,1 ml der NaOH hinzu und erwärmt im Wasserbad auf 60°. Nach Zugabe von 1,0 ± 0,05 g $Na_2S_2O_4$ erwärmt man noch 5 min auf 60°, kühlt auf 25° ab und füllt auf 100 ml auf. 10 ml dieser Lösung werden mit 50 ml Wasser, 1 ml Eisessig und 1 ml Stärkelösung versetzt und mit 0,2 n Jodlösung wird das überschüssige $Na_2S_2O_4$ zerstört. Nach Zugabe von 2,5 ml konz. HCl und 1 ml 1%iger Gelatinelösung verdünnt man zu 100 ml und polarographiert mit einem Anfangspotential von 0,3 V, und einer solchen Empfindlichkeit, daß eine Stufenhöhe von 100 bis 200 mm erhalten wird. Fehler ±1%.

3. Trimethylamin.

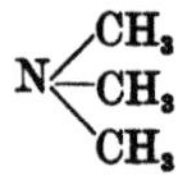

C_3H_9N: C 61,0%, H 15,2%, N 23,7%. Mol.-Gew. 59.

Eigenschaften. Farbloses Gas, das sich bei $+3°$ verflüssigt. Fischartiger, an Heringslake erinnernder Geruch. Mit dem Geruchssinn lassen sich noch 2 γ Trimethylamin nachweisen. Eine Lösung der Base riecht bei den ersten Proben nach Fisch, dann nach Monoalkylaminen und schließlich nach Ammoniak. (Geruchsumschlag nach Kauffmann und Vorländer, 1910.) In Wasser sehr leicht löslich. AgCl ist in Trimethylamin zum Unterschied von Mono- und Dimethylamin völlig unlöslich.

Derivate. *Hydrochlorid* $(CH_3)_3N \cdot HCl$. Schmilzt bei 271—275° unter Zersetzung. In siedendem Chloroform löslich.

Perjodid. $(CH_3)_3N \cdot HJ \cdot J_4$. Hexagonale Tafeln Fp. 65°. Eine Trimethylaminchlorhydratlösung scheidet noch bei einer Verdünnung von 1:50000 auf Zusatz von Jodjodkali die Perjodidverbindung ab. Durch Sättigung der Lösung mit Ammonchlorid wird die Fällbarkeitsgrenze bis auf 1:100000 erhöht.

Platinsalz. (CH₃)₃N · H₂ PtCl₆). Reguläre, orangefarbene Kristalle, die bei 190° unter Zersetzung schmelzen. In Alkohol löslicher als das Platinsalz des Dimethylamins, 0,017 g lösen sich in 100 ml siedendem absolutem Alkohol.

Pikrat. (CH₃)₃N · C₆H₂(NO₂)₃OH. Hellgelbe Prismen vom Fp. 216°. Löslich in 77 Teilen Wasser.

Pikrolonat. (CH₃)₃N · C₁₀H₈O₅N₄. Hellgelbe, rhombische Täfelchen vom Zersp. 252°. Löslich in 1121 Teilen kaltem und 166 Teilen siedendem Wasser und in 794 Teilen kaltem und 233 Teilen siedendem Alkohol.

Vorkommen: Nach KAPELLER-ADLER und VERING (1932) ist Trimethylamin in Seewasserpflanzen, nicht aber in Süßwasserpflanzen enthalten. Bei Gymnospermen bisher nicht gefunden. Die Bildung von Trimethylamin in *Chenopodium vulvaria* L. erfolgt unter der Einwirkung eines Enzyms, das aus dem Preßsaft abgetrennt werden konnte. Es zerlegt Cholin bei p_H 7,8 in Trimethylamin und Glykol. Betain, Äthanolamin, Methyl- und Dimethylaminoäthanol werden unter den gleichen Bedingungen nicht gespalten (CROMWELL, 1950). Trimethylamin entsteht durch Zersetzung von Cholin unter dem Einfluß gewisser *Clostridium*-Arten (COHEN, NISMAN und RAYNAUD, 1947; DYER und WOOD, FISH, 1947).

Mikromethode zur Trimethylaminbestimmung.
(CROMWELL 1950.)

Die Methode gründet sich auf die Beobachtung von PALUMBO (1949), daß tertiäre Amine in Gegenwart von cis-Aconitsäureanhydrid, gelöst in Essigsäureanhydrid eine rotviolette Färbung ergeben, während NH₃, primäre und sekundäre Amine nicht reagieren. Mit dieser Methode können 5—55 γ Trimethylamin erfaßt werden.

Reagentien. Cis-Aconitsäureanhydrid. Es wird erhalten aus Aconitsäure nach der Methode von R. MALACHOWSKY und MASLOWSKY (1928). Das Produkt wird durch viermaliges Umkristallisieren aus Benzol gereinigt. Eine 2,5%ige Lösung in Eisessig wird vor Gebrauch frisch bereitet. Eisessig p. A., Essigsäureanhydrid p. A., Toluol (schwefelfrei), Cholinchlorid, Betainhydrochlorid, Na₂SO₄ (wasserfrei) p. A., Kaliumcarbonat p. A.

Ausführung der Bestimmung. Eine Lösung (4 oder 5 ml), die nicht mehr als 100 γ Trimethylamin enthält, wird in einen 25 ml fassenden zylindrischen Scheidetrichter gebracht und mit 10 ml Toluol überschichtet. Nach dem Zufügen von 3 ml gesättigter K₂CO₃-Lösung wird eine Minute lang kräftig geschüttelt und 5 min stehen gelassen. Die K₂CO₃-Lösung wird abgelassen und die Toluollösung des Trimethylamins in ein Reagensglas gegossen, das mit 0,5 g wasserfreiem Na₂SO₄ beschickt ist. Das Glas wird verschlossen und 10 sec lang geschüttelt. In ein 2. Reagensglas werden 0,1 ml des Aconitsäureanhydridreagens, 2 ml Essigsäureanhydrid und 5 ml der Toluollösung in der angegebenen Reihenfolge einpipettiert. Die Lösung wird gründlich gemischt und das Glas lose verstopft in einem Wasserthermostaten bei 65° 10 min lang belassen. Die Lösung wird dann auf Zimmertemperatur abgekühlt und nach 15 min photometriert. Spektrophotometer nach ILFORD unter Benutzung eines Gelb-Grünfilters. Mit Hilfe einer Standardlösung von reinem Trimethylaminhydrochlorid wird eine Eichkurve, die eine Gerade darstellt, gewonnen, anhand derer die Trimethylaminwerte der zu untersuchenden Lösung ermittelt werden. Die Farbe ist mindestens 30 min lang stabil und bleicht langsam bei längerem Stehen aus. Andere tertiäre Amine geben rote oder violette Farben und stören die Trimethylaminbestimmung. Durch Destillation bei niedriger Temperatur und unter reduziertem Druck wird Trimethylamin von den meisten tertiären Aminen abgetrennt, die stören könnten. Toluol gibt einen schwach rosa Farbton und es ist daher notwendig, jeweils eine Kontrollbestimmung mit Toluol allein durchzuführen. Die Temperatur und Zeitangaben für die Farbentwicklung sind genau zu beachten und bei Erneuerung der

Reagentien muß jeweils eine neue Eichkurve aufgenommen werden. Die Trimethylaminlösung in Toluol muß so rasch wie möglich der Mischung des cis-Aconitanhydridreagens und des Essigsäureanhydrid zugesetzt werden, da sich bei längerem Stehen der Mischung bei Zimmertemperatur langsam eine rote Farbe entwickelt.

Bestimmung in *Chenopodium vulvaria* L. Da beim Trocknen der Pflanzen beträchtliche Trimethylaminverluste auftreten, muß für die Analyse frisches Gewebe verwendet werden. Die Blätter werden mit der Schere abgenommen, in einem Mörser unter 1 ml n HCl zu einer Paste zerrieben, mit 10 ml dest. Wasser versetzt und nach gründlichem Durchmischen wird in eine Vakuumdestillationsapparatur übergespült. Es wird 1 ml einer Antischaumlösung, bestehend aus 10 g Paraffin, Fp. 54° C in 100 ml Toluol und ein Überschuß von $Mg(OH)_2$ zugefügt. Bei vermindertem Druck und bei 40° werden in 40 min 20 ml überdestilliert. Das Destillat wird auf 25 oder 50 ml aufgefüllt. Es werden Proben von je 4 ml für die Analyse verwendet (vgl. Tab. 14).

Tabelle 14. *Trimethylamingehalt in Blättern von Chenopodium vulvaria L.* (CROMWELL, *1950*)
(γ Trimethylamin pro g Frischgewicht).

Pflanzen Nr.	Junge Pflanzen			Blühende Pflanzen		
	Intakte Blätter	Blätter ohne Drüsen	Drüsen	Intakte Blätter	Blätter ohne Drüsen	Drüsen
1	850	265	585	425	125	300
2	818	221	597	470	137	333
3	740	218	522	580	145	435
4	890	282	608	326	102	224
5	730	248	482	410	116	294
6	805	260	545	370	110	260
Durchschnitt	805	249	556	430	122	308

4. Quantitative Mikroanalyse einer Mischung von Ammoniak und von Methylaminen.
(JAMES und MARTIN 1952 b:)

Ein sehr langwieriges Verfahren hat REAY (1937) angegeben. Es beruht auf der getrennten Analyse verschiedener Anteile der Mischung nach elektivem Niederschlagen oder Zerstören der einen oder anderen Komponente. Von FUKS und RAPPOPORT (1948) wurde die Verteilungschromatographie auf einer Säule, die aus einem Gemisch aus Stärke und Calciumhydroxyd bestand, zur Analyse herangezogen. Es wurde mit n-Butanol-Wasser entwickelt. Die Amine liefen als freie Basen mit der ausfließenden Lösung aus und wurden durch kontinuierliche Titration bestimmt. Die 4 Amine wurden voneinander getrennt. Geringe Verluste entstehen infolge der Flüchtigkeit der Amine beim Auftragen auf die Säule. LAGERKVIST (1950) trennte Ammoniak und Methylamin auf Stärkesäulen mit n-Propanol—Wasser—Salzsäure als Lösungsmittel, die Positionen der Zonen wurden durch Behandeln der Eluate mit Ninhydrin bestimmt. Die wiedergefundenen Mengen wurden nicht angegeben.

Eine Trennung von Ammoniak und den 3 Methylaminen gelingt mit der Gas-Flüssigkeitschromatographie (MARTIN und SMITH, 1952) unter Verwendung von Kieselgursäulen, die mit einer Mischung von 5-Äthyl-nonan 2-ol und flüssigem Paraffin als stationäre Phase gefüllt sind, bei einer Temperatur von 78,6°. Die Amine kommen aus der Säule in der Ordnung ihrer Kochpunkte (Ammoniak —33,4°, Methylamin —6,5°, Trimethylamin +3,5°, Dimethylamin +7,4°). Bei der Anwendung des Verfahrens auf die Untersuchung von *Chenopodium vulvaria* L.

wurden die Resultate von Cromwell (1950) bestätigt. Es wurden nur Ammoniak und Trimethylamin gefunden. Es wurde eine 120 cm lange Säule, deren flüssige Phase aus Hendekanol 15% und flüssigem Paraffin (Vol./Vol.) bestand, verwendet. Flußgeschwindigkeit des Stickstoffs 6,7 ml/min. Stickstoffdruck 80 mm Hg. Es können noch 2 γ NH$_3$, 4 γ Monomethylamin, 7 γ Dimethylamin und 8 γ Trimethylamin erfaßt werden. Maximal können noch gut getrennt werden 160 γ Ammoniak, 180 γ Monomethylamin, 180 γ Dimethylamin und 220 γ Trimethylamin.

Bestimmung der Amine in *Chenopodium vulvaria* L. Die Blätter von 4 Pflanzen, etwa 100 g, werden in einem Homogenisator zerkleinert und mit 1 Liter 0,001 n HCl extrahiert. Der Extrakt wird filtriert, zentrifugiert und mit Natronlauge alkalisiert. Die Amine werden dann durch Luft in eine mit 0,001 nHCl beschickte Vorlage abgeblasen. Nach einer Stunde wird die Waschflasche entfernt, ihr Inhalt fast zur Trockne eingeengt und die Lösung auf 2 ml aufgefüllt. 0,1 ml Portionen werden für die Gasflüssigkeitsverteilungschromatographie angewandt.

Bei Modellansätzen werden nahezu die eingesetzten Aminmengen wiedergefunden. Fehlerbreite $\pm 4\%$. In der beschriebenen Weise konnte eine Reihe von aliphatischen, primären, sekundären und tertiären Aminen getrennt und quantitativ bestimmt werden. Bereitung der Chromatographiersäulen und Beschreibung der Titrationsapparatur s. James und Martin (1952).

Nach Attwood, Ford, Hoyle und Parkes (1950) wird das Gemisch der freien Amine zur Bestimmung von NH$_3$ mit 5 n HCl gegen Methylrot neutralisiert. Nach Zusatz von KCl wird eingeengt, vorsichtig mit Natriumacetat versetzt und mit 20% gefällt. Aus dem Niederschlag wird NH$_3$ durch Destillation mit NaOH in Freiheit gesetzt und in 0,1 n Säure aufgefangen.

Die Summe von Monomethylamin und NH$_3$ wird nach van Slyke mit NaNO$_2$ auf Grund des entwickelten N$_2$ in einem modifizierten Orsatschen Aparat bestimmt und Monomethylamin aus der Differenz errechnet. Dimethylamin wird mit einer 25%igen ätherischen Lösung von p-Toluolsulfonylchlorid als N-Dimethyl-p-toluolsulfonamid gefällt. Der Niederschlag wird mit H$_2$SO$_4$ hydrolysiert und Dimethylamin in einem NH$_3$-Bestimmungs-Apparat bestimmt, wobei 1 ml 0,1 n Säure 4,64 mg Dimethylamin entsprechen. Trimethylamin reagiert nicht mit HNO$_2$ und kann im Rückstand durch Redestillation und Titration bestimmt werden. Eine Korrektur für überdestilliertes NO$_2$ ist erforderlich.

5. Tetramethylammoniumhydroxyd.

(CH$_3$)$_4$N · OH kommt in *Courbonia virgata* A. Brogn. (Henry, 1948) und in *Courbonia pseudo-petalosa* (Nigeria) vor. Tetramin wird aus *Courbonia virgata* nach Henry und Grindlay (1948) in folgender Weise isoliert:

Die Base wird als Perjodid durch Zufügen eines Überschusses von Jod-Jodkaliumlösung zu dem vorher gereinigten Alkoholextrakt des pflanzlichen Materials ausgefällt. Zersetzung des Perjodids durch Erhitzen mit Wasser, Verdampfung zur Trockne und Waschen des Rückstandes mit Alkohol. Umkristallisierung aus heißem Wasser. Tetramethylammoniumjodid schmilzt bei 100°.

Tetramethylammoniumhydroxyd kann mit Silberoxyd aus den Halogensalzen freigesetzt werden und kristallisiert aus wäßrig alkoholischer Lösung als Hydrat aus. *Pentahydrat:* C$_4$H$_{13}$ON + 5 H$_2$O, sehr hygroskopische Nadeln, Fp. 62—63°. *Trihydrat* C$_4$H$_{13}$ON + 3 H$_2$O, Fp. 59—60°, geht im Vakuum in das *Monohydrat* C$_4$H$_{13}$ON + H$_2$O über, Fp. 130—135° unter Zersetzung in Trimethylamin und Methanol. *Chlorid* C$_{14}$H$_{12}$N · Cl, hygroskopische Kristalle aus abs. Alkohol. — *Pikrat* $C_4H_{12}N \cdot C_6H_2O_7N_3$, lange Nadeln, Zersetzung bei 312°. — *Chloraurat* $C_4H_{12}N \cdot AuCl_4$, lange Nadeln schwer löslich, Zersetzung bei 336°. — *Chloroplatinat* $(C_4H_{12}N)_2PtCl_6$, orangegelbe Octaeder, Zersetzung bei 278°, wenig löslich in Wasser, — *Reineckat* $(CH_3)_4N(Cr(NH_3)_2(SCN)_4H_2O$.

Isolierung aus *Courbonia virgata* neben Stachydrin s. S. 608.

6. Trimethylaminoxyd $(CH_3)_3NO$.

Bildet mit 2 Mol Wasser ein Hydrat. Hygroskopische Nadeln, Fp. 96°, süß schmeckend, leicht löslich in Methanol, unlöslich in Äther, löslich in Äthanol. Bei Sublimation des Hydrates im Vakuum bei 180° wird hygroskopisches, wasserfreies Trimethylaminoxyd erhalten. Fp. 208°.

Hydrochlorid $C_3H_9ON \cdot HCl$ aus Methanol, Fp. 205—210°. (Zers.) *Pikrat* $C_3H_9ON \cdot C_6H_3O_7N_3$, sehr geeignet zur Isolierung, da es in Äthanol und kaltem Wasser schwer löslich ist, Smp. 197°. — *Chloroplatinat* $(C_3H_9ON)_2 \cdot H_2PtCl_6$, glänzende rhombische Plättchen, Smp. 214°. — *Chloraurat* $C_3H_9ON \cdot HAuCl_4$, Fp. 250°, in kaltem Wasser schwer, in heißem leicht löslich.

Bestimmung von Trimethylaminoxyd in biologischem Material (Ronald und Jakobson, 1948). 10 g homogenisiertes Material werden mit 30—35 ml 96%igem Alkohol verrieben, die Suspension wird in einen 50 ml Kolben übergeführt, mit Alkohol aufgefüllt und nach kurzem Stehen filtriert. 10 ml des klaren Filtrates werden zur Bestimmung der Totalbasen NH_3, Mono-, Di- und Trimethylamin, mittels Wasserdampfdestillation benutzt. Ein anderer aliquoter Teil wird mit $TiCl_3$-Lösung reduziert, der $TiCl_3$-Überschuß durch Zusatz einiger Tropfen gesättigter $NaNO_3$-Lösung entfernt und wie vorher mit Wasserdampf unter Vorlage überschüssiger 0,02 n HCl destilliert.

II. Äthanolamine.

I. Colamin (Aminoäthanol) $HO \cdot CH_2 \cdot CH_2 \cdot NH_2 = C_2H_7ON$: C 39,20%, H 12,68%, N 22,97%, Mol.-Gew. 61.

Eigenschaften: Dickflüssiges farbloses Öl von schwachem Geruch, hygroskopisch, wasser- und alkoholmischbar. Wenig löslich in Äther, Benzol und Ligroin, löslich in Chloroform. Kp. 171—172°, mit Äther und mit Wasserdämpfen flüchtig. Die wäßrige Lösung reagiert stark alkalisch.

Obwohl Aminoäthanol als Phosphatidbaustein und als Zwischenprodukt der Cholinbiosynthese und der Erythrophleum-Alkaloid-Bildung auftritt, wurde es in freier Form bisher nur in seltenen Fällen nachgewiesen.

Die wäßrige Lösung reagiert stark alkalisch.

Hydrochlorid, $C_2H_7ON \cdot HCl$, hygr. Kristalle, Fp. unterhalb 100°. — *Oxalat* $C_2H_7ON \cdot C_2H_2O_4$, leicht löslich in Wasser, wenig löslich in Alkohol. — *Pikrolonat* $C_2H_7ON \cdot C_{10}H_8O_5N_4$, büschelförmig verwachsene gelbe Nadeln, Zers. bei etwa 225°, schwer löslich in Wasser, löslich in etwa 100 Teilen heißem, 400—500 Teilen kaltem Alkohol. — *Flavianat* $C_2H_7ON \cdot C_{10}H_6O_8N_2S$, Prismen, Fp. 211—212°. — Chloroaurat $C_2H_7ON \cdot HAuCl_4$, aus salzsaurer Lösung in monoklinen oder triklinen Kristallen, Fp. 186—187°, Chloroplatinat $(C_2H_7ON)_2 \cdot H_2PtCl_6 + H_2O$, goldgelbe Blättchen. — β-Glycerinphosphat, weiße wachsartige Substanz.

1. Darstellung von Colamin aus dem Phosphatid von *Lolium perenne*.
(Smith und Chibnall 1932, zit. nach Klein 1933.)

2,3 g der Phosphatidfraktion werden durch achtstündiges Kochen mit 5%iger Schwefelsäure hydrolysiert. Die Fettsäuren werden durch Ausschütteln mit Äther entfernt. Die Lösung wird auf 100 ml eingeengt. Die Schwefelsäure wird mit Baryt entfernt, das Bariumsulfat abzentrifugiert und gut mit heißem Wasser ausgewaschen. Dann wird Bleiacetat zugegeben, der geringfügige Niederschlag wird abzentrifugiert und aus der klaren Flüssigkeit das Blei mit Schwefelwasserstoff entfernt. Die so erhaltene klare Lösung wird auf 100 ml eingeengt und mit kleinen Mengen 20%iger Mercuriacetatlösung versetzt, bis eben ein bleibender orangegelber Niederschlag auftritt. Dann wird das gleiche Volumen Alkohol zugegeben und über Nacht stehen gelassen. Der Niederschlag wird abzentrifugiert, mit wenig 50%igem Alkohol gewaschen, in Wasser suspendiert, das Quecksilber mit Schwefelwasserstoff entfernt. Die erhaltene Lösung wird mit etwas Salzsäure angesäuert und im Vakuum zur Trockne gebracht. Das Chlorhydrat des Colamins wird aus dem Rückstand mit Alkohol extrahiert. Nach dem Verdampfen des Alkohols wird der Rückstand in möglichst wenig Salzsäure gelöst und mit einer konzentrierten Goldchloridlösung versetzt. Nach mehrtägigem Stehen im Exsiccator kristallisiert das Goldsalz in braunen kubischen Kristallen aus. Beim Umkristallisieren aus Wasser werden lange, organgefarbene Nadeln erhalten, die bei 189—191° schmelzen unter vorangehendem Sintern bei 187°.

2. Quantitative Bestimmung von Äthanolamin.

Das Verfahren ist zur Bestimmung von Phosphatidyl-Äthanolamin und Phosphatidylserin in Geweben entwickelt worden und gründet sich auf die Reaktion von NICOLET und SHINN (1939), die gemäß folgender Gleichung verläuft:

$$HO \cdot CH_2NH_2 + NaJO_4 = 2\ CH_2O + NH_3 + NaJO_3$$

$$HO \cdot H_2C \cdot CH(NH_2) \cdot COOH + NaJO_4 = HCOOH + CHO \cdot COOH + NH_3 + NaJO_3.$$

Das entwickelte Ammoniak wird durch Dampfdestillation abgetrennt und quantitativ bestimmt. Die Bestimmung von Äthanolamin neben Serin erfolgt in der Weise, daß in einer Probe zuerst die Summe von Äthanolamin und Serin ermittelt wird, in einer zweiten Probe wird dann durch Adsorption an Permutit das Äthanolamin vom Serin abgetrennt und dieses allein bestimmt.

Durchführung der Bestimmung. Reagentien. Perjodat nach FOLIN NaCl-Lösung 30%ig. Borsäure 12.04 g, gelöst in 150 ml n NaOH, 3,5 ml davon gemischt mit 2 ml 0,2 n mol. Perjodsäure ergibt eine Lösung von etwa p_H 9,6. Perjodsäure $(HJO_4 + 2H_2O)$ 0,2 molar; 0,005 n Schwefelsäure oder Salzsäure; 0,005 n Natriumthiosulfat; lösliche Stärke 1%ig. Alle Reagentien chemisch rein, Ammoniak frei und geschützt gegen Ammoniakaufnahme aus der Luft. Für die Perjodidabsorption wird ein 120 mm langes Glasrohr mit 7 mm äußerem Durchmesser verwendet. Das Ende ist leicht verjüngt und mit Baumwolle verstopft. Das Glas wird mit 1,5 g Perjodid beschickt am nicht verjüngten Ende mit einem Gummischlauch mit der Wasserdampfdestillationsapparatur nach PARNAS-WAGNER verbunden. Die Lösung, die etwa 0,6—3 mg Äthanolamin (oder 1—5 mg Serin) in einem Volumen von etwa 5 ml enthält, wird durch die Perjodidsäule geschickt, die dann mit wenigen ml dest. Wassers gewaschen wird. Äthanolamin kann danach mit 5 ml Natriumchloridlösung eluiert werden. Es wird 3 mal mit 1 oder 2 ml Wasser ausgewaschen. Die Ammoniakbestimmung wird wie folgt ausgeführt: In das Rezeptorgefäß der Parnas-Wagner-Apparatur werden genau 5 oder 10 ml 0,01 n H_2SO_4 eingefüllt. In die Flasche werden eingefüllt:

a) die Serin- oder Äthanolamin enthaltende Lösung,
b) ein Tropfen Paraffinöl,
c) 3,5 ml der alkalischen Boratlösung,
d) 2 ml der Perjodidlösung.

Der Hahnstopfen, der den Trichter mit dem Reaktionsgefäß verbindet, wird nach jedem Zufügen geschlossen, dann wird 7 min lang Dampf durchgeleitet. Das Rezeptorgefäß wird gesenkt, so daß die Spitze des Kühlers über dem Spiegel der vorgelegten Säule sich befindet und die Destillation 1 min lang weitergeführt. Das Kühlerende wird dann mit etwas Wasser gewaschen. Zur Vorlage werden 2 ml Kaliumjodidlösung gegeben und die Flasche verstopft. Nach 5 min wird das freigesetzte Jod auf schwach gelb mit 0,005 Natriumthiosulfat titriert (Mikrobürette). Die Titration wird nach Zufügen von 10 Tropfen Stärkelösung vervollständigt. Mit jeder Bestimmung läuft eine Leerbestimmung. Der Verbrauch von 1 ml 0,005 n Thiosulfatlösung entspricht nach Berücksichtigung des Leerwertes 0,303 mg Äthanolamin und 0,525 mg Serin. Soll Äthanolamin in Pflanzen bestimmt werden, so müssen sie zuerst hydrolysiert werden. Die Gewebe werden mit Alkohol entwässert und dann mit kochendem Alkohol in einem kontinuierlichen Extraktionsapparat extrahiert. Der Alkohol wird entfernt bei 45° unter reduziertem Druck. Der trockene Rückstand wird in Chloroform gelöst und die Lösung durch Asbest filtriert. Die Chloroformlösung wird zur Trockne verdampft und der Rückstand 3 Std. mit 5 ml 6 n HCl in Äthanol am Rückflußkühler erhitzt. Das Methanol wird dann unter reduziertem Druck bei 60° entfernt, die Entfernung des HCl sollte vollständig oder nahezu vollständig sein, da die Adsorption von Äthanolamin in saurer Lösung nicht quantitativ verläuft. Bei dem beschriebenen Vorgehen treten keine Verluste an (Serin oder) Äthanolamin auf.

3. Methylaminoäthanol: $HO \cdot CH_2 \cdot CH_2 \cdot NHCH_3$.

Methylaminoäthanol ist ein farbloses dickflüssiges Öl von schwachem Geruch, das aus der Luft Kohlensäure und Wasser anzieht. Es ist mit Wasser und Alkohol mischbar. Wenig löslich in Äther, Benzol und Ligroin, löslich in Chloroform, Kp. 171—172°, mit Ätherdämpfen flüchtiger als mit Wasserdämpfen.

Es wurde aus einer Mutante von *Neurospora crassa* isoliert. Aus 1 kg Pilz wurden 300 mg Pikrolonat erhalten (Horowitz, 1946).

Methylaminoäthanol ist ein Bestandteil des Nitrophleins und wird gewonnen nach saurer Hydrolyse dieser Substanz mit H_2SO_4. Nach Entfernen der Nitrophleinsäure wird der Rückstand mit Kaliumhydroxyd stark alkalisiert und die Mischung über freier Flamme destilliert. Das stark basische Destillat wird erschöpfend mit Äther extrahiert und die organische Lösung über Kaliumcarbonat getrocknet. Nach Zufügen von ätherischer Pikrinsäure zu einem Teil dieser Lösung scheidet sich ein kristallines Pikrat ab. Nach 2 Umkristallisationen erhält man $C_3H_9ONC_6H_3O_7N_3$, Fp. 148°. Das α-Naphthylisothiocyanat, der N-Methyl-N-oxyäthyl-N-α-naphthylthioharnstoff $C_{14}H_{16}ON_2S$ schmilzt bei 125°, (Blount, Openshaw und Todd, 1940).

4. Dimethylaminoäthanol: $HO \cdot CH_2 \cdot CH_2 \cdot N(CH_3)_2$

ist eine Flüssigkeit mit dem Kp. 135°, *Pikrat:* $C_4H_{11}ON \cdot C_6H_3O_7N_3$, derbe Nadeln mit $^1/_2 H_2O$ aus Wasser Fp. 96—97°, ziemlich löslich in Wasser und Alkohol. *Pikronolat:* $C_4H_{11}ON \cdot C_{10}H_8O_5N_4$, gelbe Nadeln aus verdünntem Alkohol. Fp. ca. 197° unter Zersetzung.

Chloroaurat $C_4H_{12}ONCl_4Au$ aus heißem Wasser, goldgelbe Nadeln, Fp. 194°.

Dimethylaminoäthanol kann als Spaltprodukt der sauren Hydrolyse des Cassaidins der Rinde von *Erythrophleum guinense* (Dalma, 1939) in Form des Goldsalzes gewonnen werden, Fp. 194°.

III. Cholin: $(CH_3)_3NOH \cdot CH_2 \cdot CH_2 \cdot OH$.

(KLEIN 1933; GUGGENHEIM 1950.)

$C_5H_{15}O_2N$: C 49,6%, H 12,4%, N 11,6%. Mol.-Gew. 121,13.

Cholin $C_5H_{15}O_2N$, farbloser Sirup, der aus der Luft CO_2 anzieht. Im Hochvakuum über P_2O_5 wird die Base als eine sehr zerfließliche Kristallmasse erhalten. Cholin ist leicht löslich in Wasser und Alkohol, unlöslich in Äther und Chloroform, wenig löslich in Amylalkohol. Die verdünnte wäßrige Lösung der Base ist beständig. Konzentrierte Lösungen zersetzen sich beim Erwärmen in Trimethylamin und Glykol. Bei der trockenen Destillation zerfällt Cholin in Trimethylamin und Glykol, daneben wird auch in geringen Mengen β-Dimethylaminoäthanol und Dimethylvinylamin gebildet. Cholinchlorid ist bis 180° beständig und zerfällt bei der Destillation fast quantitativ in Dimethylaminoäthanol und Methylchlorid. Cholin wirkt an parasympathisch inervierten Organen erregend. Diese Eigenschaft ist in seinem Acetylderivat, dem Acetylcholin, besonders stark ausgeprägt und kann zur Identifizierung von Cholin herangezogen werden (s. S. 571).

Vorkommen des Cholins. Cholin ist im ganzen Pflanzensystem, bei Phanerogamen und Kryptogamen (Algen, Moose, Pilze), bei Blütenpflanzen in allen Organen (Blätter, Stengel, Blüten, Samen, Frucht, Keimpflanzen, Wurzel, Zwiebeln, Knollen, Rinde, Holz) nachgewiesen worden, wobei allerdings bemerkt werden muß, daß eine sekundäre Entstehung von Cholin in den meisten Fällen nicht ausgeschlossen ist. Eine Zusammenstellung der bis dahin bekannten Fundstätten ist in Bd. IV 1 von J. KLEIN: Handbuch der Pflanzenanalyse, Wien 1931/33. Weitere Angaben in: Handbuch d. Drogenkunde v. FR. BERGER, 1949.

Identifizierung. Dazu sind außer dem erwähnten Acetylcholin, das noch gesondert behandelt wird, folgende Derivate geeignet:

Cholinchloroplatinat. $(C_5H_{14}O_2N \cdot Cl)_2 \cdot PtCl_4$. Aus heißen Lösungen kristallisiert es zuerst regulär und geht dann in die stabile monokline Form über. In der Regel werden bei Erkalten von heiß gesättigten alkoholischen wäßrigen Lösungen schmale Prismen erhalten, bei langsamem Verdunsten sechseckige Tafeln und dicke Prismen mit aufgesetzten Pyramiden. Als besonders charakteristische Form werden sechsseitige, dachziegelartig übereinandergeschobene kleine Plättchen oder Täfelchen angesehen. Zersetzungspunkt 234—235°. Der Dimorphismus des Cholinplatinates dient zum Nachweis des Cholins, indem die beiden Modifikationen wechselseitig ineinander übergeführt werden. Es wird auch ein Schmelzpunkt von 226° angegeben (GRÜN und LIMPÄCHER, 1926). Löslich in 5,8 Teilen Wasser, unlöslich in Alkohol.

Perchlorat. $C_5H_{14}ON \cdot ClO_4$. Smp. 273°, löslich in 29 Teilen Wasser von 15°. Beim Abdampfen einer verdünnten wäßrigen Lösung mit Salpetersäure verwandelt es sich in den Salpetersäureester der bei stärkerem Erhitzen ziemlich heftig verpufft.

Chloroaurat. $C_5H_{13}ON \cdot HAuCl_4$. Gelbe Nadeln aus heißem Alkohol, als Rohprodukt bisweilen in Würfeln. Fp. 250—255° bei raschem Erhitzen. Es sind auch höhere Schmelzpunkte angegeben. Löslich in 80 Teilen Wasser von 17,5°, leichter in heißem Wasser; auch in heißem Alkohol gut löslich.

Cholinchlorid-Zinkchlorid. $(C_5H_{14}ONCl)_2 \cdot ZnCl_2$. Beim Zusammenbringen der alkoholischen Lösungen beider Chloride. Kleine farblose Nadeln, schwer löslich in kaltem Alkohol.

Cholinchlorid-Cadmiumchlorid. $C_5H_{14}ONCl \cdot CdCl_2$. Analog erhalten wie Zinkdoppelsalz.

Cholinchlorid-Quecksilberchlorid. $C_5H_{14}ONCl \cdot 6 HgCl_2$. Säulenförmige Kristalle mit schiefer Auslöschung oft in kreuz- oder sternförmigen Gruppierungen. Fp. 244° (auch 250° angegeben). Löslich in 66 Teilen Wasser von 19,5°, sehr schwer in Alkohol. Bei 100° verliert die Verbindung an Gewicht infolge Flüchtigkeit des Sublimates.

Chloroplatinat des Benzoylcholins. Wird Cholinchlorid mit überschüssigem Benzoylchlorid 2—3 Std. im Wasserbad erwärmt, so fällt Platinchlorid aus der wäßrigen Lösung des Reaktionsproduktes das schwerlösliche Chloroplatinat des Benzoylcholins, Fp. 206°.

Phosphorwolframat. $(C_5H_{14}ON)_3 \cdot PO_4 \cdot 12 WO_3$, in Wasser nur wenig löslich, fällt aus Wasser amorph, aus verdünntem Alkohol mikrokristallinisch. Aus alkoholischer Lösung fällen Platin-, Gold- und Mercurichlorid Cholin bereits aus Lösungen von 1:2 Millionen.

Cholinperjodide. $C_5H_{14}ONJ \cdot J_5$; $C_5H_{14}ONJ \cdot J_3$. Nach STANEK (1905) fügt man zu der auf Cholin zu prüfenden Flüssigkeit eine Lösung von Jodjodkali (bereitet durch Auflösen von 153 g Jod und 100 g Jodkali in 200 ml Wasser). Das niedere Perjodid ist ein schwarzes, grünlich schimmerndes Öl, stark metallisch glänzend, in Wasser unlöslich, in Alkohol und in Kaliumjodidlösung löslich. Beim Zusammenbringen mit fein pulverisiertem Jod oder mit Kaliumtrijodidlösung geht es in das grüne kristallinische Ennea-Jodid $C_5H_{14}ON \cdot J_9$ über. Die Verbindung eignet sich zur quantitativen Bestimmung. Ist neben Cholin auch Betain anwesend, so werden die ausgefällten Perjodide durch Kochen mit Kupfer und Kupferchlorid in die Chloride übergeführt und durch Fraktionieren mit Alkohol getrennt.

Goldjodid. Zusammensetzung des Reagens wesentlich: Neutrale wäßrige Goldchloridlösung wird allmählich zu einer wäßrigen Natriumjodidlösung gegeben, und zwar solange, wie sich der entstehende Niederschlag auflöst (4 Mole NaJ, 1 Mol $AuCl_3$); aus neutraler Lösung bis 1:5000 beim Eintrocknen schwarze Prismen und Stäbchen. Erfassungsgrenze 8 γ.

Kalium-quecksilber-chlorid. Wird eine Cholinchloridlösung mit wenig Kalium-quecksilberchlorid versetzt, der Niederschlag in Alkohol aufgelöst und langsam verdunsten gelassen, so entsteht Cholin-quecksilber-jodid in Form kleiner, schwach doppelbrechender Nadeln (SCHOORL, 1918).

Quecksilberchlorid. Die Quecksilberchloridreaktion hat sich bei der Untersuchung von über 100 Pflanzenspezies auf Cholin neben der nachfolgend besprochenen Jodjodkalireaktion sehr gut bewährt (s. dazu die Einschränkung S. 559). Zur Ausführung der Reaktion wird auf den Objektträger zuerst der zu prüfende Tropfen gebracht. Dann werden mit einer Pipette 2—3 Tropfen des Reagens zugefügt (konzentrierte alkoholische oder 50%ige alkoholische, schwach salzsaure Lösung des Salzes). Der Tropfen wird sofort mit einem Deckglas bedeckt und nach 2—3 min untersucht. Die auftretenden Kristalle können von verschiedener Größe sein, charakteristisch sind quadratische Formen sowie Skelettformen, die oft gefiedert sind. Bei stärkeren Cholinkonzentrationen (bis 1:10000) kommt es vor, daß das Reaktionsprodukt amorph ausfällt. Wiederholung der Probe mit etwas verdünnterem Material liefert immer die charakteristischen Kristallformen. Treten in einer stark verunreinigten Probe Kristallformen auf, deren Zugehörigkeit zum Cholinquecksilberchloridsalz nicht ohne weiteres sicher steht, so kann durch Zufügen eines kleinen Tropfens Cholinchlorid (1:20000) zu einem neuen Probetropfen leicht festgestellt werden, ob eine Vermehrung der fraglichen Kristalle erfolgte, oder ob neben ihnen die für das Cholinquecksilberchlorid charakteristischen Formen auftreten. Erfassungsgrenze 0,4—0,8 γ.

Jodjodkali-Reagens nach STANEK (1905) gehört zu den empfindlichsten und besten Reagentien auf Cholin. In Verdünnung von 1:1000000 noch sicher nachzuweisen. Die Reaktion wird folgendermaßen ausgeführt: Auf den Objektträger kommt ein nicht zu großer Tropfen der zu untersuchenden Flüssigkeit, daneben 1 Tropfen des tiefdunkelrotbraunen Reagens. Durch das Auflegen des Deckglases werden die Tropfen so miteinander verbunden, daß sie sich nicht mischen können, sondern daß unter dem Deckglas eine Diffusionszone entsteht. In dieser Zone treten dann (meist in der stärker J-haltigen Hälfte) die charakteristischen Kristalle des Cholinjodproduktes, das vielleicht mit dem von STANEK beschriebenen Enneajodid identisch ist, auf. Die Kristalle sind, je nach der Menge des vorhandenen Cholins, verschieden große, schiefe Prismen; häufig finden sich Zwillingsbildungen, oft tafelförmig ausgebildet, immer schwach durchscheinend (im Gegensatz zu den meist auch ausfallenden Jodkristallen, die tiefschwarz und nicht durchscheinend sind). Die Kristalle sind hellgelbbraun. Diese Reaktion ist die empfindlichste auf Cholin, sie läßt sich in der angegebenen Weise ohne weiteres auch für stark verunreinigte Proben, Pflanzenextrakte und dergleichen anwenden. Die auftretenden Kristalle des Cholinjodproduktes sind nicht beständig und verwandeln sich nach kurzer Zeit in ölige Tropfen, die nach einiger Zeit verschwinden. Es ist notwendig, daß man jedes Präparat unter dem Mikroskop mindestens 5 min im Auge behält, um nach den Kristallen zu suchen, da sie sonst, zumal wenn sich die beiden Tropfen unter dem Deckglas etwas rascher mischen, leicht der Aufmerksamkeit entgehen können. Erfassungsgrenze 0,04 γ.

Kaliumwismutjodid, in salzsaurer Lösung. Bei 1:5000 feine Nädelchen und kleine Prismen. Erfassungsgrenze 4 γ.

Reineckesalz. $C_5H_{14}ON \cdot C_4H_6N_6S_4Cr$. 6seitige Tafeln, Smp. oberhalb 250°; in Wasser von 18° zu 0,002% löslich, in 10%iger HCl zu 0,03%, bei Gegenwart von überschüssigem Ammoniumreineckat beträgt die Löslichkeit nur 0,0015% (STRACK und SCHWANEBERG, 1937). Über die Bestimmung des Cholins mit Aurantiasäure s. S. 562.

1. Fällung von Cholin und Acetylcholin mit Tetraphenylbornatrium.

Nach MARQUARDT und VOGG (1952) übertrifft Tetraphenylbornatrium alle bisher bekannten Fällungsmittel für Cholin und Acetylcholin. Die Erfassungsgrenze in wäßrigen Lösungen liegt für Cholinchlorid unter 1×10^{-5} g/ml und für Acetylcholin unter 1×10^{-4} g/ml in alkoholischer Lösung für Cholinchlorid unter 1×10^{-4} g/ml und für Acetylcholinchlorid unter 5×10^{-4} g/ml. Ein weiterer

Vorteil besteht darin, daß die Fällung in schwach essigsaurer Lösung ausgeführt werden kann und somit eine Verseifung der Cholinester hintangehalten wird. Vor der Bestimmung sind Kaliumsalze, Ammoniumionen und deren organische Verbindungen zu entfernen, sofern diese Basen als echte Salze in Lösung vorliegen. Betainartige Verbindungen, Eiweiß und Aminosäuren, geben mit Tetraphenylbornatrium keine Fällung.

Die Zusammensetzung des gefällten Komplexsalzes entspricht der Formel $C_5H_{14}ON\ B\ (C_6H_5)_4$ = Tetraphenylbor-cholin bzw. $C_7H_{16}O_2N\ B\ (C_4H_5)_4$ = Tetraphenylbor-acetylcholin. In Naturprodukten ist das Acetylcholin immer von Cholin begleitet, dabei findet sich in pflanzlichem und tierischem Gewebe das Cholin immer in großem Überschuß. Keines der bekannten Fällungsmittel bewirkt eine direkte Trennung dieser beiden Substanzen. Es ergeben sich stets Mischkristalle von Cholin und Acetylcholin. Dies trifft auch für das Tetraphenylbornatrium zu. Wichtig ist jedoch, daß die Fällung des Acetylcholins auch bei einem großen Überschuß an Cholin quantitativ erfolgt. Das Vorliegen eines Esters des Cholins im Komplexsalz läßt sich qualitativ und quantitativ mittels der Ultrarotspektroskopie ermitteln. Diese Identifizierung ist auch bei einem großen Cholinüberschuß (1:100) möglich. Mit der Ultrarotspektroskopie lassen sich noch Verdünnungen des Acetylcholins von $1:5 \times 10^5$ erfassen.

Herstellung der Komplexsalze: Die 10 und 100 mg Cholinchlorid bzw. Acetylcholinchlorid enthaltenden Proben werden in etwa 50—100 ml Wasser gelöst. Es wird dabei so viel Essigsäure zugegeben, bis die Reaktion eben lackmussauer ist. Unter Umrühren läßt man zu der Reagenslösung eine etwa 0,1 Mol wäßrige Lösung von Tetraphenylbornatrium im Überschuß zutropfen. Den reichlich auftretenden Niederschlag läßt man 24 Std. im Eisschrank stehen und filtriert danach durch Porzellanfiltertiegel A_1 oder A_2. Die Fällung ist vollständig, wenn auf erneute Reagenszugabe im Filtrat keine Trübung auftritt. Der Niederschlag wird mit Wasser, das einige Tropfen Essigsäure enthält, gründlich ausgewaschen und bei 100° im Trockenschrank bis zur Gewichtskonstanz gebracht.

Physikalisch-chemische Bestimmung. Die Ultrarotspektroskopie wird am Perkin-Elmer-Spektralphotometer Modell 21 durchgeführt. Gewisse funktionelle Gruppen besitzen spezifische Absorptionsbanden, so auch die CO-Gruppe in Estern, die eine intensive Bande in der Gegend 5,7—5,8 hat. Als Lösungsmittel für die Komplexsalze eignet sich Acetonitril, da es keine CO-Gruppe besitzt. (Tetraphenylborcholin- und Acetylcholin lösen sich auch in Aceton, doch ist dieses Lösungsmittel hier nicht brauchbar, da es eine CO-Gruppe besitzt.) Die gelösten Substanzen werden in Steinsalzküvetten von 0,94 mm Schichtdicke in den Substanzstrahlengang gebracht, während sich das reine, wasserfreie Lösungsmittel im Vergleichsstrahlengang befindet.

Biologische Austestung. Praktische Durchführung s. S. 571. Zur Zerlegung der Komplexsalze wird eine abgewogene Menge in Acetonwasser (1:1) gelöst und mit überschüssigem Quecksilber (II)-Chlorid und mit Kochsalz versetzt, um die schwach saure Reaktion des Sublimats durch Komplexbildung zu unterbinden. Vom abgeschiedenen Phenylquecksilberchlorid wird abfiltriert und das klare Filtrat im Vakuum zur Trockne eingeengt. Zur biologischen Auswertung wird in 0,9%iger NaCl-Lösung aufgenommen.

2. Isolierung des Cholins.

Nach EWINS (1914) kann Cholin neben Acetylcholin wie folgt isoliert werden: 1500 ml alkoholischer Extrakt werden im Vakuum auf 400 ml eingedampft und mit wäßriger Quecksilberchloridlösung (1:16) versetzt, bis kein Niederschlag mehr entsteht. Das Filtrat wird von überschüssigem Quecksilber mit Schwefelwasserstoff befreit. Das Filtrat vom Quecksilbersulfid wird mit so viel Sodalösung versetzt, daß es schwach sauer bleibt und im Vakuum zu einem dünnen Sirup eingeengt. Dieser wird in 94%igen Alkohol eingegossen. Der entstehende Niederschlag wird nach 12stündigem Stehen abgenutscht. Das Filtrat wird abdestilliert, der Rückstand im Vakuum getrocknet und in 50 ml Methylalkohol gelöst. (Ein hierbei verbleibender Rückstand wird verworfen.) Die methylalkoholische Lösung wird mit 300 ml absolutem Alkohol gefällt. Zum Filtrat von diesem Niederschlag wird die zur Fällung erforderliche Menge alkoholischer Quecksilberchloridlösung gegeben. Nach mehrtägigem Stehen wird der Quecksilberniederschlag abfiltriert, der Niederschlag wird getrocknet, fein zerrieben und mit je 150 ml heißem Wasser viermal ausgelaugt. Der Rückstand wird durch Filtrieren entfernt. Das nach dem Erkalten nochmals filtrierte Filtrat wird im Wasserbad auf 40 ml eingeengt. Die dabei entstandene kristallinische Fällung wird nach Abkühlen abfiltriert, getrocknet, mit heißem Wasser nach feinem Zerreiben gut aufgeschlämmt und mit Schwefelwasserstoff behandelt (mehrmals wiederholen). Filtrate und Waschwässer werden durch Einleiten von Luft vom Schwefelwasserstoff befreit und die stark saure Lösung mit frisch gefälltem Silbercarbonat bis zur Chlorfreiheit des Filtrates behandelt. Der Silberüberschuß wird aus dem Filtrat durch Schwefelwasserstoff entfernt, dieser wird durch Luft verjagt. Die verbleibende Lösung wird mit Weinsäurelösung schwach angesäuert, dann im Vakuum bei 60—70° zur Trockne gebracht. Der Rückstand wird mit absolutem Alkohol erschöpfend extrahiert. Das alkoholische Filtrat wird auf 15 ml eingeengt und 2 Tage stehen gelassen; das sich ausscheidende saure weinsaure Cholin wird abfiltriert. Das alkoholische Filtrat vom weinsauren Cholin wird nach Konzentrieren bei 40—50° mit alkoholischer Platinchloridlösung versetzt. Die Fällung wird bei 30—40° mit 70%igem Alkohol ausgezogen, worin die Acetylcholinverbindung im Gegensatz zu der des Cholins gut löslich ist. Durch Abdunsten bei 30—40° wird das Acetylcholinchloroplatinat gewonnen. Der Rückstand vom Alkoholauszug, der noch bedeutende Mengen Cholinchloroplatinat enthält, wird zur Gewinnung des letzteren aus wenig Wasser unter Zusatz von Alkohol umkristallisiert.

Trennungen. In Frage kommt vor allem die Trennung vom Betain. Hierzu dient in erster Linie die verschiedene Löslichkeit der Chlorhydrate in kaltem absolutem Alkohol, worin Cholinchlorid sehr leicht, Betainchlorhydrat schwer löslich ist. Auch durch Umkristallisieren der Quecksilberverbindung läßt sich eine Reinigung erzielen, Cholinquecksilberchlorid ist in Wasser bedeutend weniger löslich als Betainquecksilberchlorid. Ist wenig Betain neben viel Cholin vorhanden, so kann die Reinigung über die Perjodide erfolgen. Betain wird durch Kaliumtrijodid nur in saurer Lösung gefällt, Cholin auch in alkalischer. Man löst die Chlorhydrate in der 30—40fachen Menge Wasser, neutralisiert mit Soda, setzt etwa 2% Natriumcarbonatlösung hinzu und fällt das Cholin mit Kaliumtrijodid. Die verschiedene Löslichkeit ihrer Quecksilberchloriddoppelsalze eignet sich zur Trennung des Cholins vom Trigonellin. In neutraler, mäßig verdünnter Lösung fällt Cholinquecksilberchlorid aus, während das Trigonellinquecksilberchlorid sich erst in stark schwefelsaurer Lösung abscheidet.

Cholin kann von Betain und Colamin getrennt werden mit Dipikrylamin (Aurantiasäure). Es sind nämlich die Aurantianate des Betains und des Colamins leichter löslich als die des Cholins. Zur Trennung des Colamins vom Cholin kann man unter Zuhilfenahme von Kalk Colamin mit Äther extrahieren, worin das Cholin sich nicht löst (THIERFELDER, 1920). Auch ist das Cholin als Chloroplatinat und in dessen Filtrat das Colamin nach VAN SLYKE bestimmt worden, doch muß bei Fällung von Cholin als Platinat in alkoholischer Lösung mit einem durchschnittlichen Verlust von 9—10% gerechnet werden. Eine Abtrennung des Colamins vom Cholin gelingt ferner durch Kupplung des Cholins mit β-Naphthalinsulfochlorid oder des Colamins mit Carbobenzoxychlorid. Auch läßt sich die verschiedene Alkohollöslichkeit der Reineckate der beiden Basen mit Nutzen verwenden. Die Abtrennung des Cholins vom Betain, bzw. Colamin mit Aurantiasäure ist besonders bequem und empfiehlt sich auch durch die vorzügliche Ausbeute an Cholin (etwa 99% aus 1%iger Lösung des Chlorids). Auffallend ist ein Unterschied in der Farbe und Kristallisation. Während Cholinaurantiat ein leuchtend scharlachrotes Pulver mikroskopischer Kristalle darstellt, sind die betreffenden Salze des Betains und Colamins bordeauxrot und bilden makroskopisch gut ausgebildete Kristalle.

3. Nachweis und Bestimmung des Cholins. (KLEIN 1933.)

Bei Behandlung von Cholin mit konzentrierter Kalilauge tritt der charakteristische Geruch des Trimethylamins nach Heringslake auf, was den Nachweis von Cholin noch in einer Verdünnung von 1:2000000 erlaubt.

Alloxanprobe: Wird nach Entfernung von Eiweißkörpern und Ammonsalzen mit einer gesättigten Lösung von Alloxan auf dem Wasserbad eingedampft, so tritt eine schöne rotviolette, auf Zusatz von Alkalien tiefblau werdende Färbung auf. Dampft man eine verdünnte Cholinperchlorat-Lösung (ca. 0,1 g in 50 ml Wasser) unter Zusatz von 2 ml reiner 65%iger Salpetersäure auf dem Wasserbade ein und löst den Rückstand in wenig heißem Wasser, so kristallisiert nach Zugabe von einigen Tropfen verdünnter Perchlorsäure das Perchlorat des Cholin-salpetersäureesters in atlasglänzenden, fast rechtwinkligen Platten, $(CH_3)_3N(HClO_4) \cdot CH_2 \cdot$ $CH_2 \cdot O \cdot NO_2$, von sehr hoher Doppelbrechung und äußerst lebhaften Polarisationsfarben (Smp. 185—186°) aus. 100 Teile Wasser lösen bei 15° nur 0,62 Teile, durch einen mäßigen Überschuß von Perchlorsäure wird die Löslichkeit noch bedeutend verringert.

Nachweis des Cholins im Gewebsschnitt. Sind in einem Pflanzenteil größere Cholinmengen vorhanden, so gelingt es leicht, durch Einlegen von Gewebsschnitten in Sublimatreagens (s. S. 560) das Cholin durch die charakteristischen Kristalle des Cholinquecksilberchlorids nachzuweisen.

Nachweis nach Extraktion. 0,5—1,0 g des frischen Pflanzenmaterials werden in kleinen Reagensgläschen mit etwa 5 ml 96%igem Alkohol übergossen. Der Alkohol enthält auf je 100 ml 1—2 ml 2 n HCl. Die Proben werden in der Kälte mindestens 24 Std. stehengelassen. Dann wird der Alkohol abgegossen und unter gelindem Erwärmen abgedunstet. Nun wird durch kurzes Umschwenken mit 1 ml Wasser aufgenommen und sofort mit 1 Tropfen dieser Flüssigkeit die Reaktion angestellt. Um einigermaßen vergleichbare Resultate zu erhalten, wird in Reihenversuchen der Abdunstungsrückstand immer mit der gleichen Menge Wasser aufgenommen. Von der mit 1 ml Wasser aufgenommenen Probe wird 1 Tropfen zur Reaktion mit Sublimatreagens verwendet. Es lassen sich aus einem Ansatz bequem 10—20 Parallelproben ausführen. (Die Methode kann nicht als einwandfrei gelten, da Cholin durch die Säure aus Lecithin abgespalten werden kann.)

4. Die Bestimmung von freiem Cholin in Pflanzen.
(STREET, KENYON und WATSON 1946.)

Cholin ist als Bestandteil des Lecithins wahrscheinlich in jeder lebenden Zelle vorhanden. SCHULZE (1912) und KLEIN und ZELLER (1930) untersuchten über 100 Pflanzen der verschiedensten Familien und fanden mit Ausnahme von 3 Flechten in allen Fällen freies Cholin. Nach DUCET (1948) trifft jedoch die Behauptung von KLEIN und LINSER (1932), daß Cholin hauptsächlich in freier Form vorkomme, nicht zu. Vielmehr liegt es nach DUCET (1948) als lipoidlösliche Verbindung, nämlich als Lecithin vor. Bei der Cholinbestimmung nach KLEIN und LINSER (1932) wurde nicht berücksichtigt, daß die in allen bisher untersuchten Pflanzen enthaltene Lecithinase vom Beginn des Welkens der Blätter an aus den pflanzlichen Phosphatiden Cholin freilegt. Neben freiem Cholin gibt es noch gebundene lösliche Formen, die auf dem Weg der Lecithinsynthese liegen und ausschließlich in jungen Organen während des Wachstums auftreten. Freies Cholin überwiegt nur in den Samen und in jungen Pflanzen bei der Keimung.

Für die quantitative Bestimmung von Cholin in Pflanzen sind zahlreiche Verfahren angegeben worden. Die Bestimmung des Cholins wurde früher mit Hilfe der Bildung von schwerlöslichem Cholinperjodid, Wismutjodid, Platinchlorid, Goldchlorid oder Reineckat durchgeführt. Mit der Perjodidmethode konnte ROMAN (1930) 0,005—5 mg Cholin in reiner wäßriger Lösung bestimmen. Fehlergrenze ±5%. Doch ist es nur sehr schwer, einheitliche Niederschläge zu erhalten und Jodverluste zu vermeiden. Dazu kommt, daß die Methode bei Pflanzen deswegen schwierig anzuwenden ist, weil Jod auch mit Alkaloiden und vielen anderen Basen Niederschläge gibt, eine Schwierigkeit, die nur teilweise durch Hinzufügen von Natriumcarbonat (SCHULZE und TRIER, 1912a; VICKERY,

1925) umgangen werden kann, auch wird die Löslichkeit des Cholin-Perjodids allgemein durch die Gegenwart von Salzen gesteigert. Bei Methoden, die Kaliumwismutjodid als Fällungsmittel verwenden, finden sich ähnliche Schwierigkeiten. STRACK und SCHWANEBERG (1937) beobachteten, daß selbst reine Basen unter bestimmten Bedingungen ein atypisches Goldsalz geben und daß· von gemischten Lösungen gemischte Salze niedergeschlagen werden, so daß Goldchlorid für das Niederschlagen oder die Bestimmung von Basen ungeeignet erscheint. Auch Platin- oder Quecksilberchlorid führt zu Salzgemischen, aus denen das Cholinderivat nur durch weitere Reaktionen und Umkristallisationen abgetrennt werden kann. Methoden, die auf die Fällung von Cholin als Reineckat sich gründen, wurden .von KAPFHAMMER und BISCHOFF (1930), BEATTIE (1936) und ENGEL (1942) angegeben. STRACK und SCHWANEBERG (1937) und NEUBERGER und SANGER (1942) geben an, daß alle basischen Substanzen Reineckate geben, die bei saurer oder manchmal bei neutraler oder schwach alkalischer Reaktion unlöslich sind; nur die quaternären Basen ohne Carboxylgruppe, wie Cholin und Methylpyridinium bilden Reineckate, die bei p_H 12—13 unlöslich sind. Nach STREET und Mitarbeiter (1946) sind jedoch die aus Pflanzenextrakten mit Reineckat bei p_H 12 erhaltenen Niederschläge so schwer· zu filtrieren, daß Zersetzung auf dem Filter erfolgt. Für die Bestimmung des Cholingehaltes des Reineckatniederschlages beschriebene Methoden (BEATTIE, 1936; JOHNSTON et al., 1939; MILLER und LOWE, 1940) sind für reines Cholinreineckat und nicht für Niederschläge, die aus Pflanzenextrakten mit Ammoniumreineckat bei neutraler Reaktion erhalten werden, anwendbar. Nach GULEWITSCH (1898) entsteht bei der Behandlung von Cholin mit konzentrierter Kalilauge oder mit feuchtem Silberoxyd Trimethylamin. Auf diese Beobachtung gründet sich das Verfahren der Cholinbestimmung nach KLEIN und ZELLER (1930) und KLEIN und LINSER (1932). Das Verfahren, welches eine 3stündige Behandlung der Pflanzenextrakte mit, 33%iger NaOH bei hoher Temperatur, bis zu 180°, einschließt, verursacht so tiefgehende Veränderungen in den Pflanzenextrakten, daß selbst Betain, eine der widerstandsfähigsten Basen, dabei teilweise zu Trimethylamin abgebaut wird, so daß es zur Cholinbestimmung ungeeignet erscheint. Nach LINTZEL und FOMIN (1931a) entsteht bei Behandlung von Cholin mit alkalischem Permanganat Trimethylamin und Glykolsäure nach der folgenden Gleichung:

$$CH_2(OH) \cdot CH_2N(CH_3)_3OH \xrightarrow{O_2} CH_2(OH) \cdot COOH + N(CH_3)_3 + H_2O.$$

Auch Acetylcholin, Neurin und Carnitin liefern bei alkalischer Oxydation Trimethylamin, nicht aber Betain, und γ-Butyrobetain. Neurin liefert bei Behandlung mit saurem Permanganat Trimethylamin, während unter diesen Bedingungen Cholin zu Betain oxydiert wird. Weder Neurin noch Carnitin sind aber bisher in höheren Pflanzen aufgefunden worden. Weitere Untersuchungen von LINTZEL und FOMIN (1931b) sowie LINTZEL und MONASTERIO (1931) waren die Basis zum Verfahren der Cholinbestimmung von STREET, KENYON und WATSON (1946), das von COTTE und KAHANE (1950) weiter entwickelt worden ist. Bei der quantitativen Bestimmung von Cholin durch oxydative Spaltung mit Permanganat in alkalischem Milieu können Verluste an Trimethylamin entstehen, durch Bildung von sekundären und primären Aminen sowie von Ammoniak. Die freiwerdenden Methylgruppen werden zu Formaldehyd oxydiert, der mit primären Aminen SCHIFFsche Basen bildet und so die Amine der maßanalytischen Bestimmung entzieht. Zur Vermeidung von Trimethylaminverlusten müssen die von COTTE und KAHANE angegebenen Oxydationsbedingungen genau eingehalten werden.

Beschreibung der Methode nach Cotte und Kahane (1953).

Cotte und Kahane bedienen sich bei der Cholinbestimmung der in Abb. 3 angegebenen Apparatur. Durch A wird der Dampfdestillationskolben C mit 10 ml 40—50%iger Natronlauge beschickt. Eventuell in der Lauge vorhandene Ammoniakspuren werden durch einen mäßigen Wasserdampfstrom entfernt. Dann wird die Dampfentwicklung unterbrochen und die auf Cholin zu analysierende Probe, z. B. roher Pflanzenextrakt durch A in den Destillationskolben gebracht. Das Tropfrohr B wird beim Vorliegen einer reinen Cholinlösung mit 0,5—1%, beim Vorliegen eines rohen Pflanzenextraktes mit 1%iger $KMnO_4$-Lösung gefüllt. Man heizt erneut zur Verdrängung der Luft aus dem Reaktionsgefäß und läßt die Permanganatlösung so langsam zutropfen, daß ein Überschuß an Oxydationsmittel (Grünfärbung der Mischung durch Manganat!) vermieden wird. Die Vorlage (D), in der das übergehende $N(CH_3)_3$ aufgefangen wird, ist mit überschüssiger 0,01 n H_2SO_4 beschickt. Indicator: Methylrot $+$ Bromkresolgrün.

Sobald eine bleibende Grünfärbung im Reaktionsgefäß auftritt, wird die Permanganatzuführung beendet. Die Destillation wird danach noch einige Minuten fortgeführt. Die gesamte Destillationsdauer beträgt etwa 15—20 min. Der Überschuß der 0,01 n Schwefelsäure wird sofort titriert mit Tashiros Indicator (s. Steward und Street, 1946) gegen

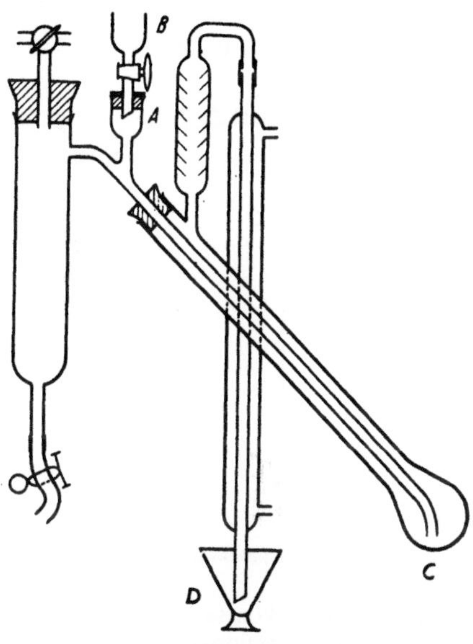

Abb. 3. Apparatur zur Cholinbestimmung nach Cotte u. Kahane.

einen Standard, der eine 0,02 n-Lösung von Trimethylamin enthält und sich in einer kohlensäurefreien Mikrobürette befindet. Der Verbrauch von 1 ml 0,01 n Schwefelsäure entspricht 1,395 mg Cholinhydrochlorid oder 1,04 mg Cholin-Ion.

Zur Trennung des Trimethylamins von anderen flüchtigen Basen, die bei der obigen Behandlung aus Naturstoffgemischen frei werden, wird die neutrale bis schwach saure Lösung mit einem großen Überschuß an Formaldehyd versetzt und so die Gesamtheit der primären Amine abgefangen. Nach Alkalisieren bei Temperaturen unter $+20°$ wird das im Pflanzenextrakt von vornherein vorhandene Trimethylamin bei der gleichen Temperatur durch Durchsaugen eines Luftstromes vor der Permanganatzufuhr aus dem Gemisch entfernt. Dabei wird die S. 566 skizzierte Apparatur (Abb. 4) verwendet. Sie besteht aus einer Waschflasche von 300 ml Inhalt (A), einer mit 2 ml Säure beschickten „Falle" (B) sowie einem mit Wasser beschickten Unterdruckgefäß (C). Die Eintauchrohre sind mit kugelförmig erweiterten durchbohrten Ansätzen (D) versehen, um das Passieren der Blasen sowie die Bindung der Amine an die Säure zu erleichtern.

Die neutrale Analysenlösung wird in das Gefäß A gebracht und 10—40 ml 40%iger Formalinlösung sowie 20—50 ml konzentrierter Natronlauge zugegeben. Man verbindet das Gefäß mit der Falle, die 2 ml 50%iger (Vol.-%) H_2SO_4 enthält, und saugt Luft mit einer Geschwindigkeit durch, die einen Unterdruck von 15—20 ml Wassersäule verursacht. Die so vorbehandelte Lösung wird dann der oben beschriebenen Permanganatoxydation zugeführt. Wird eine reine

Cholinlösung bestimmt, gibt man nach dem Einfließen der Lauge Natriumtetrathionat hinzu (Jod $+ S_2O_3Na_2$), so daß mindestens 20 mg auf 1 mg Cholinchlorid treffen. Es wird dadurch verhindert, daß Trimethylamin zu Trimethylaminoxyd weiter oxydiert wird. Bei dem Verfahren von COTTE und KAHANE erhält man Cholinausbeuten von 97—98%. Die Ester des Cholins, mit Ausnahme des Phosphorsäureesters, der unter den angewandten Bedingungen praktisch nicht angegriffen wird, liefern quantitativ Trimethylamin. Betain, Tetramethylammonium und Trimethylaminoxyd werden nicht angegriffen.

Die Verwendung von Perschwefelsäure zur Oxydation anstelle von Permanganat bietet in bezug auf die Genauigkeit der Analyse keinen Vorteil. Durch den Vergleich der Resultate, bei der Oxydation mit Permanganat mit denen, die bei Verwendung von Persulfat gewonnen werden, erhält man Hinweise auf das Vorhandensein von Substanzen wie Betain, oder den Phosphorsäureester des Cholins, für die es keine genauen Methoden der Identifizierung und der Bestimmung gibt.

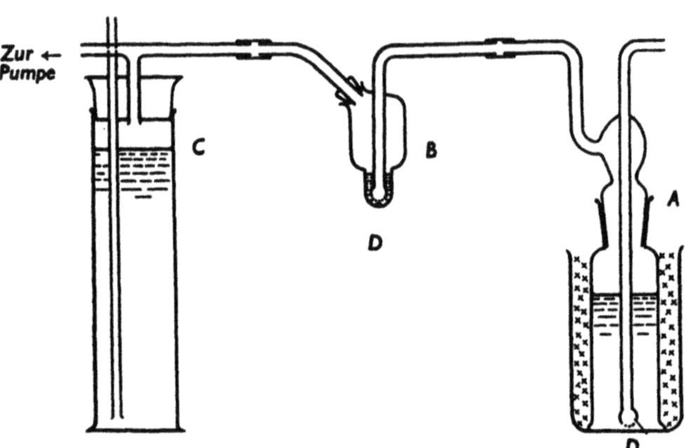

Abb. 4. Apparatur zum Abfangen des Trimethylamins.

Bei Sojabohne ist nach DUCET (1948) während der Keimung der totale Gehalt an Cholin sehr konstant bis zur Entwicklung der Cotyledonen. Der totale Cholingehalt der Cotyledonen vermindert sich, während der des Keimlings anwächst, so daß eine Wanderung des Cholins vom Cotyledon zum Pflänzchen zu verzeichnen ist. Der totale Gehalt an wasserlöslichem Cholin nimmt zwischen der 15. und der 18. Std. mit Beginn der Keimung zu. Bei Gerste und Sojabohne nimmt der Gehalt an Cholin/g Trockengewicht vom Anfang bis zum Ende des Wachstums ab. Die jungen Organe, z. B. Blätter und Früchte, haben den höchsten Gehalt an totalem Cholin. Die ausgewachsenen und gealterten Organe besitzen nur die Hälfte oder $^1/_4$ des Cholingehaltes der jungen Organe.

Die Blüten enthalten ein Maximum an Cholin im Augenblick der Befruchtung. Die Gerste und Sojabohnensamen haben ihren höchsten Cholingehalt/g Trockengewicht im Beginn ihrer Bildung. Die Cholinmenge nimmt dann bis zu einem bestimmten Gehalt ab, der bis zur Reife bestehen bleibt. Dann erfolgt ein Abfall auf den Gehalt der ruhenden Samen.

Ein Verfahren zur *Bestimmung von Cholin und Äthanolamin* gründet CHARGAFF (1937) auf die Tatsache, daß Äthanolamin mit Carbobenzoxychlorid reagiert und kristallisiert, während Cholin damit nicht reagiert. Zu einer Lösung von 1,94 g Cholinhydrochlorid und 1,70 g Äthanolaminhydrochlorid in 9 ml Wasser werden 4,7 g Natriumcarbonat und 10 ml Chloroform hinzugefügt. Die Mischung wird dann mit 4 g Carbobenzoxychlorid geschüttelt, welches in 4 Portionen zugefügt wird. In den ersten 15 min wird mit Wasser gekühlt. Nach einer Stunde wird kein CO_2 mehr entwickelt, es wird mehr Chloroform hinzugegeben, und dann werden die beiden Schichten getrennt. Die wäßrige Schicht wird viermal mit Chloroform gewaschen. Die vereinigten Chloroformextrakte werden viermal mit sehr verdünnter Salzsäure und zweimal mit Wasser gewaschen, dann mit Natriumsulfat getrocknet und im Vakuum konzentriert. Das bleibende rohe Öl wird in 5 ml Chloroform gelöst. Dann wird Petroläther hinzugefügt, bis eine Trübung sich entwickelt. Nach 12 Std. im Eisschrank sind prächtige Nadeln auskristallisiert. Nach Zufügen von Petroläther werden weitere Mengen der

Verbindung aus der Mutterlauge erhalten. Die gesamte Ausbeute entspricht 72% der Theorie. Nach einer weiteren Umkristallisierung aus Chloroform und Petroläther wird Carbobenzoxyäthylamin in längeren farblosen Nadeln erhalten, die bei 66,5° schmelzen. Formel: $C_6H_5CH_2OCONHCH_2CH_2OH$. Zur Freilegung der Base aus der Carbobenzoxyverbindung wird in 5 ml n Salzsäure suspendiert und mit Wasserstoff in Gegenwart von Paladium nach BERGMANN und ZERVAS (1932) 2 Std. hydriert. Der nach dem Einengen erhaltene Sirup wird in konzentrierter Salzsäure aufgenommen und durch Zufügen von Goldchlorid und konzentrierter Salzsäure in das Goldsalz des Äthanolamins übergeführt, das in gelben Nadeln erhalten wird. Fp. 194—195°.

APPLETON, LA DU, LEVY, MURRAY und BRODIE (1953) bestimmen Cholin durch Überführen in das Ennea-Jodid, das in Äthylendichlorid gelöst und bei 365 mμ spektrophotometrisch bestimmt wird. An den Niederschlag adsorbiertes Jod stört die Messungen nicht. Es können noch 5 γ Cholin nachgewiesen werden.

LOVERN (1950) bestimmt Cholin nach MARENZI und CARDINI (1943) durch Fällung als Reineckat, Oxydation des Chroms zu Chromat und Kolorimetrierung desselben mit Diphenylcarbazid.

5. Papierchromatographischer Nachweis des Cholins und Trennung des Cholins von anderen Aminen und von Betainen.

Mit geringem Zeit- und Arbeitsaufwand und bei geringster Materialmenge kann Cholin papierchromatographisch von Begleitstoffen abgetrennt und dann quantitativ bestimmt werden.

Die papierchromatographische Trennung von Cholin und Cholinestern gelingt nach WHITTAKER (1951) am besten unter Verwendung von Wasser, das mit n-Butanol gesättigt ist.

Die R_f-Werte bei verschiedenen Lösungsmitteln finden sich in Tab. 15.

Tabelle 15. R_f-Werte von Cholin und Cholinestern.
(Alle Verbindungen als Chloride.)

Lösungsmittel	Cholin	Acetylcholin	Propionyl-cholin	Acetyl-β-methyl-cholin	Butyryl-cholin	Benzoyl-cholin
n-Butanol[1]	0,06	0,09	0,17	0,15	0,22	0,23
Benzylalkohol, Propanol, Wasser (2:5:2)	0,25	0,33	—	0,37	0,46	0,49
Propanol, Ameisensäure, Wasser (8:1:1)[2] . . .	0,38	0,46	—	0,56	0,66	0,65

Die Methode ist geeignet zum Nachweis von Acetylcholin in Gewebsextrakten. Zur Sichtbarmachung der Flecke wird das Papier mit einer Lösung von alkalischem Hydroxylamin in Alkohol, hergestellt durch Mischen von 10%iger wäßriger Hydroxylaminlösung mit 10%iger alkoholischer Kalilauge und Abfiltrieren des ausgefallenen Kaliumchlorids, getränkt. Das Papier wird dann getrocknet und mit 3%igem ätherischem saurem Eisenchlorid besprüht. Die Carbonsäureester erscheinen als purpurrote Flecke auf gelbem Grund. Die untere Grenze der Empfindlichkeit war bei synthetischen Estern 0,5—2 γ. Auch Acetylphosphat wird auf diese Weise sichtbar gemacht (WHITTAKER, 1951). Von MALYOT und STEIN (1952) wird zur Sichtbarmachung von papierchromatographischen Flecken von

---

[1] Absteigend.   [2] Aufsteigend.

Cholinestern nach Entwickeln mit Eisessig/Pyridin/Wasser im Verhältnis 50:35:25 oder 50:30:20 mit Dipikrylamin besprüht. Sprühreagens nach Ackermann und Mauer (1943): 0,2 g Dipikrylamin in 50 ml Aceton + 50 ml bidest. Wasser. Cholin und seine Derivate geben rote Flecke auf goldgelbem Grund. Trocknung des Papiers bei Zimmertemperatur.

Chargaff, Levine und Green (1948) chromatographieren Amine und stickstoffhaltige Lipoide unter Verwendung der in Tab. 16 angegebenen Lösungsmittel. Bei Erhitzen der Chromatogramme auf 95°, und zwar 90 min lang, entstehen wohldefinierte Fluoreszenzen unter der Quarzlampe. Die primären Amine werden mit Hilfe von Ninhydrin lokalisiert. Bei Verwendung von n-Butanol—Morpholin (3:1) oder n-Butanol—Pyridin (4:1) werden die Papiere 30—60 min an der Luft getrocknet und ebenso lange bei 95°, dann wird mit einer 2%igen wäßrigen Lösung von Phosphormolybdänsäure besprüht. Nun wird das Chromatogramm in einem großen Volumen von n-Butanol 5 min lang gewaschen und ebenso lange in fließendem Brunnenwasser; schließlich wird es in eine frisch bereitete Lösung von 0,4% Stannochlorid in 3 n HCl gebracht. Es entstehen an der Stelle des Cholins gut abgegrenzte blaue Flecke. Zum Nachweis von Äthanolamin und Serin werden die Papiere getrocknet, dann mit 0,1%iger Lösung von Ninhydrin in n-Butanol besprüht und auf 95° 5 — 15 min lang erhitzt zur Entwicklung der gefärbten Flecke. Tab. 16 verzeichnet die R_f-Werte für Cholin und Äthanolamin bei Verwendung verschiedener Lösungsmittel.

Tabelle 16. *R_f-Werte von Cholin und Äthanolamin bei verschiedenen Lösungsmitteln.*

Nr. d. Lösungsmittels	Lösungsmittel	R_f-Werte	
		Cholin	Äthanolamin
1	Phenol		0,43
2	n-Phenolmorpholin (3:1)	0,30	0,51
3	n-Butanoldioxan (4:1)	0,25	0,27
4	n-Butanolpyridin (4:1)	0,24	0,25

Trimethylamin und Dimethylaminoäthanol erscheinen wegen ihrer Flüchtigkeit auf den Chromatogrammen überhaupt nicht. Monomethylaminoäthanol nur mit verminderter Intensität, wenn die Chromatogramme in alkalischem oder neutralem Lösungsmittel liefen. Wird 1- oder 2%ige Essigsäure zum Lösungsmittel gegeben, so werden die genannten Substanzen auf dem Papier zurückgehalten.

Brante (1949) macht Cholin und andere N-methylierte Alkanolamine, die papierchromatographisch getrennt worden waren, sichtbar durch direktes Sublimieren von Jod auf die getrockneten Papierchromatogramme oder durch Auftragen des Jodes in alkoholischer Lösung. Während der Sublimation oder des Besprühens entwickeln sich Flecke von brauner und gelber Farbe, die im UV-Licht am besten differenziert werden können. Die Flecke blassen schnell ab und müssen daher mit dem Bleistift umrahmt werden. Die Farbe kann durch Erneuern der Jodbehandlung wieder hervorgerufen werden. Nach dem Verschwinden des Jodes kann das gleiche Papier ohne Nachteil mit einem anderen Reagens, z. B. mit Ninhydrin, behandelt werden. Auch nach der Ninhydrinbehandlung kann das Papier mit Jod besprüht werden und man erhält auf diese Weise Resultate, als ob Ninhydrin nicht benutzt worden wäre. Man kann auch auf diese Weise Flecke, die 20—30 γ stickstoffhaltiger Verbindungen enthalten, nachweisen. Aminosäuren, Guanidinbasen, Pyrimidine und Purine werden zwar ebenfalls gefärbt, aber in den meisten Fällen viel schwächer als die Amine. Stickstoffhaltige Lösungsmittel müssen auf Grund ihrer Reaktionsbereitschaft mit Jod sehr sorgfältig vor der Jodbehandlung entfernt werden und sind deswegen zu vermeiden.

6. Mikrobiologische Cholinbestimmung.

(s. Methods of Vitamin Assay, Second Edit. 1951.)

Nach HOROWITZ und BEADLE (1943) wird eine künstlich erzeugte Mutante von *Neurospora crassa* verwendet, die cholinempfindlich ist (Nr. 34486=A.T.C.C. Nr. 9277).

Verfahren nach MÜCKE (1955).

Neurospora crassa-Test (BANDELIN, 1949). Zur quantitativen mikrobiologischen Bestimmung von Cholin kommt z. Z. nur die erwähnte röntgeninduzierte Neurospora-Mutante Nr. 34486 in Betracht.

Kulturnährmedium. Bacto-Neurospora Culture Agar (Difco) oder Testmedium (einfach stark) mit Zusatz von 0,2% Difcohefeextrakt. 0,2% Malzextrakt, 1 γ/ml Cholinchlorid und 1,5% Agar. Auch Bierwürze-Agar mit 1 γ/ml Cholinchloridzusatz ist gut brauchbar.

Stammhaltung. Der Testpilz wird auf Schrägagar bei monatlichem Überimpfen gehalten. Die beimpften Nährböden werden 3—4 Tage bei 30° C bebrütet und anschließend bei 4° C aufbewahrt.

Impflösung. Von der Dauerkultur wird eine Öse Sporenmaterial entnommen und in 10 ml sterilem Aqua bidest. suspendiert.

Testmedium (doppelt stark)

1000 ml enthalten:		
Saccharose	40	g
Ammontartrat	10	g
Ammonitrat	2	g
KH_2PO_4	2	g
$MgSO_4 \cdot 7\,H_2O$	1	g
NaCl	0,2	g
$CaCl_2$	0,2	g
Asparagin	2	g
Glutaminsäure	2	g

Spurenelemente (als Salze zugegeben):

H_3BO_3	2 ml einer	2,8 mg-%igen	Lösung	
$CuSO_4$	2 ml „	17,5 mg-%	„	„
$FeSO_4$	2 ml „	50,0 mg-%	„	„
$MnSO_4$	2 ml „	4,0 mg-%	„	„
MoO_3	2 ml „	1,5 mg-%	„	„
$ZnSO_4$	2 ml „	0,443 %	„	„

Die anorganischen Bestandteile werden gemeinsam mit der Saccharose von allen übrigen Substanzen getrennt gelöst auf 100° C erhitzt und filtriert. Nach Vereinigen beider Teilmedien wird auf p_H 5,0 eingestellt.

Vorbereitung des Testmaterials. Vom trockenen, gut zerkleinerten Testmaterial werden 50—500 mg in einen 50 ml-Erlenmeyerkolben gegeben, 10 ml 3% H_2SO_4 zugesetzt und 2 Std. bei 120° autoklaviert. Nach dem Abkühlen wird das Material quantitativ unter Nachspülen mit 5 ml Aq. bidest. in ein 50 ml-Zentrifugenröhrchen übertragen und durch Zutropfen einer gesättigten Bariumhydroxydlösung auf Kongorot neutralisiert. Das entstandene $BaSO_4$ sowie ungelöste Substratanteile werden abzentrifugiert und das Dekantat durch Papierfilter (Whatman Nr. 50) filtriert. Der Rückstand im Zentrifugenglas wird mit 3 ml Aqua bidest. versetzt und im kochenden Wasserbad mit einem Glasstab 10 min umgeruhrt. Nach Abkühlen wird nochmals zentrifugiert, filtriert, beide Filtrate vereinigt mit 1 n NaOH lackmusneutral eingestellt auf ein zweckmäßiges Volumen aufgefüllt; 1 ml soll etwa 2—5 γ/ml Cholin (als Chlorid berechnet) enthalten. Sollte im Testsubstrat Methionin in störenden Mengen enthalten sein — was erfahrungsgemäß nicht häufig ist —, ist es durch Basenaustauscher zu eliminieren. Dazu läßt man 5 ml der neutralisierten Extraktlösung langsam durch eine Permutitsäule (110 × 6 mm) laufen. Ist der Cholingehalt des Extraktes geringer als 1 γ/ml, so sind 10 ml notwendig. Mit 2 × je 5 ml 0,3%iger NaCl-Lösung wird nachgewaschen, die Filtrate werden verworfen. Schließlich wird das Cholin mit 10 ml heißer 5%iger NaCl-Lösung eluiert. Das Eluat fängt man in einem 10 ml-Maßkolben auf und ergänzt nach Abkühlen auf 10 ml. 1 ml der Lösung soll etwa 5 γ Cholin enthalten.

Durchführung der Bestimmung. In 125 ml-Erlenmeyerkölbchen pipettiert man je 10 ml Testmedium. Das Endvolumen von 20 ml je Kölbchen wird erreicht durch Zugabe von abgestuften Mengen Cholinchloridlösung bzw. Testextraktmengen und Auffüllen mit Aq. bidest.; eine Leerwertkontrolle enthält nur Testmedium und Aq. bidest. Alle Testkölbchen werden mit Glas- oder Metallkappen verschlossen, 10 min bei 115° C autoklaviert und nach

Abkühlen mit je 1 Tropfen der Sporensuspension beimpft. 72 Std. wird bei 30° C bebrütet und dann ausgewertet.

Auswertung. Gravimetrisch; beim Berechnen wendet man das „log-log"-Verfahren an (Wood, 1947).

Meßbereich. 0—100 γ-% Cholin (als Chlorid); 8—10% Fehlerbreite.

Trotz der stets gleichmäßig gut reproduzierbaren Standardkurve soll man bei jedem Testansatz stets eine Standardreihe mitlaufen lassen; man braucht dann jeweils nur einige Zwischenstufen anzusetzen, um den Verlauf der Kurve festzulegen.

Lecithin kann von *N. crassa* auch als Cholinquelle benutzt werden, doch stört es nur geringfügig bei Bestimmung des freien Cholins. Außer Methionin, das 0,2% der Aktivität des Cholins für den Testpilz besitzt, können noch folgende Cholinanalogen das *Neurospora*-Wachstum stimulieren: Acetylcholin, Dimethylaminoäthanol, Monomethylaminoäthanol, Phosphorylcholin und Arsenocholin. Die letztgenannten Substanzen finden sich jedoch in biologischem Material nur in unwesentlichen Mengen und können vernachlässigt werden.

Bei der mikrobiologischen Cholinbestimmung in pflanzlichem Material ist mit der Anwesenheit fungistatischer oder Cholin-hemmender Substanzen zu rechnen. Je nach Art des angewandten Extraktionsverfahrens werden sie in unterschiedlicher Menge in den Testextrakt eingeschleppt und können das Testergebnis stark beeinflussen. Untersuchungen an *Capsella bursa pastoris* erbrachten den Nachweis, daß in dieser Droge eine das Testergebnis beeinflussende Substanz vorkommt.

IV. Acetylcholin. $(CH_3)_3NOH \cdot CH_2 \cdot CH_2 \cdot OCO \cdot CH_3$.

Acetylcholin ist in Wasser und Alkohol leicht löslich, unlöslich in Äther. In alkalischer wäßriger Lösung zerfällt Acetylcholin in Cholin und Essigsäure. Auch starke Mineralsäuren zerstören es. Optimum der Haltbarkeit bei p_H 3,9.

Wird Acetylcholin in bicarbonatalkalischer Lösung, z. B. Lockelösung, aufgelöst, so nimmt seine pharmakologische Wirksamkeit bei Zimmertemperatur im Laufe eines Tages stark ab. In der Natur ist Acetylcholin stets von großen Mengen Cholin begleitet, welches die Beständigkeit des Wirkstoffes gegenüber Säure und Lauge, aber auch gegen den Angriff der Acetylcholinesterase schützt. Acetylcholin ist pharmakologisch hoch aktiv und erregt wie Cholin parasympathisch innervierte Organe. Seine Wirkung wird durch Behandeln der Testtiere oder -Präparate mit Atropin aufgehoben. Acetylcholin wird durch die in tierischen Organen weit verbreitete Cholinesterase in das pharmakologisch sehr viel weniger wirksame Cholin und in Essigsäure gespalten. Diese Tatsachen können zur quantitativen Bestimmung der Substanz herangezogen werden (s. S. 571).

Salze. *Chlorid* $C_7H_{16}O_2NCl$ ist sehr hygroskopisch, leicht löslich in Alkohol. Das *Bromid* $C_7H_{16}O_2NBr$ ist nicht hygroskopisch, Fp. 160—161°. Das saure *Acetylcholintartrat* ist leicht löslich in Alkohol zum Unterschied von Cholinbitartrat.

Chloroplatinat. $(C_7H_{16}O_2N)_2 \cdot PtCl_6$, anisotrope Nädelchen aus 50%igem Alkohol. Verschiedene Fp.-Angaben s. Guggenheim, S. 220. Fp. 223—224°, 256—257°, 242—244° (Zers.). Cholin-Acetylcholin-dichloroplatinat

$$C_{24}H_{60}O_6N_4Cl_{12}Pt_2 = (C_7H_{16}O_2N)_2 \cdot PtCl_6 + (C_5H_{14}ON)_2 \cdot PtCl_6,$$

isotrope Oktaeder aus Wasser, Fp. 260—261°, weniger löslich als die Komponenten.

Chloroaurat. $(C_7H_{18}O_2N) \cdot AuCl_4$, Prismen oder Nadeln. Fp.-Angaben verschieden (s. Guggenheim, S. 220) 154°, 155°, 166°, 168°. Verwandelt sich beim Schütteln mit molekularem Silber in das Chlorid.

Reineckat. $C_7H_{16}O_2N \cdot C_4H_6N_6S_4Cr$. Rhomben, Fp. 173°, löslich in Aceton, in Wasser von 21° zu 0,0183% löslich, in Alkohol zu 0,0335%.

Über eine mikrochemische Bestimmungsmethode für Acetylcholin s. S. 573.

Mitchell und Clark (1952) bestimmen Acetylcholin und Tetramethylammonium durch Behandeln der alkalischen Lösung derselben mit Bromphenolblau. Die blauen Farbstoffe werden dann mit Äthylenchlorid ausgeschüttelt und kolorimetriert. Grenze des Nachweises 2—3 γ.

1. Pharmakologische Bestimmung von Cholin und Acetylcholin.

Auf die pharmakologische Wirksamkeit des Acetylcholins gründen sich sehr empfindliche und spezifische Methoden zum Nachweis und zur quantitativen Bestimmung des Cholins und Acetylcholins. Als Testobjekte werden folgende parasympathisch innervierte Präparate verwendet. Der isolierte Blutegelrückenmuskel (MINZ, 1932), der Froschrectus (CHANG, GADDUM, 1933), FLETCHER, BEST und SOLANDT (1935) die isolierte Froschlunge (BRECHT, 1943), das isolierte Herz der Muschel Venus mercenaria (TOWER, McEACHERN, 1947), das isolierte Froschherz, der isolierte Dünndarm der Maus und des Meerschweinchens.

Zur Technik des biologischen Versuches siehe GADDUM (1936). Es sind nachweisbar am isolierten Blutegel 2-, am isolierten Kaninchenvorhof 4-, am Froschherz 10-, musculus rectus abdominalis 20-, am Blutdruck der Katze 20 γ/l. An der isolierten Froschlunge können noch sehr viel geringere Acetylcholinmengen nachgewiesen werden.

Der Nachweis am Blutegelrückenmuskel ist für Acetylcholin spezifisch, er sei daher im folgenden geschildert: Man nimmt die Längsstreifen vom vorderen Ende des Rückenhautmuskels des Blutegels; die ventrale Fläche des Muskels wird sorgfältig präpariert und gereinigt. Der Streifen wird in LOCKE-Lösung aufgehängt, die auf die 1,4fache Menge verdünnt worden ist und ungefähr 5 mg Eserin pro Liter enthält. Bewirkt diese Eserinkonzentration eine starke Kontraktion des Blutegelmuskels, so muß man eine niedrigere Konzentration anwenden. Der Blutegelstreifen wird an einem Hebel befestigt, der auf einer berußten Trommel schreibt. Die Belastung hängt von der Größe des Muskelstreifens ab und beträgt für gewöhnlich 2—3 g. Die Vergrößerung der Ausschläge ist ungefähr 5fach. Verwendet man Streifen, die durch Halbieren des Rückenhautmuskels in der Längsrichtung erhalten worden sind, so beträgt die Belastung nur etwa 1,2 g. Das Badvolumen beträgt für gewöhnlich 20 ml, doch erhält man mit einem Badvolumen von 2 ml ebenso befriedigende Ergebnisse. Die Temperatur des Bades entspricht der Zimmertemperatur und das Bad wird von einem Luftstrom durchperlt. Wird dem Bade Acetylcholin zugefügt, so beginnt die Kontraktur langsam und kann zu ihrer vollen Entwicklung mehrere Minuten beanspruchen. Wird das Acetylcholin wieder mit frischer Salzlösung ausgewaschen, so erschlafft der Muskel langsam. Dieser Vorgang läßt sich bis zu einem gewissen Grade durch vorsichtiges mechanisches Dehnen des Muskels beschleunigen. Ist die Zimmertemperatur aber sehr niedrig, unter 18° C, so muß man das Bad erwärmen, da der Muskel zu langsam oder überhaupt nicht wieder erschlafft. Der Muskel kann aber erst dann für eine zweite Gabe von Acetylcholin verwendet werden, wenn er erschlafft ist und der Schreibhebel mehrere Minuten eine horizontale Grundlinie geschrieben hat. Der Vorgang ist langsam; hat man aber gefunden, daß eine bestimmte Lösung Reaktionen hervorruft, die in ihrer Form praktisch denen des Acetylcholins entsprechen, so kann man die Bestimmung beschleunigen und ein genaues quantitatives Ergebnis erhalten, indem man einer genauen Zeiteinteilung folgt und die wirksame Lösung nur für eine genaue Zeit, 2 oder 4 min mit dem Muskel in Berührung bringt (s. CHANG und GADDUM, 1933; GUGGENHEIM, 1951).

Folgende Mikroorganismen produzieren Acetylcholin: *Claviceps purpurea* (EWINS, 1914), *Lactobacillus plantarum* (STEVENSON, ROWAT, 1947). Nach BÜLBRING, LOURIE, PARDOL (1949) synthetisieren 'Tripanosomen *(Tripanosoma rhodesience)* Acetylcholin, und zwar 2,38 bis 8,60 γ/g Feuchtgewicht. 10^{10} Tripanosomen produzieren in vitro bis zu 5,5 γ/g Acetylcholin oder 71,5 γ/g Acetontrockenpulver in 75 min bei 37°.

Der Acetylcholingehalt von Kartoffelknollen nimmt beim Lagern ab. So enthielt der Preßsaft der Serie „Bintje" 18 γ bei der Juniernte und im Januar nur noch 5—6 γ.

Unter gleichen Bedingungen auf demselben Feld gewachsene Kartoffelpflanzen 6 verschiedener Sorten wurden auf ihren Gehalt an Acetylcholin untersucht. Der Nachweis erfolgte

am Rectus abdominalis des Frosches und am Carotisdruck der Katze. Es fanden sich in den
jungen Knollen: „Chippewa" 14 γ, bei „Industrie" 6,5 γ, bei „Eigenheimer" 20 γ Acetyl-
cholin/1 ml Preßsaft. Im Januar ergaben dann die Knollen folgende Werte: „Bintje"
5—6 γ, „Chippewa" 2 γ, „Industrie" 1,5 γ, „Eigenheimer" 0,0 γ/ml Saft. Außerdem wurden
die Sorten „Ovalgelbe" und „Maibutter" untersucht. In ihnen und in den Zweigen und Blättern
aller untersuchten Sorten konnte nie Acetylcholin nachgewiesen werden.

Bei *Urtica dioica* enthielten die Würzelchen zwischen 0,45 und 1,9 γ Acetylcholin in
10 mg Gewebe. Die Rizome mehr als 3,7 γ, die höchsten Konzentrationen fanden sich in der
braunen Rinde und dem benachbarten Phloem, in dem sich 5,6—135 γ, Durchschnitt 37 γ
pro 10 mg Gewebe fanden. Die Konzentrationen des Acetylcholins in *Urtica urens* wechselt
zwischen 1:500000 und 1:20000, durchschnittliche Konzentration 1:50000. Mittlere
Konzentration im Blatt 1:80000, Cortex und Stengel 1:20000, Rinde und Wurzel 1:60000,
Würzelchen 1:40000 (s. auch Tab. 22a—d). (Emmelin und Feldberg, 1949.)

Das angezweifelte Vorkommen von Acetylcholin in der Mistel (Winterfeld, 1952) ist
einwandfrei bewiesen worden (Winterfeld und Weiss). Dagegen ist früher angegebenes
Propionylcholin (Müller, 1932) nicht vorhanden (Winterfeld und Kronenthaler, 1942).
Das Acetylcholin ist in der Pflanze und im frischen Mistelsaft wahrscheinlich am Eiweiß
oder an einem anderen Träger verankert und dadurch stabilisiert (Winterfeld, 1952).
Das Acetylcholin ist aus der Eisessiglösung des Wirkstoffextraktes durch Benzolfällung von
Viscotoin völlig abtrennbar (Winterfeld, 1952 und Winterfeld und Kronenthaler, 1942).
Acetylcholin wurde ferner nachgewiesen in: Hirtentäschelkraut, Mutterkornpilz, Kartoffeln
und ziegelrotem Rißpilz (Marquard und Mitarb., 1952). Weißdorn (Crataegus) enthält nach
Fiedler, Hildebrand und Neu (1953) Acetylcholin neben Cholin.

2. Isolierung von Acetylcholin und Cholin aus *Crataegus*.
(Fiedler, Hildebrand und Neu, 1953.)

Aus *Crataegus*-Blättern mit heißem Wasser hergestellte Auszüge werden
zunächst im Vakuum bei 40—50° stark konzentriert und durch Schütteln mit
MgO, frisch gefälltem Pb(OH)$_2$ und anschließendem sofortigem Zusatz von
Alkohol-Aceton (1:1) im Überschuß von der Hauptmenge an Gerbstoffen,
Flavonen und Schleimstoffen (die auf diese Weise in leicht filtrierbarer Form auf-
fallen) befreit. Aus dem Filtrat läßt sich durch Bleiessig bei alkalischer Reaktion
ein weiteres Flavon ausscheiden. Nach der Entfernung der Blei-Ionen mit Schwefelsäure oder sekundärem Natriumphosphat werden bei neutraler Reaktion die
organischen Lösungsmittel im Vakuum entfernt. Dann wird die Lösung stark eingeengt. Durch erneuten Zusatz von Alkohol-Aceton (1:1) werden anorganische Salze
sowie Reste von Schleim- und Eiweiß-Stoffen ausgefällt. Die organischen
Flüssigkeiten werden im Vakuum entfernt. Die so gereinigte, wasserklare Lösung
wird in der üblichen Weise mit Phosphorwolframsäure gefällt und aus dem zer-
setzten Niederschlag wird das Cholin als Reineckat zuerst fein kristallin und nach
dem Umkristallisieren aus Aceton-Wasser in den typischen sechsseitigen Tafeln
gewonnen.

Aus getrockneten und gemahlenen *Crataegus*-Früchten werden Auszüge mit
Methanol oder mit Äthanol mit 2% 2 n Salzsäure hergestellt. Der Alkohol wird
im Vakuum abdestilliert. Nach Digerieren der trockenen Rückstände mit Wasser
wird nach Entfernung vom Ungelösten mit Ammonium-Reineckat bzw. Kalignost (Fa. F. Heyl & Co., Hildesheim) gefällt. Die erhaltenen Niederschläge
werden nach dem Waschen mit eiskalter gesättigter Reineckesäurelösung in Aceton
gelöst und filtriert. Das Aceton wird abdestilliert, der Rückstand in Butanol aufgenommen, filtriert, das Butanol abdestilliert und die Behandlung erneut durchgeführt.
Durch die Behandlung der Extrakte mit organischen Lösungsmitteln konnten
die Ballaststoffe entfernt werden. Nach der Zerlegung der Reineckate waren die
Lösungen farblos.

Das Cholin wird durch Zersetzen des Reineckesalzes mit Silbersulfat in Aceton
freigemacht und das überschüssige Silber-Ion durch Natriumchlorid entfernt.
Mit Quecksilber(II)-Chlorid wird das Cholinchlorid-Quecksilberchlorid-Doppel-

salz ($C_5H_{14}ONCl \cdot 6\,HgCl_2$) gefällt, das mit dem zum Vergleich bereiteten Queck-silber-Doppelsalz aus reinem Cholinchlorid (Merck) keine Schmelzpunkt-depression ergibt. Die Filtrate von der Zerlegung der Reineckate zeigten auf Zusatz von Stanek-Reagens unter dem Mikroskop die charakteristischen Kristall-formen (an der Berührungszone beider Lösungen braune Nadeln mit einspringen-den Winkeln und Zwillingsbildungen).

Cholin wurde auch papierchromatographisch bei aufsteigender Methode mit Butanol-Dioxan und Butanol-Pyridin (jeweils 4:1 mit Wasser gesättigt) nach CHARGAFF, LEVINE und GREEN (1948) sowie Phenol-Wasser (ob. Ph.) und Butanol-Eisessig (4:1:5) nachgewiesen. Die Cholin-Flecken auf dem Papierchromato-gramm zeigen sich bereits beim Hängen an der Luft durch ihr hygroskopisches Verhalten an (Wölbung des Papiers). Die Cholin-Flecke wurden nach MUNIER und MACHEBOEUF (1949) sichtbar gemacht mit DRAGENDORFF-Reagens.

Zum Nachweis von Acetylcholin neben Cholin wurden die Filtrate von der Zerlegung der Reineckate verwendet und im Vakuum bei etwa 30° konzentriert. Das Vorliegen von Acetylcholin wurde nach FEIGL, ANGER und FREHDEN (1934) über die Eisen(III)-Salze der Acetylhydroxamsäure mikrochemisch festgestellt. Außerdem wurde der Nachweis noch papierchromatographisch nach MAR-QUARDT, SCHUMACHER und VOGG (1952) geführt, wobei mit Wasser gesättigtes Kollidin auch brauchbar war. Das besprühte Papierchromatogramm zeigte die Acetylhydroxamsäure als violette Flecke.

V. Diamine.

1. Putrescin.

Putrescin. Tetramethylendiamin, 1,4-Diaminobutan. $C_4H_{12}N_2$. C 54,6%, H 23,6% N 31,8%, Mol-Gew. 88,1.

$$\begin{array}{c} H_2C{-}CH_2{-}NH_2 \\ | \\ H_2C{-}CH_2{-}NH_2 \end{array}$$

Putrescin und Cadaverin entstehen bei der bakteriellen Zersetzung eiweiß-artiger Naturprodukte aus Ornithin und Lysin durch Einwirkung von Decarb-oxylasen von Bakterien. (Über Aminosäuredecarboxylasen in Mikroorganismen s. GALE, 1946.) Auch eine Entstehung von Putrescin aus Arginin ist möglich (ACKERMANN, 1909; HIRAI, 1936; AKASI, 1938; GALE, 1940), da unter dem Ein-fluß von Streptokokken aus Arginin zuerst Ornithin gebildet wird, das anschlie-ßend zu Putrescin decarboxyliert wird (GALE, 1940). In Mischkulturen von *Bac-terium coli* und Streptokokken verläuft die Spaltung von Arginin bei p_H 5,5 unter Bildung von Putrescin; bei p_H 4 wird durch direkte Decarboxylierung von Arginin, Agmatin gebildet (GALE, 1940).

Eigenschaften. Farblose Flüssigkeit von spermaähnlichem Geruch. Fp. 27°. Sdp. 158 bis 160°. Zieht aus der Luft Kohlendioxyd an. Leicht löslich in Wasser, wenig in Äther, mit Wasserdämpfen schwer flüchtig. Beim Destillieren mit Kalilauge beständig. Wird durch Alkaloidreagentien gefällt.

Derivate.

Chlorhydrat. $C_4H_{12}N_2 \cdot 2\,HCl$. Lange, farblose Nadeln oder Tafeln vom Zersetzungs-punkt 315°. (Mikroschmelzpunktsapparat nach KLEIN.) Nicht hygroskopisch. Sehr leicht löslich in Wasser, wenig in verdünntem Alkohol, unlöslich in abs. Alkohol und Äther. Durch seine Schwerlöslichkeit in abs. Alkohol und Äther unterscheidet es sich vom Cadaverin-chlorhydrat.

Nach BOSER und KLEIN (1932) wird die Trennung von Cadaverin und Putrescin besser durch Behandeln der Chlorhydrate mit Äther an Stelle des früher empfohlenen 96%igen Alkohols durchgeführt, da dieser etwas Putrescinchlorhydrat löst.

Bei der trockenen Destillation liefert Putrescinchlorhydrat Pyrrolidin, welches einen mit Salzsäure befeuchteten Fichtenspan rot färbt.

Dipikrat. $C_4H_{12}N_2 \cdot 2\,C_6H_3O_7N_3$. Seidenglänzende, verfilzte Nadeln, in Wasser schwer löslich. Bräunt sich bei 230° und zersetzt sich bei 250°.

Dipikrolonat. $C_4H_{12}N_2 \cdot 2\,C_{10}H_8O_5N_4$. Hellgelbe, lange schmale Prismen mit schiefer Endigung. Zersp. 263°. Sehr schwer löslich in Wasser und Alkohol. Putrescinchlorhydrat gibt noch in einer Verdünnung von 1:50000 eine Fällung mit Pikrolonsäure. Eignet sich zum mikrochemischen Nachweis von Putrescin.

Phosphorwolframat. Wenig löslich in Wasser, ziemlich löslich in Alkohol und Methanol, leicht löslich in Aceton.

Diflavianat. Zers. bei 253—254°, hydrolysiert leicht zur Monoverbindung.

Chloroplatinat. $C_4H_{12}N_2 \cdot H_2PtCl_6$, meist drusig verwachsene Nadeln oder Prismen, wenig löslich in Wasser.

Chloroaurat. $C_4H_{12}N_2 \cdot 2\,H\,AnCl_4\,2\,H_2O$, Plättchen, in Wasser wenig löslich (Unterschied vom Chloroaurat des Cadaverins).

Dibenzoylputrescin, $C_4H_{10}N_2 \cdot 2\,COC_6H_5$. Seidenglänzende Plättchen oder Nadeln vom Fp. 175—176°, unlöslich in Wasser, fast unlöslich in Äther. Wird aus alkoholischer Lösung durch Äther ausgeschieden (Unterschied vom Dibenzoylcadaverin). Schwer löslich in kaltem, leicht in heißem Alkohol. Sublimiert unzersetzt, spaltet sich mit alkoholischer Salzsäure in Diamin und Benzoesäure.

Phenylisocyanatverbindung, $C_6H_5NHCONH(CH_2)_4NHCONHC_6H_5$. In den meisten Lösungsmitteln praktisch unlöslich. Kristallisiert aus Pyridin-Aceton in zu Büscheln vereinigten Nadeln vom Fp. 240°, unlöslich in Wasser, Aceton, Benzol, Ligroin, löslich in heißem Nitrobenzol, Anilin und Pyridin.

Agmatin und Arcain können durch Bakterienenzyme zu Putrescin und Harnstoff hydrolysiert werden (LINNEWEH, 1931).

Putrescin und Cadaverin wurden in autolysierter Hefe, in Mutterkorn, Steinpilz in Kulturen von *Aspergillus orycae* (Reisschimmel) und in verdorbenem Sojabohnenmehl beobachtet (RIELÄNDER, 1908; REUTER, 1912; KUNG, 1914; YAMADA, 1926; YAMADA, 1928).

In höheren Pflanzen wurde Putrescin unter den Nebenalkaloiden von *Datura stramonium* (CIAMICIAN und RAVENNA, 1911; CROMWELL, 1943) und *Atropa belladonna* (CROMWELL, 1943; GORIS, LARSONEAU, 1921), ferner in Orangensaft aufgefunden. In Blättern und Stengeln von *Atropa belladonna* fehlte vom Mai bis zum Juni Putrescin, im August fanden sich 0,01 bzw. 0,02 g in 2 kg *Atropa belladonna* bzw. *Datura stramonium* (CROMWELL, 1943).

Gewinnung von Putrescin aus frischen Steinpilzen. (REUTER, 1912.)

5 kg Steinpilze wurden von den Hyphen abgetrennt, in einer Fleischmühle zerkleinert und mit 10 l 95%igem Alkohol übergossen. Das Material war so gut konserviert und behielt ein frisches Aussehen. Nach 15 Monaten wurde die Flüssigkeit vom Rückstand getrennt und letzterer mit 85%igem Alkohol ausgekocht. Nach Abdestillieren des Alkohols wurde die Flüssigkeit mit Bleiessig gereinigt und die Basen mit Phosphorwolframsäure gefällt. Der Phosphorwolframsäureniederschlag wurde in üblicher Weise verarbeitet. Die Lysinfraktion wurde mit Pikrinsäure neutralisiert, wobei eine Fällung entstand, welche nach einmaligem Umkristallisieren aus Wasser 0,85 g eines gelben Pulvers lieferte, welches sich bei 245° zersetzte. Das Pikrat wurde in das Chlorid übergeführt und dieses in seinen Eigenschaften und nach Überführung in Gold- und Platinsalz als Putrescindichlorhydrat erkannt.

Isolierung von Putrescin aus *Atropa belladonna* und *Datura stramonium*. (CROMWELL, 1943.)

Das Vorkommen von Putrescin in *Datura stramonium* sowie die Anwesenheit eines Diamins in Blättern von *Atropa belladonna* wurde durch CIAMICIAN (1911) wahrscheinlich gemacht. Tetramethylputrescin wurde durch E. MERCK aus *Hyoscyamus muticus* isoliert. 2 kg Pflanzenmaterial wurden fein gemahlen, mit 1 Liter Äther-Chloroform gesättigtem Wasser (4 Vol. Äther, 1 Vol. Chloroform) versetzt und 2 Std. stehengelassen. Der Brei wurde in einen Stoffbeutel gegeben, ausgepreßt und der Preßkuchen zweimal mit Wasser (250 ml) extrahiert. Die vereinigten Extrakte wurden auf 90° erhitzt, koaguliertes Eiweiß und Gewebsrückstände abzentrifugiert. Die klare Lösung wurde auf

300 ml bei vermindertem Druck konzentriert und mit Bleiacetat versetzt. Der Niederschlag wurde abzentrifugiert und das Blei aus dem Filtrat mit H_2S ausgefällt. Das Filtrat wurde weiter konzentriert, mit NaOH schwach alkalisch gemacht und die Alkaloide mit Äther-Chloroformgemisch ausgeschüttelt. Nach Entfernen der Alkaloide wurde die alkalische Lösung unter vermindertem Druck schwach erwärmt, um Äther und Ammoniak zu entfernen. Histidin und Arginin wurden mit $AgNO_3$ bei p_H 7,2 bzw. 12 ausgefällt, ebenso die Lysinfraktion mit Phosphorwolframsäure; der Lysinniederschlag wurde in Wasser suspendiert und die Basen mit Na(OH) wieder freigesetzt. Das Filtrat wurde mit verdünntem HCl schwach angesäuert und im Vakuum zur Trockne eingedampft. Der Rückstand wurde mit Soda schwach alkalisch gemacht und mit absolutem Alkohol extrahiert. Aus der alkoholischen Lösung wurden die Basen durch Zugabe einer gesättigten Lösung von $HgCl_2$ in Äthanol ausgefällt. Nach Filtrieren wurden die Niederschläge in Wasser suspendiert und von Quecksilber durch H_2S befreit. Zum Filtrat wurden einige Tropfen verdünnter Salzsäure gegeben, die Lösung auf 2 ml eingeengt und ein Überschuß absoluten Äthanols hinzugegeben. Nach einiger Zeit wurden ausgeschiedene Kristalle abfiltriert, in sehr wenig Wasser gelöst, wieder ausgefällt mit absolutem Äthanol und aus absolutem Methanol umkristallisiert. Fp der Kristalle: 289° (Putrescindihydrochlorid: Fp 290°). Der Rückstand wurde sodaalkalisch gemacht und wasserdampf-destilliert, das Destillat angesäuert und fast zur Trockne eingedampft, nach Zugabe von etwas Äthanol erfolgte Ausfällung mit Aceton. Nach nochmaliger Destillation und Ausfällung wurde Putrescin als Hydrochlorid, als Platinchlorid und als Pikrat, ferner papierchromatographisch und röntgenographisch identifiziert.

Zum Nachweis und zur Isolierung von Putrescin aus kaliumarm ernährter Gerste wurde Putrescin von anderen ninhydrinpositiven Substanzen auf einer Säule von sulfonierten Polystyrolharzen abgetrennt. Die Säulen wurden mit HCl und Ammoniumhydroxyd gewaschen und das Putrescin mit 0,2 Mol Ammoniumcarbonatlösung eluiert.

Isolierung von Putrescin aus Orangensaft (HERBST und SNELL, 1948, 1949).

5 Liter konservierten ungesüßten Orangensaftes wurden durch eine Schicht Hyflo-Super-Cel filtriert, das Filtrat (p_H 3,8) dann durch ein 3 cm weites Glasrohr mit 50 g Amberlite IR-100-H laufengelassen (50 ml pro Minute). Die Säule wurde mit 1 l dest. Wassers nachgewaschen. Nur 13% des gesamten Putrescins wurden in der ausgelaufenen Lösung vorgefunden, im Waschwasser gar nichts, wie Wachstumsteste bei Bakt. Haemophilus parainfluencae 7901 zeigten (Putrescin wurde als Wachstumsfaktor für dieses Bakterium erkannt). Die Säule wurde anschließend mit 5 Liter 2,4 n HCl gewaschen, das salzsaure Eluat mit dem Putrescin dann im Vakuum fast zur Trockne eingeengt, 50%ige NaOH zugefügt und im Wasserdampf destilliert. Das Destillat wurde mit 10%iger H_3PO_4 auf p_H 6,8 eingestellt. Nach Zugabe von Natriumpikrat fiel Putrescin-Pikrat als kristalliner Niederschlag aus. Nach mehrfachem Umkristallieren war der Schmelzpunkt F. 264°. Das Pikrat wurde ins Hydrochlorid, in die Benzoyl-Verbindung und in das Phenylisocyanat übergeführt (DUDLEY und ROSENHEIM, 1924) und Putrescin durch die einzelnen Schmelzpunkte identifiziert.

Quantitative Bestimmung des Putrescins im Wachstumstestversuch mit *Haemophilus parainfluencae* 7901. Dieser Mikroorganismus wächst auf einem reduzierten Nährboden nur in Gegenwart von Putrescin. Cadaverin, 1,3-Diaminopropan, 1,6-Diaminohexan, n-Butylamin, Pyrrolidin und Agmatin können Putrescin nicht ersetzen.

Mit Hilfe dieses Testes wurde Putrescin aus Orangensaft auf die oben geschilderte Weise isoliert.

Stammkultur: H. parainfluencae ATCC 7901. Stammkulturen werden angelegt auf sterilem Gelatineagar (Difco), dem pro 100 ml 2 ml eines sterilen Hefeextraktes zugefügt waren. Der Hefeextrakt wird in folgender Weise bereitet: 50 g Bierhefe werden in 25 ml destilliertem Wasser suspendiert und in 75 ml kochendes Wasser, enthaltend 3,4 g KH_2PO_4, eingerührt, nachdem man 20 min auf 80—85° gehalten hatte, wurde die Mischung unter Saugen durch Hyflosuper-Cel filtriert und durch Druckerhitzen stabilisiert. Die Kulturen wurden wöchentlich übertragen und 24 Std. bei 37° inkubiert, dann bei Raumtemperatur aufbewahrt.

Die Zusammensetzung des Basalmediums siehe Tab. 17. Das Impfmedium enthält Difcoproteosepepton Nr. 3 2%, 0,1% Glucose, 0,2% Natriumchlorid, 0,2% K_2HPO_4, 0,1% Difcohefeextrakt. 10 ml Proben werden in Gläser eingefüllt und autoklaviert. Unmittelbar von der Inokulierung durch Übertragung aus der Stammkultur werden 0,2 ml des filtrierten Extraktes (Seitz-Filter) aus frischer Hefe zu 10 ml dieses sterilen Mediums gegeben. Nach 10—12stündiger Inkubation bei 37° ist das Wachstum stark genug, um die Impflösung zu bereiten. Die Zellen werden durch Zentrifugieren isoliert, einmal mit 0,9%iger Natriumchloridlösung gewaschen, dann resuspendiert mit genügend steriler Salzlösung, um eine 95%ige Übertragung von einfallendem Licht zu gewährleisten (Evelin-Kolorimeter, 660 mμ Filter, 16 mm Küvette, sterile Salzlösung = 100. 0,1 ml dieser Suspension werden zu jedem der 10 ml Proben des Versuchsmediums gegeben. Durch stärkere Beimpfung wird die Empfindlichkeit des Versuches stark reduziert.

Tabelle 17. *Zusammensetzung des Basalmediums zur Kultur von Haemophilus parainfluencae 7901.* Nach Herbst u. Snell (1949).

Bestandteil	Menge pro 5 ml doppelte Stärke	Bestandteil	Menge pro 5 ml doppelte Stärke	Bestandteil	Menge pro 5 ml doppelte Stärke
	mg		mg		
D,L-Asparaginsäure	10	L-Cystin	1	Biotin	0,01 γ
L-Glutaminsäure .	10	L-Tyrosin	1	p-Aminobenzoe-	
DL-Alanin	10	Glycin	1	säure . . .	0,01 γ
L-Arginin-HCl . .	2	Glukose	10	Folinsäure . .	0,1 γ
L-Lysin-HCl . . .	2	Natriumacetat .	60	Inositol . . .	200 γ
D,L-Methionin . .	2	Adeninsulfat. . .	100 γ	Coenzym I . .	1 γ
L-Leucin	1	Guanin HCl . .	100 γ	$MgSO_4 \cdot 7\ H_2O$	1 mg
D,L-Threonin . . .	2	Uracil	100 γ	$CaCl_2 \cdot 2\ H_2O$	400 γ
D,L-Serin	2	Thiaminchlorid .	1 γ	$FeSO_4 \cdot 7\ H_2O$	135 γ
L-Prolin	1	Riboflavin . . .	1 γ	$ZnSO_4 \cdot 7\ H_2O$	4 γ
DL-Tryptophan . .	2	Nicotinsäure . .	5 γ	$CuSO_4 \cdot 5\ H_2O$	4 γ
D,L-Valin	2	Nicotinamid . .	5 γ	$CoCl_2 \cdot 6\ H_2O$	4 γ
D,L-Phenylalanin .	2	Pyridoxin-HCl .	20 γ	$MnSO_4 \cdot H_2O$	3 γ
L-Histidin	1	Ca-pantothenat .	10 γ	K_2HPO_4 . . .	15,6 mg
D,L-Isoleucin . . .	2	Cholinchlorid . .	50 γ	KH_2PO_4	1,4 mg

2. Cadaverin.

Pentamethylendiamin, 1,5-Diaminopentan.

$$HC_2 \Big\langle {}^{CH_2-CH_2-NH_2}_{CH_2-CH_2-NH_2} \qquad C_5H_{14}N_2\text{: C } 58,8\%, \text{ H } 13,7\%, \text{ N } 27,5\%, \text{ Mol-Gew. } 102.$$

Eigenschaften. Farblose, nach Piperidin riechende Flüssigkeit, erstarrt im Kältegemisch kristallinisch. Sdp. 178—179°. Bildet mit 2 Mol Wasser ein öliges Hydrat. Leicht löslich in Wasser, löslich in Alkohol, schwer löslich in Äther. Mit Wasserdämpfen flüchtig. Zieht aus der Luft Kohlendioxyd an. Wird beim Kochen mit Laugen nicht zersetzt. Gibt mit Alkaloidreagentien Niederschläge.

Derivate.

Chlorhydrat. $C_5H_{14}N_2 \cdot 2$ HCl. Zerfließliche Nadeln vom Fp. 255°. In 96%igem Alkohol löslich, in abs. Alkohol und Äther etwas löslich (Unterschied von Putrescinchlorhydrat ermöglicht Trennung von diesem). Bei der trockenen Destillation entsteht Piperidin.

Phosphorwolframat. Wenig löslich in Wasser, ziemlich löslich in Alkohol und Methanol, leicht löslich in Aceton.

Chloroaurat. $C_5H_{14}N_2 \cdot 2\,HAuCl_4$, lange stark glänzende gelbe im Exsiccator verwitternde Nadeln oder Würfel, Fp. 186—188°, leicht löslich in Wasser.

Dibenzoyl-cadaverin. $C_5H_{12}N_2(OC \cdot C_6H_5)_2$, Fp. 130°, 135°, fast unlöslich in Wasser, leicht löslich in Alkohol, sehr wenig löslich in Äther.

N,N'-Dibenzoylsulfonyl-cadaverin $C_5H_{12}N_2(O_2S \cdot C_6H_5)_2$, glänzende Kristalle aus Alkohol, Fp. 119°, leicht löslich in heißem, wenig löslich in kaltem Alkohol, leicht löslich in verdünntem Alkali.

Diphenylisocyanatverbindung des Cadaverins $C_{19}H_{24}O_2N_4$ löst sich in Pyridin etwas leichter als das entsprechende Putrescinderivat und kristallisiert im Gegensatz zu diesem aus einer Pyridin-Aceton-Mischung erst nach einiger Zeit. Die Kristallform ist ähnlich wie bei der Putrescinverbindung. Fp. 207—209°.

Dipikrat. $C_5H_{14}N_2 \cdot .2\,C_6H_3O_7$. Dünne gelbe Nadeln oder langgestreckte Tafeln. Fp. 221° unter Zers. In Wasser schwer löslich.

Dipikrolonat. $C_5H_{14}N_2 \cdot .2\,C_{10}H_8N_4O_5$. Kurze schmale Prismen, orangegelb. Zersp. 250°. Empfindlichkeit 1:30000. Wenig löslich in Wasser und Alkohol. Eignet sich zum mikrochemischen Nachweis von Cadaverin.

Styphnat. Grüngelbe, lange schmale Kristallbündel; in großen Verdünnungen, bilden sich schräg abgeschnittene Prismen. Empfindlichkeit 1:50000. Eignet sich zum mikrochemischen Nachweis von Cadaverin.

Chloroplatinat. $C_5H_{14}N_2 \cdot H_2PtCl_6$. Aus salzsaurer Lösung mit alkoholischer Platinchloridlösung. Rotgelbe Prismen, Nadelbüschel oder Oktaeder. Löslich in 113 Teilen Wasser von 12°. *Chloroaurat:* $C_5H_{14}N_2 \cdot 2\,HAuCl_4$. Lange glänzende Nadeln oder Würfel. Fp. 186—188°.

Quecksilberverbindungen bilden sich verschiedene, je nach den Bedingungen $C_5H_{14}N_2 \cdot 2\,HCl \cdot 4\,HgCl_2$, bildet bei Verwendung von überschüssigem Sublimat in salzsaurer Lösung lange farblose Nadeln vom Fp. 214—216°. In kaltem Wasser wenig löslich (in 32 Teilen bei 16°), leicht löslich in heißem Wasser. Verliert schon bei 95° HgCl. Durch die Schwerlöslichkeit in kaltem Wasser unterscheiden sich die Sublimatverbindungen des Cadaverins von denen des Putrescins.

Zur Identifizierung des Cadaverins eignet sich unter Umständen die trockene Destillation des Chlorhydrates, wobei unter Abspaltung von Ammonchlorid Piperidin entsteht.

VI. Butyl- und Amylamine.

1. Isobutylamin. $\dfrac{CH_3}{CH_3}{>}CH—CH_2 \cdot NH_2$

$C_4H_{11}N$: C 65,7%, H 15,15%, N 19,15%, Mol-Gew. 73,1.

Eigenschaften. Flüssig, noch bei —77°. Sdp. 68—69°. Mit Wasser, unter Erwärmung in allen Verhältnissen mischbar.

Derivate.

Chlorhydrat. $C_4H_{11}N \cdot HCl$. Sehr hygroskopische Schüppchen vom Fp. 177 bis 178°. 11,6 g lösen sich in 100 ml Chloroform.

Bromhydrat. $C_4H_{11}N \cdot HBr$. Blättchen vom Fp. 138°.

Chloroplatinat. $C_4H_{11}N \cdot H_2PtCl_6$. Orangefarbene Kristalle vom Zersp. 225 bis 230°.

Bromoplatinat $(C_4H_{11}N \cdot H_2) \cdot PtBr_6$. Rubinrote, monokline Prismen. Mit MAYERs-Reagens (Kaliumquecksilberjodid) entsteht noch in starker Verdünnung eine Abscheidung von der Zusammensetzung $(CH_3)_3N \cdot HgJ_2$. Löslich in einer Mischung von gleichen Teilen Essigester-Chloroform, wodurch eine Abtrennung z. B. von Dimethylamin möglich ist. Aus Alkohol lange gelbe Nadeln vom Fp. 136°.

Vorkommen. In 10 Arten von 103 verschiedenen Phanerogamen fanden Klein und Steiner (1928) in Blüten und Exhalaten geringe Mengen von Isobutylamin. Bei 14 anderen Arten Isoamylamin, die wie Trimethylamin Lockstoffe für Insekten darstellen. Isobutylamin ist auch in grünen Tabakblüten neben flüchtigen Pyrrolidinbasen beobachtet worden (Ciamician und Ravenna, 1911; Pictet und Court, 1907). — Für eine intermediäre Bildung von Isobutylamin in Pflanzen spricht auch, daß es als Baustein im Fagaramid im Spilantol, im Herculin und im Pelitorin vorhanden ist. Isobutylamin ist nach Steiner und Stein v. Kamienski (1955) in Secale cornutum enthalten.

Bei Verwendung einer außerordentlich großen Zahl von verschiedenen Bakterien verschiedenster Gruppen, die Hexonbasen in einer Nährbouillon aus tryptisch verdautem Casein und Glucose rasch decarboxylierten, wurde eine Decarboxylierung von Valin und Leucin oder irgendeiner anderen Monoaminosäure in Versuchen von Gale (1940) nicht beobachtet. Isobutylamin soll jedoch bei 4wöchiger Einwirkung von Fäulnisbakterien auf D,L-Valin gebildet werden. Aus 10g D,L-Valin wurden so von Neuberg und Karczag (1909) 0,424 g Isobutylaminchloroplatinat, Kp. 226—227°, isoliert.

Bei der Einwirkung von *Bacterium proteus* auf L-Leucin soll sich in Milchzuckerhaltiger Lösung Isobutylamin bilden (Arai, 1921).

2. Sekundäres Butylamin.

Sdp. 63°. Das aus dem Butylsenföl abgespaltene Amin ist optisch aktiv. $[\alpha]_D^{20} = +7,44°, +6,42°$. Die Salze sind linksdrehend.

Vorkommen. Als Baustein des Butylsenföls. Dieses liefert beim Erhitzen mit verdünnter Schwefelsäure freies Butylamin.

3. Isoamylamin.

$C_5H_{13}N$: C 68,87%, H 15,04%, N 16,09%. Mol.-Gew. 87,1.

Flüssigkeit vom Sdp. 95—96°, sp. Gewicht bei 18° 0,750.

Hydrochlorid. $C_5H_{13}N \cdot HCl$, bitter schmeckende, zerfl. Kristalle.

Hydrobromid. $C_5H_{13}N \cdot HBr$, Blättchen, Fp. 225°.

Saures Oxalat. $C_5H_{13}N \cdot C_2H_2O_4$, aus Alkohol und Aceton, Fp. 169°.

Neutrales Oxalat. $2 C_5H_{13}N \cdot C_2H_2O_4$, Fp. 200—207°.

Chloroplatinat. $(C_5H_{13}N)_2 \cdot H_2PtCl_6$, goldgelbe Blättchen, in heißem Wasser leicht löslich.

Chloroaurat. $C_5H_{13}N \cdot HAuCl_4$, breite, dünne Blättchen aus Wasser, Smp. 151°

L(+)-*Amylamin* (sek. Isobutylcarbinamin) siedet bei 96—97°, sp. Gewicht bei 18° 0,2550 $[\alpha]_D^{20} = -5,16°$.

Vorkommen. Die Decarboxylierung von Leucin wird durch ubiquitäre Gärungserreger wie *Bacillus casei* bewirkt (Vorkommen in Roquefortkäse). Daß auch Pilze Leucin in Isoamylamin verwandeln können, beweist das Vorkommen im Mutterkorn und *Boletus edulis*. Findet sich neben Pyrrolidinbasen in grünen Tabakblättern. Von Klein in vielen Pflanzenexhalaten nachgewiesen. Sympathicomimetisch wirksam. Bildet sich aus Leucin durch Einwirkung von Mikroorganismen in saurem Medium (Arai, 1921). Steiner und Stein v. Kamienski (1955) fanden *n-Hexylamin* ($CH_3-CH_2-CH_2-CH_2-CH_2-CH_2-NH_2$) und *n-Heptylamin* ($CH_3-CH_2-CH_2-CH_2-CH_2-CH_2-CH_2-NH_2$) neben Methyl-, Äthyl-, i-Butyl-, i-Amyl-, β-Phenyläthyl- und Trimethylamin in Secale cornutum. Nachweis dieser Substanzen mikrokristallographisch, im hängenden Tropfen mit Dinitro-α-naphthol

Nachweis: Papierchromatographisch und anhand der Pikrolonate.

Tabelle 18. *Schmelzpunkte der Pikrolonate von n-Hexyl- und Heptylamin.*

	n-Hexyl-amin in Grad	n-Heptylamin in Grad	Amin aus Mutterkorn in Grad
Pikrolonat	188—189	162	188—189
Pikrolonat + Salphen	160	142—143	160
Pikrolonat + Benzanilid	138	127—128	137—138

VII. Amine mit aromatischen Substituenten.

1. β-Phenyläthylamin. $\langle\ \rangle$—CH$_2$—CH$_2$—NH$_2$

$C_8H_{11}N$: C 79,26%, H 9,15%, N 11,59%. Mol-Gew. 121,12.

Eigenschaften. Farblose, bei 197—198° siedende, eigentümlich riechende Flüssigkeit. 1 Teil löst sich in 24 Teilen Wasser bei 20°. Leicht löslich in Alkohol und Äther. Starke einsäurige Base, die an der Luft Kohlensäure anzieht.

Derivate. Mit Phosphorwolframsäure fällt aus wäßrigen Lösungen erst bei längerem Stehen das *Phosphorwolframat* $(C_8H_{11}N)_3 \cdot H_3PO_4 \cdot 12 Wo_3$ als ein gelbliches Öl, das allmählich zu Prismen oder Nadeln erstarrt. Zur Identifizierung eignen sich folgende Derivate:

Chlorhydrat. $C_8H_{11}N \cdot HCl$. Glänzende Blättchen aus Alkohol + Äther. Fp. 217°.

Pikrat. $C_8H_{11}N \cdot C_6H_3O_7N_3$. Tetragonale Prismen vom Fp. 171 — 174°. Ziemlich löslich in warmem Wasser.

Chloroplatinat. $(C_8H_{10}N)_2 \cdot H_2PtCl_6$. Kristallisiert aus säurehaltigem Alkohol in hellgelben Blättchen oder orangefarbenen Schuppen, die sich bei 225° schwärzen und bei 246—248° unter Zersetzung schmelzen. Sehr wenig löslich in Wasser.

Chloroaurat. $C_8H_{11}N \cdot HAuCl_4$. Hellgelbe breite Nadeln vom Fp. 98 — 100°. Leicht löslich in Wasser und Alkohol.

Benzoylverbindung. $C_8H_{10}N \cdot COC_6H_5$. Fp. 114°. Isolierung aus Autolysat von *Boletus edulis* REUTHER (1912).

Vorkommen. KEIL und BARTMANN (1935) fanden Phenyläthylamin in frischen Butterpilzen *Boletus luteus* neben Putrescin und Cholin. In *Boletus elegans* wurde die Substanz vermißt. Sie ist Bestandteil von Senfölen, z. B. der Brunnenkresse. Auch im Destillat von *Viscum album* und im Extrakt der amerikanischen Mistelart *Phoradendron flavescens* kommt Phenyläthylamin vor. Die Blattknospen ("tops") von *Acacia floribunda* enthalten bis zu 0,2% eines Gemisches aus β-Phenyläthylamin und Tryptamin, Blüten bis 1%, Blattknospen von *A. pruinosa* 0,04%, Blüten von *A. longifolia* und in den Blattknospen bis 0,2%. Phenyläthylamin ist möglicherweise eine Vorstufe des Phenyläthylsenföls, das im ätherischen Öl von *Reseda odorata*, der Brunnenkresse *(Nasturtium officinale)*, Winterkresse *(Barbaraea praecox)* und *Brassica rapa* vorkommt (siehe auch RETI, 1953).

β-Phenyläthylamin entsteht bei Fäulnis von Leim, Bierhefe u. a. (Bakterienprodukt), ferner bei der Autolyse von *Boletus edulis*. Gewinnung von Phenyläthylamin, Putrescin und Cholin aus Boletus edulis s. KEIL und BARTMANN, 1935.

2. Tyramin. (p-Oxyphenyläthylamin) HO$\langle\ \rangle$—CH$_2$—CH$_2$·NH$_2$

$C_8H_{11}ON$: C 70 07%, H 8,03%, N 10,22%, Mol-Gew. 137,1.

$C_8H_{11}ON$, hexagonale Blättchen, Fp. 161°, löslich in 95 Teilen Wasser bei 15° und in 10 Teilen heißem Alkohol, ziemlich löslich in Amylalkohol, wenig löslich in Äther und Chloroform; kristallisiert aus heißem Xylol, in dem es wenig löslich ist, Sdp. 161—163° bei 2 mm, 175—179° bei 8 mm.

Derivate.

Hydrochlorid. $C_8H_{11}ON \cdot HCl$, kristallisiert aus konz. Salzsäure, Smp. 268° kurze Prismen, sehr leicht löslich in Wasser.

Pikrat. $C_8H_{11}ON \cdot C_6H_3O_7N_3$, kurze Prismen, Smp. 200°.

Chloroplatinat. $(C_8H_{11}ON)_2 \cdot H_2PtCl_6$, 6seitige Blättchen.

Monobenzoat. $C_{15}H_{15}O_2N$, hexagonale Blättchen aus Alkohol, Fp. 162°, leichter löslich als die Dibenzoylverbindung.

Dibenzoat. $C_{22}H_{19}O_3N$, Kristalle aus Alkohol, Fp. 170°.

Dicarbomethoxyderivat. $C_{12}H_{15}O_5N$, aus Methanol, Fp. 100,5°.

Vorkommen. Tyramin findet sich in Besenginster *Sarothamnus scoparius* (CORREALE und CORTESE, 1953) in Gerstenkeimlingen, in der Mariendistel (ULLMANN, 1922), in der amerikanischen Mistel *Phoradendron* (CRAWFORD und WATANABE, 1915) sowie in Mutterkorn (SCHLEMMER, 1932).

Tyramin entsteht durch Decarboxylierung von Tyrosin durch die Tyrosindecarboxylase vieler Bakterienarten (s. GALE, 1946; und WERLE, 1952). Isolierung der Tyrosindecarboxylase aus *Streptococcus haemolyticus* (s. GALE, 1940).

Tyramin wirkt am Blutdruck des narkotisierten Hundes und der Katze stark blutdrucksteigernd. Durch die Monaminoxydase wird Tyramin unter Entwicklung von Ammoniak zu p-Oxyphenylacetaldehyd abgebaut, der pharmakologisch unwirksam ist. Diese Tatsachen können zur quantitativen Bestimmung des Tyramins herangezogen werden.

Papierchromatographischer Nachweis von Tyramin in *Sarothamnus scoparius* nach CORREALE und CORTESE (1953).

Es wurden während der Winterruhe, zur Blütezeit und im Herbst Wurzeln, Stämme, Zweige und Blüten getrennt und zum Teil sofort, d. h. in ganz frischem Zustand, zum Teil nach Trocknung im Sonnenlicht oder im Thermostaten mit 5—10 Volumen 70%igem Aceton extrahiert. Nach einigem Stehen wurden die Extrakte abfiltriert, das Lösungsmittel im Vakuum vertrieben und die wäßrige Lösung mit Petroläther geschüttelt, um Fette und Pigmente zu entfernen. Dann wurde die Flüssigkeit im Vakuum eingeengt bis 1 ml derselben 2—4 g frischen Gewebes entsprach. Es wurde auf Whatman Nr. 1-Papier 24—48 Std. bei Zimmertemperatur chromatographiert. Es wurden 0,003—0,03 ml Flüssigkeit aufgetragen. Als Lösungsmittel diente n-Butanol mit n-HCl gesättigt, n-Butanol gesättigt mit Essigsäure, n-Butanol gesättigt mit Trichloressigsäure 5% und n-Butanol 8 Vol. + Methylamin (25%ig) 3 Vol. Entwicklungsreagentien: Kaliumferricyanid + NH_3-Dämpfe, Eisenchlorid, Kaliumjodid in leicht saurem Milieu, diazotierte Sulfanilsäure (Pauly-Reagens) + Natriumcarbonat 10—20%ig. α-Nitroso-β-naphthol-Lösung nach GERNGROS-VOSS und HERFELD (1933). Die letzte Reaktion fällt nur bei Monophenolen positiv aus, die Pauly-Reaktion bei Mono- und Diphenolen, die anderen Reaktionen (Oxydationsreaktionen) nur für Brenzkatechinderivate. Als reine Vergleichssubstanzen wurden Oxytyramin N-Methyloxytyramin (Epinin), Dopa, Tyramin und Tyrosin verwendet. Die R_f-Werte der gefundenen Substanzen sind in Tab. 19 wiedergegeben. Der Gehalt an Oxytyramin ist in den Zweigen am niedrigsten, nämlich 150 γ/g frischen Gewebes. Im Frühling enthalten die Zweige 250 γ/g frischen Gewebes, im Herbst 500 γ/g. Der Tyramingehalt der Extrakte aus Zweigen ist in den verschiedenen Jahreszeiten gleich dem des Oxytyramins. Die Stämme enthalten im Winter praktisch keine Phenolderivate und auch im Frühling und im Herbst sind sie erheblich ärmer als die Zweige. Tyramin ist in frischen und getrockneten Zweigen wie auch in den Blüten enthalten. Der Gehalt an Oxytyramin ist am niedrigsten im Winter (etwa 150 γ/g, gerechnet auf frisches Gewebe), nimmt im Frühling zu, auf etwa 250 γ/g, um im Herbst ein Maximum von etwa 500 γ/g zu erreichen. Der Gehalt der Zweige an Oxytyramin ist in den verschiedenen Jahreszeiten annähernd so hoch wie der Gehalt an Tyramin. Alle Mono- und Dioxyphenylderivate sind im frischen Material nachweisbar.

Sarothamnus scoparius-Samen enthält in Tyramin und Dopa nur in Spuren; Tyrosin nur in gebundener, durch Säurehydrolyse freilegbarer Form. Beim Keimen erscheinen früh die in der reifen Pflanze vorkommenden Substanzen, nämlich Tyrosin, Tyramin, Dopa-, Oxytyramin, N-Methyloxytyramin oder Epinin; Tyrosin liegt nun auch in freier Form vor. Der

Tabelle 19. *R_f-Werte von Tyramin, Oxytyramin und Methyloxytyramin.*
[Nach CORREALE und CORTESE (1953).]

Lösungsmittel.

	n-Butanol gesättigt mit Essigsäure	n-Butanol gesättigt mit Trichloressigsäure 5%	n-Butanol gesättigt mit n-HCl	n-Butanol 8 Vol. + Methylamin 25%ig 3 Vol.
Oxytyramin	0,20—0,22	0,53—0,56	0,19—0,22	—
N-Methyloxytyramin . .	0,25—0,28	0,61—0,64	0,26—0,27	—
Tyramin	—	—	0,30—0,35	0,55—0,62

Gesamtgehalt an Oxyphenylalkylaminen ist beim Keimling niedriger als bei der ausgewachsenen Pflanze. Das Verhältnis der Konzentrationen an den verschiedenen Substanzen bleibt aber unverändert. Bei rascherem Wachstum des Keimlings (in Gartenerde) nimmt der Gehalt an Oxyphenylalkylaminen stark ab. Bei *Genista spartium* ist weder im Samen noch im Keimling Brenzkatechin nachweisbar. Es wurde jedoch eine unbekannte Substanz, wahrscheinlich ein Monophenol, aufgefunden.

Tabelle 20. *Vorkommen von Tyramin, Oxytyramin und den zugehörigen Aminosäuren in Ginster nach papierchromatographischen Untersuchungen von* CORREALE *und* CORTESE *(1953).*

	Sarothamnus scoparius			Genista spartium	
	Samen	Keime I	Keime II	Samen	Keime I
Tyrosin	—	+	—	—	—
Tyramin	(Spuren)	+++	+	(Spuren)	(Spuren)
Dopa	(Spuren)	(Spuren)	—	—	—
Epinin	—	++	(Spuren)	—	—
Oxytyramin	—	++	+	—	—

Über spektrographische Untersuchungen der Inhaltsstoffe von Gerstenkeimlingen und von Mutterkorn, insbesondere über den Nachweis von Tyramin s. SCHLEMMER, 1932. Über Isolierung von Tyramin aus Semina cardui Mariac (Stechdistelkörner) s. ULLMANN, 1922.

3. Hordenin. $C_{10}H_{15}ON$ (p-Oxyphenyl-Äthyl-dimethylamin, Anhalin)

(GUGGENHEIM, 1951). $HO-\langle\bigcirc\rangle-CH_2-CH_2-N\big\langle\genfrac{}{}{0pt}{}{CH_3}{CH_3}$

Mol.-Gew. 165,23, farblose rhombische Prismen, Fp. 117—118°, sublimiert unter Zersetzung; bei 11 mm, Fp. 173—174°, leicht löslich in Alkohol, Äther und Chloroform, unlöslich in Benzol.

Hydrochlorid. $C_{10}H_{15}ON \cdot HCl$, Nadeln aus Alkohol, leicht löslich in Wasser.

Sulfat. $(C_{10}H_{15}ON)_2 \cdot H_2SO_4 + H_2O$, Nadeln, leicht löslich in Wasser, wenig löslich in Alkohol.

Pikrat. Löslich in Wasser, Fp. 139—40°.

Pikrolonat. Fp. 219—220.

Vorkommen. In Gerstenkeimlingen. Das Hordenin entsteht während der Keimung und erreicht ein Konzentrationsmaximum am 4. Keimungstag. Der Hordeningehalt beträgt in den Würzelchen etwa 0,4—0,45%, im Keimling 0,1—0,2% der Trockensubstanz. Danach nimmt das Hordenin wieder ab und ist am 25. Tag nicht mehr nachweisbar. Von 21 untersuchten Gramineen enthielten nur die Keimlinge von Gerste und der wilden Gerste Hordenin. In den Würzelchen von *Panicum miliaceum* wurden 0,86%, von *Andropogon sorghum* Brot 0,071%, von *Hordeum murinum* 8,5% des Trockengewichts gefunden. Die amerikanische Mistel *Phoradendron* enthält Hordenin neben Tyramin.

Mit Hordenin identisch erwies sich Anhalin, eine von Heffter (1894, 1896) aus der mexikanischen Kaktee *Anhalonium* isolierte Base, welcher zuerst irrtümlicherweise die Zusammensetzung $C_{10}H_{17}ON$ zugeschrieben worden war. Die Kaktee *Trichocereus candicans* enthält neben 0,5% Hordenin 0,5% Candicin (p-Oxyphenyl)-äthyl-trimethylammoniumhydroxyd.

4. Oxytyramin.

(3,4-Dioxyphenyläthylamin) $C_8H_{11}O_2N$.

Tricarbomethoxyderivat. $C_{14}H_{17}O_8N = (CH_3OOC—O)_2C_6H_3—CH_2—CH_2—NH—COO—CH_3$. Kristalle aus Methanol, Fp. 92—93°.

Tribenzoylderivat. $C_{25}H_{23}O_5N$. Fp. 141°.

Vorkommen. In Besenginster, *Genista spartium* und *Sarothamnus scoparius* (Wolfes, Kreitmair und Sieckmann, 1936 und Correale und Cortese, 1953). Mengenmäßiger Gehalt entspricht dem des Tyramins (s. dort!).

Nachweis. Aufarbeitung des Besenginsters und papierchromatographische Analyse der gewonnenen Extrakte (s. d. S. 580 bei Tyramin). Wie Tyramin, so wirkt auch Oxytyramin blutdrucksteigernd. Die pharmakologische Wirksamkeit der Substanz kann zu ihrer quantitativen Bestimmung herangezogen werden. Durch Monaminoxydase wird Oxytyramin desaminiert.

5. Tryptamin.

$$\text{(Indol)}—CH_2—CH_2 \cdot NH_2$$

(Guggenheim, 1951) $C_{10}H_{12}N_2$ (β-Indolyläthylamin). Mol.-Gew. 160,21.

Aus Alkohol + Benzol farblose Nadeln vom Fp. 118°. Wenig löslich in Wasser und in Äther. Sehr leicht löslich in Alkohol und Aceton.

Hydrochlorid, $C_{10}H_{12}N_2 \cdot HCl$, bildet farblose Prismen aus Alkohol und Äther, Fp. 156°, ist leicht löslich in Wasser, wenig löslich in Alkohol.

Pikrat, $C_{10}H_{12}N_2 \cdot C_6H_3O_7N_3$ bildet dunkelrote Kristalle, ist unlöslich in Wasser, wenig löslich in Alkohol und Chloroform, leicht löslich in Aceton, Fp. 242—243° unter Zersetzung.

Pikrolonat bildet chromgelbe Prismen aus Wasser, Fp. 231° unter Zersetzung.

Phosphorwolframat $(C_{10}H_{12}N_2)_3 \cdot H_3PO_4 \cdot 12\,WO_3$, bildet braunrote Nadeln; wenig löslich in Wasser, leicht löslich in Aceton und Alkohol, Fp. 242—243° unter Zersetzung.

Vorkommen. Blattknospen von *Acacia floribunda* und *A. pruinosa* und *A. longifolia* enthalten bis zu 0,2% eines Gemisches aus Tryptamin und β-Phenyläthylamin (Metcalfe and Sexton, 1949).

6. 5-Oxytryptamin.

Nachweis. Papierchromatographisch $C_{10}H_{12}ON_2$.

R_f-Wert in Butanol gesättigt mit nHCl 0,13—0,17. R_f-Wert in H_2O gesättigtem Butanol = 0,15. Besprühen mit Paulys-Reagens und Dinitrobenzoessigsäurereagens.

5-Oxytryptaminpikrat. $C_{10}H_{12}N_2O \cdot C_6H_3N_3O_7$, Fp. 193° (Zers.). Bei p_H 5—6 zeigt das Hydrochlorid ein Maximum der Absorption bei 275 mμ, ein 2. Maximum bei 293 mμ und ein Minimum bei 247 mμ. Lange zu Büscheln geordnete orangerote Nadeln.

Nachweis von 5-Oxytryptamin in der Juckbohne. (BOWDEN, BROWN und BEATTIE, 1954.) Die Trichome der *Mucona pruriens* waren lange dafür bekannt, daß sie ein intensives Jucken und Schmerzen auf der menschlichen Haut erzeugen. BROADBENT schrieb dies dem Vorhandensein von Histamin freisetzenden Substanzen zu. Es wurde nun festgestellt, daß in den Trichomen 5-Oxytryptamin vorkommt, und zwar nach pharmakologischen Bestimmungen 0,015%. Ein aktiver Extrakt aus Juckbohnen kann bereitet werden durch Befeuchten mit Alkohol, Extrahieren mit kaltem Wasser und Verdampfen des wäßrigen Extraktes zur Trockne unter 35°. 100 g der rohen Droge ergaben 10 g Extrakt.

Durch Fraktionieren des Rohextraktes unter pharmakologischer Kontrolle am Meerschweinchenileum, Reextraktion des rohen Materials mit kleinen Mengen Wasser, Papierchromatographie des löslichen Materials mit n-Butanol- Essigsäure-Wasser-Gemisch 100:10:30 wurde eine Fraktion erhalten, die 125mal so aktiv ist, wie der Originalextrakt. Bei absteigender Chromatographie auf Whatman-Papier Nr. 1 mit n-Butanol-Essigsäure 10:1, Wasser gesättigt, konzentrieren sich 85% der Gesamtaktivität in einer Zone mit dem R_f-Wert 0,42 der mit dem von 5-Oxytryptamin identisch ist. Identität bestand weiterhin in den Reaktionen mit ammoniakalischem Silbernitrat mit diazotierter Sulfanilsäure mit Ehrlichs-Reagens und mit Jepson und Stevens-Essigsäure-Reagens (JEPSON, STEVENS, 1953), welches für einen engen Bereich von Tryptaminderivate spezifisch ist. Wäßrige Lösungen der aktivsten Fraktion hatten die folgenden Lichtabsorptionen:

min. 247 mμ, max. 273 mμ, inf. 300 mμ. Nach HAMLIN und FISCHER (1951) sind die Daten für 5-Oxytryptamin bei p_H 5,4 min.: 250 mμ, max. 275 mμ, inf. 299 mμ.

7. N-Methyltryptamin (=Dipterin).

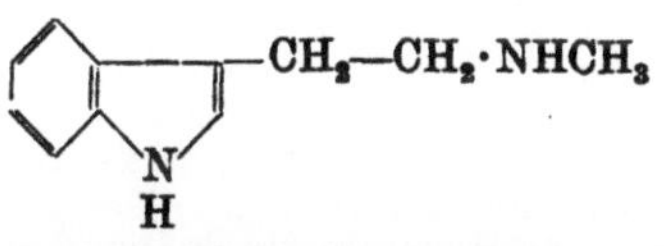

$$-CH_2-CH_2\cdot NHCH_3$$

(GUGGENHEIM, 1951)

Mol.-Gew. 174,24, Fp. 94°, bildet aus Chloroform und Petroläther sternförmige Nadeln.

Hydrochlorid. Platten, Fp. 180°.

Pikrat. Rote Nadeln, Fp. 191°.

Vorkommen. In Chenopodiaceen, *Girgensohnia diptera*, *Arthrophytum leptocladum* (YURASHEWSKY, 1939, 1940, 1941) (MANSKE, 1949).

8. Gramin.

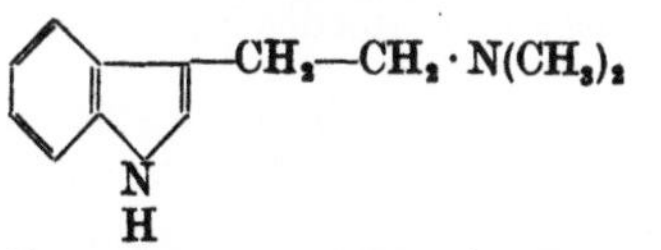

$$-CH_2-CH_2\cdot N(CH_3)_2$$

(GUGGENHEIM, 1951) $C_{11}H_{14}N_2$

(Donaxin; 3-Dimethylaminoäthylindol) Mol.-Gew. 174,24, Fp. 134°, wenig löslich in Wasser, unlöslich in Petroläther.

Pikrat. $C_{11}H_{14}N_2 \cdot C_6H_3O_7N_3$, bildet gelbe Nadeln, aus heißem Wasser, Fp. 140—42°.

Chloroplatinat. $C_{11}H_{14}N_2H_2 \cdot PtCl_6$.

Perchlorat. $C_{11}H_{14}N_2 \cdot HClO_4$, Fp. 148—50°.

Jodmethylat. $C_{12}H_{17}N_2J$, aus Alkohol, Fp. 175—76°.

Nachweis. Mit 4-Dimethylamino-benzaldehyd entsteht nach Unterschichten mit konzentrierter Schwefelsäure eine Grünfärbung, mit diazotierter Sulfanilsäure schwache Rotfärbung. Bei Zinkstaubdestillation entsteht Skatol. In vorgereinigten Extrakten gelingt die quantitative Bestimmung von Gramin durch Ermittlung des Extinktionskoeffizienten bei $\lambda = 280$ mμ.

Vorkommen. In den Blättern von *Arundo donax* L., (Mazza und Stolfi, 1931) und verschiedenen Gerstensippen (v. Euler und Erdtmann, 1935).

Isolierung. Zerkleinerte frische Gerstenblätter werden mit salzsaurem Alkohol ausgezogen und bei saurer Reaktion von in Äther löslichen Bestandteilen befreit. Bei alkalischer Reaktion wird die Graminbase in Äther übergeführt.

Ausbeute. 0,11°/$_{00}$ vom Gewicht des frischen Ausgangsmaterials. Man kann auch das Gramin den zerriebenen Blättern durch Perkolieren mit Äthylenchlorid bei ammoniakalischer Reaktion entziehen, es dann aus dem Äthylenchlorid mit verdünnter Salzsäure extrahieren und schließlich bei ammoniakalischer Reaktion wieder in Äther überführen.

VIII. Histamin.

$$HC \!=\! C\!-\!CH_2 \cdot CH_2 \cdot NH_2$$

$$| \qquad |$$

$$HN \quad N$$

$$\underset{\displaystyle H}{\underset{\displaystyle C}{\diagdown \diagup}}$$

Imidazolyl-äthylamin $C_5H_9N_3$: C 54,05%, H 8,15%, N 37,8%, Mol.-Gew. 111,1.

Eigenschaften. Kristallisiert in farblosen, zerfließlichen Platten, sehr leicht löslich in Wasser und Alkohol, leicht in heißem Chloroform, unlöslich in Äther. Fp. 83—84°, siedet bei 18 mm bei 210° ohne wesentliche Zersetzung. Wäßrige Lösung reagiert alkalisch. In unsteriler Lösung nicht haltbar, da Histamin durch Diaminoxydase haltige Luftkeime zerstört wird. Histamin ist durch eine außerordentlich starke pharmakologische Aktivität ausgezeichnet. Es wirkt an der Katze und am Hund schon in γ-Mengen stark blutdrucksenkend und kontrahiert den isolierten Meerschweinchendarm. Durch Behandeln der Testtiere oder -objekte mit Antihistaminica wird die pharmakologische Wirkung des Histamins aufgehoben. Durch Diaminoxydase, ein auch in pflanzlichen Geweben vorkommendes Ferment, wird Histamin oxydativ desaminiert, wodurch es seine pharmakologische Aktivität verliert. Unter Berücksichtigung dieser Tatsachen wird von der pharmakologischen Mengenbestimmung des Histamins sehr viel Gebrauch gemacht.

Derivate.

Dichlorhydrat. $C_5H_9N_3 \cdot 2$ HCl. In Wasser und Alkohol leicht löslich. Fp. 240°.

Monopikrat. $C_5H_9N_3 \cdot C_6H_3O_7N_3$. Nadeln vom Fp. 262—264°.

Dipikrolonat. $C_5H_9N_3 \cdot 2$ $C_{10}H_8O_5N_4$, Fp. 262—264°.

Dipikrat. Tiefgelbe rhombische Prismen oder Blättchen. Styphnat: Sechsseitige Säulen, Empfindlichkeit 1:20000.

Platinjodidverbindung. Mit großer Konstanz treten innerhalb der Konzentrationen 1:10000 und 1:20000 schwarze geweihartige Bildungen auf, die eine eindeutige Erkennung des Histamins erlauben.

Vorkommen. Überall da, wo histidinhaltiges organisches Material durch Bakterien infiziert wird, kann Histamin entstehen. Ein Vorkommen in Pflanzen ist nur dort gesichert, wo eine bakterielle Bildung ausgeschlossen wurde. Zur Histaminbildung sind vor allem die Coli- und Clostridien-Arten befähigt, durch ihren Besitz an Histidindecarboxylase, die in schwach saurem Medium vermehrt gebildet wird und deren p_H-Optimum bei fast allen Bakterienarten im sauren Bereich liegt. Histamin kommt aber auch in sterilem Pflanzenmaterial vor, in welchem eine Entstehung auf Grund seiner bakteriellen Fäulnis ausgeschlossen erscheint. Der Mechanismus seiner Entstehung ist noch nicht geklärt, d. h., es ist noch nicht entschieden, ob das Histamin z. B. durch fermentative Decarboxylierung von Histidin oder durch reduktive Aminierung von Imidazylacetaldehyd entsteht. Es steht jedoch fest, daß der Histamingehalt von jungen Spinatpflanzen, deren Wurzeln in eine histidinhaltige Nährlösung tauchten, mehr Histamin enthalten als Pflanzen, in deren Nährlösung Histidin fehlte. Im tierischen Organismus wird Histamin auch in gebundener Form, z. B. als Acetyl-Histamin angetroffen. Ein konjungiertes Histamin ist in Pflanzen bisher nicht beschrieben worden. — Histamin ist in hoher Konzentration in der Flüssigkeit der Nesselhaare von Brennesselarten enthalten. Es wurde auch im Spinat- und Tomatensaft nachgewiesen. HOLTZ und JANISCH (1937). Histamin wurde ferner in Hefeextrakten und in Baumwollstaub festgestellt, wobei allerdings eine bakterielle Entstehung wahrscheinlich ist (Vorkommen von Histamin s. Tab. 26, WERLE und RAUB, 1944).

1. Nachweis und quantitative Bestimmung.

Pharmakologische Bestimmung. Histamin kann auf Grund seiner hohen pharmakologischen Aktivität in kleinsten Mengen bis herab zu 0,01 γ/ml bestimmt werden. Am besten eignet sich der isolierte Meerschweinchendünndarm, dann aber auch die Bestimmung am Blutdruck der Katze. Da die Reaktionen der biologischen Testobjekte für Histamin nicht spezifisch sind, müssen bei alleiniger pharmakologischer Bestimmung zur Sicherung des Befundes Kontrollmessungen durchgeführt werden, z. B. in der Weise, daß man feststellt, ob anhand einer Standardlösung sich bei Verwendung verschiedenartiger Testobjekte, z. B. am Blutdruck und Meerschweinchendarm die gleichen Histaminwerte ergeben. Ferner kann man durch Vorbehandeln des Versuchsobjektes mit Antihistaminkörpern feststellen, ob der beobachtete pharmakologische Effekt durch Histamin verursacht ist oder nicht. Die in Pflanzenextrakten nachgewiesenen Histaminmengen sind meist zur chemischen Identifizierung zu gering und werden indirekt identifiziert, durch Messung an verschiedenen Versuchstieren durch Aufhebung der pharmakologischen Wirkung mit Antihistaminsubstanzen und durch Abbau durch Histaminase, z. B. aus Niere oder Darmschleimhaut, WERLE und RAUB (1948). — Zur pharmakologischen Bestimmung ihres Histamingehaltes müssen die Gewebe extrahiert werden, z. B. durch Homogenisieren in einer 5%igen Trichloressigsäurelösung. Es wird vom Ungelösten abfiltriert oder abzentrifugiert und der Trichloressigsäureextrakt entweder nach der Methode von BARSOUM und GADDUM (1935) in der von ANREP und Mitarbeitern (1939) modifizierten Form ausgeäthert und dann mit n-HCl 1 Std. lang gekocht, dann wird im Vakuum zur Trockne verdampft, die Salzsäure durch 3maliges Eintragen und Verdampfen von Alkohol im Vakuum entfernt, dann wird das Histamin aus dem Rückstand in mit Kaliumchlorid gesättigtem abs. Alkohol aufgenommn. Der Alkohol wird nach dem Abfiltrieren des Ungelösten im Vakuum verdampft, der Rückstand in eine geeignete Menge Wasser aufgenommen und die Lösung neutralisiert. Nach CODE (1937) kann man die Trichloressigsäure durch 3stündiges Verkochen des Extraktes mit konzentrierter HCl entfernen und im übrigen wie oben angegeben verfahren.

BORN, VANE und FILPOT (1952) machen Gebrauch von der raschen Diffusion von Histamin durch Cellophan; bei Dialyse von Blut gegen das gleiche Volumen Wasser ist schon nach einer Stunde völliger Konzentrationsausgleich von innen nach außen erreicht. EMMELIN (1945) ultrafiltriert das Histamin unter gleichzeitigem Zentrifugieren.

Die Vorbereitung histaminhaltiger Extrakte zum biologischen, papierchromatographischen und chemischen Nachweis kann noch auf anderen Wegen erfolgen. McINTIRE, ROTH und SHAW (1947) adsorbieren Histamin aus vorgereinigten Extrakten an Baumwollsäuresuccinat Der wäßrige histaminhaltige Extrakt wird zuerst mit 2 n-Butanol extrahiert, so daß praktisch alles Histamin in *einer* Extraktion in das Butanol übergeführt wird. Aus dem Butanol wird das Histamin durch Baumwollsäuresuccinat, einen Kationenaustauscher, aufgenommen und aus diesem mit einem kleinen Volumen verdünnter Salzsäure wieder eluiert und mit Natriumhydroxyd neutralisiert zu einer isotonischen für den biologischen Versuch geeigneten Lösung.

Baumwollsäuresuccinat wird in folgender Weise gewonnen: Geschmolzenes Natriumacetat (5 g) und Bernsteinsäureanhydrid (40 g) werden in 300 ml Eisessig gelöst, 10 g Baumwolle werden in dieser Lösung suspendiert. Eine Trockenpistole wird mit der Flasche verbunden und die Temperatur 48 Std. auf 100° gehalten. Die teilweise veresterte Baumwolle wird abfiltriert, gut gewaschen mit Wasser, verdünnter Salzsäure und Wasser, anschließend mit Alkohol. Nach Lubschetz (1950) wird Histamin aus dem butanolischen Extrakt durch Amberlite IRC 50 aufgenommen. Der Amberlite wird mit der Butanollösung geschüttelt. Schon Spuren von anorganischen Salzen verhindern die Adsorption von Histamin an Baumwollsäuresuccinat und an Amberlite, so daß auf eine scharfe Abtrennung der wäßrigen Lösung von Butanol zu achten ist. Zur quantitativen Elution des Histamins läßt man die eluierende Säure 15 min lang mit dem Amberlite in Berührung.

Anrep (1944) adsorbiert das Histamin an aktivierte Tierkohle und eluiert mit salzsäurehaltigem Alkohol; Roberts und Adam (1950) adsorbieren an den Kationenaustauscher Decalso und eluieren mit Ammoniumhydroxyd. Spez. Gewicht 0,880 und waschen nach mit analysenreinem Chloroform, das mit Ammoniakgas gesättigt ist. Das Eluat wird im Vakuum zur Trockne verdampft und der Rückstand mit absolutem Alkohol, der 3% Salzsäure enthält, extrahiert. Vom Ungelösten wird abzentrifugiert. Es wird wiederum zur Trockne verdampft, dann nach dem Durchfeuchten mit absolutem Alkohol wiederum zur Trockne verdampft. Dieser Rückstand kann dann in Wasser aufgenommen und biologisch am Meerschweinchendarm bestimmt werden. *Permutit* (0,5—1,0 g) kann aus 20 ml einer wäßrigen Lösung 10—1000 γ Histamin quantitativ adsorbieren. Histidin, β-Imidazolyl-4(5)-Milchsäure, Imidazol, Tyramin, Tyrosin und Phenol werden nicht aufgenommen. Das Histamin wird mit konzentrierter Kochsalzlösung eluiert (Schwartz und Riegert, 1936).

2. Farb- und Fällungsreaktionen für Histamin.

Die älteste und früher gebräuchlichste Methode zum kolorimetrischen Histaminnachweis ist die Diazoreaktion nach Pauly (1904), mit diazotierter Sulfanilsäure, die mit Histamin aber auch mit Histidin eine kirschrote Färbung ergibt. Die Diazoreaktion wurde von Koessler und Hanke (1919) zur quantitativen Histaminbestimmung in biologischem Material verwendet.

Ausführung der Diazoreaktion nach Maciag und Schöntal (1938). 2 ml einer wäßrigen Histamindihydrochlorid-Lösung werden mit 1 ml gesättigter Na_2CO_3-Lösung 0,5 ml frischem Diazoreagens (5 g Sulfanilsäure + 50 ml konz. HCl + 1000 ml Wasser) gemischt mit 2 Teilen einer 0,5%igen $NaNO_2$-Lösung versetzt. Es tritt kirschrote Färbung auf. Unmittelbar darauf werden 1,5 ml 95%igen Alkohols zugefügt. Diese Menge Alkohol reicht aus, um den Farbstoff in Lösung zu halten. Die Farbe ist noch in einer Verdünnung von 1:100000 zu erkennen. Auch Tyrosin und andere Phenolderivate kuppeln mit Diazobenzolsulfosäure unter Bildung einer rotbraunen Färbung. Man kann ihr Kupplungsvermögen jedoch durch vorsichtige Benzoylierung unter Beibehaltung der Reaktionsfähigkeit der Imidazolverbindungen ausschalten. Die photometrische

Bestimmung mittels der Diazoreaktion mit Sulfanilsäure wurde von Havinga (1947), von Barac (1950) verfeinert. Am besten wird nach Havinga, Sekkles und Strengers (1947) ein photoelektrisches Photometer verwendet. Die Messungen werden anhand von diazotierten Histaminstandardlösungen bei 5000 Å durchgeführt. Das Lambert-Beersche Gesetz ist genau erfüllt. Die Grenze der Bestimmungsmöglichkeit liegt bei 1 mg Histamindichlorhydrat pro Liter. Bei Verwendung größerer Küvetten können noch größere Histaminverdünnungen quantitativ bestimmt werden.

Reagentien. 0,9 g Sulfanilsäure in 100 ml bidestilliertem Wasser enthaltend 9 ml 37%iger Salzsäure. — Natriumnitrit 7 g/100 ml. — Natriumcarbonat 18,3 g/100 ml. — Zur Kupplung läßt sich auch anstelle der Diazobenzolsulfosäure diazotierte p-Nitranilinsäure verwenden, wodurch eine Erfassungsgrenze von 0,5 γ Histamin/ml erreicht wird (Rosenthal und Tabor, 1948).

3. Kolorimetrische Bestimmung (Lowry, 1954).

Die Farbreaktionen erreichen aber bei weitem nicht die Empfindlichkeit der biologischen Methoden, mit deren Hilfe Messungen bis zu Konzentrationen von 10^{-8} molar möglich sind. Eine die biologischen Verfahren an Empfindlichkeit noch übertreffende Methode gründet sich auf die Reaktion der Aminogruppe des Histamins mit Dinitrofluorbenzol (D N F B) nach Sanger (1945). Das von McIntire, White und Sproull (1950) und neuerdings von Lowry (1954) entwickelte fluoreszenzphotometrische Verfahren erfaßt noch 0,01—0,02 γ Histamin mit einer Genauigkeit von 25% bei Vorliegen einer 4×10^{-10} mol. Lösung und einer Genauigkeit von 5—10% bei Vorliegen von 0,1 γ Histamin oder mehr.

Das Verfahren besteht aus folgenden Schritten:

1. Extraktion des Histamins aus dem biologischen Material mit Trichloressigsäure.

2. 50fache Anreicherung des Histamins und Trennung von vielen anderen Aminen durch Adsorption an Decalso.

3. Bildung des gefärbten Histaminderivates mit DNFB.

4. Weitere Konzentrierung (8fach) und Trennung dieses Derivates von DNFB-Derivaten anderer übriggebliebener Amine durch Extraktion mit 2 Octanon und anschließende Überführung in eine kleine Menge Salzsäure.

5. Messung der Farbintensität in Mikroküvetten bei einer Wellenlänge von 360 mμ.

Reagentien. 1. Wasser, bidestilliert in Glas.

2. Trichloressigsäure (TCA), redestilliert.

3. Natriumacetat, 4 mol.

4. Kaliumbromid, 40 g auf 100 ml.

5. Karbonatpuffer (p$_H$ 10): 21 g (0,2 mol.) Na$_2$CO$_3$ wasserfrei und 8,4 g (0,1 mol.) NaHCO$_3$ auf 1 l.

6. Natrium-Diäthyldithiocarbamat (Destillation Products Industries, Rochester, N. Y.) 12,5 mg-%ige Lösung, bereitet aus einer Standardlösung (1,25%ig). Beide Lösungen sollen bei 5° C aufbewahrt werden.

7. 2,4-Dinitrofluorbenzol (DNFB) (Jasonols Chem. Corporation, 1085 Myrtle Avenue, Brooklyn 6, N. Y.) 2 % (Volumen) in absolutem Alkohol.

8. 2-Octanon. Wenn eine Trübung beim letzten Schritt der Analyse eintritt, kann diese sehr verringert werden durch Redestillation und/oder Auswaschen des Octanols mit Säure.

9. Salzsäure 2 n.

10. Histaminstandardlösung, 1—100 γ Histaminbase (1,66 bis 166 γ Histamindihydrochlorid) per ml in 0,01 n HCl. Diese Standardlösungen sind über sehr lange Zeit haltbar, wenn sie bei —20° C eingefroren werden.

11. Decalso (The Permutit Company, New York, N. Y.). Decalso wird gekocht mit 3% Essigsäure, in H$_2$O gewaschen und getrocknet. Der 40- bis 60-Maschenanteil wird benutzt.

Apparate. (Abb. 5) Thermostat mit einer Temperatur zwischen 75 und 80°. Er soll die Elutionsflüssigkeit warm halten.

2. **Adsorptionssäule (Abb. 5) (A. S. Aloe Company, St. Louis).** Diese besteht aus einem oberen Reservoir, einer langen Röhre, um einen genügenden Druck zu erreichen, einem dritten engeren Teil zur Aufnahme des Adsorbens, dieser Teil muß sehr eng sein, um die Menge des benötigten Adsorbens und der Elutionsflüssigkeit sehr klein zu halten, einer Verengerung, die den Sand und das Adsorbens zurückhält und einem spitz zulaufenden Ende. Eine befriedigende Elution wird nur erreicht, wenn das dritte Segment (Adsorbens) vollständig in den Wärmeschrank hineinreicht, Säulen, die zu lang sind, können hochgezogen und durch eine Gummimanschette befestigt werden. Mit Hilfe von fließendem bidestilliertem Wasser wird eine einige mm dicke Schicht von säuregewaschenem Sand (40—60 Maschen) in die Röhre eingebracht und dann eine Säule aus Decalso von 3 ± 0,3 cm Höhe. Diese Angaben gelten für die durch leichtes Aufstoßen fest gepackte Röhre. Die Höhe der Decalsoschicht ist entscheidender als das Gewicht. Der Durchfluß durch die so zubereitete Röhre soll 0,1—0,5 ml in der Minute betragen. Luftblasen über dem adsorbierenden Material, die den Durchfluß verhindern oder verlangsamen, können mit einem Stahldraht entfernt werden. Wenn Sandkörnchen die Verengung verlegen, können sie mit einem sehr feinen Draht, der durch die untere Öffnung eingeführt wird, entfernt werden.

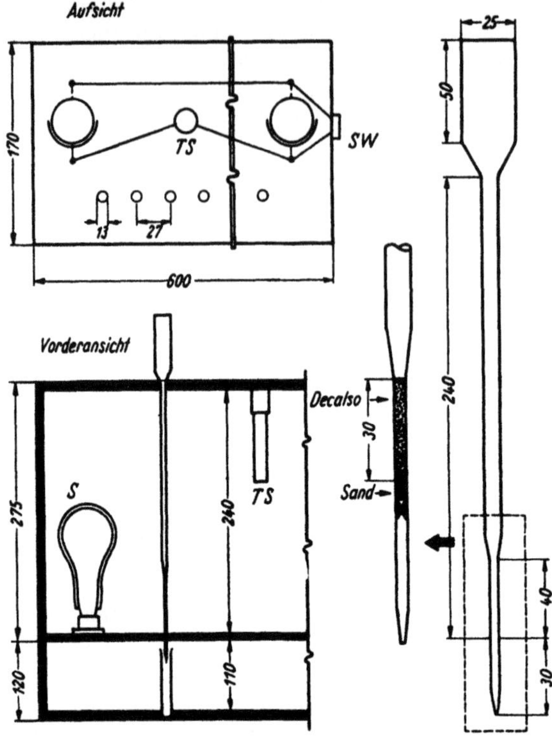

Abb. 5. Adsorptionssäule und Wärmeschrank. Die Zahlen bedeuten Abmessungen in mm. Innerer Durchmesser des mittleren Abschnittes der Säule 8 mm; Decalso-Abschnitt 1,8 mm; Verengerung 0,3 mm. *TS* Thermostat, *SW* Schalter, *S* 100 Watt-Birne, teilweise abgeschirmt durch Aluminiumfolie, um eine stellenweise Überhitzung zu vermeiden.

3. **Rührapparat für schnelles** Mischen (und Extraktion) von Lösungen (Bessey et al., 1946). Dieser besteht aus einem leicht gebogenen Stab, der im Futter eines mit ziemlich hoher Geschwindigkeit laufenden Motors (2000 Umdrehungen/min oder mehr) befestigt wird. Der Stab läuft waagerecht. Wenn ein Röhrchen im Winkel gegen den sich drehenden Stab gehalten wird, wird der Inhalt heftig durchgequirlt.

4. **Mikroküvetten,** 1—1,4 mm lichte Weite mit 10 mm Lichtdurchfall (Pyrocell Mfg., Co., 507 East 84th Street, New York, N. Y.) werden im Beckmann-Spektrophotometer benutzt, das eingerichtet wird, um die Absorption kleinster Proben von 25—35 Kubikmillimeter (Lowry und Bessey) (1946), messen zu können.

5. **Röhrchen mit Säure** gereinigt zum Auffangen des Eluates und aller bei der Durchführung der Histaminbestimmung anfallenden Flüssigkeiten. Sie werden gespült, mit konzentrierter Salpetersäure, die 1:1 mit Wasser verdünnt ist, dann ausgespült, mit destilliertem Wasser gekocht und schließlich mit bidestilliertem Wasser gespült.

6. **Lange Levy-Pipetten** (Levy, 1936) (Microchemical Specialities Company, Berkeley, California; The Carlsberg Laboratory, Copenhagen, Denmark. Mit diesen werden alle Volumina unter 1 ml gemessen. Mit Hilfe ausgezogener Spezialpipetten, 25 cm lang, mit einer Kapazität von 0,5 ml, wird die KBr-Elutionsflüssigkeit auf die Decalsoschicht in der Röhre gebracht.

Das Verfahren ist zur Histaminbestimmung im Blutplasma ausgearbeitet, dürfte aber unter Weglassung der die Blutgerinnung betreffenden Angaben auch auf pflanzliches Material anwendbar sein.

Adsorption und Elution. Zu einem abgemessenen Extraktvolumen (z. B. 10 cm³) wird $^1/_{10}$ Volumen 4 mol. Natriumacetat gegeben, um das p_H auf ungefähr 4 zu bringen. Mit ein wenig Wasser wird diese Mischung in die Decalso-Säule gespült. Nachdem die Flüssigkeit durch die Säule gelaufen ist, mit einer Geschwindigkeit, die nicht größer sein soll als 0,5 ml pro Minute, wird die Säule mit 5 ml Wasser gewaschen. Die letzten Tropfen des Waschwassers werden herausgeblasen. Die Säule wird in den Wärmeschrank gebracht, dann werden 0,5 ml 40%iges KBr in die Säule, gerade oberhalb des Decalso, einpipettiert. Das Eluat wird in einer Röhre aufgefangen (6 ml, 11 × 100 mm, Nr. 9450 Arthur H.homas, Co., Philadelphia), die ungefähr 40 ml Natriumdiäthylthiocarbamat (0,0008 mol.) enthält. Die letzten Tropfen des Eluates werden ausgeblasen.

Farbentwicklung und Extraktion. Zu jedem Eluat werden 0,25 ml Carbonatpuffer und 20 ml 2%iges DNFB gegeben (kurz vorher mit dem Mischapparat gemischt). Die Röhrchen werden in einem Wasserbad von konstanter Temperatur bei 60° C 30 min lang inkubiert. Nach der Zugabe von 0,4 ml Methyl-n-hexyl-keton wird jedes Röhrchen heftig gemischt und zentrifugiert. Ein aliquoter Teil des Ketons (0,30 oder 0,35 ml) wird in ein 1 ml-Röhrchen (7 × 70 mm, Kimble Nr. 45060-S181, A. S. Aloe Co., St. Louis) gebracht, das 50 ml 2 n HCl enthält. Diese Mischung wird 10 sec lang gerührt, dann 5 min lang bei 3000 Umdrehungen zentrifugiert.

Ablesung und Standardisierung. Die Ketonschicht wird sorgfältig abgesaugt und verworfen. Die Absorption der Säureschicht wird bei 360 mμ in einer Mikroküvette abgelesen. Die Prozedur kann eine Nacht lang oder länger unterbrochen werden, entweder wenn das Histamin an der Decalsosäule adsorbiert ist, oder wenn das Histamin in der letzten Säurelösung ist. In letzterem Fall wird die Ketonschicht nicht entfernt, um das Verdampfen zu verhindern, und die Probe wird eingefroren.

Standardlösungen, die 0,05—0,2 γ Histamin enthalten, und Leerproben, jede mit einem Volumen von 5% TCA, gleich dem Volumen des benutzten Extraktes, werden durch den ganzen Prozeß mitgeführt. Der photometrische Wert der Leerproben ist gewöhnlich so niedrig, wie der von 0,02 γ Histamin. Bei großen Histaminmengen ergibt eine entsprechende Verdünnung mit 2 n HCl befriedigende Ergebnisse bis hinauf zu 10 γ Histamin.

4. Bestimmung des Histamins als Isotopen-Derivat.

Eine sehr empfindliche quantitative Bestimmung des Histamins in Form eines Isotopenderivates stammt von SCHAYER, KOBAYASHI und SMILEY (1954). Mit p-Jod (J 131)-phenylsulfanyl-chlorid (Pipsylchlorid) in Dioxan gelöst, setzt sich Histamindihydrochlorid noch in stärkster Verdünnung zu 93—96% zu Dipipsylhistamin um, das anhand von Träger-Pipsylhistamin, also nicht markiertem Pipsylhistamin, auf Grund seiner Radioaktivität quantitativ bestimmt werden kann.

Dipipsylhistamin. C: 31,7, H: 2,37, N: 6,53%.

Das bisher nur bei tierischem Material verwendete Verfahren geht aus von Trichloressigsäureextrakten, die auch bei pflanzlichem Material das Histamin zuverlässig erfassen.

5. Fällungsreaktionen für Histamin.

Nach ZEIDLER (1952) wird Histamin auch durch Tetraphenylbornatrium (Kalignost[1]) gefällt. Die Löslichkeit in Wasser beträgt 0,01%; Fp. 135°.

Zur Freisetzung des Histamins aus der Tetraphenylborverbindung wird mit Salzsäure angesäuert, das freie Tetraphenylbor wird ausgeäthert.

Beispiel: 0,08 g Tetraphenyl-bor-histamin werden mit 1 ml starker HCl versetzt. Dann wird mit Äther mehrmals ausgeschüttelt. Der Äther wird jeweils mittels einer Kapillarpipette abgesaugt. Nach Trocknen des wäßrigen Anteils i. V. bleiben 0,02 g Histamindichlorid zurück. Ber. 0,02 g. Das Fällungsmittel muß stets frisch gelöst werden und die Lösung des zu fällenden Histamins muß neutral oder schwach sauer sein.

Fällung von Histamin und anderen Aminen mit Tetraphenylbornatrium (Kalignost).

Außer Cholin und Acetylcholin (s. S. 560) werden auch Histamin und weitere Amine durch Kalignost (s. Tab. 23) gefällt.

---

[1] Als „Kalignost" von der Firma Heyl und Co., Hildesheim, zu beziehen.

Nach Galat und Friedmann (1949) wird Histamin aus seinen Lösungen durch Bis-o-dichlorbenzolsulfonsäure (Vickery, 1942) gefällt. Mehrere Stunden nach Zugabe der Säure scheiden sich farblose Kristalle des Bis-o-dichlorbenzolsulfonates ab. Sie werden abfiltriert und aus kochendem Wasser unter Zugabe von aktivierter Tierkohle umkristallisiert. Das bei 100° getrocknete Produkt $C_{17}H_{17}O_6Cl_4N_3S_2$ schmilzt bei 225—227° (korr.). Die Überführung des Bisdichlorsulfonates in Histamindihydrochlorid erfolgt durch Lösen in kochendem n-Butanol, Sättigung dieser Lösung mit HCl und Abkühlung auf Zimmertemperatur. Die freie Base wird durch Zersetzen des Bisdichlorsulfonates mit Bariumhydroxyd gewonnen. Nach Ackermann (1934) kann Histamin auch durch 2,6-Dijodphenolsulfonsäure gefällt werden. Weitere Fällungsmöglichkeiten siehe allgemeiner Teil.

Tabelle 21. *Tetraphenylbor-Verbindungen einiger biogener Amine.*

Cadaverin	wasserlöslich in %	Fp.
Cadaverin	0,031	164°
Histamin	0,01	135°
Putrescin	0,027	196°
Tetramethylammonium- hydroxyd	0,05	200° (Zers.)

6. Papierchromatographischer Nachweis von Histamin.

Die vorbereiteten Extrakte können mit verschiedenen Lösungsmitteln chromatographiert werden. Bremner und Kenten (1951), Steiner und Stein von Kamienski (1953) verwenden Butanol-Eisessig. R_f-Wert 0,15—0,25. Das Histamin der Chromatogramme kann auch kolorimetrisch quantitativ bestimmt werden (Dent, 1947).

Bei der Chromatographie mit Mineralsäure-(HCl oder H_2SO_4)haltigen Lösungsmitteln kann beim Trocknen des Papiers Verkohlung eintreten. Es ist dann zweckmäßig, das Papier vorher mit einer 10%igen Ammoniumhydroxydlösung zu sättigen (Williams und Cirby, 1948). Zur Papierchromatographie des Histamins siehe ferner Dent (1948), Urbach (1948), Urbach (1949), Penau, Saias und Andretti (1952) und West und Riley (1954).

7. Vorkommen von Histamin.

Die Haare von Brennesselarten *Urtica urens* und *Urtica dioica* enthalten nach Emmelin und Feldberg (1949) neben Acetylcholin auch Histamin und eine dritte, noch nicht identifizierte Substanz. In *Laportea gigas* wurde Acetylcholin und Histamin durch Lindigkeit und Jung (1953) nachgewiesen. Acetylcholin wurde am eserinisierten Froschrektus bestimmt und identifiziert auf Grund seiner Empfindlichkeit gegenüber Alkali und Cholinesterase sowie seiner Kochbeständigkeit in saurer Lösung, Wirkung am isolierten Meerschweinchendarm, Aufhebung der Wirkung durch Atropinisieren des Versuchsobjektes und Resistenz der Darmwirkung gegenüber einer Antihistaminsubstanz. Die Histaminbestimmung wurde am atropinisierten Meerschweinchendarm vorgenommen. Die Wirkung der als Histamin angesprochenen Substanz wurde durch Vorbehandlung des Versuchsobjektes mit Antihistaminsubstanz aufgehoben. Die größeren Stengelhaare enthalten bei *U. dioica* und *U. urens* mehr Histamin als die kleineren Blätterhaare. Bei *U. dioica* besteht kein Unterschied in bezug auf die Größe der Haare.

Histamin kommt auch in Mariendistelkörnern *Silybum marianum* L. vor (s. Berger, 1950).

Tabelle 22a. *I. Acetylcholin- und Histamingehalt der Haare von Urtica dioica.*

Art der Haare	Zahl der Haare für die Bestimmung verwendet	Acetylcholin in γ pro Haar	Histamin in γ pro Haar
Stengel	30	0,067	—
Blätter	30	0,046	—
Stengel	500	0,040	0,01
Blätter	500	0,049	0,0035
Stengel (kleine Haare) . .	100	0,013	

Tabelle 22b. *Acetylcholin und Histamin in γ pro 10 mg Blattgewebe.* *(Haare entfernt.)*

U. urens:									
Acetylcholin . . .	0,09	0,14	0,05	0,02	0,27	0,18	—	—	—
Histamin	1,0	1,0	0,83	—	—	0,19	0,5	—	—
U. dioica:									
Acetylcholin . . .	0,095	0,095	0,67	0,005	0,35	0,12	0,21	0,3	0,65
Histamin	0,2	0,17	0,2	0,17	0,063	—	—	—	—

Tabelle 22c. *Verteilung von Acetylcholin und Histamin in äußeren und inneren Schichten der Stiele.*

	Gehalt in γ pro mg Gewebe			
	Acetylcholin		Histamin	
Pflanze	Assimilierende Rinde	Stiel	Assimilierende Rinde	Stiel
U. urens . .	0,45	0,06	1,1	0,23
U. urens . .	0,5	0,06	0,8	0,17
U. dioica .	0,26	0,11	0,03	0,013
U. dioica .	0,07	0,02	0,05	0,01
U. dioica .	0,5	0,14	—	—

Tabelle 22d. *Verteilung von Acetylcholin und Histamin in dem unterirdischen Gewebe von Urtica urens.*

Exp. Nr.	Acetylcholin in γ pro 10 mg Gewebe Hauptwurzel				Histamin in γ pro 10 mg Gewebe Hauptwurzel		
	Würzelchen	ganze	Rinde	Stiel	Würzelchen	Rinde	Stiel
1	0,55	—	0,09	0,01	1,1	0,38	0,06
2	0,19	—	0,05	—	0,55	0,7	—
3	—	—	0,24	0,06	—	2,7	0,65
4	0,14	—	0,08	—	—	1,0	
5	0,13	0,08	—	—	—	—	—
6	0,13	0,03	—	—	—	—	—
7	—	—	0,03	0,02	—	—	—
8	—	—	0,54	0,01	—	—	—
Durchschnitt	0,23	0,06	0,17	0,025	0,8	1,2	0,36

Tabelle 23a. *Histamingehalt in γ pro g Organ Keimpflanze.*
(WERLE und RAUB, 1948.)

Pflanze Nr.	Alter der Keimlinge in Tagen	Wurzel	Stengel	Keimblatt	1. Blatt	2. Blatt
1	30	19	106	32	139	
2	43	20	142		40,9	
3	51	32	179		47,5	67,5

Tabelle 23 b. *Histamingehalt in γ pro g Organ ausgewachsener, überwinterter Pflanze.*

Pflanze	Nr.	Wurzel	Stengel	Blatt, frischgrün	Blatt, welkend	Blüte
Spinat ♂	1	61	22	52	123	598
Spinat ♂	2	80	113	291		1378
Spinat ♀	3	84	6	14	54	858
G. Heinrich	1	117[3]				
G. Heinrich	2		39	6		192
G. Heinrich	3		30	117[1]		81
G. Heinrich	4	212[3]		30[2]		

Tabelle 24. *Vorkommen von Histamin in verschiedenen Pflanzenfamilien.*
(Werle und Raub, 1948.)

Untersuchte Pflanze	Histamingehalt in γ pro g frischen Pflanzengewebes	
	von Blättern	von anderen Pflanzenteilen
Urticaceae		
Urtica dioica	11,3*[4]; 43,6	
Urtica urens	120*; 20*; 122 143,5	
Urera	30*; 34,3*	
Laportea	50*	Frucht: 2,2*; 3*
Piperaceae		
Piper nigrum	0; 0	
Chenopodiaceae		
Spinacia	siehe Tab. 23 b 14,4—292	Samen 12,6; 27,3; 31,4
		Samen geschält 15,5
		Samenschale allein 50,4
		Keimling 6 Tage alt 465
		Keimling 8 Tage alt 60,5; 385
		Keimling 18 Tage alt 180
		Keimlinge gleicher Saat über, längeren Zeitraum unters.e s. Abb. 6. Werte für dis. übrigen Pflanzenorgane Tab. 23a u. b.
Beta vulgar. var. rapa	5,2	
Beta trigyna	4,2	
Chenopodium bonus	45; 52	
Henricus	40,7	Samen 0*; 0*
	siehe Tab. 23 b 64—118	Wurzel 22,5*
		Keimpflanze 16,5*
		versch. Org. s. Tab. 23 b
Chenopodium album	6*	unreif. Frucht 120; 20,5
Chenopodium quinoa		unreife Frucht 20,8
Kochia childsii		Stengel + Blätter + unreife Früchte 5,2*; 10,5
Salsola Kali		Stengel + Blätter + unreife Früchte 13,4; 14,2
Ranunculaceae		
Delphinium		Blüte: 2,9
Sarraceniaceae		
Sarracenia	Schlauchblatt 10,2*[5]; 2,6	
Nepenthaceae		
Nepenthes	Blattkanne 13,8*; 4,1 Blatt ohne Kanne 1,7	

[1] Jüngere, dicht unter der Blüte stehende Blätter.

[2] Ältere Blätter, tiefer am Sproß stehend.

[3] Die Bestimmung des Histamingehaltes wurde zu Ende der Vegetationsperiode durchgeführt.

[4] Die mit * versehenen Zahlen geben nicht den Histamingehalt des Pflanzengewebes, getestet an Aufarbeitungen, sondern den einfachen Wasserextrakt der betreffenden Pflanzenteile an, die in einigen Fällen zum Vergleich in die Tabelle eingefügt sind, jedoch nicht von derselben untersuchten Probe stammen.

[5] Siehe Fußnote 4, S. 593.

Tabelle 24. (Fortsetzung.)

Untersuchte Pflanze	Histamingehalt in γ pro g frischen Pflanzengewebes	
	von Blättern	von anderen Pflanzenteilen
Papaveraceae Chelidonium majus	24*; 5,7*; 40,5*; 10,6; 107	
Fumariaceae Corydalis glauca	0	
Droseraceae Drosera	Blatt mit Drüsenköpfchen 1,2*; 6,7 Blatt ohne Drüsenköpf. 6,1	
Sterculaceae Theobroma cacao	0	
Geraniaceae „Geranie"	0	
Rutaceae Ruta graveolescens	0	
Citrus vulgaris	17,3	
Aquifoliaceae Ilex	0	
Crassulaceae Sempervivum	0	
Saxifragaceae Saxifraga sarmentosa	0	
Mimosaceae Mimosa	2,7*; 13,3*; 8,3	
Caesalpinaceae Tamarindus indica	0	
Papilionaceae Genista palida	0	
Trifolium pratense	3*; 4,5*	
Trifolium repens	4* 6,2; 13,1	
Onagraceae Fuchsia	0	
Araliaceae Hedera helix	0	
Primulaceae Primula obconica	0	
Cyclamen	0,9* 6,2	Blüte 1,3*; 4,2 Knolle 2,7* Wurzel 0,6*
Labiatae Lamium album	20* 20,2	
Salvia	2,9	
Orobanchaceae Orobanche		Sproß 0
Lentibulariaceae Pinguicula	2—10	
Plantaginaceae Plantago lanceolata	3,6* 0,7; 6,2	

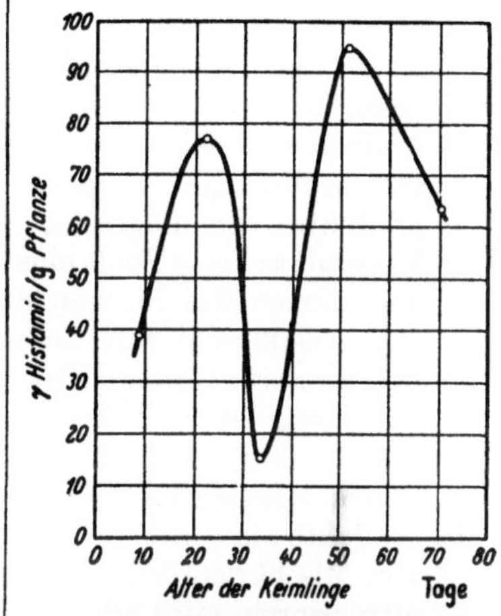

Abb. 6. Histamingehalt des Spinats, untersucht vom Keimling bis zur fruchtenden Pflanze. — Die Ordinate gibt den Histamingehalt in γ pro g frischer Pflanze, die Abszisse das Alter der Pflanzen in Tagen an. (Werle und Raub, 1948.)

8. Mikrochemischer Nachweis von Putrescin, Cadaverin, Phenyläthylamin, Tyramin und Histamin.

Klein und Boser (1932) geben mikrochemische Reaktionen an, die eine Erkennung von Putrescin, Cadaverin, Histamin, Tyramin und Phenyläthylamin nebeneinander erlauben. Von den untersuchten Reagentien geben die meisten jodhaltigen Verbindungen mit diesen Aminen keine Fällungen. Fast alle anderen Fällungsmittel bilden mit den Aminen zwar Reaktionsprodukte, die aber wenig charakteristisch sind. Phosphorwolframsäure eignet sich nur zum Nachweis von Tyramin in ziemlich konzentrierter Lösung.

Tabelle 25.

Reagens	Erfassungsgrenze				
	Putrescin	Cadaverin	Histamin	Phenyl-äthylamin	Tyramin
Goldbromid			1 γ		4 γ
Goldjodid		4 γ			
Platinjodid			2 γ	4 γ	
Pikrinsäure			2 γ	40 γ	40 γ
Pikrolonsäure	4/5 γ	4/3 γ			40 γ
Trinitroresorcin	4/5 γ	4/5 γ	2 γ		
Trinitrobenzoesäure . . .		4 γ			
Phosphorwolframsäure . .					40 γ

Zu den Reagentien bemerken die Autoren noch folgendes: Die Edelmetallsalze, insbesondere Platin- und Goldjodid, und ihre Reaktionsprodukte mit Aminen sind sehr polymorph. Geringe Schwankungen der äußeren Kristallisationsbedingungen verändern die Kristallform des Reagens und die des Reaktionsproduktes. Im folgenden sind nun Reaktionen beschrieben, die sich entweder konstant auf *eine* Form einstellen (Goldbromid mit Tyramin) oder bei denen neben Variationen der Formen doch eine typische Form immer wieder durchschlägt (z. B. Goldjodid mit Cadaverin immer wieder die Tendenz zur Bildung des Sechsersterns, auch Platinjodid mit Histamin stets geweihartige Formen). Ungünstig ist, daß die Reaktionsprodukte ziemlich labil sind und zur Bildung von Mischkristallen neigen. Auch bleiben in Substanzgemischen einzelne Reaktionen sehr leicht aus. Bessere Ergebnisse liefern Nitrokörper. Ihre Reaktionsprodukte sind fast immer formtypisch und formkonstant. Tendenz zur Bildung von Mischkristallen konnten Klein und Boser nur bei der Reaktion von Trinitroresorcin mit Cadaverin und Putrescin beobachten.

Tabelle 26.

Reagens	Lösung der reinen Substanz					Gemische									
						aliphat.		cyclische			aliphat. u. cyclische.				
	Pu	Ca	Hi	Phe	Ty	Pu	Ca	Hi	Phe	Ty	Pu	Ca	Hi	Phe	Ty
Goldbromid	−	−	+	−	+	−	−	−	−	−	−	−	−	−	+
Goldjodid	−	+	−	−	−	−	+	−	−	−	−	−	−	−	−
Platinjodid	−	−	+	+	−	−	−	+	−	−	−	−	+	−	−
Pikrinsäure	−	−	+	+	+	−	−	+	+	+	−	−	+	+	+
Pikrolonsäure	+	+	−	−	+	+	+	−	−	+	+	+	−	−	−
Trinitroresorcin	+	+	+	−	−	M	M	+	−	−	−	+	+	−	−
Trinitrobenzoesäure	−	+	−	−	−	−	+	−	−	−	−	+	−	−	−
Phosphorwolframsäure . . .	−	−	−	−	+	−	−	−	−	+	−	−	−	−	+

Die Reaktionsprodukte der einzelnen Amine mit den oben angeführten Reagentien sind also typisch, auch im Gemisch mit anderen Aminen. Eine Trennung der Amine vor ihrer Bestimmung ist demnach nicht nötig.

Der mikrochemische Nachweis der im Mutterkorn enthaltenen Basen gelang Klein und Boser (1932) auf folgende Weise: 50 g Mutterkorn werden 3 Tage in 1 l Chloroform belassen. Nach Entfernung des Chloroforms wird dem an der Luft getrockneten Material die Amine durch Extraktion mit 1 l 80%igem Alkohol entzogen. Der Extrakt wird schwach schwefelsauer gemacht und verdampft. Der Rückstand wird in Wasser aufgenommen. Nötigenfalls wird mit Carboraffin entfärbt. Die Vorbehandlung mit Chloroform entfernt Begleitstoffe, welche die Fällung der Amine durch verschiedene Reagentien verhindern.

Nachweis von Putrescin mit Pikrolonsäure und Trinitroresorcin, Cadaverin mit Pikrolonsäure und Trinitroresorcin, Histamin mit Pikrinsäure und Trinitroresorcin, Tyramin mit Pikrinsäure und Goldbromid, Phenyläthylamin konnte nicht nachgewiesen werden.

F. Betaine.

I. Allgemeines.

Betaine entstehen aus Aminosäuren durch vollkommene Methylierung der NH_2-Gruppe. Sie sind nach BREDIG (1898) als Zwitterionen aufzufassen mit der allgemeinen Formel $(CH_3)_3N^+R$—C—O⁻, wobei R der Rest einer Aminosäure

bedeutet.

In Pflanzen sind bisher folgende Betaine nachgewiesen worden:

$$H_2N\text{—}CH_2\text{—}COOH \longrightarrow (CH_3)_3N^+\text{—}CH_2COO^-$$

Glykokoll → Glykokollbetain

Histidin → Hercynin / Ergothionin

Tryptophan → Hypaphorin

Prolin → Stachydrin

Cxyprolin → Betonicin, Turicin

Nicotinsäure → Trigonellin

Die freien Betaine reagieren neutral und sind in Wasser, meist auch in Alkohol, leicht löslich. Sie besitzen gewöhnlich 1, in seltenen Fällen auch 2 (Ergothionin) oder 3 (Butyrobetain) Mole Kristallwasser, das sie beim Erwärmen abgeben. Bei höherer Temperatur lagern sich die Betaine in die isomeren Methylester der tertiären Aminosäuren um (Willstätter, 1904). γ-Butyrobetain zersetzt sich in Butyrolakton und Trimethylamin. Hydrochloride und andere mineralsaure Salze der Betaine reagieren sauer. Aus neutraler und saurer Lösung werden die Betaine durch Phosphorwolframsäure, Phosphormolybdänsäure, Kaliumperjodid, Kaliumwismutjodid, Kaliumquecksilberjodid und Goldchlorid, aus alkalischer Lösung mit Platinchlorid und mit Quecksilberchlorid gefällt. Bei dem allgemeinen Verfahren zur Aufteilung der basischen Extraktivstoffe nach Kutscher, Kossel und Ackermann gelangen die Betaine in die Lysin-, Cholin-, Betainfraktion, welche nach Ausfällung der mit Silbernitrat und Baryt abscheidbaren Basen in der barytalkalischen Mutterlauge verbleibt. Überschüssiges Silber wird daraus mit Salzsäure, Baryt mit Schwefelsäure entfernt und die Basen bei schwefelsaurer Reaktion mit einer konzentrierten wäßrigen Lösung von Phosphorwolframsäure als 2. Phosphorwolframatniederschlag abgeschieden. Der Niederschlag wird mit Baryt zerlegt, das Filtrat von abgeschiedenen Kristallen, die aus Kaliumcarbonat bestehen, abfiltriert. Andere anorganische Salze, im wesentlichen Kaliumchlorid, werden nach dem Ansäuern mit Salzsäure mit Alkohol ausgefällt, das Filtrat wird zur Trockne verdampft. Nach Aufnehmen des Rückstandes in wenig Wasser fällt alkoholische Pikrinsäure Lysin zusammen mit Betain als schwer lösliches Pikrat aus der konzentrierten Lösung aus. Da Lysinpikrat in überschüssiger Pikrinsäure löslich ist, ist ein Überschuß zu vermeiden. Beim Umkristallisieren des Gemisches von Lysin- und Betainpikrat aus heißem Wasser wird nur Lysinpikrat abgeschieden. Die quaternären Ammoniumbasen der Mutterlauge fallen nach Entfernung der Pikrinsäure und nach Überführung in Alkohol mit alkoholischer Sublimatlösung aus, während noch vorhandenes Lysin in Lösung bleibt.

1. Mikrochemischer Nachweis der Betaine.
(Klein, Krisch, Pollauf und Soos 1931.)

1. Jodjodkalium. Reagens nach Stanek bestehend aus 200 ml Wasser, 100 g Kaliumjodid und 153 g Jod. Die zu untersuchende Lösung wird mit einem Tropfen des Reagens versetzt und sofort beobachtet. Bei Gegenwart von Cholin bilden sich an der Berührungszone von Reagens und Probetropfen sofort schöne, sechsseitige Prismen von dunkelbrauner Farbe. Empfindlichkeit 1:10000000. Glykokoll-Betain, Nicotinsäure und Trigonellin geben keine vom freien Jod unterscheidbaren Produkte. Stachydrincarbonat gibt unregelmäßig aneinander gelagerte Nadeln und Spindeln, das Chlorid gibt rechtwinklige rotbraune Platten. Nachweis wenig empfindlich.

Nach Krimberg (1938) ist es eine gemeinsame Eigenschaft aller Betaine, daß sie in freiem Zustand bzw. als Carbonate in äthylalkoholischer Lösung 2 —3 Moleküle Sublimat addieren. Die Sublimatverbindungen der freien Betaine sind sehr kristallisationsfähig und ändern ihre Zusammensetzung beim Umkristallisieren aus Wasser nicht. Auch ihre Schmelzpunkte sind scharf und konstant.

Das Cholinprodukt unterscheidet sich sehr deutlich vom Trigonellin und der Nicotinsäure, die sich voneinander nicht abtrennen lassen.

Cholin. Sehr viele kleine, aus je vier kurzen Prismen zusammengesetzte Aggregate. Empfindlichkeit 1:20000.

Glykokollbetain. Kein charakteristisches Produkt. Manchmal einige quadratische Platten.

Nicotinsäure. Büschel- und besenartige Aggregate aus farblosen Nadeln. Empfindlichkeit 1:2000.

2. Platinchlorid. Der zu untersuchenden Lösung wird ein Tropfen einer 5%igen wäßrigen Lösung von Platinchlorid zugefügt. Die Reaktion mit Platinchlorid ist für Nicotinsäure charakteristisch, die auf Grund der Kristallform und der Empfindlichkeit neben anderen Körpern gut erkennbar ist.

Cholin. Am Rande wenige gelbe, vierseitige Platten. Empfindlichkeit 1:200, beim Eintrocknen noch bei geringeren Konzentrationen positiv.

Glykokollbetain verhält sich wie Cholin.

Nicotinsäure. Viele kleine gelbe Würfel. Empfindlichkeit 1:2000.

Trigonellin. Große, gelbe, vierseitige Platten am Rand. Empfindlichkeit 1:2000, beim Eintrocknen bis 1:5000.

Stachydrin. In wäßriger Lösung keine Fällung. Wird die Reaktion in alkoholischer Lösung ausgeführt, werden gelbe Nadeln erhalten, die beim Abdunsten des Alkohols Wasser anziehen und zerfließen.

3. Platinjodid. Reagens: 5% iges wäßriges Platinchlorid wird mit 5% igem wäßrigem Kaliumjodid im Verhältnis 1:2 oder 1:3 gemischt und 1 Tropfen dieser Lösung der zu untersuchenden Lösung zugefügt. Charakteristische Produkte geben das Trigonellin und das Stachydrin, die auf Grund der Kristall- und Aggregatform leicht von anderen Körpern zu unterscheiden sind. Sie lassen sich auch voneinander unterscheiden, wenn sie nicht in allzu geringer Menge vorhanden sind.

Cholin. Amorphe, schwarze Fällung. Am Rande wenige, sehr kleine, schwarze vierseitige Platten, nicht charakteristisch.

Glykokollbetain. Kleine schwarze rechteckige Platten. Empfindlichkeit 1:2000.

Nicotinsäure. Fällung von kleinen schwarzen Drusen. Am Rande dunkelbraune, vier- oder sechsseitige Platten. Empfindlichkeit 1:20000.

Trigonellin. Am Rande des Präparates kleine schwarze vier- und sechsseitige Platten; Büschel aus unregelmäßigen Kristallen. Empfindlichkeit 1:20000.

Stachydrin. Rotbraune und dunkelbraune, quadratische rechteckige und sechsseitige Platten. Rosetten aus kleinen Prismen. Empfindlichkeit 1:20000.

4. Goldchlorid. Zu einem Tropfen der zu untersuchenden Probe wird 1 Tropfen einer 5% igen wäßrigen Lösung von Goldchlorid zugesetzt.

Goldchlorid gibt mit Nicotinsäure, Trigonellin, Betain und Cholin charakteristische Produkte. Nicotinsäure läßt sich ebenso wie das Trigonellin von allen anderen leicht unterscheiden, während Cholin, Betain und Stachydrin voneinander nicht zu unterscheiden sind.

5. Goldjodid. Reagens: 5% ige wäßrige Lösung von Goldchlorid mit 5% iger wäßriger Lösung von Kaliumjodid im Verhältnis 1:2 oder 1:3 gemischt.

Goldjodid gibt mit Trigonellin ein sehr charakteristisches Produkt, durch das sich das Trigonellin von allen anderen leicht unterscheiden läßt.

Cholin. Kleine schwarze, rechteckige Platten, wenig Nadelkreuze. Empfindlichkeit 1:2000.

Glykokollbetain. Sehr kleine, schwarze rechteckige Platten. Empfindlichkeit 1:2000.

Nicotinsäure. Kein charakteristisches Produkt.

Trigonellin. Büschel aus langen, rotbraunen Kristallfäden, daneben einige Rosetten aus rotbraunen, rhombischen Platten. Empfindlichkeit 1:20000.

Stachydrin. Sehr kleine, rotbraune, rechteckige und sechseckige Platten, zum Teil zu Rosetten vereinigt. Empfindlichkeit 1:2000.

6. Goldbromid. Reagens: 5% ige wäßrige Goldchloridlösung und 5% ige wäßrige Natriumbromidlösung im Verhältnis 1:1, 1:2 bzw. 1:3 mischen.

Die Reaktion gestattet den sicheren Nachweis von Trigonellin und gibt Anhaltspunkte für das Vorhandensein von Betain und besonders von Stachydrin.

Cholin. Sehr kleine rotbraune Prismen mit vier- oder sechsseitigem Umriß, daneben kleine, rotbraune, dreieckige Platten. Empfindlichkeit 1:1000.

Glykokollbetain. Rosetten aus je vier oder sechs rotbraunen Platten oder Würfeln, daneben, besonders am Rande des Präparates, kleine rotbraune, quadratische Platten. Empfindlichkeit 1:20000.

Nicotinsäure. Am Rande des Präparates rotbraune Prismen mit schiefer Endfläche. Empfindlichkeit 1:20000.

Trigonellin. Wirre Büschel aus goldbraunen Fäden. Empfindlichkeit 1 : 20 000.

Stachydrin. Sehr kleine, quadratische und regelmäßige sechseckige, rotbraune Platten, daneben aus drei oder vier Platten zusammengesetzte Rosetten. Empfindlichkeit 1 : 20 000.

7. Kaliumwismutjodid. Reagens: 18 ml Wasser, 3 ml wäßrige Salzsäure, 7 g Kaliumjodid und 1,5 g basisches Wismutnitrat bis zur Lösung aufkochen. 1,5 g Jod zufügen und mit Wasser auf das doppelte Volumen verdünnen.

Sehr charakteristische Fällung mit Trigonellin, Stachydrin und Nicotinsäure. Stachydrin läßt sich gut neben Trigonellin erkennen.

Cholin. Kleine schwarze Nadeln und dunkelbraune rechteckige Platten. Empfindlichkeit 1 : 20 000.

Glykokollbetain. Rotbraune, vierseitige, schiefwinklige Platten und Büschel aus braunen Fäden. Empfindlichkeit 1 : 2000.

Nicotinsäure. Schwarze und dunkelbraune Doppelbüschel und Drusen aus Nadeln, Kristallfäden und Rosetten aus vierseitigen, schiefwinkligen Platten. Empfindlichkeit 1 : 20 000.

Trigonellin. Drusen und Rosetten aus rotbraunen und leuchtend roten Prismen mit schiefer Endfläche, darunter breitere, gelbbraune sechsseitige lange, schmale Platten. Empfindlichkeit 1 : 20 000.

Bei Reaktionen mit pflanzlichen Extrakten treten hauptsächlich letztere Formen auf, die auch oft sehr klein und nadelförmig sein können.

Stachydrin. Vierteilige und sechsteilige Rosetten aus schwarzen, vierkantigen, schiefwinkligen Prismen, daneben charakteristische T- und F-förmige Aggregate aus solchen Kristallen. Die Reaktion ist für Stachydrin sehr charakteristisch. Daneben sechsstrahlige, schwarze und rotbraune Sternaggregate. Empfindlichkeit 1 : 20 000.

2. Nachweis der Betaine in Pflanzen.

Zum Nachweis in der Pflanze werden die Betaine extrahiert. Bei alkaloidführenden Pflanzen werden die Alkaloide durch Vorextraktion, z. B. mit Chloroform, entfernt. KLEIN (1931) stellt die Extrakte folgendermaßen her: Etwa 2 g des lufttrockenen Pflanzenmaterials werden gut pulverisiert und mit etwa 20 ml 1%iger Salzsäure 24 Std. stehen gelassen. Die Lösung wird abgesaugt und auf dem Wasserbad auf 5 ml eingedampft. Der dabei ausfallende dunkle Niederschlag wird abfiltriert, dem Filtrat wird etwas Tierkohle zugesetzt; es wird kurz aufgekocht und dann erkalten lassen. Nach abermaligem Filtrieren wird ein Tropfen des Extraktes für die Mikroreaktion verwendet. Bei Samen, die mit 1%iger Salzsäure stark quellen, wird zur Extraktion zweckmäßig Alkohol verwendet. Bei fettreichem Material wird mit Chloroform vorextrahiert. Bei Anwesenheit von Begleitsubstanzen können die Mikroreaktionen etwas anders ausfallen als mit den reinen Vergleichssubstanzen. Kaliumwismutjodid z. B. gibt häufig, besonders anfänglich, nur amorphe Fällungen. Es ist vorteilhaft, die Präparate längere Zeit stehen zu lassen, und sie ab und zu zu beobachten, da vielfach Umlagerungen eintreten. Durch Erhitzen der Präparate kann der Niederschlag umkristallisiert werden, und man erhält beim Erkalten in der Regel viel schönere Kristalle als nach etwa vierstündigem Stehen der Präparate. Beim Auswerten solcher Beobachtungen muß man sehr vorsichtig sein. Glykokollbetain gibt beim Erhitzen mit Kaliumwismutjodid ähnliche sechsstrahlige Sterne, wie sie für das Stachydrin charakteristisch sind. Beim Stehenlassen des Präparates ohne Erwärmen wurden beim Glykokollbetain derartige Kristalle nie beobachtet. Zur Unterscheidung von Glykokollbetain und Stachydrin kann die verschiedene Löslichkeit der Kaliumwismutjodidverbindungen in Salzsäure herangezogen werden. Stachydrin-Kaliumwismutjodid ist in Salzsäure unlöslich, während die entsprechende Betainverbindung darin löslich ist. Die Produkte des Cholins und des Trigonellins sind ebenfalls unlöslich. Wird die Reaktion bei gleichzeitigem Zusatz von etwas Salzsäure ausgeführt, so fallen nur die Produkte des Cholins, Trigonellins und des Stachydrins aus, während Betain keine Fällung gibt.

. Als die charakteristischsten und brauchbarsten Reaktionen erwiesen sich bei der Untersuchung pflanzlicher Extrakte jene mit Goldbromid und Kaliumwismutjodid.

II. Spezieller Teil.

1. Betain.

(Glykokollbetain) $C_5H_{11}O_2N$: C 51,3%, H 9,4%, N 12,0%, Mol.-Gewicht 117,1.

Eigenschaften: Kristallisiert mit 1 Mol. Wasser in zerfl. Kristallen, bei 100° wird das Kristallwasser abgegeben. Löslich in Wasser und Alkohol, aus letzterem in großen an der Luft leicht zerfließlichen Kristallen erhältlich. Aus der alkoholischen Lösung kann es mit Äther in Blättchen gefällt werden. Bei 14,3° lösen sich in 100 g Wasser 157,1 g wasserfreies Betain, bei 18,3° lösen 100 g Äthylalkohol 8,6 g. 100 g Methylalkohol bei 21,1° 54,36 g.

Betain schmilzt wasserfrei bei 293° unter Umwandlung in den isomeren Dimethylaminoessigsäure-methylester.

$$(CH_3)_3\overset{+}{N}CH_2CO\overset{-}{O} \longrightarrow (CH_3)_2NCH_2COOOCH_3.$$

Unter 135° sind beide Isomeren beständig, zwischen 135° und 293° ist das Betain die beständigere Form, über 293° ist Betain nicht existenzfähig. Über 290° erhitzt, zersetzt es sich teilweise in CO_2 und Trimethylamin. Beim längeren Erhitzen mit Natronlauge liefert Betain, Glykolsäure und Trimethylamin. Beim Erhitzen im Rohr auf 270—280° entstehen CO_2, Glykolsäure, Trimethylamin, Tetramethylammoniumhydroxyd. Bei der Gärung bilden sich Trimethylamin, Essigsäure, Propionsäure und Buttersäure.

Salze.

Hydrochlorid. $C_5H_{11}O_2N \cdot HCl$, Tafeln, Fp. 227° (Zers.), leicht löslich in Wasser, sehr wenig löslich in Alkohol.

Hydrojodid. $C_5H_{11}O_2N \cdot HJ$, nicht hygroskopische Nadeln, Fp. 188—190°, sehr leicht löslich in heißem, wenig löslich in kaltem Alkohol. Glykokollbetain bildet Doppelsalze mit Kaliumjodid von der Zusammensetzung $C_5H_{11}O_2N \cdot KJ \cdot 2H_2O$ und $(C_5H_{11}O_2N)_2 \cdot KJ \cdot 2H_2O$.

Perjodid. $C_5H_{11}O_2N \cdot HJ \cdot J_5$, verliert an der Luft Jod.

Perchlorat. $C_5H_{11}O_2N \cdot HClO_4$, viel leichter löslich als Cholinperchlorat, bei 19° lösen sich 17,73 Teile in 100 Teilen Wasser.

Pikrat. $C_5H_{11}O_2N \cdot C_6H_3O_7N_3$, Nadeln, Fp. 180—181°, wenig löslich in Alkohol, leichtlöslich in Wasser.

Pikrolonat. $C_5H_{11}O_2N \cdot C_{10}H_8O_5N_4$, Nadeln, Zersetzung bei 200°, leichtlöslich in Wasser und Alkohol.

Chloroaurat. $C_5H_{11}O_2N \cdot HAuCl_4$, dimorph, a) rhombisches System aus warmer 0,5 bis 1%iger HCl, Blättchen, Prismen und Platten, Fp. 248—250°, zur Identifizierung am besten geeignet. Schmelzpunkt und Goldgehalt des Betainchloroaurats können infolge der Bildung basischer Salze weitgehend variieren, b) reguläres System $C_5H_{11}O_2N \cdot HAuCl_4 + 1\frac{1}{2} H_2O$, aus warmer 5%iger wäßriger Lösung bei Gegenwart eines geringen Überschusses von Goldchlorid sternförmige Aggregate. Fp. 200—209° je nach der Art des Erhitzens. Beim Umkristallisieren aus Wasser erhält man ein blaßgelbes Salz mit niedrigem Gehalt an Gold, wahrscheinlich vermengt mit einem wasserhaltigen Salz. Nach BLOKER (1936) existieren die von FISCHER und WILLSTÄTTER angenommenen sogenannten modifizierten Goldsalze des Glykokollbetains mit $1\frac{1}{2}$ Mol bzw. 2 Mol H_2O nicht. Es handelt sich vielmehr um Chloroaurate, die auf 1 Mol Goldchlorid mehr als 1 Mol Betain enthalten. Die Bildung solcher basischer Chloroaurate läßt sich vermeiden, wenn man bei der Darstellung mit einem Überschuß an Goldchlorid in salzsaurer Lösung arbeitet.

Chloroplatinat. $(C_5H_{11}O_2N)_2 \cdot H_2PtCl_6 \cdot 4H_2O$ aus Wasser große rhombische Tafeln mit abgestumpften Ecken, Fp. 242°, unlöslich in Alkohol, aus heißem Wasser in gelben Prismen mit verschiedenem Wassergehalt. — Die aus heißem Wasser wasserfrei auskristallisierenden Nadeln verwandeln sich in der Mutterlauge in 4seitige Tafeln mit $4H_2O$. Doppelsalz mit *Mercurichlorid* $(C_5H_{11}O_2N \cdot HCl)_2 \cdot HgCl_2$, Täfelchen, in Wasser ziemlich leicht löslich, in Alkohol schwer löslich.

Phosphorwolframat. $(C_5H_{11}O_2N)_3 \cdot H_3PO_4 \cdot 12WO_3$ aus heißem Wasser oder verdünntem Alkohol große Rhomboeder, sehr leicht löslich in Aceton, wenig löslich in Wasser.

Reineckat. $C_5H_{11}O_2N \cdot C_4H_7N_6S_4Cr$, flache langgestreckte Prismen bei 152° Sintern, bei 154° zähe dunkelviolette Schmelze, sehr wenig löslich in Wasser, wenig löslich in Alkohol, leicht löslich in Aceton und Dioxan.

Vorkommen. Betain findet sich in vielen Pflanzenarten und ist in allen Organen (Wurzeln, Blätter, Blüten, Pollen, Früchten, Samen) nachweisbar. In verhältnismäßig hohen Konzentrationen ist es bei Chenopodiaceen und Gramineen, ferner in *Lycium chinense* Mill. *(Solanaceae)* und *Vicia*-Arten enthalten.

Quantitative Betainbestimmung nach Cromwell (1953).

Nach Trocknen bei 80° in einem Wärmeschrank mit guter Entlüftung werden 0,1—1 g des Pflanzenmaterials in die Extraktionshülse einer Mikro-Soxhlet-Apparatur eingewogen und mit Methanol 1 Std. lang extrahiert. Das Methanol wird auf einem Wasserbad unter Zusatz von Siedesteinen bei 80° verdampft. Der Rückstand wird in 5 ml dest. Wasser aufgenommen, dann werden 0,5—2 g Aktivkohle in Pulverform je nach der Menge des eingewogenen Pflanzenmaterials hinzugefügt. Die Lösung wird fast zum Kochen gebracht und durch einen porösen Glasfiltertiegel, dessen Boden mit einer dünnen Schicht Gooch-Asbest belegt ist, filtriert. Das farblose Filtrat wird in ein Reagenzglas gegossen und der Kolben 3mal mit 2 ml heißem Wasser ausgewaschen, das durch den Filtertiegel gegeben wird. Die vereinigten Filtrate werden in eine Abdampfschale übergeführt und auf dem Wasserbad auf etwa 3 ml eingedampft. Die konzentrierte Lösung wird in das Reagenzglas zurückgegossen und die Schale 2mal mit ·~ 1 ml Wasser ausgewaschen. Der p_H-Wert der Lösung wird durch Zufügen von 0,05 ml konzentrierter Salzsäure auf etwa 1 eingestellt. Dann werden 10 ml einer gesättigten auf p_H 1 eingestellten Lösung von Ammoniumreineckatlösung hinzugefügt, und das Reagenzglas wird 1 Std. bei 4° C gehalten. Die Reineckatfällung wird durch einen abgekühlten Glasfiltertiegel mittlerer Porengröße filtriert und 4mal mit 2 ml wassergesättigtem Äther gewaschen. Die Unterseite des Tiegels wird getrocknet und die Fällung in 5 ml 74%igem Aceton (Vol./Vol.) aufgelöst. Das Reineckation wird dann entfernt durch Zufügen von 5 ml Silbernitratreagens (0,1 n Silbernitrat und Natriumnitrat). Der durch Silbernitrat erzeugte Niederschlag wird abzentrifugiert und die klare überstehende Flüssigkeit in einen Siedekolben gegeben. Der Niederschlag wird 2mal mit je 5 ml Silbernitratreagens und 2 mal mit je 5 ml destilliertem Wasser gewaschen. Die Waschwässer werden mit einigen Tonsiedesteinchen in den Siedekolben gegeben. Der Kolben wird zur Entfernung des Acetons gelinde erwärmt und dann werden 4 ml einer 0,2n NaOH langsam zugegeben. Anschließend wird auf etwa 5 ml eingeengt. Nach dem Abkühlen werden 0,15 ml konzentrierte Salzsäure zugegeben und die Silberchloridfällung wird durch einen mit Gooch-Asbest bedeckten Glasfiltertiegel filtriert. Die Silberchloridfällung wird 3mal mit je 2 ml Wasser gewaschen. Die vereinigten Filtrate und Waschwässer werden in einer kleinen Porzellanschale auf dem Wasserbad auf etwa 3 ml eingeengt und in das Glas zurückgegeben. Die Schale wird 2mal mit 1 ml Wasser ausgewaschen und diese Waschwässer dem Inhalt des Reagenzglases zugegeben. Das Gesamtvolumen beträgt dann etwa 5 ml. Der p_H-Wert wird auf den Wert 1 gebracht, dann werden 10 ml gesättigte Ammoniumreineckatlösung zugefügt und das Reagenzglas 1 Std. bei 4° aufbewahrt. Die Fällung wird in einem gekühlten Glasfiltertiegel mittlerer Porengröße abfiltriert und mit 3 ml Portionen von wassergesättigtem Äther gewaschen, bis die Waschwässer säurefrei sind. Das Betain-Reineckat wird in 75%igem Aceton (Vol./Vol.) aufgelöst und in ein konisches Zentrifugenglas von 10 ml gefüllt. Nach Zufügen von 5 ml Silbernitrat wird vom Silberreineckat abzentrifugiert, die überstehende Flüssigkeit sorgfältig dekantiert, in ein konisches 100 ml-Gefäß gespült und das Silberreineckat von Betainnitrat freigewaschen durch 5 ml-Portionen von destilliertem Wasser bis zu einem Gesamtvolumen von 40 ml (Filtrat + Waschwässer). Zu den vereinigten Filtraten und Waschwässern werden 4 Tropfen einer 0,1%igen Lösung von Methylrot in Äthanol gegeben und dann wird die Lösung auf p_H 6 mit 0,02 n-Diphenylguanidin in 70%igem Äthanol titriert. 1 Milliäquivalent Betainnitrat = 1 Milliäquivalent Diphenylguanidin. Eine Standardlösung von 0,05 n-Diphenylguanidin in 70%igem Alkohol wird aus umkristallisiertem Diphenylguanidin, das 2mal aus kochendem Toluol kristallisiert war, bereitet. Bei der Entfernung der störenden Substanzen ist darauf zu achten, daß durch den Überschuß an Silbernitrat nach der Entfernung des Reineckations als Silberreineckat Silberoxyd entsteht. Die Menge der zugefügten Natronlauge ist sorgfältig so zu bemessen, daß ein kleiner Überschuß an Silbernitrat in Lösung bleibt. Wenn Guanin und Adenin anwesend sind, werden sie mit den Betainen ausgefällt, oder sie werden als unlösliche Silberverbindungen entfernt, wenn der rohe Reineckatniederschlag in wäßrigem Aceton aufgelöst und mit Silbernitrat behandelt wird. Andere Betaine, wie z. B. Trigonellin, Stachydrin und Betonicin, werden durch die Behandlung mit Silberoxyd nicht verändert. Wenn sie anwesend sind, dann geben sie verhältnismäßig unlösliche Niederschläge mit Ammoniumreineckat bei p_H 1. Bei der Entfernung der Reineckationen mit Silbernitrat geben die Nitrate dieser Betaine saure Lösungen und werden als Betaine bestimmt. Man muß daher vorsichtig sein bei der Berechnung der Ergebnisse, wenn die Gegenwart von anderen Betainen bei der Voruntersuchung der Pflanzenextrakte durch Papierchromatographie festgestellt wurde. Der Schmelzpunkt und der Chromgehalt dieser gereinigten Betain-Reineckate wird als Kriterium für ihre Reinheit benutzt. Die Methode ergibt die befriedigendsten Ergebnisse, wenn die Probe des Pflanzenmaterials 1—10 mg Betain enthält. Es wurden 99—101% wiedergefunden, wenn bekannte Mengen von Betain zu Gewebsextrakten und zu Mischungen von Aminosäuren und Purinen zugefügt wurden.

Bestimmung von Betain in *Beta vulgaris* L. und anderen Pflanzen.

Das Pflanzenmaterial wird so bald wie möglich nach dem Ernten getrocknet und bis zur Analyse in einem Exsiccator aufbewahrt. Wenn nichts anderes angegeben, wurden alle Blätter des Schößlings jeder Pflanze zur Analyse verwendet. Tab. 27 verzeichnet die Betainwerte (Mittelwerte von je 2 Bestimmungen) für *Beta vulgaris, Atriplex patula* L. und *Atriplex hortensis* L. bei fortschreitender Entwicklung.

Tab. 28 enthält Betainwerte von einer Reihe von Pflanzenarten. Es werden die Resultate von STANEK (1906, 1911) bestätigt, wonach die Blätter von jungen Keimlingen einen niedrigeren Betaingehalt haben als die Blätter älterer Pflanzen. Der Betaingehalt des ganzen Blattsystems junger Pflanzen nimmt über die Vegetationsperiode zu und erreicht zur Blütezeit ein Maximum.

Darüber hinaus enthalten junge Keimlingsblätter von *B. vulgaris* viel weniger Betain als junge wachsende Blätter, die aus der Krone einer reifen Wurzel aussprießen.

Ein Konzentrationsgefälle ist feststellbar während des Wachstums der untersuchten Pflanzen, in dem der Betaingehalt von der Wurzel zu den Blättern stetig zunimmt. Im Gegensatz dazu haben die unentfalteten Blätter des Zentrums der Krone der reifen Wurzel mehr Betain als die älteren Blätter der Peripherie der Krone. Blätter der Gattung *Atriplex (A. halimus* und *A. canescens)* haben einen bemerkenswerten hohen Betaingehalt.

Nach einer Untersuchung von CROMWELL und RENNIE (1953) wird das Betain von den Wurzeln zu den Blättern transportiert. Wenn das Betain im intermediären Stoffwechsel nicht verwertet wird und die Produktion in den Wurzeln weitergeht, so ist eine allmähliche Anhäufung von Betain in dem Schößlingssystem während der Wachstumsperiode zu erwarten. So ist die hohe Konzentration von Betain in den jungen Blättern, die aus dem Zentrum der Krone einer reifen Wurzel aussprießen, durch eine Überführung von Betain aus einer relativ großen Masse von Wurzelgewebe zu erklären, das im Ruhezustand große Mengen Betain enthält. Der hohe Betaingehalt in den Blättern der Immergrünart könnte erklärt werden als Resultat einer langsamen Anhäufung über eine längere Zeitperiode.

Nach DAVIES und DOWDEN (1936) beträgt der mittlere Betaingehalt in Melasseschnitzeln 1,8%, in der Trockensubstanz von Rübenköpfen 1,6%. Ferner in Gerste 0,06%. Andere Quellen, die Betain enthalten, sind: virginischer Tabak, Mutterkornpilz, stinkender Gänsefuß, Sojabohne, Arnica, echter Eibisch in der Wurzel, Weizenkleie (0,35%; NOTTBOHM und MAYER, 1935).

Isolierung von Betain aus dem Guayulestrauch.
(MURRAY und WALTER 1945.)

Trockene entblätterte Guayulestauden (*Parthenium argentatum* Gray) (2586 g) werden in der Mühle zerkleinert, mit 15 l kaltem dest. Wasser mechanisch verrührt und dann filtriert. Jede Extraktion wird wiederholt. Die vereinigten Filtrate werden in der Hitze eingeengt, dann mit 100 ml konzentrierter HCl versetzt. Der entstehende Niederschlag wird abfiltriert und verworfen. Das Filtrat wird weiter konzentriert, die ausfallenden Niederschläge werden abfiltriert und verworfen. Der Rückstand wird mit warmem Methanol gewaschen und die

Tabelle 27. *Betaingehalt der Blätter von Beta vulgaris während verschiedener Wachstumsstadien.*

Alter (Tage)		g Betain (pro 100 g) Trockengewicht		
		Blätter	Blattstiele	Wurzeln
Beta vulgaris	25	0,25	—	—
	33	0,36	—	1,27
	44	0,48	—	—
	56	0,64	—	—
	62	0,84	—	1,15
	75	0,87	—	—
	81	1,02	—	—
	96	1,32	1,09	—
	132	1,36	1,65	0,79
	140	2,20	—	
	167	2,55	1,08	0,18
		(äußere Blätter)	(obere Hälfte)	
		3,34	0,99	
		(mittlere Blätter)	(untere Hälfte)	
		4,85		
		(innere Blätter)		
Atriplex patula	30	1,32		
	60	1,71		
	80	2,64		
		(Pflanzenkeime)		
	90	2,48		
		(Pflanzen in Blüte)		
	120	1,84		
		(früchtetragende Pflanzen)		
Atriplex hortensis	39	0,78		
	48	1,04		
	107	0,99		
		(alte Blätter)		
		1,26		
		(junge Blätter)		

Tabelle 28. *Betaingehalt verschiedener Pflanzenarten.*

Pflanzen		g Betain (pro 100 g Trockengewicht)
Spinacea oleracia L.	Blätter (35 Tage)	0,27
Chenopodium vulvaria L.	Blätter (reif)	1,86
Chenopodium bonus-henricus L. . .	Blätter (reif)	2,23
Atriplex canescens James	Blätter (reif)	3,27
Atriplex halimus L.	Blätter (reif)	4,90
Amaranthus caudatus L.	Blätter (reif)	0,54
Lycium chinense Mill	Blätter (reif)	1,95
Vicia faba L.	Cotyledonen	0,22
(Keimlinge, im Dunkeln gewachsen)	Schoten	1,47
	Wurzeln	1,10
(*Triticum vulgare* Vill. var. Atle)		
Etiolierte Keimlinge (8 Tage) . . .	Coleoptilen	0,83
	Wurzeln	0,42
Beta vulgaris L.		
Etiolierte Blätter		5,00

Waschwässer werden mit dem Filtrat vereinigt. Das Filtrat wird mehrere Male mit Kohle geklärt und weiter eingeengt, bis das Betainhydrochlorid auskristallisiert. Die Mischung wird gekühlt, das Betainhydrochlorid abfiltriert, mit kaltem absolutem Äthanol gewaschen und aus 95%igem Alkohol umkristallisiert. Es werden 12 g oder etwa 0,4% Betainhydrochlorid erhalten.

Bestimmung von Betain neben Cholin.
(BANDELIN und PANKRATZ 1953.)

Cholin wird quantitativ als Reineckat aus alkalischer Lösung gefällt, Betain aus saurer Lösung.

Reagentien. Ammoniumreineckat. Man schüttelt 1 g Ammoniumreineckat mit 50 ml Wasser und filtriert 3n HCl; Na_3PO_4 25 %ig. — Verdünnte Reineckat-Phosphat-Lösung: 2 ml einer gesättigten Ammoniumreineckatlösung in einem Liter 5%igem Trinatriumphosphat. — Aceton und 75%iges Aceton.

Durchführung. 5 ml der zu untersuchenden Lösung (enthaltend 3—8 mg Cholin bzw. Betain) werden in einem 15 ml-Zentrifugenglas mit 1 ml Na_3PO_4 und 5 ml Ammoniumreineckatlösung versetzt. Man kühlt in Eiswasser und zentrifugiert den Niederschlag ab, der mit 2 ml der Reineckatphosphatlösung ausgewaschen wird. Das ausgefällte Cholinreineckat wird dann mit Aceton zu genau 10 ml gelöst und die Extinktion bei 525 mμ gemessen. Die vereinigten Zentrifugate und Waschflüssigkeiten dienen zur Betainbestimmung. Man versetzt sie in einem Zentrifugenglas mit 1 ml HCl, stellt in Eiswasser und zentrifugiert das ausgefallene Reineckat ab. Dieses wird in 75%igem Aceton zu 10 ml gelöst. Man photometriert dann die Lösung bei 525 mμ.

Papierchromatographische Bestimmung von Betain neben Trigonellin, Hordenin und Cholin.
(MUNIER und MACHEBOEUF 1949, 1951.)

Es wird aufsteigend chromatographiert. Die Art des Lösungsmittels richtet sich nach der Basizität der zu chromatographierenden Betaine und Alkaloide. Bei schwachen Basen wird in alkalischem, bei starken in saurem Medium entwickelt (Ameisensäure, Essigsäure, und andere Säuren als Zusätze zur organischen Lösungsmittelphase). Betain- und Trigonellinflecke werden sichtbar gemacht mit Hilfe eines DRAGENDORFF-Reagens folgender Zusammensetzung: 1. 2,5 Teile Bismutum subnitricum, 20 Teile Wasser, 5 Teile konz. HCl. 2. Eisessig, 5 Teile, 3. 4 g Kaliumjodid, zuvor in 10 ml Wasser in kleinen Anteilen gelöst. Man mischt, es erscheint ein orangefarbener Niederschlag, über dem eine gelb gefärbte Flüssigkeit sich befindet. Man filtriert und bewahrt das Filtrat in einer dunklen Flasche auf. Diese Lösung wird im Augenblick der Verwendung in folgenden Verhältnissen verdünnt: 5 ml der Stammlösung, 10 ml Essigsäure und Auffüllen mit Wasser auf 100 ml. Es können Mengen von etwa 20 γ Trigonellin oder Betain festgestellt werden. Ephedrin, Trigonellin und Hordenin werden mit Kaliumjodoplatinat nachgewiesen. Dazu wird 1 ml einer wäßrigen 10%igen Lösung von $PtCl_4$ zu 25 ml einer Lösung von 44% Kaliumjodid in Wasser gegeben, dann wird auf 50 ml mit Wasser aufgefüllt. Mit dieser Lösung können Flecke die mindestens 20 γ der erwähnten Substanzen enthalten, sichtbar gemacht werden. Cholin kann mit Phosphormolybdänsäure-reagens nachgewiesen werden. Man gibt 0,5 g Phosphormolybdänsäure und 3 Tropfen Essigsäure mit Wasser auf 100 ml. Man legt das Papier eine Minute lang in dieses Reagens, dann 2 min lang in fließendes

Wasser und gibt es schließlich in eine Stannochloridlösung, welche die Phosphormolybdänsäure zu Molybdänblau reduziert. Die Stannochloridlösung muß frisch bereitet werden.

1g Stannochlorid ($SnCl_2 \cdot 2\ H_2O$) 10 ml reine Essigsäure, 90 ml Wasser. Man läßt das Reagens 30 sec lang auf dem Papier reagieren, dann wird gewaschen und getrocknet.

Betain, Trigonellin, Kaffein, Theobromin und Nicotinsäureamid ergeben bei dieser Behandlung keine Farbflecke. Wenn auf „Durieux" Papier Nr. 122, für langsame Filtration aufsteigend chromatographiert wurde, so erhielt man die in Tab. 29 verzeichneten R_f-Werte.

2. Hypaphorin.

$C_{14}H_{18}O_2N_2$, Mol. Gewicht 264,1. $C = 68,62\%$, $H = 7,37\%$, $N = 11,39\%$.

Bei 238° C färbt es sich braun, Schmelzpunkt bei 255° C unter Zersetzung. $[\alpha]_D^{20} = +91°$ bis $+93°$ in 3%iger Lösung. Hypaphorin ist sehr leicht löslich in Wasser, leicht löslich in Alkohol; unlöslich in anderen organischen Lösungsmitteln, wie Äther u. a. Aus Wasser kristallisiert es in großen, 2 Mol Wasser enthaltenden farblosen Kristallen, die zu langnadeligen Bündeln vereinigt sind. Das Kristallwasser wird im Exsiccator abgegeben. Durch wäßrige Kalilauge wird Hypaphorin zersetzt unter Bildung von Trimethylamin und Indol. Es reduziert Goldchlorid, Kaliumpermanganat und Eisensalze. Die Spaltung durch Alkalien erfolgt leichter als bei Tryptophan. Pharmakologisch ist Hypaphorin wenig wirksam. Vom Kaninchen werden 0,5—1,0 g i.v. ohne merkliche Symptome vertragen; es wird unverändert im Harn ausgeschieden. Frösche zeigen nach 12—15 mg eine mehrere Tage andauernde Reflexerregbarkeit und Tetanus

Tabelle 29. *R_f-Werte von Trigonellin, Hordenin, Cholin und Betain bei Verwendung von Lösungsmitteln verschiedener Zusammensetzung.*

Saure Lösungsmittel.

n-Butanol	100	100	100	80
Essigsäure	4	10	20	20
Wasser	24	30	50	100
R_f-Werte bei 20°.				
Trigonellin	0,08	0,13	0,25	0,30
Hordenin	0,53	0,52	0,62	0,67
Betain	0,14	0,17	0,27	0,33
Cholin	0,71	0,65	0,74	0,675

Alkalische Lösungsmittel.

n-Butanol	90	100	100	80	—	—	—
Eucalyptol	—	—	—	—	—	50	50
Isoamylalkohol	—	—	—	—	100	—	—
Äthanol	10	—	—	—	—	—	—
Glykolmonochlorhydrin	—	—	—	—	—	20	50
Pyridin	—	—	—	20	—	—	—
Ammoniak bei 22° Bé	1	2	—	—	2	3	5
verdünnt 1:10 (Ammoniak)	—	—	24	—	—	—	—
Wasser	97	16	—	25	8	2	13
R_f-Werte bei 20°.							
Trigonellin	0,09	0,102	0,06	0,10	0,016	0,00	0,17
Cholin	0,01	0,148	1,09	0,108	0,025	0,00	0,235
Betain	0,107	1,14	0,08	0,106	0,04	0,116	—

(ROMBURG, 1911; BARGER, 1911; CHEN, CHEN, 1933). Die Synthese des Hypaphorins gelingt durch erschöpfende Methylierung des Tryptophans (ROMBURG, 1911).

Derivate:

Nitrat. $C_{14}H_{18}O_2N_2 \cdot HNO_3$. Schwer löslich in Wasser (1:170 bei Zimmertemperatur). Der Schmelzpunkt liegt bei 215—220° C. $[\alpha]_D^{20} = +94,70°$ einer Lösung von 0,299 g Nitrat in 18 ml wäßrigem Ammoniak. Nach ROMBURG und BARGER (1911) hat das Nitrat einen Schmelzpunkt von 220—222° C.

Hydrochlorid. Lange, prismenförmige Kristalle, aus Wasser umkristallisierbar. Fp. 234—235° $[\alpha]_D^{20} = 89,2°$. KLEIN gibt als Fp. 227° C an.

Hydrobromid. $C_{14}H_{18}N_2O_2 \cdot HBr$. Fp. 227° C.

Flavianat. $C_{14}H_{18}N_2O_2 \cdot C_{10}H_6O_9N_2S$. DEULOFEU, HUGE und MAZZOCCO (1939); geeignet zur Isolierung: 0,15 g Hypaphorinhydrochlorid werden in 25 ml Wasser aufgelöst und eine Lösung von 0,4 g Flaviansäure hinzugefügt. Es scheiden sich beinahe augenblicklich feine organgegelbe Nadeln aus, die aus Wasser umkristallisiert werden können. Manchmal erhält man auch rote, rhombische Kristalle, die mit den gelben jedoch nicht identisch sind. Fp. 235° C.

Jodmethylat. $C_{14}H_{18}O_2N_2 \cdot CH_3J$. Glitzernde Platten aus heißem Wasser, löslich in 200 Teilen Wasser bei 18° C. ·

Nachweis. Hypaphorin gibt mit Kaliumdichromat in konzentrierter Schwefelsäure eine intensive Violettfärbung. Mit Glyoxylsäure und Schwefelsäure gibt es die ADAMKIEWICZsche Reaktion, unterscheidet sich aber vom Tryptophan dadurch, daß es durch Ferrichlorid nur sehr schlecht zu β-Indolaldehyd oxydiert wird und ninhydrinnegativ ist.

Isolierung von Hypaphorin aus Pflanzenmaterial.

1000 g Samen von *Erythrina variegata* var. *orientalis* (L.) werden fein pulverisiert und durch Extraktion mit Äther entfettet. Das vom Äther befreite Samenpulver wird mit Alkohol perkoliert, der Alkohol im Vakuum vollständig entfernt und der sirupöse Rückstand in Wasser gelöst. Die filtrierte wäßrige Lösung wird mit Phosphormolybdänsäure versetzt, bis keine Fällung mehr entsteht. Der mit phosphormolybdänsäurehaltigem Wasser ausgewaschene Niederschlag wird in noch feuchtem Zustand mit Soda vermischt und auf dem Wasserbad getrocknet. Der Rückstand wird mehrmals mit absolutem Alkohol ausgekocht. Die alkoholischen Lösungen werden vereinigt und konzentriert. Beim Erkalten kristallisiert das rohe Hypaphorin aus, das aus Alkohol unter Zusatz von Tierkohle in farblosen, zu Büscheln vereinigten langen Prismen erhalten wird (MARANON und SANTOS, 1932). 10,0 g Samen von *Erythrina crista galli* werden gemahlen mit 10,0 g Calciumhydroxyd gemischt, getrocknet und im Soxhlet mit Alkohol extrahiert. Der halbfeste Niederschlag, der beim Abkühlen erhalten wird, wird entfernt und das Filtrat auf dem Wasserbad zur Trockne eingedampft. Nach Zufügen von n HCl zum Rückstand bilden sich bald Kristalle aus. Diese werden gesammelt und mit wenig Alkohol gewaschen. Ein weiterer Anteil wird nach dem Eindampfen der Mutterlauge erhalten. Die gesamte Ausbeute an Hydrochlorid beträgt 1,86 g vom Fp. 227—229° C (DEULOFEU, HUGE und MAZZOCCO, 1939). 980 g Pulver von Samen der Arten *Erythrina Folkersii, E. fusca, E. velutina* oder *E. macrophylla* werden mit 1500 ml Petroläther im Soxhlet entfettet (6 Std.). Nach dem Abdestillieren des Lösungsmittels werden 77 g der gelben Fettfraktion erhalten. Die Extraktion des Samenpulvers wird 30 Std. lang mit 1500 ml-Portionen Methanol fortgesetzt. Dieser Extrakt wird dann durch Destillieren konzentriert und schließlich werden die letzten Reste Methanol bei 30° C in einem Vakuum von 18 mm Hg entfernt. Der Rückstand, bestehend aus 150 g eines gummiartigen Produkts, wird in 700 ml Wasser gelöst, mit 14 ml konzentrierter

HCl behandelt und noch von rückständigen emulgierten Öltropfen durch Extraktion mit Petroläther und anschließend mit Chloroform befreit. Die klare wäßrige Lösung wird eben alkalisch gemacht und mit 100 ml Chloroform 10 mal extrahiert. Das Hypaphorin kann aus der alkalischen wäßrigen Lösung durch Behandeln mit Salzsäure isoliert werden. (Folkers und Koniuszy, 1939.)

Vorkommen. Hypaphorin wurde von Greshoff in *Erythrina Hypaphorus* Boerl entdeckt. Es ist zu 3% des Trockengewichts in dieser Droge enthalten. Von Maranon und Santos (1932) wurde es in *Erythrina variegata* var. *orientalis* nachgewiesen. Rao und Mitarbeiter (1938) isolierten es aus *E. indica*, Deulofeu, Huge und Mazzacco (1939) aus *E. cristagalli* und Folkers und Koniuszy (1939, 1940) stellten sein Vorkommen in *E. sandwiscensis, E. sucumbrans, E. glauca, E. Folkersii, E. fusca, E. velutina* und *E. velutina* var. *auriantica, E. Grisebachii* und *E. macrophylla* DC fest. Schließlich berichten auch noch Gentile und Labriola (1942) über ein Vorkommen in *E. falcata* und *E. Dominguezzi.* Hier erscheint das Hypaphorin in den Samen, in der Rinde und in geringen Mengen in den Blättern. Nach Klein findet es sich möglicherweise auch in der Familie der *Chenopodiaceae*, in *Beta vulgaris* var. *rapa*; es könnte die Muttersubstanz des beim Kochen der Melasse mit Erdalkalien sich bildenden Indols sein.

3. Stachydrin.

(Prolinbetain, Methylbetain der Hygrinsäure.) $C_7H_{13}C_2N$.

Eigenschaften. Stachydrin bildet farblose, durchsichtige, zerfließliche Kristalle, leicht löslich in Wasser und Alkohol, schwer löslich in siedendem, unlöslich in kaltem Chloroform und Äther. Das Kristallwasser wird zum Teil schon über Schwefelsäure abgegeben. Fp. 235°.

Aus *Stachys*-Knollen isoliertes Stachydrin ist optisch inaktiv. Das aus *Galeopsis grandiflora* und *Citrus aurantium* stammende Präparate sowie aus l-Prolin hergestelltes Stachydrin ist optisch aktiv: $[\alpha]_D = -26,5°$ (salzsaures Salz in 5%iger wäßriger Lösung).

Bleibt es längere Zeit mit Basen oder Säuren in Berührung, so sinkt die spezifische Drehung ab bis z. B. auf $-9°$; beim Kochen mit Baryt wird das Stachydrin racemisiert. Die Lösungen reagieren neutral; Geschmack unangenehm süßlich. Das salzsaure Salz kristallisiert in großen wasserfreien Prismen, leicht löslich in Wasser und auch in kaltem Alkohol. Fp. 235° unter Zersetzung.

Mercurichlorid erzeugt in wäßriger Lösung aus Stachydrin erst auf Zusatz von Salzsäure eine Fällung. Das Chlorid gibt mit Gerbsäure keine Fällung, mit Pikrinsäure nur bei sehr hoher Konzentration. Phosphorwolframsäure fällt Stachydrin aus 5%iger schwefelsaurer Lösung fast quantitativ aus; mit Phosphomolybdänsäure entsteht ein gelblicher Niederschlag. Das Staneksche Reagens aus Perjodid fällt bei alkalischer Reaktion beträchtliche Mengen Stachydrin mit dem Cholin aus. Beim Erhitzen mit konzentriertem Alkali entweicht Dimethylamin. Durch Permanganat tritt in der Kälte keine Oxydation ein. Bei der Destillation unter stark vermindertem Druck geht das Stachydrin in den isomeren Hygrinsäuremethylester über, und zwar zu etwa $^1/_3$ der angewandten Menge. Daneben entsteht Trimethylamin.

Nachweis und Bestimmung. Salzsaures Stachydrin gibt beim Erhitzen Dämpfe, die einen mit HCl befeuchteten Fichtenspan rot färben. Ferner zeigt das Chloroaurat unter dem Mikroskop eine charakteristische Form, vierseitige Blättchen von rhombischem Habitus. Eine ähnliche Form hat auch das normale Trigonellinaurat, doch lassen sich die beiden Verbindungen leicht durch den Schmelzpunkt unterscheiden.

Zur quantitativen Bestimmung eignet sich die Fällung mit Phosphorwolfram-säure, wobei sich etwa 4,7% der Bestimmung entziehen, und der Niederschlag mit Mercurichlorid, wobei 2,7% der Fällung entgehen.

Zur Identifizierung eignen sich folgende Derivate:

Chloroaurat. $C_7H_{13}O_2N \cdot HAuCl_4$. Entsteht beim Zusatz von Goldchlorid zur Lösung des salzsauren Stachydrins als gelber Niederschlag, der nach einiger Zeit kristallinisch wird: vierseitige Blättchen von rhombischem Habitus, die in kaltem Wasser sehr schwer, auch in heißem Wasser nicht leicht löslich sind. Fp. 225° unter Zersetzung bei raschem Erhitzen.

Pikrat: $C_{13}H_{16}N_4O_9$, Nadeln, in Wasser ziemlich löslich. Fp. 195—196°.

Chloroplatinat $(C_7H_{13}ON_2)_2 \cdot H_2PtCl_6$. Sehr leicht löslich in Wasser und verdünntem Alkohol, schwer löslich in 80%igem Alkohol, unlöslich in absolutem Alkohol. Kristallisiert aus Alkohol in Nadeln, aus Wasser in großen orangeroten rhombischen Kristallen mit 2 Mol Wasser oder in unscheinbaren Formen mit 4 Mol Wasser.

Saures Oxalat. $(C_7H_{13}O_2N \cdot C_2H_2O_4)$, Nadeln, unlöslich in kaltem absolutem Alkohol, Chloroform und Äther. Sehr schwer löslich in warmem 95%igem Alkohol. Fp. 105—107°.

Hydrochlorid. $C_7H_{13}O_2N \cdot HCl$, große Prismen, Zersetzung bei 235°, leicht löslich in Wasser, 1 Teil löst sich in 12,7 Teilen kaltem Alkohol.

Mercurichlorid fällt nur das Hydrochlorid, aber nicht die freie Base, am besten in alkoholischer Lösung.

Chloroaurat des Methylesters. $C_6H_{12}N \cdot COOCH_3 \cdot HAuCl_4$. Schwer löslich in Wasser. Fp. 85°.

Phosphorwolframat. $(C_7H_{13}O_2N)_3 \cdot H_3PO_4 \cdot 12 WO_3$, Rhomboeder, leicht löslich in Aceton.

Chloroaurat des Äthylesters. $C_6H_{12}N \cdot COOC_2H_5 \cdot HAuCl_4$. Schwer löslich in Wasser, Smp. 59—60°, zersetzt sich bei 241—244°.

Pikrat des Äthylesters. Nadeln, Smp. 94—96°.

Jodwasserstoffsaures Salz des Äthylesters. Fp. 88—89°.

Isolierung von cis- und trans-3-Oxystachydrin aus den Schalen von *Courbonia virgata*.

(CORNFORTH und HENRY 1952.)

280 g der lufttrockenen und gemahlenen Schalen der reifen Frucht werden bei Zimmertemperatur mit 95%igem Äthanol zuerst perkoliert und dann über Nacht digeriert. Die vereinigten Extrakte (1800 ml) werden verdampft und der Rückstand wird in 110 ml Wasser aufgenommen. Nach Ansäuern mit 3 Tropfen 2 n Schwefelsäure und Extraktion mit Petroläther, zur Entfernung von Fett, wird die Lösung durch gelindes Erwärmen eingeengt und schließlich über konzentrierter Schwefelsäure im Exsiccator belassen. Die völlige Durchkristallisation wird beschleunigt durch öfteres Durchbrechen der Kruste an der Oberfläche. Die Kristalle vom Schmelzpunkt 250° (u. Zers.) bestehen hauptsächlich aus dem Trans-isomeren. Es werden 4 Portionen, nämlich 6,1; 3,4; 6,5 und 4,8 g, erhalten. Zum Waschen und Sammeln der Kristalle werden kleine Portionen Äthanol verwendet. Eine nochmalige Extraktion der Schalen mit kochendem 95%igem Äthanol (500 ml) während 3½ Std. ergibt eine Ausbeute von 5,1 g 3-Oxystachydrin. Die rohen Kristalle werden aus Wasser wie oben beschrieben umkristallisiert. Nach zwei solchen Behandlungen wird die endgültige Reinigung

erreicht durch Eingießen einer konzentrierten wäßrigen Lösung in kaltes Äthanol. Farblose Prismen Fp. 250° (unter Verpuffen) $[\alpha]_D^{20} + 10,0°$.

Pikrat. $C_7H_{13}O_2N$, $C_6H_3O_7N_3$, gelb aus Äthanol, Fp. 160°.

Hydrochlorid. $C_7H_{13}O_2N$, HCl (aus Äthanol Nadeln), Fp. 196° (Zers.); nicht hygroskopisch.

Isolierung von cis-3-Oxystachydrin.

Die Mutterlauge aus der letzten Umkristallisation des trans-3-Oxystachydrins a wird über Schwefelsäure getrocknet. Nach 10 Tagen wird die viskose kristallhaltige Schicht ohne Zusatz von Äthanol durch Absaugen so gut wie möglich von der Mutterlauge befreit. Die unreinen Kristalle werden in 10 ml Wasser gelöst und die Lösung nach Behandlung mit Aktivkohle (0,3 g) und Filtrieren bei Raumtemperatur auf ein kleines Volumen eingeengt. Der Rückstand wird in einem Kühlschrank abgekühlt und mit 5 ml Äthanol verrührt. Die entstehenden großen Kristalle werden gesammelt, gewaschen und mit 10 ml kaltem Äthanol luftgetrocknet. Ausbeute 2,46 g. Fp. 210—212° (Zers.) nach Trocknen bei 110°. Die Kristalle werden in 5 ml Wasser gelöst, dann in 25 ml kaltes Äthanol gegossen und im Kühlschrank belassen. Am nächsten Tag werden die farblosen Prismen vom cis-3-Oxystachydrin-Monohydrat (1,97 g) abgesaugt. Fp. 209—210° (Zers.) $[\alpha]_D^{22} + 53°$.

Hydrochlorid aus absolutem Alkohol, Fp. 201—202° (Zers.).

In ähnlicher Weise wurden aus 810 g entfetteten Kernen 3—4 g 3-Oxystachydrin gewonnen.

In Wurzeln fehlten cis- und trans-3-Oxystachydrin.

l-Stachydrinäthylestersalze wurden in der Wurzel von *C. virgata* durch Henry und King (1950) nachgewiesen.

Cornforth und Henry (1952) isolierten Stachydrin in ähnlicher Weise aus *Capparis tomentosa* Cam. in Form des Perjodids. Es wurden aus 0,5 kg Samenschalen 3,18 g, aus 1 kg Fruchtfleisch 2,92 g Stachydrin-Perjodid erhalten.

Die Isolierung von (—)Stachydrin und Tetramin aus *Courbonia virgata*.
(Henry und King 1950.)

Das Verfahren beruht auf der Fällung der Base in Form ihres Perjodids durch Zufügung eines Überschusses an *Wagners*-Reagens (Jodjodkaliumlösung) zum vorgereinigten äthanolischen Extrakt des pflanzlichen Materials. Zersetzung des Perjodids durch Erhitzen mit Wasser, Verdampfen zur Trockne und Waschen des Rückstandes mit Äthanol, um leicht lösliche Jodide und unzersetzte Perjodide zu entfernen. Dann wird aus heißem Wasser umkristallisiert. Tetramethylammoniumperjodid schmilzt im reinen Zustand bei 100° und seine Zersetzung zum Jodid durch Erhitzen mit Wasser auf dem Wasserbad schreitet nur langsam vorwärts, da es in heißem Wasser ziemlich löslich ist. Der Perjodidniederschlag der aus dem gereinigten Extrakt aus *C. virgata* gebildet wird, enthält außer Tetramethylammoniumperjodid die Perjodide anderer basischer Bestandteile. In der Wurzel von *C. virgata* kommt nach Henry und King (1950) neben Stachydrinäthylester Di- und Trimethylamin vor. Das Stachydrin wird in Form des Perjodids isoliert.

Isolierungsverfahren. 5 kg lufttrockener Wurzeln von *C. virgata*, die nach dem Zermahlen ein 30 Maschensieb passiert haben, werden 8 Std. lang mit Alkohol auf dem Wasserbad in 1 kg-Portionen (pro kg mit 2 l rektifiziertem Alkohol) am Rückflußkühler extrahiert.

Der Alkohol, welcher nach der Extraktion sauer war, wird abfiltriert und der Rückstand mit Äthanol gewaschen. In der gleichen Weise noch einmal extrahiert und dann abfiltriert. Der Alkohol, der für die 2. Extraktion verwendet wird, dient jeweils zur 1. Extraktion des nächsten Wurzelanteils. Der Alkohol wird auf dem Wasserbad verdampft. Der dickflüssige Rückstand wird mit 2,5 l destilliertem Wasser verdünnt und von Lipoidmaterial durch Behandeln mit geschmolzenem Paraffinwachs, und durch einige Extraktionen der kalt filtrierten Lösung, zuerst mit Leuchtpetroleum und mit Chloroform vollkommen befreit. Die noch intensiv gefärbte Lösung wird mit Ammoniak alkalisiert und danach erschöpfend mit Chloroform und dann mit Äther extrahiert. Diese Extrakte werden verworfen. Die verbleibende Mutterlauge wird mit neutralem Bleiacetat versetzt bis die Lösung sauer wird, dann mit basischem Bleiacetat, von dem große Mengen benötigt werden, bis keine weitere Fällung mehr auftritt. Das Filtrat wird mit dem 4fachen Volumen Alkohol versetzt, wodurch ein weiterer dicker Niederschlag entsteht, er wird abfiltriert und das Filtrat auf etwa 250 ml eingeengt, um das Äthanol zu entfernen, dann vom Bleiacetat mit Hilfe von Schwefelwasserstoff nach starkem Verdünnen befreit. Das noch intensiv gefärbte Filtrat wird auf 250 ml eingeengt und mit 75 g Ammoniumjodid gelöst, in 100 ml Wasser versetzt. Die Lösung wird im Eisschrank über Nacht aufbewahrt. Diese Behandlung entfernt den Hauptanteil des Tetramins als Jodid. Es wurden 71 g Tetraminjodid erhalten.

Die Mutterlauge wird in einen 1,5 l-Becher übertragen und auf 60° erhitzt. Es wird eine Schicht Leuchtpetroleum zugegeben, dann ein Überschuß an Jodlösung (850 ml 25%iges Jod in 18% Ammoniumjodid), die langsam unter Rühren zugefügt wird. Das niedergeschlagene Perjodid, welches sich als eine flüssige 1 cm dicke Schicht am Boden des Bechers sammelt, wird mit einer etwa 1,5 cm dicken Schicht von Chloroform bedeckt, in welchem es sich nur wenig löst und der Überstand der wäßrigen Schicht vollständig abgehebert. Der Becher wird dann mit Wasser gefüllt und die wäßrige Schicht wieder abgehoben. Dieser Prozeß wird öfter wiederholt, um eine vollständige Entfernung der wasserlöslichen Verunreinigungen, (Ammoniumjodid, Glucose und Fructose, Kaliumsalze, Essigsäure usw.) zu erzielen. Das Perjodid wird so vollständig wie möglich mit Hilfe von rektifiziertem Alkohol gesammelt, der dann auf dem Wasserbad entfernt wird, so daß eine bewegliche 1 cm dicke Schicht in dem 1,5 l-Becher verbleibt.

Das Perjodid wird nun noch 2mal auf dem Wasserbad mit 1 l Wasser behandelt, um die letzten Spuren von wasserlöslicher Substanz zu entfernen. Der wäßrige Extrakt wird verworfen. Die Zersetzung der Perjodide zu den Jodiden der Basen erfolgt nun durch langsames Erhitzen auf dem Wasserbad unter heftigem Rühren und gelegentlichem Abdekantieren des wäßrigen Extraktes in Abdampfschalen und Verdampfen zu einem kleinen Volumen. Der Rückstand wird zu einer bröckligen Masse getrocknet, die zum Teil in Äthanol gelöst wird unter Hinterlassen eines festen Rückstandes, der noch einen großen Anteil verfügbaren Jods enthält. Die Lösung, die den Hauptanteil der leichter zersetzbaren Perjodide enthält, wird mit gereinigter Tierkohle behandelt, abfiltriert und auf dem Wasserbad mit Hilfe eines Föns bis fast zur Trockne verdampft. Der Rückstand wird mit absolutem Äthanol extrahiert, um den Hauptteil des verbleibenden Tetraminjodids abzutrennen, der eine niedrige Löslichkeit hat. Die Alkohollösung färbt sich beim Aufbewahren intensiv dunkel. Der Alkohol wird verdampft. Bei Behandlung des Rückstandes mit Wasser resultiert eine nahezu farblose Lösung und es verbleibt eine Kugel von öligem Perjodid. Nach einigen Tagen bilden sich in der filtrierten wäßrigen Lösung, die nun nahezu braun ist, große Kristalle eines Perjodids. Die Kristallisation wird 16 Tage belassen, die Kristalle werden gesammelt, mit eiskaltem Wasser gewaschen und getrocknet. Ausbeute 1,09 g. Die Kristalle riechen nur leicht nach Jod. Die Mutterlauge dieser Kristalle wird bei Raumtemperatur beiseitegestellt, wonach sich weitere Kristalle bilden. Nach 20 Tagen wird eine weitere Ausbeute von 1,53 g erhalten. Die Analyse ergab, daß beide Anteile identisch sind. Es handelte sich um Stachydrinäthylestertrijodid. Es wurde noch eine dritte und vierte Kristallisation erhalten, die aber sehr stark verunreinigt war durch ein flüssiges Perjodid. Das oben erwähnte flüssige Perjodid wurde in heißem Äthanol (z. B. 100 ml) gelöst, beim Erkalten trat Kristallisation ein. Die Kristalle wurden gesammelt mit kaltem Äthanol gewaschen und bis zur Gewichtskonstanz getrocknet. Ausbeute 3,73 g. Das Filtrat wurde konzentriert und dann bei Zimmertemperatur belassen, bis das Äthanol vollkommen verdampft war. Auch diese Kristallgruppe war mit Stachydrinäthylestertrijodid identisch (0,5 g). Das Trijodid wurde in 50%igem Äthanol gelöst und mit einem Überschuß von Silberpulver behandelt. Die farblose Lösung wurde dann zu einem kleinen Volumen konzentriert (1 ml) und mit einem Überschuß an gesättigter wäßriger Natriumpikratlösung behandelt. Das Pikrat kristallisierte rasch, Fp. 101°. $C_9H_{18}O_2N \cdot C_6H_2O_7N_3$. Das mit Silber behandelte Perjodid, also der freie Stachydrinäthylester, ergab ein $[\alpha]_D^{18}$ von —4,4°, bezogen auf das Jodid.

Der Stachydrinäthylester wurde zur weiteren Kennzeichnung in das Aurichlorid $(C_9H_{18}O_2NAuCl_4$; Fp. 61—62°) und in das Chloroplatinat, Fp. 195° (unter Zers.) übergeführt.

Trennung von Cholin und Stachydrin. (SCHULZE u. TRIER, 1910.) Zur Trennung des Stachydrins vom Cholin benutzt man, falls nicht schon durch Umkristallisieren aus Wasser oder Alkohol das Ziel erreicht wird, die Fällbarkeit des Cholins durch Kaliumtrijodid in alkalischer Lösung; dabei kann jedoch, wie oben angegeben, eine beträchtliche Menge Stachydrin ausgefällt werden. Man kann die Trennung auch auf den Umstand gründen, daß eine wäßrige Lösung von freiem Stachydrin auf Zusatz von Mercurichlorid keine Fällung gibt; erst nach Zusatz von Salzsäure scheidet sich ein Niederschlag aus. Enthält die in oben beschriebener Weise erhaltene Mutterlauge neben Stachyrin Cholin, so wird unter schwachem Erwärmen so viel Silberoxyd zugesetzt, bis die Flüssigkeit neutrale Reaktion annimmt. Vom Chlorsilber abfiltrieren, eine geringe Menge gelösten Silbers mit Hilfe von Schwefelwasserstoff entfernen, die Lösung, in welcher neben freiem Stachydrin sich Cholinchlorid vorfindet, auf ein geringes Volumen eindunsten und sodann wäßrige Mercurichloridlösung hinzufügen. Den entstehenden Niederschlag nach dem Abfiltrieren und Auswaschen durch Schwefelwasserstoff zersetzen, das im Filtrat vom Quecksilbersulfid sich vorfindende Chlorid ist Cholinchlorid. Die Methode liefert keine ganz sicheren Resultate, die Trennung ist nicht quantitativ. Der Phosphowolframsäureniederschlag enthält Cholin nur in kleiner Menge. Beim Auskristallisieren des salzsauren Stachydrins bleibt das Cholinchlorid wegen seiner Leichtlöslichkeit in der Mutterlauge.

4. Ergothionein.

(Thionin, Thiasin, Sympectothion.)

$C_9H_{15}O_2N_3S$: C 54,78%, H 7,67%, N 21,32%, Mol-Gewicht 229,1.

Eigenschaften. Kristallisiert aus Wasser in farblosen monoklinen Nadeln mit 2 Mol Kristallwasser. Dieses wird über Schwefelsäure abgegeben und beim Stehen an der Luft wieder aufgenommen. Sehr leicht löslich in heißem Wasser, löslich in 8,6 Teilen Wasser von 20°. Ziemlich leicht löslich in verdünntem Alkohol, kaum löslich in heißem Methanol und Aceton. Unlöslich in Äther und Chloroform. Fp. 290° unter Zersetzung. $[\alpha]_D^{20} = +110°$. In frischem Zustand geruchlos, nach längerem Aufbewahren riecht es unangenehm. Nach dem Schmelzen mit Alkali und Ansäuern tritt Schwefelwasserstoff auf. Der Schwefel wird aber beim Kochen mit Kalilauge nicht abgespalten. Beim Erwärmen mit Chloroform und Kalilauge Grünfärbung, die auf Zusatz von Säure nach Blau umschlägt. Goldchlorid färbt Ergothionein blutrot. Mit 9 Mol Eisenchlorid erhält man unter Abspaltung des Schwefels in guter Ausbeute das Histidinbetain.

Nachweis und Bestimmung. Nach HUNTER (1939) läßt sich Ergothionein mit diazotierter Sulfanilsäure noch in Verdünnungen von 1:5 Millionen nachweisen. Zur kolorimetrischen Bestimmung eignet sich die Rotfärbung mit Diazobenzolsulfosäure unter Benutzung einer $1^0/_{00}$igen Lösung von Ergothionein oder Phenolsulfophthalein als Standard (Empfindlichkeitsgrenze 1:5 Millionen). Das Verfahren wurde zuerst von HUNTER (1949), später durch MELVILLE und LUBSCHEZ (1953) sowie von BALDRIDGE (1953) verbessert.

BALDRIDGE und Mitarbeiter erniedrigen die Kupplungszeit mit gepulvertem Diazoreagens von 45 auf 30 sec und messen die Extinktion bei 420—540 mμ, was eine Ausschaltung des mit dem Diazoreagens kuppelnden Histidins ermöglicht. MELVILLE und Mitarbeiter beseitigen störende Fremdstoffe durch Verwendung des Kunstharzaustauschers Amberlite IRA 410.

Beide Verfahren sind bisher nur für Blut und tierisches Gewebe ausgearbeitet worden. Zum Schutz des Ergothioneins wird vor der Fällung der Eiweißkörper mit Trichloressigsäure, Glutathion und Natriumhydrogensulfit zugesetzt. Das Filtrat wird mit Chloroform ausgeschüttelt und mit dem Austauscher behandelt. Danach wird mit diazotierter Sulfanilsäure in starker Natronlauge gekuppelt und bei 540 mμ gegen eine Vergleichslösung gemessen. Die beiden Verfahren dürften auch bei der Bestimmung des Ergothioneins bei pflanzlichem Material anwendbar sein.

Zur Identifizierung sind folgende Derivate geeignet:

Chlorhydrat $C_9H_{15}N_3S \cdot HCl \cdot 2 H_2O$. Rhombische Kristalle, die bei 105° das Kristallwasser abgeben. Fp. 250°. Sehr leicht löslich in Wasser und Methanol, leicht löslich in verdünntem Alkohol $[\alpha]_D^{20} = +88,5°$. Gibt mit wenig Silbernitrat einen käsigen, kompliziert zusammengesetzten Niederschlag.

Sulfat. $(C_9H_{15}O_2N_3S)_2 \cdot H_2SO_4 \cdot 2 H_2O$, löslich in 7 Teilen Wasser von 10°. $[\alpha]_D^{20} = +87,4°$. Fp. 265° unter Zersetzung.

Phosphat. $C_9H_{15}O_2N_3S \cdot H_3PO_4$. Löslich in 20 Teilen Wasser von 19° $[\alpha]_D^{20} = +83,8°$.

Quecksilberdoppelsalz. $C_9H_{15}O_2N_3S \cdot HCl \cdot HgCl_2$. Löslich in 180 Teilen kaltem Wasser, bei Gegenwart von Sublimat fast unlöslich.

Chloroplatinat, orangerot, ziemlich löslich in Wasser. *Jodide.* Bei allmählichem Zusatz von Jodjodkali zu einer 10%igen Ergothioneinchlorhydratlösung bildet sich zuerst ein schwarzer Niederschlag, der sich wieder auflöst, dann gelbe Nadeln, welche nach dem Filtrieren mit mehr Jod versetzt einen orangegelben Niederschlag geben. Die vereinigten Niederschläge geben aus 60%igem Alkohol umkristallisiert orangegelbe Nadeln, die leicht durch Wasser zersetzt werden und anscheinend ein Gemisch verschiedener Verbindungen darstellen.

Trijodid. $C_9H_{15}O_2N_3SJ_3 \cdot 2H_2O$, entsteht aus der Lösung eines Salzes mit einem geringen Überschuß von Jodjodkali. Schwarzbrauner schwerlöslicher Niederschlag.

Gewinnung aus Mutterkorn nach EAGLES (1928).

1 kg fein gemahlenes Mutterkorn wird in einem Extraktionsapparat 4 Std. mit 90%igem Alkohol (300 ml Alkohol/100 g Droge) extrahiert. Der Rückstand wird 2mal mit Alkohol ausgekocht, durch Leinwand filtriert und das Filtrat mit dem Hauptextrakt vereinigt. Der Alkohol wird abdestilliert, dann wird Wasser zugesetzt, Fette und Harze werden abfiltriert. Das Volumen der Lösung beträgt nunmehr etwa 2 l. Durch vorsichtigen Zusatz von Bariumhydroxyd können beträchtliche Mengen von Farbstoffen entfernt werden. Nun fügt man basisches Bleiacetat hinzu, bis kein Niederschlag mehr entsteht, zentrifugiert und entfernt das überschüssige Blei mit Schwefelsäure. Die resultierende Lösung wird mit 2,5 n-NaOH alkalisiert, dann werden die Alkaloide mit Chloroform ausgeschüttelt.

Die Lösung wird mit Essigsäure angesäuert und mit einer warmen 8%igen Sublimatlösung versetzt, bis kein Niederschlag mehr entsteht. Der Sublimatniederschlag wird gut mit einer 0,25%igen Sublimatlösung gewaschen, in 100 ml Wasser suspendiert und mit Schwefelwasserstoff zersetzt. Das Filtrat wird durch Durchleiten von Luft von Schwefelwasserstoff befreit, und das Ergothionein mit Bleiacetat-NaOH gefällt. Zur vollständigen Ausfällung des Ergothioneins wurden 61 ml 20%ige Bleizuckerlösung, 27,5 ml 2,5 n NaOH und 8 ml 10%ige Kochsalzlösung benutzt. Nach der Zersetzung des Bleisalzes mit Schwefelsäure wird mit Phosphorwolframsäure gefällt, und das freie Ergothionein in bekannter Weise isoliert. Ausbeute 0,65 g Ergothionein aus 1 kg Mutterkorn.

Papierchromatographischer Nachweis von Ergothionein.
(BALDRIDGE und LEWIS 1953.)

BALDRIDGE und LEWIS (1953) verwenden zur aufsteigenden Papierchromatographie eine proteinfreie Ergothioneinlösung in 5%iger wäßriger Natriumcitratlösung. Davon werden aliquote Teile auf Papierstreifen aufgetragen. Als Lösungsmittel dienten Mischungen von Benzol, Butanol, Methanol und Wasser 1:1:2:1 oder Butanol, Eisessig, Wasser 5:3:3. Die Flecke auf den Chromatogrammen wurden durch Besprühen der getrockneten Streifen mit einer frisch bereiteten Mischung von gleichen Teilen Natriumacetatlösung und diazotierter Sulfanilsäurelösung sichtbar gemacht (HUNTER, 1939). Die gelbe Farbe der Ergothioneinflecke schlägt nach Betupfen mit 10 n Natronlauge nach Magenta um.

Ergothioneinbestimmung nach MELLVILLE und LUBSCHEZ (1953).

Die Methode gründet sich auf die Diazoreaktion mit Ergothionein und ist für die Ergothioneinbestimmung im Blut ausgearbeitet. Auf pflanzliches Material bisher nicht angewandt. Es werden Substanzen, welche die Diazoreaktion stören und die auch in pflanzlichem Gewebe vorkommen können, mit Hilfe von Austauschern eliminiert. Histidin, Tyrosin, Thiolhistidin und Carnosin geben gelbe bis orangerote Färbung. Zu niedrige Werte für Ergothionein ergeben sich in Gegenwart von Eisen-, Kupfer- oder Ammoniumionen, Jodid, Fluorid, Sulfit oder Schwefel, Ascorbinsäure, Cystein oder Glutathion, Kreatinin, Alkoholen, Phenolen oder Ketonen. Es besteht eine direkte Proportionalität zwischen Ergothioneinkonzentration und Farbstärke. Ein günstiger Konzentrationsbereich ist 15—20 γ Ergothionein in 3 ml. Zu Gewebsextrakten gegebenes Ergothionein wird praktisch quantitativ wiedergefunden.

Reagentien.

1. Trichloressigsäure. In einer Allglas-Apparatur redestilliert. 35%ige Lösung in Wasser. Ein Überschuß von Trichloressigsäure soll vermieden werden, weil sehr viel Austauschharz verwendet werden müßte, um den Überschuß zu entfernen. Die Lösung wird im Eisschrank aufbewahrt und jede Woche erneuert.

2. Chloroform. 6mal mit dem gleichen Volumen von destilliertem Wasser gewaschen und in einer dunklen Glasflasche aufbewahrt. Wird am Tag der Benutzung noch einmal mit Wasser gewaschen.

3. Gepulvertes Natriumhydrosulfit. Glutathion, wäßrige Lösung 25 mg/ml unmittelbar vor Verwendung bereitet. Einige Proben von Glutathion enthalten Verunreinigungen, welche die Diazoreaktion stören.

4. Ionenaustauschharz Amberlite IRA 410, aufbereitet durch 3 Std. langes Schütteln von 500 g des handelsfeuchten Chloridsalzes mit 2 l 5%iger Na$_2$CO$_3$-Lösung. Waschen mit destilliertem Wasser bis zur Alkalifreiheit und Sammeln auf einem Trichter unter Saugen. Das feuchte Material wird in einer dichtschließenden Flasche aufbewahrt.

5. Alkalische Pufferlösung. 200 g wasserfreies Natriumcarbonat, analysenrein und 57 g Natriumcitratdihydrat werden in etwa 800 ml Wasser unter Erwärmen gelöst. Die Lösung wird gekühlt, filtriert und auf 1 l verdünnt. Die Lösung wird in Polyäthylenbehältern aufbewahrt und wird wieder filtriert, wenn Sedimente gebildet worden sind.

6. Sulfanilsäure. Analysenrein 0,2%ige Lösung in 1%iger Salzsäure.

7. Natriumnitrat. Analysenrein 4%ige Lösung in Wasser.

8. Diazotierte Sulfanilsäurelösung. 1 Vol. der kalten Natriumnitritlösung wird zu 10 Volumen kalter Sulfanilsäurelösung hinzugegeben. Die Mischung wird im Eisschrank aufbewahrt. Wird jeweils frisch am Tage der Verwendung bereitet. 19 n Natriumhydroxyd bereitet aus Natriumhydroxyd analysenrein, gefiltert durch ein gesintertes Glas und in Polyäthylenflaschen aufbewahrt. Die Lösung sollte filtriert werden, wenn Wolkenbildung sich entwickelt.

Ein 3 ml-Anteil, der mit Chloroform und dem Austauscher vorbehandelten Ergothioneinlösung wird in ein einfaches Reagenzglas von 18 mm Durchmesser gegeben, das sich in einem Eisbad befindet. 0,5 ml alkalische Pufferlösung werden zugegeben und die Lösung durchgemischt. Nun werden 0,5 ml diazotierter Sulfanilsäure zugefügt, es wird durchgemischt und 45 sec im Eisbad belassen. Dann wird 1 ml 19 n Natronlauge rasch zugegeben und wieder gut gemischt. Die Lösung bleibt nun bei Zimmertemperatur stehen, bis sie von Gasblasen frei wird, dann wird sie in einem Spektrophotometer bei 540 mμ bestimmt. Nicht bei 510 mμ wie nach Hunter (1939), da die Lichtabsorption der störenden Substanzen wie Histidin und Tyrosin, nicht aber die des Ergothioneins bei den größeren Wellenlängen geringer ist. Die Extinktion eines gleichartigen Ansatzes, bei dem 3 ml der zu bestimmenden Lösung durch Wasser ersetzt sind, wird vom Wert des Hauptversuchs abgezogen.

5. Hercynin (Histidinbetain).

$C_9H_{15}O_2N_3$: C 54,78%, H 7,67%, N 21,31%, Mol-Gew. 197,12.

Eigenschaften. Das freie Betain ist noch nicht beschrieben worden $[\alpha]_D^{20} = +41,1°$ (in Gegenwart von 8 Mol HCl), Chlorid löslich in Wasser und Alkohol. Nitrat bildet große durchsichtige Platten und Oktaeder.

Derivate.

Chloroaurat. $C_9H_{15}O_2N_3 \cdot 2$ HAuCl$_4$. Sehr schwer löslich in Wasser, kristallisiert aus heißer verdünnter Salzsäure in langen orangegelben Spießen. Fp. 184°. In Gegenwart von verdünnter Salzsäure und überschüssigem Goldchlorid bilden sich Salze anderer Zusammensetzung.

Monopikrat. $C_9H_{15}O_2N_3 \cdot C_6H_3O_7N_3 + H_2O$, wird in feinen weichen Nädelchen erhalten, wenn eine Lösung der Base mit Pikrinsäure neutralisiert wird oder wenn eine neutrale Lösung mit Natriumpikrat umgesetzt wird. Das Kristallwasser wird bei 105° abgegeben. Fp. 201°.

Dipikrat. $C_9H_{15}O_2N_3 \cdot 2$ $C_6H_3O_7N_3 + 2$ H_2O, Fp. 123° aus dem Monopikrat durch Zusatz einer gesättigten Pikrinsäurelösung. Aus dem salzsauren Salz des Hercynins durch Eindampfen mit Salzsäure und Zufügen von Natriumpikratlösung zu dem mit Wasser aufgenommenen, von überschüssiger Salzsäure befreiten Salz. Löslich in 25 Teilen Wasser. Das wasserfreie Dipikrat verliert das Kristallwasser bei 105°, schmilzt aus Alkohol umkristallisiert bei 213—214°.

Pikrolonat bildet lange, dünne, orangegelbe Nadeln, Fp. 229—230°.

Vorkommen. Bisher nur in Pilzen gefunden, im Handelspräparat der wasserlöslichen Extraktstoffe von Champignons (KUTSCHER, 1912) im alkoholischen, Extrakt von *Boletus edulis* (REUTER, 1912), vielleicht auch im Fliegenpilz (KÜNG, 1913). Kann durch Entschwefeln von Ergothionein entstehen.

Isolierung aus der Pflanze. Im Gegensatz zu anderen Imidazolverbindungen gelangt das Hercynin bei der Aufteilung des Phosphorwolframsäureniederschlages nicht in die Histidinfraktion. KUTSCHER (1912) fand es ausschließlich unter den mit Silbernitrat und Baryt nicht fällbaren Basen (Lysinfraktion), während REUTER es in der Argininfraktion nachwies. Bei der Isolierung aus Fliegenpilzen verfuhr KÜNG (1913) folgendermaßen: Zerkleinertes Material wurde dreimal in der Kälte mit 70—80%igem Alkohol ausgezogen und der Rückstand jedesmal scharf abgepreßt. Die dunkelbraun gefärbten Extrakte wurden vereinigt, in Wasser aufgenommen, im Scheidetrichter von dem an der Oberfläche schwimmenden Fett, Lecithin und Cholesterin getrennt und mit Äther geschüttelt, um diese Stoffe möglichst vollständig zu entfernen. Aus der Argininfraktion wurde nach Entfernung des Silbers und des beigemischten Baryts und nachfolgender Abscheidung mit Phosphorwolframsäure aus dem Niederschlag ein nur schwach alkalisch reagierender Sirup erhalten, der auf Zusatz von alkoholischer Pikrinsäurelösung eine gelbbraun gefärbte Fällung lieferte, die in heißem Wasser gelöst wurde und beim Erkalten zuerst ein flockiges Pikrat abschied, das nach längerem Stehen in schöne Drusen vereinigte glänzende Blättchen bildete, die nach Umkristallisieren aus Wasser bei 201—202° schmolzen. Das Präparat stimmte sowohl im Aussehen als auch im Zersetzungspunkt mit dem von REUTER beschriebenen Hercyninmonopikrat überein. Das Pikrat wurde durch Zusatz von Salzsäure in das Chlorid übergeführt und die Pikrinsäure mit Äther entfernt. Goldchlorid erzeugte eine Fällung, welche bei 118—120° unzersetzt schmolz, während REUTER für das entsprechende Hercyninsalz 184, KUTSCHER 180—182° angeben. Es ist danach nicht erwiesen, ob tatsächlich Hercynin vorlag.

6. Betonicin und Turicin.

Betain der 4-Oxypyrrolidin-2-carbonsäure (γ-Oxyprolinbetain). $C_7H_{13}O_3N$: C 52,79%, H 8,23%, N 8,81%, Mol-Gewicht 159.

Das Vorhandensein zweier asymmetrischer Kohlenstoffatome im 4-Oxyprolin schafft die Möglichkeit von 4 stereoisomeren Betainen. Im Betonicin und Turicin liegen Diastereoisomere vor, entstanden durch Racemisierung eines der beiden asymmetrischen Kohlenstoffatome, wahrscheinlich des C(2). Die beiden Basen finden sich nebeneinander im Kraut von *Betonica officinalis* zu etwa 1% des Trockengewichts, ferner neben Trigonellin in Blättern und Stengeln von *Stachys silvatica*, Betonicin neben Trigonellin in der Jackbohne.

Betonicin und Turicin sind die optischen Antipoden des γ-Oxyprolinbetains. Betonicin dreht die Ebene des polarisierten Lichtes nach links, Turicin nach rechts. Sie entstehen bei der Methylierung der γ-Oxy-hygrinsäure und Oxyprolin nebeneinander.

Betonicin. Derivate. *Hydrochlorid* $C_7H_{13}O_3N \cdot HCl$, feine Nadeln oder Prismen aus absolutem Alkohol. Sehr leicht löslich in Wasser, löslich in heißem, wenig in kaltem absolutem Alkohol. Reagiert sauer und ist nicht hygroskopisch. Smp. 232° unter Zersetzung. $[\alpha]_D^{20} = -24{,}8°$.

Chloroaurat. $C_7H_{13}O_3N \cdot HAuCl_4$ aus Wasser gelbe, zu fächerartigen Drusen vereinigte Kristallblättchen. In kaltem Wasser schwer löslich. Aus verdünnten Lösungen aber nicht fällbar. Zersetzungspunkt 230—232° (auch 242° angegeben).

Chloroplatinat. $(C_7H_{13}NO_3)_2H_2PtCl_6 \cdot 2\ H_2O$. Dichromatische Prismen aus Wasser. Fp. 225—226° unter Zersetzung.

Turicin. *Isolierung aus der Pflanze.* Wie unter Betonicin beschrieben. Die beiden Betaine lassen sich isolieren, indem man sie erst frei, dann die in salzsaure Salze verwandelten Fraktionen mit Alkohol behandelt. Beim Digerieren der freien Betaine mit Alkohol wird hauptsächlich Betonicin gelöst, während Turicin zurückbleibt. Die salzsauren Salze verhalten sich umgekehrt, man erhält somit reines salzsaures Turicin, indem man den ungelösten Teil der freien Basen mit überschüssiger Salzsäure zur Trockne eindunstet und mit wenig absolutem Alkohol auszieht.

Eigenschaften. Kristallisiert aus siedendem Alkohol in seidenglänzenden flachen Prismen, bei rascher Abkühlung in feinen glänzenden Nadeln. Nicht hygroskopisch, verwittert sehr rasch im Exsiccator; wird weiß und undurchsichtig. Sehr leicht löslich in Wasser, schmeckt süß, schmilzt wasserfrei unter Zersetzung bei 260°, wasserhaltig bei etwa 249° und zersetzt sich dann bei 256° $[\alpha]_D^{20} = +36{,}26°$. Gibt in Dampfform starke Fichtenspanreaktion. Fällt aus der wäßrigen Lösung mit Phosphorwolframsäure.

Turicinchlorhydrat. Feine, glänzende Nadeln oder sechsseitige Tafeln aus absolutem Alkohol, zersetzen sich bei 224°, reagieren sauer. $[\alpha]_D^{20} = +24{,}65°$.

Chloroaurat. $C_7H_{13}O_3N \cdot HAuCl_4$, gelbe Prismen, oft zu glänzenden Drusen vereinigt. Fp. 230—232° unter Zersetzung. Fällt nur aus sehr konzentrierter Lösung.

Chloroplatinat $(C_7H_{13}NO_3)_2H_2PtCl_6 + H_2O$. Fp. 223° unter Zersetzung.

Isolierung von Betonicin aus *Canavalia ensiformis*.
(Ackermann und Appel 1939.)

20 kg Samen von *Canavalia ensiformis* wurden zermahlen und nach Entfettung durch Benzol mit 50% Alkohol dreimal in der Siedehitze extrahiert. Die Flüssigkeit von ungefähr 120 Litern wurde im Vakuum vom Alkohol befreit und auf etwa 15 Liter gebracht. Zu der schwach phosphorsauren Flüssigkeit

wurden 570 ml 40%ige Tanninlösung gegeben und aus dem Filtrat des entstandenen Niederschlags das Tannin mit Baryt und nachfolgende Behandlung mit Bleioxyd bei Gegenwart von Schwefelsäure beseitigt. Die schwach schwefelsaure Flüssigkeit wurde nach dem Einengen mit 8 Litern 40%iger Flaviansäure gefällt und der dicke Niederschlag am anderen Tage abgesaugt. Er enthielt Canavanin vermischt mit etwas Arginin. Nach Beseitigung der Flaviansäure wurde aus der starken alkalischen Flavianmutterlauge das Arginin als Kupfernitratsalz ausgefällt. Die Testlösung wurde in bekannter Weise in eine Lösung der freien Basen übergeführt und bei Gegenwart von 5% H_2SO_4 mit Phosphorwolframsäure gefällt. Die nach Zersetzung des Niederschlages gewonnene Basenlösung wurde nach dem Einengen mit überschüssigem Kupfercarbonat gekocht. Nach dem Abfiltrieren des Niederschlages (Canein und Kitagin) wurde nach Beseitigung des Kupfers mit H_2S aus seiner Lösung, die curcumaalkalisch reagierte, mit Silbernitrat ohne Barytzusatz gefällt, was etwa einer Histidinfraktion entspricht. Der Niederschlag wurde mit H_2S vom Silber befreit und aufgeteilt in eine mit wäßrigem $HgSO_4 + H_2SO_4$ (Hopkins) fällbare Fraktion in eine zweite, die erst mit diesem Fällungsmittel bei Gegenwart von Alkohol fiel und eine dritte, die auch unter diesen letzten Bedingungen nicht niederzuschlagen war. In dem mit $HgSO_4$ auch bei Gegenwart von Alkohol nicht fällbaren Anteil war nach Beseitigung des Alkohols und des Quecksilbers in der Lösung der Basen noch Salpetersäure zugegen, die von der Zersetzung der Histidinfraktion stammte. Die Mutterlauge des Desaminocanavaninhalbnitrates gab beim Fällen mit Goldchloridwasserstoffsäure in salzsaurer Lösung ein Goldsalz, das sich nach mehrfachem Umkristallisieren aus Salzsäure als Trigonellinchloroaurat erwies. Aus der Lysinfraktion wurde durch Reineckefällung Cholin gewonnen. Das Filtrat des ammoniakalischen Reineckatsalzes wurde entsprechend dem Vorgehen von Strack und Schwaneperg (1936) mit Schwefelsäure angesäuert, worauf ein neuer Niederschlag entstand, der nach Beseitigung der Reineckesäure in eine Lösung der Chloride verwandelt wurde. Sie lieferten neben etwas Trigonellin 4,20 g Betonicin-Chlorid. Die Lösung des Chlorides reagiert wie die eines Betains kongosauer und hat die spezifische Drehung $[\alpha]_D^{20} = -24,1°$, während Kühn und Trier (1913) $[\alpha]_D^{15} = -24,79°$ angeben.

Das rechtsdrehende Turicin fand sich nicht.

Isolierung von Turicin aus Stachys silvatica.

Am besten werden ältere Pflanzen von *Stachys silvatica* verwendet. Sie werden mit verdünntem Alkohol extrahiert und dann zuerst von den durch Bleiessig fällbaren Bestandteilen befreit. Aus der erhaltenen Lösung werden die durch Silbernitrat und Baryt fällbaren Basen entfernt. Die übriger Basen werden wieder durch Phosphorwolframsäure gefällt und in salzsaure Salze übergeführt. Turicin und Betonicin lassen sich jetzt trennen, indem man sie erst freisetzt, dann in die salzsauren Salze verwandelt und diese mit Alkohol behandelt. Beim Extrahieren der freien Betaine mit Alkohol wird hauptsächlich Betonicin gelöst. Turicin bleibt zurück. Die salzsauren Salze verhalten sich umgekehrt. Man erhält demnach reines salzsaures Turicin, indem man den ungelösten Teil der freien Basen mit überschüssiger Salzsäure zur Trockne eindunstet und mit wenig absolutem Alkohol auszieht.

7. Trigonellin.

(Methylbetain der Nicotinsäure, Coffearin.)

$C_7H_7O_2N$: C 62,02%, H 5,15%, N 10,22%, Mol-Gewicht 137.

Eigenschaften. Kristallisiert in farblosen Prismen mit 1 Mol Kristallwasser; wird bei 100° wasserfrei. In Wasser und Alkohol sehr leicht löslich. In Äther unlöslich, schwer löslich in warmem Chloroform. Geschmack unangenehm süßlich. Schmilzt bei 130° im Kristallwasser. Wasserfrei schmilzt es gegen 220° unter vorangehender Braunfärbung und Zersetzung. Fällbar durch die meisten Alkaloid-fällungsmittel. Zum Teil auch durch Gerbsäure fällbar, leicht löslich im Überschuß des Fällungsmittels. Im normalen Analysengang findet es sich in der Lysinfraktion.

Nachweis. Zum Nachweis eignet sich das charakteristische Aussehen des Chlorhydrats und vor allem der Chloroaurate. Beim Erhitzen tritt der typische Geruch des Pyridins auf.

Chlorhydrat. $C_7H_7O_2N \cdot HCl$. Kristallisiert in flachen, stark glänzenden, rechtwinklig begrenzten Tafeln, die in Wasser sehr leicht, in kaltem absolutem Alkohol sehr schwer löslich sind. Löslichkeit in absolutem Alkohol 1:344. Schmilzt unter Zersetzung bei 260°.

Chloroaurate. Bei der Fällung des salzsauren Salzes mit Goldchloridlösung scheint zunächst ein wasserhaltiges Goldsalz zu entstehen, das keinen scharfen Schmelzpunkt und keine konstante Zusammensetzung besitzt. Aus verdünnter Salzsäure mit überschüssigem Goldchlorid umkristallisiert erhält man das normale

Chloroaurat. $C_7H_7NO_2 \cdot AuCl_4$. In kaltem Wasser schwerlösliche Blätt-chen oder flache Prismen, die bei 197—198° ohne Zersetzung schmelzen. Kristallisiert man die Fällung nur aus Wasser um, so erhält man ein basisches Chloraurat der Zusammensetzung $(C_7H_7NO_2)_4 \cdot 3\,AuCl_4$. Dieses Salz enthält 37,7% Au, manchmal wurde auch ein höherer Au-Gehalt gefunden. Kristallisiert in feinen Nadeln, die sich in kaltem Wasser schwer lösen und bei 185—186° ohne Zersetzung schmelzen. Eignet sich zur Bestimmung.

Chloroplatinate. $(C_7H_7NO_2)_2 \cdot PtCl_6$. Es sind Platinsalze mit 4 Mol, mit 1 Mol und ohne Kristallwasser beschrieben worden. Das Chloroplatinat mit 1 Mol H_2O schmilzt unter Zersetzung bei 225—226°, NOTTBOHM und MAYER (1931). Die Chloroplatinate lösen sich leicht in Wasser, sind aber in Alkohol schwer löslich. Am häufigsten scheint die Verbindung mit 1 Mol Wasser aufzutreten. Wasserfrei bildet die Verbindung derbe Prismen.

Pikrat. $C_7H_7NO_2 \cdot C_6H_3N_3O_7$. Glänzende Prismen, in Wasser und Methanol leicht, in absolutem Äthylalkohol schwerer löslich. Fast unlöslich in Äther. Fp. 198—200°.

Bestimmung des Trigonellins im rohen Kaffee.
(NOTTBOHM und MAYER 1931.)

20 g feinpulverisierte Kaffeebohnen werden 3 Std. lang mit Chloroform im Soxhlet extrahiert. Das anhaftende Chloroform wird im Vakuum abgesaugt, dann wird das Pulver 3 Std. mit 96%igem Alkohol extrahiert. (Wechseln des Alkohols nach 1 Std.) Die vereinigten alkoholischen Extrakte werden mit 8 ml Bleiessig versetzt, der Niederschlag wird abzentrifugiert und mit Alkohol ausgewaschen. Man entfernt das Blei durch Einleiten von Schwefelwasserstoff, fügt zum Filtrat 5 ml 25%ige Salzsäure und dampft auf etwa 50 ml ein. Von der dabei auftreten-den öligen Ausscheidung wird abgetrennt, und die Lösung nach Behandeln mit etwas Tierkohle zur Trockne verdampft. Das Abdampfen wird mit der gleichen Menge Salzsäure noch 2—3 mal wiederholt, bis der gesamte Zucker zerstört ist. Der Rückstand darf nicht klebrig sein. Er wird in heißem Wasser aufgeschwemmt, nach Zusatz von etwas Tierkohle aufgekocht, filtriert und auf 5 ml (oder weniger) eingeengt. Aus der klaren Lösung wird das Trigonellin als Jodverbindung aus-gefällt, indem man 2—3 Tropfen Salzsäure und 11 ml 0,1 n Jodlösung zusetzt. Der sofort auftretende braune flockige Niederschlag kristallisiert in der Regel nach etwa 10 min. Man nutscht auf einem Zuckerbestimmungsröhrchen nach

Allihn ab, wäscht mit wenigen Millilitern eiskaltem Wasser aus, löst den Rückstand in warmem Alkohol, verdünnt stark mit Wasser und titriert mit 0,1 n Thiosulfatlösung. Die Trigonellinjodverbindung enthält auf 1 Mol Trigonellin etwa 3 Atome Jod. Zur Berechnung muß man sich mit einem Titer begnügen, der durch Verwendung von reinem Trigonellin unter den gleichen Bedingungen zu ermitteln ist.

Isolierung aus Kaffee.
(Nottbohm und Mayer 1931.)

Den rohen gemahlenen Kaffee mit Äther entfetten. Coffein durch Extraktion mit Chloroform entfernen. Mit Wasser kochen, mit Baryt bis zur schwach alkalischen Reaktion versetzen und einige Stunden unter häufigem Rühren stehen lassen. Durch Leinwand filtrieren, Filtrat mit Bleiacetat versetzen, abermals filtrieren und überschüssiges Barium und Blei mit Schwefelsäure entfernen. Filtrat dieser Fällung nötigenfalls zur Entfernung von Coffein mit Chloroform behandeln und Lösung weiter eindampfen. Den sirupösen Rückstand in verdünnter Schwefelsäure lösen und mit Phosphorwolframsäure versetzen. Niederschlag mit 5%iger Schwefelsäure waschen und in bekannter Weise mit Baryt versetzen. Die abfiltrierte Lösung vorsichtig eindampfen und Rückstand mit Alkohol behandeln. Den so erhaltenen kristallinischen Körper in Salzsäure lösen. Beim Eindunsten erhält man das Chlorhydrat des Trigonellins in charakteristischen, stark glänzenden Platten, die in Alkohol schwer löslich sind.

Bestimmung von Trigonellin in pflanzlichem Material.

Bestimmung nach Kodicek und Wang (1941). Prinzip: Bei Hydrolyse von Trigonellin mit Alkali in Alkohol wird Methylamin abgespalten und dabei der verbleibende Pyridinring geöffnet. Dieser kuppelt mit aromatischen Aminen unter Bildung von Farbstoffen. Für die Trigonellinbestimmung ist am besten geeignet Benzidin, das mit dem Hydrolyseprodukt ein orangerotes Derivat ergibt. Die Probe scheint für Trigonellin spezifisch zu sein. Von vorgelegten Trigonellinmengen werden $95 \pm 8\%$ wiedergefunden. Pyridoxin, Nicotinsäure, Nipekotinsäure und Nicotin ergeben negative Proben. Nur N-Methylpyridiniumhydroxyd gibt eine ähnliche Farbreaktion.

Zur Analyse von pflanzlichem Material verwendet man z. B. 10 ml Extrakt und versetzt mit 1 ml konzentrierter HCl, 0,4 ml 10%iger Bariumchloridlösung und 0,5 g Tierkohle. Es wird einige Minuten geschüttelt, dann wird filtriert. 2 ml-Portionen des Filtrates werden in 250 ml-Gefäße gegeben. Zu einem der Gefäße gibt man 0,5 ml einer Standardtrigonellinlösung, die 50 γ pro ml enthält.

Dann fügt man 4,2 ml 40%iges Natriumhydroxyd zu jeder Flasche und erhitzt auf dem Wasserbad $^1/_2$ Std. lang. Nun wird sorgfältig neutralisiert, auf 30 ml aufgefüllt und abzentrifugiert; 10 ml Anteile des Zentrifugates werden in Reagenzgläser gegeben. Zum Kontrollversuch wird 1 ml 5%iger HCl, zum Hauptversuch 1 ml 1%iges Benzidin in 5%iger Salzsäure gegeben. Man läßt die Farbe 1 Std. sich entwickeln und mißt dann in einem Pulfrich-Photometer mit Filter S 50.

In Kaffee und Tee wurden 7,5 mg/g und 0,13 mg/g Trigonellin gefunden.

Bestimmung nach Fox (1943). Das Verfahren von Kodicek und Wang wurde durch Fox (1943) verfeinert, und zwar in folgender Weise: Er ersetzte den Äthylalkohol durch Methylalkohol, verwendet Dianisidin zur Farbstoffbildung, und kolorimetriert mit Filter 520. Das Verfahren wird nur durch Glucose gestört. Nicotinsäure, α-Pyrollidin, Nicotin, Coffein, Betain, Cholin, Thymol, Pyridoxinbetain und Methylnicotinsäure stören die Reaktion nicht. Auf der

Methode von Kodicek und Wang beruht auch das Verfahren zur Trigonellinbestimmung von Sarret (1943).

Roggen (1943) verwendet zur Bestimmung des Trigonellins den nach Aufspaltung mit Alkali entstehenden Dialdehyd, der mit Anilin kondensiert wird und kolorimetriert $1^1/_2$ Std. später die gelbe Farbe mit Filter S 450. Es können auf diese Weise 10 γ Trigonellin/ml erfaßt werden. Die Farbreaktion störende Substanzen können durch Oxydieren mit Permanganat, gegen das Trigonellin beständig ist, beseitigt werden. Trigonellin kann dann weiterhin gereinigt werden durch Adsorption an Lloyds Reagens.

Die geschilderten Verfahren sind meist zur Trigonellinbestimmung im Urin verwendet worden. Erfahrungen über Bestimmungen in pflanzlichem Material wurden von Kodicek und Wang selbst mitgeteilt. Raffaele (1939) weist Trigonellin nach durch Kochen mit einigen Tropfen 10%iger Natronlauge und Zugabe von einigen Tropfen 2%iger alkalischer 2,4-Dinitrochlorbenzollösung zu der entstehenden gelben Flüssigkeit, die beim Erhitzen in Tiefrot umschlägt. Erhitzen von Trigonellin mit wäßriger Resorcinlösung in salzsaurer Lösung und Alkalisieren mit Natronlauge ergibt eine tiefrote Färbung.

Moores und Greninger (1951) bestimmen Trigonellin z. B. im Kaffee (grünen oder gerösteten) durch Perkolieren mit 50%igem Äthanol und Adsorption des Extraktes an eine Celite Nr. 545/Filtrolsäule, Verhältnis 2:1, durch die zuvor 2 n H_2SO_4 gesaugt wurde. Sie eluieren mit 3%igem Ammoniak das Trigonellin und bestimmen den Trigonellingehalt spektrophotometrisch vor und nach Fällung des Trigonellins durch Zinkferrocyanid. Bei geröstetem Kaffee werden die störenden Röstprodukte im Eluat vor der Bestimmung des Trigonellins durch milde Oxydation mit $KMnO_4$-Sulfit entfernt.

Kogan, Dicardo und Maynard (1953) bestimmen Trigonellin im Kaffee neben Coffein nach Abtrennung der Chlorogensäure mittels Magnesiumoxyd. Als Solvens diente dabei 65%iger wäßriger Isopropylalkohol, 2 n Salzsäure. Trigonellin und Coffein werden im Chromatogramm an der UV-Fluoreszenz erkannt, ihre R_f-Werte betragen 0,59 bzw. 0,78. Vorher nicht quantitativ abgetrennte Chlorogensäure fluoresziert ebenfalls im UV-Licht, stört jedoch in kleinen Mengen nicht, da ihr R_f-Wert bei 0,86 liegt und sich die Flecke nicht überlappen.

Trigonellinbestimmung im Kaffee nach Moores und Greninger (1951).

Reagentien und Apparate. Schwefelsäure 0,05 n; Magnesiumoxyd schwer, U. S. P. Lösungsmittel: 65% salzsaurer Isopropylalkohol (in bezug auf HCl 2 n), muß gelegentlich erneuert werden, da durch Verdampfen das Verhältnis von Alkohol zu Säure sich ändert. Salzsäure 0,1 n, Whatman Filterpapier Nr. 1, Bogen zu 15:37 cm, UV-Lampe, Beckman-Spektrophotometer Modell D. V. mit Quarzküvetten.

Vorgang. 5—10 g getrockneter, gemahlener, gerösteter oder grüner Kaffee werden in einen tarierten 1 Liter-Erlenmeyerkolben gegeben. Dazu werden 200 ml destilliertes Wasser gefügt und 50 ml 0,05 n-H_2SO_4. Die Mischung wird 20 min lang auf einer Heizplatte gelind gekocht. Dann werden 25 g Magnesiumoxyd zugegeben und weitere 20 min gekocht. Es wird auf Zimmertemperatur abgekühlt und das verdampfte Wasser ersetzt. Es wird gründlich gemischt und durch Whatman Nr. 17-Papier filtriert. Unter Benutzung einer graduierten Pipette werden 0,02 ml Anteile des Filtrates auf Whatman Nr. 1-Papier im Abstand von etwa 8 cm aufgetragen. Nach dem Trocknen des Papiers werden weitere 0,02 ml-Portionen auf die gleichen Flecke gegeben. Das Papier wird nach jeder Auftragung wieder getrocknet. Es wird 18—24 Std. chromatographiert. Nach Trocknen an der Luft wird der Bogen vor einer UV-Lampe im verdunkelten Zimmer betrachtet. Die Frontlinien und die fluoreszierenden Flecke werden markiert. Die Coffeinflecke haben einen R_f-Wert von etwa 0,78, die Trigonellinflecke von 0,59.

Die Felder werden ausgeschnitten und getrennt in Reagenzgläser gegeben. Dann wird von der freien Fläche zwischen den fluoreszierenden Flecken ein Feld der gleichen Größe und des gleichen R_f-Wertes ausgeschnitten und als Leerversuch in ein Reagenzglas gegeben. Es werden 5 ml 0,1 n-Salzsäure in jedes Glas gegeben und bei Zimmertemperatur mindestens 4 Std. belassen.

Dann werden die Gläser leicht geschüttelt und das Eluat in die Quarzküvetten zur Ablesung im Beckman-Spektrophotometer übergeführt.

Die Absorption von Coffeinlösungen wird bei 272 mμ, die von Trigonellin bei 265 mμ abgelesen.

Papierchromatographische Bestimmung von Trigonellin und Cholin nach MUNIER und MACHEBOEUF (1951).

Durchführung, Lösungsmittel, R_f-Werte und Sichtbarmachung der Flecke s. S. 603.

Eine sehr gute Trennung von Trigonellin und Cholin wird erreicht, wenn in saurem Milieu (100 Teile Butanol, 20 Teile Essigsäure und 50 Teile Wasser) entwickelt wird. R_f-Wert von Cholin 0,33, für Trigonellin 0,20. Hordeninflecke werden sichtbar gemacht mit Joddampf.

Literatur.

ACKERMANN, D.: Hoppe-Seylers Z. **60**, 482 (1909). — ACKERMANN, D.: Hoppe-Seylers Z. **225**, 46 (1934). — ACKERMANN, D., u. W. APPEL: Hoppe-Seylers Z. **262**, 103 (1939). — ACKERMANN, D., u. H. MAUER: Hoppe-Seylers Z. **279**, 114 (1943). — AKASI, S.: Acta Scholae med. Univ. Kioto **22**, 433 (1938). — ALEXANDER, B.: J. Biol. Chem. **209**, 795 (1954). — ALEXANDER, B., G. LANDWEHR and A. G. SELIGMANN: J. Biol. Chem. **160**, 51 (1945). — AMATSU, H., and M. TSUDJI: Acta scholae med. Univ. Kioto II, 447 (1918); Biochem. Z. **122**, 251 (1921). — ANREP, G. V., M. S. AYADI, G. S. BARSOUM, J. R. SMITH and M. M. TALAAT: J. Physiol. **103**, 155 (1944). — ANREP, G. V., G. S. BARSOUM, M. TALAAT and E. WIENINGER: J. Physiol. **95**, 476 (1939). — APPLETON, H. D., and B. N. LA DU jr., B. B. LEVY, J. MURRAY, B. B. BRODIE: J. Biol. Chem. **205**, 803(1953). — ARAI, M.: Biochem. **122**, 251 (1921). — ATTWOOD, C. W., I. A. U. FORD, E. R. HOYLE and D. W. PARKES: J. Soc. Chem. Industr. **69**, 181 (1950).

BAKER, W., J. B. HARBORNE and W. D. OLLIS: J. Chem. Soc. (London) **1952**, 315. — BALDRIDGE, R. C., and H. B. LEWIS: J. Biol. Chem. **202**, 169 (1953). — BANDELIN, F. J.: J. Amer. Pharmaceut. Assoc. **38**, 304(1949). — BARAC, G.: Bull. Soc. chim. Biol. **32**, 287 (1950). — BARGER, G.: J. Chem. Soc. London **99**, 2068 (1911). — BARGER, G., and F. R. TUTIN: Biochem. J. **12**, 402 (1918). — BARSOUM, G. S., and J. H. GADDUM: J. Physiol. **85**, 1 (1935). — J. Sci. Technol. B. **31**, 1 (1950). — BEATTIE, F. J. R.: Biochem. J. **30**, 1554 (1936). — BECKER, M.: Biochem. Z. **288**, 348 (1936). — BENEDICT, S. R.: J. Biol. Chem. **51**, 187 (1922). — BERGER, F.: Handbuch der Drogenkunde, Bd./1, 327 (1949). — BERGMANN, M., u. L. ZERVAS: Hoppe-Seylers Z. **173**, 80 (1928). — BESSEY, O. A., O. H. LOWRY and M. J. BROCK: J. Biol. Chem. **164**, 321 (1946). — BESSEY, O. A., O. H. LOWRY, M. J. BROCK and J. L. LOPEZ: J. Biol. Chem. **166**, 177 (1946). — BINDER, H.: Angew. Chemie **66**, 268 (1954). — BILLMANN, J. H., and O. E. MAHONI: J. Amer. Chem. Soc. **61**, 2340 (1939). — BISSET, S. K.: Biochem. J. **58**, 225 (1954). — BLOCK, R. J.: Analyt. Chem. **22**, 1327 (1950). — BLOOD, J. W., and H. T. CRANFIELD: Analyst **61**, 289 (1935). — BLOUNT, B. K., H. T. OPENSHAW and A. R. TODD: J. Chem. Soc. (London) **1940**, 1, 286. — BLUMRICH, K. G., u. G. BANDEL: Angew. Chem. **54**, 374 (1941). — BOARDMAN, N. K., and S. M. PARTRIDGE: Nature **172**, 209 (1953). — BORN, G. V., J. R. VANE and E. J. FILPOT: Brit. J. Pharmacol. **7**, 298 (1952). — BOSER, D., u. G. KLEIN: Arch. d. Pharmaz. **270**, 374 (1932). — BOURER, A., J. L. MONGAR and H. O. SCHILD: Brit. J. Pharmacol. and Chemotherap. **9**, 24 (1954). — BOWDEN, K., B. G. BROWN and J. BATTY: Nature **174**, 925 (1954). — BRANTE, G.: Nature **163**, 651 (1949). — BRECHT, K.: Pflügers

Arch. **246**, 553 (1943). — Bredig, G.: Z. f. Physik und Chemie **13**, 323 (1898). — Bredig, G., u. K. Winkelblech: Z. f. Elektrochemie **6**, 33 (1900). — Bregoff, H. M., E. Roberts and C. C. Delwiche: J. Biol. Chem. **205**, 565 (1953). — Bremner, J. H., and R. H. Kenten: Biochem. J. **49**, 651 (1951). — Broadbent, J. L.: Brit. J. Pharm. 8, 263—270 (1953). — Brown, E. L., and N. Campbell: J. Chem. Soc. (London) **1937**, 1699. — Buchler, C. A., and J. D. Calfee: Ind. Eng. Chem. Analyt. Edit. **6**, 351 (1934). — Bülbring, E., E. M. Lourie and U. Pardoe: Brit. J. Pharmacol. **4**, 290 (1949).

Chang, H. C., and J. H. Gaddum: J. Physiol. **79**, 255 (1933); — Chargaff, E.: J. Biol. Chem. **118**, 417—419 (1937).—Chargaff, E., and E. Vischer: J. Biol. Chem. **76**, 703 (1948). — Chargaff, E., C. Levine and Ch. Green: J. of Biol. Chem. **175**, 67 (1948). — Chen, K, K., and A. L. Chen: J. Amer. Pharmaceut. Assoc. **22**, 813 (1933). — Christensen, O.T.: J. Pract. Chem. **45**, 356 (1892).—Ciamician, G., e. C. Ravenna: Atti d. Reale accad. dei Lincei Roma **20**, I, 614 (1911). — Code, F. J.: J. Physiol. **89**, 257; **90**, 349 (1937). — Cohen, G. N., B. Nishman et M. Raynaud: C. r. Acad, Sci. Paris **225**, 647 (1947). — Colombo, E.: Bull. chim. Farnac. **89**, 129—31 (1950). — Conway, E. J., and J. Byrne: Biochem. J. **27**, 419 (1933). — Cornforth, J. W., and A. J. Henry: J. Chem. Soc. **1952**, 597. — Correale, P., u. I. Cortese: Naturwissenschaften **40**, 57 (1953). — Cotte, J., et E. Kahane: Bull. Soc. chim. Biol. **17**, 639 (1950). — Cotte, E., et E. Kahane: Bull. Soc. chim. Biol. France S. 151 (1953). — Craig, C., G. Gould, C. Shedlowsky and A. Jakobs: J. Biol. Chem. **125**, 289 (1938). — Crawford, A. C., and W. K. Watanabe: J. Biol. Chem. **19**, 303 (1914); **24**, 169 (1915). — Cromwell, B. T.: Biochem. J. **37**, 717, 722 (1943); **45**, 84 (1949); **46**, 578 (1950). — Cromwell, B. T., and S. D. Rennie: Biochem. J. **55**, 189 (1953).

Dalma, G.: Helv. Chim. Acta **22**, 1497 (1939). — Davies, D. F., K. M. Wolfe and H. M. Perry: J. of Lab. and Clin. Med. **41**, 802 (1953). — Davies, W. L., and H. C. Dowden: J. Soc. chim. Ind. Chem. & Ind. **75**, Trans. **175—179**, 26/6 (1936). — Dent, C. E.: Biochem. J. **43**, 169 (1948). — Detsch, M. G., Eggleton and P. Eggleton: Biochem. J. **32**, 203 (1938). — Deulofeu, V., E. Hug and P. Mazzocco: J. Chem. Soc. London **1939**, 1841. — Dihlmann, W.: Biochem. Z. **325**, 295 (1954); Zorn und W. Dihlmann: Dtsch. Gesundheitswesen **50**, 1522 (1953). — Drummond, J. C.: Biochem. J. **12**, 5 (1918). — Ducet, G.: Anal. Chem. Acta **2**, 839—843 (1948). — Dudley, H. W., and Rosenheim: Biochem. J. **18**, 1263 (1924). — Duquenois, P., et M. Faller: Bull. Soc. chim. France **1939**, 998—1008. — Dyer, F. E., A. J. Wood: J. Fish. Res. Bd. **7**, 17 (1947).

Eagles, B. A.: J. Amer. Chem. Soc. **50**, 1386 (1928). — Eggert, A. H.: J. Bacter. **37**, 205 (1939). — Eggert, A. H., R. J. Litteru and J. V. Deutsch: J. Bacter. **37**, 187 (1939). — Ehrismann, O., u. O. E. Werle: Biochem. Z. **318**, 560 (1947). — Emmelin, N. J.: Acta Physiol. **9**, 378 (1945). — Emmelin, N., and W. Feldberg: New Phytologist **48**, 143 (1949). — Engel, R. W.: J. Biol. Chem. **144**, 701 (1942). — Fngeland, R.: Hoppe-Seylers Z. **57**, 49 (1908). — English, F. L.: Analyt. Chem. **23**, 344 (1951). — Erspamer, V., u. G. Falconieri: Naturwissenschaften **39**, 431 (1953). — Euler, U. v. S., Erdtmann: Liebigs Ann. **520**, 1 (1935). — Ewins, A. J.: Biochem. J. **8**, 44 (1914).

Feigl, F., u. H. E. Feigl: Microchimica Acta **1954**, 85. — Fiedler, U., G. Hildebrand und R. Neu: Arzneimittelf. **3**, 436 (1953). — Fischer, E.: Ber. dtsch. chem. Ges. **35**, 1593 (1902). — Fitzpatrick, W. H.: Science **109**, 469 (1949). — Flatcher, J. P., Ch. H. Best and O. Mc. K. Solandt: Biochem. J. **29**, 2278 (1935). — Folkers, K., and F. Koniuszy: J. Am. Chem. Soc. **61**, 1232 (1939); **62**, 436, 1940.—Foster, G. L., and C. Schmidt: Proc. Soc. Exp. Biol. and Med. **22**, 19 (1921). — Fox, S. W.: J. Biol. Chem. **147**, 645 (1943).— Francois, M.: C. r. Acad. Sci. Paris **144**, 567, 857 (1907). — François, M., C. E. M. Pugh and J. H. Quastel: Biochem. J. **282** (1937). — Franzen, H., and A. Schneider: Biochem. J. **116**, 195 (1921). — Fritz, J. R.: Analyt. Chem. **22**, 1028 (1950). — Fuks, N. A., et M. A. Rappoport: C. r. Acad. Sci. U. R. S. S. **60**, 7, 1219 (1948).

Gaddum, J. H.: Gefäßerweiternde Stoffe der Gewebe. Leipzig 1936, Georg Thieme Verlag: — Galat, A., and H. L. Friedmann: J. Am. Chem. Soc. **71**, 3976 (1949). — Gale, E. F.: Biochem. J. **34**, 853 (1940); **34**, 846 (1940); **34**, 392 (1940). — Gale, E. F.: Advances in Enzymology **6**, 1 (1946). — Gaind, K. N., and S. Dutt: Bull. Acad. Agra Oudh. Allahabad **3**, 79 (1933). — Gebauer-Fuelnegg, E. u. A. J. Kendall: Ber. dtsch. chem. Ges. **64**, 1067 (1931). — Gentile, R., A. und R. Labriola: J. org. Chem. **7**, 136—39 (1942). — Gerngross, O., K. Voss und H. Herfeld: Ber. dtsch. chem. Ges. **66**, 43 (1933). — Glick, D.: J. Biol. Chem. **156**, 643 (1944). — Goris, A., et A. Larsonneau: J. Pharmacie et Chim. **23**, 475 (1921). — Gregor, R. G., and W. V. Thorpe: Biochem. J. **27**, 1394 (1933). — Grün, A., u. R. Limpächer: Ber. dtsch. chem. Ges. **59**, 1345 (1926). — Guggenheim, M.: Die biogenen Amine. Basel: S. Karger (1951). — Gulewitsch, W.: Hoppe-Seylers Z. **24**, 513 (1942).

Hamlin, K. E., and F. E. Fischer: J. Am. Chem. Soc. **73**, 5007 (1951). — Havinga, E., L. Seekles et T. Strengers: Rec. Trav. chim. Pays-Bas **66**, 605 (1947). — Heath, H., C. Rimington and C. E. Stearle: Biochem. J. **50**, 530 (1952). — Hefter R,:

Ber. dtsch. chem. Ges. 27, 1975 (1894); 29, 216 (1896). — Henry, A. J.: Brit. J. Pharmacol. 3, 187 (1948). — Henry, A. J., and F. Grindlay: J. Soc. Chem. Ind. 68, 9 (1949). — Henry, A. J., and H. King: J. Chem. Soc. 1950, 2866. — Hein, F., u. Fr. A. Segitz: Z. f. anal. Chem. 72, 119 (1927). — Herbst, R. M., and H. T. Clarke: J. Biol. Chem. 104, 769 (1934). — Herbst, E. H., and E. E. Snell: J. Biol. Chem. 176, 989 (1948); 181, 47 (1949). — Hess, W. C., E. Work and M. X. Sullivan: Lancet 1, 652 (1949). — Hirai, K.: Biochem. Z. 283, 390 (1936). — Holtz, P., u. H. Janisch: Arch. exper. Path. und Pharmakol. 187, 336 (1937); 186, 684 (1937). — Horowitz, N. H., and G. W. Beadle: J. Biol. Chem. 150, 325 (1943). — Hunter, G.: Canad. J. Res. Sect. E 27, 230 (1939). — Husemann, A., u. W. Marme: Liebigs Ann. 240, 239 (1864).

Jakobi, H. P., C. A. Baumann and W. J. Meck: J. Biol. Chem. 138, 571 (1941). — James, A. T., and G. Martin: Biochem. J. 50, 679 (1952a); 52, 238 (1952b). — Jaminet, F. R.: J. Pharm. Belg. 1, 23 (1953). — Jepson, J. B., and B. J. Stevens: Nature 172, 772 (1953). — Johnston, C. C. G., J. L. Irvin and C. Walton: J. Biol. Chem. 131, 425 (1939). — Jones, J. H., and J. Assn: Official Agr. Chem. 27, 462 (1944); Chem. Abstr. 38, 6276 (1944). — McIntre, F. C., L. W. Roth and J. L. Shaw: J. Biol. Chem. 170, 537 (1947). — McIntire, F. C., F. B. White and M. Sproull: Arch. Biochem. and Biophysics 29, 376 (1950).

Kahane, E., et J. Levy: Arch. des Sciences Physiologiques VII, 45a, 62 (1953). — Kapeller-Adler, A., and F. Vering: Biochem. J. 243, 292 (1931). — Kapfhammer, J., u. C. Bischoff: Hoppe-Seylers Z. 179 (1930). — Kaufmann, M., u. D. Vorlaender: Ber. dtsch. chem. Ges. 43, 2755 (1910). — Keen, T., and J. R. Fritz: Analyt. Chem. 25, 564 (1952). — Keil, W., u. H. Bartmann: Biochem. Z. 280, 61/64 (1935). — Kiesel, A.: Hoppe-Seylers Z. 53, 215 (1907). — Klein, G.: Handbuch der Pflanzenanalyse IV/1, 237 (1931). — Klein, G., u. D. Boser: Arch. d. Pharm. 270, 374 (1932). — Klein, G., M. Krisch, G. Pollauf u. G. Soos: Österr. Bot. Z. 80, 273 (1931). — Klein, G., u. H. Linser: Hoppe-Seylers Z. 209, 75 (1932). — Klein, G., u. G. Steiner: Jahrb. wiss. Botanik 68, 602 (1928). — Klein, G., u. A. Zeller: Österr. Bot. Z. 79, 40 (1930). — Kodicek, E., and Y. L. Wang: Nature 2, 23—24 (1941). — Koessler, K. K., and M. T. Hanke: J. Biol. Chem. 39, 521 (1919). — Kogan, L., F. Dicardo and W. E. Maynard: Analyt. Chem. 45, 1118 (1953). — Kolthoff, J. M., and H. Bendix: Industr. Engn. Chem. Anal. Ed. 2, 95 (1939). — Kossel, A., u. F. Weiss: Hoppe-Seylers Z. 68, 165 (1910). — Kossel, A., u. R. E. Gross: Hoppe-Seylers Z. 165, 967 (1924). — Krimberg, R.: Biochem. Z. 297, 261/269 (1938). — Küng, A., u. G. Trier: Hoppe-Seylers Z. 85, 214 (1913). — Kuhn, A., u. A. Gerhardt: Ber. dtsch. pharm. Ges. 281, 378—798 (1943). — Küng, H. P.: Hoppe-Seylers Z. 91, 241 (1914). — Kung, H. P., and Wei-Yuan-Huang: J. Am. Chem. Soc. 71, 1836 (1949). — Kutscher, F.: Zbl. f. Physiol. 54, 775 (1910); /26, 569 (1912).

Lagerkvist, U.: Acta Chem. Scand. 4, 543 (1950). — Langley, W. D., and A. J. Albrecht: J. Biol. Chem. 108, 729 (1935). — Lecher, H., u. Holschneider: Ber. dtsch. chem. Ges. 57, 755 (1924). — Lesure, J.: J. Pharmacie et Chim. 56, 23 (1937). — Levene, P. A., and T. Ingvaldsen: J. Biol. Chem. 43, 355 (1920). — Levy, M.: C. r. Trav. Lab. Carlsberg, Serie chim. 21, 101 (1936). — Lindigkeit, R., u. F. Jung: Pharmazie 8, 78—79 (1953). — Linneweh, W.: Hoppe-Seylers Z. 200, 115 (1931); 202, 1 (1931). — Lintzel, W., u. S. Fomin: Biochem. Z. 238, 438 (1931). — Lintzel, W., u. G. Monasterio: Biochem. Z. 241, 273 (1931). — Lovern, J. A.: Chem. and Ind. 707, 21/10 (1950). — Lowry, O. H., H. T. Graham, F. B. Harris, M. K. Priebat, A. R. Marks and R. U. Bregman: J. of Pharmacology and exper. Ther. 112, 116 (1954). — Lubschez, R.: J. Biol. Chem. 183, 731 (1950).

Mactag, A., u. R. Schoental: Mikrochemie 24, 243—250 (1938). — Malachowsky, R., u. Maslowsky: Ber. dtsch. chem. Ges. 61, 2521 (1928). — Malyoth, G., u. H. W. Stein: Biochem. Z. 322, 165—67 (1951). — Maranon, J., u. J. K. Santos: The Phillipina J. of Science 48, 564 (1932). — Marenzi, A. D., u. C. E. Cardini: J. Biol. Chem. 147, 363 (1943). — Manske, H. F.: Ann. Rev. Biochem. 13, 545 (1944). — Marquardt, P., u. G. Vogg: Hoppe-Seylers Z. 291 (1952). — Martin, W. H., M. J. Pelczar u. P. A. Hansen: Science 116, 483 (1952). — Mazza u. Stolfi: Arch. sci. biol. 16, 183 (1931). — Merck & Co. Inc. Rahway, N. J., V. St. A, (E. P. 522 225 v. 9/12. 1938 ausg. 11/7. 1940. A. Prior. 16/12. 1937 u. 27/7. 1938) C. 1941, I, 1323. — Merz, R. W., u. K. F. Bergner: Arch. pharmaz. Ges. 278, 49—70 (1940). — Metcalfe u. Sexton: Biochem. J. 45, 143 (1949). — Melville, D. B., u. R. Lubschetz: J. Biol. Chem. 200, 275 (1953). — Miller, C. C., and A. J. Lowe: J. Chem. Soc. 1940, 1263. — Minz, B.: Arch. exper. Path. und Pharm. 168, 292 (1932). — Mitchell, R., Hawkins and Smith: J. Am. Chem. Soc. 66, 782 (1944). — Mitchell, R., and B. B. Clark: Proc. Soc. Biol. and Med. 81, 105, 9 (1952). — Moores, R. G., and D. M. Greninger: Analyt. Chem. 23, 327 (1951). — Mücke, D.: Einführung in mikrobiologische Bestimmungsverfahren VEB. Georg Thieme Leipzig 1955. — Müller, H. E.: Z. physiol. Chem. 209, 207 (1932); 268, 245 (1941); 269, 31 (1941). — Müller, H. E., u. H. Sievers: Hoppe-Seylers Z. 89, 37 (1929); 92, 513 (1932). — Munier, R.: Bull. Soc. chim. Biol. 33, 862 (1951); Chem. Zbl. 1952 II, S. 5627, 1. — Munier, R., et M. Ma-

CHEBOEUF: Bull. Soc. chim. Biol. **81**, 5, 1144 (1949). — MURRAY, C. W., and E. W. WALTER: J. Amer. Chem. Soc. **67**, 1422 (1945).

NEUBERG, C., u. L. KARCZAG: Biochem. Z. **18**, 435 (1909). — NEUBERG, K.: Biochem. Z. **29**, 434 (1910). — NEUBERGER, A., and S. SANGER: Biochem. J. **86**, 662 (1942). — NICOLET, B. H., and L. A. SHINN: J. Am. chem. Soc. **61**, 1615 (1939). — NOTTBOHM, F. E., u. F. MAYER: Z. f. Unters. Lebensmittel **61**, 202 (1931); **69**, 289 (1935).

OPFER-SCHAUM, R.: Microchimica Acta **82**, 148 (1944). — ORMSBY, A. A., and S. JOHNSON: J. Biol. Chem. **187**, 711 (1950).

PALUMBO, M.: Farm. sci. Pavia **8**, 675 (1948), Chem. Abstr. **48**, 3115 (1949). — PARTRIDGE, M. S.: Biochem. J. **42**, 238 (1948). — PARTRIDGE, S. M.: Analyt. Chem. **77**, 955 (1952). — PAULY, H.: Hoppe-Seylers Z. **42**, 508 (1904); **94**, 284 (1915). — PENAU, H., E. SAIS et C. ANDRETTI: Ann. pharmac. franc. **10**, 514/29 (1952). — PETERS, R. A.: Biochem. J. **54**, 1852 (1933). — PICTET, A., u. G. COURT: Ber. dtsch. chem. Ges. **40**, 3771 (1907). — POLUEKTOFF, N. S.: Kalic USSR, **2**, Nr. 10, 44 (1933). — PRATT, A. E.: J. Biol. Chem. **67**, 351 (1926).

RAFFAELE, G.: Pharmacista italiano Suppl. 7, 46—47 (1939). — GENTILE, R. A., and R. LABRIOLA: J. Am. Chem. Soc. **61**, 1232 (1939); **61**, 3052 (1939); **62**, 436 (1940). — RAO, P., R. O. C. SURYBACRAS, T. R. VENKATA and T. R. SESHADRI: Progr. Ind. Acta Sciencia A 7, 179 (1938). — REAY, G. A.: Rep. Food Invest. Board (London) **1987**, 69. — REDEMANN, C. E., and C. NIERMANN: J. Am. Chem. Soc. **62**, 590 (1940). — RETI, L.: The Alkaloids III. Acad. Press. Inc. 1953; Biochem. Z. **280**, 61 (1936). — REUTER, C.: Hoppe-Seylers Z. **78**, 167 (1912). — RIELÄNDER, G.: Sitzgsb. Ges. Naturwiss. Marburg 5. 8. 1908, 173. — ROBERTS, M., and H. M. ADAM: Brit. J. Pharmacol. and Chemotherp. **5**, 526 (1950). — ROCHE, J., W. FELIX, Y. ROBIN et N. v. THOAI: C. r. Acad. Sci. Paris **288**, 1688 (1951). — ROGGEN, J. C.: Ber. wiss. Biol. **60**, 260 (1943). — ROMAN, W.: Biochem. Z. **219**, 218 (1930). — ROMBURG, H.: Kon. Akad. Wetensch. Amsterdam 1258 (1911). — RONALD and JAKOBSOHN: J. Soc. Chem. Ind. **66**, 160 (1947). — ROSENTHAL, S. M., and H. TABOR: J. of Pharmacology and exper. Ther. **92**, 425 (1948).

SAMUELSON, G.: J. of Pharmacy and Pharmacology V, Nr. 4, 239 (1953). — SANGER, F.: Biochem. J. **39**, 507 (1945). — SARRET, H. P.: J. Biol. Chem. **150**, 159—64 (1943). — SASAKI, S.: Bull. Agr. chem. Soc. Japan **9**, 87 (1933). — SCHAYER, R. W., Y. KOBAYASHI and R. L. SMILEY: J. Biol. Chem. **212**, 593 (1955). — SCHLEMMER, F.: Arch. Pharmazie 270 (1932) — SCHÖNBERG, A., and W. I. AWAD: J. Chem. Soc. (London) **1949**, 766. — SCHOORL, N.: Pharm. Weekbl. **864** (1918). — SCHULZE, E. und G. TRIER: Hoppe-Seylers Z. f. physiol. Chem. **67**, 60 (1910). — SCHULTZE, E., u. G. TRIER: Hoppe-Seylers Z. **60**, 155 (909); **76**, 258 (1912); **81**, 53 (1912). — SCHWARTZ u. RIEGERT: C. r. Soc. Biol. **128**, 219 (1936). — SCHWYZER, R.: Acta Chem. Scand. **6**, 219 (1952). — SLYKE, VAN D. D., R. T. DILLON, D. A. MC FAYDEN and P. HAMILTON: J. Biol. Chem. **141**, 627 (1941). — SMITH, I., and A. C. CHIBNALL: Biochem. J. **26**, 1345 (1932). — SNEATH, P. H. A.: Nature **175**, 818 (1955). — STANEK, V.: Hoppe-Seylers Z. **46**, 290 (1905); **48**, 334 (1905); **72**, 402 (1911). — STANEK, V.: Hoppe-Seylers Z. **48**, 334 (1906); **72**, 402 (1911). — STANLEY, E. L., and R. V. SAVACOOL: Analyt. Chem. **21**, 312 (1949). — STEINER, M.: Handbuch der Pflanzenanalyse IV/2,1608(1953). — STEINER, M., u. E. STEIN v. KAMIENSKI: Naturwissenschaften **40**, 483 (1953). — STEINER, M., E. STEIN v. KAMIENSKI: Naturwissenschaften **42**, 345 (1955). — STEVENSON, E., and ROWATT: J. Gen. Microbiol. **1**, 279 (1947). — STEWARD, F. C., and H. E. STREET: Plant. Physiol. **21**, 155 (1946). — STRACK, E., u. H. SCHWANEBERG: Hoppe-Seylers Z. **245**, 11 (1937); **245**, 11 (1936). — STREET, H. E., A. E. KENYON u. G. M. WATSON: Biochem. Z. **40**, 869—74 (1946); Ann. Appl. Biol. **33**, I(1946).

THIERFELDER, H. S., u. O. SCHULTZE: Hoppe-Seylers Z. **96**, 296 (1915). — TOWER, A. A., and MCEACHERN: Can. J. Res. **26**, 183 (1947). — THIES, H., u. F. W. REUTHER: Naturwiss. **41**, 230 (1954).

ULLMANN, A.: Biochem. Z. **128**, 4026 (1922). — URBACH, K. F.: Proc. Soc. Exper. Biol. and Med. N. Y. **68**, 430 (1948); **70**, 146 (1949).

VICKERY, H. B.: J. Biol. Chem. **65**, 81 (1925); **143**, 77 (1942).

WACEK, A., u. H. LÖFFLER: Mikrochemie 18, (N. F. 12) 277—82 (1935). — WAGNER, CH. D., R. H. BROWN and E. D. PETERS: J. Am. Chem. Soc. **69**, 2609 (1947). — WALKER, H. G., and R. ERLANDSEN: Analyt. Chem. **28**, 1309(1951). — WANAG, G. u. A. LODE: Ber. dtsch. chem. Ges. **69**, 1066 (1936); **70**, 547 (1937). — WANAG, G.: J. Anal. Chem. **113**, 21 (1938). — WANAG, G.: Z. f. anal. Chem. **119**, 413—417 (1940). — WANAG, G., u. A. DOMBROWSKY: Ber. dtsch. chem. Ges. **75**, I, 82 (1942). — WELSH, L. H.: Bull. Johns Hopkins Hosp. **83**, 568 (1948). — WERLE, E.: Biochem. Z. **809**, 61 (1941); Angew. Chemie **64**, 311 (1952). — WERLE, E., u. A. RAUB: Biochem. Z. **818**, 538 (1944). — WEST, G. B., and

J. F. Riley: Nature 174, 882 (1954). — Whittaker, V. P., u. S. Wijesundera: Biochem. Z. 49, XIV (1951). — Wieland, Th., C. Fischer u. F. Moewus: Liebigs Ann. 561, 47 (1948). — Wiley, J., and Sons: Organic Synthetics Coll. Vol. I, p. 757. New York, 1943. — Williams, R. J., and H. F. Kirby: Science 107, 481 (1948). — Willstätter, R.: Ber. dtsch. chem. Ges. 35, 270, 584 (1901). — Wilson, K. W., F. E. Anderson and R. W. Donohoe: Analyt. Chem. 23, 1032 (1951). — Winterfeld, K., u. A. Kronenthaler: A. d. Pharm. 280, 103 (1942). — Winterfeld, K., u. E. Weiss: Pharmazeutische Ztg. 1952, Seite 573. — Wolfes, O., H. Kreitmair u. W. Sieckmann: Mercks Jahresber. 111, 15 (1936). — Wood, E. C.: Analyst 72, 84 (1947).

Yamada, and Ishida: J. Agric. Chem. Soc. Japan 2, Nr. 7, Seite 1 (1926). — Yoshimura, K.: Biochem. Z. 28, 16 (1910). — Yoshima, K.: Biochem. Z. 274, 408 (1934). — Yurashewsky: J. Gen. Chem. (USSR) 9, 595 (1939); 10, 781 (1940); 11, 157 (1941).

Zeidler, L.: Hoppe-Seylers Z. 291, 177 (1952). — Zincke, Th., u. Fr. Farr: Liebigs Ann. 391, 57 (1912).

Pantothensäure und Coenzym A.

Von

E. Werle.

Mit 3 Abbildungen.

Pantothensäure ist ein von jeder lebenden Zelle benötigter Wirkstoff von Vitamincharakter und daher in Bakterien, Pilzen, Algen und höheren Pflanzen enthalten. Ihre außerordentliche Bedeutung hat die Pantothensäure als Bestandteil des Coenzyms A, das zur Übertragung der Acetylgruppe, aber auch von anderen Säureresten befähigt ist und in diesem Zusammenhang eine zentrale Stellung im Stoffwechsel der Fette, Eiweiße und Kohlenhydrate und anderer funktionell wichtiger Substanzen einnimmt (Übersicht bei LIPMANN, 1954, LYNEN 1951, 1952—1953, WIELAND, 1954). Die Pantothensäure liegt daher in den Zellen der Pflanzen zum größten Teil nicht in freier, sondern in gebundener Form vor (NOVELLI, KAPLAN und LIPMANN, 1949), und muß zur quantitativen Bestimmung aus diesen Bindungen freigelegt werden. Außer im Coenzym A findet sich die Pantothensäure in Zwischenstufen der Synthese, die von der Pantothensäure zum Coenzym A führen. Von diesen ist der sog. *Lactobacillus bulgaricus-Faktor* (L. B. F.), ein Wuchsstoff für den *Lactobacillus bulgaricus*, am besten untersucht (SNELL und BROWN, 1953). Die Beziehungen des L. B. F. zur Pantothensäure und zum Coenzym A sind in dem folgenden Schema wiedergegeben, das auch Hinweise enthält, die erkennen lassen, wie die Pantothensäure aus ihren Bindungen gelöst werden kann (SNELL und BROWN, 1953). (Zur Struktur von Coenzym A s. auch J. BADDILEY, 1955.) Der L. B. F. ist in Pflanzenmaterial (WILLIAMS, HOFF-JORGENSEN und SNELL, 1949) und in Bakterien-Kulturfiltraten (RASMUSSEN, SMILEY, ANDERSON, VAN LANEN, WILLIAMS und SNELL, 1951; LONG, WILLIAMS, 1951; VITUCCI, BOHONOS, WIELAND, LEFEMINE, HUTCHINGS, 1951) enthalten.

Coenzym A und seine Vorstufen.

Coenzym A

$$\text{HOCH}_2\text{C(CH}_3)_2\text{-CHOH-CO-NHCH}_2\text{CH}_2\text{CO-NHCH}_2\text{CH}_2\text{SH} \rightarrow \text{HOCH}_2\text{C(CH}_3)_2\text{-CHOH-CO-NHCH}_2\text{CH}_2\text{CO-NHCH}_2\text{CH}_2\text{SR}_1$$

L.B.F.

L.B.F.—1A

$R_1 =$ eine Alkylgruppe von niedrigem Molekulargewicht

Leberenzym

$$2 \text{ Pantethein} \xrightarrow[+ 2\text{H}]{- 2\text{H}} \text{Pantethin} \rightarrow \begin{array}{c} \text{R--S} \\ | \\ \text{R--S} \end{array}$$

R = L.B.F. Rest.

$$\text{HOCH}_2\text{C(CH}_3)_2\text{-CHOH-CO-NHCH}_2\text{CH}_2\text{COOH} + \text{NH}_2\text{CH}_2\text{CH}_2\text{SH}$$

Cysteamin

Pantothensäure

$$\text{HOCH}_2\text{C(CH}_3)_2\text{-CH}_2\text{COOH} + \text{NH}_2\text{CH}_2\text{CH}_2\text{COOH}$$

β-Alanin

Pantoinsäure

Bakterien-Kulturfiltrate sind die besten Quellen für seine Gewinnung. Besonders ausgiebig sind die von *Ashbya gossypii* und *B. subtilis* (VITUCCI, BOHONOS, WIELAND, LEFEMINE, HUTCHINGS, BROWN, WILLIAMS und SNELL, 1953).

A. Pantothensäure.

Pantothensäure $C_9H_{17}N_5O$, Mol.-Gew. 219,2.

Konstitution: d-α-Dioxy-β-β-dimethylbutyryl-β-alanin. Blaßgelbes visköses Öl, das auch bei tiefer Temperatur nicht kristallisiert, äußerst leicht löslich in Wasser und Alkohol (mit saurer Reaktion) $\alpha_{26}^D = + 37,5°$. Beständig gegen Licht und Sauerstoff, jedoch in freier Form und alkalischer Lösung (pH höher als 8) empfindlich gegen Erhitzen.

Salze: Natriumpantothenat $C_9H_{16}NO_5Na$ (Fp. 120—31°) $\alpha_{20}^D = + 26°$. Feine weiße verfilzte Nadeln oder Knollen, hygroskopisch, leicht löslich in Wasser (reagiert alkalisch), unlöslich in Äther; thermostabil.

Calciumpantothenat: $C_{18}H_{32}N_2O_{10}Ca$, Mol.-Gew. 476,5 (Fp. 195—196°) $\alpha_{26}^D = + 24,3°$, empfindlich gegen Säure, Alkali und Hitze. In Wasser, Äthylacetat und Eisessig leicht löslich, in Äther und Amylalkohol wenig löslich. In Benzol und Chloroform unlöslich.

Bestimmungsmethoden: (SNELL, E. E.: Vitamin Methods, Vol. I. New York: Acad. Press. Inc. 1951) D. MÜCKE Einführung in mikrobiologische Bestimmungsverfahren, VEB Thieme Leipzig 1955.

Die Pantothensäure kann chemisch, fermentchemisch, mikrobiologisch und im Wachstumstest an höheren Tieren (Ratte oder Hühnchen) bestimmt werden.

Mikrobiologisch wird nur die freie Pantothensäure, im Wachstumstest dagegen die Gesamtpantothensäure, also auch die gebundene ohne vorherige

Hydrolyse erfaßt. Von den im folgenden zu beschreibenden Bestimmungsmethoden für Pantothensäure werden nur diejenigen genau wiedergegeben, die zur Bestimmung in natürlichen Materialien geeignet sind.

I. Mikrobiologische Bestimmung.

Zur Pantothensäurebestimmung sind eine Reihe von Mikroorganismen so *Proteus morganii* (Pelczar und Porter, 1941) *Lactobacillus arabinosus* (Skeggs und Wright, 1944), *Lactobacillus helveticus* (Strong und Neal, 1943), *Lactobacillus* casei, *Neurospora* Mutanten und Hefe *(Saccharomyces carlsbergensis)* und andere herangezogen worden (Hoag, Sarret und Cheldelin, 1945). Die besten Resultate wurden mit *Lactobacillus arabinosus* erzielt.

1. Behandlung des zu untersuchenden Materials mit einer Mischung von alkalischer Phosphatase und Leberenzym.

Die Pantothensäure wird nur zum geringen Teil bei der Gewebsautolyse (Farrer, 1951), aber vollständig durch Zusatz von gewissen Enzymen freigelegt, am besten durch eine Mischung eines Leberenzympräparates und einer a'kalischen Phosphatase. Auch ein als „Mylase P" bezeichnetes käufliches Fermentpräparat (Wallerstein Laboratories, New York) ist zur Freilegung der Pantothensäure vielfach verwendet worden (Buskirk, Bergdahl und Delor, 1948).

Gewinnung der alkalischen Phosphatase aus Kalbsdünndarm nach Schmidt und Thannhauser, 1943. Es werden die ersten 2 m des Dünndarms vom Pylorus abwärts verwendet. Die Mucosa wird abgeschabt, in Wasser bei pH 9 verrührt und unter Zusatz von Trypsin unter Toluol 1—2 Wochen bei 37° C verdaut. Aus dem Verdauungsansatz wird die Phosphatase durch 0,9 Sättigung mit $(NH_4)_2SO_4$ niedergeschlagen. Nach dem Abzentrifugieren wird der Niederschlag in Wasser aufgenommen. Das Ferment wird nochmals durch 0,8 Sättigung mit Ammoniumsulfat niedergeschlagen, in Wasser aufgenommen und dann gegen Ammoniumacetatpuffer von pH 8,8 dialysiert. Weitere Verunreinigungen werden durch Adsorption an Aluminiumhydroxyd abgetrennt, Trypsin wird durch Adsorption an Kaolin entfernt. Dann wird das Enzym aus der Lösung mit eiskaltem Aceton gefällt, der Niederschlag in Wasser aufgenommen und die Fällung mit Aceton wiederholt. Das Enzym wird in Wasser aufgelöst und gegen Ammoniumacetatpuffer dialysiert. Die wäßrige Lösung ist monatelang im Eisschrank unverändert haltbar. Die Aktivität der erhaltenen Lösung ist verschieden hoch. Ein von Neilands und Strong (1948) verwendetes Präparat hatte 1150 Phosphataseeinheiten/mg Stickstoff; 1484 Einheiten/ml. Das Präparat kann auch von Armour Comp. Chicago bezogen werden. Eine 2%ige Lösung dieses Fermentpulvers enthält 20—30 Phosphataseeinheiten/ml.

Bereitung des Leberenzyms (Kaplan und Lipmann, 1948). Die Lebern frisch getöteter 1 Monate alter Küken werden abgekühlt und mit 20 Vol. Aceton von 0° 1 min lang homogenisiert. Das Aceton wird abgesaugt, wobei der Niederschlag nicht lufttrocken werden darf. Das Aceton wird durch Waschen des Niederschlages mit kaltem peroxydfreiem Äther entfernt und das ätherfeuchte Material im Vakuumexsiccator in der Kälte getrocknet. Nach dem Pulverisieren wird Bindegewebe mit Hilfe eines Siebes entfernt. Das Präparat ist in der Kälte mindestens 6 Wochen ohne Verlust haltbar. Zum Gebrauch wird es mit dem 10fachen seines Gewichts kalter, frisch bereiteter 0,02 molarer $NaHCO_3$-Lösung extrahiert, vom Ungelösten wird abzentrifugiert; das Enzym wird in 1 oder 2 ml

Portionen eingefroren und unmittelbar vor Anwendung bei Raumtemperatur aufgetaut. Durch Behandlung des Fermentes mit Ionenaustauschern (Dowex 1) werden gebundene Formen der Pantothensäure entfernt, wodurch der Korrekturwert
für das Fermentpräparat bei der Pantothensäurebestimmung stark herabgesetzt
wird (NOVELLI und SCHMELZ, 1951). Zur Freilegung der Pantothensäure geben
NEILANDS und STRONG (1948) zu 1—40 mg des zu untersuchenden homogenisierten
Pflanzenmaterials 0,05 ml Leberenzymlösung, 0,2 ml der Darmphosphataselösung,
0,05 ml 0,1 mol. Bicarbonatlösung und füllen auf 1 ml auf. Die Gläser werden bei
37° 4 Std. lang inkubiert. Das zu untersuchende Material soll etwa 5 γ Pantothensäure enthalten. Ist störendes Fettmaterial vorhanden, so wird in der oben
beschriebenen Weise weiter verfahren. In einer Kontrollbestimmung wird jeweils
der Pantothensäuregehalt der zugesetzten Enzymlösung bestimmt. Ob die
Mischung des Leberenzyms und der alkalischen Phosphatase in rohen Pflanzenhomogenaten die ganze vorhandene Pantothensäuremenge, wie sie z. B. der
Kükenversuch ergibt, freilegt, ist noch nicht entschieden. Jedenfalls stimmen
nach HEGSTEDT und LIPMANN (1948) die Werte für den Pantothensäuregehalt
von Coenzym A-Präparaten, die nach Einwirkung der Phosphatase und des Leberenzyms im mikrobiologischen Test erhalten wurden, überein mit denjenigen, die
der Kükenversuch bei i.peritonealer Verabreichung des Coenzyms ergeben hatte.

Nach HARRISON (1949) ist es in gewissen Fällen möglich, die Pantothensäure-
ausbeute durch Behandeln mit kaltem Alkali vor der Enzymeinwirkung um
200—400% zu erhöhen. Die höchsten Werte (z. B. für Malz) wurden nach folgendem Verfahren erhalten: 1 g Material wird 30 min lang in 0,25 ml 0,1 mol. Natriumacetatlösung mit Dampf behandelt. Dann wird abgekühlt, 5 ml 10 n Natriumhydroxydlösung hinzugefügt und 5 min bei Zimmertemperatur belassen. Dann
wird mit Schwefelsäure auf pH4,5 eingestellt und 0,5 g Papain und 0,5 g Takadiastase von bekanntem Pantothensäuregehalt werden zugefügt. Es wird nun 3 Tage
bei 37° unter Toluol inkubiert, dann wird abfiltriert, auf pH6,5 eingestellt und auf
ein bestimmtes Volumen aufgefüllt. Nach dieser Aufbereitung ergab die acidi-
metrische mikrobiologische Bestimmung der Pantothensäure mit *Lactobacillus
arabinosus* einer Malzprobe 13 γ/g gegenüber 5 γ/g, bei üblicher Behandlung. Zur
Freilegung der Pantothensäure geben BUSKIRK, BERGDAHL und DELOR (1948)
folgende Vorschrift: In einen 50 ml Kolben werden zu 0,4 g des fein gepulverten
Materials 0,4 g „Mylase P" (die Menge hängt ab von der Aktivität) gegeben, 10 ml
2%iger Essigsäure und 1 ml n Natronlauge, wodurch auf pH 4,2—4,5 gepuffert
wird. Liegen Flüssigkeiten oder Suspensionen vor, so muß jedes organische
Lösungsmittel wie z. B. Alkohol entfernt werden. Aliquote Anteile, die etwa 0,4 g
des Materials enthalten, werden auf 10 ml verdünnt oder konzentriert, mit 0,2 ml
Eisessig, 1 ml n NaOH und 0,4 g „Mylase P" versetzt. Die Mischung wird
2—3 Std. bei 50° belassen. Dann werden die Proben mit dest. Wasser auf 250 ml
aufgefüllt und durch Whatmann-Papier Nr. 2 filtriert. Ein aliquoter Teil des
klaren Filtrates wird so verdünnt, daß die Flüssigkeit 0,02—0,2 γ Pantothensäure
pro ml enthält. Zur Berechnung der Resultate ist es notwendig, die Pantothensäure, die in „Mylase P" enthalten ist, zu berücksichtigen. Meist ist ihr Wert
vernachlässigbar gering, muß aber in Rechnung gestellt werden, wenn der Gehalt
des zu untersuchenden Präparates selbst sehr niedrig ist. Das Wachstum der zur
Pantothensäurebestimmung benützten Bakterien, insbesondere von *L. arabinosus*
wird stark beeinflußt durch Fettsäuren, Lezithin usw. in dem zu bestimmenden
Material; es kann in Abhängigkeit von der Konzentration dieser Stoffe in Gegenwart suboptimaler Mengen von Pantothensäure stark angeregt oder gehemmt werden. Um diese Effekte auszuschließen, muß das verdaute Medium bei pH 4,5—4,8
durch Papier filtriert werden. Meist genügt die Filtration allein.

2. Bestimmung der Pantothensäure mit *Lactobacillus arabinosus* 17—5 ATCC Nr. 8014[1].

(Skeggs und Wright 1944[2]).

Vorbemerkung zur Technik der mikrobiologischen Wirkstoffbestimmung. Mikrobiologische Bestimmungen können nur mit Organismen ausgeführt werden, deren Ansprüche auf die verschiedenen Nahrungsbestandteile auch bei längerer Züchtung sich nicht ändern. Die für den Test verwendete Nährlösung muß alle für das Wachstum des betreffenden Mikroorganismus notwendigen Stoffe in optimaler Konzentration enthalten mit Ausnahme des zu bestimmenden Vitamins, so daß weitere Zugaben das Wachstum nicht beschleunigen können. Die Zugabe steigender Mengen der zu bestimmenden Substanz muß in einem bestimmten Konzentrationsbereich einen gut abgestuften Wachstumseffekt hervorrufen, so daß eine steile Standardkurve resultiert.

Ausführung der Bestimmung. Aus der Stammkultur des zu verwendenden Mikroorganismus wird eine Impflösung hergestellt, die je nach Vorschrift 18 bis 24 Std. bebrütet wird. Sodann wird eine größere Anzahl von Versuchsgläsern mit dem gleichen Volumen Nährlösung (die Nährlösung enthält alle für das Wachstum des betreffenden Mikroorganismus notwendigen Bestandteile mit Ausnahme des zu bestimmenden Vitamins) und mit abgestuften Mengen des reinen zu bestimmenden Vitamins bzw. des zu analysierenden Extraktes beschickt. Es werden jeweils mindestens 3 Haupt- und 3 Leerversuche angesetzt. Alle Ansätze werden auf das gleiche Endvolumen aufgefüllt, im Autoklaven sterilisiert und nach dem Abkühlen bei Zimmertemperatur beimpft. Die Vorschriften für die Bereitung der Impflösung und der Menge des zu verwendenden Impfgutes (je 1 Tropfen) müssen genau eingehalten werden. Nach ein- bis mehrtägiger Bebrütung kann bei Bakterien oder Hefe der Trübungswert der Suspension oder bei Milchsäurebakterien die gebildete Milchsäure titriert werden. Bei Pilzen wird das Trockengewicht des Mycels bestimmt. Die quantitative Auswertung wird anhand des Vergleichs der Wachstumswirkung von Standard und Extraktdosen vorgenommen (s. Abb. 1).

Behandlung der Stammkultur. L. arabinosus wird auf Schrägagar folgender Zusammensetzung gezüchtet: 1% Glucose, 1% Hefeextrakt (Difco), 1,5% Agar. Die Stammkultur wird monatlich weiter geimpft, unmittelbar nach der Beimpfung wird die Kultur 24 Std. bei 37° bebrütet und bis zur nächsten Überimpfung bei 4° aufbewahrt.

Bereitung von Extrakten des auf Pantothensäure zu untersuchenden Materials. 1 g des zerkleinerten Materials wird in 40 ml 0,5% Acetatpuffer von pH 4,5 aufgenommen, mit je 20 g Papain und Takadiastase versetzt und während 24 Std. bei 37° unter Toluol belassen. Zur Inaktivierung der zugesetzten Enzyme wird der Extrakt anschließend 30 min lang strömendem Dampf ausgesetzt, dann auf pH 6,8 eingestellt, und so verdünnt, daß pro ml etwa 0,02 γ Pantothensäure enthalten sind. Fetthaltiges Material (z. B. Weizenkeime) wird vor der Neutralisation mit Äther ausgeschüttelt.

Bereitung der Impflösung. Als Impflösung dient Pantothensäurenährlösung, welcher pro ml 0,2 γ Pantothensäure zugesetzt worden sind. Die Lösung wird 10 min bei 115° sterilisiert. Die Nährlösung hat folgende Zusammensetzung: Glucose 20 g, Natriumacetat (wasserhaltig) 10 g, Natriumchlorid 5 g, Peptonlösung 200 ml, Caseinhydrolysat 25 ml, 1-Cystinlösung 25 ml, Adenin + Uracil-Lösung 10 ml. Lactoflavinlösung 4 ml, Ca-Pantothenatlösung 1 ml, Pyridoxin-

---

[1] Reine Kulturen können bezogen werden von der American Type Culture Collection (ATCC 2029 M. Street, N. W. Washington 6, D. C.).

[2] Dargestellt nach O. Wiss, Ztschr. f. Vitamin-, Hormon- und Fermentforschg. 4, 191 (1951).

lösung 1 ml, Nicotinsäureamid 1 ml, Biotinlösung 4 ml, p-Aminobenzoesäurelösung 1 ml, Folsäurelösung (1 : 5 verdünnt) 0,5 ml, Salzlösung A 5 ml, Salzlösung B 5 ml. Die frisch bereitete Lösung wird auf 100° erhitzt, filtriert und mit 40%iger Natronlauge auf pH 6,5 eingestellt. Es wird angeimpft und 24 Std. bei 33° inkubiert, die Zellen werden auszentrifugiert und in steriler physiologischer Salzlösung resuspendiert. Ein Anteil dieser dicken Suspension wird mit steriler Salzlösung wieder weiter verdünnt, um eine leicht getrübte Suspension zu erhalten. Davon wird 1 Tropfen verwendet zur Impfung in jedes mit dem Grundmedium beschickte Versuchsglas.

Bereitung der für die Pantothensäurebestimmung notwendigen Reagentien.

l-Cystinlösung. 2 g l-Cystin werden in etwa 50 ml Wasser suspendiert und gekocht. Durch Zusatz von 2,5 ml konzentrierter Salzsäure und durch längeres Erhitzen geht das Cystin in Lösung. Das Volumen wird auf 500 ml mit Wasser ergänzt. Haltbarkeit unter Toluol bei Zimmertemperatur: einige Monate.

dl-Tryptophanlösung: 2 g dl-Tryptophan werden mit etwa 50 ml Wasser zum Sieden erhitzt und tropfenweise bis zur vollständigen Lösung mit konzentrierter Salzsäure versetzt. Mit Wasser auf 500 ml ergänzen. Haltbarkeit unter Toluol im Eisschrank: mehrere Monate.

Adenin + Guanin + Uracil-Lösung: Je 50 mg Adenin, Guanin und Uracil werden in 50 ml Wasser suspendiert und nach Zusatz von 15—20 Tropfen konzentrierter Salzsäure bis zur vollständigen Lösung auf dem siedenden Wasserbad gehalten. Nach Abkühlen auf Zimmertemperatur wird das Volumen mit Wasser auf 50 ml ergänzt. Haltbarkeit im Eisschrank unter Toluol: 14 Tage.

10 mg Nicotinsäure werden in 100 ml Wasser gelöst.

10 mg Pyridoxin werden in je 100 ml Wasser gelöst.

Haltbarkeit im Eisschrank: 8 Tage.

Lactoflavinlösung (Vitamin B_2). Es werden genau 25 mg Lactoflavin eingewogen, im Litermeßkolben in wenig Wasser suspendiert, mit 1 ml Eisessig versetzt und auf 1000 ml Wasser aufgefüllt. Die Lösung wird unter Toluol im Eisschrank im Dunkeln aufbewahrt und ist 14 Tage haltbar.

Aneurinlösung (Vitamin B_1). 10 mg Aneurin werden in wenig Wasser gelöst und das Volumen mit 0,2% Salzsäure auf 100 ml ergänzt. Haltbarkeit im Eisschrank: einige Wochen.

Biotinlösung. Der Inhalt einer Ampulle (25 γ Biotin) wird mit 1 ml Salzlösung A versetzt und auf 250 ml mit Wasser verdünnt. Haltbarkeit im Eisschrank unter Toluol: etwa 3 Monate.

p-Aminobenzoesäurelösung. 10 mg p-Aminobenzoesäure werden in 100 ml Wasser suspendiert und zur vollständigen Lösung 3—5 Tropfen Eisessig zugegeben. Haltbarkeit unter Toluol im Eisschrank: 8 Tage.

Salzlösung A. 25 g KH_2PO_4 und 25 g K_2HPO_4 werden in 250 ml Wasser gelöst. Haltbarkeit im Eisschrank unbeschränkt.

Salzlösung B. 10 g $MgSO_4 \cdot 7 H_2O$, 0,5 g NaCl, 0,5 g $FeSO_4 . 7 H_2O$ und 0,5 g $MnSO_4 \cdot 4 H_2O$ werden in 250 ml Wasser gelöst. Um ein Ausfallen von Salzen zu verhindern, werden 5 Tropfen konzentrierter Salzsäure zugegeben. Haltbarkeit im Eisschrank unbeschränkt.

Versuchsansätze. Reagenzgläser von etwa 2 cm Durchmesser und mindestens 15 cm Länge werden mit 5 ml Nährlösung beschickt, dann wird der zu untersuchende Extrakt bzw. der Vergleichsstandard zugegeben und mit Wasser auf 10 ml aufgefüllt. Die zugesetzten Standardlösungen enthalten 0,01; 0,02; 0,04; 0,06; 0,08; 0,1; 0,12; 0,15 γ Calciumpantothenat. In den Hauptversuchen werden 1,0, 2,0 und 4,0 ml Extrakt zugegeben. Sind die Extrakte gefärbt oder getrübt, so werden gleichwertige Versuchsansätze mitgeführt, die jedoch nicht beimpft werden. Starke Färbung oder starke Trübung macht die Bestimmung unmöglich. Die Ansätze werden während 15 min im strömenden Dampf sterilisiert, dann mit 1 Tropfen der bereiteten Impfsuspension beimpft und 3 Tage bei 37° bebrütet.

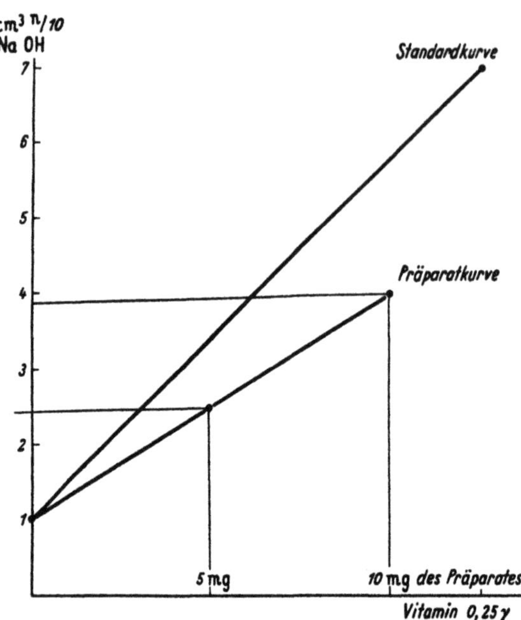

Abb. 1. Auswertung der Pantothensäure-Bestimmung mit Lactobacillus arabinosus. (WISS, 1951).

Auswertung. Zur Auswertung wird die Standardkurve auf Millimeterpapier aufgetragen. Auf der Abszisse werden die Standarddosen auf der Ordinate die Mittelwerte des korrespondierenden Trübungswertes, Titrationswertes oder Mycel-Gewichts aufgetragen. Bei gekrümmtem Verlauf der Kurve wird der Gehalt der einzelnen Dosen direkt auf der Kurve abgelesen. Bei einer einwandfreien Bestimmung sollten die Werte innerhalb der Standardkurve liegen und der errechnete Gehalt der einzelnen Dosen nicht mehr als 10% voneinander abweichen. Eine im Bereich der Extraktdosen gerade Standardkurve ermöglicht eine genaue Auswertung. Der gerade verlaufende Teil der Kurve wird bis zur Ordinate verlängert, die 3 Extraktdosen werden eingetragen und ebenfalls durch eine Gerade verbunden. Gemeinsame oder nur wenig voneinander abweichende Schnittpunkte der beiden Kurven mit der Ordinate sind Voraussetzungen für ein genaues Ergebnis (s. Abb. 1).

3. Pantothensäurebestimmung mit Hilfe von *Saccharomyces Carlsbergensis* 4228, ATCC Nr 9080[1].

Methoden zur Pantothensäurebestimmung mit Hilfe von Hefe hatten sich früher deswegen nicht eingebürgert, weil β-Alanin den gleichen stimulierenden Effekt ausübt, wie Pa-S. Dieser Effekt wird sehr verstärkt durch Entfernen von Asparagin aus dem Nährmedium und wird weitgehend unterdrückt durch Zufügung von Asparagin (WILLIAMS und ROHRMANN, 1936) und (WILLIAMS und SAUNDERS, 1934)[1]. In den heutigen Verfahren zur Pa-S-Bestimmung mit *Saccharomyces carlsbergensis* wird die Wirkung von β-Alanin für alle praktischen Zwecke völlig eliminiert durch hohe Konzentrationen von Asparagin. Zugefügte Pantothensäure wird zu 97—116% wiedergefunden, mit einem Durchschnitt von 103,6%. Die Reproduzierbarkeit der Methode ist ausgezeichnet. So wurde in 10 getrennten Versuchen mit einer getrockneten Hefe im Durchschnitt 166,3 γ Pantothensäure/g mit einer durchschnittlichen Abweichung von 3,2% und einer Standardabweichung von 3,8% wiedergefunden.

Durchführung der Pantothensäurebestimmung mit Hefe nach ATKIN, WILLIAMS, SCHULTZ und FREI (1944). Die Hefe wird auf Schrägkulturen auf Difco-Malzagar vorrätig gehalten und wird monatlich übertragen. Nach der Überimpfung wird die Kultur 24 Std. bei 30° im Kühlschrank aufbewahrt. Jeweils einen Tag vor dem Ansatz einer Messung wird eine frische Überimpfung durchgeführt und die Hefe 24 Std. inkubiert. Die Hefe wird dann vom Schrägagar in ein Glas mit steriler Salzlösung übergeführt, bis die Lichtabsorption eine Konzentration von 3 mg Hefe/ml anzeigt. Die Kalibrierung wird mit Hilfe einer Suspension feuchter Bäckerhefe vorgenommen. 10 ml dieser Suspension werden in einem Erlenmeyer-Kolben zu 90 ml steriler Salzlösung gegeben. Diese Suspension enthält 0,3 mg feuchter Hefe/ml und ist für den Versuch fertig. Das Grundmedium für die Pa-Messungen wird durch Mischung einer Reihe von Stammlösungen, welche die folgenden Bestandteile enthält, bereitet:

1. Zucker- und Salzlösung: 1 l enthält 200 g reinste Dextrose, 2,2 g KH_2PO_4, 1,7 g KCl, 0,5 g $MgSO_4 \cdot 7 H_2O$, 0,5 g $CaCl_2 \cdot 2H_2O$, 0,01 g $MnSO_4$, 0,01 g $FeCl_3$.

2. Kaliumcitratpuffer: 1 l enthält 100 g Kaliumcitratmonohydrat und 20 g Citronensäuremonohydrat.

3. Thiaminlösung: 10 γ/ml.

4. Inositlösung: 1 mg/ml.

5. Pyridoxinhydrochloridlösung: 10 γ/ml.

6. Biotinlösung: 0,8 γ/ml.

7. Ammonsulfatlösung: 150 mg/ml.

8. Asparaginlösung: 30 mg (in der Wärme gelöst).

Die Lösungen werden je 30 min lang an 3 aufeinanderfolgenden Tagen im strömenden Dampf sterilisiert und können bei Zimmertemperatur, jedoch vor Licht geschützt, aufbewahrt werden. Um 100 ml des Grundmediums herzustellen, die für 20 Reagenzgläser genügen, werden von den Stammlösungen folgende Mengen gemischt:

50 ml Zucker- und Salzlösungen, 10 ml Kaliumcitratpuffer, 5 ml Inositlösung, 5 ml Ammonsulfatlösung, 5 ml Thiaminlösung, 5 ml Pyridoxinlösung, 5 ml Biotin-lösung, 12,5 ml Asparaginsäurelösung, das Gemisch wird mit Wasser auf 100 ml aufgefüllt. Es ist nicht nötig, die Mischung für jeden Versuch frisch herzustellen. Sie kann bei einigen Grad unter Null 3 Monate lang ohne Nachteil aufbewahrt werden. Die gegenüber der endgültigen Versuchslösung doppelt konzentrierte Lösung enthält pro 100 ml folgende Substanzen (siehe untenstehende Tab.).

Durchführung der Messung. 5 ml des Basalmediums plus die auf ihren Panto-thensäuregehalt zu untersuchende Lösung oder ein aliquoter Teil des Pantothen-säurestandards werden in eine Reihe von Reagenzröhrchen (25 × 250 mm) gegeben und mit Wasser auf 9,0 ml aufgefüllt. Geeignete Mengen von Ca-Pantothenat um eine Standardkurve aufzustellen sind: 0; 50; 100; 150; 200; 300 und 400 mγ. Von der zu untersuchenden Lösung werden Volumina verwendet, die Pantothensäure-mengen enthalten, die etwa denen der Vergleichsansätze entsprechen. Die Gläser werden mit Wattepfropfen verschlossen 10 min lang bedampft, abgekühlt und mit 1 ml der Impflösung beimpft. Die Gläser werden dann bei 30° 16—18 Std. geschüttelt. Das Wachstum in jedem Glas wird dann turbidimetrisch im photoelektrischen Kolorimeter bestimmt. Das Ergebnis wird anhand der Standardkurve berechnet.

Glucose	10	g
Asparagin	075	mg
Biotin	4	γ
Inosit	5	mg
Pridoxinhydrochlorid	50	γ
Thiaminchlorid	50	γ
Ammoniumsulfat	750	γ
Kaliumcitrat (Monohydrat)	1,0	g
Citronensäure (Monohydrat)	200	mg
KH_2PO_4	100	mg
KCl	85	mg
$CaCl_2 \cdot 2 H_2O$	25	mg
$MgSO_4 \cdot 7 H_2O$	25	mg
$MnCl_2$	500	γ
$FeCl_3$	500	γ

Pantothensäurebestimmung mit einer Pantothensäure-empfindlichen *Neurospora*-Mutante[1].

Der Test ist zwar nicht ganz so empfindlich, wie der Bakterientest mit *Lactobacillus arabinosus*, aber leichter und mit geringerer Vorbereitungszeit durchzuführen. Die Grundnährlösung für die *Neurospora*-Mutante hat folgende Zusammensetzung:

Dest. Wasser 1000 ml; Saccharose 30 g; Ammoniumtartrat 10 g; K_2HPO_4 5 g; $MgSO_4 \cdot 7 H_2O$ 1 g; NaCl 0,2 g; $CaCl_2$ 0,2 g; $FeCl_3$ 10 mg; $ZnSO_4 \cdot 7 H_2O$ 2 mg; Biotin 8 γ. In dieser Nährlösung erfolgte ohne Zusatz von Pantothensäure kein Wachstum. Mit steigenden Zugaben (2—10 γ) Pa-S je ml wurden Myceltrockengewichtsernten bis zu 80 mg/10 ml bei viertägiger Kultur in 50 ml Kolben bei 20—25° erzielt. Die Impfung erfolgte mit einer Sporensuspension, welche durch Abschwemmen einer Stammkultur hergestellt wurde, in obige Grundnährlösung.

Die Stammkultur wurde auf folgendem Substrat herangezogen: Maltose 3,8 g, Bactopepton 0,8 g, Bactomalt 0,2 g, Agar 2,5 g, Glucose 0,5 g, Wasser 100 ml, Pantothensäure 2 Tropfen einer 0,01%igen Lösung.

Zur Pa-S-Bestimmung bei pflanzlichen Extrakten wurde für jede Versuchserie eine eigene Testkurve durch Auswertung einer Serie mit Zusätzen bekannter Mengen von Pantothensäure in der geschilderten Weise als Bezugsbasis aufgestellt. Die zu prüfenden Extrakte müssen auf das Vorhandensein von Hemmstoffen untersucht werden, was durch Zusatz bekannter Mengen von Pantothensäure erfolgt. Es wird empfohlen, die zu analysierende Pa-S-Lösung vor dem Ansatz des Testversuches mit Absorptionskohle zu reinigen.

II. Biologische Bestimmung von Pantothensäure mit Hilfe des Wachstumstestes bei Ratten und Küken.

(Dargestellt nach C. I. Bliss und P. György in Vitamin Methods II., S. 223 ff. Acad. Press. Inc. Publ. New York 1951, dort ausführliche Literaturangaben.)

Erhalten junge Ratten eine Nahrung, in der Pantothensäure fehlt, die aber sonst ausreichend ist, so tritt nach Ablauf von etwa 3 Wochen Wachstumsstillstand ein (Gewicht 50—65 g). Mit der zweiten Woche beginnend wird bei den Tieren der Pelz rauh und dünn und es entsteht eine Dermatitis. In der dritten Woche beobachtet man starke nasale Sekretion, entzündlich veränderte Partien um die Schnauze mit blutig verbackenen Tasthaaren bei den meisten Tieren. Pigmentierte Ratten-Stämme ergrauen an den pigmentierten Stellen des Pelzes (Achromotrichie). Bei Küken entwickelt sich bei Pantothensäuremangel eine typische Dermatitis: Um den Schnabel und um die Augen wird die Haut rauh. Gleichzeitig wird die Wachstumsgeschwindigkeit stark vermindert. Für quantitative Bestimmungen der Pantothensäure kann bei Ratten und Hühnchen das Verhalten der Dermatitis oder der Wachstumsgeschwindigkeit bei Zufütterung des zu untersuchenden Materials herangezogen werden. Meist wird der Wachstumstest verwendet, weshalb nur dieser im folgenden beschrieben wird. Die Tiere, die in den Versuch genommen werden sollen, erhalten eine Diät der folgenden Zusammensetzung (siehe vorstehende Tab.).

Basaldiät für Ratten und Küken.

	Ratten	Küken
Rohrzucker	69	
Glucose	—	60,5
Casein.	20	18
Gelatine.	—	10
Pflanzliches Öl . . .	6,7	5
Salzmischung . . .	4	6
Cystin.	0,2	0,3
Cholinchlorid. . . .	0,1	0,2

---

[1] Villela, G. G.: Bull. Soc. Chim. biol. (Paris) **29**, 763—776 (1947).

Die Salzmischung ist wie folgt zusammengesetzt (siehe nachstehende Tab.).
Die Salze werden gemischt und gepulvert, Kaliumjodid wird mit Natriumchlorid verrieben und erst bei Gebrauch den übrigen Salzen zugefügt. Die Diät muß durch Vitamine ergänzt werden, da sie vitaminfrei ist. Die Vitamine werden in Form einer Mischung von kristallinen, wasserlöslichen Vitaminen der Diät zugefügt und durch öllösliche Vitamine ergänzt, die mit der Diät gemischt oder wöchentlich oder halbwöchentlich direkt verabreicht werden. In der folgenden Tab. ist eine geeignete Vitaminzusammensetzung für Ratten und Küken angegeben.

	Gramm		Gramm
K_2HPO_4	322	$MnSO_4 \cdot 4 H_2O$	5,1
$CaCO_3$	300	KJ	0,8
$NaCl$	167	$CuSO_4 \cdot 5 H_2O$	0,3
$MgSO_4 \cdot 7 H_2O$	102	$ZnCl_2$	0,25
$CaHPO_4 \cdot 2 H_2O$	75	$CoCl_2 \cdot 6 H_2O$	0,05
$Fe(C_6H_5O_7) \cdot 6 H_2O$	27,5		
			1000,00

Bei Ratten gestalten sich die Pantothensäurebestimmungen wie folgt.

Junge Ratten erhalten die Mangeldiät und werden nach etwa 3 Wochen gewichtskonstant. Sie werden rascher pantothensäurearm, wenn schon die säugenden Mütter auf der angegebenen Basaldiät gehalten wurden. Um eine hohe Sterblichkeit zu vermeiden, sollten die jungen Ratten erst im Alter von 14—16 Tagen in den Versuch genommen werden. Sobald die Tiere gewichtskonstant geworden sind, werden sie in Einzelkäfige gebracht und willkürlich auf Gruppen verteilt, bei denen der Nahrung eine bekannte Menge der zu bestimmenden Lösung zugesetzt wird. Das Wachstum der Ratten

Tägliche Gabe in Milligramm.	
Thiamin HCl	0,03
Riboflavin	0,5
Niacin	1,0
Pyridoxin HCl	0,03
Inosit	1,0
Cholinchlorid	3,0
Weizenkeimlingsöl. . .	500,0
Haliveröl	wöchentl. pro Ratte 2 Tropfen

ist dann eine lineare Funktion, des Logarithmus der aufgenommenen Pantothensäuredosis im Bereich von 10—100 γ bei täglicher Zufuhr.

Für die Pantothensäurebestimmung an Küken werden 1 Tag alte Küken auf einer käuflichen Ausgangsdiät 4—7 Tage gehalten, um Unterschiede des anfänglichen Vitamingehaltes auszugleichen. Dann erhalten sie die oben angegebene Basaldiät 3—4 Wochen lang. Im prophylaktischen Testversuch erhalten 2 oder mehr Gruppen bekannte Pantothensäuremengen, 2 oder mehr Gruppen von Küken erhalten verschiedene Mengen des Materials, dessen Pantothensäuregehalt bestimmt werden soll. Das Gewicht der Tiere nimmt mehr oder weniger genau linear mit dem Logarithmus der Pantothensäuredosis im Bereich von 0,2—1 mg Pantothensäure pro 100 g Diät zu.

III. Chemische Bestimmung der Pantothensäure[1].

Chemische Bestimmungen von Pantothensäure gründen sich auf Farbreaktionen der Produkte der hydrolytischen Spaltung des Vitamins. Hydrolyse in alkalischem Medium führt zur Bildung von α-γ-Dioxy-β-β-dimethylbuttersäure (Pantoinsäure s. Schema, S. 624) und zu β-Alanin. Bei der Bestimmungsmethode von SZALKOWSKY, MADER und FREDIANI (1950) wird das gebildete β-Alanin mit Kaliumpermanganat in Gegenwart von Kaliumbromid behandelt, und dann mit 2,4-Dinitrophenylhydrazin zu einem unlöslichen Hydrazon kondensiert. Nach

---

[1] Dargestellt nach P. GYÖRGY, Chemical Methods of Vitamin Assay. In: Vitamin Methods Vol. II, und zwar S. 662, Acad. Press. Inc. New York, 1951.

Auflösung dieser Verbindung in Pyridin und Verdünnung mit Natriumhydroxyd wird die resultierende blaue Farbe bei 570 mμ gemessen. Niacin, Niacinamid, Thiamin und Vitamin B$_6$, die die Bestimmung stören können, werden durch Chromatographieren der wäßrigen Pantothensäureprobe an Aluminiumoxydsäulen entfernt. Vitamin C, das ebenfalls die Bestimmung stören kann, konnte durch die chromatographische Behandlung noch nicht abgetrennt werden. Wenn die Spaltung des Pantothenats unterhalb von pH 4,7 erfolgt, so wird Pantoyllakton anstatt Pantoinsäure gebildet (Frost, 1943). Auf dieser Tatsache haben Woolish und Schmall (1950) eine kolorimetrische Bestimmung von Pantothensäure gegründet. Das Pantoyllakton reagiert mit Hydroxylamin bei alkalischer Reaktion und bildet Pantoyllaktonhydroxamsäure. Nach dem Ansäuern und nach dem Zufügen von Eisenchlorid entsteht eine purpurrote Farbe, die bei 500 mμ gemessen wird. (Reaktion für Ester und Laktone nach Feigl, Anger und Frehden, 1934.)

Die Methode wurde allerdings nur für reine d-Calciumpantothenatlösungen und zur Bestimmung von Pantothensäure in Vitamintabletten verwendet. Andere Vitamine, wie Thiamin, Riboflavin, Niacinamid, Thiamin und Folsäure stören vor oder nach Hydrolyse die Reaktion nicht. Pyridoxin gibt eine leicht braune Färbung der gleichen Intensität vor und nach der Hydrolyse. Ascorbinsäure jedoch stört die Reaktion stark. Pantoinsäure wird als Pantoyllakton nach saurer Hydrolyse mit entfernt, aber nicht in der unhydrolysierten Probe.

Durchführung der Analyse nach Woolish und Schmall. Folgende Reagentien werden benötigt: Salzsäure 0,1 n bis 1 n und 2 n. 2,4-Dinitrophenol: 1%ig in 95% Äthanol. Ferrichlorid: 2%ige wäßrige Lösung. Eine wäßrige Lösung von reiner d-Pantothensäure enthaltend 2 mg per ml. Acetatpuffer: 25 ml von 2,5 mol. Natriumacetat + 40 ml 1 n Schwefelsäure (pH ungefähr 4,2).

Hydrolyse und Kolorimetrie. Die Hydrolyse von Pantothenat zu Pantoyllakton erfolgt durch dreistündiges Erhitzen in saurer Lösung auf 80°, oder 1,5 Std. auf 100°. Lösungen von Pantoin oder Pantothenat (4 ml, enthaltend etwa 3 mg), die keine störenden Substanzen enthalten, werden in einem 50 ml fassenden Kolben mit 3 ml n HCl versetzt. Der Kolben wird vor dem Erhitzen lose verstopft. Nach der Hydrolyse und nach dem Abkühlen auf Zimmertemperatur werden zu jedem Kolben 2 ml der Hydroxylamin-Natriumhydroxydlösung zugegeben und dann 5 ml 1 n NaOH. Nach 5 min werden 3 Tropfen 2,4-Dinitrophenolindikator zugegeben und die Lösung sorgfältig mit n HCl, in der Höhe des Endpunktes mit 0,1 n HCl titriert, bis ein farbloser Endpunkt erreicht ist. Das Volumen wird dann auf 50 ml mit Wasser verdünnt. Aliquote Teile dieser Testlösung und 2% Eisenchlorid im Verhältnis 5:1 werden in ein Kolorimeterglas pippetiert, z. B. 5 ml und 1 ml. Die Ablesungen werden dann an einem photoelektrischen Kolorimeter mit 500 mμ Filter innerhalb 45 sec bis 1 min nach dem Zufügen des Ferrichlorids durchgeführt. Das Instrument wird vorher mit einem vollständigen Leeransatz auf die Absorption Null eingestellt. Die Leeransätze enthalten die gleichen Anteile der Testlösungen wie die Hauptansätze, werden jedoch nicht hydrolysiert. Ein 5,00 ml-Anteil des U.S.P.-Standard von d-Calciumpantothenat (3 mg) wird hydrolysiert und in genau der gleichen Weise wie die zu prüfende Lösung kolorimetriert. Ein unhydrolysierter Leerversuch wird auch für den Standard abgelesen. Zur Berechnung des Pantothenats gegen eine entsprechende Standardlösung dient folgende Formel:

$$\frac{T - T_B}{S - S_B} \times \frac{C}{D} = \text{mg } d\text{-Calciumpantothenat/Einheit der Probe.}$$

T = Absorption der hydrolysierten Testlösung.
T_B = Absorption der unhydrolysierten Testlösung (Leerversuch).
S = Absorption der hydrolisierten Standardlösung.
S_B = Absorption der unhydrolysierten Standardlösung (Leerversuch).

C = mg d-Calciumpantothenat oder Pantoin in dem 5 ml Anteil der Standardlösung, der zur Hydrolyse genommen wurde.

D = Verdünnungsfaktor. (Aliquoter Anteil des Eluates, der zur Hydrolyse verwendet wurde.) Wenn d-Panthenol gegen einen Standard von Calciumpantothenat gemessen wurde, so muß das nach obiger Formel errechnete Resultat mit 0,8613 multipliziert werden, um von mg Pantothenat auf mg Panthenol zu kommen.

Das Hydroxylamin-Natriumhydroxydreagens muß vor dem Natriumhydroxyd zugefügt werden. Wird die Reihenfolge umgekehrt, so wird einiges Pantoyllakton zu Pantoinsäure umgewandelt, bevor Hydroxamsäure gebildet werden konnte. Die Intensität der Farbe ist bei 8° größer als bei 20°, zeigt aber geringere Änderungen zwischen 20 und 35°. Die Farbenintensität mit einer gegebenen Menge Pantoyllakton nimmt ab, wenn das Volumen, zu dem die Hydroxylamin-Natriumhydroxydlösung zugegeben wird, zunimmt. Es sollte das gleiche Volumen für Test- und Standardlösung verwendet werden.

Nach CROKAERT (1941), und FRAME, RUSSEL und WILHELMI (1943) kann eine chemische Bestimmungsmethode auch auf die braungelbe Farbreaktion gegründet werden, die nach Hydrolyse der Pantothensäure zwischen dem freigesetzten β-Alanin und Naphthochinonsulfonat sich abspielt. Das Verfahren ist auf pflanzliches Material bisher nicht angewandt worden. Die Pantothensäurebestimmung wird durch die Gegenwart von Substanzen mit freien Aminogruppen gestört. Man kann die Pantothensäure an Tonerde durch Adsorption bei pH 3,3 von den neutralen und vorwiegend basischen Aminosäuren, die nicht zurückgehalten werden, abtrennen. Weiter ist es möglich, durch Durchsaugen des auf Pantothensäure zu untersuchenden Materials durch eine saure Dowex-50-Säule Pigmente und andere Begleitsubstanzen abzutrennen (CROKAERT, MOORE und BIGWOOD, 1951).

Obige Tabelle verzeichnet den Pantothensäuregehalt von Pflanzenmaterial vor und nach der fermentativen Freilegung der Pantothensäure, nach mikrobiologischen Messungen sowie

Tabelle 1. *Gehalt von Pflanzenmaterial an Pantothensäure (γ/g). (Aus V. Schmidt: Z. f. Vitamin-, Hormon und Fermentforschg. 1951—1952, S. 326.)*

Art	Mikrobiol. unbehandelt	Mikrobiol. nach Fermentbehandlung	Kükenwachstum
Broccoli (Blätter und Blüten)			87
Kartoffel	3—4	—	6,5
Roggenmehl	9,0		
Weizen	11,0		
Weizenkleie			24
Weizenmehl	3,1—5,0		
Hafergrütze	13,0		
Reis	3,5		10
Reiskleie			22
Hefe	100,0	400	200
Bete	1,3—1,7		
Kohl	1,4—1,8		
Lauch			1,2
Blumenkohl	5—10—9		
Bohnen	1,2		
Erbsen	13		21
Karotten	2—2,5		2
Salat	1—1,1		
Spinat	1—1,8	4	
Tomate	1—4		
Erdbeeren	2—2,6		
Melone	3,1		
Apfelsine	2—3,5		0,7
Äpfel	0,5—0,6		
Rosinen	0,9		
Stachelbeeren	2		
Heidelbeeren	1,5		
Holunderbeeren	1,5		
Rhabarber	0,6		
Champignons	8—25		
Zucchini Kürbis			35
Wirsingkohl			30
Taro-Wurzel			28
Alfalfa (Luzerne)			25
Zwiebel			9
Batate			13
Banane			4

nach Bestimmung am Kükenwachstumstest nach Freilegung der Pantothensäure mit Clarase und Papain.

Tabelle 2. *Gehalt von Gräsern an Pantothensäure*
(γ/g Trockenmaterial)[1].

Gräser	Gehalt an Pantothensäure[2]
Achillea millefolium (Schafgarbe)	5,1
Plantago lanceolata (Engblättriger Wegerich) . . .	.5,6
Poterium sanguisorba (Becherblume)	11,7
Cichorium intybus (Cichorie)	18,8
Trifolium hybridum (alsike)	22,1
Lotus corniculatus (Klee)	20,2
Medicago sativa (Luzerne)	43,4
Onobrychis sativa (Hahnenkopf)	15,4
Lolium perenne (Raigras)	17,4
Phleum pratense (Timotheus)	9,1
Dactylis glomerata (Hahnenfuß)	4,7
Poa trivialis (Rauhstieliges Wiesengras)	11,6
Cynosurus cristatus (Geschopfter Hundeschwanz)	15,5
Festuca pratensis (Schwingel)	5,1
F. rubra susp. fallax (Wiederkäuerschwingel) . .	15,5
F. elatior (Großer Schwingel).	7,7
F. fermina rubra (Roter Schwingel)	12,3
Molinia coerula (Windhalm)	4,6
Eriphorum vaginatum (Ziehmoos).	7,4
Juncus squarrosus (Stuhlhalm)	9,7

B. Coenzym A.

Die Coenzym A-Bestimmungen gründen sich alle auf die katalytischen Fähigkeiten, die das Coenzym A in Verbindung mit spezifischen Fermentproteinen hat. Es kann daher jede Fermentreaktion herangezogen werden, die primär oder sekundär Coenzym A-abhängig ist.

1. Coenzym A-Bestimmung nach KAPLAN und LIPMANN (1948).

Die Geschwindigkeit der Acetylierung von Sulfanilamid durch ein Ferment aus Taubenleber in Gegenwart von Acetat und Adenosintriphosphat ist ausschließlich eine Funktion der Coenzym A-Konzentration. Bei der praktischen Durchführung der Reaktion werden noch Cystein und Citrat dem Versuchsansatz zugegeben. Cystein reaktiviert und stabilisiert das acetylierende Enzym, Citrat entionisiert das Calcium des Versuchsansatzes und verhindert so die Spaltung von A. T. P. durch die Adenosin-Triphosphatase. Taubenleberextrakt verliert bei der Autolyse seine Fähigkeit zur Acetylierung von Sulfonamid vollkommen (LIPMANN, 1945). Er gewinnt sie wieder durch Zufügen eines coenzymhaltigen Extraktes. Die Acetylierung ist so auf ein Apoenzym und ein Coenzym-System zurückzuführen mit Coenzym A als erstem und Adenosintriphosphat als zweitem Cofaktor. Die Leichtigkeit, mit der das Coenzym A durch Autolyse zerstört wird, bietet eine bequeme Methode der Bereitung des Apoenzyms, mit Hilfe dessen wiederum Coenzym A-Bestimmungen durchgeführt werden können.

Gewinnung eines Trockenpulvers aus Taubenleber. Man verwendet ein Acetontrockenpulver, dessen wichtiger Vorteil gegenüber frischem Extrakt ein niedriger Gehalt an Adenosintriphosphatase ist. Für die Bereitung des Präparates werden Tauben durch Dekapitierung getötet und so vollkommen wie möglich ausgeblutet.

---

[1] Lit.: THOMAS, B., u. H. F. WALKER: Empire J. of Exp. Agric. **17**, No. 67, 170 (1949).
[2] Mikrobiologisch bestimmt nach E. C. BARTON-WRIGHT: Analyst. **70**, 283 (1945).

Die Lebern werden rasch herausgenommen und bis zur Verwendung eingefroren. Die gefrorenen Lebern werden in Stücke zerschnitten und in einem gekühlten Homogenisator mit dem 20fachen ihres Gewichtes an kaltem Aceton 2 min lang homogenisiert. Das Homogenat wird rasch auf einem Büchnertrichter abfiltriert. Der Rückstand wird auf dem Trichter mit Aceton und Äther (peroxydfrei!) gewaschen. Die Homogenisierung wird im Kühlschrank ausgeführt. Werden gut gekühltes Aceton und Äther verwendet, so ist Filtration unter Kühlung nicht notwendig. Das Pulver wird weiter im Vakuumexsiccator über Phosphorpentoxyd getrocknet. Bindegewebe wird mit Hilfe eines Siebes entfernt. Es werden 15 bis 20 g eines rosa Pulvers aus 100 g frischer Leber von 10—12 Tauben erhalten.

Die rosa Farbe ist ein Hinweis auf gute Aktivität des Präparates und stammt von undenaturiertem Hämoglobin. Ist das Hämoglobin denaturiert, so entsteht ein bräunliches Acetonpulver, das unweigerlich nur eine geringe Aktivität aufweist.

Bereitung des Apoenzyms der Acetylierung aus Taubenleber-Trockenpulver. Je 10 g Leberpulver werden mit 100 ml 0,02 mol Na-Bicarbonatlösung sorgfältig verrieben. Der unlösliche Rückstand wird durch Zentrifugieren entfernt. Zwischen 60—75 ml einer dunkelroten, leicht opalisierenden Flüssigkeit werden erhalten. Sie wird über Nacht in einer Tiefkühlkammer gehalten. Zur Inaktivierung wird dieses Material aufgetaut auf Raumtemperatur gebracht und 4 Std. belassen. Das Ferment wird dabei zu 95—100% inaktiviert. Nach dem Altern wird der Extrakt wieder zentrifugiert und der Niederschlag verworfen. Der gealterte Extrakt hält sich im Tiefkühler unbegrenzt[1]). Der gealterte rezentrifugierte Extrakt soll eine rötliche wasserklare Lösung sein. Das für die Inaktivierung hauptsächlich verantwortliche Ferment ist unstabil und seine Aktivität nimmt während der Inkubierung bei Zimmertemperatur stark ab. Dieser Umstand ist von Vorteil für die Verwendung des Extraktes zur Bestimmung von Coenzym A.

Ausführung der Acetylierung. In Reagenzgläser von 1—1,3 cm Durchmesser werden einpipettiert:

1. Die auf Coenzym A zu untersuchende Lösung. Das Volumen sollte 0,3 ml nicht übersteigen.
2. 0,3 ml Natriumacetat-ATP-Sulfanilamid-Lösung (s. weiter unten!).
3. 0,08 ml frisch bereiteter molarer Natriumcarbonatlösung.
4. 0,1 ml 0,1 mol. Cysteinhydrochlorid.
5. 0,25 ml des gealterten Enzyms.
6. Wasser zum Auffüllen.

Bereitung der Acetat-ATP-Sulfanilamidlösung: 10 ml 0,004 M Sulfanilamid, 2,5 ml m Natriumacetat, 8,0 ml annähernd 0,05 ml Kaliumadenosintriphosphat, 10 ml 0,2 n Natriumacetat werden gemischt. Der pH-Wert der Mischung sollte etwa 7 sein. Größere Mengen dieser Mischung werden in Portionen von je 45 ml eingefroren und erst jeweils im Augenblick der Anwendung aufgetaut.

Der Leerversuch enthält alle Komponenten mit Ausnahme des Coenzyms A; die Gläser werden verstopft, durchgemischt und 2 Std. im Wasserbad bei 37° inkubiert. Die Reaktion wird unterbrochen durch Zufügen von 4 ml 5%iger Trichloressigsäure. Nach dem Zentrifugieren werden aliquote Teile von 1 ml für

---

[1] Es ist wichtig, die Extrakte vor dem Altern zu frieren, da durch das Ausfrieren eine große Menge von Adenosintriphosphatase entfernt wird. Eine Wiederholung der Zentrifugierung sollte nach dem Altern durchgeführt werden, sonst verliert das Präparat viel seiner Aktivität. Ein Taubenleberextrakt, welcher zuerst zentrifugiert wurde, und dann alterte, acetylierte 16 γ Sulfanilamid, während der gleiche Extrakt nach Altern und nachfolgendem Zentrifugieren unter gleichen Bedingungen 44,2 γ/ml acetylierte.

die Sulfanilamidbestimmung nach (Bratton, und Marshall, 1938, und 1939) (Ausführung der Bestimmung s. weiter unten) verwendet.

Es wird in einem Klett-Fotokolorimeter mit Filter 54 abgelesen. Die Differenz der Sulfanilamidgehalte der Haupt- und Leerversuche entspricht der acetylierten Amidmenge.

Coenzym A-Einheit nach Lipmann. Als Coenzym A-Einheit wird nach Lipmann diejenige Menge Coenzym A bezeichnet, die das beschriebene System zur Hälfte seiner maximalen Aktivität aktiviert. Die Coenzym A-Einheit enthält 0,7γ gebundene Pantothensäure und ist unabhängig von der Reinheit des Co-A-Präparates. Bei der Coenzym A-Bestimmung wird ein gutes Enzympräparat 75—85% des zugeführten Sulfanilamids acetylieren. Die besten Ablesungen werden erhalten in der Gegend der Standardkurve, bei der 40% des Sulfanilamids während der Inkubation verschwinden.

Standard-Coenzym A-Lösung. Eine gut wirksame Coenzymlösung wird bereitet durch rasches Aufkochen von frischer Tauben- oder Kaninchenleber mit 3 Volumen Wasser. Um ein Coenzym A-Trockenpräparat zu erhalten, wird diese Lösung mit Salzsäure auf pH 2 gebracht und dann mit dem 10fachen Volumen Aceton gefällt. Der Niederschlag wird mit Aceton gewaschen und mit Äther getrocknet. Das Präparat ist hygroskopisch. In einem Exsiccator bei

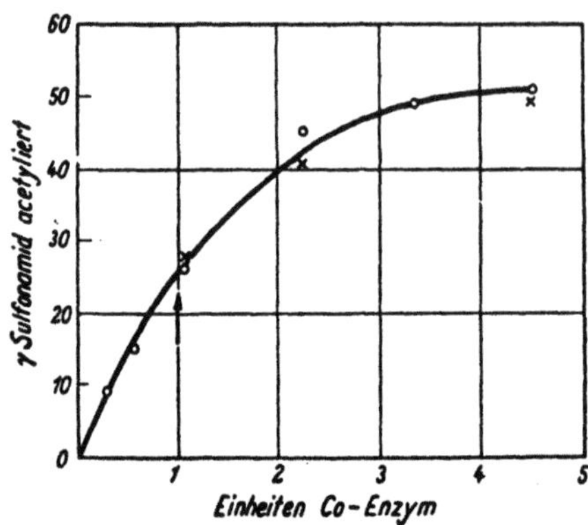

Abb. 2. Konzentrations-Aktivitätskurve für CoA-Präparate verschiedener Reinheit. Der Pfeil gibt den Punkt einer Einheit auf der Kurve an. Die Ordinate die Menge acetylierten Sulfanilamids —O— Meßpunkte für rohes CoA 0,25 E/mg und — + — Meßpunkte für gereinigtes Coenzym 130 E/mg. [J. Biol. Chem. 174, 41 (1948)]. Die Kurve läßt erkennen, daß das Ergebnis der Coenzym A-Bestimmung vom Reinheitsgrad des Coenzyms A praktisch unabhängig ist.

niedriger Temperatur aufbewahrt, bleibt es stabil. Solche Coenzym A-Präparate enthalten 0,25—0,5 Einheiten Coenzym A pro mg Trockensubstanz.

Dieses Präparat wird als Bezugs-Coenzym A verwendet. Mit ihm wird eine Aktivitäts-Konzentrationskurve angelegt und die Menge, die der Einheit entspricht, wird notiert. Man erhält die in Abb. 2 wiedergegebene Konzentrations-Aktivitätskurve für Coenzym A.

Zur Bestimmung unbekannter Coenzym A-Mengen werden Ansätze in 2 Konzentrationen bestimmt, etwas oberhalb und etwas unterhalb von dem Punkt, der einer Einheit entspricht. Bei jeder Bestimmung werden die Einheiten und die Sättigungspunkte mit dem Bezugs-Coenzym A festgelegt. Eine genauere Coenzym A-Bestimmung wird in der Weise ausgeführt, daß man mit einem Bezugs-Coenzym A für jede Extraktprobe über einen weiten Konzentrationsbereich eine Standardkurve anlegt, von der der Coenzym A-Gehalt der unbekannten Probe direkt abgelesen wird.

Sulfanilamidbestimmung nach Bratton und Marshall. Sulfanilamid wird hierbei durch Diazotieren und Kuppeln in saurer Lösung mit N-(1-Naphthyl)-äthylendiamindihydrochlorid bestimmt. Dazu werden benötigt:

1. 0,1%ige Natriumnitratlösung.

2. Eine wäßrige Lösung von N-Naphthyläthylendiamindihydrochlorid enthaltend 100 mg/100 ml. Diese Lösung sollte in einer dunkel gefärbten Flasche gehalten werden.

3. Eine Saponinlösung mit 0,5 g Saponin/l.

4. 4 nHCl

5. Eine Lösung von Ammoniumsulfat enthaltend 0,5 g/100 g.

6. Eine Stammlösung von Sulfanilamid in Wasser 200 mg/l enthaltend.

Diese Lösung kann mehrere Monate im Eisschrank aufbewahrt werden. Aus ihr werden Standardlösungen bereitet, indem man 5; 2,5 und 1 ml der Stammlösung mit 18 ml einer 15%igen Trichloressigsäure-Lösung versetzt und mit Wasser auf 100 ml auffüllt. Absorptionsmaximum zwischen 540 und 550 mμ.

Bei einer Modifikation dieser Coenzym A-Bestimmung, die von F. LYNEN und D. REINWEIN stammt, wird die umständliche Entwicklung des Azofarbstoffes, wie die der Originalmethode von KAPLAN und LIPMANN, umgangen. Es wird an Stelle des farblosen Sulfanilamids gelbes p-Nitroanilin benützt, welches dann bei der Acetylierung in das nahezu farblose p-Nitroacetanilid übergeht. Die Acetylierung kann deshalb kolorimetrisch bei 405 mμ verfolgt werden.

Ausführung des Testes: In 1 cm Küvetten gibt man:

0,75 ml einer Reaktionsmischung bestehend aus: 35 Teilen 0,4 mol Kaliumphosphatpuffer pH 6,8, 0,4 Teilen mol-MgCl$_2$, 0,3 Teilen 0,35 mol Natrium-thioglykolat, 0,5 Teilen 0,0034 ml p-Nitroanilin, 2 Teilen 0,1 mol Natriumcitrat, 0,25 Teilen mol Natriumacetat und 0,5 Teilen 0,1 mol K-ATP.

x ml der zu testendenCoA-Lösung, (0,85—x) ml Wasser.

Man startet durch Zugabe von 0,1 ml Taubenleberextrakt, bereitet nach KAPLAN und LIPMANN, und verfolgt die Extinktionsabnahme der Lösung während einstündiger Inkubation im Thermostaten bei 25°, $\lambda = 405$ mμ.

Um den CoA-Gehalt angeben zu können, muß man auch hier unter Einsatz eines CoA-Standardpräparates eine

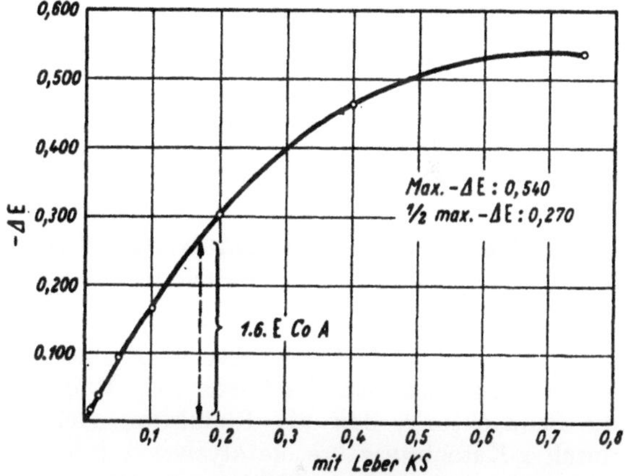

Abb. 3. Coenzym A-Eichkurve. CoA-Bestimmung. Abhängigkeit von der Leberkochsaftkonzentration. 60 min bei 25° inkubiert. (HILZ, 1953.)

Eichkurve aufstellen. Die halbe Extinktionsabnahme entspricht dann, da das Volumen des Ansatzes 1,6 ml ausmacht, 1,6 CoA-Einheiten. Der CoA-Gehalt der unbekannten Lösung kann aus dieser Eichkurve abgelesen werden (Abb. 3.) Es ist zweckmäßig, Mengen zwischen 0,5 und 2 CoA-Einheiten, einzusetzen, da die Kurve in diesem Gebiet annähernd linearen Verlauf zeigt.

Eine weitere CoA-Bestimmung bedient sich des Enzyms der Acetataktivierung, welches die Reaktion (a) katalysiert. Beim Test wird das gebildete Acetyl-CoA durch Hydroxylamin weiter zu Acethydroxamsäure umgesetzt, deren Menge mit Eisen III Salz kolorimetrisch gemessen wird. [Reaktion (b)]:

a) ATP + Acetat + CoA $\longrightarrow$ PP + Acetyl-CoA

b) Acetyl-CoA + Hydroxylamin $\longrightarrow$ CoA + Acethydroxamsäure.

Da die Reaktion b CoA zurückliefert, welches bei Reaktion a verbraucht wurde, wirkt Coenzym A auch in diesem Test katalytisch.

Der Testansatz, der sich in spitzen Zentrifugengläsern befindet, enthält bei einem Gesamtvolumen von 0,5 ml:

$5\mu M$ ATP, $5\mu M$ Acetat, $5\mu M$ MgSO$_4$, $5\mu M$ Cystein-HCl, $100\mu M$ NH$_2$OH, (pH 7,3)[1], 0,02—0,04 ml Enzymlösung und variierte Mengen Coenzym A.

Das Ganze wird bei 37° 1 Std. lang inkubiert, dann werden 2 ml einer Lösung, bestehend aus 2 Teilen 2,5 n Salzsäure (1 Volumen konz. Salzsäure $+$ 3 Vol. H$_2$O), 2 Teilen 12%iger Trichloressigsäure und 2 Teilen 5%iger FeCl$_3$-Lösung in 0,1 n HCl zugegeben. Das denaturierte Eiweiß wird abzentrifugiert und die klare, überstehende Lösung im Photometer Eppendorf bei $\lambda = 546$ mμ (Schichtdicke 1 cm) kolorimetriert.

Ein für diesen Test geeignetes Enzympräparat kann aus Herzmuskel (Beinert, Green, Hele, Hift, von Korff und Ramakrishnan, 1953) oder besser aus Hefe (Jones, Black, Flynn und Lipmann, 1953; Hilz, 1953) gewonnen werden. Man setzt von der Lösung des Acetat aktivierenden Enzyms so viel ein, daß unter den Bedingungen der Bestimmung und bei Sättigung des Systems mit CoA eine Extinktion von 0,6—0,7 (entsprechend einer Synthese von 6—7 Molen Acethydroxamsäure) gemessen wird.

Eine weitere für fermentative Studien geeignete Bestimmung des Coenzyms A gründet sich auf die Phosphotransacetylase-Reaktion. Die Transacetylase, die aus *Clostridium kluyveri* erhalten werden kann (Stadtmann, Novelli und Lipmann, 1951), katalysiert folgende reversible Reaktion:

$$CH_3COOPO_3 + HS—CoA \rightleftharpoons CH_3CO—S—CoA + HPO_4''$$

Die unter Standardbedingungen entstandene Acetyl-CoA-Menge kann über die sog. Arsenolyse-Reaktion bestimmt werden.

$$Acetyl—CoA + Arsenat \longrightarrow Acetyl—Arsenat + CoA—SH$$

Rasche spektrophotometrische Bestimmung des Coenzyms A nach Korff (1953).

Eine rasche und spezifische Methode zur Bestimmung katalytischer Mengen von Coenzym A (0,5—1,5 Lipmann-Einheiten), die sich bei der Kontrolle der verschiedenen Schritte der Coenzym A-Reinigung bewährt hat, gründet sich auf die Beobachtung, daß die Reduktion von Diphosphopyridinnucleotid (DPN) durch α-Ketoglutarsäure, katalysiert durch eine lösliche Oxydase und ein Hilfsenzym (Deacylase) CoA-abhängig ist wie es die folgenden Gleichungen veranschaulichen:

1. α-Ketoglutarat $+$ CoA $+$ DPN $\xrightarrow{\text{Oxydase}}$ Succinyl-CoA $+$ DPNH $+$ CO$_2$ $+$ H$^+$

2. Succinyl-CoA $\xrightarrow{\text{Deacylase}}$ Succinat $+$ CoA.

Es wird dabei die Geschwindigkeit der DPN-Reduktion gemessen. Lösliche α-Ketoglutarsäureoxydase wird nach Sanadi, Littlefield und Bock (1952), Succinyl-CoA-Deacylase nach Gergeley, Hele und Ramakrishnan (1952) bereitet.

Ausführung der Bestimmung. Es werden folgende Ansätze fertiggestellt:

Cystein (0,1 mol., pH 7) 0,1 ml; CoA-Lösung (enthaltend 15 Einheiten CoA/mg, pH 7) 0,1 ml; Glycinat (0,1 mol., pH 9) 0,1 ml; α-Ketoglutarat (0,2 mol., pH 7) 0,1 ml; Deacylase 0,1 ml, α-Ketoglutarsäureoxydase 0,01 ml; mit Wasser auf 2,9 ml aufgefüllt. Eine bekannte CoA-Menge wird jeweils in der gleichen Weise angesetzt. Die Geschwindigkeit der DPN-Reduktion/Einheit CoA variiert nämlich nicht nur von einem Enzympräparat zum anderen, sondern auch bis zu einem gewissen Grad aus noch nicht ganz bekannten Gründen in aufeinanderfolgenden Analysen desselben Enzympräparates. Ein geeigneter CoA-Standard wird aus

---

[1] Bereitung der Hydroxylaminlösung: 1 Vol. wäßriger 28%iger NH$_2$OH · HCl-Lösung wird im Eisbad mit 1 Vol. 4 n KOH versetzt und damit pH 7—7,3 eingestellt.

relativ reinem Coenzym A (BEINERT, VON KORFF, GREEN, BUYSKE, HANDSCHUH-MACHER, HIGGINS und STRONG, 1952) hergestellt, welches erwiesenermaßen Pantothensäure nur in Form von CoA enthält (d. h. mikrobiologische Bestimmung des Coenzym A-Gehaltes muß den gleichen Wert ergeben wie die enzymatische).

Nach der Mischung werden die Ansätze 5 min lang bei 30° inkubiert (Gefäßdurchmesser 1 cm) und danach die Absorption im Beckman-Apparat bei 340 mμ gegen dest. Wasser gemessen. Die Reaktion wird durch Zufügen von 0,1 ml 1%iger DPN-Lösung zu jedem Ansatz in Gang gesetzt. Die Zellen werden bedeckt, die Ansätze durchgemischt und die Absorption bei 340 mμ im zeitlichen Abstand von je 1 min 3—5 min lang abgelesen, mit einem Abstand von 10 sec zwischen den einzelnen Ansätzen. Die durchschnittliche Zunahme der Absorption/min wird für jede Probe ermittelt. Die Menge von Coenzym A in der zu bestimmenden Probe wird dann durch einfaches Proportionieren berechnet. Um den Einfluß von Leerversuchen und der Abweichung von einer vollständig linearen Reaktion auf ein Minimum herabzudrücken, sollte die unbekannte Probe in einer Höhe angesetzt werden, die eine Geschwindigkeit der DPN-Reaktion erzeugt, die bis auf $\pm$ 20% derjenigen ist, die durch den Standard bewirkt wird. Adenosindiphosphat, Adenosintriphosphat und anorganisches Pyrophosphat oder Kupfer sollten in dem Versuchssystem nicht zugegen sein, da diese Substanzen

Tabelle 3. *Coenzym A-Gehalt von Mikroorganismen.* [Nach N. O. KAPLAN und F. LIPMANN, J. Biol. Chem. 174, 37, (1948)].

Mikroorganismus	Coenzym A Einheit./g Trockengewicht
Proteus morganii	572
Lactobacillus arabinosus .	150
Lactobacillus delbrueckii .	40
Luftgetrocknete Hefe . .	72
Escherichia coli	320
Propionsäurebakterien . .	330
Clostridium butylicum . .	2000

Tabelle 4. *Coenzym A in höheren Pflanzen.*

	Coenzym A Einheit /g Frischgewicht
Spinat	0,74
Tomaten	1,3
gefrorene Erbsen	4,5
Weizenkeimlinge	30

die Reaktion hemmen. Pyridin, das oft zur Elution von Coenzym A aus Tierkohle verwendet wird, stört nicht ernstlich, wenn seine Konzentration in der Reaktionsmischung 0,3% nicht übersteigt. Die mit der Methode VON KORFFs erhaltenen Werte stimmen zwar vorzüglich mit denen, die mit anderen Verfahren erhalten wurden überein, doch dürfte das Bestimmungsverfahren VON KORFFs nur für relativ reine Coenzymlösungen, also nicht zur Bestimmung von Coenzym A-Gehalten von Pflanzenhomogenaten geeignet sein.

Literatur.

ATKIN, L., W. L. WILLIAMS, A. S. SCHULZ and C. N. FREY: Ind. Eng. Chem. Anal. Ed. 16, 67 (1944).

BADDILEY, J.: Advances in Enzymology 16, 1 (1955).

BEINERT, H., D. E. GREEN, P. HELE, H. HILFT, R. W. v. KORFF and C. V. RAMAKRISHNAN: J. Biol. Chem. 203, 35 (1953). — BEINERT, H., R. W. v. KORFF, D. E. BUYSKE, D. A. HANDSCHUHMACHER, R. E. HIGGINS and F. M. STRONG: J. Amer. Chem. Soc. 74, 854 (1952). — BRATTON, A. C., and E. K. MARSHALL: J. Biol. Chem. 128 537 (1939); 122, 263 (1938). — BUSKIRK, H. H., A. M. BERGDAHL and R. A. DELOR: J. Biol. Chem. 172, 671 (1948).

CROKAERT, R.: Bull. Soc. chim. XXXII (1941). — CROKAERT, R., S. MOORE et E. J. BIGWOOD: Bull. Soc. Chim. biol. 33, 1209—1213 (1951).

FARRER, K. T. H.: J. of Exper. Biol. a. Med. Sci 29, 285 (1951). — FEIGL, F., V. ANGER u. O. FREHDEN: Mikrochemie 15, 9, (1934). — FRAME, E. G., J. A. RUSSEL and A. E. WILHELMI: J. Biol. Chem. 149, 255 (1943). — FROST, D. V.: Ind. Eng. Chem. Anal. Edit. 15, 306 (1943).

Gergeley, J., P. Hele and C. V. Ramakrishnan: J. Biol. Chem. 198, 323 (1952).

Harrison, J. S.: Nature (London) 163, 798 (1949). — Hegstedt, D. M., and F. Lipmann: J. Biol. Chem. 174, 89 (1948). — Hilz, H.: Chem. Dissertation Univ. München 1953. — Hoag, H. R., and H. P. Sarret, F. H. Cheldelin: Ind. Eng. Chem. Anal. Ed. 17, 60 (1945).

Jones, M. E., S. Black, R. M. Flynn and F. Lipmann: Biochem. et Biophys. Acta 12, 1941 (1953).

Kaplan, N. O., and F. Lipmann: J. Biol. Chem. 174, 37 (1948). — Korrff, R. W. v.: J. Biol. Chem. 200, 401 (1953).

Long, C. L., and W. L. Williams: J. Bact. 61, 195 (1951). — Lipmann, F.: J. Biol. Chem. 160, 173 (1945). — Lipmann, F., N. O. Kaplan, G. D. Novelli, L. C. Tuttle and B. Guirard: J. Biol. Chem. 167, 869 (1947). — Lipmann, F.: The metabolic function of pantothenic acid in: "The Vitamins", Vol. II, 598 New York. Acad. Press. Inc. 1954:

Lynen, F., u. D. Reinwein: Private Mitteilung. —

Lynen, F., E. Reichert und L. Rueff: Ann. 574, 1 (1951).

Lynen, F.: The Harvey Lectures Acad. Press New York 1952—1953, p. 210.

Lynen, F.: Fed. Proc. 12, 683 (1953).

Neilands, J.B., and F.M. Strong: Arch. of Biochem. 19, 287 (1948). — Novelli, G. D., N. O. Kaplan and F. Lipmann: J. Biol. Chem. 177, 97 (1949). — Novelli, G. D., and F. J. Schmelz jr.: J. Biol. Chem. 192, 181 (1951).

Pelczar, M. J., and J. R. Porter: J. Biol. Chem. 139, 11 (1941).

Rasmussen, R. A., K. L. Smiley, J. G. Anderson, J. M. van Lanen, W.L. Williams and E. E. Snell: Proc. Soc. Exptl. Biol. Med. 73, 658 (1950).

Sanadi, D. R., J. W. Littlefield and R. M. Bock: J. Biol. Chem. 197, 851 (1952). — Schmidt, G., and S. J. Thannhauser: J. Biol. Chem. 149, 369 (1943). — Skeggs, H. R., and L. D. Wright: J. Biol. Chem. 156, 21 (1944). — Snell, E. E., and G. M. Brown: Adv. in Enzymology 14, 49 (1953). — Stadtmann, E. R., G. D. Novelli and F. Lipmann: J. Biol. Chem. 191, 365 (1951). — Stadtmann, E. R.: J. Biol. Chem. 196, 535 (1952). — Strong, F. M., and A. L. Neal: Ind. Eng. Chem. Anal. Edit. 15, 654 (1943). — Szalkowsky, C., and W. Mader, H. Frediani: Anal. Chem. 22, 369 (1950).

Vitucci, J. C., N. Bohonos, V. P. Wieland, D. V. Lefemine and B. L. Hutchings: Arch. Biochem. and Biophys. 34, 409 (1951).— Vitucci, J. C., N. Bohonos, O. P. Wieland, D. V. Lefemine, B. L. Hutchings, G. M. Brown, W. L. Williams and E. E. Snell: J. Am. Chem. Soc. 75, 6203 (1953).

Wieland, O.: Ärztl. Forschg. 10. Juli (1954). — Williams, W. L., E. Hoff-Jorgensen and E. E. Snell: J. Biol. Chem. 177, 933 (1949). — Williams, R. H., and E. Rohrmann: J. Amer. Chem. Soc. 58, 695 (1936). — Williams, R. J., and D. H. Saunders: Biochem. J. 28, 1887 (1934). — Wiss, O.: Z. f. Vitamin-, Hormon- und Fermentforschg. Bd. 4, 191, (1951). — Woolish, E. G., and H. Schmall: Analyt. Chem. 22, 1033 (1950).

Riboflavin, Folic Acid and Biotin.

By

F. M. Strong.

A. Riboflavin.

Riboflavin occurs in nature both in the free form, and as the coenzymes riboflavin adenine dinucleotide[1] (FAD) and riboflavin mononucleotide[1] (FMN, riboflavin-5'-phosphate). Little information is available as to the proportion of the total riboflavin which exists in these various forms in plant tissues. In the rat, BESSEY, LOWRY and LOVE (1949) found 70—90% of the total in the form of FAD and only negligible amounts of free riboflavin. The same situation probably holds true for many plant tissues, provided they have not been damaged or allowed to autolyze. For example, in the early days of riboflavin investigations KUHN, WAGNER-JAUREGG and KALTSCHMITT (1934) observed that over half the riboflavin in spinach leaves was in a bound state, i. e., not diffusable. As is now well known, the flavin coenzymes are more or less firmly bound to specific proteins as essential components of the various yellow enzymes, and most of the riboflavin of intact, metabolically active tissues no doubt exists in these combinations.

I. Distribution in Plants.

The total riboflavin concentration of a number of representative plant tissues is given in Table 1. Most information of this sort has been collected in connection with analysis of foods for nutritional studies, and there seem to have been few systematic attempts to ascertain the distribution of riboflavin throughout the various parts of entire plants. The available information up to 1948 has been reviewed by BONNER and BONNER. The vitamin is synthesized by growing root tips (BONNER, 1942), in seeds during germination (BURKHOLDER and McVEIGH, 1945), and presumably in green leaves and other plant tissues. As might be expected, the concentration roughly parallels metabolic activity of the tissue, and is higher in leaves, embryos, and young actively-growing shoots than in stems and storage tissues such as tubers and the endosperm of cereal grains.

II. Principles of Detection and Estimation.

Riboflavin has been determined by biological, microbiological, and fluorometric methods. The first two depend on the essential nature of riboflavin for growth of higher animals such as the chick and rat, which have been used for assay purposes, and for the growth and fermentative activity of various microorganisms, particularly the lactic acid bacteria. The animal assay is almost never used at present.

The physical methods are based on the intense yellow-green fluorescence of the vitamin (or its photolysis product, lumiflavin) produced by irradiation with visible light of wave length ca. 445 mμ. The specificity of the assay has been

---

[1] Although FAD and FMN are not true nucleotides, the above names are well established and will probably continue to be used.

Table 1. *Distribution of Riboflavin in Representative Plant Tissues.*

Plant	Part Analyzed	Riboflavin content[1] µg./g.	References
Cereals			
Corn	Whole seed	1.4	Andrews et al. (1942)
	Seed germinated 5 days, whole seedling	4.3	Burkholder (1943)
	Cobs	0.8—2.6	Hall et al. (1952)
	Tassels	6.5—9.4	Van Lanen et al. (1946)
Oat	Whole seed	1.3	Andrews et al. (1942)
	Seed germinated 5 days, whole seedling	11.6	Burkholder (1943)
	Endosperm	1.6	McVeigh (1944)
	Root of embryo	13.7	McVeigh (1944)
	Coleoptile of embryo	20.0	McVeigh (1944)
Sorgum	Whole seed	1.3	Hubbard et al. (1950)
	Endosperm	0.9	Hubbard et al. (1950)
	Germ	3.9	Hubbard et al. (1950)
	Bran	4.0	Hubbard et al. (1950)
Wheat	Whole seed	1.1—2.2	Andrews et al. (1942)
	Endosperm (patent flour)	0.34	Andrews et al. (1942)
	Bran	2.8	Andrews et al. (1942)
	Seed germinated 5 days whole seedling	5.4	Burkholder (1943)
Flowers			
Cucurbita mexicana	Entire flower, dried	25.0	Cravioto et al. (1945)
Brassica campestris	Entire flower, dried	1.5	Cravioto et al. (1945)
Agave atrovirens	Entire flower, dried	5.6	Cravioto et al. (1945)
Yucca aloifolia	Entire flower, dried	9.0	Cravioto et al. (1945)
Fruits			
Avocado	Edible portion of fruit	4.9	Cravioto et al. (1945)
Blackberries	Whole fruit	2.7	Watt et al. (1950)
Grapefruit	Juice	2.0	Watt et al. (1950)
Grape	Juice	2.6	Watt et al. (1950)
Mango	Fruit	2.4	Cravioto et al. (1945)
Peach	Edible portion of fruit	3.8	Watt et al. (1950)
Tomato	Whole fruit	6.8	Watt et al. (1950)
Grasses etc.			
Alfalfa	Aerial parts (alfalfa hay)	13.6	Morrison (1948)
	Leaf meal	19.2	Morrison (1948)
Bluegrass	Aerial parts (hay)	9.9	Morrison (1948)
Pea	Vines (from canning factory)	29.0	Morrison (1948)
Timothy	Aerial parts (hay)	9.0	Morrison (1948)
Legume seeds, and nuts			
Almonds	Shelled nuts	6.7	Watt et al. (1950)
Beans, kidney	Whole seed	2.4	Watt et al. (1950)
Peanuts	Shelled nuts	1.3	Watt et al. (1950)
Peas	Mature, dry seeds	2.8	Watt et al. (1950)
Pecans	Shelled nuts	1.1	Watt et al. (1950)
Soybeans	Mature, dry seeds	3.1	Watt et al. (1950)
Walnuts	Shelled nuts	1.3	Watt et al. (1950)
Vegetables			
Asparagus	Shoots	27.2	Watt et al. (1950)
Beans, green	Immature seed pods, with seeds	10.0	Watt et al. (1950)
Broccoli	Flower stalks	21.0	Watt et al. (1950)
Carrot	Roots	5.0	Watt et al. (1950)

[1] Dry basis.

Table 1. (Continued.)

Plant	Part Analyzed	Riboflavin content[1] µg./g.	References
Chard	Leaves	20.0	WATT et al. (1950)
Cucumber	Fruit	8.7	CRAVIOTO et al. (1945)
Lettuce	Leaves	15.4	WATT et al. (1950)
Parsley	Leaves and stems	17.5	WATT et al. (1950)
Potato	Tubers	1.8	WATT et al. (1950)
Miscellaneous			
Forsythia	Buds	8.4	BURKHOLDER and McVEIGH (1945)
Horse chestnut	Buds	2.1	BURKHOLDER and McVEIGH (1945)
	Leafy shoots	15.0	BURKHOLDER and McVEIGH (1945)
	Mature leaves	9.4	BURKHOLDER and McVEIGH (1945)
Tomato	Apex of plant	33.7	BONNER and BONNER (1948)
	Top (younger) leaves	30—37	BONNER and BONNER (1948)
	Bottom (older) leaves	10—20	BONNER and BONNER (1948)
	Top 10 cm. of stem	6.2	BONNER and BONNER (1948)
	Bottom 10 cm. of stem	2.8	BONNER and BONNER (1948)
	Roots	10.3	BONNER and BONNER (1948)
White oak	Buds	3.0	BURKHOLDER and McVEIGH (1945)
White pine	Needles	7.6	BURKHOLDER and McVEIGH (1945)
Yeast	Dried baker's	55.0	WATT et al. (1950)

improved by use of suitable filters for both the inciting and fluorescent light, removal of interfering pigments by oxidation, and use of pure riboflavin as an internal standard to compensate for errors arising from the effect of various extraneous materials on the fluorescence.

Both the fluorometric and microbiological methods have been developed into reliable analytical procedures, which nicely supplement each other. Since some samples are best handled by one procedure and some by the other, both will be described.

Purely qualitative methods for detection of riboflavin have received little attention, although the characteristic yellow-green fluorescence has been employed in cytochemical work for localization and identification of the vitamin in cells and tissues (for literature see GLICK, 1949).

III. Quantitative Determination of Riboflavin.

1. Fluorometric Method.

The procedure described is taken mainly from that adopted by the Association of Official Agricultural Chemists (1949), and from the detailed directions given by the Association of Vitamin Chemists (1951). Many additional details and descriptions of alternative methods are given by STILLER in GYÖRGY (1950). All operations must be carried out in dim light, preferably with a red safety lamp, or in red or amber glassware (see DE MEERE and BROWN, 1944; OTTES and ROBERTS, 1949). Contact with materials which may contribute fluorescing

---

[1] Dry basis.

impurities, such as cork, rubber, stopcock grease, etc., should be avoided. To prevent destruction of riboflavin, solutions must be kept below pH 7.0 at all times.

Reagents.

1. Riboflavin stock solution. Prepare a solution of pure dry riboflavin in 0.02 N acetic acid containing 25 μg. per ml.[1], and store under toluene in a brown glass bottle in the refrigerator.

2. Riboflavin standard solution. Dilute an aliquot of the stock solution with water to a concentration of 0.5 μg. per ml. This standard solution is freshly prepared on the day it is to be used.

3. Hydrochloric acid and sodium hydroxide solutions. 10 N, 1 N, and 0.1 N solutions of each.

4. Glacial acetic acid, and 0.02 N acetic acid.

5. Potassium permanganate solution. A 3% solution in water, freshly prepared each week.

6. 3% hydrogen peroxide solution.

7. Solid sodium hydrosulfite ($Na_2S_2O_4$).

Apparatus. An electronic photofluorometer is required which is suitable for measuring the fluorescence of riboflavin solutions containing ca. 0.05—0.1 μg. per ml. If the instrument available has a higher or lower sensitivity, appropriate changes in the concentration of the riboflavin standard solution and in the amount of sample taken for analysis should be made so that the riboflavin concentration in the solutions finally measured falls within the range of the instrument. The instrument should be equipped with appropriate filters to screen both the incident and fluorescent light. For example, Hoffer, Alcock and Geddes (1944) recommend Corning[2] filters 511 and 038 for the incident beam and filter 351 for the fluorescent beam. According to the manufacturer, the wavelengths which pass through the 511 plus 038 filter combination are about 410 to 460 mμ, while transmission through the 351 filter is zero below 510 mμ and 80—90% above 540 mμ. Any filters passing approximately these wavelengths would undoubtedly be satisfactory.

Preparation of Test Solution. A weight of sample estimated to contain 5—10 μg. of riboflavin and 50 ml. of 0.1 N HCl are mixed thoroughly so that all lumps are dispersed, and the mixture is autoclaved at 15 pounds per square inch pressure (121° C) for thirty minutes[3].

The pH of the cooled sample is raised to 6.0 by careful addition of sodium hydroxide with constant agitation, and immediately lowered to 4.5 with 1 N HCl. The mixture is then diluted with water to 100 ml. and filtered or centrifuged. To 50 ml. of the filtrate 1 N HCl is added dropwise until no further precipitate forms, and this is followed by a volume of 1 N NaOH equal to that of the 1 N HCl used. The mixture is diluted to 100 ml. with water and filtered again if necessary. This product is designated the *test solution*. This solution should contain about 0.025—0.05 μg. of riboflavin per ml. If more dilute, it should be evaporated under vacuum to approximately this concentration.

---

[1] Higher concentrations may partially crystallize out on standing in the refrigerator (Hanson and Weiss, 1945).

[2] Corning Glass Works, Corning, New York, USA. The designation of these filters has now been changed as follows: 511 = 5113, 038 = 3389, and 351 = 3486.

[3] The volume of the acid solution in ml. should be at least ten times the dry weight of the sample in grams. Certain low potency samples may therefore require more than the indicated 50 ml. of acid. For samples which contain appreciable amounts of basic substances or have significant buffering capacity sufficient additional acid should be added to compensate for these materials. This is added as a concentrated solution to avoid excessive dilution.

Removal of Interfering Pigments. Place 10 ml. portions of the test solution, prepared as described above, in each of four test tubes (or fluorometer cuvettes). To each of two of these tubes add 1 ml. of riboflavin standard solution (0.5 μg. per ml.), and to each of the others add 1 ml. of water. Now add to each tube with mixing, first 1 ml. of glacial acetic acid, then 1 ml. of 3% potassium permanganate solution and finally, exactly two minutes after addition of the permanganate, 1 ml. of 3% hydrogen peroxide. The permanganate color should be discharged within 10 seconds. Shake or tap the tubes to remove air bubbles if any are present. The solutions are now ready for measurement of their fluorescence.

Fluorometry. The linearity of response of the fluorometer to varying concentrations of known fluorescent substances should be verified, if necessary, as described by LOOFBOUROW and HARRIS (1942).

Measure the fluorescence (expressed in terms of galvanometer deflections) of the sample solutions containing 1 ml. of added riboflavin solution, and designate the average of the two results "reading A". Similarly measure the sample solutions containing 1 ml. of added water, and designate the average "reading B". Add to two or more tubes approximately 20 mg. of solid, powdered sodium hydrosulfite and measure the fluorescence within 10 seconds to obtain "reading C". All tubes should give the same reading after this treatment, which reduces the riboflavin present to the non-fluorescent leuco form.

The riboflavin content of the test solution may then be calculated as follows: —

$$\frac{B-C}{A-B} \times \frac{0.5}{10} = \mu g. \text{ per ml.}$$

Calculate the riboflavin content of the original sample on the basis of the aliquots taken and dilutions made during the analysis.

Sources of Error. The fluorometric procedure is unreliable on samples which contain high concentrations of iron unless the iron is removed, e. g., by precipitation as the phosphate at pH 4.5—6.6. Likewise highly pigmented materials such as those which have become caramelized by heating are unsuitable. In general, interfering substances must be removed, and this is best accomplished by precipitation of proteins by pH adjustment, followed by permanganate oxidation. If these measures are insufficient, the microbiological method should be used. Adsorption on and elution from Florisil columns has been suggested as a means of removing impurities, but experience has shown that this procedure causes additional difficulties, and that it can better be omitted (ANDREWS, 1944; HOFFER et al., 1944). Destruction of some riboflavin may occur during the permanganate oxidation procedure if the time or concentration limits stated above are exceeded.

2. Microbiological Method.

The assay procedure described below is essentially that given by SNELL (1950), although details have also been taken from the Association of Vitamin Chemists (1951). Much information and background material is also presented by BARTON-WRIGHT (1952), and by STRONG (1947). These references also describe the general techniques for carrying out microbiological assays, and should be consulted by those analysts who are unfamiliar with this type of work. (Cf. also the chapter on "Principles of Bioassays", Vol. I of this handbook.)

The principle of the microbiological assay, in brief, is that the growth and acid production of certain lactobacilli are, within definite limits, proportional to the amount of riboflavin available to the culture. In practice a riboflavin-free basal medium is prepared which is sufficiently complete with respect to other required nutrients so that the bacterial response is limited only by riboflavin. This medium is distributed among a series of test tubes, graded amounts of sample extract or standard riboflavin solution added, and the sterilized tubes inoculated with the test organism and incubated. Comparison of the response to sample extracts with that from known amounts of pure riboflavin permits estimation of the sample potency.

Basal Medium. Although many modifications of the original medium of Snell and Strong (1939) have been suggested, none appears to afford more reliable assay results. Since the utility of the original medium has been verified by widespread use, it is again recommended in the present procedure.

A basal medium stock solution is first prepared which has exactly twice the concentration desired in the final assay tubes. This arrangement leaves room for addition of sample solutions as needed. The composition of this stock solution is as follows:

photolyzed peptone solution	100 ml.
cystine hydrochloride solution	100 ml.
yeast supplement solution	10 ml.
salt solution A	5 ml.
salt solution B	5 ml.
anhydrous glucose	10 g.

The solutions are mixed, the glucose dissolved in the mixture, the pH adjusted to 6.6—6.8, and the solution diluted to 500 ml. with water. This quantity is sufficient for 100 tubes[1].

The stock solutions needed for the basal medium may be preserved satisfactorily for at least a month if kept in the dark in the presence of about 0.1% chloroform and 0.5% toluene. Natural mixtures such as the yeast and peptone solutions should be refrigerated to retard microbial growth. Other stock solutions may be stored at room temperature so long as microbial growth is not evident.

Reagents and Stock Solutions.

Photolyzed Peptone Solution. To a solution of 40 g. of commercial peptone (e. g., "Bacto-Peptone", Difco Laboratories, Detroit, Michigan) in 250 ml. of water is added 20 g. of sodium hydroxide dissolved in 250 ml. of water. The solution is placed in the light (window sill) in a plain glass bottle for 24 hrs. The pH is then adjusted to 6.6—6.8 by addition of glacial acetic acid (ca. 27—29 ml.), 7 g. of anhydrous sodium acetate added, and the mixture diluted to 800 ml. with water.

Cystine Hydrochloride Solution. One gram of cystine is dissolved in a little concentrated hydrochloric acid (4—5 ml.) and the solution diluted to 1000 ml. with water.

Yeast Supplement Solution. To 25 g. of commercial dried yeast extract (e. g., Bacto Yeast Extract, Difco Laboratories, Detroit, Michigan) dissolved in 125 ml. of water is added a solution of 38 g. of basic lead acetate in 125 ml. of water[2]. The solutions are mixed well, adjusted to pH 9—10 with concentrated ammonium hydroxide, and the mixture filtered with suction. The filtrate is brought to pH 6.5 with glacial acetic acid, and excess lead removed with hydrogen sulfide. After filtration, the solution is concentrated *in vacuo* about 20% to remove dissolved hydrogen sulfide, then diluted with water to 250 ml.

---

[1] The complete medium is available in dehydrated form from the Difco Laboratories, Detroit, Michigan, USA (see "Difco Manual", 7th ed., 1953). Dehydrated media are also available for the biotin and folic acid microbiological assays.

[2] If dried yeast extract is not available, 500 g. of fresh baker's yeast (starch free) is mixed with 5 l. of water, heated 2 hrs. in flowing steam, then autoclaved at 15 pounds pressure for 15 min., and filtered. After concentrating the filtrate to 125 ml. *in vacuo*, it may be precipitated with lead as described above.

Inorganic Salt Solutions. Solution A consists of 25 g. of K_2HPO_4 and 25 g. of KH_2PO_4 in 250 ml. of water. Solution B contains 10 g. of $MgSO_4 \cdot 7H_2O$, 0.5 g. of NaCl, 0.5 g. of $FeSO_4 \cdot 7H_2O$, and 0.5 g. of $MnSO_4 \cdot H_2O$ in 250 ml. of water.

Riboflavin Standard Solution. Prepare a stock solution containing 25 μg. per ml. of pure, dry riboflavin in 0.02 N acetic acid. This solution may be stored in the refrigerator under toluene in a brown glass bottle for at least a month. The standard solution is prepared by diluting an aliquot of the stock solution with water to a concentration of 0.1 μg. per ml. Fresh standard solution is prepared for each assay.

Cultures and Inoculum. The medium used for carrying stock cultures of the test organism is prepared as follows:—

Agar	1.5 g.
Glucose, anhydrous	1.0 g.
Dried yeast extract	0.75 g.
Bacteriological peptone (e. g., Difco Bacto)	0.75 g.
KH_2PO_4	0.2 g.
Tomato juice preparation[1]	10 ml.
"Tween 80", 10% solution in 95% ethanol[2]	1 ml.

The ingredients are mixed with enough water to make a total volume of 80—90 ml., the mixture is adjusted to pH 6.8, warmed to bring all solids into solution, and diluted to 100 ml. While still warm enough to remain liquid, 10 ml. portions are distributed among a number of test tubes, which are plugged with cotton, sterilized at 15 pounds for 15—20 minutes, and cooled in an upright position. The finished tubes are stored in the refrigerator until needed. The liquid medium for growing inoculum has the same composition except that the agar is omitted.

The assay organism, *Lactobacillus casei*, is obtainable from the American Type Culture Collection, 2029 M Street, N. W., Washington, D. C., under number 7469. When first obtained, a transfer is made into the liquid inoculum medium, and the tube incubated at 37° $\pm$ 1° C for 12—24 hrs. Subcultures are made in this medium at daily intervales, until the organism is so active that heavy growth is evident after only 6—8 hrs. incubation. Stab cultures are then made in the agar medium, and after being incubated 16—24 hrs. at 37° are held in the refrigerator until needed. These stock cultures are transferred at weekly intervals.

Inoculum is grown by transferring cells from a stock culture to a tube containing 10 ml. of the sterilized liquid inoculum medium, and incubating for 6—8 hrs. If heavy growth does not occur in this time, the organism is activated by daily transfers as described above. This procedure avoids the loss of response to added known vitamines sometimes encountered in microbiological assay work (NYMON and GORTNER, 1946; JONES and MORRIS, 1949). When a satisfactory culture has been obtained, the cells are centrifuged, the supernatant decanted, and the cells resuspended in 20 ml. of sterile, 0.9% sodium chloride solution. One drop of this suspension is used to inoculate each assay tube.

Preparation of Test Solution. An amount of the finely ground or homogenized sample sufficient to contain an estimated 10 μg. of riboflavin is suspended in 50 ml. of 0.1 N hydrochloric acid, and the mixture autoclaved at 15 lbs. pressure (121° C) for 15 minutes. Cool, adjust to pH 4.5 with 2.5 N sodium acetate solution,

---

[1] Commercial canned tomato juice is centrifuged or filtered with the help of a diatomaceous filter aid, and the clear filtrate used.

[2] Obtainable from Atlas Powder Company, Wilmington 99, Delaware, USA. It is probable that the use of this ingredient is not absolutely essential.

dilute to 100 ml., and filter[1]. This procedure removes interfering materials especially fatty acids (Strong and Carpenter, 1942). An aliquot of the filtrate estimated to contain 5 μg. of riboflavin is then adjusted to pH 6.6—6.8 and diluted to 100 ml[2].

Assay Procedure. A series of lipless test tubes (16 or 18 × 150 mm.) held in a metal rack is used for the assay. To duplicate tubes are added amounts of riboflavin standard solution (0.1 μg. per ml.) sufficient to supply 0.0, 0.05, 0.10, 0.15, 0.20, and 0.25 μg. of the vitamin. To other tubes are added aliquots of the test solution, also in duplicate, in such amounts as to provide three or more levels of riboflavin estimated to fall in the range 0.05—0.20 μg. per tube. The volume in each tube is then brought to 5 ml. with distilled water, and to each tube is added 5 ml. of the basal medium stock solution. The tubes are then plugged with non-adsorbent cotton, or covered with metal or glass caps, and the rack plus tubes sterilized by autoclaving at 15 pounds pressure for 10—15 minutes. After cooling, the tubes are inoculated, and placed at 37° ± 1° in the dark for 72 hours.

At the end of the incubation period the bacterial response is measured by titration of the lactic acid produced in each tube. Alternatively the bacterial growth may be estimated after 16—18 hrs. incubation by turbidity measurements with a photoelectric colorimeter or turbidimeter provided turbidity is not contributed by the sample. In general the titrimetric procedure is the more reliable.

Calculation of Results. A standard curve is prepared by plotting the titration data from the tubes of the standard series against the corresponding amounts of pure riboflavin present. The blank titration should require the equivalent of about 0.5—1.0 ml. of 0.1 N sodium hydroxide, but in any event must be below 2.0 ml. for the assay to be valid. The curve should then extend in a linear or nearly linear fashion to a titration of 8—10 ml. of 0.1 N alkali at about 0.25 μg. of riboflavin. Titration values on replicate tubes should agree within 0.2 ml.

The riboflavin content of the sample tubes is estimated by interpolation on the standard curve. At least two and preferably three of the levels tested should fall in the range of 0.05—0.25 μg. of riboflavin per tube. If these requirements are fulfilled, the results from each level tested are divided by the milliliters of test solution used to arrive at the riboflavin content per ml. of the test solution. If these values agree within ± 10% of the mean, and show no consistent upward or downward variation with increasing levels tested, all are averaged to obtain the final result.

Sources of Error. A variety of materials are known to enhance or diminish the response of lactic acid bacteria to riboflavin and hence to influence the assay results (Strong, 1947). Perhaps the most common are fatty acids, which must be removed if present in appreciable amounts. Filtration at pH 4.5 serves very well in most cases, although ether extraction is required for some highfat samples (Strong and Carpenter, 1942). One of the most common and troublesome difficulties is failure of riboflavin values obtained from different levels of samples to agree, but rather to show a consistent "drift" in one direction or the other. The cause and cure of such drifts have been considered at length by Snell (1950). This difficulty will probably not be encountered with most samples, if the recommended procedure is followed.

Slight variations in incubator temperatures have been shown to cause significant errors (Price and Graves, 1944), and some authors therefore prefer to use a water bath for incubation. Certain foods both of animal and plant origin have

---

[1] More complete precipitation may result if the pH is first brought to 5—6 and then adjusted to 4.5.

[2] Larger volumes of extractant may be required for some samples, see p. 646.

been reported to show an apparent increase in riboflavin content on processing (HINMAN et al., 1946), and to the author's knowledge no adequate explanation for the observed discrepancies has been advanced.

Modified Methods. The microbiological determination of riboflavin has been scaled downward 50-fold to produce an ultra-micro method (LOWRY and BESSEY, 1944). The assay is carried out in a total volume of 0.2 ml., and permits the determination of 0.5—2.0 millimicrograms of the vitamin with a probable error of about 3%.

KORNBERG, LANGDON and CHELDELIN (1948) have proposed the use of *Leuconostoc mesenteroides* (American Type Culture Collection, No. 10,100) as an assay organism for riboflavin. Since this organism is fifty times more sensitive to riboflavin than is *L. casei*, smaller amounts of sample are required and the disturbing effects of extraneous substances in the sample correspondingly diminished.

3. Other Flavins.

The riboflavin nucleotides, FAD and FMN as well as free riboflavin, may be determined fluorometrically by the procedure of BESSEY et al. (1949). This method is quicker and more sensitive than the enzymatic methods previously used (KLEIN and KOHN, 1940). However, it does not appear to have been applied to plant tissues to date. For detailed directions the original publication should be consulted.

B. Pteroylglutamic Acid and Related Compounds.

1. State of Combination in Living Organisms.

Pteroylglutamic acid, I (folic acid, or folacin), apparently represents one stage in the biosynthesis of 5-formyl-5,6,7,8-tetrahydropteroyl-L-glutamic acid, II (folinic acid, "Leucovorin", or "citrovorum factor"), which is, or is more closely related to, the physiologically active form of this vitamin (BAUMANN, 1953).

Pteroylglutamic acid, I.

Leucovorin, II.

Synthetic Leucovorin has recently been resolved (COSULICH, SMITH and BROQUIST, 1952), and the levo-rotary isomer shown to be very probably identical with natural "citrovorum factor" isolated by KERESZTESY and SILVERMAN (1951).

Pteroylglutamic acid exists in biological materials partially in combination with additional glutamic acid, specifically as pteroyltriglutamate (teropterin) and pteroylheptaglutamate ("vitamin B_c conjugate"), and the possible existence of still other forms in nature is recognized. The "citrovorum factor" also exists to a large extent as a conjugate in biological materials (Couch and Doctor, 1953; Chang, 1953). Pteroic acid itself, and 10-formylpteroic acid, III (rhizopterin, or the "*Streptococcus lactis* R factor") do not possess typical folic acid activity for

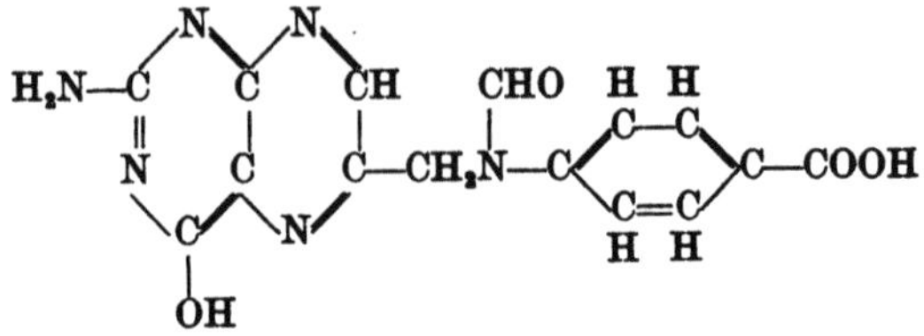

Rhizopterin, III.

higher animals, although the latter is found in nature. The glutamic acid derivative of rhizopterin (10-formyl-pteroylglutamic acid, formyl folic acid), however, does possess the typical vitamin activity. This compound has been prepared synthetically, and may also have natural occurrence.

The complex group of substances related to pteroylglutamic acid has been the subject of numerous reviews (for literature see Williams, Eakin, Beerstecker and Shive, 1950). For purposes of the present discussion the expression "folic acid activity" will be used to designate the biological effect of any, or any combination of, the active members of the group which may occur in natural materials. For convenience the following abbreviations will be used: CF = natural "citrovorum factor" (levo rotatory 5-formyl-5,6,7,8-tetrahydropteroyl-L-glutamic acid); PGA = pteroyl-L-glutamic acid.

Distribution in Plants. It now seems probable that intact tissues contain most of their folic acid activity in the form of CF (Chang, 1953), mostly in a bound form (Couch et al., 1953; Hill and Scott, 1951). However, this situation has only recently come to light, and nearly all the available analytical data are expressed in terms of PGA. Representative values are listed in Table 2. It may be seen that the metabolically more active tissues, especially leaves, are relatively high in folic acid activity.

2. Principle of Detection and Estimation.

Compounds of the folic acid group have been determined almost exclusively by microbiological methods to date. Most of the data in the literature have been obtained with the aid of *Lactobacillus casei* or *Streptococcus faecalis* as the test organisms. Samples have commonly been subjected to a hydrolytic procedure intended to convert the polyglutamates into PGA so as to measure all the members of the complex insofar as possible. Because of the sensitivity of these substances to chemical hydrolysis, enzymatic methods must be employed, and "conjugase" preparations from such sources as chick pancreas or hog kidney have been used. It has been observed, also, that conjugase inhibitors are present in some samples (Mims, Swendseid and Bird, 1947). Recently, since the discovery of CF, this substance has been determined by microbiological assay based on the use of *Leuconostoc citrovorum*. As pointed out above, it is probable that much of the total folic acid activity of biological materials exists in this form, and it is, therefore, necessary to select an assay method which will include CF. *L. casei* and *S. faecalis* are both suitable from this standpoint as they respond fully to both

CF and PGA. However, it is generally agreed that *L. casei*, although more sensitive to PGA, is also more susceptible to disturbance from various extraneous substances, and *S. faecalis* is usually preferred. An *S. faecalis* method has been adopted as an official procedure by the Association of Official Agricultural Chemists (1950), and a modified version is recommended by BARTON-WRIGHT (1952) on the basis of the studies of JONES and MORRIS (1949). The procedure described below is very similar to that of the Association of Vitamin Chemists (1951).

Table 2. *Distribution of Pteroylglutamic Acid in Plants.*

Plant	Part analyzed	Pteroyl-glutamic acid content[1] µg./g.	Reference
Alfalfa	Aerial parts (hay)	3.00	FAGER et al. (1949)
Almond	Shelled nuts	1.04	TOEPFER et al. (1951)
Asparagus	Shoots (edible portion)	12.0	FAGER et al. (1949)
Avocado	Edible portion of fruit	1.35	TOEPFER et al. (1951)
Bean (maturing)	Upper large leaves	9.5	FAGER et al. (1949)
	Lower leaves	8.6	
	Small beans	10.2	
	Seeds (older beans)	5.7	
	Pods (older beans)	4.2	
Beet	Roots (edible portion)	2.4	FAGER et al. (1949)
Brewers yeast	Dried yeast	21.8	TOEPFER et al. (1951)
Broccoli	Flower stalks	3.33	TOEPFER et al. (1951)
Carrot	Roots	0.55	TOEPFER et al. (1951)
Forsythia	Buds	1.47	BURKHOLDER and McVEIGH (1945)
Green bean	Immature pods and seeds (edible portion)	6.2	FAGER et al. (1949)
Hemlock	Needles	3.31	BURKHOLDER and McVEIGH (1945)
Horse chestnut	Buds	0.13	BURKHOLDER and McVEIGH (1945)
	Leafy shoots	3.79	
	Mature leaves	0.81	
Leaf lettuce	Aerial parts (edible portion)	22.3	FAGER et al. (1949)
Oat	Whole, mature seeds	0.40	FAGER et al. (1949)
Orange	Juice	0.36	TOEPFER et al. (1951)
Parsley	Aerial parts (edible portion)	12.0	FAGER et al. (1949)
Pea	Immature seeds	1.14	FAGER et al. (1949)
Peach	Edible portion of fruit	1.35	TOEPFER et al. (1951)
Peanut	Shelled nuts	1.26	TOEPFER et al. (1951)
Potato	Tubers	0.18—0.93	TOEPFER et al. (1951)
Soybean	Whole, mature seeds	5.3	FAGER et al. (1949)
Spinach	Aerial parts (edible portion)	22.3	FAGER et al. (1949)
Timothy	Aerial parts (hay)	3.00	FAGER et al. (1949)
Tobacco (15 inches tall)	Stem tip and small leaves	11.6	FAGER et al. (1949)
	Upper large leaves	11.1	
	Lower leaves	2.6	
	Stem	1.5	
	Midrib from central leaves	1.4	
	Central leaves, minus midrib	8.5	
	Roots	3.1	
Walnut	Shelled nuts	1.04	TOEPFER et al. (1951)
Wheat	Whole, mature seeds	5.3	FAGER et al. (1949)
White oak	Buds	0.25	BURKHOLDER and McVEIGH (1945)
White pine	Needles	2.13	BURKHOLDER and McVEIGH (1945)

[1] Dry basis. *S. faecalis* method.

3. Quantitative Determination of Folic Acid Activity with
Streptococcus Faecalis.

Basal Medium. The composition of the basal medium is given in Table 3.

Table 3. *Basal Medium for Determination of Folic Acid Activity with Streptococcus Faecalis.*

Component	Amount[1]	Component	Amount[1]
	grams		milligrams
Sodium citrate dihydrate . .	35	$MgSO_4 \cdot 7\,H_2O$	200
Glucose, anhyd.	20	$MnSO_4 \cdot H_2O$	50
Hydrolyzed casein	5	NaCl	[2]
K_2HPO_4	3.5	$FeSO_4 \cdot 7\,H_2O$	10
	milligrams		micrograms
L-Cystine	400	Pyridoxine	1250
Glycine	200	Riboflavin	500
DL-Alanine	200	Niacin	500
Asparagine	200	Pyridoxal	500
DL-Tryptophan	200	Ca pantothenate	400
Glutamic acid	20	Thiamine	200
Xanthine	10	p-Aminobenzoic acid	100
Adenine.	10	Biotin	10
Guanine	10		
Uracil	10		

Reagents and Stock Solutions.

Acid-hydrolyzed Casein Solution. Reflux 100 g. of "vitamin-free" casein with 500 ml. of approximately 20% hydrochloric acid for 8—10 hrs., concentrate *in vacuo* to a thick paste, dissolve in 1—2 volumes of water, and reconcentrate. Dissolve the paste in water, adjust to pH 3.5 with sodium hydroxide solution, dilute to ca. 600 ml. and decolorize with 20—40 g. of activated charcoal[3] (e. g., Norite A, or Darco G-60) at 80°. The filtrate from the charcoal should be pale amber to colorless; if otherwise, repeat the treatment. This removes folic acid activity and thereby makes possible low blank values in the assay. Finally adjust to pH 6.6—6.8 and dilute to 1 liter.

Cystine-tryptophan Solution. Suspend 4.0 g. of L-cystine and 2.0 g. of DL-tryptophan (or 1.0 g. of L-tryptophan) in 700 ml. of water and add ca. 12 ml. of 20% hydrochloric acid with stirring at 70—80° until the amino acids dissolve. Cool and dilute with water to 1 liter. L-Tryptophan from natural sources may contain sufficient folic acid activity to produce a high blank. Such activity may be removed with charcoal at pH 3 as described above[4].

Adenine-guanine-uracil Solution. Dissolve 0.1 g. each of adenine sulfate, guanine hydrochloride, and uracil in 5 ml. of 20% hydrochloric acid with heating, adding water if necessary, and dilute with water to 500 ml.

Xanthine Solution. Dissolve 0.1 g. of xanthine in 100 ml. of ca. *N* ammonium hydroxide solution (warming if necessary), and dilute with water to 500 ml.

Asparagine Solution. Prepare a 2% solution of L-asparagine monohydrate in water.

DL-*Alanine Solution.* Dissolve 2 g. of DL-alanine in water and dilute to 100 ml.

---

[1] Amount for 1 l. of final medium, or 500 ml. of the double strength basal medium stock solution.

[2] Supplied in adequate amounts by the hydrolyzed casein solution.

[3] More complete removal of folic acid activity may often be obtained if the charcoal is heated 2 hrs. at 100° before use.

[4] Some workers prefer to weigh out solid cystine and tryptophan as needed, because this solution tends to become discolored on storage and high blanks result.

Salts Solution. Dissolve 20 g. of $MgSO_4 \cdot 7 H_2O$, 1 g. of $FeSO_4 \cdot 7 H_2O$, and 5 g. of $MnSO_4 \cdot H_2O$ in water, add 1 ml. of concentrated hydrochloric acid and dilute to 1000 ml. with water.

Vitamin Solution. Dissolve in 20% ethanol 20 mg. of thiamine hydrochloride, 50 mg. of riboflavin, 50 mg. of nicotinic acid, 125 mg. of pyridoxine hydrochloride, 50 mg. of calcium pantothenate, 10 mg. of p-aminobenzoic acid, and 1 mg. of biotin, and dilute with the same solvent to 1 liter. Protect from light.

Pyridoxal Solution. Dissolve 50 mg. of pyridoxal hydrochloride in 50% ethanol and dilute to 1 liter.

Basal Medium Stock Solution. This solution, which is exactly twice the concentration of the final medium in the assay tubes, is prepared as follows from the stock solutions described:

Sodium citrate dihydrate	35 g.
Glucose, anhydrous	20 g.
Dipotassium phosphate (K_2HPO_4)	3.5 g.
Cystine-tryptophan solution	100 ml.
Acid hydrolyzed casein solution	50 ml.
Adenine-guanine-uracil solution	50 ml.
Xanthine solution	10 ml.
Asparagine solution	10 ml.
Vitamin solution	10 ml.
Pyridoxal solution	10 ml.
DL-Alanine solution	10 ml.
Salts solution	10 ml.
Glycine	200 mg.
Glutamic acid	20 mg.

The various components are mixed with enough water to make about 450 ml., adjusted to pH 6.8, and diluted to 500 ml. This is sufficient medium for 100 tubes.

Pteroylglutamic Acid Solutions. A stock solution is prepared by dissolving 100 mg. of pure PGA in a 0.01 N solution of sodium hydroxide in 20% ethanol (by volume), adjusting to pH 7—8 with dilute hydrochloric acid, and diluting with 20% ethanol to 1 liter. From this a working standard solution is prepared fresh for each assay by diluting a suitable aliquot with water to a concentration of 0.002 μg. per ml.

Cultures and Inoculum. The test organism is *Streptococcus faecalis*, strain R, obtainable from the American Type Culture Collection, as number 8043. The organism is carried as stab cultures in the medium described on p. 649 under the microbiological determination of riboflavin. Inoculum is also prepared as described under the riboflavin method.

Preparation of Test Solution. Avoid exposure of folic acid solutions or sample extracts to direct sunlight, although light from an ordinary incandescent bulb is permissible (JONES and MORRIS, 1949). An amount of the homogenized or finely divided sample estimated to contain 1—2 μg. of PGA, or the equivalent in the form of other compounds with folic acid activity, is placed in a test tube or small flask, 5 ml. of a 1.25% aqueous solution of sodium acetate added, and the mixture heated at 100° (water bath) for 5 minutes. After cooling, 2 ml. of enzyme preparation[1] and 0.5—1.0 ml. of toluene are added, and the mixture incubated at 37° for 24 hrs. Place the tube in a 100° water bath for 5—10 minutes, cool, adjust to pH 6.6—6.8, and dilute to such a volume that the solution contains about 0.5 to

---

[1] Disperse fresh hog kidney tissue in three times its weight of water (e. g., with a Waring Blendor or equivalent mechanical device), centrifuge and filter the supernatant liquid through a diatomaceous filter aid. Keep frozen in 2 ml. portions until used.

2.0 mμg. of PGA (or the equivalent) per ml. Filter a sufficient portion for assay. Enzyme blanks must be run and suitable corrections made to allow for the folic acid activity of the hog kidney preparation.

Chicken pancreas is also a suitable source of conjugase for determination of folic acid activity. The fresh tissue is ground or homogenized with 5 volumes of acetone (e. g., in a Waring Blendor), the insoluble material filtered off, rewashed with acetone, and air dried. This acetone powder is stored in the cold, and is stable for several months. For use, a sufficient quantity to provide 20 mg. per gram of sample is uniformly suspended in a little water, and substituted for the hog kidney preparation in the above procedure. In this case, also, the sample must be dispersed in 5 ml. of 0.2 M phosphate buffer pH 7, rather than 1.25% sodium acetate solution.

Assay Procedure. Except as otherwise indicated the assay procedure is the same as that for the microbiological determination of riboflavin. The standard curve is obtained from a series of duplicate tubes containing 0.0, 0.5, 1.0, 1.5, 2.0, 2.5, 3.0 and 4.0 ml., respectively, of the working standard solution (0, 1, 2, 3, 4, 5, 6 and 8 mμg., respectively, of PGA). The standard curve should extend from a blank value of ca. 2 ml. of 0.1 N alkali or less to ca. 12 ml. at 8 mμg. of PGA. The test solution is set up in duplicate at four or more levels estimated to fall in the range corresponding to 1—6 mμg. of PGA. Calculate the results from not less than three sets of sample tubes which show folic acid activity values agreeing with each other to within ± 10% of the average.

The bacterial response may also be estimated from turbidity measurements after 16—18 hrs. provided there is no interference from the sample.

Sources of Error. The greatest uncertainty in the above method relates to the treatment of the sample to liberate the bound forms of PGA and CF. It is never certain that the enzymatic release is complete, and in doubtful cases or when dealing with previously untested materials, it is well to investigate the effect of varying the amount of enzyme preparation used (cf. Toepfer et al., 1951).

False positive results may be obtained because the response of *S. faecalis* is not strictly specific. Thus it is able to use pteroic acid and formylpteroic acid (rhizopterin) in place of PGA, whereas higher animals are not. Likewise, high levels of thymine or thymidine show full folic acid activity for *S. faecalis*. These and other possible sources of error are discussed at length by Snell (1950). An extended discussion of the reliability and reproduceability of the assay is also given by Toepfer et al. (1951).

Other Analytical Methods. Folic acid activity is also commonly estimated microbiologically with *L. casei* as the test organism (Association of Vitamin Chemists, 1951; Toepfer et al., 1951). Animal assay methods (Bliss and György, 1951) are now largely of historical interest only.

C. Biotin.

Much of the biotin in yeast, seeds, nuts, and animal products exists as a tightly-bound, water-insoluble complex, whereas vegetables, green plant materials, and fruits contain a water-extractable form (Lampen, Bahler and Peterson, 1942; Wright, 1947). The water-soluble form may be either free biotin or a combination with other still unknown components (Bowden and Peterson, 1948). At least one such substance has been isolated and characterized, viz. biocytin, a peptide-like compound of biotin and lysine (Wright et al., 1951). A brief survey of the occurrence of biotin in plant materials is given in Table 4.

Table 4. *Biotin Content of Representative Plant Tissues.*

Plant	Part analyzed	Biotin content[1] μg./g.	Reference
Alder	Buds	0.76	BURKHOLDER and McVEIGH (1945)
Alfalfa	Aerial parts	0.28	LAMPEN et al. (1942)
Corn	Cobs	0.04	HALL et al. (1952)
Forsythia	Buds	0.11	BURKHOLDER and McVEIGH (1945)
Hemlock	Needles	0.17	BURKHOLDER and McVEIGH (1945)
Horse chestnut	Buds	0.52	BURKHOLDER and McVEIGH (1945)
	Leafy shoots	2.26	
Oat	Entire plants, 24—45 days after planting	1.5—2.3	KOHLER (1944)
Orange	Juice	0.04—0.10	KREHL and COWGILL (1950)
Pea	Immature seeds	0.11	LAMPEN et al. (1942)
Peanut	Raw nuts	0.20	LAMPEN et al. (1942)
Sorghum	Whole grain	0.09—0.46	TANNER et al. (1947)
	Whole grain	0.20	HUBBARD et al. (1950)
	Endosperm	0.11	HUBBARD et al. (1950)
	Germ	0.57	
	Bran	0.35	
Soybean	Dry, whole seeds	0.63—0.94	BURKHOLDER and McVEIGH
	Sprouted seeds	1.08—1.79	(1945)
Spinach	Aerial parts	0.48	LAMPEN et al. (1942)
Walnut, English	Shelled nuts	0.37	LAMPEN et al. (1942)
Wheat	Whole seeds	0.07	LAMPEN et al. (1942)
Yeast, brewer's	Dried whole cells	0.83	LAMPEN et al. (1942)

As might be expected from its exceedingly great biological activity, the concentration of biotin in natural materials is very low.

I. Microbiological Determination of Biotin.

The analysis of biological samples for their biotin content is carried out almost exclusively by microbiological methods, chiefly with yeast and lactic acid bacteria. Most authorities appear to agree that *Lactobacillus arabinosus* is the assay organism of choice.

Reagents and Stock Solutions. The following stock solutions used in the folic acid assay may also be used for the biotin method:— hydrolyzed casein solution, cystine-tryptophan solution, and adenine-guanine-uracil solution. However, in preparing the casein hydrolysate, 15 ml. of 30% hydrogen peroxide is used instead of the decolorizing charcoal. After standing 24 hrs. at room temperature, the pH is brought from 3.5 to 7.0 with sodium hydroxide, 10 g. of powdered manganese dioxide added, and the mixture stirred until evolution of oxygen ceases. It is then filtered and diluted to 1 liter. This treatment removes traces of biotin which otherwise would produce high blanks. The charcoal treatment which is used to remove nicotinic and folic acids is unnecessary when the solution is to be used for biotin assay.

The basal medium for biotin assay also contains the salt solutions A and B described under the microbiological method for riboflavin. The composition of the final medium is given in Table 5.

---

[1] Dry basis.

Table 5. *Basal Medium for Biotin Assay with Lactobacillus Arabinosus.*

Ingredient	Amount present[1] grams	Ingredient	Amount present[1] milligrams
Glucose, anhydrous	20.00	Adenine sulfate	5
Sodium acetate, anhydrous	6.00	Guanine hydrochloride . . .	5
Hydrolyzed casein	5.00	Uracil	5
DL-Tryptophan	0.10	Pyridoxine hydrochloride . .	2
L-Cystine	0.20	Calcium pantothenate. . . .	1
K_2HPO_4	0.50	Thiamine chloride	1
KH_2PO_4	0.50	Riboflavin	1
$MgSO_4 \cdot 7\,H_2O$	0.20	Nicotinic acid	1
NaCl	0.01	p-Aminobenzoic acid	1
$FeSO_4 \cdot 7\,H_2O$	0.01		
$MnSO_4 \cdot H_2O$	0.01		

Vitamin Solution for Biotin Assay. Dissolve in 20% ethanol 40 mg. of pyridoxine hydrochloride and 20 mg. each of riboflavin, nicotinic acid, calcium pantothenate, thiamine chloride, and p-aminobenzoic acid, and dilute to 1 liter with the same solvent.

Basal Medium Stock Solution. Using the stock solutions listed, assemble the basal medium stock solution as follows:—

Hydrolyzed casein solution	50 ml.
Cystine-tryptophan solution	50 ml.
Vitamin solution	50 ml.
Adenine-guanine-uracil solution	25 ml.
Salt solution A	5 ml.
Salt solution B	5 ml.
Glucose, anhydrous	20 g.
Sodium acetate, anhydrous	6 g.

Adjust to pH 6.6—6.8 and dilute to 500 ml. This is sufficient double strength medium for 100 tubes.

Biotin Solutions. Prepare a solution of 25.0 mg. of D-biotin (free acid) in 500 ml. of 50% (by volume) ethanol. For a working standard solution a suitable small aliquot of this stock solution is diluted with 50% ethanol to a concentration of 0.2 mμg. of biotin per ml. This working standard is freshly prepared for each assay as needed.

Cultures and Inoculum. The test organism, *Lactobacillus arabinosus* 17—5 is obtainable from the American Type Culture Collection, under number 8014. The stock cultures are maintained and inoculum grown in the same media and by the same procedures as those described under the determination of riboflavin (p. 648), except that a few drops of the inoculum so prepared is further diluted with 10 ml. of sterile 0.9% saline before use. This very dilute inoculum should show barely visible turbidity on swirling. *L. arabinosus* produces somewhat less copious growth than *L. casei* or *S. faecalis* on the inoculum medium, but may be used, nevertheless, after 6—8 hrs. incubation, if the culture is an active one.

Preparation of Test Solutions. An amount of the sample estimated to contain approximately 200 mμg. of biotin is autoclaved with 25 ml. of 4 N sulfuric acid at 15 pounds pressure for 1—2 hrs., cooled, diluted to 100 ml., and a portion filtered. A 10 ml. aliquot of the filtrate is diluted to 60—70 ml., adjusted to pH 6.8, and then diluted accurately to 100 ml. This solution should contain about 0.2 mμg. of biotin per ml.

---

[1] For 1 liter of final medium or 500 ml. of double strength basal medium stock solution.

Assay Procedure. The assay is conducted in the manner described for riboflavin and folic acid. The standard curve is prepared from the following levels of pure biotin: 0.0, 0.1, 0.2, 0.3, 0.4, 0.5, 0.6 and 0.8 mμg. per tube. The bacterial response is measured either by turbidity after 16—18 hrs. incubation, or by titration after 72 hrs. incubation. Results from various levels of the test solution are converted into biotin values by interpolation on the standard curve, the various levels compared, and the final biotin content of the test solution and original sample calculated in the usual manner.

Sources of Error. Fatty acids such as oleic if present in the assay tubes to the extent of about 100—300 μg. per tube will partially or wholly replace biotin for *L. arabinosus* especially in the presence of water-soluble fatty acid derivatives such as "Tween 40" or "Tween 80" (WILLIAMS, BROQUIST and SNELL, 1947). The great dilution of the sample, and the filtration of the test solution while still strongly acid, are sufficient to reduce the quantity of fatty acids present in most cases to safe levels. However, certain high-fat samples may require a preliminary ether extraction to obviate this source of error.

There is no positive assurance that the acidic hydrolysis recommended will quantitatively release biotin from all samples, but more drastic treatment should not be routinely employed as some destruction of biotin has been reported to occur when higher acid concentrations and (or) longer heating periods were used (WRIGHT, 1947). Such destruction apparently is more pronounced with plant materials than with samples of animal origin (PETERSON, McDANIEL and McCOY, 1940).

References.

ANDREWS, J. S.: Cereal Chem. **21**, 398 (1944). — ANDREWS, J. S., H. M. BOYD and D. E. TERRY: Cereal Chem. **19**, 55 (1942). — Association of Official Agricultural Chemists: J. Assoc. Offic.Agr. Chemists **82**, 108 (1949); Methods of Analysis, 7th ed., Washington 4, D. C., p. 784, 1950. — Association of Vitamin Chemists: Methods of Vitamin Assay, 2nd ed., New York: Interscience Publishers, Inc. 1951.

BARTON-WRIGHT, E. C.: The Microbiological Assay of the Vitamin B Complex and Amino Acids. London: Pitman Publishing Corporation 1952. — BAUMANN, C. A.: J. Amer. Dietet. Assoc. **29**, 548 (1953). — BESSEY, O. A., O. H. LOWRY and R. H. LOVE: J. Biol. Chem. **180**, 755 (1949). — BLISS, C. I., and P. GYÖRGY: In GYÖRGY "Vitamin Methods", vol. 2, p. 232. New York: Academic Press 1951. — BONNER, J.: Botan. Gaz. **103**, 581 (1942). — BONNER, J., and H. BONNER: Vitamins and Hormones **6**, 225—275 (1948). — BOWDEN, J. P., and W. H. PETERSON: J. Biol. Chem. **178**, 533 (1948). — BURKHOLDER, P. R.: Science (Lancaster, Pa.) **97**, 562 (1943). — BURKHOLDER, P. R., and I. McVEIGH: Plant Physiol **20**, 276, 301 (1945).

CHANG, S. C.: J. Biol. Chem. **200**, 827 (1953). — COSULICH, D. B., J. M. SMITH and H. P. BROQUIST: J. Amer. Chem. Soc. **74**, 4215 (1952). — COUCH, V. M., and J. R. DOCTOR: J. Biol. Chem. **200**, 223 (1953). — CRAVIOTO B., R. F. F. LOCKHART, R. K. ANDERSON, F. DE P. MIRANDA, R. S. HARRIS, E. AGUILAR, E. W. TAPIA, H. S. LOCKHART, M. K. NUTTER and L. P. GUILD: J. Nutrition **29**, 317—329 (1945).

DE MERRE, L. J., and W. S. BROWN: Arch. Biochem. **5**, 181 (1944).

FAGER, E. E. C., O. E. OLSON, R. H. BURRIS and C. A. ELVEHJEM: Food Research **14**, 1 (1949). — FLYNN, L. M.: J. Assoc. Offic. Agr. Chemists **33**, 633 (1950).

GALSTON, A. W., and R. S. BAKER: Amer. J. Botany **36**, 773 (1949). — GLICK, D.: Techniques of Histo- and Cytochemistry, pp. 104, 109. New York: Interscience Publishers, Inc. 1949.

HALL, H. H., J. J. CURTIS and M. C. SHEKLETON: Cereal Chem. **29**, 156 (1952). — HANSON, S. W. F., and A. F. WEISS: Analyst **70**, 48 (1945). — HILL, C. H., and M. L. SCOTT: Federation Proc. **10**, 197 (1951). — HINMAN, W. F., R. E. TUCKER, L. M. JANS and E. Cr. HALLIDAY: Ind. Eng. Chem. (Anal. Ed.) **18**, 296 (1946). — HOFFER, A., A. W. ALCOCK and W. F. GEDDES: Cereal Chem. **21**, 515 (1944). — HUBBARD, J. E., H. H. HALL and F. R. EARLE: Cereal Chem. **27**, 415 (1950).

JONES, A., and S. MORRIS: Analyst. **74**, 29 (1949).

KERESZTESY, J. C., and M. SILVERMANN: J. Amer. Chem. Soc. **73**, 5510 (1951). — KLEIN, J. R., and H. I. KOHN: J. Biol. Chem. **136**, 177 (1940). — KOHLER, G. O.: J. Biol. Chem. **152**,

215 (1944). — Kornberg, H. A., R. S. Langdon and V. H. Cheldelin: Analyt. Chem. **20**, 81 (1948). — Krehl, W. A., and G. W. Cowgill: Food Research **15**, 179 (1950). — Kuhn, R., T. Wagner-Jauregg and H. Kaltschmitt: Ber. dtsch. chem. Ges. **67**, 1453 (1934).

Lampen, J. O., G. P. Bahler and W. H. Peterson: J. Nutrition **23**, 11 (1942). — Loofbourow, J. R. and R. S. Harris: Cereal Chem. **19**, 151 (1942). — Lowry, O. H., and O. A. Bessey: J. Biol. Chem. **155**, 71 (1944).

McVeigh, I.: Bull. Torrey Botan. Club **71**, 438 (1944). — Mims, V., M. E. Swendseid and O. D. Bird: J. Biol. Chem. **170**, 367 (1947). — Morrison, F. B.: Feeds and Feeding, 21st ed., p. 1164, Ithaca. New York: Morrison Publishing Co. 1948.

Nymon, M. C., and W. A. Gortner: J. Biol. Chem. **163**, 277 (1946).

Ottes, R. T., and F. R. Roberts: J. Assoc. Offic. Agr. Chemists **32**, 797 (1949).

Peterson, W. H., L. E. McDaniel and E. McCoy: J. Biol. Chem. **133**, lxxv (1940). — Price, S. A., and H. C. H. Graves: Nature (London) **153**, 461 (1944).

Snell, E. E.: In P. György, Vitamin Methods, vol. 1, pp. 327—360. New York: Academic Press. 1950. — Snell, E. E., and F. M. Strong: Ind. Eng. Chem. (Anal. Ed.) **11**, 346 (1939). — Stiller, E. T.: In P. György, "Vitamin Methods", vol. 1, pp. 102—125. New York: Academic Press 1950. — Strong, F. M.: Biological Symposia **12**, 143—165 (1947). — Strong, F. M., and L. E. Carpenter: Ind. Eng. Chem. (Anal. Ed.) **14**, 909 (1942).

Tanner, F. W., S. E. Pfeiffer and J. J. Curtis: Cereal Chem. **24**, 268 (1947). — Toepfer, E. W., E. G. Zook, M. L. Orr and L. R. Richardson: "Folic Acid Content of Foods", U. S. Dept. of Agriculture, Handbook No. 29. Washington, D. C.: Government Printing Office 1951.

Van Lanen, J. M., F. W. Tanner and S. E. Pfeiffer: Cereal Chem. **23**, 428 (1946).

Watt, B., and A. L. Merrill: United States Department of Agriculture, Handbook No. 8, Composition of Foods — Raw, Processed, Prepared. Washington D. C.: U. S. Government Printing Office 1950. — Williams, W. L., H. P. Broquist and E. E. Snell: J. Biol. Chem. **170**, 619 (1947). — Williams, R. J., R. E. Eakin, E. Beerstecker and W. Shive: "Biochemistry of the B Vitamins". New York: Reinhold Publishing Corporation 1950. — Wright, L. D.: Biological Symposia **12**, 290 (1947). — Wright, L. D., E. L. Cresson, H. R. Skeggs, R. L. Peck, D. E. Wolf, T. R. Wood J. Valiant, and K. Folkers: Science **114**, 635 (1951).

Melanins.

By

M. Thomas.

Melanins have been defined comprehensively as the dark brown or black pigments which accumulate in certain parts of plants and animals. On chemical grounds this definition is not satisfactory for the higher plants, as there exist strong reasons for supposing that whereas some of the dark pigments in these plants are indolic compounds, others do not contain nitrogen. Examples of non-nitrogenous pigments (see p. 662) are first the phytomelanes of the *Compositae*, and secondly Japanese lac and a whole list of other oxidative products of nitrogen-free phenols, seen in functioning specialized organs such as bud-scale leaves, or making their appearance when cells are injured. In the present article, even the blackest of these non-nitrogenous pigments will not be regarded as a melanin. The use of the term will be confined to dark pigments which are nitrogenous because they are indole derivatives. The definition now to be given will cover such authentic melanins as have been extracted from animal organs or produced by enzyme action *in vitro*. A naturally occurring melanin is a dark polymeric indole derivative of high molecular weight which is produced by a series of reactions (section F) involving tyrosinase, oxygen and, as organic substrates, either tyrosine or dihydroxyphenylalanine ("dopa").

As we shall see in section C there is strong presumptive evidence of the occurrence in vascular plants and fungi of melanins as so defined. None of these presumptive melanins has, however, yet been isolated and analysed to prove that it contains nitrogen in amounts comparable with those found in authentic melanins (section A). Nor have many observations and experiments been made to compare the properties of presumptive plant melanins with those of authentic melanins (section B). No attempts appear to have been made as yet to detect, by chromatographic or spectrophotometric methods, the appearance of dopachrome or other melanin precursors (section F) in cells which are destined to produce dark pigments during their normal development. We still require, therefore, decisive proof that indolic melanins accumulate as "secondary plant products." (as defined by PAECH, 1950) in the normal ontogeny of any plant.

In their admirable review LERNER and FITZPATRICK (1950) state why animal melanins have been so widely studied. One of the probable reasons for the neglect of plant melanins as objects of study is that, as a rule, no obvious function can be assigned to them. Further the occurrence of dark brown or black pigments has not often been used as a diagnostic feature; but specific appellations such as *nigricans* and *niger* indicate that they have not been completely overlooked. So far, however, there has been no major study on the inheritance of melanic pigmentation in vascular plants. In contrast we may note the statement by GINSBURG (1944) that melanic pigmentation in the mammal has been subjected to a more extensive genetic analysis than any other single character yet studied in this group. The elucidation of problems of inheritance has also been the incentive for recent experimental work on presumptive melanins in fungi (section B). For vascular plants it is probable that a much wider exploratory examination of the occurrence of presumptive melanins (section C), and of the factors which promote or prevent their appearance (section D and E), will be necessary before genetical problems can be clearly defined. The opening sections of this article have been written to put the reader in touch with literature describing some of the methods, relating mainly to investigations on animal melanism, which may conceivably be adapted for use in such exploratory experimentation on higher plants. If it is conceded that melanogenesis *in vivo* follows the same path as that which occurs *in vitro* it follows that some knowledge of the methods developed by chemists in their attempts to elucidate the nature of intermediate stages, and the molecular structure of melanin precursors, should prove of value in the wider study of melanism in living plants. The present is a period of renewed enquiry into these matters. Accordingly in the final section of this article, reference is made to some of the older and more modern literature concerning the chemistry of melanins and melanogenesis.

A. Notes on the Isolation and Analysis of Dark Pigments.

No melanin, whether a naturally occurring pigment or produced *in vitro* by the action of tyrosinase or by chemical means, has yet been isolated as a single chemical compound of definite composition. Partially purified dark products from animal sources (but none from plants) have been analysed, and the resulting figures compared with those obtained by the analysis of melanins produced *in vitro* by the action on tyrosine or dopa of tyrosinase from plant or animal sources. For native presumptive melanins pronounced differences have been found, and even for tyrosinase and dopa melanin (see p. 673) the results have been variable. One example must suffice. Panizzi and Nicolaus (1952) extracted, purified and analysed sepia melanin, obtaining the following values: C 57.49, H 2.91, N 10.55, Ash 0.47. Oxygen values are omitted. They also purified a biosynthetic melanin prepared by the action of potato tyrosinase on tyrosine. Analysis gave the following values: C 62.85, H 2.85, N 7.85 (Dumas), 8.82 (Kjeldahl). In addition to differences in the percentage nitrogen content, they found a qualitative difference between sepia and tyrosinase melanins. By oxidation of the sepia melanin, but not of tyrosinase melanin, they obtained a volatile compound which gave a red colour with pinewood steeped in hydrochloric acid, and identified this compound as pyrrole tricarboxylic acid. Panizzi and Nicolaus cite analytical values obtained by other workers. These and other records (see for example values quoted by Lerner and Fitzpatrick, 1950) support the conclusion that the composition of natural melanin may be variable. It has been alleged that some contain sulphur. There is good evidence that melanoproteins exist, and Sizer (1947) reported that tyrosinase can oxidize, *in vitro*, a purified protein to yield a dark product. However all the presumptive melanins from animal sources and authentic tyrosinase melanins have had one property in common; as a rule they contain more than 8% nitrogen.

At least one generalization can therefore be made with reasonable confidence, on the basis of the definition at the beginning of this article. A purified black pigment obtained from plants, which on analysis is found not to contain nitrogen, is not a melanin.

Thus phytomelanes, which occur in the fruits of *Tagetes patula* and *Helianthus annuus* and in other *Compositae*, are not melanins. For example, de Vries (1948) isolated the dark product from *Tagetes* fruits and found that "its atomic ratio deviated only slightly from $C_x (H_2O)_y$". No further analytical results for plant products can be cited yet. The dark lac described by Bertrand (1894, 1895 and 1896) was probably free from nitrogen, since it was formed *in vitro* by the action of an extracted enzyme laccase on the phenolic substance laccol. Presumably many of the brown substances formed in specialized organs, or injury, result from oxidation of non-nitrogenous phenols.

The present writer has been interested in the dark markings on the involucral bracts of *Senecio* spp. and of certain other plants belonging to the *Compositae*. In his experiments on *Senecio vulgaris* he has failed so far to find any evidence of the presence of tyrosinase (but see p. 668). The pigment, therefore, is probably not a melanin. It might well be a phytomelane. Dr. K. B. BLACKBURN (Newcastle) has observed dark granules in the involucral cells. A desirable experiment would be to isolate enough of the pigment to determine whether or not it contains nitrogen. The absence of nitrogen would be decisive proof that this dark pigment is not a melanin.

We conclude that it is highly desirable for attempts to be made to isolate and analyse a range of dark brown or black pigments which occur in vascular plants. Their resistance to the action of most organic solvents and hot strong acids will serve to separate melanins from many cell components. Such solubility as they may show in alkali may then be used to separate them from cellulose residues. The addition of acid to the alkaline solution will lead to precipitation of melanin. As an example of an attempt along these lines to isolate and estimate quantitatively an animal melanin the experiments of EINSELE (1937) on the hair of the house-mouse may be cited. Doubtless for plant pigments special purification processes may have to be elaborated for each particular separation.

B. Some Properties which have been Used in Attempts to Identify Pigments as Melanins.

A quotation from JACOBSON and MILLOTT (1953) may serve to introduce a section dealing with tests which have been used to strengthen a presumption that a certain dark pigment is a melanin. In their report on melanogenesis in the coelomic fluid of an echinoid these authors made the following statement. — "Although chemical tests specific for melanin are non-existent, it is possible to identify, at least provisionally, a naturally occurring brown or black pigment as a melanin, provided that it shows certain characteristics, namely resistance to solvents, bleaching when subjected to the action of oxidants, and the capacity to reduce directly ammoniacal solutions of silver nitrate." In this statement importance should be attached to the qualifying adverb "provisionally". There are slight differences in the solubilities of melanins from different sources, but it is broadly true to say that they are insoluble in the most commonly used organic solvents, although some dissolve slightly in pyridine, ethylene chlorohydrin and certain other solvents. Unquestionably melanin, when first formed by the action *in vitro* of tyrosinase on tyrosine, exists in colloid aqueous solution. Moreover Dr. K. B. BLACKBURN (Newcastle) has observed the presence of dark pigments in the vacuole of living cells of the flower of *Vicia faba*. The cell sap was probably slightly acid. It is likely, however, that the concentration of dissolved melanin was low. Much remains to be described, especially for plant melanins. It appears to be established, however, that many animal melanins are insoluble even in hot strong acids but are dissolved by alkalies (see for example, EINSELE, 1937). However, even hot alkaline solutions do not always effect complete solution. These are properties which have been exploited in attempts to obtain purified specimens (see above). The bleaching action of bromine water, hydrogen peroxide, potassium permanganate and other oxidants is described by LISON (1936). He also gives a detailed account of MASSON's method of performing the "argentaffine" reaction, i. e. of bringing about the reduction of ammoniacal silver nitrate. He points out that the reaction is not a specific test.

FIGGE (1939) prepared a tyrosinase melanin *in vitro*, and showed that the black solution was reduced by sodium hydrosulphite to give a light brown colour. Oxidation by potassium ferricyanide led to darkening again. The degree of colour change was measured by a photoelectric colorimeter. Figge (1940) applied the test to a naturally occurring animal melanin, and concluded that it was reversibly oxidized, although there was interference by a contaminating substance. Like other previous and later workers he saw a "general similarity of behaviour of melanins, but much variability in details".

In the 1920's Gallerani in Italy and Bloch and Schaaf in Switzerland made spectrophotometric studies of melanin solutions. More recently such studies have been made in relation to genetical problems. For example Daniel (1938) reported that the absorption spectra of alkaline solutions of melanins from mice of different colours were qualitatively identical, and concluded that colour differences were not owing to difference in chemical composition. Baker and Andrews (1944) plotted regression coefficients of the logarithm of optical density against wave length and obtained curves for dark pigments from guinea pigs which were similar to those for melanins from mice and horse, and for dopa melanin, but the possibility of qualitative differences was admitted. Ginsburg (1944) performed similar experiments and concluded that "Dopa melanin in any stage of oxidation or however derived, tyrosine melanin at any stage of oxidation, potato melanin and dissolved pigment from black or sepia guinea pigs have similar spectra". This was considered to be compatible with the notions that all these pigments are closely related substances, and that lighter and darker shades result from quantitative variations and not from qualitative differences.

An inference of doubtful validity is that a dark pigment is probably a melanin if it has a similar absorption spectrum to that of an authentic melanin. Fox and Gray (1950) suspected the presence of melanin in one type of a certain strain of *Neurospora crassa*, which turned dark as the mycelium aged. They did not extract the pigment and examine its properties, but they extracted an enzyme from the mycelium which converted tyrosine into a dark solution. They measured the absorption spectrum of the dark solution at wave lengths ranging from 470 to 610 mμ and found that the slope of the absorption curve was closely similar to that reported by Spiegel-Adolf (1937) for "melanin". We should note, however, that Spiegel-Adolf's "melanins" were dark products obtained by the irradiation not only of tyrosine, but of tryptophane and of phenyl alanine. However, she noted that the slopes of her curves for these "photosynthetic melanins" were similar to those described in earlier literature for genuine melanins. Fox and Gray stated that the pigment derived in their experiment was "decolourized by potassium permanganate, as is melanin, so that there is little doubt about its identity". This reliance on a single bleaching test for confirmation illustrates the inadequacy of present analytical methods.

Schaeffer (1953) described a black mutant of *Neurospora crassa*, which he stated arose from its parent by a single gene mutation. He obtained 6-day old cultures, grown in a solution containing tyrosine but of low sulphur content. Horowitz and Shen (1952) had previously found that the development of tyrosinase activity in *Neurospora* is enhanced in such a medium. In Schaeffer's experiments the medium darkened as well as the mycelium. He made experiments on the dark pigments which, it should be noted, unlike the pigment investigated by Fox and Gray, had been produced by the living mycelium, and not *in vitro* by an extracted enzyme. Schaeffer treated the wet mycelium with 0.5 normal sodium hydroxide for 1—2 hours at room temperature, obtaining a deep brown solution. (In a parallel experiment with the parent strain of the black mutant, the alkaline extract was almost colourless.) By boiling under a reflux condenser, he increased the amount of pigment extracted. The alkaline solution was cooled, and the pigment was precipitated either by acidification or by saturation with ammonium sulphate i. e. it behaved like a melanin prepared from tyrosine *in vitro*. SCHAEFFER then carried out spectrophotometric measurements of alkaline solutions in wave lengths between 400 and 620 mμ, and evaluated the slopes of the logarithm of the optical density. The values obtained did not accord well with those reported by FOX and GRAY (1950) for the melanin produced *in vitro* by extracted tyrosinase. SCHAEFFER pointed out that the solvent conditions were different. SCHAEFFER also compared his results with those obtained by BAKER and ANDREWS (1944), for the pigment from guinea pigs. The slopes were different, but SCHAEFFER pointed out that the preparations of BAKER and ANDREWS contained hydrolysed keratin.

Clearly, before reliance can be placed on conclusions drawn from similarities or dissimilarities of such slopes, further extensive work must be done to standardize the conditions under which the spectrophotometric measurements are made. Even if conditions are satisfactorily standardized, there may still remain some doubt about identifying a pigment as a melanin from the slope of the absorption curve. MASON (1948) has stated why he considers such an identification is not convincing. He pointed out that all the solutions of authentic and presumptive melanins so far studied have shown a general absorption in the visible spectrum, without specific features. In the absence of such features he does not believe chemical characterization of the dark compounds to be justifiable.

One preliminary experiment performed by Mr. J. M. A. BROWN (Newcastle) may be mentioned here. He compared the absorption spectrum of dark solutions of two presumptive melanins from higher plants and a presumptive phytomelane from sunflower fruits, and found no distinguishing features. If further work confirms his finding, the conclusion would be that dark colloid solutions of substances which have very different chemical constitutions may have similar absorption spectra.

Possibly, for the present, we should suspend judgment concerning the value of the spectrophotometer as a tool in the identification of a dark pigment as a melanin, remembering, however, that it may, in the future, prove as invaluable an aid in following the early stages of melanogenesis *in vivo* as it has already proved in the investigation of these stages *in vitro* (see MASON, 1948, and p. 672 below).

C. The Detection of Tyrosinase and its Substrates in Plants Producing Presumptive Melanin.

All experiments on enzyme and substrates concerned with melanin formation in plants derive from those of BERTRAND (1896). From the organs of some higher plants (e. g. dahlia tubers) and the fungus *Russula nigricans* he obtained an enzyme which, in air, acted on tyrosine *in vitro* to give successively red and black products. Accordingly he named the enzyme tyrosinase and performed experiments to show that it was distinct from laccase. Both enzymes were soluble in water and were precipitated from aqueous solutions by the addition of ethyl alcohol.

At the present day the preparation of highly active enzyme, whether it be called tyrosinase or polyphenol oxidase (see KERTÉSZ, 1952), is usually based on methods employed in the separation of purified enzymes as copper proteids from potato by KUBOWITZ, F. (1938) and from mushroom by KEILIN and MANN (1938). KERTÉSZ describes a simplified method, and refers to work by other authors, especially in the U.S.A. in the 1940's.

BERTRAND (1896) detected the presence of tyrosine in plants containing tyrosinase, and prepared a crystalline specimen of this amino acid from dahlia. General acceptance has been given to his conclusion that colouring *via* red to black in air of extracted plant juices is owing to the oxidation by tyrosinase usually of tyrosine. There is, however, an alternative substrate in plants, namely dihydroxyphenylalanine ("dopa"). The isolation of dopa from *Vicia faba* was described dy GUGGENHEIM (1913) and from *Stizolobium* spp. by MILLER (1920).

By two dimensional paper chromatography Mr. R. F. Jones (Newcastle) has obtained some evidence that both tyrosine and dopa are present in extracts obtained from *Vicia faba*. Chromatographic methods should prove useful in ascertaining if one both of these tyrosinase substrates are present in organs capable of producing a black pigment. It is noted, in passing, that aqueous solutions of dopa are oxidized spontaneously in air to give melanins (see Miller, 1920). The oxidation is, however, much accelerated by tyrosinase.

A clear extension of Bertrand's conclusion is that rapid discolourations *via* red to black brought about by injuring a tissue (e. g. that of potato by grinding, or by exposure to chloroform vapour) strongly suggest the presence of tyrosinase and tyrosine or dopa. (If dopa is present, the darkening may be owing, in some measure, to non-enzymic oxidation.) The evidence is strengthened if the enzyme is extracted free from substrate and then shown to act *in vitro* on a solution of tyrosine. An enzyme which oxidizes tyrosine will also act on dopa.

It can be argued that for a plant organ which gives positive reactions in all these tests, a dark pigment, produced under natural conditions when the organ is injured or during the final stages of senescence, is probably a melanin as defined at the opening of this article. A long list of examples could be given: a few must suffice. The leaves of some species of willow (e. g. *Salix nigricans* Smith), and broom (*Cytisus nigricans* Linn.) turn black on injury. Striking black colours in dying or dead cells are seen in the autumn in the pods of broom against a background of green healthy shoots, and in the upper surface of leaves of white poplar, contrasting with the down covered under surface.

Presumptive melanins occurring in dead cells of pericarps, testas, and scale leaves of winter buds (e. g. those of ash) or inflorescences (e. g. those of some willows), can be regarded as "secondary plant products" produced in normal development, i. e. independently of external injurious agencies. For the vascular cryptogams we may add to this list the dark pigments seen in the leaf stalks of *Adiantum* and certain species of *Asplenium* and in the leaf sheaths of some species of *Equisetum*. Dr. Blackburn has observed that cell walls become brown in the melanic parts of these plants. Pigment formation may begin before cell contents have completely disintegrated. There are not many published records of the appearance of presumptive melanins as secondary plant products in still turgid functioning organs. The black pigment in the wing petals of *Vicia faba* is the best known example.

Cogent but not decisive evidence concerning the identity of dark pigments sometimes formed in fungal cultures, is that the degree of pigmentation has been shown to relate to the tyrosinase activity of mycelial extracts (for *Glomerella cingulata* see Markert, 1950, for *Neurospora* see Fox and Gray, 1950, Horowitz and Shen, 1952, and Schaeffer, 1953). Horowitz and Shen (1952) give clear accounts of methods for measuring this activity. They used a Warburg apparatus for determining the oxygen uptake resulting from the addition to the mycelial extract of tyrosine (or dopa) dissolved in $M/60$ sodium hydroxide. In addition, with a photoelectric colorimeter (blue filter No. 42), they determined the time taken for the formation of the red intermediate compound in melanin production after tyrosine (or dopa) was added to the extract. They stated that "Proportionality between the enzyme concentration and the maximal rate of colour development obtains within the period of formation of the red intermediate". They found that readings were erratic when melanin began to precipitate and judged them to be valueless for a quantitative comparison.

As far as the present writer knows no attempt has yet been made to isolate, purify, and analyse a dark pigment produced by a fungal mycelium.

D. Inhibitors and the Possible Nature of Dopa-Oxidase and Tyrosinase.

BLOCH (1917) observed the deposit of melanin granules in the cytoplasm of slightly pigmented cells of human skin, when frozen sections containing such cells were placed in a dilute solution of dopa. He showed that this "dopa reaction" required oxygen, was inhibited by hydrogen sulphide and hydrogen cyanide, and, indeed, had the general properties of an enzymic oxidation. It did not take place after heating the tissue to 100° C. The enzyme specificity appeared to be confined to L-dopa. Tyrosine was not attacked. The enzyme was called dopa oxidase, and for many years it was believed to be distinct from the tyrosinase of plants, mealworms etc. (It was realized that tyrosinase can convert dopa as well as tyrosine into melanin.) Mammalian pigmentation was attributed to the activity of dopa oxidase in spite of the fact that free dopa was never found in the tissue of these animals (see LERNER and FITZPATRICK, 1950).

FIGGE (1941) considered the possibility that tyrosinase and tyrosine might be present in a cell without melanin formation if a strong inhibitor were present which prevented the conversion of tyrosine into dopa. If it did not completely suppress the enzymic conversion of dopa into melanin, dopa oxidase activity could be shown. FIGGE, obtained evidence that small amounts of glutathione may act in this way. He therefore obtained support for his idea that the limited specificity of dopa oxidase is a resultant of the combined activities of tyrosinase and an inhibitor.

It is now known (see LERNER and FITZPATRICK, 1950) that a water soluble, dialysable, heat stable inhibitor of plant tyrosinase may occur in human skin, and that there is a direct relationship between the inhibitory power of epidermal extracts and the amount of sulphydryl present in these extracts. It is evidently possible that a mammalian skin may contain a tyrosinase but show no power of melanin formation because of the presence of components containing sulphydryl or other inhibiting groups.

Doubts about attributing mammalian pigmentation to dopa oxidase rather than to tyrosinase increased when LERNER, FITZPATRICK, CALKINS, and SUMMERSON (1948) separated cytoplasmic particles from mouse melanoma and showed that, after an induction period, the particles oxidized tyrosine, with an oxygen uptake which they measured. They concluded that the distinction between tyrosinase and dopa oxidase is not valid. They obtained evidence, however, that dopa can shorten the induction period in the oxidation of tyrosine.

VAN DUIJN (1953) has repeated some of BLOCH's experiments on sections of mammalian skin. Her main conclusions appear to be that it is premature to identify the "dopa factor" with any known enzyme, and mammalian melanogenesis with the tyrosinase-tyrosine reaction. Her conclusions and those of other workers mentioned in this section will certainly require consideration in the light of those of KERTÉSZ (1952) in which he casts doubt on the existence of an enzyme which acts directly on tyrosine. Accordingly he regards the name "tyrosinase" as most inappropriate, but uses it in order to attach his findings to those of earlier workers. His chief conclusion is that tyrosinase is compounded of an enzymic copper proteid, one or more orthodihydroxyphenols, and certain free metallic ions (e. g. copper ions). The enzyme acts on the o-phenols to give o-quinones, which can then carry out a range of cellular oxidations. One of the slower oxidations is that of tyrosine to dopa, but this oxidation is greatly accelerated by certain metallic ions. Dopa would give dopaquinone and, therefore, melanins in the absence of these ions. These oxidations are described in section F.

NELSON and DAWSON (1944) and LERNER and FITZPATRICK (1950) discuss experimental evidence and other views relating to the composition of tyrosinase, for example that it is compounded of two enzymic proteins, a monophenolase (cresolase) and a polyphenolase (catecholase), or, alternatively, of a single protein possessing two active catalytic centres.

On the basis of all these views any substance which tends to reconvert o-quinones into o-phenols would retard or inhibit so-called tyrosinase action, as judged by experiments on pigment formation or oxygen uptake on the addition of tyrosine. In addition to compounds containing sulphydryl groups, it is known that ascorbic acid can act as an inhibitor. Moreover SCHROEDER (1934) reported that the enzymic oxidation of dopa *in vivo* could be inhibited by ascorbic acid.

E. The Infrequent Appearance of Melanins as Secondary Plant Products.

Although presumptive melanins occur in plants more frequently than has been realized, the fact remains that plants which do not produce black pigments in functioning organs greatly outnumber those which do. Moreover in the latter melanin formation is very localized. For example Dr. Blackburn observed that in the dark streaks of the vexillium petal of *Vicia faba* melanic and non-melanic cells lie side by side.

Let us accept that the appearance of an authentic melanin means that during normal ontogeny, tyrosinase and its natural substrate (presumably free or bound tyrosine or dopa) come together in the absence of effective amounts of an inhibitor (e. g. ascorbic acid or a compound containing sulphydryl), and react with an absorption of oxygen to yield melanin by the series of reactions to be described in section F.

Except by the slow autoxidation of such dopa as may be present, melanin will not appear under any conditions in plant cells lacking tyrosinase, whatever meaning we may attach to this word. The present writer's preliminary experiments seemed to indicate that tyrosinase may be absent from all parts of some plants (e. g. *Senecio vulgaris*, but see below). He has, however, so far failed to demonstrate that in plants containing tyrosinase, there is a differentiation into parts containing and parts not containing the enzyme. For example, tyrosinase was present in all tested parts of *Vicia faba*. The enzyme was not confined to the cells which produce melanin as a secondary plant product.

Failure of cells which contain tyrosinase to produce melanin may be owing to absence of free substrate. Chromatographic methods (see. p. 666) should prove useful in investigations on the distribution of tyrosine and dopa. It appears to be probable that in plant organs which change colour quickly on injury one or both of these substrates are present. For *Vicia faba* we have seen that this is the case. In most of the healthy cells of such plants it may be that substrate and oxygen do not have simultaneous access to enzyme. Melanin formation will then only take place when there is physical disorganization of the protoplasm. Alternatively tyrosinase activity may be inhibited by the presence of some strong reducing agent like ascorbic acid or a sulphydryl compound. On physical disorganization these reducing agents may become oxidized, with the removal of the inhibition. Paech (1950) has given evidence for and discussed this type of effect for ascorbic acid.

Horowitz and Shen (1952) found that the development of tyrosinase activity by a certain wild type of *Neurospora* depended on sulphur nutrition. Grown in media deficient in sulphur, mycelia gave extracts which showed strong tyrosinase activity, whereas extracts from mycelia grown in "sulphur sufficient media" showed only weak activity. The difference was found to be owing in part to the presence in the latter of a dialysable inhibitor, which might well be a compound containing an —SH group.

It appears therefore that in order to prove the absence of tyrosinase from a plant extract we must show that the extract does not contain an inhibitor. For this reason the present writer does not consider his evidence of the absence of tyrosinase from *Senecio vulgaris*, is quite decisive as yet. Mr. J. M. A. Brown (Newcastle) has demonstrated that extracts from this plant strongly retard the oxygen uptake shown when tyrosine is added to potato slices or extracts of potato tissue. The inhibiting substances may well be ascorbic acid. However, under conditions which should remove ascorbic acid from extracts of *Senecio vulgaris*, these extracts have not shown tyrosinase activity. For vascular plants no results parallel to those of Horowitz and Shen's for *Neurospora* can be cited yet.

F. Intermediate Stages in Melanogenesis in Vitro and Suggestions Concerning the Chemical Structure of Melanins.

Raper's conclusions on the chemistry of melanogenesis (see Evans and Raper, 1937 and Raper, 1938) are summarized in the accompanying scheme. His conclusions were based on copious and convincing evidence which he and his colleagues at Manchester obtained by standard chemical and biochemical techniques. A few illustrations of the kind of evidence obtained are outlined below. Details concerning experimental methods will be found in the papers cited. Broadly similar results have been obtained with tyrosinase preparations from animals (e. g. mealworms) and plants (e. g. mushrooms), and by the chemical oxidation of dopa.

RAPER (1926) allowed a tyrosinase preparation to act on tyrosine (I) in a well-oxygenated solution for a relatively short time. He isolated a product, representing not more than 3% of the tyrosine oxidized, which he identified as dopa (II), by analysis and by comparing its properties with those of a specimen of dopa prepared from *Vicia faba*, using the method described by GUGGENHEIM (1913). Both substances, when acted on by tyrosinase, gave red solutions, which had similar properties. For example, under anaerobic conditions at pH 8 they were decolourized. Quantitative measurement of the nitrogen distribution in the colourless solutions, showed that 77% of the original amino-N of tyrosine was no longer in the amino form (see below). When the colourless solutions were aerated, they blackened immediately. Since tyrosine, when acted on by tyrosinase, yielded dopa, and since both tyrosine and dopa were oxidized to yield red and black pigments, it was inferred that dopa is the first intermediate in the production of melanin from tyrosine. It was considered that the red substance was another intermediate and that in the red solution other precursors of melanin were present in which the nitrogen was no longer in the amino form.

RAPER (1927) produced a red solution by the action of tyrosinase on tyrosine. He allowed the red solution to decolourize *in vacuo* or in the presence of sulphurous acid. After employing a methylation technique on the decolourized solutions he succeeded in isolating two crystalline products which were identical in all respects with eynthetic preparations of 5,6-dimethoxy-indole and 5,6-dimethoxy-indole-2-carboxylic acid. Obviously he had obtained strong evidence that at least one of the corresponding ortho-dihydroxy-indoles (VII and VI respectively) is a precursor of melanin. By demonstrating the presence of the indole ring in these compounds, he accounted for the disappearance of amino-N at some stage during the conversion of tyrosine into melanin precursors. (In the above scheme, cyclisation is indicated as occurring in the conversion, by intramolecular rearrangement of III into IV.) By showing that the red substance obtained by the action of tyrosinase on dopa prepared from *Vicia faba* could also be converted into 5,6-dimethoxy-indole, RAPER obtained further evidence that dopa is the first product formed during the oxidation of tyrosine.

On the basis of this work RAPER argued that the most probable structure of the red substance is that of the orthoquinone V portrayed above, and that substance III and IV are likely precursors. It should be noted, however, that none of these substances was isolated. Nor indeed were the free phenolic substances VI and VII isolated. Moreover at this time these substances had not been synthesized (see below). Accordingly RAPER could do no more than make suggestions about the chemical mechanism of melanin formation *in vitro* when tyrosinase acts on tyrosine or dopa. He believed that tyrosinase melanin is produced mainly by the oxidation of VII and to a lesser extent by that of the carboxylic acid VI. He considered it unlikely that tyrosinase melanin is always identical with melanin pigments in cells, since the latter could be produced by the action of tyrosinase on tyrosine combined in protein molecules. Evidence of the production of melanoproteins *in vitro* has since been given by SIZER (1947, 1948). He cited evidence that the destructive distillation of natural melanins yields compounds which give the pine-shaving reaction, suggesting that they are cyclic compounds containing nitrogen.

Using a modified form of Haldane's blood-gas apparatus, Dulière and Raper (1930) measured the oxygen uptake during the oxidation to melanin of tyrosine and of dopa. They found that five atoms of oxygen were consumed in converting one molecule of tyrosine into melanin, and four consumed in converting one of dopa into melanin. This accords with the view indicated in Raper's scheme that three atoms of oxygen are involved in the conversion of tyrosine into 5,6-dihydroxyindole, and two atoms in the oxidative polymerization of this cylic compound into melanin. At Manchester, the evolution of carbon dioxide in the conversion of VI into VII or directly into melanin does not appear to have been measured. (For more modern work see below.)

Evans and Raper (1937) thought that the only conversions requiring the presence of tyrosinase were (i) tyrosine (I) to dopa (II), and (ii) dopa (II) to dopa quinone (III). The experimental evidence indicated that all the other reactions could occur spontaneously, but would proceed faster in the presence of tyrosinase. A noteworthy point is that melanin formation from dopa is faster than from tyrosine.

Actually, dopa can undergo autoxidation in air to yield melanin at pH 8 (see for example, Clemo, Duxbury and Swan, 1952). Further we note that, if the views held by Kertész (see p. 667) are correct, Evans and Raper's conclusion that the conversion of tyrosine into dopa is enzymic may require modification. These matters of detail are fundamental in relation to genetical problems, but do not affect the arguments in the remainder of this section.

Another point about dopa may have significance in relation to enzymic systems in plants. Evans and Raper (1937) obtained evidence that the addition of ascorbic acid to the tyrosinase-tyrosine system increases the accumulation of dopa by the reduction of dopa-quinone. This effect would tend to inhibit melanin formation (see p. 667). Subsequently, using the Warburg apparatus, Robinson and Nelson (1944) found that the addition of ascorbic acid occasioned a reduction in oxygen uptake attributable to tyrosinase action. We recall that compounds containing an —SH group may also have a retarding effect.

A few additional remarks about the red pigment are desirable at this stage. In 1931 Mazza and Stolfti isolated a red pigment (hallochrome) from *Halla parthenopoea* Costa, a marine worm containing tyrosinase. They concluded that the hallochrome was identical with the red pigment obtained by the oxidation of dopa. Evans and Raper (1937) agreed with this conclusion. They regarded hallochrome as 2-carboxy-2,3-dihydro-indole-5,6-quinone (substance V in the above scheme). For example, they prepared specimens of hallochrome from worms obtained from Naples, and compared its properties with those of the red pigment formed by the action of potato tyrosinase on tyrosine. They stated that reduction in aqueous solution of each of these pigments with palladium and hydrogen leads to the production of substance IV. Accordingly for a number of years the red substance V formed as an intermediate in melanin production *in vitro* was called hallochrome. However, Mason (1948) has questioned the conclusion that the red pigment produced by tyrosinase and hallochrome are identical. He found that the maximum light absorption of the former occurs at 475 mμ, whereas that of hallochrome is at 539 mμ. According to Bu'Lock and Harley-Mason (1951) the red pigment formed by tyrosinase is now often referred to as dopachrome. Presumably this is the red pigment described by Bertrand (see p. 665) as a melanin precursor, and the one whose rate of production was taken by Horowitz and Shen (see p. 666) as a measure of tyrosinase activity.

In more recent times most of Raper's ideas have been confirmed, and suggestive evidence has been obtained relating to melanin formation. In 1938 Raper wrote

"The processes taking place in the last stage of the reaction, which consists in the conversion of 5:6-dihydroxy-indole into melanin, are at present obscure." The obscurity remains but, perhaps, it is not so profound now.

CLEMO and WEISS (1945), sought for a compound readily produced from 5,6-dihydroxy indole and convertible by a non-enzymic oxidation into melanin. They suggested that 5,6:5',6'-tetrahydroxy-indigo, possibly as the corresponding quinone or semi-quinone, might be such a compound. With a view to testing their suggestion they attempted to prepare this indigo compound, but without success. The ideas behind their experiments were pioneering, but their suggestion that melanins have an indigoid structure is not favoured at present (see e. g. HARLEY-MASON, 1948).

BEER, CLARK, KHORANA and ROBERTSON (1948) reported the synthesis of 5:6-dihydroxy indole (VII) and related compounds. They showed that this synthetic dihydroxy indole readily undergoes oxidative polymerization to yield melanins. By experiments with purified specimens RAPER's conclusion that VII is a melanin precursor, based on work with solutions of mixtures of substances, had been confirmed. Melanin was also formed from the 2-methyl derivative. We shall return to the significance of this fact later.

BEER, McGRATH, ROBERTSON and WOODIER (1949) reported the synthesis of 2-carboxy-5,6-dihydroxy-indole (VI). Over 24 hours it was only slowly oxidized to give a brown solution, with no precipitation of melanin. RAPER had regarded this substance as well as VII as a melanin precursor, but these experiments on the synthetic acid do not support the view that it is an immediate precursor. The action of tyrosinase on the synthetic product VI has not yet been reported. However, by 1949, MASON's work, which will now be described,

RAPER's Scheme (see text).

I, L-parahydroxyphenyl-alanine (tyrosine); II, L-3,4-dihydroxyphenyl-alanine (dopa); III, quinone of II; IV, 2-carboxy-5,6-dihydroxydihydro-indole (leuco body); V, 2-carboxy-2,3-dihydro-indole-5,6-quinone (red substance); VI, 2-carboxy-5,6-dihydroxy-indole; VII, 5,6-dihydroxy-indole.

had made it possible that it is the quinone of VII which acts as the building stone for melanin construction.

Mason (1948) made a spectrophotometric study of the oxidation of dopa by silver oxide and by tyrosinase. He recognized three chromophoric phases. In the first, dopa was converted into the red pigment dopachrome with maximum absorption spectra at 305 and 475 mμ. The second phase started after dopachrome had rearranged to yield 5,6-dihydroxy-indole. In this phase a substance (a "purple pigment") was produced with maximum absorption at 300 and 540 mμ. As in parallel experiments the gentle oxidation by tyrosinase of a preparation containing 5,6-dihydroxy-indole gave the same pigment, it was concluded that the purple pigment is indole-5,6-quinone (VIII).

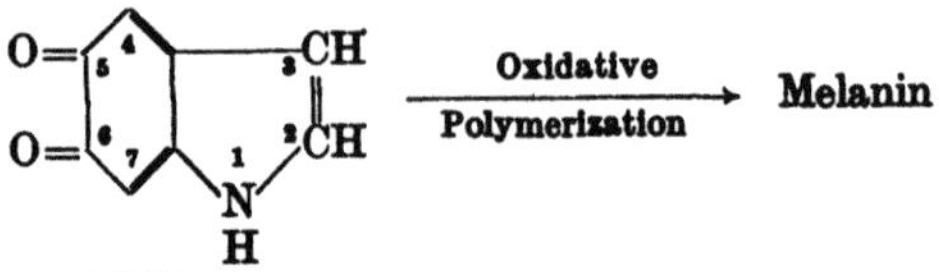

$$O=\underset{\underset{\displaystyle \text{VIII}}{}}{\left\langle \begin{matrix} 5 & 4 \\ 6 & 7 \end{matrix} \right.} \quad \underset{\overset{N}{H}}{\left. \begin{matrix} 3 \\ 1 \end{matrix} \right\rangle} \begin{matrix} {}_3CH \\ {}_2CH \end{matrix} \xrightarrow[\text{Polymerization}]{\text{Oxidative}} \text{Melanin}$$

Clearly Mason not only provided confirmatory evidence of Raper's scheme, but extended it. Indeed he suggested that VIII, having quinone, amine and pyrrole functions, may be capable of undergoing coupling with itself to yield a polymer or quinonoid oxidative product, viz dopa melanin. Melanogenesis provided the third chromophoric phase. Light absorption was general without peaks, except possibly in the ultra violet. Up to the present no advances in knowledge have come from spectrophotometric measurements of solutions containing melanin (see p. 665).

At Cambridge, in considering the possible structure of the melanin polymer, Bu'Lock and Harley-Mason (1951) performed model experiments on reactions between quinones and indoles. The object was to consider the possible ways in which the putative intermediate indole-5,6-quinone (see Mason above) might undergo self-condensation. They showed that indoles can react at the 3-position with quinones. The 3-indolyl quinones obtained in this way are blue-violet to black substances, possessing similar properties.

The conclusions drawn from this and later Cambridge work and experiments at other centres (e. g. Liverpool, see de Beer et al., 1948, 1949) have been briefly stated by Cromartie and Harley-Mason (1953). Seven methylated derivatives of 5,6-dihydroxy-indole have been prepared, viz. all the five possible mono-methylated derivatives and also the 2,3- und the 4,7-dimethyl derivatives. The oxidation by molecular oxygen of these derivatives and of 5,6-dihydroxy-indole has been subjected to comparative spectroscopic study at pH 6.85. For example it was found that insoluble pigments resembling melanin were obtained by the autoxidation of 5,6-dihydroxy indole and of 1-, 2-, 4- and 7-methyl derivatives. Tyrosinase had little effect on the rate of oxidation. The other derivatives gave blue-violet or orange-red solutions. The evidence pointed to the conclusion that for melanin formation from substituted 5,6-dihydroxy-indoles, the 3- and either the 4- or the 7-position must be free. This conclusion supports the view put forward by Bu'Lock and Mason (1951) that melanin is produced by repeated condensations of indole-5,6-quinone, involving reaction between the cationoid 4- or 7-position of one molecule, and the anionoid 3-position of another. Two repeating units which form the "backbone" of the molecule of melanin are represented in IX.

The suggested empirical formula for the repeating unit is $C_8H_3O_2N$ (see below). The authors point out that extensive conjugation through long chains of varying length would account for the intense light absorption displayed by melanin over a wide range of wave lengths. When the 1- and 2-positions remain free cross linkages may occur subsequently. The authors suggest that the existence of such cross links could account for the extreme insolubility of melanin.

Several analyses (see e. g. p. 662) have been made of tyrosinase melanin and dopa melanin. CLEMO and WEISS (1945) cite three, in which the N found was 8.1%, 8.3% and 8.65%. In the repeating unit $C_8H_3O_2N$, the calculated N content is 8.6% approximately. BU'LOCK and HARLEY-MASON point out that the empirical formula they propose is not very different from the formula $C_8H_5O_3N$ and $C_{16}H_{10}O_5N_2$ arrived at by other workers after the analysis of melanin preparations. The former could be compounded of $(C_8H_3O_2N + H_2O)$, and the latter of $[(C_8H_3O_2N)_2 + H_2O + 2H]$. Moreover they note that, for the production of their proposed repeating unit from tyrosine, five atoms of oxygen would be required. An equation may be written,

$$C_9H_{11}O_3N + 5O = C_8H_3O_2N + CO_2 + 4H_2O.$$

As regards oxygen consumption this equation conforms with the experimental finding of DULIÈRE and RAPER (p. 670).

Using a modified BARCROFT apparatus, MASON and WRIGHT (1949) measured not only the oxygen uptake when dopa was oxidized by tyrosinase under certain selected conditions of pH, buffer nature and concentration, and enzyme concentration, but also the carbon dioxide evolved at pH values between 5.1 and 5.5. These experiments appear to be the first in which carbon dioxide evolution during tyrosinase action has been studied. The results obtained were in fair accord with predictions that could be made from RAPER's scheme especially if one accepts MASON's development of the scheme, namely that the quinone (VIII) of the diphenol (VII) and not the quinone of the carboxylic acid (VI) is the repeating unit in melanin. Until oxygen uptake had proceeded to an extent equivalent to two atoms of oxygen absorbed per molecule of dopa oxidized, it was too fast to be measured accurately. This phase was related to the conversion of dopa (II) into the red substance (V) (old name, hallochrome, newer name, dopachrome). The authors consider that there was no evolution of carbon dioxide during this phase. Subsequently this gas was liberated during the conversion of V into melanin whilst oxygen uptake continued, but now at a measurable rate which indicated that two atoms of oxygen were absorbed for every molecule of carbon dioxide evolved. This finding is in agreement with the view that decarboxylation without oxygen uptake occurs in the conversion of V to dihydroxy indole (VII), and that two oxygen atoms are involved in the oxidative polymerization of VII into melanin. One of these oxygen atoms would be the hydrogen acceptor in the oxidation of VII into its quinone (VIII), and the other the acceptor in the oxidative condensation of this quinone.

Mason and Wright do not claim to have established the main quantitative relations in gaseous exchange during melanin formation. Indeed they tabulate their results to emphasize the variability they found in the number of atoms of oxygen consumed and molecules of carbon dioxide evolved per molecule of dopa oxidized over a prolonged period. The oxygen consumption varied from 2.9 to 4.6 atoms, whereas Raper's scheme requires 4, and the carbon dioxide evolved varied from 0.6 to 1.0 molecules, whereas the scheme requires one molecule. They realize that melanins are ill-defined substances in which autoxidations and decompositions (e. g. ring fission, see below) may occur subsequent to their formation. They write "Dopa melanin cannot be regarded as a homogeneous reaction product, the composition of which is independent of conditions of preparation."

The experimental results obtained by Clemo, Duxbury and Swan (1952) provide further evidence that there still exist many complexities which remain to be eludicated. They studied melanin formation when mushroom tyrosinase, acted on D,L-(carboxy-^{14} C)-tyrosine and when 3,4-dihydroxyphenyl-D,L-(carboxy-^{14} C)-alanine underwent autoxidation at pH 8. They had previously synthesized both of these substrates. In the first place they found that the aerobic oxidation of a solution of dopa at pH 8 led to an evolution of carbon dioxide at least equal to that which could theoretically result from the complete decarboxylation of this amino acid. The yield of melanin, however, was only 60% of the weight of dopa used. Because they used carboxyl-labelled tyrosine and dopa, by measuring the radio activity of the carbon dioxide evolved, they were able to show that only half of this carbon dioxide arises from the carboxyl group. They concluded that the remainder comes from "one or more of the other eight carbon atoms of the amino acid molecule". Clearly this finding means that, as regards gaseous exchange, a modification of or addition to the scheme portrayed on p. 671 is required. Moreover they demonstrated that the aerobic oxidation of 5,6-dihydroxy-indole (VII) at pH 8 leads to melanin production with the evolution of carbon dioxide. The authors state that they have no proof that the residues of tyrosine, dopa or VII which evolve the non-carboxyl carbon dioxide are incorporated into the melanin molecule. Cromartie and Harley-Mason (1953) suggest that the evolution of carbon dioxide might result from ring-fission after the melanin polymer is constructed (cf. Mason and Wright's views stated above about the decomposition of melanins). Swan and Wright (1954) have obtained evidence that, if hydrogen peroxide is produced during the course of melanin formation, as it might well be during the production of the intermediate orthoquinones, carbon dioxide could be evolved by secondary reactions. For example when hydrogen peroxide was added to a solution in which 5,6-dihydroxy-indole was being converted into melanin, carbon dioxide evolution was increased to about twice the normal value and less melanin was formed. Evidence that hydrogen peroxide might be produced during the normal conversion of the dihydroxy-indole into melanin was obtained when they found that in the presence of catalase the normal amount of carbon dioxide evolved was halved. The results of another set of experiments suggested that the evolution of gas resulted from ring fission, occurring as a side-reaction, and not as an essential step in melanin formation. In agreement with Cromartie and Harley-Mason (1953) they do not favour a recent attempt to revive a theory "in which oxidation of 5:6-dihydroxy indole to pyrrole-2:3-diacetic acid is postulated as an essential feature of the reaction sequence".

The most favoured view held at present is, therefore, that in the conversion of dopachrome to tyrosinase melanin, the main sequence of changes leads to the oxidative condensation of VIII, without the production by ring fission of pyrrole intermediates. Related to this conversion there is a liberation of one molecule of carbon dioxide for every two atoms of oxygen absorbed. In addition however, side reactions may occur involving further oxygen absorption and carbon dioxide production. The nature of these side reactions does not concern us here, but the recognition of their existence enables confidence to be retained in the scheme on p. 671, and in its more recent development, in spite of the occasional departure of values for oxygen consumption and carbon dioxide evolution from those required by the scheme.

In conclusion we state the provisional hypothesis that plants of any kind which produce melanins as secondary plant products (for probable examples see p. 666) do so by a sequence of reactions involving compounds 1 to IX inclusive, with a gas exchange quotient $\left(\dfrac{CO_2}{O_2}\right)$ of about 0.4 (see equation on p. 673). The detection of these intermediate substances and the measurement of gaseous exchanges during melanogenesis in living plants are among the problems calling for investigation by modern methods.

Thanks are due to Dr. G. A. Swan (Newcastle) for calling the writer's attention to some of the work described in Section VI and for reading the first draft of this section. Thanks are also extended to Dr. E. Nicolai (Leeds) for translating an Italian paper, and to the Newcastle colleagues mentioned in the text for carrying out certain preliminary observations and experiments at the writer's request.

References.

Baker, M. R., and A. C. Andrews: Genetics 29 104 (1944). — Beer, R. J. S., K. Clarke, H. G. Khorana and A. Robertson: J. Chem. Soc. 1948, 2223. — Beer, R. J. S., L. McGrath, A. Robertson and A. B. Woodier: J. Chem. Soc. 1949, 2061. — Bertrand, G.: C. R. Acad. Sci. (Paris) 118, 1215 (1894); 120, 266 (1895); (a) C. R. Acad. Sci. (Paris) 122, 1132 (1896); (b) C. R. Acad. Sci. (Paris) 122, 1215 (1896). — Bloch, B.: Hoppe-Seylers Z. 98, 226 (1917). — Bu'Lock, J. D., and J. Harley-Mason: (a) J. Chem. Soc. 1951, 703; (b) J. Chem. Soc. 1951, 2248.

Clemo, G. R., F. K. Duxbury and G. A. Swan: J. Chem. Soc. 1952, 3464. — Clemo, G. R., and J. Weiss: J. Chem. Soc. 1945, 702. — Cromartie, R. I. T., and J. Harley-Mason: Chem. Ind. Rev. 1953, 972.

Daniel, J.: J. Genet. 36, 139 (1938). — van Duijn, P.: J. Histochem. and Cytochem. 1, 143 (1953). — Dulière, W. L., and H. S. Raper: Biochem. J. 24, 239 (1930).

Einsele, W.: J. Genet. 34, 1 (1937). — Evans, W. C., and H. S. Raper: Biochem. J. 31, 2162 (1937). — Figge, F. H. J.: Proc. Soc. Exp. Biol. (N. Y.) 41, 127 (1939); 44, 293 (1940); 46, 269 (1941). — Fox, A. S., and W. D. Gray: Proc. Nat. Acad. Sci. (Wash.) 36, 538 (1950).

Ginsburg, B.: Genetics 29, 176 (1944). — Guggenheim, M.: Hoppe-Seylers Z. 88, 276 (1913).

Harley-Mason, F.: J. Chem. Soc. 1948, 1244. — Horowitz, N. H., and San-Chian Shen: J. Biol. Chem. 197, 513 (1952).

Jacobson, F. W., and N. Millott: Proc. Roy. Soc. B, 141, 231 (1953).

Keilin, D., and T. Mann: Proc. Roy. Soc. B 125, 187 (1938). — Kertész, D.: Biochim. et Biophys. Acta 9, 170 (1952). — Kubowitz, F.: Biochem. Z. 299, 32 (1938).

Lerner, A. B., and T. B. Fitzpatrick: Physiol. Rev. 30, 91 (1950). — Lerner, A. B., T. B. Fitzpatrick, E. Calkins and W. H. Summerson: J. Biol. Chem. 178, 185 (1948). — Lison, L.: "Histochemie Animale". Paris: Gauthier-Villars 1936.

Markert, C. L.: Genetics 50, 60 (1950). — Mason, H. S.: J. Biol. Chem. 172, 83 (1948). — Mason, H. S., and C. I. Wright: J. Biol. Chem. 180, 235 (1949). — Miller, E. R.: J. Biol. Chem. 44, 481 (1920).

Nelson, J. M., and C. R. Dawson: Advances in Enzymology 4, 99 (1944).

Paech, K.: "Biochemie und Physiologie der sekundären Pflanzenstoffe". Berlin-Göttingen Heidelberg: Springer-Verlag 1950. — Panizzi, L., and R. Nicolaus: Gazetta 82, 435 (1952).

Raper, H. S.: Biochem. J. 20, 735 (1926); 21, 89 (1927); J. Chem. Soc. 1938, 125. — Robinson, E. S., and J. M. Nelson: Arch. Biochem. 4, 111 (1944).

Schaeffer, P.: Arch. Biochem. 47, 359 (1953). — Schroeder, H.: Klin. Wschr. 1934, 533. — Sizer, I. W.: J. Biol. Chem. 169, 303 (1947). — Sizer, I. W.: Science 108, 335 (1948). — Spiegel-Adolf, M.: Biochem. J. 31, 1303 (1937). — Swan, G. A., and D. Wright: J. Chem. Soc. 1954, 381.

de Vries, M. A.: "Over de vorming van Phytomelaan bij Tagetes patula L. en enige andere, Composieten", Leiden: Buurman 1948.

Blausäure-Verbindungen.

Von

P. Seifert.

Mit 3 Abbildungen.

Seit VAUQUELIN (1803), veranlaßt durch eine Untersuchung von SCHRADER, Cyanwasserstoff erstmalig aus bitteren Mandeln und Aprikosenkernen und BERGEMANN (1812) aus der Rinde von *Prunus padus* isolieren konnten, ist eine große Zahl von Pflanzen (über 360 Arten aus 148 Gattungen und 41 Familien) bekannt geworden, die Cyanwasserstoff oder Blausäure leicht abspalten. Der Cyanwasserstoff (HCN) kommt in den Pflanzen gebunden in Form von Blausäure-(Cyanhydrin- oder Oxynitril-)Glykosiden vor. Freies HCN tritt nach BRUNSWIK (1921), PAECH u. a. höchstens in Spuren als Produkt der prämortalen enzymatischen Spaltung auf.

Der im Prinzip gleichbleibende Aufbau eines derartigen „cyanogenen" Glykosids sei an dem Beispiel des am längsten bekannten Blausäureglykosids, dem Amygdalin, das sich in vielen Prunoideen und Pomoideen findet, aufgezeigt. Im Amygdalin ist das Cyanhydrin oder Oxynitril des Benzaldehyds glykosidisch mit dem Disaccharid Gentiobiose verbunden:

$$\text{Benzaldehyd} \ (C_6H_5{-}CH{=}O) + HCN = \text{Cyanhydrin} \ (C_6H_5{-}CH(OH){-}CN) + C_{12}H_{22}O_{11} = \text{Amygdalin} \ (C_6H_5{-}CH(O{-}C_{12}H_{21}O_{10}){-}CN)$$

Benzaldehyd — Cyanhydrin = Oxynitril des Benzaldehyds = Mandelsäurenitril — Gentiobiose — Amygdalin = Mandelsäurenitril-gentiobiosid

Als Aldehydkomponente kommt außer Benzaldehyd auch p-Oxybenzaldehyd, z. B. im Dhurrin vor. Auch Cyanhydrine von Ketonen, wie Aceton im Linamarin, ein Trioxyketon im Gynocardin, sowie Cyanhydrine von Zuckern, Maltose im Lotusin, können als Grundbausteine der cyanogenen Glykoside auftreten.

Als glykosidisch gebundene Komponente finden sich zumeist Mono- und Disaccharide, wie Gentiobiose, im Amygdalin. Die meisten cyanogenen Glykoside sind echte Glucoside, in denen Glucose glykosidisch gebunden vorkommt, so im Dhurrin, Gynocardin, Linamarin, Prunasin und in seinen Stereoisomeren Sambunigrin und Prulaurasin sowie im Hiptagin. Im Vicianin liegt ein Disaccharid aus Glucose und Arabinose, die Vicianose, vor. Im Lotusin ist ein Nichtsaccharid, das Lotoflavin, mit dem Maltose-Cyanhydrin glykosidisch verbunden. Die Bindung ist offensichtlich eine vorwiegend β-glykosidische.

Der Gehalt cyanogener Pflanzen an einzelnen Blausäureglykosiden ist sehr unterschiedlich, und auch die verschiedenen Organe der gleichen Pflanze enthalten ungleich viel. Selbst der HCN-Gehalt gleichartiger Blätter variiert. Nach ARMSTRONG besteht jedoch kein ausgesprochener Unterschied im Glykosid-Gehalt zwischen Sonnen- und Schattenblättern. Der Gehalt der beiden Hälften eines Blattes ist identisch. Die Blattflächen haben im allgemeinen weniger HCN als die Mittelnerven (ROSENTHALER). Je nach der Herkunft wechselt der Gehalt an glykosidischem HCN in den Wurzeln von *Manihot utilissima* zwischen 0,065 und 0,015%. Die Rinde der Wurzel enthält 60% des HCN (A. CORREIA 1947). Hier dürfte die Bodenbeschaffenheit eine entscheidende Rolle spielen. WESTER konnte bei Düngung mit K, P und N eine deutliche Steigerung des Glykosidgehaltes beobachten. Mit dem Alter und fortschreitender Jahreszeit sowie bei Chlorose nimmt der HCN-Gehalt ab. Reife Stengel, vergilbte Blätter und reife Kapselsubstanz von Flachs enthalten kein Linamarin mehr. Der Gehalt von Linamarin in Leinsamen wird jedoch durch Lagerung kaum verändert (LUEDTKE, 1952). Eine Neubildung von Blausäureglykosiden findet offensichtlich nur in den chlorophyllhaltigen Pflanzenteilen statt. Seit BRIDEL im Samen von Kirschlorbeer Amygdalin, in den Blättern dagegen Prulaurasin nachweisen konnte, ist anzunehmen, daß auch mehrere Glykoside von Cyanhydrinen gleichzeitig in einer Pflanze vorkommen können. Der cyanogene Charakter ist nach DILLEMANN (1953) teilweise dominant vererbbar.

Nach der Annahme von TREUB sind HCN und die cyanogenen Glykoside als eine Etappe des N auf dem Wege von anorganischem Nitrat-N über Amide zum Eiweiß aufzufassen. Die Cyanogenese geht der Proteinbildung parallel und steigert sich nach Verletzungen. Der von IWANOFF und OSNIZKAJA (zit. nach PAECH 1950) festgestellte Stoffwechsel der HCN-Glykoside würde ebenfalls hierfür sprechen. Es wäre somit auch die Zunahme des HCN beim Keimen von Samen, beim Austreiben der Knospen und die Tatsache, daß junge Blätter mehr HCN enthalten als ausgewachsene, verständlich. Nach neueren Vorstellungen (PAECH) stellt aber der Umsatz der HCN-Verbindungen nur einen Seitenweg sowohl der Assimilation wie auch der Dissimilation des Eiweißes dar.

Die Blausäureglykoside bildenden Pflanzen verfügen auch über spezifische Enzyme, die sowohl die Assimilation wie auch die Dissimilation der Glykoside steuern. In den Säften und den kalt hergestellten wäßrigen Auszügen sind Substrat und Enzym nebeneinander vorhanden, so daß die enzymatische Zersetzung von selbst in Gang kommt. Das bekannte Emulsin oder Mandelenzym ist ein kompliziertes Enzymgemenge. Es enthält Amygdalase, die Amygdalin in L-Mandelsäurenitril und Gentiobiose spaltet, und Oxynitrilase, die Mandelsäurenitril in HCN und Benzaldehyd aufspaltet. Nur das natürlich vorkommende L-Mandelsäurenitril wird angegriffen. Aus racemischem (D,L) Amygdalin kann daher D-Mandelsäurenitril unzersetzt erhalten werden. Wird Emulsin auf 65° erhitzt, so bleibt allein die Oxynitrilase wirksam. Auf Amygdalin wirksame Fermente kommen übrigens sehr verbreitet im Pflanzenreich vor, so in Moosen, Bakterien, Myxomyceten, Rost- und Hutpilzen u. a. (CZAPEK). Prunase ist ebenfalls sehr verbreitet. Sie läßt Amygdalin unverändert. Sie findet sich zusammen mit Linase in Linaceen. In älteren Pflanzen überwiegt jedoch die Prunase. In vivo sind Enzym und Substrat in der Pflanze räumlich voneinander getrennt. Bei der Narkose (PAECH) oder der Zerstörung des Gewebes kommen Enzym und Glykosid zusammen, was eine Hydrolyse zur Folge hat. Wie der Glykosidgehalt, so wechselt auch die Enzymaktivität der einzelnen Pflanzen und Pflanzenteile unter verschiedenen Umweltbedingungen und in verschiedenen Entwicklungsstadien (WARTH; CHARLTON; COUCH u. BRIESE)[1]. Dies ist von nicht unerheblicher Bedeutung für die analytische Bestimmung von HCN, sofern enzymatisch hydrolysiert wird.

Da die Blausäureglykoside beim Welken der Pflanzen durch deren Eigenenzyme eine Zersetzung erfahren, muß bei quantitativen Untersuchungen für eine baldige Verarbeitung Sorge getragen werden. Durch schnelles Einbringen der zerkleinerten Pflanzenteile in siedendes Wasser oder Alkohol, denen vorteilhaft zur Absättigung von Pflanzensäuren $CaCO_3$ beigegeben wird, werden die Enzyme unwirksam gemacht. Hierfür hat sich nach BRIEGER auch siedendes Chloroform gut bewährt.

Alle bekannten Blausäureglykoside sind in Wasser und Alkohol, wenn auch mit unterschiedlichen Löslichkeitskoeffizienten, löslich. In Äther und Petroläther sind sie unlöslich, so daß bei der Isolierung eine Abtrennung von Fett- und Lipoidsubstanzen einfach ist. Ein Teil von ihnen löst sich außerdem in Essigester so weitgehend, daß sie mit Hilfe dieses Lösungsmittels von den in Wasser zwar löslichen, in Essigester aber unlöslichen störenden Beimengungen, vor allem von Sacchariden, getrennt werden können.

Über die weitere *Isolierung* und *Reindarstellung* der cyanogenen Glykoside läßt sich wenig Allgemeines sagen, da jedes Glykosid wegen seines unterschiedlichen Aufbaues und seiner dadurch bedingten Unterschiede in den chemischen und physicochemischen Eigenschaften ein individuelles Vorgehen erfordert. Hier muß auf die in der Literatur des öfteren mitgeteilten Verfahren, die heute noch die gleichen sind, verwiesen werden (Zusammenstellung bei ROSENTHALER in G. KLEIN: Hb. der Pflanzenanalyse, Bd. III [2. Hälfte], S. 1045—1056, Wien 1932). Tab. 1 bringt eine Zusammenstellung der bisher isolierten und identifizierten cyanogenen Glykoside, ihres chemischen Aufbaues und einiger physicochemischer Eigenschaften.

---

[1] Die β-Glykosidasen, die für die Abspaltung der glykosidisch gebundenen Komponente der Blausäureglykoside (resp. für die Herstellung der glykosidischen Bindung) verantwortlich sind, lassen sich nach einer persönlichen, bisher unveröffentlichten Mitteilung von Herrn Dr. H. REZNIK, Heidelberg, mit Indikan histochemisch nachweisen.

Tabelle 1. *Zusammenstellung der bisher bekannten und in ihrer chemischen Zusammensetzung aufgeklärten Blausäure-Glykoside.*

| Name | Ausgangsmaterial | Fp | $[\alpha]_D/t$ | kristallisiert aus | Chemischer Aufbau | | durch Emulsin spaltbar | Um-rechnungs-faktor Glykosid: HCN |
					Carbonyl-Komponente oder Cyanhydrin resp. Oxynitril von	glykosidische Komponente		
A. Glucoside								
Dhurrin	*Sorghum vulgare*	Zers. >200°		H_2O, Alkohol	p-Oxybenzaldehyd	Glucose	+	11.522
Gynocardin	Samen von *Gynocardia odorata*	162°	$\frac{+72,5}{21°}$	H_2O, Aceton	Trioxyketon $C_5H_5(OH)_3CO$	Glucose	+	12.333
Hiptagin	Wurzelrinde von *Hiptage Madablota*	110°	$\frac{+\,3,5}{20°}$	Alkohol-Wasser 1:1	Hiptagenin $C_4H_4N_2O_4$	Glucose		11.333
Linamarin	Keimlinge des Leins	141°	$\frac{-27,4}{15°}$	Alkohol	Aceton	Glucose		9.148
Prulaurasin	Kirschlorbeerblätter	122°	$\frac{-54,6}{20°}$	Essigester	Benzaldehyd	Glucose	+	10.926
Prunasin	Hefe	139°	$\frac{-27}{20°}$	Essigester, Chloroform	Benzaldehyd	Glucose	+	10.926
Sambunigrin	Blätter von *Sambucus nigra* u. Kirschlorbeer	152°	$\frac{-76,1}{20°}$	Essigester .	Benzaldehyd	Glucose	+	10.926
B. Glykoside								
Amygdalin	Bittere Mandeln	etwa 210°	$\frac{-41,96}{20°}$	H_2O, Alkohol	Benzaldehyd	Gentiobiose	+	16.926
Corynocarpin	Samen von *Corynocarpus laevigata*	140°		H_2O, Alkohol				
Karakin	dto.	122°		H_2O, Alkohol	Maltose	Lotoflavin		18.000
Lotusin	*Lotus arabicus*	Zers.		Alkohol	Benzaldehyd			23.593
Vicianin	Samen von *Vicia angustifolia*	160°	$\frac{-20,7}{16-18°}$	H_2O		Vicianose (Disaccharid aus Glucose u. Arabinose)		15.815

I. Qualitativer Nachweis von Blausäure.

1. Freisetzung der Blausäure.

Außer durch die Eigenenzyme der jeweiligen Ausgangspflanze findet eine Zersetzung der HCN-Glykoside in die einzelnen Komponenten durch verdünnte Säure statt. Starke Salzsäure liefert nach Abspaltung der glykosidischen Gruppe die aus dem Oxynitril hervorgehende Oxysäure, z. B. Mandelsäure aus dem Amygdalin. Mit starken Alkalien dagegen entsteht (ohne Abspaltung der glykosidischen Komponente) das Alkalisalz der jeweiligen Glykosidooxysäure.

Wegen der Langwierigkeit und Schwierigkeit, die einzelnen Glykoside der Blausäure zu isolieren oder als solche zu identifizieren, scheiden diese selbst für die routinemäßige analytische Bestimmung aus. Ebenso sind die Carbonylkomponente und die glykosidische Komponente für den gleichen Zweck ungeeignet, da sie von anderen Stoffwechselprodukten der Pflanze nicht genügend differenziert sind und außerdem von Glykosid zu Glykosid wechseln können. Nur in sehr beschränktem Maße kann bei Kenntnis des Oxynitrils der enzymolytische Index nach BOURQUELOT (ROSENTHALER, 1931) dazu dienen, die Identifizierung eines cyanogenen Glykosids zu führen. Für eine qualitative und quantitative Bestimmung der cyanogenen Glykoside kann am ehesten die für sie spezifische Komponente, der Cyanwasserstoff, herangezogen werden. Es ist nach den bisherigen Erfahrungen anzunehmen, daß die Blausäure in den cyanogenen Glykosiden eine stabile Fixierung gefunden hat. Abgesehen von Alkaloid Ricinin ist es bis heute nicht sicher erwiesen, ob es im Pflanzenreich noch andere HCN-Verbindungen gibt. Der qualitative HCN-Nachweis kann somit für das Vorhandensein von cyanogenem Glykosid, seine quantitative Bestimmung für die Menge desselben genommen werden.

Die Freisetzung des HCN aus den Blausäureglykosiden ist nach folgenden Verfahren möglich:

1. Autoenzym-Verfahren,
2. Säurehydrolytisches Verfahren,
3. Enzymatisches sog. Gesamt-HCN-Verfahren (WINKLER), bei dem Enzympräparate anderer Pflanzen (meist Emulsin) zusätzlich zur Spaltung Verwendung finden,
4. Kombination von Enzym- und Säure-Verfahren.

Die nach dem Autoenzym-Verfahren in Freiheit gesetzten Blausäuremengen sind beträchtlichen Schwankungen unterworfen, weil die Enzymaktivitäten der Pflanzen erheblich differieren können. Diese Verfahren sind daher für quantitative Untersuchungen nicht verwertbar. In Gegenwart von Sublimat ($HgCl_2$) dagegen haben COUCH und BRIESE mit dem Autoenzym-Verfahren theoretische Ausbeuten erhalten, jedoch nur dann, wenn die Proben über Wochen reagieren konnten. Die sehr lange Reaktionszeit macht dieses Vorgehen für routinemäßiges analytisches Arbeiten zwar ungeeignet, aber als Standardisierungsmethode (WINKLER) wäre dieses Vorgehen zu empfehlen. Auch die zusätzliche Verwendung von Enzympräparaten führt in den meisten Fällen nicht zu 100%igen Ausbeuten an HCN. Dies liegt in der Eigenart enzymatischer Reaktionen allgemein begründet[1]. Befriedigende Ergebnisse lassen sich jedoch nach dem Vorgehen von BISHOP erzielen, der das Untersuchungsmaterial mit kaltem Wasser ansetzt, langsam auf 100° erwärmt, das entstehende HCN aus dem Reaktionsgemisch entfernt, wieder auf 35° abkühlt und mit Emulsin den Rest des Glykdosis (nach dem Autor 5—9%) zersetzt.

___

[1] Eine gute Zerkleinerung des Materials ist für quantitative Untersuchungen unerläßlich. Nach LUEDTKE (1952) ist Zerreiben im Mörser sehr zu empfehlen.

Eine allzu lange Fermentierung ist nach Luedtke (1952) nicht anzuraten. Für den Linamarin-Nachweis im Flachs fand er eine Fermentierungsdauer von 2—3 Std. bei 40° C und einem p_H-Wert von 6,0 als optimal (s. a. Dillemann 1951).

Die säurehydrolytischen Verfahren erfassen den HCN-Gehalt ebenfalls nicht quantitativ, wohl liefern sie in bezug auf den Gesamt-HCN-Gehalt (die HCN-Glykoside repräsentierend) relativ konstantere Ausbeuten. Dies hat seine Ursache darin, daß ein Teil des Cyanhydrins zum Amid und weiter zur Oxycarbonsäure und NH_3 hydrolysiert wird. Mit gutem Erfolg wird die Kombination von Enzym- und Säure-Verfahren verwendet, wobei man zuerst die enzymatische Zersetzung durchführt und die Zersetzung mit Säure anschließt. Bei der enzymatischen Zersetzung kann man sich die Beobachtung von Mirande zunutze machen, wonach die enzymatische Zersetzung der Blausäureglykoside durch Dämpfe von Chloroform, Äther, Chloräthyl, CS_2 und Hg beschleunigt wird, weil dadurch Enzym und Substrat nach Abtöten der Zellen miteinander in Kontakt treten können.

Die Inaktivierung der Autoenzyme durch Eintauchen in siedendes Wasser und auch Alkohol ist immer mit kleinen Verlusten an HCN verbunden, worauf schon Treub aufmerksam machte. Der Verlust beträgt nach Bishop etwa 6%[1]. Völlig ungeeignet für quantitative Untersuchungen ist die Inaktivierung mit Wasserdampf, da ein großer Teil des Glykosids hierbei bereits zersetzt wird. Bei Verwendung von siedendem Chloroform (Brieger 1931) hält sich der HCN-Verlust offensichtlich ebenfalls in erträglichen Grenzen.

Cyanwasserstoff ist ein Gas und kann daher leicht aus dem Reaktionsgemisch entfernt werden. Es wird im allgemeinen in alkalischer Lösung aufgefangen, wo sich die Alkalicyanide bilden. In diesen alkalischen Lösungen erfolgt zumeist auch die eigentliche Bestimmung. Das Auffangen in gutgekühlter wäßriger Lösung ist für quantitative Untersuchungen nicht zu empfehlen, da mit Verlusten gerechnet werden muß. HCN läßt sich außerdem in Sublimatlösung absorbieren ($2\,HCN + HgCl_2 = Hg(CN)_2 + 2\,HCl$). Das dabei entstehende $Hg(CN)_2$ bleibt in Lösung. Dieses Vorgehen ist aber unpraktisch, da die Blausäure zum Nachweis aus dieser Lösung erst wieder in Freiheit gesetzt werden muß, was sehr umständlich und auch mit Verlusten verbunden ist.

Die Destillationsmethode[2] ist zur quantitativen Isolierung des HCN nicht brauchbar, da wahrscheinlich durch die Carbonylkomponente des Glykosids Verluste auftreten. Im Modellversuch konnte Bishop aus Cyanidlösungen nach dem Ansäuern 95—97% HCN wiedergewinnen. Nach der Zugabe von Benzaldehyd trat jedoch ein Defizit von 30—40% ein. Sehr gute Ergebnisse liefert dagegen das *Aerations-Verfahren*, wobei das in Freiheit gesetzte HCN mit Hilfe eines Luftstromes — hierzu kann auch ein inertes Gas, wie z. B. Stickstoff, Verwendung finden — aus dem Reaktionsgemisch ausgetrieben wird.

2. Nachweis-Methoden.

Peche hat den *histochemischen Nachweis* der Blausäure mit Mercuronitrat führen können. Hierbei entsteht neben $Hg(CN)_2$ metallisches Hg an den Stellen der Umsetzung, das als dunkler Niederschlag im mikroskopischen Bilde erscheint. HCN wurde dabei besonders stark an den Chloroplasten der Palisadenzellen von *Prunus laurocerasus* beobachtet.

---

[1] Die Eigenlöslichkeit der Blausäureglykoside in Alkohol und Wasser darf ebenfalls nicht übersehen werden.

[2] Dies gilt auch für die Wasserdampf-Destillation (Luedtke, 1952).

Die analytischen Blausäure-Verfahren[1] sind mit Ausnahme der Reaktion mit Silber-Ion ausschließlich kolorimetrische Verfahren. Die folgenden Prinzipien sind für den *qualitativen Blausäure-Nachweis* herangezogen worden:

1. Silber-Reaktion,
2. Berliner-Blau-Reaktion,
3. Rhodanid-Reaktion,
4. Nitroprussid-Reaktion,
5. Pikrinsäure-Reaktion,
6. Phenolphthalein-Reaktion und ähnliche Reaktionen,
7. Methämoglobin-Reaktion,
8. Jod-Reaktion.

Die Prinzipien 4—8 werden nur noch relativ wenig benutzt. Die Phenolphthalein-Reaktion (WEEHUIZEN-THIERY) führt in Gegenwart von Cu(II)-Ionen durch Oxydation zu Phenolphthalein, das in alkalischer Lösung eine Rotfärbung zeigt. Auf ganz ähnlicher Reaktion beruht die SCHOENBEIN-Reaktion von alkoholischem Guajakharz-Extrakt (1:10) und Cu(II), die zu einer Blaufärbung führt. Mit Aloin entsteht eine Rotfärbung (KLUNGE). Hydrocoerulignon-Reagens ($CuSO_4$ + Essigsäure + Hydrocoerulignon in Wasser bei 50° digeriert und filtriert) ergibt mit HCN einen Niederschlag von purpurglänzendem Coerulignon, bei wenig HCN nur eine Rotfärbung (MOIR). Benzidin gibt unter den gleichen Bedingungen einen blauen Niederschlag resp. eine Blaufärbung der Lösung (MOIR, PERTUSI-GASTALDI). Diese Reaktionen sind auch als Papierreaktionen durchführbar. Sie sind aber nicht spezifisch, da sie auch von solchen Pflanzenstoffen gegeben werden, die oxydierend wirken. Sie sind daher nur als Vorproben zu verwerten. Da sie andererseits sehr empfindlich sind, ist ihr negativer Ausfall für die Abwesenheit von HCN beweisend.

Die Methämoglobin-Reaktion (KOBERT) wird kaum noch zum Blausäure-Nachweis verwandt. Sie beruht darauf, daß bereits geringe Mengen von HCN in Lösungen von Methämoglobin (aus Hämoglobin-Lösungen mit etwas rotem Blutlaugensalz leicht zu erzeugen) eine hellrote Färbung ergeben, die auch beim Erhitzen auf Körpertemperatur bestehen bleibt.

Die Jodstärke-Reaktion beruht darauf, daß HCN mit freiem Jod dieses reduziert ($2\,J$ + + HCN = JCN + HJ), was eine Entfärbung der blauen Lösung (Jod-Stärke) zur Folge hat. Diese Reaktion ist wie die Reaktion mit Methaemoglobin von großer Empfindlichkeit und Spezifität.

Erwärmt man die alkalische HCN-Lösung mit einem Tropfen Pikrinsäurelösung (2%ig), so tritt eine Rotfärbung auf (isopurpursaures Alkali). Die Reaktion ist aber wenig empfindlich (1:3000). Mit reduzierend wirkenden Substanzen entsteht die gleiche Reaktion. Sie läßt sich nach GUIGNARD auch als Papierreaktion durchführen. Das Papier wird hierzu mit einer Lösung von 1 g Pikrinsäure, 10 g Soda in 100 ml Wasser getränkt. In einer HCN-haltigen Atmosphäre färbt sich das Papier rot. Sie kann im Reagensglas so ausgeführt werden, daß einige Tropfen Chloroform (MIRANDE) am Boden und darüber das Pflanzenmaterial eingebracht werden. Nach Einführung eines Streifens GUIGNARD-Papiers (mit dem Korken eingeklemmt) wird längere Zeit verkorkt gehalten.

Die Nitroprussid-Reaktion wird nach VAN GIFFEN so angestellt, daß man in der zu untersuchenden Lösung etwas $NaNO_2$ auflöst, 2—3 Tropfen $FeCl_3$-Lösung hinzufügt, schüttelt und mit verdünnter Schwefelsäure ansäuert. Man erhitzt zum Sieden, fällt das überschüssige Fe mit NH_3 aus, filtriert, dampft ein, nimmt mit Wasser auf, kühlt mit Eis und fügt dann 1 Tropfen Schwefelammonium zu, wodurch bei Gegenwart von HCN eine violettrote Färbung entsteht, die in Blau, Grün und Gelb übergeht. Bei geringen Mengen von HCN entsteht sofort eine blaugrüne Färbung. Die Empfindlichkeit beträgt 1:300000.

Die Nitroprussid-Reaktion kommt bezüglich ihrer Spezifität bereits der Rhodanid- und Berliner-Blau-Reaktion nahe. Die *Rhodanid-Reaktion* geht bereits auf LIEBIG zurück. Die alkalische HCN-Lösung wird im einfachsten Falle zu diesem Zweck mit einigen Tropfen einer gelben Lösung von Schwefelammonium

---

[1] Die Geschichte der Blausäureglykoside ist nunmehr etwa 150 Jahre alt. Die Aufklärung der meisten heute bekannten Glykoside fällt bereits in die klassischen Anfänge der modernen Chemie, ebenso die ihrer Analytik. Mit den Publikationen der zwanziger Jahre stand das analytische Gebäude im allgemeinen bereits festgefügt, und es sind seitdem wenige prinzipielle Neuerungen hinzugekommen.

zusammen auf dem Wasserbad zur Trockne verdampft. Der nunmehr Rhodanid ($NaCN + S = NaSCN$)- enthaltende Rückstand wird mit verdünnter Salzsäure aufgenommen und mit einigen Tropfen $FeCl_3$-Lösung versetzt, wobei sofort eine blutrote Färbung der Lösung resultiert ($Fe(SCN)_3$), wenn HCN zugegen war. Die Empfindlichkeit ist außerordentlich groß (1:4000000). Eine nur schwache Rotfärbung muß kritisch beurteilt werden, da andere N-haltige Substanzen unter Umständen bei den gleichen Bedingungen (Überhitzung) Spuren von HCN liefern können (Rosenthaler).

Nach Kolthof kann die Rhodanid-Reaktion auch so ausgeführt werden, daß man die alkalische HCN-Lösung (5 ml) mit 1 ml Natriumtetrathionat-Lösung (1%ig) und 5 Tropfen konz. Ammoniak-Lösung 5 min im Wasserbad auf 50—55° erwärmt, dann mit verdünnter Salzsäure ansäuert und einige Tropfen $FeCl_3$-Lösung (5%ig) zusetzt.

Die *Silber-Reaktion* wird zur qualitativen HCN-Bestimmung so ausgeführt, daß zu einer trüben Mischung von 2 ml Ammoniak-Lösung (10%ig), 1 Tropfen Jodkali-Lösung (5%ig), 20 ml Wasser und 1 Tropfen $AgNO_3$-Lösung (1%ig) die alkalische HCN-Lösung zugesetzt wird. Ist Blausäure vorhanden, so wird die von AgJ trübe (oder besser opake) Lösung geklärt, weil sich die lösliche Ag-Cyan-Komplexverbindung bildet:

$$AgJ + 2\,NaCN = Na[Ag(CN)_2] + NaJ.$$
unlöslich   löslich

Die Reaktion geht bereits auf Deniges zurück. Sie ist für die Praxis von einer ausreichenden Empfindlichkeit und absoluter Spezifität.

Häufiger als die Silber- und Rhodanid-Reaktion, die beide in der beschriebenen methodischen Durchführung eine vorangehende Isolierung des HCN in Alkali fordern, ist die *Berliner-Blau-Reaktion* — Lander, Walden und Treub haben den hierauf beruhenden Methoden bereits den Vorzug gegeben — zur qualitativen Bestimmung des HCN herangezogen worden. Sie kann auch direkt, d. h. ohne vorherige Überführung in Alkali durchgeführt werden. Sie beruht darauf, daß sich komplexes Ferrocyanid ($Na_4[Fe(CN)_6]$) bildet, das mit Fe(III)-Ionen Berliner-Blau (Ferriferrocyanid) ergibt. Die auf dieser Reaktion aufbauenden Verfahren sind vielfach modifiziert worden. In der einfachen Ausführung von Lockemann ist sie immer noch zu empfehlen.

Filtrierpapier wird mit Natronlauge (10%ig) befeuchtet und über der Öffnung eines Reagensglases befestigt, welches die mit verdünnter Schwefelsäure versetzten, zu untersuchenden Pflanzenteile enthält. Man erhitzt bis zum Sieden und gibt nach einiger Zeit auf die feuchte Stelle einige Tropfen $FeSO_4$-Lösung (5%ig) und nach einigen weiteren Minuten einige Tropfen Salzsäure (10%ig). Es tritt nur dann eine Blaufärbung des Papieres auf, wenn HCN freigemacht wurde.

In der Modifikation nach Druce, Lande und Walden sowie nach Vorlaender kann HCN in der alkalischen Lösung in kleinsten Mengen nachgewiesen werden (0,00002 g in 1 ml). Man geht dabei folgendermaßen vor: Die alkalische HCN-Lösung wird mit einigen Tropfen 5%iger $FeSO_4$-Lösung 1—2 min erwärmt. Vom entstehenden Niederschlag wird abfiltriert, das Filtrat mit verdünnter Salzsäure angesäuert und mit 3—10 Tropfen einer frischen kaltgesättigten Lösung von $FeCl_2$ oder $FeSO_4$ versetzt. Unter spontaner Oxydation tritt Berliner-Blau auf, wenn HCN vorlag.

Den auf der Silber-, Rhodanid- und Berliner-Blau-Reaktion beruhenden Methoden zum qualitativen HCN-Nachweis kommt zweifellos die größte Spezifität zu. Die Gefahr einer Störung der Rhodanid-Reaktion durch andere N-haltige Pflanzenstoffe (s. vorher) ist praktisch ohne Bedeutung, wenn das HCN nach dem Aerationsverfahren in Alkali übergeführt wird. Die Rhodanid-Reaktion dürfte demnach besonders geeignet sein, Spuren von HCN nachzuweisen. Auch die Berliner-Blau-Reaktion ist hierzu noch von einer ausreichenden Empfindlichkeit, während die Silber-Reaktion in der beschriebenen Ausführung etwas weniger empfindlich ist (s. dagegen mikrochemischer Nachweis). Es ist immer zu empfehlen, zur qualitativen Blausäure-Bestimmung wenigstens zwei verschiedene Prinzipien zu verwenden.

Nach BRUNSWIK (1921) kann der *mikrochemisch-qualitative Blausäure-Nachweis* mit der Silber- und Benzidin-Reaktion folgendermaßen geführt werden: In eine kleine Glaskammer (Objektträger mit Glasring verkittet) wird wenig Pflanzenmaterial mit Wasser und 1 Tropfen Äther oder Chloroform befeuchtet. Den Verschluß nach oben bildet ein Deckgläschen mit einem Tropfen 1%iger $AgNO_3$-Lösung, die mit Methylenblau schwach angefärbt wird. Wird HCN aus dem Untersuchungsmaterial freigemacht, so entstehen in dem Tropfen Körnchen, Nadeln, Ranken, Drusen und Sphärite von AgCN, die vom Methylenblau deutlich blaugrün gefärbt sind. Die Empfindlichkeit der Methode beträgt 0,06 γ HCN.

Wird das Deckglas in gleicher Weise mit einem Tropfen Benzidin-Reagens (Gemisch von 1 ml 3%iger Cupriacetat-Lösung, 10 ml 5%iger Benzidinacetat-Lösung und 16 ml Wasser) versehen, so entstehen blaue Nadeln oder Körnchen. Die Empfindlichkeit dieser Methode beträgt 0,02 γ HCN.

Ist nur mit äußerst kleinen Mengen von Blausäure zu rechnen, dann verfährt man so, daß man mindestens 1 kg Pflanzenmaterial aufbereitet. Dieses wird in möglichst frischem Zustand zerkleinert und mit Wasser befeuchtet. Dazu verwendet man am besten einen Langhals-Rundkolben mit Waschflaschen-Einsatz. Nach mehreren Stunden unter Verschluß wird mit verdünnter Schwefelsäure angesäuert (nach BISHOP kann man auch ganz langsam auf 100° erwärmen, abkühlen lassen und dann ansäuern). Nach dem Ansäuern verbindet man sofort über eine Waschflasche mit KOH oder NaOH (10—20 ml 10%ig) mit der Wasserstrahlpumpe. Die durch das Untersuchungsmaterial gesaugte Luft wird vorher durch zwei Waschflaschen mit $HgCl_2$ (1:20)-Lösung und NaOH (10%ig) von CO_2 und eventuellen HCN-Beimengungen der Raumluft befreit. Es wird nun über mehrere Stunden ein starker Luftstrom durch das System gesaugt. Schäumen kann durch Octyl- oder Caprylalkohol vermieden werden (einige Tropfen). Der HCN-Nachweis kann nunmehr in der alkalischen Lösung geführt werden, am besten zuerst mit der Rhodanid-Reaktion, sodann mit der Berliner-Blau-Reaktion und mit der Silber-Reaktion. Der positive Ausfall von mindestens zwei Reaktionen zeigt mit Sicherheit Blausäure im Untersuchungsmaterial an.

II. Quantitative Methoden.

Für die quantitative Bestimmung der Blausäure kommen praktisch nur die Silber- und Berliner-Blau-Reaktion in Frage. Die Titration mit Jodlösung nach FORDOS und GELIS bis zur Gelbfärbung (HCN + J_2 = JCN + HJ) sowie die Titration mit Brom (HCN + Br_2 = BrCN + HBr), wobei der in äquivalenter Menge entstehende Bromwasserstoff mit Lauge titriert wird, sind nicht spezifisch und empfindlich genug. Die *Silber-Reaktion* wurde bisher fast ausschließlich angewandt. Die *gravimetrische Bestimmung* (Ausfällung als AgCN und Wägung des getrockneten Niederschlages oder des durch Glühen erhaltenen Silbers) ist zwar sehr exakt, aber umständlicher und weniger empfindlich als die *titrimetrischen Silber-Verfahren*, die deswegen allgemein bevorzugt werden. Die meisten quantitativen HCN-Bestimmungen von Pflanzen sind mit dem letztgenannten Verfahren durchgeführt worden. Man unterscheidet eine Titration in saurer von einer solchen in alkalischer Lösung. Für die Vornahme der erstgenannten wird zuerst mit NH_3 und $AgNO_3$ im Überschuß versetzt, das Filtrat mit HNO_3 angesäuert und mit Ammoniumrhodanid und $FeCl_3$ als Indikator das überschüssige Silber zurücktitriert (VOLHARD).

Für die Titration in alkalischer Lösung benutzt man KJ als Indikator. Sobald ein Mol von 2 Mol HCN in AgCN übergeführt wurde, entsteht ein bleibender Niederschlag von AgJ, da durch das 2. Mol HCN das zuerst gebildete AgCN in das lösliche Komplexsalz $K[Ag(CN)_2]$ übergeführt wird. Der erste überschüssige Tropfen von $AgNO_3$-Lösung erzeugt eine bleibende Trübung. Dieses bereits auf LIEBIG-DENIGES zurückgehende Prinzip wurde von den meisten Autoren mit Erfolg für die HCN-Bestimmung in Pflanzenmaterial verwandt (TREUB, ROSENTHALER, BISHOP, WINKLER u. v. a.). BISHOP hat empfohlen, für die Titration zur besseren Erkennung der ersten AgJ-Opalescenz einen dunklen Hintergrund zu wählen, wodurch er eine Empfindlichkeit von 0,00001 g HCN mit einer Fehlerbreite von 1—3% erreichen konnte (Umschlag deutlich nach 1 Tropfen n/100 $AgNO_3$). Dies befähigte ihn, die HCN-Bestimmung an dem Material eines einzigen Blattes durchzuführen.

Die HCN-Titration mit $AgNO_3$ in alkalischer Lösung. Man versetzt einen aliquoten Teil der alkalischen HCN-Lösung mit 2 ml NH_3-Lösung (10%ig) oder NH_4Cl-Lösung (20%ig) bis zur deutlich ammoniakalischen Reaktion, setzt ein Körnchen Jodkali oder 3 Tropfen einer 5%igen Jodkali-Lösung hinzu und titriert mit einer n/50 oder n/100 $AgNO_3$-Lösung (3,398 resp. 1,699 g auf 1000), bis gegen einen dunklen Hintergrund der erste Tropfen der $AgNO_3$-Lösung eine bleibende Opalescenz verursacht. 1 ml n/100 $AgNO_3$-Lösung entspricht 0,54 mg HCN oder z. B. 0,54 · 11,522 Dhurrin.

Die kolorimetrische Bestimmung der Blausäure kann nach BERL und DELPY vorgenommen werden. Ein aliquoter Teil der alkalischen HCN-Lösung wird mit einem Überschuß von Ferrosulfat (5 ml einer 3%igen Lösung) und 3—5 Tropfen einer 5%igen $FeCl_3$-Lösung 15 min lang gekocht. Nach dem Erkalten wird mit 10%iger Salzsäure angesäuert und mindestens 5 Std. stehen gelassen. Je nach der Menge der zu erwartenden Blausäure wird auf 50 oder 100 ml aufgefüllt und im Kolorimeter gegen eine unter gleichen Bedingungen mit bestimmten HCN-Mengen erhaltene Suspension von Berliner-Blau verglichen.

Die kolimetrische Bestimmung der Blausäure ist auch in einfacher Form nach einer *Testfleckenmethode* durchzuführen, die auf GETTLER und GOLDBAUM zurückgeht und deren Brauchbarkeit SEIFERT für die Blausäure-Analytik bestätigen konnte. Sie ist für HCN absolut spezifisch und von einer derartigen Empfindlichkeit, daß Bruchteile von einem Mikrogramm noch einwandfrei ermittelt werden können.

Das Prinzip der Methode besteht darin, daß aus der HCN-haltigen Lösung das HCN-Gas durch Ansäuern mit H_2SO_4 in Freiheit gesetzt wird, mit einem Luftstrom oder dem Strom eines inerten Gases, wie Stickstoff, ein mit Ferrosulfat und NaOH präpariertes Testpapier passiert, wo es absorbiert wird. Beim Behandeln des Testpapieres mit HCl entsteht auf diesem ein Fleck von Berliner-Blau dessen Intensität der in Freiheit gesetzten HCN-Menge direkt proportional ist. Durch Vergleich des nach dem Trocknen erhaltenen Testflecks mit einer Standardreihe von Testflecken, die unter gleichen Bedingungen von bekannten HCN-Mengen gegeben wurden, läßt sich mit einer für die praktischen Belange ausreichenden Genauigkeit der HCN-Gehalt des Substrates ermitteln. Die Methode ist von einer überraschenden Einfachheit und eignet sich besonders für serienmäßige Routine-Bestimmungen, da sie schnell durchführbar ist.

Herstellung des Testpapieres. Man löst 5 g Ferrosulfat (krist. p. a. Merck) in 50 ml Wasser und filtriert, falls ein unlöslicher Rückstand bleibt. In diese

Lösung taucht man ein Blatt Filterpapier (Schleicher & Schüll 589³). Nach 5 min wird es herausgenommen und an einer Klammer freihängend an der Luft getrocknet. Das eben lufttrockene Papier wird dann einen kurzen Augenblick in eine 20%ige Natronlauge eingetaucht und wieder getrocknet (nicht dem direkten Sonnenlicht aussetzen). Aus dem trocken schmutzig grün aussehenden Papier schneidet man runde Scheiben von 2,5 cm Durchmesser. Diese gebrauchsfertigen Testpapiere lassen sich über einen Zeitraum von einigen Monaten verwenden, wenn sie kühl und im Dunkeln aufbewahrt werden.

Die Apparatur. Die Bestimmung wird in einer Apparatur vorgenommen, wie sie die Abb. 1 zeigt. A, B und C sind 50 cm³ fassende Reagenzgläser mit einer lichten Weite von 2—2,5 cm mit Waschflascheneinsätzen. In A wird

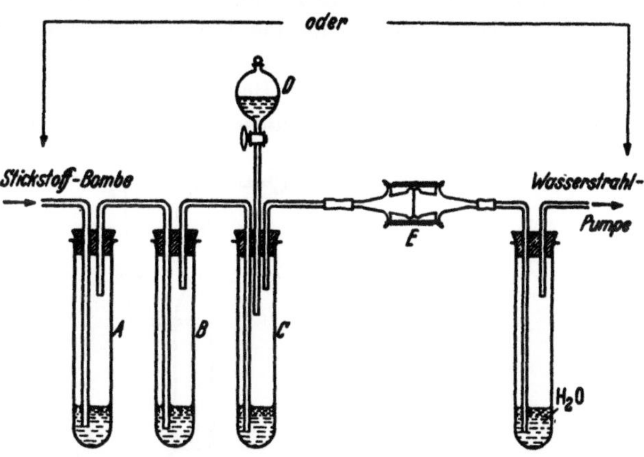

Abb. 1. Testflecken-Apparatur zur HCN-Bestimmung.

die durchgesaugte Luft resp. der durchgedrückte Stickstoff aus der Stahlflasche mit Reduzierventil von eventuellen HCN-Beimengungen durch 10 ml $HgCl_2$-Lösung (10%ig) und in B durch 10 ml 10%ige NaOH von CO_2 befreit. In C wird die zu untersuchende alkalische HCN-Lösung in aliquoter Menge (1—5 ml) eingefüllt, durch den Einfülltrichter D mit 5 ml 10%iger H_2SO_4 angesäuert. Das HCN-haltige Gas passiert in E das Testpapier. Es wird gehalten von sog. Glasflanschen, die aufeinander eingeschliffen das Testpapier zwischen sich aufnehmen (Abb. 2). Ein Flanschen-Durchmesser (lichte Weite) von 10 mm wird zumeist ausreichen. Eine lichte Weite von 5 mm kann die Empfindlichkeit noch um ein Beträchtliches steigern.

Die Bestimmung. In E wird ein Testpapier eingelegt. Man kann aus

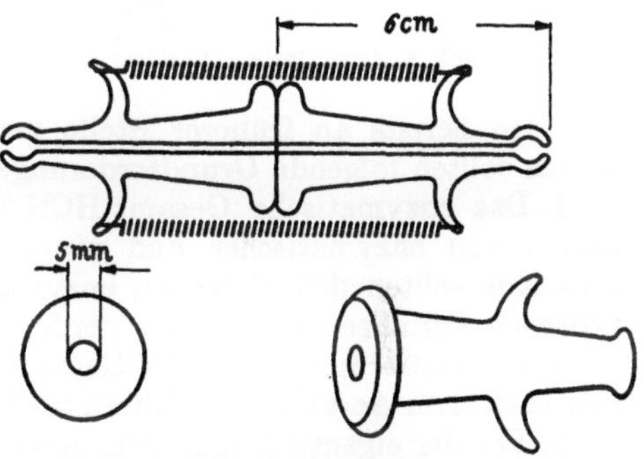

Abb. 2 Glasflanschen.

einer dünnen Kapillare ein kleines Tröpfchen Wasser zum leichten Anfeuchten auftropfen, was sich für eine gleichmäßige Ausbildung des Testflecks als günstig erwiesen hat. D wird mit 10 ml 10%iger H_2SO_4 gefüllt. Dann wird die zu untersuchende Lösung (1—5 ml je nach der zu erwartenden HCN-Menge) eingefüllt. C wird in ein auf 80—90 C° erwärmtes Becherglas mit Wasser bis zur Höhe des Flüssigkeitsspiegels eingetaucht. Nun wird mit Hilfe einer Wasserstrahlpumpe (im einfachsten Fall genügt der Unterdruck einer mindestens 2 l fassenden Mariotteschen Flasche) ein lebhafter Luftstrom durchgesaugt, oder aus einer Gasbombe Stickstoff durch das System gedrückt, und die Apparatur auf Dichtigkeit geprüft. Sodann läßt man die H_2SO_4 aus D in C einlaufen und reguliert den Luft- oder Stickstoffstrom so ein, daß in der Sekunde etwa 2 Blasen durchgehen. Nach 15 min wird das Testpapier herausgenommen, in 20 ml 10%iger HCl solange gebadet, bis das Papier bis auf den Berliner-Blau-Fleck vollkommen weiß geworden ist. Das Testpapier wird dann noch einige Minuten in fließendem Leitungswasser gewässert, getrocknet und mit der Standard-Testfleckenreihe verglichen, die unter völlig gleichen Bedingungen mit bekannten Mengen von analysenreinem Cyankalium erhalten wurden. Bei einem Flanschendurchmesser von 10 mm ist der Bereich von 0,1 bis etwa 5 γ als optimal anzusehen. Der Berliner-Blau-Fleck ist für das Vorliegen von HCN beweisend, seine Intensität ein quantitativer Ausdruck für seine Menge.

Weniger die eigentliche Bestimmung der Blausäure als ihre *quantitative Isolierung aus dem zu untersuchenden Pflanzenmaterial* bereitet Schwierigkeiten.

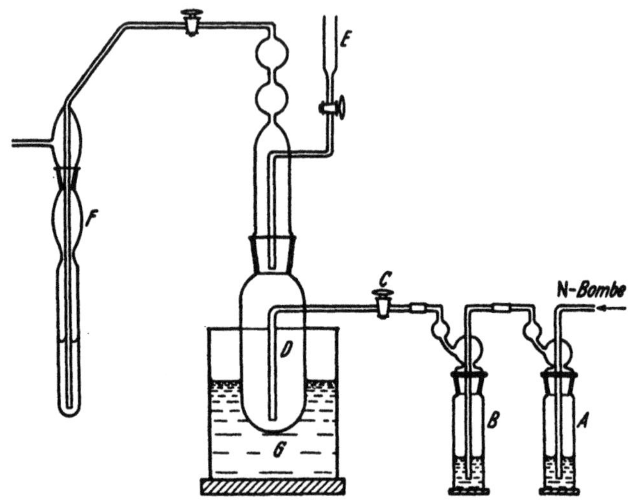

Abb. 3. Apparatur zur Zersetzung der HCN-Glykoside und Isolierung der Blausäure.

Aus den bereits an früherer Stelle gemachten grundsätzlichen Ausführungen heraus sollten folgende Grundforderungen eingehalten werden:

1. Das enzymatische Gesamt-HCN-Verfahren (Winkler) oder die Kombination von enzymatischer und säurehydrolytischer Spaltung der Blausäureglykoside sollten den einfachen enzymatischen oder ausschließlich säurehydrolytischen Verfahren vorgezogen werden.

2. Die Entfernung des HCN-Gases aus dem Reaktionsgemisch sollte einheitlich nach dem Aerations-Verfahren vorgenommen werden.

3. Vor der eigentlichen Bestimmung der Blausäure sollte diese in alkalischer Lösung absorbiert werden.

Außerdem sollte die Tatsache Berücksichtigung finden, daß die Zersetzung durch Enzyme nur in einem bestimmten p_H-Bereich optimal ist. Die Spaltung von Amygdalin ist bei p_H 6,0, die von Prunasin bei p_H 4,9 und die von Linamarin bei p_H 6,0, allgemein zwischen p_H 5,0 und 6,0 optimal.

In Anlehnung an die apparativen Prinzipien von Brunswik (1923) Luedtke (1952) u. a. ist die Schliffapparatur der Abb. 3 für die Zersetzung der Blausäureglykoside und die Isolierung der in Freiheit gesetzten Blausäure zu empfehlen. Die Waschflasche *A* enthält 20 ml 2%ige Sublimatlösung zur Absorption eventueller HCN-Beimengungen der durchgesaugten Luft oder des durchgedrückten Stickstoffs. Waschflasche *B* enthält 20 ml 10%iger NaOH zur Absorption von CO_2. Über den Hahn *C* kann der Gas-Durchgang reguliert werden. *D* nimmt das zu untersuchende Pflanzenmaterial auf. *E* ist ein Einfülltrichter, über den Enzympräparationen oder Säure zum Substrat gegeben werden können, ohne daß die Apparatur zu dem Zweck geöffnet werden muß. F ist das Absorptionsgefäß für HCN, das mit 10 oder 20 ml NaOH (0,1—1,0 normal) beschickt wird. G ist ein Becherglas mit Wasser, das erhitzt werden kann. Für Mikrobestimmungen ist zu empfehlen, die Ausmaße der Apparatur entsprechend klein zu halten.

Zur Zersetzung der Blausäureglykoside kann man folgendermaßen vorgehen: Man bringt das zu untersuchende Pflanzenmaterial in gut zerkleinerter Form in bestimmter Menge je nach der zu erwartenden HCN-Menge mit Wasser (5 g Trockenmaterial mit 50 ml, 10—20 g Frischmaterial mit 25—30 ml [WINKLER]) zusammen, setzt Pufferlösung (für Linamarin nach LUEDTKE (1952) 48 ml Citratpuffer[1]), 2—3 Tropfen Capryl- oder Octylalkohol zur Vermeidung von Schaumbildung hinzu, mischt gut durch und läßt mindestens 3 Std. oder über Nacht bei geschlossener Apparatur autolysieren[2]. Dann wird unter langsamer Erwärmung des Wasserbades auf 100° C (BISHOP) 2—3 Std. durchlüftet. Darauf läßt man auf 35° abkühlen und setzt aus dem Einfülltrichter eine Lösung oder Suspension von 0,1 g Emulsin in 20 ml Wasser hinzu. Man läßt nach guter Durchmischung erneut 3 Std. bei geschlossener Apparatur stehen und aerisiert unter langsamem Erwärmen ein zweites Mal.

Anstelle der Zugabe von Emulsin[3] kann man nach der 1. Aeration aus dem Einfülltrichter auch 30%ige H_2SO_4 (bei 5 g Trockenmaterial 25 ml, bei 10 g Frischmaterial 20 ml) zusetzen. Man läßt bei Raumtemperatur 15 min stehen und durchlüftet unter Erwärmung des Wasserbades auf 100° C 2—3 Std.

Überall dort, wo es angängig ist, sollte der Arbeitsweise von DUNSTAN und HENRY der Vorzug gegeben werden. Die Forscher isolierten das Blausäureglykosid zuerst durch mehrstündige Extraktion des zu untersuchenden Pflanzenmaterials mit Alkohol im Soxleth. Der Alkohol wird (möglichst unter vermindertem Druck) abgedampft und im Extraktionsrückstand wird die Blausäure durch Erhitzen mit H_2SO_4, was mit der Apparatur nach Abb. 3 vorgenommen werden kann, in Freiheit gesetzt. Der die cyanogenen Glykoside enthaltende Extraktionsrückstand kann auch direkt in die Testfleckenapparatur zur HCN-Bestimmung überführt werden.

Literatur.

ALVES CORREIA: Bragantia 7, 15 (1947); nach Ref. in Chem. Abstr. 1951, 11814. — ARMSTRONG: Proc. Roy. Soc. (London) 82, 558 (1910). — BERL u. DELPY: Ber. dtsch. chem. Ges. 43, 1430 (1910). — BISHOP: Biochemic. J. 21, 1162 (1927). — BRIDEL: J. Pharm. et Chim. (7) 12, 249 (1915). — BRIEGER: In G. KLEIN, Handbuch der Pflanzenanalyse, Bd. I, S. 520, Wien 1931. — BRUNSWIK: Sitzgsber. Akad. Wien 130 I, 405 (1921); Österr. bot. Z. 72, 58 (1923). — CHARLTON: Mem. Dep. Agric. (India) 9, 1 (1926); zit. nach WINKLER. — COUCH and BRIESE: J. Agricult. Res. 57, 81 (1938); 62, 493 (1941). J. Wash. Acad. Sci. 29, 146 (1030); 30, 413 (1940); zit. nach WINKLER. — CZAPEK: Biochemie der Pflanzen, Bd. III, S. 205—220, Jena 1921. — DILLEMANN: C. r. Acad. Sci. (Paris) 232, 1961 (1951) u. Diss. Paris 1953. — DRUCE, LANDE u. WALDEN: Z. Unters. Nahrungs- u. Genußmittel 23, 399 (1912). — DUNSTAN and HENRY: Proc. Roy. Soc. (London) B 72, 285 (1903). — FINNEMORE and WILLIAMS: Austral. J. Pharmacol. N. s. 16, 40 (1935). — FORDOS et GELIS: J. Pharmac. et Chim. [3] 23, 48 (1853). — GETTLER and GOLDBAUM: Analyt. Chem. 19, 270 (1947). — GUIGNARD:

---

[1] 21,01 g Citronensäure $+$ 8,0 g NaOH ad 1000, davon 30 ml $+$ 18 ml 0,1 n-NaOH-Lösung.

[2] Inaktiviertes Frischmaterial wird nach ROSENTHALER mit einem wäßrigen kalt hergestellten Extrakt der gleichen Frischpflanze, der durch längere Aeration absolut von HCN befreit wurde, angesetzt.

[3] Überall dort, wo Emulsin nicht wirksam ist, wird das spezifische Enzym, z. B. Linamarase, oder aber der gut durchlüftete kalt hergestellte wäßrige Auszug der zu untersuchenden Pflanze selbst verwendet.

C. r. Acad. Sci. (Paris) 141, 16; 142, 552; 145, 1115. Klunge: Zit. n. Posenthaler. — Kobert: Chem. Zbl. 1901 I, 51. — Kolthof: Z. analyt. Chem. 63, 188 (1923). — Lockemann: Ber. dtsch. chem. Ges. 34, 2127 (1910).— Luedtke: Biochem. Z. 322, 310 (1952); 323, 428 (1953). — Mirande: C. r. Acad. Sci. (Paris) 149, 140 (1909). — Moir: Proc. Chem. Soc. 1910, 115. — Paech: Biochemie und Physiologie der sekundären Pflanzenstoffe. S. 250—252, Berlin-Göttingen-Heidelberg, Springer-Verlag 1950. — Peche: Sitzgsber. Wien. Akad. Wiss. 121, 33 (1912). — Pertusi-Gastaidi: Zit. n. Rosenthaler. — Rosenthaler: Nachweis organischer Verbindungen. S. 491 bis 499. Stuttgart 1914 u. Blausäureglucoside in G. Klein: Handbuch der Pflanzenanalyse, Bd. III (2), Wien 1931. — Seifert: Dtsch. Z. gerichtl. Med. 41, 441 (1952). — Van Giffen: Pharmaz. Weekbl. 41, 1043 (1910). — Volhard: Ann. Chem. 190, 47 (1878). — Vorlaender: Zit. nach Rosenthaler. — Warth: Mem. Dep. Agric. 7—8, 1—29 (1923—1926); zit. nach Winkler. — Weehuizen-Thiery: Pharmaz. Weekbl. 42, 271 (1905). — Wester: Ber. pharmaz. Ges. 24, 123 (1914). — Winkler: J. Assoc. Agricult. Chemists 34, 541 (1951).

Senföle, Lauchöle und andere schwefelhaltige Pflanzenstoffe.

Von

A. Stoll und E. Jucker.

Mit 2 Abbildungen.

A. Einleitung.

Die Bedeutung des Schwefels im Aufbau von Pflanzenstoffen ist schon vor Jahrzehnten erkannt worden (vgl. z. B. MULDER, 1846; GMELIN-KRAUT, 1870); trotzdem sind unsere Kenntnisse über seine Rolle im Stoffwechsel der höheren Pflanzen und über den Gesamtumsatz der Schwefelverbindungen in der pflanzlichen Zelle noch verhältnismäßig lückenhaft. Gerade in den letzten Jahren sind aber auf diesem Gebiet beträchtliche Fortschritte erzielt worden. Es kann in diesem Kapitel auf die interessanten Befunde in der Phytochemie des Schwefels nur gelegentlich hingewiesen werden, da hier in erster Linie von der Analyse und Bestimmung der pflanzlichen Schwefelverbindungen die Rede sein soll.

Der Nachweis und die Bestimmung einzelner schwefelhaltiger Naturstoffe erfolgt nach spezifischen analytischen Methoden, die weiter unten an Beispielen erörtert werden; einzig der Nachweis des Schwefels und die Bestimmung seines prozentualen Anteils im Molekül wird bei allen Schwefelverbindungen gleich ausgeführt:

Zur *qualitativen* Schwefelbestimmung wird die zu untersuchende organische Verbindung durch Glühen mit Kalium oder Natrium aufgeschlossen, und der Glührückstand wird noch heiß in Wasser gelöst. Diese alkalische Lösung wird nach Filtration mit Nitroprussidnatrium versetzt, wonach bei Vorliegen von Schwefel eine rote Färbung auftritt.

Ein anderer Nachweis des Schwefels wird von ROSENTHALER (1937) beschrieben. Das Verfahren beruht auf der Oxydation des Schwefels organischer Verbindungen zu schwefliger Säure, auf die man Jodsäure einwirken läßt. Dabei bildet sich zunächst freies Jod, das mit Hilfe der Jodstärkereaktion erkannt wird. Liegt sehr viel Schwefel vor, so ist theoretisch die weitere Reduktion des Jods zu Jodwasserstoff möglich, was ein Verschwinden der Jod-anzeigenden Blaufärbung zur Folge hätte. Dieser Fall tritt aber praktisch nie ein.

Die *quantitative* Schwefelbestimmung wird hauptsächlich auf drei Arten ausgeführt:

1. Katalytische Oxydation des Schwefels zu Sulfat und gravimetrische Bestimmung als Bariumsulfat: Oxydation des Schwefels erfolgt durch Verbrennen der Substanz im *Perlenrohr*. Um die bei der Verbrennung ebenfalls entstehende schweflige Säure zu Schwefelsäure zu oxydieren, werden die Verbrennungsprodukte in wäßrige Perhydrol-Lösung geleitet, und die Sulfationen darin durch Bariumionen erfaßt. Bei organischen Verbindungen, die keinen Stickstoff und kein Halogen enthalten, kann die entstandene Schwefelsäure mit Lauge direkt titriert werden.

2. Oxydation nach CARIUS und Bestimmung der SO_4-Ionen als Bariumsulfat: Die Oxydtion der schwefelhaltigen Verbindungen nach CARIUS wird durch

Erhitzen der Substanzprobe im Rohr mit rauchender Salpetersäure bei Gegenwart von Bariumchlorid vorgenommen, wobei Bariumsulfat gebildet wird. Diese Methode kann indessen infolge Abscheidung von Kieselsäure aus dem Glas des Bombenrohres zu Fehlresultaten führen.

3. Katalytische Reduktion zu Sulfid, das jodometrisch bestimmt wird: Bei dieser Methode wird eine Probe des schwefelhaltigen Präparates mit Platin als Katalysator hydriert, wobei der Schwefel zu Sulfid reduziert, und dieses jodometrisch titriert wird. Nach dieser Methode lassen sich nur Verbindungen, die weder Arsen, noch Phosphor oder Metallkomplexe enthalten, bestimmen.

Neuerdings wurde von Bürger (1941) ein neues, einfaches Verfahren für die Schwefelbestimmung entwickelt: Die Substanzprobe wird in einem geschlossenen Röhrchen durch schmelzendes Kali zersetzt, wobei sich Kaliumsulfid bildet. Nach Beendigung der Reaktion wird das zerschnittene Röhrchen mit Wasser versetzt, und die Lösung nach Abstumpfen mit Bicarbonat jodometrisch titriert. Diese Methoden sind von Pregl-Roth (1949) ausführlich beschrieben worden.

B. Schwefel und einfache Schwefelverbindungen.

Elementarer Schwefel. Elementarer Schwefel wurde bisher im Pflanzenreich nicht aufgefunden. Manche Bakterien und Pilze speichern hingegen Polysulfid-schwefel auf. Verschiedene Pflanzen können elementaren Schwefel (z. B. aus Schädlingsbekämpfungsmitteln) mit Hilfe einer oxydoreduktiven Umsetzung mit geeigneten, in diesen Pflanzen enthaltenen Thiolverbindungen, z. B. Glutathion, umsetzen, wobei das Disulfid entsteht, nach der Gleichung:

$$2 \text{ HS-Glutathion} + \text{S} \rightarrow \text{Glutathion-S-S-Glutathion} + H_2S$$

(Kendall und Nord, 1926; Nord, 1927).

Schwefelwasserstoff. Schwefelwasserstoff ist als Hemmstoff und Destruktor lebenswichtiger Redoxasen für die meisten niederen und höheren Pflanzen ein starkes Gift. Es gibt indessen Bakterien, die Schwefelwasserstoff für ihren Stoff-wechsel benötigen. So leben z. B. Schwefelbakterien des Schwarzen Meeres in Schwefelwasserstoff enthaltendem Wasser (Karaoglanov, 1942) und gewinnen die zur Aufrechterhaltung ihrer Lebensfunktionen nötigen Kalorien offenbar durch Oxydation des Schwefelwasserstoffs. Weiterhin seien die farblosen anaeroben Spirillen, die farbstoffhaltigen Schwefelpurpurbakterien (van Niel, 1931) und die grünen Schwefelbakterien erwähnt, von denen die letztgenannten den Schwefelwasserstoff bis zum elementaren Schwefel oxydieren, während die Schwefelpurpurbakterien diese Oxydation bis zur Sulfatstufe weiterführen.

Schwefelkohlenstoff. Farblose, stark lichtbrechende Flüssigkeit vom Kp. 47°. Schwefelkohlenstoff findet sich nur in geringen Mengen im Öl des schwarzen Senfs, wahrscheinlich als Zersetzungsprodukt des Allylsenföls (Gildemeister und Hoffmann, 1928). Er wird im Pflanzenreich auch als Stoffwechselprodukt angetroffen. Im javanischen Hutpilz *Schizophyllum lobatum* wurden beträchtliche Mengen Schwefelkohlenstoff nachgewiesen (Czapek, 1921).

Mercaptane.

Methylmercaptan. CH₃SH.

Widerlich nach faulem Kohl riechendes Gas, Kp. 6°. Methylmercaptan findet sich in Spuren in frischen Wurzeln von *Raphanus sativus* L., in den Blättern von verschiedenen *Lasianthus*-Arten: *Lasianthus lucidus* Bl., *L. purpureus* Bl., *L. stercorarius* Bl., *L. bracteolatus* Miqu. (Koolhaas, 1931). Geringe Spuren eines Mercaptans wurden auch im Bärlauchöl *(Allium ursinum L.)* nachgewiesen (Gildemeister und Hoffmann, 1928, 1929).

Zum *Nachweis* dient das schwerlösliche Quecksilbermercaptid $(CH_3S)_2Hg$, das beim Umsatz mit Quecksilberoxyd oder durch Einleiten von Methylmercaptan in eine wäßrige Quecksilbercyanidlösung entsteht. Mikroskopische Prismen, die bei 175° schmelzen (Zers.). Ein von NENCKI ausgearbeitetes und von NAKAMURA (1925) verbessertes Verfahren erlaubt den sicheren Nachweis von Methylmercaptan: Zerriebenes Wurzelmaterial wird nach Ansäuern mit Oxalsäure der Vakuumdestillation unterworfen, wobei das Destillat in eine alkoholische Quecksilbercyanidlösung eingeleitet wird. Der entstehende gelbliche Niederschlag wird getrocknet und mit heißem Aceton extrahiert, worauf man erkalten läßt. Das farblose Quecksilber-methylmercaptid kristallisiert nach einiger Zeit in glänzenden Kristallen vom Smp. 172° (unkorr.).

KOOLHAAS (1931) beschreibt eine Modifikation dieses Verfahrens, wonach frisches pflanzliches Material zerrieben und mit Wasserdampf destilliert wird. Die Dämpfe werden in 40%iger Kalilauge aufgefangen, und das Mercaptan wird aus dieser Lösung durch Eintropfenlassen von 10%iger Schwefelsäure in Freiheit gesetzt. Während dieser Operation wird ein langsamer Luftstrom durch die Apparatur geleitet, womit das Methylmercaptan (nach Passieren eines Kühlsystems) in ein mit fast gesättigter methanolischer Quecksilbercyanid-Lösung beschicktes U-Rohr übergeführt wird. Der entstandene Niederschlag wird abfiltriert, mit Methanol gewaschen und bei 40° über Calciumoxyd im Vakuum bis zur Gewichtskonstanz getrocknet. Nach dem Umkristallisieren dieses Niederschlags aus Aceton werden Quecksilber- und Schwefelgehalt analytisch bestimmt. Die im U-Rohr erzeugte Fällung kann auf Reinheit geprüft werden, indem aus einer Probe in einer geeigneten Apparatur mit Salzsäure das Mercaptan freigesetzt und jodometrisch bestimmt wird. Das Mercaptan wird durch einen schwachen Luftstrom aus der sauren, gelinde siedenden Lösung in eine Vorlage, die 0,1 n Jodlösung enthält, übergetrieben, dort quantitativ absorbiert und zum Disulfid oxydiert. Das unverbrauchte Jod wird zurücktitriert und daraus der Mercaptangehalt berechnet.

Sehr oft handelt es sich darum, Methylmercaptan in *Spuren* nachzuweisen. Die Methode von DENIGÈS, 1920 (vgl. CZAPEK) führt rasch und sicher zum Ziel: Eine Probe des zu prüfenden Materials wird in einigen ml reiner konz. Schwefelsäure gelöst und mit einer kleinen Menge einer 1%igen Lösung von Isatin in konz. Schwefelsäure versetzt. Ist ein Mercaptan vorhanden, so tritt Grünfärbung ein. Man kann die Empfindlichkeit dieser Reaktion noch steigern, indem man das mit der Isatin-Schwefelsäure benetzte, kugelförmig geschmolzene Ende eines Glasstabs in den Dampfraum, wo sich gasförmiges Mercaptan befindet, einführt. Die bei Vorhandensein von Mercaptan sehr schnell auftretende Grünfärbung wird noch deutlicher, wenn man das gefärbte Ende des Glasstabs in etwas konz. Schwefelsäure eintaucht. Durch mehrfache Wiederholung dieser Operation wird die Farbe in der Schwefelsäure stark intensiviert. Diese Farbreaktion fällt nur bei Mercaptanen, nicht aber bei Sulfiden positiv aus.

Da dieser Mercaptan-Nachweis durch Aldehyde oder höhere Alkohole gestört wird, muß man bei deren Anwesenheit folgendermaßen vorgehen: Man versetzt das zu untersuchende pflanzliche Material in wäßriger Lösung oder in Suspension mit wenig Kali- oder Natronlauge, schüttelt einige Augenblicke, verdünnt mit Wasser und gibt Nitroprussidnatrium zu. Eine hierbei auftretende rotviolette Färbung ist nicht auf H_2S zurückzuführen, falls sich nach Zusatz alkalischer Bleilösung kein Bleisulfid abscheidet. Zeigt indessen die Reaktion mit alkalischer Bleilösung H_2S an, dann ist die Reaktion nochmals auszuführen, wobei die Lauge durch alkalische Bleilösung ersetzt, und das Nitroprussidnatrium erst nachher zugegeben wird, sonst verhindert die schwarze Färbung des Bleisulfids die Beobachtung der rotvioletten Färbung, die vom Mercaptan herrührt.

2,2'-Dithiolisobuttersäure. $C_4H_8O_2S_2$ (HS—CH$_2$)$_2$CH—COOH.

Farblose Kristalle aus Petroläther vom Smp. 61—62°, wurde von Jansen (1948) aus *Asparagus* über eine nicht kristallisierende Disulfidverbindung isoliert. Im frischen Pflanzensaft von *Asparagus* scheint ein Gleichgewicht zwischen einer Sulfhydryl- und einer Disulfidgruppe (oxydierte Form) zu bestehen. In der Luft ausgesetztem Pflanzensaft ist jedoch nur noch die Disulfidgruppe nachweisbar, woraus hervorgeht, daß in der Pflanze die Dithiolisobuttersäure als solche vorkommt.

C. Schwefelsäureester, Sulfonsäuren, Sulfite, Thioäther.

I. Schwefelsäureester.

Schwefelsäureester wurden hauptsächlich in Membranen verschiedener Algen festgestellt. Sie sind außerdem ständig in der Hefe anzutreffen.

Cholinschwefelsäureester. $C_5H_{13}O_4NS(CH_3)_3N^+CH_2—CH_2—OSO_3^-$ (Woolley und Peterson, 1937) wurde im Mycel von *Aspergillus sydowi* gefunden.

Agar-Agar, der bekannteste, seit vielen Jahren zur Bereitung von Nährböden benutzte Schwefelsäureester, ist ein Polygalaktosid-schwefelsäureester mit einer wahrscheinlich aus 9 D-Galaktopyranoseresten aufgebauten Ketteneinheit (Jones und Peat, 1942) und findet sich in verschiedenen Rotalgen und Tangen, z. B. in *Gigartina spinosa*, *Gelidium corneum*, *Spirida filamentosa* (Cioglia, 1940) und in *Phyllophora nervosa* (Korenzwit, 1938).

Carrageenschleim enthält Schwefelsäureester. Auch in *Gigartina stellata*, *Chondrus crispus* und in *Agardhiella tenera* wurden Schwefelsäureester gefunden. Die Verbreitung solcher Ester in höheren Pflanzen ist neuerdings wieder eingehender studiert worden (Meyer und Chaffee, 1940), doch sind die Kenntnisse auf diesem Gebiet immer noch lückenhaft.

II. Sulfonsäuren.

Untersuchungen über die Verbreitung von Sulfonsäuren im Pflanzenreich sind erst in allerneuster Zeit unternommen worden.

Die Sulfoessigsäure, HO$_3$S—CH$_2$—COOH, kommt als Bestandteil von Alkaloidestern in *Erythrina glauca* Willd., *E. pallida Briton*, *E. pallida Rose* und *E. poeppigiana* (Walp.) vor (Folkers, Koniuszy und Shavel jr., 1944). Die beiden Sulfosäureester, die auf den Frosch ähnlich wie Curare wirken, besitzen folgende Zusammensetzung und Eigenschaften:

Erysothiovin. $C_{20}H_{23}O_7NS \cdot 2\,H_2O$, Smp. 187°, $[\alpha]_D^{25} = +208°$ (in Alkohol).

Erysothiopin. $C_{19}H_{21}O_7NS \cdot 2\,H_2O$, Smp. 168—169°, $[\alpha]_D^{25} = +194°$ (in Alkohol).

Sulfite sind wiederholt in pflanzlichen Zellen, vor allem im Phloem der Wurzeln und Sprosse beobachtet worden (Nightingale, Schermerhorn und Robbins, 1932).

III. Thioäther, aliphatische Di- und Polysulfide und Sulfoxyde (Lauchöle).

Thioäther sind in der Natur nicht sehr verbreitet; ihr Vorkommen beschränkt sich fast ausschließlich auf Lauchgewächse, wo sie als sog. *Lauchöle* vorliegen. Sie zeichnen sich durch einen widerlichen, stark anhaftenden Geruch aus und dürften mit den Senfölen in genetischem Zusammenhang stehen. Es wird angenommen, daß die Lauchöle in der lebenden Pflanze als Glucoside vorliegen. Über die Funktion der Thioäther im Pflanzenorganismus ist noch wenig bekannt, doch scheint es, daß sie bei Redoxumsetzungen eine Rolle spielen.

Die Charakterisierung der Thioäther kann z. B. durch Überführung in Sulfone geschehen. Einige Thioäther geben mit verschiedenen Schwermetallsalzen, z. B. mit Quecksilberchlorid, Platinchlorid und Goldchlorid schwerlösliche, zur Identifizierung gut geeignete Additionsverbindungen.

Für den Nachweis und die Bestimmung müssen die Thioäther und Dialkyldisulfide zuerst aus dem pflanzlichen Material isoliert werden, was z. B. durch Destillation mit Wasserdampf geschehen kann. Die Zerlegung des dabei erhaltenen ätherischen Öls in die einzelnen Bestandteile erfolgt durch fraktionierte Destillation, wobei bereits die Siedepunkte und die Elementaranalyse über die Zusammensetzung der einzelnen Fraktionen Anhaltspunkte geben können. Die eindeutige Identifizierung erfolgt durch Überführung in kristallisierte und wohldefinierte Additionsprodukte.

1. **Dimethylsulfid.** $CH_3—S—CH_3$.

Leicht flüchtiges, ätherisch und zugleich nach Meerrettich riechendes Öl vom Kp. 38°, $D_{21} = 0,846$, reagiert mit Mercurichlorid in alkoholischer Lösung unter Bildung eines weißen voluminösen Niederschlags der Bruttozusammensetzung $[(CH_3)_2S]_2 \cdot [HgCl_2]_3$. Diese komplexe Verbindung kristallisiert in Nadeln, die sich am Licht schwach färben und bei raschem Erhitzen bei 150—151° schmelzen. Dimethylsulfid liefert mit Platinchlorid ein gelbes Kristallpulver der Zusammensetzung $[(CH_3)_2S]_2PtCl_4$, Smp. 218° (Zers.). Methyljodid vereinigt sich mit Dimethylsulfid zu einem Trimethylsulfoniumjodid $(CH_3)_3SJ$, das in kaltem Äthanol sehr schwer löslich ist; Prismen aus Wasser, Zers. 215°. Beim Behandeln mit konz. Salpetersäure erhält man das Nitrat des Dimethyl-sulfoxyds, während rauchende Salpetersäure bei höherer Temperatur das Dimethylsulfon, $CH_3—SO_2—CH_3$ (Smp. 108°), liefert. Dimethylsulfid kann aus amerikanischem Pfefferminzöl isoliert werden (GILDEMEISTER und HOFFMANN, 1928).

2. **Dimethyldisulfid.** $CH_3—S—S—CH_3$.

Kp. 112°, ist wahrscheinlich als Oxydationsprodukt des primär durch Methylierung von H_2S von *Schizophyllum commune* gebildeten CH_3SH anzusehen. Es wurde ferner im Pfefferminzöl und Geraniumöl aufgefunden.

3. **Divinylsulfid.** $CH_2=CH—S—CH=CH_2$.

Farbloses Öl, von ätherischem Geruch, an Diallylsulfid erinnernd, Kp. 101°, $D = 0,9125$. Divinylsulfid ist in Wasser wenig löslich, mischt sich hingegen mit Alkohol oder Äther in jedem Verhältnis. Konz. Schwefelsäure zersetzt es unter Rotfärbung und Entwicklung unerträglich riechender Gase. Die Reaktion mit konz. Salpetersäure ist äußerst heftig, unter Umständen explosionsartig unter Entzündung. An Divinylsulfid kann bei vorsichtigem Vorgehen unter Luftausschluß und Erwärmung Brom angelagert werden, wobei ein farbloses, dickflüssiges Öl vom Kp. 195° entsteht. Einwirkung von trockenem Silberoxyd bei 30° führt zum Austausch des Schwefels gegen Sauerstoff unter Bildung des Vinyläthers. Löst man Divinylsulfid in der gleichen Menge Alkohol und versetzt mit einer alkoholischen Lösung von Silbernitrat, so bildet sich ein weißer Niederschlag der Zusammensetzung $(C_2H_3)_2S \cdot (AgNO_3)_2$ vom Smp. 87°. Die analoge Reaktion mit Quecksilberchlorid führt zu einer kompliziert zusammengesetzten Verbindung $C_8H_{12}Cl_4S_2Hg_2$ vom Smp. 91°. Verreibt man sie mit Kaliumrhodanid, so bildet sich Vinylsenföl, $CH_2=CH—NCS$. Ein weiteres charakteristisches Derivat des Divinylsulfids erhält man beim Versetzen seiner Lösung in Alkohol (1:1) mit einer alkoholischen Lösung von Platinchlorid im Überschuß und vorsichtigem Verdünnen mit Wasser. Nach mehrtägigem Stehen kristallisiert eine gelbe flockige Verbindung aus, die zur Entfernung von Verunreinigungen mehrmals mit warmem Äthanol behandelt wird. In reinem Zustand ist dieses Derivat ein feurig gelbes Pulver der Zusammensetzung $C_{12}H_{18}Cl_8S_3Pt_2$, Smp. 93°.

Divinylsulfid wird nach der Vorschrift von Semmler (1887) wie folgt isoliert: Die zu untersuchenden Pflanzenteile von z. B. *Allium ursinum* L. werden mit Wasserdampf destilliert, und das dunkelbraun gefärbte, stark lichtbrechende, knoblauchähnlich riechende Öl fraktioniert destilliert. Die zwischen 90 und 107° siedende Fraktion wird gesondert abgefangen, getrocknet und bleibt während mehreren Tagen über Kalium stehen, bis die Gasentwicklung aufgehört hat. Nach dem Abgießen vom Kalium wird das ätherische Öl nochmals fraktioniert destilliert, wobei die zwischen 99 und 103° siedende Fraktion in der Hauptsache aus Divinylsulfid besteht. Bereits die Elementaranalyse dieser Fraktion liefert einen Hinweis auf ihre Zusammensetzung. Der exakte Nachweis erfolgt jedoch wie oben beschrieben durch Überführung in definierte Derivate.

4. **Diallylsulfid.** $CH_2 = CH - CH_2 - S - CH_2 - CH = CH_2$.

Farbloses, nach Knoblauch riechendes Öl, Kp. 140°, $D_{27} = 0{,}887$, in Wasser nur wenig löslich, wurde zusammen mit Allylsenföl im ätherischen Öl von *Cochlearia armoracia* L. (Hubatka, 1843) und in den Wurzeln verschiedener Akazienarten aufgefunden.

Löst man Diallylsulfid in der gleichen Menge Alkohol und versetzt mit einer alkoholischen Lösung von Silbernitrat, so bildet sich ein weißer Niederschlag von in Wasser leicht löslichen Nadeln der Zusammensetzung $C_6H_{10}S \cdot (AgNO_3)_2$.

Neuere Untersuchungen deuten darauf hin, daß Diallylsulfid im Knoblauchöl überhaupt nicht vorkommt, dieses hingegen in der Hauptsache aus Polysulfiden, wie z. B. Diallyldisulfid, Diallyltrisulfid, Diallyltetrasulfid und Propyl-allyl-disulfid besteht.

Diese Polysulfide sind Zersetzungs- und Umwandlungsprodukte der Knoblauch-Muttersubstanz Alliin (s. unter III, 18) und werden bei der Destillation der ganzen Pflanze *(Allium sativum L.)* mit Wasserdampf als gelb gefärbtes, ätherisches Öl von intensivem Knoblauchgeruch erhalten. Verschiedene Reagentien, wie Quecksilberchlorid, Platinchlorid und Aurichlorid geben in alkoholischer Lösung weiße bis gelbe Fällungen. Die Zerlegung des Knoblauchöls in die einzelnen Fraktionen erfolgt nach Semmler (in Gildemeister und Hoffmann, 1929) durch fraktionierte Vakuumdestillation, wobei die bei 16 mm zwischen 65 und 125° übergehende Fraktion aufgefangen wird. Diese liefert nach wiederholter Fraktionierung die oben beschriebenen Allyl-polysulfide. Die am tiefsten siedende Fraktion (16 mm bis 70°) besteht aus Propyl-allyl-disulfid. Die nächste Fraktion (16 mm, 70—84°) macht die Hauptmenge (60%) des Knoblauchöls aus und besteht aus Diallyl-disulfid. Die nächsthöhere Fraktion (16 mm, 112—122°) ist ein gelbes Öl, zur Hauptsache Diallyl-trisulfid; der nach dem Abdestillieren der zwischen 65 und 125° siedenden Knoblauchölanteile hinterbleibende Rückstand enthält hauptsächlich Diallyl-tetrasulfid.

5. **Diallyl-disulfid.** $CH_2 = CH - CH_2 - S - S - CH_2 - CH = CH_2$.

Nach Knoblauch riechende Flüssigkeit, Kp. 78—80°/16 mm, $D_{15} = 1{,}010$, ist der Hauptbestandteil des Knoblauchöls *(Allium sativum L.)* neben anderen Polysulfiden (Wertheim, 1844, 1845; Semmler, 1892), und entsteht wohl aus dem Sulfoxyd Allicin (s. unter III, 17), das sich seinerseits vom Alliin (s. unter III, 18) ableitet.

6. **Propyl-allyl-disulfid.** $CH_3 - CH_2 - CH_2 - S - S - CH_2 - CH = CH_2$.

Kp. 66—69°/16 mm, $D_{15} = 1{,}0231$. Nach Küchenzwiebeln riechende Flüssigkeit, kommt im Knoblauchöl neben anderen Alkylsulfiden vor (Semmler, 1892).

7. Isobutyl-propenyl-disulfid.

$$CH_3-CH_2-\overset{\displaystyle CH_3}{\overset{|}{CH}}-S-S-CH=CH-CH_3$$

Unangenehm riechendes Öl, Kp. 82—84°, $[\alpha]_D = -17,62°$, wurde aus den ätherischen Ölen von *Asa foetida* isoliert.

8. Diallyl-trisulfid. $CH_2=CH-CH_2-S-S-S-CH_2-CH=CH_2$.

Kp. 112—122°/16 mm, $d_{15} = 1,0845$. Nach Knoblauch riechende Flüssigkeit, kommt im Knoblauchöl neben anderen verwandten Produkten vor (WERTHEIM, 1844, 1845; SEMMLER, 1892).

9. Diallyl-tetrasulfid. $CH_2=CH-CH_2-S-S-S-S-CH_2-CH=CH_2$.

Flüssigkeit von penetrantem Geruch, nicht unzersetzt destillierbar, kommt im Knoblauchöl neben anderen schwefelhaltigen Produkten vor (SEMMLER, 1892).

10. Djenkolsäure, $CH_2(S-CH_2-\overset{\displaystyle }{\underset{\underset{\displaystyle NH_2}{|}}{CH}}-COOH)_2$, ist das L-Cysteinthioacetal des

Formaldehyds. Djenkolsäure kristallisiert in Nadeln vom Smp. 300—350° (Zers.), $[\alpha]_D^{26} = -44,5°$ (in 1%iger HCl); $[\alpha]_D^{20} = -25°$ (in 1%iger HCl). Sie wurde in *Pithecolobium lobatum* (Djenkolbohne) aufgefunden (VAN VEEN und HYMAN, 1935). Die Strukturaufklärung gelang DU VIGNEAUD und PATTERSON (1936), und die Synthese ARMSTRONG und DU VIGNEAUD (1947).

11. Methionol (γ-Methylmercapto-propanol). $CH_3-S-CH_2-CH_2-CH_2OH$.

Kp. 99—101°/23 mm, bildet einen Bestandteil des ätherischen Öls von jap. Maggi „Shoyu" (AKABORI und KANEKO, 1936).

12. β-Methyl-thiopropionsäure-methylester,

$$CH_3-S-CH_2-CH_2-COOCH_3,$$

wurde aus dem ätherischen Öl des Ananassaftes *(Ananas sativus* Lindl.) isoliert (HAAGEN-SMIT, KIRCHNER, DEASY und PRATER, 1945).

13. Adenyl-D-thiomethylribose. $C_{11}H_{15}O_3N_5S$.

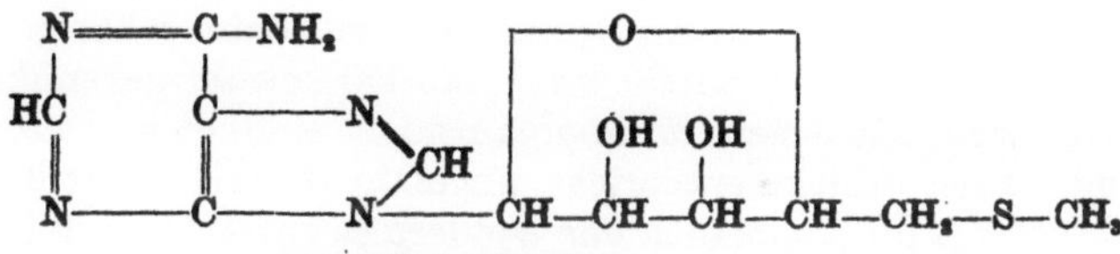

vermutliche Struktur

Smp. 212°, findet sich in der Hefe (SUZUKI, ODAKE und MORI, 1924; SUZUKI und MORI, 1925; LEVENE und SOBOTKA, 1925). WENDT (1942) gelang es, die Konstitution aufzuklären, und er schrieb ihr die obenstehende Strukturformel zu.

14. Uscharin. $C_{31}H_{41}O_3NS$.

Smp. 265° (Zers.), $[\alpha]_D = +29,0°$ (in $CHCl_3$), von HESSE, REICHENEDER und EYSENBACH (1939) aus dem frischen Milchsaft von *Calotropis procera* R. Br. und *Calotropis gigantea*, die ein afrikanisches Pfeilgift liefern, isoliert. Die Konstitution der Verbindung ist noch nicht sichergestellt. Es handelt sich wahrscheinlich um einen Thioäther. Kurzes Erwärmen mit Säure spaltet den Stickstoff als NH_3 und den Schwefel als flüchtige organische Verbindung ab, wobei Uscharidin, $C_{29}H_{38}O_9$, entsteht.

Neueste Untersuchungen von HESSE und GAMPP (1952) zeigen, daß man Uscharin mit Mercurosalzen schon bei Zimmertemperatur hydrolytisch spalten kann, wobei Uscharidin, Ammoniak und dimerer Mercaptoacetaldehyd erhalten wurden. Da bei dieser Spaltung gleichzeitig eine Carbonylgruppe freigelegt wird, schließen die Autoren auf das Vorhandensein eines Thiazolringes.

15. Terthienyl. $C_{12}H_8S_3$.

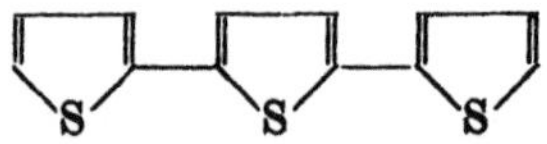

Smp. 94—95,5° aus Methanol, wurde von Zechmeister und Sease (1947) aus den gelben Blumenblättern der „Lemon" oder aus der afrikanischen Varietät von *Tagetes erecta* L. isoliert. Die Lösungen oder Adsorbate zeigen eine intensive tintenblaue Fluoreszenz.

16. Gliotoxin. $C_{13}H_{14}N_2O_4S_2$.

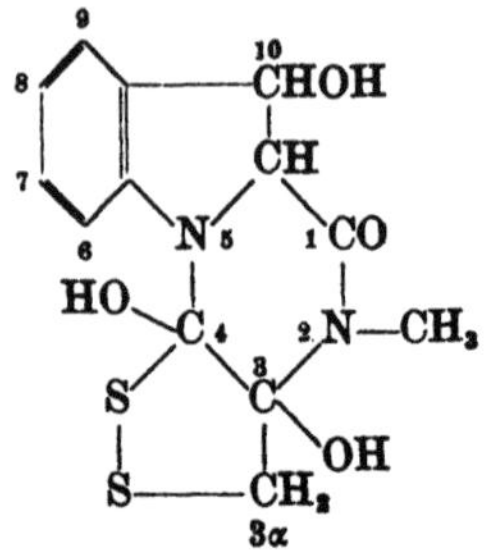

Smp. 221° (Zers.) aus Methanol, $[\alpha]_D^{25} = -290°$ (in Äthanol); $[\alpha]_D^{25} = -270°$ (in Pyridin). Gliotoxin ist ein Antibiotikum, das aus den Kulturfiltraten von *Aspergillus fumigatus*, ferner aus *Gliocladium fimbriatum* und *Trichoderma lignorum* isoliert wurde (Weindling et al., 1932, 1934, 1936, 1937, 1938, 1941); über die Konstitution der Verbindung vgl. Johnson, Bruce und Dutcher (1943).

17. Allicin (Diallyl-disulfid-oxyd).

$$CH_2{=}CH{-}CH_2{-}\overset{\overset{O}{\uparrow}}{S}{-}S{-}CH_2{-}CH{=}CH_2.$$

Farbloses, unbeständiges Öl, das nicht ohne Zersetzung destilliert werden kann. Allicin wurde von Cavallito und Bailey (1944) und Cavallito, Buck und Suter (1944) aus dem Knoblauch isoliert. Seine wäßrige Lösung zeigt noch in Verdünnungen von 1:85000 bis 1:125000 sowohl gegenüber Gram-positiven als auch Gram-negativen Mikroorganismen eine gleich starke antibakterielle Wirkung. 1 mg Allicin entspricht ungefähr 15 OE Penicillin. Allicin kommt nicht als solches im Knoblauch vor, sondern es entsteht nach Stoll und Seebeck (1948a, 1949a) durch enzymatische Spaltung der genuinen Muttersubstanz Alliin (s. folgender Abschnitt, vgl. auch den Beitrag von Skinner in Bd. 3 dieses Handbuches).

18. Alliin [(+)-S-Allyl-L-cystein-sulfoxyd].

$$C_6H_{11}O_3NS \cdot 1/2\,H_2O \qquad CH_2{=}CH{-}CH_2{-}\overset{\overset{O}{\uparrow}}{S}{-}CH_2{-}\overset{\overset{NH_2}{|}}{CH}{-}COOH \cdot 1/2\,H_2O$$

Kristallisiert aus verdünntem Alkohol oder Aceton in äußerst feinen, langen, farblosen, oft zu Büscheln vereinigten, geruchlosen Nadeln, Smp. 163—165° (Zers.), $[\alpha]_D^{21} = +62,8°$ (in Wasser).

Da das Alliin eine der am eingehendsten untersuchten schwefelhaltigen Substanzen des Pflanzenreiches ist und außerdem mit Alliin eine ganz neuartige enzymatische Spaltung, die der quantitativen Bestimmung dieses Stoffes zugrunde liegt, entdeckt und studiert werden konnte, so soll unten über die Untersuchungen von Alliin ausführlicher berichtet werden.

19. Schwefelhaltige Aminosäuren.

Tierisches und pflanzliches Eiweiß enthält ständig je nach Herkunft in wechselnder Menge auch schwefelhaltige Aminosäuren. Cystin, Cystein, Glutathion und Methionin sind die wichtigsten Vertreter dieser Gruppe; sie werden im Abschnitt „Aminosäuren" (S. 8, 9, 29) eingehend behandelt. Hier seien nur einige Literaturangaben über wichtige Bestimmungsmethoden der Aminosäuren zusammengestellt: ANDREWS und ANDREWS (1937); BAERNSTEIN (1936); BENISCHKE (1948); BROWN (1942); DENIGÈS (1938); DIEMAIR und KOCH (1940); HESS und SULLIVAN (1945); KOLB und TOENNIES (1940); LAVINE (1935); McFARREN (1951); MIRSKY und ANSON (1935); SCHORMÜLLER und BALLSCHMIETER (1951); SURMATIS und WILLARD (1937); WEIDINGER (1937); WOJAHN (1952); ZEILE und OETZEL (1949).

20. Sulfide unbekannter Konstitution.

Solche sind aus einer Reihe von Pflanzen isoliert worden: So gewann SEMMLER (s. bei GILDEMEISTER, 1929) aus dem Öl der Küchenzwiebel *(Allium cepa* L.*)* ein Disulfid der Zusammensetzung $C_6H_{12}S_2$ vom Kp. 75—83°/10 mm. Mit dem Propyl-allyl-disulfid aus dem Knoblauchöl scheint es nicht identisch zu sein, da die Siedepunkte verschieden sind. Durch Zinkstaub-Destillation wird es in ein bei 130° destillierendes Sulfid, $C_6H_{12}S$, durch nascierenden Wasserstoff (durch Kalium im Öl erzeugt) in ein gesättigtes Disulfid $C_6H_{14}S_2$ (Kp. 68—69°/10 mm) verwandelt.

Das Zwiebelöl wird durch Wasserdampfdestillation der ganzen Pflanze als dunkelbraunes, ziemlich dünnflüssiges Rohöl in einer Ausbeute von 0,046% erhalten ($D_{8,7} = 1,0410$, $[\alpha]_D = -5°$) und zersetzt sich bei der Destillation unter gewöhnlichem Druck. Unter 10 mm Druck geht es hingegen fast vollständig zwischen 64 und 125° über. Die I. Fraktion (bis 100°) bildet ein hellgelbes Öl von deutlichem, nicht unangenehmem Zwiebelgeruch. Sie ist fast ebenso stark linksdrehend wie das Rohöl ($[\alpha]_D = -4°$, $D_{12} = 1,0234$). Die II. Fraktion, die den Hauptanteil ausmacht, siedet zwischen 100 und 125°, ist dunkelgelb und spezifisch schwerer ($D = 1,0385$). Durch erneute fraktionierte Destillation der beiden Fraktionen wurde im wesentlichen das gleiche oben erwähnte Disulfid der Zusammensetzung $C_6H_{12}S_2$ isoliert. In der II. Fraktion (100—125°) ist daneben noch ein Sulfid mit höherem Kohlenstoffgehalt und höherer Dichte enthalten, das möglicherweise mit einer der höher siedenden Verbindungen aus dem Öl von *Asa foetida* identisch ist. Der Destillationsrückstand des Zwiebelöls ist dunkelbraungelb und von widerlichem Geruch. Er enthält ein Polysulfid mit höherem Schwefelgehalt als das oben erwähnte Disulfid, denn mit Zinkstaub entsteht daraus dasselbe Monosulfid wie aus diesem. Alle Fraktionen des Zwiebelöls geben mit einer alkoholischen Lösung von $HgCl_2$ weiße, mit Platin- bzw. Goldchlorid gelbe Niederschläge.

Aus dem Öl von *Asa foetida* isolierte SEMMLER (s. bei GILDEMEISTER u. HOFFMANN, 1928):

1. ein Disulfid $C_7H_{14}S_2$, Kp. 83—84°/9 mm, Kp. 210—212°. Beim längeren Erhitzen mit Zinkstaub auf 130—150° und darauffolgendem schnellem Abdestillieren wird daraus ein Monosulfid $C_7H_{14}S$ erhalten. Das Disulfid gibt mit alkoholischem $HgCl_2$ einen weißen, durch Wasserzusatz sich vermehrenden Niederschlag, aus dem man durch Auskochen mit Alkohol und Erkaltenlassen des alkoholischen Auszuges schöne Nadeln der Zusammensetzung $C_7H_{14}S_2 \cdot [HgCl_2]_2$ gewinnt.

2. ein Disulfid $C_8H_{16}S_2$, Kp. 92—96°/9 mm;
3. ein Disulfid $C_{10}H_{18}S_2$, Kp. 112—115°/9 mm, beide in geringer Menge;
4. ein Disulfid $C_{11}H_{20}S_2$, Kp. 126—127°/9 mm.

Die Bestimmung von Aneurin (Thiamin, Vitamin B_1) ist in Bd. 4, S. 345, diejenige von Biotin in Bd 4., S. 656 beschrieben.

Weitere lauchartig riechende Stoffe, die chemisch fast unerforscht sind, kommen bei Leguminosen vor. Eine flüchtige schwefelhaltige, stickstofffreie Verbindung ist in der Rinde von *Scorodophleus zenkeri*, eine ähnliche Substanz in den Samen von *Acacia farnesiana* Willd., sowie in den Wurzeln und Zweigen anderer *Acacia*-Arten beobachtet worden (Czapek, 1921).

IV. Die Bestimmung von Alliin.

Stoll und Seebeck (1947, 1948a) gelang es, aus *Allium sativum* L. und *Allium ursinum* L. das bis dahin unbekannte Alliin rein und kristallisiert darzustellen, als schwefelhaltige Aminosäure zu charakterisieren, und schließlich auch totalsynthetisch aufzubauen.

Um eine gute Ausbeute an Alliin zu erhalten, ist es notwendig, von einer guten, frischen Droge (Zwiebel) auszugehen. Es hat sich gezeigt, daß der Gehalt des Knoblauchs an Alliin mit seinem Schwefelgehalt parallelgeht und stark von der Provenienz der Droge abhängt. Um eine enzymatische Spaltung des Alliins während der Zerkleinerung und der Extraktion des Knoblauchs zu verhüten, wird das Ausgangsmaterial tiefgekühlt, bei tiefer Temperatur zerkleinert und mit Methyl- oder Äthylalkohol, dessen Wassergehalt 15—20% nicht übersteigen darf, extrahiert. Nach Entfernung des Alkohols im Vakuum bei niedriger Temperatur wird die verbleibende wäßrige sirupöse Lösung mit Äther von fettigen Bestandteilen befreit. Die Abtrennung von weiteren Begleitstoffen, vor allem von Kohlenhydraten, kann durch fraktioniertes Ausfällen mit Alkohol aus konzentrierten wäßrigen Lösungen und Digerieren mit Methanol, oder durch Überführen des Alliins in verflüssigtes Phenol, aus dem das Alliin mit Äther ausgefällt wird, erreicht werden. Zusatz von Methanol zu einer konzentrierten wäßrigen Lösung, die das Alliin bereits stark angereichert enthält, bewirkt dessen erste Kristallisation in farblosen Nadeln, die beim Umkristallisieren aus wasserhaltigem Aceton bereits das reine Produkt liefern. Die Konstitution des Alliins (s. untenstehende Formel) konnte ebenfalls ermittelt werden (Stoll und Seebeck, 1948a).

Alliin wird von der *Alliinase* (Stoll und Seebeck, 1949a), dem spezifischen Enzym des Knoblauchs, in das antibakteriell wirkende Allicin (Cavallito und Bailey, 1944), Brenztraubensäure und Ammoniak gespalten nach folgender Gleichung:

$$
\begin{array}{ccccc}
 & & CH_2 & & CH_2 \quad CH_2 \\
 & & \| & & \| \quad\quad \| \\
CH_2 & & CH & & CH \quad CH \\
\| & 2\ & | & \xrightarrow{-H_2O} & | \quad\quad | \\
CH & & CH_2 & & CH_2 \quad CH_2 \\
| & & | & & | \quad\quad\quad | \\
CH_2 & & H\!-\!S\!=\!O & & O\!=\!S\!\rule{1cm}{0.4pt}\!S \\
| & & & & \\
2\ S\!\rightarrow\!O & \longrightarrow & & & \text{Allicin (Cavallito)} \\
| & & & & \\
CH_2 & & CH_2 & & CH_3 \\
| & & \| & & \| \\
H_2N\!-\!CH & 2\ H_2N\!-\!C & \xrightarrow{+2H_2O} & 2\ C\!=\!O & +2\,NH_3 \\
| & | & & | & \\
COOH & COOH & & COOH & \\
\text{Alliin} & & & &
\end{array}
$$

Die bei dieser quantitativ verlaufenden Reaktion gebildete Brenztraubensäure kann als 2,4-Dinitro-phenylhydrazon bestimmt werden. Andererseits erlaubt die

Ermittlung des gebildeten Ammoniaks auf einfachste Weise die quantitative Bestimmung der ursprünglich vorhandenen Alliinmenge. Im folgenden wird die Gewinnung einer gereinigten Enzymlösung und die vergleichende quantitative Bestimmung von Ammoniak und Brenztraubensäure nach STOLL und SEEBECK (1949a) beschrieben.

Die Gewinnung einer gereinigten Enzymlösung. 100 g frische Knoblauchzwiebeln werden mit Trockeneis fein gemahlen und dann mit 400 ml Wasser versetzt. Der dünne Brei wird unter ständigem Umrühren im Thermostaten auf 37° C erwärmt und bei dieser Temperatur noch weitere 20 min gerührt. Man nutscht nun von festen Bestandteilen ab und filtriert die trübe Lösung durch eine mit Talk gedichtete Nutsche, wobei 425 ml einer hellgelben, klaren Lösung vom p_H 6,2 erhalten werden (= rohe Enzymlösung). Der Zusatz von 21 ml 10%iger Essigsäure unter Umrühren bewirkt eine gallertige Fällung, die abzentrifugiert und in 150 ml Wasser suspendiert wird. Beim Versetzen mit 10%igem Ammoniak, bis ein p_H von 6,4 erreicht ist, geht ein großer Teil der Suspension in Lösung. Man filtriert von ungelösten Flocken ab, säuert die klare Lösung mit 10%iger Essigsäure auf p_H 4,0 an und zentrifugiert den ausgefallenen Niederschlag ab. Das in 400 ml 1/15-m. Phosphatpuffer vom p_H 6,4 unter Zusatz von etwas Toluol wieder aufgelöste Enzym wurde bei allen Versuchen, bei denen im folgenden nicht ausdrücklich ein anderes Enzympräparat erwähnt wird, verwendet.

Vergleichende quantitative Bestimmung von Ammoniak und Brenztraubensäure. a) 30 ml Enzymlösung werden im Thermostaten auf 37° erwärmt und mit 20 ml einer ebenfalls erwärmten wäßrigen Lösung von 300 mg Alliin versetzt. Die enzymatische Reaktion wird nach 90 sec durch Zusatz von 10 ml 2 n-Schwefelsäure unterbrochen. In 20 ml, das sind $^2/_5$ der Lösung, bestimmt man den Gehalt an Ammoniak nach FOLIN (1902). Den Rest der Lösung ($^3/_5$) versetzt man mit 3 ml einer 20%igen wäßrigen Lösung von Trichloressigsäure und zentrifugiert den entstandenen Niederschlag ab. Er wird in 10 ml Wasser suspendiert und erneut zentrifugiert. Die vereinigten Flüssigkeiten werden mit 30 ml einer heißen 1%igen Lösung von 2,4-Dinitrophenylhydrazin in 2 n-Salzsäure versetzt. Nach einer Stunde filtriert man den kristallinen Nie-

<table>
<tr><td colspan="6" align="center">Tabelle 1.</td></tr>
<tr><td>t
in Minuten</td><td>mg NH$_3$</td><td>Spaltung in %</td><td>mg DPH[1]</td><td>Spaltung in %</td></tr>
<tr><td>a) 1,5</td><td>5,30</td><td>58,0</td><td>165,0</td><td>57,3</td></tr>
<tr><td>b) 4</td><td>6,39</td><td>69,9</td><td>191,0</td><td>66,4</td></tr>
</table>

derschlag ab, wäscht mit verdünnter Salzsäure und Wasser nach und trocknet bei 110°. Eine Probe liefert nach dem Umkristallisieren aus Methanol das reine 2,4-Dinitrophenyl-hydrazon der Brenztraubensäure.

b) Ein Vergleichsversuch mit einer anderen Enzymlösung wird z. B. nach 4 min unterbrochen.

Die folgende Zusammenstellung der Ergebnisse aus zwei Versuchen zeigt eine relativ gute Übereinstimmung der Spaltungswerte, wie sie sich aus der Bestimmung des gebildeten NH$_3$ und der Brenztraubensäure in äquimolaren Mengen ergaben (Tab. 1).

STOLL und SEEBECK (1949a) haben die Wirkung und die Eigenschaften der Alliinase eingehend untersucht. Da über Enzyme, die mit schwefelhaltigen Naturprodukten assoziiert sind und deren Aufspaltung herbeiführren, zur Zeit noch sehr wenig bekannt ist, so seien die Ergebnisse dieser enzymatischen Untersuchung von STOLL und SEEBECK in extenso wiedergegeben:

Zeitlicher Verlauf der Spaltung von Alliin mit Alliinase. 15 ml Enzymlösung und 5 ml Wasser

<table>
<tr><td colspan="5" align="center">Tabelle 2.</td></tr>
<tr><td>t
in Minuten</td><td>0,1 n-NaOH
ml</td><td>Differenz</td><td>mg NH$_3$</td><td>Spaltung in %</td></tr>
<tr><td>0</td><td>19,62</td><td>—</td><td>—</td><td>—</td></tr>
<tr><td>0,5</td><td>18,14</td><td>1,48</td><td>2,51</td><td>27,5</td></tr>
<tr><td>1</td><td>17,04</td><td>2,58</td><td>4,38</td><td>48,0</td></tr>
<tr><td>2</td><td>15,21</td><td>4,41</td><td>7,49</td><td>81,9</td></tr>
<tr><td>4</td><td>14,53</td><td>5,09</td><td>8,65</td><td>94,7</td></tr>
</table>

werden im Thermostaten auf 37° erwärmt und mit 10 ml einer 1%igen wäßrigen Lösung von Alliin (=100 mg Alliin) von derselben Temperatur versetzt. Nach t min fügt man 5 ml 2 n-Schwefelsäure zu, um die Wirkung des Enzyms zu unterbrechen, und bestimmt sogleich den Gehalt an Ammoniak nach FOLIN (Tab. 2).

---

[1] DPH = 2,4-Dinitrophenylhydrazon der Brenztraubensäure.

Tabelle 3.

Alter in Tagen	0,1 n-NaOH ml	Differenz	mg NH₃	Spaltung in %
0	15,32	4,54	7,77	84,4
3	15,22	4,64	7,88	86,1
7	16,91	2,95	5,01	54,8
10	18,98	0,88	1,49	16,3
14	19,50	—	—	—

Nullwert: 19,86 ml 0,1 n-Natronlauge.

Haltbarkeit der Enzymlösung. Zur Bestimmung der Haltbarkeit der Alliinase bewahrte man eine frischhergestellte Lösung unter etwas Toluol bei 3° C auf und entnahm in verschiedenen Zeitabschnitten Proben, um ihre Wirksamkeit festzustellen.

Ansatz: 15 ml Enzymlösung und 5 ml Wasser werden auf 37° erwärmt und mit 10 ml einer 1%igen wäßrigen Alliinlösung von derselben Temperatur versetzt. Nach 2 min wird die enzymatische Reaktion durch Hinzufügen von 5 ml 2 n-Schwefelsäure unterbrochen, und das gebildete Ammoniak bestimmt.

Bei einer 10 Tage lang bei 3° C aufbewahrten Alliinaselösung ist die Wirksamkeit auf weniger als ¹/₅ des Anfangswertes zurückgegangen (Tab. 3).

Über den Einfluß des pH auf die Alliinasewirkung. Je 15 ml der rohen Enzymlösung werden mit 0,8 ml 10%iger Essigsäure versetzt, und die ausgeschiedenen Niederschläge abzentrifugiert. Sie werden in je 15 ml nachfolgender Pufferlösungen suspendiert bzw. gelöst:

für pH 3: 1/10 Mol Citratpuffer,
für pH 4: 1/10 Mol Citratpuffer,
für pH 5: 1/15 Mol Phosphatpuffer,
für pH 6: 1/15 Mol Phosphatpuffer,
für pH 7: 1/15 Mol Phosphatpuffer,
für pH 8: 1/10 Mol Boratpuffer,
für pH 9: 1/10 Mol Boratpuffer.

Dann verdünnt man mit je 5 ml Wasser, erwärmt im Thermostaten auf 37° und versetzt mit je 10 ml einer 1%igen wäßrigen Alliinlösung von derselben

Tabelle 4.

pH	0,1 n-NaOH ml	Differenz	mg NH₃	Spaltung in %
3	19,61	—	—	—
4	18,23	1,45	2,46	27,0
5	15,06	4,62	7,85	85,8
6	15,31	4,37	7,43	81,5
7	15,21	4,47	7,59	83,1
8	15,21	4,47	7,59	83,1
9	19,24	0,44	0,74	8,1

Nullwert: 19,68 ml 0,1 n-Natronlauge.

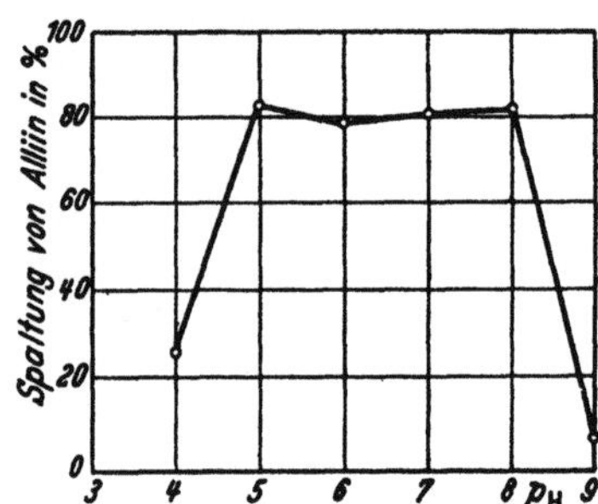

Abb. 1. Über den Einfluß des pₕ-Wertes auf die Alliinasewirkung.

Temperatur. Unterbrechung der Reaktion mit 5 ml 2 n-Schwefelsäure nach 2 min und Bestimmung des gebildeten Ammoniaks.

Wie die letzte Kolonne der Tabelle 4 und die graphische Darstellung (Abb. 1) zeigen, ist die Alliinasewirkung innerhalb relativ weiter Grenzen vom pH unabhängig.

Temperaturabhängigkeit der Alliinasewirkung. 15 ml Enzymlösung und 5 ml Wasser werden bei $t°$ mit 10 ml einer 1%igen wäßrigen Alliinlösung von derselben Temperatur versetzt. Nach 2 min unterbricht man die Reaktion mit 5 ml 2 n-Schwefelsäure und bestimmt das gebildete Ammoniak.

Als interessantes Ergebnis dieser Messungen (Tab. 5 und Abb. 2) erscheint die Beobachtung, daß die Alliinase selbst bei 0° in 2 min $^1/_5$ des Substrates zu spalten vermag.

Tabelle 5.

t in °C	0,1 n-NaOH ml	Differenz	mg NH$_3$	Spaltung in %
0	18,74	1,12	1,90	20,8
20	15,97	3,89	6,61	72,3
30	15,45	4,41	7,50	82,1
37	15,22	4,64	7,88	86,1
45	15,62	4,24	7,20	78,9
60	19,24	0,62	1,05	11,1

Nullwert: 19,86 ml 0,1 n-Natronlauge.

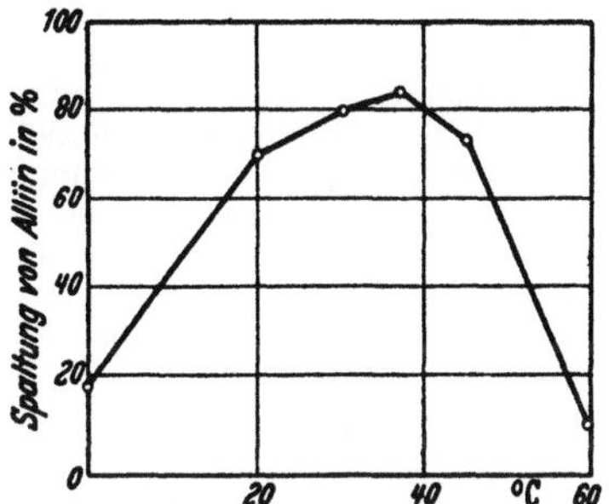

Abb. 2. Temperaturabhängigkeit der Alliinasewirkung.

Inaktivierung der Alliinase durch Erhitzen. 15 ml Enzymlösung werden im Dampfbad während 30 min erhitzt. Die trübe Lösung verdünnt man nach dem Erkalten mit 5 ml Wasser, bringt sie im Thermostaten auf 37° und vermengt sie mit 10 ml einer 1%igen wäßrigen Alliinlösung von derselben Temperatur. Nach 5 min werden 5 ml 2 n-Schwefelsäure zugesetzt, und das gebildete Ammoniak wird bestimmt.

Tabelle 6.

	0,1 n-NaOH ml	Differenz	mg NH$_3$	Spaltung in %
Nullwert . . .	19,96	—	—	—
Versuch . . .	19,90	—	—	—
Kontrollversuch	15,60	4,36	7,40	81,0

Die erhitzte Lösung zeigt keine Wirkung mehr.

Für die Gewinnung des Alliinase-Talk-Präparates werden 120 ml Enzymlösung mit 20 g Talk unter starkem Umrühren mit so viel 10%iger Essigsäure versetzt, bis der isoelektrische Punkt der Alliinase vom pH 4,0 erreicht ist. Die abfiltrierte Talkschicht wird mit 50 ml 1/150 m Citratpuffer vom pH 4,0 gewaschen und im Vakuum über konzentrierter Schwefelsäure getrocknet. In frischem Zustand zeigt das Präparat eine gute enzymatische Wirksamkeit, wie im Beispiel f) der folgenden Versuchsreihe (Tab. 7) gezeigt wird.

Schädigung der Alliinase durch organische Lösungsmittel. a) 2,5 g eines frischen Alliinase-Talk-Präparates werden 5 min lang mit 15 ml Alkohol verrührt, abgenutscht, mit 5 ml Äthylalkohol nachgewaschen und im Vakuum getrocknet. Das so behandelte Alliinase-Talk-Präparat suspendiert man in 15 ml $^1/_{15}$ mol Phosphatpuffer pH 6,4 und 5 ml Wasser, erwärmt die Lösung auf 37° und versetzt mit 10 ml einer 1%igen wäßrigen Alliinlösung von derselben Temperatur. Nach 2 min gibt man 5 ml 2 n-Schwefelsäure zu und bestimmt den Gehalt an Ammoniak. Wie im Beispiel a) mit Alkohol, wird im Beispiel b) mit Chloroform, c) mit Aceton, d) mit Essigester und e) mit Methanol behandelt, während im Beispiel f) das Alliinase-Talk-Präparat unbehandelt zur Wirkung kommt.

Tabelle 7.

	0,1 n-NaOH ml	Differenz	mg NH$_3$	Spaltung in %
a) Äthylalkohol	19,65	0,25	—	—
b) Chloroform .	19,25	0,65	1,10	12,1
c) Aceton . . .	19,63	0,27	—	—
d) Essigester .	19,71	0,19	—	—
e) Methanol . .	19,70	0,20	—	—
f) unbehandelt	15,87	4,03	6,86	75,0

Nullwert: 19,90 ml 0,1 n-Natronlauge.

Von den in die Untersuchung einbezogenen Lösungsmitteln läßt nur Chloroform eine geringe Wirksamkeit des Enzyms bestehen.

Die Synthese des natürlichen Alliins und seiner drei Isomeren wurde von Stoll und Seebeck (1950, 1951) durchgeführt.

L-Cystein wurde in Gegenwart von Natronlauge mit Allylbromid zu S-Allyl-L-cystein (Desoxo-alliin) kondensiert und mit Hydrogenperoxydlösung in das am Schwefelatom racemische S-Allyl-L-cystein-sulfoxyd übergeführt. Durch fraktionierte Kristallisation gelang es, diese synthetische Verbindung in die beiden in bezug auf das Schwefelatom optisch aktiven Isomeren zu trennen. Das dabei erhaltene (+)-S-Allyl-L-cystein-sulfoxyd erwies sich als identisch mit dem natürlichen Alliin. Damit ist also nicht nur die Totalsynthese verwirklicht, sondern zugleich auch die Konfiguration des natürlichen Alliins aufgeklärt. Das bei der fraktionierten Kristallisation ebenfalls isolierte (—)-S-Allyl-L-cystein-sulfoxyd ist in bezug auf die Konfiguration des Schwefelatoms der Antipode des natürlichen Alliins.

Auch die beiden optisch aktiven Isomeren des Alliins, die sich vom D-Cystein ableiten und am Schwefelatom (+)- oder (—)-Konfiguration besitzen, sind hergestellt worden (Stoll und Seebeck, 1950, 1951). Ausgehend von D,L-Cystein wurde das S-Allyl-D,L-cystein hergestellt, das mit 90%iger Ameisensäure und Acetanhydrid in das S-Allyl-D,L-N-formyl-cystein übergeführt wurde. Durch fraktionierte Kristallisation seines Brucinsalzes gelang es, das S-Allyl-D-cystein zu erhalten, das durch Behandlung mit Hydrogenperoxydlösung in S-Allyl-D-cystein-sulfoxyd übergeführt werden konnte. Diese in bezug auf das Schwefelatom racemische Verbindung konnte, wie in der oben angeführten entsprechenden L-Form, durch fraktionierte Kristallisation in (+)-S-Allyl-D-cystein-sulfoxyd und (—)-S-Allyl-D-cystein-sulfoxyd aufgetrennt werden. Somit sind nun alle optischen Isomeren des Alliins bekannt.

In der folgenden Tabelle sind einige Eigenschaften von Alliin und seinen drei künstlich hergestellten Isomeren zusammengestellt.

Tabelle 8. *Eigenschaften der 4 isomeren Alliine.*

Nr.	Substanz	Schmelzpunkt	Spez. Drehung in H_2O	Krist. in
I	Alliin natürlich = (+)-S-Allyl-L-cystein-sulfoxyd	163—165° 164—166°	+62,8° +63,2°	Nadeln aus Wasser/Aceton
II	(—)-S-Allyl-L-cystein-sulfoxyd	160—161,5°	—60,7°	Nadeln aus verd. Methanol
III	(+)-S-Allyl-D-cystein-sulfoxyd	164—166°	+64,7°	Nadeln aus 70 bis 80% Methanol
IV	(—)-S-Allyl-D-cystein-sulfoxyd	167—168°	—59,2°	Nadeln aus verd. Aceton

An diesen vier Isomeren und analog gebauten synthetischen Derivaten konnte die sehr ausgeprägte konfigurative Spezifität der Alliinase studiert werden (Stoll und Seebeck, 1949b), wobei es sich z. B. zeigte, daß Derivate des D-(+)-Cysteins von dem Enzym nicht abgebaut werden können.

D. Thioaldehyde.

Vertreter dieser Körperklasse sind in der Natur erst in der allerneusten Zeit entdeckt worden. Es ist nicht anzunehmen, daß diese reaktionsfähigen Körper in der Pflanze als solche vorliegen. Wahrscheinlich bilden sie sich erst im Laufe des Aufarbeitungsprozesses. Über die Bedeutung der Thioaldehyde oder ihrer Muttersubstanzen für die Pflanze ist noch nichts Genaueres bekannt. Es wird angenommen, daß sie Zwischenprodukte für den Aufbau von Thiolen bilden.

Thiopropionaldehyd, $CH_3—CH_2—CHS$, wurde von Kohmann (1947) bei der Wasserdampfdestillation (Vac.) von Zwiebeln als tränenerzeugende Verbindung nachgewiesen. Thiopropionaldehyd ist nicht stabil; besonders unter der Einwirkung von Säuren polymerisiert er sich rasch.

E. Thioharnstoffe und L-5-Vinyl-2-thiooxazolidon.

Verbindungen dieser Körperklasse sind u. a. in mehreren Arten der Familie der Cruciferen ziemlich verbreitet.

Thioharnstoff. NH_2—CS—NH_2.

Smp. 182°. KLEIN und FARKASS (1930) gelang es, Thioharnstoff in *Cytisus*-Arten mikrochemisch mittels Palladiumchlorid, das mit Thioharnstoff charakteristische schwarze Kristalle bildet, sowie mit Xanthydrol (Reaktionsprodukt Dixanthyl-thioharnstoff) eindeutig nachzuweisen. Er liegt in der Natur wahrscheinlich als Ureid vor.

Ergothionein (Thionin, Thiasin, Sympectothion), $C_9H_{15}O_2N_3S$,

$$CH=C-CH_2-CH-CO^-$$
$$\underset{N}{|}\qquad \underset{NH}{|}\qquad \underset{N^+(CH_3)_3}{|}$$
$$\underset{C}{\diagdown\diagup}$$
$$|$$
$$SH$$

kristallisiert aus Wasser in farblosen, monoklinen Nadeln mit 2 Mol H_2O, Smp. 290° (Zers.), $[\alpha]_D = +110°$, wurde erstmals von TANRET (1909) aus *Secale cornutum* isoliert. Die Extraktion wurde später von EAGLES (1928) und von PIRIE (1933) verbessert. Ergothionein ist auch synthetisch hergestellt worden (HEAT, LAWSON und RIMINGTON, 1950). Es ist ein kompliziert gebautes Derivat des Thioharnstoffs, das gleichzeitig ein Betain ist.

Zur Bestimmung wird Ergothionein vorteilhafterweise zunächst in möglichst reinem Zustande isoliert und dann mit diazotierter Sulfanilsäure umgesetzt. Die Ausführung des Verfahrens wird von HUNTER (1949) beschrieben.

L-5-Vinyl-2-thiooxazolidon.

$$H_2C———NH$$
$$CH_2=CH\cdot HC\qquad CS$$
$$\diagdown_O\diagup$$

Smp. 50°, farblose Kristalle, $[\alpha]_D^{21} = -71°$ (aus 2%igem Methanol), wurde von ASTWOOD, GREER und ETTLINGER (1949) aus *Brassica napobrassica* isoliert, und dessen Struktur wurde durch Synthese sichergestellt. Die Verbindung stellt einen antithyreoiden Faktor dar.

F. Senföle und Senfölglykoside.

Senföle sind Ester der Isothiocyansäure der allgemeinen Formel RNCS. Sie unterscheiden sich somit ganz wesentlich von den unter C. III. beschriebenen Lauchölen, die in ihrem Molekül keinen Stickstoff enthalten. Die flüchtigen Senföle zeichnen sich durch scharfen Geruch aus, sie reizen die Schleimhäute und erzeugen auf der Haut Blasen. In der Pflanze liegen sie vorwiegend als Glykoside vor und werden erst im Laufe des Aufarbeitungsprozesses oder nach enzymatischem Abbau als zuckerfreie Verbindungen erhalten.

In reinem Zustand sind die meisten Senföle farblose Flüssigkeiten; einige sind bei Zimmertemperatur kristallin. Die einfacheren Verbindungen lassen sich im Vakuum unzersetzt destillieren. Die Struktur der Senföle ergibt sich einerseits aus der Synthese: Umsetzung von primären Alkylaminen mit Schwefelkohlenstoff und darauffolgende Spaltung der so entstandenen Dithiocarbaminate RNH—CS—SH · RNH_2 durch Erhitzen mit wäßriger $HgCl_2$-Lösung, andererseits durch Spaltung mittels Wasser, verdünnten Säuren oder Alkalien, wobei die entsprechenden Alkylamine wieder erhalten werden: $RNCS + 2 H_2O = H_2S + CO_2 + RNH_2$.

Der *Nachweis* der Senföle kann entweder durch Überführung in geeignete Derivate oder durch Spaltung zum entsprechenden Amin erfolgen. Geeignete Derivate lassen sich z. B. wie folgt herstellen:

1. Umsetzung von Senfölen mit Ammoniak (meist in alkoholischer Lösung) führt gemäß der Gleichung $RNCS + NH_3 = RHN—CS—NH_2$ zu Monoalkyl-thioharnstoffen.

2. Umsatz von Senfölen mit Phenylhydrazin in warmem Äthanol führt zu Phenyl-alkyl-thiosemicarbaziden:

$$RNCS + H_2N—NH—C_6H_5 = RHN—CS—NH—NH—C_6H_5,$$

die als gut kristallisierbare Verbindungen zum Nachweis von Senfölen dienen können.

3. Beim Umsatz von Senfölen mit Borneolnatrium entstehen nach Hydrolyse der primär gebildeten Natriumverbindung Alkyl-thiocarbaminsäure-bornylester der Formel $C_{10}H_{18}O—CS—NHR$, die sich ebenfalls gut für den Nachweis von Senfölen eignen.

Zum mikrochemischen Nachweis von Senfölen hat PIETSCHMANN (1924) die beiden folgenden Verfahren, die gute orientierende Resultate liefern, vorgeschlagen:

1. Ein Tropfen einer 0,00002%igen Lösung von Allylsenföl gibt auf dem Objektträger (am besten nach Zusatz von ein wenig Methanol) mit freiem Phenylhydrazin einen aus feinen dünnen Nadeln bestehenden Niederschlag. Bei größerer Konzentration entstehen außer längeren, manchmal verzweigten Nadeln längere, 6seitig zugespitzte, unter dem Mikroskop weiß erscheinende Plättchen, die in Methanol unlöslich, in Wasser, Glycerin und Chloralhydrat löslich sind.

2. Das aus der zu untersuchenden Pflanze erhaltene Wasserdampfdestillat wird mit der halben Menge konz. Ammoniak versetzt und bleibt während 12 Std. in gut verschlossenem Gefäß stehen. Anschließend engt man im Becherglas auf dem Dampfbad bis auf einen kleinen Rest ein, bringt von dieser Flüssigkeit 1—2 Tropfen auf einen Objektträger und läßt bei Zimmertemperatur oder bei schwach erhöhter Temperatur verdunsten. Bei sehr verdünnten Lösungen kann diese Operation auf dem gleichen Objektträger wiederholt werden. Beim Verdunsten dieser Lösung entstehen Kristalle von Alkyl-thioharnstoff, deren Form stark von der Konzentration der Lösung und von den Verunreinigungen abhängt. Fügt man zu diesen Kristallen 1 Tropfen Silbernitratlösung (hergestellt aus 0,3 g Silbernitrat, 25 Tropfen verd. Salpetersäure und 100 ml Wasser), so erhält man lange, feine, meist verzweigte bündel- oder büschelförmig angeordnete Nadeln. Die Kristallisation tritt bei sehr starker Verdünnung erst beim Verdunsten der Lösung ein.

Ein anderes Verfahren zum mikrochemischen Nachweis flüchtiger Senföle wurde von ROSENTHALER (1926) entwickelt: Das zu untersuchende pflanzliche Material (z. B. Senfmehl) wird in der MOLISCHschen Gaskammer mit wenig Wasser befeuchtet, mit einem Deckglas bedeckt, an dessen Unterseite 1 Tropfen einer mit Salzsäure angesäuerten 1%igen Kaliumpermanganat-Lösung hängt, und nach 10 min mit einem Mikrobrenner erwärmt. Nach weiteren 5 min wird das Deckglas abgenommen, die Permanganatlösung mit 1 Tropfen Salzsäure versetzt und sowohl Permanganat als auch Braunstein durch leichtes Erwärmen über dem Mikrobrenner zum Verschwinden gebracht. Gibt man zu der so erhaltenen Lösung 1 Tropfen Bariumchloridlösung, so tritt bei Anwesenheit von Senföl Trübung durch Bariumsulfat ein. Dieser Nachweis eignet sich gut bei allen flüchtigen Schwefelverbindungen, die durch Permanganat zu Schwefelsäure oxydiert werden. Für den eindeutigen Nachweis empfiehlt es sich, eine Parallelbestimmung nach der oben beschriebenen Methode mit Phenylhydrazin vorzunehmen.

Vor kurzem wurde von BÖHME (1949) ein verhältnismäßig einfaches und zuverlässiges Verfahren zur *quantitativen* Bestimmung von Senfölen mitgeteilt; es beruht auf der Umsetzung der Senföle mit Wasserstoffperoxyd in alkalischer Lösung:

$$\ce{>C=S} + 2\,OH' + 4\,H_2O_2 = \ce{>C=O} + 5\,H_2O + SO_4''$$

Die nach dieser Gleichung gebildeten Sulfat-Ionen können entweder gravimetrisch bestimmt oder titriert werden; beide Wege führen zu durchaus befriedigenden Resultaten. SCHULTZ u. a. (1952, 1953) haben Senfölglykosid-Drogen, sowohl Frischpflanzen als auch Cruciferensamen, über deren Inhaltsstoffe noch wenig bekannt war, mit Hilfe der *Papierchromatographie* untersucht. Damit gelang es, auf einfache Weise die Anwesenheit und Zahl der Senfölglykoside zu ermitteln und zu wenigen Gruppen zusammenzufassen, wobei manche Unsicherheiten in der bisherigen Zuteilung beseitigt werden konnten.

1. Rhodanwasserstoff. HSCN.

Farblose Kristalle, die sich oberhalb 0° rasch zersetzen. Rhodanwasserstoff besitzt einen scharfen ätzenden Geruch und ist in wäßriger Lösung stark sauer. Verdünnte wäßrige Lösungen sind haltbar. Durch Schwefelwasserstoff tritt Spaltung in Ammoniak und Schwefelkohlenstoff ein. Diese Reaktion ist wahrscheinlich für die Entstehung von Spuren von Schwefelkohlenstoff in Pflanzen verantwortlich.

Geringe Mengen von Rhodanwasserstoff bzw. seiner Salze kommen in manchen pflanzlichen Organen, z. B. in den Samen der Hülsenfrüchte (BEILSTEIN, 1921) vor. Der frisch ausgepreßte Saft der Zwiebel von *Allium cepa* L. enthält die freie Säure in nicht unerheblicher Menge. Sie wurde auch in verschiedenen Cruciferensamen nachgewiesen (CZAPEK, 1921).

Zum Nachweis von Rhodanwasserstoff dienen in erster Linie Ferrisalze, die damit eine blutrote Färbung geben. Dieses blutrote Komplexsalz läßt sich mit Äther ausschütteln. Sehr charakteristisch ist auch die smaragdgrüne Färbung, die beim Umsatz von Rhodanwasserstoff oder von Rhodaniden mit Kupfersulfat entsteht.

Die quantitative Bestimmung kann entweder gravimetrisch oder titrimetrisch erfolgen. Bei der gravimetrischen Bestimmung oxydiert man das zu untersuchende Rhodanid mit einem Gemisch von konz. Salpetersäure und Bromwasser oder durch Wasserstoffperoxyd in Gegenwart von Ammoniak und ermittelt die entstandene Schwefelsäure als Bariumsulfat. Diese Methode wird durch das Vorliegen von Halogen- und Cyan-Ionen nicht gestört. Die titrimetrische Bestimmung von Rhodaniden wird nach VOLHARD so durchgeführt, daß man die zu untersuchende Lösung mit einem Überschuß an 0,1 n-Silbernitratlösung versetzt, mit Salpetersäure ansäuert, Eisen-ammoniak-alaun zufügt, und den Überschuß der Silber-Ionen mit 0,1 n-Rhodankaliumlösung zurücktitriert.

Ein weiteres Verfahren wurde von THIEL (1902) und von RUPP (1905) beschrieben. Dabei wird zur Rhodanidlösung bei Gegenwart von Bicarbonat-Ionen 0,1 n-Jodlösung im Überschuß zugesetzt, nach 4 stündigem Stehen bei Zimmertemperatur (in einer Flasche mit lose sitzendem Glasstopfen) mit Salzsäure angesäuert und der Jodüberschuß mit Thiosulfat sofort zurücktitriert.

Der mikrochemische Nachweis kann nach DENIGÈS ähnlich wie beim Schwefelkohlenstoff erfolgen. Einige ml der auf Rhodanid-Ionen zu prüfenden Flüssigkeit werden mit dem gleichen bis doppelten Volumen einer Mercurisulfatlösung versetzt, umgeschüttelt, wenn nötig filtriert und während 1—5 min zum schwachen Sieden erhitzt. Bei Gegenwart eines Rhodanids bildet sich eine Trübung bzw. eine kristalline Fällung. Unter dem Mikroskop lassen sich meist strahlig gruppierte Prismen, bei starker Verdünnung auch rautenförmig gekreuzte Kriställchen von Dithio-trimercurosulfat erkennen. Diese Reaktion ist sehr empfindlich: Noch 0,25 mg Rhodanwasserstoffsäure im Liter lassen sich damit eindeutig nachweisen.

2. Allylsenföl (Allyl-isothiocyanat, Iso-thiocyanallyl). $CH_2=CH-CH_2-NCS$.

Kp. 150,7°, D_{15} = 1,020—1,025. Farbloses, mit der Zeit gelblich werdendes Öl von sehr stechendem, zu Tränen reizendem Geruch, auf der Haut Blasen erzeugend. Die Verbindung ist in Wasser nur sehr wenig, in organischen Lösungsmitteln meist in jedem Verhältnis löslich. Die Einwirkung von konz. Schwefelsäure führt unter stürmischer Reaktion zu COS und schwefelsaurem Allylamin, $(C_3H_5NH_2)_2 \cdot H_2SO_4$. Bei der Einwirkung von verdünnter Salzsäure bei höherer Temperatur entsteht Allylaminhydrochlorid, $C_3H_5NH_2 \cdot HCl$, Smp. 105—110° aus abs. Alkohol.

Allylsenföl kommt in den Pflanzen hauptsächlich glykosidisch gebunden als Sinigrin vor. Dieses Glykosid ist bisher in den folgenden Pflanzen gefunden worden:

In den Samen und Keimlingen von *Sinapis nigra* L. (Will, 1844), in geringer Konzentration in den Wurzeln von *Cochlearia armoracia* L. (Hubatka, 1843; Stoll und Seebeck, 1948b), in *Alliaria officinalis* D. C. (Wertheim, 1845), in der Rinde und im Holz der Wurzeln und Stengel von *Armoracia lapathifolia* Gilib. (Pietschmann, 1924), in *Capsella bursa pastoris* (Bodinus, 1920), in *Iberis amara* L., *Lepidium campestre* R. Br. L., *Draba* L., *Thlaspi arvense* L. (Pless, 1846), in *Brassica oleracea botrytis* L. (Schneider und Lohmann, 1912), in *Isatis tinctoria* L. und anderen Cruciferen (Pietschmann, 1924), im besonderen *Brassica*-Arten (Grimme, 1912).

Die Gewinnung von Allylsenföl, schlechthin als Senföl bezeichnet, erfolgt z. B. aus Samen von schwarzem Senf. Diese werden zunächst durch starkes Pressen von Öl befreit, die zerkleinerten Preßkuchen anschließend durch Stehenlassen in lauwarmem Wasser vergoren und das dabei gebildete Schwefel-haltige Öl durch Wasserdampf abgetrieben. Auf diese Weise erhält man 0,5—1% Allylsenföl, berechnet auf die Menge der angesetzten Samen. Die Bildung des Senföls durch enzymatische Spaltung seines Glykosids Sinigrin (s. weiter unten) ist von Nebenreaktionen begleitet. Diese sind die Ursache für die Bildung zweier im natürlichen Senföl nie ganz fehlender Substanzen. Durch längere Berührung mit Wasser oder mit dem Kupfer der Destillationsblase wird nämlich das Senföl unter Schwefelabscheidung in Allylcyanid, C_3H_5CN, übergeführt, dessen Menge unter Umständen im Verhältnis zur Senfölfraktion bedeutend sein kann; daneben entsteht noch Schwefelkohlenstoff. Außer diesen beiden Verunreinigungen kann Senföl natürlichen oder synthetischen Ursprungs erhebliche Mengen des isomeren Allylrhodanids C_3H_5SCN enthalten. Dieses lagert sich mit der Zeit in Allylsenföl um. Um reines Allylsenföl zu erhalten, ist man deshalb genötigt, Präparate, die während längerer Zeit (z. B. 2 Monate) gestanden haben, fraktioniert zu destillieren.

Der Nachweis des Allylsenföls geschieht am besten durch Einwirkung von Ammoniak in alkoholischer Lösung unter Bildung des gut kristallisierenden Allylthioharnstoffs, $C_3H_5NH-CS-NH_2$: Rhombische Prismen, die bei 74° schmelzen. Als weitere für die Identifizierung geeignete Derivate seien noch das mit Phenylhydrazin in Alkohol entstehende Phenyl-allyl-thiosemicarbazid

C_3H_5NH—CS—NH—NH—C_6H_5 vom Smp. 118° und der Allyl-thiocarbaminsäure-bornylester (s. S. 704), $C_{10}H_{18}O$—CH—NH—C_3H_5 vom Smp. 59—60° erwähnt. Eine Farbreaktion, die sich ebenfalls zum Nachweis eignet, besteht im Umsatz von Allylsenföl in alkoholischer Lösung mit Phloroglucin und konz. HCl, wobei blaßrote Färbung eintritt, deren Intensität durch Erwärmen gesteigert werden kann.

Die quantitative Bestimmung erfolgt gravimetrisch durch Wägung des oben beschriebenen Allylthioharnstoffs oder besser noch durch Umsetzung des Senföls mit ammoniakalischer Silbernitratlösung, wobei der intermediär entstehende Allylthioharnstoff unter Bildung von Silbersulfid zerfällt. Dieses Verfahren kann titrimetrisch oder gravimetrisch durchgeführt werden:

a) Titrimetrisch: Von einer Lösung, bestehend aus 1 g Senföl in 49 g Äthanol werden etwa 5 g (genau eingewogen) in einem 100 ml Meßkolben mit 50 ml 0,1 n-Silbernitratlösung und 10 ml Ammoniak (D = 0,960) versetzt. Dann wird der Kolben mit einem 1 m langen Steigrohr versehen und während 1 Std. im siedenden Wasserbad gehalten. Man kühlt den Kolbeninhalt auf Zimmertemperatur ab, füllt mit Wasser bis zur Marke auf, schüttelt kräftig durch und filtriert ab. 50,0 ml des Filtrats werden mit 6 ml Salpetersäure (D = 1,153) und etwas Ferriammoniumsulfatlösung versetzt und mit 0,1 n-Rhodanammoniumlösung bis zur bleibenden Rotfärbung titriert. Die doppelte Anzahl der verbrauchten ml Rhodanidlösung wird von 50 subtrahiert. Diese Zahl ergibt die in Reaktion getretene Silbermenge. Wenn a die Menge des angewandten Senföls in g, b die Anzahl der verbrauchten ml Silberlösung darstellt, so erhält man den Prozentgehalt an Allylsenföl gemäß der Gleichung:

$$\frac{b \cdot 24{,}78}{a}$$

Eine weitere Methode (Hypojoditverfahren) zur Bestimmung von Senfölen wird von WOJAHN (1952) angegeben. Unter Verbrauch von 8 Äquivalenten Jod wird der Schwefel im Allylisothiocyanat quantitativ zu Sulfat oxydiert. Zweckmäßig führt man die Senföle vorerst durch Einwirkung von Ammoniak in Thioharnstoffe über. Dieses Verfahren eignet sich z. B. auch zur Bestimmung des Gehaltes an Allylsenföl in der Droge.

b) Gravimetrisch: Die gravimetrische Bestimmung erfolgt zunächst wie bei der soeben beschriebenen titrimetrischen Analyse, mit dem Unterschied, daß keine eingestellte Silberlösung angewendet werden muß. Nach beendeter Reaktion läßt man den Silbersulfidniederschlag gut absitzen und sammelt ihn durch Filtration der heißen Flüssigkeit auf einem vorher mit Ammoniak, dann heißem Wasser, Alkohol und Äther gewaschenen, getrockneten und gewogenen Filter, wäscht ihn mit heißem Wasser gut aus, spült mit Äthanol und dann mit Äther gut nach und trocknet den so behandelten Niederschlag bei etwa 80° bis zur Gewichtskonstanz. Wenn a das Gewicht des angewandten Senföls, b das Gewicht des Silbersulfids bedeutet, so beträgt der Prozentgehalt an Allylsenföl

$$\frac{b \cdot 39{,}995}{a}$$

Zur Bestimmung des Senföls in Samen werden 5 g davon in einem Kolben von etwa 300 ml mit 100 ml Wasser von 20—25° übergossen. Der verschlossene Kolben bleibt unter wiederholtem Umschwenken 2 Std. lang stehen, worauf man den Kolbeninhalt sorgfältig abdestilliert, die zuerst übergehenden 40—50 ml Destillat in einem Meßkölbchen von 100 ml Inhalt, das mit 10 ml konz. Ammoniak beschickt ist, auffängt und mit 20 ml 0,1 n-Silbernitratlösung versetzt. Diese Mischung wird nun 1 Std. lang auf dem Wasserbad erhitzt, wobei als Kühlzapfen ein in den Kölbchenhals hineinragender Glastrichter dient. Nach dem Abkühlen wird mit Wasser bis zur Marke aufgefüllt, filtriert, und 50 ml des klaren Filtrats nach Zusatz von 6 ml Salpetersäure und 5 ml Ferriammoniumsulfatlösung mit 0,1-n Ammoniumrhodanidlösung titriert. Aus der Anzahl der verbrauchten ml Silberlösung errechnet sich die Menge des in 5 g Samenmaterial vorhandenen Senföls, da 1 ml Silberlösung 0,004956 g Allylsenföl entspricht.

Eine einfache volumetrische Bestimmung des Senföls in Spiritus sinapis teilen Kaiser und Fürst (1953) mit.

Eine Bestimmung des Allylsenföls im Senfmehl wird ferner von Gros und Pichon (1934) angegeben.

3. Sinigrin. $CH_2=CH-CH_2-N=C-S-C_6H_{11}O_5 \cdot H_2O$
$$\underset{OSO_3K}{|}$$

Sinigrin ist das Glucosid des Allylsenföls. Smp. 127—128° mit 1 H_2O (Zers.); wasserfrei, Smp. 179°, $[\alpha]_D = -17,42°$ (in 82—83%igem Alkohol). Es ist in Wasser sehr leicht löslich, schwer löslich in konz. Alkohol, unlöslich in Äther, Benzol usw. Sinigrin wurde aus den Samen von *Sinapis nigra* L., aus den Rhizomen von *Cochlearia armoracia* L. (Stoll und Seebeck, 1948b) sowie aus vielen Cruciferen isoliert (Grimme, 1912). (Vgl. vorhergehender Abschnitt 2, S. 706.)

Schultz, Gmelin und Keller (1953) berichten über die Reindarstellung des Sinigrins mittels Ionenaustauscher aus *Sinapis nigra*.

4. Sek. D-Butylsenföl (sek. D-Butylisothiocyanat). $CH_3-CH_2-CH-NCS$
$$\underset{CH_3}{|}$$

Kp. 159—163°, $[\alpha]_D = +66,22°$ (in 94%igem Alkohol), findet sich als Hauptbestandteil im ätherischen Öl von *Cochlearia officinalis* (Gadamer, 1899; Hofmann, 1869; Ter Meulen, 1900), in den Samen von *Cardamine amara* L. (Kuntze, 1907), in *Cardamine hirsuta* L., *Cardamine pratensis* L. und *Cochlearia danica* L. (Blanksma, 1914). In den Pflanzen liegt es als Glykosid Gluco-cochlearin vor.

Bei der Reduktion mit Zinkstaub und Schwefelsäure in alkoholisch-wäßriger Lösung entsteht sek. D-Butylamin, dessen Platinat schöne, gelbrote Prismen vom Smp. 204—210° bildet. Der Umsatz des Senföls mit Ammoniak bei 100° führt zum sek. D-Butylthioharnstoff vom Smp. 137°, $[\alpha]_D = +33,97°$ (in gesättigter wäßriger Lösung). Beim Umsatz mit Phenylhydrazin entsteht ein Thiosemicarbazid vom Smp. 135°, $[\alpha]_D = +20°$ (in Äthanol).

Zur Isolierung des sek. D-Butylsenföls verwendet man das sorgfältig zerkleinerte Kraut der *Cochlearia officinalis* L., das mit Wasser angerührt wird und über Nacht stehen bleibt. Bei Verwendung von trockenem Kraut muß außer Wasser noch $^1/_5$ gepulvertes Senfmehl zugegeben werden, da im trockenen *Cochlearia*-Kraut das glykosidabbauende Ferment nicht mehr wirksam ist. Nach dem Stehen über Nacht destilliert man mit Wasserdampf und isoliert das ätherische Öl auf übliche Weise. Aus trockenem Kraut gewinnt man 0,175—0,305% sek. D-Butylsenföl. Das Rohöl siedet zwischen 150 und 162°; es hat einen scharfen, senfölähnlichen, nicht unangenehmen Geruch. Die Gehaltsbestimmung erfolgt analog wie beim Allylsenföl.

5. Glucocochlearin.

Glucocochlearin ist das Glykosid des sek. D-Butylsenföls; sein Vorkommen deckt sich daher mit den unter 4. für das freie sek. D-Butylsenföl gemachten Angaben. Das Glykosid liegt indessen bis heute noch nicht in reinem Zustand vor.

6. Crotonylsenföl (Allyl-methylisothiocyanat, Allyl-methylsenföl).
$CH_2=CH-CH_2-CH_2-NCS,$

Stark lichtbrechende Flüssigkeit von senfölartigem Geruch, Kp. 174° (geringe Zers.), findet sich hauptsächlich in den Samen von *Brassica napus* L. (SJOLLEMA, 1901; TER MEULEN, 1905), in *Brassica oleracea* L. *campestris chinoleifera* Vieh. (Chines. Colza) (VIEHOEVER, CLEVENGER und EWING, 1920) und in den Samen des indischen Senfs an Zucker gebunden im Glucosid Gluconapin vor.

Um das Crotonylsenföl zu isolieren, werden die Samen zusammen mit weißem Senfmehl als Enzymmaterial mit Wasser angeteigt und bleiben zur fermentativen Abspaltung des Zuckerrestes bei Zimmertemperatur stehen. Dann wird mit Wasserdampf destilliert. Vorteilhafter ist es jedoch, die gepulverten Samen zur Entfernung des Öls zunächst abzupressen, dann in heißes Wasser zu geben und nach dem Erkalten die nötige Fermentmenge in Form von weißem Senfmehl beizugeben. Nach dem Stehen über Nacht wird das aus dem Gluconapin abgespaltene Crotonylsenföl auf dem Wasserbad im Teilvakuum mit Wasserdampf abdestilliert, und das ätherische Öl aus dem Destillat ausgeäthert. Die Ausbeute beträgt bis 0,8% vom Gewicht der verwendeten Samen.

Der Nachweis und die quantitative Bestimmung des Crotonylsenföls erfolgt auf analoge Weise, wie unter 2. (S. 706) für Allylsenföl beschrieben wurde.

7. Gluconapin. $CH_2=CH-CH_2-CH_2-N=C-S-C_6H_{11}O_5$

$$OSO_3K$$

Über das Vorkommen von Gluconapin siehe unter 6. dieses Kapitels. Es ist das Glucosid von Crotonylsenföl, das bei der enzymatischen Spaltung neben Glucose und $KHSO_4$ erhalten wird.

8. Butylsulfid-crotonylsenföl. $C_4H_9-S-CH=CH-CH_2CH_2NCS$ (?).

HEIDUSCHKA und ZWERGAL (1932) führen den scharfen Geschmacksstoff des Rettichs (*Raphanus sativus niger* und *R. sativus albus*) auf eine Verbindung dieser Art zurück.

Zur Herstellung von Butylsulfid-crotonylsenföl wird der Rettich in feine Scheiben geschnitten und mit Wasserdampf destilliert. Das wäßrige Destillat schüttelt man mit Äther oder Petroläther aus und verdampft das Lösungsmittel auf dem Wasserbad. Die Rektifikation des öligen Rohproduktes im Vakuum liefert außer einer salbenartigen Substanz (s. weiter unten) ein hellgelbes Öl vom Kp. 140—142°/20 mm. Dieses besitzt den scharfen Geruch und Geschmack des frischen Rettichs. Die Elementaranalyse deutet auf die Bruttozusammensetzung $C_9H_{15}NS_3$ oder $C_9H_{17}NS_2$. Die Doppelbindung ist nicht sicher erwiesen. Der Gesamtgehalt des frischen Rettichs an diesem Senföl beträgt 0,01—0,02%.

9. Cheirolin (Methylsulfon-n-propyl-isothiocyanat), $CH_3SO_2CH_2CH_2CH_2NCS$ weicht in seiner Zusammensetzung von den übrigen Senfölen insofern wesentlich ab, als es eine Sulfongruppe besitzt. Rhombisch pyramidale Prismen aus Äther, Smp. 47—48°, Kp. etwa 200°/3 mm.

Cheirolin findet sich im Samen von *Cheiranthus cheiri* L. (SCHNEIDER, 1908, 1909; SCHNEIDER und LOHMANN, 1912; SCHNEIDER und SCHÜTZ, 1913), *Erysimum arkansanum* Nutt. und *E. nanum compactum aureum* (SCHNEIDER, 1910), wo es in Form des Glykosids Glucocheirolin vorliegt.

Darstellung des Cheirolins: Gemahlene Samen von *Cheiranthus cheiri* L. werden zuerst mit Äther entölt, dann zur Zersetzung des Glucocheirolins mit 5%iger Sodalösung versetzt und mit Äther ausgeschüttelt. Nach dem Abdunsten des Äthers verbleibt ein braun gefärbtes Öl, das in 0,5%iger Schwefelsäure von 50—60° aufgenommen wird; es werden dazu pro kg verarbeiteter Samen 600 ml Säure benötigt. Das von der Säure nicht gelöste Öl wird abgetrennt und noch zweimal mit je 200 ml 0,5%iger 50—60° warmer Schwefelsäure ausgeschüttelt. Die vereinigten, noch mindestens 40° warmen Schwefelsäureauszüge gießt man zur Klärung durch ein doppeltes Filter und salzt das Cheirolin in einem geräumigen Scheidetrichter mit festem Ammonsulfat aus. Das sich abscheidende Öl wird mit Äther aufgenommen und der ätherische Auszug mit Pottasche getrocknet. Sollte ein Teil des Cheirolins auskristallisieren, so bringt man die Kristalle durch gelindes Erwärmen wieder in Lösung und filtriert von der Pottasche noch warm ab. Die ätherische Lösung wird nun eingeengt und das zurückbleibende Öl mit einer kleinen Menge Cheirolin angeimpft, wenn die Kristallisation nicht von selbst einsetzt. Cheirolin kristallisiert dabei in großen rechteckigen Tafeln, die mit einem Briefcouvert Ähnlichkeit haben. Zur weiteren Reinigung kristallisiert man Cheirolin aus Äther, bei größeren Mengen aus der gleichen Gewichtsmenge Methanol um. Die Ausbeute beträgt 1,6—1,7% berechnet auf die trockenen Goldlacksamen. Aus Samen von *Erysimum arkansanum* gewinnt man die Verbindung in einer Ausbeute von 1,3%.

Bei Behandlung des Cheirolins mit $^1/_2$ Mol Mercurioxyd in warmem Wasser entsteht der Di-(γ-methylsulfonpropyl)-thioharnstoff $(C_4H_9O_2S—NH)_2CS$ vom Smp. etwa 125°. Eine erneute Behandlung mit Quecksilberoxyd führt den Thioharnstoff in Di-(γ-methylsulfonpropyl)-harnstoff $(C_4H_9O_2S—NH)_2CO$ vom Smp. 172° über. Weitere Derivate des Cheirolins können durch Umsatz mit alkoholischem Ammoniak erhalten werden, wobei ein Thioharnstoff vom Smp. 116° entsteht. Der Umsatz von Cheirolin mit Anilin in Äthanol liefert einen Phenylthioharnstoff vom Smp. 136°.

10. **Glucocheirolin,** $CH_3—SO_2—CH_2—CH_2—CH_2—N{=}C—S—C_6H_{11}O_5 \cdot H_2O$
$$\underset{\displaystyle OSO_3K}{\big|}$$

kristallisiert in farblosen Nädelchen, Smp. 158—160°, $[\alpha]_D^{20} = +21,5°$; es ist in Wasser sehr leicht löslich und gänzlich geschmacklos. Vorkommen wie Cheirolin (s. unter 9.). Glucocheirolin liefert bei der enzymatischen Spaltung Cheirolin, Glucose und $KHSO_4$ (Schneider und Lohmann, 1912; Schneider und Schütz, 1913).

11. **Erysolin (Methylsulfon-n-butylisothiocyanat).**
$CH_3—SO_2—CH_2—CH_2—CH_2—CH_2—NCS.$

Prismen aus Äther, Smp. 59—60°, wurde aus den Samen von *Erysimum perowskianum* Fisch. et Mey. (Schneider und Kaufmann, 1912) isoliert. In den Pflanzen bildet es einen Bestandteil des Glucosids Glucoerysolin.

Erysolin reizt die Schleimhäute stark. Die Verseifung liefert das dem Erysolir zugrunde liegende Amin, δ-Amino-butylmethylsulfon. Die Oxydation mit rote₁ rauchender Salpetersäure führt wie beim Cheirolin zur Bildung von Methylsulfon säure. Beim Umsatz mit äthanolischem Ammoniak entsteht ein Thioharnstof vom Smp. 143—144°. Die Isolierung des Erysolins wird analog derjenigen de Cheirolins (s. unter 9.) ausgeführt. Zur Erhöhung der Ausbeute bleibt das ent fettete Samenmehl vor der Ätherextraktion einige Stunden bei 35° mit Wasse stehen, wobei der Zuckerrest enzymatisch abgespalten wird.

12. **Glucoerysolin.** $CH_3—SO_2—CH_2—CH_2—CH_2—CH_2—N{=}C—S—C_6H_{11}O_5$
$$\underset{\displaystyle OSO_3K}{\big|}$$

Glucoerysolin ist das Glucosid des Erysolins und findet sich als dessen Mutte₁ substanz in den Samen von *Erysimum perowskianum* Fisch. et Mey. (Schneide und Kaufmann, 1912).

13. **Benzylsenföl (Benzylisothiocyanat).** $C_6H_5CH_2NCS.$

Farblose Flüssigkeit von scharfem Kressengeruch. In Wasser unlöslich, löslic in den gebräuchlichen organischen Lösungsmitteln, Kp.243°, Kp.140—141°/17 m₁ $D_{15} = 1,1246.$

Benzylsenföl wurde aus ätherischen Ölen von *Tropaeolum majus* L. (GADAMER, 1899), aus dem ätherischen Öl von *Lepidium sativum* L. (GADAMER, 1899), aus den Wurzeln von *Sisymbrium alliarium* Scop., *Isatis tinctoria* und aus *Salvadora oleoides* Den. *(S. persica* L.) isoliert. Es kommt auch in den Blättern von *Cardamine pratensis* L., in den Samen von *Raphanus sativus* L. *niger* und *R. sativus* L. *radicula* vor. In den Pflanzen liegt es zum Teil als Glucosid (Glucotropaeolin) vor.

Zum Nachweis des Benzylsenföls dient der Benzylthioharnstoff vom Smp. 162°, der beim Umsatz mit äthanolischem Ammoniak entsteht. Charakteristisch ist auch das Phenyl-benzyl-thiosemicarbazid vom Smp. 158°, das sich aus Benzylsenföl und Phenylhydrazin bildet.

Als Ausgangsmaterial für die Gewinnung des Benzylsenföls dient Kapuzinerkresse *(Tropaeolum majus* L.), die ihren scharfen Geruch dem Benzylsenföl verdankt. Durch fermentative Spaltung des Glucotropaeolins wird das Benzylsenföl freigesetzt. Damit diese Zuckerabspaltung quantitativ vor sich geht, muß das Kressenkraut vor der Fermentierung sorgfältig zerkleinert werden und der Krautbrei einige Stunden stehenbleiben, bevor er mit Äther extrahiert wird. An Stelle der Ätherextraktion kann man den Kressenkrautbrei nach beendeter Fermentation einer Wasserdampfdestillation unterwerfen und das Benzylsenföl aus dem Destillat mit Äther aufnehmen. Aus frischem Kraut gewinnt man Benzylsenföl in einer Ausbeute von etwa 0,03% (berechnet auf rohes Öl).

14. **Glucotropaeolin,** C_6H_5—CH_2—$N=C$—S—$C_6H_{11}O_5$

$$\overset{|}{O}SO_3K \qquad (\text{GADAMER})$$

findet sich in den Samen und im Kraut von *Tropaeolum majus* L. (GADAMER, 1899) und *Lepidium sativum* L. (GADAMER, 1899).

Durch fermentative Spaltung wird daraus Benzylsenföl, Glucose und $KHSO_4$ erhalten (GADAMER, 1899). GADAMER nahm für das Glucosid die vorstehend angegebene Konstitution an.

15. **Phenyläthylsenföl** (β-Phenäthylsenföl, β-Phenäthyl-isothiocyanat). C_6H_5—CH_2—CH_2—NCS, dickflüssiges Öl vom Kp. 141—142°/13mm, $D_{15} = 1,0997$. Riecht retticartig.

Phenyläthylsenföl wurde aus den Wurzeln von *Reseda lutea* und *R. luteola*, aus dem ätherischen Öl von *Reseda odorata* L. (BERTRAM und WALBAUM, 1894), aus *Nasturtium officinale* R. Br. (GADAMER, 1899) sowie aus dem Kraut von *Barbaraea praecox* R. Br. (GADAMER, 1899), aus *Brassica rapa var. rapifera* Metzg. (KUNTZE, 1907) und aus dem Rhizomen von *Cochlearia armoracia* L. neben dem Hauptbestandteil Allylsenföl und Phenylpropylsenföl aufgefunden (HEIDUSCHKA und ZWERGAL, 1932).

Ein charakteristisches Derivat ist der Phenyläthylthioharnstoff, der beim Umsatz des Senföls mit alkoholischem Ammoniak entsteht und bei 137° schmilzt. Die Entschwefelung dieses Thioharnstoffs mit Silbernitrat und Barytwasser führt zum Phenyläthylharnstoff: lange, feine Nadeln vom Smp. 111—112°. Die Behandlung des Phenäthylsenföls mit konz. HCl im Bombenrohr führt zum Hydrochlorid des β-Phenyläthylamins, Blättchen aus Äthanol vom Smp. 217°.

Bei der Darstellung des Phenäthylsenföls geht man vom Kraut der sorgfältig zerkleinerten Brunnenkresse *(Nasturtium officinale)* oder der perennierenden amerikanischen Winterkresse *(Barbaraea praecox)* aus. Der Krautbrei wird nach einigem Stehen bei gelinder Wärme mit Wasserdampf destilliert und das Destillat wie unter 13. für Benzylsenföl beschrieben mit Äther ausgezogen. Ebenso gut können für die Isolierung des Senföls frische zerkleinerte Resedawurzeln dienen, die bei der Destillation 0,014—0,035% Öl ergeben, das zur Hauptsache aus Phenyläthylsenföl besteht. Ein anderes Verfahren geht von den Schalen der Wasser- und Stoppelrüben von *Brassica rapa var. rapifera* Metzg. aus, die in sehr geringer Menge β-Phenyläthylsenföl enthalten.

16. **Gluconasturtiin,** $C_6H_5—CH_2—CH_2—N=C—S—C_6H_{11}O_5$

$$\underset{\displaystyle OSO_3K}{\big|}$$

wurde aus dem Samen von *Nasturtium officinale* R. Br. (GADAMER, 1899) und aus anderen Cruciferen, z. B. *Cardamine pratensis* L., *Cochlearia armoracia* L., isoliert. Bei der Hydrolyse liefert das Glucosid β-Phenyläthylsenföl.

17. **p-Oxybenzylsenföl** (p-Oxybenzyl-isothiocyanat, Sinalbinsenföl).

$$HO—\underset{}{\boxed{}}—CH_2—NCS$$

Gelbes Öl, schmeckt sehr scharf und zersetzt sich beim Erhitzen. Es liegt als Glucosid Sinalbin im weißen Senf *Sinapis alba* L. vor und wird daraus durch enzymatische Spaltung mittels Myrosinase erhalten (WILL und LAUBENHEIMER, 1879; GADAMER, 1897 und SALKOWSKI, 1889).

Ein charakteristisches Derivat ist der Phenyl-p-oxybenzylthioharnstoff vom Smp. 170—171° (gelbe Nadeln aus verdünntem Äthanol), der bei der Einwirkung von Anilin auf das in Äther gelöste Senföl entsteht. Die Entschwefelung mit Quecksilberoxyd führt zum Phenyl-p-oxybenzylharnstoff vom Smp. 140—142° (farblose Nadeln aus Äthanol). Der Umsatz des Senföls mit Phenylhydrazin in äthanolischer Lösung ergibt Phenyl-p-oxybenzyl-thiosemicarbazid vom Smp. 124° (farblose Blättchen aus wäßrigem Äthanol).

Zur Darstellung des p-Oxybenzylsenföls werden weiße Senfsamen gemahlen, mit Schwefelkohlenstoff entölt, mit Wasser zu einem Brei angerührt und nach der enzymatischen Hydrolyse des Sinalbins abgepreßt. Die Flüssigkeit und der zerkleinerte Preßkuchen werden nun mit Äther ausgezogen. Nach dem Abdestillieren des Äthers hinterbleibt das rohe Senföl.

18. **Sinalbin.**

$$HO—\underset{}{\boxed{}}—CH_2—N=C—S—C_6H_{11}O_5$$

Smp. 84° (wasserfrei 138—140°), $[\alpha]_D = —8°$ (—8,23°), wurde aus dem Samen von *Sinapis alba* L. isoliert (ROBIQUET und BOUTRON-CHARLARD, 1831).

Die enzymatische Spaltung des Glykosids in Sinalbinsenföl, Sinapinsulfat und Glucose wurde von WILL und LAUBENHEIMER (1879) und von GADAMER (1897, 1897) aufgeklärt. SALKOWSKI (1889) konnte das Sinalbinsenföl als p-Oxybenzylsenföl identifizieren.

19. **Sulforaphen** (Raphanin) Methyl-[4-thioisocyan-buten-(1)-yl-(1)]sulfoxyd

$$CH_3—S—CH=CH—CH_2—CH_2—NCS$$
$$\downarrow$$
$$O$$

Farbloses Öl, Kp. 125—130°/0,015 mm, $[\alpha]_D^{18} = —108°$ (in $CHCl_3$).

PLESS (1846), BERTRAM und WALBAUM (1894), GADAMER (1899), MOREIGNE (1896) sowie HEIDUSCHKA und ZWERGAL (1932) konnten aus Rettichöl und -knollen schwefelhaltige Öle isolieren. IVÁNOVICS und HORVÁTH (1947) gewannen aus einem wäßrigen Auszug von Rettichsamen ein leicht gelbliches Öl vom Kp. 135°/0,06 mm, dem sie die Formel $C_{17}H_{26}N_3O_3S_5$ oder $C_{17}H_{26}N_3O_4S_5$ zuerteilten.

Schmid und Karrer (1948a) konnten das Sulforaphen aus den Samen von *Raphanus sativus var. alba* L., das darin als Glykosid gebunden ist, über eine Silberverbindung und durch Hochvakuumdestillation und Verteilungschromatographie an feuchtem Kieselsäuregel (aus Äther-Chloroformlösung) isolieren und dessen Konstitution durch Abbaureaktionen beweisen. Die optische Aktivität ist durch die Asymmetrie der Sulfoxydgruppe bedingt. Sulforaphen bildet durch Umsatz mit alkoholischem NH_3 einen Thioharnstoff, $CH_3—S—CH=CH—CH_2—CH_2—NH—CS—NH_2$, mit Ani-

$$CH_3—\overset{\uparrow}{S}—CH=CH—CH_2—CH_2—NH—CS—NH_2$$

lin einen Phenylthioharnstoff,

$$CH_3—\overset{\uparrow}{S}—CH=CH—CH_2—CH_2—NH—CS—NHC_6H_5,$$

mit Anisidin einen p-Anisyl-thioharnstoff, die alle gut kristallisieren. Oxydation mit rauchender Salpetersäure ergibt Methansulfonsäure, $CH_3—SO_3H$, die als kristallisiertes Bariumsalz isoliert werden kann. Wird Sulforaphen mit 3%iger alkoholischer Natronlauge erwärmt und nach dem Abkühlen mit Natriumnitroprussidlösung versetzt, so entsteht eine intensiv rot-violette Färbung.

Die quantitative Bestimmung des labilen Schwefels kann auf folgende Weise durchgeführt werden:

a. Titrimetrisch: 153,9 mg Sulforaphen werden in 10 ml 10% NH_3 und 50,00 ml 0,1 n-$AgNO_3$-Lösung $1^1/_2$—2 Std. auf dem Wasserbad erwärmt, worauf man auf Zimmertemperatur abkühlt, mit Wasser auf 100 ml auffüllt, durch ein trockenes Filter filtriert und, im Filtrat nach dem Ansäuern die unverbrauchten Ag-Ionen durch Titration mit Kaliumrhodanid bestimmt.

b. Gravimetrisch wird der Schwefelgehalt auf folgende Weise ermittelt: 145,0 mg Sulforaphen werden in 10 ml 10% NH_3 und 50,00 ml 0,1 n-$AgNO_3$-Lösung $1^1/_2$—2 Std. auf dem Wasserbad erwärmt. Das ausgeschiedene AgS wird abfiltriert, bei 100° bis zur Gewichtskonstanz getrocknet und gewogen (Schmid und Karrer 1948a).

Sulforaphen besitzt gegenüber verschiedenen Mikroorganismen eine schwache antibiotische Wirkung.

20. **Glykosid des Sulforaphens.** $C_{12}H_{20}O_{10}NS_3K$.

Vermutliche Struktur: $CH_3—S—CH=CH—CH_2—CH_2—N=C—S—C_6H_{11}O_5$

$$CH_3—\overset{\uparrow}{S}—CH=CH—CH_2—CH_2—N=C—\underset{\underset{OSO_3K}{|}}{S}—C_6H_{11}O_5$$

Dieses Glykosid kommt in *Raphanus sativus var. alba* L. vor, konnte aber bis heute noch nicht isoliert werden. Seine hydrolytische Spaltung liefert das von Schmid und Karrer (1948a) isolierte Sulforaphen.

Neben dem Glykosid des Sulforaphens kommen in *Raphanus sativus var alba* L. noch weitere schwefelhaltige Verbindungen vor.

21. **4-Methyl-sulfoxyd-buten-(3)-yl-cyanid.**

$$CH_3—\overset{\uparrow}{S}—CH=CH—CH_2—CH_2—CN$$

Farbloses, fast geruchloses Öl, Kp. 120—125°/0,001 mm, $[\alpha]_D^{13} = -196°$ (in abs. Alkohol). Die Verbindung ist in Wasser und Alkohol leicht, weniger in Äther und sehr schwer in Petroläther löslich. Das als Glykosid gebundene 4-Methyl-sulfoxyd-buten-(3)-yl-cyanid wurde von Schmid und Karrer (1948b) aus *Raphanus sativus var. alba* L. isoliert und seine Konstitution aufgeklärt.

22. **Das Glykosid des 4-Methyl-sulfoxyd-buten-(3)-yl-cyanids** kommt in *Raphanus sativus var. alba* L. vor, konnte aber bis heute noch nicht als solches gefaßt werden (Schmid und Karrer, 1948b).

23. **5-Methylsulfoxyd-amylen-(4)-yl-cyanid,**

$$H_3C—\overset{\uparrow}{S}—CH=CH—CH_2—CH_2—CH_2—CN \qquad (?)$$

Tabelle 9. *Vorkommen der schwefelhaltigen Naturstoffe*[1].

Pflanze	Inhaltsstoffe
Acacia farnesiana Willd. und andere Akazienarten	lauchartig riechende Stoffe unbekannter Konstitution (*30*)
Alliaria officinalis s. unter: *Sisymbrium alliaria* Scop.	
Allium cepa L.	Disulfid unbekannter Konstitution (*44*) Thiopropionaldehyd (*71*) Rhodanwasserstoff (*29*)
Allium sativum L.	Diallyldisulfid (*44 117, 148, 149,*) Propyl-allyldisulfid (*44, 117*) Diallyltrisulfid (*44, 117, 148, 149*) Diallyltetrasulfid (*44, 117*) Diallyldisulfid-oxyd (Allicin)(*19,20*) Alliin (*123*)
Allium ursinum L	Methylmercaptan (*43, 44*) Divinylsulfid (*44, 116*) Alliin (*123*)
Ananas sativus Lindl.	β-Methyl-thiopropionsäure-methyl-ester (*49*)
Armoracia lapathifolia Gilib. s. unter *Cochlearia armoracia* L.	
Asa foetida	Disulfide: $C_7H_{14}S_2$; $C_8H_{16}S_2$; $C_{10}H_{18}S_2$; $C_{11}H_{20}S_2$ (*30*)
Asarum canadense L.	schwefelhaltige, stark antibiotische Verb. $C_{21}H_{20}O_8N_2S$ (*21*)
Aspergillus sydowi (Mycel)	Cholinschwefelsäureester (*156*)
Atractylis gummifera L.	Atractylsäure (*1, 2, 157, 158*)
Barbaraea praecox R. Br.	Phenyläthylsenföl (*42*)
Brassica	
B. juncea Hook. et Thoms.	Crotonylsenföl (*69*)
Sinapis juncea L. [= *Brassica juncea* (L.)]	Allylsenföl (*44*)
B. napobrassica	L-5-Vinyl-2-thiooxazolidon (*7*)

Pflanze	Inhaltsstoffe
B. napus L.	Crotonylsenföl (*120, 133*)
B. nigra (L.) Koch	Allylsenföl (*151*)
Sinapis nigra L. (= *Brassica nigra* Koch)	Schwefelkohlenstoff (*43*)
B. rapa L.	Allylsenföl (*47*)
B. rapa rapifera Metzg.	Phenyläthylsenföl (*47, 75*)
Brassica oleracea L. *asparagoides* (Dalech.)	Allylsenföl (*47*)
B. oleracea L. *botrytis*	Allylsenföl (*111*)
B. oleracea L. *capitata alba et rubra*	Allylsenföl (*47*)
B. oleracea L. *chinoleifera* Vieh.	Crotonylsenföl (*137*)
B. oleracea L. *crispa DC*	Allylsenföl (*47*)
B. oleracea L. *gemmifera DC*	Allylsenföl (*47*)
B. oleracea L. *gongylodes*	Allylsenföl (*47*)
B. oleracea L. *quercifolia DC*	Allylsenföl (*47*)
B. oleracea L. *sabauda*	Allylsenföl (*47*)
B. oleracea L. *acephala vulgaris*	Allylsenföl (*47*)
Calotropis gigantea (Dryand. in Ait.)	Uscharin (*54*)
C. procera (Dryand in Ait.)	Uscharin (*54*)
Capsella bursa pastoris	Allylsenföl (*15*)
Cardamine amara L.	D-sek. Butylsenföl (*75*)
C. hirsuta L.	D-sek. Butylsenföl (*14*)
Cardamine pratensis L.	D-sek. Butylsenföl (*14*) Benzylsenföl (*11*)
Cheiranthus cheiri L.	Cheirolin (*106, 107, 110, 112*)
Cochlearia armoracia L.	Diallylsulfid (*57*) Allylsenföl (*57, 124*) Phenyläthylsenföl (*51*) Phenylpropylsenföl (*51*) Allylsenföl (*91*)
Armoracia lapathifolia Gilib. (= *Cochlearia armoracia* L.)	
Cochlearia danica L.	D-sek. Butylsenföl (*14*)
C. officinalis L.	D-sek. Butylsenföl (*39, 56, 132*)
Cruciferensamen	Rhodanwasserstoff (*30*)

[1] Es sei außerdem auch auf WEHMER: „Die Pflanzenstoffe" (1929, 1931, 1935) (*138, 139*) verwiesen.

Tabelle 9. (Fortsetzung.)

Pflanze	Inhaltsstoffe	Pflanze	Inhaltsstoffe
Cytisus-Arten	Thioharnstoff (*70*)	*Raphanus sativus* L. *albus DC*	Butylsulfid-crotonylsenföl (*51*)
Diplotaxis tenuifolia DC	Allylsenföl (*69, 94*)	*R. sativus* var. *alba* L.	Sulforaphen (*104*)
Erysimum arkansanum Nutt. (= *E. asperum DC*)	Cheirolin (Methyl-sulfon-n-propyl-isothiocyanat) (*108*)		4-Methyl-sulfoxyd-buten-(3)-yl cyanid (*105*)
E. nanum Boiss. et Hohen.	Cheirolin (*43, 44, 69*)	*R. sativus* L.	Methylmercaptan (*73*)
E. perowskianum Fisch. et Mey.	Erysolin (Methyl-sulfon-n-butyl-isothiocyanat) (*109*)	*R. sativus* L. *niger*	Butylsulfidcrotonylsenföl (*51*)
			Benzylsenföl (*11*)
Erythrina glauca Willd.	Erysothiovin, Erysothiopin (Sulfonessigsäureester) (*36*)	*R. sativus* L. *radicula*	Benzylsenföl (*11*)
		Reseda lutea	Phenyläthylsenföl (*11*)
Erythrina pallida Britton u. Rose	Erysothiovin, Erysothiopin (Sulfonessigsäureester) (*36*)	*R. luteola* L.	Phenyläthylsenföl (*11*)
		R. odorata L.	Phenyläthylsenföl (*12*)
E. poeppigiana (Walp.)	Erysothiovin, Erysothiopin (Sulfonessigsäureester) (*36*)	Reunionöl	Dimethylsulfid (*43, 44*)
		Salvadora oleoides Decne. (= *S. persica* L.)	Benzylsenföl (*40, 69*)
Gelidium corneum	Agar-Agar (*11, 27*)		
Geraniumöl (afrik.)	Dimethylsulfid (*43, 44*)	*Schizophyllum commune* Nutt.	Dimethylsulfid (*25*)
Gigartina spinosa	Agar-Agar (*11, 27*)		Dimethyldisulfid (*25*)
Gliocladium fimbriatum	Gliotoxin (*143, 145*)	*Sch. lobatum* ?	Schwefelkohlenstoff (*30*)
Hefe	Adenyl-D-thiomethyl-ribose (*78, 129, 130*)	*Scorodophloeus zenkeri* Harms	lauchartig riechende Stoffe unbekannter Struktur (*30*)
Hülsenfrüchte	Rhodanwasserstoff (*9*)	*Secale cornutum*	Ergothionein (Thionin, Thiasin, Sympectothion) (*34, 92, 131*)
Iberis amara L.	Allylsenföl (*93*)		
Isatis tinctoria L.	Allylsenföl (*91*) / Benzylsenföl (*91*)	„*Shoyu*", jap. Maggi	Methionol (γ-Methyl-mercapto-propanol) (*3*)
Lasianthus bracteolatus Miqu.	Methylmercaptan (*11*)	*Sinapis alba* L.	p-Oxybenzylsenföl (*37, 38, 103, 152*)
L. lucidus Bl.	Methylmercaptan (*73*)	*S. arvensis* L.	Allylsenföl (*69*)
L. purpureus Bl.	Methylmercaptan (*73*)	*S. chinensis* L. Mant. (*Brassica urbaniana* O.-E. Schultz)	Allylsenföl (*69*)
L. sativus L.	Benzylsenföl (*69*)		
Lasianthus stercorarius Bl.	Methylmercaptan (*73*)	*S. dissecta* (Lagasca)	Allylsenföl (*69*)
Leguminosen (diverse)	lauchartig riechende Stoffe unbekannter Struktur (*30*)	*S. juncea* L. s. unter *Brassica juncea*	
		S. nigra L. s. unter *Brassica nigra* Koch	
Lepidium campestre R. Br. (in Ait.)	Allylsenföl (*93*)	*Sisymbrium alliaria* Scop.	Benzylsenföl (*11*)
L. campestre R. Br. (in Ait.)	Benzylsenföl (*69*)	*Alliaria officinalis* (= *Sisymbrium alliaria* Scop.)	Allylsenföl (*52, 148*)
Lepidium draba L.	Allylsenföl (*93*)		
L. latifolium L.	Benzylsenföl (*69*)	*Sisymbrium officinale* Scop.	Allylsenföl (*11*)
L. ruderale L.	Allylsenföl (*93*)	*Spirida filamentosa*	Agar-Agar (*11, 27*)
L. rurale ?	Benzylsenföl (*69*)	*Tagetes erecta* L.	Terthienyl (*159*)
L. sativum L.	Benzylsenföl (*42*)	*Thlaspi arvense* L.	Allylsenföl (*93*)
Nasturtium officinale R. Br. (in Ait.)	Phenyläthylsenföl (*41, 42*)	*Trichoderma lignorum*	Gliotoxin (*141, 142, 143, 146*)
Pfefferminzöl (amerik.)	Dimethylsulfid (*43*)	*Tropaeolum majus* L.	Benzylsenföl (*40, 42*)
Phyllophora nervosa	Agar-Agar (*62, 74*)		
Pithecolobium lobatum Benth.in Hook.	Djenkolsäure (*136*)		

wurde von Zwergal (1951) als wasserdampfflüchtiges, helles Öl von kraut- bis radieschenähnlichem Geschmack aus *Brassica oleracea* L. *var. gongylodes* L. auf folgende Weise isoliert: Das schwach angesäuerte Wasserdampfdestillat von Kohlrabischnitzeln wurde ausgeäthert, der Äther abdestilliert und der Rückstand unter vermindertem Druck in 4 Fraktionen zerlegt. Aus der höchstsiedenden Fraktion konnte das 5-Methylsulfoxyd-amylen-(4)-yl-cyanid gewonnen werden.

Daß diese Verbindung kein Senföl ist, geht daraus hervor, daß sie mit alkoholischer $AgNO_3$-Lösung keine Fällung gibt; es ist somit keine Isothiocyangruppe vorhanden. Hingegen läßt sich eine Nitrilgruppe nachweisen; denn beim Verkochen mit 2 n-Lauge wird Ammoniak abgespalten. Die Jodzahl (155) entspricht einer Doppelbindung.

24. Phenylpropylsenföl. $C_6H_5-CH_2-CH_2-CH_2-CNS$ (?).

Kp. 163—166°/15 mm, hellgelbes Öl, riecht nach Senföl. Phenylpropylsenföl wurde von Heiduschka und Zwergal (1932) aus *Cochlearia armoracia* L. isoliert.

25. Das Glykosid des Phenylpropylsenföls kommt in *Cochlearia armoracia* L. vor (Heiduschka und Zwergal, 1932). Bis heute konnte aber nur das Produkt seiner hydrolytischen Spaltung, das Phenylpropylsenföl, isoliert werden.

26. Atractylsäure.

Smp. 173°, $[\alpha]_D = -64°$. Sie ist der giftige Bestandteil der Wurzeln von *Atractylis gummifera* L. Ihre Hydrolyse mit Säure liefert Glucose, Valeriansäure, Kaliumbisulfat und Atractyligenin, $C_{14}H_{22}O_4$, dessen Konstitution noch nicht aufgeklärt werden konnte (Wunschendorff und Braudel, 1931; Ajello, 1933: Wunschendorff und Valier, 1934; Ajello, 1934).

27. Antibiotische Verbindung aus *Asarum canadense*.

Cavallito und Bailey (1946) isolierten aus *Asarum canadense* L. eine schwefelhaltige, stark antibiotisch wirksame Verbindung der Zusammensetzung $C_{21}H_{20}O_8N_2S$, die aus 95%igem Alkohol kristallisiert und sich oberhalb 160° zersetzt. Sie wirkt auf *Streptococcus aureus* etwa gleich stark wie Penicillin. Die Struktur dieses Antibiotikums ist noch unbekannt.

Literatur.

1. Ajello, T.: Gazz. chim. ital. **63**, 289 (1933); Chem. Zentralbl. **1933** II, 2399. — 2. Ajello, T.: Gazz. chim. ital. **64**, 59 (1934); Chem. Zentralbl. **1934** I, 3861. — 3. Akabori, S., and T. Kaneko: Proc. Imp. Acad. (Tokyo) **12**, 131 (1936). — 4. Andrews, J. C., and K. Grandall Andrews: J. Biol. Chem. **118**, 555 (1937); Z. anal. Chem. **124**, 452 (1942). — 5. Armstrong, M. D., and V. du Vigneaud: J. Biol. Chem. **168**, 373 (1947). — 6. Arnold, A., and C. A. Elvehjem: J. Nutrition **15**, 403 (1938). — 7. Astwood, E. B., M. A. Greer and M. G. Ettlinger: Science (Lancaster, Pa.) **109**, 631 (1949).

8. Baernstein, H. D.: J. Biol. Chem. **115**, 25 (1936); Z. anal. Chem. **115**, 240 (1938). — 9. Beilstein: Handbuch der organischen Chemie, 4. Aufl., Bd. 3, 1921. — 10. Benischke, H.: Z. Pflanzenernährung und Bodenkunde **41**, 42 (1948); Z. anal. Chem. **129**, 468 (1949). — 11. Bersin, Th.: Advances in Enzymology, Vol. X. New York: Interscience Publ. 1950. — 12. Bertram, J., und H. Walbaum: J. pr. Chem. [2] **50**, 555 (1894). — 13. Bicknell, F., and F. Prescott: The Vitamins in Medicine. New York: Grune and Stratton 1947. — 14. Blanksma, J. J.: Pharm. Weekbl. **51**, 1383 (1914). — 15. Bodinus: Apoth.-Ztg. **35**, 183 (1920). — 16. Böhme, H.: Z. anal. Chem. **129**, 129 (1949). — 17. Brown, W. L.: J. Biol. Chem. **142**, 299 (1942). — 18. Bürger, K.: Z. angew. Chem. **54**, 479 (1941).

19. Cavallito, C. J., and J. H. Bailey: J. Amer. Chem. Soc. **66**, 1950 (1944). — 20. Cavallito, C. J., J. S. Buck and C. M. Suter: J. Amer. Chem. Soc. **66**, 1952 (1944). — 21. Cavallito, C. J., and J. H. Bailey: J. Amer. Chem. Soc. **68**, 489 (1946). — 22. Cerecedo, L. R., and D. J. Hennessy: J. Amer. Chem. Soc. **59**, 1617 (1937). — 23. Cerecedo, L. R., and F. J. Kaszuba: J. Amer. Chem. Soc. **59**, 1619 (1937). — 24. Cerecedo, L. R., and J. J. Thornton: J. Amer. Chem. Soc. **59**, 1621 (1937). — 25. Challenger, F., and Ph. T. Charlton: J. Chem. Soc. **1947**, 424. — 26. Chase, E. F.: Diss. Columbia 1928. — 27. Cioglia, L.: Chem. Zbl. **1940** II, 916. — 28. Cook, E. F.: J. Amer. Pharm. Assoc. **28**, 267 (1939). —

29. Czapek, F.: Biochemie der Pflanzen, 2. Aufl., Bd. 2, S. 54. Jena: G. Fischer 1920. — 30. Czapek, F.: Biochemie der Pflanzen, 2. Aufl., Bd. 3. Jena: G. Fischer 1921.

31. Denigès, G.: Bull. Trav. Soc. pharm. Bordeaux 76, 180 (1938); Z. anal. Chem. 127, 50 (1944). — 32. Diemair, W., u. J. Koch: Z. anal. Chem. 119, 94 (1940). — 33. Du Vigneaud, V., and W. I. Patterson: J. Biol. Chem. 114, 533 (1936).

34. Eagles, B. A.: J. Amer. Chem. Soc. 50, 1386 (1928).

Folin, O.: Z. physiol. Chem. 37, 161 (1902). — 36. Folkers, K., F. Koniuszy and J. Shavel jr.: J. Amer. Chem. Soc. 66, 1083 (1944)

37. Gadamer, J.: Arch. Pharm. 235, 83 (1897). — 38. Gadamer, J.: Ber. dtsch. chem. Ges. 30, 2328, 2330 (1897). — 39. Gadamer, J.: Arch. Pharm. 237, 92 (1899). — 40. Gadamer, J.: Arch. Pharm. 237, 111 (1899). — 41. Gadamer, J.: Arch. Pharm. 237, 507 (1899). — 42. Gadamer, J.: Ber. dtsch. chem. Ges. 32, 2335 (1899). — 43. Gildemeister, E., u. F. Hoffmann: Die ätherischen Öle, 3. Aufl., Bd. 1. Leipzig: L. Staakmann 1928. — 44. Gildemeister, E., u. F. Hoffmann: Die ätherischen Öle, 3. Aufl., Bd. 2. Leipzig: L. Staakmann 1929. — 45. Gmelin-Kraut: Handbuch der organischen Chemie, Bd. VII, S. 2198, 1870. — 46. Greene, R. D., and A. Black: J. Amer. Chem. Soc. 59, 1395 (1935). — 47. Grimme, C.: Pharm. Zentralhalle 53, 733 (1912). — 48. Gros, R., et G. Pichon: J. Pharm. Chim. [8] 19, 249 (1934); Z. anal. Chem. 103, 449 (1935).

49. Haagen-Smit, A. J., J. G. Kirchner, C. L. Deasy and A. N. Prater: J. Amer. Chem. Soc. 67, 1651 (1945). — 50. Heat, H., A. Lawson and C. Rimington: Nature (Lond.) 166, 106 (1950). — 51. Heiduschka, A., u. A. Zwergal: J. prakt. Chem. 132, 201 (1932). — 52. Hérissey, H., u. H. R. Boivin: Chem. Zbl. 1929 I, 358. — 53. Hess, W. C., and M. X. Sullivan: Ind. Eng. Chem. (Anal. Ed.) 17, 717 (1945). — 54. Hesse, G., F. Reicheneder u. H. Eysenbach: Liebigs Ann. 537, 67 (1939). — 55. Hesse, G., u. H. W. Gampp: Ber. dtsch. chem. Ges. 85, 933 (1952). — 56. Hofmann, A. W.: Ber. dtsch. chem. Ges. 2, 102 (1869). — 57. Hubatka, C.: Liebigs Ann. 47, 153 (1843). — 58. Hunter, G.: Canad, J. Res. (Sect. E) 27, 230 (1949); Z. ar al. Chem. 133, 461 (1951).

59. Ivánovicz, G., and St. Horváth: Nature (Lond.) 160, 297 (1947).

60. Jansen, E. F.: J. Biol. Chem. 176, 657 (1948). — 61. Johnson, J. R., W. F. Bruce and J. D. Dutcher: J. Amer. Chem. Soc. 65, 2005 (1943). — 62. Jones, W. G. M., and S. Peat: J. Chem. Soc. 1942, 225.

63. Kaiser, H., u. E. Fürst: Dtsch. Apoth.-Ztg. 50, 1734 (1953); Z. anal. Chem. 113, 380 (1938). — 64. Karaoglanov, Z.: Österr. Chem.-Ztg. 45, 154 (1942). — 65. Kendall, E. C., and F. F. Nord: J. Biol. Chem. 69, 295 (1926). — 66. Kinnersley, H. W., J. R. O'Brien and R. A. Peters: Biochem. J. 27, 232 (1933). — 67. Kinnersley, H. W., and R. A. Peters: Biochem. J. 28, 667 (1934). — 68. Kinnersley, H. W., and R. A. Peters: Biochem. J. 32, 1516 (1938). — 69. Klein, G.: Handbuch der Pflanzenanalyse, Bd. 3, 2. Teil. Wien: Julius Springer 1932. — 70. Klein, G., u. E. Farkass: Österr. bot. Z. 79, 115 (1930). — 71. Kohmann, E. F.: Science (Lancaster, Pa.) 106, 625 (1947). — 72. Kolb, J. J., and G. Toennies: Ind. Eng. Chem. (Anal. Ed.) 12, 723 (1940). — 73. Koolhaas, D. R.: Biochem. Z. 230, 446 (1931). — 74. Korenzwit, A.: Chem. J. Ser. B (Russ.) 11, 331 (1938); Chem. Zbl. 1939 I, 2097. — 75. Kuntze, M.: Arch. Pharm. 245, 657 (1907).

76. Lavine, Th. F.: J. Biol. Chem. 109, 141 (1935); Z. anal. Chem. 111, 439 (1938). — 77. Lenormant, H., et M. Hérisson: C. r. Acad. Sci. (Paris) 232, 815 (1951). — 78. Levene, P. A., and H. Sobotka: J. Biol. Chem. 65, 551 (1925).

Malyoth, G., u. H. W. Stein: Biochem. Z. 323, 265 (1952). — 80. McFarren, F.: Anal. Chem. 23, 168 (1951); Z. anal. Chem. 136, 363 (1952). — 81. Meyer, K., and E. Chaffee: Proc. Soc. Exp. Biol. Med. 43, 487 (1940). — 82. Mirsky, A. E., and M. L. Anson: J. gen. Physiol. 18, 307 (1935); Z. anal. Chem. 108, 130 (1937). — 83. Moreigne, H.: J. Pharm. Chim. [6] 4, 10 (1896). — 84. Moreigne, H.: Bull. soc. chim. France [3] 15, 797 (1896). — 85. Mulder, G. J.: Chem. Zbl. 17, 833 (1846).

86. Naiman, B.: Science (Lancaster, Pa.) 85, 290 (1937). — 87. Nakamura, N.: Biochem. Z. 164, 31 (1925). — 88. Nightingale, G. T., L. G. Schermerhorn and W. R. Robbins: Plant Physiol. 7, 565 (1932). — 89. Nord, F. F.: J. Phys. Chem. 31, 867 (1927). —

90. Odake, S.: J. Agr. Chem. Soc. (Jap.) 10, 409 (1934).

91. Pietschmann, A.: Mikrochemie 2, 33 (1924). — 92. Pirie, N. W.: Biochem. J. 27, 204 (1933). — 93. Pless, F.: Liebigs Ann. 58, 36 (1846). — 94. Pottiez, Ch.: Chem. Zbl. 1922 II, 1195. — 95. Prebluda, H. J., and E. V. McCollum: J. Biol. Chem. 127, 495 (1939). — 96. Pregl, F.: Quantitative organische Mikroanalyse, 6. Aufl., bearb. von H. Roth. Wien: Springer-Verlag 1949.

97. Raybin, H. W.: Science (Lancaster, Pa.) 88, 35 (1938). — 98. Robiquet et Boutron-Charlard: J. de Pharm. [II] 17, 279 (1831). — 99. Rosenberg, H. R.: Chemistry and Physiology of the Vitamins. New York 1945. — 100. Rosenthaler, L.: Pharm. Acta Helv. 1, 72 (1926). — 101. Rosenthaler, L.: Z. anal. Chem. 108, 24 (1937). — 102. Rupp, E.:

Arch. Pharm. **243**, 460 (1905); vgl. auch J. Houben: Die Methoden der organischen Chemie, Bd. 1, S. 240. Leipzig: G. Thieme 1921.

103. Salkowski, H.: Ber. dtsch. chem. Ges. **22**, 2137 (1889). — 104. Schmid, H., u. P. Karrer: (a) Helv. Chim. Acta **31**, 1017 (1948). — 105. Schmid, H., u. P. Karrer: (b) Helv. Chim. Acta **31**, 1087 (1948). — 106. Schneider, W.: Ber. dtsch. chem. Ges. **41**, 446 (1908). — 107. Schneider, W.: Ber. dtsch. chem. Ges. **42**, 3416 (1909). — 108. Schneider, W.: Liebigs Ann. **375**, 207 (1910). — 109. Schneider, W., u. H. Kaufmann: Liebigs Ann. **392**, 1 (1912). — 110. Schneider, W., u. W. Lohmann: Ber. dtsch. chem. Ges. **45**, 2954 (1912). — 11. Schneider, W., u. W. Lohmann: Ber. dtsch. chem. Ges. **45**, 2955 (1912). — 112. Schneider, W., u. L. A. Schütz: Ber. dtsch. chem. Ges. **46**, 2634 (1913). — 113. Schormüller, J., u. H. Ballschmieter: Z. anal. Chem. **132**, 1 (1951). — 114. Schultz, O.-E., R. Gmelin u. A. Keller: Z. Naturforschg. 8 b, 14 (1953). — 115. Schultz, O.-E., u. R. Gmelin: Z. Naturforschg. 7 b, 500 (1952); 8 b, 151 (1953). — 116. Semmler, F. W.: Liebigs Ann. 241, 90 (1887).— 117. Semmler, F. W.: Arch. Pharm. **230**, 434 (1892). — 118. Sherman, H. C., and S. L. Smith: The Vitamins. Chemical Catalog Co. 1931. — 119. Sherman, H. C., and A. Spohn: J. Amer. Chem. Soc. **45**, 2719 (1923). — 120. Sjollema, B.: Rec. Trav. Chim. (Pays-Bas) **20**, 237 (1901). — 121. Spruyt: Chem. Weekbl. **27**, 298 (1930). — 122. Stoll, A., u. E. Seebeck: Experientia (Basel) **3**, 114 (1947). — 123. Stoll, A., u. E. Seebeck: (a) Helv. Chim. Acta **31**, 189 (1948). — 124. Stoll, A., u. E. Seebeck: (b) Helv. Chim. Acta **31**, 1432. (1948). — 125. Stoll, A., u. E. Seebeck: (a) Helv. Chim. Acta **32**, 197 (1949). — 126. Stoll, A., u. E. Seebeck: (b) Helv. Chim. Acta **32**, 866 (1949). — 127. Stoll, A., u. E. Seebeck: Experientia (Basel) **4**, 330 (1950); Helv. Chim. Acta **34**, 481 (1951). — 128. Surmatis, J. D., u. M. L. Willard: Mikrochemie **21**, 167 (1937); Z. anal. Chem. **115**, 444 (1938). — 129. Suzuki, U., S. Odake u. T. Mori: Biochem. Z. **154**, 278 (1924). — 130. Suzuki, U., u. T. Mori: Biochem. Z. **162**, 413 (1925).

131. Tanret, C.: C. r. Acad. Sci. (Paris) **149**, 222 (1909). — 132. Ter Meulen, H.: Rec. Trav. Chim. (Pays-Bas) **19**, 37 (1900). — 133. Ter Meulen, H.: Rec. Trav. Chim. (Pays-Bas) **24**, 481 (1905). — 134. Thiel, A.: Ber. dtsch. chem. Ges. **35**, 2766 (1902); vgl. auch J. Houben: Die Methoden der organischen Chemie, Bd. 1, S. 240. Leipzig: Georg Thieme 1921.

135. van Niel, C. B.: Arch. Mikrobiol. **3**, 1 (1931). — 136. van Veen, A. G., et A. J. Hyman: Rec. Trav. Chim. (Pays-Bas) **54**, 493 (1935). — 137. Viehoever, A., J. F. Clevenger and C. O. Ewing: J. Agr. Res. **20**, 117 (1920).

138. Wehmer, C.: Die Pflanzenstoffe, 2. Aufl., Bd. 1. Jena: G. Fischer 1929. — 139. Wehmer, C.: Die Pflanzenstoffe, 2 Aufl., Bd. 2. Jena: G. Fischer 1931; Erg.-Bd. 1930—34. Jena: G. Fischer 1935. — 140. Weidinger, A.: Rec. Trav. Chim. (Pays-Bas) **56**, 562 (1937); Z. anal. Chem. **120**, 351 (1940). — 141. Weindling, R.: Phytopathology **22**, 837 (1932). — 142. Weindling, R.: Phytopathology **24**, 1153 (1934). — 143. Weindling, R.: Phytopathology **27**, 1175 (1937).—144. Weindling, R.: Bot. Rev. **4**, 475 (1938). — 145. Weindling, R.: Phytopathology **31**, 991 (1941). — 146. Weindling, R., and O. H. Emerson: Phytopathology **26**, 1068 (1936). — 147. Wendt, G.: Z. physiol. Chem. **272**, 152 (1942). — 148. Wertheim, Th.: Liebigs Ann. **51**, 289 (1844). — 149. Wertheim, Th.: Liebigs Ann. **55**, 297 (1845). — 150. Westenbrink, H. G. K., u. E. P. Setyn-Parvé: Int. Z. Vit. Forschg. **21**, 461 (1949). — 151. Will, H.: Liebigs Ann. **52**, 1 (1844). — 152. Will, H., u. A. Laubenheimer: Liebigs Ann. **199**, 150 (1879). — 153. Williams, R. R., R. E. Waterman and J. C. Keresztesy: J. Amer. Chem. Soc. **56**, 1187 (1934). — 154. Wojahn, H.: Pharm. Zentralhalle **91**, 326 (1952). — 155. Wojahn, H.: Arch. Pharm. **285**, 86 (1952). — 156. Woolley, D. W., and W. H. Peterson: J. Biol. Chem. **122**, 213 (1937).—157. Wunschendorff, H., et P. Braudel: Bull. Soc. Chim. Biol. **13**, 758, 764 (1931); Chem. Zbl. **1932** II, 70. — 158. Wunschendorff, H., et P. Valier: Bull. Soc. Chim. Biol. **16**, 74, 80 (1934); Chem. Zbl. **1934** I, 3752.

159. Zechmeister, L., and J. W. Sease: J. Amer. Chem. Soc. **69**, 273 (1947). — 160. Zeile, K., u. M. Oetzel: Z. physiol. Chem. **284**, 1 (1949); Z. anal. Chem. **131**, 390 (1951). — 161. Zwergal, A.: Die Pharmazie **6**, 245 (1951).

Sachverzeichnis.

(Deutsch—Englisch)

Wegen allgemeiner Stichworte wie Extraktion, Abtrennung, Reinigung usw. der einzelnen Stoffgruppen vergleiche man auch das Inhaltsverzeichnis am Anfang dieses Bandes.

cis-, trans-, n-, D-, L- und ähnliche Isomere sind unter dem Anfangsbuchstaben der Verbindung und nicht unter dem Präfix eingeordnet.

Alle iso-Verbindungen finden sich unter Iso-.

Ä, Ö, Ü sind wie Ae, Oe, Ue eingereiht.

Bei gleicher Schreibweise in beiden Sprachen sind die Verbindungen jeweils einfach aufgeführt.

Subject Index.

(English — German)

For general terms such as extraction, isolation, purification etc. of various groups of compounds cf. also the table of contents at the beginning of the volume.

cis-, trans-, n-, D-, L- and similar isomers are listed according to the first letter of the following word.

All iso-compounds are to be found under Iso-.

Ä, Ö, Ü are taken as Ae, Oe, Ue.

Where English and German spelling of a word is identical, the italicised (German) entry is omitted.

Dimethyldisulfide, *Dimethyldisulfid* 693, 715.
Dimethylsulfide, *Dimethylsulfid* 693, 715.
Dinitrobenzoic acid, *Dinitrobenzoesäure* 525.
Dinitro-α-naphthol 534.
Dinucleoside monophosphates, *Dinucleosid-monophosphate* 287.
Dinucleotides, *Dinucleotide* 287.
Dioscorine, *Dioscorin* 498, 500.
4:6 Dioxy-2,5-diaminopyrimidine, *4,6-Dioxy-2,5-diaminopyrimidin* 299.
2:6-Dioxy-5-methyl-pyrimidine, *2,6-Dioxy-5-methylpyrimidin* 255.
β,β-Diphenylethylamine, R_f value, *β,β-Diphenyläthylamin, R_f-Wert* 540.
4-Diphenyl isothiocyanate and amines, *4-Diphenyl-isothiocyanat u. Amine* 527.
Dipeptidase 79.
Diphosphates, cyclic, *Diphosphate, cyclische* 288.
Diphosphopyridinenucleotide, *Diphospho-pyridinnucleotid* 320.
Dipropylamine, R_f value, *Dipropylamin, R_f-Wert* 538.
Dipterin-N-methyltryptamine, *Dipterin-N-methyltryptamin* 584.
Disaccharides, *Disaccharide* 676.
DISCHE reagent, DISCHE-*Reagens* 269.
Disulfides, aliphatic, *Disulfide, aliphatische* 692, 714.
2,2'-Dithioisobutyric acid, *2,2'-Dithioiso-buttersäure* 692.
Divicine, *Divicin* 249, 299.
Divinyl sulfide, *Divinylsulfid* 693, 714.
Djenkolic acid, *Djenkolsäure* 12, 695, 715.
DNA, s. Deoxyribonucleic acid, *DNA, s. Desoxyribonucleinsäure.*
DPN, s. also Co-dehydrogenases, *DPN, s. auch Codehydrasen* 320.
—, estimation, *Bestimmung* 326.
—, fermentation test, *Gärtest* 330.
—, preparation, *Darstellung* 339.
—, Thunberg method, *Thunberg-Technik* 328.
DPN-H₂, calibration curve for photometr. determ., *Eichkurve für photometrische Bestimmung* 337.
n-Dodecylamine, R_f value, *n-Dodecylamin, R_f-Wert* 542.
Domesticine, *Domesticin* 382, 384.
Dopa 581.
— -Oxidase, *Dopa-Oxydase* 667.
Double-wedge trough in spectroscopy, *Doppelkeiltrog in der Spektroskopie* 209, 212, 213, 218, 243.

Edestin 83, 85, 90, 91, 93.
Eisenin 6.
Electrophoresis apparatus, *Elektrophorese-Apparatur* 279.
—, power supply, *Energieversorgung* 279, 280.
Electrophoresis, buffer solutions, *Elektrophorese, Pufferlösungen* 280.
— in borate, *in Borat* 273.

Electrophoretic uniformity of proteins, *Elektrophoretische Einheitlichkeit v. Proteinen* 44.
Emetamine, *Emetamin* 429.
Emetine, *Emetin* 428.
Emulsin 670, 687.
Endosperm protein, *Endosperm-Eiweiß* 72.
Enzymes, *Enzyme* 33, 57, 77, 97, 288.
—, action, control, *Enzymwirkung, Regulierung* 43.
—, functions, *Funktionen* 60.
—, glycosidic, *glykosidspaltende* 677.
— in resting seed, *in ruhenden Samen* 79.
— in potato tuber, *in Kartoffelknollen* 63.
Ephedra alkaloids, *Ephedra-Alkaloide* 418.
Ephedrine, *Ephedrin* 419.
—, diliturate, *Diliturat* 529.
—, isolation from *Ephedra, Isolierung aus Ephedra* 419.
—, qualitative tests, *qual. Tests* 420.
—, quantitative estimation, *quant. Bestimmg.* 420.
—, R_f value, *R_f-Wert* 540.
L-nor-Ephedrine, L-*nor-Ephedrin* 419.
ERDMANN's reagent, ERDMANNs *Reagens* 374.
Ergocornine, *Ergocornin* 411.
Ergocorninine, *Ergocorninin* 411.
Ergocristine, *Ergocristin* 411.
Ergocryptine, *Ergocryptin* 411.
Ergometrine, *Ergometrin* 410.
Ergometrinine, *Ergometrinin* 410.
Ergosine, *Ergosin* 411.
Ergosinine, *Ergosinin* 411.
Ergot alkaloids, *Mutterkornalkaloide* 7, 409 to 416.
—, detection and estimation, *Nachweis u. Bestimmg.* 413.
—, isolation, *Isolierung* 411.
—, paper chromatogr., *Papierchromatogr.* 414.
Ergotamine, *Ergotamin* 410.
Ergotaminine, *Ergotaminin* 410.
Ergothionein 595, 610—612, 703, 715.
Erysolin 710, 715.
Erysothiopin 692.
Erysothiovin 692, 715.
Eseramine, *Eseramin* 391.
Ethanolamine, *Äthanolamin* 524, 556, 557.
—, diliturate, *Diliturat* 529.
—, R_f value, *R_f-Wert* 538, 540, 568.
γ-Ethylamide of L-glutamic acid, *γ-Äthylamid von L-Glutaminsäure* 12.
Ethylamine, *Äthylamin* 535.
—, absorpt. spectr. of ninhydrin complex, *Absorpt. Spektr. v. Ninhydrin-Komplex* 544.
—, diliturate, *Diliturat* 529.
—, rel. absorpt., *relative Absorption* 545.
—, R_f value, *R_f-Wert* 540, 542.
Excelsin 83, 93.

Ferriheme chloride, *Ferrihäm-chlorid* 199.
Ferriheme hydroxide, *Ferrihäm-hydroxyd* 199.
Ferrihemochromogen, *Ferrihämochromogen* 199.